Index of Study Strategies

About the Author

George Woodbury,
College of the Sequoias

George Woodbury earned a bachelor's degree in mathematics from the University of California, Santa Barbara, in 1990 and a master's degree in mathematics from California State University, Northridge, in 1994. He currently teaches at College of the Sequoias in Visalia, California—just outside of Fresno. George has been honored as an instructor by his students as well as his colleagues. Aside from teaching and writing, he served as the department chair for College of the Sequoias' math/engineering division from 1999–2004. Additionally, he is the author of *Elementary Algebra* and *Intermediate Algebra*.

Elementary and Intermediate Algebra

Third Edition

George Woodbury
College of the Sequoias

Addison-Wesley

Boston Columbus Indianapolis New York San Francisco Upper Saddle River

Amsterdam Cape Town Dubai London Madrid Milan Munich Paris Montreal Toronto

Delhi Mexico City Sao Paulo Sydney Hong Kong Seoul Singapore Taipei Tokyo

Editorial Director: Christine Hoag
Editor in Chief: Maureen O'Connor
Acquisitions Editor: Dawn Giovaniello
Executive Project Manager: Kari Heen
Senior Content Editor: Lauren Morse
Editorial Assistant: Chelsea Pingree
Media Producer: Carl Cottrell
Software Development: Mary Durnwald, Tanya Farber, and Eileen Moore
Production Project Manager: Beth Houston
Senior Design Supervisor: Andrea Nix
Cover Designer: Christina Gleason
Text Designer: Tamara Newnam
Production Services: PreMedia Global
Executive Marketing Manager: Michelle Renda
Associate Marketing Manager: Tracy Rabinowitz
Marketing Coordinator: Alicia Frankel
Prepress Services Buyer: Caroline Fell
Manufacturing Buyer: Linda Cox

Library of Congress Cataloging-in-Publication Data

Woodbury, George, 1967-
Elementary and intermediate algebra / George Woodbury. -- 3rd ed.
p. cm.
Includes index.
ISBN-13: 978-0-321-66548-5 (student ed.)
ISBN-10: 0-321-66548-1 (student ed.)
ISBN-13: 978-0-321-66584-3 (instructor ed.)
ISBN-10: 0-321-66584-8 (instructor ed.)
1. Algebra--Textbooks. I. Title.
QA152.3.W66 2012
512.9--dc22 2010002670

10 9 8 7 6 5 4

Prentice Hall
is an imprint of

www.pearsonhighered.com

ISBN-10: 0-321-66548-1
ISBN-13: 978-0-321-66548-5

To Tina, Dylan, and Alycia
You make everything meaningful and worthwhile—
yesterday, today, and tomorrow.

Contents

Preface

Dear Instructors,

Developmental mathematics is our students' gateway to academic success and a better life. I wrote this textbook with my own students and their success in mind. I hope you find that this textbook provides you with the necessary tools to help your students learn and increase their understanding of math.

I wrote this textbook as a combined algebra textbook from the ground up, rather than piecing together two separate elementary algebra and intermediate algebra textbooks. My goal was to create one book that would take students through elementary and intermediate algebra in a seamless fashion, eliminating much of the overlap between the two courses. The centerpiece of this strategy is Chapter 8: A Transition. This chapter reviews essential elementary algebra topics while quickly extending them to new intermediate algebra topics.

My approach to functions is "Early and Often." By introducing functions in Chapter 3, and including them in nearly every subsequent chapter, I give elementary algebra students plenty of time to get accustomed to function notation, evaluating functions, and graphing functions before the difficult topics of composition of functions and inverse functions.

I have been using MyMathLab® in my classes for over ten years, and I have personally witnessed the power of MyMathLab® to help students learn and understand mathematics. In fact, when I first decided to pursue writing this textbook, I chose Pearson as my publisher because I was such a strong advocate of incorporating MyMathLab® into developmental math classes.

As I have traveled throughout the country and have had the chance to speak with instructors, it has become clear to me that while it is quite easy for instructors to get started with MyMathLab®, many instructors needed to develop a strategy to effectively incorporate MyMathLab® into their classes to promote learning and understanding. Common questions include the following:

- *How long should a homework assignment be?*
- *Is homework sufficient, or should I incorporate quizzes?*
- *What portion of the overall grade should come from MyMathLab®?*
- *How do I incorporate MyMathLab® into a traditional course? Into an online course?*

To answer those and other questions, I have created a manual for instructors that focuses on strategies for successfully incorporating MyMathLab® into a course. In addition, I address many practical how-to questions. The manual is intended to help new instructors get started with MyMathLab® while at the same time helping those instructors who are experienced with MyMathLab® to use it in a more effective manner.

If you have questions or want to explore MyMathLab® further, feel free to visit my website: www.georgewoodbury.com. There you will find many helpful articles. You also can access my blog and e-mail me through the contact page or get in touch with me via Twitter or Facebook.

Best of luck this semester!

George Woodbury

NEW TO THIS EDITION

Responses from instructors and students have led to adjustments in the coverage and distribution of certain topics and encouraged expansion of the book's examples, exercises, and updated applications.

Content and Organization

- A new section covers basic statistics (Section 1.6).
- Dimensional analysis is introduced in Section 2.4 (Applications Involving Percentages; Ratio and Proportion), and additional coverage can be found online.
- The presentation of topics and objectives in Section 7.1 (Rational Expressions and Functions) has been reorganized.
- A general strategy for solving quadratic equations was added to Chapter 10 (Section 10.2, The Quadratic Formula).
- The presentation of topics in Section 10.4 (Graphing Quadratic Equations) was revised to clarify graphing quadratic equations in standard form.
- Starting in Section 11.3 (Quadratic Functions), a new approach to graphing that focuses on shifts and transitions instead of a point-plotting method is used throughout Chapters 11–13.
- Sections 12.1 (Exponential Functions) and 12.2 (Logarithmic Functions) focus on the basic graphs of $f(x) = b^x$ and $f(x) = \log_b x$. The graphs of $f(x) = b^{x-h} + k$ and $f(x) = \log_b (x - h) + k$ are then covered completely in Section 12.6 (Graphing Exponential and Logarithmic Functions).
- The Chapter Summaries have been expanded to a two-column procedure/example format.

Examples, Exercises, and Applications

- Additional Mixed Practice problems have been added throughout the text.
- Section 3.1 (The Rectangular Coordinate System; Equations in Two Variables) now has an additional example and exercises that use real-world data for plotting ordered pairs.
- The factoring exercises in Section 6.5 (Factoring Polynomials: A General Strategy) have been restructured.
- The coverage of factoring polynomials in Section 8.4 (Review of Factoring; Quadratic Equations and Rational Equations) has been expanded with additional examples and a General Factoring Strategy.
- An example that uses systems of two linear equations in two unknowns to solve real-world problems was added to Section 8.5 (Systems of Equations, Two Equations in Two Unknowns, Three Equations in Three Unknowns).
- New application problems on Body Surface Area (BSA) and distance to the horizon were added to Section 9.5 (Radical Equations and Applications of Radical Equations).

Resources for the Student and Instructor

- **George Woodbury's Guide to MyMathLab®** provides instructors with helpful ways to make the most out of their MyMathLab® experience. New and experienced users alike will benefit from George Woodbury's tips for implementing the many useful features available through MyMathLab®.
- The new **Guide to Skills and Concepts**, specifically designed for the Woodbury series, includes additional exercises and resources for every section of the text to help students make the transition from acquiring skills to learning concepts.

George Woodbury's Approach

The Transition from Elementary to Intermediate Algebra

This text was written as a combined book from the outset; it is not a merging of separate elementary and intermediate algebra texts. Chapter 8 (page 419) is representative of the author's direct approach to teaching elementary and intermediate algebra with purpose and consistency. Serving as the transition between the two courses, this chapter is designed to begin the intermediate algebra course by **reviewing and extending essential elementary algebra concepts** in order to **introduce new intermediate algebra topics.** Each section in Chapter 8 includes a review of one of the key topics in elementary algebra coupled with the introduction to an extension of that topic at the intermediate algebra level.

Early-and-Often Approach to Graphing and Functions

Woodbury introduces the primary algebraic concepts of graphing and functions early in the text (Chapter 3) and then consistently incorporates them throughout the text, providing optimal opportunity for their use and review. By introducing functions and graphing early, the text helps students become comfortable with reading and interpreting graphs and function notation. Working with these topics throughout the text establishes a basis for understanding that better prepares students for future math courses.

Practice Makes Perfect!

Examples Based on his experiences in the classroom, George Woodbury has included an abundance of clearly and completely worked-out examples.

Quick Checks The opportunity for practice shouldn't be designated only for the exercise sets. Every example in this text is immediately followed by a Quick Check exercise, allowing students to practice what they have learned and to assess their understanding of newly learned concepts. Answers to the Quick Check exercises are provided in the back of the book.

Exercises Woodbury's text provides more exercises than most other algebra texts, allowing students ample opportunity to develop their skills and increase their understanding. The exercise sets are filled with traditional skill-and drill exercises as well as unique exercise types that require thoughtful and creative responses.

Types of Exercises

Vocabulary Exercises: Each exercise set starts out with a series of exercises that check students' understanding of the basic vocabulary covered in the preceding section (page 77).

Mixed Practice Exercises: Mixed Practice exercises (the number of which has been increased in this edition) are provided as appropriate throughout the book to give students an opportunity to practice multiple types of problems in one setting. In these exercises, students are to determine the correct method used to solve a problem, thereby reducing their tendency to simply memorize steps to solve the problems for each objective (page 151).

Writing in Mathematics Exercises: Asking students to explain their answer in written form is an important skill that often leads to a higher level of understanding as acknowledged by the AMATYC Standards. At relevant points in each chapter, students also may be invited to write **Solutions Manual Exercises or Newsletter Exercises.** Solutions Manual exercises require students to solve a problem completely with step-by-step explanations as if they were writing their own solutions manual. Newsletter Exercises can be used to encourage students to be creative in their mathematical writing. Students are asked to explain a mathematical topic, and their explanation should be in the form of a short, visually appealing article that could be published in a newsletter that is read by people who are interested in learning mathematics (pages 79 and 110).

Quick Review Exercises: Appearing once per chapter, Quick Review Exercises are a short selection of review exercises aimed at helping students maintain the skills they learned previously and preparing them for upcoming concepts (page 101).

Applying Skills and Solving Problems Problem solving is a skill that is required daily in the real world and in mathematics. Based on George Pólya's text *How to Solve It*, George Woodbury presents a **six-step problem-solving strategy** in Chapter 2 that lays the foundation for solving applied problems. He then expands on this problem-solving strategy throughout the text by incorporating **hundreds of applied problems** on topics such as motion, geometry, and mixture. Interesting themes in the applied problems include investing and saving money, understanding sports statistics, landscaping, owning a home, and using a cell phone.

Building Your Study Strategies Woodbury introduces a Study Strategy in each chapter opener. The strategy is revisited and expanded upon prior to each section's exercise set in Building Your Study Strategy boxes and then again at the end of the chapter. These helpful Study Strategies outline good study habits and ask students to apply these skills as they progress through the textbook. Study Strategy topics include Study Groups, Using Your Textbook, Test Taking, and Overcoming Math Anxiety (pages 118, 125, and 192).

Mathematicians in History These activities provide a structured opportunity for students to learn about the rich and diverse history of mathematics. These short research projects, which ask students to investigate the life of a prominent mathematician, can be assigned as independent work or used as a collaborative learning activity (page 117).

Classroom Examples Having in-class practice problems at your fingertips is extremely helpful whether you are a new or experienced instructor. These instructor examples, called Classroom Examples, are included in the margins of the Annotated Instructor's Edition (page 64).

The optional **Using Your Calculator** feature is presented throughout the text, giving students guided calculator instruction (with screen shots as appropriate) to complement the material being covered (page 74).

A Word of Caution This feature, located throughout the text, help students avoid misconceptions by pointing out errors that students often make (page 93).

End-of-Chapter Content Each chapter concludes with a newly expanded **Chapter Summary**, a summary of the chapter's **Study Strategies, Chapter Review Exercises**, and a **Chapter Test**. Together these are an excellent resource for extra practice and test preparation. **Full solutions** to highlighted Chapter Review exercises are provided at the back of the text as yet another way for students to assess their understanding and check their work. A set of **Cumulative Review** exercises can be found after Chapters 4, 7, 11, and 14. These exercises are strategically placed to help students review for midterm and final exams.

Overview of Supplements

The supplements available to students and instructors are designed to provide the extra support needed to help students be successful. As you can see from the following list of supplements, all areas of support are covered—from tutoring help (Pearson Tutor Center) to guided solutions (video lectures and solutions manuals) to help in being a better math student. These additional supplements will help students master the skills, gain confidence in their mathematical abilities, and move on to the next course.

Student Supplements

Student's Solutions Manual

- Worked-out solutions for the odd-numbered section-level exercises
- Solutions to all problems in the Chapter Review, Chapter Test, and Cumulative Review exercises

ISBNs: 0-321-71562-4, 978-0-321-71562-3

Guide to Skills and Concepts

Includes the following resources for each section of the text to help students make the transition from acquiring skills to learning concepts:

- Learning objectives
- Vocabulary terms with fill-in-the-blank exercises
- Reading Ahead writing exercises
- Warm-up exercises
- Guided examples
- Extra practice exercises

ISBNs: 0-321-71563-2, 978-0-321-71563-0

Video Resources on DVD featuring Chapter Test Prep Videos

- Short lectures for each section of the text presented by author George Woodbury and Mark Tom
- Complete set of digitized videos on DVD-ROMs for student to use at home or on campus
- Ideal for distance learning or supplemental instruction
- Video lectures that include optional English captions
- Students can watch instructors work through step-by-step solutions to all the Chapter Test exercises from the textbook. Chapter Test Prep Videos are also available on YouTube™ (search using WoodburyElemIntAlg).
- Also available via MyMathLab®

ISBNs: 0-321-74542-6, 978-0-321-74542-2

Instructor Supplements

Annotated Instructor's Edition

- Answers to all exercises in the textbook
- Teaching Tips and Classroom Examples

ISBNs: 0-321-66584-8, 978-0-321-66584-3

NEW! George Woodbury's Guide to MyMathLab®

- Helpful tips for getting the most out of MyMathLab®, including quick-start guides and general how-to instructions, strategies for successfully incorporating MyMathLab® into a course, and more

ISBNs: 0-321-65353-X, 978-0-321-65353-6

Instructor's Resource Manual with Tests

- Two free-response tests per chapter and two multiple-choice tests per chapter
- Two free-response and two multiple-choice final exams
- Resources to help both new and adjunct faculty with course preparation and classroom management by offering helpful teaching tips correlated to the sections of the text
- Short quizzes for every section that can be used in class, for individual practice or for group work
- Full answers to Guide to Skills and Concepts
- Available in MyMathLab® and on the Instructor's Resource Center

Instructor's Solutions Manual

- Worked-out solutions to all section-level exercises
- Solutions to all Quick Check, Chapter Review, Chapter Test, and Cumulative Review exercises
- Available in MyMathLab® and on the Instructor's Resource Center

TestGen®

TestGen® (www.pearsoned.com/testgen) enables instructors to build, edit, print, and administer tests using a computerized bank of questions developed to cover all of the objectives of the text. TestGen® is algorithmically based, allowing instructors to create multiple but equivalent versions of the same question or test with the click of a button. Instructors also can modify test bank questions or add new questions. The software and test bank are available for download from Pearson Education's online catalog.

PowerPoint® Slides

- Key concepts and definitions from the text
- Available in MyMathLab® and on the Instructor's Resource Center

MyMathLab® Online Course (access code required)

MyMathLab® is a text-specific, easily customizable online course that integrates interactive multimedia instruction with textbook content. MyMathLab® provides the tools you need to deliver all or a portion of your course online, whether your students are working in a lab setting or from home.

- **Interactive homework exercises**, correlated to your textbook at the objective level, are algorithmically generated for unlimited practice and mastery. Most exercises are free-response and provide guided solutions, sample problems, and tutorial learning aids for extra help.
- **Personalized homework** assignments can be designed to meet the needs of your class. MyMathLab® tailors the assignment for each student based on his or her test or quiz scores. Each student receives a homework assignment that contains only the problems he or she still needs to master.
- **Personalized Study Plan**, generated when students complete a test, a quiz, or homework, indicates which topics have been mastered and links to tutorial exercises for topics students have not mastered. You can customize the Study Plan so that the topics available match your course content.
- **Multimedia learning aids**, (for example, video lectures and podcasts, animations, and a complete multimedia textbook) help students independently improve their understanding and performance. You can assign these multimedia learning aids as homework to help your students grasp the concepts.
- **Homework and Test Manager** lets you assign homework, quizzes, and tests that are automatically graded. Select just the right mix of questions from the MyMathLab® exercise bank, instructor-created custom exercises, and/or TestGen® test items.
- **Gradebook**, designed specifically for mathematics and statistics, automatically tracks students' results, lets you stay on top of student performance, and gives you control over how to calculate final grades. You also can add off-line (paper-and-pencil) grades to the gradebook.
- **MathXL® Exercise Builder** allows you to create static and algorithmic exercises for your online assignments. You can use the library of sample exercises as an easy starting point, or you can edit any course-related exercise.
- **Pearson Tutor Center** (www.pearsontutorservices.com) access is automatically included with MyMathLab®. The Tutor Center is staffed by qualified math instructors who provide textbook-specific tutoring for students via toll-free phone, fax, e-mail, and interactive Web sessions.

MathXL® Online Course (access code required)

MathXL® is a powerful online homework, tutorial, and assessment system that accompanies Pearson Education's textbooks in mathematics and statistics. With MathXL®, instructors can:

- Create, edit, and assign online homework and tests using algorithmically generated exercises correlated at the objective level to the textbook.
- Create and assign their own online exercises and import TestGen® tests for added flexibility.
- Maintain records of all student work tracked in MathXL's online gradebook.

With MathXL®, students can:

- Take chapter tests in MathXL® and receive personalized study plans and/or personalized homework assignments based on their test results.

- Use the study plan and/or the homework to link directly to tutorial exercises for the objectives they need to study.
- Access supplemental animations and video clips directly from selected exercises.

MathXL® is available to qualified adopters. For more information, visit the website at www.mathxl.com or contact your Pearson representative.

Acknowledgments

Writing this textbook was a monumental task, and I would like to take this opportunity to thank everyone who helped me along the way. The following reviewers provided thoughtful suggestions and were instrumental in the development of *Elementary and Intermediate Algebra*, Third Edition.

Frederick Adkins *Indiana University of Pennsylvania*
Jan Archibald *Ventura County Community College District*
Madelaine Bates *Bronx Community College*
Scott Beslin *Nicholls State University*
Norma Bisulca *University of Maine at Augusta*
Kevin Bodden *Lewis and Clark Community College*
Megan Bomer *Illinois Central College*
Shirley Brown *Weatherford College*
Dr. Brett J. Butler *Horry-Georgetown Technical College*
Linh Tran *Changaris Jefferson Community and Technical College*
Ivette Chuca El *Paso Community College*
Theodore Cluver *California State University, Chico*
Yong S. Colen *Indiana University of Pennsylvania*
Victor Cornell *Phoenix College*
Douglas Culler *Midlands Technical College*
Benjamin Donath *Buena Vista University*
Barbara Duncan *Hillsborough Community College*
Guangwei Fan *Maryville University*
Gail Small Ferrell *Truckee Meadows Community College*
Donna Flint *South Dakota State University*
Dolen Freeouf *Southeast Community College*
Deborah Donovan Fries *Wor-Wic Community College*
Matthew Gardner *North Hennepin Community College*
Wendy Grooms *Howard Payne University*
Pauline Hall *Iowa State University*
William S. Harmon *El Paso Community College*
Rosanne Harris *The College of Staten Island, CUNY*
Brian Heck *Nicholls State University*
Devin Henson *Midlands Technical College*
Nicole Holden Kennebec *Valley Community College*
Gary Hughes Iowa *Western Community College*
Marilyn Jacobi *Gateway Community College*
Rose Jenkins *Midlands Technical College*
Sheryl Johnson *Johnston Community College*
Brian Karasek *South Mountain Community College*
Kimberly Kuster-Smith *Hudson Valley Community College*
Maria Ortiz Kelly *Reedley College*
Gary S Kersting *North Central Michigan College*
Harriet Kiser *Georgia Highlands College*
Kathy Kopelousos *Lewis and Clark Community College*
Tamela Kostos *McHenry County College*
Kimberly Kuster-Smith *Hudson Valley Community College*
Betty J. Larson *South Dakota State University*
R. Warren Lemerich *Laramie County Community College*
Mickey Levendusky *Pima County Community College, Downtown*
Janna Liberant *SUNY/Rockland Community College*
R. Scott Linder *Ohio Wesleyan University*
DeAnna McAleer, Ed.D. *University of Maine at Augusta*
Timothy McKenna *University of Michigan–Dearborn*
Andrew McKintosh *Glendale Community College*
Monica Meissen *Clarke College*
James M. Meyer *University of Wisconsin – Green Bay*
Erika Miller *Towson University*
Chris Milner *Clark College*
Deborah Mixson-Brookshire *Kennesaw State University*
Gary Motta *Lassen Community College*
Carol Murphy *San Diego Miramar College*
Sanjivendra ("Scotty") Nath *McHenry County College*
Dana Onstad *Midlands Technical College*

Gail Opalinski *University of Alaska, Anchorage*
JoAnn Paderi *Glendale Community College*
Lourdes Pajo *Pikes Peak Community College*
Ramakrishna Polepeddi *Westwood College, Denver North Campus*
Sharonda Burns Ragland *ECPI College of Technology*
Kim Rescorla *Eastern Michigan University*
Daniel Schaal *South Dakota State University*
Kathryn G. Shafer, Ph.D. *Bethel College*
Pavel Solin *The University of Texas at El Paso*
Stephen M. Son *Dyersburg State Community College*
Dina Spain *Horry-Georgetown Technical College*
Fereja Tahir *Illinois Central College*
Linda Tansil *Southeast Missouri State University*
Jane H. Theiling *Dyersburg State Community College*
Mary Lou Townsend *Wor-Wic Community College*
Linda Tucker *Rose State College*
Gary B. Turner *Rochester College*
Vivian Turner *Rochester College*
Giovanni Viglino *Ramapo College of New Jersey*
Cindie Wade *St. Clair County Community College*
Patrick Ward *Illinois Central College*
Gary Wardall *University of Wisconsin – Green Bay*
C Timothy Weier *North Central Michigan College*
Floyd Wouters *University of Wisconsin Oshkosh*

I have truly enjoyed working with the team at Pearson Education. I do owe special thanks to my editor, Dawn Giovaniello, as well as Lauren Morse, Chelsea Pingree, Mary St. Thomas, Jon Wooding, Beth Houston, Michelle Renda, Tracy Rabinowitz, and Carl Cottrell. Thanks are due to Greg Tobin and Maureen O'Connor for believing in my vision and taking a chance on me and to Susan Winslow and Jenny Crum for getting this all started. Stephanie Logan Collier's assistance during the production process was invaluable, and Gary Williams, Carrie Green, and Irene Duranczyk deserve credit for their help as accuracy checkers.

Thanks to Jared Burch, Chris Keen, Vineta Harper, Mark Tom, Don Rose, and Ross Rueger, my colleagues at College of the Sequoias, who have provided great advice along the way and frequently listened to my ideas. I also would like to thank my students for keeping my fires burning. It truly is all about the students.

Most importantly, thanks to my wife Tina and our wonderful children Dylan and Alycia. They are truly my greatest blessing, and I love them more than words can say. The process of writing a textbook is long and difficult, and they have been supportive and understanding at every turn.

Finally, this book is dedicated to my nephew Pat Slade and to the memory of my wife's grandmother Miriam Spaulding. Pat is one of the strongest men I know, and his journey is always foremost in our thoughts. We are forever in debt to Miriam—she showed us the value of hard work and empathy, and we miss her greatly.

George Woodbury

CHAPTER 1

Review of Real Numbers

This chapter reviews properties of real numbers and arithmetic that are necessary for success in algebra. The chapter also introduces several algebraic properties.

STUDY STRATEGY

Study Groups Throughout this book, study strategies will help you learn and be successful in this course. This chapter will focus on getting involved in a study group.

Working with a study group is an excellent way to learn mathematics, improve your confidence and level of interest, and improve your performance on quizzes and tests. When working with a group, you will be able to work through questions about the material you are studying. Also, by being able to explain how to solve a particular problem to another person in your group, you will increase your ability to retain this knowledge.

We will revisit this study strategy throughout this chapter so you can incorporate it into your study habits. See the end of Section 1.1 for tips on how to get a study group started.

1.1 Integers, Opposites, and Absolute Value

OBJECTIVES

1. **Graph whole numbers on a number line.**
2. **Determine which is the greater of two whole numbers.**
3. **Graph integers on a number line.**
4. **Find the opposite of an integer.**
5. **Determine which is the greater of two integers.**
6. **Find the absolute value of an integer.**

A **set** is a collection of objects, such as the set consisting of the numbers 1, 4, 9, and 16. This set can be written as $\{1, 4, 9, 16\}$. The braces, $\{\ \}$, are used to denote a set, and the values listed inside are said to be **elements**, or members, of the set. A set with no elements is called the **empty set** or **null set**. A **subset** of a set is a collection of some or all of the elements of the set. For example, $\{1, 9\}$ is a subset of the set $\{1, 4, 9, 16\}$. A subset also can be an empty set.

Whole Numbers

Objective 1 Graph whole numbers on a number line. For the most part, this text deals with the set of real numbers. The set of real numbers is made up of the set of rational numbers and the set of irrational numbers.

Rational Numbers

A **rational number** is a number that can be expressed as a fraction, such as $\frac{3}{4}$ and $\frac{2}{9}$. Decimal numbers that terminate, such as 2.57, and decimal numbers that repeat, such as 0.444 . . . , are also rational numbers.

Irrational Numbers

An **irrational number** is a number that cannot be expressed as a fraction, but instead is a decimal number that does not terminate or repeat. The number π is an example of an irrational number: $\pi = 3.14159\ldots$.

One subset of the set of real numbers is the set of natural numbers.

Natural Numbers

The set of **natural numbers** is the set $\{1, 2, 3, \ldots\}$.

If we include the number 0 with the set of natural numbers, we have the set of **whole numbers**.

Whole Numbers

The set of **whole numbers** is the set $\{0, 1, 2, 3, \ldots\}$. This set can be displayed on a number line as follows:

The arrow on the right-hand side of the number line indicates that the values continue to increase in this direction. There is no largest whole number, but we say that the values approach infinity (∞).

To graph any particular number on a number line, we place a point, or dot, at that location on the number line.

EXAMPLE 1 Graph the number 6 on a number line.

SOLUTION To graph any number on a number line, place a point at that number's location.

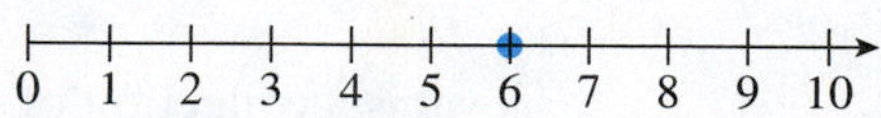

Quick Check 1

Graph the number 4 on a number line.

Inequalities

Objective 2 Determine which is the greater of two whole numbers. When comparing two whole numbers a and b, we say that a is **greater than** b, denoted $a > b$, if the number a is to the right of the number b on the number line. The number a is **less than** b, denoted $a < b$, if a is to the left of b on the number line. The statements $a > b$ and $a < b$ are called **inequalities**.

EXAMPLE 2 Write the appropriate symbol, $<$ or $>$, between the following: 6 _____ 4

SOLUTION Let's take a look at the two values graphed on a number line.

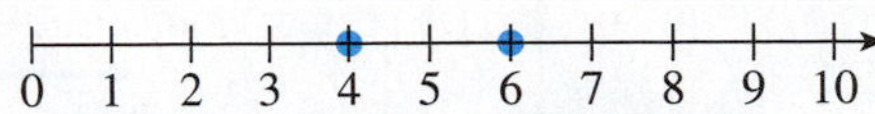

Because the number 6 is to the right of the number 4 on the number line, 6 is greater than 4. So $6 > 4$.

EXAMPLE 3 Write the appropriate symbol, $<$ or $>$, between the following: 2 _____ 5

SOLUTION Because 2 is to the left of 5 on the number line, $2 < 5$.

Quick Check 2

Write the appropriate symbol, $<$ or $>$, between the following:

a) 8 _____ 3

b) 19 _____ 23

Integers

Objective 3 Graph integers on a number line. Another important subset of the real numbers is the set of integers.

Integers

The set of **integers** is the set $\{\ldots, -3, -2, -1, 0, 1, 2, 3, \ldots\}$. We can display the set of integers on a number line as follows:

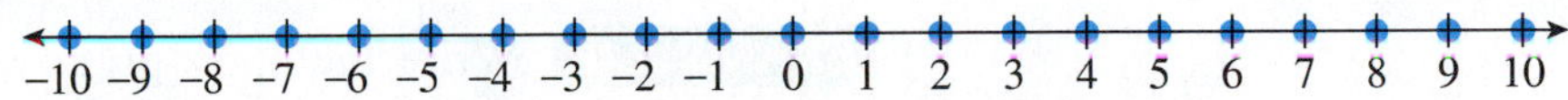

The arrow on the left side indicates that the values continue to decrease in this direction, and they are said to approach negative infinity $(-\infty)$.

Opposites

Objective 4 Find the opposite of an integer. The set of integers is the set of whole numbers together with the opposites of the natural numbers. The **opposite** of a number is a number on the other side of 0 on the number line and the same distance from 0 as that number. We denote the opposite of a real number a as $-a$. For example, -5 and 5 are opposites because both are 5 units away from 0 and one is to the left of 0 while the other is to the right of 0.

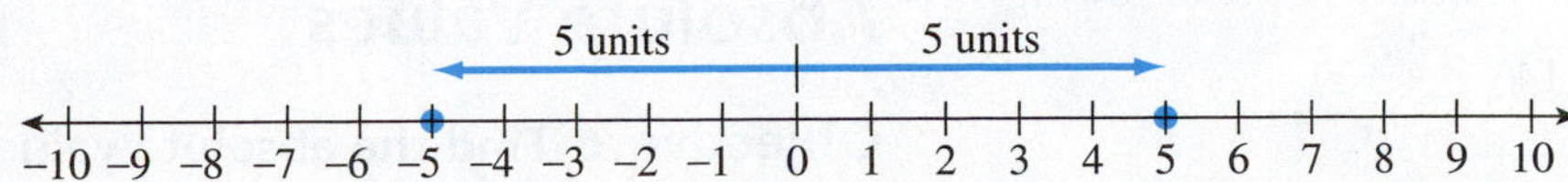

Numbers to the left of 0 on the number line are called **negative numbers**. Negative numbers represent a quantity less than 0. For example, if you have written checks that the balance in your checking account cannot cover, your balance will be a negative number. A temperature that is below 0° F, a golf score that is below par, and an elevation that is below sea level are other examples of quantities that can be represented by negative numbers.

EXAMPLE 4 What is the opposite of 7?

SOLUTION The opposite of 7 is −7 because −7 also is 7 units away from 0 but is on the opposite side of 0.

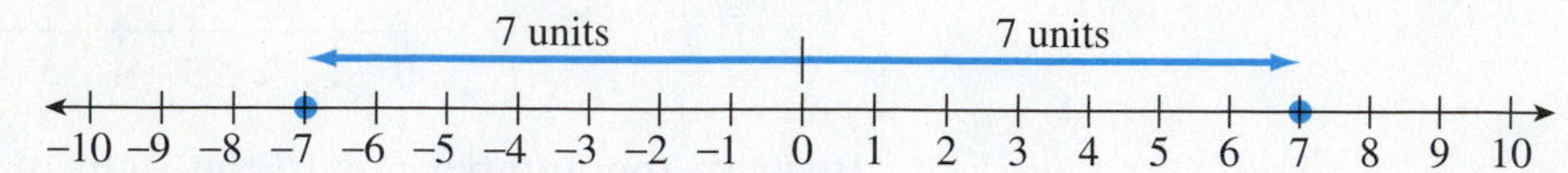

EXAMPLE 5 What is the opposite of −6?

SOLUTION The opposite of −6 is 6.

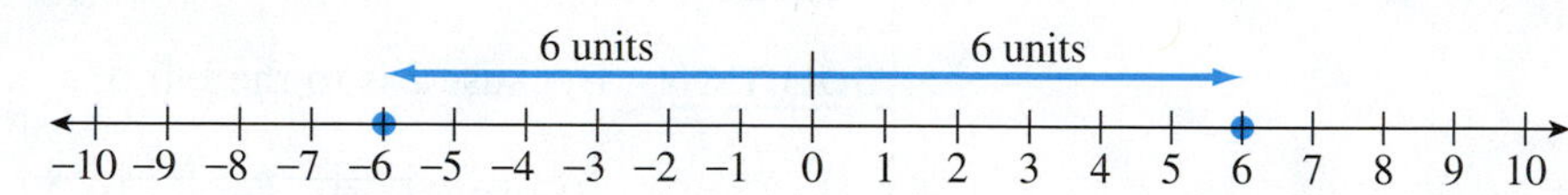

Quick Check 3

Find the opposite of the given integer.

a) −13 **b)** 8

The opposite of 0 is 0 itself. Zero is the only number that is its own opposite.

Inequalities with Integers

Objective 5 Determine which is the greater of two integers. Inequalities for integers follow the same guidelines as they do for whole numbers. If we are given two integers a and b, the number that is greater is the number that is to the right on the number line.

EXAMPLE 6 Write the appropriate symbol, $<$ or $>$, between the following: −3 ____ 5

SOLUTION Looking at the number line, we can see that −3 is to the left of 5; so $-3 < 5$.

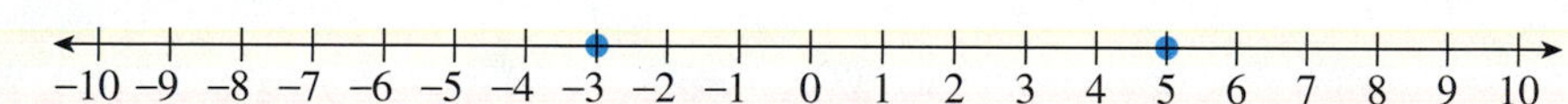

EXAMPLE 7 Write the appropriate symbol, $<$ or $>$, between the following: −2 ____ −7

SOLUTION On the number line, −2 is to the right of −7; so $-2 > -7$.

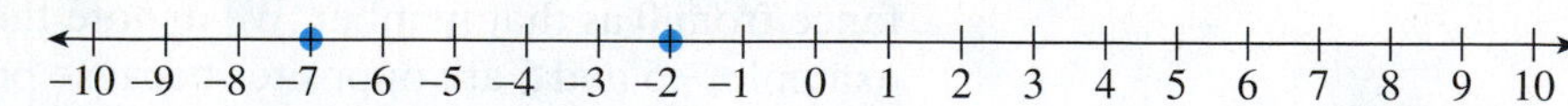

Quick Check 4

Write the appropriate symbol, $<$ or $>$, between the following:

a) −14 ____ −11
b) 6 ____ −20

Absolute Values

Objective 6 Find the absolute value of an integer.

Absolute Value

The **absolute value** of a number a, denoted $|a|$, is the distance between a and 0 on the number line.

Distance cannot be negative, so the absolute value of a number a is always 0 or higher.

EXAMPLE 8 Find the absolute value of 6.

SOLUTION The number 6 is 6 units away from 0 on the number line, so $|6| = 6$.

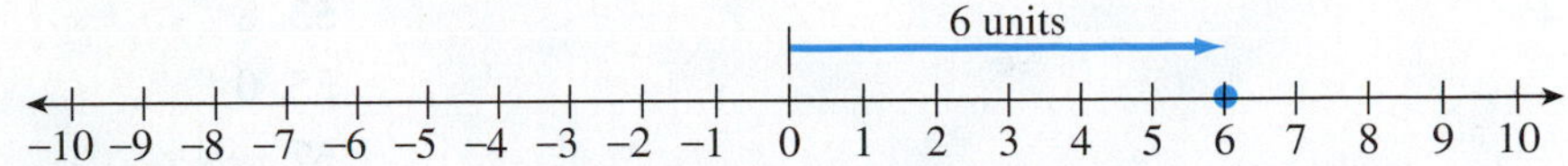

EXAMPLE 9 Find the absolute value of −4.

SOLUTION The number −4 is 4 units away from 0 on the number line, so $|-4| = 4$.

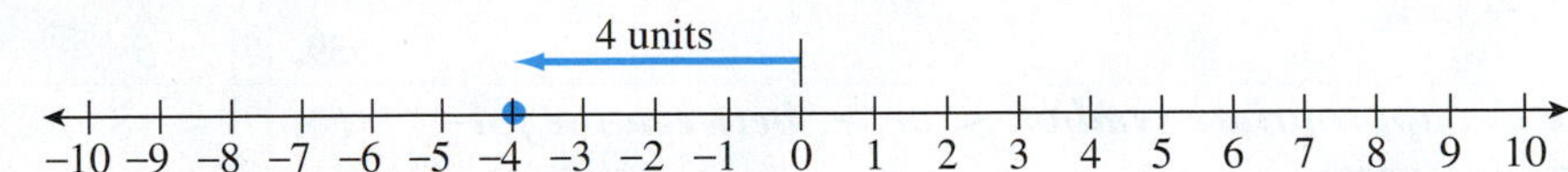

Quick Check 5

Find the absolute value of −9.

BUILDING YOUR STUDY STRATEGY

Study Groups, 1 With Whom to Work? To form a study group, you must begin with this question: With whom do I want to work? Look for students who are serious about learning, who are prepared for each class, and who ask intelligent questions during class.

Look for students with whom you believe you can get along. You are about to spend a great deal of time working with this group, sometimes under stressful conditions.

If you take advantage of tutorial services provided by your college, keep an eye out for classmates who do the same. There is a strong chance that classmates who use the tutoring center are serious about learning mathematics and earning good grades.

Exercises 1.1

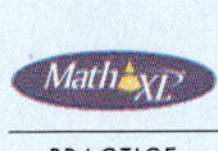

Vocabulary

1. A set with no elements is called the ________.
2. A number m is ______ than another number n if it is located to the left of n on a number line.
3. The arrow on the right side of a number line indicates that the values approach ________.
4. $c > d$ if c is located to the ________ of d on a number line.

Graph the following whole numbers on a number line.

5. 7
6. 3
7. 6
8. 9
9. 2
10. 13

Write the appropriate symbol, < or >, between the following whole numbers.

11. 3 ___ 13
12. 7 ___ 9
13. 8 ___ 6
14. 12 ___ 5
15. 45 ___ 42
16. 33 ___ 37

Graph the following integers on a number line.

17. −4
18. −7
19. 5

20. -9

21. -12

22. 4

Find the opposite of the following integers.

23. -7 **24.** 5
25. 22 **26.** -13
27. 0 **28.** -39

Write the appropriate symbol, < or >, between the following integers.

29. -7 ___ -9 **30.** -5 ___ -2
31. -13 ___ -11 **32.** -8 ___ -14
33. -16 ___ 0 **34.** 5 ___ -3
35. 9 ___ -14 **36.** -10 ___ 6

Find the following absolute values.

37. $|-15|$ **38.** $|9|$
39. $|0|$ **40.** $|-6|$
41. $|-7|$ **42.** $|-12|$
43. $-|7|$ **44.** $-|12|$
45. $-|-29|$ **46.** $-|-8|$

Write the appropriate symbol, < or >, between the following integers.

47. $|-7|$ ___ 4 **48.** $|-17|$ ___ $|13|$
49. -16 ___ $|-16|$ **50.** $|8|$ ___ $|-19|$
51. $-|-24|$ ___ $|-47|$ **52.** $|-8|$ ___ $-|8|$

Identify whether the given number is a member of the following sets of numbers: A. natural numbers, B. whole numbers, C. integers, D. real numbers.

53. 8 **54.** -6
55. 0 **56.** 3.14
57. -9 **58.** 20

Find the missing number if possible. There may be more than one number that works, so find as many as possible. There may be no number that works.

59. $|?| = 5$
60. $|?| = 18$
61. $|?| = -7$
62. $|?| = 0$
63. $|?| - 8 = 6$
64. $4 \cdot |?| + 3 = 27$

Writing in Mathematics

Answer in complete sentences.

65. A fellow student tells you that to find the absolute value of any number, make the number positive. Is this always true? Explain in your own words.

66. True or false: The opposite of the opposite of a number is the number itself.

67. If the opposite of a nonzero integer is equal to the absolute value of that integer, is the integer positive or negative? Explain your reasoning.

68. If an integer is less than its opposite, is the integer positive or negative? Explain your reasoning.

1.2 Operations with Integers

OBJECTIVES

1. Add integers.
2. Subtract integers.
3. Multiply integers.
4. Divide integers.

Addition and Subtraction of Integers

Objective 1 Add integers. Using the number line can help us learn how to add and subtract integers. Suppose we are trying to add the integers 3 and -7, which could be written as $3 + (-7)$. On a number line, we will start at 0 and move 3 units in the positive, or right, direction. Adding -7 tells us to move 7 units in the negative, or left, direction.

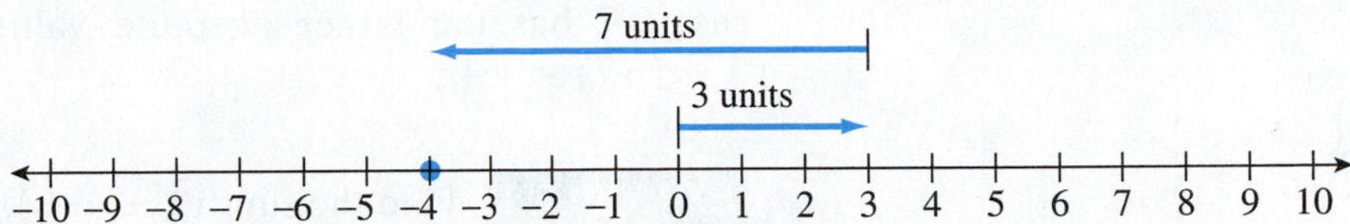

Ending up at -4 tells us that $3 + (-7) = -4$.

We can use a similar approach to verify an important property of opposites: the sum of two opposites is equal to 0.

Sum of Two Opposites

For any real number a, $a + (-a) = 0$.

Suppose that we want to add the opposites 4 and -4. Using the number line, we begin at 0 and move 4 units to the right. We then move 4 units to the left, ending at 0. So $4 + (-4) = 0$.

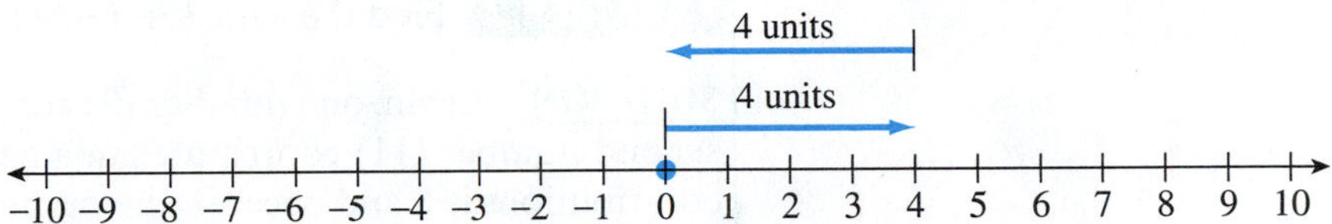

We also can see that $3 + (-7) = -4$ through the use of manipulatives, which are hands-on tools used to demonstrate mathematical properties. Suppose we had a bag of green and red candies. Let each piece of green candy represent a positive 1 and each piece of red candy represent a negative 1. To add $3 + (-7)$, we begin by combining 3 green candies (positive 3) with 7 red candies (negative 7). Combining 1 red candy with 1 green candy has a net result of 0, as the sum of two opposites is equal to 0. So each time we make a pair of a green candy and a red candy, these two candies cancel each other's effect and can be discarded. After doing this, we are left with 4 red candies. The answer is -4.

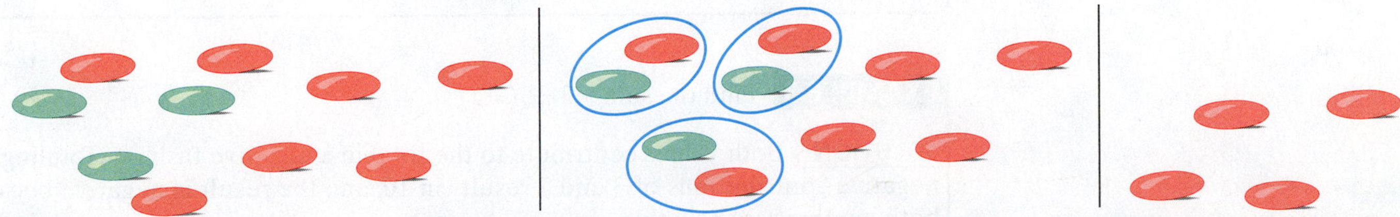

Now we will examine another technique for finding the sum of a positive number and a negative number. In the sum $3 + (-7)$, the number 3 contributes to the sum in a positive fashion while the number -7 contributes to the sum in a negative fashion. The two numbers contribute to the sum in an opposite manner. We can think of the sum as the difference between these two contributions.

Adding a Positive Number and a Negative Number

1. **Take the absolute value of each number and find the difference between these two absolute values.** This is the difference between the two numbers' contributions to the sum.
2. **Note that the sign of the result is the same as the sign of the number that has the largest absolute value.**

For the sum $3 + (-7)$, we begin by taking the absolute value of each number: $|3| = 3$, $|-7| = 7$. The difference between the absolute values is 4. The sign of the sum is the same as the sign of the number that has the larger absolute value. In this case, -7 has the larger absolute value, so the result is negative. Therefore, $3 + (-7) = -4$.

EXAMPLE 1 Find the sum $12 + (-8)$.

SOLUTION

$\lvert 12\rvert = 12; \quad \lvert 8\rvert = 8$	Find the absolute value of each number.
$12 - 8 = 4$	The difference between the absolute values is 4.
$12 + (-8) = 4$	Because the number with the larger absolute value is positive, the result is positive.

Quick Check 1

Find the sum $14 + (-6)$.

Notice in the previous example that $12 + (-8)$ is equivalent to $12 - 8$, which also equals 4. What the two expressions have in common is that there is one number (12) contributes to the total in a positive fashion and a second number (8) that contributes to the total in a negative fashion.

EXAMPLE 2 Find the sum $3 + (-11)$.

SOLUTION Again, one number (3) contributes to the total in a positive way and a second number (11) contributes in a negative way. The difference between their contributions is 8 and because the number making the larger contribution is negative, the result is -8.

$$3 + (-11) = -8$$

Quick Check 2

Find the sum $4 + (-17)$.

Note that $-11 + 3$ also equals -8. The rules for adding a positive integer and a negative integer still apply when the first number is negative and the second number is positive.

Adding Two (or More) Negative Numbers

1. Total the negative contributions of each number.
2. Note that the sign of the result is negative.

EXAMPLE 3 Find the sum $-3 + (-7)$.

SOLUTION Both values contribute to the total in a negative fashion. Totaling the negative contributions of 3 and 7 results in 10, and the result is negative because both numbers are negative.

$$-3 + (-7) = -10$$

Quick Check 3

Find the sum $-2 + (-9)$.

Objective 2 Subtract integers. To subtract a negative integer from another integer, we use the following property:

Subtraction of Real Numbers

For any real numbers a and b, $a - b = a + (-b)$.

This property says that adding the opposite of b to a is the same as subtracting b from a. Suppose we are subtracting a negative integer, as in the example $-8 - (-19)$. The property for subtraction of real numbers says that subtracting -19 is the same as adding its opposite (19); so we convert this subtraction to $-8 + 19$. Remember that subtracting a negative number is equivalent to adding a positive number.

EXAMPLE 4 Subtract $6 - (-27)$.

SOLUTION

$6 - (-27) = 6 + 27$ Subtracting -27 is the same as adding 27.

$= 33$ Add.

Quick Check 4

Subtract $11 - (-7)$.

General Strategy for Adding/Subtracting Integers

- Rewrite "double signs." *Adding a negative number,* $4 + (-5)$, *can be rewritten as subtracting a positive number,* $4 - 5$. *Subtracting a negative number,* $-2 - (-7)$, *can be rewritten as adding a positive number,* $-2 + 7$.
- Look at each integer and determine whether it is contributing positively or negatively to the total.
- Add any integers contributing positively to the total, resulting in a single positive integer. In a similar fashion, add all integers that are contributing to the total negatively, resulting in a single negative integer. Finish by finding the sum of these two integers.

Rather than saying to add or subtract, the directions for a problem may state to "simplify" a numerical expression. To **simplify** an expression means to perform all arithmetic operations.

EXAMPLE 5 Simplify $17 - (-11) - 6 + (-13) - (-21) + 3$.

SOLUTION Begin by working on the *double signs.* This produces the following:

$17 - (-11) - 6 + (-13) - (-21) + 3$

$= 17 + 11 - 6 - 13 + 21 + 3$ Rewrite double signs.

$= 52 - 19$ The four integers that contribute in a positive fashion (17, 11, 21, and 3) total 52. The two integers that contribute in a negative fashion total -19.

$= 33$ Subtract.

Quick Check 5

Simplify $14 - 9 - (-22) - 6 + (-30) + 5$.

Using Your Calculator When using your calculator, you must be able to distinguish between the subtraction key and the key for a negative number. On the TI-84, the subtraction key [−] is listed above the addition key on the right side of the calculator, while the negative key [(-)] is located to the left of the [ENTER] key at the bottom of the calculator. Here are two ways to simplify the expression from the previous example using the TI-84.

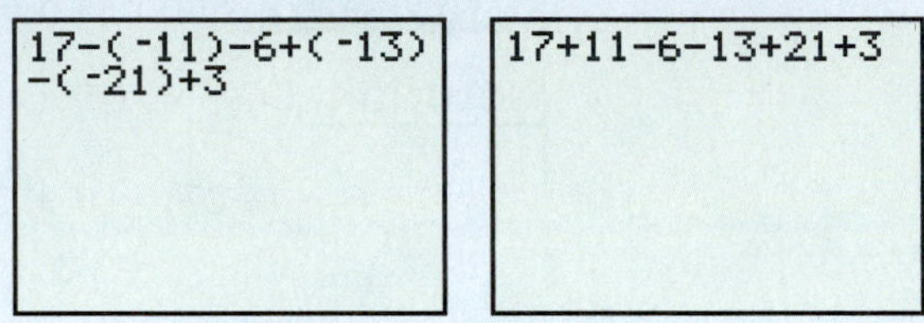

In either case, pressing the [ENTER] key produces the result 33.

Multiplication and Division of Integers

Objective 3 Multiply integers. The result obtained when multiplying two numbers is called the **product** of the two numbers. The numbers that are multiplied are called **factors**. When we multiply two positive integers, their product also is a positive integer. For example, the product of the two positive integers 4 and 7 is the positive integer 28. This can be written as $4 \cdot 7 = 28$. The product $4 \cdot 7$ also can be written as 4(7) or (4)(7).

The product $4 \cdot 7$ is another way to represent the repeated addition of 7 four times.

$$\begin{aligned} 4 \cdot 7 &= 7 + 7 + 7 + 7 \\ &= 28 \end{aligned}$$

This concept can be used to show that the product of a positive integer and a negative integer is a negative integer. Suppose we want to multiply 4 by −7. We can rewrite this as −7 being added four times, or $(-7) + (-7) + (-7) + (-7)$. From our work earlier in this section, we know that this total is −28; so $4(-7) = -28$. Anytime we multiply a positive integer and by a negative integer, the result is negative. So $(-7)(4)$ also is equal to −28.

Products of Integers

(Positive) · (Negative) = Negative
(Negative) · (Positive) = Negative

EXAMPLE 6 Multiply $5(-8)$.

SOLUTION We begin by multiplying 5 and 8, which equals 40. The next step is to determine the sign of the result. Whenever we multiply a positive integer by a negative integer, the result is negative.

$$5(-8) = -40$$

Quick Check 6
Multiply $10(-6)$.

A WORD OF CAUTION Note the difference between $5 - 8$ (a subtraction) and $5(-8)$ (a multiplication). A set of parentheses *without* a sign in front of them is used to imply multiplication.

Product	Result
$(3)(-5)$	-15
$(2)(-5)$	-10
$(1)(-5)$	-5
$(0)(-5)$	0
$(-1)(-5)$	?
$(-2)(-5)$	?

The product of two negative integers is a positive integer. Let's try to understand why this is true by considering the example $(-2)(-5)$. Examine the table to the left, which shows the products of some integers and -5.

Notice the pattern in the table. Each time the integer multiplied by -5 decreases by 1, the product increases by 5. As we go from $0(-5)$ to $(-1)(-5)$, the product should increase by 5. So $(-1)(-5) = 5$ and, by the same reasoning, $(-2)(-5) = 10$.

Product of Two Negative Integers

$$(\text{Negative}) \cdot (\text{Negative}) = \text{Positive}$$

EXAMPLE 7 Multiply $(-9)(-8)$.

SOLUTION We begin by multiplying 9 and 8, which equals 72. The next step is to determine the sign of the result. Whenever we multiply a negative integer by a negative integer, the result is positive.

$$(-9)(-8) = 72$$

Quick Check 7

Multiply $(-7)(-9)$.

Products of Integers

- If a product contains an *odd number* of negative factors, the result is negative.
- If a product contains an *even number* of negative factors, the result is positive.

The main idea behind this principle is that every two negative factors multiply to be positive. If there are three negative factors, the product of the first two is a positive number. Multiplying this positive product by the third negative factor produces a negative product.

EXAMPLE 8 Multiply $7(-2)(-5)(-3)$.

SOLUTION Because there are three negative factors, the product will be negative.

$$7(-2)(-5)(-3) = -210$$

Quick Check 8

Multiply $-4(-10)(5)(-2)$.

Using Your Calculator Here is how the screen looks when you are using the TI-84 to multiply the expression in the previous example:

```
7(-2)(-5)(-3)
            -210
```

Notice that parentheses can be used to indicate multiplication without using the [×] key.

Before continuing on to division, let's consider multiplication by 0. Any real number multiplied by 0 is 0; this is the multiplication property of 0.

Multiplication Property of 0

For any real number x,

$$0 \cdot x = 0$$
$$x \cdot 0 = 0.$$

Objective 4 Divide integers. When dividing one number called the **dividend** by another number called the **divisor**, the result obtained is called the **quotient** of the two numbers:

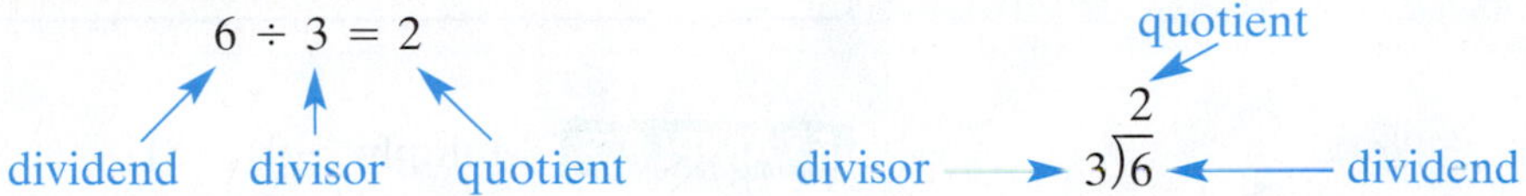

The statement "6 divided by 3 is equal to 2" is true because the product of the quotient and the divisor, $2 \cdot 3$, is equal to the dividend 6.

$$6 \div 3 = 2 \longleftrightarrow 2 \cdot 3 = 6$$

When we divide two integers that have the same sign (both positive or both negative), the quotient is positive. When we divide two integers that have different signs (one negative, one positive), the quotient is negative. Note that this is consistent with the rules for multiplication.

Quotients of Integers

(Positive) ÷ (Positive) = Positive (Positive) ÷ (Negative) = Negative

(Negative) ÷ (Negative) = Positive (Negative) ÷ (Positive) = Negative

EXAMPLE 9 Divide $(-54) \div (-6)$.

SOLUTION When we divide a negative number by another negative number, the result is positive.

$$(-54) \div (-6) = 9 \quad \text{Note that } -6 \cdot 9 = -54.$$

EXAMPLE 10 Divide $(-33) \div 11$.

SOLUTION When we divide a negative number by a positive number, the result is negative.

$$(-33) \div 11 = -3$$

Quick Check 9

Divide $72 \div (-8)$.

Whenever 0 is divided by any integer (except 0), the quotient is 0. For example, $0 \div 16 = 0$. We can check that this quotient is correct by multiplying the quotient by the divisor. Because $0 \cdot 16 = 0$, the quotient is correct.

Division by Zero

Whenever an integer is divided by 0, the quotient is said to be **undefined**.

Use the word **undefined** to state that an operation cannot be performed or is meaningless. For example, $41 \div 0$ is undefined. Suppose there was a real number a for which $41 \div 0 = a$. In that case, the product $a \cdot 0$ would be equal to 41. Because the product of 0 and any real number is equal to 0, such a number a does not exist.

BUILDING YOUR STUDY SKILLS

Study Groups, 2 **When to Meet** Once you have formed a study group, determine where and when to meet. It is a good idea to meet at least twice a week for at least an hour per session. Consider a location where quiet discussion is allowed, such as the library or tutorial center.

Some groups like to meet the hour before class, using the study group as a way to prepare for class. Other groups prefer to meet the hour after class, allowing them to go over material while it is fresh in their minds. Another suggestion is to meet at a time when your instructor is holding office hours.

Exercises 1.2

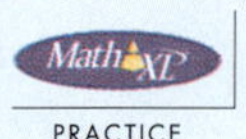
PRACTICE

WATCH

DOWNLOAD
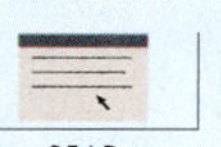
READ

REVIEW

Vocabulary

1. When finding the sum of a positive integer and a negative integer, the sign of the result is determined by the sign of the integer with the ________ absolute value.
2. The sum of two negative integers is a(n) ________ integer.
3. Subtracting a negative integer can be rewritten as adding a(n) ________ integer.
4. The product of a positive integer and a negative integer is a(n)________ integer.
5. The product of a negative integer and a negative integer is a(n) ________ integer.
6. If a product contains a(n) ________ number of negative integers, the product is negative.
7. In a division problem, the number you divide by is called the ________.
8. Division by 0 results in a quotient that is ________.

Add.

9. $8 + (-13)$
10. $16 + (-11)$
11. $6 + (-33)$
12. $52 + (-87)$
13. $(-4) + 5$
14. $-9 + 2$
15. $-14 + 22$
16. $(-35) + 50$
17. $-5 + (-6)$
18. $-9 + (-9)$

Subtract.

19. $8 - 6$
20. $13 - 9$
21. $5 - 11$
22. $4 - 12$
23. $(-5) - 3$
24. $(-9) - 6$
25. $-9 - 13$
26. $-47 - 16$
27. $36 - (-25)$
28. $64 - (-19)$
29. $-42 - (-33)$
30. $-27 - (-60)$

Simplify.

31. $8 - 13 - 6$
32. $-6 + 12 - 20$
33. $-9 + 7 - 4$
34. $-5 - 8 + 23$
35. $6 - (-16) + 5$
36. $18 - 21 - (-62)$
37. $4 + (-15) - 13 - (-25)$
38. $-13 + (-12) - (-1) - 29$
39. A mother with \$30 in her purse paid \$22 for her family to go to a movie. How much money did she have remaining?
40. A student had \$60 in his checking account prior to writing an \$85 check to the bookstore for books and supplies. What is his account's new balance?
41. The temperature at 6 A.M. in Fargo, North Dakota, was $-8°$C. By 3 P.M., the temperature had risen by 12° C. What was the temperature at 3 P.M.?
42. If a golfer completes a round at 3 strokes under par, her score is denoted -3. A professional golfer had rounds of -4, -2, 3, and -6 in a recent tournament. What was her total score for this tournament?
43. Dylan drove from a town located 400 feet below sea level to another town located 1750 feet above sea level. What was the change in elevation traveling from one town to another?
44. After withdrawing \$80 from her bank using an ATM card, Alycia had \$374 remaining in her savings account.

How much money did Alycia have in her account prior to withdrawing the money?

Multiply.

45. $7(-6)$
46. $-4(9)$
47. $-8 \cdot 5$
48. $-6(-8)$
49. $-15(-12)$
50. $-11 \cdot 17$
51. $82(-1)$
52. $-1 \cdot 19$
53. $-6 \cdot 0$
54. $0(-240)$
55. $-6(-3)(5)$
56. $-2(-4)(-8)$
57. $5 \cdot 3(-2)(-6)$
58. $-7 \cdot 2(-7)(-2)$

Divide if possible.

59. $45 \div (-5)$
60. $56 \div (-7)$
61. $-36 \div 6$
62. $-91 \div 13$
63. $-32 \div (-8)$
64. $-75 \div (-5)$
65. $126 \div (-9)$
66. $-420 \div 14$
67. $0 \div (-13)$
68. $0 \div 11$
69. $29 \div 0$
70. $-15 \div 0$

Mixed Practice, 71–82

Simplify.

71. $-11(-12)$
72. $126 \div (-6)$
73. $5 - 13$
74. $5(-13)$
75. $17 - (-11) - 49$
76. $8(-7)(-6)$
77. $-432 \div 3$
78. $-5 \cdot 17$
79. $9(-24)$
80. $9 + (-24)$
81. $5 \cdot 3(-17)(-29)(0)$
82. $-16 + (-11) - 42 - (-58)$

83. A group of 4 friends went out to dinner. If each person paid \$23, what was the total bill?

84. Three friends decided to start investing in stocks together. In the first year, they lost a total of \$13,500. How much did each person lose?

85. Tina owns 400 shares of a stock that dropped in value by \$3 per share last month. She also owns 500 shares of a stock that went up by \$2 per share last month. What is Tina's net income on these two stocks for last month?

86. Mario took over as the CEO for a company that lost \$20 million dollars in 2007. The company lost three times as much in 2008. The company went on to lose \$13 million more in 2009 than it had lost in 2008. How much money did Mario's company lose in 2009?

87. When a certain integer is added to -34, the result is -15. What is that integer?

88. Thirty-five less than a certain integer is -13. What is that integer?

89. When a certain integer is divided by -8, the result is 16. What is that integer?

90. When a certain integer is multiplied by -4 and that product is added to 22, the result is -110. What is that integer?

True or False (If false, give an example that shows why the statement is false.)

91. The sum of two integers is always an integer.
92. The difference of two integers is always an integer.
93. The sum of two whole numbers is always a whole number.
94. The difference of two whole numbers is always a whole number.

Writing in Mathematics

Explain each of the following in your own words.

95. Explain why subtracting a negative integer from another integer is the same as adding the opposite of that integer to it. Use the example $11 - (-5)$ in your explanation.

96. Explain why a positive integer times a negative integer produces a negative integer.

97. Explain why a negative integer times another negative integer produces a positive integer.

98. Explain why $7 \div 0$ is undefined.

1.3 Fractions

OBJECTIVES

1. **Find the factor set of a natural number.**
2. **Determine whether a natural number is prime.**
3. **Find the prime factorization of a natural number.**
4. **Simplify a fraction to lowest terms.**
5. **Change an improper fraction to a mixed number.**
6. **Change a mixed number to an improper fraction.**

Factors

Objective 1 Find the factor set of a natural number. To factor a natural number, express it as the product of two natural numbers. For example, one way to factor 12 is to rewrite it as $3 \cdot 4$. In this example, 3 and 4 are said to be factors of 12. The collection of all factors of a natural number is called its **factor set**. The factor set of 12 can be written as $\{1, 2, 3, 4, 6, 12\}$ because $1 \cdot 12 = 12, 2 \cdot 6 = 12$, and $3 \cdot 4 = 12$.

EXAMPLE 1 Write the factor set for 18.

SOLUTION Because 18 can be factored as $1 \cdot 18, 2 \cdot 9$, and $3 \cdot 6$, its factor set is $\{1, 2, 3, 6, 9, 18\}$.

Quick Check 1

Write the factor set for 36.

Prime Numbers

Objective 2 Determine whether a natural number is prime.

Prime Numbers

A natural number is **prime** if it is greater than 1 and its only two factors are 1 and itself.

For instance, the number 13 is prime because its only two factors are 1 and 13. The first 10 prime numbers are 2, 3, 5, 7, 11, 13, 17, 19, 23, and 29. The number 8 is not prime because it has factors other than 1 and 8, namely, 2 and 4. A natural number greater than 1 that is not prime is called a **composite** number. The number 1 is considered to be neither prime nor composite.

EXAMPLE 2 Determine whether the following numbers are prime or composite:

a) 26 **b)** 37

SOLUTION

a) The factor set for 26 is $\{1, 2, 13, 26\}$. Because 26 has factors other than 1 and itself, it is a composite number.

b) Because the number 37 has no factors other than 1 and itself, it is a prime number.

Quick Check 2

Determine whether the following numbers are prime or composite.

a) 57
b) 47
c) 48

Prime Factorization

Objective 3 Find the prime factorization of a natural number. When we rewrite a natural number as a product of prime factors, we obtain the **prime factorization** of the number. The prime factorization of 12 is $2 \cdot 2 \cdot 3$ because 2 and 3 are prime numbers and $2 \cdot 2 \cdot 3 = 12$. A **factor tree** is a useful tool for finding the prime factorization of a number. Here is an example of a factor tree for 72.

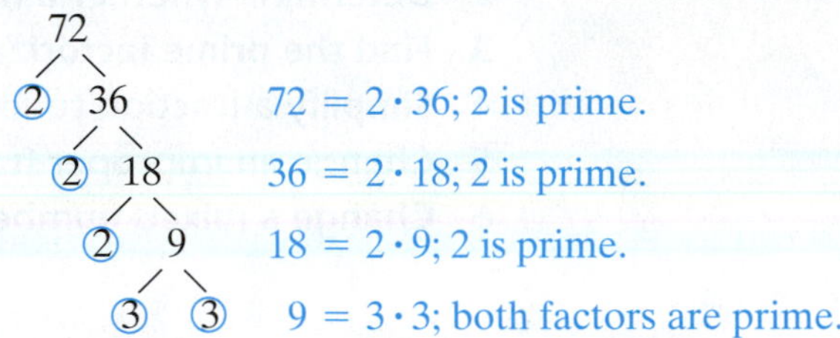

The prime factorization of 72 is $2 \cdot 2 \cdot 2 \cdot 3 \cdot 3$. We could have begun by rewriting 72 as $8 \cdot 9$ and then factored those two numbers. The process for creating a factor tree for a natural number is not unique, although the prime factorization for the number is unique.

EXAMPLE 3 Find the prime factorization of 60.

SOLUTION

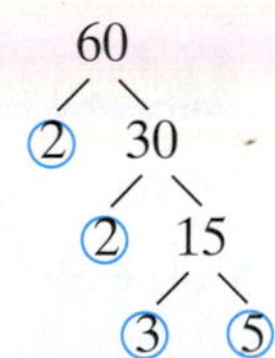

The prime factorization is $2 \cdot 2 \cdot 3 \cdot 5$.

Quick Check 3

Find the prime factorization of 63.

Fractions

Objective 4 Simplify a fraction to lowest terms. Recall from Section 1.1 that a rational number is a real number that can be written as the quotient (or ratio) of two integers, the second of which is not zero. An irrational number is a real number that cannot be written this way, such as the number π.

Rational numbers are often expressed using fraction notation such as $\frac{3}{7}$. Whole numbers such as 7 can be written as a fraction whose denominator is 1: $\frac{7}{1}$. The number on the top of the fraction is called the **numerator**, and the number on the bottom of the fraction is called the **denominator**.

$$\frac{\text{numerator}}{\text{denominator}}$$

If the numerator and denominator do not have any common factors other than 1, the fraction is said to be in **lowest terms**.

To simplify a fraction to lowest terms, begin by finding the prime factorization of both the numerator and denominator. Then divide the numerator and the denominator by their common factors.

EXAMPLE 4 Simplify $\frac{18}{30}$ to lowest terms.

SOLUTION

$\frac{18}{30} = \frac{2 \cdot 3 \cdot 3}{2 \cdot 3 \cdot 5}$ Find the prime factorization of the numerator and denominator. $18 = 2 \cdot 3 \cdot 3$, $30 = 2 \cdot 3 \cdot 5$.

$= \frac{\overset{1}{\cancel{2}} \cdot \overset{1}{\cancel{3}} \cdot 3}{\underset{1}{\cancel{2}} \cdot \underset{1}{\cancel{3}} \cdot 5}$ Divide out common factors.

$= \frac{3}{5}$ Simplify.

EXAMPLE 5 Simplify $\frac{4}{24}$ to lowest terms.

SOLUTION

$\frac{4}{24} = \frac{2 \cdot 2}{2 \cdot 2 \cdot 2 \cdot 3}$ Find the prime factorization of the numerator and denominator.

$\frac{\overset{1}{\cancel{2}} \cdot \overset{1}{\cancel{2}}}{\underset{1}{\cancel{2}} \cdot \underset{1}{\cancel{2}} \cdot 2 \cdot 3}$ Divide out common factors.

$= \frac{1}{6}$ Simplify.

Quick Check 4

Simplify to lowest terms:

a) $\frac{45}{210}$ **b)** $\frac{24}{384}$

A WORD OF CAUTION It is customary to leave off the denominator of a fraction if it is equal to 1. For example, rather than writing $\frac{19}{1}$, we usually write 19. However, we cannot omit a numerator that is equal to 1. In the previous example, it would have been a mistake if we had written 6 instead of $\frac{1}{6}$.

Mixed Numbers and Improper Fractions

Objective 5 Change an improper fraction to a mixed number. An **improper fraction** is a fraction whose numerator is greater than or equal to its denominator, such as $\frac{7}{4}$, $\frac{400}{150}$, $\frac{8}{8}$, and $\frac{35}{7}$. (In contrast, a proper fraction's numerator is smaller than its denominator.) An improper fraction is often converted to a **mixed number**, which is the sum of a whole number and a proper fraction. For example, the improper fraction $\frac{14}{3}$ can be represented by the mixed number $4\frac{2}{3}$, which is equivalent to $4 + \frac{2}{3}$.

To convert an improper fraction to a mixed number, begin by dividing the denominator into the numerator. The quotient is the whole number portion of the mixed number. The remainder becomes the numerator of the fractional part, while the denominator of the fractional part is the same as the denominator of the improper function.

$$\begin{array}{r} 4 \\ 3\overline{)14} \\ -12 \\ \hline 2 \end{array}$$

Whole-number portion of mixed number: 4

Denominator of fractional portion of mixed number: 3

Numerator of fractional portion of mixed number: 2

EXAMPLE 6 Convert the improper fraction $\frac{71}{9}$ to a mixed number.

SOLUTION Begin by dividing 9 into 71, which divides in 7 times with a remainder of 8.

$$\begin{array}{r} 7 \\ 9\overline{)71} \\ \underline{-63} \\ 8 \end{array}$$

The mixed number for $\frac{71}{9}$ is $7\frac{8}{9}$.

Quick Check 5

Convert the improper fraction $\frac{121}{13}$ to a mixed number.

Objective 6 Change a mixed number to an improper fraction. Often we have to convert a mixed number such as $2\frac{7}{15}$ into an improper fraction before proceeding with arithmetic operations.

Rewriting a Mixed Number as an Improper Fraction

- Multiply the whole number part of the mixed number by the denominator of the fractional part of the mixed number.
- Add this product to the numerator of the fractional part of the mixed number.
- The sum is the numerator of the improper fraction. The denominator stays the same.

$2\frac{7}{15}$ — Multiply; Add product to numerator

$$2\frac{7}{15} = \frac{37}{15}$$

EXAMPLE 7 Convert the mixed number $5\frac{4}{7}$ to an improper fraction.

SOLUTION Begin by multiplying $5 \cdot 7 = 35$. Add this product to 4 to produce a numerator of 39.

$$5\frac{4}{7} = \frac{39}{7}$$

Quick Check 6

Convert the mixed number $8\frac{1}{6}$ to an improper fraction.

BUILDING YOUR STUDY STRATEGY

Study Groups, 3 Where to Meet

- Some study groups prefer to meet off campus in the evening. One good place to meet is at a coffee shop with tables large enough to accommodate everyone, provided that the surrounding noise is not too distracting.
- Some groups take advantage of study rooms at public libraries.
- Other groups like to meet at members' homes. This typically provides a comfortable, relaxing atmosphere in which to work.

Exercises 1.3

MyMathLab

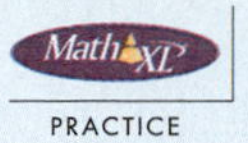
PRACTICE

WATCH

DOWNLOAD

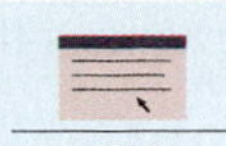
READ

REVIEW

Vocabulary

1. The collection of all factors of a natural number is called its ________.
2. A natural number greater than 1 is ________ if its only factors are 1 and itself.
3. A natural number greater than 1 that is not prime is called a(n) ________ number.
4. Define the prime factorization of a natural number.
5. The numerator of a fraction is the number written on the ________ of the fraction.
6. The denominator of a fraction is the number written on the ________ of the fraction.
7. A fraction is in lowest terms if its numerator and denominator contain no ________ other than 1.
8. A fraction whose numerator is less than its denominator is called a(n) ________ fraction.
9. A fraction whose numerator is greater than or equal to its denominator is called a(n) ________ fraction.
10. An improper fraction can be rewritten as a whole number or as a(n) ________.

11. Is 7 a factor of 247?
12. Is 13 a factor of 273?
13. Is 6 a factor of 4836?
14. Is 9 a factor of 32,057?
15. Is 15 a factor of 2835?
16. Is 103 a factor of 1754?

Write the factor set for the following numbers.

17. 48
18. 60
19. 27
20. 15
21. 20
22. 16
23. 81
24. 64
25. 31
26. 103
27. 91
28. 143

Write the prime factorization of the following numbers. (If the number is prime, state this.)

29. 18
30. 20
31. 42
32. 36
33. 39
34. 50
35. 27
36. 32
37. 125
38. 49
39. 29
40. 76
41. 99
42. 90
43. 31
44. 209
45. 120
46. 109

Simplify the following fractions to lowest terms.

47. $\frac{10}{16}$
48. $\frac{35}{42}$
49. $\frac{9}{45}$
50. $\frac{38}{2}$
51. $\frac{168}{378}$
52. $\frac{60}{84}$
53. $\frac{27}{64}$
54. $\frac{66}{154}$
55. $\frac{160}{176}$
56. $\frac{56}{45}$
57. $\frac{49}{91}$
58. $\frac{72}{140}$

Convert the following mixed numbers to improper fractions.

59. $3\frac{4}{5}$
60. $7\frac{2}{9}$
61. $2\frac{16}{17}$
62. $6\frac{5}{14}$
63. $13\frac{8}{11}$
64. $17\frac{16}{33}$

Convert the following improper fractions to whole numbers or mixed numbers.

65. $\frac{39}{5}$
66. $\frac{56}{8}$
67. $\frac{101}{7}$
68. $\frac{12}{4}$
69. $\frac{141}{19}$
70. $\frac{109}{8}$

71. List four fractions that are equivalent to $\frac{3}{4}$.

72. List four fractions that are equivalent to $1\frac{2}{3}$.

73. List four whole numbers that have at least three different prime factors.

74. List four whole numbers greater than 100 that are prime.

Writing in Mathematics

Answer in complete sentences.

75. Describe a real-world situation involving fractions. Describe a real-world situation involving mixed numbers.

76. Describe a situation in which you should convert an improper fraction to a mixed number.

1.4 Operations with Fractions

OBJECTIVES

1. Multiply fractions and mixed numbers.
2. Divide fractions and mixed numbers.
3. Add and subtract fractions and mixed numbers with the same denominator.
4. Find the least common multiple (LCM) of two natural numbers.
5. Add and subtract fractions and mixed numbers with different denominators.

Multiplying Fractions

Objective 1 Multiply fractions and mixed numbers. To multiply fractions, we multiply the numerators together and multiply the denominators together. When multiplying fractions, we may simplify any individual fraction, as well as divide out a common factor from a numerator and a different denominator. Dividing out a common factor in this fashion is often referred to as **cross-canceling**.

EXAMPLE 1 Multiply $\frac{4}{11} \cdot \frac{5}{6}$.

SOLUTION The first numerator (4) and the second denominator (6) have a common factor of 2 that we can eliminate through division.

$$\frac{4}{11} \cdot \frac{5}{6} = \frac{\overset{2}{\cancel{4}}}{11} \cdot \frac{5}{\underset{3}{\cancel{6}}} \qquad \text{Divide out the common factor 2.}$$

$$= \frac{2}{11} \cdot \frac{5}{3} \qquad \text{Simplify.}$$

$$= \frac{10}{33} \qquad \text{Multiply the two numerators and the two denominators.}$$

Quick Check 1

Multiply $\frac{10}{63} \cdot \frac{9}{16}$.

EXAMPLE 2 Multiply $3\frac{1}{7} \cdot \frac{14}{55}$.

SOLUTION When multiplying a mixed number by another number, convert the mixed number to an improper fraction before proceeding.

$$3\frac{1}{7} \cdot \frac{14}{55} = \frac{22}{7} \cdot \frac{14}{55} \quad \text{Convert } 3\frac{1}{7} \text{ to the improper fraction } \frac{22}{7}.$$

$$= \frac{\overset{2}{\cancel{22}}}{\underset{1}{\cancel{7}}} \cdot \frac{\overset{2}{\cancel{14}}}{\underset{5}{\cancel{55}}} \quad \text{Divide out the common factors 11 and 7.}$$

$$= \frac{4}{5} \quad \text{Multiply.}$$

Quick Check 2

Multiply $2\frac{2}{3} \cdot 8\frac{5}{8}$.

Dividing Fractions

Objective 2 Divide fractions and mixed numbers.

Reciprocal

When we invert a fraction such as $\frac{3}{5}$ to $\frac{5}{3}$, the resulting fraction is called the **reciprocal** of the original fraction.

Consider the fraction $\frac{a}{b}$, where a and b are nonzero real numbers. The reciprocal of this fraction is $\frac{b}{a}$. Notice that if we multiply a fraction by its reciprocal, such as $\frac{a}{b} \cdot \frac{b}{a}$, the result is 1. This property will be important in Chapter 2.

Reciprocal Property

For any nonzero real numbers a and b, $\frac{a}{b} \cdot \frac{b}{a} = 1$.

To divide a number by a fraction, invert the divisor and then multiply.

EXAMPLE 3 Divide $\frac{16}{25} \div \frac{22}{15}$.

SOLUTION

$$\frac{16}{25} \div \frac{22}{15} = \frac{16}{25} \cdot \frac{15}{22} \quad \text{Invert the divisor and multiply.}$$

$$= \frac{\overset{8}{\cancel{16}}}{\underset{5}{\cancel{25}}} \cdot \frac{\overset{3}{\cancel{15}}}{\underset{11}{\cancel{22}}} \quad \text{Divide out the common factors 2 and 5.}$$

$$= \frac{24}{55} \quad \text{Multiply.}$$

Quick Check 3

Divide $\frac{12}{25} \div \frac{63}{10}$.

A WORD OF CAUTION When dividing a number by a fraction, we must invert the divisor (not the dividend) before dividing out a common factor from a numerator and a denominator.

When performing a division involving a mixed number, begin by rewriting the mixed number as an improper fraction.

EXAMPLE 4 Divide $2\frac{5}{8} \div 3\frac{3}{10}$.

SOLUTION Begin by rewriting each mixed number as an improper fraction.

$$2\frac{5}{8} \div 3\frac{3}{10} = \frac{21}{8} \div \frac{33}{10} \quad \text{Rewrite each mixed number as an improper fraction.}$$

$$= \frac{21}{8} \cdot \frac{10}{33} \quad \text{Invert the divisor and multiply.}$$

$$= \frac{\overset{7}{\cancel{21}}}{\underset{4}{\cancel{8}}} \cdot \frac{\overset{5}{\cancel{10}}}{\underset{11}{\cancel{33}}} \quad \text{Divide out common factors.}$$

$$= \frac{35}{44} \quad \text{Multiply.}$$

Quick Check 4

Divide $\frac{20}{21} \div 2\frac{2}{3}$.

Adding and Subtracting Fractions

Objective 3 Add and subtract fractions and mixed numbers with the same denominator. To add or subtract fractions that have the same denominator, we add or subtract the numerators, placing the result over the common denominator. Make sure you simplify the result to lowest terms.

EXAMPLE 5 Subtract $\frac{3}{8} - \frac{9}{8}$.

SOLUTION The two denominators are the same (8), so we subtract the numerators. When we subtract $3 - 9$, the result is -6. Although we may leave the negative sign in the numerator, it often appears in front of the fraction itself.

$$\frac{3}{8} - \frac{9}{8} = -\frac{6}{8} \quad \text{Subtract the numerators.}$$

$$= -\frac{3}{4} \quad \text{Simplify to lowest terms.}$$

Quick Check 5

Subtract $\frac{17}{20} - \frac{5}{20}$.

When performing an addition involving a mixed number, begin by rewriting the mixed number as an improper fraction.

EXAMPLE 6 Add $3\frac{5}{12} + 2\frac{11}{12}$.

SOLUTION Begin by rewriting each mixed number as an improper fraction.

$$3\frac{5}{12} + 2\frac{11}{12} = \frac{41}{12} + \frac{35}{12} \quad \text{Rewrite } 3\frac{5}{12} \text{ and } 2\frac{11}{12} \text{ as improper fractions.}$$

$$= \frac{76}{12} \quad \text{Add the numerators.}$$

$$= \frac{19}{3} \quad \text{Simplify to lowest terms.}$$

$$= 6\frac{1}{3} \quad \text{Rewrite as a mixed number.}$$

It is not necessary to rewrite the result as a mixed number, but this is often done when you perform arithmetic operations on mixed numbers.

Quick Check 6

Add $6\frac{1}{8} + 5\frac{5}{8}$.

Objective 4 Find the least common multiple (LCM) of two natural numbers. Two fractions are said to be **equivalent fractions** if they have the same numerical value and both can be simplified to the same fraction when simplified to lowest terms. To add or subtract two fractions with different denominators, we must first convert them to equivalent fractions with the same denominator. To do this, we find the **least common multiple (LCM)** of the two denominators. This is the smallest number that is a multiple of both denominators. For example, the LCM of 4 and 6 is 12 because 12 is the smallest multiple of both 4 and 6.

To find the LCM for two numbers, begin by factoring them into their prime factorizations.

Finding the LCM of Two or More Numbers

- Find the prime factorization of each number.
- Find the common factors of the numbers.
- Multiply the common factors by the remaining factors of the numbers.

EXAMPLE 7 Find the LCM of 24 and 30.

SOLUTION Begin with the prime factorizations of 24 and 30.

$$24 = 2 \cdot 2 \cdot 2 \cdot 3$$
$$30 = 2 \cdot 3 \cdot 5$$
$$24 = \textcircled{2} \cdot 2 \cdot 2 \cdot \textcircled{3}$$
$$30 = \textcircled{2} \cdot \textcircled{3} \cdot 5$$

The common factors are 2 and 3. Additional factors are a pair of 2's as well as a 5. So to find the LCM, multiply the common factors (2 and 3) by the additional factors (2, 2, and 5).

$$2 \cdot 3 \cdot 2 \cdot 2 \cdot 5 = 120$$

The least common multiple of 24 and 30 is 120.

Quick Check 7

Find the least common multiple of 18 and 42.

Another technique for finding the LCM for two numbers is to start listing the multiples of the larger number until we find a multiple that also is a multiple of the smaller number. For example, the first few multiples of 6 are

$$6\text{: } 6, 12, 18, 24, 30, \ldots$$

The first multiple listed that also is a multiple of 4 is 12, so the LCM of 4 and 6 is 12.

Objective 5 Add and subtract fractions and mixed numbers with different denominators. When adding or subtracting two fractions that do not have the same denominator, we first find a common denominator by finding the LCM of the two denominators. Then convert each fraction to an equivalent fraction whose denominator is that common denominator. Once we rewrite the two fractions so they have the same denominator, we can add (or subtract) as done previously in this section.

Adding or Subtracting Fractions with Different Denominators

- Find the LCM of the denominators.
- Rewrite each fraction as an equivalent fraction whose denominator is the LCM of the original denominators.
- Add or subtract the numerators, placing the result over the common denominator.
- Simplify to lowest terms if possible.

EXAMPLE 8 Add $\frac{5}{12} + \frac{9}{14}$.

SOLUTION The prime factorization of 12 is $2 \cdot 2 \cdot 3$, and the prime factorization of 14 is $2 \cdot 7$. The two denominators have a common factor of 2. If we multiply this common factor by the other factors of these two numbers, 2, 3, and 7, we see that the LCM of these two denominators is 84. Begin by rewriting each fraction as an equivalent fraction whose denominator is 84. Multiply the first fraction by $\frac{7}{7}$ and the second fraction by $\frac{6}{6}$. Because $\frac{7}{7}$ and $\frac{6}{6}$ are both equal to 1, we do not change the value of either fraction.

$$\frac{5}{12} + \frac{9}{14} = \frac{5}{12} \cdot \frac{7}{7} + \frac{9}{14} \cdot \frac{6}{6}$$ Multiply the first fraction's numerator and denominator by 7. Multiply the second fraction's numerator and denominator by 6.

$$= \frac{35}{84} + \frac{54}{84}$$ Multiply.

$$= \frac{89}{84}$$ Add.

This fraction is already in lowest terms.

Quick Check 8

Add $\frac{4}{9} + \frac{2}{15}$.

When performing an addition or a subtraction involving a mixed number, we can begin by rewriting the mixed number as an improper fraction.

EXAMPLE 9 Subtract $4\frac{1}{3} - \frac{3}{4}$.

SOLUTION Begin by rewriting $4\frac{1}{3}$ as an improper fraction.

$$4\frac{1}{3} - \frac{3}{4} = \frac{13}{3} - \frac{3}{4}$$ Rewrite $4\frac{1}{3}$ as an improper fraction.

$$= \frac{13}{3} \cdot \frac{4}{4} - \frac{3}{4} \cdot \frac{3}{3}$$ The LCM of the denominators is 12. Multiply the first fraction by $\frac{4}{4}$. Multiply the second fraction by $\frac{3}{3}$.

$$= \frac{52}{12} - \frac{9}{12}$$ Multiply.

$$= \frac{43}{12}$$ Subtract.

$$= 3\frac{7}{12}$$ Rewrite as a mixed number.

Quick Check 9

Subtract $5\frac{1}{5} - 3\frac{5}{6}$.

BUILDING YOUR STUDY STRATEGY

Study Groups, 4 Going Over Homework A study group can go over homework assignments together. It is important that each group member work on the assignment before arriving at the study session. If you struggled with a problem or could not do it at all, ask for help or suggestions from your group members.

If there was a problem that you seem to understand better than the members of your group do, share your knowledge; explaining how to do a certain problem increases your chances of retaining that knowledge until the exam and beyond.

At the end of each session, quickly review what the group accomplished.

Exercises 1.4

MyMathLab

PRACTICE

WATCH

DOWNLOAD

READ

REVIEW

Vocabulary

1. Before multiplying by a mixed number, convert it to a(n) ______________.
2. When a fraction is inverted, the result is called its ______________.
3. Explain how to divide by a fraction.
4. When fractions are added or subtracted, they must have the same ______________.
5. The smallest number that is a multiple of two numbers is called their ______________.
6. A board is cut into two pieces that measure $4\frac{1}{6}$ feet and $3\frac{3}{8}$ feet, respectively. Which operation will give the length of the original board?

   a) $4\frac{1}{6} + 3\frac{3}{8}$ **b)** $4\frac{1}{6} - 3\frac{3}{8}$

   c) $4\frac{1}{6} \cdot 3\frac{3}{8}$ **d)** $4\frac{1}{6} \div 3\frac{3}{8}$

Multiply. Your answer should be in lowest terms.

7. $\frac{3}{8} \cdot \frac{4}{27}$ **8.** $\frac{6}{35} \cdot \frac{25}{29}$

9. $\frac{20}{21} \cdot \left(-\frac{77}{90}\right)$ **10.** $-\frac{9}{30} \cdot \frac{28}{42}$

11. $4\frac{2}{7} \cdot \frac{14}{25}$ **12.** $-3\frac{5}{9} \cdot 2\frac{1}{6}$

13. $5 \cdot 6\frac{3}{10}$ **14.** $8 \cdot \frac{7}{12}$

15. $\frac{2}{3} \cdot \frac{8}{9}$ **16.** $-\frac{12}{35}\left(-\frac{14}{99}\right)$

Divide. Your answer should be in lowest terms.

17. $\frac{6}{25} \div \frac{8}{45}$ **18.** $\frac{15}{32} \div \frac{9}{20}$

19. $-\frac{22}{56} \div \frac{33}{147}$ **20.** $-\frac{24}{91} \div \left(-\frac{9}{39}\right)$

21. $\frac{17}{40} \div \frac{1}{2}$ **22.** $\frac{7}{30} \div \left(-\frac{1}{5}\right)$

23. $\frac{4}{11} \div 3\frac{1}{5}$ **24.** $3\frac{3}{8} \div 6$

25. $-2\frac{4}{5} \div 6\frac{2}{3}$ **26.** $7\frac{1}{9} \div 13\frac{3}{8}$

Add or subtract.

27. $\frac{7}{15} + \frac{4}{15}$ **28.** $\frac{1}{8} + \frac{5}{8}$

29. $\frac{5}{9} + \frac{4}{9}$ **30.** $\frac{3}{10} + \frac{9}{10}$

31. $\frac{2}{5} - \frac{4}{5}$ **32.** $\frac{17}{18} - \frac{5}{18}$

33. $\frac{9}{16} - \frac{3}{16}$ **34.** $\frac{13}{42} - \frac{29}{42}$

Find the LCM of the given numbers.

35. 10, 15 **36.** 8, 12

37. 12, 42 **38.** 9, 30

39. 16, 80 **40.** 16, 27

41. 8, 10, 14 **42.** 20, 35, 50

Simplify. Your answer should be in lowest terms.

43. $\frac{4}{5} + \frac{3}{4}$ **44.** $\frac{4}{7} + \frac{1}{4}$

45. $\frac{7}{10} + \frac{5}{8}$ **46.** $\frac{3}{4} + \frac{5}{6}$

47. $6\frac{1}{5} + 5$ **48.** $3 + 8\frac{3}{7}$

49. $6\frac{2}{3} + 5\frac{1}{6}$ **50.** $11\frac{4}{9} + 5\frac{1}{3}$

51. $\frac{2}{3} - \frac{7}{15}$ **52.** $\frac{1}{2} - \frac{7}{9}$

53. $\frac{3}{4} - \frac{2}{7}$ **54.** $\frac{5}{8} - \frac{5}{6}$

55. $7\frac{1}{2} - 3\frac{1}{4}$ **56.** $12\frac{2}{3} - 6\frac{2}{5}$

57. $12\frac{3}{10} - 9$ **58.** $6 - 4\frac{3}{4}$

59. $-\frac{5}{9} - \frac{7}{12}$ **60.** $-\frac{9}{10} - \frac{11}{14}$

61. $-\frac{9}{16} + \frac{5}{24}$ **62.** $-\frac{3}{8} + \frac{13}{24}$

63. $\frac{6}{7} - \left(-\frac{8}{15}\right)$ **64.** $\frac{1}{12} - \left(-\frac{19}{30}\right)$

65. $-\frac{4}{15} + \left(-\frac{13}{18}\right)$ **66.** $-\frac{17}{24} + \left(-\frac{25}{42}\right)$

67. $\frac{3}{16} + \frac{9}{20} - \frac{11}{12}$

68. $\frac{10}{21} - \frac{13}{18} + \frac{8}{15}$

Mixed Practice, 69–88

Simplify.

69. $\frac{8}{9} \cdot \frac{3}{5}$

70. $\frac{3}{4} + \frac{7}{10}$

71. $\frac{7}{30} \div \frac{35}{48}$

72. $\frac{12}{35} \cdot \frac{14}{27}$

73. $\frac{1}{6} - \frac{7}{8}$

74. $\frac{7}{24} - \frac{29}{40}$

75. $\frac{19}{30} + \frac{11}{18}$

76. $3\frac{1}{5} \cdot 4\frac{3}{8}$

77. $13 \div \frac{1}{8}$

78. $\frac{3}{5} - \frac{2}{3} - \frac{7}{10}$

79. $3\frac{4}{7} + 6\frac{3}{5} - 8$

80. $12\frac{1}{3} + 7\frac{1}{6} - 5\frac{1}{2}$

81. $\frac{15}{56} + \left(-\frac{16}{21}\right)$

82. $\frac{7}{12} - \left(-\frac{23}{30}\right)$

83. $-\frac{9}{13} + \frac{19}{36}$

84. $-\frac{3}{8} - \frac{81}{100}$

Find the missing number.

85. $\frac{11}{24} + \frac{5}{?} = \frac{13}{12}$

86. $\frac{?}{10} - \frac{1}{3} = \frac{1}{6}$

87. $\frac{10}{21} \cdot \frac{?}{75} = \frac{4}{45}$

88. $\frac{11}{40} + \frac{?}{40} = \frac{9}{10}$

89. Bruce is fixing a special dinner for his girlfriend. The three recipes he is preparing call for $\frac{1}{2}$ cup, $\frac{3}{4}$ cup, and $\frac{1}{3}$ cup of flour, respectively. In total, how much flour does Bruce need to make these three recipes?

90. Sue gave birth to twins. One of the babies weighed $4\frac{7}{8}$ pounds at birth, and the other baby weighed $5\frac{1}{4}$ pounds. Find the total weight of the twins at birth.

91. A chemist has $\frac{23}{40}$ fluid ounce of a solution. If she needs $\frac{1}{8}$ fluid ounce of the solution for an experiment, how much of the solution will remain?

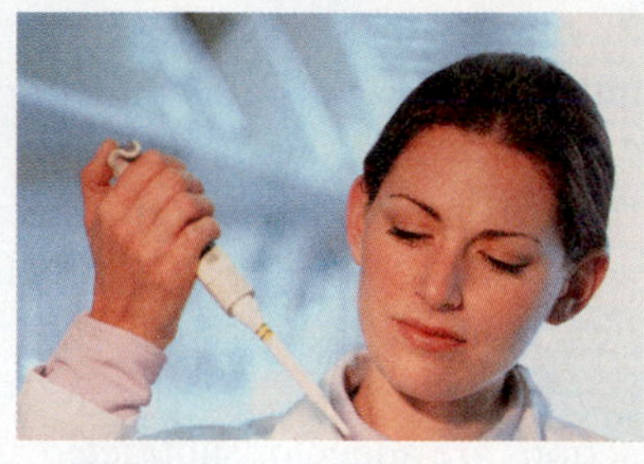

92. A popular weed spray concentrate recommends using $1\frac{1}{4}$ tablespoons of concentrate for each quart of water. How much concentrate needs to be mixed with 6 quarts of water?

93. A board that is $4\frac{1}{5}$ feet long needs to be cut into 6 pieces of equal length. How long will each piece be?

94. Ross makes a batch of hot sauce that will be poured into bottles that hold $5\frac{3}{4}$ fluid ounces. If Ross has 115 fluid ounces of hot sauce, how many bottles can he fill?

95. A craftsperson is making a rectangular picture frame. Each of two sides will be $\frac{5}{6}$ of a foot long, while each of the other two sides will be $\frac{2}{3}$ of a foot long. If the craftsperson has one board that is $2\frac{3}{4}$ feet long, is this enough to make the picture frame? Explain.

96. A pancake recipe calls for $1\frac{1}{3}$ cups of whole wheat flour to make 12 pancakes. How much flour is needed to make 48 pancakes?

97. A pancake recipe calls for $1\frac{1}{3}$ cups of whole wheat flour to make 12 pancakes. How much flour is needed to make 6 pancakes?

Writing in Mathematics

Answer in complete sentences.

98. Explain, using your own words, the difference between dividing a number in half and dividing a number by one-half.

99. Explain why you cannot divide a common factor of 2 from the numbers 4 and 6 in the expression $\frac{4}{7} \cdot \frac{6}{11}$.

100. Explain why it would be a bad idea to rewrite fractions with a common denominator before multiplying them.

1.5 Decimals and Percents

OBJECTIVES

1. **Perform arithmetic operations with decimals.**
2. **Rewrite a fraction as a decimal number.**
3. **Rewrite a decimal number as a fraction.**
4. **Rewrite a fraction as a percent.**
5. **Rewrite a decimal as a percent.**
6. **Rewrite a percent as a fraction.**
7. **Rewrite a percent as a decimal.**

Decimals

Rational numbers also can be represented using **decimal notation**. The decimal 0.23 is equivalent to the fraction $\frac{23}{100}$, or twenty-three hundredths. The digit 2 is in the tenths place, and the digit 3 is in the hundredths place. The following chart shows several place values for decimals:

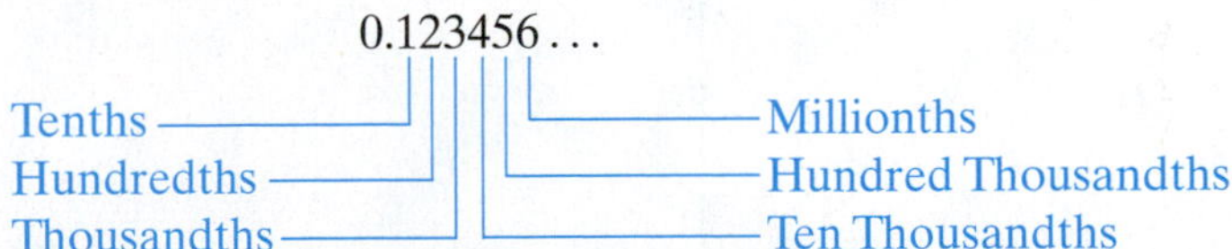

Objective 1 Perform arithmetic operations with decimals. Here is a brief summary of arithmetic operations using decimals.

- To add or subtract two decimal numbers, align the decimal points and add or subtract as you would with integers.

$$3.96 + 12.072$$

$$\begin{array}{r} 3.96 \\ +\ 12.072 \\ \hline 16.032 \end{array}$$

- To multiply two decimal numbers, multiply them as you would integers. The total number of decimal places in the two factors shows how many decimal places are in the product.

$$-2.09 \cdot 3.1$$

In this example, the two factors have a total of three decimal places, so the product must have three decimal places. Multiply these two numbers as if they were 209 and 31 and then insert the decimal point in the appropriate place, leaving three digits to the right of the decimal point.

$$\begin{array}{r} -209 \\ \times\ 31 \\ \hline -6479 \end{array} \qquad \begin{array}{r} -2.09 \\ \times\ 3.1 \\ \hline -6.479 \end{array}$$

3 places

- To divide two decimal numbers, move the decimal point in the divisor to the right so that it becomes an integer. Then move the decimal point in the other number (dividend) to the right by the same number of spaces. The decimal point in the answer will be aligned with this new location of the decimal point in the dividend.

$8.24 \div 0.4$

$0.4\overline{)8.24}$ Begin by moving *each* decimal point one place to the right.

$0.4\overline{)8.24}$

$$\begin{array}{r} 20.6 \\ 4\overline{)82.4} \end{array}$$

Then perform the division.

Rewriting Fractions as Decimals and Decimals as Fractions

Objective 2 Rewrite a fraction as a decimal number. To rewrite any fraction as a decimal, we divide its numerator by its denominator. The fraction line is simply another way to write "÷".

EXAMPLE 1 Rewrite the fraction $\frac{5}{8}$ as a decimal.

SOLUTION To rewrite this fraction as a decimal, divide 5 by 8.

Now begin the division, adding a decimal point after the 5 and 0's to the end of the dividend until there is no remainder.

$$\begin{array}{r} 0.625 \\ 8\overline{)5.000} \\ \underline{-4.8} \\ 20 \\ \underline{-16} \\ 40 \\ \underline{-40} \\ 0 \end{array}$$

$$\frac{5}{8} = 0.625$$

Quick Check 1

Rewrite the fraction $\frac{3}{4}$ as a decimal.

EXAMPLE 2 Rewrite the fraction $\frac{23}{30}$ as a decimal.

SOLUTION When we divide 23 by 30, the result is a decimal that does not terminate (0.76666 . . .). The pattern continues repeating the digit 6 forever. This is an example of a **repeating decimal**. We may place a bar over the repeating digit(s) to denote a repeating decimal.

$$\frac{23}{30} = 0.7\overline{6}$$

Quick Check 2

Rewrite the fraction $\frac{5}{18}$ as a decimal.

Objective 3 Rewrite a decimal number as a fraction. Suppose we want to rewrite a decimal number such as 0.48 as a fraction. This decimal is read as "forty-eight hundredths" and is equivalent to the fraction $\frac{48}{100}$. Simplifying this fraction shows that $0.48 = \frac{12}{25}$.

EXAMPLE 3 Rewrite the decimal 0.164 as a fraction in lowest terms.

SOLUTION This decimal ends in the thousandths place, so start with a fraction of $\frac{164}{1000}$.

$$0.164 = \frac{164}{1000} \quad \text{Rewrite as a fraction whose denominator is 1000.}$$

$$= \frac{41}{250} \quad \text{Simplify to lowest terms.}$$

Quick Check 3

Rewrite the decimal 0.425 as a fraction in lowest terms.

Using Your Calculator To rewrite a decimal as a fraction using the TI-84, key the decimal, press the MATH key, and select option 1. The following screens show how to rewrite the decimal 0.164 as a fraction using the TI-84, as in Example 3.

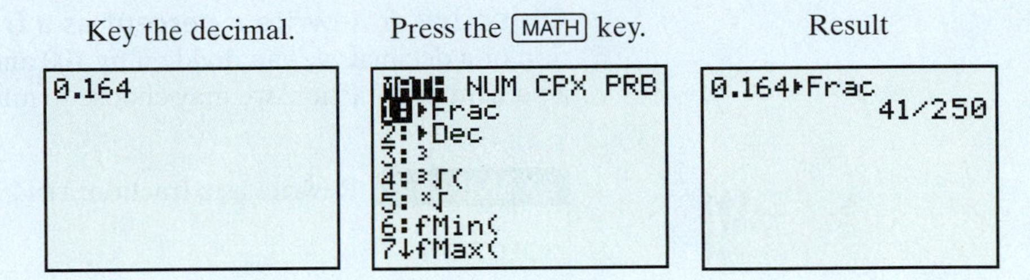

Percents

Objective 4 Rewrite a fraction as a percent. **Percents** (%) are used to represent numbers as parts of 100. One percent, which can be written as 1%, is equivalent to 1 part of 100, or $\frac{1}{100}$, or 0.01. The fraction $\frac{27}{100}$ is equivalent to 27%. Percents, decimals, and fractions are all ways to write a rational number. We will learn to convert back and forth between percents and fractions as well as between percents and decimals.

Rewriting Fractions and Decimals as Percents

To rewrite a fraction or a decimal as a percent, we multiply it by 100%. Because 100% is equal to 1 this will not change the value of the fraction or decimal number.

EXAMPLE 4 Rewrite as a percent: a) $\frac{2}{5}$, b) $\frac{3}{8}$.

SOLUTION

a) Begin by multiplying by 100% and simplifying.

$$\frac{2}{\cancel{5}_{1}} \cdot \cancel{100}^{20}\% = 40\%$$

b) Again, multiply by 100%. Occasionally, as in this example, we will end up with an improper fraction, which can be changed to a mixed number.

$$\frac{3}{\cancel{8}_{2}} \cdot \cancel{100}^{25}\% = \frac{75}{2}\%, \text{ which can be rewritten as } 37\frac{1}{2}\%.$$

Quick Check 4

a) Rewrite $\frac{7}{10}$ as a percent.

b) Rewrite $\frac{21}{40}$ as a percent.

Objective 5 Rewrite a decimal as a percent.

EXAMPLE 5 Rewrite 0.3 as a percent.

SOLUTION When we multiply a decimal by 100, the result is the same as moving the decimal point two places to the right.

$$0.30$$

$$0.3 \cdot 100\% = 30\%$$

Quick Check 5

Rewrite 0.42 as a percent.

Rewriting Percents as Fractions and Decimals

Objective 6 Rewrite a percent as a fraction. To rewrite a percent as a fraction or a decimal, we can divide it by 100 and omit the percent sign. When rewriting a percent as a fraction, we may choose to multiply by $\frac{1}{100}$ rather than dividing by 100.

EXAMPLE 6 Rewrite as a fraction: a) 44%, b) $16\frac{2}{3}\%$.

SOLUTION

a) Begin by multiplying by $\frac{1}{100}$ and omitting the percent sign.

$$\overset{11}{\cancel{44}} \cdot \frac{1}{\underset{25}{\cancel{100}}} = \frac{11}{25}$$

b) Rewrite the mixed number $16\frac{2}{3}$ as an improper fraction $\left(\frac{50}{3}\right)$, multiply by $\frac{1}{100}$, and simplify.

$$\frac{\overset{1}{\cancel{50}}}{3} \cdot \frac{1}{\underset{2}{\cancel{100}}} = \frac{1}{6}$$

Quick Check 6

a) Rewrite 35% as a fraction.
b) Rewrite $11\frac{2}{3}\%$ as a fraction.

Objective 7 Rewrite a percent as a decimal. In the next example, we will rewrite percents as decimals rather than as fractions.

EXAMPLE 7 Rewrite as a decimal: a) 56%, b) 143%.

SOLUTION

a) Begin by dropping the percent sign and dividing by 100. Keep in mind that dividing a decimal number by 100 is the same as moving the decimal point two places to the left.

$$.56$$

$$56 \div 100 = 0.56$$

b) When a percent is greater than 100%, its equivalent decimal must be greater than 1.

$$143 \div 100 = 1.43$$

Quick Check 7

a) Rewrite 8% as a decimal. **b)** Rewrite 240% as a decimal.

BUILDING YOUR STUDY STRATEGY

Study Groups, 5 Three Questions Another way to structure a group study session is to have each member bring a list of three questions to the meeting. The questions can be about specific homework problems or about topics or procedures that have been covered in class. Once the members have asked their questions, the group should attempt to come up with answers that each member understands. If the group cannot answer a question, see your instructor at the beginning of the next class or during office hours, asking him or her for an explanation.

Exercises 1.5

MyMathLab

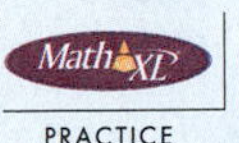
PRACTICE

WATCH

DOWNLOAD
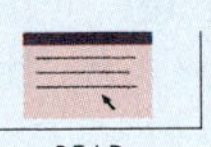
READ

REVIEW

Vocabulary

1. The first place to the right of a decimal point is the ___________ place.
2. The third place to the right of a decimal point is the ___________ place.
3. To rewrite a fraction as a decimal divide the ___________ by the ___________.
4. To rewrite a fraction as a percent ___________ it by 100%.
5. To rewrite a percent as a fraction ___________ it by 100 and omit the percent sign.
6. Percents are used to represent numbers as parts of ___________.

Simplify the following decimal expressions.

7. $4.23 + 3.62$
8. $13.89 - 2.54$
9. $-7(5.2)$
10. $69.54 \div 6$
11. $8.4 - 3.7$
12. $-7.9 + (-4.5)$
13. $13.568 \div 0.4$
14. $3.6(4.7)$
15. $-2.2 \cdot 3.65$
16. $6.2 - 15.9$
17. $13.47 - (-21.562)$
18. $5.283 \div 0.25$
19. $-6.3(3.9)(-2.25)$
20. $-4.84 \div (-0.016)$
21. $37.278 + 56.722$
22. $109.309 - 27.46 - 52.3716$

Rewrite the following fractions as decimal numbers.

23. $\frac{9}{10}$
24. $\frac{2}{5}$
25. $-\frac{23}{8}$
26. $\frac{59}{4}$
27. $\frac{13}{25}$
28. $-\frac{11}{16}$
29. $24\frac{29}{50}$
30. $7\frac{3}{20}$

Rewrite the following decimal numbers as fractions in lowest terms.

31. 0.2
32. 0.5
33. 0.85
34. 0.36
35. −0.74
36. −0.56
37. 0.375
38. 0.204
39. The normal body temperature for humans is 98.6° F. If Melody has a temperature that is 2.8° F above normal, what is her temperature?
40. Paul spent the following amounts on gifts for his wife's birthday: \$32.95, \$16.99, \$47.50, \$12.37, and \$285. How much did Paul spend on gifts for his wife?

41. The balance of Carie's checking account is \$427.36. If she writes checks for \$19.95, \$34.40, and \$148.68, what will her new balance be?

42. At the close of the stock market on Tuesday, the price for one share of Google was \$426.17. Over the next three days, the stock went down by \$9.63, up by \$14.08, and down by \$7.84. What was the price of the stock at the end of Friday's session?

43. An office manager bought 12 cases of paper. If each case cost \$21.47, what was the total cost for the 12 cases?

44. Jean gives Chris a \$20 bill and tells him to go to the grocery store and buy as many hot dogs as he can. If each package of hot dogs costs \$2.65, how many packages can Chris buy? How much change will Chris have?

Rewrite as percents.

45. $\frac{3}{4}$ **46.** $\frac{3}{5}$

47. $\frac{4}{5}$ **48.** $\frac{5}{8}$

49. $\frac{7}{8}$ **50.** $\frac{11}{12}$

51. $\frac{27}{4}$ **52.** $\frac{12}{5}$

53. 0.4 **54.** 0.6

55. 0.15 **56.** 0.87

57. 0.09 **58.** 0.03

59. 3.2 **60.** 2.75

Rewrite as fractions.

61. 84% **62.** 80%

63. 7% **64.** 2%

65. $11\frac{1}{9}\%$ **66.** $18\frac{2}{11}\%$

67. 520% **68.** 275%

Rewrite as decimals.

69. 54% **70.** 71%

71. 16% **72.** 29%

73. 7% **74.** 9%

75. 0.3% **76.** 61.3%

77. 400% **78.** 320%

79. Find three fractions that are equivalent to 0.375.

80. Find three fractions that are equivalent to 0.4.

Complete the following table.

	Fraction	Decimal	Percent
81.		0.2	
82.	$\frac{7}{40}$		
83.			32%
84.			45%
85.	$\frac{13}{8}$		
86.		0.64	

Writing in Mathematics

Answer in complete sentences.

87. Stock prices at the New York Stock Exchange used to be reported as fractions. Now prices are reported as decimals. Do you think this was a good idea? Explain.

88. Describe a real-world application involving decimals.

1.6 Basic Statistics

OBJECTIVES

1. Calculate basic statistics for a set of data.
2. Construct a histogram for a set of data.

In today's data-driven society, we often see graphs presenting data or information on television news as well as in newspapers, magazines, and online. This section focuses on the calculation of statistics used to describe a set of data as well as the creation of a histogram to represent a set of data.

Objective 1 Calculate basic statistics for a set of data. There are two basic types of statistics—those that describe the typical value for a set of data and those that describe how varied the values are. Statistics that describe the typical value for a set of data are often called **measures of center**, while statistics that describe how varied a set of data is are often called **measures of spread**.

Mean

The **mean** of a set of data is one measure of center for a set of data. To calculate the mean for a set of data, we simply add all of the values and divide by how many values there are. The mean is the arithmetic average of the set of data.

$$\text{Mean} = \frac{\text{Sum of All Values}}{\text{Number of Values}}$$

EXAMPLE 1 Eight students were asked how far they drive to school each day. Here are the results, in miles: 17, 8, 30, 1, 2, 5, 15, 10. Calculate the mean for this set of data.

SOLUTION The total of these eight values is 88.

$$\text{Mean} = \frac{88}{8} = 11$$

The mean mileage for these students is 11 miles.

Quick Check 1

Five families were asked how much they spend on groceries each month. Here are the results: $850, $1020, $970, $635, $795. Calculate the mean for this set of data.

Median

The **median** of a set of data is another measure of center for a set of data, often used for types of data that have unusually high or low values, such as home prices and family income. To find the median for a set of data, we begin by writing the values in ascending order. If a set of data has an odd number of values, the single value in the middle of the set of data is the median. If a set of data has an even number of values, the median is the average of the two center values.

EXAMPLE 2 Find the median of the given set of values.

a) 37, 16, 59, 18, 30, 4, 75, 46, 62

b) 65, 72, 74, 81, 71, 83, 89, 82, 53, 48, 77, 65

SOLUTION

a) Begin by rewriting the values in ascending order: 4, 16, 18, 30, 37, 46, 59, 62, 75. Because there is an odd number of values (9), the median is the single value in the center of the list.

4 16 18 30 37 46 59 62 75

↑ Median

The median is 37.

b) Here are the values in ascending order: 48, 53, 65, 65, 71, 72, 74, 77, 81, 82, 83, 89. Because there is an even number of values (12), the median is the average of the two values in the center of the list.

48 53 65 65 71 (72 74) 77 81 82 83 89

$$\text{Median} = \frac{72 + 74}{2} = \frac{146}{2} = 73$$

The median is 73.

Quick Check 2

Find the median of the given set of values.

a) 54, 21, 39, 16, 7, 75
b) 51, 3, 29, 60, 62, 25, 43, 102, 14

Mode and Midrange

The **mode** and **midrange** are two other measures of center for a set of data. The mode is the value that is repeated most often.

- If there are no repeated values, the set of data has no mode. The set 1, 2, 3, 4, 5 has no mode because no value is repeated.
- A set of data can have more than one mode if two or more values are repeated the same number of times. The set 5, 5, 7, 8, 8 has two modes—5 and 8.

The midrange is the average of the set's minimum value and maximum value.

$$\text{Midrange} = \frac{\text{Minimum Value} + \text{Maximum Value}}{2}$$

EXAMPLE 3 During a medical trial, the LDL cholesterol levels of 16 adult males were measured. Here are the results.

104 122 115 90 116 88 167 105 154 129 81 157 143 122 106 87

Find the mode and the midrange for this data.

SOLUTION Only one value, 122, has been repeated; so the mode is 122. To find the midrange, identify the maximum and minimum values for this set of data. The smallest value in this set is 81, and the largest value is 167.

$$\text{Midrange} = \frac{81 + 167}{2} = \frac{248}{2} = 124$$

The midrange is 124.

Quick Check 3

Here are the heights, in inches, of 10 adult females.

56	65	66	66	65	66	65	60	61	65

Find the mode and the midrange for this data.

Range

The **range** is a measure of spread for a set of data, showing how varied the values are. To find the range of a set of values, we subtract the minimum value in the set from the maximum value in the set.

$$\text{Range} = \text{Maximum Value} - \text{Minimum Value}$$

EXAMPLE 4 Here are the test scores of 9 math students.

65	75	96	91	78	81	73	92	61

Find the range of these test scores.

SOLUTION To find the range, identify the maximum and minimum values for this set of data. The maximum value in this set is 96, and the minimum value is 61.

$$\text{Range} = 96 - 61 = 35$$

The range is 35 points.

Quick Check 4

Here are the heights, in centimeters, of 8 adult males.

165	168	174	179	182	159	171	180

Find the range for this data.

Histogram

Objective 2 Construct a histogram for a set of data.

A **histogram** is a graph that can be used to show how a set of data is distributed, giving an idea of where the data values are centered as well as how they are dispersed. To construct a histogram, we begin with a **frequency distribution**, which divides the data into groups, called **classes**, and lists how many times each class is represented in the set of data. Here is a frequency distribution showing the ages of U.S. Presidents at inauguration.

Age	Frequency
40 to 44	2
45 to 49	7
50 to 54	13
55 to 59	12
60 to 64	7
65 to 69	3

This frequency distribution shows that two presidents were between 40 and 44 years old when they were inaugurated, seven presidents were between the ages of 45 and 49, and so on.

To construct a histogram, we begin by drawing two axes as shown. Mark the beginning of each class at the bottom of the graph on the horizontal axis, including the value that would be the lower limit of the next class. In the example, that will be the numbers 40, 45, 50, 55, 60, 65, and 70. On the vertical axis, mark the frequencies. Make sure your axis goes at least to the highest frequency in the frequency distribution.

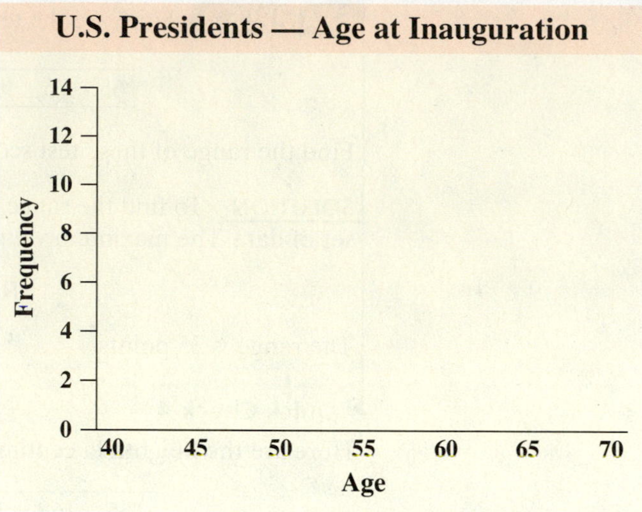

A bar is drawn above each class, and the height of the bar is determined by the frequency of that class. The first bar, from 40 to 45, should have a height of 2; the second bar should have a height of 7; and so on. Here is the histogram.

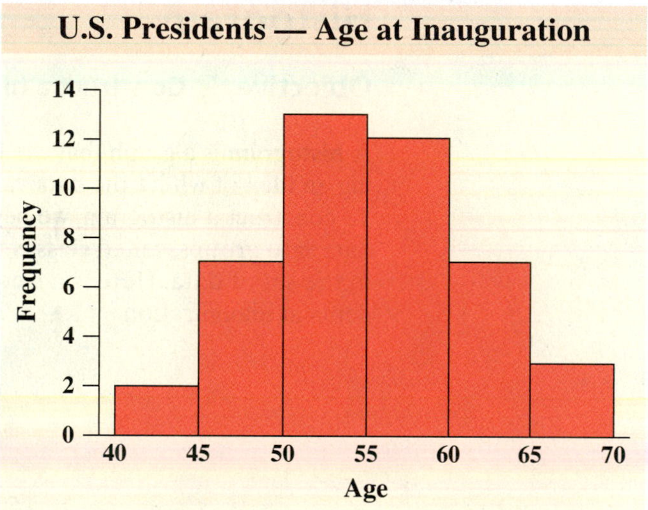

From this histogram, we can see that the bulk of the values are between 50 and 60, which is the center of the values.

EXAMPLE 5 Here is a frequency distribution showing the ages of 30 students enrolled in an online algebra class.

Age	Frequency
17 to 20	17
21 to 24	6
25 to 28	2
29 to 32	1
33 to 36	3
37 to 40	1

Draw a histogram for the frequency distribution.

SOLUTION On the horizontal axis, we begin the labels at 17 and increase by 4 until we reach 41. On the vertical axis, we must have labels that reach at least 17, which is the largest frequency.

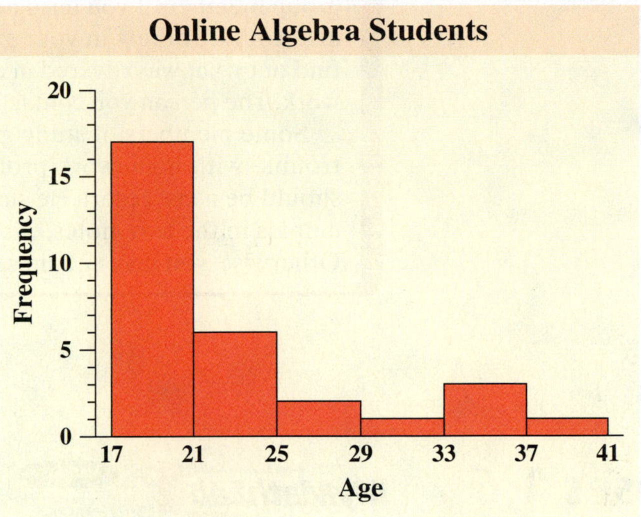

Quick Check 5

Here is a frequency distribution showing the ages of 86 passengers on a cruise.

Ages	Frequency
25 to 34	3
35 to 44	7
45 to 54	6
55 to 64	15
65 to 74	35
75 to 84	20

Draw a histogram for the frequency distribution.

BUILDING YOUR STUDY STRATEGY

Study Groups, 6 Keeping in Touch It is important for group members to share phone numbers and e-mail addresses. This will allow you to contact other group members if you misplace the homework assignment for that day. You can also contact others in your group if you have to miss class. This allows you to find out what was covered in class and which problems were assigned for homework. The person you contact can also give you advice on certain problems.

Some members of study groups agree to call each other if they are having trouble with homework problems. Calling a group member for this purpose should be a last resort. Be sure that you have used all available resources (examples in the text, notes, etc.) and have given the problem your fullest effort. Otherwise, you may end up calling for help on all of the problems.

Exercises 1.6

Vocabulary

1. Statistics that describe the typical value for a set of data are often called measures of _______.
2. Statistics that describe how varied a set of data is are often called measures of _______.
3. The _______ of a set of data is the sum of all of the values divided by the number of values.
4. The _______ of a set of data is the value in the center of the data once the values are arranged in ascending order.
5. The _______ of a set of data is the value that is repeated most often.
6. The _______ of a set of data is the average of its minimum and maximum values.
7. The _______ of a set of data is the difference between its maximum and minimum values.
8. A(n) _______ is a graph that can be used to show how a set of data is distributed.

Find the mean for the given values.

9. 63, 98, 21, 42, 71
10. 84, 37, 29, 46, 15, 65
11. 5, 17, 21, 35, 42, 59, 89, 106
12. 30, 70, 74, 82, 95, 113, 128, 140

Find the median for the given values.

13. 97, 76, 22, 103, 80, 45, 66
14. 87, 3, 20, 62, 55, 73, 101, 49, 75
15. 68, 47, 32, 90, 85, 40, 83, 39, 50, 77
16. 123, 304, 290, 175, 260, 209, 321, 275

Find the mode, if it exists, for the given values.

17. 70, 56, 63, 35, 56, 63, 36, 56, 19
18. 80, 50, 70, 80, 50, 70, 80, 40, 60
19. 7, 41, 32, 56, 41, 19, 8, 32, 25
20. 61, 47, 47, 17, 29, 16, 25, 92, 16
21. 5, 35, 89, 106, 42, 17, 59, 21
22. 88, 45, 6, 99, 32, 75, 16, 100, 42

For the given values, find the midrange and the range.

23. 70, 140, 87, 62, 196, 125, 155
24. 93, 47, 28, 80, 94, 60, 93
25. 406, 354, 509, 427, 516, 379
26. 165, 82, 97, 155, 79, 203, 121, 99

For Exercises 27–34, find the a) mean, b) median, c) mode, d) midrange, and e) range.

27. 22, 13, 16, 30, 32, 19, 24, 30, 21
28. 59, 41, 46, 62, 41, 50, 65
29. 48, 45, 63, 36, 50, 38, 73, 63
30. 98, 84, 44, 40, 50, 82, 43, 84, 46, 70

31. 63, 86, 76, 85, 59, 71, 34, 44, 30, 67, 44, 77, 50, 83, 76

32. 48, 91, 38, 101, 93, 66, 31, 57, 84, 47, 73, 41, 86, 90, 96, 86, 62

33. 196, 295, 213, 69, 371, 77, 253, 210, 298, 210, 426, 327, 270, 323, 262, 70, 459, 481, 278, 192

34. 257, 50, 23, 223, 125, 249, 197, 191, 99, 194, 239, 227, 192, 96, 50, 147, 259, 296

Find the mean for the given set of data.

35. IQ of 12 college students

95 82 104 119 118 126 82 96 116 85 90 90

36. Red Sox home runs (2001–2008)

2001	2002	2003	2004	2005	2006	2007	2008
198	177	238	222	199	192	166	173

37. Systolic blood pressure of seven 60-year-old men

133 112 142 154 102 139 149

38. Systolic blood pressure of nine 60-year-old women

119 160 121 92 109 95 114 112 122

39. Weight (in ounces) of 10 newborn baby girls

101 110 125 120 106 113 102 108 132 135

40. Starting salary for 5 bachelor's degrees

Chemical Engineering	Computer Science	Mathematics	Political Science	English
$61,800	$54,200	$43,500	$39,400	$36,700

(Source: PayScale.com)

Find the median for the given set of data.

41. Number of Facebook friends for 6 college math instructors

243 18 21 152 93 125

42. Serum glucose level (mg/dL) of 8 people

90 91 94 122 113 142 59 92

43. Math test scores of 9 members of a study group

80 96 100 89 74 96 95 98 87

44. Pregnancy duration (days) for 11 women

267 255 263 261 265 273 264 267 268 275 273

45. Number of hours spent studying last week by 10 college students

25 12 17 3 20 20 16 34 1 9

46. Number of hours spent working last week by 10 college students

8 4 20 0 12 32 8 40 20 16

For exercises 47–52, find the a) mean, b) median, c) mode, d) midrange, and e) range.

47. Room rate at 10 Las Vegas hotels (Valentine's Day, 2010)

$219 $259 $127 $199 $259 $169
$219 $229 $299 $199

48. Touchdown passes thrown by Joe Montana by year (16 seasons)

1 15 19 17 26 28 27 8 31 18 26 26 0 2 13 16

49. Time (in seconds) of the 8 songs on Bruce Springsteen's "Born to Run"

289 191 180 390 271 270 198 574

50. Number of calories in 16 different brands of beer

188 166 163 165 149 209 135 150 96 145 170
124 158 110 314 94

51. Cell phone minutes used by 14 families last month

636 754 662 884 1346 659 1006 1357
1129 904 1747 1336 1234 388

52. Systolic blood pressure of thirteen 65-year-old smokers (mmHg)

110 118 137 127 134 163 129 102
102 136 150 130 113

Find the missing value x that satisfies the given condition for the set of values.

53. Mean: 75

80 86 100 81 30 57 90 x

54. Mean: 82.5

43 96 90 x 81 104 111 72 66 89

55. Median: 101.5

112 98 121 72 x 65

56. Median: 93

97 x 81 100 104 88 121 79

57. Range: 84; midrange: 52

66 25 94 37 x 85 42

58. Range: 47; midrange: 35.5

30 29 58 12 16 45 x 50

Construct a histogram for the given frequency distribution.

59. Average 2007 SAT math score for the 50 states and Washington, D.C.
(*Source: The College Board*)

Average Score	Frequency
460 to 479	2
480 to 499	6
500 to 519	14
520 to 539	6
540 to 559	6
560 to 579	8
580 to 599	5
600 to 619	4

60. Scores of 40 students on an algebra exam

Score	Frequency
40 to 49	1
50 to 59	3
60 to 69	4
70 to 79	7
80 to 89	14
90 to 99	11

61. High school graduation rates for the 50 states (2005)

Graduation Rate	Frequency
45.0% to 49.9%	1
50.0% to 54.9%	2
55.0% to 59.9%	2
60.0% to 64.9%	4
65.0% to 69.9%	7
70.0% to 74.9%	16
75.0% to 79.9%	13
80.0% to 84.9%	5

62. Daily caloric intake of 60 participants in a health study

Calories	Frequency
1400 to 1599	1
1600 to 1799	6
1800 to 1999	15
2000 to 2199	21
2200 to 2399	9
2400 to 2599	8

Complete the frequency distribution for the given data and use it to construct a histogram.

63. Number of heads in 1000 coin flips, repeated by 60 students

471	505	515	503	506	507	471	522	548	478	514	490
511	463	515	514	490	485	467	531	487	482	500	506
504	492	522	497	508	499	515	499	516	495	499	496
510	520	509	500	488	512	501	506	488	497	498	503
496	488	522	505	517	497	500	502	472	525	477	506

Number of Heads	Frequency
450 to 469	
470 to 489	
490 to 509	
510 to 529	
530 to 549	

64. Starting salaries of 50 Certified Public Accountants (CPAs)

\$52,500	\$57,700	\$55,100	\$60,400	\$61,900
\$57,700	\$52,600	\$57,200	\$52,700	\$59,200
\$58,300	\$58,800	\$61,400	\$56,600	\$56,900
\$60,300	\$57,700	\$55,400	\$56,200	\$59,300
\$59,000	\$58,300	\$52,300	\$56,000	\$59,400
\$60,100	\$53,400	\$54,700	\$55,600	\$62,800
\$62,300	\$55,000	\$56,300	\$57,200	\$59,600
\$56,200	\$62,300	\$55,900	\$50,100	\$64,500
\$55,900	\$56,600	\$57,200	\$59,600	\$60,300
\$55,900	\$63,400	\$62,100	\$55,800	\$63,100

Salary	Frequency
\$50,000 to \$52,499	
\$52,500 to \$54,999	
\$55,000 to \$57,499	
\$57,500 to \$59,999	
\$60,000 to \$62,499	
\$62,500 to \$64,999	

65. IQs of 36 college students

104	114	88	96	105	120	92	107	134	110	95	133
132	119	115	116	129	114	100	95	105	140	110	131
97	113	104	120	95	128	128	106	96	114	127	133

IQ	Frequency
85 to 94	
95 to 104	
105 to 114	
115 to 124	
125 to 134	
135 to 144	

66. Blood glucose level (mg/dL) of 40 women participating in a clinical study

63	141	84	108	93	64	94	80	95	100
90	115	96	89	68	114	130	111	86	75
85	93	116	102	72	87	95	105	109	74
115	120	119	101	92	100	84	148	97	100

Blood Glucose Level	Frequency
60 to 74	
75 to 89	
90 to 104	
105 to 119	
120 to 134	
135 to 149	

Use the given histogram to create a frequency distribution.

67.

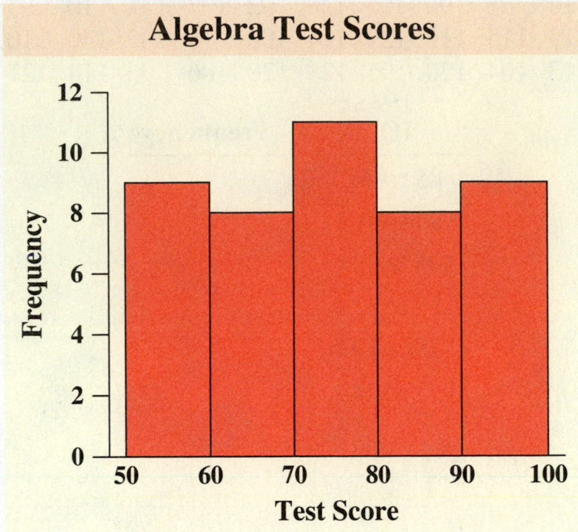

68.

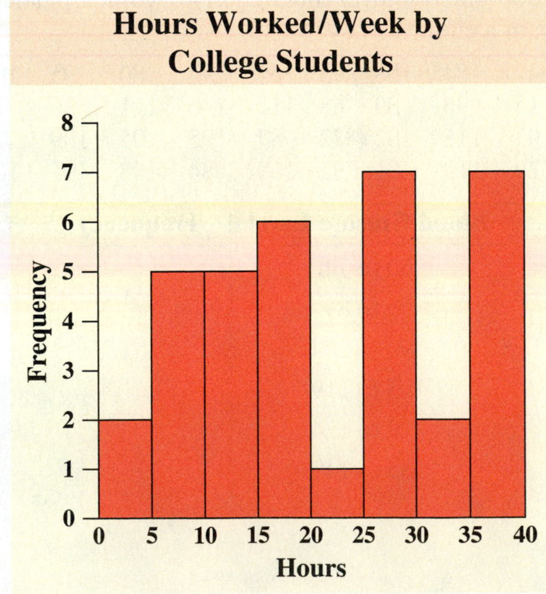

69.

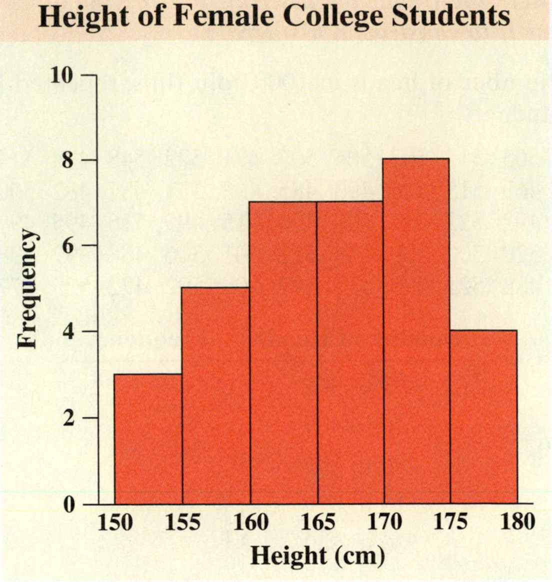

70.

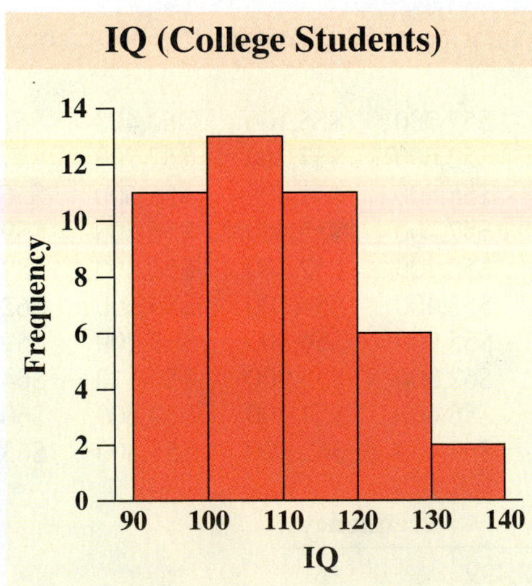

1.7 Exponents and Order of Operations

OBJECTIVES

1. Simplify exponents.
2. Use the order of operations to simplify arithmetic expressions.

Exponents

Objective 1 Simplify exponents. The same number used repeatedly as a factor can be represented using **exponential notation**. For example, $3 \cdot 3 \cdot 3 \cdot 3 \cdot 3$ can be written as 3^5.

Base, Exponent

For the expression 3^5, the number being multiplied (3) is called the **base**. The **exponent** (5) tells how many times the base is used as a factor.

We read 3^5 as *three raised to the fifth power* or simply *three to the fifth power.* When raising a base to the second power, we usually say that the base is being **squared**. When raising a base to the third power, we usually say that the base is being **cubed**. Exponents of 4 or higher do not have a special name.

EXAMPLE 1 Simplify 2^7.

SOLUTION In this example, 2 is a factor seven times.

$$\begin{aligned} 2^7 &= 2 \cdot 2 \cdot 2 \cdot 2 \cdot 2 \cdot 2 \cdot 2 && \text{Write 2 as a factor seven times.} \\ &= 128 && \text{Multiply.} \end{aligned}$$

Quick Check 1

Simplify 4^3.

Using Your Calculator Many calculators have a key that can be used to simplify expressions with exponents. When using the TI-84, use the [^] key. Other calculators may have a key that is labeled [y^x] or [x^y]. Here is the screen you should see when using the TI-84 to simplify the expression in Example 1.

```
2^7
                128
```

EXAMPLE 2 Simplify $\left(\frac{2}{3}\right)^3$.

SOLUTION When the base is a fraction, the same rules apply. Use $\frac{2}{3}$ as a factor three times.

$$\begin{aligned} \left(\frac{2}{3}\right)^3 &= \frac{2}{3} \cdot \frac{2}{3} \cdot \frac{2}{3} && \text{Write as a product.} \\ &= \frac{8}{27} && \text{Multiply.} \end{aligned}$$

Quick Check 2

Simplify $\left(\frac{1}{8}\right)^4$.

Consider the expression $(-3)^2$. The base is -3; so we multiply $(-3) \cdot (-3)$, and the result is positive 9. However, in the expression -2^4, the negative sign is not included in a set of parentheses with the 2. So the base of this expression is 2, not -2. We will use 2 as a factor 4 times and then take the opposite of the result: $-2^4 = -(2 \cdot 2 \cdot 2 \cdot 2) = -16$.

Order of Operations

Objective 2 Use the order of operations to simplify arithmetic expressions. Suppose we were asked to simplify the expression $2 + 4 \cdot 3$. We could obtain two different results depending on whether we performed the addition or the multiplication first. Performing the addition first would give us $6 \cdot 3 = 18$, while performing the multiplication first would give us $2 + 12 = 14$. Only one result is correct. The **order of operations agreement** is a standard order in which arithmetic operations are performed, ensuring a single correct answer.

Order of Operations

1. **Remove grouping symbols.** Begin by simplifying all expressions within parentheses, brackets, and absolute value bars. Also perform any operations in the numerator or denominator of a fraction. This is done by following Steps 2–4, presented next.
2. **Perform any operations involving exponents.** After all grouping symbols have been removed from the expression, simplify any exponential expressions.
3. **Multiply and divide.** These two operations have equal priority. Perform multiplications or divisions in the order they appear from left to right.
4. **Add and subtract.** At this point, the only remaining operations should be additions and subtractions. Again, these operations are of equal priority, and we perform them in the order they appear from left to right. We also can use the strategy for totaling integers from Section 1.2.

Considering this information, when we simplify $2 + 4 \cdot 3$, the correct result is 14 because multiplication takes precedence over the addition.

$$\begin{aligned} 2 + 4 \cdot 3 &= 2 + 12 && \text{Multiply } 4 \cdot 3. \\ &= 14 && \text{Add.} \end{aligned}$$

EXAMPLE 3 Simplify $4 + 3 \cdot 5 - 2^6$.

SOLUTION In this example, the operation with the highest priority is 2^6, because simplifying exponents takes precedence over addition or multiplication. Note that the base is 2, not -2, because the negative sign is not grouped with 2 inside a set of parentheses.

$$\begin{aligned} 4 + 3 \cdot 5 - 2^6 &= 4 + 3 \cdot 5 - 64 && \text{Raise 2 to the 6th power.} \\ &= 4 + 15 - 64 && \text{Multiply } 3 \cdot 5. \\ &= -45 && \text{Simplify.} \end{aligned}$$

Quick Check 3
Simplify $-2 \cdot 7 + 11 - 3^4$.

EXAMPLE 4 Simplify $(-5)^2 - 4(-2)(6)$.

SOLUTION Begin by squaring negative 5, which equals 25.

$$\begin{aligned} (-5)^2 - 4(-2)(6) &= 25 - 4(-2)(6) && \text{Square } -5. \\ &= 25 - (-48) && \text{Multiply } 4(-2)(6). \\ &= 25 + 48 && \text{Write as a sum, eliminating the double signs.} \\ &= 73 \end{aligned}$$

Quick Check 4
Simplify $9^2 - 4(-2)(-10)$.

A WORD OF CAUTION When we square a negative number such as $(-5)^2$ in the previous example, the result is a positive number.

$$(-5)^2 = (-5)(-5) = 25$$

EXAMPLE 5 Simplify $8 \div 2 + 3(7 - 4 \cdot 5)$.

SOLUTION The first step is to simplify the expression inside the set of parentheses. Here the multiplication takes precedence over the subtraction. Once we have simplified the expression inside the parentheses, we proceed to multiply and divide. We finish by subtracting.

$$\begin{aligned} 8 \div 2 + 3(7 - 4 \cdot 5) &= 8 \div 2 + 3(7 - 20) && \text{Multiply } 4 \cdot 5. \\ &= 8 \div 2 + 3(-13) && \text{Subtract } 7 - 20. \\ &= 4 + 3(-13) && \text{Divide } 8 \div 2. \\ &= 4 - 39 && \text{Multiply } 3(-13). \\ &= -35 && \text{Subtract.} \end{aligned}$$

Quick Check 5

Simplify $20 \div 5 \cdot 10(3 \cdot 6 - 9)$.

Using Your Calculator When using your calculator to simplify an expression using the order of operations, you may want to perform one operation at a time. However, if you are careful to enter all of the parentheses, you can enter the entire expression at one time. Here is how to simplify the expression in Example 5 using the TI-84.

```
8/2+3(7-4*5)
                -35
```

Occasionally, an expression will have one set of grouping symbols inside another set, such as the expression $3[7 - 4(9 - 3)] + 5 \cdot 4$. This is called **nesting** grouping symbols. We begin by simplifying the innermost set of grouping symbols and work our way out from there.

EXAMPLE 6 Simplify $3[7 - 4(9 - 3)] + 5 \cdot 4$.

SOLUTION We begin by simplifying the expression inside the set of parentheses. Once we do that, we simplify the expression inside the square brackets.

$$\begin{aligned} & 3[7 - 4(9 - 3)] + 5 \cdot 4 && \text{Subtract } 9 - 3 \text{ to simplify the expression inside} \\ &= 3[7 - 4 \cdot 6] + 5 \cdot 4 && \text{the parentheses. Then turn our attention to simplifying the expression inside the square brackets.} \\ &= 3[7 - 24] + 5 \cdot 4 && \text{Multiply } 4 \cdot 6. \\ &= 3[-17] + 5 \cdot 4 && \text{Subtract } 7 - 24. \\ &= -51 + 20 && \text{Multiply } 3[-17] \text{ and } 5 \cdot 4. \\ &= -31 && \text{Simplify.} \end{aligned}$$

Quick Check 6

Simplify $4[3^2 + 5(2 - 8)]$.

BUILDING YOUR STUDY STRATEGY

Study Groups, 7 Productive Group Members For any group or team to be effective, it must be made up of members who are committed to giving their full effort to success. Here are some pointers for being a productive study group member:

- **Arrive at each session fully prepared.** Make sure you have completed all required homework assignments. Bring a list of any questions you have.
- **Stay focused during study sessions.** Do not spend your time socializing.
- **Be open-minded.** During study sessions, you may be told that you are wrong. Keep in mind that the goal is to learn mathematics, not always to be correct initially.
- **Consider the feelings of others.** When a person has made a mistake, be supportive and encouraging.
- **Know when to speak and when to listen.** A study group works best when it is a collaborative body.

Exercises 1.7

MyMathLab

PRACTICE

WATCH

DOWNLOAD
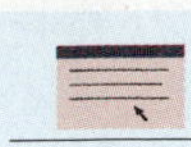
READ

REVIEW

Vocabulary

1. In exponential notation, the factor being multiplied repeatedly is called the ________.
2. In exponential notation, the exponent tells us how many times the ______ is repeated as a(n) ________.
3. When a base is ________, it is raised to the second power.
4. When a base is _______, it is raised to the third power.
5. Symbols such as ( ) and [ ] are called __________ symbols.
6. When simplifying arithmetic expressions, once all grouping symbols have been removed, we perform any operations involving ____________.

Rewrite the given expression using exponential notation.

7. $2 \cdot 2 \cdot 2$

8. $9 \cdot 9 \cdot 9 \cdot 9 \cdot 9$

9. $(-2)(-2)(-2)(-2)$

10. $(-6)(-6)(-6)(-6)(-6)$

11. $-3 \cdot 3 \cdot 3 \cdot 3 \cdot 3$

12. $-8 \cdot 8 \cdot 8 \cdot 8$

13. *five to the third power*

14. *seven squared*

Simplify the given expression.

15. 3^4

16. 4^3

17. 7^5

18. 2^9

19. 10^6

20. 10^5

21. 1^{723}

22. 0^{2364}

23. $\left(\frac{3}{4}\right)^3$

24. $\left(\frac{1}{5}\right)^5$

25. 0.2^5

26. 0.5^4

27. $(-3)^4$

28. -3^4

29. -2^7

30. $(-2)^7$

31. $2^3 \cdot 5^2$

32. $9^2 \cdot 4^4$

33. $-6^2 \cdot (-2)^2$

34. $-3^2 \cdot (-8)^2$

Simplify the given expression.

35. $9 + 3 \cdot 4$

36. $8 - 2 \cdot 7$

37. $20 \div 5 \cdot 2^2$

38. $80 - 16 \div 2^3$

39. $8 \cdot 5 - 9 \cdot 7$

40. $8(5 - 9 \cdot 7)$

41. $3.2 + 2.8(-6.3)$

42. $(3.2 + 2.8) - 6.3$

43. $17.1 - 8.58 \div 3.9$

44. $4.7 \cdot 13.9 - 3.6^2$

45. $(-3)^2 - 4(-5)(-4)$

46. $(-6)^2 - 4(2)(7)$

47. $3^2 + 4^2$

48. $(3 + 4)^2$

49. $-4 - 5(7 - 3 \cdot 6)$

50. $3 - 6(5^2 - 4 + 2 \cdot 7)$

51. $\frac{1}{2} + \frac{1}{2} \cdot \frac{4}{7}$

52. $\frac{3}{5} \cdot \frac{2}{3} + \frac{15}{7} \cdot \frac{21}{20}$

53. $\frac{2}{9} \div \frac{5}{3} \cdot \frac{33}{50}$

54. $4 \cdot \frac{3}{8} - \left(\frac{7}{6}\right)^2$

55. $\frac{1}{9}\left(\frac{19}{28} - \frac{1}{4}\right)$

56. $\frac{8}{25} \div \left(\frac{4}{15} - \frac{1}{3} + \frac{3}{5}\right) \cdot \frac{7}{18}$

57. $\frac{2^2 - 9}{3^3 - 7}$

58. $\frac{3 + 5 \cdot 7 - 4 \cdot 2}{1 + 2 \cdot 17}$

59. $\frac{(3 + 5) \cdot 6 - 8}{1 + 3^2 + 2}$

60. $\frac{10^3 + 9^3}{1^3 + 12^3}$

61. $3 - [4(5 - 6 \cdot 7)]$

62. $18 + [9 - (4 - 5 \cdot 8)] \div 3^2$

63. $-4 \cdot 9 - |-7(3 + 5)| - 2^3$

64. $-6^2 + 2|(7 - 49) \div (2 \cdot 3)|$

65. $9 - |3^2 + 2^3 - 10 \cdot 9|$

66. $|-4^2 + 19| - |(-7)^2 + 5(10)|$

67. $-21(|4 - 3 \cdot 9| + |8(13 - 4)|)$

68. $(6 \cdot 5 + 40 \div 20)(|8^2 - 2 \cdot 17| - |10^2 - 2 \cdot 51|)$

Construct an "order-of-operations problem" of your own that involves at least four numbers and produces the given result. Answers will vary. Examples are shown.

69. 41

70. 0

71. −13

72. −19

The size of a computer's memory is measured by the number of bytes that it can store. The following table lists the number of bytes in commonly used storage units.

1 kilobyte (KB)	2^{10} bytes
1 megabyte (MB)	2^{20} bytes
1 gigabyte (GB)	2^{30} bytes
1 terabyte (TB)	2^{40} bytes

Calculate the number of bytes in each of the following.

73. 1 kilobyte

74. 1 megabyte

75. 1 gigabyte

76. 1 terabyte

Find the missing number.

77. $5^? = 125$

78. $6^? = 7776$

79. $?^3 = 729$

80. $?^4 = 2401$

81. $\left(\frac{3}{?}\right)^4 = \frac{81}{4096}$

82. $\left(\frac{2}{5}\right)^? = \frac{64}{15{,}625}$

Insert the arithmetic operation signs +, −, •, and ÷ between the values to produce the desired results. (You may use parentheses as well.)

83. 3 5 9 8 = 14

84. 4 7 2 6 = 14

85. 3 7 9 2 = 33

86. 8 2 3 14 4 7 = 2

Writing in Mathematics

Answer in complete sentences.

87. Explain the difference between $(-2)^6$ and -2^6.

88. ***Newsletter*** Write a newsletter explaining how to use the order of operations to simplify an expression.

1.8 Introduction to Algebra

OBJECTIVES

1. Build variable expressions.
2. Evaluate algebraic expressions.
3. Use the commutative, associative, and distributive properties of real numbers.
4. Identify terms and their coefficients.
5. Simplify variable expressions.

Variables

The number of students absent from a particular English class changes from day to day, as does the closing price for a share of Microsoft stock and the daily high temperature in Providence, Rhode Island. Quantities that change, or vary, are often represented by variables. A **variable** is a letter or symbol that is used to represent a quantity that changes or that has an unknown value.

Variable Expressions

Objective 1 Build variable expressions. Suppose a buffet restaurant charges \$7 per person to eat. To determine the bill for a family to eat in the restaurant, we multiply the number of people by \$7. The number of people can change from family to family, so we can represent this quantity by a variable such as x. The bill for a family with x people can be written as $7 \cdot x$, or simply $7x$. This expression for the bill, $7x$, is known as a variable expression. A **variable expression** is a combination of one or more variables with numbers or arithmetic operations. When the operation is multiplication such as $7 \cdot x$, we often omit the multiplication dot. Here are other examples of variable expressions.

$$3a + 5 \qquad x^2 + 3x - 10 \qquad \frac{y + 5}{y - 3}$$

EXAMPLE 1 Write an algebraic expression for *the sum of a number and 17.*

SOLUTION Choose a variable to represent the unknown number. Let x represent the number. The expression is $x + 17$.

Other terms to look for that suggest addition are *plus, increased by, more than,* and *total.*

Quick Check 1

Write an algebraic expression for *9 more than a number.*

EXAMPLE 2 Write an algebraic expression for *five less than a number.*

SOLUTION Let x represent the number. The expression is $x - 5$.

Be careful with the order of subtraction when the expression *less than* is used. *Five less than a number* says that we need to subtract 5 from that number. A common error is to write the subtraction in the opposite order.

Other terms that suggest subtraction are *difference, minus,* and *decreased by.*

Quick Check 2

Write an algebraic expression for *a number decreased by 25.*

A WORD OF CAUTION *Five less than a number* is written as $x - 5$, not $5 - x$.

EXAMPLE 3 Write an algebraic expression for *the product of 3 and two different numbers.*

SOLUTION Because we are building an expression involving two different unknown numbers, we need to introduce two variables. Let x and y represent the two numbers. The expression is $3xy$.

Other terms that suggest multiplication are *times, multiplied by, of,* and *twice.*

Quick Check 3

Write an algebraic expression for *twice a number.*

EXAMPLE 4 Four friends decide to rent a fishing boat for the day. Assuming that all 4 friends decide to split the cost of renting the boat evenly, write a variable expression for the amount each friend will pay.

SOLUTION Let c represent the cost of the boat. The expression is $c \div 4$ or $\frac{c}{4}$.

Other terms that suggest division are *quotient, divided by,* and *ratio.*

Quick Check 4

Write an algebraic expression for *the quotient of a number and 20.*

Here are some phrases that translate to $x + 5$, $x - 10$, $8x$, and $\frac{x}{6}$:

$x + 5$	$x - 10$
The sum of a number and 5 A number plus 5 A number increased by 5 5 more than a number The total of a number and 5	10 less than a number The difference of a number and 10 A number minus 10 A number decreased by 10
$8x$	**$\frac{x}{6}$**
The product of 8 and a number 8 times a number 8 multiplied by a number	The quotient of a number and 6 A number divided by 6 The ratio of a number and 6

Evaluating Variable Expressions

Objective 2 Evaluate algebraic expressions. We often will have to evaluate variable expressions for particular values of variables. To do this, we substitute the appropriate numerical value for each variable and then simplify the resulting expression using the order of operations.

EXAMPLE 5 Evaluate $2x - 7$ for $x = 6$.

SOLUTION The first step in evaluating a variable expression is to rewrite the expression, replacing each variable with a set of parentheses. For example, rewrite $2x - 7$ as $2(\ ) - 7$. Then we can substitute the appropriate value for each variable and simplify.

$$\begin{aligned} &2x - 7 && \\ &2(6) - 7 && \text{Substitute 6 for } x. \\ &= 12 - 7 && \text{Multiply.} \\ &= 5 && \text{Subtract.} \end{aligned}$$

The expression $2x - 7$ is equal to 5 for $x = 6$.

Quick Check 5

Evaluate $5x + 2$ for $x = 11$.

EXAMPLE 6 Evaluate $x^2 - 5x + 6$ for $x = -5$.

SOLUTION

$$\begin{aligned} &x^2 - 5x + 6 && \\ &(-5)^2 - 5(-5) + 6 && \text{Substitute } -5 \text{ for } x. \\ &= 25 - 5(-5) + 6 && \text{Square } -5. \\ &= 25 + 25 + 6 && \text{Multiply.} \\ &= 56 && \text{Add.} \end{aligned}$$

Quick Check 6

Evaluate $x^2 - 13x - 40$ for $x = -8$.

EXAMPLE 7 Evaluate $b^2 - 4ac$ for $a = 3$, $b = -2$, and $c = -10$.

SOLUTION

$$\begin{aligned} &b^2 - 4ac && \\ &(-2)^2 - 4(3)(-10) && \text{Substitute 3 for } a, -2 \text{ for } b, \text{ and } -10 \text{ for } c. \\ &= 4 - 4(3)(-10) && \text{Square negative 2.} \\ &= 4 - (-120) && \text{Multiply.} \\ &= 4 + 120 && \text{Rewrite without double signs.} \\ &= 124 && \text{Add.} \end{aligned}$$

Quick Check 7

Evaluate $b^2 - 4ac$ for $a = -1$, $b = 5$, and $c = 18$.

Properties of Real Numbers

Objective 3 Use the commutative, associative, and distributive properties of real numbers. Now we examine three properties of real numbers. The first is the **commutative property**, which is used for addition and multiplication.

Commutative Property

For all real numbers a and b,

$$a + b = b + a \quad \text{and} \quad a \cdot b = b \cdot a.$$

This property states that changing the order of the numbers a sum or a product does not change the result. For example, $3 + 9 = 9 + 3$ and $7 \cdot 8 = 8 \cdot 7$. Note that this property does not work for subtraction or division. Changing the order of the numbers in subtraction or division generally changes the result.

The second property is the **associative property**, which also holds for addition and multiplication.

Associative Property

For all real numbers a, b, and c,

$$(a + b) + c = a + (b + c) \quad \text{and} \quad (ab)c = a(bc).$$

This property states that changing the grouping of the numbers in a sum or a product does not change the result. Notice that $(2 + 7) + 3$ is equal to $2 + (7 + 3)$.

$$\begin{aligned} &(2 + 7) + 3 \\ &= 9 + 3 \\ &= 12 \end{aligned} \qquad \begin{aligned} &2 + (7 + 3) \\ &= 2 + 10 \\ &= 12 \end{aligned}$$

Also notice that $5(12x)$ can be rewritten using the associative property as $(5 \cdot 12)x$, which is equal to $60x$.

The third property of real numbers is the **distributive property**.

Distributive Property

For all real numbers a, b, and c,

$$a(b + c) = ab + ac.$$

This property says that we can distribute the factor outside the parentheses to each number being added in the parentheses, perform the multiplications, and then add. (This property also holds true when the operation inside the parentheses is subtraction.) Consider the expression $3(5 + 4)$. The order of operations says that this is equal to $3 \cdot 9$, or 27. Here is how to simplify the expression using the distributive property.

$$\begin{aligned} 3(5 + 4) &= 3 \cdot 5 + 3 \cdot 4 && \text{Apply the distributive property.} \\ &= 15 + 12 && \text{Multiply } 3 \cdot 5 \text{ and } 3 \cdot 4. \\ &= 27 && \text{Add.} \end{aligned}$$

Either way we get the same result.

EXAMPLE 8 Simplify $2(4 + 3x)$ using the distributive property.

SOLUTION

$$\begin{aligned} 2(4 + 3x) &= 2 \cdot 4 + 2 \cdot 3x && \text{Distribute the 2.} \\ &= 8 + 6x && \text{Multiply. Recall from the associative property that } 2 \cdot 3x \text{ equals } 6x. \end{aligned}$$

Quick Check 8

Simplify $7(5x - 4)$ using the distributive property.

A WORD OF CAUTION The expression $8 + 6x$ is not equal to $14x$.

EXAMPLE 9 Simplify $7(2 + 3a - 4b)$ using the distributive property.

Quick Check 9

Simplify $12(x - 2y + 3z)$ using the distributive property.

SOLUTION When more than two terms are inside the parentheses, distribute the factor to each term. Also, it is a good idea to distribute the factor and multiply mentally.

$$7(2 + 3a - 4b) = 14 + 21a - 28b$$

EXAMPLE 10 Simplify $-4(2x - 5)$ using the distributive property.

SOLUTION When the factor outside the parentheses is negative, the negative number must be distributed to each term inside the parentheses. This will change the sign of each term inside the parentheses.

$$\begin{aligned} -4(2x - 5) &= (-4) \cdot 2x - (-4) \cdot 5 && \text{Distribute the } -4. \\ &= -8x - (-20) && \text{Multiply.} \\ &= -8x + 20 && \text{Rewrite without double signs.} \end{aligned}$$

Quick Check 10

Simplify $-6(4x + 11)$ using the distributive property.

Simplifying Variable Expressions

Objective 4 Identify terms and their coefficients. In an algebraic expression, a **term** is a number, a variable, or a product of a number and variables. Terms in an algebraic expression are separated by addition. The expression $7x - 5y + 3$, has three terms: $7x$, $-5y$, and 3. The numerical factor of a term is its **coefficient**. The coefficients of these terms are 7, −5, and 3.

EXAMPLE 11 For the expression $-5x + y + 3xy - 19$, determine the number of terms, list them, and state the coefficient for each term.

SOLUTION This expression has four terms: $-5x$, y, $3xy$, and -19.

What is the coefficient for the second term? Although y does not appear to have a coefficient, its coefficient is 1. This is because y is the same as $1 \cdot y$. So the four coefficients are −5, 1, 3, and −19.

Quick Check 11

For the expression $x^3 - x^2 + 23x - 59$, determine the number of terms, list them, and state the coefficient for each term.

Objective 5 Simplify variable expressions. Two terms that have the same variable factors with the same exponents, or that are both constants, are called **like terms**. Consider the expression $9x + 8y + 6x - 3y + 8z - 5$. There are two sets of like terms in this expression: $9x$ and $6x$ as well as $8y$ and $-3y$. There are no like terms for $8z$ because no other term has z as its sole variable factor. Similarly, there are no like terms for the constant term −5.

Combining Like Terms

When simplifying variable expressions, we can combine like terms into a single term with the same variable part by adding or subtracting the coefficients of the like terms.

EXAMPLE 12 Simplify $4x + 11x$ by combining like terms.

SOLUTION These two terms are like terms because they both have the same variable factors. We can simply add the two coefficients to produce the expression $15x$.

$$4x + 11x = 15x$$

Quick Check 12

Simplify $3x + 7y + y - 5x$ by combining like terms.

A general strategy for simplifying algebraic expressions is to begin by applying the distributive property. We can then combine any like terms.

EXAMPLE 13 Simplify $8(3x - 5) - 4x + 7$.

SOLUTION The first step is to use the distributive property by distributing the 8 to each term inside the parentheses. Then we will be able to combine like terms.

$$8(3x - 5) - 4x + 7 = 24x - 40 - 4x + 7$$ Distribute the 8.

$$= 20x - 33$$ Combine like terms.

Quick Check 13

Simplify $5(2x + 3) + 9x - 8$.

EXAMPLE 14 Simplify $5(9 - 7x) - 10(3x + 4)$.

SOLUTION We must make sure that we distribute the *negative* 10 into the second set of parentheses.

$$5(9 - 7x) - 10(3x + 4) = 45 - 35x - 30x - 40$$ Distribute the 5 to each term in the first set of parentheses and distribute the -10 to each term in the second set of parentheses.

$$= -65x + 5$$ Combine like terms.

Usually we write the simplified result with variable terms preceding constant terms, but it also is correct to write $5 - 65x$ because of the commutative property.

Quick Check 14

Simplify $3(2x - 7) - 8(x - 9)$.

A WORD OF CAUTION If a factor in front of a set of parentheses is negative or has a subtraction sign in front of it, we must distribute the negative sign along with the factor.

BUILDING YOUR STUDY STRATEGY

Study Groups, 8 Dealing with Unproductive Group Members Occasionally, a group will have a member who is not productive or is even disruptive. You should not allow one person to prevent your group from being successful. If you have a group member who is not contributing to the group in a positive way, talk to that person one-on-one discreetly.

Try to find solutions that will be acceptable to the group and the group member in question. Seek advice from your instructor. Undoubtedly, your instructor has seen a similar situation and may have valuable advice for you.

Try to be professional about the situation rather than adversarial. Remember, you will continue to see this person in class each day and you want to stay on good terms.

Exercises 1.8

PRACTICE

WATCH

DOWNLOAD

READ

REVIEW

Vocabulary

1. A(n) __________ is a letter or symbol used to represent a quantity that changes or has an unknown value.
2. A combination of one or more variables with numbers and/or arithmetic operations is a(n) __________ expression.
3. To __________ a variable expression, substitute the appropriate numerical value for each variable and then simplify the resulting expression using the order of operations.
4. For any real numbers a and b, the __________ property states that $a + b = b + a$ and $a \cdot b = b \cdot a$.
5. For any real numbers a, b, and c, the __________ property states that $(a + b) + c = a + (b + c)$ and $(a \cdot b) \cdot c = a \cdot (b \cdot c)$.
6. For any real numbers a, b, and c, the __________ property states that $a(b + c) = ab + ac$.
7. A(n) __________ is a number, a variable, or a product of numbers and variables.
8. A(n) __________ is the numerical factor of a term.
9. Two terms that have the same variable factors, or that are constant terms, are called __________.
10. To combine two like terms, add their __________.

Build a variable expression for the following phrases.

11. A number increased by 15
12. A number decreased by 33
13. Twenty-four less than a number
14. Forty-one more than a number
15. Three times a number
16. A number divided by 8
17. Nineteen more than twice a number
18. Seven less than 4 times a number
19. The sum of two different numbers
20. One number divided by another
21. Seven times the difference of two numbers
22. Half the sum of a number and 25
23. A college charges $325 per credit for tuition. Letting c represent the number of credits that a student is taking, build a variable expression for the student's tuition.
24. The admission charge for a particular amusement park is $54.95 per person. Letting p represent the number of people that attended the amusement park yesterday, build a variable expression for the total admission charges the park collected yesterday.
25. A professional baseball player is appearing at a baseball card convention. The promoter agreed to pay the player a flat fee of $25,000 plus $22 per autograph signed. Letting a represent the number of autographs signed, build a variable expression for the amount of money the player will be paid.
26. Jim Rockford, a private investigator from the 1970's TV show *The Rockford Files,* charged $200 per day plus expenses to take a case. Letting d represent the number of days Rockford worked on a case and assuming that he had $425 in expenses, build a variable expression for the amount of money he would charge for the case.

Write the given expression using words.

27. $x - 9$
28. $x + 16$
29. $7x$
30. $6x + 5$
31. $8x - 10$
32. $2x - 7$

Evaluate the following algebraic expressions under the given conditions.

33. $8x + 31$ for $x = 6$
34. $27 - 4x$ for $x = 9$
35. $7a - 13b$ for $a = 8$ and $b = 11$
36. $9m + 10n$ for $m = 15$ and $n = -5$
37. $2(3x + 8)$ for $x = 5$
38. $6x + 16$ for $x = 5$
39. $x^2 + 9x + 18$ for $x = 5$
40. $a^2 - 7a - 30$ for $a = 3$
41. $y^2 + 4y - 17$ for $y = -3$

42. $x^2 - 5x - 9$ for $x = -4$

43. $m^2 - 9$ for $m = 3$

44. $5 - x^2$ for $x = -2$

45. $b^2 - 4ac$ for $a = 2, b = 5$, and $c = -3$

46. $b^2 - 4ac$ for $a = -5, b = 7$, and $c = -4$

47. $b^2 - 4ac$ for $a = -1, b = -2$, and $c = 10$

48. $b^2 - 4ac$ for $a = 15, b = -6$, and $c = 9$

49. $5(x + h) - 17$ for $x = -3$ and $h = 0.01$

50. $-3(x + h) + 4$ for $x = -6$ and $h = 0.001$

51. $(x + h)^2 - 5(x + h) - 14$ for $x = -3$ and $h = 0.1$

52. $(x + h)^2 + 4(x + h) - 28$ for $x = 6$ and $h = 1$

53. The commutative property works for addition but does not work in general for subtraction.

a) Give an example of two numbers a and b such that $a - b \neq b - a$.

b) Can you find two numbers a and b such that $a - b = b - a$?

54. The commutative property works for multiplication but does not work in general for division.

a) Give an example of two numbers a and b such that $\frac{a}{b} \neq \frac{b}{a}$.

b) Can you find two numbers a and b such that $\frac{a}{b} = \frac{b}{a}$?

Simplify where possible.

55. $3(x - 9)$

56. $4(2x + 5)$

57. $5(3 - 7x)$

58. $12(5x - 7)$

59. $-2(5x + 7)$

60. $-3(8x + 3)$

61. $-6(9x - 5)$

62. $-4(-10x + 1)$

63. $7x + 9x$

64. $14x + 3x$

65. $3x - 8x$

66. $22a - 15a$

67. $3x - 7 + 4x + 11$

68. $15x + 32 - 19x - 57$

69. $5x - 3y - 7x - 19y$

70. $15m - 11n - 6m + 22n$

71. $3(2x - 5) + 4x + 7$

72. $8 - 19k + 6(3k - 5)$

73. $2(4x - 9) - 11$

74. $3(2a + 11b) - 13a$

75. $6y - 5(3y - 17)$

76. $5 - 9(3 - 4x)$

77. $3(4z - 7) + 9(2z + 3)$

78. $-6(5x - 9) - 7(13 - 12x)$

79. $-2(5a + 4b - 13c) + 3b$

80. $-7(-2x + 3y - 17z) - 5(3x - 11)$

For the following expressions:
a) Determine the number of terms.
b) Write down each term.
c) Write down the coefficient for each term.
Make sure you simplify each expression before answering.

81. $5x^3 + 3x^2 - 7x - 15$

82. $-3a^2 - 7a - 10$

83. $3x - 17$

84. $9x^4 - 10x^3 + 13x^2 - 17x + 329$

85. $5(3x^2 - 7x + 11) - 6x$

86. $4(3a - 5b - 7c - 11) - 2(6b - 5c)$

87. $5(-7a - 3b + 5c) - 3(6b - 9c - 23)$

88. $2(-4x + 7y) - (3x + 9y)$

Writing in Mathematics

Answer in complete sentences.

89. Give an example of a real-world situation that can be described by the variable expression $60x$. Explain why the expression fits your situation.

90. Give an example of a real-world situation that can be described by the variable expression $20x + 35$. Explain why the expression fits your situation.

91. ***Solutions Manual*** Write a solutions manual page for the following problem.
Simplify $4(2x - 5) - 3(3x - 8) - 7x$.

CHAPTER 1 SUMMARY

Section 1.1 Integers, Opposites, and Absolute Value

Inequalities, pp. 2–4

The number a is greater than b $(a > b)$ if a is to the right of the number b on the number line.

The number a is less than b $(a < b)$ if a is to the left of b on the number line.

Place the correct sign, $<$ or $>$, between the two integers: 7 ______ 3

$7 > 3$ because 7 is to the right of 3 on the number line.

Opposites, pp. 3–4

Two numbers are opposites if they are on different sides of 0 on the number line and are the same distance from 0.

The opposite of 4: -4, The opposite of -12: 12

Absolute Values, pp. 4–5

The absolute value of a number a, denoted $|a|$, is the distance between a and 0 on the number line.

$|-15| = 15, |9| = 9$

Section 1.2 Operations with Integers

Adding a Positive Integer and a Negative Integer, pp. 6–8

Find the difference between the absolute values of the two integers.

The sign of the result is the same as the sign of the number that has the largest absolute value.

Add: $-13 + 8$

$$|-13| = 13, |8| = 8$$
$$13 - 8 = 5$$
$$-13 + 8 = -5$$

Adding Two Integers with the Same Sign, p. 8

Add the absolute values of the two integers.

The sign of the result is the same as the sign of the two integers.

Add: $-25 + (-18)$

$$|-25| = 25, |-18| = 18$$
$$25 + 18 = 43$$
$$-25 + (-18) = -43$$

General Strategy for Adding/Subtracting Integers, p. 9

- Simplify double signs.
- Determine whether each integer is contributing positively or negatively to the total.
- Add all integers that are contributing positively to the total. Add all integers that are contributing negatively to the total.
- Finish by finding the sum of these two totals.

Simplify: $9 + (-17) - 16 + 4 - (-11)$

$$\begin{aligned} &9 + (-17) - 16 + 4 - (-11) \\ &= 9 - 17 - 16 + 4 + 11 \\ &= 24 - 33 \\ &= -9 \end{aligned}$$

Section 1.3 Fractions

Factor Set, p. 15

The collection of all factors of a natural number is called its factor set.

Find the factor set of 12.

$12 = 1 \cdot 12,\ 2 \cdot 6,\ 3 \cdot 4$

Factor set: $\{1, 2, 3, 4, 6, 12\}$

Prime Numbers, pp. 15–16

A natural number is prime if it is greater than 1 and its only two factors are 1 and itself.

Is 39 prime?

No, because $3 \cdot 13 = 39$.

Prime Factorization, p. 16

To find the prime factorization of a natural number, rewrite the number as a product of prime factors.

Find the prime factorization of 120.

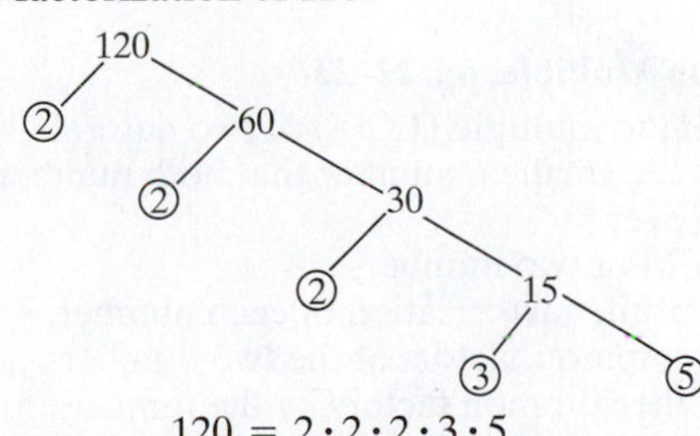

$120 = 2 \cdot 2 \cdot 2 \cdot 3 \cdot 5$

Lowest Terms, pp. 16–17

A fraction is in lowest terms if its numerator and denominator do not have any common factors other than 1.

Write in lowest terms: $\frac{12}{42}$

$$\frac{12}{42} = \frac{2\cdot 2\cdot 3}{2\cdot 3\cdot 7} = \frac{\overset{1}{\cancel{2}}\cdot 2\cdot \overset{1}{\cancel{3}}}{\underset{1}{\cancel{2}}\cdot \underset{1}{\cancel{3}}\cdot 7} = \frac{2}{7}$$

Rewriting an Improper Fraction as a Mixed Number, pp. 17–18

To rewrite an improper fraction as a mixed number, divide the numerator by the denominator.
The quotient is the whole number part of the mixed number, and the remainder is the numerator of the fraction part.

Express as a mixed number: $\frac{53}{8}$

$$\begin{array}{r} 6 \\ 8\overline{)53} \\ -48 \\ \hline 5 \end{array}$$

$$\frac{53}{8} = 6\frac{5}{8}$$

Rewriting a Mixed Number as an Improper Fraction, p. 18

To change a mixed number to an improper fraction, multiply the whole number part of the mixed number by the denominator of the fraction part.
Add the numerator to this product to find the numerator of the improper fraction.

Express as an improper fraction: $7\frac{2}{9}$

$$7\cdot 9 = 63$$
$$63 + 2 = 65$$
$$7\frac{2}{9} = \frac{65}{9}$$

Section 1.4 Operations with Fractions

Multiplying Fractions, pp. 20–21

To multiply two fractions, divide out any factors that are common to a numerator and a denominator.
Multiply the two numerators and multiply the two denominators.

Multiply: $\frac{3}{10}\cdot\frac{8}{9}$

$$\frac{\overset{1}{\cancel{3}}}{\underset{5}{\cancel{10}}}\cdot\frac{\overset{4}{\cancel{8}}}{\underset{3}{\cancel{9}}} = \frac{4}{15}$$

Dividing Fractions, pp. 21–22

To divide a fraction by another fraction, invert the divisor and multiply the resulting fractions.

Divide: $\frac{35}{12} \div \frac{21}{20}$

$$\frac{35}{12} \div \frac{21}{20} = \frac{35}{12}\cdot\frac{20}{21}$$
$$= \frac{\overset{5}{\cancel{35}}}{\underset{3}{\cancel{12}}}\cdot\frac{\overset{5}{\cancel{20}}}{\underset{3}{\cancel{21}}}$$
$$= \frac{25}{9}$$

Adding and Subtracting Fractions with the Same Denominator, p. 22

To add or subtract fractions that have the same denominator, add or subtract the numerators, placing the result over the common denominator.

Add: $\frac{7}{8} + \frac{5}{8}$

$$\frac{7}{8} + \frac{5}{8} = \frac{12}{8}$$
$$= \frac{3}{2}$$

Least Common Multiple, pp. 22–23

The least common multiple (LCM) of two natural numbers is the smallest number that both numbers divide into evenly.
To find the LCM of two numbers:

- Find the prime factorization of each number.
- Find the common factors of the two numbers.
- Multiply the common factors by the remaining factors of the two numbers.

Find the LCM of 36 and 54.

$$36 = 2\cdot 2\cdot 3\cdot 3$$
$$54 = 2\cdot 3\cdot 3\cdot 3$$
$$\text{LCM: } 2\cdot 2\cdot 3\cdot 3\cdot 3 = 108$$

Adding and Subtracting Fractions with Different Denominators, pp. 23–24

- Find the LCM of the denominators.
- Rewrite each fraction as an equivalent fraction whose denominator is the LCM of the original denominators.
- Add or subtract the numerators, placing the result over the common denominator.

Add: $\frac{3}{8} + \frac{7}{12}$

LCM: 24

$$\frac{3}{8} + \frac{7}{12} = \frac{3}{8}\cdot\frac{3}{3} + \frac{7}{12}\cdot\frac{2}{2} = \frac{9}{24} + \frac{14}{24} = \frac{23}{24}$$

Section 1.5 Decimals and Percents

Addition/Subtraction with Decimals, p. 27

Align the decimal points and add or subtract as you would with integers.

Add: 9.62 + 4.583

$$\begin{array}{r} \overset{1\,1}{9.620} \\ +\ 4.583 \\ \hline 14.203 \end{array}$$

Multiplication with Decimals, p. 27

Multiply two decimal numbers as you would integers. The number of decimal places in the product is the total number of decimal places in the two factors.

Multiply: $3.47 \cdot 5.2$

$$\begin{array}{r} 347 \\ \times\ 52 \\ \hline 694 \\ 17350 \\ \hline 18044 \end{array}$$

$3.47 \cdot 5.2 = 18.044$

Division with Decimals, p. 28

Move the decimal point in the divisor to the right so that it becomes an integer. Move the decimal point in the dividend to the right by the same number of spaces. The decimal point in the answer will be aligned with this new location of the decimal point in the dividend.

Divide: $9.568 \div 2.3$

$9.568 \div 2.3 \rightarrow 95.68 \div 23$

$$\begin{array}{r} 4.16 \\ 23\overline{)95.68} \\ -92 \\ \hline 36 \\ -23 \\ \hline 138 \\ -138 \\ \hline 0 \end{array}$$

Rewriting Fractions as Decimals, p. 28

To rewrite any fraction as a decimal, divide its numerator by its denominator.

Rewrite as a decimal: $\frac{8}{25}$

$$\frac{8}{25} = 8 \div 25 = 0.32$$

Rewriting Decimals as Fractions, pp. 28–29

Write the decimal as a whole number in the numerator. The denominator of the fraction can be found by determining the place value of the last decimal place.

Rewrite as a fraction: 0.74

$$0.74 = \frac{74}{100} = \frac{37}{50}$$

Section 1.6 Basic Statistics

Mean, p. 33

The mean of a set of data is the arithmetic average of the values.

Find the mean of 7, 12, 16, 25, 40.

$$\text{Mean} = \frac{7 + 12 + 16 + 25 + 40}{5} = \frac{100}{5} = 20$$

Median, pp. 33–34

To find the median for a set of data, begin by writing the values in ascending order.

If there is an odd number of values, the single value in the middle of the set of data is the median.

If there is an even number of values, the median is the average of the two center values.

7, 12, 16, 25, 40

Median = 16

13, 18, 20, 33, 40, 59

$$\text{Median} = \frac{20 + 33}{2} = \frac{53}{2} = 26.5$$

Mode, p. 34

The mode of a set of data is the value that is repeated most often.

12, 7, 19, 3, 7, 11, 16, 12, 7, 15

Mode = 7

Midrange, p. 34

The midrange of a set of data is the average of the set's minimum value and maximum value.

13, 18, 20, 33, 40, 59

$$\text{Midrange} = \frac{13 + 59}{2} = \frac{72}{2} = 36$$

Range, pp. 34–35

To find the range of a set of values, subtract the minimum value in the set from the maximum value in the set.

13, 18, 20, 33, 40, 59

Range = 59 − 13 = 46

Histogram, pp. 35–37

A histogram is a graph that can be used to show how a set of data is distributed, giving an idea of where the data values are centered as well as how they are dispersed.

Age	Frequency
18 to 21	21
22 to 25	7
26 to 29	4
30 to 33	2
34 to 37	1

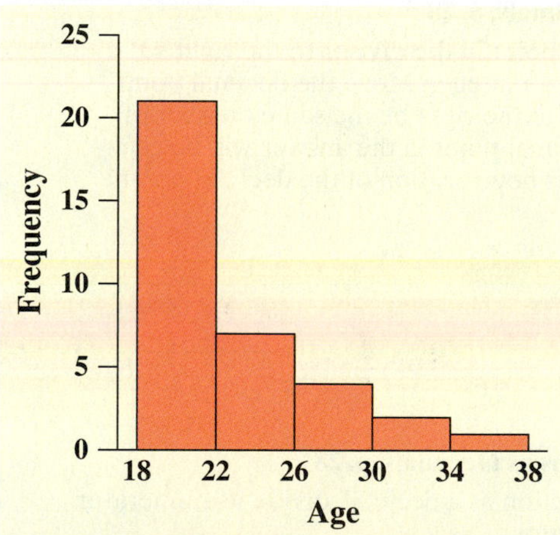

Section 1.7 Exponents and Order of Operations

Exponents, p. 43

Using the number x repeatedly as a factor y times can be represented using exponential notation as x^y. The base is x, and the exponent is y.

Simplify: 4^3

$$4^3 = 4 \cdot 4 \cdot 4 = 64$$

Order of Operations, pp. 44–45

The order of operations is a standard order in which arithmetic operations are performed, ensuring a single correct answer.

1. Remove grouping symbols.
2. Perform any operations involving exponents.
3. Perform any multiplication or division from left to right.
4. Perform any addition or subtraction from left to right.

Simplify: $8 \div (3 \cdot 5 - 19) - 7^2$

$$\begin{aligned} 8 \div (3 \cdot 5 - 19) - 7^2 &= 8 \div (15 - 19) - 7^2 \\ &= 8 \div (-4) - 7^2 \\ &= 8 \div (-4) - 49 \\ &= -2 - 49 \\ &= -51 \end{aligned}$$

Section 1.8 Introduction to Algebra

Evaluating Variable Expressions, p. 49

To evaluate a variable expression for a particular value of a variable, substitute the value for the variable and simplify the resulting expression using the order of operations.

Evaluate $x^2 - 5x - 32$ for $x = -8$.

$$(-8)^2 - 5(-8) - 32$$
$$= 64 - 5(-8) - 32$$
$$= 64 + 40 - 32$$
$$= 72$$

Distributive Property, pp. 50–51

For all real numbers a, b, and c, $a(b + c) = ab + ac$.

Simplify: $7(3x - 2)$

$$7(3x - 2) = 7 \cdot 3x - 7 \cdot 2$$
$$= 21x - 14$$

Combining Like Terms, pp. 51–52

When simplifying variable expressions, we can combine like terms into a single term by adding or subtracting the coefficients of the like terms.

Simplify: $5x - 9 - 2x - 18$

$$5x - 9 - 2x - 18 = 3x - 27$$

SUMMARY OF CHAPTER 1 STUDY STRATEGIES

Creating a study group is one of the best ways to help you learn mathematics and improve your performance on quizzes and exams.

- Sometimes a concept will be easier for you to understand if one of your peers explains it to you. One of your group members may have some insight into a particular topic, and that person's explanation may be just the trick to turn on the "lightbulb" in your head.
- When you explain a certain topic to a fellow student or show a student how to solve a particular problem, you increase your chances of retaining this knowledge.
- You also will find it helpful to have a support group—students who can support one another when times are difficult.

CHAPTER 1 REVIEW

***Write the appropriate symbol, $<$ or $>$, between the following integers.* [1.1]**

1. -10 ___ -7 **2.** 3 ___ -12

***Find the following absolute values.* [1.1]**

3. $|8|$ **4.** $|-13|$

***Find the opposite of the following integers.* [1.1]**

5. -6 **6.** 12

***Add or subtract.* [1.2]**

7. $9 + (-16)$ **8.** $-10 + 7$

9. $-22 - 19$ **10.** $4 - (-23)$

***Simplify.* [1.2]**

11. $3 - 16 - 24$ **12.** $8 - (-19) - 7$

***Multiply or divide.* [1.2]**

13. $9(-6)$ **14.** $-144 \div (-9)$

***Write the factor set for the following numbers.* [1.3]**

15. 42

16. 108

***Write the prime factorization of the following numbers. (If prime, state this.)* [1.3]**

17. 32 **18.** 60

***Simplify the following fractions to lowest terms.* [1.3]**

19. $\frac{24}{42}$ **20.** $\frac{9}{72}$

21. $\frac{40}{234}$ **22.** $\frac{98}{7}$

***Rewrite the following mixed numbers as improper fractions.* [1.3]**

23. $5\frac{2}{3}$ **24.** $12\frac{7}{25}$

***Rewrite the following improper fractions as mixed numbers.* [1.3]**

25. $\frac{38}{10}$ **26.** $\frac{55}{6}$

***Multiply or divide.* [1.4]**

27. $\frac{9}{16} \cdot \frac{28}{75}$ **28.** $5\frac{1}{2} \cdot \frac{14}{33}$

29. $\frac{8}{15} \div \frac{20}{21}$ **30.** $3\frac{1}{5} \div 7\frac{1}{2}$

***Add or subtract.* [1.4]**

31. $\frac{5}{8} + \frac{7}{12}$ **32.** $\frac{13}{20} + \frac{5}{6}$

33. $\frac{11}{18} - \frac{4}{9}$ **34.** $\frac{7}{36} - \frac{13}{42}$

***Simplify.* [1.4]**

35. $\frac{30}{49} \cdot \frac{35}{66}$ **36.** $\frac{7}{16} - \frac{11}{48}$

37. $\frac{9}{40} \div \frac{39}{35}$ **38.** $\frac{7}{20} + \frac{1}{15}$

***Simplify the following decimal expressions.* [1.5]**

39. $8.7 + 3.92$

40. $24.308 - 15.49$

41. $8.4 \cdot 3.6$

42. $40.92 \div 4.65$

***Rewrite the following fractions as decimal numbers.* [1.5]**

43. $\frac{8}{25}$ **44.** $\frac{15}{16}$

***Rewrite the following decimal numbers as fractions in lowest terms.* [1.5]**

45. 0.75 **46.** 0.28

***Rewrite as a percent.* [1.5]**

47. $\frac{2}{5}$ **48.** $\frac{7}{25}$

49. 0.9 **50.** 0.45

***Rewrite as a fraction.* [1.5]**

51. 30% **52.** 55%

***Rewrite as a decimal.* [1.5]**

53. 90% **54.** 4%

Worked-out solutions to Review Exercises marked with ● can be found on page AN-3.

55. Jeff bought a book at a yard sale for \$23. If he had \$40 prior to buying the book, how much money did Jeff have after buying the book? [1.2]

56. Gray had \$78 in his checking account prior to writing a \$125 check to the bookstore for books and supplies. What is the new balance of his account? [1.2]

57. Three investors plan to start a new company. If start-up costs are \$37,800, how much will each person have to invest? [1.2]

58. If one recipe calls for $1\frac{1}{2}$ cups of flour and a second recipe calls for $2\frac{2}{3}$ cups of flour, how much flour is needed to make both recipes? [1.4]

59. Find the mean. Height (inches) of basketball players: 77, 71, 83, 80, 73, 74, 73, 81 [1.6]

60. Find the median. Pulse of 40-year-old men during exercise: 113, 127, 133, 128, 108, 126, 117, 129, 128, 121 [1.6]

61. Find the mode, midrange, and range. Exam scores of study group members: 96, 88, 95, 79, 88, 99 [1.6]

62. Construct a histogram for the given frequency distribution. [1.6]

25-Year-Old Females Systolic Blood Pressure (mmHg)	Frequency
80 to 89	2
90 to 99	8
100 to 109	12
110 to 119	19
120 to 129	6
130 to 139	3

Simplify the given expression. **[1.7]**

63. 4^3

64. $\left(\frac{2}{5}\right)^3$

65. -2^6

66. $3^5 \cdot 5^2$

Simplify the given expression. **[1.7]**

67. $5 + 8 \cdot 4$

68. $25 - 15 \div 5$

69. $3 + 13 \cdot 5 - 20$

70. $(3 + 13) \cdot 5 - 20$

71. $54 - 27 \div 3^2$

72. $\frac{3}{4} + \frac{1}{4} \cdot \frac{12}{25}$

Build a variable expression for the following phrases. **[1.8]**

73. The sum of a number and 14

74. A number decreased by 20

75. Eight less than twice a number

76. Nine more than 6 times a number

77. A coffeehouse charges \$3.55 for a cup of coffee. Letting c represent the number of cups of coffee a coffeehouse sells on a particular day, build a variable expression for the revenue from coffee sales. [1.8]

78. A rental company rents moving vans for \$20 plus \$0.15 per mile. Letting m represent the number of miles, build a variable expression for the cost to rent a moving van from this rental company. [1.8]

Evaluate the following algebraic expressions under the given conditions. **[1.8]**

79. $3x + 17$ for $x = 9$

80. $9 - 8x$ for $x = -2$

81. $10a - 4b$ for $a = 2$ and $b = -9$

82. $(8x - 9)(2x - 11)$ for $x = 4$

83. $x^2 - 7x - 30$ for $x = -3$

84. $b^2 - 4ac$ for $a = -1, b = -8,$ and $c = 5$

Simplify. **[1.8]**

85. $5(x + 7)$

86. $6x + 21x$

87. $8x - 25 - 3x + 17$

88. $8y - 6(4y - 21)$

89. $15 - 23k + 7(4k - 9)$

90. $-8(2x + 25) - (103 - 19x)$

For the following expressions:
a) Determine the number of terms.
b) Write down each term.
c) Write down the coefficient for each term. **[1.8]**

91. $x^3 - 4x^2 - 10x + 41$

92. $-x^2 + 5x - 30$

CHAPTER 1 TEST

For Extra Help

CHAPTER Test Prep VIDEOS

Step-by-step test solutions are found on the Chapter Test Prep Videos available via the Video Resources on DVD, in MyMathLab, and on YouTube (search "WoodburyElemIntAlg" and click on "Channels").

Write the appropriate symbol, < or >, between the following integers.

1. -15 ___ -18

Find the following absolute value.

2. $|-17|$

Simplify.

3. $7 + (-13)$

4. $-7(-9)$

5. Write the factor set for 45.

6. Write the prime factorization of 108. (If the number is prime, state this.)

7. Simplify $\frac{60}{84}$ to lowest terms.

8. Rewrite $\frac{67}{18}$ as a mixed number.

Simplify.

9. $\frac{11}{63} \cdot \frac{15}{44}$

10. $\frac{2}{9} \div \frac{8}{21}$

11. $\frac{3}{5} + \frac{11}{12}$

12. $\frac{5}{24} - \frac{4}{9}$

13. Simplify 8.05(2.27).

14. Rewrite 0.36 as a fraction. Your answer should be in lowest terms.

15. Eleanor bought 13 computers for \$499 each. What was the total cost for the 13 computers?

16. After Lindsay made a deposit of \$407.83 in her checking account, the balance was \$1203.34. What was the balance before the deposit?

17. Rewrite 72% as a fraction.

18. Rewrite 6% as a decimal.

19. Find the mean and median for the given set of values.

43 46 71 95 85 27 37 8 44 26 34 85 91 79 89 20

Simplify the given expression.

20. $16 - 8 \cdot 5$

21. $-9 + 4 \cdot 13 - 6 \cdot 3$

22. $\frac{4^2 + 3^2}{3 + 4 \cdot 13}$

23. Build a variable expression for the expression *seven less than four times a number.*

24. A landscaper charges \$50 plus \$20 per hour for yard maintenance. Letting h represent the number of hours spent working on a particular yard, build a variable expression for the landscaper's charge.

Evaluate the following algebraic expressions under the given conditions.

25. $16 - 5x$ for $x = -9$

26. $x^2 + 6x - 17$ for $x = -8$

Simplify.

27. $5(2x - 13)$

28. $7y - 8(2y - 30)$

Mathematicians in History

Srinivasa Ramanujan was a self-taught Indian mathematician, viewed by many to be one of the greatest mathematical geniuses in history. As a student, he became so totally immersed in his work with mathematics that he ignored his other subjects and failed his college exams. Ramanujan's life is chronicled in the biography *The Man Who Knew Infinity: A Life of the Genius Ramanujan* by Robert Kanigel.

Write a one-page summary (*or* make a poster) of the life of Srinivasa Ramanujan and his accomplishments.

Interesting issues:

- Where and when was Srinivasa Ramanujan born?
- At age 16, Ramanujan borrowed a mathematics book that strongly influenced his life as a mathematician. What was the name of the book?
- Ramanujan got married on July 14, 1909. The marriage was arranged by his mother. How old was his bride at the time?
- What jobs did Ramanujan hold in India?
- Which renowned mathematician invited Ramanujan to England in 1914?
- Ramanujan's health in England was poor. What was the cause of his poor health?
- What is the significance of the taxicab number 1729?
- What were the circumstances that led to Ramanujan's death? How old was Ramanujan when he died?

CHAPTER

2 Linear Equations

In this chapter, we will learn to solve linear equations and investigate applications of this type of equation. The chapter also discusses applications involving percents and proportions. The chapter concludes with a section on linear inequalities.

STUDY STRATEGY

Using Your Textbook In this chapter, we will focus on how to get the most out of your textbook. Students who treat their books solely as a source of homework exercises are turning their backs on one of their best resources.

Throughout this chapter, we will revisit this study strategy and help you incorporate it into your study habits.

2.1 Introduction to Linear Equations

OBJECTIVES

1. **Identify linear equations.**
2. **Determine whether a value is a solution of an equation.**
3. **Solve linear equations using the multiplication property of equality.**
4. **Solve linear equations using the addition property of equality.**
5. **Solve applied problems using the multiplication property of equality or the addition property of equality.**

Linear Equations

Objective 1 Identify linear equations. An **equation** is a mathematical statement of equality between two expressions. It is a statement that asserts that the value of the expression on the left side of the equation is equal to the value of the expression on the right side. Here are a few examples of equations.

$$2x = 8 \qquad x + 17 = 20 \qquad 3x - 8 = 2x + 6 \qquad 5(2x - 9) + 3 = 7(3x + 16)$$

All of these are examples of linear equations. A **linear equation** in one variable has a single variable, and the exponent for that variable is 1. For example, if the variable in a linear equation is x, the equation cannot have terms containing x^2 or x^3. The variable in a linear equation cannot appear in a denominator either. Here are some examples of equations that are not linear equations.

$x^2 - 5x - 6 = 0$ Variable is squared.

$m^3 - m^2 + 7m = 7$ Variable has exponents greater than 1.

$\dfrac{5x + 3}{x - 2} = -9$ Variable is in denominator.

Solutions of Equations

Objective 2 Determine whether a value is a solution of an equation. A **solution** of an equation is a value that when substituted for the variable in the equation, produces a true statement, such as $5 = 5$.

EXAMPLE 1 Is $x = 3$ a solution of $9x - 7 = 20$?

SOLUTION To check whether a particular value is a solution of an equation, we substitute that value for the variable in the equation. If after simplifying both sides of the equation we have a true mathematical statement, the value is a solution.

$$9x - 7 = 20$$

$$9(3) - 7 = 20 \quad \text{Substitute 3 for } x.$$

$$27 - 7 = 20 \quad \text{Multiply.}$$

$$20 = 20 \quad \text{Subtract.}$$

Because 20 is equal to 20, we know that $x = 3$ is a solution.

EXAMPLE 2 Is $x = -2$ a solution of $3 - 4x = -5$?

SOLUTION Again, substitute for x and simplify both sides of the equation.

$$3 - 4x = -5$$

$$3 - 4(-2) = -5 \quad \text{Substitute } -2 \text{ for } x.$$

$$3 + 8 = -5 \quad \text{Multiply.}$$

$$11 = -5 \quad \text{Add.}$$

This statement, $11 = -5$, is not true because 11 is not equal to -5. Therefore, $x = -2$ is not a solution of the equation.

Quick Check 1

Is $x = -7$ a solution of $4x + 23 = -5$?

The set of all solutions of an equation is called its **solution set**. The process of finding an equation's solution set is called **solving the equation**. When we find all of the solutions to an equation, we write those values using set notation inside braces { }.

When solving a linear equation, our goal is to convert it to an equivalent equation that has the variable isolated on one side with a number on the other side (for example, $x = 3$). The value on the opposite side of the equation from the variable after it has been isolated is the solution of the equation.

Multiplication Property of Equality

Objective 3 Solve linear equations using the multiplication property of equality. The first tool for solving linear equations is the **multiplication property of equality**. It says that for any equation, if we multiply both sides of the equation by the same nonzero number, both sides remain equal to each other.

Multiplication Property of Equality

For any algebraic expressions A and B and any nonzero number n,

$$\text{if } A = B, \text{ then } n \cdot A = n \cdot B.$$

Think of a scale that holds two weights in balance. If we double the amount of weight on each side of the scale, will the scale still be balanced? Of course it will.

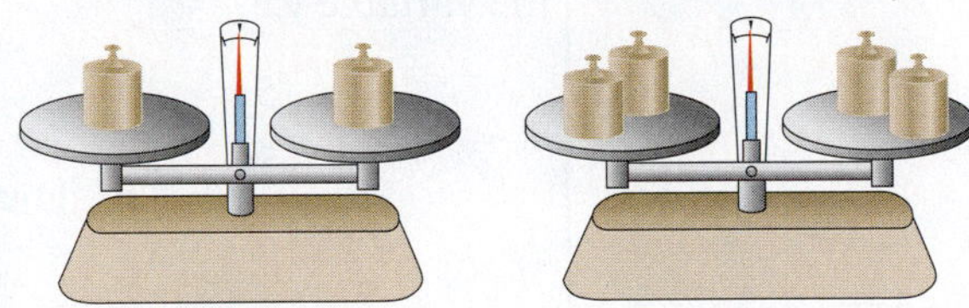

Think of an equation as a scale and the expressions on each side as the weights that are balanced. Multiplying both sides by the same nonzero number leaves both sides still balanced and equal to each other.

Why is it important to multiply both sides of an equation by a *nonzero* number? Multiplying both sides of an equation by 0 can take an equation that is false and make it true. For example, we know that $5 = 3$ is false. If we multiply both sides of that equation by 0, the resulting equation is $0 = 0$, which is true.

EXAMPLE 3 Solve $\frac{x}{3} = 4$ using the multiplication property of equality.

SOLUTION The goal when solving this linear equation is to isolate the variable x on one side of the equation. We begin by multiplying both sides of the equation by 3. The expression $\frac{x}{3}$ is equivalent to $\frac{1}{3}x$; so when we multiply by 3, we are multiplying by the reciprocal of $\frac{1}{3}$. The product of reciprocals is equal to 1, so the resulting expression on the left side of the equation is $1x$, or x.

Another way to think of $\frac{x}{3}$ is as $x \div 3$. Division and multiplication are inverse operations, and we solve the equation by multiplying both sides of the equation by 3. This will "undo" dividing by 3.

$$\frac{x}{3} = 4$$

$$3 \cdot \frac{x}{3} = 3 \cdot 4 \quad \text{Multiply both sides by 3.}$$

$$\overset{1}{\cancel{3}} \cdot \frac{x}{\underset{1}{\cancel{3}}} = 3 \cdot 4 \quad \text{Divide out common factors.}$$

$$x = 12 \quad \text{Multiply.}$$

The solution that we found is $x = 12$. Before moving on, we must check this value to ensure that we made no mistakes and that it is a solution.

Check

$$\frac{x}{3} = 4$$

$$\frac{12}{3} = 4 \quad \text{Substitute 12 for } x.$$

$$4 = 4 \quad \text{Divide.}$$

This is a true statement; so $x = 12$ is a solution, and the solution set is $\{12\}$.

Quick Check 2

Solve $\frac{x}{8} = -2$ using the multiplication property of equality.

The multiplication property of equality also allows us to divide both sides of an equation by the same nonzero number without affecting the equality of the two sides. This is because dividing both sides of an equation by a nonzero number, n, is equivalent to multiplying both sides of the equation by the reciprocal of the number, $\frac{1}{n}$.

EXAMPLE 4 Solve $5y = 40$ using the multiplication property of equality.

SOLUTION Begin by dividing both sides of the equation by 5, which will isolate the variable y.

$$5y = 40$$

$$\frac{5y}{5} = \frac{40}{5} \quad \text{Divide both sides by 5.}$$

$$\frac{\overset{1}{\cancel{5}}y}{\underset{1}{\cancel{5}}} = \frac{40}{5} \quad \text{Divide out common factors on the left side.}$$

$$y = 8 \quad \text{Simplify.}$$

Again, check the solution before writing it in solution set notation.

Check

$$5y = 40$$

$$5(8) = 40 \quad \text{Substitute 8 for } y.$$

$$40 = 40 \quad \text{Multiply.}$$

$y = 8$ is indeed a solution, and the solution set is $\{8\}$.

Quick Check 3

Solve $4a = 56$ using the multiplication property of equality.

If the coefficient of the variable term is negative, we must divide both sides of the equation by that negative number.

EXAMPLE 5 Solve $-7n = -56$ using the multiplication property of equality.

SOLUTION In this example, the coefficient of the variable term is -7. To solve this equation, we need to divide both sides by -7.

$$-7n = -56$$

$$\frac{-7n}{-7} = \frac{-56}{-7} \quad \text{Divide both sides by } -7.$$

$$n = 8 \quad \text{Simplify.}$$

The check of this solution is left to the reader. The solution set is $\{8\}$.

Quick Check 4

Solve $-9a = 144$ using the multiplication property of equality.

A WORD OF CAUTION When dividing both sides of an equation by a negative number, keep in mind that this will change the sign of the number on the other side of the equation.

Consider the equation $-x = 16$. The coefficient of the variable term is -1. To solve this equation, we can multiply both sides of the equation by -1 or divide both sides by -1. Either way, the solution is $x = -16$. We also could have solved the equation by inspection by reading the equation $-x = 16$ as *the opposite of x is 16.* If the opposite of x is 16, we know that x must be equal to -16.

We will now learn how to solve an equation in which the coefficient of the variable term is a fraction.

EXAMPLE 6 Solve $\frac{3}{8}a = -\frac{5}{2}$ using the multiplication property of equality.

SOLUTION In this example, we have a variable multiplied by a fraction. In such a case, we can multiply both sides of the equation by the reciprocal of the fraction. When we multiply a fraction by its reciprocal, the result is 1. This will leave the variable isolated.

$$\frac{3}{8}a = -\frac{5}{2}$$

$$\frac{\overset{1}{\cancel{8}}}{\underset{1}{\cancel{3}}} \cdot \frac{\overset{1}{\cancel{3}}}{\underset{1}{\cancel{8}}}a = \frac{\overset{4}{\cancel{8}}}{3}\left(-\frac{5}{\underset{1}{\cancel{2}}}\right)$$ Multiply by the reciprocal of the fraction $\frac{3}{8}$ and divide out common factors.

$$a = -\frac{20}{3}$$ Simplify.

The check of this solution is left to the reader. The solution set is $\{-\frac{20}{3}\}$.

Quick Check 5

Solve $\frac{9}{16}x = \frac{21}{8}$ using the multiplication property of equality.

Addition Property of Equality

Objective 4 **Solve linear equations using the addition property of equality.** The **addition property of equality** says that we can add the same number to both sides of an equation or subtract the same number from both sides of an equation without affecting the equality of the two sides.

Addition Property of Equality

For any algebraic expressions A and B and any number n,

if $A = B$, then $A + n = B + n$

and $A - n = B - n$.

This property helps us solve equations in which a number is added to or subtracted from a variable on one side of an equation.

EXAMPLE 7 Solve $x + 4 = 11$ using the addition property of equality.

SOLUTION In this example, the number 4 is being added to the variable x. To isolate the variable, use the addition property of equality to subtract 4 from both sides of the equation.

$$x + 4 = 11$$

$$x + 4 - 4 = 11 - 4$$ Subtract 4 from both sides.

$$x = 7$$ Simplify.

To check this solution, substitute 7 for x in the original equation.

Check

$$x + 4 = 11$$

$$7 + 4 = 11$$ Substitute 7 for x.

$$11 = 11$$ Add.

The statement is true, so the solution set is $\{7\}$.

Quick Check 6

Solve $a + 22 = -8$ using the addition property of equality.

EXAMPLE 8 Solve $13 = y - 9$ using the addition property of equality.

SOLUTION When a value is subtracted from a variable, isolate the variable by adding the value to both sides of the equation. In this example, add 9 to both sides.

$$13 = y - 9$$
$$13 + 9 = y - 9 + 9 \quad \text{Add 9 to both sides.}$$
$$22 = y \quad \text{Add.}$$

Check

$$13 = y - 9$$
$$13 = 22 - 9 \quad \text{Substitute 22 for } y.$$
$$13 = 13 \quad \text{Subtract.}$$

This is a true statement, so the solution set is $\{22\}$.

Quick Check 7

Solve $-13 = x - 28$ using the addition property of equality.

In the next section, we will learn how to solve equations requiring us to use both the multiplication and addition properties of equality.

Applications

Objective 5 Solve applied problems using the multiplication property of equality or the addition property of equality. This section concludes with an example of an applied problem that can be solved with a linear equation.

EXAMPLE 9 Admission to the county fair is \$8 per person; so the admission price for a group of x people can be represented by $8x$. If a Cub Scout group paid a total of \$208 for admission to the county fair, how many people were in the group?

SOLUTION We can express this relationship in an equation using the idea that the total cost of admission is equal to \$208. Because the total cost for x people can be represented by $8x$, the equation is $8x = 208$.

$$8x = 208$$
$$\frac{8x}{8} = \frac{208}{8} \quad \text{Divide both sides by 8.}$$
$$x = 26 \quad \text{Divide.}$$

We can verify that this value checks as a solution. Whenever we work on an applied problem, we should write the solution as a complete sentence. There were 26 people in the Cub Scout group at the county fair.

Quick Check 8

Josh spent a total of \$26.50 to take his date to a movie. This left him with only \$38.50 in his pocket. How much money did he have with him before going to the movie?

BUILDING YOUR STUDY STRATEGY

Using Your Textbook, 1 Reading Ahead One effective way to use your textbook is to read a section in the text before it is covered in class. This will give you an idea about the main concepts covered in the section, and your instructor can clarify these concepts in class.

When reading ahead, you should scan the section. Look for definitions that are introduced; as well as procedures that are developed. Pay close attention to the examples. If you find a step in the examples that you do not understand, ask your instructor about it in class.

Exercises 2.1

MyMathLab

PRACTICE

WATCH

DOWNLOAD

READ

REVIEW

Vocabulary

1. A(n) ________ is a mathematical statement of equality between two expressions.

2. A(n) ________ of an equation is a value that when substituted for the variable in the equation, produces a true statement.

3. The ___________ of an equation is the set of all solutions to that equation.

4. State the multiplication property of equality.

5. State the addition property of equality.

6. Freebird's Pizza charges \$12 for a pizza. If the bill for an office pizza party is \$168, which equation can be used to determine the number of pizzas that were ordered?

a) $x + 12 = 168$ **b)** $x - 12 = 168$

c) $12x = 168$ **d)** $\frac{x}{12} = 168$

Is the given equation a linear equation? If not, explain.

7. $5x^2 - 7x = 3x + 8$

8. $4x - 9 = 17$

9. $3x - 5(2x + 3) = 8 - x$

10. $\frac{7}{x^2} - \frac{5}{x} - 13 = 0$

11. $y = 5$

12. $x^4 - 1 = 0$

13. $\frac{3x}{11} - \frac{5}{4} = \frac{2x}{7}$

14. $x \cdot 4 - 1 = 0$

15. $x + \frac{3}{x} + 18 = 0$

16. $3(4x - 9) + 7(2x + 5) = 15$

Check to determine whether the given value is a solution of the equation.

17. $x = 7,\quad 5x - 9 = 26$

18. $x = 3,\quad 2x - 11 = x + 2$

19. $a = 8,\quad 3 - 2a = a + 2a - 11$

20. $m = 4,\quad 15 - 8m = 3m - 29$

21. $z = \frac{1}{4},\quad \frac{2}{3}z + \frac{11}{6} = 2$

22. $t = \frac{5}{3},\quad \frac{1}{10}t + \frac{1}{3} = \frac{1}{2}$

23. $m = 3.4,\quad 3m - 2 = 2m + 0.4$

24. $a = -2.5,\quad 4a - 6 = 9 + 10a$

Solve using the multiplication property of equality.

25. $7x = -91$ **26.** $9a = 72$

27. $6y = 84$ **28.** $11x = -1331$

29. $8b = 22$ **30.** $20z = -35$

31. $5a = 0$ **32.** $0 = 12x$

33. $-5t = 35$ **34.** $-11x = -44$

35. $-2x = -28$ **36.** $-9h = 54$

37. $-t = 45$ **38.** $-y = -31$

39. $\frac{x}{3} = 7$ **40.** $\frac{x}{4} = 3$

41. $-\frac{t}{8} = 12$ **42.** $-\frac{g}{13} = -7$

43. $\frac{7}{12}x = \frac{14}{3}$ **44.** $\frac{3}{4}x = -\frac{9}{2}$

45. $-\frac{2}{5}x = 4$ **46.** $-\frac{5}{6}x = 15$

47. $3.2x = 6.944$ **48.** $-4.7x = 15.04$

Solve using the addition property of equality.

49. $a + 9 = 16$ **50.** $b + 4 = 28$

51. $x + 11 = 3$ **52.** $x + 5 = -18$

53. $n - 13 = 30$ **54.** $n - 9 = -23$

55. $a + 3.2 = 5.7$

56. $x - 4.9 = -11.2$

57. $12 = x + 3$

58. $4 = m + 10$ **59.** $b - 7 = 13$

60. $a - 9 = 99$ **61.** $t - 7 = -4$

62. $n - 3 = -18$ **63.** $-4 + x = 19$

64. $-15 + x = -7$ **65.** $x + 9 = 0$

66. $b - 17 = 0$ **67.** $9 + a = 5$

68. $7 + b = -22$

69. $a + 5 + 6 = 7$

70. $x + 4 - 9 = -3$

Mixed Practice, 71–88

Solve the equation.

71. $60 = 5a$

72. $-m = -15$

73. $x - 27 = -11$

74. $\frac{3}{14}x = \frac{5}{2}$

75. $-\frac{b}{10} = -3$

76. $b + 39 = 30$

77. $x + 24 = -17$

78. $\frac{x}{5} = 13$

79. $11 = -\frac{n}{7}$

80. $47 = y + 35$

81. $0 = x + 56$

82. $x - 38 = -57$

83. $-t = 18$

84. $\frac{t}{15} = -7$

85. $\frac{4}{9}x = -\frac{14}{15}$

86. $x - 3 = -9$

87. $0 = 45m$

88. $40 = -6m$

Provide an equation that has the given solution.

89. $x = 7$

90. $x = -13$

91. $n = \frac{5}{2}$

92. $m = -\frac{3}{10}$

Set up a linear equation and solve it for the following problems.

93. Zoe has only nickels in her pocket. If she has \$1.35 in her pocket, how many nickels does she have?

94. Fruit smoothies were sold at a campus fund-raiser for \$3.25. If total sales were \$217.75, how many smoothies were sold?

95. A local garage band, the Grease Monkeys, held a rent party. They charged \$3 per person for admission, with

the proceeds used to pay the rent. If the rent is \$425 and they ended up with an extra \$52 after paying the rent, how many people came to see them play?

96. Ross organized a tour of a local winery. Attendees paid Ross \$12 to go on the tour, plus another \$5 for lunch. If Ross collected \$493, how many people came on the tour?

97. An insurance company hired 8 new employees. This brought the total to 174 employees. How many employees did the company have before these 8 people were hired?

98. Geena scored 17 points lower than Jared on the last math exam. If Geena's score was 65, find Jared's score.

99. As a cold front was moving in, the temperature in Visalia dropped by 19°F in a two-hour period. If the temperature dropped to 37°F, what was the temperature before the cold front moved in?

100. A company was forced to lay off 37 workers due to an economic downturn. If the company now has 144 employees, how many workers were employed before the layoffs?

Writing in Mathematics

Answer in complete sentences.

101. Explain why the equation $0x = 15$ cannot be solved.

102. Find a value for x such that the expression $x + 21$ is less than -39.

2.2 Solving Linear Equations: A General Strategy

OBJECTIVES

1. Solve linear equations using both the multiplication property of equality and the addition property of equality.
2. Solve linear equations containing fractions.
3. Solve linear equations using the five-step general strategy.
4. Identify linear equations with no solution.
5. Identify linear equations with infinitely many solutions.
6. Solve literal equations for a specified variable.

In the previous section, we solved equations that required only one operation to isolate the variable. In this section, we will learn how to solve equations requiring the use of both the multiplication and addition properties of equality.

Solving Linear Equations

Objective 1 Solve linear equations using both the multiplication property of equality and the addition property of equality. Suppose we needed to solve the equation $4x - 7 = 17$. Should we divide both sides by 4 first? Should we add 7 to both sides first? We refer to the order of operations. This says that in the expression $4x - 7$, we multiply 4 by x and then subtract 7 from the result. To isolate the variable x, we undo these operations in the opposite order. We first add 7 to both sides to undo the subtraction and then divide both sides by 4 to undo the multiplication.

Solution	**Check**
$4x - 7 = 17$	$4(6) - 7 = 17$
$4x - 7 + 7 = 17 + 7$	$24 - 7 = 17$
$4x = 24$	$17 = 17$
$\frac{4x}{4} = \frac{24}{4}$	
$x = 6$	

Because the solution $x = 6$ checks, the solution set is $\{6\}$.

EXAMPLE 1 Solve the equation $3x + 41 = 8$.

SOLUTION To solve this equation, begin by subtracting 41 from both sides. This will isolate the term $3x$. Then divide both sides of the equation by 3 to isolate the variable x.

$$3x + 41 = 8$$

$$3x + 41 - 41 = 8 - 41 \quad \text{Subtract 41 from both sides to isolate } 3x.$$

$$3x = -33 \quad \text{Subtract.}$$

$$\frac{3x}{3} = -\frac{33}{3} \quad \text{Divide both sides by 3 to isolate } x.$$

$$x = -11 \quad \text{Simplify.}$$

Now check the solution.

$$3x + 41 = 8$$

$$3(-11) + 41 = 8 \quad \text{Substitute } -11 \text{ for } x.$$

$$-33 + 41 = 8 \quad \text{Multiply.}$$

$$8 = 8 \quad \text{Simplify.}$$

Because $x = -11$ produced a true statement, the solution set is $\{-11\}$.

Quick Check 1

Solve the equation $5x - 2 = 33$.

EXAMPLE 2 Solve the equation $-8x - 19 = 13$.

SOLUTION

$$-8x - 19 = 13$$

$$-8x - 19 + 19 = 13 + 19 \quad \text{Add 19 to both sides to isolate } -8x.$$

$$-8x = 32 \quad \text{Add.}$$

$$\frac{-8x}{-8} = \frac{32}{-8} \quad \text{Divide by } -8 \text{ to isolate } x.$$

$$x = -4 \quad \text{Simplify.}$$

Now check the solution.

$$-8x - 19 = 13$$
$$-8(-4) - 19 = 13 \quad \text{Substitute } -4 \text{ for } x.$$
$$32 - 19 = 13 \quad \text{Multiply.}$$
$$13 = 13 \quad \text{Simplify.}$$

The solution set is $\{-4\}$.

Quick Check 2

Solve the equation $6 - 4x = 38$.

Solving Linear Equations Containing Fractions

Objective 2 Solve linear equations containing fractions. If an equation contains fractions, we may find it helpful to convert it to an equivalent equation that does not contain fractions before we solve it. This can be done by multiplying both sides of the equation by the LCM of the denominators.

EXAMPLE 3 Solve the equation $\frac{2}{3}x - \frac{5}{6} = \frac{1}{2}$.

SOLUTION This equation contains three fractions, and the denominators are 3, 6, and 2. The LCM of these denominators is 6, so begin by multiplying both sides of the equation by 6.

$$\frac{2}{3}x - \frac{5}{6} = \frac{1}{2}$$
$$6\left(\frac{2}{3}x - \frac{5}{6}\right) = 6\left(\frac{1}{2}\right) \quad \text{Multiply both sides by the LCM 6.}$$
$$\overset{2}{\cancel{6}} \cdot \frac{2}{\underset{1}{\cancel{3}}}x - \overset{1}{\cancel{6}} \cdot \frac{5}{\underset{1}{\cancel{6}}} = \overset{3}{\cancel{6}} \cdot \frac{1}{\underset{1}{\cancel{2}}} \quad \text{Distribute and divide out common factors.}$$
$$4x - 5 = 3 \quad \text{Multiply.}$$
$$4x - 5 + 5 = 3 + 5 \quad \text{Add 5 to both sides to isolate } 4x.$$
$$4x = 8 \quad \text{Add.}$$
$$\frac{4x}{4} = \frac{8}{4} \quad \text{Divide by 4 to isolate } x.$$
$$x = 2 \quad \text{Simplify.}$$

Now check the solution.

$$\frac{2}{3}x - \frac{5}{6} = \frac{1}{2}$$
$$\frac{2}{3}(2) - \frac{5}{6} = \frac{1}{2} \quad \text{Substitute 2 for } x.$$
$$\frac{4}{3} - \frac{5}{6} = \frac{1}{2} \quad \text{Simplify.}$$
$$\frac{8}{6} - \frac{5}{6} = \frac{1}{2} \quad \text{Rewrite } \frac{4}{3} \text{ as } \frac{8}{6}.$$
$$\frac{3}{6} = \frac{1}{2} \quad \text{Subtract.}$$
$$\frac{1}{2} = \frac{1}{2} \quad \text{Simplify.}$$

The solution set is $\{2\}$.

Quick Check 3

Solve the equation $\frac{2}{7}x + \frac{1}{2} = \frac{4}{3}$.

A WORD OF CAUTION When multiplying both sides of an equation by the LCM of the denominators, make sure you multiply each term by the LCM, including any terms that do not contain fractions.

A General Strategy for Solving Linear Equations

Objective 3 Solve linear equations using the five-step general strategy. Now we will examine a process that can be used to solve any linear equation. This process works not only for types of equations we have already learned to solve, but also for more complicated equations such as $7x + 4 = 3x - 20$ and $5(2x - 9) + 3x = 7(3 - 8x)$.

Solving Linear Equations

1. **Simplify each side of the equation completely.**
   - Use the distributive property to clear any parentheses.
   - If there are fractions in the equation, multiply both sides of the equation by the LCM of the denominators to clear the fractions from the equation.
   - Combine any like terms that are on the same side of the equation. After you have completed this step, the equation should contain, at most, one variable term and one constant term on each side.
2. **Collect all variable terms on one side of the equation.** If there is a variable term on each side of the equation, use the addition property of equality to place both variable terms on the same side of the equation.
3. **Collect all constant terms on the other side of the equation.** If there is a constant term on each side of the equation, use the addition property of equality to isolate the variable term.
4. **Divide both sides of the equation by the coefficient of the variable term.** At this point the equation should be of the form $ax = b$. Use the multiplication property of equality to find the solution.
5. **Check your solution.** Check that the value creates a true equation when substituted for the variable in the original equation.

In the next example, we will solve equations that have variable terms and constant terms on both sides of the equation.

EXAMPLE 4 Solve the equation $3x + 8 = 7x - 6$.

SOLUTION In this equation, there are no parentheses or fractions to clear and there are no like terms to be combined. Begin by gathering the variable terms on one side of the equation.

$$3x + 8 = 7x - 6$$

$$3x + 8 - 3x = 7x - 6 - 3x$$ Subtract $3x$ from both sides to gather the variable terms on the right side of the equation.

$$8 = 4x - 6$$ Subtract.

$$8 + 6 = 4x - 6 + 6$$ Add 6 to both sides to isolate $4x$.

$$14 = 4x$$ Add.

$$\frac{14}{4} = \frac{4x}{4}$$ Divide both sides by 4 to isolate x.

$$\frac{7}{2} = x$$ Simplify.

The check of this solution is left to the reader. The solution set is $\{\frac{7}{2}\}$.

Quick Check 4

Solve the equation $6x + 19 = 3x - 8$.

Using Your Calculator You can use a calculator to check the solutions to an equation. Substitute the value for the variable and simplify the expressions on each side of the equation. Here is how the check of the solution $x = \frac{7}{2}$ would look on the TI-84.

```
3(7/2)+8
              18.5
7(7/2)-6
              18.5
```

Because each expression is equal to 18.5 when $x = \frac{7}{2}$, this solution checks.

In the next example, we will solve an equation containing like terms on the same side of the equation.

EXAMPLE 5 Solve the equation $13x - 43 - 9x = 6x + 37 + 5x - 66$.

SOLUTION This equation has like terms on each side of the equation, so begin by combining these like terms.

$$13x - 43 - 9x = 6x + 37 + 5x - 66$$

$$4x - 43 = 11x - 29 \quad \text{Combine like terms on each side.}$$

$$4x - 43 - 4x = 11x - 29 - 4x \quad \text{Subtract } 4x \text{ from both sides to gather variable terms on the right side of the equation.}$$

$$-43 = 7x - 29 \quad \text{Simplify.}$$

$$-43 + 29 = 7x - 29 + 29 \quad \text{Add 29 to both sides to isolate } 7x.$$

$$-14 = 7x \quad \text{Simplify.}$$

$$-\frac{14}{7} = \frac{7x}{7} \quad \text{Divide both sides by 7 to isolate } x.$$

$$-2 = x \quad \text{Divide.}$$

The check of this solution is left to the reader. The solution set is $\{-2\}$.

Quick Check 5

Solve the equation $6x - 9 + 4x = 3x + 13 - 8$.

EXAMPLE 6 Solve the equation $6(3x - 8) + 14 = x - 3(x - 5) + 1$.

SOLUTION Begin by distributing the 6 on the left side of the equation and the -3 on the right side of the equation. Then combine like terms before solving.

$$6(3x - 8) + 14 = x - 3(x - 5) + 1$$

$$18x - 48 + 14 = x - 3x + 15 + 1 \quad \text{Distribute 6 and } -3.$$

$$18x - 34 = -2x + 16 \quad \text{Combine like terms.}$$

$$18x - 34 + 2x = -2x + 16 + 2x \quad \text{Add } 2x \text{ to both sides to gather variable terms on the left side of the equation.}$$

$$20x - 34 = 16 \quad \text{Simplify.}$$

$$20x - 34 + 34 = 16 + 34 \quad \text{Add 34 to both sides to isolate } 20x.$$

$$20x = 50 \quad \text{Simplify.}$$

$$\frac{20x}{20} = \frac{50}{20} \quad \text{Divide by 20 to isolate } x.$$

$$x = \frac{5}{2} \quad \text{Simplify.}$$

The check of this solution is left to the reader. The solution set is $\{\frac{5}{2}\}$.

Quick Check 6

Solve the equation $3(2x - 7) + x = 2(x - 9) - 18$.

In the next example, we will clear the equation of fractions before solving.

EXAMPLE 7 Solve the equation $\frac{3}{5}x - \frac{2}{3} = 2x + \frac{5}{6}$.

SOLUTION Begin by finding the LCM of the three denominators (5, 3, and 6), which is 30. Then multiply both sides of the equation by 30 to clear the equation of fractions.

$$\frac{3}{5}x - \frac{2}{3} = 2x + \frac{5}{6}$$

$$30\left(\frac{3}{5}x - \frac{2}{3}\right) = 30\left(2x + \frac{5}{6}\right) \quad \text{Multiply both sides by LCM (30).}$$

$$\overset{6}{\cancel{30}} \cdot \frac{3}{\underset{1}{\cancel{5}}}x - \overset{10}{\cancel{30}} \cdot \frac{2}{\underset{1}{\cancel{3}}} = 30 \cdot 2x + \overset{5}{\cancel{30}} \cdot \frac{5}{\underset{1}{\cancel{6}}} \quad \text{Distribute and divide out common factors.}$$

$$18x - 20 = 60x + 25 \quad \text{Multiply.}$$

$$18x - 20 - 18x = 60x + 25 - 18x \quad \text{Subtract } 18x \text{ from both sides to gather variable terms on the right side of the equation.}$$

$$-20 = 42x + 25 \quad \text{Simplify.}$$

$$-20 - 25 = 42x + 25 - 25 \quad \text{Subtract 25 from both sides to isolate } 42x.$$

$$-45 = 42x \quad \text{Simplify.}$$

$$-\frac{45}{42} = \frac{42x}{42} \quad \text{Divide both sides by 42 to isolate } x.$$

$$-\frac{15}{14} = x \quad \text{Simplify.}$$

The check of this solution is left to the reader. The solution set is $\left\{-\frac{15}{14}\right\}$.

Quick Check 7

Solve the equation $\frac{1}{10}x - \frac{2}{5} = \frac{1}{20}x - \frac{7}{10}$.

Contradictions and Identities

Objective 4 Identify linear equations with no solution. Each equation we have solved to this point has had exactly one solution. This will not always be the case. Now we consider two special types of equations: contradictions and identities.

A **contradiction** is an equation that is always false regardless of the value substituted for the variable. A contradiction has no solution, so its solution set is the empty set { }. The empty set also is known as the **null set** and is denoted by the symbol $\varnothing$.

EXAMPLE 8 Solve the equation $3(x + 1) + 2(x + 4) = 5x + 6$.

SOLUTION Begin solving this equation by distributing on the left side of the equation.

$$3(x + 1) + 2(x + 4) = 5x + 6$$

$$3x + 3 + 2x + 8 = 5x + 6 \quad \text{Distribute.}$$

$$5x + 11 = 5x + 6 \quad \text{Combine like terms.}$$

$$5x + 11 - 5x = 5x + 6 - 5x \quad \text{Subtract } 5x \text{ from both sides.}$$

$$11 = 6 \quad \text{Simplify.}$$

We are left with an equation that is a false statement because 11 is never equal to 6. This equation is a contradiction and has no solutions. Its solution set is $\varnothing$.

Quick Check 8

Solve the equation $3x - 4 = 3x + 4$.

Objective 5 Identify linear equations with infinitely many solutions. An **identity** is an equation that is always true. If we substitute any real number for the variable in an identity, it will produce a true statement. The solution set for an identity is the set of all real numbers. We denote the set of all real numbers as $\mathbb{R}$. An identity has infinitely many solutions rather than a single solution.

EXAMPLE 9 Solve the equation $9x - 5 = 2(4x - 1) + x - 3$.

SOLUTION Begin to solve this equation by distributing on the right side of the equation.

$$9x - 5 = 2(4x - 1) + x - 3$$
$$9x - 5 = 8x - 2 + x - 3 \quad \text{Distribute.}$$
$$9x - 5 = 9x - 5 \quad \text{Combine like terms.}$$
$$9x - 5 - 9x = 9x - 5 - 9x \quad \text{Subtract } 9x \text{ from both sides.}$$
$$-5 = -5 \quad \text{Simplify.}$$

Because -5 is always equal to -5 regardless of the value of x, this equation is an identity. Its solution set is the set of all real numbers $\mathbb{R}$.

Quick Check 9

Solve the equation $5a + 4 = 4 + 5a$.

Literal Equations

Objective 6 Solve literal equations for a specified variable. A **literal equation** is an equation that contains two or more variables. Literal equations are often used in real-world applications involving many unknowns, such as geometry problems.

Perimeter of a Rectangle

The **perimeter of a rectangle** is a measure of the distance around the rectangle. The formula for the perimeter (P) of a rectangle with length L and width W is $P = 2L + 2W$.

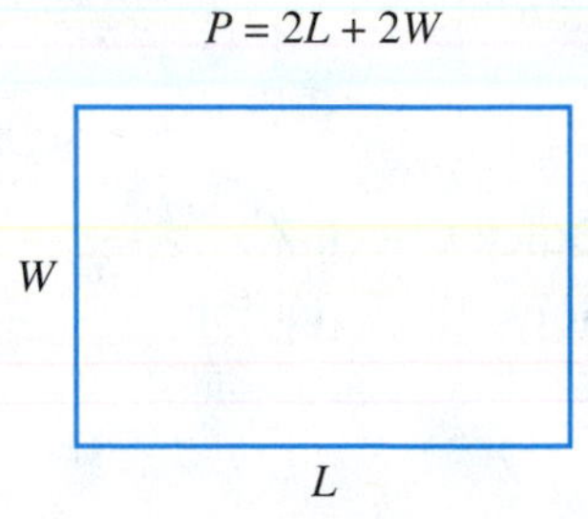

The equation $P = 2L + 2W$ is a literal equation with three variables. We may be asked to solve literal equations for one variable in terms of the other variables in the equation. The equation $P = 2L + 2W$ is solved for the variable P. If we solved for the width W in terms of the length L and the perimeter P, we would have a formula for the width of a rectangle if we knew its length and perimeter.

We solve literal equations by isolating the specified variable. We use the same general strategy that we used to solve linear equations. We treat the other variables in the equation as if they were constants.

EXAMPLE 10 Solve the literal equation $P = 2L + 2W$ (perimeter of a rectangle) for W.

SOLUTION Gather all terms containing the variable W on one side of the equation and gather all other terms on the other side. This can be done by subtracting $2L$ from both sides.

$$P = 2L + 2W$$
$$P - 2L = 2L + 2W - 2L \quad \text{Subtract } 2L \text{ to isolate } 2W.$$
$$P - 2L = 2W \quad \text{Simplify.}$$
$$\frac{P - 2L}{2} = \frac{2W}{2} \quad \text{Divide by 2 to isolate } W.$$
$$\frac{P - 2L}{2} = W$$

We usually rewrite the equation so that the variable we solved for appears on the left side: $W = \frac{P - 2L}{2}$.

Quick Check 10

Solve the literal equation $x + 2y = 5$ for y.

EXAMPLE 11 Solve the literal equation $A = \frac{1}{2}bh$ (area of a triangle) for h.

SOLUTION Because this equation has a fraction, begin by multiplying both sides by 2.

$$A = \frac{1}{2}bh$$

$$2 \cdot A = 2 \cdot \frac{1}{2}bh \quad \text{Multiply both sides by 2 to clear fractions.}$$

$$2 \cdot A = \overset{1}{\cancel{2}} \cdot \frac{1}{\underset{1}{\cancel{2}}}bh \quad \text{Divide out common factors.}$$

$$2A = bh \quad \text{Multiply.}$$

$$\frac{2A}{b} = \frac{bh}{b} \quad \text{Divide both sides by } b \text{ to isolate } h.$$

$$\frac{2A}{b} = h \quad \text{or} \quad h = \frac{2A}{b}$$

Quick Check 11

Solve the literal equation $\frac{1}{5}xy = z$ for x.

Some geometry formulas contain the symbol π, such as $C = 2\pi r$ and $A = \pi r^2$. The symbol π is the Greek letter pi. It is used to represent an irrational number that is approximately equal to 3.14. When solving literal equations containing π, do not replace it with its approximate value.

BUILDING YOUR STUDY STRATEGY

Using Your Textbook, 2 Quick Check Exercises The Quick Check exercises following most examples in this text are similar to the examples in the book. After reading through and possibly reworking an example, try the corresponding Quick Check exercise. You can then decide whether you need more practice.

Exercises 2.2

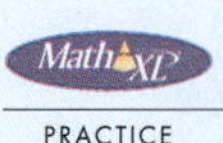

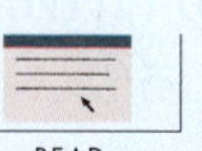

Vocabulary

1. To clear fractions from an equation, multiply both sides of the equation by the ______ of the denominators.
2. An equation that is never true is a(n) ____________.
3. The solution set to a contradiction can be denoted ∅, which represents ____________.
4. An equation that is always true is a(n) _________.
5. To solve the equation $2x - 9 = 7$, the best first step is to ______.
   a) divide both sides of the equation by 2
   b) add 9 to both sides of the equation
   c) subtract 7 from both sides of the equation
6. To solve the equation $3(4x + 5) = 11$, the best first step is to ______.
   a) distribute 3 on the left side of the equation
   b) divide both sides of the equation by 4
   c) subtract 5 from both sides of the equation

Solve.

7. $5x + 31 = 16$
8. $2x + 9 = 31$
9. $23 - 4x = 9$
10. $33 - 6x = -24$
11. $6 = 2a + 20$
12. $-29 = 6b - 17$
13. $9x + 24 = 24$
14. $-4x + 30 = -34$
15. $16.2x - 43.8 = 48.54$
16. $-9.5x - 72.35 = 130$
17. $7a + 11 = 5a - 9$

18. $3m - 11 = 9m + 5$

19. $16 - 4x = 2x + 61$

20. $6n + 14 = 27 + 6n$

21. $5x - 9 = -9 + 5x$

22. $x - 7 = 11 - 3x$

23. $16 - 5x = 2x - 5$

24. $-31 + 11n = 4n + 60$

25. $3.2x + 8.3 = 1.3x + 19.7$

26. $4x - 29.2 = 7.5x - 6.8$

27. $3x + 8 + x = 2x - 9 + 13$

28. $3x + 14 + 8x - 90 = 4x + 11 + x + 27$

29. $x + (x + 1) + (x + 2) = 378$

30. $x + (x + 2) + (x + 4) = 447$

31. $2L + 2(L - 5) = 94$

32. $2(w + 7) + 2w = 74$

33. $2(2w - 3) + 2w = 102$

34. $2L + 2(3L - 50) = 68$

35. $4(2x - 3) - 5x = 3(x + 4)$

36. $2(5x - 3) + 4(x - 8) = 11$

37. $13(3x + 4) - 7(2x - 5) = 5x - 13$

38. $3(2x - 8) - 5(15 - 6x) = 9(4x - 11)$

39. $0.06x + 0.03(4000 - x) = 156$

40. $0.11x + 0.05(7500 - x) = 675$

41. $0.17x - 0.4(x + 1700) = -2566$

42. $0.09x - 0.02(x - 2100) = 322$

43. $0.48x + 0.72(120 - x) = 0.66(120)$

44. $0.3x + 0.55(95 - x) = 0.45(95)$

45. $\frac{1}{4}x - \frac{1}{3} = \frac{5}{12}$

46. $\frac{3}{11}x + \frac{5}{2} = 8$

47. $\frac{x}{12} - \frac{11}{6} = \frac{5}{4}$

48. $\frac{2}{5} - \frac{3}{8}x = -\frac{11}{10}$

49. $\frac{1}{2}x - 3 = \frac{11}{5} - \frac{3}{4}x$

50. $\frac{3}{5} - \frac{5}{6}x = 2x + \frac{4}{3}$

51. $\frac{2}{9}x + \frac{3}{4} = \frac{1}{6}x - \frac{5}{3}$

52. $\frac{4}{7}x - \frac{3}{2} = x - \frac{5}{4}$

53. Write a linear equation that is a contradiction.

54. Write a linear equation that is an identity.

55. Write an equation that has infinitely many solutions.

56. Write an equation that has no solutions.

57. Write an equation whose single solution is negative.

58. Write an equation whose single solution is a fraction.

Solve the following literal equations for the specified variable.

59. $5x + y = -2$ for y

60. $-6x + y = 9$ for y

61. $7x + 2y = 4$ for y

62. $9x + 4y = 20$ for y

63. $-4x + 3y = 10$ for y

64. $-8x + 5y = -6$ for y

65. $P = a + b + c$ for b

66. $P = a + b + 2c$ for c

67. $d = r \cdot t$ for t

68. $d = r \cdot t$ for r

69. $C = 2\pi r$ for r

70. $S = 2\pi rh$ for r

To convert a Celsius temperature (C) to a Fahrenheit temperature (F), use the formula $F = \frac{9}{5}C + 32$.

71. If the temperature outside is 68°F, find the Celsius temperature.

72. If the normal body temperature for a person is 98.6°F, find the Celsius equivalent.

73. A number is tripled and then added to 64. If the result is 325, find the number.

74. If two-thirds of a number is added to 48, the result is 74. Find the number.

75. It costs \$200 to rent a booth at a craft fair. Tina wants to sell homemade kites at the fair. It costs Tina \$4 in material to make each kite.

   a) If she has \$520 available to buy material and pay expenses, how many kites can she make to sell at the craft fair after paying the \$200 rental charge?

   b) If she is able to sell all of the kites, how much will she need to charge for each kite in order to break even?

   c) How much should she charge for each kite in order to make a profit of \$500?

76. A churro is a Mexican dessert pastry. Dan is able to buy churros for 75 cents each, which he plans to sell at a high school baseball game. Dan must donate \$50 to the high school team to be allowed to sell the churros at the game.

   a) If Dan has \$110 to invest in this venture, how many churros will he be able to buy after paying \$50 to the team?

   b) If Dan charges \$2 for each churro, how many must he sell to break even?

   c) If Dan charges \$2 for each churro and he is able to sell all of his churros, what will his profit be?

77. Find the missing value such that $x = 3$ is a solution of $5x - ? = 11$.

78. Find the missing value such that $x = -2$ is a solution of $3(2x - 5) + ? = 5(3 - x)$.

79. Find the missing value such that $x = -\frac{3}{2}$ is a solution of $2x + 9 = 6x + ?$.

80. Find the missing value such that $x = \frac{2}{5}$ is a solution of $4(2x + 7) = 7(3x + ?)$.

Mixed Practice, 81–92

Solve. (If the equation is a literal equation, solve for x.)

81. $14x = -105$

82. $\frac{1}{6}x + \frac{3}{4} = \frac{5}{9}$

83. $27x - 16 - 6x = 14 + 21x - 30$

84. $2x - 3y + 4z = 0$

85. $-5n + 11 - 4n + 9 = n - 45$

86. $a + 43 = 16$

87. $11x - 8y = 34$

88. $2(3w + 7) + 2w = 86$

89. $\frac{2}{5}x - \frac{3}{8} = 3x - \frac{7}{10}$

90. $5t + 72 + 8t = 17t - 28 + 11t$

91. $0.16x + 0.07(3000 - x) = 372$

92. $9n - 17 - 5n - 32 = 23 + 4n + 5$

Writing in Mathematics

Answer in complete sentences.

93. Solve the linear equation $14 = 2x - 9$, showing all of the steps. Next to this, show all of the steps necessary to solve the literal equation $y = 2x - 9$ for x. Write a paragraph explaining how these two processes are similar and how they are different.

94. Write a word problem whose solution can be found from the equation $4.75x + 225 = 795$.

95. ***Newsletter*** Write a newsletter explaining the steps for solving linear equations.

2.3 Problem Solving; Applications of Linear Equations

OBJECTIVES

1. Understand the six steps for solving applied problems.
2. Solve problems involving an unknown number.
3. Solve problems involving geometric formulas.
4. Solve problems involving consecutive integers.
5. Solve problems involving motion.
6. Solve other applied problems.

Introduction to Problem Solving

Although the ability to perform mathematical computations and abstract algebraic manipulations is valuable, one of the most important skills developed in math classes is the skill of problem solving. Every day we are faced with making important decisions and predictions, and the thought process required in decision making is similar to the process of solving mathematical problems.

When faced with a problem to solve in the real world, we first take inventory of the facts that we know. We also determine exactly what it is we are trying to figure out. Then we develop a plan for taking what we already know and using it to figure out how to solve the problem. Once we have solved the problem, we reflect on the route we took to solve it to make sure that route was a logical way to solve the problem. After examining the solution for correctness and practicality, we may finish by presenting the solution to others for their consideration, information, or approval.

An example of one such real-world problem is determining how early to leave for school on the first day of classes. Suppose your first class is at 9 A.M. The problem to solve is figuring out what time to leave your house so that you will not be late. Think about the facts you know: it is normally a 20-minute drive to school, and the classroom is a 10-minute walk from the parking lot. Add information specific to the first day of classes: traffic near the school will be more congested than normal, and it will take longer to find a parking space in the parking lot. Based on past experience, you figure an extra 15 minutes for traffic and another 10 minutes for parking. Adding up all of these times tells you that you need 55 minutes to get to your class. Adding an extra 15 minutes just to be sure, you decide that you need to leave home by 7:50 A.M.

Objective 1 Understand the six steps for solving applied problems. In the previous example, we identified the problem to be solved, gathered the facts, and used them to find a solution to the problem. Solving applied math problems requires that we follow a similar procedure. George Pólya, this chapter's Mathematician in History, was a leader in the study of problem solving. Here is a general plan for solving applied math problems that is based on the work of Pólya's text *How to Solve It.*

Solving Applied Problems

1. **Read the problem.** This step is often overlooked, but misreading or misinterpreting the problem essentially guarantees an incorrect solution. Quickly read the problem once to get a rough idea of what is going on, then read it more carefully a second time to gather all of the important information.
2. **List all of the important information.** Identify all known quantities presented in the problem and determine which quantities you need to find. Creating a table to hold this information or a drawing to represent the problem is a good idea.
3. **Assign a variable to the unknown quantity.** If there is more than one unknown quantity, express each one in terms of the same variable, if possible.
4. **Find an equation relating the known values to the variable expressions representing the unknown quantities.** Sometimes this equation can be translated directly from the wording of the problem. Other times the equation will depend on general facts that are known outside the statement of the problem, such as geometry formulas.
5. **Solve the equation.** Solving the equation is not the end of the problem, as the value of the variable often is not what you were originally looking for. If you were asked for the length and width of a rectangle and you replied "x equals 4," you would not have answered the question. Check your solution to the equation.
6. **Present the solution.** Take the value of the variable from the solution to the equation and use it to figure out all unknown quantities. Check these values to ensure that they make sense in the context of the problem. For example, the length of a rectangle cannot be negative. Finally, present your solution in a complete sentence using the proper units.

Now we will put this strategy to use. All of the applied problems in this section lead to linear equations.

Objective 2 Solve problems involving an unknown number.

EXAMPLE 1 Seven more than twice a number is 39. Find the number.

SOLUTION This is not a very exciting problem, but it provides a good opportunity to practice setting up applied problems.

There is one unknown in this problem. Let's use n to represent the *unknown number.*

Unknown
Number: n

Now translate the sentence *Seven more than twice a number is 39* into an equation. In this case, twice a number is $2n$; so 7 more than twice a number is $2n + 7$.

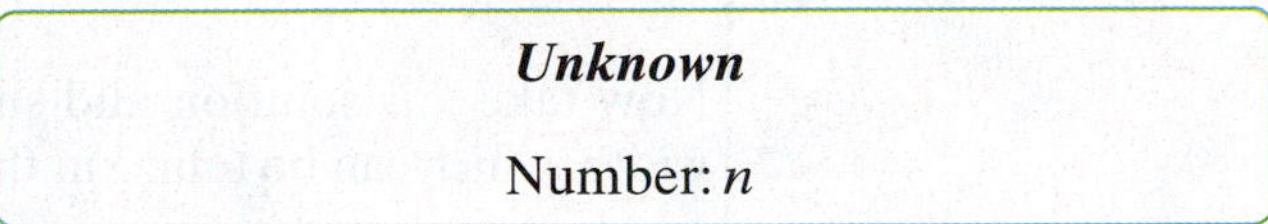

Now solve the equation.

$$2n + 7 = 39$$
$$2n = 32 \quad \text{Subtract 7 from both sides.}$$
$$n = 16 \quad \text{Divide both sides by 2.}$$

The solution of the equation is $n = 16$. Now refer back to the table containing the unknown quantity. Because the unknown number is n, the solution to the problem is the same as the solution of the equation. The number is 16.

It is a good idea to check the solution. Twice the number is 32, and 7 more than that is 39.

Quick Check 1

Three less than 4 times a number is 29. Find the number.

Geometry Problems

Objective 3 Solve problems involving geometric formulas. The next example involves the perimeter of a rectangle, which is a measure of the distance around the outside of the rectangle. The perimeter P of a rectangle whose length is L and whose width is W is given by the formula $P = 2L + 2W$.

EXAMPLE 2 Mark's vegetable garden is in the shape of a rectangle, and he can enclose it with 96 feet of fencing. If the length of his garden is 16 feet more than 3 times the width, find the dimensions of his garden.

SOLUTION The unknown quantities are the length and the width. Because the length is given in terms of the width, pick a variable to represent the width and write the length in terms of this variable. Let w represent the width of the rectangle. The length is 16 feet more than 3 times the width, so express the length as $3w + 16$. One piece of information provided is that the perimeter is 96 feet. This information can be recorded in the following table or drawing:

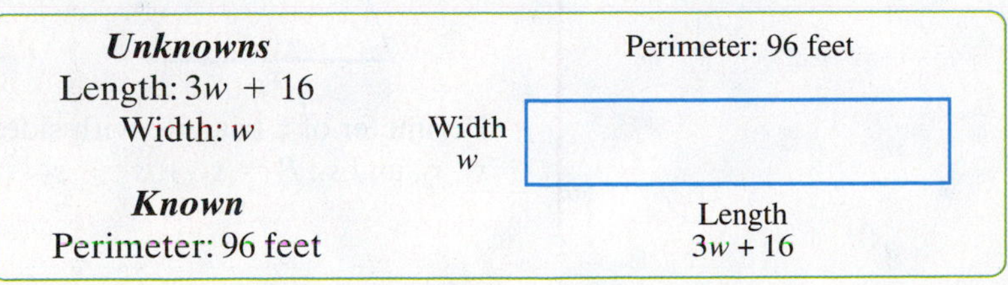

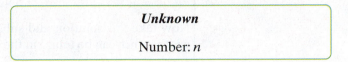

Start with the formula for perimeter and substitute the appropriate values and expressions.

$P = 2L + 2W$

$96 = 2(3w + 16) + 2w$ Substitute 96 for P, $3w + 16$ for L, and w for W.

$96 = 6w + 32 + 2w$ Distribute.

$96 = 8w + 32$ Combine like terms.

$64 = 8w$ Subtract 32 from both sides.

$8 = w$ Divide both sides by 8.

Now take this solution and substitute 8 for w in the expressions for length and width, which can be found in the table that lists the unknowns.

Length: $3w + 16 = 3(8) + 16 = 40$
Width: $w = 8$

The length of Mark's garden is 40 feet, and the width is 8 feet. It checks that the perimeter of this rectangle is 96 feet.

Quick Check 2

A rectangular living room has a perimeter of 80 feet. The length of the room is 8 feet less than twice the width of the room. Find the dimensions of the room.

Here is a summary of useful formulas and definitions from geometry.

Geometry Formulas and Definitions

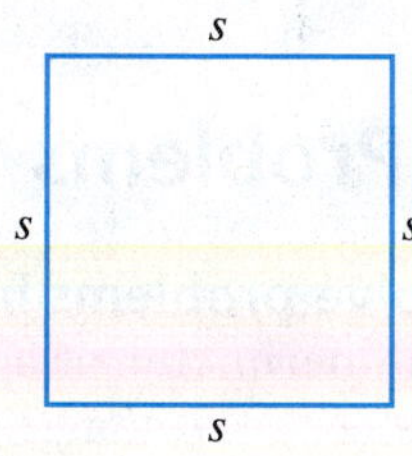

- Perimeter of a square with side s: $P = 4s$
- Area of a square with side s: $A = s^2$

- Perimeter of a rectangle with length L and width W: $P = 2L + 2W$
- Area of a rectangle with length L and width W: $A = LW$

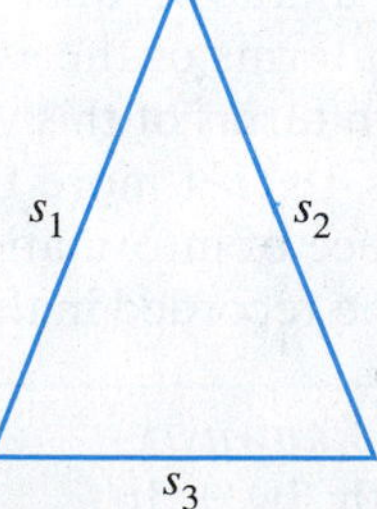

- Perimeter of a triangle with sides s_1, s_2, and s_3: $P = s_1 + s_2 + s_3$

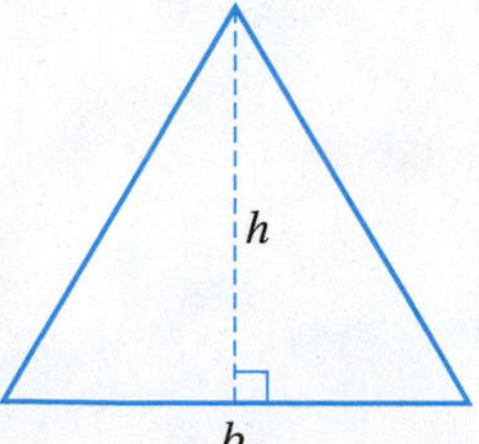

- Area of a triangle with base b and height h: $A = \frac{1}{2}bh$

- An **equilateral triangle** is a triangle that has three equal sides.
- An **isosceles triangle** is a triangle that has at least two equal sides.

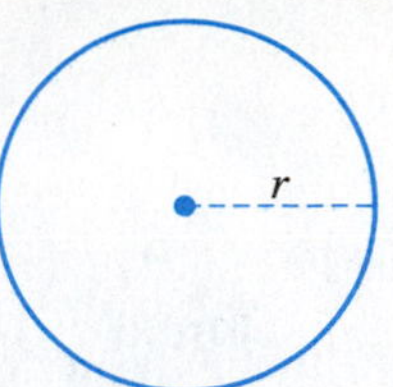

- **Circumference** of a circle with radius r: $C = 2\pi r$
- Area of a circle with radius r: $A = \pi r^2$

EXAMPLE 3 An isosceles triangle has a perimeter of 30 centimeters. The third side is 3 centimeters shorter than each of the sides that have equal lengths. Find the lengths of the three sides.

SOLUTION In this problem, there are three unknowns, which are the lengths of the three sides. Because the triangle is an isosceles triangle, the first two sides have equal lengths. Represent the length of each side using the variable x. The third side is 3 centimeters shorter than the other two sides, and its length can be represented by $x - 3$. Here is a summary of this information, together with the known perimeter.

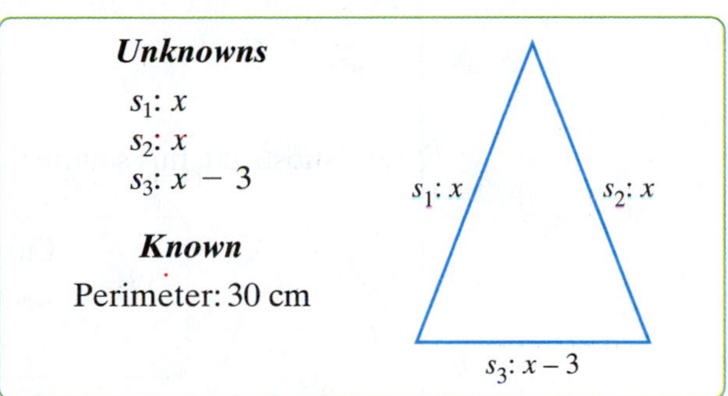

Now set up the equation and solve for x.

$$P = s_1 + s_2 + s_3$$

$30 = x + x + (x - 3)$ Substitute 30 for P, x for s_1 and s_2, and $x - 3$ for s_3.

$30 = 3x - 3$ Combine like terms.

$33 = 3x$ Add 3 to both sides.

$11 = x$ Divide both sides by 3.

Looking back to the table, substitute 11 for x.

s_1: $x = 11$

s_2: $x = 11$

s_3: $x - 3 = 11 - 3 = 8$

The three sides are 11 centimeters, 11 centimeters, and 8 centimeters. The perimeter of this triangle checks to be 30 centimeters.

Quick Check 3

The perimeter of a triangle is 112 inches. The longest side is 8 inches longer than 3 times the shortest side. The other side of the triangle is 20 inches longer than twice the shortest side. Find the lengths of the three sides.

Consecutive Integers

Objective 4 Solve problems involving consecutive integers. The following example involves consecutive integers, which are integers that are 1 unit apart from each other on the number line.

EXAMPLE 4 The sum of three consecutive integers is 87. Find them.

SOLUTION In this problem, the three consecutive integers are the three unknowns. Let x represent the first integer. Because consecutive integers are 1 unit apart from each other, let $x + 1$ represent the second integer. Adding another 1, let $x + 2$ represent the third integer. Here is a table of the unknowns.

Unknowns
First: x
Second: $x + 1$
Third: $x + 2$

We know that the sum of these three integers is 87, which leads to the equation.

$$x + (x + 1) + (x + 2) = 87$$
$$3x + 3 = 87 \quad \text{Combine like terms.}$$
$$3x = 84 \quad \text{Subtract 3 from both sides.}$$
$$x = 28 \quad \text{Divide both sides by 3.}$$

Now substitute this solution to find the three unknowns.

First: $x = 28$
Second: $x + 1 = 28 + 1 = 29$
Third: $x + 2 = 28 + 2 = 30$

The three integers are 28, 29, and 30. The three integers add up to 87.

Quick Check 4
The sum of three consecutive integers is 213. Find them.

Suppose a problem involves consecutive *odd* integers rather than consecutive integers. Consider any string of consecutive odd integers (for example, 15, 17, 19, 21, 23, ...). How far apart are these consecutive odd integers? Each odd integer is 2 away from the previous one. When working with consecutive odd integers (or consecutive even integers for that matter), let x represent the first integer and add 2 to find the next consecutive integer of that type. Here is a table showing the pattern of unknowns for consecutive integers, consecutive odd integers, and consecutive even integers.

Consecutive Integers	Consecutive Odd Integers	Consecutive Even Integers
First: x	First: x	First: x
Second: $x + 1$	Second: $x + 2$	Second: $x + 2$
Third: $x + 2$	Third: $x + 4$	Third: $x + 4$
⋮	⋮	⋮

Motion Problems

Objective 5 Solve problems involving motion.

Distance Formula

When an object such as a car moves at a constant rate of speed, r, for an amount of time, t, the distance traveled, d, is given by the formula $d = r \cdot t$.

Suppose a person drove 238 miles in 3.5 hours. To determine the car's rate of speed for this trip, we can substitute 238 for d and 3.5 for t in the equation $d = r \cdot t$.

$$d = r \cdot t$$
$$238 = r \cdot 3.5 \quad \text{Substitute 238 for } d \text{ and 3.5 for } t.$$
$$\frac{238}{3.5} = \frac{r \cdot 3.5}{3.5} \quad \text{Divide both sides by 3.5.}$$
$$68 = r \quad \text{Simplify.}$$

The car's rate of speed for this trip was 68 mph.

EXAMPLE 5 Tina drove her car at a rate of 60 mph from her home to Rochester. Her mother, Linda, made the same trip at a rate of 80 mph, and it took her 2 hours less than it took Tina to make the trip. How far is Tina's home from Rochester?

SOLUTION For this problem, begin by setting up a table showing the relevant information. Let t represent the time it took Tina to make the trip. Because it took Linda 2 hours less to make the trip, represent her time by $t - 2$. Multiply each person's rate of speed by the time she traveled to find the distance she traveled.

	Rate	Time	Distance ($d = r \cdot t$)
Tina	60	t	$60t$
Linda	80	$t - 2$	$80(t - 2)$

Because both trips were exactly the same distance, we get the equation by setting the expression for Tina's distance equal to the expression for Linda's distance and then solve for t.

$$60t = 80(t - 2)$$
$$60t = 80t - 160 \quad \text{Distribute.}$$
$$-20t = -160 \quad \text{Subtract } 80t \text{ from both sides.}$$
$$t = 8 \quad \text{Divide both sides by } -20.$$

At this point, we must be careful to answer the appropriate question. We were asked to find the distance traveled. We can substitute 8 for t in either of the expressions for distance. Using the expression for the distance traveled by Tina, $60t = 60(8) = 480$. Using the expression for the distance traveled by Linda, $80(t - 2)$, produces the same result. Tina's home is 480 miles from Rochester.

Quick Check 5

Jake drove at a rate of 85 mph for a certain amount of time. His brother Elwood then took over as the driver. Elwood drove at a rate of 90 mph for 3 hours longer than Jake had driven. If the two brothers drove a total of 970 miles, how long did Elwood drive?

Other Problems

Objective 6 Solve other applied problems.

EXAMPLE 6 One number is 5 more than twice another number. If the sum of the two numbers is 80, find the two numbers.

SOLUTION In this problem, there are two unknown numbers. Let one of the numbers be x. Because one number is 5 more than twice the other number, we can represent the other number as $2x + 5$.

Unknown
First: x
Second: $2x + 5$

Finally, we know that the sum of these two numbers is 80, which leads to the equation $x + 2x + 5 = 80$.

$$x + 2x + 5 = 80$$
$$3x + 5 = 80 \quad \text{Combine like terms.}$$
$$3x = 75 \quad \text{Subtract 5 from both sides.}$$
$$x = 25 \quad \text{Divide both sides by 3.}$$

Now substitute this solution to find the two unknown numbers.

First: $x = 25$
Second: $2x + 5 = 2(25) + 5 = 50 + 5 = 55$

The two numbers are 25 and 55. The sum of these two numbers is 80.

Quick Check 6

A number is 1 more than five times another number. If the sum of the two numbers is 103, find the two numbers.

This section concludes with a problem involving coins.

EXAMPLE 7 Ernie has nickels and dimes in his pocket. He has 5 more dimes than nickels. If Ernie has \$2.30 in his pocket, how many nickels and dimes does he have?

SOLUTION We know that the amount of money that Ernie has in nickels and dimes totals \$2.30. Because the number of dimes is given in terms of the number of nickels, let n represent the number of nickels. Because we know that Ernie has 5 more dimes than nickels, we can represent the number of dimes by $n + 5$. A table can be quite helpful in organizing this information, as well as in finding the equation.

	Number of Coins	Value of Coin	Amount of Money
Nickels	n	0.05	$0.05n$
Dimes	$n + 5$	0.10	$0.10(n + 5)$

The amount of money that Ernie has in nickels and dimes is equal to $2.30.

$$0.05n + 0.10(n + 5) = 2.30$$
$$0.05n + 0.10n + 0.50 = 2.30 \quad \text{Distribute.}$$
$$0.15n + 0.50 = 2.30 \quad \text{Combine like terms.}$$
$$0.15n = 1.80 \quad \text{Subtract 0.50 from both sides.}$$
$$n = 12 \quad \text{Divide both sides by 0.15.}$$

The number of nickels is 12. The number of dimes $(n + 5)$ is $12 + 5$, or 17. Ernie has 12 nickels and 17 dimes. You are responsible for verifying that these coins are worth a total of $2.30.

Quick Check 7

Bert has a pocketful of nickels and quarters. He has 13 more nickels than quarters. If Bert has $3.35 in his pocket, how many nickels and quarters does he have?

BUILDING YOUR STUDY STRATEGY

Using Your Textbook, 3 Supplementing Notes and Using Examples When you are rewriting your classroom notes, your textbook can be helpful in supplementing those notes. If your instructor gave you a set of definitions in class, look up the corresponding definitions in the textbook.

You also should keep the textbook handy while you are working on your homework. If you are stuck on a particular problem, look through the section for an example. It may give you an idea about how to proceed.

Exercises 2.3

PRACTICE

WATCH

DOWNLOAD

READ

REVIEW

Vocabulary

1. Which equation can be used to solve *7 less than twice a number is 23?*
   a) $2(n - 7) = 23$ **b)** $2n - 7 = 23$
   c) $7 - 2n = 23$
2. State the formula for the perimeter of a rectangle.
3. If a rectangle has a perimeter of 80 feet and its length is 8 feet less that 3 times its width, which equation can be used to find the dimensions of the rectangle?
   a) $2(W - 8) + 2(3W) = 80$
   b) $2(3W - 8) + 2W = 80$
   c) $2(L) + 2(3L - 8) = 80$
4. Consecutive integers are integers that are ______ unit(s) apart on a number line.
5. If you let x represent the first of three consecutive integers, you can represent the other two integers by ________ and ________.
6. If you let x represent the first of three consecutive even integers, you can represent the other two integers by ________ and ________.
7. If you let x represent the first of three consecutive odd integers, you can represent the other two integers by ________ and ________.
8. Charles has 7 fewer $5 bills than $10 bills. The total value of these bills is $265. Which equation can be used to determine how many of each type of bill Charles has?
   a) $5(n - 7) + 10n = 265$
   b) $5n - 7 + 10n = 265$
   c) $5n + 10(n - 7) = 265$
   d) $5n + 10n - 7 = 265$

9. Five more than twice a number is 97. Find the number.
10. Eighteen more than 7 times a number is 109. Find the number.
11. Eight less than 5 times a number is 717. Find the number.
12. Nine less than 3 times a number is 42. Find the number.
13. When 3.2 is added to 6 times a number, the result is 56.6. Find the number.

14. When 17.2 is subtracted from twice a number, the result is 11.4. Find the number.

15. The width of a rectangle is 5 meters less than its length, and the perimeter is 26 meters. Find the length and width of the rectangle.

16. A rectangle has a perimeter of 62 feet. If its length is 9 feet longer than its width, find the dimensions of the rectangle.

17. A rectangular patio has a perimeter of 56 feet. If the length of the patio is 2 feet less than twice the width, find the dimensions of the patio.

18. A rectangular sheet of paper has a perimeter of 39 inches. The length of the paper is 6 inches less than twice its width. Find the dimensions of the sheet of paper.

19. A rectangle has a perimeter of 130 centimeters. If the length of the rectangle is 4 times its width, find the dimensions of the rectangle.

20. A homeowner has a rectangular garden in her backyard. The length of the garden is 9 times as long as the width, and the perimeter is 160 feet. Find the dimensions of the garden.

21. A square has a perimeter of 220 feet. What is the length of one of its sides?

22. The bases in a Major League Baseball infield are set out in the shape of a square. When Kevin Youkilis hits a home run, he has to run 360 feet to make it all the way around the bases. How far is the distance from home plate to first base?

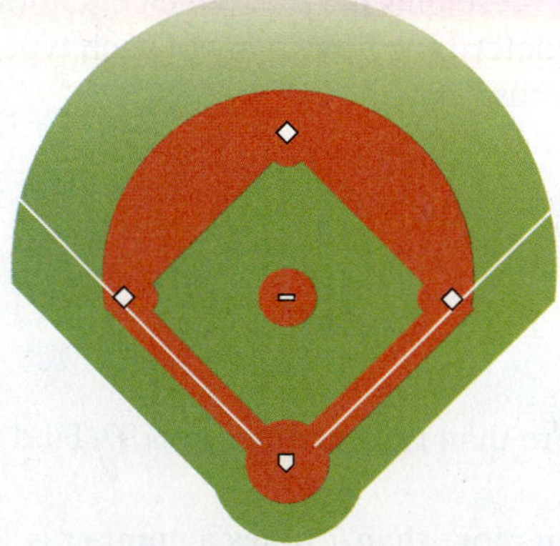

23. An equilateral triangle has a perimeter of 135 centimeters. Find the length of each side of this triangle.

24. The third side of an isosceles triangle is 2 inches shorter than each of the other two sides. If the perimeter of this triangle is 67 inches, find the length of each side.

25. One side of a triangle is 5 inches longer than the side that is the base of the triangle. The third side is 7 inches longer than the base. If the perimeter is 36 inches, find the length of each side.

26. Two sides of a triangle are 3 feet and 6 feet longer than the third side of the triangle. If the perimeter of the triangle is 36 feet, find the length of each side.

27. A circle has a circumference of 69.08 inches. Use $\pi \approx 3.14$ to find the radius of the circle.

28. A circle has a circumference of 48π inches. Find its radius.

Two positive angles are said to be **complementary** if their measures add up to 90°.

29. Angles A and B are complementary angles. If the measure of angle A is 30° more than the measure of angle B, find the measures of the two angles.

30. Angles A and B are complementary angles. Angle A is 15° less than angle B. Find the measures of the two angles.

31. An angle is 10° more than 3 times its complementary angle. Find the measures of the two angles.

32. An angle is 6° less than twice its complementary angle. Find the measures of the two angles.

Two positive angles are said to be **supplementary** if their measures add up to 180°.

33. An angle is 12° more than 3 times its supplementary angle. Find the measures of the two angles.

34. An angle is 39° less than twice its supplementary angle. Find the measures of the two angles.

The measures of the three angles inside a triangle total 180°.

35. One angle in a triangle is 10° more than the smallest angle in the triangle, while the other angle is 20° more than the smallest angle. Find the measures of the three angles.

36. A triangle contains three angles: A, B, and C. Angle B is twice angle A, and angle C is 20° more than angle A. Find the measures of the three angles.

37. If the perimeter of a rectangle is 104 inches and the length of the rectangle is 6 inches more than its width, find the area of the rectangle.

38. If the perimeter of a rectangle is 76 inches and the width of the rectangle is 4 inches less than its length, find the area of the rectangle.

39. A rectangle has a perimeter of 46 inches. The length of the rectangle is 7 inches shorter than twice its width. If a square's side is as long as the length of this rectangle, find the perimeter of the square.

40. A rectangular pen with a perimeter of 200 feet is divided into four square pens by building three fences inside the rectangular pen as follows:

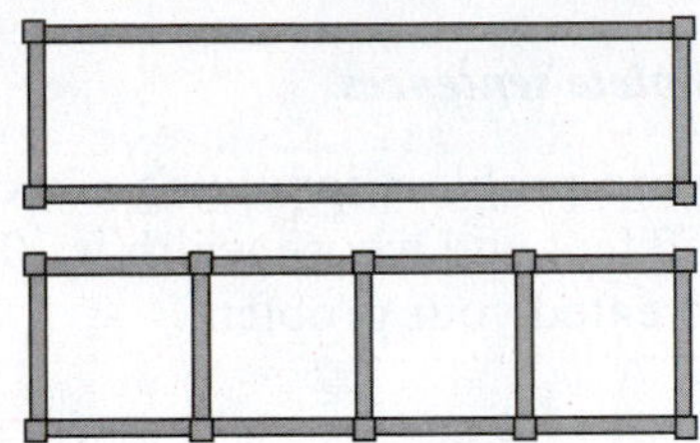

a) What are the original dimensions of the rectangular pen?

b) What is the total length of the three sections of fencing that were added inside the rectangular pen?

c) If the additional fencing cost $2.85 per foot, find the cost of the fencing required to make the four rectangular pens.

41. The sum of three consecutive integers is 459. Find the three integers.

42. The sum of three consecutive integers is 186. Find the three integers.

43. The sum of three consecutive even integers is 246. Find the integers.

44. The sum of five consecutive odd integers is 755. Find the integers.

45. The sum of four consecutive odd integers is 304. Find the integers.

46. The sum of four consecutive even integers is 508. Find the integers.

47. The smallest of three consecutive integers is 18 less than the sum of the two larger integers. Find the three integers.

48. There are three consecutive odd integers, and the sum of the smaller two integers is 49 less than 3 times the largest integer. Find the three odd integers.

49. A. J. drove his race car at a rate of 125 miles per hour for 4 hours. How far did he drive?

50. Angela made the 300-mile drive to Las Vegas in 6 hours. What was her average rate for the trip?

51. If Mario drives at a rate of 68 miles per hour, how long will it take him to drive 374 miles?

52. A bullet train travels 240 kilometers per hour. How far can it travel in 45 minutes?

53. Janet swam a 50-meter race in 28 seconds. What was her speed in meters per second?

54. In 1927, Charles Lindbergh flew his airplane, *Spirit of St. Louis,* 3500 miles from New York City to Paris in 33.5 hours. What was his speed for the trip, to the nearest tenth of a mile per hour?

55. Susan averaged 80 miles per hour on the way to Phoenix to make a sales call. On the way home, she averaged 70 miles per hour and it took her 1 hour longer to drive home than it did to drive to Phoenix.

a) What was the total driving time for Susan's trip?

b) How far does Susan live from Phoenix?

56. Geoffrey drove 60 miles per hour on the way to visit his parents. On the way back to school, he drove 75 miles per hour and it took him 1 hour less than it did to drive to his parents' home.

a) What was the total driving time for Geoffrey's trip?

b) How far is Geoffrey's school from his parents' home?

57. Don started driving east at a rate of 60 miles per hour. If Dennis leaves the same place 2 hours later heading east at a rate of 80 miles per hour, how long will it take him to catch up to Don?

58. Maxine started driving south at a rate of 50 miles per hour. One hour later Claudia started riding her motorcycle south at a rate of 70 miles per hour. How long will it take Claudia to catch Maxine?

59. Lance heads directly east at a constant rate of 25 miles per hour. One hour later, Levi leaves the same spot heading directly west at a constant rate of 20 miles per hour. How long after Levi leaves will the two cyclists be 250 miles apart?

60. This morning a train left the station heading directly north at 50 miles per hour. Two hours later a bus left the same station heading directly south at 60 miles per hour. How long after the bus leaves the station will the bus and the train be 650 miles apart?

61. One number is 17 more than another. If the sum of the two numbers is 71, find the two numbers.

62. One number is 20 more than another. If the sum of the two numbers is 106, find the two numbers.

63. One number is 8 less than another number. If the sum of the two numbers is 96, find the two numbers.

64. One number is 27 less than another number. If the sum of the two numbers is 153, find the two numbers.

65. One number is 5 more than another number. If the smaller number is doubled and added to 3 times the larger number, the sum is 80. Find the two numbers.

66. One number is 11 more than another number. If 3 times the smaller number is added to 5 times the larger number, the sum is 191. Find the two numbers.

67. Chris has a jar with dimes and quarters in it. The jar has 7 more quarters in it than dimes. If the total value of the coins is $5.60, how many quarters are in the jar?

68. Kay's change purse has nickels and dimes in it. The number of nickels is 5 more than twice the number of dimes. If the total value of the coins is $4.05, how many nickels are in Kay's purse?

69. The film department held a fund-raiser by showing the movie π. Admission was $4 for students and $7 for nonstudents. The number of students who attended was 10 more than 4 times the number of nonstudents who attended. If the department raised $500, how many students attended the movie?

70. For a matinee, a movie theater charges $4.50 for children and $6.75 for adults. At today's matinee, there are 20 more children than adults and the total receipts are $405. How many children are at today's matinee?

Writing in Mathematics

Answer in complete sentences.

71. Write a word problem involving a rectangle whose length is 50 feet and whose width is 20 feet. Explain how you created your problem.

72. Write a word problem whose solution is *There were 70 children and 40 adults in attendance.* Explain how you created your problem.

2.4 Applications Involving Percents; Ratio and Proportion

OBJECTIVES

1. Use the basic percent equation to find an unknown amount, base, or percent.
2. Solve applied problems using the basic percent equation.
3. Solve applied problems involving percent increase or percent decrease.
4. Solve problems involving interest.
5. Solve mixture problems.
6. Solve for variables in proportions.
7. Solve applied problems involving proportions.

The Basic Percent Equation

Objective 1 Use the basic percent equation to find an unknown amount, base, or percent. We know that 40 is one-half of 80. Because the fraction $\frac{1}{2}$ is the same as 50%, we also can say that 40 is 50% of 80. In this example, the number 40 is referred to as the **amount** and the number 80 is referred to as the **base**. The basic equation relating these two quantities reflects that the amount is a percentage of the base, or

$$\text{Amount} = \text{Percent} \cdot \text{Base}.$$

In this equation, it is important to express the percent as a decimal or a fraction rather than a percent. Obviously, it is crucial to identify the amount, the percent, and the base correctly. It is a good idea to write the information in the following form:

_______ is _______ % of _______
(Amount) (Percent) (Base)

EXAMPLE 1 What number is 40% of 45?

SOLUTION Letting n represent the unknown number, write the information as follows:

$$\underset{\text{(Amount)}}{\underline{\quad n \quad}} \text{ is } \underset{\text{(Percent)}}{\underline{\quad 40 \quad}} \text{ \% of } \underset{\text{(Base)}}{\underline{\quad 45 \quad}}$$

We see that the amount is n; the percent is 40%, or 0.4; and the base is 45.

Now translate this information to an equation, making sure the percent is written as a decimal rather than a percent.

$$n = 0.4(45)$$
$$n = 18 \quad \text{Multiply.}$$

We find that 18 is 40% of 45.

Quick Check 1

25% of 44 is what number?

A WORD OF CAUTION When using the basic percent equation, make sure you rewrite the percent as a decimal or a fraction before solving for an unknown amount or base.

EXAMPLE 2 Eight percent of what number is 12?

SOLUTION Letting n represent the unknown number, rewrite the information as follows:

$$\underset{\text{(Amount)}}{\underline{\quad 12 \quad}} \text{ is } \underset{\text{(Percent)}}{\underline{\quad 8 \quad}} \text{ \% of } \underset{\text{(Base)}}{\underline{\quad n \quad}}$$

The amount is 12, the percent is 8%, and the base is unknown. Make sure you write the percent as a decimal before working with the equation.

$$12 = 0.08n$$
$$\frac{12}{0.08} = \frac{0.08n}{0.08} \quad \text{Divide both sides by 0.08.}$$
$$150 = n \quad \text{Simplify.}$$

Eight percent of 150 is 12.

Quick Check 2

39 is 60% of what number?

EXAMPLE 3 What percent of 75 is 39?

SOLUTION In this problem, the percent is unknown. Letting p represent the percent, write the information as follows:

$$\underset{\text{(Amount)}}{\underline{\quad 39 \quad}} \text{ is } \underset{\text{(Percent)}}{\underline{\quad p \quad}} \text{ \% of } \underset{\text{(Base)}}{\underline{\quad 75 \quad}}$$

Now translate directly to the basic percent equation and solve it.

$$39 = p \cdot 75$$
$$\frac{39}{75} = \frac{p \cdot 75}{75} \quad \text{Divide both sides by 75.}$$
$$0.52 = p \quad \text{Simplify.}$$

0.52 is equivalent to 52%, so 39 is 52% of 75.

Quick Check 3

56 is what percent of 80?

A WORD OF CAUTION After using the basic percent equation to solve for an unknown percent, make sure you rewrite the solution as a percent by multiplying by 100%.

When using the basic percent equation to solve for an unknown percent, keep in mind that the base is not necessarily the larger of the two given numbers. For example, if we are trying to determine what percent of 80 is 200, the base is the smaller number (80).

Applications

Objective 2 Solve applied problems using the basic percent equation. Now we will use the basic percent equation to solve applied problems. We will continue to use the strategy for solving applied problems developed in Section 2.3.

EXAMPLE 4 There are 7107 female students at a certain community college. If 60% of all students at the college are female, what is the total enrollment of the college?

SOLUTION We know that 60% of the students are female, so the number of female students (7107) is 60% of the total enrollment. Letting n represent the unknown total enrollment, write the information as follows:

$$\underset{\text{(Amount)}}{\underline{\ \ 7107\ \ }} \text{ is } \underset{\text{(Percent)}}{\underline{\ \ 60\ \ }} \text{ \% of } \underset{\text{(Base)}}{\underline{\ \ n\ \ }}$$

Translate this sentence to an equation and solve.

$$7107 = 0.6n$$
$$11{,}845 = n \qquad \text{Divide both sides by 0.6.}$$

There are 11,845 students at the college.

Quick Check 4

Of the 500 children who attend an elementary school, 28% buy their lunch at school. How many children buy their lunch at this school?

Percent Increase and Percent Decrease

Objective 3 Solve applied problems involving percent increase or percent decrease. **Percent increase** is a measure of how much a quantity has increased from its original value. The amount of increase is the amount in the basic percent equation, while the original value is the base.

$$\text{Amount of Increase} = \text{Percent} \cdot \text{Original}$$

Percent decrease is a similar measure of how much a quantity has decreased from its original value.

EXAMPLE 5 An art collector bought a lithograph for \$2500. After three years, the lithograph was valued at \$3800. Find the percent increase in the value of the lithograph over the three years.

SOLUTION The amount of increase in value is \$1300 because $3800 - 2500 = 1300$. We need to determine what percent of the original value is the increase in value. Letting p represent the unknown percent, write the information as follows:

$$\underset{\text{(Amount)}}{\underline{\ \ 1300\ \ }} \text{ is } \underset{\text{(Percent)}}{\underline{\ \ p\ \ }} \text{ \% of } \underset{\text{(Base)}}{\underline{\ \ 2500\ \ }}$$

This translates to the equation $1300 = p \cdot 2500$, which we can solve.

$$1300 = p \cdot 2500$$
$$0.52 = p \qquad \text{Divide both sides by 2500.}$$

Converting this decimal result to a percent, we see that the lithograph increased in value by 52%.

Quick Check 5

Lee bought a new car for \$30,000 three years ago, and now it is worth \$13,800. Find the percent decrease in the value of the car.

A WORD OF CAUTION When solving problems involving percent increase or percent decrease, remember that the base is always the *original amount*.

Percents are involved in many business applications. One such problem is determining the sale price of an item after a percent discount has been applied.

EXAMPLE 6 A department store is having a "20% off" sale, so all prices have been reduced by 20% of the original price. If a robe was originally priced at $37.50, what is the sale price after the 20% discount?

SOLUTION Begin by finding the amount by which the original price has been discounted. Let d represent the amount of the discount. Because the amount of the discount is 20% of the original price ($37.50), we can write the information as follows:

$$\underset{\text{(Amount)}}{\underline{\quad d \quad}} \text{ is } \underset{\text{(Percent)}}{\underline{\quad 20 \quad}} \text{ \% of } \underset{\text{(Base)}}{\underline{\quad \$37.50 \quad}}$$

This translates to the equation $d = 0.2(37.50)$, which we solve for d.

$$d = 0.2(37.50)$$
$$d = 7.50 \qquad \text{Multiply.}$$

The discount is $7.50, so the sale price is $37.50 − $7.50, or $30.00.

Quick Check 6

A department store bought a shipment of MP3 players for $42 each wholesale. If the store marks the MP3 players up by 45%, what is the selling price of an MP3 player?

Interest

Objective 4 Solve problems involving interest. **Interest** is a fee a borrower pays for the privilege of borrowing a sum of money. Banks also pay interest to customers who deposit money in their institutions. Simple interest is calculated as a percentage of the **principal**, which is the amount borrowed or deposited. If r is the annual interest rate, the simple interest (I) owed on a principal (P) is given by the formula $I = P \cdot r \cdot t$, where t is time in years.

If an investor deposited $3000 in an account that paid 4% annual interest for one year, we could determine how much money would be in the account at the end of one year by substituting 3000 for P, 0.04 for r, and 1 for t in the formula $I = P \cdot r \cdot t$.

$$I = P \cdot r \cdot t$$
$$I = 3000(0.04)(1) \qquad \text{Substitute 3000 for } P\text{, 0.04 for } r\text{, and 1 for } t.$$
$$I = 120 \qquad \text{Multiply.}$$

The interest earned in one year is $120. There is $3000 plus the $120 in interest, or $3120, in the account after one year.

EXAMPLE 7 Martha invested some money in an account that paid 6% annual interest. Her friend Stuart found an account that paid 7% interest, and he invested $5000 more in this account than Martha did in her account. If the pair earned a total of $1650 in interest from the two accounts in one year, how much did Martha invest? How much did Stuart invest?

SOLUTION Letting x represent the amount Martha invested at 6%, use $x + 5000$ to represent the amount Stuart invested at 7%. The equation for this problem comes from the fact that the interest earned in each account must add up to $1650. Use a table to display the important information for this problem.

Investor	Principal (P)	Rate (r)	Time (t)	Interest $I = P \cdot r \cdot t$
Martha	x	0.06	1	$0.06x$
Stuart	$x + 5000$	0.07	1	$0.07(x + 5000)$
Total				1650

The interest from the first account is $0.06x$, and the interest from the second account is $0.07(x + 5000)$; so their sum must be \$1650. Essentially, we can find this equation in the final column in the table.

$$\begin{aligned} 0.06x + 0.07(x + 5000) &= 1650 \\ 0.06x + 0.07x + 350 &= 1650 && \text{Distribute.} \\ 0.13x + 350 &= 1650 && \text{Combine like terms.} \\ 0.13x &= 1300 && \text{Subtract 350 from both sides.} \\ x &= 10{,}000 && \text{Divide both sides by 0.13.} \end{aligned}$$

Because $x = 10{,}000$, the amount Martha invested at 6% interest (x) is \$10,000 and the amount Stuart invested at 7% interest $(x + 5000)$ is \$15,000. You may verify that Martha earned \$600 in interest and Stuart earned \$1050 in interest, which totals \$1650.

Quick Check 7

Mark invested money in two accounts. One account paid 3% annual interest, and the other account paid 5% annual interest. He invested \$2000 more in the account that paid 5% interest than in the account that paid 3% interest. If Mark earned a total of \$500 in interest from the two accounts in one year, how much did he invest in each account?

EXAMPLE 8 Donald loaned a total of \$13,500 to two borrowers, Carolyn and George. Carolyn paid 8% annual interest, while George paid 12%. If Donald earned a total of \$1320 in interest from the two borrowers in one year, how much did Carolyn borrow? How much did George borrow?

SOLUTION Letting x represent the amount that Carolyn borrowed, use $13{,}500 - x$ to represent the amount George borrowed. (By subtracting the amount Carolyn borrowed from \$13,500, we are left with the amount George borrowed.) The equation for this problem comes from the fact that the interest earned by Donald must add up to \$1320. Again, use a table to display the important information for this problem.

Borrower	Principal (P)	Rate (r)	Time (t)	Interest $I = P \cdot r$
Carolyn	x	0.08	1	$0.08x$
George	$13{,}500 - x$	0.12	1	$0.12(13{,}500 - x)$
Total	13,500			1320

The interest paid by Carolyn is $0.08x$, and the interest paid by George is $0.12(13{,}500 - x)$; so their sum must be \$1320. Again, we can find this equation in the final column in the table.

$$\begin{aligned} 0.08x + 0.12(13{,}500 - x) &= 1320 \\ 0.08x + 1620 - 0.12x &= 1320 && \text{Distribute.} \\ -0.04x + 1620 &= 1320 && \text{Combine like terms.} \\ -0.04x &= -300 && \text{Subtract 1620 from both sides.} \\ x &= 7500 && \text{Divide both sides by } -0.04. \end{aligned}$$

Because $x = 7500$, the amount Carolyn borrowed is \$7500 and the amount George borrowed $(13{,}500 - x)$ is \$6000. You may verify that Carolyn paid \$600 in interest and George paid \$720 in interest, which totals \$1320.

Quick Check 8

Patsy invested a total of \$1900 in two accounts. One account paid 4% annual interest, and the other account paid 5% annual interest. If Patsy earned a total of \$86.50 in interest from the two accounts, how much did she invest in each account?

Mixture Problems

Objective 5 Solve mixture problems. The next example is a **mixture problem**. In this problem, we will be mixing two solutions that have different concentrations of alcohol to produce a mixture whose concentration of alcohol is somewhere between the two individual solutions. We will determine how much of each solution should be used to produce a mixture of the desired specifications.

EXAMPLE 9 A chemist has two solutions. The first is 20% alcohol, and the second is 30% alcohol. How many milliliters (mL) of each solution should she use if she wants to make 40 mL of a solution that is 24% alcohol?

SOLUTION Let x represent the volume of the 20% alcohol solution in milliliters. We need to express the volume of the 30% solution in terms of x as well. Because the two quantities must add up to 40 mL, the volume of the 30% solution can be represented by $40 - x$.

The equation that solves this problem is based on the volume of alcohol in each solution as well as the volume of alcohol in the mixture. When we add the volume of alcohol that comes from the 20% solution to the volume of alcohol that comes from the 30% solution, it should equal the volume of alcohol in the mixture.

Because we want to end up with 40 mL of a solution that is 24% alcohol, this solution should contain $0.24(40) = 9.6$ mL of alcohol. The following table, which is similar to the one used in the previous example involving interest, presents all of the information:

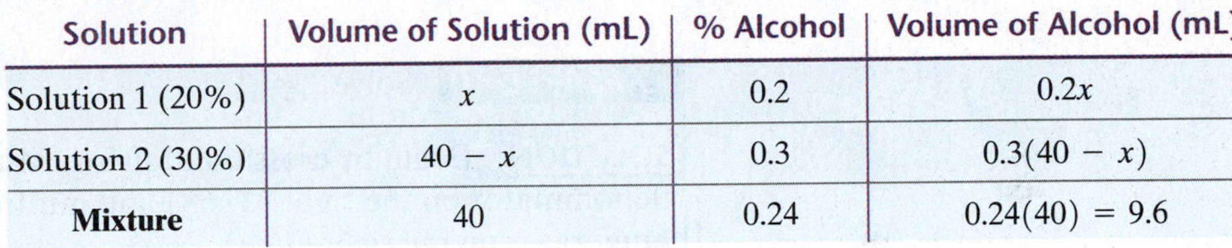

Solution	Volume of Solution (mL)	% Alcohol	Volume of Alcohol (mL)
Solution 1 (20%)	x	0.2	$0.2x$
Solution 2 (30%)	$40 - x$	0.3	$0.3(40 - x)$
Mixture	40	0.24	$0.24(40) = 9.6$

$$\text{Alcohol from Solution 1} + \text{Alcohol from Solution 2} = \text{Alcohol in Mixture}$$
$$0.2x + 0.3(40 - x) = 9.6$$

Now solve the equation for x.

$$0.2x + 0.3(40 - x) = 9.6$$
$$0.2x + 12 - 0.3x = 9.6 \quad \text{Distribute.}$$
$$-0.1x + 12 = 9.6 \quad \text{Combine like terms.}$$
$$-0.1x = -2.4 \quad \text{Subtract 12 from both sides.}$$
$$x = 24 \quad \text{Divide both sides by } -0.1.$$

Because x represented the volume of the 20% alcohol solution that the chemist should use, she should use 24 mL of the 20% solution. Subtracting this amount from 40 mL tells us that she should use 16 mL of the 30% alcohol solution.

Quick Check 9

Marie had one solution that was 30% alcohol and a second solution that was 42% alcohol. How many milliliters of each solution should be mixed to make 80 milliliters of a solution that is 39% alcohol?

Ratio and Proportion

A **ratio** is a comparison of two quantities using division and is usually written as a fraction. Suppose a first-grade student has 20 minutes to eat her lunch, and this is followed by a 30-minute recess. The ratio of the time for lunch to the time for recess is $\frac{20}{30}$, which can be simplified to $\frac{2}{3}$. This means that for every 2 minutes of eating time, there are 3 minutes of recess time.

Objective 6 Solve for variables in proportions. Ratios are important tools for solving some applied problems involving proportions. A **proportion** is a statement of equality between two ratios. In other words, a proportion is an equation in which two ratios are equal to each other. Here are a few examples of proportions.

$$\frac{5}{8} = \frac{n}{56} \qquad \frac{8}{11} = \frac{5}{n} \qquad \frac{9}{2} = \frac{4x + 5}{x}$$

The **cross products** of a proportion are the two products obtained when we multiply the numerator of one fraction by the denominator of the other fraction. The two cross products for the proportion $\frac{a}{b} = \frac{c}{d}$ are $a \cdot d$ and $b \cdot c$.

When we solve a proportion, we are looking for the value of the variable that produces a true statement. We use the fact that the cross products of a proportion are equal if the proportion is indeed true.

$$\text{If } \frac{a}{b} = \frac{c}{d}, \text{ then } a \cdot d = b \cdot c.$$

Consider the two equal fractions $\frac{3}{6}$ and $\frac{5}{10}$. If we write these two fractions in the form of an equation $\frac{3}{6} = \frac{5}{10}$, both cross products ($3 \cdot 10$ and $6 \cdot 5$) are equal to 30. This holds true for any proportion.

To solve a proportion, we multiply each numerator by the denominator on the other side of the equation and set the products equal to each other. We then solve the resulting equation for the variable in the problem. This process is called **cross multiplying** and will be demonstrated in the next example.

EXAMPLE 10 Solve $\frac{3}{4} = \frac{n}{68}$.

SOLUTION Begin by cross multiplying. Multiply the numerator on the left by the denominator on the right ($3 \cdot 68$) and multiply the denominator on the left by the numerator on the right ($4 \cdot n$).

Then write an equation that states that the two products are equal to each other and solve the equation.

$3 \cdot 68$ $\quad$ $4 \cdot n$

$$\frac{3}{4} = \frac{n}{68}$$

$$\frac{3}{4} = \frac{n}{68}$$

$$3 \cdot 68 = 4 \cdot n \quad \text{Cross multiply.}$$

$$204 = 4n \quad \text{Multiply.}$$

$$51 = n \quad \text{Divide both sides by 4.}$$

The solution set is $\{51\}$.

Quick Check 10

Solve $\frac{3}{17} = \frac{n}{187}$.

A WORD OF CAUTION When solving a proportion, cross multiply. *Do not "cross-cancel."*

Applications of Proportions

Objective 7 Solve applied problems involving proportions. Now we turn our attention to applied problems involving proportions.

EXAMPLE 11 Studies show that 1 out of every 9 people is left-handed. In a sample of 216 people, how many would be left-handed according to these studies?

SOLUTION Begin by setting up a ratio based on the known information. Because 1 person out of every 9 is left-handed, the ratio is $\frac{1}{9}$. Notice that the numerator represents the number of left-handed people while the denominator represents the number of all people. When setting up a proportion, keep this ordering consistent. This given ratio will be the left side of the proportion. To set up the right side of the proportion, write a second ratio relating the unknown quantity to the given quantity. The unknown quantity is the number of left-handed people in the group of 216 people. Let n represent the number of left-handed people. The proportion then is $\frac{1}{9} = \frac{n}{216}$. Again, make sure the ratio on the left side $\left(\frac{\text{left-handed}}{\text{total}}\right)$ is consistent with the ratio on the right side. Thus,

$$\frac{1}{9} = \frac{n}{216}$$

$$216 = 9n \quad \text{Cross multiply.}$$

$$24 = n \quad \text{Divide both sides by 9.}$$

According to these studies, there should be 24 left-handed people in the group of 216 people.

Quick Check 11

Studies show that 4 out of 5 dentists recommend sugarless gum for their patients who chew gum. In a group of 85 dentists, according to these studies, how many would recommend sugarless gum to their patients?

Another use of proportions is for **dimensional analysis**: the process of converting from one unit to another. Here is a set of unit conversions that will allow us to convert from the English system of measurement to the metric system of measurement.

Conversions

Length

1 kilometer (km) ≈ 0.62 mile (mi)	1 mi ≈ 1.61 km
1 meter (m) ≈ 3.28 feet (ft)	1 ft ≈ 0.305 m
1 centimeter (cm) ≈ 0.39 inch (in.)	1 in. ≈ 2.54 cm

Volume

1 liter (L) ≈ 0.264 gallons (gal)	1 gal ≈ 3.785 L

Mass

1 kilogram (kg) ≈ 2.2 pounds (lb)	1 lb ≈ 0.454 kg
1 gram (g) ≈ 0.035 ounce (oz)	1 oz ≈ 28.35 g

EXAMPLE 12 Convert 80 meters to feet.

SOLUTION To convert from meters to feet, we have a choice of two conversion factors: 1 m ≈ 3.28 ft or 1 ft ≈ 0.305 m. (Note that depending on which conversion factor is used, answers may vary slightly.) Use 1 m ≈ 3.28 ft to set up a proportion, letting n represent the number of feet in 80 m.

$$\frac{1}{3.28} = \frac{80}{n} \quad \text{Set up the proportion with meters in the numerator and feet in the denominator.}$$

$$n = 262.4 \quad \text{Cross multiply.}$$

There are 262.4 ft in 80 m. (If we had used 1 ft ≈ 0.305 m as the conversion factor, the answer would have been 262.3 feet.)

Quick Check 12

Use the fact that 1 lb ≈ 0.454 kg to convert 175 pounds to kilograms.

BUILDING YOUR STUDY STRATEGY

Using Your Textbook, 4 Creating Note Cards Your textbook can be used to create a series of note cards for studying new terms and procedures or difficult problems. Each time you find a new term in the textbook, write the term on one side of a note card and the definition on the other side. Collect all of these cards and cycle through them, reading the term and trying to recite the definition by memory. You can do the same thing for each new procedure introduced in the textbook.

If you are struggling with a particular type of problem, find an example that has been worked out in the text. Write the problem on the front of the card and write the complete solution provided on the back of the card. As you cycle through the cards, try to solve the problem and then check the solution provided on the back of the card.

Some students have trouble recalling only the first step for solving particular problems. If you are in this group, you can write a series of problems on the front of some note cards and the first step on the back of the cards. You can then cycle through these cards in the same way.

Exercises 2.4

PRACTICE

WATCH

DOWNLOAD
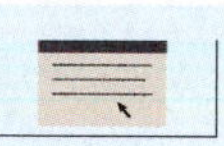
READ

REVIEW

Vocabulary

1. State the basic percent equation.
2. In a percent problem, the ______ is a percentage of the base.
3. Percent increase is a measure of how much a quantity has increased from its ____________.
4. Simple interest is calculated as a percentage of the ________.
5. State the formula for calculating simple interest.
6. A(n) ______ is a comparison of two quantities using division and is usually written as a fraction.
7. A(n) __________ is a statement of equality between two ratios.
8. If $\frac{a}{b} = \frac{c}{d}$, then $a \cdot d =$ ____.

9. Forty-five percent of 80 is what number?
10. What percent of 130 is 91?
11. Forty percent of what number is 92?
12. Fifty-seven is 6% of what number?
13. What percent of 256 is 224?
14. What number is 37% of 94?
15. What is $37\frac{1}{2}$% of 104?
16. Fifteen is what percent of 36?
17. What is 240% of 68?
18. What percent of 24 is 108?
19. Fifty-five is 125% of what number?
20. What is 600% of 53?
21. Thirty percent of the M&Ms in a bowl are brown. If there are 280 M&Ms in the bowl, how many are brown?
22. A doctor helped 320 women deliver their babies last year. Sixteen of these women had twins. What percent of the doctor's patients had twins?
23. Forty percent of the registered voters in a certain precinct are registered as Democrats. If 410 of the registered voters in the precinct are registered as Democrats, how many registered voters are in the precinct?
24. An online retailer adds a 15% charge to all items for shipping and handling. If a shipping-and-handling fee of $96 is added to the cost of a computer, what was the original price of the computer?
25. A certain brand of rum is 40% alcohol. How many milliliters of alcohol are there in a 750-milliliter bottle of rum?
26. A certificate of deposit (CD) pays 3.08% interest. Find the amount of interest earned in one year on a $5000 CD.
27. Marlana invested $3500 in the stock market. After one year, her portfolio had increased in value by 17%. What is the new value of her portfolio?
28. Kristy bought shares of a stock valued at $48.24. If the stock price dropped to $30.15, find the percent decrease in the value of the stock.

29. The price of a gallon of milk increased by 26 cents. This represented a price increase of 8%. What was the original price of the milk?

30. A bookstore charges its customers 32% over the wholesale cost of the book. How much does the bookstore charge for a book that has a wholesale price of $64?

31. A community college has 1600 parking spots on campus. When it builds its new library, the college will lose 120 of its parking spots. What will be the percent decrease in the number of available parking spots?

32. A community college had its budget slashed by $3,101,000. This represents a cut of 7% of its total budget from last year. What was last year's budget?

33. Last year the average SAT math score for students at a high school was 500.

a) This year the average math score decreased by 20%. Find the new average score.

b) By how many points must the average math score increase to get back to last year's average?

c) By what percent do scores need to increase next year to bring the average score back to 500?

34. During 2008, an auto plant produced 25% fewer cars than it had in 2007. Production fell by another 20% during 2009. By what percent will production have to increase in 2010 to make the same number of cars as the plant produced in 2007?

35. Tina invested $30,000 in a stock. In the first year, the stock increased in value by 10%. In the second year, the stock decreased in value by 20%. What percentage gain is required in the third year for Tina's stock to return to its original value? (Round to the nearest tenth of a percent.)

36. Khalid invested $100,000 in a mutual fund. In the first year, the mutual fund decreased in value by 25%. In the second year, the mutual fund increased in value by 20%. What percentage gain is required in the third year for Khalid's mutual fund to return to its original value? (Round to the nearest tenth of a percent.)

37. In 2007, Donald's stock portfolio decreased in value by 25%. In 2008, his stock portfolio decreased in value by another 40%. What percentage gain is required in 2009 for Donald's stock portfolio to return to the value it had at the beginning of 2007? (Round to the nearest tenth of a percent.)

38. In 2007, Giada's retirement fund decreased in value by 8%. In 2008, her retirement fund decreased in value by another 10%. What percentage gain is required in 2009 for Giada's retirement fund to return to the value it had at the beginning of 2007? (Round to the nearest tenth of a percent.)

39. Janet invested her savings in two accounts. One of the accounts paid 2% interest, and the other paid 5% interest. Janet put twice as much money in the account that paid 5% as she put in the account that paid 2%. If she earned a total of $84 in interest from the two accounts in one year, find the amount invested in each account.

40. One of Jaleel's mutual funds made a 10% profit last year, while the other mutual fund made an 8% profit. Jaleel had invested $1400 more in the fund that made an 8% profit than he did in the fund that made a 10% profit. If he made a $715 profit last year, how much was invested in each fund?

41. Kamiran deposits a total of $6500 in savings accounts at two different banks. The first bank pays 4% interest, while the second bank pays 5% interest. If he earns a total of $300 in interest in one year, how much was deposited at each bank?

42. Dianne was given $35,000 when she retired. She invested some at 7% interest and the rest at 9% interest. If she earned $2910 in interest in one year, how much was invested in each account?

43. Aurora invested $8000 in two stocks. One stock decreased in value by 8%; the other, by 20%. If she lost a total of $1000 on these two stocks, how much was invested in the stock that decreased in value by 8%?

44. Lebron invested $20,000 in two stocks. The first stock decreased in value by 3%, and the other decreased in value by 32%. If he lost a total of $5675 on these two stocks, how much was invested in the stock that decreased in value by 32%?

45. Marquis invested $4800 in two mutual funds. Last year one of the funds went up by 6% and the other fund went down by 5%. If his portfolio increased in value by $90 last year, how much did Marquis invest in the fund that earned a 6% profit?

46. Veronica received a $50,000 insurance settlement. She invested some of the money in a bank CD that paid 3.5% annual interest and invested the rest in a stock. In one year, the stock decreased in value by 30%. If Veronica lost $1600 from her $50,000 investment, how much did she put in the bank CD?

47. A mechanic has two antifreeze solutions. One is 70% antifreeze, and the other is 40% antifreeze. How much of each solution should be mixed to make 60 gallons of a solution that is 52% antifreeze?

48. A chemist has two saline solutions. One is 2% salt, and the other is 6% salt. How much of each solution should be mixed to make 2 liters of a saline solution that is 3% salt?

49. A dairyfarmer has some milk that is 5% butterfat as well as some lowfat milk that is 2% butterfat. How much of each needs to be mixed together to make 1000 gallons of milk that is 3.2% butterfat?

50. Patti has two solutions. One is 27% acid, and the other is 39% acid. How much of each solution should be mixed together to make 9 liters of a solution that is 29% acid?

51. A chemist has two solutions. One is 27% salt, and the other is 43% salt. How much of each solution should be mixed together to make 44 milliliters of a solution that is 39% salt?

52. A chemist has two solutions. One is 6% acid and the other is 9% acid. How much of each solution should be mixed together to make 72 milliliters of a solution that is 8.5% acid?

53. A chemist has two solutions. One is 40% alcohol, and the other is pure alcohol. How much of each solution should be mixed together to make 1.6 liters of a solution that is 52% alcohol?

54. A chemist has 60 milliliters of a solution that is 65% alcohol. She plans to add pure alcohol until the solution is 79% alcohol. How much pure alcohol should she add to this solution?

55. Paula has 400 milliliters of a solution that is 80% alcohol. She plans to dilute it so that it is only 50% alcohol. How much water must she add to do this?

56. A bartender has rum that is 40% alcohol. How much rum and how much cola need to be mixed together to make 5 liters of rum and cola that is 16% alcohol?

Solve the proportion.

57. $\frac{2}{3} = \frac{n}{48}$

58. $\frac{5}{8} = \frac{n}{32}$

59. $\frac{20}{n} = \frac{45}{81}$

60. $\frac{22}{30} = \frac{55}{n}$

61. $\frac{3}{4} = \frac{n}{109}$

62. $\frac{7}{10} = \frac{n}{86}$

63. $\frac{13.8}{n} = \frac{2}{9}$

64. $\frac{7.2}{n} = \frac{5}{8}$

65. $\frac{3}{4} = \frac{n + 12}{28}$

66. $\frac{5}{9} = \frac{3n - 2}{18}$

67. $\frac{2n + 15}{18} = \frac{5n + 13}{24}$

68. $\frac{n + 10}{40} = \frac{1 - n}{48}$

Set up a proportion and solve it for the following problems.

69. At a certain community college, 3 out of every 5 students are female. If the college has 9200 students, how many are female?

70. A day care center has a policy that there will be at least 4 staff members present for every 18 children. How many staff members are necessary to accommodate 63 children?

71. The directions for a powdered plant food state that 3 tablespoons of the food need to be mixed with 2 gallons of water. How many tablespoons need to be mixed with 15 gallons of water?

72. Eight out of every 9 people are right-handed. If a factory has 352 right-handed workers, how many left-handed workers are there at the factory?

73. If 5 out of every 6 teachers in a city meet the minimum qualifications to be teaching and the city has 864 teachers, how many do not meet the minimum qualifications?

74. A survey showed that 7 out of every 10 registered voters in a county are in favor of a bond measure to raise money for a new college campus. If there are 140,000 registered voters in the county, how many are not in favor of the bond measure?

Convert the given quantity to the desired unit. Round to the nearest tenth if necessary.

75. 60 cm to in.

76. 150 mi to km

77. 15.2 L to gal

78. 6 gal to L

79. 80 kg to lb

80. 35 oz to g

Writing in Mathematics

Answer in complete sentences.

81. Write a word problem for the following table and equation. Explain how you created the problem.

	Principal (P)	Rate (r)	Time (t)	Interest $I = P \cdot r \cdot t$
	x	0.03	1	$0.03x$
	$x + 10{,}000$	0.05	1	$0.05(x + 10{,}000)$
Total				2100

Equation: $0.03x + 0.05(x + 10{,}000) = 2100$

82. Write a word problem for the following table and equation. Explain how you created the problem.

Solution	Amount of Solution (ml)	% Alcohol	Amount of Alcohol (ml)
Solution 1	x	0.38	$0.38x$
Solution 2	$60 - x$	0.5	$0.5(60 - x)$
Mixture	60	0.4	24

Equation: $0.38x + 0.5(60 - x) = 24$

Quick Review Exercises

Section 2.4

Solve.

1. $4x + 17 = 41$
2. $5x - 19 = 3x - 51$
3. $2x + 2(x + 8) = 68$
4. $x + 3(x + 2) = 2(x + 4) + 30$

2.5 Linear Inequalities

OBJECTIVES

1. Present the solutions of an inequality on a number line.
2. Present the solutions of an inequality using interval notation.
3. Solve linear inequalities.
4. Solve compound inequalities involving the union of two linear inequalities.
5. Solve compound inequalities involving the intersection of two linear inequalities.
6. Solve applied problems using linear inequalities.

Suppose you went to a doctor and she told you that your temperature was normal. This would mean that your temperature was 98.6° F. However, if the doctor told you that you had a fever, could you tell what your temperature was? No, all you would know is that it was above 98.6° F. If you let the variable t represent your temperature, this relationship could be written as $t > 98.6$. The expression $t > 98.6$ is an example of an inequality. An inequality is a mathematical statement comparing two quantities using the symbols $<$ (*less than*) or $>$ (*greater than*). It states that one quantity is smaller than (or larger than) the other quantity.

Two other symbols that may be used in an inequality are $\leq$ and $\geq$. The symbol $\leq$ is read as *less than or equal to,* and the symbol $\geq$ is read as *greater than or equal to.* The inequality $x \leq 7$ is used to represent all real numbers that are less than 7 or are equal to 7. Inequalities involving the symbols $\leq$ or $\geq$ are often called **weak inequalities**, while inequalities involving the symbols $<$ or $>$ are called **strict inequalities** because they involve numbers that are strictly less than (or greater than) a given value.

Presenting Solutions of Inequalities Using a Number Line

Objective 1 Present the solutions of an inequality on a number line.

Linear Inequality

A **linear inequality** is an inequality containing linear expressions.

A linear inequality often has infinitely many solutions, and it is not possible to list every solution. Consider the inequality $x < 3$. There are infinitely many solutions of this inequality, but the one thing the solutions share is that they are all to the left of 3 on the number line. Values located to the right of 3 on the number line are greater than 3. All shaded numbers to the left are solutions of the inequality and can be represented on a number line as follows:

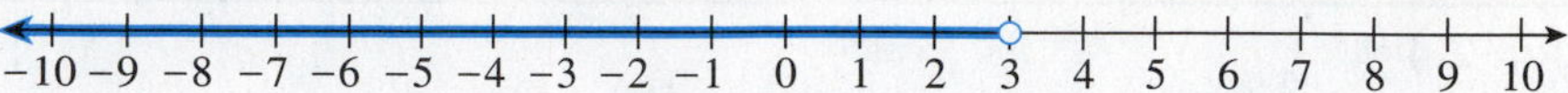

An **endpoint** of an inequality is a point on a number line that separates values that are solutions from values that are not. The open circle at $x = 3$ tells us that the endpoint is not included in the solution because 3 is not less than 3. If the inequality were $x \le 3$, we would fill in the circle at $x = 3$.

Here are three more inequalities with their solutions graphed on a number line:

Inequality	Solution
$x \le -2$	−10 −9 −8 −7 −6 −5 −4 −3 −2 −1 0 1 2 3 4 5 6 7 8 9 10
$x > 5$	−10 −9 −8 −7 −6 −5 −4 −3 −2 −1 0 1 2 3 4 5 6 7 8 9 10
$x \ge -4$	−10 −9 −8 −7 −6 −5 −4 −3 −2 −1 0 1 2 3 4 5 6 7 8 9 10

Saying that a is less than b is equivalent to saying that b is greater than a. In other words, the inequality $a < b$ is equivalent to the inequality $b > a$. This is an important piece of information, as we will want to rewrite an inequality so that the variable is on the left side of the inequality before we graph it on a number line.

EXAMPLE 1 Graph the solutions of the inequality $1 > x$ on a number line.

SOLUTION To avoid confusion about which direction should be shaded on the number line, before graphing the solutions, rewrite the inequality so that the variable is on the left side. Saying that 1 is greater than x is the same as saying that x is less than 1, or $x < 1$. The values that are less than 1 are to the left of 1 on the number line.

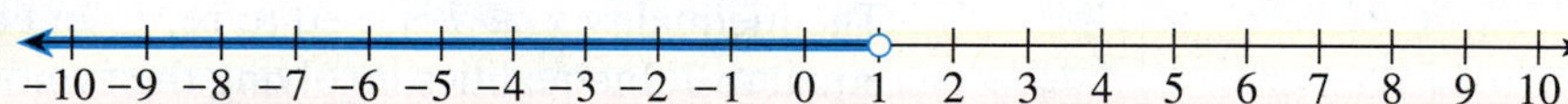

Quick Check 1

Graph the solutions of the inequality $8 < x$ on a number line.

Interval Notation

Objective 2 Present the solutions of an inequality using interval notation. Another way to present the solutions of an inequality is by using **interval notation**. A range of values on a number line, such as the solutions of the inequality $x \ge 4$, is called an **interval**. Inequalities typically have one or more intervals as their solutions. Interval notation presents an interval that is a solution by listing its left and right endpoints. We use parentheses around the endpoints if the endpoints are not included as solutions, and we use brackets if the endpoints are included as solutions.

When an interval continues on indefinitely to the right on a number line, we will use the symbol ∞ (infinity) in place of the right endpoint and follow it with a parenthesis. Let's look again at the number line associated with the solutions of the inequality $x \ge -4$.

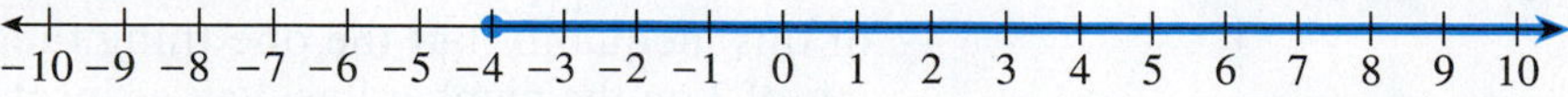

The solutions begin at −4 and include any number that is greater than −4. The interval is bounded by −4 on the left side and extends without bound on the right side, which can be written in interval notation as $[-4, \infty)$. The endpoint −4

is included in the interval, so we write a square bracket in front of it because it is a solution. We always write a parenthesis after ∞, because it is not an actual number that ends the interval at that point.

If the inequality had been $x > -4$ instead of $x \geq -4$, we would have written a parenthesis rather than a square bracket before -4. In other words, the solutions of the inequality $x > -4$ can be expressed in interval notation as $(-4, \infty)$.

For intervals that continue indefinitely to the left on a number line, we use $-\infty$ in place of the left endpoint.

Here is a summary of different inequalities with the solutions presented on a number line and in interval notation.

Inequality	Number Line	Interval Notation
$x > 4$	−10 −9 −8 −7 −6 −5 −4 −3 −2 −1 0 1 2 3 4 5 6 7 8 9 10	$(4, \infty)$
$x \geq 4$	−10 −9 −8 −7 −6 −5 −4 −3 −2 −1 0 1 2 3 4 5 6 7 8 9 10	$[4, \infty)$
$x < 4$	−10 −9 −8 −7 −6 −5 −4 −3 −2 −1 0 1 2 3 4 5 6 7 8 9 10	$(-\infty, 4)$
$x \leq 4$	−10 −9 −8 −7 −6 −5 −4 −3 −2 −1 0 1 2 3 4 5 6 7 8 9 10	$(-\infty, 4]$

Solving Linear Inequalities

Objective 3 Solve linear inequalities. Solving a linear inequality such as $7x - 11 \leq -32$ is similar to solving a linear equation. In fact, there is only one difference between the two procedures:

> Whenever you multiply both sides of an inequality by a negative number or divide both sides by a negative number, the direction of the inequality changes.

Why is this? Consider the inequality $3 < 5$, which is a true statement. If we multiply each side by -2, is the left side $(-2 \cdot 3 = -6)$ still less than the right side $(-2 \cdot 5 = -10)$? No, $-6 > -10$ because -6 is located to the right of -10 on the number line. Changing the direction of the inequality after multiplying (or dividing) by a negative number produces an inequality that is still a true statement.

EXAMPLE 2 Solve the inequality $7x - 11 \leq -32$. Present your solutions on a number line and in interval notation.

SOLUTION Think about the steps you would take to solve the equation $7x - 11 = -32$. You follow the same steps when solving this inequality.

$$7x - 11 \leq -32$$

$$7x \leq -21 \quad \text{Add 11 to both sides.}$$

$$x \leq -3 \quad \text{Divide both sides by 7.}$$

Now present the solution on a number line.

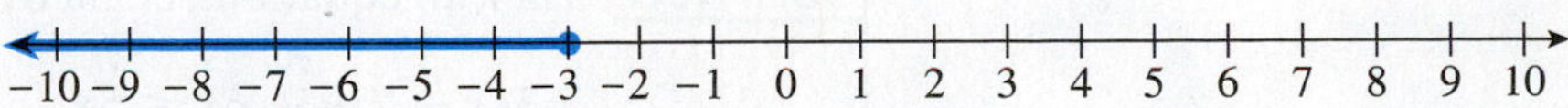

This can be expressed in interval notation as $(-\infty, -3]$.

Quick Check 2

Solve the inequality $4x + 3 < 31$. Present your solutions on a number line and in interval notation.

EXAMPLE 3 Solve the inequality $-2x - 5 > 9$. Present your solutions on a number line and in interval notation.

SOLUTION Begin by solving the inequality.

$$-2x - 5 > 9$$

$$-2x > 14 \quad \text{Add 5 to both sides.}$$

$$\frac{-2x}{-2} < \frac{14}{-2} \quad \text{Divide both sides by } -2 \text{ to isolate } x. \text{ Notice that the direction of the inequality must change because you are dividing by a negative number.}$$

$$x < -7 \quad \text{Simplify.}$$

Now present the solutions.

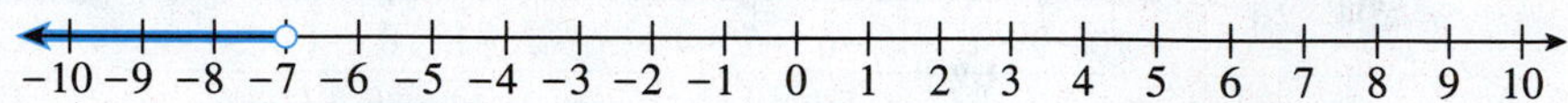

In interval notation, this is $(-\infty, -7)$.

Quick Check 3

Solve the inequality $7 - 4x \leq -13$. Present your solutions on a number line and in interval notation.

EXAMPLE 4 Solve the inequality $5x - 12 > 8 - 3x$. Present your solutions on a number line and in interval notation.

SOLUTION Collect all variable terms on one side of the inequality and all constant terms on the other side.

$$5x - 12 > 8 - 3x$$

$$8x - 12 > 8 \quad \text{Add } 3x \text{ to both sides.}$$

$$8x > 20 \quad \text{Add 12 to both sides.}$$

$$x > \frac{5}{2} \quad \text{Divide both sides by 8 and simplify.}$$

Here is the solution on a number line.

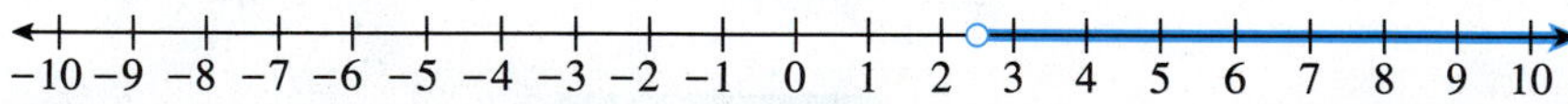

In interval notation, this is written as $\left(\frac{5}{2}, \infty\right)$.

Quick Check 4

Solve the inequality $3x + 8 < x + 2$. Present your solutions on a number line and in interval notation.

EXAMPLE 5 Solve the inequality $3(7 - 2x) + 6x \leq 5x - 4$. Present your solutions on a number line and in interval notation.

SOLUTION As with equations, begin by simplifying each side completely.

$$3(7 - 2x) + 6x \leq 5x - 4$$

$$21 - 6x + 6x \leq 5x - 4 \quad \text{Distribute.}$$

$$21 \leq 5x - 4 \quad \text{Combine like terms.}$$

$$25 \leq 5x \quad \text{Add 4 to both sides.}$$

$$5 \leq x \quad \text{Divide both sides by 5.}$$

We can rewrite this solution as $x \geq 5$, which will help when you display the solutions on a number line.

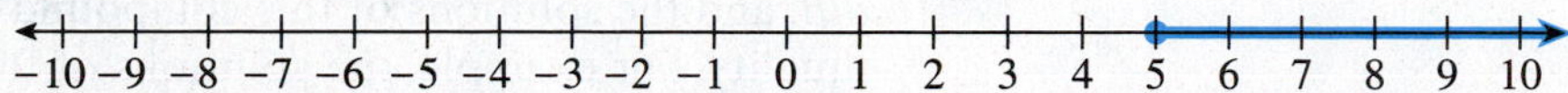

This can be expressed in interval notation as $[5, \infty)$.

Quick Check 5

Solve the inequality $(11x - 3) - (2x - 13) \geq 4(2x + 1)$. Present your solutions on a number line and in interval notation.

Compound Inequalities

Objective 4 Solve compound inequalities involving the union of two linear inequalities. A **compound inequality** is made up of two or more individual inequalities. One type of compound inequality involves the word *or*. In this type of inequality, we are looking for real numbers that are solutions of one inequality or the other. For example, the solutions of the compound inequality $x < 2$ or $x > 4$ are real numbers that are less than 2 or greater than 4. The solutions of this compound inequality can be displayed on a number line as follows:

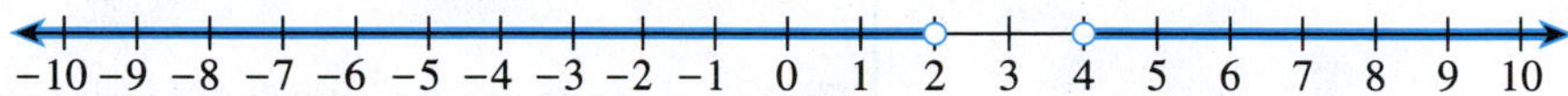

To express the solutions using interval notation, we write both intervals with the symbol for **union** ($\cup$) between them. The interval notation for $x < 2$ or $x > 4$ is $(-\infty, 2) \cup (4, \infty)$.

To solve a compound inequality involving *or*, we simply solve each inequality separately.

EXAMPLE 6 Solve $9x + 20 < -34$ or $5 + 4x \geq 19$.

SOLUTION Solve each inequality separately. Begin with $9x + 20 < -34$.

$$9x + 20 < -34$$
$$9x < -54 \quad \text{Subtract 20 from both sides.}$$
$$x < -6 \quad \text{Divide both sides by 9.}$$

Now solve the second inequality, $5 + 4x \geq 19$.

$$5 + 4x \geq 19$$
$$4x \geq 14 \quad \text{Subtract 5 from both sides.}$$
$$x \geq \frac{7}{2} \quad \text{Divide both sides by 4 and simplify.}$$

Combine the solutions on a single number line.

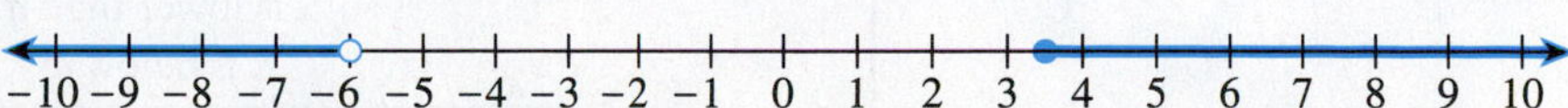

The solutions can be expressed in interval notation as $(-\infty, -6) \cup \left[\frac{7}{2}, \infty\right)$.

Quick Check 6

Solve $3x - 2 \leq 10$ or $4x - 13 \geq 15$.

Objective 5 Solve compound inequalities involving the intersection of two linear inequalities. Another type of compound inequality involves the word *and*, and the solutions of this compound inequality must be solutions of each inequality. For example, the solutions of the inequality $x > -5$ and $x < 1$ are real numbers that are, at the same time, greater than -5 and less than 1. This type of inequality is generally written in the condensed form $-5 < x < 1$. You can think of the solutions as being between -5 and 1.

To solve a compound inequality of this type, we need to isolate the variable in the middle part of the inequality between two real numbers. A key aspect of the approach is to work on all three parts of the inequality at once.

EXAMPLE 7 Solve $-5 \leq 2x - 1 < 9$.

SOLUTION We are trying to isolate x in the middle of this inequality. The first step is to add 1 to all three parts of this inequality, after which we will divide by 2.

$$-5 \leq 2x - 1 < 9$$

$$-4 \leq 2x < 10 \quad \text{Add 1 to each part of the inequality.}$$

$$-2 \leq x < 5 \quad \text{Divide each part of the inequality by 2.}$$

The solutions are presented on the following number line:

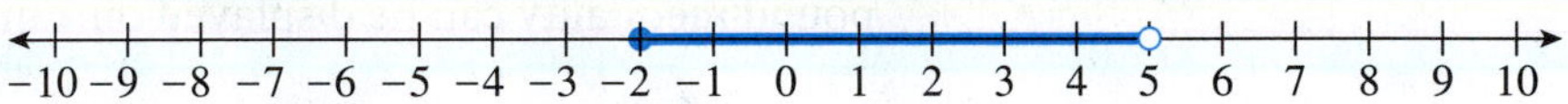

These solutions can be expressed in interval notation as $[-2, 5)$.

Quick Check 7

Solve $-11 < 4x + 1 < 7$.

Applications

Objective 6 Solve applied problems using linear inequalities. Many applied problems involve inequalities rather than equations. Here are some key phrases and their translations into inequalities.

Key Phrases for Inequalities

Phrase	Inequality
x is greater than a x is more than a x is higher than a x is above a	$x > a$
x is at least a x is a or higher	$x \geq a$
x is less than a x is lower than a x is below a	$x < a$
x is at most a x is a or lower	$x \leq a$
x is between a and b, exclusive x is more than a but less than b	$a < x < b$
x is between a and b, inclusive x is at least a but no more than b	$a \leq x \leq b$

EXAMPLE 8 A sign next to an amusement park ride says that you must be at least 48″ tall to get on the ride. Set up an inequality that shows the heights of people who can get on the ride.

SOLUTION Let h represent the height of a person. If a person must be at least 48″ tall, his or her height must be 48″ or above. In other words, the person's height must be greater than or equal to 48″. The inequality is $h \geq 48$.

A WORD OF CAUTION The phrase *at least* means "greater than or equal to" ($\geq$) and does not mean "less than."

EXAMPLE 9 An instructor tells her students that on the final exam, they need a score of 70 or higher out of a possible 100 to pass. Set up an inequality that shows the scores that are not passing.

SOLUTION Let s represent a student's score. Because a score of 70 or higher will pass, any score lower than 70 will not pass. This inequality can be written as $s < 70$. If we happen to know that the lowest possible score is 0, we also could write the inequality as $0 \leq s < 70$.

Quick Check 8

The Brainiac Club has a bylaw stating that a person must have an IQ of at least 130 to be admitted to the club. Set up an inequality that shows the IQs of people who are unable to join the club.

Now we will set up and solve applied problems involving inequalities.

EXAMPLE 10 If a student averages 90 or higher on the five tests given in a math class, the student will earn a grade of A. Sean's scores on the first four tests are 82, 87, 93, and 92. What score on the fifth test will give Sean an A?

SOLUTION To find the average of five test scores, add the five scores and divide by 5. Let x represent the score of the fifth test. The average can then be expressed as $\frac{82 + 87 + 93 + 92 + x}{5}$. We are interested in which scores on the fifth test give an average that is 90 or higher.

$$\frac{82 + 87 + 93 + 92 + x}{5} \geq 90$$

$$\frac{354 + x}{5} \geq 90 \quad \text{Simplify the numerator.}$$

$$5 \cdot \frac{354 + x}{5} \geq 5 \cdot 90 \quad \text{Multiply both sides by 5 to clear the fraction.}$$

$$354 + x \geq 450 \quad \text{Simplify.}$$

$$x \geq 96 \quad \text{Subtract 354 from both sides.}$$

Sean must score at least 96 on the fifth test to earn an A.

Quick Check 9

If a student averages lower than 70 on the four tests given in a math class, the student will fail the class. Bobby's scores on the first three tests are 74, 78, and 80. What scores on the fourth test will result in Bobby failing the class?

BUILDING YOUR STUDY STRATEGY

Using Your Textbook, 5 A Word of Caution/Using Your Calculator Two features of this textbook you may find helpful are labeled "A Word of Caution" and "Using Your Calculator." Each Word of Caution box warns you about common errors for certain problems and tells you how to avoid repeating the same mistake. Look through the section for this feature before attempting the homework exercises.

Using Your Calculator shows how you can use the Texas Instruments TI-84 calculator to help with selected examples in the text. If you own one of these calculators, look for this feature before beginning your homework.

Exercises 2.5

PRACTICE

WATCH

DOWNLOAD
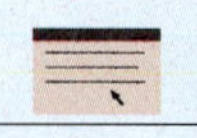
READ

REVIEW

Vocabulary

1. A(n) __________ is an inequality containing linear expressions.
2. When graphing an inequality, use a(n) ________ circle to indicate an endpoint that is not included as a solution.
3. You can express the solutions of an inequality using a number line or ________ notation.
4. When solving a linear inequality, you change the direction of the inequality whenever you multiply or divide both sides of the inequality by a(n) ________ number.
5. An inequality that is composed of two or more individual inequalities is called a(n) ________ inequality.
6. Which inequality can be associated with the statement *The person's height is at least 80 inches*?

   a) $x \le 80$ **b)** $x < 80$ **c)** $x \ge 80$ **d)** $x > 80$

Graph each inequality on a number line and present it in interval notation.

7. $x < 3$

8. $x > -6$

9. $x \ge -1$

10. $x \le 13$

11. $-2 < x < 8$

12. $3 < x \le 12$

13. $x > \frac{9}{2}$

14. $x > 5.5$

15. $x > 8$ or $x \le 2$

16. $x < -5$ or $x > 0$

Write an inequality associated with the given graph or interval notation.

17. −10 −8 −6 −4 −2 0 2 4 6 8 10

18. −10 −8 −6 −4 −2 0 2 4 6 8 10

19. −10 −8 −6 −4 −2 0 2 4 6 8 10

20. −10 −8 −6 −4 −2 0 2 4 6 8 10

21. $(-\infty, 2) \cup (9, \infty)$

22. $(-3, -2)$

23. $(-\infty, -4)$

24. $[-8, \infty)$

Solve. Present your solution on a number line and in interval notation.

25. $x + 9 < 5$

26. $x - 7 > 3$

27. $2x > 12$

28. $4x \geq -20$

29. $-5x > 15$

30. $-8x < -24$

31. $3x \geq -10.5$

32. $2x \leq 9.4$

33. $3x + 2 < 14$

34. $2x - 7 > 11$

35. $-2x + 9 \leq 29$

36. $7 - 4x > 19$

37. $8 < 3x - 13$

38. $1 \geq 5x + 16$

39. $\frac{2}{3}x > \frac{10}{9}$

40. $\frac{5}{9}x - \frac{4}{3} < 7$

41. $5x + 13 > 2x - 11$

42. $-2x + 51 < 9 - 8x$

43. $5(x + 3) + 2 > 3(x + 4) - 6$

44. $5(2x - 3) - 3 < 2(2x + 9) - 4$

45. $2x < -8$ or $x + 7 > 5$

46. $x - 3 \leq -5$ or $6x > -6$

47. $2x - 7 \leq -3$ or $2x - 7 \geq 3$

48. $3x - 6 < -9$ or $3x - 6 > 9$

49. $-2x + 17 < 5$ or $10 - x > 7$

50. $-4x + 13 \leq 9$ or $6 - 5x \geq 21$

51. $3x + 5 < 17$ or $2x - 9 < 5$

52. $7x - 1 \geq 13$ or $4x + 5 \geq -7$

53. $\frac{1}{4}x + \frac{7}{24} \leq \frac{2}{3}$ or $\frac{1}{4}x + \frac{9}{20} \geq \frac{6}{5}$

54. $\frac{1}{8}x - \frac{1}{5} < \frac{1}{12}x - \frac{3}{40}$ or $\frac{1}{2}x - \frac{3}{20} > \frac{1}{4}x + \frac{11}{10}$

55. $-4 < x - 3 < 1$

56. $2 \leq x + 7 \leq 5$

57. $-2 < 5x - 7 < 28$

58. $18 \leq 4x - 10 \leq 50$

59. $6 < 3x + 15 \leq 33$

60. $-12 \leq 4x - 18 < 7$

61. $\frac{1}{5} \leq \frac{1}{2}x - \frac{1}{3} \leq \frac{7}{4}$

62. $-\frac{3}{10} < \frac{1}{5}x - \frac{1}{2} < \frac{3}{5}$

63. In your own words, explain why the inequality $8 < 9x + 7 < 3$ has no solutions.

64. In your own words, explain why solving the compound inequality $x + 7 < 9$ or $x + 7 < 13$ is the same as solving the inequality $x + 7 < 13$ only.

65. Adrian needs to score at least 2 goals in the final game of the season to set a new scoring record. Write an inequality that shows the number of goals that will set a new record.

66. If Eddie sells fewer than 7 cars this week, he will be fired. Write an inequality that shows the number of cars that will result in Eddie being unemployed.

67. A certain type of shrub is tolerant to 30° F, which means that it can survive at temperatures down to 30° F. Write an inequality that shows the temperatures at which the shrub cannot survive.

68. To be eligible for growth funding from the state, a community college must have 11,200 or more students this semester. Write an inequality that shows the number of students that makes the college eligible for growth funding.

For 69–76, write an inequality for the given situation and solve it.

69. John is paid to get signatures on petitions. He gets paid 10 cents for each signature. How many signatures does he need to gather today to earn at least $30?

70. Carlo attends a charity wine-tasting festival. There is a $10 admission fee. In addition, there is a $4 charge for each variety tasted. If Carlo brings $40 with him, how many varieties can he taste?

71. Angelica is going out of town on business. Her company will reimburse her up to $85 for a rental car. She rents a car for $19.95 plus $0.07 per mile. How many miles can Angelica drive without going over the amount her company will reimburse?

72. An elementary school's booster club is holding a fundraiser by selling cookie dough. Each tub of cookie dough sells for $11, and the booster club gets to keep half of the proceeds. If the booster club wants to raise at least $12,000, how many tubs of cookie dough does it need to sell?

73. Charles told his son Harry that he should give at least 15% of his earnings to charity. If Harry earned $37,500 last year, how much should he have given to charity?

74. Karina is looking for a formal dress to wear to a charity event, and she has $400 to spend. If the store charges 7% sales tax on each dress, what price range should Karina be considering?

75. Students in a real estate class must have an average score of at least 80 on their exams in order to pass. If Jacqui has scored 92, 93, 85, and 96 on the first four exams, what scores on the fifth exam would allow her to pass the course?

76. Students with an average test score below 70 after the third test will be sent an Early Alert warning. Robert scored 62 and 59 on the first two tests. What scores on the third exam will save Robert from receiving an Early Alert warning?

Writing in Mathematics

Answer in complete sentences.

77. In your own words, explain why you must change the direction of an inequality when dividing both sides of that inequality by a negative number.

78. ***Solutions Manual*** Write a solutions manual page for the following problem:

Solve. $-10 < 3x + 2 \leq 29$

CHAPTER 2 SUMMARY

Section 2.1 Introduction to Linear Equations

Linear Equations, pp. 63–64

An equation is a mathematical statement of equality between two expressions.

A linear equation has a single variable, and that variable does not have an exponent that is greater than 1.

$3x = 18$

Solution of an Equation, p. 64

A value that when substituted for the variable in the equation produces a true statement

Is $x = -3$ a solution of $4x - 8 = 3x - 11$?

$$\begin{aligned} 4(-3) - 8 &= 3(-3) - 11 \\ -12 - 8 &= -9 - 11 \\ -20 &= -20 \end{aligned}$$

$x = -3$ is a solution.

Multiplication Property of Equality, pp. 64–67

For any algebraic expressions A and B and any nonzero number n, if $A = B$, then $n \cdot A = n \cdot B$.

Solve $-5x = 40$.

$$\begin{aligned} -5x &= 40 \\ \frac{-5x}{-5} &= \frac{40}{-5} \\ x &= -8 \\ \{-8\} \end{aligned}$$

Addition Property of Equality, pp. 67–68

For any algebraic expressions A and B and any number n, if $A = B$, then $A + n = B + n$ and $A - n = B - n$.

Solve $x - 6 = -13$.

$$\begin{aligned} x - 6 &= -13 \\ x - 6 + 6 &= -13 + 6 \\ x &= -7 \\ \{-7\} \end{aligned}$$

Section 2.2 Solving Linear Equations: A General Strategy

Solving Linear Equations, pp. 71–75

1. Simplify each side of the equation completely.
2. Collect all variable terms on one side of the equation.
3. Collect all constant terms on the other side of the equation.
4. Divide both sides of the equation by the coefficient of the variable term.
5. Check your solution.

Solve $2(4x + 5) - 5x = x + 6$.

$$\begin{aligned} 2(4x + 5) - 5x &= x + 6 \\ 8x + 10 - 5x &= x + 6 \\ 3x + 10 &= x + 6 \\ 2x + 10 &= 6 \\ 2x &= -4 \\ x &= -2 \\ \{-2\} \end{aligned}$$

Contradictions, p. 75

A contradiction is an equation that has no solution, so its solution set is the empty set { }.

The empty set also is known as the null set and is denoted by the symbol $\varnothing$.

Solve $2(x + 3) + 7 = 2x + 1$.

$$\begin{aligned} 2(x + 3) + 7 &= 2x + 1 \\ 2x + 6 + 7 &= 2x + 1 \\ 2x + 13 &= 2x + 1 \\ 13 &= 1 \\ \varnothing \end{aligned}$$

Identities, pp. 75–76

An identity is an equation that is always true. The solution set for an identity is the set of all real numbers, denoted $\mathbb{R}$. An identity has infinitely many solutions.

Solve $3x + 8 - 9 = 2x + 7 + x - 8$.

$$\begin{aligned} 3x + 8 - 9 &= 2x + 7 + x - 8 \\ 3x - 1 &= 3x - 1 \\ -1 &= -1 \\ \mathbb{R} \end{aligned}$$

Literal Equations, pp. 76–77

A literal equation is an equation that contains two or more variables.

To solve a literal equation for a particular variable, write it in terms of the other variable(s).

Solve $3x + 2y = 9$ for y.

$$\begin{aligned} 3x + 2y &= 9 \\ 2y &= -3x + 9 \\ y &= \frac{-3x + 9}{2} \end{aligned}$$

Section 2.3 Problem Solving; Applications of Linear Equations

Solving Applied Problems, pp. 80–87

1. Read the problem.
2. List all of the important information.
3. Assign a variable to the unknown quantity.
4. Find an equation relating the known values to the variable expressions representing the unknown quantities.
5. Solve the equation.
6. Present the solution.

Seven less than 3 times a number is 59. Find the number.

Unknown Number: x

$$3x - 7 = 59$$
$$3x = 66$$
$$x = 22$$

The number is 22.

Geometric Formulas and Definitions, pp. 81–83

- Perimeter of a triangle with sides s_1, s_2, and s_3: $P = s_1 + s_2 + s_3$
- Perimeter of a square with side s: $P = 4s$
- Perimeter of a rectangle with length L and width W: $P = 2L + 2W$
- Circumference of a circle with radius r: $C = 2\pi r$

The length of a rectangle is 5 inches more than twice its width. If the perimeter is 46 inches, find the dimensions of the rectangle.

Unknowns Length: $2x + 5$; width: x

$$2(2x + 5) + 2x = 46$$
$$4x + 10 + 2x = 46$$
$$6x + 10 = 46$$
$$6x = 36$$
$$x = 6$$

Length: $2x + 5 = 2(6) + 5 = 12 + 5 = 17$

Width: $x = 6$

The length is 17 inches, and the width is 6 inches.

Distance, pp. 84–85

When an object such as a car moves at a constant rate of speed (r) for an amount of time (t) the distance traveled (d) is given by the formula $d = r \cdot t$.

If a person drives at a constant speed of 65 mph, how long will it take to drive 338 miles?

Unknown Time driving: t

$$338 = 65 \cdot t$$
$$\frac{338}{65} = t$$
$$5.2 = t$$

It will take 5.2 hours.

Section 2.4 Applications Involving Percents; Ratio and Proportion

Basic Percent Equation, pp. 90–92

The amount is a percentage of the base.

Amount = Percent · Base

The percent must be expressed as a decimal or a fraction rather than as a percent.

What percent of 68 is 17?

$$\underline{17} \text{ is } \underline{n}\ \% \text{ of } \underline{68}.$$
$$17 = n \cdot 68$$
$$\frac{17}{68} = n$$
$$0.25 = n$$

17 is 25% of 68.

Percent Increase/Decrease, pp. 92–93

Percent increase is a measure of how much a quantity has increased from its original value. Percent decrease is a similar measure of how much a quantity has decreased from its original value. The amount is the amount of increase/decrease, and the base is the original value.

A painting was purchased for \$540 and sold a year later for \$648. Find the percent increase in the value of the painting.

Amount of increase: \$108; base: \$540

$$\underline{108} \text{ is } \underline{n}\ \% \text{ of } \underline{540}.$$
$$108 = n \cdot 540$$
$$\frac{108}{540} = n$$
$$0.2 = n$$

The painting's value increased by 20%.

Interest, pp. 93–94

If r is the interest rate, the simple interest (I) owed on a principal (P) is given by the formula $I = P \cdot r \cdot t$, where t is time in years.

Mark invested a total of $10,000 in two accounts. One account paid 4% annual interest, and the other paid 5% annual interest. In the first year, Mark earned $460 in interest. How much did he invest in each account?

Acct.	P	r	t	$I = P \cdot r \cdot t$
First	x	0.04	1	$0.04x$
Second	$10{,}000 - x$	0.05	1	$0.05(10{,}000 - x)$
Total	$10,000			$460

$$\begin{aligned} 0.04x + 0.05(10{,}000 - x) &= 460 \\ 0.04x + 500 - 0.05x &= 460 \\ -0.01x + 500 &= 460 \\ -0.01x &= -40 \\ x &= 4000 \end{aligned}$$

Mark invested $4000 at 4% and $6000 at 5%.

Ratio, p. 95

A ratio is a comparison of two quantities using division and is usually written as a fraction.

$$\frac{63 \text{ miles}}{2 \text{ hours}}$$

Proportions, p. 95

A proportion is a statement of equality between two ratios.

$$\frac{n}{40} = \frac{17}{24}$$

Cross Products of a Proportion, pp. 96–97

If $\frac{a}{b} = \frac{c}{d}$, then $a \cdot d = b \cdot c$.

Solve $\frac{n}{12} = \frac{48}{64}$.

$$\begin{aligned} 64n &= 12 \cdot 48 \\ 64n &= 576 \\ n &= \frac{576}{64} \\ n &= 9 \end{aligned}$$

Section 2.5 Linear Inequalities

Linear Inequalities, p. 101

A linear inequality is an inequality containing linear expressions.

$$2x - 3 \ge 9$$

Solutions: Number Line, pp. 101–102

The solutions to a linear inequality can be represented on a number line.
If the endpoint is a solution, you use a closed circle on the number line.
If the endpoint is not a solution, you use an open circle.

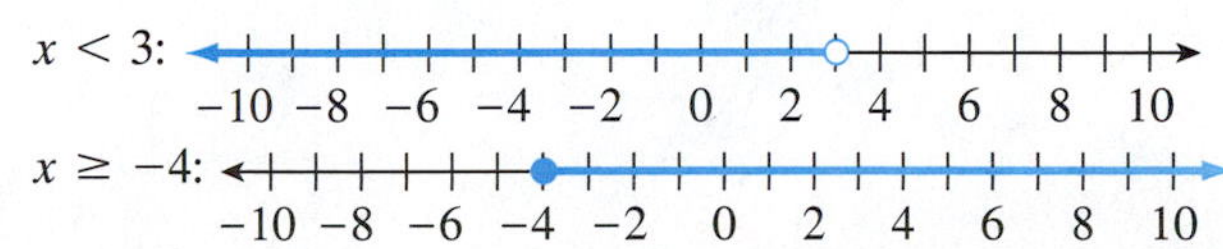

Solutions: Interval Notation, pp. 102–103

Interval notation presents an interval by listing its left and right endpoints. Parentheses are used around endpoints not included in the interval, and square brackets are used when the endpoints are included in the interval.

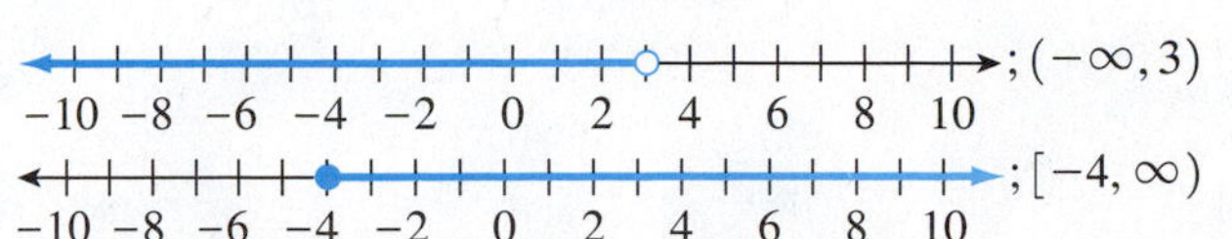

Solving Linear Inequalities, pp. 103–105

Solving a linear inequality is similar to solving a linear equation except that multiplying or dividing both sides of an inequality by a negative number changes the direction of the inequality.

Solve $-3x + 2 \le -16$.

$$\begin{aligned} -3x + 2 &\le -16 \\ -3x &\le -18 \\ \frac{-3x}{-3} &\ge \frac{-18}{-3} \\ x &\ge 6 \end{aligned}$$

−10 −8 −6 −4 −2 0 2 4 6 8 10 , $[6, \infty)$

Compound Inequalities, pp. 105–106

A compound inequality is made up of two or more individual inequalities.

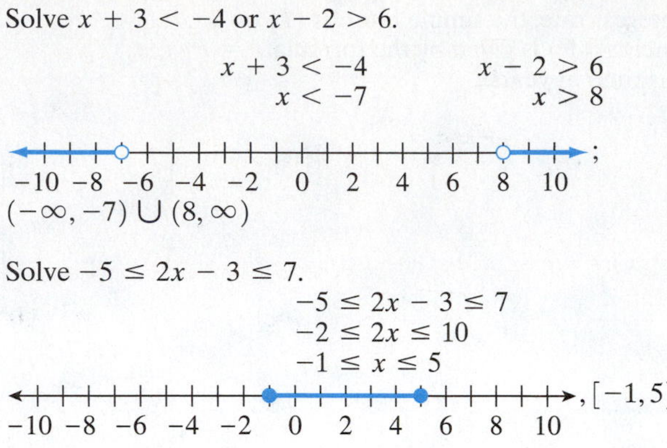

Solve $x + 3 < -4$ or $x - 2 > 6$.

$$x + 3 < -4 \qquad x - 2 > 6$$
$$x < -7 \qquad x > 8$$

$(-\infty, -7) \cup (8, \infty)$

Solve $-5 \le 2x - 3 \le 7$.

$$-5 \le 2x - 3 \le 7$$
$$-2 \le 2x \le 10$$
$$-1 \le x \le 5$$

$[-1, 5]$

SUMMARY OF CHAPTER 2 STUDY STRATEGIES

Using your textbook as more than a source of the homework exercises and answers to the odd problems will help you learn mathematics.

- Read the section the night before your instructor covers it to familiarize yourself with the material.
- Reread the section after it is covered in class. After you read through an example, attempt the Quick Check exercise that follows it.
- Use the textbook to create a series of note cards for each section.
- In each section, look for the feature labeled "A Word of Caution." This feature points out common mistakes and offers advice to help you avoid making the same mistakes.
- Finally, the feature labeled "Using Your Calculator" will show you how to use the TI-84 calculator to help you solve the problems in that section.

CHAPTER 2 REVIEW

***Determine whether the given value is a solution of the equation.* [2.1]**

1. $x = 7, 4x + 9 = 19$

2. $a = -9, (4a - 15) - (2a + 7) = -2(11 - a)$

***Solve using the multiplication property of equality.* [2.1]**

3. $2x = 32$

4. $-9x = 54$

***Solve using the addition property of equality.* [2.1]**

5. $x - 3 = -9$

6. $a + 14 = 8$

***For each of the following problems, set up a linear equation and solve it.* [2.1]**

7. A local club charged \$8 per person to see a "Battle of the Bands." After paying the winning band a prize of \$250, the club made a profit of \$694. How many people came to see the "Battle of the Bands?"

8. Thanks to proper diet and exercise, Randy's cholesterol dropped 23 points in the last two months. If his new cholesterol level is 189 points, what was his cholesterol level two months ago?

***Solve.* [2.2]**

9. $5x - 8 = 27$

10. $-11 = 3a + 16$

11. $-4x + 15 = 7$

12. $51 - 2x = -21$

13. $\frac{x}{6} - \frac{5}{12} = \frac{5}{4}$

14. $\frac{1}{5}x - \frac{1}{2} = \frac{5}{4}$

15. $4x - 19 = 5x + 14$

16. $2m + 23 = 17 - 8m$

17. $2n - 11 + 3n + 4 = 7n - 39$

18. $3(2x - 7) - (x - 9) = 2x - 33$

***Solve the following literal equations for the specified variable.* [2.2]**

19. $7x + y = 8$ for y

20. $15x + 7y = 30$ for y

21. $C = \pi \cdot d$ for d

22. $P = 3a + b + 2c$ for a

23. One number is 26 more than another number. If the sum of the two numbers is 98, find the two numbers. [2.3]

24. The width of a rectangle is 7 feet less than its length, and the perimeter is 50 feet. Find the length and the width of the rectangle. [2.3]

25. The sum of three consecutive even integers is 204. Find the three even integers. [2.3]

26. If Jason drives at a speed of 72 miles per hour, how long will it take him to drive 342 miles? [2.3]

27. Alycia's piggy bank has dimes and quarters in it. The number of dimes is 4 less than 3 times the number of quarters. If the total value of the coins is \$7.85, how many dimes are in the piggy bank? [2.3]

28. Twenty percent of 95 is what number? [2.4]

29. What percent of 136 is 51? [2.4]

30. Thirty-two percent of what number is 1360? [2.4]

31. Fifty-five percent of the students at a college are female. If the college has 10,520 students, how many are female? [2.4]

32. In a group of 320 college graduates, 204 decided to pursue a master's degree. What percent of these students decided to pursue a master's degree? [2.4]

33. Jerry's bank account earned \$553.50 interest last year. This account pays 4.5% interest. How much money did Jerry have in the account at the start of the year? [2.4]

34. A store is discounting all of its prices by 20%. If a sweater's original price was \$59.95, what is the price after the discount? [2.4]

35. Karl deposits a total of \$2000 in savings accounts at two banks. The first bank pays 6% interest, while the second bank pays 5% interest. If he earns a total of \$117 interest in the first year, how much did he deposit at each bank? [2.4]

36. Merle has two solutions. One is 17% alcohol, and the other is 42% alcohol. How much of each solution should be mixed to make 80 milliliters of a solution that is 32% alcohol? [2.4]

***Solve the proportion.* [2.4]**

37. $\frac{n}{6} = \frac{18}{27}$

38. $\frac{4}{9} = \frac{n}{72}$

39. $\frac{32}{13} = \frac{192}{n}$

40. $\frac{38}{n} = \frac{57}{75}$

***Set up a proportion and solve it to solve the following problems.* [2.4]**

41. At a community college, 3 out of every 7 students plan to transfer to a four-year university. If the college has 14,952 students, how many plan to transfer to a four-year university?

Worked-out solutions to Review Exercises marked with can be found on page AN-4.

42. Six tablespoons of a concentrated weed killer must be mixed with 1 gallon of water before spraying. How many tablespoons of the weed killer need to be mixed with $2\frac{1}{2}$ gallons of water? [2.4]

Solve. Present your solution on a number line and in interval notation. [2.5]

43. $x + 5 < 2$

44. $x - 3 > -8$

45. $2x - 7 \geq 3$

46. $3x + 4 \leq -8$

47. $3x < -18$ or $x + 3 > 10$

48. $x + 5 \leq -3$ or $4x \geq -12$

49. $-21 \leq 3x - 6 \leq 24$

50. $-11 < 2x - 9 < 11$

51. Erin needs to make at least 7 sales this week to qualify for a bonus. Write an inequality that shows the number of sales that will qualify Erin to receive a bonus. [2.5]

52. A company is trying to raise capital and offers 2 million shares of stock for sale. If the company needs to raise at least $69.5 million, what price does it need to receive for the stock? [2.5]

53. If Steve's test average is at least 70 but lower than 80, his final grade will be a C. His scores on the first four tests were 58, 72, 66, and 75. What scores on the fifth exam will give him a C for the class? (Assume that the highest possible score is 100.) [2.5]

CHAPTER 2 TEST

For Extra Help

Step-by-step test solutions are found on the Chapter Test Prep Videos available via the Video Resources on DVD, in MyMathLab, and on YouTube (search "WoodburyElemIntAlg" and click on "Channels").

1. Solve $-8x = 56$ using the multiplication property of equality.

2. Solve $x + 17 = -22$ using the addition property of equality.

Solve.

3. $7x - 20 = 36$

4. $72 = 30 - 8a$

5. $3x + 34 = 5x - 21$

6. $3(5x - 16) - 9x = x - 63$

7. $\frac{1}{8}x - \frac{2}{3} = \frac{5}{6}$

8. $\frac{n}{100} = \frac{22}{40}$

9. Solve the literal equation $4x + 3y = 28$ for y.

10. What number is 16% of 175?

11. What percent of 660 is 99?

Solve. Present your solution on a number line and in interval notation.

12. $6x + 8 \leq -7$

13. $x + 4 \leq -5$ or $2x + 3 \geq 0$

14. $-23 < 3x + 7 < 23$

15. The length of a rectangle is 3 feet less than twice its width, and the perimeter is 72 feet. Find the length and the width of the rectangle.

16. Barry's wallet contains $5 and $10 bills. He has 7 more $5 bills than $10 bills. If Barry has $185 in his wallet, how many $5 bills are in the wallet?

17. Ernesto invested a total of $10,000 in two mutual funds. In the first year, one account earned a 4% profit and the other earned an 11% profit. If Ernesto's total profit was $862, how much did he invest in the mutual fund that earned an 11% profit?

18. A chemist has two solutions. One is 13% salt, and the other is 21% salt. How much of each solution should be mixed to make 100 milliliters of a solution that is 18% salt?

19. At a certain university, 2 out of every 9 professors earned their degree outside the United States. If the university has 396 professors, how many of them earned their degree outside the United States?

20. Wayne earns a commission of $500 for every car he sells. How many cars does he need to sell to earn at least $17,000 in commissions this month?

Mathematicians in History

George Pólya was a Hungarian mathematician who spent much of his career on problem solving and wrote the landmark text *How to Solve It*. Alan Schoenfeld has said of this book, "For mathematics education and the world of problem solving it marked a line of demarcation between two eras, problem solving before and after Pólya."

Write a one-page summary (*or* make a poster) of the life of George Pólya and his accomplishments. Also look up Pólya's most famous quotes and list your favorite.

Interesting issues:

- Where and when was George Pólya born?
- What circumstances led to Pólya leaving Göttingen after Christmas in 1913?
- The political situation in Europe in 1940 forced Pólya to leave Zürich for the United States. At which university did Pólya work after arriving in the United States?
- When was Pólya's monumental book *How to Solve It* published?
- Summarize Pólya's strategy for solving problems.
- At which American university did Pólya spend most of his career?
- What was Pólya's mnemonic for the first fourteen digits of π?
- How old was Pólya when he taught his last class? What was the subject?

CHAPTER 3

Graphing Linear Equations

In this chapter, we will examine linear equations in two variables. Equations involving two variables relate two unknown quantities to each other, such as a person's height and weight or the number of items sold by a company and the profit the company made. The primary focus is on graphing linear equations, which is the technique used to display the solutions to an equation in two variables.

STUDY STRATEGY

Making the Best Use of Your Resources In this chapter, we will focus on how to get the most out of your available resources. What works best for one student may not work for another student. Consider all of the different resources presented in this chapter, and use the ones you believe will help you.

Throughout this chapter, we will revisit this study strategy and help you incorporate it into your study habits.

3.1 The Rectangular Coordinate System; Equations in Two Variables

OBJECTIVES

1. Determine whether an ordered pair is a solution of an equation in two variables.
2. Plot ordered pairs on a rectangular coordinate plane.
3. Complete ordered pairs for a linear equation in two variables.
4. Graph linear equations in two variables by plotting points.

Equations in Two Variables and Their Solutions; Ordered Pairs

Objective 1 Determine whether an ordered pair is a solution of an equation in two variables. In this section, we will begin to examine equations in two variables, which relate two different quantities to each other. Here are some examples of equations in two variables.

$$y = 4x + 7 \qquad 2a + 3b = -6$$

$$y = 3x^2 - 2x - 9 \qquad t = \frac{4n + 9}{3n - 7}$$

As in the previous chapter, we will be looking for solutions of these equations. A **solution of an equation in two variables** is a pair of values that when substituted for the two variables produce a true statement. Consider the equation $3x + 4y = 10$. One solution of this equation is $x = 2$ and $y = 1$. We can see that this is a solution by substituting 2 for x and 1 for y into the equation.

$$\begin{aligned} 3x + 4y &= 10 \\ 3(2) + 4(1) &= 10 && \text{Substitute 2 for } x \text{ and 1 for } y. \\ 6 + 4 &= 10 && \text{Multiply.} \\ 10 &= 10 && \text{Add.} \end{aligned}$$

Because this produces a true statement, together the pair of values form a solution. Solutions to this type of equation are written as **ordered pairs** (x, y). In the example, the ordered pair $(2, 1)$ is a solution. We call it an ordered pair because the two values are listed in the specific order of x first and y second. The values are often referred to as **coordinates**.

EXAMPLE 1 Is the ordered pair $(-3, 2)$ a solution of the equation $y = 3x + 11$?

SOLUTION To determine whether this ordered pair is a solution, substitute -3 for x and 2 for y.

$$\begin{aligned} y &= 3x + 11 \\ (2) &= 3(-3) + 11 && \text{Substitute } -3 \text{ for } x \text{ and 2 for } y. \\ 2 &= -9 + 11 && \text{Multiply.} \\ 2 &= 2 && \text{Simplify.} \end{aligned}$$

Because this is a true statement, the ordered pair $(-3, 2)$ is a solution of the equation $y = 3x + 11$.

EXAMPLE 2 Is $(4, 0)$ a solution of the equation $y - 2x = 8$?

SOLUTION Again, check to see whether this ordered pair is a solution by substituting the given values for x and y. Be careful to substitute 4 for x and 0 for y, not vice versa.

$$\begin{aligned} y - 2x &= 8 \\ (0) - 2(4) &= 8 && \text{Substitute 4 for } x \text{ and 0 for } y. \\ 0 - 8 &= 8 && \text{Multiply.} \\ -8 &= 8 && \text{Simplify.} \end{aligned}$$

Because this is a false statement, $(4, 0)$ is not a solution of the equation $y - 2x = 8$.

Quick Check 1

Is $(0, -6)$ a solution of the equation $3x + 2y = 12$?

The Rectangular Coordinate Plane

Objective 2 Plot ordered pairs on a rectangular coordinate plane. Ordered pairs can be displayed graphically using the **rectangular coordinate plane**. The rectangular coordinate plane also is known as the **Cartesian plane**, named after the famous French philosopher and mathematician René Descartes.

The rectangular coordinate plane is made up of two number lines, known as **axes**. The axes are drawn at right angles to each other, intersecting at 0 on each axis. The accompanying graph shows how it looks.

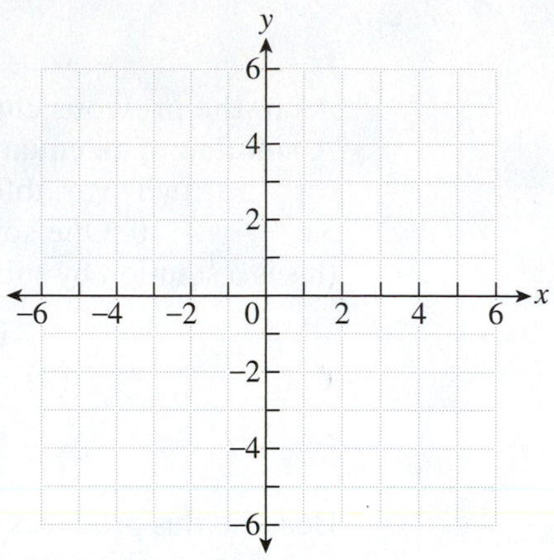

x-axis, *y*-axis, origin

The **horizontal axis,** or ***x*-axis,** is used to show the first value of an ordered pair. The **vertical axis,** or ***y*-axis,** is used to show the second value of an ordered pair. The point where the two axes cross is called the **origin** and represents the ordered pair $(0, 0)$.

Positive values of x are found to the right of the origin, and negative values of x are found to the left of the origin. Positive values of y are found above the origin, and negative values of y are found below the origin.

We can use the rectangular coordinate plane to represent ordered pairs that are solutions of an equation. To plot an ordered pair (x, y) on a rectangular coordinate plane, we begin at the origin. The x-coordinate tells us how far to the left or right of the origin to move. The y-coordinate then tells us how far up or down to move from there.

Suppose we wanted to plot the ordered pair $(-3, 2)$ on a rectangular coordinate plane. The x-coordinate is -3, which tells us to move 3 units to the left of the origin. The y-coordinate is 2, so we move up 2 units to find the location of the ordered pair and place a point at this location. An ordered pair is often referred to as a **point** when it is plotted on a graph.

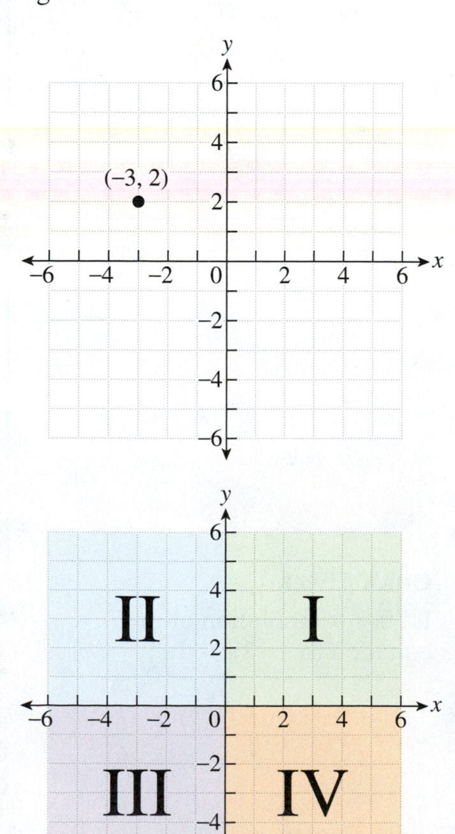

One final note about the rectangular coordinate plane is that the two axes divide the plane into four sections called **quadrants**. The four quadrants are labeled I, II, III, and IV. A point that lies on one of the axes is not considered to be in one of the quadrants.

EXAMPLE 3 Plot the ordered pairs $(5, 1)$, $(4, -2)$, $(-2, -4)$, and $(-4, 0)$ on a rectangular coordinate plane.

SOLUTION The four ordered pairs are plotted on the following graph:

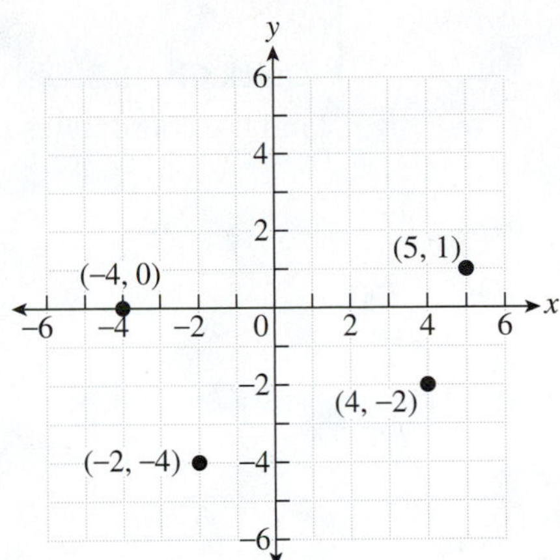

Quick Check 2

Plot the ordered pairs $(3, 8)$, $(5, -4)$, $(-3, 6)$, and $(0, 7)$ on a rectangular coordinate plane.

Choosing the appropriate scale for a graph is an important skill. To plot the ordered pair $(\frac{7}{2}, -4)$ on a rectangular coordinate plane, we could use a scale of $\frac{1}{2}$ on the axes. A rectangular plane with axes showing values ranging from -10 to 10 is not large enough to plot the ordered pair $(-45, 25)$. A good choice is to label the axes showing values that range from -50 to 50. Because this is a wide range of values, it makes sense to increase the scale. Because each coordinate is a multiple of 5, we use a scale of 5 units.

It is important to be able to read the coordinates of an ordered pair from a graph.

EXAMPLE 4 Find the coordinates of points A through G on the following graph:

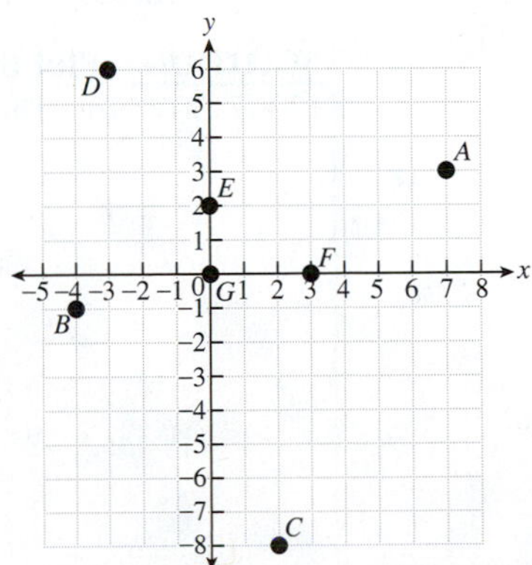

SOLUTION To find the x-coordinate of a point, determine how far the point is to the left or right of the origin. Points to the right of the origin have positive x-coordinates, and points to the left of the origin have negative x-coordinates. Points on the y-axis have an x-coordinate of 0.

Next, determine the y-coordinate of the point by measuring how far the point is above or below the origin. If the point is above the origin, the y-coordinate is positive. If the point is below the origin, the y-coordinate is negative. Points on the x-axis have a y-coordinate of 0.

Here is a table containing the coordinates of each point.

Point	A	B	C	D	E	F	G
Coordinates	$(7, 3)$	$(-4, -1)$	$(2, -8)$	$(-3, 6)$	$(0, 2)$	$(3, 0)$	$(0, 0)$

Quick Check 3

Find the coordinates of points A through E on the following graph:

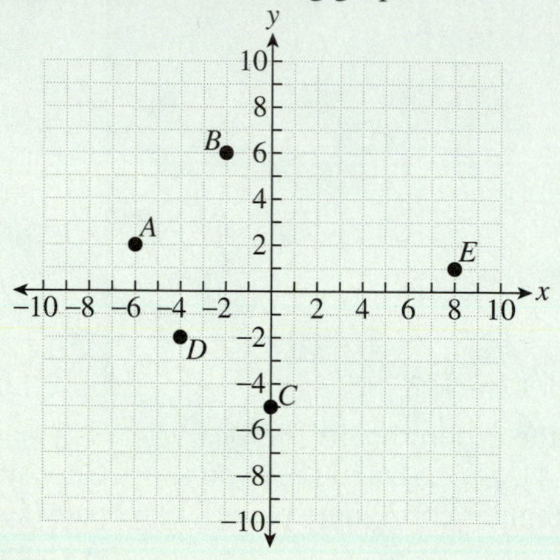

EXAMPLE 5 Plot the ordered pairs associated with the given data. (Let the year be x and the median household income be y.)

Year	1967	1977	1987	1997	2007
Median Household Income ($1000's)	7	14	26	37	50

(Source: U.S. Census Bureau)

SOLUTION Plot the ordered pairs $(1967, 7)$, $(1977, 14)$, $(1987, 26)$, $(1997, 37)$, and $(2007, 50)$. Note that we focus on x values from 1967 to 2007.

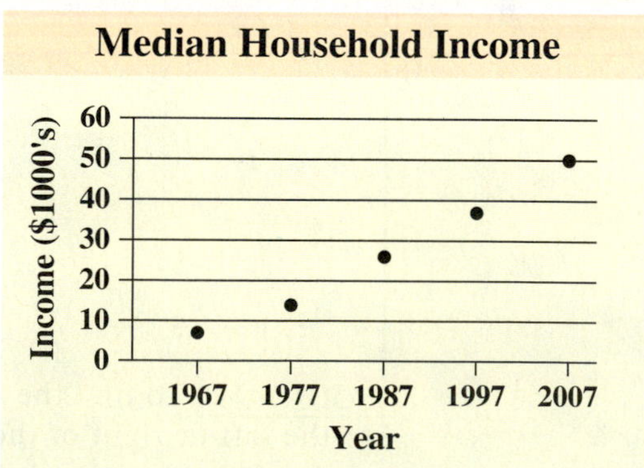

Quick Check 4

Plot the ordered pairs associated with the given data. (Let the year be x and the number of subscribers be y.)

Year	2004	2005	2006	2007	2008
Subscribers (millions)	2.6	4.2	6.3	7.5	9.4

(Source: Netflix)

Linear Equations in Two Variables

Objective 3 **Complete ordered pairs for a linear equation in two variables.**

Standard Form of a Linear Equation

A linear equation in two variables is an equation that can be written in the form $Ax + By = C$, where A, B, and C are real numbers and A and B are not both 0. This is called the **standard form** of a linear equation in two variables.

A linear equation has multiple solutions. To find an ordered pair that is a solution of a linear equation, we can select an arbitrary value for one of the variables, substitute this value into the equation, and solve for the remaining variable. The next example provides practice finding solutions of linear equations in two variables.

EXAMPLE 6 Find the missing coordinate such that the ordered pair $(2, \_)$ is a solution of the equation $y = 5x - 4$.

SOLUTION We are given an x-coordinate of 2, so we substitute this for x in the equation.

$y = 5(2) - 4$ Substitute 2 for x and solve for y.

$y = 10 - 4$ Multiply.

$y = 6$ Subtract.

The ordered pair $(2, 6)$ is a solution of this equation.

Quick Check 5

Find the missing coordinate such that the ordered pair $(3, \_)$ is a solution of the equation $y = -3x + 5$.

The solutions of any linear equation in two variables follow a pattern. Now we will begin to explore this pattern.

EXAMPLE 7 Complete the following table of ordered pairs with solutions of the linear equation $x + y = 5$.

x	y
3	
	−2
0	

Plot the solutions on a rectangular coordinate plane.

SOLUTION The first ordered pair has an x-coordinate of 3. Substituting this for x results in the equation $(3) + y = 5$, which has a solution of $y = 2$. $(3, 2)$ is a solution.

The second ordered pair has a y-coordinate of −2. Substituting for y, this produces the equation $x + (-2) = 5$. The solution of this equation is $x = 7$, so $(7, -2)$ is a solution.

The third ordered pair has an x-coordinate of 0. When substituting this for x, we get the equation $(0) + y = 5$. This equation has a solution of $y = 5$, so $(0, 5)$ is a solution. The completed table follows.

x	y
3	2
7	−2
0	5

Quick Check 6

Complete the table of ordered pairs with solutions of the linear equation $x - y = 6$.

x	y
4	
	3
−1	

Plot the solutions on a rectangular coordinate plane.

When we plot these three points on a rectangular coordinate plane, it looks like the following graph. Notice that the three points appear to fall on a straight line.

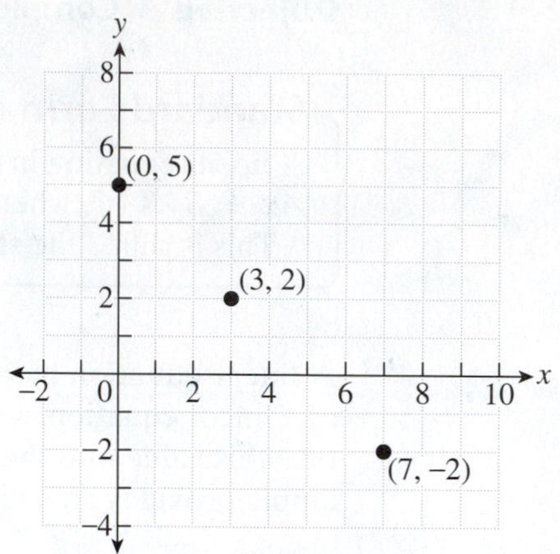

Graphing Linear Equations by Plotting Points

Objective 4 Graph linear equations in two variables by plotting points. In the previous example, notice that the three points appear to be in line with each other. We can draw a straight line that passes through each of the three points. In addition to the three points on the graph, every point that lies on the line is a solution of the equation. Every ordered pair that is a solution must lie on this line.

We can sketch a graph of the solutions of a linear equation by plotting ordered pairs that are solutions and then drawing the straight line through these points. The line we sketch continues indefinitely in both directions. Although only two points are needed to determine a line, we will find three points. The third point is used as a check to make sure the first two points are accurate. We will pick three arbitrary values for one of the variables and find the other coordinates.

EXAMPLE 8 Graph the linear equation $y = 3x - 4$.

SOLUTION When the equation has y isolated on one side of the equal sign, it is a good idea to select values for x and find the corresponding y-values. We find three points on the line by letting $x = 0$, $x = 1$, and $x = 2$. These values were chosen arbitrarily. If we start with other values, the ordered pairs will still be on the same line.

$x = 0$	$x = 1$	$x = 2$
$y = 3(0) - 4$	$y = 3(1) - 4$	$y = 3(2) - 4$
$y = 0 - 4$	$y = 3 - 4$	$y = 6 - 4$
$y = -4$	$y = -1$	$y = 2$
$(0, -4)$	$(1, -1)$	$(2, 2)$

Quick Check 7

Graph the linear equation $y = 2x + 1$.

Plot these three points on a rectangular coordinate plane and draw a line through them.

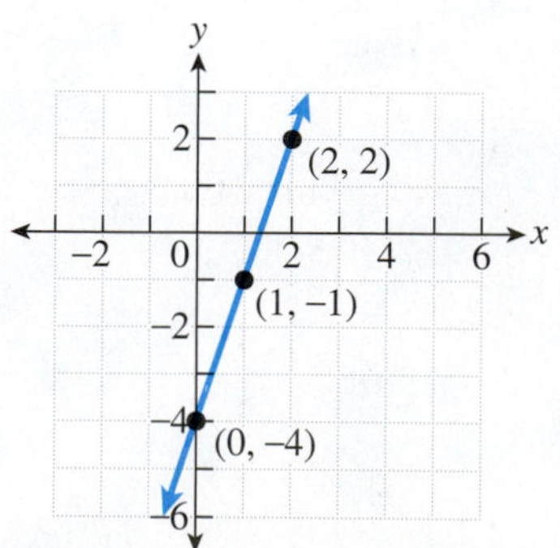

BUILDING YOUR STUDY STRATEGY

Using Your Resources, 1 Your Instructor One of the most important resources available to you is your instructor. If you have a question during the lecture, ask it. Your question is probably on the minds of other students as well.

Find out where your instructor's office is located and when office hours are held. Visiting your instructor during office hours is a great way to get one-on-one help. Some questions are difficult for the instructor to answer completely during class and are more easily handled in his or her office.

Exercises 3.1

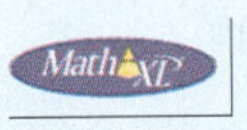
PRACTICE

WATCH

DOWNLOAD

READ

REVIEW

Vocabulary

1. The standard form of a linear equation in two variables is ____________.
2. Solutions to an equation in two variables are written as a(n) ____________.
3. The ______ is used to show the first coordinate of an ordered pair on a rectangular coordinate plane.
4. The ______ is used to show the second coordinate of an ordered pair on a rectangular coordinate plane.
5. The point at $(0, 0)$ is called the ______.
6. A(n) _____ on a rectangular coordinate plane is used to show the solutions of a linear equation in two variables.

Is the ordered pair a solution of the given equation?

7. $(7, 6)$, $5x + 3y = 53$
8. $(2, 9)$, $2x + 9y = 85$
9. $(4, -3)$, $6x - 2y = 18$
10. $(9, 5)$, $2x + 7y = 53$
11. $(10, 9)$, $y = \frac{3}{2}x - 6$
12. $(8, 7)$, $y = -\frac{1}{4}x + 5$
13. $(3, 2)$, $-2x + 5y = 4$
14. $(5, -2)$, $-x + 7y = -9$
15. $(-3, 0)$, $-4x + y = 12$
16. $(0, 4)$, $2x - 5y = -20$

Plot the points on a rectangular coordinate plane.

17. $(5, 2)$, $(3, 4)$, $(2, -6)$, $(-4, 1)$
18. $(-3, -5)$, $(-5, 3)$, $(4, -3)$, $(-2, -1)$
19. $(0, 4)$, $(-2, 0)$, $(-4, 2)$, $(2, 4)$
20. $(-4, 0)$, $(-5, 0)$, $(0, -3)$, $(0, 2)$
21. $\left(\frac{7}{2}, 5\right)$, $\left(-2, \frac{13}{4}\right)$, $\left(-\frac{23}{8}, -1\right)$, $\left(\frac{3}{2}, -\frac{5}{2}\right)$
22. $(-70, 30)$, $(20, -90)$, $(45, 25)$, $(-50, -68)$

Find the coordinates of the labeled points A, B, C, and D.

23.

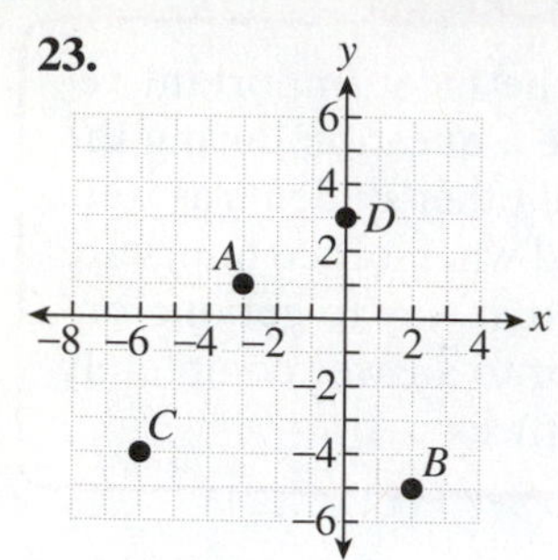

24.

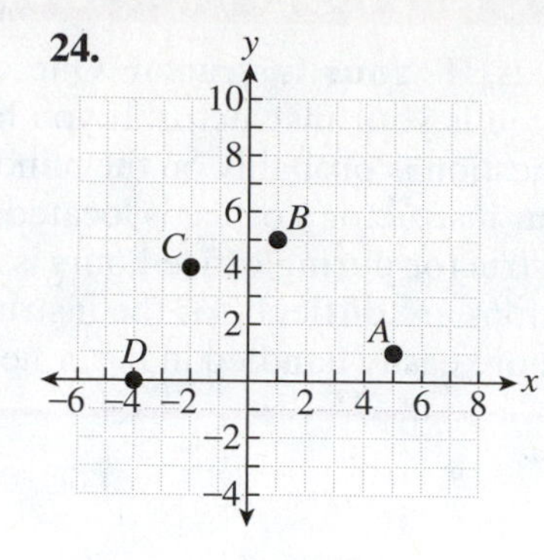

25. **26.**

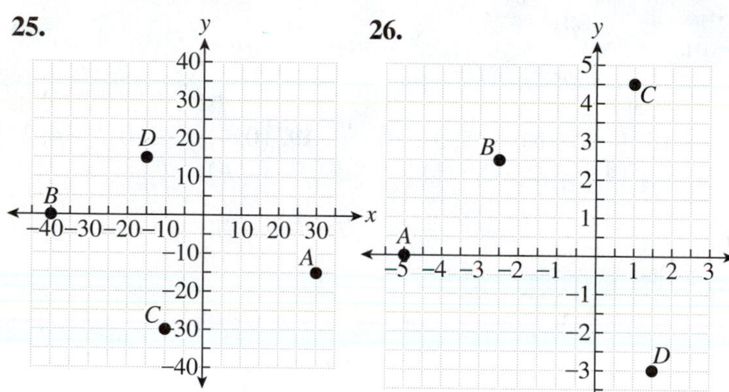

In which quadrant (I, II, III, or IV) is the given point?

27. $(5, -1)$

28. $(-2, 8)$

29. $(3, 4)$

30. $(-6, -4)$

31. $(2, 5)$

32. $(-3, -3)$

33. $\left(\frac{1}{2}, -\frac{3}{5}\right)$

34. $(-700, 1000)$

Plot the ordered pairs associated with the given data. (Let the first row be x and the second row be y.)

35. Number of identity theft complaints filed. (*Source*: *Federal Trade Commission*)

Year	2000	2002	2004	2006	2008
Complaints (1000's)	31	162	247	246	314

36. Median home prices in California from 2003–2008. (*Source*: *California Association of Realtors*)

Year	2003	2004	2005	2006	2007	2008
Median Price ($1000's)	$373	$451	$524	$557	$558	$381

37. Number of Starbucks stores worldwide from 2003–2008. (*Source*: *Starbucks*)

Year	2003	2004	2005	2006	2007	2008
Stores	7225	8569	10,241	12,440	15,011	16,680

38. U.S. cell phone subscribers, in millions, for selected years. (*Source*: *Cellular Telecommunications & Internet Associates*)

Year	1987	1992	1997	2002	2007
Subscribers (millions)	1	11	55	141	250

39. Estimated number of bariatric (weight-loss) surgeries performed from 2002–2008. (*Source*: *American Society for Metabolic and Bariatric Surgery*)

Year	2002	2003	2004	2005	2006	2007	2008
Surgeries (thousands)	63	103	141	171	178	205	220

40. Total value of online retail sales in the United States, in billions, from 2000–2007. (*Source*: *U.S. Census Bureau*)

Year	2000	2001	2002	2003	2004	2005	2006	2007
Sales	\$24	\$31	\$41	\$54	\$67	\$84	\$108	\$128

Using the given coordinate, find the other coordinate that makes the ordered pair a solution of the equation.

41. $3x - 2y = 8, (2, \_\_)$

42. $x + 4y = 7, (\_\_, 3)$

43. $7x - 4y = -11, (\_\_, -6)$

44. $5x + 4y = 12, (-4, \_\_)$

45. $y = 2x - 7, (9, \_\_)$

46. $y = -5x + 4, (3, \_\_)$

47. $7x - 11y = -22, (0, \_\_)$

48. $4x + 3y = -20, (\_\_, 0)$

49. $3x + \frac{1}{2}y = -2, (\_\_, 8)$

50. $\frac{1}{5}x + 3y = 20, (10, \_\_)$

51. $6x + 2y = 9, (\frac{2}{3}, \_\_)$

52. $3x + 4y = 15, (\_\_, \frac{7}{2})$

Complete each table with ordered pairs that are solutions.

53. $5x - 3y = -12$

x	y
−3	
0	
3	

54. $-2x + 4y = -16$

x	y
−2	
0	
2	

55. $y = 4x - 9$

x	y
0	
1	
$\frac{3}{2}$	

56. $y = -3x + 5$

x	y
0	
$\frac{1}{3}$	
2	

57. $y = \frac{2}{3}x - 2$

x	y
−3	
0	
6	

58. $y = \frac{7}{5}x + 1$

x	y
−5	
0	
5	

59. Find three equations that have the ordered pair (5, 2) as a solution.

60. Find three equations that have the ordered pair (3, −4) as a solution.

61. Find three equations that have the ordered pair (0, 6) as a solution.

62. Find three equations that have the ordered pair (−8, 0) as a solution.

Find three ordered pairs that are solutions of the given linear equation and use them to graph the equation.

63. $y = 4x + 1$

64. $y = 2x - 3$

65. $y = -2x + 7$

66. $y = -3x + 6$

67. $y = -2x$

68. $y = 4x$

69. $y = \frac{1}{4}x + 3$

70. $y = \frac{1}{2}x - 3$

71. $y = -\frac{3}{5}x + 6$

72. $y = -\frac{5}{2}x + 4$

73. $3x + y = 9$

74. $2x + 3y = 6$

Writing in Mathematics

Answer in complete sentences.

75. In your own words, explain why the order of the coordinates in an ordered pair is important. Is there a difference between plotting the ordered pair (a, b) and the ordered pair (b, a)?

76. If the ordered pair (a, b) is in quadrant II, in which quadrant is (b, a)? In which quadrant is $(-a, b)$? Explain how you determined your answers.

3.2 Graphing Linear Equations and Their Intercepts

OBJECTIVES

1. Find the x- and y-intercepts of a line from its graph.
2. Find the x- and y-intercepts of a line from its equation.
3. Graph linear equations using their intercepts.
4. Graph linear equations that pass through the origin.
5. Graph horizontal lines.
6. Graph vertical lines.
7. Interpret the graph of an applied linear equation.

In the previous section, we learned to use a rectangular coordinate plane to display the solutions of an equation in two variables. We also learned that all of the solutions of a linear equation in two variables are on a straight line. In this section, we will learn to graph the line associated with a linear equation in two variables in a more systematic fashion.

Intercepts

Objective 1 Find the x- and y-intercepts of a line from its graph.

A point at which a graph crosses the x-axis is called an ***x*-intercept,** and a point at which a graph crosses the y-axis is called a ***y*-intercept.**

EXAMPLE 1 Find the coordinates of any x-intercepts and y-intercepts.

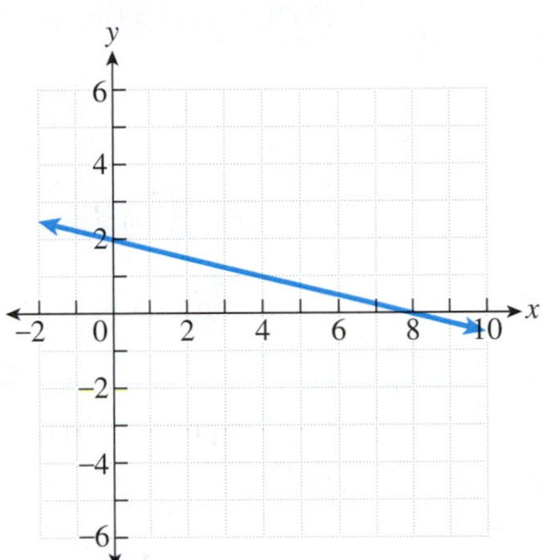

SOLUTION An x-intercept is a point at which the graph crosses the x-axis. This graph crosses the x-axis at the point $(8, 0)$, so the x-intercept is the point $(8, 0)$.

This graph crosses the y-axis at the point $(0, 2)$, so the y-intercept is $(0, 2)$.

Quick Check 1

Find the coordinates of any x-intercepts and y-intercepts.

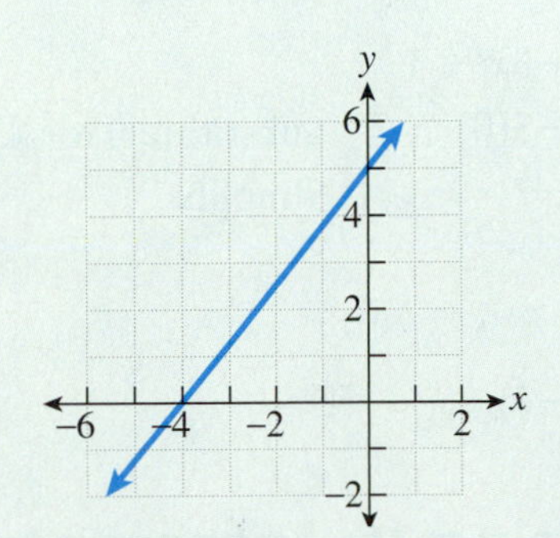

Some lines do not have both an x- and y-intercept.

This horizontal line does not cross the x-axis, so it does not have an x-intercept. Its y-intercept is the point $(0, -6)$.

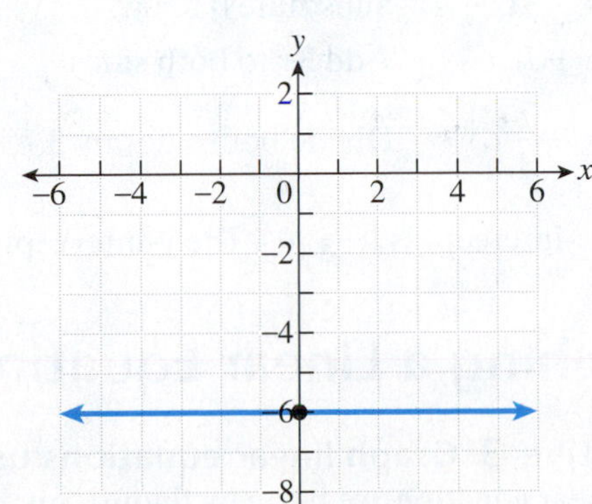

The x-intercept of this vertical line is the point $(4, 0)$. This graph does not cross the y-axis, so it does not have a y-intercept.

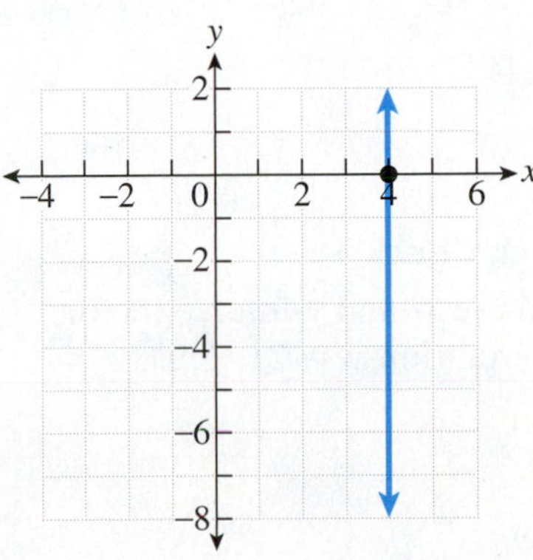

Objective 2 Find the x- and y-intercepts of a line from its equation. Any point that lies on the x-axis has a y-coordinate of 0. Similarly, any point that lies on the y-axis has an x-coordinate of 0. We can use these facts to find the intercepts of an equation's graph algebraically. To find an x-intercept, we substitute 0 for y and solve for x. To find a y-intercept, we substitute 0 for x and solve for y.

Finding Intercepts

To Find an x-Intercept
Substitute 0 for y and solve the resulting equation for x. The ordered pair will be of the form $(x, 0)$.

To Find a y-Intercept
Substitute 0 for x and solve the resulting equation for y. The ordered pair will be of the form $(0, y)$.

EXAMPLE 2 Find the x- and y-intercepts for the equation $5x - 2y = 10$.

SOLUTION Begin with the x-intercept by substituting 0 for y and solving for x.

$$5x - 2y = 10$$
$$5x - 2(0) = 10 \quad \text{Substitute 0 for } y.$$
$$5x = 10 \quad \text{Simplify the left side of the equation.}$$
$$x = 2 \quad \text{Divide both sides by 5.}$$

The x-intercept is $(2, 0)$. To find the y-intercept, substitute 0 for x and solve for y.

$$5x - 2y = 10$$
$$5(0) - 2y = 10 \quad \text{Substitute 0 for } x.$$
$$-2y = 10 \quad \text{Simplify the left side of the equation.}$$
$$y = -5 \quad \text{Divide both sides by } -2.$$

The y-intercept is $(0, -5)$.

Quick Check 2
Find the x- and y-intercepts for the equation $-3x + 4y = -12$.

EXAMPLE 3 Find the x- and y-intercepts for the equation $y = -3x - 7$.

SOLUTION

x-intercept

$$y = -3x - 7$$
$$0 = -3x - 7 \quad \text{Substitute 0 for } y.$$
$$3x = -7 \quad \text{Add } 3x \text{ to both sides.}$$
$$x = -\frac{7}{3} \quad \text{Divide both sides by 3.}$$

y-intercept

$$y = -3x - 7$$
$$y = -3(0) - 7 \quad \text{Substitute 0 for } x.$$
$$y = -7 \quad \text{Simplify.}$$

The x-intercept is $\left(-\frac{7}{3}, 0\right)$. The y-intercept is $(0, -7)$.

Quick Check 3
Find the x- and y-intercepts for the equation $y = 2x - 9$.

Graphing a Linear Equation Using Its Intercepts

Objective 3 Graph linear equations using their intercepts. When graphing a linear equation, we begin by finding any intercepts. This often gives us two points for the graph. Although only two points are necessary to graph a straight line, we plot at least three points to make sure the first two points are accurate. In other words, we use a third point as a "check" for the first two points we find. We find this third point by selecting a value for x, substituting it into the equation, and solving for y.

Graphing a Line by Using Its Intercepts

- Find the x-intercept by substituting 0 for y and solving for x.
- Find the y-intercept by substituting 0 for x and solving for y.
- Select a value of x, substitute it into the equation, and solve for y to find the coordinates of a third point on the line.
- Graph the line that passes through these three points.

EXAMPLE 4 Graph $3x + y = 6$. Label any intercepts.

SOLUTION Begin by finding the x- and y-intercepts.

***x*-intercept**

$3x + (0) = 6$ Substitute 0 for y.
$3x = 6$ Simplify.
$x = 2$ Divide both sides by 3.

***y*-intercept**

$3(0) + y = 6$ Substitute 0 for x.
$y = 6$ Simplify.

The x-intercept is $(2, 0)$. The y-intercept is $(0, 6)$.

Before looking for a third point, let's plot the two intercepts on a graph. This can help us select a value to substitute for x when finding a third point. It is a good idea to select a value for x that is somewhat near the x-coordinates of the other two points. For example, $x = 1$ is a good choice for this example. If we choose a value of x that is too far from the x-coordinates of the other two points, we may have to change the scale of the axes or extend them to show the third point.

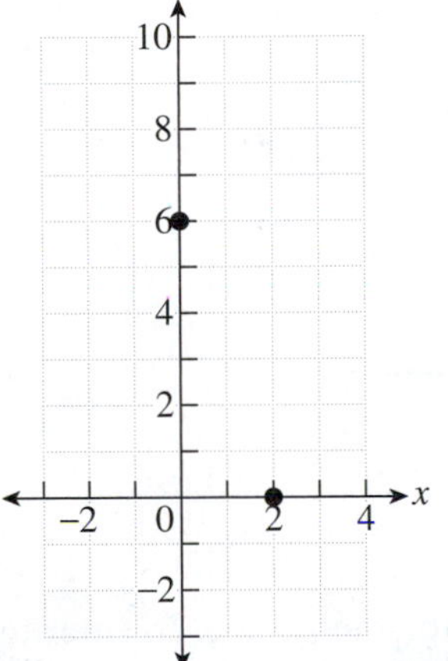

$3(1) + y = 6$ Substitute 1 for x.
$3 + y = 6$ Simplify the left side of the equation.
$y = 3$ Subtract 3.

The point $(1, 3)$ is the third point. Keep in mind that the third point can vary with our choice for x, but the three points still should be on the same line. After plotting the third point, we finish by drawing a straight line that passes through all three points.

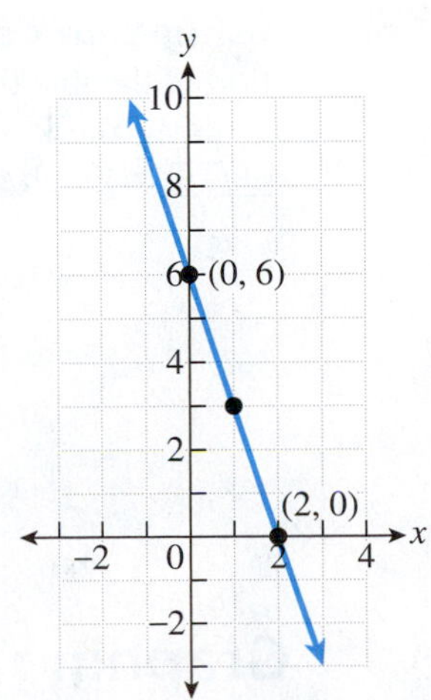

Quick Check 4

Graph $2x - 3y = 12$. Label any intercepts.

EXAMPLE 5 Graph $y = 2x + 5$. Label any intercepts.

SOLUTION Begin by finding the x- and y-intercepts.

x-intercept $(y = 0)$	y-intercept $(x = 0)$
$0 = 2x + 5$	$y = 2(0) + 5$
$-5 = 2x$	$y = 5$
$-\frac{5}{2} = x$	$(0, 5)$
$\left(-\frac{5}{2}, 0\right)$	

Choosing $x = 1$ to find the third point is a good choice: It is close to the two intercepts, and it will be easy to substitute 1 for x in the equation.

$y = 2(1) + 5$ Substitute 1 for x in the equation.

$y = 7$ Simplify.

The third point is $(1, 7)$. Here is the graph.

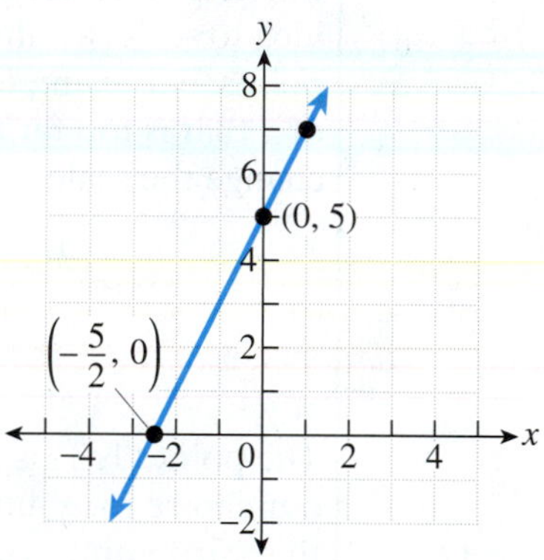

Quick Check 5

Graph $y = -3x - 9$. Label any intercepts.

Using Your Calculator You can use the TI-84 to graph a line. To enter the equation of the line that you graphed in Example 5, press the [Y=] key. Enter $2x + 5$ next to Y_1 using the key labeled [X,T,Θ,n] to key the variable x. Then press the key labeled [GRAPH] to graph the line.

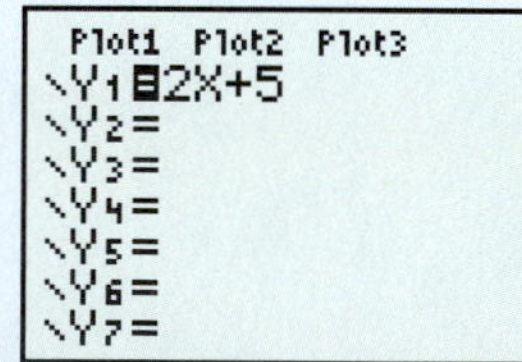

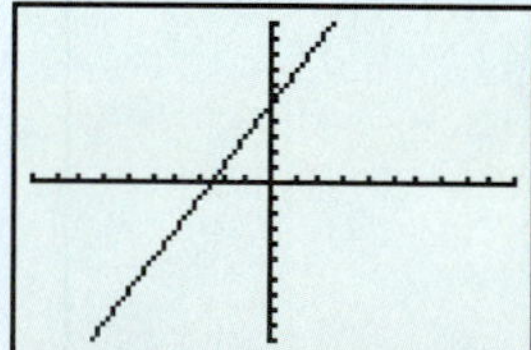

Graphing Linear Equations That Pass through the Origin

Objective 4 **Graph linear equations that pass through the origin.** Some lines do not have two intercepts. For example, a line that passes through the origin has its x- and y-intercepts at the same point. In this case, we need to find two additional points.

Graphing a Line Passing through the Origin

- Determine that both the x- and y-intercepts are at the origin.
- Find a second point on the line by selecting a value of x and substituting it into the equation. Solve the equation for y.
- Find a third point on the line by selecting another value of x and substituting it into the equation. Solve the equation for y.
- Graph the line that passes through these three points.

EXAMPLE 6 Graph $2x - 8y = 0$. Label any intercepts.

SOLUTION Let's start by finding the x-intercept.

$$2x - 8(0) = 0 \quad \text{Substitute 0 for } y.$$
$$2x = 0 \quad \text{Simplify.}$$
$$x = 0 \quad \text{Divide both sides by 2.}$$

The x-intercept is at the origin $(0, 0)$. This point also is the y-intercept, which we can verify by substituting 0 for x and solving for y. We need to find two more points before we graph the line. Suppose we choose $x = 1$.

$$2(1) - 8y = 0 \quad \text{Substitute 1 for } x.$$
$$2 - 8y = 0 \quad \text{Simplify.}$$
$$2 = 8y \quad \text{Add } 8y.$$
$$\frac{1}{4} = y \quad \text{Divide both sides by 8 and simplify.}$$

The point $\left(1, \frac{1}{4}\right)$ is on the line. However, it may not be easy to put this point on the graph because of its fractional y-coordinate. We can try to avoid this problem by solving the equation for y before choosing a value for x.

$$2x - 8y = 0$$
$$2x = 8y \quad \text{Add } 8y \text{ to both sides.}$$
$$\frac{1}{4}x = y \quad \text{or} \quad y = \frac{1}{4}x \quad \text{Divide both sides by 8 and simplify.}$$

If we choose a value for x that is a multiple of 4, we will have a y-coordinate that is an integer. We will use $x = 4$ and $x = -4$, although many other choices would work.

$x = 4$	$x = -4$
$y = \frac{1}{4}(4)$	$y = \frac{1}{4}(-4)$
$y = 1$	$y = -1$
$(4, 1)$	$(-4, -1)$

Quick Check 6

Graph $y = 3x$. Label any intercepts.

Here is the graph.

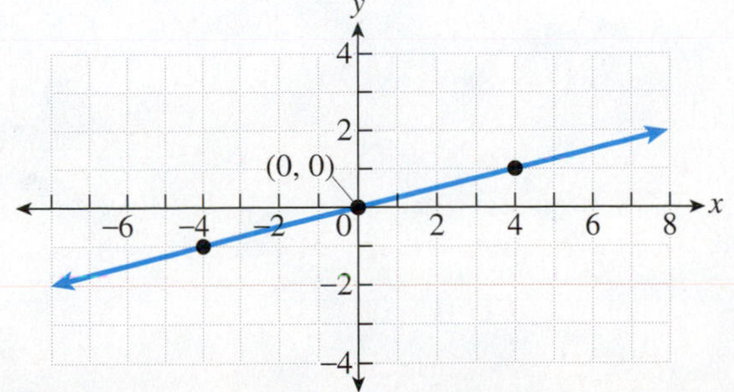

Horizontal Lines

Objective 5 Graph horizontal lines. Can we draw a line that does not cross the x-axis? Yes; here is an example.

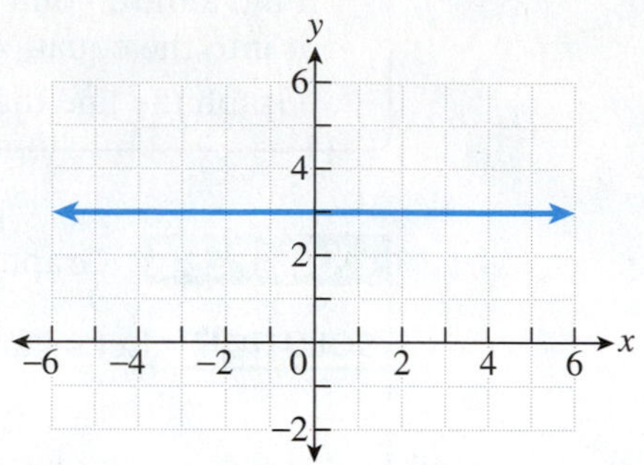

What is the equation of the line in this graph? Notice that each point has the same y-coordinate (3). The equation for the line is $y = 3$. The equation for a horizontal line is always of the form $y = b$, where b is a real number. To graph the line $y = b$, plot the y-intercept $(0, b)$ on the y-axis and draw a horizontal line through this point.

EXAMPLE 7 Graph $y = -2$. Label any intercepts.

SOLUTION This graph will be a horizontal line, as its equation is of the form $y = b$. If an equation does not have a term containing the variable x, its graph will be a horizontal line.

Begin by plotting the point $(0, -2)$, which is the y-intercept. Then draw a horizontal line through this point.

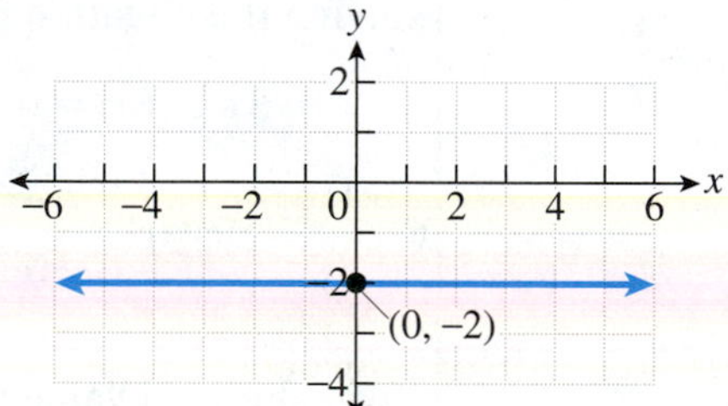

Quick Check 7

Graph $y = 7$. Label any intercepts.

Vertical Lines

Objective 6 Graph vertical lines. Vertical lines have equations of the form $x = a$, where a is a real number. Just as the equation of a horizontal line does not have any terms containing the variable x, the equation of a vertical line does not have any terms containing the variable y. We begin to graph an equation of this form by plotting its x-intercept at $(a, 0)$. Then we draw a vertical line through this point.

EXAMPLE 8 Graph $x = 6$. Label any intercepts.

SOLUTION The x-intercept for this line is at $(6, 0)$. Plot this point and draw a vertical line through it.

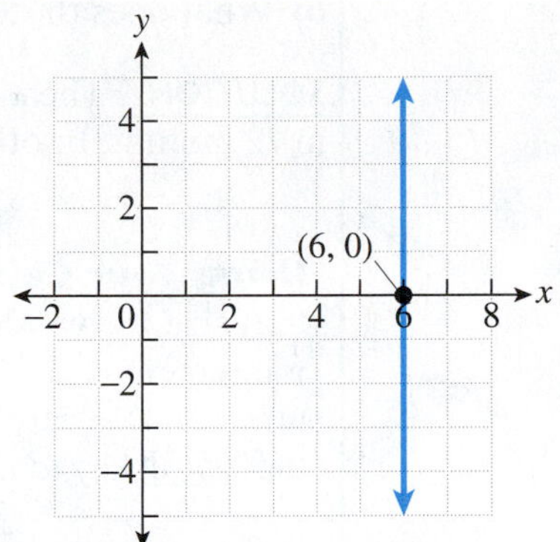

Quick Check 8

Graph $x = -15$. Label any intercepts.

Horizontal Lines	**Vertical Lines**
Equation: $y = b$	Equation: $x = a$
• Plot the y-intercept at $(0, b)$.	• Plot the x-intercept at $(a, 0)$.
• Graph the horizontal line passing through this point.	• Graph the vertical line passing through this point.

Applications of the Graphs of Linear Equations

Objective 7 Interpret the graph of an applied linear equation. The intercepts of a graph have important interpretations in the context of an applied problem. The y-intercept often provides an initial condition, such as the amount of money owed on a loan or the fixed costs to attend college before the number of units that a student is taking are added in. The x-intercept often provides a break-even point, such as when a business goes from a loss to a profit.

EXAMPLE 9 Chelsea borrowed \$3600 from her friend Mack, promising to pay him \$50 per month until she had paid off the loan. The amount of money (y) that Chelsea owes Mack after x months have passed is given by the equation $y = 3600 - 50x$.

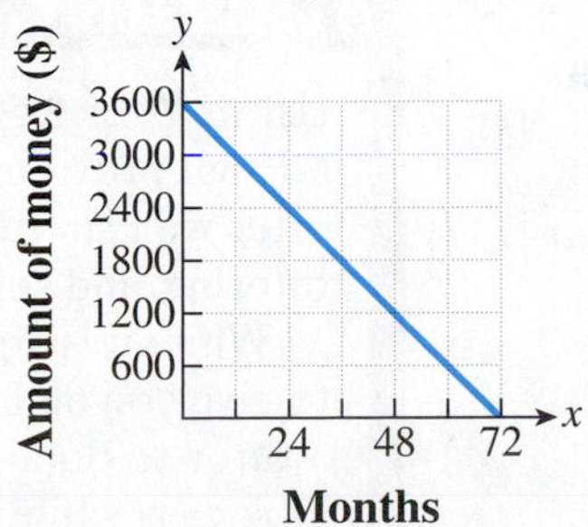

Use the graph of this equation to answer the following questions. (Because the number of months cannot be negative, the graph begins at $x = 0$.)

a) What does the y-intercept represent in this situation?

SOLUTION The y-intercept is at $(0, 3600)$. It tells us that zero months after borrowing the money from Mack, Chelsea owes him \$3600.

b) What does the x-intercept represent in this situation?

SOLUTION The x-intercept is at $(72, 0)$. It tells us that Chelsea will owe Mack \$0 in 72 months. In other words, her debt will be paid.

Using Your Calculator When graphing a line using the TI-84, you may need to resize the window. In Example 9, the x-intercept is at the point $(72, 0)$ and the y-intercept is at $(0, 3600)$. The graph should show these important points. After entering $3600 - 50x$ for Y_1, press the WINDOW key. Fill in the screen as follows and press the GRAPH key to display the graph.

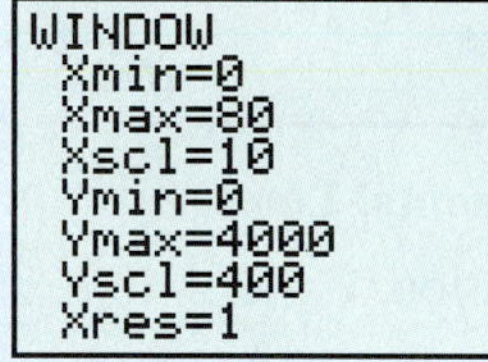

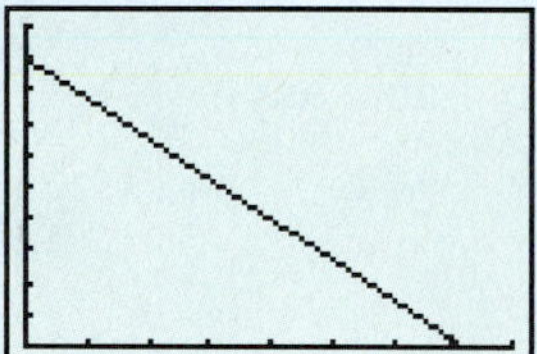

Quick Check 9

Nancy invests \$10,000 in a new business and expects to make a \$2000 profit each month. Nancy's net profit (y) after x months have passed is given by the equation $y = 2000x - 10{,}000$.

Use the graph of this equation to answer the following questions. (Because the number of months cannot be negative, the graph begins at $x = 0$.)

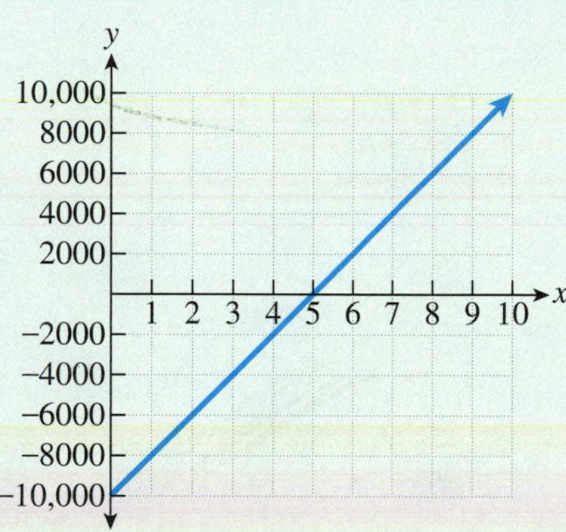

a) What does the y-intercept represent in this situation?

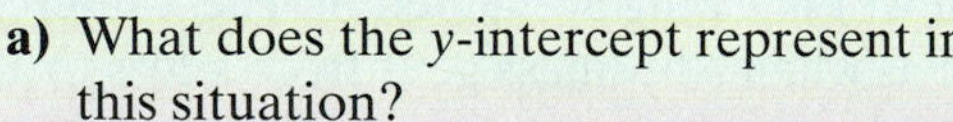

b) What does the x-intercept represent in this situation?

BUILDING YOUR STUDY STRATEGY

Using Your Resources, 2 Tutors Most college campuses have a tutorial center that provides free tutoring in math and other subjects. Some tutorial centers offer walk-in tutoring, others schedule appointments for one-on-one or group tutoring, and some combine these approaches.

While it is a good idea to ask a tutor if you are doing a problem correctly or to give you an idea of a starting point for a problem, it is not a good idea to ask a tutor to do a problem for you. You learn by doing, not by watching. It may look easy while the tutor is working a problem, but this will not help you do the next problem. If you do not understand a tutor's answer or directions, do not be afraid to ask for further explanation.

Exercises 3.2

MyMathLab PRACTICE WATCH DOWNLOAD READ REVIEW

Vocabulary

1. A point at which a graph crosses the x-axis is called a(n) ____________.
2. A point at which a graph crosses the y-axis is called a(n) ____________.
3. A(n) ____________ line has no y-intercept.
4. A(n) ____________ line has no x-intercept.
5. State the procedure for finding an x-intercept.
6. State the procedure for finding an y-intercept.

Find the x- and y-intercepts from the graph.

7.

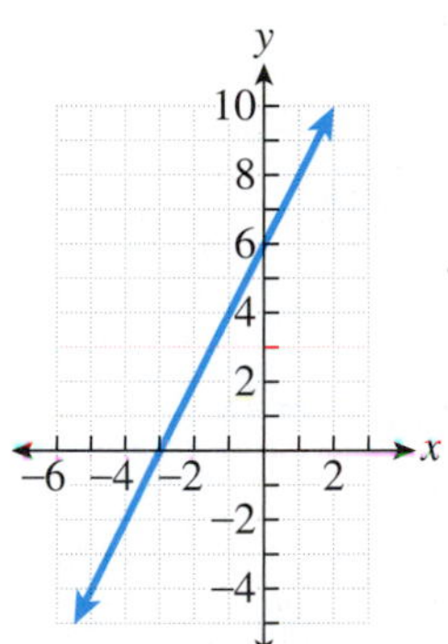

8.

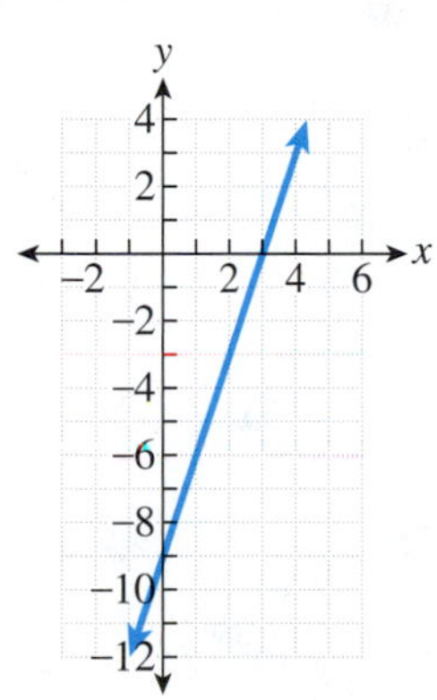

9.

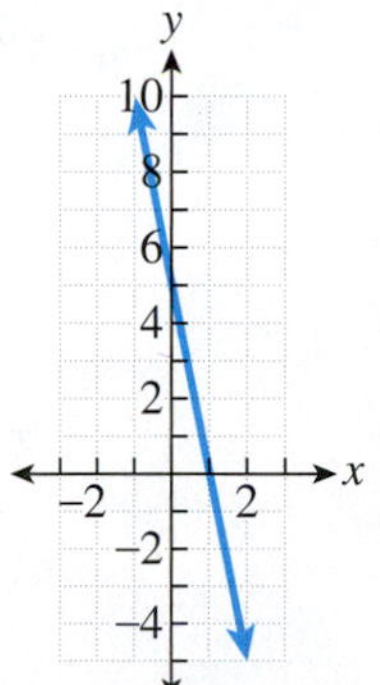

10.

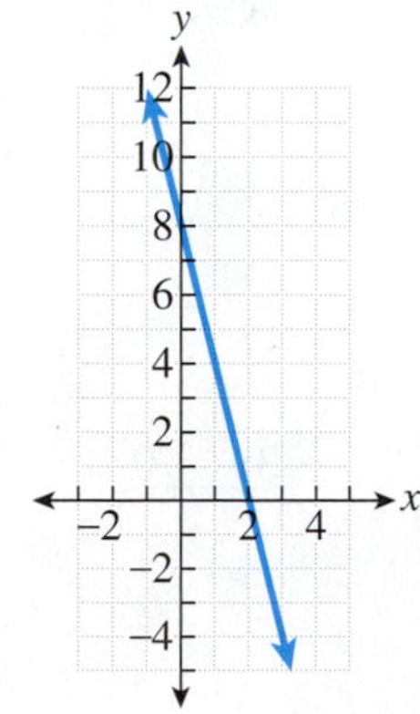

11.

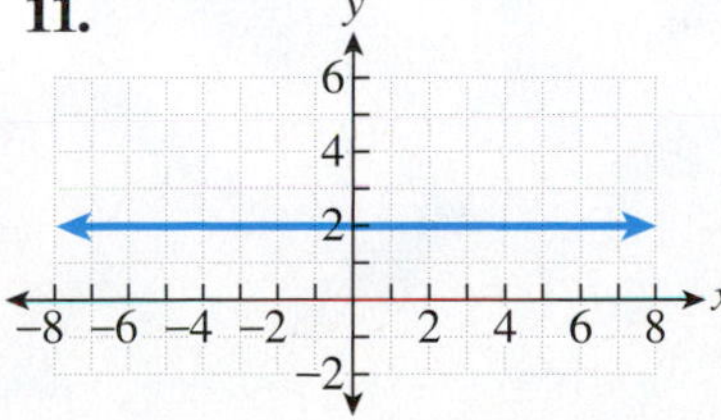

12.

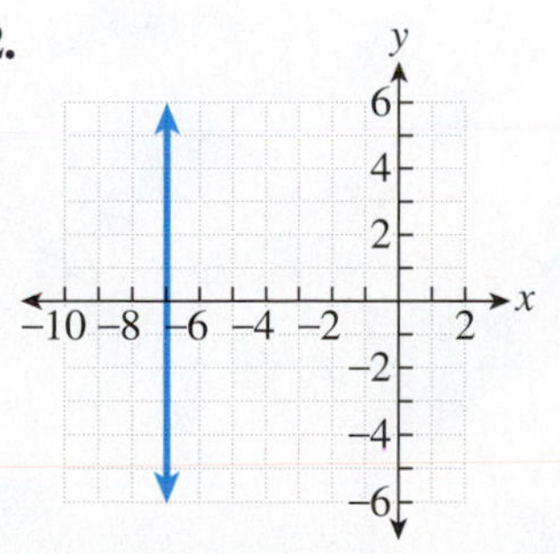

Find the x- and y-intercepts.

13. $x + y = 6$
14. $x + y = -3$
15. $x - y = -8$
16. $x - y = 4$
17. $3x + 2y = 6$
18. $2x + 5y = -10$
19. $-3x + 5y = -30$
20. $2x - 6y = 24$
21. $\frac{3}{5}x + \frac{1}{2}y = 3$
22. $\frac{2}{9}x - \frac{1}{12}y = \frac{2}{3}$
23. $y = -x + 8$
24. $y = x - 5$
25. $y = 3x - 10$
26. $y = -2x + 5$
27. $y = -2x$
28. $y = 3x$
29. $y = 2$
30. $x = 10$
31. Find three equations that have an x-intercept at $(4, 0)$.
32. Find three equations that have an x-intercept at $(-7, 0)$.
33. Find three equations that have a y-intercept at $(0, -10)$.
34. Find three equations that have a y-intercept at $(0, 2)$.

Find the intercepts and then graph the line.

35. $x + y = 4$

36. $x + y = -2$

37. $x - y = -3$

38. $x - y = 6$

39. $2x - 5y = 10$

40. $-4x + 2y = 8$

41. $3x + 4y = -24$

42. $6x + 5y = -30$

43. $-4x + 7y = 14$

44. $3x - 4y = 18$

45. $y = x - 4$

46. $y = x + 6$

47. $y = -2x + 7$

48. $y = 4x - 2$

49. $y = 5x$

50. $y = -4x$

51. $2x - 3y = 0$

52. $3x + 5y = 0$

53. $y = 4$

54. $y = -6$

55. $x = 3$

56. $x = -8$

57. A lawyer bought a new copy machine for her office in 2010, paying $12,000. For tax purposes, the copy machine depreciates at $1500 per year. If you let x represent the number of years after 2010, the value of the copier (y) after x years is given by the equation $y = 12{,}000 - 1500x$.

a) Find the y-intercept of this equation. In your own words, explain what this intercept signifies.

b) Find the x-intercept of this equation. In your own words, explain what this intercept signifies.

c) In 2013, what will be the value of the copy machine?

58. Jerry was ordered to perform 400 hours of community service as a reading tutor. Jerry spends eight hours each Saturday teaching people to read. The equation that tells the number of hours (y) that remain on the sentence after x Saturdays is $y = 400 - 8x$.

a) Find the y-intercept of this equation. In your own words, explain what this intercept signifies.

b) Find the x-intercept of this equation. In your own words, explain what this intercept signifies.

c) After 12 Saturdays, how many hours of community service does Jerry have remaining?

59. A public golf course charges a $300 annual fee to belong to its club. Members of the club pay $24 to play a round of golf. The equation that gives the cost (y) to belong to the club and to play x rounds of golf per year is $y = 300 + 24x$.

a) Find the y-intercept of this equation. In your own words, explain what this intercept signifies.

b) In your own words, explain why this equation has no x-intercept in the context of this problem.

c) Find the total cost for a club member who plays 50 rounds of golf.

60. In addition to paying $185 per unit, a community college student pays a registration fee of $200 per semester. The equation that gives the cost (y) to take x units is $y = 185x + 200$.

a) Find the y-intercept of this equation. In the context of this problem, is this cost possible? Explain.

b) In your own words, explain why this equation has no x-intercept in the context of this problem.

c) A full load for a student is 12 units. How much will a full-time student pay per semester?

Writing in Mathematics

Answer in complete sentences.

61. ***Solutions Manual*** Write a solutions manual page for the following problem:

Find the x- and y-intercepts and then graph the line $3x - 4y = 18$.

Quick Review Exercises

Section 3.2

Solve for y.

1. $2x + y = 8$
2. $4x - y = 6$
3. $-8x + 2y = 10$
4. $4x + 6y = 12$

3.3 Slope of a Line

OBJECTIVES

1. Understand the slope of a line.
2. Find the slope of a line from its graph.
3. Find the slope of a line passing through two points using the slope formula.
4. Find the slopes of horizontal and vertical lines.
5. Find the slope and y-intercept of a line from its equation.
6. Find the equation of a line given its slope and y-intercept.
7. Graph a line using its slope and y-intercept.
8. Interpret the slope and y-intercept in real-world applications.

Slope of a Line

Objective 1 Understand the slope of a line. Here are four different lines that pass through the point $(0, 2)$.

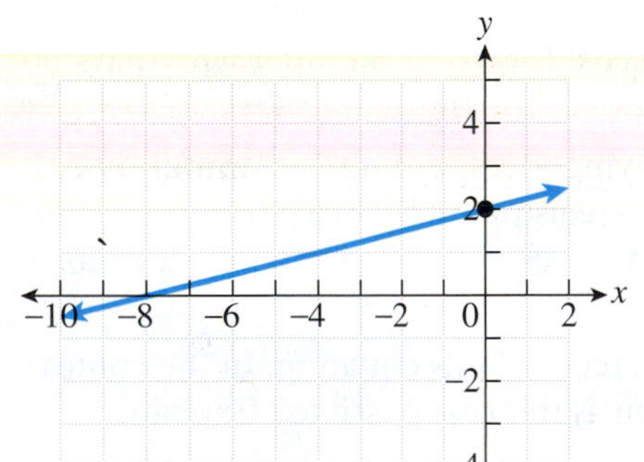

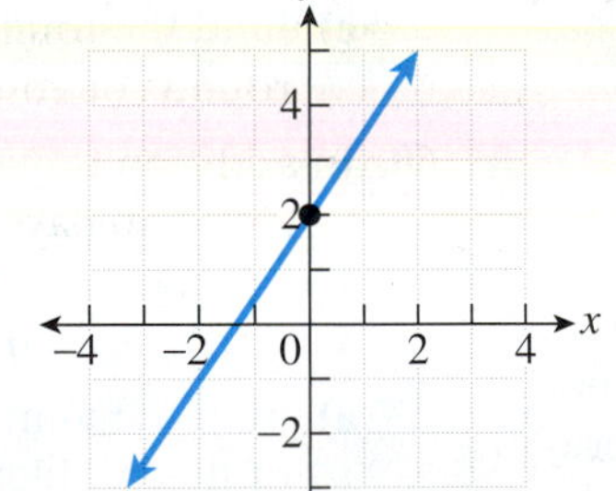

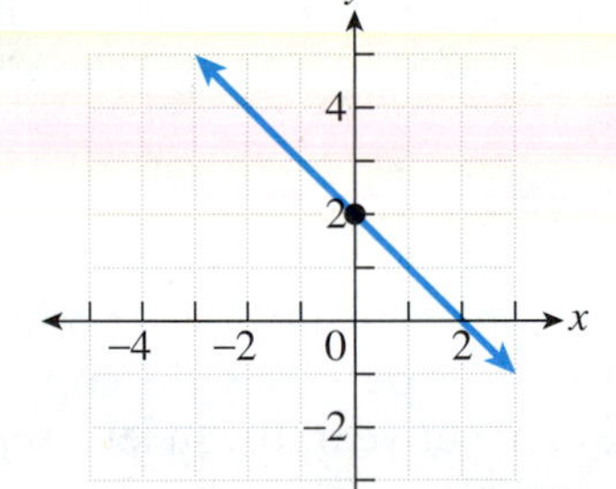

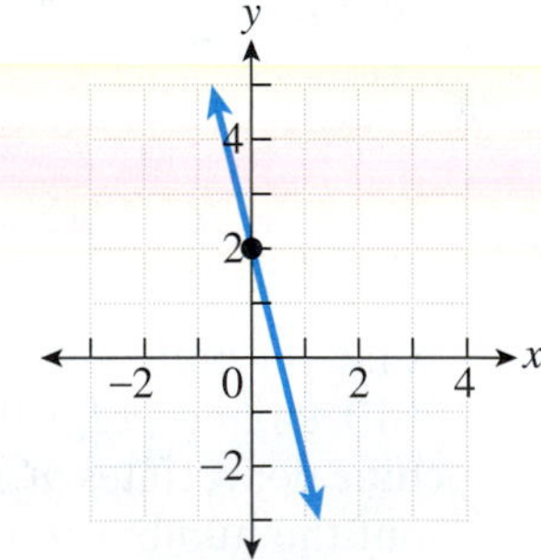

Notice that some of the lines are rising to the right, while others are falling to the right. Also, some are rising or falling more steeply than others. The characteristic that distinguishes these lines is their slope.

Slope

The **slope** of a line is a measure of how steeply a line rises or falls as it moves to the right. We use the letter m to represent the slope of a line.

If a line rises as it moves to the right, its slope is positive. If a line falls as it moves to the right, its slope is negative.

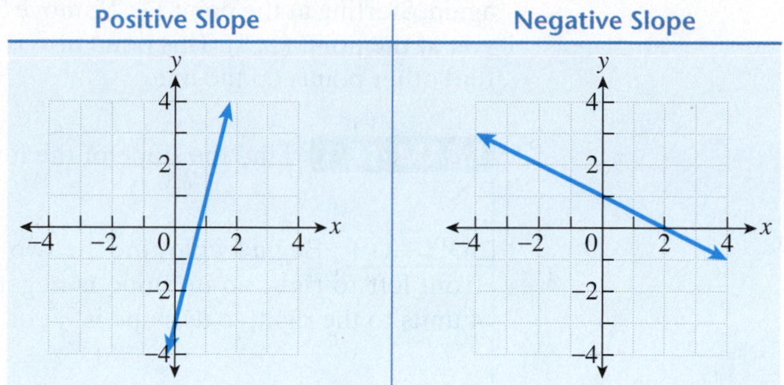

To find the slope of a line, we begin by selecting two points that are on the line. As we move from the point on the left to the point on the right, we measure how much the line rises or falls. The slope is equal to this vertical distance divided by the distance traveled from left to right. This is often referred to as "rise over run," or as the change in y divided by the change in x. Sometimes the change in y is written as Δy and the change in x is written as Δx. The Greek letter Δ (delta) is often used to represent change in a quantity. The slope of a line is represented as

$$m = \frac{\text{rise}}{\text{run}}, \quad \text{or} \quad m = \frac{\Delta y}{\Delta x}.$$

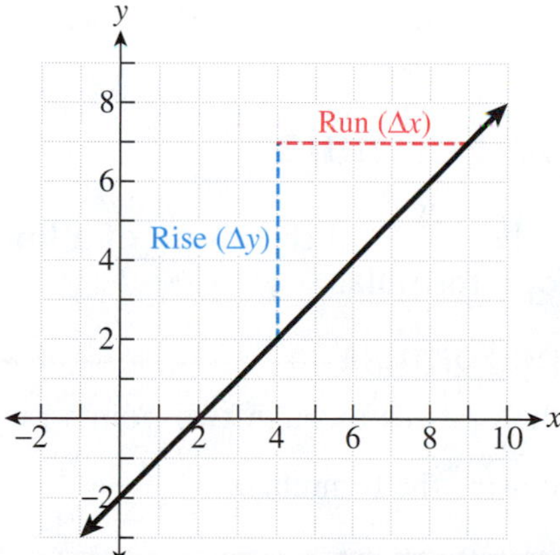

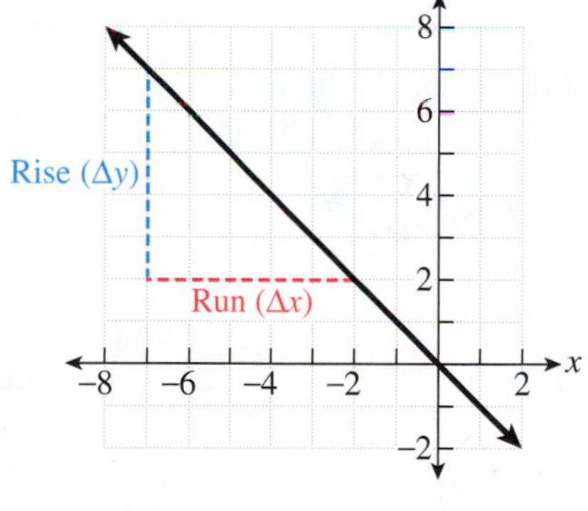

Finding the Slope of a Line from Its Graph

Objective 2 **Find the slope of a line from its graph.** Consider the following line that passes through the points $(1, 2)$ and $(3, 8)$. To move from the point on the left to the point on the right, the line rises by 6 units as it moves 2 units to the right.

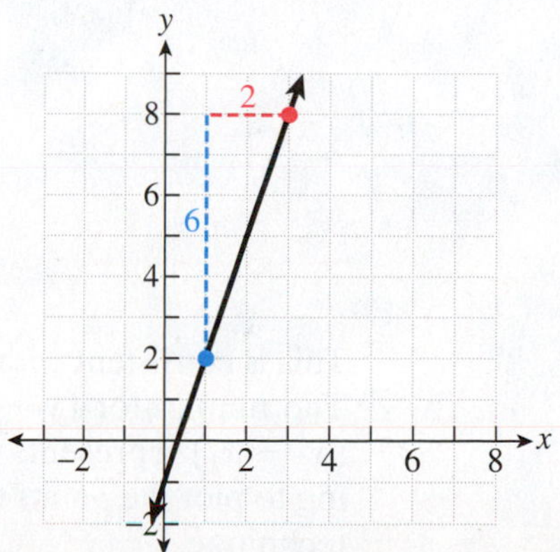

The slope m is $\frac{6}{2}$, or 3. When $m = 3$, every time the line moves up by 3 units (y increases by 3), it moves 1 unit to the right (x increases by 1). Look at the graph again. Starting at the point $(1, 2)$, move 3 units up and 1 unit to the right. This places you at the point $(2, 5)$. This point also is on the line. We can continue this pattern to find other points on the line.

EXAMPLE 1 Find the slope of the line that passes through the points $(4, 9)$ and $(8, 1)$.

SOLUTION Begin by plotting the two points on a graph. Notice that this line falls from left to right, so its slope is negative. The line drops by 8 units as it moves 4 units to the right, so its slope is $\frac{-8}{4}$, or -2.

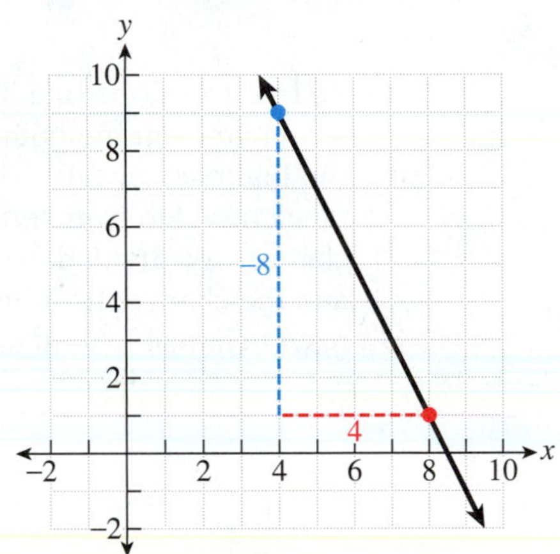

Quick Check 1

Find the slope of the line that passes through $(-2, 5)$ and $(4, -7)$.

Slope Formula

Objective 3 Find the slope of a line passing through two points using the slope formula.

Slope Formula

If a line passes through two points (x_1, y_1) and (x_2, y_2), we can calculate its slope using the formula $m = \dfrac{y_2 - y_1}{x_2 - x_1}$.

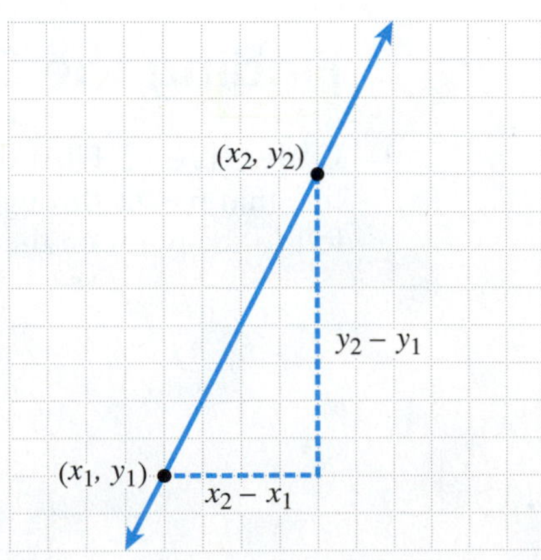

This is consistent with the technique we used to find slope in the previous example. The numerator $(y_2 - y_1)$ represents the vertical change, while the denominator $(x_2 - x_1)$ represents the horizontal change. Using this formula saves us from having to plot the points on a graph, but keep in mind that you can continue to use that technique.

We have seen that the slope of a line that passes through the points (1, 2) and (3, 8) is 3. Now we will use the formula $m = \frac{y_2 - y_1}{x_2 - x_1}$ to find the slope of the line through these two points. We will use the point (1, 2) as (x_1, y_1) and the point (3, 8) as (x_2, y_2).

$$\begin{aligned} m &= \frac{8 - 2}{3 - 1} \\ &= \frac{6}{2} \\ &= 3 \end{aligned}$$

Notice that we get the same result as before. We could change the ordering of the points and still calculate the same slope. In other words, we could use the point (3, 8) as (x_1, y_1) and the point (1, 2) as (x_2, y_2) and still find that $m = 3$. We can subtract in either order as long as we are consistent. Whichever order we use to subtract the y-coordinates, we must subtract the x-coordinates in the same order.

EXAMPLE 2 Use the formula $m = \frac{y_2 - y_1}{x_2 - x_1}$ to find the slope of the line that passes through the points (1, 7) and (4, 3).

SOLUTION One way to think of the numerator in this formula is *the second y-coordinate minus the first y-coordinate,* which, in the example, is $3 - 7$. Using a similar approach for the x-coordinates in the denominator, we begin with a denominator of $4 - 1$.

$$\begin{aligned} m &= \frac{3 - 7}{4 - 1} && \text{Substitute into the formula.} \\ &= \frac{-4}{3} && \text{Simplify numerator and denominator.} \end{aligned}$$

The slope of the line that passes through these two points is $-\frac{4}{3}$. This line falls 4 units for every 3 units it moves to the right.

Quick Check 2

Use the formula $m = \frac{y_2 - y_1}{x_2 - x_1}$ to find the slope of the line that passes through the points (1, 2) and (5, 14).

A WORD OF CAUTION It does not matter which point is labeled as (x_1, y_1) and which point is labeled as (x_2, y_2). It *is* important that the order in which the y-coordinates are subtracted in the numerator is the order in which the x-coordinates are subtracted in the denominator.

EXAMPLE 3 Use the formula $m = \frac{y_2 - y_1}{x_2 - x_1}$ to find the slope of the line that passes through the points $(-5, -2)$ and $(-1, 4)$.

SOLUTION In this example, we learn to apply the formula to points that have negative x- or y-coordinates.

$$\begin{aligned} m &= \frac{4 - (-2)}{-1 - (-5)} && \text{Substitute into the formula.} \\ &= \frac{4 + 2}{-1 + 5} && \text{Eliminate double signs.} \\ &= \frac{6}{4} && \text{Simplify numerator and denominator.} \\ &= \frac{3}{2} && \text{Simplify.} \end{aligned}$$

The slope of this line is $\frac{3}{2}$. The line rises by 3 units for every 2 units it moves to the right.

Quick Check 3

Use the formula $m = \frac{y_2 - y_1}{x_2 - x_1}$ to find the slope of the line that passes through the points $(-3, 5)$ and $(1, -5)$.

Horizontal and Vertical Lines

Objective 4 Find the slopes of horizontal and vertical lines. Let's look at the horizontal line $y = 3$ that passes through the points $(1, 3)$ and $(4, 3)$.

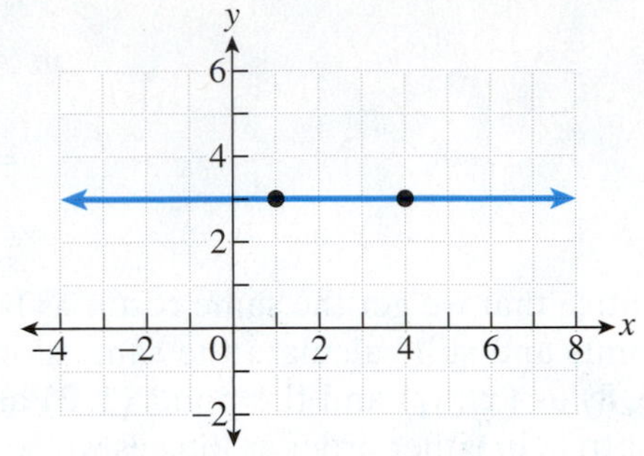

What is its slope?

$$m = \frac{3 - 3}{4 - 1} = \frac{0}{3} = 0$$

The slope of this line is 0. The same is true for any horizontal line. The vertical change between any two points on a horizontal line is always equal to 0, and when we divide 0 by any nonzero number, the result is equal to 0. Thus, the slope of any horizontal line is equal to 0.

Here is a brief summary of the properties of horizontal lines.

Horizontal Lines

Equation: $y = b$, where b is a real number
y-intercept: $(0, b)$
Slope: $m = 0$

Look at the following vertical line:

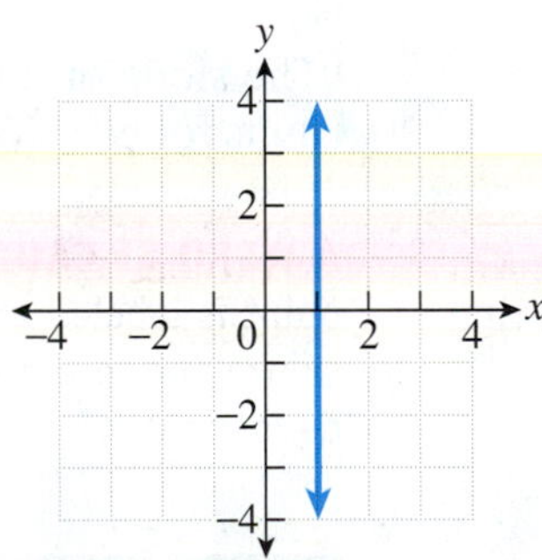

Notice that each point on this line has an x-coordinate of $x = 1$. We can attempt to find the slope of this line by selecting any two points on the line, such as $(1, 2)$ and $(1, 5)$, and using the slope formula.

$$m = \frac{5 - 2}{1 - 1} = \frac{3}{0}$$

The fraction $\frac{3}{0}$ is undefined. Thus, the slope of a vertical line is said to be undefined.

Here is a brief summary of the properties of vertical lines.

Vertical Lines

Equation: $x = a$, where a is a real number
x-intercept: $(a, 0)$
Slope: Undefined

EXAMPLE 4 Find the slope, if it exists, of the line $x = -2$.

SOLUTION This line is a vertical line. The slope of a vertical line is undefined, so this line has undefined slope.

Quick Check 4

Find the slope, if it exists, of the line $x = 5$.

EXAMPLE 5 Find the slope, if it exists, of the line $y = 4$.

SOLUTION This line is a horizontal line. The slope of any horizontal line is 0, so the slope of this line is 0.

Quick Check 5

Find the slope, if it exists, of the line $y = -2$.

A WORD OF CAUTION Avoid stating that a line has "no slope." This is a vague phrase; some may take it to mean that the slope is equal to 0, while others may interpret it as meaning that the slope is undefined.

Slope–Intercept Form of a Line

Objective 5 Find the slope and y-intercept of a line from its equation.

If we solve a given equation for y so that it is in the form $y = mx + b$, this is the **slope–intercept form** of a line. The number being multiplied by x is the slope m, while b represents the y-coordinate of the y-intercept.

Slope–Intercept Form of a Line

$y = mx + b$

m: Slope of the line

b: y-coordinate of the y-intercept

To verify that the y-intercept is $(0, b)$, substitute 0 for x in the equation $y = mx + b$.

$$y = mx + b$$
$$y = m(0) + b$$
$$y = b$$

x	$y = 2x + 5$	(x, y)
0	$y = 2(0) + 5 = 5$	$(0, 5)$
1	$y = 2(1) + 5 = 7$	$(1, 7)$
2	$y = 2(2) + 5 = 9$	$(2, 9)$
3	$y = 2(3) + 5 = 11$	$(3, 11)$

To verify that the slope is the coefficient of the x-term, let's look at the equation $y = 2x + 5$ by creating a table of values, as shown to the left.

Notice that the y-values increase by 2 as the x-values increase by 1; so the slope is 2.

EXAMPLE 6 Find the slope and y-intercept of the line $y = -3x + 7$.

SOLUTION Because this equation is already solved for y, we can read the slope and the y-intercept directly from the equation. The slope is -3, which is the coefficient of the term containing x in the equation. The y-intercept is $(0, 7)$ because 7 is the constant in the equation.

Quick Check 6

Find the slope and the y-intercept of the line $y = \frac{5}{2}x - 6$.

EXAMPLE 7 Find the slope and y-intercept of the line $2x + 2y = 11$.

SOLUTION Begin by solving for y.

$$2x + 2y = 11$$
$$2y = -2x + 11 \quad \text{Subtract } 2x \text{ from both sides.}$$
$$\frac{2y}{2} = \frac{-2x}{2} + \frac{11}{2} \quad \text{Divide each term by 2.}$$
$$y = -x + \frac{11}{2} \quad \text{Simplify.}$$

The slope of this line is -1. When we see $-x$ in the equation, we need to remember that this is the same as $-1x$. The y-intercept is $\left(0, \frac{11}{2}\right)$.

Quick Check 7

Find the slope and the y-intercept of the line $6x + 2y = 10$.

A WORD OF CAUTION We cannot determine the slope and y-intercept of a line from its equation unless the equation has already been solved for y first.

Objective 6 Find the equation of a line given its slope and y-intercept.

EXAMPLE 8 Find the equation of a line that has a slope of 2 and a y-intercept of $(0, -15)$.

SOLUTION We will use the slope–intercept form $(y = mx + b)$ to help us find the equation of this line. Because the slope is 2, we can replace m with 2. Also, because the y-intercept is $(0, -15)$, we can replace b with -15. The equation of this line is $y = 2x - 15$.

Quick Check 8

Find the equation of a line that has a slope of $\frac{1}{2}$ and a y-intercept of $\left(0, -\frac{4}{7}\right)$.

Graphing a Line Using Its Slope and y-Intercept

Objective 7 Graph a line using its slope and y-intercept. Once we have an equation in slope–intercept form, we can use this information to graph the line. We can begin by plotting the y-intercept as the first point. We can then use the slope of the line to find a second point.

Graphing a Line Using Its Slope and y-Intercept

- Plot the y-intercept $(0, b)$
- Use the slope m to find another point on the line.
- Graph the line that passes through these two points.

EXAMPLE 9 Graph the line $y = -4x + 8$ using its slope and y-intercept.

SOLUTION We will start at the y-intercept, which is $(0, 8)$. Because the slope is -4, the line moves down 4 units as it moves 1 unit to the right. This gives us a second point at $(1, 4)$. Here is the graph.

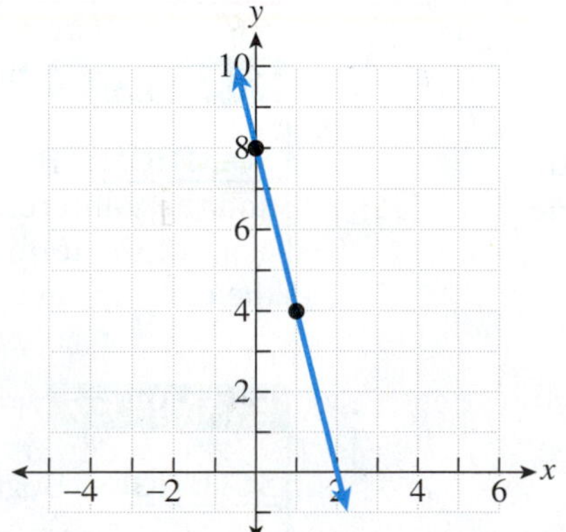

Quick Check 9

Graph the line $y = 2x + 6$ using its slope and y-intercept.

EXAMPLE 10 Graph the line $y = \frac{1}{2}x - 3$ using its slope and y-intercept.

SOLUTION We will start by plotting the y-intercept at $(0, -3)$. The slope is $\frac{1}{2}$, which tells us that the line rises by 1 unit as it moves 2 units to the right. This will give us a second point at $(2, -2)$.

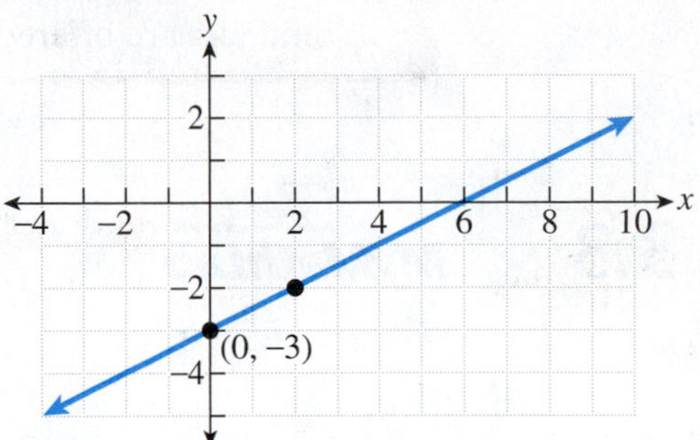

Quick Check 10

Graph the line $y = -\frac{2}{3}x + 4$ using its slope and y-intercept.

We have two techniques for graphing lines that are not horizontal or vertical lines. We can graph a line by finding its x- and y-intercepts or by using the y-intercept and the slope of the line. There are certain times when one technique is more efficient to apply than the other, and knowing which technique to use for a particular linear equation can save time and prevent errors. In general, if the equation is already solved for y, such as $y = 3x + 7$ or $y = -4x + 5$, graphing the line using the y-intercept and the slope is a good choice. If the equation is not already solved for y, such as $2x + 3y = 12$ or $5x - 2y = -10$, consider finding both the x- and y-intercepts and then drawing the line that passes through these two points.

Applications

Objective 8 Interpret the slope and y-intercept in real-world applications. The concept of slope becomes more important when we apply it to real-world situations.

EXAMPLE 11 The number of Americans (y), in millions, who have cell phones can be approximated by the equation $y = 13x + 77$, where x is the number of years after 1999. Interpret the slope and y-intercept of this line. (*Source: Cellular Telecommunications Industry Association*)

SOLUTION The slope of this line is 13, which tells us that each year we can expect an additional 13 million Americans to have cell phones. (The number of Americans who have cell phones is increasing because the slope is positive.) The y-intercept at $(0, 77)$ tells us that approximately 77 million Americans had cell phones in 1999.

Quick Check 11

The number of women (y) accepted to medical school in a given year can be approximated by the equation $y = 218x + 7485$, where x is the number of years after 1997. Interpret the slope and y-intercept of this line. (*Source: Association of American Medical Colleges*)

BUILDING YOUR STUDY STRATEGY

Using Your Resources, 3 **Study Skills Courses** Some colleges run short-term study skills courses or seminars, and these courses can be helpful to students who are learning mathematics. These courses cover everything from taking notes and taking tests to overcoming math anxiety. To find out whether study skill courses or seminars are offered at your school, talk to an academic counselor.

Exercises 3.3

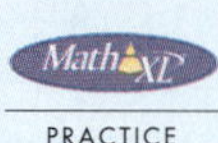
PRACTICE

WATCH

DOWNLOAD
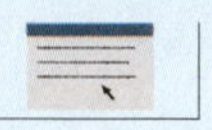
READ

REVIEW

Vocabulary

1. The ________ of a line is a measure of how steeply a line rises or falls as it moves to the right.
2. A line that rises from left to right has ________ slope.
3. A line that falls from left to right has ________ slope.
4. State the formula for the slope of a line that passes through two points.
5. A(n) ________ line has a slope of 0.
6. The slope of a(n) ________ line is undefined.
7. The ________ form of the equation of a line is $y = mx + b$.
8. So that the slope of a line can be determined from its equation, the equation must be solved for ________.

Determine whether the given line has a positive or negative slope.

9.

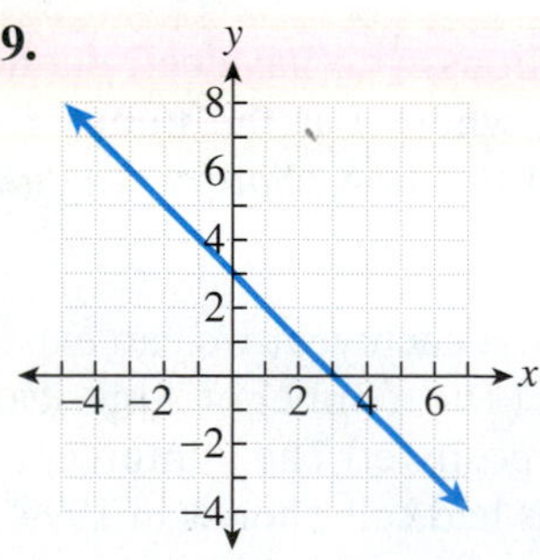

10.

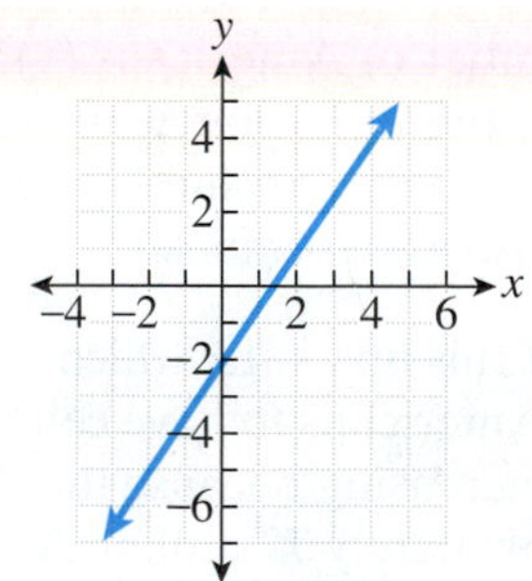

11.

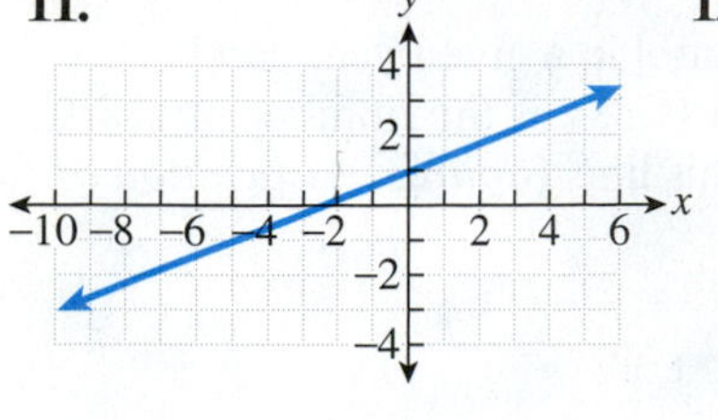

12.

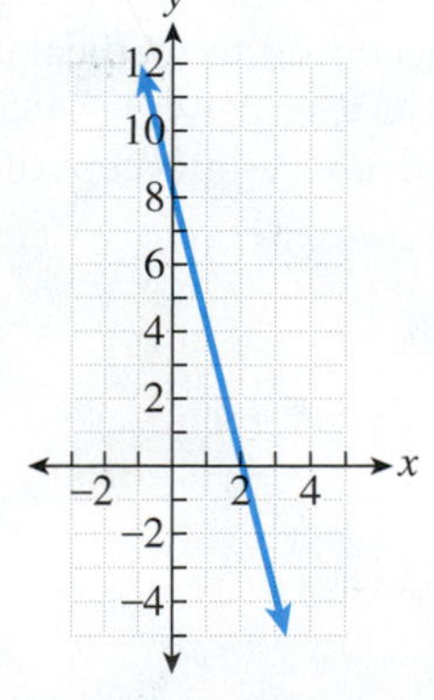

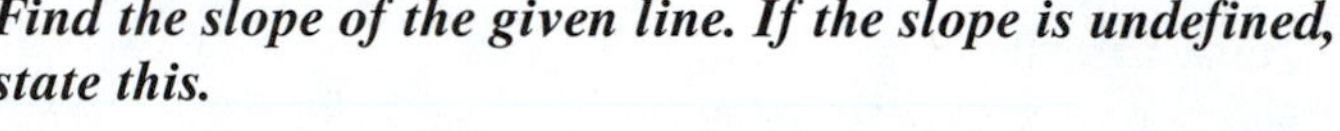

Find the slope of the given line. If the slope is undefined, state this.

13.

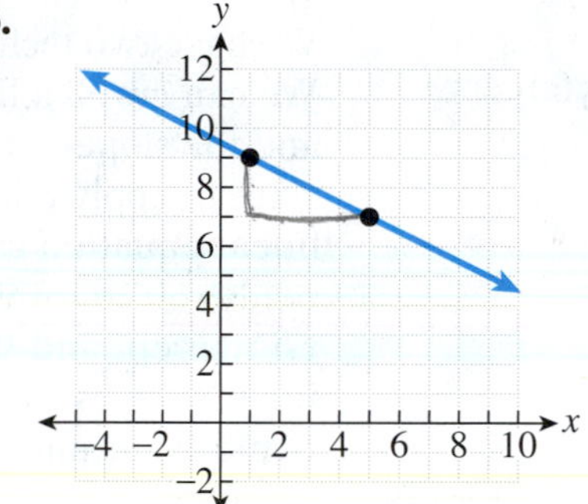

14.

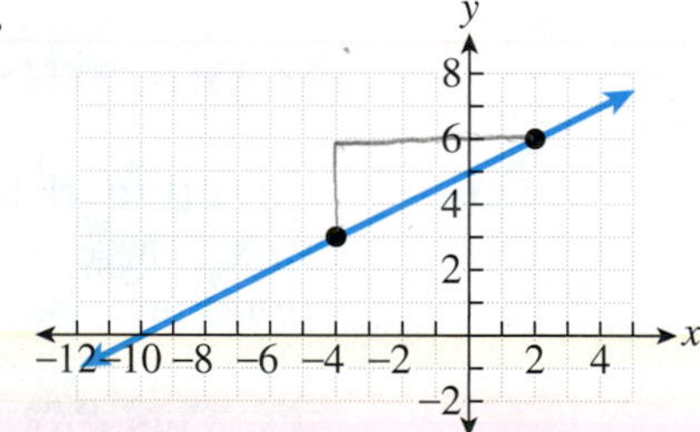

15.

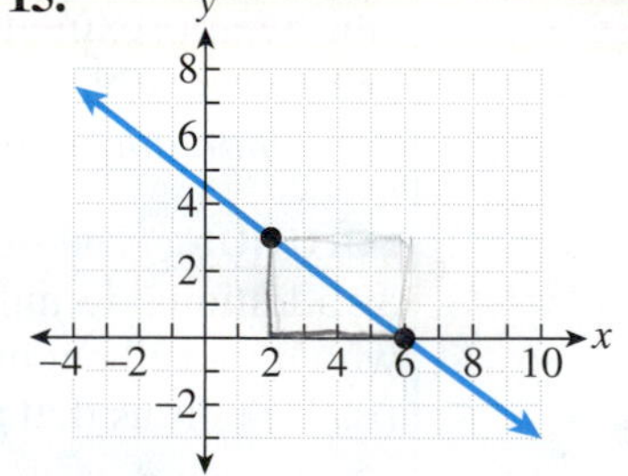

16.

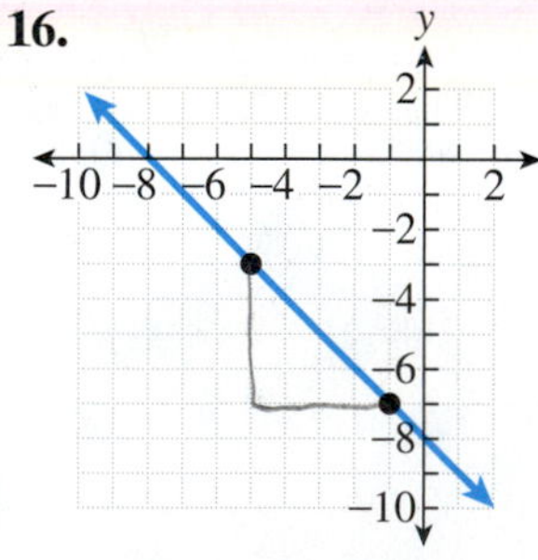

17.

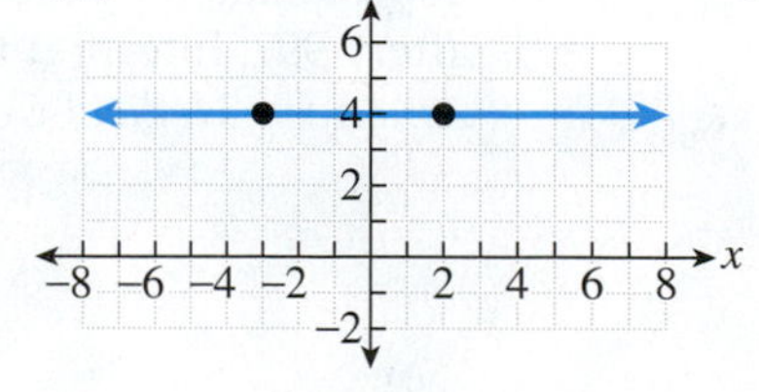

18.

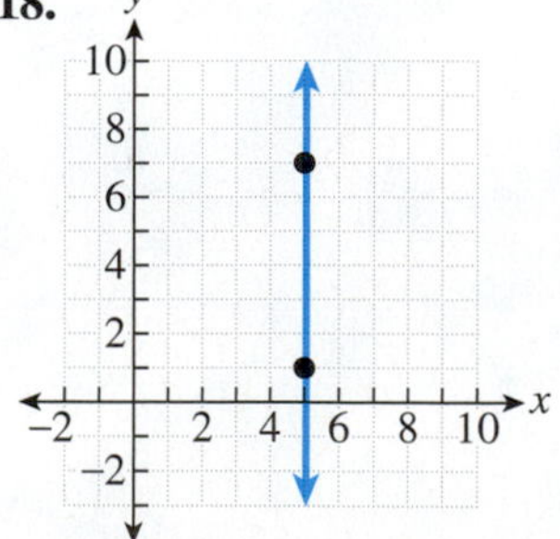

Find the slope of a line that passes through the given two points. If the slope is undefined, state this.

19. (3, 5) and (4, 7)

20. (2, 6) and (3, 1)

21. (−2, 3) and (−5, −6)

22. (5, −4) and (−1, 5)

23. (13, −6) and (7, 2)

24. (−11, −4) and (−7, 6)

25. (−32, 15) and (−24, −3)

26. (46, 33) and (19, 24)

27. (3, 6) and (−4, 6)

28. (−2, −6) and (2, 6)

29. (0, 4) and (0, 7)

30. (0, 0) and (4, 7)

Graph. Label any intercepts and determine the slope of the line. If the slope is undefined, state this.

31. $y = 4$

32. $x = -2$

33. $x = 3$

34. $y = -\frac{17}{5}$

35. $x = 8$

36. $y = -3$

Determine the equation of the given line as well as the slope of the line. If the slope is undefined, state this.

37.

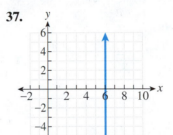

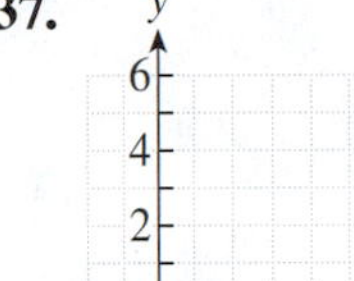

38.

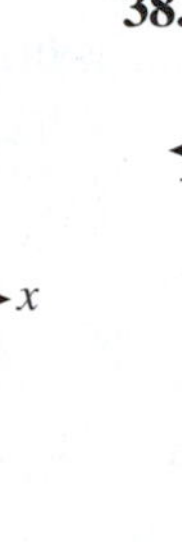

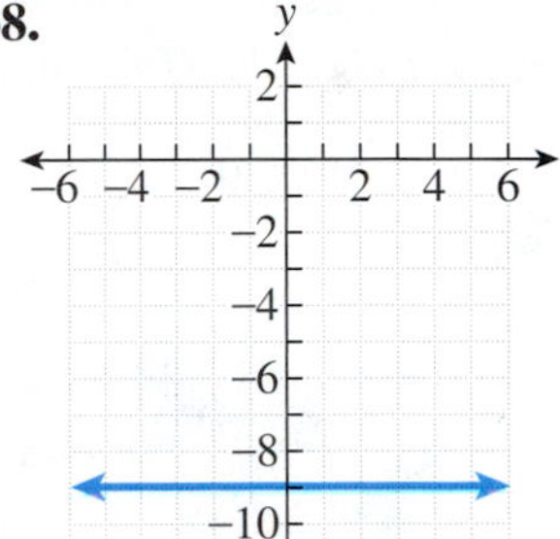

39.

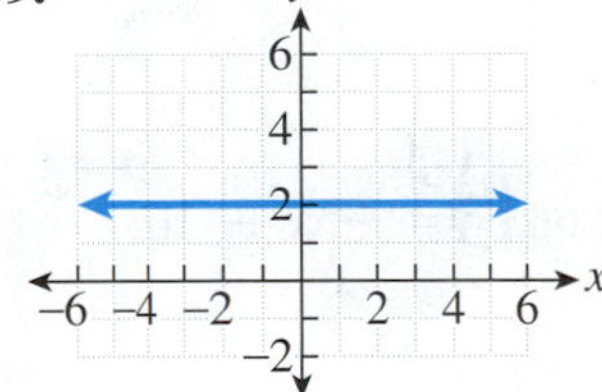

40.

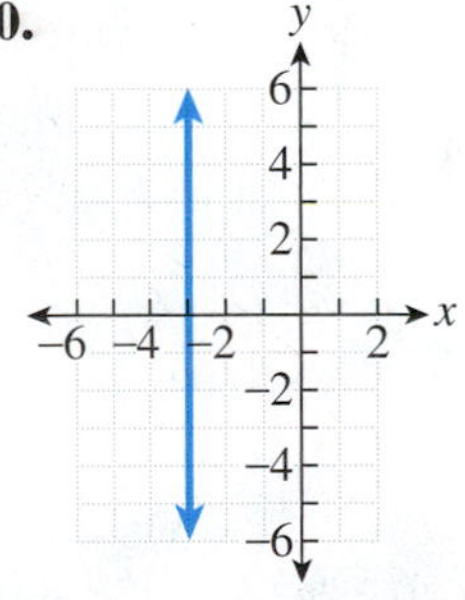

Find the slope and the y-intercept of the given line.

41. $y = 6x - 7$

42. $y = 4x + 11$

43. $y = -2x + 3$

44. $y = -5x - 8$

45. $6x + 4y = -10$

46. $2x + 3y = -12$

47. $5x - 8y = 10$

48. $8x - 6y = 12$

49. $x - 5y = 8$

50. $x + 4y = 14$

Find the equation of a line with the given slope and y-intercept.

51. Slope -2, y-intercept $(0, 5)$
52. Slope 4, y-intercept $(0, 3)$
53. Slope 3, y-intercept $(0, -6)$
54. Slope -5, y-intercept $(0, -2)$
55. Slope 0, y-intercept $(0, -4)$
56. Slope 0, y-intercept $(0, 1)$

Graph using the slope and y-intercept.

57. $y = 3x + 6$

58. $y = 2x - 8$

59. $y = -2x - 6$

60. $y = -5x + 10$

61. $y = x - 3$

62. $y = -x - 1$

63. $y = 4x$

64. $y = -2x$

65. $y = \frac{7}{2}x - 7$

66. $y = -\frac{5}{4}x + 10$

67. $3y = -9x - 6$

68. $2y = 4x + 10$

69. $-5x + 4y = 8$

70. $2x + 5y = -25$

Mixed Practice, 71–88

Graph using the most efficient technique, finding the x- and y-intercepts or using the slope and y-intercept.

71. $y = -x + 2$

72. $y = -\frac{1}{4}x + 2$

73. $y = -2x + 8$

74. $y = -3x + 7$

75. $y = 6x - 9$

76. $y = -2$

77. $y = -x - 5$

78. $y = -5x$

79. $3x - 2y = -12$

80. $2x + 5y = 15$

81. $y = \frac{4}{5}x$

82. $x = \frac{9}{2}$

83. $y = 4$

84. $x = -3$

85. $3x + y = 1$

86. $y = \frac{2}{5}x + 1$

87. $5x - y = -5$

88. $-8x + 2y = -10$

Graph the two lines and find the coordinates of the point of intersection.

89. $y = 3x - 6$ and $y = -x - 2$

90. $5x - 4y = 20$ and $y = \frac{1}{2}x + 1$

91. The value y of Dave's car x years after 2008 is given by the equation $y = -3000x + 25{,}000$.

a) Find the slope of the equation. In your own words, explain what this slope signifies.

b) Find the y-intercept of the equation. In your own words, explain what this y-intercept signifies.

c) Use this equation to predict the value of the car in 2014.

92. The number of nursing students enrolled in accelerated beaccelereate programs x years after 2004 can be approximated by the equation $y = 1200x + 6090$. (Based on 2004–2008 data. *Source: American Association of Colleges of Nursing*)

a) Find the slope of the equation. In your own words, explain what this slope signifies.

b) Find the y-intercept of the equation. In your own words, explain what this y-intercept signifies.

c) Use this equation to predict the number of nursing students enrolled in accelerated baccalaureate programs in 2016.

93. The **pitch** of a roof is measure of its slope, dividing the vertical rise by the horizontal span.

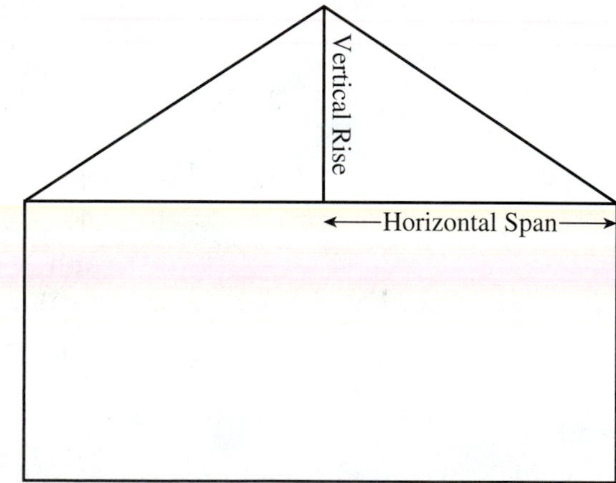

Find the pitch of the roof.

a)

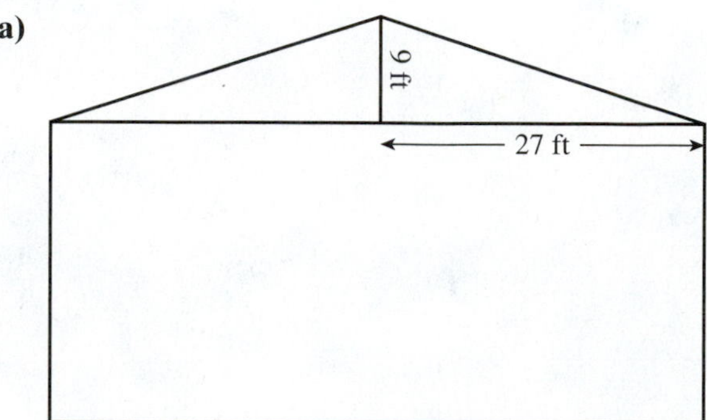

b)

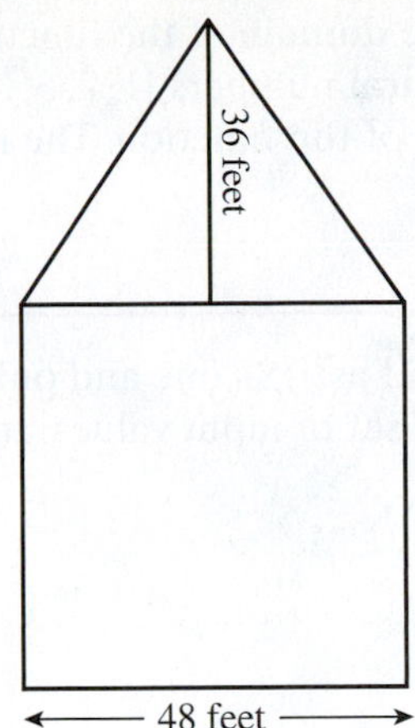

94. A contractor is building a square house that is 72 feet wide. If he wants the pitch of the roof to be $\frac{1}{3}$, what will be the vertical rise of the roof?

95. On a 1000-foot stretch of highway through the mountains, the road rises by a total of 50 feet. Find the slope of the road.

96. George walked 1 mile (5280 feet) on a treadmill. The grade was set to 4%, which means that the slope of the treadmill is 4%, or 0.04. Over the course of his workout, how many feet did George climb (vertically)?

Writing in Mathematics

Answer in complete sentences.

97. The governor of a state predicts that unemployment rates will increase for the next five years. Would an equation relating unemployment rate (y) to the number of years after the statement was made (x) have a positive or negative slope? Explain.

98. Consider the equations $y = -\frac{3}{8}x + 7$ and $20x - 8y = 320$. For each equation, do you believe that finding the x- and y-intercepts or using the slope and y-intercept is the most efficient way to graph the equation? Explain your choices.

99. ***Solutions Manual*** Write a solutions manual page for the following problem:

Find the slope of the line that passes through the points $(-2, 6)$ *and* $(4, -8)$.

3.4 Linear Functions

OBJECTIVES

1. **Define *function, domain,* and *range.***
2. **Evaluate functions.**
3. **Graph linear functions.**
4. **Interpret the graph of a linear function.**
5. **Determine the domain and range of a function from its graph.**

A coffeehouse sells coffee for $2 per cup. We know that it would cost $4 to buy 2 cups, $6 for 3 cups, $8 for 4 cups, and so on. We also know that the general formula for the cost of x cups in dollars is $2 \cdot x$. The cost depends on the number of cups bought. We say that the cost is a function of the number of cups bought.

Functions; Domain and Range

Objective 1 Define *function, domain,* and *range.* A **relation** is a rule that takes an input value from one set and assigns a particular output value from another set to it. A relation for which each input value is assigned one and only one output value is called a **function**. In the example about coffee, the input value is the number of cups of coffee bought and the output value is the cost. For each number of cups bought, there is only one possible cost. The rule for determining the cost is to multiply the number of cups by $2.

Cups	1	2	3	4	...	x	...
	↓	↓	↓	↓	...	↓	
Cost	$2	$4	$6	$8		$2 · x	

The set of input values for a function is called the **domain** of the function. The domain for the example about coffee is the set of natural numbers {1, 2, 3, . . . }. The set of output values for a function is called the **range** of the function. The range of the function in the coffee example is {2, 4, 6, . . . }.

Function, Domain, and Range

A **function** is a relation that takes an input value and assigns one and only one output value to it. The **domain** of the function is the set of input values, and the **range** is the set of output values.

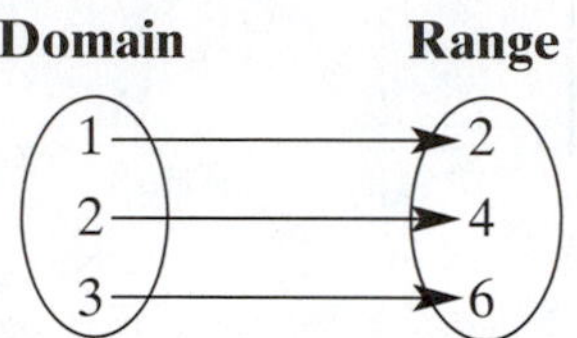

If an input value can be associated with more than one output value, the relation is not a function. For example, a relation that took a month of the year as its input and listed the people at your school who were born that month as its output would not be a function because each month has more than one person born in that month.

EXAMPLE 1 A mail-order company is selling holiday ornaments for \$4 each. There is an additional \$4.95 charge per order for shipping and handling. Find the function for the total cost of an order as well as the domain and range of the function.

SOLUTION To determine the cost of an order, we begin by multiplying the number of ornaments by \$4. To this we still need to add \$4.95 for shipping and handling.

Function: If n is the number of ornaments ordered, the cost is $4n + 4.95$.
Domain: Set of possible number of ornaments ordered {1, 2, 3, . . . }
Range: Set of possible costs of the orders {\$8.95, \$12.95, \$16.95, . . . }

Quick Check 1

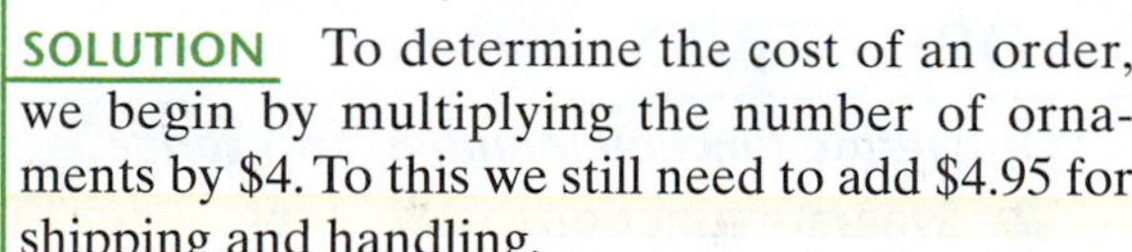

A rental agency rents small moving trucks for \$19.95 plus \$0.15 per mile. Find the function for the total cost to rent a truck as well as the domain and range of the function.

Function Notation

The perimeter of a square with side x can be found using the formula $P = 4x$. This is a function that also can be expressed as $P(x) = 4x$ using function notation. **Function notation** is a way to present the output value of a function for the input x. The notation on the left side, $P(x)$, tells us the name of the function, P, as well as the input variable, x. $P(x)$ tells us that P is a function of x and is read as *P of x*. The parentheses on the left side are used to identify the input variable, not to indicate multiplication. Although we used P as the name of the function (P for perimeter), we could have used any letter. The letters f and g are frequently used for function names. The expression on the right side, $4x$, is the formula for the function.

$P(x) = 4x$ ↑	$P(x) = 4x$ ↑	$P(x) = 4x$ ↑
The variable inside the parentheses is the input variable for the function.	$P(x)$ is the output value of the function P when the input value is x.	The expression on the right side of the equation is the formula for this function.

Evaluating Functions

Objective 2 Evaluate functions. Suppose we wanted to find the perimeter of a square with a side of 3 inches. We are looking to evaluate the perimeter function, $P(x) = 4x$, for an input of 3, or, in other words, $P(3)$. Finding the output value of a function for a particular value of x is called **evaluating** the function. To evaluate a function for a particular value of the variable, we substitute that value for the variable in the function's formula and then simplify the resulting expression. To evaluate $P(3)$, we substitute 3 for x in the formula and simplify the resulting expression.

$$\begin{aligned} P(x) &= 4x \\ P(3) &= 4(3) \\ &= 12 \end{aligned}$$

Because $P(3) = 12$, the perimeter is 12 inches.

EXAMPLE 2 Let $f(x) = 3x + 7$. Find $f(4)$.

SOLUTION We need to replace x in the function's formula by 4 and simplify the resulting expression.

$$\begin{aligned} f(4) &= 3(4) + 7 && \text{Substitute 4 for } x. \\ &= 19 && \text{Simplify.} \end{aligned}$$

$f(4) = 19$. This means that when the input is $x = 4$, the output of the function is 19.

EXAMPLE 3 Let $g(x) = \frac{2}{3}x - 8$. Find $g(-9)$.

SOLUTION In this example, we need to replace x with -9. As the functions become more complicated, we should use parentheses when substituting the input value.

$$\begin{aligned} g(-9) &= \frac{2}{3}(-9) - 8 && \text{Substitute } -9 \text{ for } x. \\ &= -6 - 8 && \text{Multiply.} \\ &= -14 && \text{Simplify.} \end{aligned}$$

Quick Check 2

Let $g(x) = 2x + 9$. Find $g(-6)$.

EXAMPLE 4 Let $g(x) = 8 - 3x$. Find $g(a + 3)$.

SOLUTION In this example, we are substituting a variable expression for x in the function. After we replace x with $a + 3$, we need to simplify the resulting variable expression.

$$\begin{aligned} g(a + 3) &= 8 - 3(a + 3) && \text{Replace } x \text{ with } a + 3. \\ &= 8 - 3a - 9 && \text{Distribute } -3. \\ &= -3a - 1 && \text{Combine like terms.} \end{aligned}$$

Quick Check 3

Let $g(x) = 7x - 12$. Find $g(a + 8)$.

Linear Functions and Their Graphs

Objective 3 Graph linear functions.

A **linear function** is a function of the form $f(x) = mx + b$, where m and b are real numbers.

Some examples of linear functions are $f(x) = x - 9$, $f(x) = 5x$, $f(x) = 3x + 11$, and $f(x) = 6$. We now turn our attention to graphing linear functions. We graph any function $f(x)$ by plotting points of the form $(x, f(x))$. The output value of the function $f(x)$ is treated as the variable y was when we were graphing linear equations in two variables. When we graph a function $f(x)$, the vertical axis is used to represent the output values of the function.

We can begin to graph a linear function by finding the y-intercept. As with a linear equation that is in slope–intercept form, the y-intercept for the graph of a linear function $f(x) = mx + b$ is the point $(0, b)$. For example, the y-intercept for the graph of the function $f(x) = 5x - 8$ is the point $(0, -8)$. In general, to find the y-intercept of any function $f(x)$, we can find $f(0)$. After plotting the y-intercept, we can use the slope m to find other points.

EXAMPLE 5 Graph the linear function $f(x) = \frac{1}{2}x + 2$.

Quick Check 4

Graph the linear function $f(x) = \frac{3}{4}x - 6$.

SOLUTION We can start with the y-intercept, which is $(0, 2)$. The slope of the line is $\frac{1}{2}$; so beginning at the point $(0, 2)$, we move 1 unit up and 2 units to the right. This leads to a second point at $(2, 3)$.

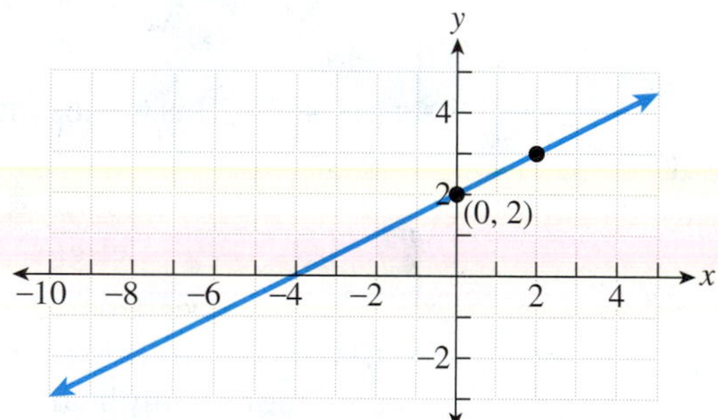

EXAMPLE 6 Graph the linear function $f(x) = -3$.

Quick Check 5

Graph the linear function $f(x) = 4$.

SOLUTION This function is known as a **constant function**. The function is constantly equal to -3, regardless of the input value x. Its graph is a horizontal line with a y-intercept at $(0, -3)$.

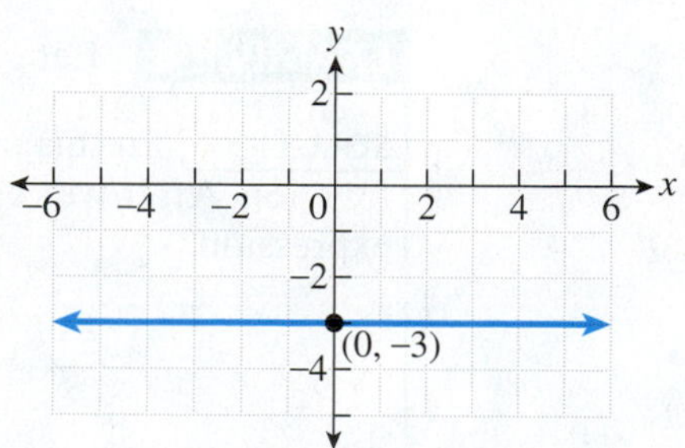

Interpreting the Graph of a Linear Function

Objective 4 Interpret the graph of a linear function. Here is the graph of a function $f(x)$.

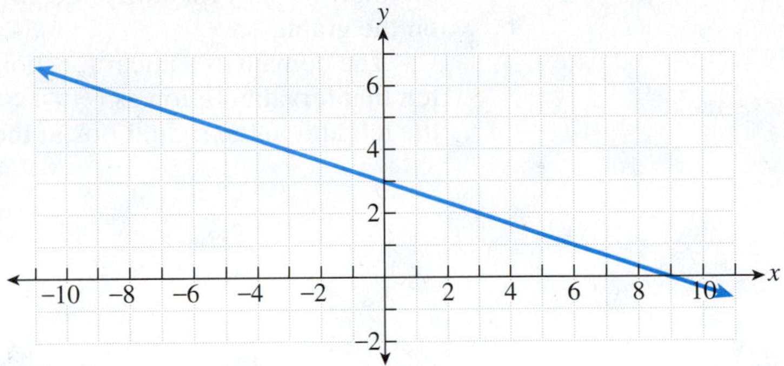

The ability to read and interpret a graph is an important skill. This line has an x-intercept at the point $(9, 0)$, so $f(9) = 0$. The y-intercept is at the point $(0, 3)$, so $f(0) = 3$.

Suppose we wanted to find $f(3)$ for this particular function. We can do so by finding a point on the line that has an x-coordinate of 3. The y-coordinate of this point is $f(3)$. In this case, $f(3) = 2$.

We also can use this graph to solve the equation $f(x) = 6$. Look for the point on the graph that has a y-coordinate of 6. The x-coordinate of this point is -9, so $x = -9$ is the solution of the equation $f(x) = 6$.

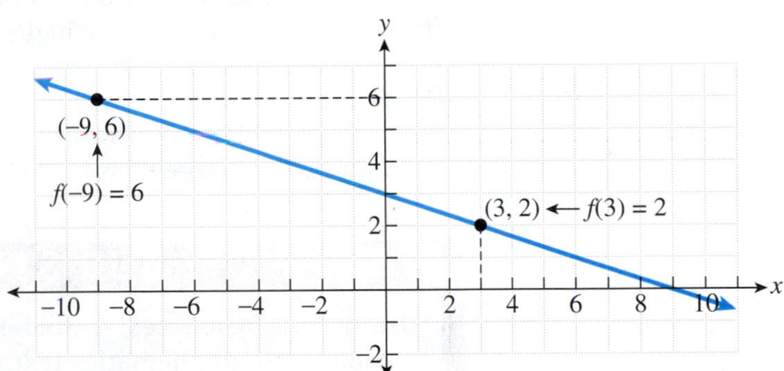

EXAMPLE 7 Consider the graph of the function $f(x)$ to the left.

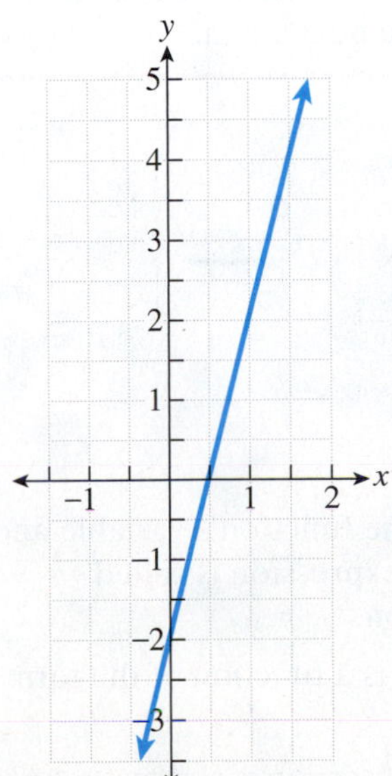

a) Find $f\left(1\frac{1}{2}\right)$.

SOLUTION We are looking for a point on the line that has an x-coordinate of $1\frac{1}{2}$. The point is $\left(1\frac{1}{2}, 4\right)$, so $f\left(1\frac{1}{2}\right) = 4$.

b) Find a value x such that $f(x) = 2$.

SOLUTION We are looking for a point on the line that has a y-coordinate of 2, and this point is $(1, 2)$. The value that satisfies the equation $f(x) = 2$ is $x = 1$.

Quick Check 6

Consider the following graph of a function:

a) Find $f(1)$.
b) Find a value x such that $f(x) = -2$.

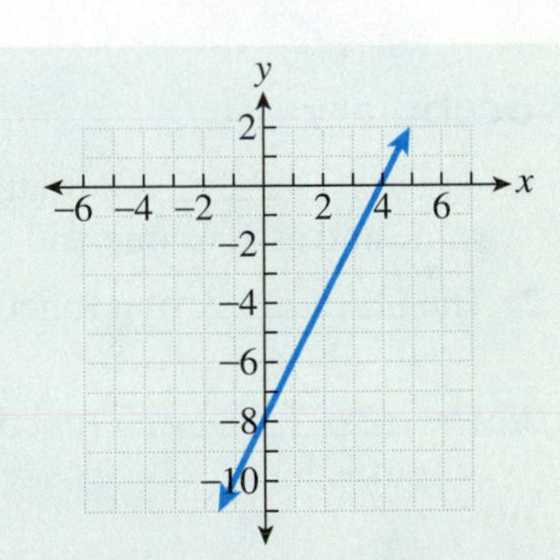

Objective 5 Determine the domain and range of a function from its graph. The domain and range of a function also can be read from a graph. Recall that the domain of a function is the set of all input values. This corresponds to all of the x-coordinates of the points on the graph. The domain is the interval of values on the x-axis for which the graph exists. We read the domain from left to right on the graph.

The domain of a linear function is the set of all real numbers, which can be written in interval notation as $(-\infty, \infty)$. The graphs of linear functions continue on to the left and to the right. Look at the following three graphs of linear functions.

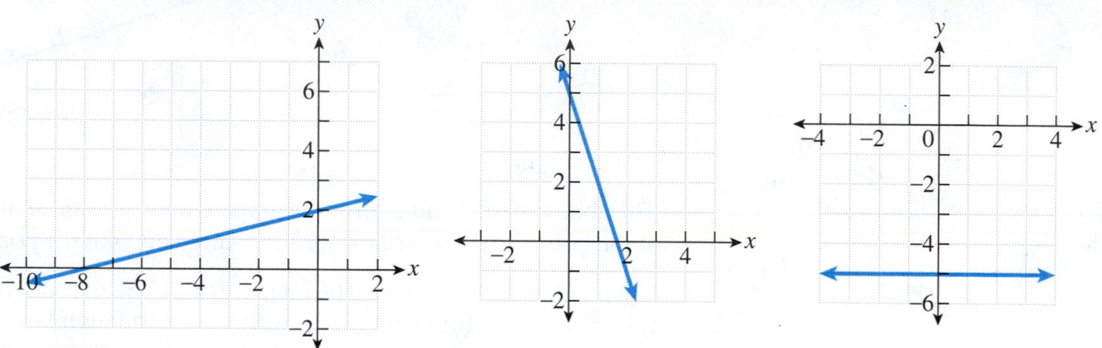

Each line continues to the left as well as to the right. This tells us that the graph exists for all values of x in the interval $(-\infty, \infty)$.

The range of a function can be read vertically from a graph. The range goes from the lowest point to the highest point on the graph. All linear functions have $(-\infty, \infty)$ as their range, except constant functions.

Because a constant function of the form $f(x) = c$, where c is a real number, has only one possible output value, its range is $\{c\}$. For example, the range of the constant function $f(x) = -5$ is $\{-5\}$.

BUILDING YOUR STUDY STRATEGY

Using Your Resources, 4 Student's Solutions Manual The student's solutions manual for a mathematics textbook can be a valuable resource when used properly, but it can be detrimental to learning when used improperly. You should refer to the solutions manual to check your work or to see where you went wrong when finding the solutions.

However, if you do nothing except essentially copy the solutions manual onto your paper, you will likely have difficulty on the next exam.

Exercises 3.4

PRACTICE

WATCH

DOWNLOAD
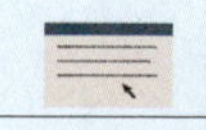
READ

REVIEW

Vocabulary

1. A(n) __________ is a relation that takes one input value and assigns one and only one output value to it.
2. The __________ of a function is the set of all possible input values.
3. The __________ of a function is the set of all possible output values.
4. Substituting a value for the function's variable and simplifying the resulting expression is called __________ the function.
5. A(n) __________ function is a function of the form $f(x) = mx + b$.
6. The __________ of a linear function is always $(-\infty, \infty)$.

The following table contains the names of the nine American League All-Stars from 2008, along with the player's AL team.

Name	Team
Cliff Lee (P)	Cleveland Indians
Joe Mauer (C)	Minnesota Twins
Kevin Youkilis (1B)	Boston Red Sox
Dustin Pedroia (2B)	Boston Red Sox
Derek Jeter (SS)	New York Yankees
Alex Rodriguez (3B)	New York Yankees
Manny Ramirez (LF)	Boston Red Sox
Josh Hamilton (CF)	Texas Rangers
Ichiro Suzuki (RF)	Seattle Mariners

7. Would a relation that took a player's name as an input and listed his AL team as an output be a function? Why or why not?

8. Could a function be defined in the opposite direction, with the name of the AL team as the input and the All-Star player's name as the output? Why or why not?

9. Would a relation that took a person as an input and listed that person's birth mother as an output be a function? Why or why not?

10. Could a function be defined with a woman as an input and her child as an output? Why or why not?

For Exercises 11–14, determine whether a function exists with:
a) Set A as the input and set B as the output.
b) Set B as the input and set A as the output.

11. High Temperatures on December 16

Set *A* City	Set *B* High Temp.
Boston, MA	39° F
Orlando, FL	75° F
Providence, RI	39° F
Rochester, NY	26° F
Visalia, CA	57° F

12.

Set A Person	Set *B* Last 4 Digits of SSN
Jenny Crum	1234
Maureen O'Connor	5283
Dona Kenly	6405
Michelle Renda	5555
Lauren Morse	9200

13.

Set *A* Person	Set *B* Birthday
Greg Erb	December 15
Karen Guardino	December 17
Jolene Lehr	October 15
Siméon Poisson	June 21
Lindsay Skay	May 28
Sharon Smith	June 21

14.

Set *A* Competitive Eater	Set *B* Hot Dogs Eaten in 12 Minutes
Joey Chestnut	66
Takeru Kobayashi	63
Pat Bertoletti	49
Tim Janus	43
Sonya Thomas	39

For the given set of ordered pairs, determine whether a function could be defined for which the input would be an x-coordinate and the output would be the corresponding y-coordinate. If a function cannot be defined in this manner, explain why.

15. $\{(-2, 4), (-1, 1), (0, 0), (1, 1), (2, 4), (3, 9)\}$

16. $\{(2, -2), (1, -1), (0, 0), (1, 1), (2, 2)\}$

17. $\{(5, 3), (2, 7), (-4, -6), (5, -2), (0, 4)\}$

18. $\{(-6, 3), (-2, 3), (1, 3), (5, 3), (11, 3)\}$

19. $\{(2, -5), (2, -1), (2, 0), (2, 3), (2, 5)\}$

20. $\{(1, 1), (2, 2), (3, 3), (4, 4), (5, 5)\}$

21. A Celsius temperature can be converted to a Fahrenheit temperature by multiplying it by $\frac{9}{5}$ and then adding 32.

a) Create a function $F(x)$ that converts a Celsius temperature x to a Fahrenheit temperature.

b) Use the function $F(x)$ from part a to convert the following Celsius temperatures to Fahrenheit temperatures.

0° C 100° C 30° C −10° C −40° C

22. To convert a Fahrenheit temperature to a Celsius temperature, subtract 32 and then multiply that difference by $\frac{5}{9}$. Create a function $C(x)$ that converts a Fahrenheit temperature x to a Celsius temperature.

23. A college student takes a summer job selling newspaper subscriptions door-to-door. She is paid \$36 for a four-hour shift. She also earns \$7 for each subscription sold.

a) Create a function $f(x)$ for the amount she earns on a shift during which she sells x subscriptions.

b) Use the function $f(x)$ to determine how much she earns on a shift during which she sells 12 subscriptions.

24. A cell phone carrier offers a plan with a \$29.99 monthly fee and charges \$0.40 per minute for each minute above 300 minutes for the month.

a) Create a function $f(x)$ for the amount a person pays in a month if he or she uses x minutes above 300 minutes that month.

b) Use the function $f(x)$ to determine the monthly bill for a subscriber who used 850 minutes last month.

25. Create a linear function whose graph has a slope of 4 and a y-intercept at $(0, 3)$.

26. Create a linear function whose graph has a slope of 5 and a y-intercept at $(0, -9)$.

27. Create a linear function whose graph has a slope of -3 and a y-intercept at $(0, -4)$.

28. Create a linear function whose graph has a slope of $-\frac{1}{2}$ and a y-intercept at $(0, \frac{2}{3})$.

29. Create a linear function whose graph has a slope of 0 and a y-intercept at $(0, 6)$.

30. Create a linear function whose graph has a slope of 0 and a y-intercept at $(0, 0)$.

Evaluate the given function.

31. $g(x) = x - 9, g(-13)$

32. $h(x) = 4x, h(-9)$

33. $f(x) = -8x + 3, f(10)$

34. $f(x) = 6x + 7, f(18)$

35. $f(x) = -\dfrac{2}{5}x + 9, f(-10)$

36. $f(x) = \dfrac{3}{4}x + 6, f(-20)$

37. $f(x) = 9x - 25, f(0)$

38. $f(x) = 6x + 13, f(0)$

39. $g(x) = 3x - 1, g\left(\dfrac{2}{3}\right)$

40. $g(x) = 5x + 7, g\left(\dfrac{9}{5}\right)$

41. $f(x) = 3x + 4, f(a)$

42. $f(x) = 2x - 3, f(b)$

43. $f(x) = 7x - 2, f(a + 3)$

44. $f(x) = 5x + 9, f(a - 7)$

45. $f(x) = 16 - 3x, f(2a - 5)$

46. $f(x) = 5 - 6x, f(3a - 4)$

47. $f(x) = 6x + 4, f(x + h)$

48. $f(x) = 3x + 11, f(x + h)$

Graph the linear function.

49. $f(x) = 6x - 6$

50. $f(x) = 2x + 6$

51. $f(x) = -3x + 3$

52. $f(x) = -x - 8$

53. $f(x) = \dfrac{4}{3}x + 4$

54. $f(x) = \dfrac{2}{5}x - 2$

55. $f(x) = \dfrac{8}{3}x$

56. $f(x) = -\dfrac{3}{7}x$

57. $f(x) = 4$

58. $f(x) = -6$

59. Refer to the graph of the function $f(x)$.
a) Find $f(5)$.
b) Find a value a such that $f(a) = -8$.
c) Find the domain and range.

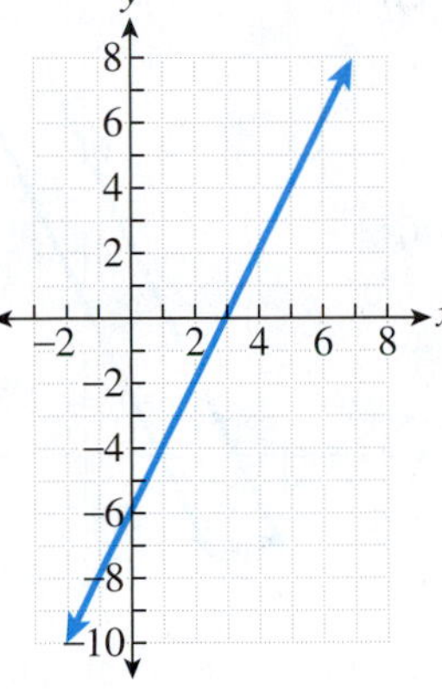

60. Refer to the graph of the function $g(x)$.
a) Find $g(-4)$.
b) Find a value a such that $g(a) = -2$.
c) Find the domain and range.

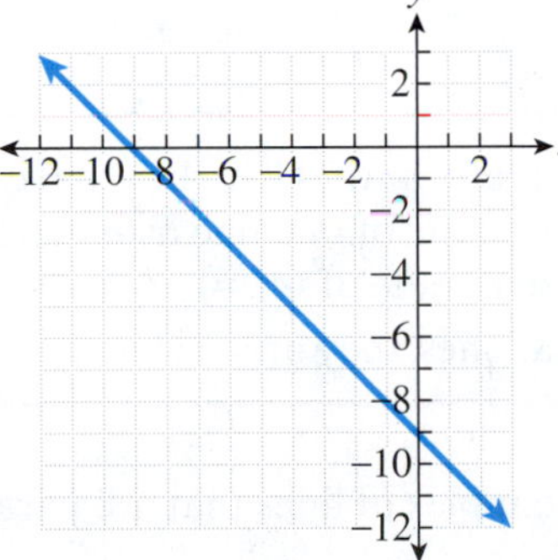

61. Refer to the graph of the function $f(x)$.
a) Find $f(2)$.
b) Find a value a such that $f(a) = 9$.
c) Find the domain and range.

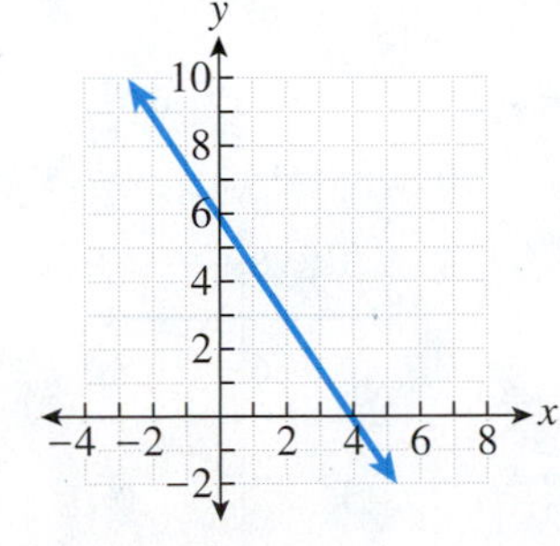

62. Refer to the graph of the funciton $f(x)$.
a) Find $f(-1)$.
b) Find a value a such that $f(a) = -8$.
c) Find the domain and range.

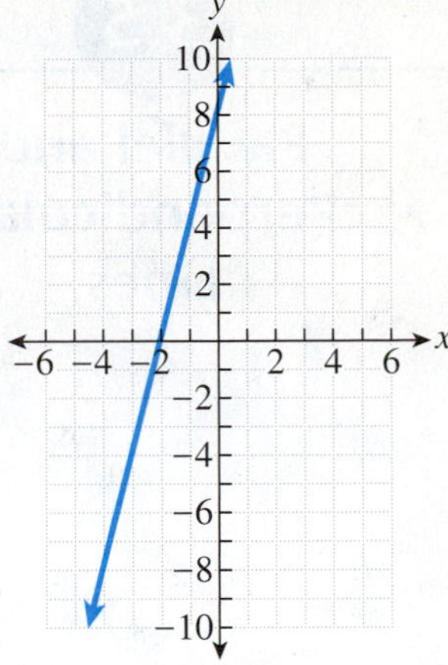

63. a) Find $f(4)$.
b) Find the domain and range.

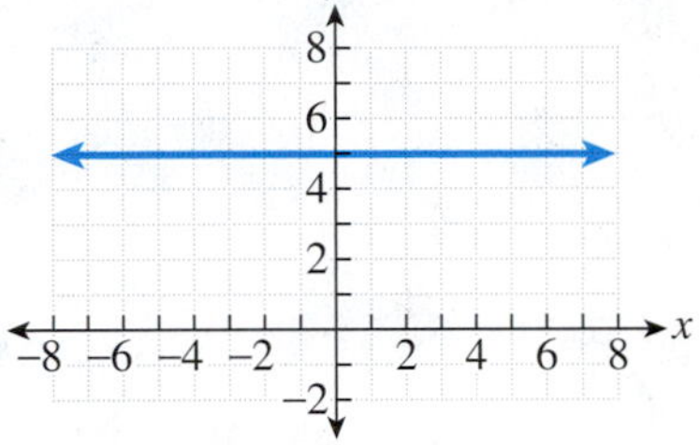

64. a) Find $f(-3)$.
b) Find the domain and range.

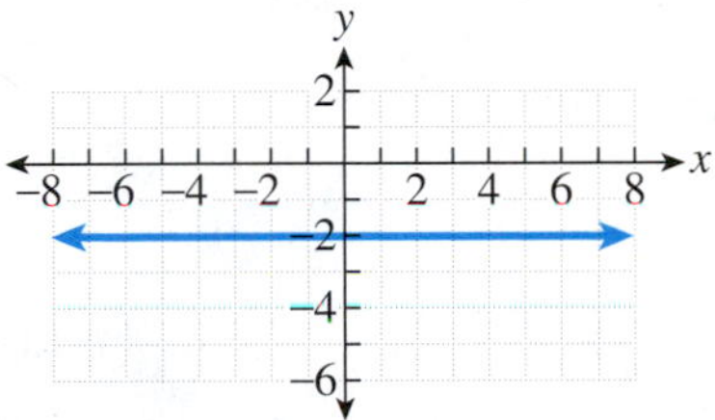

Writing in Mathematics

Answer in complete sentences.

65. Give an example of two sets A and B and a rule that is a function from set A to set B. Explain why your rule meets the definition of a function. Give another rule that would not be a function from set B to set A. Explain why your rule does not meet the definition of a function.

Quick Review Exercises

Section 3.4

Find the slope of the given line.

1. $6x + 2y = 17$

2. $5x - 4y = -16$

3. $x - 5y = 10$

4. $14x + 10y = 30$

3.5 Parallel and Perpendicular Lines

OBJECTIVES

1 Determine whether two lines are parallel.
2 Determine whether two lines are perpendicular.

Parallel Lines

Objective 1 Determine whether two lines are parallel. In this section, we will examine the relationship between two lines. Consider the following pair of lines:

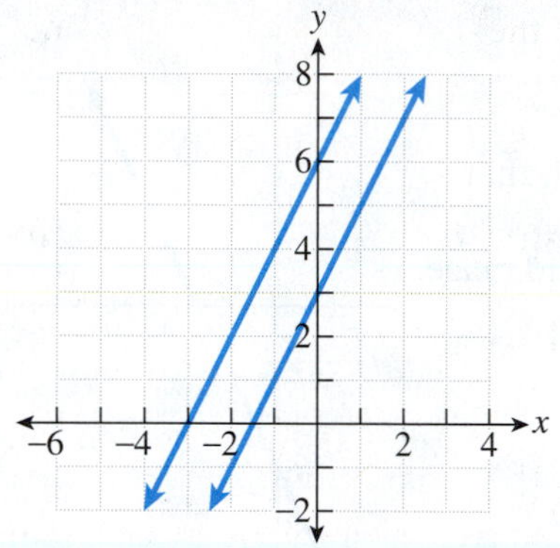

Notice that these two lines do not intersect. The lines have the same slope and are called **parallel lines**.

Parallel Lines

- Two nonvertical lines are **parallel** if they have the same slope. In other words, if we denote the slope of one line as m_1 and the slope of the other line as m_2, the two lines are parallel if $m_1 = m_2$.
- If two lines are vertical lines, they are parallel.

Following are some examples of lines that are parallel:

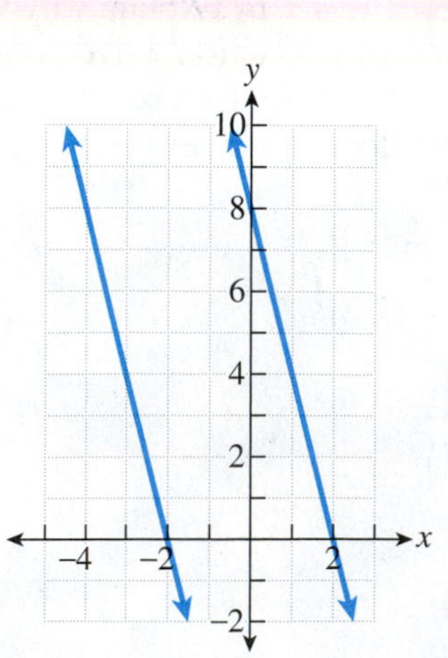

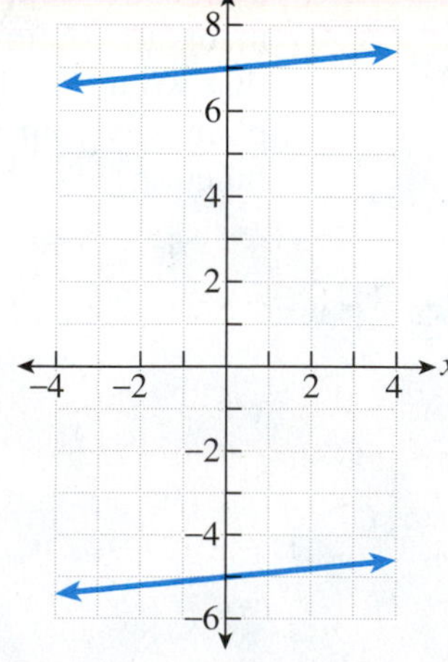

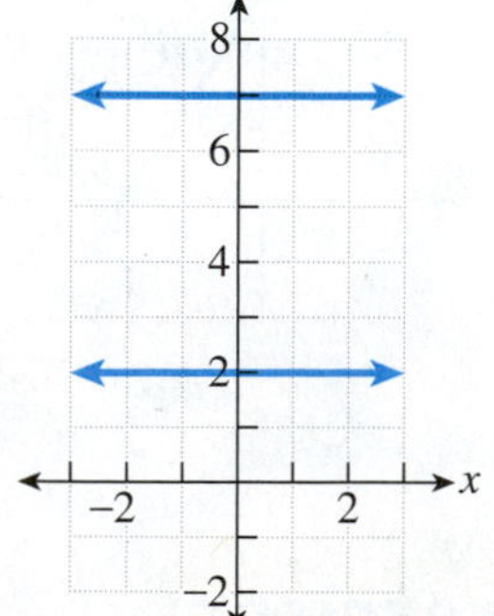

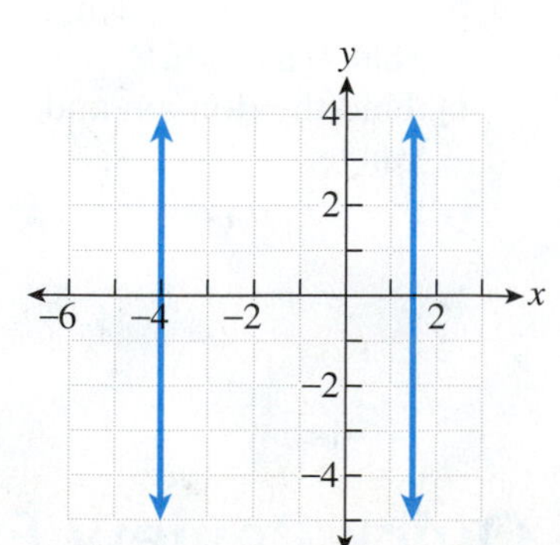

EXAMPLE 1 Are the two lines $y = 2x - 5$ and $y = 3x + 1$ parallel?

SOLUTION These two equations are in slope–intercept form, so we can see that the slope of the first line is 2 and the slope of the second line is 3. Because the slopes are not equal, these two lines are not parallel.

Quick Check 1

Are the two lines $y = 4x + 3$ and $y = -4x$ parallel?

EXAMPLE 2 Are the two lines $6x + 3y = -9$ and $y = -2x + 2$ parallel?

SOLUTION To find the slope of the first line, solve the equation for y.

$$6x + 3y = -9$$
$$3y = -6x - 9 \quad \text{Subtract } 6x \text{ from both sides.}$$
$$y = -2x - 3 \quad \text{Divide by 3.}$$

The slope of the first line is -2. Because the second line is already in slope–intercept form, we see that the slope of that line also is -2. Because the two slopes are equal, the two lines are parallel.

Quick Check 2

Are the two lines $8x + 6y = 36$ and $y = -\frac{4}{3}x + 5$ parallel?

Any two horizontal lines, such as $y = 4$ and $y = 1$, are parallel to each other because their slopes are 0. Any two vertical lines, such as $x = 2$ and $x = -1$, are, by definition, parallel to each other as well.

EXAMPLE 3 Find the slope of a line that is parallel to the line $4x + 3y = 8$.

SOLUTION For a line to be parallel to $4x + 3y = 8$, it must have the same slope as this line. To find the slope of this line, we solve the equation for y.

$$4x + 3y = 8$$
$$3y = -4x + 8 \quad \text{Subtract } 4x \text{ from both sides.}$$
$$y = -\frac{4}{3}x + \frac{8}{3} \quad \text{Divide by 3.}$$

The slope of the line $4x + 3y = 8$ is $-\frac{4}{3}$. A line that is parallel to $4x + 3y = 8$ has a slope of $-\frac{4}{3}$.

Quick Check 3

Find the slope of a line that is parallel to the line $2x - 9y = 36$.

Perpendicular Lines

Objective 2 Determine whether two lines are perpendicular.

Two distinct lines that are not parallel intersect at one point. **Perpendicular lines** are one special type of intersecting lines. Here is an example of two lines that are perpendicular. Perpendicular lines intersect at right angles. Notice that one of these lines has a positive slope and the other line has a negative slope.

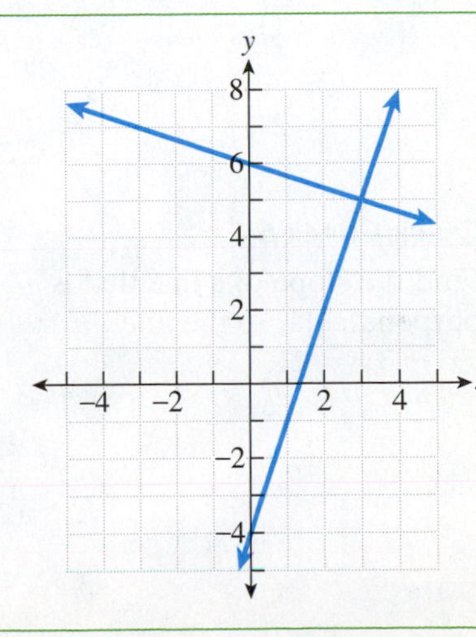

Perpendicular Lines

- Two nonvertical lines are **perpendicular** if their slopes are **negative reciprocals.** In other words, if we denote the slope of one line as m_1 and the slope of the other line as m_2, the two lines are perpendicular if $m_1 = -\frac{1}{m_2}$. (This is equivalent to saying that two lines are perpendicular if the product of their slopes is -1.)
- A vertical line is perpendicular to a horizontal line.

Two numbers are **negative reciprocals** if they are reciprocals that have opposite signs, such as 5 and $-\frac{1}{5}$, -2 and $\frac{1}{2}$, and $-\frac{3}{10}$ and $\frac{10}{3}$.

EXAMPLE 4 Are the two lines $y = 4x + 5$ and $y = -4x + 5$ perpendicular?

SOLUTION The slopes of these two lines are 4 and -4, respectively. While the signs of these two slopes are opposite, the slopes are not reciprocals. Therefore, the two lines are not perpendicular.

Quick Check 4

Are the two lines $y = x + 7$ and $y = -x + 3$ perpendicular?

EXAMPLE 5 Are the two lines $y = 3x + 7$ and $2x + 6y = 5$ perpendicular?

SOLUTION The slope of the first line is 3. To find the slope of the second line, we solve the equation for y.

$$2x + 6y = 5$$

$$6y = -2x + 5 \quad \text{Subtract } 2x.$$

$$y = -\frac{2}{6}x + \frac{5}{6} \quad \text{Divide by 6.}$$

$$y = -\frac{1}{3}x + \frac{5}{6} \quad \text{Simplify.}$$

The slope of the second line is $-\frac{1}{3}$. The two slopes are negative reciprocals, so the lines are perpendicular.

Quick Check 5

Are the two lines $y = \frac{2}{3}x + 5$ and $6x + 4y = 12$ perpendicular?

Any vertical line, such as $x = 2$, is perpendicular to any horizontal line, such as $y = -2$.

EXAMPLE 6 Find the slope of a line that is perpendicular to the line $5x + 2y = 8$.

SOLUTION We begin by finding the slope of this line. We do this by solving the equation $5x + 2y = 8$ for y.

$$5x + 2y = 8$$

$$2y = -5x + 8 \quad \text{Subtract } 5x \text{ from both sides.}$$

$$y = -\frac{5}{2}x + \frac{8}{2} \quad \text{Divide by 2.}$$

$$y = -\frac{5}{2}x + 4 \quad \text{Simplify.}$$

The slope of this line is $-\frac{5}{2}$. To find the slope of a line perpendicular to this line, we take the reciprocal of this slope and change its sign from negative to positive. A line is perpendicular to $5x + 2y = 8$ if it has a slope of $\frac{2}{5}$.

Quick Check 6

Find the slope of a line that is perpendicular to the line $7x - 3y = 21$.

Determining Whether Two Lines Are Parallel, Perpendicular, or Neither

To summarize, nonvertical lines are parallel if and only if they have the same slope and are perpendicular if and only if their slopes are negative reciprocals. Horizontal lines are parallel to other horizontal lines, and vertical lines are parallel to other vertical lines. Finally, a horizontal line and a vertical line are perpendicular to each other. These concepts are summarized in the following table:

If . . .	the lines are parallel if . . .	the lines are perpendicular if . . .
two nonvertical lines have slopes m_1 and m_2	$m_1 = m_2$	$m_1 = -\frac{1}{m_2}$
one of the two lines is horizontal	the other line is horizontal	the other line is vertical
one of the two lines is vertical	the other line is vertical	the other line is horizontal

EXAMPLE 7 Are the two lines $y = 6x + 2$ and $y = \frac{1}{6}x - 3$ parallel, perpendicular, or neither?

SOLUTION The slope of the first line is 6, and the slope of the second line is $\frac{1}{6}$. The slopes are not equal, so the lines are not parallel.

The slopes are not negative reciprocals, so the lines are not perpendicular either.

The two lines are neither parallel nor perpendicular.

Quick Check 7

Are the two lines $y = 2x - 5$ and $y = -2x$ parallel, perpendicular, or neither?

EXAMPLE 8 Are the two lines $-3x + y = 7$ and $-6x + 2y = -4$ parallel, perpendicular, or neither?

SOLUTION Find the slope of each line by solving each equation for y.

$$-3x + y = 7$$
$$y = 3x + 7 \quad \text{Add } 3x \text{ to both sides.}$$

The slope of the first line is 3. Now find the slope of the second line.

$$-6x + 2y = -4$$
$$2y = 6x - 4 \quad \text{Add } 6x \text{ to both sides.}$$
$$y = 3x - 2 \quad \text{Divide by 2.}$$

The slope of the second line also is 3.

Because the two slopes are equal, the lines are parallel.

Quick Check 8

Are the two lines $y = 4x + 3$ and $8x - 2y = 12$ parallel, perpendicular, or neither?

EXAMPLE 9 Are the two lines $10x + 2y = 0$ and $x - 5y = 3$ parallel, perpendicular, or neither?

SOLUTION Begin by finding the slope of each line.

$$10x + 2y = 0$$
$$2y = -10x \quad \text{Subtract } 10x \text{ from both sides.}$$
$$y = -5x \quad \text{Divide by 2.}$$

The slope of the first line is -5. Now find the slope of the second line.

$$x - 5y = 3$$

$$-5y = -x + 3 \quad \text{Subtract } x \text{ from both sides.}$$

$$y = \frac{1}{5}x - \frac{3}{5} \quad \text{Divide by } -5.$$

The slope of the second line is $\frac{1}{5}$. Because the two slopes are negative reciprocals, the lines are perpendicular.

Quick Check 9

Are the two lines $y = -\frac{2}{5}x + 9$ and $5x - 2y = -8$ parallel, perpendicular, or neither?

BUILDING YOUR STUDY STRATEGY

Using Your Resources, 5 Classmates Some frequently overlooked resources are the other students in your class.

- If you have a quick question about a particular problem, a classmate can help you.
- If your notes are incomplete for a certain day, a classmate may allow you to use his or her notes to fill in the holes in your own notes.
- If you are forced to miss class, a quick call to a classmate can help you find out what was covered in class and what is the homework assignment.

Exercises 3.5

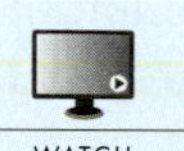

Vocabulary

1. Two nonvertical lines are ________ if they have the same slope.
2. Two nonvertical lines are ___________ if their slopes are negative reciprocals.
3. A vertical line is parallel to a(n) _______ line.
4. A vertical line is perpendicular to a(n) ________ line.
5. Two _______ lines do not intersect.
6. Two ___________ lines intersect at right angles.

Are the two given lines parallel?

7. $y = 3x + 5, y = 3x - 2$
8. $y = 5x - 7, y = -5x + 3$
9. $4x + 2y = 9, 3y = 6x + 7$
10. $x + 3y = -4, 3x + 9y = 8$
11. $y = 6, y = -6$
12. $x = 2, x = 7$

Are the two given lines perpendicular?

13. $y = 4x, y = \frac{1}{4}x - 3$
14. $y = -\frac{3}{2}x + 2, y = \frac{2}{3}x + 1$
15. $15x + 3y = 11, x - 5y = -4$
16. $x + y = 6, x - y = -3$
17. $x = 3, y = 4$
18. $y = -7, y = \frac{1}{7}$

Are the two given lines parallel, perpendicular, or neither?

19. $y = 6x - 2, y = -6x + 5$
20. $y = 7x - 9, y = \frac{1}{7}x + 3$
21. $y = 4x + 3, y = -\frac{1}{4}x - 6$
22. $y = 8x - 16, y = 8x - 1$

23. $8x + 6y = 12, 12x + 9y = -27$
24. $15x - 10y = 20, 6x + 9y = -45$
25. $5x - 4y = 44, 10x + 8y = -32$
26. $3y = 7x - 6, 21x - 9y = 63$
27. $y = -\frac{8}{7}, 8y - 7 = 0$
28. $x = \frac{3}{4}, x = -\frac{4}{3}$
29. $x = 5, y = -2$
30. $x + y = 16, y = x - 7$

Are the lines associated with the given functions parallel, perpendicular, or neither?

31. $f(x) = 7x - 5, g(x) = 7x + 3$
32. $f(x) = -2x - 5, g(x) = -\frac{1}{2}x + 4$
33. $f(x) = 6, g(x) = 6x + 4$
34. $f(x) = 4x + 9, g(x) = -\frac{1}{4}x + 3$
35. $f(x) = x + 7, g(x) = x + 2$
36. $f(x) = 5, g(x) = -5$

Find the slope of a line that is parallel to the given line. If the slope is undefined, state this.

37. $y = -6x + 7$
38. $y = 3x + 2$
39. $y = 5x$
40. $y = -3$
41. $12x + 4y = 8$
42. $2x + 6y = -7$

Find the slope of a line that is perpendicular to the given line. If the slope is undefined, state this.

43. $y = 8x - 7$
44. $y = -\frac{1}{5}x + 2$
45. $10x + 6y = 30$
46. $8x - 14y = -28$
47. $12x + 21y = 33$
48. $4x - 7 = 0$
49. Find the slope of a line that is parallel to the line $Ax + By = C$. (A, B, and C are real numbers, $A \neq 0$, and $B \neq 0$.)
50. Find the slope of a line that is perpendicular to the line $Ax + By = C$. (A, B, and C are real numbers, $A \neq 0$, and $B \neq 0$.)
51. Is the line that passes through $(3, 7)$ and $(5, -1)$ parallel to the line $y = -4x + 11$?
52. Is the line that passes through $(-6, 2)$ and $(4, 8)$ parallel to the line $-3x + 5y = 15$?
53. Is the line that passes through $(4, 7)$ and $(-1, 9)$ perpendicular to the line $-10x + 4y = 8$?
54. Is the line that passes through $(-3, -8)$ and $(2, 7)$ perpendicular to the line that passes through $(-5, 3)$ and $(7, 7)$?

Writing in Mathematics

Answer in complete sentences.

55. Describe three real-world examples of parallel lines.
56. Describe three real-world examples of perpendicular lines.
57. Explain the process for determining whether two lines of the form $Ax + By = C$ are parallel.
58. Explain the process for determining whether two lines of the form $Ax + By = C$ are perpendicular.

Quick Review Exercises

Section 3.5

Find the slope of a line that passes through the given two points. If the slope is undefined, state this.

1. $(5, 7)$ and $(8, 1)$
2. $(-4, 2)$ and $(2, 10)$
3. $(-9, -5)$ and $(-7, 5)$
4. $(6, 0)$ and $(-2, -6)$

3.6 Equations of Lines

OBJECTIVES

1. **Find the equation of a line using the point–slope form.**
2. **Find the equation of a line given two points on the line.**
3. **Find a linear equation to describe real data.**
4. **Find the equation of a parallel or perpendicular line.**

We are already familiar with the slope–intercept form of the equation of a line: $y = mx + b$. When an equation is written in this form, we know both the slope of the line (m) and the y-coordinate of the y-intercept (b). This form is convenient for graphing lines. In this section, we will look at another form of the equation of a line. We also will learn how to find the equation of a line if we know the slope of a line and any point through which the line passes.

Point–Slope Form of the Equation of a Line

Objective 1 Find the equation of a line using the point–slope form. In Section 3.3, we learned that if we know the slope of a line and its y-intercept, we can write the equation of the line using the slope–intercept form of a line $y = mx + b$, where m is the slope of the line and b is the y-coordinate of the y-intercept. If we know the slope of a line and the coordinates of any point on that line, not just the y-intercept, we can write the equation of the line using the point–slope form of an equation.

Point–Slope Form of the Equation of a Line

The **point–slope form** of the equation of a line with slope m that passes through the point (x_1, y_1) is

$$y - y_1 = m(x - x_1).$$

This form can be derived directly from the slope formula $m = \frac{y_2 - y_1}{x_2 - x_1}$. If we let (x, y) represent an arbitrary point on the line, this formula becomes $m = \frac{y - y_1}{x - x_1}$. Multiplying both sides of that equation by $(x - x_1)$ produces the point–slope form of the equation of a line.

If a line has slope 2 and passes through $(3, 1)$, we find its equation by substituting 2 for m, 3 for x_1, and 1 for y_1.

$$y_1 = 1 \qquad m = 2 \qquad x_1 = 3$$

$$y - y_1 = m(x - x_1)$$
$$y - 1 = 2(x - 3)$$

The equation of the line is $y - 1 = 2(x - 3)$. This equation can be converted to slope–intercept form or to standard form.

EXAMPLE 1 Find the equation of a line with slope 4 that passes through the point $(-2, 5)$. Write the equation in slope–intercept form.

SOLUTION Substitute 4 for m, -2 for x_1, and 5 for y_1 in the point–slope form. Then solve the equation for y to write the equation in slope–intercept form.

$y - 5 = 4(x - (-2))$ Substitute into point–slope form.

$y - 5 = 4(x + 2)$ Eliminate double signs.

$y - 5 = 4x + 8$ Distribute.

$y = 4x + 13$ Add 5 to isolate y.

The equation for a line with slope 4 that passes through $(-2, 5)$ is $y = 4x + 13$. Below is the graph of the line, showing that it passes through the point $(-2, 5)$.

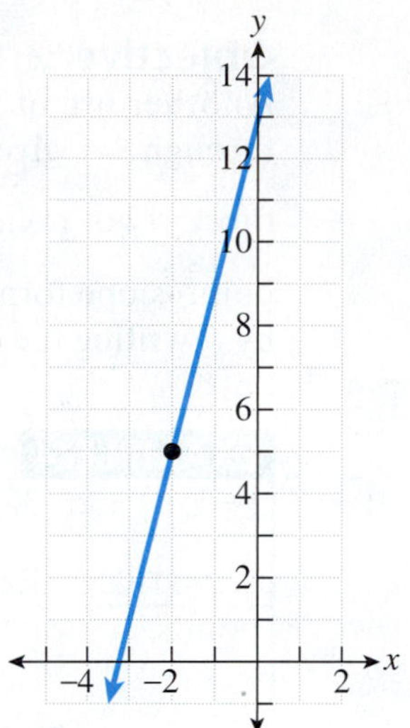

Quick Check 1

Find the equation of a line with slope 2 that passes through the point $(1, 5)$. Write the equation in slope–intercept form.

EXAMPLE 2 Find the equation of a line with slope -5 that passes through the point $(4, -3)$. Write the equation in slope–intercept form.

SOLUTION Begin by substituting -5 for m, 4 for x_1, and -3 for y_1 in the point–slope form. After substituting, solve the equation for y.

$$y - (-3) = -5(x - 4) \quad \text{Substitute into point–slope form.}$$
$$y + 3 = -5(x - 4) \quad \text{Simplify left side.}$$
$$y + 3 = -5x + 20 \quad \text{Distribute } -5.$$
$$y = -5x + 17 \quad \text{Subtract 3 to isolate } y.$$

Quick Check 2

Find the equation of a line with slope -3 that passes through the point $(4, -6)$. Write the equation in slope–intercept form.

The equation for a line with slope -5 that passes through $(4, -3)$ is $y = -5x + 17$.

In the previous example, we converted the equation from point–slope form to slope–intercept form. This is a good idea in general. We use the point–slope form because it is a convenient form for finding the equation of a line if we know its slope and the coordinates of a point on the line. We convert the equation to slope–intercept form because it is easier to graph a line when the equation is in this form.

EXAMPLE 3 The line associated with the linear function $f(x)$ has a slope of 2. If $f(-3) = -7$, find the function $f(x)$.

SOLUTION A linear function is of the form $f(x) = mx + b$. In this example, we know that $m = 2$, so $f(x) = 2x + b$. We will now use the fact that $f(-3) = -7$ to find b.

$$f(-3) = -7$$
$$2(-3) + b = -7 \quad \text{Substitute } -3 \text{ for } x \text{ in the function } f(x).$$
$$-6 + b = -7 \quad \text{Multiply.}$$
$$b = -1 \quad \text{Add 6.}$$

Replace m with 2 and b with -1. The function is $f(x) = 2x - 1$.

Quick Check 3

The line associated with the linear function $f(x)$ has a slope of -4. If $f(-2) = 13$, find the function $f(x)$.

The previous example also could have been solved by using the point–slope form with the point $(-3, -7)$.

Finding the Equation of a Line Given Two Points on the Line

Objective 2 Find the equation of a line given two points on the line. Another use of the point–slope form is to find the equation of a line that passes through two given points. We begin by finding the slope of the line passing through those two points using the slope formula $m = \frac{y_2 - y_1}{x_2 - x_1}$. Then we use the point–slope form with this slope and either of the points we were given. We finish by rewriting the equation in slope–intercept form.

EXAMPLE 4 Find the equation of a line that passes through the two points $(-1, 6)$ and $(3, -2)$.

SOLUTION Begin by finding the slope of the line that passes through these two points.

$$m = \frac{-2 - 6}{3 - (-1)} \quad \text{Substitute into the formula } m = \frac{y_2 - y_1}{x_2 - x_1}.$$

$$= \frac{-8}{4} \quad \text{Simplify numerator and denominator.}$$

$$= -2 \quad \text{Simplify.}$$

Now substitute into the point–slope form using either of the points with $m = -2$. Use $(3, -2)$:

$$y - (-2) = -2(x - 3) \quad \text{Substitute in point–slope form.}$$

$$y + 2 = -2x + 6 \quad \text{Simplify.}$$

$$y = -2x + 4 \quad \text{Solve for } y.$$

The equation of the line that passes through these two points is $y = -2x + 4$. If we had used the point $(-1, 6)$ instead of $(3, -2)$, the result would have been the same.

At the right is the graph of the line passing through the points $(-1, 6)$ and $(3, -2)$.

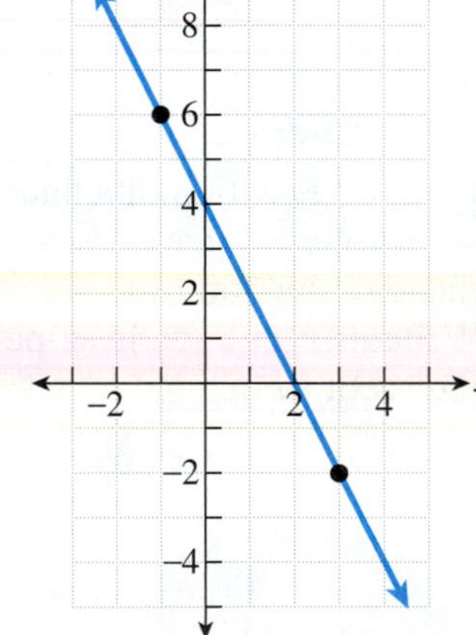

Quick Check 4

Find the equation of a line that passes through the two points $(-2, 7)$ and $(6, -5)$.

EXAMPLE 5 Find the equation of a line that passes through the two points $(8, 2)$ and $(8, 9)$.

SOLUTION Begin by attempting to find the slope of the line that passes through these two points.

$$m = \frac{9 - 2}{8 - 8} \quad \text{Substitute into the slope formula.}$$

$$= \frac{7}{0} \quad \text{Simplify numerator and denominator.}$$

The slope is undefined, so this line is a vertical line. (We could have discovered this by plotting the two points on a graph.) The equation for this line is $x = 8$.

Quick Check 5

Find the equation of a line that passes through the two points $(-6, 4)$ and $(3, 4)$.

Finding a Linear Equation to Describe Linear Data

Objective 3 Find a linear equation to describe real data.

EXAMPLE 6 In 2005, approximately 8.9 million U.S. households had a net worth of at least $1 million. By 2007, that number had increased to 9.9 million U.S. households. Find a linear equation that describes the number (y) of U.S. millionaire households (in millions) x years after 2005. (*Source: TNS Financial Services, Affluent Market Research Program*)

SOLUTION Because x represents the number of years after 2005, $x = 0$ for 2005 and $x = 2$ for 2007. This tells us that two points on the line are (0, 8.9) and (2, 9.9). We begin by calculating the slope of the line.

$$m = \frac{9.9 - 8.9}{2 - 0} \quad \text{Substitute into the slope formula.}$$

$$= \frac{1}{2} \quad \text{Simplify numerator and denominator.}$$

$$= 0.5 \quad \text{Divide.}$$

The slope is 0.5, which tells us that the number of U.S. millionaires increases by 0.5 million per year. In this example, we know that the y-intercept is (0, 8.9), so we can write the equation directly in slope–intercept form. The equation is $y = 0.5x + 8.9$. (If we did not know the y-intercept of the line, we would find the equation by substituting into the point–slope form.)

Quick Check 6

In 2003, 3620 master's degrees were awarded in Mathematics and Statistics. In 2007, this total increased to 4884. Find a linear equation that tells the number (y) of master's degrees awarded in Mathematics and Statistics x years after 2003. (*Source:* Digest of Education Statistics, *U.S. Department of Education*)

Finding the Equation of a Parallel or Perpendicular Line

Objective 4 Find the equation of a parallel or perpendicular line. To find the equation of a line, we must know the slope of the line and the coordinates of a point on the line. In the previous examples, we were given the slope or we calculated the slope using two points on the line. Sometimes the slope of the line is given in terms of another line. We could be given the equation of a line either parallel or perpendicular to the line for which we are trying to find the equation.

EXAMPLE 7 Find the equation of a line that is parallel to the line $y = -\frac{3}{4}x + 15$ and that passes through $(-8, 5)$.

SOLUTION Because the line is parallel to $y = -\frac{3}{4}x + 15$, its slope must be $-\frac{3}{4}$. Substitute this slope, along with the point $(-8, 5)$, into the point–slope form to find the equation of this line.

$$y - 5 = -\frac{3}{4}[x - (-8)]$$ Substitute into point–slope form.

$$y - 5 = -\frac{3}{4}(x + 8)$$ Eliminate double signs.

$$y - 5 = -\frac{3}{4}x - \frac{3}{\underset{1}{\cancel{4}}} \cdot \overset{2}{\cancel{8}}$$ Distribute and divide out common factors.

$$y - 5 = -\frac{3}{4}x - 6$$ Simplify.

$$y = -\frac{3}{4}x - 1$$ Add 5.

The equation of the line parallel to $y = -\frac{3}{4}x + 15$ and passing through $(-8, 5)$ is $y = -\frac{3}{4}x - 1$.

Quick Check 7

Find the equation of a line that is parallel to the line $y = \frac{2}{5}x - 9$ and that passes through $(5, 4)$.

EXAMPLE 8 Find the equation of a line that is perpendicular to the line $y = -\frac{1}{2}x + 7$ and that passes through $(5, 3)$.

SOLUTION The slope of the line $y = -\frac{1}{2}x + 7$ is $-\frac{1}{2}$, so the slope of a perpendicular line must be the negative reciprocal of $-\frac{1}{2}$, or 2. Now substitute into the point–slope form.

$$y - 3 = 2(x - 5)$$ Substitute into point–slope form.

$$y - 3 = 2x - 10$$ Distribute.

$$y = 2x - 7$$ Add 3.

The equation of the line is $y = 2x - 7$.

Quick Check 8

Find the equation of a line that is perpendicular to the line $9x - 12y = 4$ and that passes through $(6, -2)$.

If we are given . . .	We find the equation by . . .
The slope and the y-intercept	Substituting m and b into the slope–intercept form $y = mx + b$
The slope and a point on the line	Substituting m and the coordinates of the point into the point–slope form $y - y_1 = m(x - x_1)$
Two points on the line	Calculating m using the slope formula $m = \frac{y_2 - y_1}{x_2 - x_1}$ and then substituting the slope and the coordinates of one of the points into the point–slope form $y - y_1 = m(x - x_1)$
A point on the line and the equation of a parallel line	Substituting the slope of that line and the coordinates of the point into the point–slope form $y - y_1 = m(x - x_1)$
A point on the line and the equation of a perpendicular line	Substituting the negative reciprocal of the slope of that line and the coordinates of the point into the point–slope form $y - y_1 = m(x - x_1)$

Keep in mind that if the slope of the line is undefined, the line is vertical and its equation is of the form $x = a$, where a is the x-coordinate of the given point. We do not use the point–slope form or the slope–intercept form to find the equation of a vertical line.

BUILDING YOUR STUDY STRATEGY

Using Your Resources, 6 Internet Resources Consider using the Internet as a resource to conduct further research on topics you are learning. Many websites have alternative explanations, examples, and practice problems. If you find a site that helps you with a particular topic, check that site when you are researching another topic. If you have any questions about what you find, ask your instructor.

Exercises 3.6

MyMathLab PRACTICE WATCH DOWNLOAD READ REVIEW

Vocabulary

1. The point–slope form of the equation of a line with slope m that passes through the point (x_1, y_1) is ______.
2. State the procedure for finding the equation of a line that passes through the points (x_1, y_1) and (x_2, y_2).

Write the following equations in slope–intercept form.

3. $-6x + 2y = 10$
4. $3x + 6y = 15$
5. $x - 5y = 10$
6. $-12x - 16y = 10$
7. $7x - y = 3$
8. $6x + 4y = 0$

Find the equation of a line with the given slope and y-intercept.

9. Slope -3, y-intercept $(0, 5)$
10. Slope 2, y-intercept $(0, -4)$
11. Slope $\frac{2}{3}$, y-intercept $(0, -3)$
12. Slope 0, y-intercept $(0, 9)$
13. Slope 5, y-intercept $(0, 8)$
14. Slope $-\frac{3}{5}$, y-intercept $(0, 2)$

Find the slope–intercept form of the equation of a line with the given slope that passes through the given point.

15. Slope 3, through $(7, 4)$
16. Slope -2, through $(1, 6)$
17. Slope -4, through $(6, -2)$
18. Slope -1, through $(-6, -3)$
19. Slope $\frac{3}{2}$, through $(-6, -9)$
20. Slope $\frac{2}{5}$, through $(-5, -4)$
21. Slope 0, through $(8, 5)$
22. Undefined slope, through $(8, 5)$
23. Find a linear function $f(x)$ with slope -2 such that $f(-4) = 23$.
24. Find a linear function $f(x)$ with slope 5 such that $f(3) = 12$.
25. Find a linear function $f(x)$ with slope $\frac{3}{5}$ such that $f(-15) = -17$.
26. Find a linear function $f(x)$ with slope 0 such that $f(316) = 228$.

Find the slope–intercept form of the equation of a line that passes through the given points.

27. $(2, -3), (7, 2)$
28. $(4, 6), (8, 10)$
29. $(-3, -3), (1, 9)$
30. $(-6, 9), (-2, -3)$
31. $(-10, -12), (15, 18)$
32. $(6, -9), (-2, 3)$
33. $(-2, -9), (-2, -3)$
34. $(-4, 8), (2, 8)$

For Exercises 35–40, find the slope–intercept form of the equation of a line whose x-intercept and y-intercept are given.

	x-intercept	y-intercept
35.	(6, 0)	(0, 2)
36.	(−9, 0)	(0, 3)
37.	(4, 0)	(0, −4)
38.	(−10, 0)	(0, −2)
39.	(6, 0)	(0, −6)
40.	(10, 0)	(0, 4)

41. Jamie sells newspaper subscriptions door-to-door to help pay her tuition. She is paid a certain salary each night plus a commission on each sale she makes. On Monday, she sold 3 subscriptions and was paid \$66. On Tuesday, she sold 8 subscriptions and was paid \$116.

a) Find a linear equation that calculates Jamie's pay on a night she sells x subscriptions.

b) How much will Jamie be paid on a night she makes no sales?

42. Luis contracted with a landscaper to install a brick patio in his backyard. The landscaper charged \$10,000. Luis paid the landscaper a deposit the first month and agreed to make a fixed monthly payment until the balance was paid in full. After three months, Luis owed \$6800. After seven months, Luis still owed \$5200.

a) Find a linear equation for Luis's balance (y) after x months.

b) How much was the deposit that Luis paid?

c) How many months will it take to pay off the entire balance?

43. Members of a racquetball club pay an annual membership fee. In addition, they pay a fee each time they play racquetball. Last year Jay played racquetball 76 times and paid the club a total of \$1126. Last year Sammy played racquetball 35 times and paid the club a total of \$613.50.

a) Find a linear equation for the amount paid to the club by a member who plays racquetball x times a year.

b) What is the annual membership fee at this club?

c) What is the fee that is charged each time a member plays racquetball?

44. A computer repairperson charges a service fee in addition to an hourly rate to fix a computer. To fix Frank's computer, the repair person took two hours and charged a total of \$225 (service fee plus hourly rate for two hours of work). Bundy's computer had more problems and took six hours to fix. The charge to Bundy was \$505.

a) Find a linear equation for the charge (y) for a repair that takes x hours to perform.

b) How much is the repairperson's service fee?

c) How much is the repairperson's hourly rate?

45. A banquet hall hosts wedding receptions. There is a charge to rent the hall, in addition to a per person charge for the meal. Chrissy had 120 guests at her wedding and paid a total of \$4100. Adaeze had 175 guests at her wedding and paid a total of \$5750.

a) Find a linear equation for the total charge (y) to host a wedding with x guests.

b) What is the charge to rent the hall?

c) What is the charge for each person's dinner?

d) What would the charge be to host a wedding with 200 guests?

46. Masaru has a small business in which he caters sushi parties. He charges a flat party fee in addition to a charge for each person to cover the food costs. He charges a total of \$319.50 for a party with 10 guests and \$618.75 for a party with 25 guests.

a) Find a linear equation for the total charge (y) for a party with x guests.

b) What is the flat party fee?

c) What is the charge for each person's food?

d) What would the charge be for a party with 16 guests?

Find the slope–intercept form of the equation of a line that is parallel to the given line and that passes through the given point.

47. Parallel to $y = -2x + 13$, through $(-6, -3)$

48. Parallel to $y = 4x - 11$, through $(3, 8)$

49. Parallel to $-3x + y = 9$, through $(4, -6)$

50. Parallel to $10x + 2y = 40$, through $(-2, 9)$

51. Parallel to $y = 5$, through $(2, 7)$

52. Parallel to $x = -5$, through $(-6, -4)$

Find the slope–intercept form of the equation of a line that is perpendicular to the given line and that passes through the given point.

53. Perpendicular to $y = 3x - 7$, through $(9, 4)$

54. Perpendicular to $y = -\frac{1}{2}x + 5$, through $(1, -3)$

55. Perpendicular to $4x + 5y = 9$, through $(-4, 2)$

56. Perpendicular to $3x - 2y = 8$, through $(-6, -7)$

57. Perpendicular to $y = 3$, through $(-5, -3)$

58. Perpendicular to $x = -8$, through $(6, 1)$

Find the slope–intercept form of the equation of a line that is parallel to the graphed line and that passes through the point plotted on the graph. (Begin by finding the slope of the graphed line.)

59.

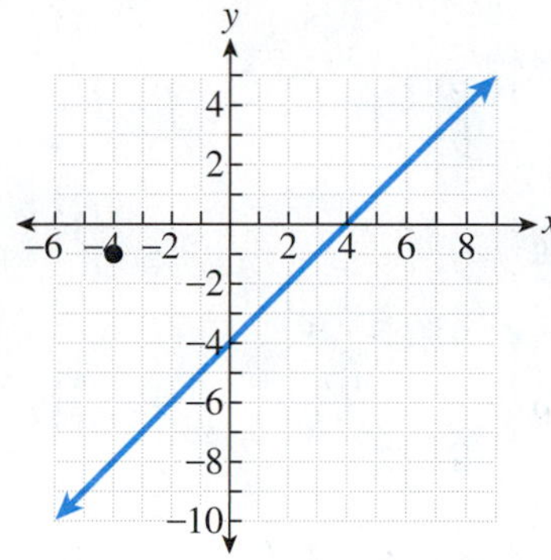

60.

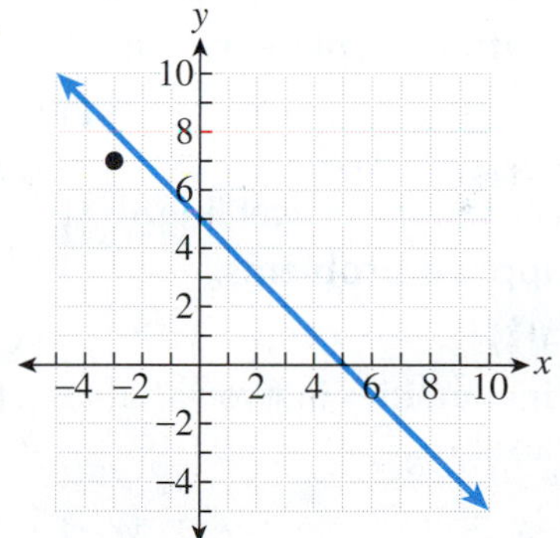

Find the slope–intercept form of the equation of a line that is perpendicular to the graphed line and that passes through the point plotted on the graph. (Begin by finding the slope of the graphed line.)

61.

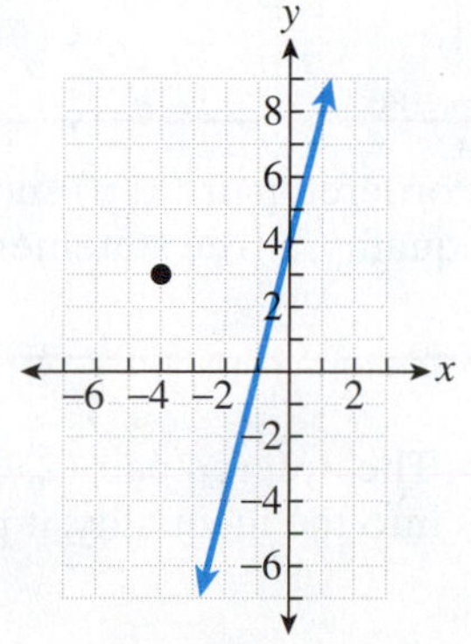

62.

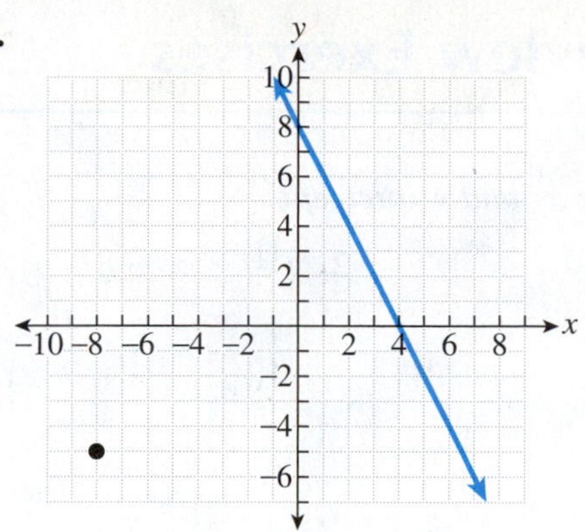

Mixed Practice, 63–72

Find the equation of a line that meets the given conditions.

63. Slope -3, passes through $(4, -8)$

64. Perpendicular to $6x - 2y = 10$, passes through $(6, -3)$

65. Parallel to $5x - 3y = -3$, passes through $(-9, 8)$

66. Passes through $(-2, 7)$ and $(4, -8)$

67. Slope 0, passes through $(13, 11)$

68. Parallel to $3x + 4y = 19$, passes through $(-12, -7)$

69. Perpendicular to $x + 4y = 8$, passes through $(-2, 13)$

70. Passes through $(-6, 14)$ and $(-6, -39)$

71. Passes through $(1, -23)$ and $(5, 33)$

72. Slope $\frac{4}{5}$, passes through $(-15, -11)$

Writing in Mathematics

Answer in complete sentences.

73. Explain why the equation of a vertical line has the form $x = a$ and the equation of a horizontal line has the form $y = b$.

74. Explain why a horizontal line has a slope of 0 and a vertical line has undefined slope.

Quick Review Exercises

Section 3.6

Graph. Label any x- and y-intercepts.

1. $5x - 2y = 10$

2. $y = -3x - 8$

3. $y = \frac{2}{3}x - 4$

4. $y = -8$

3.7 Linear Inequalities

OBJECTIVES

1. **Determine whether an ordered pair is a solution of a linear inequality in two variables.**
2. **Graph a linear inequality in two variables.**
3. **Graph a linear inequality involving a horizontal or vertical line.**
4. **Graph linear inequalities associated with applied problems.**

In this section, we will learn how to solve **linear inequalities** in two variables. Here are some examples of linear inequalities in two variables.

$$2x + 3y \leq 6 \qquad 5x - 4y \geq -8 \qquad -x + 9y < -18 \qquad -3x - 4y > 7$$

Solutions of Linear Inequalities in Two Variables

Objective 1 Determine whether an ordered pair is a solution of a linear inequality in two variables.

Solutions of Linear Inequalities

A solution of a linear inequality in two variables is an ordered pair (x, y) such that when the coordinates are substituted into the inequality, a true statement results.

For example, consider the linear inequality $2x + 3y \leq 6$. The ordered pair $(2, 0)$ is a solution because when we substitute these coordinates into the inequality, it produces the following result:

$$\begin{aligned} 2x + 3y &\leq 6 \\ 2(2) + 3(0) &\leq 6 \\ 4 + 0 &\leq 6 \\ 4 &\leq 6 \end{aligned}$$

The last inequality is a true statement, so $(2, 0)$ is a solution. The ordered pair $(3, 4)$ is not a solution because $2(3) + 3(4)$ is not less than or equal to 6. Any ordered pair (x, y) for which $2x + 3y$ evaluates to be less than or equal to 6 is a solution of this inequality, and there are infinitely many solutions to this inequality. We will display our solutions on a graph.

Graphing Linear Inequalities in Two Variables

Objective 2 Graph a linear inequality in two variables. Ordered pairs that are solutions of the inequality $2x + 3y \le 6$ are one of two types: ordered pairs (x, y) for which $2x + 3y = 6$ are solutions and ordered pairs (x, y) for which $2x + 3y < 6$ are also solutions. Points satisfying $2x + 3y = 6$ lie on a line, so we begin by graphing this line. Because the equation is in standard form, a quick way to graph this line is by finding its x- and y-intercepts.

x-intercept $(y = 0)$	y-intercept $(x = 0)$
$2x + 3(0) = 6$	$2(0) + 3y = 6$
$2x + 0 = 6$	$0 + 3y = 6$
$2x = 6$	$3y = 6$
$x = 3$	$y = 2$
$(3, 0)$	$(0, 2)$

Here is the graph of the line:

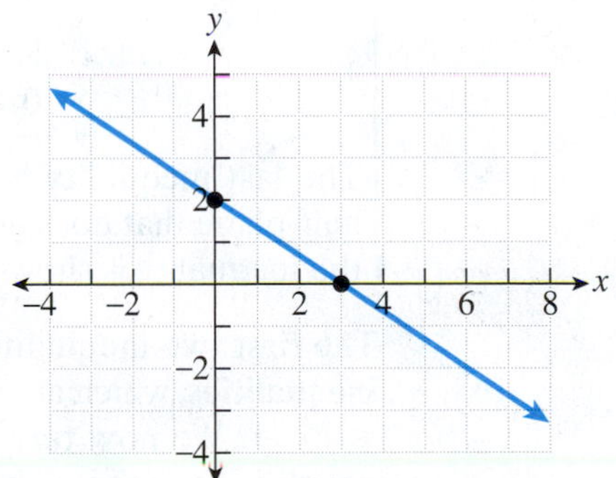

This line divides the plane into two half-planes and is the dividing line between ordered pairs for which $2x + 3y < 6$ and ordered pairs for which $2x + 3y > 6$. To finish graphing the solutions, we must determine which half-plane contains the ordered pairs for which $2x + 3y < 6$. To do this, we use a **test point,** which is a point that is not on the graph of the line whose coordinates are used for determining which half-plane contains the solutions of the inequality. We substitute the test point's coordinates into the original inequality. If the resulting inequality is true, this point and all other points on the same side of the line are solutions, and we shade that half-plane. If the resulting inequality is false, the solutions are on the other side of the line, and we shade that half-plane instead. A wise choice for the test point is the origin $(0, 0)$ if it is not on the line that has been graphed because its coordinates are easy to work with when substituting into the inequality. Because $(0, 0)$ is not on the line, we will use it as a test point.

Test Point: $(0, 0)$

$$2(0) + 3(0) \le 6$$
$$0 + 0 \le 6$$
$$0 \le 6$$

Because the last line is a true inequality, (0, 0) is a solution, and we shade the half-plane containing this point.

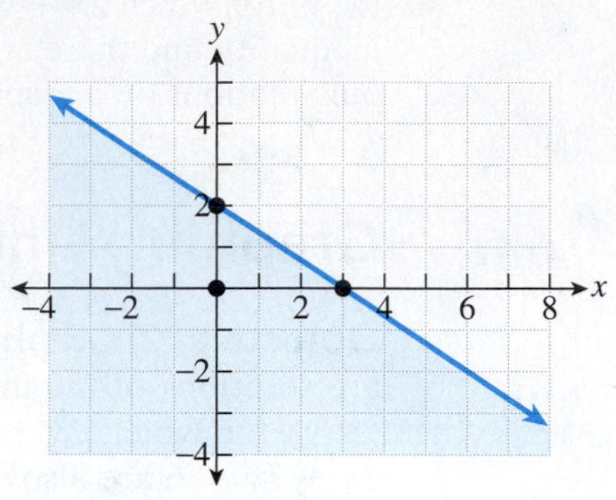

EXAMPLE 1 Graph $y \geq 4x + 3$.

SOLUTION Begin by graphing the line $y = 4x + 3$. This equation is in slope–intercept form, so we can graph it by plotting its y-intercept (0, 3) and using the slope (up 4 units, 1 unit to the right) to find other points on the line.

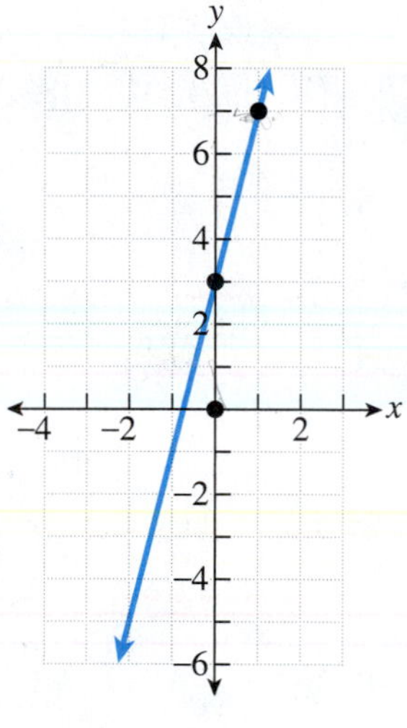

Because the line does not pass through the origin, we can use (0, 0) as a test point.

Test Point: (0, 0)

$0 \geq 4(0) + 3$

$0 \geq 0 + 3$

$0 \geq 3$

The last inequality is false, so the solutions are in the half-plane that does not contain the origin. The graph of the inequality is shown at the right.

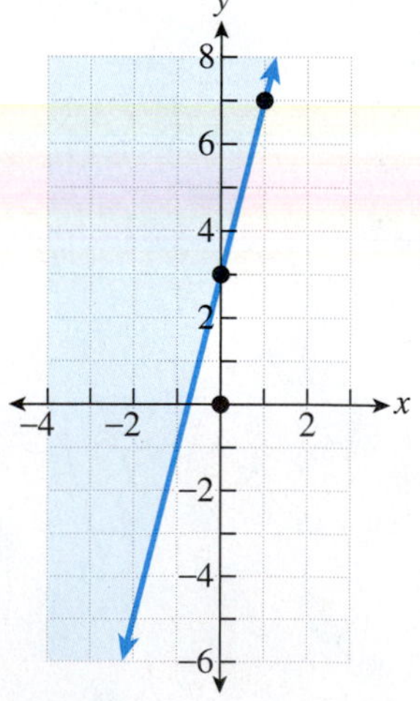

Quick Check 1

Graph $y \geq 2x - 6$.

The first two inequalities we graphed were **weak linear inequalities,** which are inequalities involving the symbols $\leq$ or $\geq$. We now turn our attention to **strict linear inequalities,** which involve the symbol < or >. An example of a strict linear inequality is $x - 5y < 5$. Ordered pairs for which $x - 5y = 5$ are not solutions to this inequality, so the points on the line are not included as solutions. We denote this on the graph by graphing the line as a dashed or broken line. We still pick a test point and shade the appropriate half-plane as we did in the previous examples.

EXAMPLE 2 Graph $x - 5y < 5$.

SOLUTION Begin by graphing the line $x - 5y = 5$ as a dashed line. This equation is in standard form, so we can graph the line by finding its intercepts.

***x*-intercept ($y = 0$)**	***y*-intercept ($x = 0$)**
$x - 5(0) = 5$	$0 - 5y = 5$
$x - 0 = 5$	$0 - 5y = 5$
$x = 5$	$-5y = 5$
	$y = -1$
(5, 0)	(0, −1)

At the right is the graph of the line, with an x-intercept at $(5, 0)$ and a y-intercept at $(0, -1)$. Notice that the line is a dashed line.

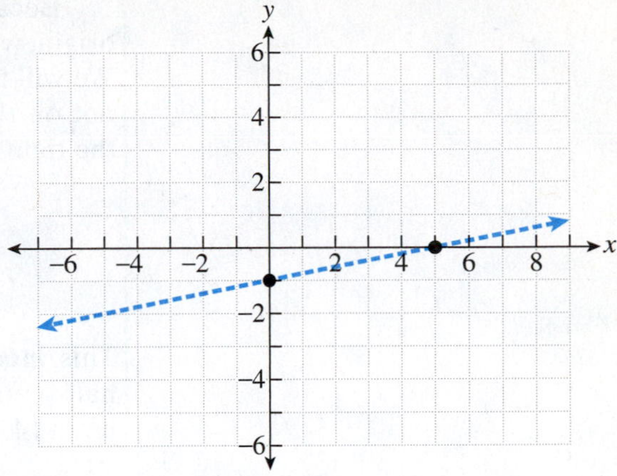

Because the line does not pass through the origin, we will use $(0, 0)$ as a test point.

Test Point: $(0, 0)$

$$0 - 5(0) < 5$$
$$0 - 0 < 5$$
$$0 < 5$$

This inequality is true, so the origin is a solution. We shade the half-plane containing the origin.

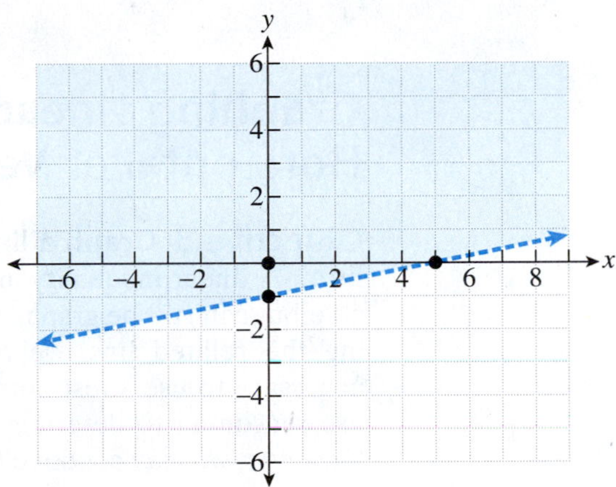

Quick Check 2

Graph $x + 2y < 6$.

EXAMPLE 3 Graph $y > 3x$.

SOLUTION This is a strict inequality, so we begin by graphing the line $y = 3x$ as a dashed line.

The equation is in slope–intercept form, with a slope of 3 and a y-intercept at $(0, 0)$. After plotting the y-intercept, we can use the slope to find a second point at $(1, 3)$. The graph of the line is shown below.

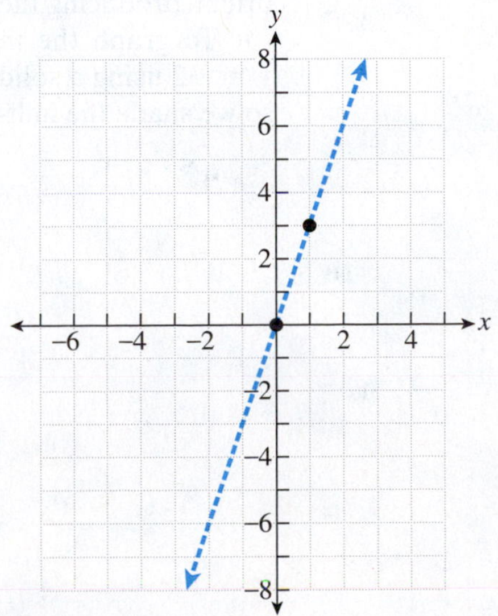

Because the line passes through the origin, we cannot use $(0, 0)$ as a test point. We will try to choose a point that is clearly not on the line, such as $(4, 0)$, which is to the right of the line.

Test Point: $(4, 0)$

$0 > 3(4)$

$0 > 12$

This inequality is false, so we shade the half-plane that does not contain the test point $(4, 0)$. At the right is the graph.

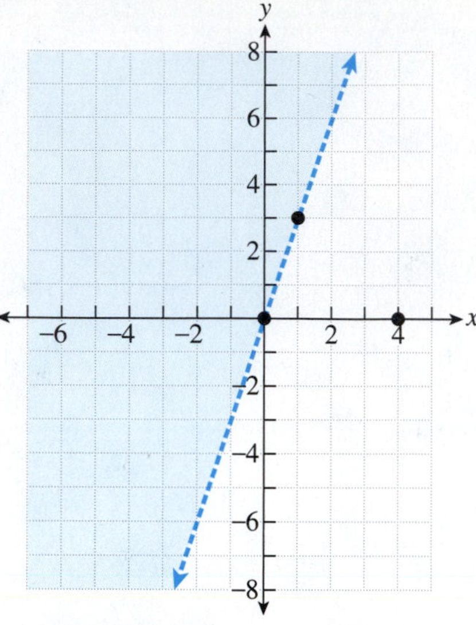

Quick Check 3

Graph $y \leq -\frac{4}{3}x$.

Graphing Linear Inequalities Involving Horizontal or Vertical Lines

Objective 3 Graph a linear inequality involving a horizontal or vertical line. A linear inequality involving a horizontal or vertical line has only one variable but can still be graphed on a plane rather than on a number line. After graphing the related line, we find that it is not necessary to use a test point. Instead, we can use reasoning to determine where to shade. However, we may continue to use test points if we choose.

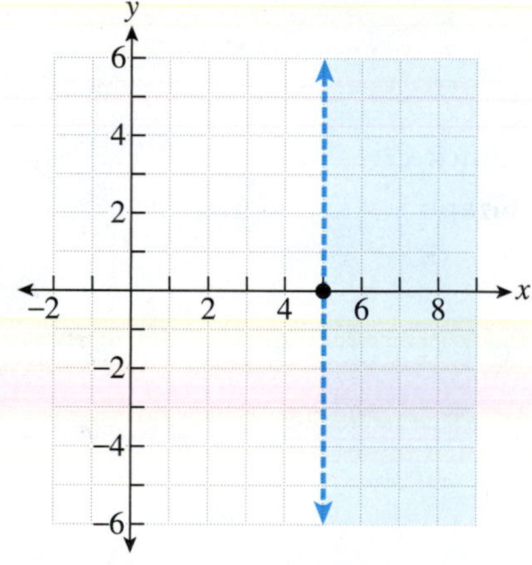

To graph the inequality $x > 5$, we begin by graphing the vertical line $x = 5$ using a dashed line. The values of x that are greater than 5 are to the right of this line, so we shade the half-plane to the right of $x = 5$. If we had used the origin as a test point, the resulting inequality $(0 > 5)$ would be false; so we would shade the half-plane that does not contain the origin, producing the same graph.

To graph the inequality $y \leq -2$, we begin by graphing the horizontal line $y = -2$ using a solid line. The values of y that are less than -2 are below this line, so we shade the half-plane below the line.

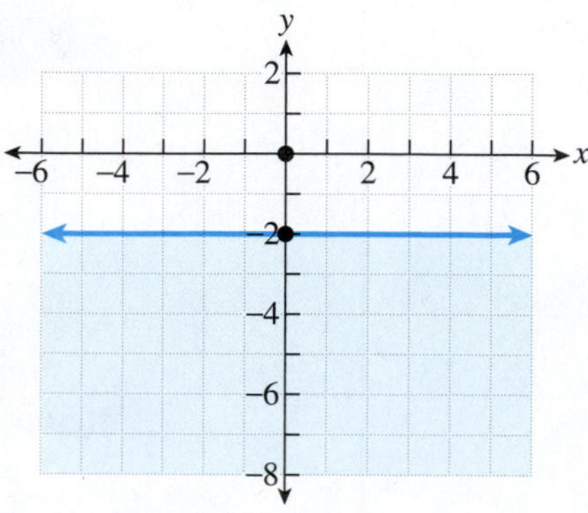

Here is a brief summary of how to graph linear inequalities in two variables.

Graphing Linear Inequalities in Two Variables

- Graph the line related to the inequality, which is found by replacing the inequality symbol with an equal sign.
- If the inequality symbol includes equality ($\leq$ or $\geq$), graph the line as a solid line.
- If the inequality symbol does not include equality ($<$ or $>$), graph the line as a dashed, or broken, line.
- Select a test point that is not on the line. Use the origin if possible. Substitute the coordinates of the point into the original inequality. If the resulting inequality is true, shade the region on the side of the line that contains the test point. If the resulting inequality is false, shade the region on the side of the line that does not contain the test point.

Applications

Objective 4 Graph linear inequalities associated with applied problems. We turn our attention to an application problem to end the section.

EXAMPLE 4 A movie theater has 120 seats. The number of adults and children admitted cannot exceed the number of seats. Set up and graph the appropriate inequality.

SOLUTION There are two unknowns in this problem: the number of adults and the number of children. Let x represent the number of adults and y represent the number of children. Because the total number of adults and children cannot exceed 120, we know that $x + y \leq 120$. (We also know that $x \geq 0$ and $y \geq 0$ because we cannot have a negative number of children or adults. Therefore, we are restricted to the first quadrant, the positive x-axis and the positive y-axis.)

To graph this inequality, begin by graphing $x + y = 120$ as a solid line. Because this equation is in standard form, graph the line using its x-intercept $(120, 0)$ and its y-intercept $(0, 120)$.

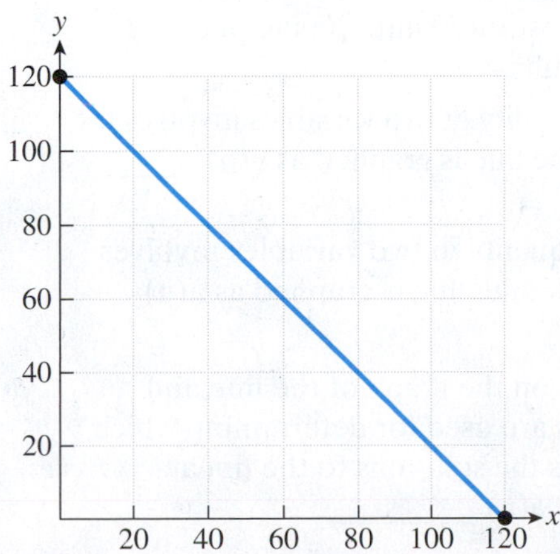

The origin is not on this line, so we can use $(0, 0)$ as a test point.

$$x + y \leq 120$$

$$0 + 0 \leq 120 \quad \text{Substitute 0 for } x \text{ and 0 for } y.$$

$$0 \leq 120 \quad \text{True.}$$

Quick Check 4

To make a fruit salad, Irv needs a total of at least 60 pieces of fruit. If Irv decides to buy only apples and pears, set up and graph the appropriate inequality.

This is a true statement, so we shade on the side of the line that contains the origin.

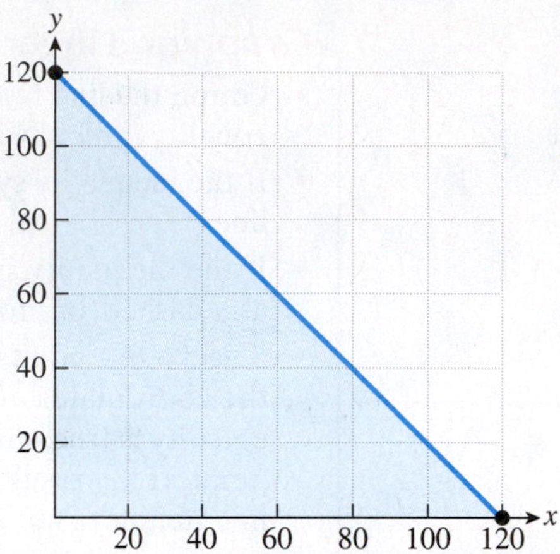

BUILDING YOUR STUDY STRATEGY

Using Your Resources, 7 MyMathLab This textbook has an online resource called MyMathLab.com. At this site, you can access video clips of lectures for each section in the book. These are useful for topics you are struggling with or for catching up if you missed a class. You also can find practice tutorial exercises that provide feedback. At this website, you also have access to sample tests, the Student's Solutions Manual, and other supplementary information.

Exercises 3.7

Vocabulary

1. A(n) ________ to a linear inequality in two variables is an ordered pair (x, y) such that when the coordinates are substituted into the inequality, a true statement results.
2. When a linear inequality in two variables involves the symbols $\le$ or $\ge$, the line is graphed as a(n) _____ line.
3. When a linear inequality in two variables involves the symbols $<$ or $>$, the line is graphed as a(n) _______ line.
4. A point that is not on the graph of the line and whose coordinates are used for determining which half-plane contains the solutions to the linear inequality is called a(n) ________.

Determine whether the ordered pair is a solution to the given linear inequality.

5. $5x + 3y \le 22$
   a) $(0, 0)$ **b)** $(8, -4)$
   c) $(2, 4)$ **d)** $(-3, 9)$

6. $2x - 4y > 6$
   a) $(0, 0)$ **b)** $(2, -3)$
   c) $(7, 2)$ **d)** $(5, 1)$

7. $y < 6x - 11$
   a) $(5, 8)$ **b)** $(0, 0)$
   c) $(2, 1)$ **d)** $(-4, -13)$

8. $x - 7y \ge -3$
   a) $(4, 1)$ **b)** $(9, 2)$
   c) $(0, 0)$ **d)** $(-16, -2)$

9. $y < 7$
   a) $(0, 0)$ **b)** $(8, 6)$
   c) $(-3, 10)$ **d)** $(6, 7)$

10. $x \le -2$
    a) $(0, 0)$ **b)** $(-1, 3)$
    c) $(-2, 19)$ **d)** $(-5, -4)$

Complete the solution of the linear inequality by shading the appropriate region.

11. $4x + y \geq 7$

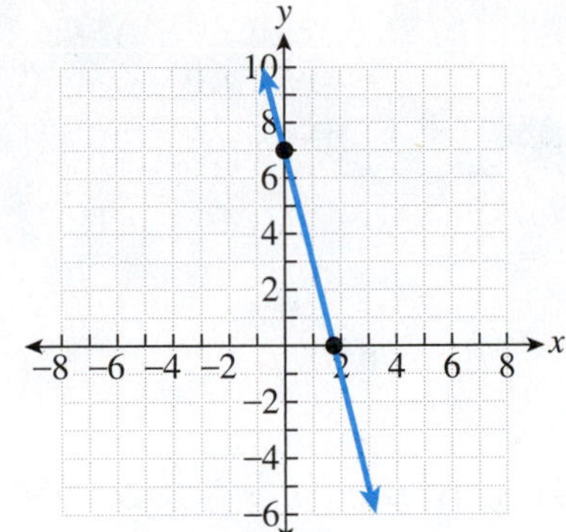

12. $-3x + 2y \leq -6$

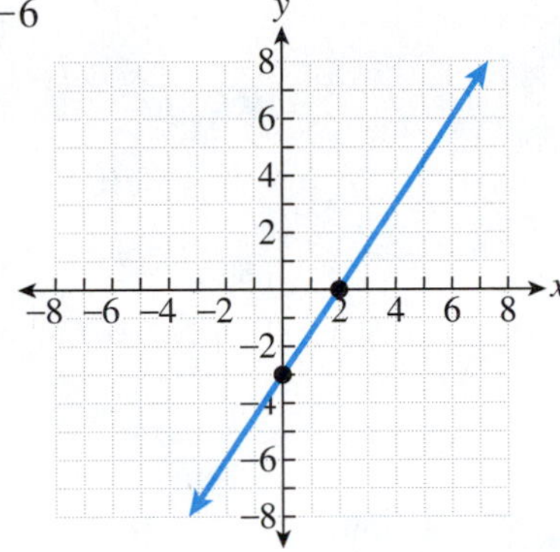

13. $2x + 8y < -4$

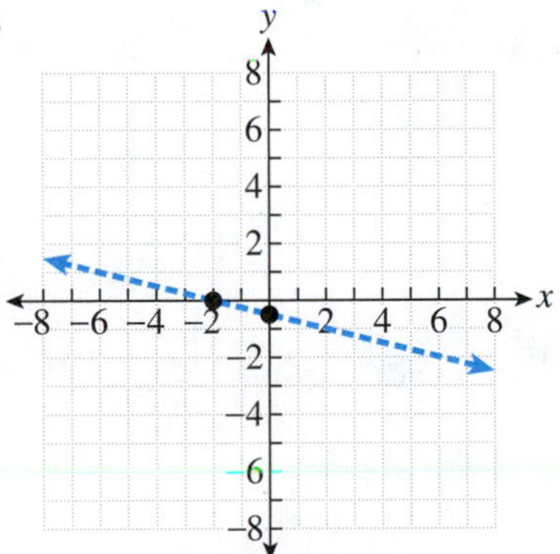

14. $10x - 2y > 0$

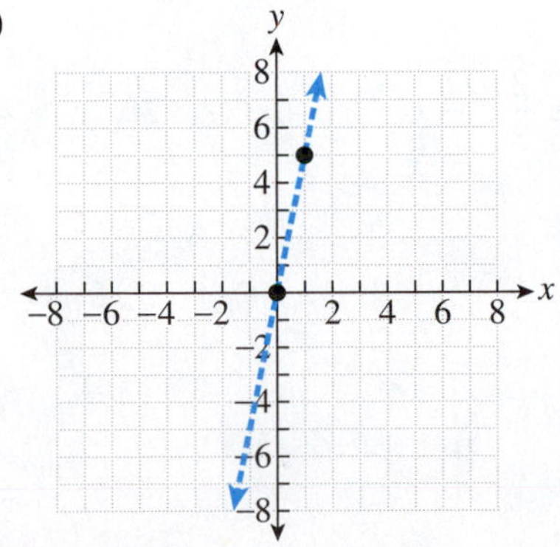

Which graph, A or B, represents the solution to the linear inequality?

15. $3x + 4y \geq 24$

A)

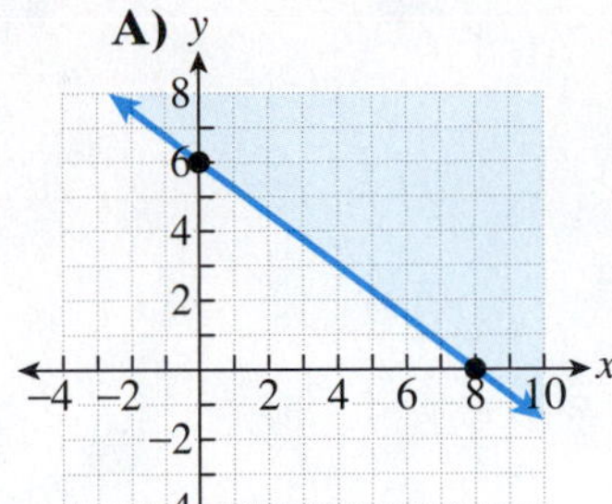

B)

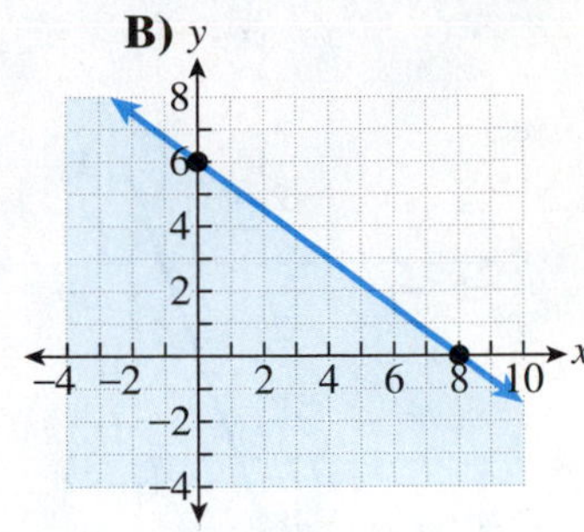

16. $y \leq \frac{1}{8}x$

A)

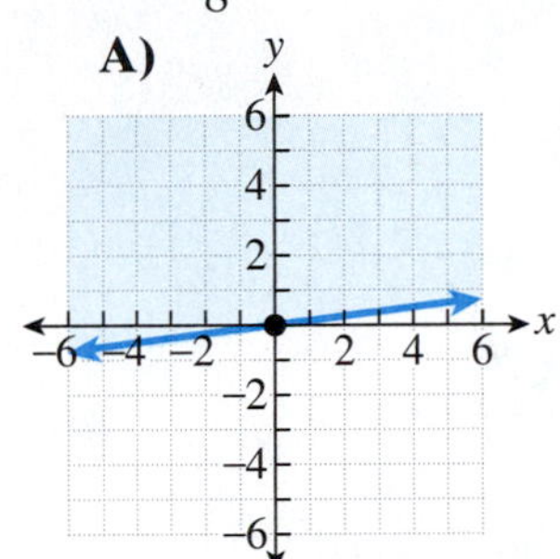

B)

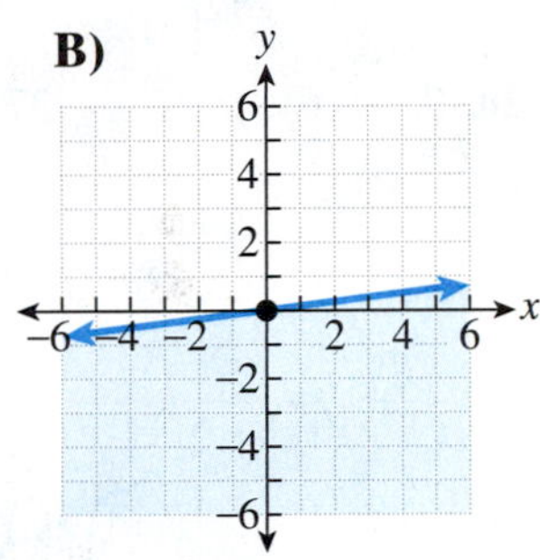

17. $3x - 2y < -18$

A)

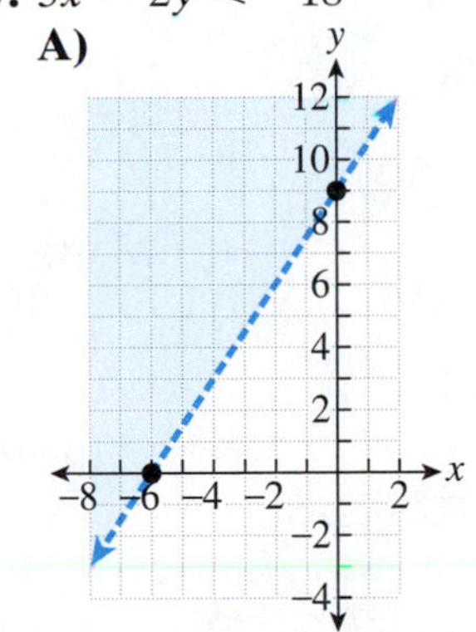

B)

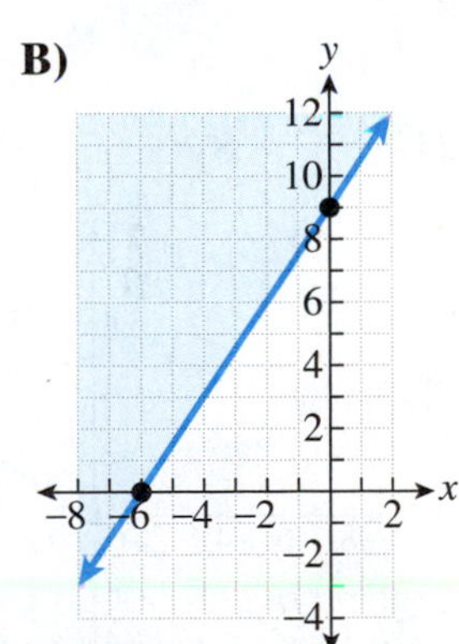

18. $5x - 4y > 20$

A)

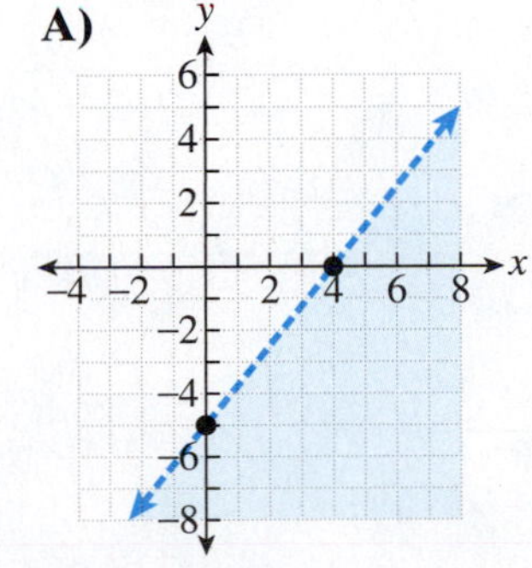

B)

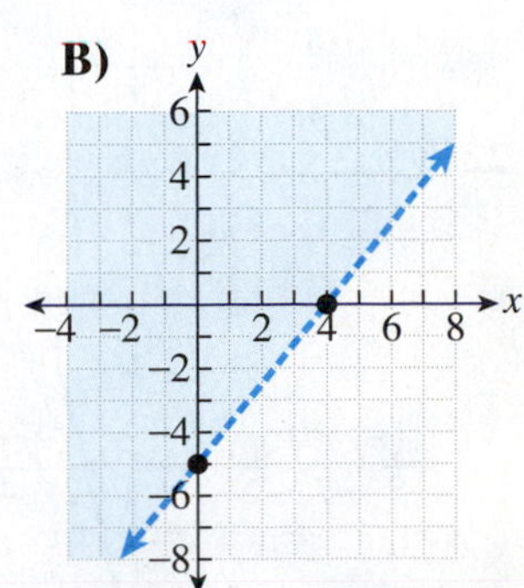

Determine the missing inequality sign (<, >, ≤, or ≥) for the linear inequality based on the given graph.

19. $8x - 3y$ ______ 24

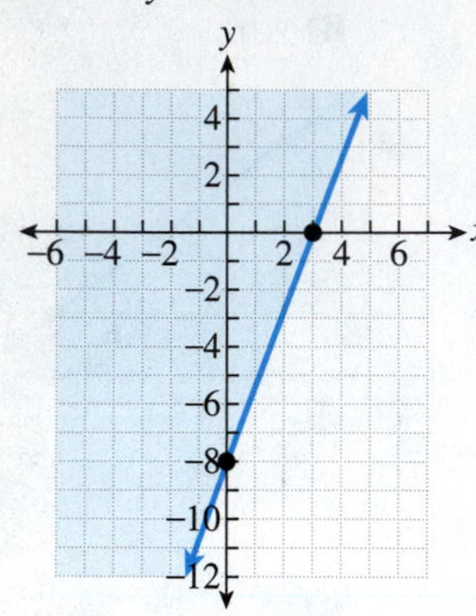

20. $3x + 6y$ ______ -27

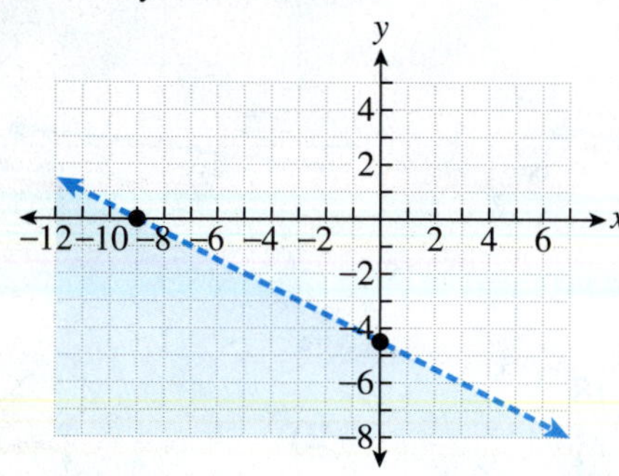

21. $-5x + 8y$ ______ 0

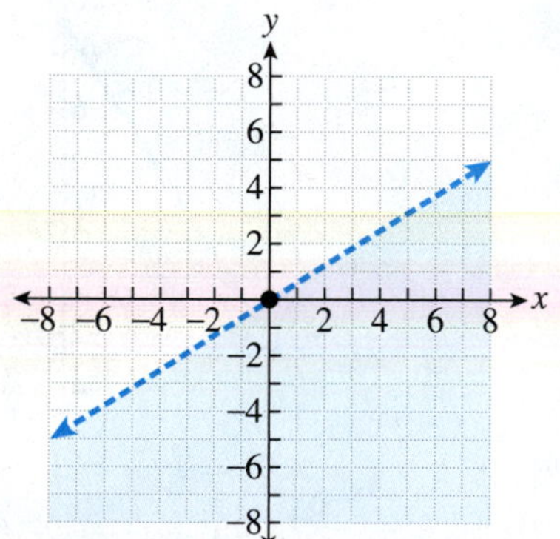

22. y ______ $4x - 6$

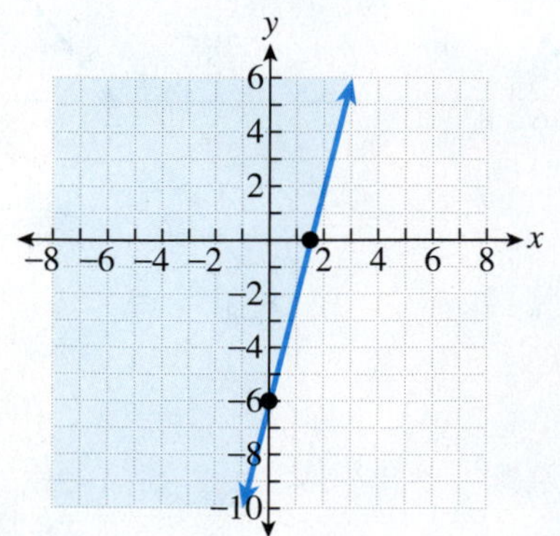

Graph the linear inequality.

23. $6x - 5y < -30$

24. $2x + 4y \leq 16$

25. $-9x + 4y \leq 0$

26. $y < -\frac{1}{4}x$

27. $y > 3$

28. $x \geq 5$

29. $y \geq \frac{2}{3}x - 6$

30. $y > 4x - 8$

31. $y \leq -2x - 4$

32. $y < -5$

33. $y > -\frac{3}{2}x - \frac{3}{2}$

34. $y \geq \frac{3}{5}x$

35. $x \leq -2$

36. $y < x$

Determine the linear inequality associated with the given solution. (Find the equation of the line. Then rewrite the equation as an inequality with the appropriate inequality sign: <, >, ≤, or ≥.)

37.

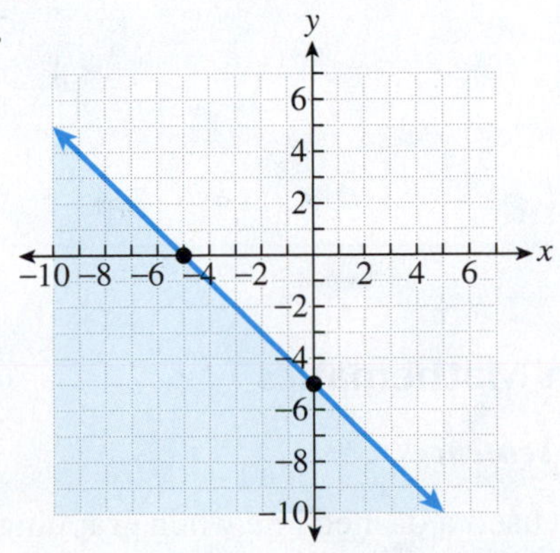

38.

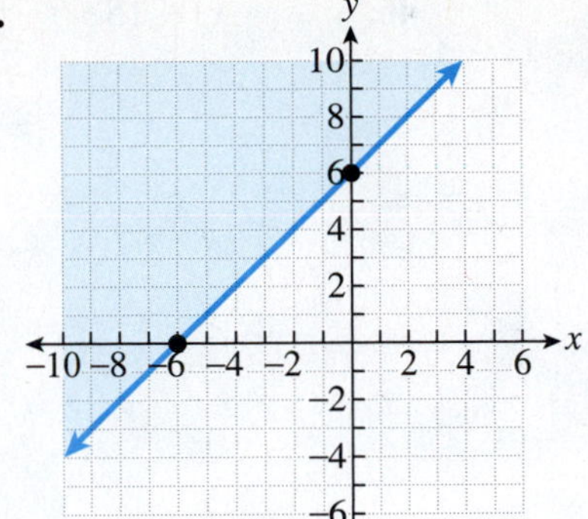

39.

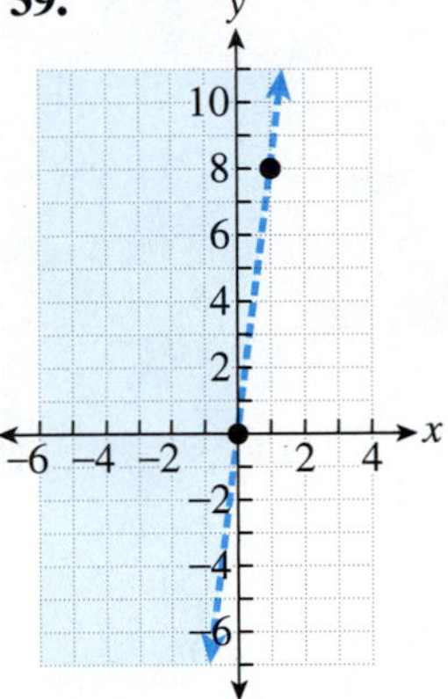

40.

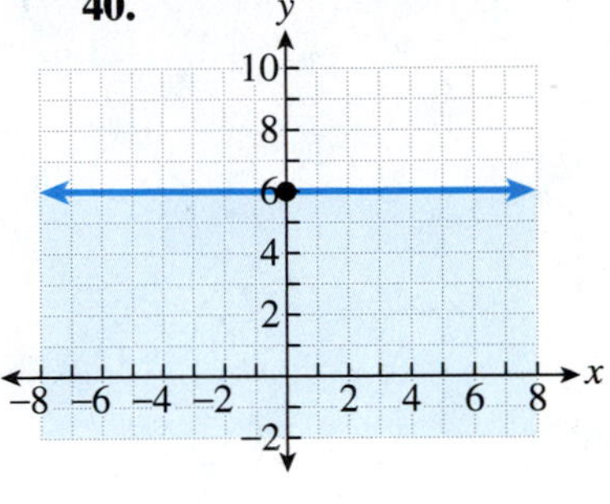

Mixed Practice, 41–50

Graph.

41. $-x + 3y \geq 9$

42. $y = \frac{3}{4}x + 5$

43. $4x - 3y = 24$

44. $x = -8$

45. $5x + 4y = 0$

46. $y \geq -x + 15$

47. $y = -3x + 8$

48. $7x - 2y > -14$

49. $y > 2x - 30$

50. $x + 3y = -9$

51. More than 50 people attended a charity basketball game. Some of the people were faculty members, and the rest were students. Set up and graph an inequality involving the number of faculty members and the number of students in attendance.

52. An elevator has a warning posted inside the car that the maximum capacity is 1200 pounds. Suppose the average weight of an adult is 160 pounds and the average weight of a child is 60 pounds.

a) Set up and graph an inequality involving the number of adults and the number of children that can safely ride the elevator.

b) Looking at the graph from part a, can two adults and 16 children ride the elevator at the same time?

Find the region that contains ordered pairs that are solutions to both inequalities. Graph each inequality separately; then shade the region that the two graphs have in common.

53. $x + y < 3$ and
$y > x - 7$

54. $2x + 3y \leq 18$ and
$2x + y \geq 8$

Writing in Mathematics

Answer in complete sentences.

55. Explain why we used a dashed line when graphing inequalities involving the symbols $<$ and $>$ and a solid line when graphing inequalities involving the symbols $\leq$ and $\geq$.

CHAPTER 3 SUMMARY

Section 3.1 The Rectangular Coordinate System; Equations in Two Variables

Ordered Pairs, pp. 118–119

An ordered pair, (x, y), is a pair of values listed in the specific order of x first and y second. The values are often referred to as coordinates.

$(3, 4)$

$(0, -8)$

Solutions of an Equation in Two Variables, pp. 118–119

An ordered pair (x, y) is a solution of an equation in two variables if it produces a true equation when its coordinates are substituted for x and y.

Is $(4, -3)$ a solution of $2x + 3y = -1$?

$$2(4) + 3(-3) = -1$$
$$8 - 9 = -1$$
$$-1 = -1$$

$(4, -3)$ is a solution.

Plotting Ordered Pairs, pp. 119–122

To plot an ordered pair (x, y) on a rectangular coordinate plane, begin at the origin. The x-coordinate tells how far to the left or right of the origin to move. The y-coordinate then tells how far up or down to move from there.

Plot $(3, 5)$, $(6, -2)$, $(-4, 7)$, and $(0, -8)$ on a rectangular plane.

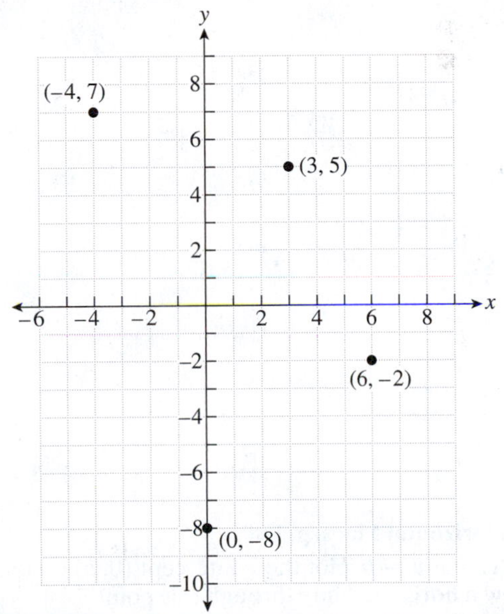

Graphing a Linear Equation by Plotting Points, p. 124

Plot ordered pairs that are solutions and draw the straight line that goes through them.

Arbitrarily select three values for x and find the y-coordinates associated with them to find the ordered pairs to plot.

Graph $y = 3x - 2$.

$x = 0$	$y = 3(0) - 2 = -2$	$(0, -2)$
$x = 1$	$y = 3(1) - 2 = 1$	$(1, 1)$
$x = 2$	$y = 3(2) - 2 = 4$	$(2, 4)$

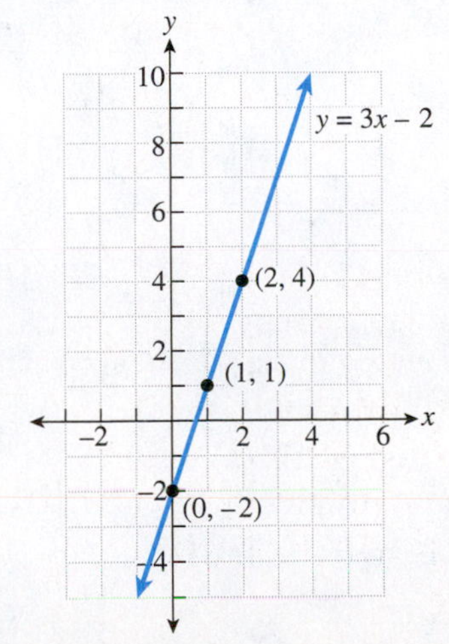

Section 3.2 Graphing Linear Equations and Their Intercepts

Intercepts, pp. 129–130

A point at which a graph crosses the x-axis is called an x-intercept, and a point at which a graph crosses the y-axis is called a y-intercept.
To find an x-intercept, substitute 0 for y and solve for x.
To find a y-intercept, substitute 0 for x and solve for y.

Find the intercepts: $3x - 4y = -24$

x-intercept	y-intercept
$3x - 4(0) = -24$ $3x = -24$ $x = -8$	$3(0) - 4y = -24$ $-4y = -24$ $y = 6$
$(-8, 0)$	$(0, 6)$

Graphing a Linear Equation Using Its Intercepts, pp. 130–133

Find the x-intercept and the y-intercept and plot them on the plane.
A third point with an arbitrarily chosen value of x can be used as a check point.

Graph: $x + 2y = 8$

x-intercept	y-intercept
$x + 2(0) = 8$ $x = 8$	$(0) + 2y = 8$ $2y = 8$ $y = 4$
$(8, 0)$	$(0, 4)$

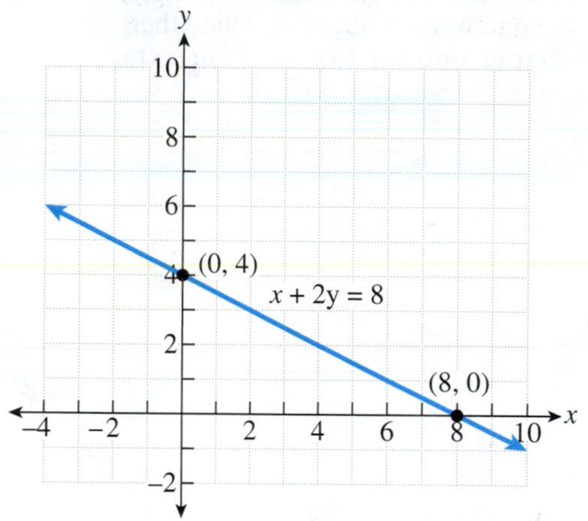

Graphing Horizontal Lines, p. 134

To graph the line $y = b$, plot the y-intercept $(0, b)$ on the y-axis and draw a horizontal line through this point.

Graph: $y = 7$

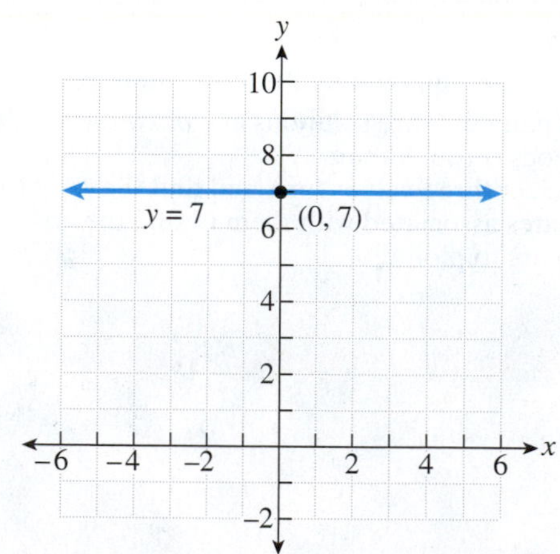

Graphing Vertical Lines, pp. 134–135

To graph the line $x = a$, plot the x-intercept $(a, 0)$ on the x-axis and draw a vertical line through this point.

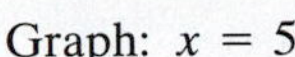

Graph: $x = 5$

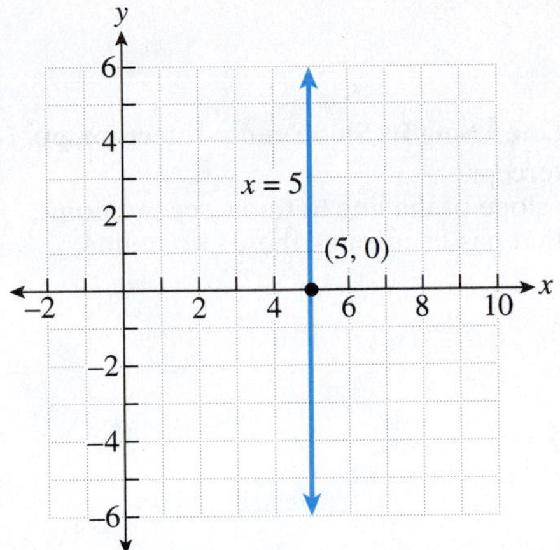

Section 3.3 Slope of a Line

Slope, pp. 140–142

The slope of a line, m, is a measure of how steeply a line rises or falls as it moves to the right.

If a line rises as it moves to the right, its slope is positive.

If a line falls as it moves to the right, its slope is negative.

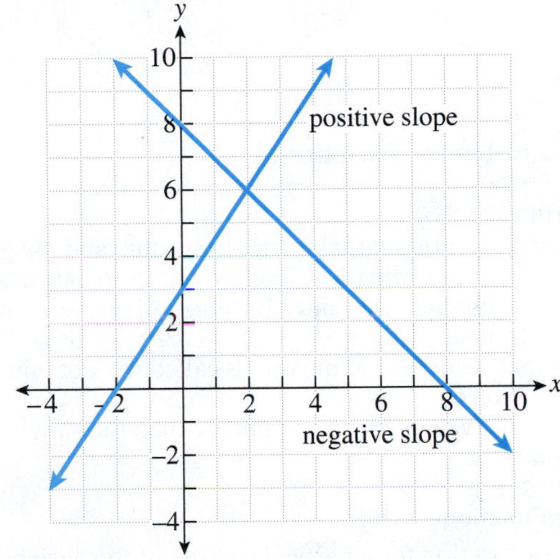

Slope Formula, pp. 142–143

If a line passes through two points (x_1, y_1) and (x_2, y_2), you can calculate its slope using the formula $m = \dfrac{y_2 - y_1}{x_2 - x_1}$.

Find the slope of the line that passes through the two points: $(4, -7), (6, 3)$

$$m = \frac{3 - (-7)}{6 - 4} = \frac{10}{2} = 5$$

Slope–Intercept Form of a Line, pp. 145–146

The equation of a line is in slope–intercept form if it is written in the form $y = mx + b$. The coefficient of x is the slope m, while b represents the y-coordinate of the y-intercept.

Find the slope and y-intercept of the line: $4x + 5y = 15$

$$4x + 5y = 15$$
$$5y = -4x + 15$$
$$\frac{5y}{5} = \frac{-4x}{5} + \frac{15}{5}$$
$$y = -\frac{4}{5}x + 3$$

The slope is $-\frac{4}{5}$, and the y-intercept is $(0, 3)$.

Slope of a Horizontal or Vertical Line, pp. 144–145

The slope of a horizontal line is 0. The slope of a vertical line is undefined.

Find the slope of the line: $y = -2$
The line is horizontal, so $m = 0$.

Find the slope of the line: $x = 6$
The line is vertical, so its slope is undefined.

Graphing a Line Using Its Slope and y-Intercept, pp. 146–147

Plot the y-intercept.
Then use the slope of the line to find a second point.
Draw a line that passes through these two points.

Graph: $y = 4x + 2$

y-intercept: $(0, 2)$
Slope: 4 (up 4 units, 1 to the right)

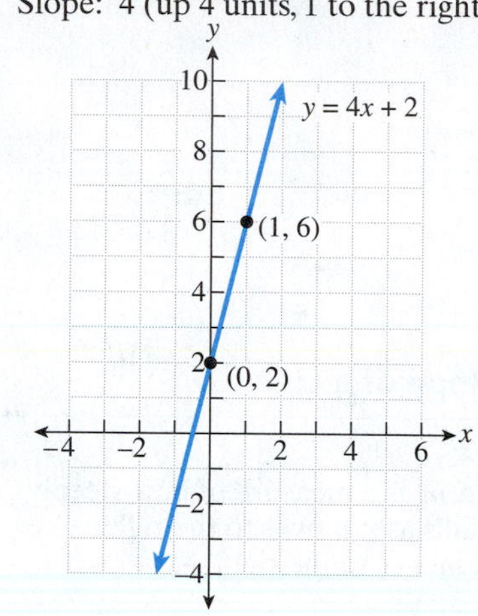

Section 3.4 Linear Functions

Functions, pp. 153–155

A function is a relation that takes an input value and assigns a particular output value to it. For a relation to define a function, each input value must be assigned one and only one output value.
The set of input values for a function is called the domain of the function.
The set of output values for a function is called the range of the function.

Is a relation whose input value is a student at your college and whose output is the student's ID number a function?

Yes, because each student has only one ID number.

Evaluating Functions, p. 155

To evaluate a function for a particular value of the variable, substitute that value for the variable in the function's formula, then simplify the resulting expression.

For $f(x) = 7x + 23$, find $f(-11)$.

$$\begin{aligned} f(-11) &= 7(-11) + 23 \\ &= -77 + 23 \\ &= -54 \end{aligned}$$

Graphing Linear Functions, p. 156

A linear function is a function of the form $f(x) = mx + b$, where m and b are real numbers.
To graph a linear function, begin by plotting its y-intercept at $(0, b)$. Then use the slope m to find other points.

Graph: $f(x) = -\frac{2}{3}x + 5$

y-intercept: $(0, 5)$
Slope: $-\frac{2}{3}$ (down 2 units, 3 to the right)

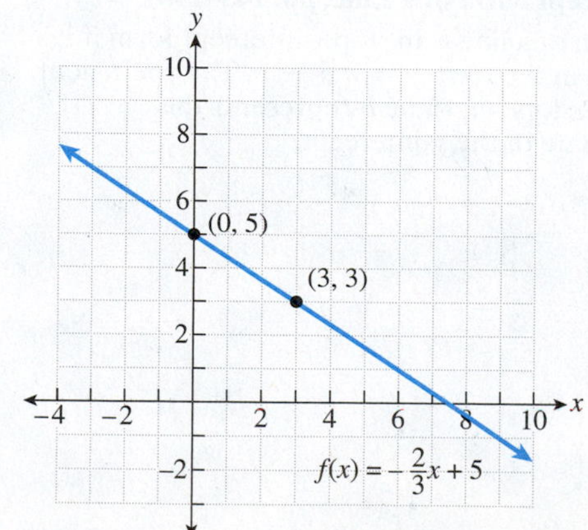

Constant Functions, p. 156

A function of the form $f(x) = b$ is called a constant function. The graph of a constant function is a horizontal line whose y-intercept is the point $(0, b)$.

Graph: $f(x) = -6$

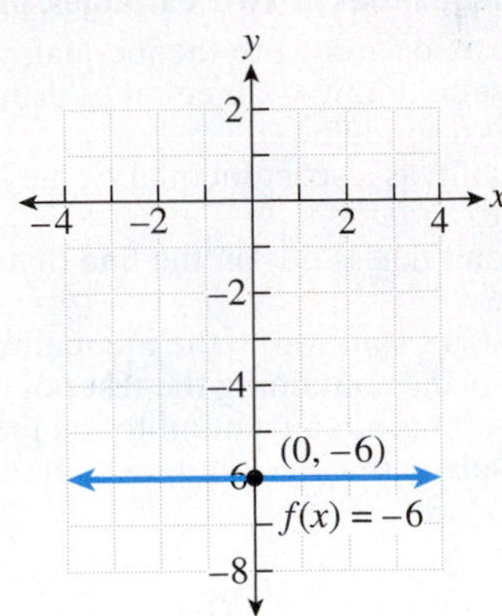

Section 3.5 Parallel and Perpendicular Lines

Parallel Lines, pp. 162–163

Two nonvertical lines are parallel if they have the same slope.

If two lines are vertical, they are parallel.

Are the two lines parallel?

$y = -4x + 7, 8x + 2y = 10$

The slope of the first line is -4.

$$\begin{aligned} 8x + 2y &= 10 \\ 2y &= -8x + 10 \\ y &= -4x + 5 \end{aligned}$$

The slope of the second line also is -4.

The lines are parallel.

Perpendicular Lines, pp. 163–164

Two nonvertical lines are perpendicular if their slopes are negative reciprocals.

A vertical line is perpendicular to a horizontal line.

Are the two lines perpendicular?

$y = -\frac{2}{5}x + 3, -5x + 2y = 24$

The slope of the first line is $-\frac{2}{5}$.

$$\begin{aligned} -5x + 2y &= 24 \\ 2y &= 5x + 24 \\ y &= \frac{5}{2}x + 12 \end{aligned}$$

The slope of the second line is $\frac{5}{2}$.

The slopes are negative reciprocals, so the lines are perpendicular.

Section 3.6 Equations of Lines

Point–Slope Form of a Line, pp. 168–169

The point–slope form of the equation of a line with slope m that passes through the point (x_1, y_1) is $y - y_1 = m(x - x_1)$.

This form of a line is used to find the equation of a line if its slope and the coordinates of a point on the line are known.

Find the equation of a line whose slope is 3 that passes through the point $(2, -17)$.

$$\begin{aligned} y - y_1 &= m(x - x_1) \\ y - (-17) &= 3(x - 2) \\ y + 17 &= 3x - 6 \\ y &= 3x - 23 \end{aligned}$$

Finding the Equation of a Line Given Two Points on the Line, p. 170

Begin by finding the slope of the line passing through the two points using the slope formula $m = \frac{y_2 - y_1}{x_2 - x_1}$.

Then use the point–slope form with this slope and either of the two points given.

Finish by rewriting the equation in slope–intercept form.

Find the equation of a line that passes through the points $(3, 4)$ and $(5, 12)$.

$$\begin{aligned} m &= \frac{12 - 4}{5 - 3} \\ &= \frac{8}{2} \\ &= 4 \\ y - y_1 &= m(x - x_1) \\ y - 4 &= 4(x - 3) \\ y - 4 &= 4x - 12 \\ y &= 4x - 8 \end{aligned}$$

Section 3.7 Linear Inequalities

Graphing Linear Inequalities in Two Variables, pp. 177–181

1. Graph the line associated with the inequality.
   If the inequality is a weak inequality, using $\leq$ or $\geq$, the line is a solid line.
   If the inequality is a strict inequality, the line is a dashed line.
2. Select a test point that is not on the line that has been graphed.
3. If the test point is a solution to the inequality, shade the half-plane containing the test point.
   If the test point is not a solution to the inequality, shade the half-plane that does not contain the test point.

Graph: $y \geq 2x - 9$

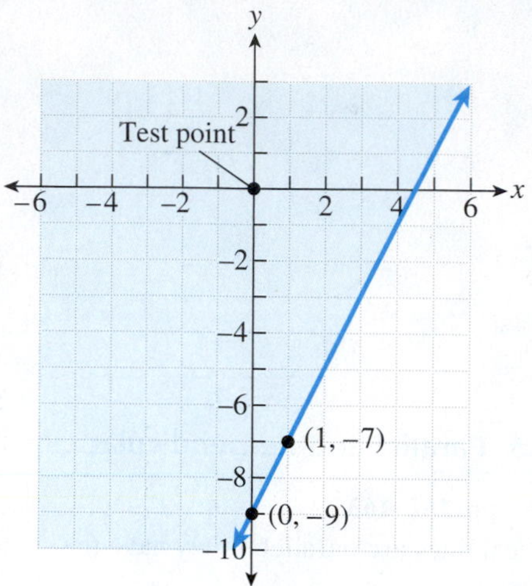

SUMMARY OF CHAPTER 3 STUDY STRATEGIES

You have a great number of resources at your disposal to help you learn mathematics, and your success may depend on how well you use them.

- Your instructor is a valuable resource. He or she can answer your questions and provide advice. In addition to being available during class and office hours, some instructors take questions from students before class begins or after class ends.
- The tutorial center is another valuable resource. Keep in mind that your goals are to understand the material and to solve the problems yourself.
- Many colleges offer short courses or seminars in study skills. Ask an academic counselor whether these courses are available and whether you could benefit from them.
- Other resources discussed in this chapter that you may want to use are the Student's Solutions Manual, your classmates, and the Internet.
- Finally, the online supplement to this textbook at MyMathLab.com can be helpful. This site provides video clips, tutorial exercises, and more.

CHAPTER 3 REVIEW

Find the coordinates of the labeled points A, B, C, and D. **[3.1]**

1.

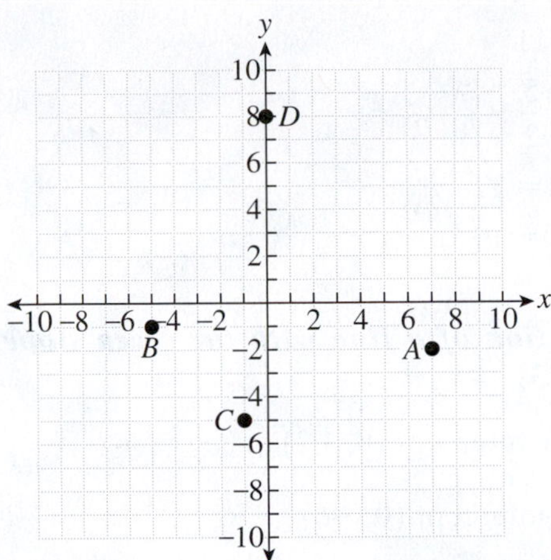

In which quadrant (I, II, III, or IV) is the given point located? **[3.1]**

2. $(-3, -3)$

3. $(-9, 4)$

Is the ordered pair a solution to the given equation? **[3.1]**

4. $(5, -2)$, $4x - 2y = 16$

5. $\left(\frac{7}{2}, \frac{3}{5}\right)$, $4x + 5y = 17$

6. $(-6, -4)$, $x - 3y = 6$

Find the x-intercept and y-intercept if possible. **[3.2]**

7. $2x - 8y = -40$

8. $5x - y = -15$

9. $3x - 7y = 0$

10. $4x - 7y = 14$

Find the x- and y-intercepts and use them to graph the equation. **[3.2]**

11. $-3x + 2y = 12$

12. $x - 4y = -8$

13. $2x + 3y = 0$

14. $4x + 3y = -6$

15. $y = -\frac{3}{2}x + 6$

16. $y = 2x - 4$

Find the slope of the line that passes through the given points. **[3.3]**

17. $(-5, 3)$ and $(-3, 9)$

18. $(6, -2)$ and $(3, 10)$

19. $(3, -8)$ and $(8, -3)$

20. $(2, -7)$ and $(7, -7)$

Worked-out solutions to Review Exercises marked with ◯ can be found on page AN-11.

***Find the slope and the y-intercept of the given line.* [3.3]**

21. $y = -2x + 7$

22. $y = 4x - 6$

23. $y = -\frac{2}{3}x$

24. $4x - 2y = 9$

25. $5x + 3y = 18$

26. $y = -7$

***Graph using the slope and y-intercept.* [3.3]**

27. $y = 3x - 6$

28. $y = -5x - 5$

29. $y = \frac{2}{5}x + 2$

30. $y = x$

31. $y = -\frac{7}{3}x + 3$

32. $y = \frac{1}{2}x + \frac{3}{2}$

***Are the two given lines parallel, perpendicular, or neither?* [3.5]**

33. $y = 2x - 9$
$y = -2x + 9$

34. $4x + y = 11$
$y = \frac{1}{4}x + \frac{5}{2}$

35. $12x - 9y = 17$
$-8x + 6y = 10$

***Find the equation of a line with the given slope and y-intercept.* [3.3]**

36. Slope -4, y-intercept $(0, -2)$

37. Slope $\frac{2}{5}$, y-intercept $(0, -6)$

38. Monique's vacation fund has \$720 in it. She plans to add \$50 to this fund each month. [3.4]

a) Create a function $f(x)$ for the amount of money in Monique's vacation fund after x months.

b) Use the function from part a to determine how much money will be in the vacation fund after 11 months.

c) Use the function from part a to determine how many months it will take Monique to save enough money for a trip to Hawaii that will cost \$2750.

***Evaluate the given function.* [3.4]**

39. $f(x) = 9x + 7, f(-2)$

40. $g(x) = 3 - 8x, g(-5)$

41. $f(x) = 3x + 7, f(5a - 1)$

42. Consider the following graph of a function $f(x)$. [3.4]

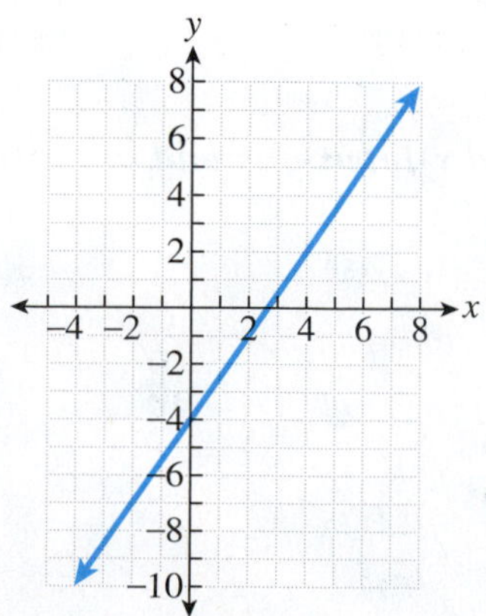

a) Find $f(-2)$.

b) Find all values x for which $f(x) = 5$.

c) Find the domain of $f(x)$.

d) Find the range of $f(x)$.

***Find the slope–intercept form of the equation of a line with the given slope that passes through the given point.* [3.6]**

43. Slope 1, through $(4, -2)$

44. Slope -5, through $(1, 3)$

45. Slope $-\frac{3}{2}$, through $(-4, 9)$

***Find the slope–intercept form of the equation of a line that passes through the two given points.* [3.6]**

46. $(-2, 1)$ and $(2, -7)$

47. $(-6, -5)$ and $(2, 7)$

48. $(5, 3)$ and $(-9, 3)$

***Find the equation of the line that has been graphed.* [3.6]**

49.

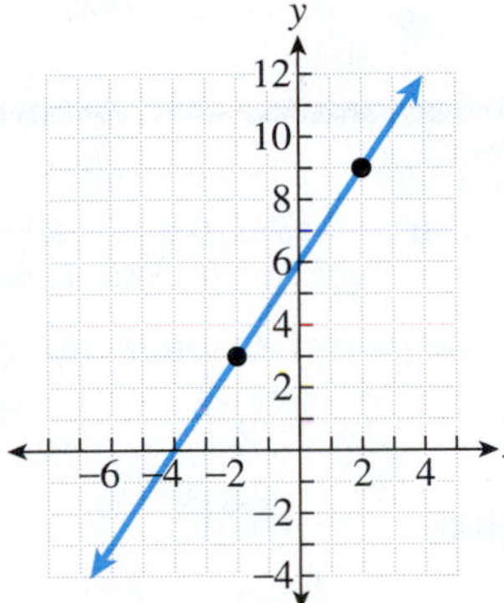

50.

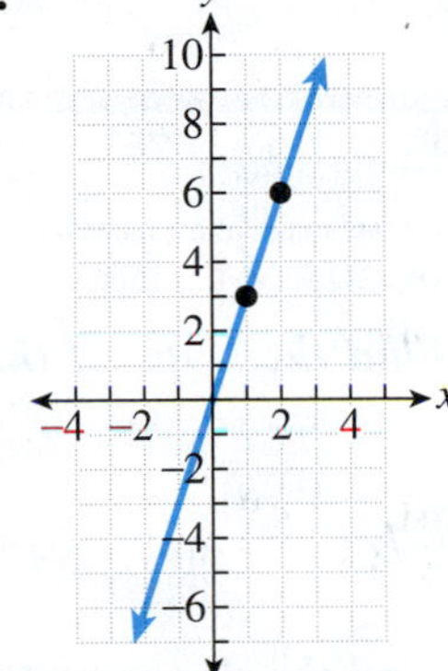

***Graph.* [3.2/3.3]**

51. $y = 4x + 8$

52. $y = -\frac{1}{2}x + 3$

53. $y = -5x + 2$

54. $x = 5$

55. $3x - 4y = 24$

56. $y = -6$

Graph the inequality on a plane. [3.7]

57. $3x + y < -9$

58. $2x + 7y \geq 14$

59. $y \leq -4$

60. $y > \frac{3}{2}x + 5$

CHAPTER 3 TEST

Step-by-step test solutions are found on the Chapter Test Prep Videos available via the Video Resources on DVD, in MyMathLab, and on YouTube (search "WoodburyElemIntAlg" and click on "Channels").

In which quadrant (I, II, III, or IV) is the given point located?

1. $(-3, 7)$

Is the ordered pair a solution to the given equation?

2. $(4, -2)$, $-3x - 7y = 2$

Find the x-intercept and y-intercept if possible.

3. $5x - 6y = -15$

4. $y = -\frac{7}{2}x + 21$

Find the x- and y-intercepts and use them to graph the equation.

5. $4x - y = 6$

6. $y = \frac{2}{3}x - 4$

Find the slope of the line that passes through the given points.

7. $(2, 8)$ and $(-3, 9)$

Find the slope and the y-intercept of the given line.

8. $y = 7x - 8$

9. $3x + 5y = 45$

Graph using the slope and y-intercept.

10. $y = -2x + 3$

11. $y = -\frac{6}{5}x + 5$

Are the two given lines parallel, perpendicular, or neither?

12. $4y = 8x - 11$
$2x - y = 44$

Find the equation of a line with the given slope and y-intercept.

13. Slope -6, y-intercept $(0, 5)$

Evaluate the given function.

14. $f(x) = 7 - 2x, f(-3)$

15. Find the slope–intercept form of the equation of a line with a slope of $-\frac{5}{2}$ that passes through the point $(-2, 9)$.

16. Find the slope–intercept form of the equation of a line that passes through the points $(-6, 4)$ and $(-1, -6)$.

Graph.

17. $y = -4x + 6$

18. $-4x + 5y = 20$

19. $y = 2$

20. Graph the inequality $x - 6y < -3$ on a plane.

Mathematicians in History

René Descartes was an early 17th-century philosopher who made important contributions to mathematics. Cartesian geometry resulted from his application of algebra to geometry. The rectangular coordinate plane also is called the Cartesian plane in his honor. Descartes once said, "Mathematics is a more powerful instrument of knowledge than any other that has been bequeathed to us by human agency."

Write a one-page summary (*or* make a poster) of the life of René Descartes and his accomplishments. Also look up Descartes' most famous quotes and list your favorite.

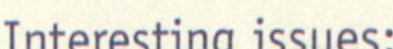

Interesting issues:

- Where and when was René Descartes born?
- How old was Descartes when he enrolled at the Jesuit College at La Flèche?
- Descartes's first major treatise on physics was *Le Monde, ou Traité de la Lumière.* Why did he choose not to publish his results?
- Descartes' most famous quote, in Latin, is "Cogito, ergo sum." What is the English translation of this quote?
- How did a fly help Descartes come up with the idea for the rectangular coordinate system?
- Queen Christina of Sweden invited Descartes to Sweden in 1649, where he died shortly thereafter. Describe the circumstances that led to his death.

CHAPTER 4

Systems of Equations

In this chapter, we will learn to set up and solve systems of equations. A system of equations is a set of two or more associated equations containing two or more variables. We will learn to solve systems of linear equations in two variables using three methods: graphing, the substitution method, and the addition method. We also will learn to set up systems of equations to solve applied problems. We will finish the chapter by learning to solve systems of linear inequalities in two variables.

STUDY STRATEGY

Doing Your Homework Doing a homework assignment should not be viewed as just another requirement. Homework exercises are assigned to help you learn mathematics. In this chapter, we will discuss how you should do your homework to get the most out of your effort.

4.1 Systems of Linear Equations; Solving Systems by Graphing

OBJECTIVES

1. **Determine whether an ordered pair is a solution of a system of equations.**
2. **Identify the solution of a system of linear equations from a graph.**
3. **Solve a system of linear equations graphically.**
4. **Identify systems of linear equations with no solution or infinitely many solutions.**

Systems of Linear Equations and Their Solutions

A **system of linear equations** consists of two or more linear equations. Here are some examples.

$$\begin{aligned} 5x + 4y &= 20 \\ 2x - y &= 6 \end{aligned} \qquad \begin{aligned} y &= \frac{3}{5}x \\ y &= -4x + 3 \end{aligned} \qquad \begin{aligned} x + y &= 11 \\ y &= 5 \end{aligned}$$

Objective 1 Determine whether an ordered pair is a solution of a system of equations. The equations in a system are examined together, and we are looking for the solution(s) that the equations have in common.

Solution of a System of Linear Equations

A **solution of a system of linear equations** is an ordered pair (x, y) that is a solution of each equation in the system.

EXAMPLE 1 Is the ordered pair $(2, -3)$ a solution of the given system of equations?

$$5x + y = 7$$
$$2x - 3y = 13$$

SOLUTION To determine whether the ordered pair $(2, -3)$ is a solution, substitute 2 for x and -3 for y in each equation. If the resulting equations are true, the ordered pair is a solution of the system. Otherwise, it is not.

$$\begin{aligned} 5x + y &= 7 \\ 5(2) + (-3) &= 7 \\ 10 - 3 &= 7 \\ 7 &= 7 \end{aligned} \qquad \begin{aligned} 2x - 3y &= 13 \\ 2(2) - 3(-3) &= 13 \\ 4 + 9 &= 13 \\ 13 &= 13 \end{aligned}$$

Because $(2, -3)$ is a solution of each equation, it is a solution of the system of equations.

EXAMPLE 2 Is the ordered pair $(-5, -1)$ a solution of the given system of equations?

$$y = 2x + 9$$
$$y = 14 - 3x$$

SOLUTION Again, begin by substituting -5 for x and -1 for y in each equation.

$$\begin{aligned} y &= 2x + 9 \\ (-1) &= 2(-5) + 9 \\ -1 &= -10 + 9 \\ -1 &= -1 \end{aligned} \qquad \begin{aligned} y &= 14 - 3x \\ (-1) &= 14 - 3(-5) \\ -1 &= 14 + 15 \\ -1 &= 29 \end{aligned}$$

Although $(-5, -1)$ is a solution of the equation $y = 2x + 9$, it is not a solution of the equation $y = 14 - 3x$. Therefore, $(-5, -1)$ is not a solution of the system of equations.

Quick Check 1

Is the ordered pair $(-6, 3)$ a solution of the given system of equations?

$$5x - 2y = -36$$
$$-x + 7y = 27$$

Objective 2 Identify the solution of a system of linear equations from a graph. Systems of two linear equations can be solved graphically. Because an ordered pair that is a solution of a system of two equations must be a solution of each equation, we know that this ordered pair is on the graph of each equation. To solve a system of two linear equations, we graph each line and look for points of intersection. Often this will lead to a single solution. Here are some graphical examples of systems of equations that have the ordered pair $(3, 1)$ as a solution.

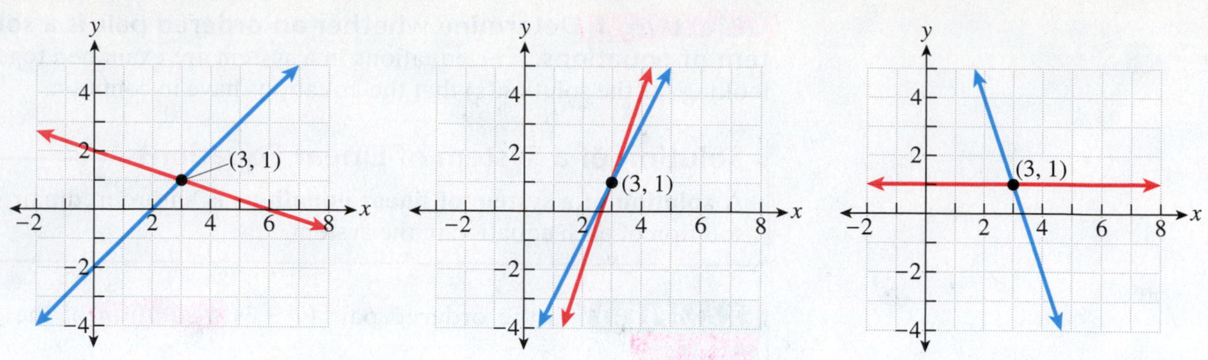

EXAMPLE 3 Find the solution of the system of equations plotted on the graph.

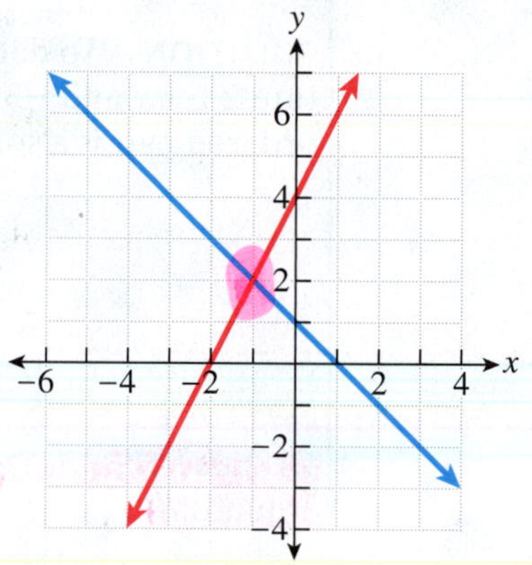

SOLUTION The two lines intersect at the point $(-1, 2)$, so the solution of the system of equations is $(-1, 2)$.

Quick Check 2

Find the solution of the system of equations plotted on the graph.

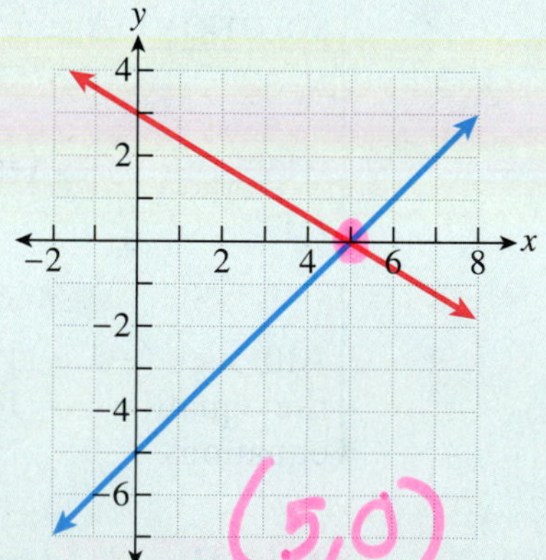

Solving Linear Systems of Equations by Graphing

Objective 3 Solve a system of linear equations graphically. Now we will solve a system of linear equations by graphing the lines and finding their point of intersection.

EXAMPLE 4 Solve the system by graphing.

$$y = 5x - 9$$
$$y = -\frac{7}{2}x + 8$$

SOLUTION Both equations are in slope–intercept form, so we will graph the lines by plotting the y-intercept and then using the slope of the line to find another point on the line.

The first line has a y-intercept at $(0, -9)$. From that point, we can find a second point by moving up 5 units and 1 unit to the right, because the slope is 5.

The y-intercept of the second line is $(0, 8)$. Because the slope of this line is $-\frac{7}{2}$, we can move down 7 units and 2 units to the right from this point to find a second point on the line. Both graphs are shown on the same set of axes at the right.

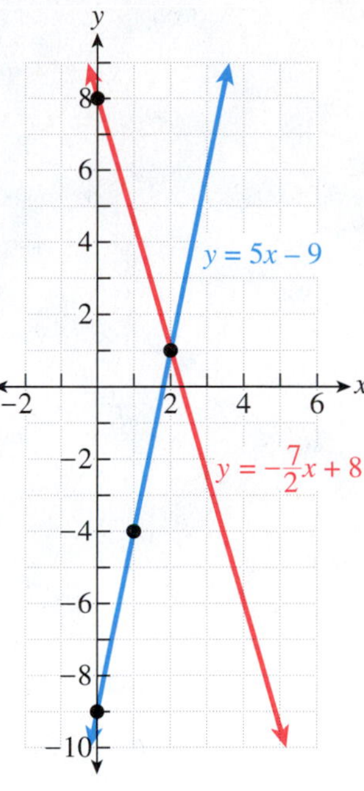

Examining the graph, we can see that the two lines intersect at the point $(2, 1)$. The solution of the system of equations is the ordered pair $(2, 1)$. We could check that this ordered pair is a solution by substituting its coordinates into both equations. The check is left to the reader.

Quick Check 3

Solve the system by graphing.

$$y = 3x - 2$$
$$y = 5x - 6$$

Using Your Calculator To solve a system of linear equations using the TI–84, you must graph each equation and then look for points of intersection. To graph the equations from Example 4, begin by tapping the [Y=] key. Next to Y_1 key $5x - 9$, and next to Y_2 key $-\frac{7}{2}x + 8$. Tap the [GRAPH] key to display the graph.

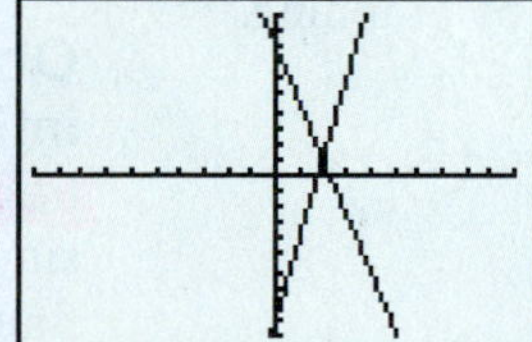

To find the points of intersection, press [2nd][TRACE] to access the CALC menu and select option **5: intersect.** You should see the following screen:

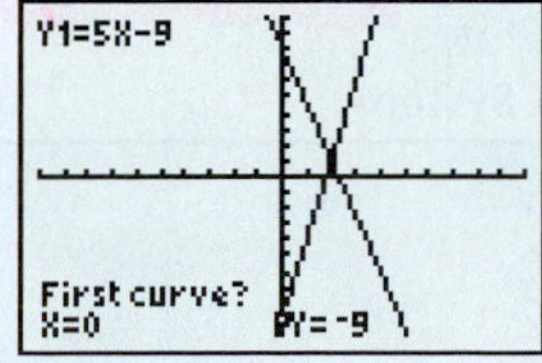

When prompted for the first curve, select Y_1 by tapping [ENTER]. When prompted for the second curve, select Y_2 by tapping [ENTER].

Now you must make an initial guess for one of the solutions. In the next screen shot, the TI–84 asks if you would like $(0, 8)$ to be your guess. Accept this guess by tapping [ENTER]. The solution is the point $(2, 1)$.

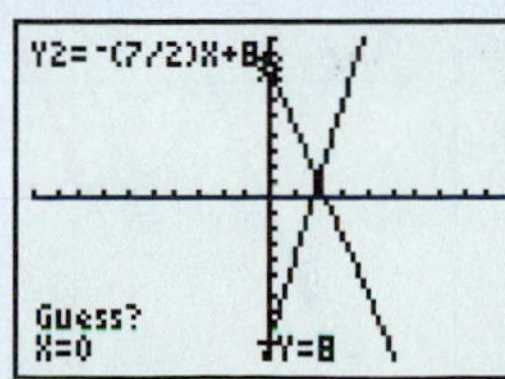

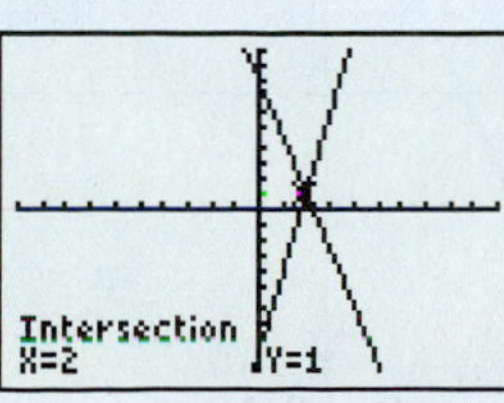

EXAMPLE 5 Solve the system by graphing.

$$2x - 3y = 12$$
$$x + y = 1$$

SOLUTION Both equations are in general form, so we will graph them using their x- and y-intercepts.

$2x - 3y = 12$		$x + y = 1$	
x-intercept ($y = 0$)	y-intercept ($x = 0$)	x-intercept ($y = 0$)	y-intercept ($x = 0$)
$2x - 3(0) = 12$ $2x = 12$ $x = 6$ $(6, 0)$	$2(0) - 3y = 12$ $-3y = 12$ $y = -4$ $(0, -4)$	$x + (0) = 1$ $x = 1$ $(1, 0)$	$(0) + y = 1$ $y = 1$ $(0, 1)$

Here are the graphs of each line on the same set of axes.

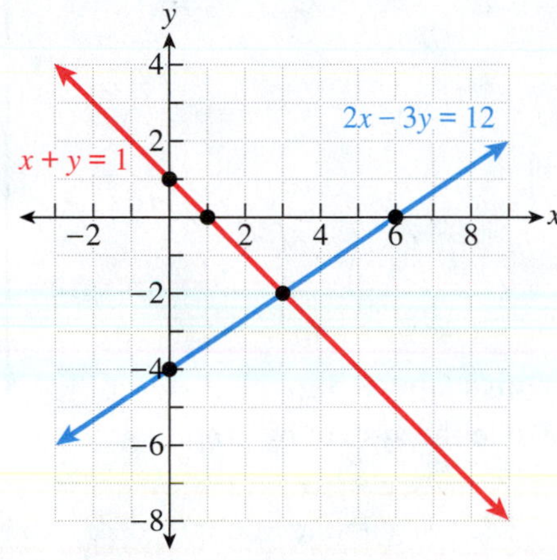

The two lines intersect at the point $(3, -2)$, so the solution of this set of equations is the ordered pair $(3, -2)$. The check is left to the reader.

Quick Check 4

Solve the system by graphing.

$$x + 5y = 5$$
$$x - y = -7$$

Objective 4 Identify systems of linear equations with no solution or infinitely many solutions. A system of two linear equations that has a single solution is called an **independent system.** Not every system of linear equations has a single solution. Suppose the graphs of the two equations are parallel lines. Parallel lines do not intersect, so such a system has no solution. A system that has no solution is called an **inconsistent system.**

Occasionally, the two equations in a system will have identical graphs. In this case, each point on the line is a solution of the system. Such a system is called a **dependent system** and has infinitely many solutions.

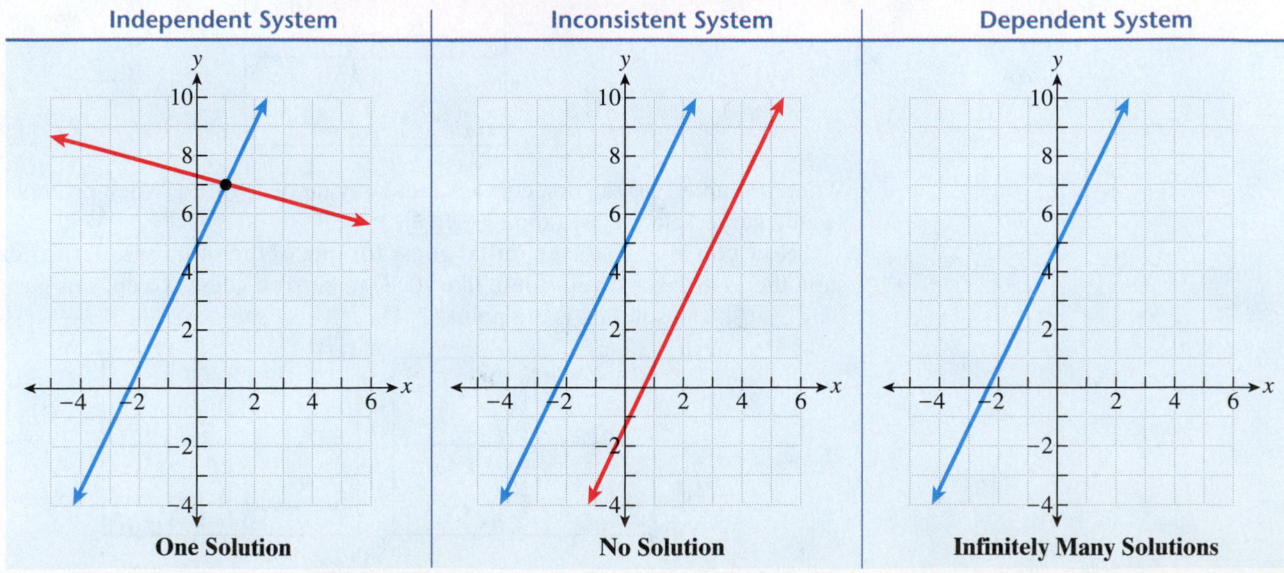

EXAMPLE 6 Solve the system by graphing.

$$y = -3x + 5$$
$$y = -3x - 3$$

SOLUTION Both equations are in slope–intercept form, so we can graph each line by plotting its y-intercept and then using the slope to find a second point on the line. The first line has a y-intercept at $(0, 5)$, and the slope is -3. The second line has a y-intercept at $(0, -3)$, and the slope also is -3. The graphs are at the right.

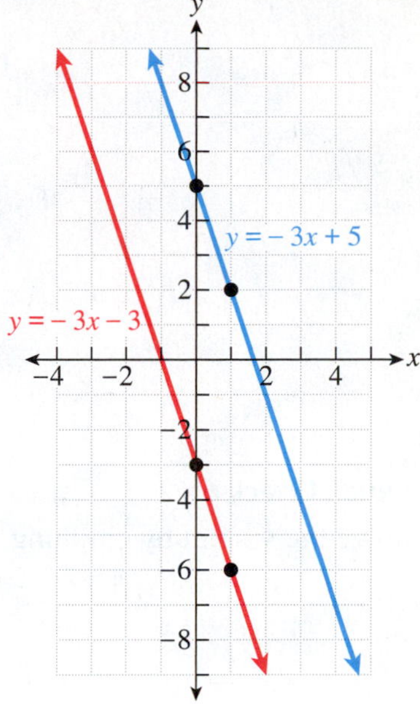

The two lines are distinct parallel lines because they have the same slope but different y-intercepts. The system is an inconsistent system and has no solution. In general, if we know that the two lines have the same slope but different y-intercepts, we know that the system is an inconsistent system and has no solution. However, if the two lines have the same slope *and* the *same* y-intercept, the system is a dependent system.

Quick Check 5

Solve the system by graphing.

$$y = -4x + 6$$
$$12x + 3y = 15$$

EXAMPLE 7 Solve the system by graphing.

$$y = 2x - 4$$
$$4x - 2y = 8$$

SOLUTION The first line is in slope–intercept form, so we can graph it by first plotting its y-intercept at $(0, -4)$. Using its slope of 2, we see that the line also passes through the point $(1, -2)$ as shown in the graph below.

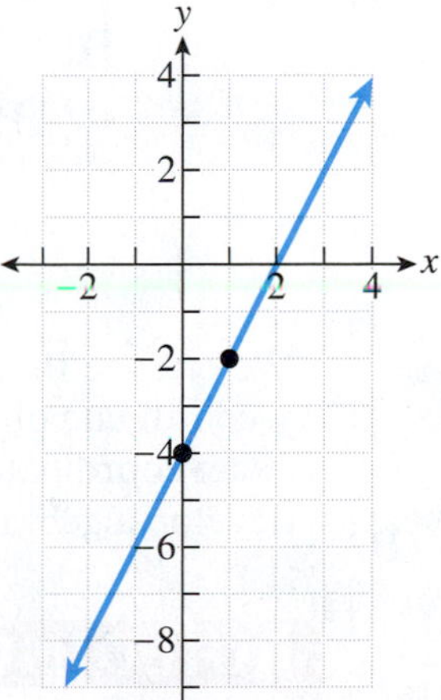

To graph the second line, we find the x- and y-intercepts, as the equation is in general form.

x-intercept $(y = 0)$	*y*-intercept $(x = 0)$
$4x - 2(0) = 8$	$4(0) - 2y = 8$
$4x = 8$	$-2y = 8$
$x = 2$	$y = -4$
$(2, 0)$	$(0, -4)$

After plotting these intercepts on the graph, we see that the two lines are exactly the same. This system is a dependent system. We can display the solutions of a dependent system by showing the graph of either equation. Every point on the line is a solution to the system of equations. To state the solution set, we write the general form for a point (x, y) that is on the line. To do this, we solve one of the equations for y in terms of x and replace y in the ordered pair (x, y) with this expression. In this example, we already know that $y = 2x - 4$; so our solutions are of the form $(x, 2x - 4)$.

$$y = 2x - 4$$
$$\downarrow$$
$$(x, y)$$
$$\downarrow$$
$$(x, 2x - 4)$$

Notice that if we rewrite the equation $4x - 2y = 8$ in slope–intercept form, it is exactly the same as the first equation. Equations in a dependent linear system of equations are multiples of each other.

Quick Check 6

Solve the system by graphing.

$$4x + 6y = 12$$
$$y = -\tfrac{2}{3}x + 2$$

Solving systems of equations by graphing does have a major drawback. Often we will not be able to accurately determine the coordinates of the solution of an independent system. Consider the graph of the following system:

$$y = 2x - 6$$
$$y = \frac{2}{5}x + 1$$

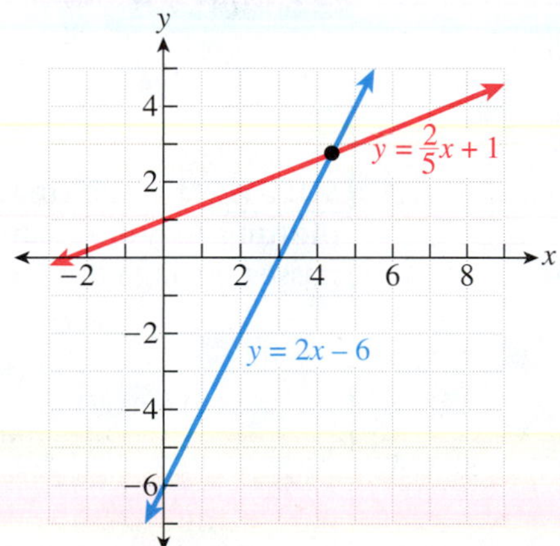

We can see that the x-coordinate of this solution is between 4 and 5 and that the y-coordinate of this solution is between 2 and 3. However, we cannot determine the exact coordinates of this solution from the graph. In the next two sections, we will develop algebraic techniques for finding solutions of linear systems of equations.

BUILDING YOUR STUDY STRATEGY

Doing Your Homework, 1 **Review First** Before beginning any homework assignment, it is a good idea to review. Start by going over your notes from class as a reminder of what types of problems your instructor covered and of how your instructor solved them. After reviewing your notes, keep them handy for further reference as you work through the homework assignment.

Next, you should review the appropriate section in the text. Pay particular attention to the examples. Look for the feature labeled "A Word of Caution" for advice on avoiding common errors. As you proceed through the homework assignment, refer to the section in the text as needed.

Exercises 4.1

MyMathLab

PRACTICE

WATCH

DOWNLOAD
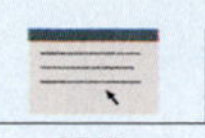
READ

REVIEW

Vocabulary

1. A(n) ______________ consists of two or more linear equations.
2. Give an example of a system of two linear equations in two unknowns.
3. An ordered pair (x, y) is a(n) __________ of a system of two linear equations if it is a solution of each equation in the system.
4. A system of two linear equations is a(n) __________ system if it has exactly one solution.
5. A system of two linear equations is a(n) __________ system if it has no solution.
6. A system of two linear equations is a(n) __________ system if it has infinitely many solutions.

Is the ordered pair a solution of the given system of equations?

7. $(5, 2)$, $y = 2x - 8$
$3x + 4y = 23$

8. $(3, 9)$, $x = 5y - 42$
$7x + 2y = 39$

9. $(-6, 1)$, $3x + 2y = -16$
$2x - 5y = -7$

10. $(-7, -8)$, $x + y = -15$
$6x - 5y = -2$

11. $(-4, -10)$, $3x - 7y = 58$
$-x + 4y = -36$

12. $(5, -5)$, $4x + 9y = -25$
$3x - y = 10$

13. $(8, -6)$, $\frac{3}{4}x - \frac{2}{3}y = 2$
$\frac{1}{8}x + 2y = -11$

14. $(9, 10)$, $\frac{2}{3}x + 5y = 56$
$7x - \frac{8}{5}y = 47$

15. $\left(\frac{3}{2}, \frac{5}{6}\right)$, $8x - 12y = 2$
$5x + 9y = 15$

16. $\left(\frac{2}{3}, \frac{15}{4}\right)$, $3x + 8y = 32$
$9x - 4y = 21$

Find the solution of the system of equations on each graph. If there is no solution, state this.

17.

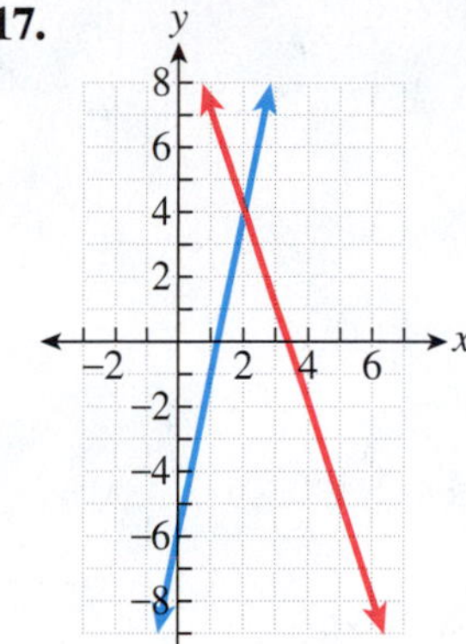

18.

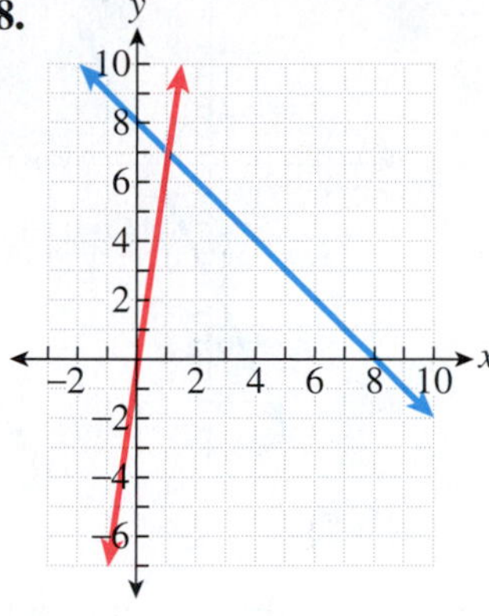

19.

20.

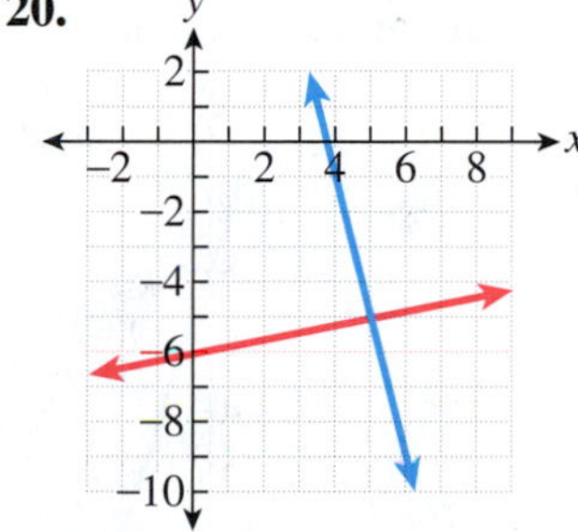

Solve the system by graphing. If the system is inconsistent and has no solution, state this. If the system is dependent, write the form of the solution for any real number x.

21. $y = x - 1$
$3x - y = 9$

22. $y = 3x - 2$
$x - y = -4$

23. $y = -2x + 8$
$y = \frac{3}{5}x - 5$

24. $4x + 2y = -8$
$y = \frac{3}{4}x + 7$

25. $y = 2x + 4$
$6x - 3y = -12$

26. $y = -5x + 6$
$3x - 4y = -24$

27. $y = \frac{3}{2}x + 3$
$7x - 2y = -14$

28. $y = -3x + 3$
$12x + 4y = 12$

29. $5x - 3y = 15$
$2x - y = 4$

30. $7x + 3y = 0$
$x + y = 4$

31. $4x - 5y = -20$
$-4x + 7y = 28$

32. $-2x + y = 10$
$2x + 3y = 6$

33. $y = 5x - 4$
$10x - 2y = 6$

34. $4x + 5y = 10$
$y = -\frac{3}{5}x + 3$

35. $9x - 2y = -18$
$y = \frac{1}{2}x - 7$

36. $y = -\frac{2}{7}x + 3$
$4x + 14y = 6$

37. Draw the graph of a linear equation that, along with the given line, forms an inconsistent system.

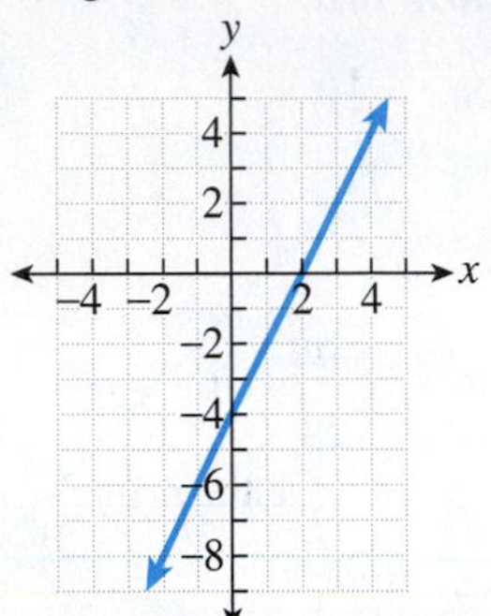

38. Draw the graph of a linear equation that, along with the given line, forms a dependent system.

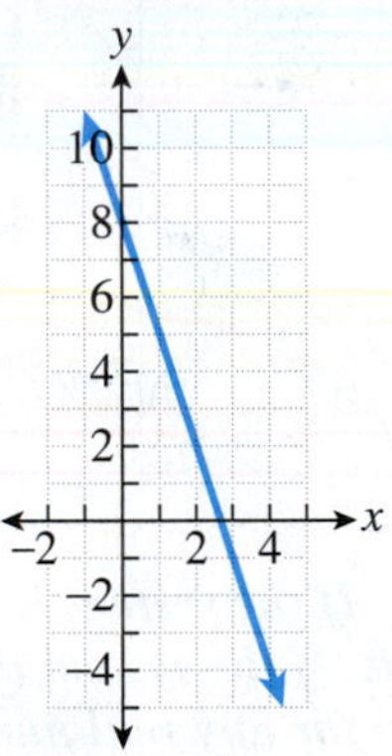

39. Draw the graph of a linear equation that, along with the given line, forms a system of equations whose single solution is the ordered pair (5, 1).

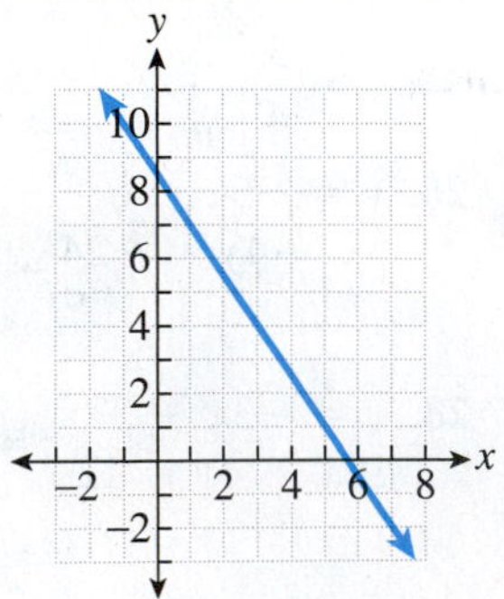

40. Draw the graphs of two linear equations that form an inconsistent system.

41. Draw the graphs of two linear equations that form a dependent system.

42. Draw the graphs of two linear equations that form a system of equations whose single solution is the ordered pair (−3, 2).

Writing in Mathematics

Answer in complete sentences.

43. Explain why the solution of an independent system of equations is the ordered pair that is the point of intersection for the two lines.

44. When solving a system of equations by graphing, you may not be able to accurately determine the coordinates of the solution of an independent system. Explain.

Quick Review Exercises

Section 4.1

Solve.

1. $8x + 3(2x - 5) = 27$
2. $7x - 5(3x + 9) = -13$
3. $2(y + 6) - 3y = 17$
4. $-3(6y + 1) + 7y = 74$

4.2

Solving Systems of Equations by Using the Substitution Method

OBJECTIVES

1. **Solve systems of linear equations by using the substitution method.**
2. **Use the substitution method to identify inconsistent systems of equations.**
3. **Use the substitution method to identify dependent systems of equations.**
4. **Solve systems of equations without variable terms with a coefficient of 1 or −1 by the substitution method.**
5. **Solve applied problems by using the substitution method.**

Solving systems of equations by graphing each equation works well when the solution has integer coordinates. However, when the coordinates are not integers, exact solutions are difficult to find when graphing by hand. In the next two sections, we will examine algebraic methods for solving systems of linear equations. The goal of each method is to combine the two equations with two variables into a single equation with only one variable. After solving for this variable, we will substitute the value of that variable into one of the original equations and solve for the other variable.

The Substitution Method

Objective 1 Solve systems of linear equations by using the substitution method. Consider the following system of equations:

$$y = x - 5$$
$$2x + 3y = 20$$

Looking at the first equation, $y = x - 5$, we see that y is equal to the expression $x - 5$. We can use this fact to replace y with $x - 5$ in the second equation $2x + 3y = 20$. We are substituting the expression $x - 5$ for y, because the two expressions are equivalent. The first example will solve this system through use of the substitution method.

The Substitution Method

To solve a system of equations by using the substitution method, we solve one of the equations for one of the variables and then substitute that expression for the variable in the other equation.

EXAMPLE 1 Solve the system by substitution.

$$y = x - 5$$
$$2x + 3y = 20$$

SOLUTION Because the first equation is solved for y in terms of x, we can substitute $x - 5$ for y in the second equation.

$$y = x - 5$$
$$\downarrow$$
$$2x + 3y = 20$$

This will produce a single equation with only one variable, x.

$2x + 3y = 20$	
$2x + 3(x - 5) = 20$	Substitute $x - 5$ for y.
$2x + 3x - 15 = 20$	Distribute.
$5x - 15 = 20$	Combine like terms.
$5x = 35$	Add 15.
$x = 7$	Divide both sides by 5.

We are now halfway to the solution. We know that the x-coordinate of the solution is 7, and we now need to find the y-coordinate.

A WORD OF CAUTION The solution of a system of equations in two variables is an *ordered pair* (x, y), not a single value.

We can find the y-coordinate by substituting 7 for x in either of the original equations. It is easier to substitute the value into the equation $y = x - 5$, because y is already isolated in this equation. In general, it is a good idea to substitute back into the equation that was used to make the original substitution.

$$
\begin{aligned}
y &= x - 5 \\
y &= (7) - 5 && \text{Substitute 7 for } x. \\
y &= 2 && \text{Subtract.}
\end{aligned}
$$

The solution to this system is $(7, 2)$. This solution can be checked by substituting 7 for x and 2 for y in the equation $2x + 3y = 20$. The check is left to the reader.

Quick Check 1

Solve the system by substitution.

$$
\begin{aligned}
y &= x - 3 \\
7x - 4y &= 27
\end{aligned}
$$

In the first example, one of the equations was already solved for y. To use the substitution method, we often have to solve one of the equations for x or y. When choosing which equation to work with or which variable to solve for, look for an equation that has an x or y term with a coefficient of 1 or -1 ($x, -x, y$, or $-y$). When we find one of those terms, solving for that variable can be done easily through addition or subtraction.

EXAMPLE 2 Solve the system by substitution.

$$
\begin{aligned}
x + 2y &= 7 \\
4x - 3y &= -16
\end{aligned}
$$

SOLUTION The first equation can easily be solved for x by subtracting $2y$ from both sides of the equation. This produces the equation $x = 7 - 2y$. We can now substitute the expression $7 - 2y$ for x in the second equation $4x - 3y = -16$.

A WORD OF CAUTION Do not substitute $7 - 2y$ into the equation it came from, $x + 2y = 7$, as the resulting equation will be $7 = 7$. Substitute it into the other equation in the system. Never substitute back into the same equation.

$$
\begin{aligned}
4x - 3y &= -16 \\
4(7 - 2y) - 3y &= -16 && \text{Substitute } 7 - 2y \text{ for } x. \\
28 - 8y - 3y &= -16 && \text{Distribute 4.} \\
28 - 11y &= -16 && \text{Combine like terms.} \\
-11y &= -44 && \text{Subtract 28 from both sides.} \\
y &= 4 && \text{Divide both sides by } -11.
\end{aligned}
$$

Because the y-coordinate of the solution is 4, we can substitute this value for y into the equation $x = 7 - 2y$ to find the x-coordinate.

$$
\begin{aligned}
x &= 7 - 2(4) && \text{Substitute 4 for } y. \\
x &= -1 && \text{Simplify.}
\end{aligned}
$$

Quick Check 2

Solve the system by substitution.

$$
\begin{aligned}
x - y &= -4 \\
3x + 5y &= 36
\end{aligned}
$$

The solution of this system of equations is $(-1, 4)$. We could check this solution by substituting -1 for x and 4 for y in the equation $4x - 3y = -16$. This is left to the reader. Although the first coordinate we solved for was y, keep in mind that the ordered pair must be written in the order (x, y).

Inconsistent Systems

Objective 2 Use the substitution method to identify inconsistent systems of equations. Recall from the previous section that a linear system of equations with no solution is called an inconsistent system. The two lines associated with the equations of an inconsistent system are parallel lines. The next example shows how to determine that a system is an inconsistent system while using the substitution method.

EXAMPLE 3 Solve the system by substitution.

$$x = 3y - 7$$
$$3x - 9y = 18$$

SOLUTION Because the first equation is solved for x, substitute $3y - 7$ for x in the second equation.

$$\begin{aligned} 3x - 9y &= 18 \\ 3(3y - 7) - 9y &= 18 \quad \text{Substitute } 3y - 7 \text{ for } x. \\ 9y - 21 - 9y &= 18 \quad \text{Distribute.} \\ -21 &= 18 \quad \text{Combine like terms.} \end{aligned}$$

The resulting equation is false because -21 is not equal to 18. Because this equation can never be true, the system of equations has no solution ($\varnothing$) and is an inconsistent system.

Quick Check 3

Solve the system by substitution.

$$-2x + y = 3$$
$$8x - 4y = 10$$

Dependent Systems

Objective 3 Use the substitution method to identify dependent systems of equations. A dependent system of linear equations has two equations with identical graphs and is a system with infinitely many solutions. When using the substitution method to solve a dependent system, we obtain an identity, such as $3 = 3$, after making the substitution.

EXAMPLE 4 Solve the system by substitution.

$$4x - y = 3$$
$$-8x + 2y = -6$$

SOLUTION Solve the first equation for y.

$$\begin{aligned} 4x - y &= 3 \\ -y &= -4x + 3 \quad \text{Subtract } 4x. \\ y &= 4x - 3 \quad \text{Divide both sides by } -1. \end{aligned}$$

Now substitute $4x - 3$ for y in the second equation.

$$\begin{aligned} -8x + 2y &= -6 \\ -8x + 2(4x - 3) &= -6 \quad \text{Substitute } 4x - 3 \text{ for } y. \\ -8x + 8x - 6 &= -6 \quad \text{Distribute 2.} \\ -6 &= -6 \quad \text{Combine like terms.} \end{aligned}$$

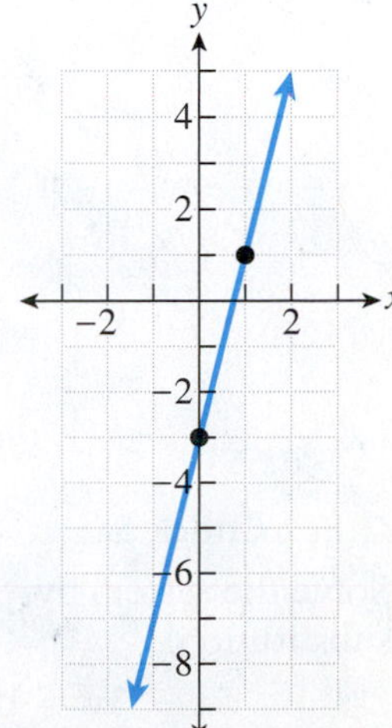

This equation is an identity that is true for all values of x. This system of equations is a dependent system with infinitely many solutions.

Because we have already shown that $y = 4x - 3$, the ordered pairs that are solutions have the form $(x, 4x - 3)$. The solutions are shown in the graph at the right, which is the graph of the line $y = 4x - 3$.

Quick Check 4

Solve the system by substitution.

$$x - 6y = 4$$
$$3x - 18y = 12$$

Objective 4 Solve systems of equations without variable terms with a coefficient of 1 or −1 by the substitution method. All of the examples in this section thus far had at least one equation containing a variable term with a coefficient of 1 or −1. Such equations are easy to solve for one variable in terms of the other. In the next example, we will learn how to use the substitution method when this is not the case.

EXAMPLE 5 Solve the system by substitution.

$$3x + 2y = -2$$
$$-2x + 7y = 43$$

SOLUTION There are no variable terms with a coefficient of 1 or −1 (x, $-x$, y, or $-y$). We will solve the first equation for y, although this selection is completely arbitrary. We could solve either equation for either variable and still find the same solution of the system.

$$3x + 2y = -2$$
$$2y = -3x - 2 \quad \text{Subtract } 3x.$$
$$y = -\frac{3}{2}x - 1 \quad \text{Divide both sides by 2.}$$

Now substitute $-\frac{3}{2}x - 1$ for y in the second equation.

$$-2x + 7y = 43$$
$$-2x + 7\left(-\frac{3}{2}x - 1\right) = 43 \quad \text{Substitute } -\tfrac{3}{2}x - 1 \text{ for } y.$$
$$-2x - \frac{21}{2}x - 7 = 43 \quad \text{Distribute 7.}$$
$$2\left(-2x - \frac{21}{2}x - 7\right) = 2 \cdot 43 \quad \text{Multiply both sides by 2 to clear fractions.}$$
$$2(-2x) - \overset{1}{\cancel{2}}\left(\frac{21}{\underset{1}{\cancel{2}}}x\right) - 2 \cdot 7 = 2 \cdot 43 \quad \text{Distribute and divide out common factors.}$$
$$-4x - 21x - 14 = 86 \quad \text{Multiply.}$$
$$-25x - 14 = 86 \quad \text{Combine like terms.}$$
$$-25x = 100 \quad \text{Add 14.}$$
$$x = -4 \quad \text{Divide both sides by } -25.$$

We know that the x-coordinate of the solution is −4. We will substitute this value for x in the equation $y = -\frac{3}{2}x - 1$.

$$y = -\frac{3}{2}(-4) - 1 \quad \text{Substitute } -4 \text{ for } x.$$
$$y = 5 \quad \text{Simplify.}$$

The solution to this system is $(-4, 5)$.

Quick Check 5

Solve the system by substitution.

$$5x - 3y = 17$$
$$2x + 4y = -14$$

Before turning our attention to an application of linear systems of equations, we present the following general strategy for solving these systems:

Solving a System of Equations by Using the Substitution Method

1. **Solve one of the equations for either variable.** If an equation has a variable term whose coefficient is 1 or -1, try to solve that equation for that variable.
2. **Substitute this expression for the variable in the other equation.** At this point, we have an equation with only one variable.
3. **Solve this equation.** This gives us one coordinate of the ordered-pair solution. (If the equation is an identity, the system is dependent. If the equation is a contradiction, the system is inconsistent.)
4. **Substitute this value for the variable into the equation from Step 1.** After simplifying, this will give us the other coordinate of the solution.
5. **Check the solution.**

Applications

Objective 5 Solve applied problems by using the substitution method.

A college performed a play. There were 75 people in the audience. If admission was \$6 for adults and \$2 for children and the total box office receipts were \$310, how many adults and how many children attended the play?

SOLUTION There are two unknowns in this problem: the number of adults and the number of children. We will let A represent the number of adults and C represent the number of children.

Because there are two variables, we need a system of two equations. The first equation comes from the total attendance of 75 people. Because all of these people are either adults or children, the number of adults (A) plus the number of children (C) must equal 75. The first equation is $A + C = 75$.

The second equation comes from the amount of money paid. The amount of money paid by adults (\$$6A$) plus the amount of money paid by children (\$$2C$) must equal the total amount of money paid. The second equation is $6A + 2C = 310$. Here is the system of equations we need to solve.

$$\begin{aligned} A + C &= 75 \\ 6A + 2C &= 310 \end{aligned}$$

Before we solve this system of equations, let's look at a table that contains all of the pertinent information for this problem.

	Number of Attendees	Cost per Ticket (\$)	Money Paid (\$)
Adults	A	6	$6A$
Children	C	2	$2C$
Total	75		310

Notice that the first equation in the system $A + C = 75$ can be found in the first column of the table labeled "Number of Attendees". The second equation in the system $6A + 2C = 310$ can be found in the last column in the table labeled "Money Paid (\$)".

The first equation in this system can be solved for A or C; we will solve it for C. This produces the equation $C = 75 - A$, so the expression $75 - A$ can be substituted for C in the second equation.

$$6A + 2C = 310$$
$$6A + 2(75 - A) = 310 \quad \text{Substitute } 75 - A \text{ for } C.$$
$$6A + 150 - 2A = 310 \quad \text{Distribute.}$$
$$4A + 150 = 310 \quad \text{Combine like terms.}$$
$$4A = 160 \quad \text{Subtract 150.}$$
$$A = 40 \quad \text{Divide both sides by 4.}$$

There were 40 adults. We now substitute this value for A in the equation $C = 75 - A$.

$$C = 75 - (40) \quad \text{Substitute 40 for } A.$$
$$C = 35 \quad \text{Subtract.}$$

There were 40 adults and 35 children at the play.

We should check the solution for accuracy. If there were 40 adults and 35 children, that is a total of 75 people. Also, each of the 40 adults paid \$6, which is \$240. Each of the 35 children paid \$2, which is another \$70. The total paid is \$310, so the solution checks.

Quick Check 6

A small business owner bought 15 new computers for his company. He paid \$700 for each desktop computer and \$1000 for each laptop computer. If he paid \$11,700 for the computers, how many desktop computers and how many laptop computers did he buy?

BUILDING YOUR STUDY STRATEGY

Doing Your Homework, 2 Neat and Complete Two important words to keep in mind when working on your homework exercises are *neat* and *complete*. When your homework is neat, you can check your work more easily and your homework will be easier to understand when you review prior to an exam.

Be as complete as you can when working on the homework exercises. By listing each step, you are increasing your chances of being able to remember all of the steps necessary to solve a similar problem on an exam or a quiz. When you review, you may have difficulty remembering how to solve a problem if you did not write down all of the necessary steps.

Exercises 4.2

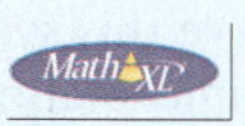
PRACTICE

WATCH

DOWNLOAD

READ

REVIEW

Vocabulary

1. To solve a system of equations by substitution, begin by solving ______________________________.
2. Once one of the equations has been solved for one of the variables, that expression is __________ for that variable in the other equation.
3. If substitution results in an equation that is an identity, the system is a(n) __________ system.
4. If substitution results in an equation that is a contradiction, the system is a(n) __________ system.

Solve the system by substitution. If the system is inconsistent and has no solution, state this. If the system is dependent, write the form of the solution for any real number x.

5. $y = 3x - 5$
   $x + 3y = 15$
6. $x = 4y + 3$
   $2x + 5y = -7$
7. $x = 2 - 5y$
   $7x + 10y = 89$
8. $x = 1 - 4y$
   $4x + 2y = -17$

9. $y = 5x - 10$
$-10x + 2y = 20$

10. $y = 2x + 7$
$6x - 3y = -21$

11. $18x + 6y = -12$
$y = -3x - 2$

12. $x = -\frac{2}{3}y + \frac{3}{4}$
$12x + 8y = 6$

13. $y = 11 - 5x$
$-7x - 2y = 14$

14. $x = 2y - 15$
$3x + 14y = 5$

15. $y = 2.2x - 6.8$
$3x + 2y = 16$

16. $x = 4.7y - 40.6$
$5x - 2y = -31$

17. $x = \frac{2}{3}y - 4$
$6x + 7y = 75$

18. $y = \frac{3}{4}x - \frac{7}{2}$
$5x - 4y = 2$

19. $y = 4x - 3$
$y = -2x - 21$

20. $y = 5x + 2$
$y = 9x - 6$

21. $y = \frac{5}{6}x + 3$
$y = 3x - \frac{7}{2}$

22. $y = \frac{3}{2}x - 9$
$y = -\frac{7}{4}x + 4$

23. $x + 3y = 19$
$3x - 4y = -21$

24. $2x - y = -15$
$5x + 6y = 5$

25. $x + y = -6$
$4x - 3y = 46$

26. $x - y = 1$
$-2x + 7y = -42$

27. $5x + 2y = -44$
$3x + y = -26$

28. $3x + 12y = 10$
$x + 4y = 7$

29. $2x - y = -1$
$8x + 2y = 17$

30. $9x - 6y = 19$
$-x + 8y = 4$

31. $x = 4y - 5$
$2x + 2y = 90$

32. $y = 2x - 17$
$2x + 2y = 38$

33. $x + y = 40$
$0.10x + 0.25y = 8.50$

34. $x + y = 200$
$6.75x + 4.25y = 1150$

35. $x + y = 5000$
$0.04x + 0.07y = 260$

36. $x + y = 9000$
$0.06x + 0.15y = 1215$

37. $y = x + 100$
$0.02x + 0.025y = 22.75$

38. $y = x - 120$
$0.015x + 0.04y = 15$

39. $x + y = 80$
$0.18x + 0.33y = 0.24(80)$

40. $x + y = 50$
$0.25x + 0.75y = 0.60(50)$

41. $6x + y = 21$
$-18x - 3y = -63$

42. $10x - y = -23$
$-20x + 2y = -46$

43. $x + 2.8y = 3.4$
$2x - 9y = -37$

44. $-3.4x + y = -30.8$
$2.1x + 3.2y = 57.2$

45. $\frac{3}{2}x + y = -7$
$-3x + 2y = 34$

46. $x + \frac{3}{8}y = 24$
$16x - 11y = -24$

47. $4x + 2y = 10$
$5x - 4y = 6$

48. $9x + 3y = 36$
$7x - 6y = 78$

49. $3x + 5y = 14$
$6x - 5y = -32$

50. $9x + 4y = 23$
$5x - 2y = -2$

51. A college basketball team charges \$5 admission to the general public for its games, while students at the college pay only \$2. If the attendance for last night's

game was 1200 people and the total receipts were \$5250, how many of the people in attendance were students and how many were not students?

52. Kenny walks away from a blackjack table with a total of \$62 in \$1 and \$5 chips. If Kenny has 26 chips, how many are \$1 chips and how many are \$5 chips?

53. A sorority set up a charity dinner to help raise money for a local children's support group. The sorority sold chicken dinners for \$8 and steak dinners for \$10. If 142 people ate dinner at the fund-raiser and the sorority raised \$1326, how many people ordered chicken and how many people ordered steak?

54. Jeannie has some \$10 bills and some \$20 bills. If she has 273 bills worth a total of \$4370, how many of the bills are \$10 bills and how many are \$20 bills?

55. The following equations give the median age (y) for men and women at their first marriage in a particular year, where x represents the number of years after 1995.

Men: $y = 0.05x + 26.7$
Women: $y = 0.10x + 24.6$

If the current trend continues, in what year will the median age at first marriage be equal for men and women? (*Source: U.S. Census Bureau*)

56. The following equations give the number of college faculty (y) by gender in a particular year, where x represents the number of years after 1987.

Men: $y = 10{,}045x + 502{,}183$
Women: $y = 16{,}475x + 247{,}693$

If the current trend continues, in what year will the number of male and female faculty members be equal? (*Source: U.S. Department of Education*)

Writing in Mathematics

Answer in complete sentences.

57. In your own words, explain how to solve a system of linear equations using the substitution method.

58. When using the substitution method, how can you tell that a system of equations is inconsistent and has no solutions? How can you tell that a system of equations is dependent and has infinitely many solutions? Explain your answer.

59. ***Solutions Manual*** Write a solutions manual page for the following problem.

Solve the system of equations by the substitution method.

$$y = 3x - 7$$
$$5x - 7y = -15$$

4.3 Solving Systems of Equations by Using the Addition Method

OBJECTIVES

1. **Solve systems of equations by using the addition method.**
2. **Use the addition method to identify inconsistent and dependent systems of equations.**
3. **Solve systems of equations with coefficients that are fractions or decimals by using the addition method.**
4. **Solve applied problems by using the addition method.**

The Addition Method

Objective 1 Solve systems of equations by using the addition method. In this section, we will examine the **addition method** for solving systems of linear equations. (This method also is known as the **elimination method.**)

The Addition Method

The addition method is an algebraic alternative to the substitution method. The two equations in a system of linear equations will be combined into a single equation with a single variable by adding them together.

In the last section, we saw that solving a system of equations by substitution is fairly easy when one of the equations contains a variable term with a coefficient of 1 or -1. The substitution method can become quite tedious for solving a system when this is not the case.

As with the substitution method, the goal of the addition method is to combine the two given equations into a single equation with only one variable. This method is based on the addition property of equality from Chapter 2, which says that when we add the same number to both sides of an equation, the equation remains true. (For any real numbers a, b, and c, if $a = b$, then $a + c = b + c$.) This property can be extended to cover adding equal expressions to both sides of an equation: if $a = b$ and $c = d$, then $a + c = b + d$.

Consider the following system:

$$\begin{aligned} 3x + 2y &= 20 \\ 5x - 2y &= -4 \end{aligned}$$

If we add the left side of the first equation $(3x + 2y)$ to the left side of the second equation $(5x - 2y)$, this expression will be equal to the total of the two numbers on the right side of the equations. Here is the result of this addition.

$$\begin{aligned} 3x + 2y &= 20 \\ 5x - 2y &= -4 \\ \hline 8x \qquad &= 16 \end{aligned}$$

Notice that when we added these two equations, the two terms containing y were opposites and summed to zero, leaving an equation containing only the variable x. Next, we solve this equation for x and substitute this value for x in either of the two original equations to find y. The first example will walk through this entire process.

EXAMPLE 1 Solve the system by addition.

$$\begin{aligned} 3x + 2y &= 20 \\ 5x - 2y &= -4 \end{aligned}$$

SOLUTION Because the two terms containing y are opposites, we can use addition to eliminate y.

$$\begin{aligned} 3x + 2y &= 20 \\ 5x - 2y &= -4 && \text{Add the two equations.} \\ \hline 8x \qquad &= 16 \\ x &= 2 && \text{Divide both sides by 8.} \end{aligned}$$

The x-coordinate of the solution is 2. We now substitute this value for x in the first original equation. (We could have chosen the other equation, because it will produce exactly the same solution.)

$$\begin{aligned} 3x + 2y &= 20 \\ 3(2) + 2y &= 20 && \text{Substitute 2 for } x. \\ 6 + 2y &= 20 && \text{Multiply.} \\ 2y &= 14 && \text{Subtract 6.} \\ y &= 7 && \text{Divide both sides by 2.} \end{aligned}$$

The solution of this system is the ordered pair (2, 7). We could check this solution by substituting 2 for x and 7 for y in the equation $5x - 2y = -4$. This is left to the reader.

Quick Check 1

Solve the system by addition.

$$\begin{aligned} 4x + 3y &= 31 \\ -4x + 5y &= -23 \end{aligned}$$

If the two equations in a system do not contain a pair of opposite variable terms, we can still use the addition method. We must first multiply both sides of one or both equations by a constant(s) in such a way that two of the variable terms become opposites. Then we proceed as in the first example.

EXAMPLE 2 Solve the system by addition.

$$2x - 3y = -16$$
$$-6x + 5y = 32$$

SOLUTION Adding these two equations at this point would not be helpful, as the sum would still contain both variables. If we multiply the first equation by 3, the terms containing x will be opposites. This allows us to proceed with the addition method.

$$\begin{aligned} 2x - 3y &= -16 \\ -6x + 5y &= 32 \end{aligned} \xrightarrow{\text{Multiply by 3}} \begin{aligned} 6x - 9y &= -48 \\ -6x + 5y &= 32 \end{aligned}$$

$$\begin{aligned} 6x - 9y &= -48 \\ -6x + 5y &= 32 & &\text{Add.} \\ \hline -4y &= -16 \\ y &= 4 & &\text{Divide both sides by } -4. \end{aligned}$$

A WORD OF CAUTION Be sure to multiply *both* sides of the equation by the same number. Do not multiply just the side containing the variable terms.

Now substitute 4 for y in the equation $2x - 3y = -16$ and solve for x.

$$\begin{aligned} 2x - 3(4) &= -16 & &\text{Substitute 4 for } y. \\ 2x - 12 &= -16 & &\text{Multiply.} \\ 2x &= -4 & &\text{Add 12.} \\ x &= -2 & &\text{Divide both sides by 2.} \end{aligned}$$

The solution of this system is $(-2, 4)$. (Be sure to write the x-coordinate first in the ordered pair even though we solved for y first.)

Quick Check 2

Solve the system by addition.

$$2x - 9y = 33$$
$$4x + 3y = 3$$

Before continuing, let's outline the basic strategy for using the addition method to solve systems of linear equations.

Using the Addition Method

1. **Write each equation in standard form ($Ax + By = C$).** For each equation, gather all variable terms on the left side and all constants on the right side.
2. **Multiply one or both equations by the appropriate constant(s).** The goal of this step is to make the terms containing x or the terms containing y opposites.
3. **Add the two equations together.** At this point, we should have one equation containing a single variable.
4. **Solve the resulting equation.** This will give us one of the coordinates of the solution.
5. **Substitute this value for the appropriate variable in either of the original equations and solve for the other variable.** Choose the equation that you believe will be easier to solve. When we solve this equation, we find the other coordinate of the solution.
6. **Write the solution as an ordered pair.**
7. **Check your solution.**

EXAMPLE 3 Solve the system by addition.

$$5x + 8y - 22 = 0$$
$$7x = 6y + 5$$

SOLUTION To use the addition method, we must write each equation in standard form.

$$5x + 8y = 22$$
$$7x - 6y = 5$$

Then we must decide which variable to eliminate. A wise choice for this system is to eliminate y, as the two coefficients already have opposite signs. If we multiply the first equation by 3 and the second equation by 4, the terms containing y will be $24y$ and $-24y$. Note that 24 is the LCM of 8 and 6.

$$5x + 8y = 22 \xrightarrow{\text{Multiply by 3}} 15x + 24y = 66$$
$$7x - 6y = 5 \xrightarrow{\text{Multiply by 4}} 28x - 24y = 20$$

$$\begin{aligned} 15x + 24y &= 66 \\ \underline{28x - 24y} &\underline{= 20} \quad \text{Add to eliminate } y. \\ 43x \qquad &= 86 \\ x &= 2 \quad \text{Divide both sides by 43.} \end{aligned}$$

The x-coordinate of the solution is 2. Substituting 2 for x in the equation $5x + 8y = 22$ will allow us to find the y-coordinate.

$$\begin{aligned} 5(2) + 8y &= 22 \quad \text{Substitute 2 for } x. \\ 10 + 8y &= 22 \quad \text{Multiply.} \\ 8y &= 12 \quad \text{Subtract 10.} \\ y &= \frac{12}{8} \quad \text{Divide both sides by 8.} \\ y &= \frac{3}{2} \quad \text{Simplify.} \end{aligned}$$

The solution to this system is $\left(2, \frac{3}{2}\right)$. The check is left to the reader.

Quick Check 3

Solve the system by addition.

$$3x + 2y = -26$$
$$5y = 2x + 11$$

Objective 2 Use the addition method to identify inconsistent and dependent systems of equations. Recall from the previous section that when solving an inconsistent system (no solution), we obtain an equation that is a contradiction, such as $0 = -6$. When solving a dependent system (infinitely many solutions), we end up with an equation that is an identity, such as $7 = 7$. The same is true when applying the addition method.

EXAMPLE 4 Solve the system by addition.

$$3x + 5y = 15$$
$$6x + 10y = 30$$

SOLUTION Suppose we chose to eliminate the variable x. We can do so by multiplying the first equation by -2. This produces the following system:

$$3x + 5y = 15 \xrightarrow{\text{Multiply by } -2} -6x - 10y = -30$$
$$6x + 10y = 30 \qquad\qquad 6x + 10y = 30$$

$$\begin{aligned} -6x - 10y &= -30 \\ \underline{6x + 10y} &\underline{= \quad 30} \quad \text{Add to eliminate } x. \\ 0 &= \quad 0 \end{aligned}$$

Because the resulting equation is an identity, this is a dependent system with infinitely many solutions. To determine the form of the solutions, we can solve the equation $3x + 5y = 15$ for y.

$$3x + 5y = 15$$
$$5y = -3x + 15 \quad \text{Subtract } 3x \text{ to isolate } 5y.$$
$$y = -\frac{3}{5}x + 3 \quad \text{Divide both sides by 5 and simplify.}$$

The ordered pairs that are solutions have the form $(x, -\frac{3}{5}x + 3)$. We also could display the solutions by graphing the line $3x + 5y = 15$, or $y = -\frac{3}{5}x + 3$.

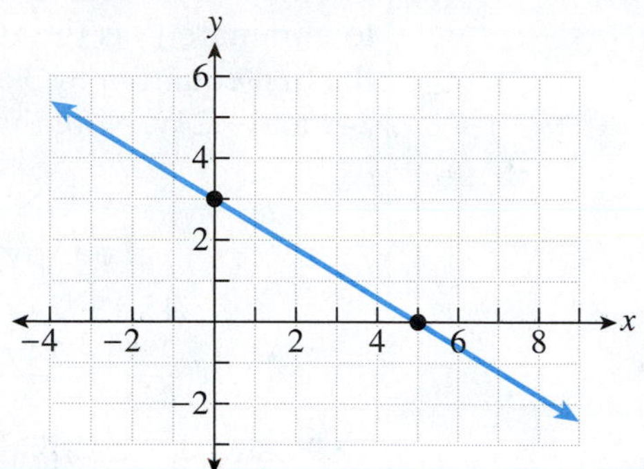

Quick Check 4

Solve the system by addition.

$$2x - 3y = 12$$
$$-4x + 6y = -24$$

EXAMPLE 5 Solve the system by addition.

$$6x - 3y = 7$$
$$-2x + y = -3$$

SOLUTION To eliminate the variable y, multiply the second equation by 3. The variable term containing y in the second equation would then be $3y$, which is the opposite of $-3y$ in the first equation.

$$\begin{array}{rcl} 6x - 3y = 7 & & 6x - 3y = 7 \\ -2x + y = -3 & \xrightarrow{\text{Multiply by 3}} & -6x + 3y = -9 \end{array}$$

$$\begin{array}{r} 6x - 3y = 7 \\ -6x + 3y = -9 \\ \hline 0 = -2 \end{array} \quad \text{Add to eliminate } y.$$

Because the resulting equation is false, this is an inconsistent system with no solution.

Quick Check 5

Solve the system by addition.

$$3x = 4y + 11$$
$$9x - 12y = 30$$

Systems of Equations with Coefficients That Are Fractions or Decimals

Objective 3 Solve systems of equations with coefficients that are fractions or decimals by using the addition method. If one or both equations in a system of linear equations contain fractions or decimals, we can clear the fractions or decimals first and then solve the resulting system.

EXAMPLE 6 Solve the system by addition.

$$\frac{5}{4}x + \frac{1}{3}y = 3$$
$$\frac{3}{8}x - \frac{5}{6}y = \frac{13}{2}$$

SOLUTION The LCM of the denominators in the first equation is 12; so if we multiply both sides of the first equation by 12, we can clear the equation of fractions.

$$12\left(\frac{5}{4}x + \frac{1}{3}y\right) = 12 \cdot 3 \quad \text{Multiply both sides by 12.}$$
$$\overset{3}{\cancel{12}} \cdot \frac{5}{\underset{1}{\cancel{4}}}x + \overset{4}{\cancel{12}} \cdot \frac{1}{\underset{1}{\cancel{3}}}y = 12 \cdot 3 \quad \text{Distribute and divide out common factors.}$$
$$15x + 4y = 36 \quad \text{Simplify.}$$

Multiplying both sides of the second equation by 24 will clear the fractions from that equation.

$$24\left(\frac{3}{8}x - \frac{5}{6}y\right) = 24 \cdot \frac{13}{2} \quad \text{Multiply both sides by 24.}$$
$$\overset{3}{\cancel{24}} \cdot \frac{3}{\underset{1}{\cancel{8}}}x - \overset{4}{\cancel{24}} \cdot \frac{5}{\underset{1}{\cancel{6}}}y = \overset{12}{\cancel{24}} \cdot \frac{13}{\underset{1}{\cancel{2}}} \quad \text{Distribute and divide out common factors.}$$
$$9x - 20y = 156 \quad \text{Simplify.}$$

Here is the resulting system of equations.

$$15x + 4y = 36$$
$$9x - 20y = 156$$

Because the coefficient of the term containing y in the second equation is a multiple of the coefficient containing y in the first equation, we will eliminate y. This can be done by multiplying the first equation by 5.

$$\begin{aligned} 15x + 4y &= 36 \\ 9x - 20y &= 156 \end{aligned} \quad \xrightarrow{\text{Multiply by 5}} \quad \begin{aligned} 75x + 20y &= 180 \\ 9x - 20y &= 156 \end{aligned}$$

$$\begin{aligned} 75x + 20y &= 180 \\ 9x - 20y &= 156 && \text{Add to eliminate } y. \\ \hline 84x \qquad &= 336 \\ x &= 4 && \text{Divide both sides by 84.} \end{aligned}$$

The x-coordinate of the solution is 4. To solve for y, we can substitute 4 for x in either of the original equations, or we can substitute into either of the equations that was created by clearing the fractions. We will use the equation $15x + 4y = 36$.

$$\begin{aligned} 15(4) + 4y &= 36 && \text{Substitute 4 for } x. \\ 60 + 4y &= 36 && \text{Multiply.} \\ 4y &= -24 && \text{Subtract 60.} \\ y &= -6 && \text{Divide both sides by 4.} \end{aligned}$$

The solution of this system is the ordered pair $(4, -6)$.

Quick Check 6

Solve the system by addition.

$$-\frac{2}{3}x + \frac{5}{2}y = 14$$
$$\frac{5}{12}x + \frac{3}{4}y = \frac{1}{2}$$

If one or both equations in a system of linear equations contain decimals, we can clear the decimals by multiplying by the appropriate power of 10 and then solve the resulting system.

EXAMPLE 7 Solve the system by addition.

$$0.3x + 2y = 7.5$$
$$0.04x + 0.07y = 0.41$$

SOLUTION If we multiply both sides of the first equation by 10, we can clear the equation of decimals.

$$10(0.3x + 2y) = 10(7.5) \quad \text{Multiply both sides by 10.}$$
$$3x + 20y = 75 \quad \text{Distribute and multiply.}$$

For the second equation, multiply both sides by 100 to clear the decimals.

$$100(0.04x + 0.07y) = 100(0.41) \quad \text{Multiply both sides by 100.}$$
$$4x + 7y = 41 \quad \text{Distribute and multiply.}$$

Here is the resulting system of equations.

$$3x + 20y = 75$$
$$4x + 7y = 41$$

Eliminate x by multiplying the first equation by 4 and the second equation by -3.

$$3x + 20y = 75 \xrightarrow{\text{Multiply by 4}} 12x + 80y = 300$$
$$4x + 7y = 41 \xrightarrow{\text{Multiply by } -3} -12x - 21y = -123$$

$$\begin{aligned} 12x + 80y &= 300 \\ -12x - 21y &= -123 \quad \text{Add to eliminate } x. \\ \hline 59y &= 177 \\ y &= 3 \quad \text{Divide both sides by 59.} \end{aligned}$$

The y-coordinate of the solution is 3. To solve for x, substitute 3 for y in either of the original equations or in either of the equations that was created by clearing the decimals. The equation $3x + 20y = 75$ is a good choice, as it will be easier to solve for x in an equation that does not contain decimals.

$$3x + 20(3) = 75 \quad \text{Substitute 3 for } y.$$
$$3x + 60 = 75 \quad \text{Multiply.}$$
$$3x = 15 \quad \text{Subtract 60.}$$
$$x = 5 \quad \text{Divide both sides by 3.}$$

The solution to this system is the ordered pair $(5, 3)$.

Quick Check 7

Solve the system by addition.

$$0.03x + 0.05y = 29.5$$
$$x + y = 700$$

Which method is more efficient to use: substitution or addition?

The substitution and addition methods for solving a system of equations will solve any system of linear equations. There are certain times when one method is more efficient to apply than the other, and knowing which method to choose for a particular system of equations can save time and prevent errors. In general, if one of the equations in the system is already solved for one of the variables, such as $y = 3x$ or $x = -4y + 5$, using the substitution method is a good choice. The substitution method also works well when one of the equations contains a variable term with a coefficient of 1 or -1 (such as $x + 6y = 17$ or $7x - y = 14$). Otherwise, consider using the addition method.

Applications

Objective 4 Solve applied problems by using the addition method. We conclude this section with an application of solving systems of equations using the addition method.

EXAMPLE 8 Dylan has \$3.20 in change in his pocket. Each coin is either a dime or a quarter. If Dylan has 17 coins in his pocket, how many are dimes and how many are quarters?

SOLUTION There are two unknowns in this problem: the number of dimes and the number of quarters. We will let d represent the number of dimes and q represent the number of quarters.

The first equation in the system comes from the fact that Dylan has 17 coins. Because all of these coins are either dimes or quarters, the number of dimes (d) plus the number of quarters (q) must equal 17. The first equation is $d + q = 17$.

The second equation comes from the amount of money Dylan has in his pocket. The amount of money in dimes (\$0.10d) plus the amount of money in quarters (\$0.25q) must equal the total amount of money he has. The second equation is $0.10d + 0.25q = 3.20$.

	Number of Coins	Value per Coin (\$)	Money in Pocket (\$)
Dimes	d	0.10	$0.10d$
Quarters	q	0.25	$0.25q$
Total	17		3.20

Here is the system of equations that we need to solve.

$$\begin{aligned} d + \quad q &= 17 \\ 0.10d + 0.25q &= 3.20 \end{aligned}$$

Multiplying both sides of the second equation by 100 will clear the equation of decimals, producing the following system:

$$\begin{aligned} d + \quad q &= 17 \\ 10d + 25q &= 320 \end{aligned}$$

Now we can eliminate d by multiplying the first equation by -10.

$$\begin{aligned} d + \quad q &= 17 \xrightarrow{\text{Multiply by } -10} & -10d - 10q &= -170 \\ 10d + 25q &= 320 & 10d + 25q &= 320 \end{aligned}$$

$$\begin{aligned} -10d - 10q &= -170 \\ 10d + 25q &= 320 \quad \text{Add to eliminate } d. \\ \hline 15q &= 150 \\ q &= 10 \quad \text{Divide both sides by 15.} \end{aligned}$$

There are 10 quarters in Dylan's pocket. To find the number of dimes he has, we can substitute 10 for q in the equation $d + q = 17$ and solve for d.

$$\begin{aligned} d + (10) &= 17 \quad \text{Substitute 10 for } q. \\ d &= 7 \quad \text{Subtract 10.} \end{aligned}$$

Dylan has 7 dimes and 10 quarters in his pocket. You can verify that the total amount of money in his pocket is \$3.20.

Quick Check 8

Erica has \$5.35 in change in her pocket. Each coin is either a nickel or a dime. If Erica has 75 coins in her pocket, how many are nickels and how many are dimes?

BUILDING YOUR STUDY STRATEGY

Doing Your Homework, 3 Difficult Problems Eventually, there will be a homework exercise that you cannot answer correctly. Rather than giving up, here are some options to consider.

- Review your class notes. A similar problem may have been discussed in class.
- Review the related section in the text. You may be able to find a similar example, or there may be a Word of Caution warning you about typical errors on that type of problem.
- Call someone in your study group. A member of your study group may be able to help you figure out what to do.

If you still cannot solve the problem, move on and try the next problem. Ask your instructor about the problem at your next class.

Exercises 4.3

PRACTICE

WATCH

DOWNLOAD

READ

REVIEW

Vocabulary

1. If an equation in a system of equations contains fractions, you can clear the fractions by multiplying both sides of the equation by the ______ of the denominators.
2. If an equation in a system of equations contains decimals, you can clear the decimals by multiplying both sides of the equation by the appropriate power of ______.
3. If the addition method results in an equation that is an identity, the system is a(n) __________ system.
4. If the addition method results in an equation that is a contradiction, the system is a(n) __________ system.

Solve the system by addition. If the system is inconsistent and has no solution, state this. If the system is dependent, write the form of the solution for any real number x.

5. $5x + 4y = -56$
   $3x - 4y = -8$
6. $7x + 5y = 39$
   $-7x + 2y = -53$
7. $-3x + 6y = 36$
   $3x - 11y = -51$
8. $9x - 8y = 25$
   $-3x + 8y = 29$
9. $4x + 3y = 1$
   $13x - 6y = 82$
10. $8x + 7y = -26$
    $-4x + 9y = 138$
11. $2x - 4y = 11$
    $-4x + 8y = -22$
12. $8x - 3y = 34$
    $-2x + 9y = 8$
13. $x + 4y = 11$
    $3x + 16y = 37$
14. $10x + 2y = 44$
    $3x + 8y = 65$
15. $12x + 9y = -69$
    $3x + 5y = -31$
16. $2x + 3y = 13$
    $10x + 15y = 26$
17. $5x + 6y = -9$
    $-x + 9y = -39$
18. $8x + 3y = -33$
    $9x - y = -59$
19. $7x + 4y = 85$
    $11x - 6y = -63$
20. $8x + 5y = -87$
    $-10x + 9y = -59$
21. $-8x + 5y = -28$
    $3x + 8y = -29$
22. $5x - 5y = 45$
    $-4x + 4y = -36$
23. $4x - 3y = -6$
    $6x + 5y = 29$
24. $-7x + 4y = -25$
    $5x + 14y = 60$
25. $-6x + 9y = 17$
    $-4x + 6y = 12$
26. $3x - 11y = 43$
    $2x - 7y = 27$
27. $5x + 7y = 16$
    $4x + 8y = 14$

28. $4x + 6y = -20$
$6x + 13y = -48$

29. $4y = 3x - 13$
$6x + 5y = 52$

30. $7x = -18 - 4y$
$3x - 2y = -30$

31. $5x = 3y + 21$
$-5x + 4y = -23$

32. $5y = 2x + 13$
$4x + 3y = 65$

33. $\frac{1}{2}x + \frac{2}{5}y = \frac{7}{5}$
$\frac{1}{4}x - \frac{1}{6}y = \frac{1}{3}$

34. $\frac{1}{4}x + \frac{1}{2}y = \frac{11}{4}$
$-\frac{1}{9}x + \frac{1}{3}y = \frac{4}{9}$

35. $\frac{2}{3}x + \frac{1}{6}y = 4$
$\frac{5}{3}x - y = -7$

36. $\frac{1}{7}x + \frac{1}{28}y = \frac{1}{2}$
$\frac{1}{3}x - \frac{2}{9}y = -\frac{23}{18}$

37. $x + y = 3600$
$0.03x + 0.07y = 144$

38. $x + y = 6000$
$0.04x + 0.06y = 330$

39. $x + y = 120$
$8.50x + 5.75y = 827.50$

40. $x + y = 100$
$0.25x + 0.17y = 23.40$

41. $x + y = 100$
$0.36x + 0.48y = 0.45(100)$

42. $x + y = 60$
$0.42x + 0.22y = 0.36(60)$

43. A man bought roses and carnations for his wife on Valentine's Day. Each rose cost \$6, and each carnation cost \$4. If the man spent \$68 for a total of 15 roses and carnations, how many of each type of flower did he buy?

44. For matinees, a movie theater charges \$7 per adult and \$4 per child. If 84 people attended a matinee and paid a total of \$432, how many adults and how many children were there?

45. A family purchased 7 trees and 4 shrubs from a local nursery and paid a total of \$312. Another family purchased 4 trees and 15 shrubs for a total of \$280. How much does the nursery charge for each tree and each shrub?

46. A large group at Maxine's Pizzeria purchased 6 pizzas and 4 pitchers of soda for \$114. Another group bought 9 pizzas and 7 pitchers of soda for \$175.50. How much does the pizzeria charge for each pizza and each pitcher of soda?

47. Alycia opens her piggy bank to find only dimes and nickels inside. She counts the coins and finds that she has 81 coins totaling \$6.50. How many dimes and how many nickels were in the piggy bank?

48. Charlotte has 24 postage stamps. Some of the stamps are 28¢ stamps, and the rest are 44¢ stamps. If the total value of the stamps is \$9.12, how many of each type of stamp does Charlotte have?

49. Jessica has a handful of nickels and quarters worth a total of \$6.95. If she has seven more nickels than quarters, how many of each type of coin does she have?

50. The cash box at a bake sale contains nickels, dimes, and quarters. The number of quarters is four more than three times the number of dimes. The number of nickels is the same as the number of dimes. If the cash box contains \$16.30 in coins, how many of each type of coin are in the box?

Mixed Practice, 51–62

Solve by substitution or addition.

51. $5x - 2y = 47$
$3x + 4y = 23$

52. $x = 4y + 12$
$3x + 2y = -34$

53. $y = 2x + 20$
$8x - 5y = -86$

54. $7x - 4y = -81$
$4x + 5y = -141$

55. $y = \frac{3}{4}x + 1$
$3x + 8y = 116$

56. $\frac{2}{3}x - \frac{1}{2}y = 6$
$\frac{5}{12}x + \frac{3}{8}y = 1$

57. $y = 4x - 8$
$-12x + 3y = -24$

58. $14x + 21y = 28$
$12x + 18y = -24$

59. $13x - 8y = 20$
$9x + 10y = -126$

60. $x = -2y + 1$
$-3x + 17y = -72$

61. $y = 6x - 14$
$4x + 5y = -19$

62. $2x - 6y = 13$
$8x + 9y = \frac{115}{2}$

Writing in Mathematics

63. ***Newsletter*** Write a newsletter explaining how to solve a system of two linear equations by using the addition method.

4.4 Applications of Systems of Equations

OBJECTIVES

1. Solve applied problems involving systems of equations by using the substitution method or the addition method.
2. Solve geometry problems by using a system of equations.
3. Solve interest problems by using a system of equations.
4. Solve mixture problems by using a system of equations.
5. Solve motion problems by using a system of equations.

Objective 1 Solve applied problems involving systems of equations by using the substitution method or the addition method. This section focuses on applications involving systems of linear equations with two variables. We will use both the substitution and addition methods to solve these systems. When trying to solve a system, use the method you believe will be more efficient.

We begin to solve an applied problem by identifying the two unknown quantities and choosing a variable to represent each quantity. We then need to find two equations that relate these quantities.

EXAMPLE 1 Ross and Carolyn combined their DVD collection when they got married. Ross had 42 more DVDs than Carolyn had. Together, they have 148 DVDs. How many DVDs did each person have?

SOLUTION The two unknown quantities in this problem are the number of DVDs that Ross had and the number of DVDs that Carolyn had. We will let r represent the number of Ross's DVDs and c represent the number of Carolyn's DVDs. We choose the variables r and c as a reminder that they stand for ***R***oss's DVDs (r) and ***C***arolyn's DVDs (c). One sentence in the problem tells us that Ross had 42 more DVDs than Carolyn had. This translates to the equation $r = c + 42$. We need a second equation, and this can be found from the sentence that tells us that together Ross and Carolyn have 148 DVDs. This translates to the equation $r + c = 148$. Here is the system of equations we need to solve.

$$r = c + 42$$
$$r + c = 148$$

We will solve this system of equations using the substitution method because the first equation has already been solved for r. We will substitute the expression $c + 42$ for the variable r in the second equation.

$$r + c = 148$$
$$(c + 42) + c = 148 \quad \text{Substitute } c + 42 \text{ for } r.$$
$$2c + 42 = 148 \quad \text{Combine like terms.}$$
$$2c = 106 \quad \text{Subtract 42.}$$
$$c = 53 \quad \text{Divide both sides by 2.}$$

Carolyn had 53 DVDs. To find out how many DVDs Ross had, we can substitute 53 for c in the equation $r = c + 42$.

$$r = (53) + 42 \quad \text{Substitute 53 for } c.$$
$$r = 95 \quad \text{Add.}$$

Ross had 95 DVDs, and Carolyn had 53 DVDs. Combined, this is a total of 148 DVDs.

Quick Check 1

At a community college, there are 68 instructors who teach math or English. The number of English instructors is 12 more than the number of math instructors. How many English instructors are there?

The setup for the previous example can be used for any problem in which we know the sum and difference of two quantities. The same system of equations would be used to find two numbers whose difference was 42 and whose sum was 148.

Geometry Problems

Objective 2 Solve geometry problems by using a system of equations. In Section 2.3, we solved problems involving the perimeter of a rectangle. We will now use a system of two equations in two variables to solve the same type of problem.

EXAMPLE 2 The length of a rectangle is 5 feet more than twice its width. If the perimeter of the rectangle is 124 feet, find the length and width of the rectangle.

SOLUTION The two unknowns are the length and the width of the rectangle. We will use the variables l and w for the length and width, respectively.

We are told that the length is 5 feet more than twice the width of the rectangle. We express this relationship in an equation as $l = 2w + 5$. The second equation in the system comes from the fact that the perimeter is 124 feet. Because the perimeter of a rectangle is equal to twice the length plus twice the width, the second equation can be written as $2l + 2w = 124$. Here is the system we are solving.

$$l = 2w + 5$$
$$2l + 2w = 124$$

We will use the substitution method to solve this system, as the first equation is already solved for l. We begin by substituting the expression $2w + 5$ for l in the equation $2l + 2w = 124$.

$$2(2w + 5) + 2w = 124 \quad \text{Substitute } 2w + 5 \text{ for } l.$$
$$4w + 10 + 2w = 124 \quad \text{Distribute.}$$
$$6w + 10 = 124 \quad \text{Combine like terms.}$$
$$6w = 114 \quad \text{Subtract 10.}$$
$$w = 19 \quad \text{Divide both sides by 6.}$$

The width of the rectangle is 19 feet. To solve for the length, substitute 19 for w in the equation $l = 2w + 5$.

$$l = 2(19) + 5 \quad \text{Substitute 19 for } w.$$
$$l = 43 \quad \text{Simplify.}$$

The length of the rectangle is 43 feet, and the width is 19 feet.

Quick Check 2

A rectangular concrete slab is being poured for a new house. The perimeter of the slab is 230 feet, and the length of the slab is 35 feet longer than the width. Find the dimensions of the concrete slab.

Interest Problems

Objective 3 Solve interest problems by using a system of equations. For certain problems, it's easier to set up a system of equations in two variables than to try to express both unknown quantities in terms of the same variable. For example, the mixture and interest problems covered in Section 2.4 are often easier to solve when working with a system of equations in two variables. We will begin with a problem involving interest.

EXAMPLE 3 Serena invested \$8000 in two certificates of deposit (CDs). Some of the money was deposited in a CD that paid 7% annual interest, and the rest was deposited in a CD that paid 6% annual interest. If Serena earned \$530 in interest in the first year, how much did she invest in each CD?

SOLUTION The two unknowns are the amount invested at 7% and the amount invested at 6%. We will let x represent the amount invested at 7% and y represent the amount invested at 6%. To determine the interest earned, we multiply the principal (the amount of money invested) by the interest rate by the time in years.

If the first account earns 7% interest, we can represent the interest earned in one year by $0.07x$. In a similar fashion, we can represent the interest earned in one year from the investment at 6% as $0.06y$. The following table summarizes this information:

Account	Principal	Interest Rate	Time (years)	Interest Earned
CD 1	x	0.07	1	$0.07x$
CD 2	y	0.06	1	$0.06y$
Total	8000			530

The first equation in the system comes from the fact that the amount invested at 7% plus the amount invested at 6% is equal to the total amount invested. As an equation, this can be represented as $x + y = 8000$. This equation can be found in the table in the column labeled "Principal." The second equation comes from the amount of interest earned and can be found in the column labeled "Interest Earned." We know that Serena earned $0.07x$ in interest from the first CD and $0.06y$ in interest from the second CD. Because the total amount of interest earned comes from these two CDs, the second equation in the system is $0.07x + 0.06y = 530$. Here is the system we must solve.

$$\begin{aligned} x + y &= 8000 \\ 0.07x + 0.06y &= 530 \end{aligned}$$

We will clear the second equation of decimals by multiplying both sides of the equation by 100.

$$\begin{aligned} x + \quad y &= 8000 \\ 0.07x + 0.06y &= 530 \end{aligned} \xrightarrow{\text{Multiply by 100}} \begin{aligned} x + y &= 8000 \\ 7x + 6y &= 53{,}000 \end{aligned}$$

To solve this system of equations, we can use the addition method. (The substitution method would work just as well.) If we multiply the first equation by -6, the variable terms containing y will be opposites.

$$\begin{aligned} x + y &= 8000 \\ 7x + 6y &= 53{,}000 \end{aligned} \quad \xrightarrow{\text{Multiply by } -6} \quad \begin{aligned} -6x - 6y &= -48{,}000 \\ 7x + 6y &= 53{,}000 \end{aligned}$$

$$\begin{aligned} -6x - 6y &= -48{,}000 \\ 7x + 6y &= 53{,}000 \quad \text{Add to eliminate } y. \\ \hline x \quad &= 5{,}000 \end{aligned}$$

Because x represented the amount invested at 7% interest, we know that Serena invested \$5000 at 7%. To find the amount invested at 6%, we substitute 5000 for x in the original equation $x + y = 8000$.

$$\begin{aligned} (5000) + y &= 8000 \quad \text{Substitute 5000 for } x. \\ y &= 3000 \quad \text{Subtract 5000.} \end{aligned}$$

Serena invested \$5000 at 7% interest and \$3000 at 6% interest.

Quick Check 3

Annika invested \$4200 in two certificates of deposit (CDs). Some of the money was deposited in a CD that paid 5% annual interest, and the rest was deposited in a CD that paid 4% annual interest. If Annika earned \$200 in interest in the first year, how much did she invest in each CD?

Mixture Problems

Objective 4 Solve mixture problems by using a system of equations.

EXAMPLE 4 Gunther works at a coffee shop that sells Kona coffee beans for \$40 per pound and Colombian coffee beans for \$15 per pound. The coffee shop also sells a blend of Kona coffee and Colombian coffee for \$20 per pound. If Gunther's boss tells him to make 30 pounds of this blend, how much Kona coffee and Colombian coffee must be mixed together?

SOLUTION The two unknowns are the number of pounds of Kona coffee (k) and the number of pounds of Colombian coffee (c) that need to be mixed together to form the blend. Because Gunther needs to make 30 pounds of coffee with a value of \$20 per pound, the total value of this mixture can be found by multiplying 30 by 20. The total cost of this blend is \$600. The following table summarizes this information:

Coffee	Pounds	Cost per Pound	Total Cost
Kona	k	40	$40k$
Colombian	c	15	$15c$
Blend	30	20	600

The first equation in the system comes from the fact that the weight of the Kona coffee plus the weight of the Colombian coffee must equal 30 pounds. As an equation, this can be represented by $k + c = 30$. This equation can be found in the table in the column labeled "Pounds." The second equation comes from the total cost of the coffee. We know that the total cost is \$600, so the second equation in the system is $40k + 15c = 600$. This equation can be found in the column labeled "Total Cost." Here is the system we must solve.

$$\begin{aligned} k + c &= 30 \\ 40k + 15c &= 600 \end{aligned}$$

To solve this system of equations, we can use the addition method. If we multiply the first equation by -15, the variable terms containing c will be opposites.

$$\begin{aligned} k + \quad c &= 30 & \xrightarrow{\text{Multiply by } -15} \quad -15k - 15c &= -450 \\ 40k + 15c &= 600 & 40k + 15c &= \quad 600 \end{aligned}$$

$$\begin{aligned} -15k - 15c &= -450 \\ 40k + 15c &= \quad 600 && \text{Add to eliminate } c. \\ \hline 25k \qquad &= \quad 150 \\ k &= \quad 6 && \text{Divide both sides by 25.} \end{aligned}$$

We know that Gunther must use 6 pounds of Kona coffee. To determine how much Colombian coffee Gunther must use, we substitute 6 for k in the equation $k + c = 30$.

$$\begin{aligned} (6) + c &= 30 && \text{Substitute 6 for } k. \\ c &= 24 && \text{Subtract 6.} \end{aligned}$$

Gunther must use 6 pounds of Kona coffee and 24 pounds of Colombian coffee.

Quick Check 4

Hannibal is making 48 pounds of a mixture of almonds and cashews that his nut shop will sell for \$6 per pound. If the nut shop sells almonds for \$5 per pound and cashews for \$8 per pound, how many pounds of almonds and cashews should Hannibal mix together?

The previous example is a mixture problem in which two items of different values or costs are combined into a mixture with a specific cost. In other mixture problems, two liquid solutions of different concentration levels are mixed together to form a single solution with a given strength, as in the next example.

EXAMPLE 5 Woody has one solution that is 28% alcohol and a second solution that is 46% alcohol. He wants to combine these solutions to make 45 liters of a solution that is 40% alcohol. How many liters of each original solution need to be used?

SOLUTION The two unknowns are the volume of the 28% alcohol solution and the volume of the 46% alcohol solution that need to be mixed together to make 45 liters of a solution that is 40% alcohol.

Solution	Volume of Solution	% Concentration of Alcohol	Volume of Alcohol
28%	x	0.28	$0.28x$
46%	y	0.46	$0.46y$
Mixture (40%)	45	0.40	18

The first equation in the system comes from the fact that the volume of the two solutions must equal 45 liters. As an equation, this can be represented by $x + y = 45$. This equation can be found in the table in the column labeled "Volume of Solution." The second equation comes from the total volume of alcohol, under the column labeled "Volume of Alcohol." We know that the total volume of alcohol is 18 liters (40% of the 45 liters in the mixture is alcohol), so the second equation in the system is $0.28x + 0.46y = 18$. Here is the system we must solve.

$$\begin{aligned} x + y &= 45 \\ 0.28x + 0.46y &= 18 \end{aligned}$$

We begin by clearing the second equation of decimals. This can be done by multiplying both sides of the second equation by 100.

$$\begin{aligned} x + \quad y &= 45 \\ 0.28x + 0.46y &= 18 \end{aligned} \xrightarrow{\text{Multiply by 100}} \begin{aligned} x + \quad y &= 45 \\ 28x + 46y &= 1800 \end{aligned}$$

To solve this system of equations, we can use the addition method. If we multiply the first equation by -28, the variable terms containing x will be opposites.

$$\begin{aligned} x + \quad y &= 45 \\ 28x + 46y &= 1800 \end{aligned} \xrightarrow{\text{Multiply by } -28} \begin{aligned} -28x - 28y &= -1260 \\ 28x + 46y &= 1800 \end{aligned}$$

$$\begin{aligned} -28x - 28y &= -1260 \\ 28x + 46y &= 1800 \quad \text{Add to eliminate } x. \\ \hline 18y &= 540 \\ y &= 30 \quad \text{Divide both sides by 18.} \end{aligned}$$

We know that Woody must use 30 liters of the solution that is 46% alcohol. To determine how much of the 28% alcohol solution Woody must use, we substitute 30 for y in the equation $x + y = 45$.

$$\begin{aligned} x + (30) &= 45 \quad \text{Substitute 30 for } y. \\ x &= 15 \quad \text{Subtract 30.} \end{aligned}$$

Woody must use 15 liters of the 28% alcohol solution and 30 liters of the 46% alcohol solution.

Quick Check 5

A chemist has one solution that is 60% acid and a second solution that is 50% acid. She wants to combine the solutions to make a new solution that is 54% acid. If she needs to make 400 milliliters of this new solution, how many milliliters of the two solutions should be mixed?

Motion Problems

Objective 5 Solve motion problems by using a system of equations. Now we will turn our attention to motion problems, which were introduced in Section 2.3. Recall that the equation for motion problems is $d = r \cdot t$, where r is the rate of speed, t is the time, and d is the distance traveled. For example, if a car is traveling at a speed of 50 miles per hour for 3 hours, the distance traveled is $50 \cdot 3$, or 150, miles.

EXAMPLE 6 Efrain is finishing his training for the Shaver Lake Triathlon. Yesterday he focused on running and biking, spending a total of 5 hours on the two activities. Efrain runs at a speed of 10 miles per hour and rides his bike at a speed of 30 miles per hour. If Efrain covered a total of 120 miles yesterday, how much time did he spend running and how much time did he spend riding?

SOLUTION The two unknowns in this problem are the amount of time Efrain ran and the amount of time he rode his bike. We will let x represent the time Efrain ran and y represent the amount of time he rode his bike. All of the information is displayed in the following table:

	Rate	Time	Distance ($d = r \cdot t$)
Running	10	x	$10x$
Biking	30	y	$30y$
Total		5	120

We know that Efrain spent a total of 5 hours on these two activities, so $x + y = 5$. We also know that the total distance traveled was 120 miles. This leads to the equation $10x + 30y = 120$. Here is the system of equations we must solve.

$$\begin{aligned} x + y &= 5 \\ 10x + 30y &= 120 \end{aligned}$$

We can solve the first equation for x ($x = 5 - y$), so we will use the substitution method to solve this system of equations. We substitute the expression $5 - y$ for the variable x in the equation $10x + 30y = 120$.

$$10(5 - y) + 30y = 120 \quad \text{Substitute } 5 - y \text{ for } x.$$
$$50 - 10y + 30y = 120 \quad \text{Distribute.}$$
$$50 + 20y = 120 \quad \text{Combine like terms.}$$
$$20y = 70 \quad \text{Subtract 50.}$$
$$y = \frac{7}{2} \quad \text{Divide both sides by 20 and simplify.}$$

Writing the fraction $\frac{7}{2}$ as a mixed number, we find that Efrain spent $3\frac{1}{2}$ hours riding his bike. To find out how long he was running, we substitute $3\frac{1}{2}$ for y in the equation $x = 5 - y$ and solve for x.

$$x = 5 - 3\frac{1}{2} \quad \text{Substitute } 3\frac{1}{2} \text{ for } y.$$
$$x = 1\frac{1}{2} \quad \text{Subtract.}$$

Efrain spent $1\frac{1}{2}$ hours running and $3\frac{1}{2}$ hours riding his bike.

Quick Check 6

On a trip to Las Vegas, Fong drove at an average speed of 70 miles per hour except for a period when he was driving through a construction zone. In the construction zone, Fong's average speed was 40 miles per hour. If it took Fong 5 hours to drive 335 miles, how long was he in the construction zone?

Suppose a boat travels at a speed of 15 miles per hour in still water. The boat will travel faster than 15 miles per hour while heading downstream because the speed of the current is pushing the boat forward as well. The boat will travel slower than 15 miles per hour while heading upstream because the speed of the current is pushing the boat back. If the speed of the current is 3 miles per hour, this boat will travel 18 miles per hour downstream $(15 + 3)$ and 12 miles per hour upstream $(15 - 3)$. The speed of an airplane is affected in a similar fashion depending on whether it is flying with or against the wind.

EXAMPLE 7 Juana's motorboat can take her downstream from her campsite to the nearest store, which is 27 miles away, in 1 hour. Returning upstream from the store to the campsite takes 3 hours. How fast is Juana's boat in still water, and what is the speed of the current?

SOLUTION The two unknowns in this problem are the speed of the boat in still water and the speed of the current. We will let b represent the speed of the boat in still water, and we will let c represent the speed of the current. The information for this problem can be summarized in the following table:

	Rate	Time	Distance $(d = r \cdot t)$
Downstream	$b + c$	1	$1(b + c) = b + c$
Upstream	$b - c$	3	$3(b - c) = 3b - 3c$

Because the distance in each direction is 27 miles, we know that both $b + c$ and $3b - 3c$ are equal to 27. Here is the system we need to solve.

$$b + c = 27$$
$$3b - 3c = 27$$

We will solve this system of equations by using the addition method. Multiplying both sides of the first equation by 3 will help us eliminate the variable c.

$$\begin{aligned} b + c &= 27 \xrightarrow{\text{Multiply by 3}} & 3b + 3c &= 81 \\ 3b - 3c &= 27 & 3b - 3c &= 27 \end{aligned}$$

$$\begin{aligned} 3b + 3c &= 81 \\ 3b - 3c &= 27 \quad \text{Add to eliminate } c. \\ \hline 6b &= 108 \\ b &= 18 \quad \text{Divide both sides by 6.} \end{aligned}$$

The speed of the boat in still water is 18 miles per hour. To find the speed of the current, we substitute 18 for b in the original equation $b + c = 27$.

$$\begin{aligned} (18) + c &= 27 \quad \text{Substitute 18 for } b. \\ c &= 9 \quad \text{Subtract 18.} \end{aligned}$$

The speed of the current is 9 miles per hour.

Quick Check 7

An airplane traveling with a tailwind can make a 2250-mile trip in $4\frac{1}{2}$ hours. However, traveling into the same wind would take 5 hours to fly 2250 miles. What is the speed of the plane in calm air, and what is the speed of the wind?

BUILDING YOUR STUDY STRATEGY

Doing Your Homework, 4 **Homework Diary** Once you finish the last exercise of a homework assignment, it is not necessarily time to stop. Use a homework diary to try summarizing what you just accomplished. In this diary, you can list the types of problems you solved, the types of problems you struggled with, and important notes about these problems. When you begin preparing for a quiz or an exam, this diary will help you to decide where to focus your attention. If you are using note cards as a study aid, this would be a good time to create a set of study cards for this section.

Exercises 4.4

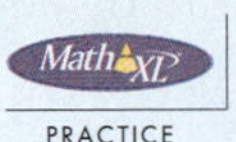
PRACTICE

WATCH

DOWNLOAD

READ

REVIEW

Vocabulary

1. State the formula for the perimeter of a rectangle.

2. State the formula for calculating simple interest.

3. State the formula for calculating distance.

4. A chemist is mixing a solution that is 35% alcohol with a second solution that is 55% alcohol to create 80 milliliters of a solution that is 40% alcohol. Which system of equations can be used to solve this problem?

   a) $x + y = 40$
   $0.35x + 0.55y = 80$

   b) $x + y = 80$
   $0.35x + 0.55y = 0.40$

   c) $x + y = 80$
   $0.35x + 0.55y = 0.40(80)$

5. A mathematics instructor is teaching an elementary algebra class and a statistics class. She has 17 more students in her statistics class than she has in her elementary algebra class. The two classes have a total of 93 students. How many students does she have in each class?

6. A hardcover math book is 289 pages longer than a paperback math book. If the total number of pages in these two books is 1691, how many pages are in each book?

7. Andre's father is 28 years older than Andre. If you added Andre's age to his father's age, the total would be 98. How old is Andre? How old is his father?

8. Maggie's ma tells everyone that she is 14 years younger than she actually is. If you add her actual age to the age that she tells people she is, the total is 122. What is the actual age of Maggie's ma (as reported by Bob Dylan)?

9. Two baseball players hit 79 home runs combined last season. The first player hit 7 more home runs than twice the number of home runs the second player hit. How many home runs did each player hit?

10. A piece of wire 240 centimeters long was cut into two pieces. The longer of the two pieces is 25 centimeters longer than four times the length of the shorter piece. How long is each piece?

11. In football, a team gets 3 points when its kicker makes a field goal attempt and 1 point when he makes an extra point attempt. Last season a team scored 77 points from a total of 47 successful kicks. How many field goals and how many extra points did the team's kicker make last year?

12. A salesperson earns a $10 commission for each printer he sells and a $15 commission for each computer he sells. Last month the salesperson made 42 sales that resulted in a total commission of $590. How many printers and how many computers did he sell?

13. On an exam, students earn 5 points for each correct response and lose 2 points for each incorrect response. Arturo answered 27 questions on the exam and earned a score of 93 points. How many of his responses were correct?

14. In Pai-Gow poker, a player earns a profit of $9.50 for each hand won, loses $10 for each hand lost, and breaks even in case of a tie. Moriko played 82 hands and tied 46 times. Of the remaining hands, she made a profit of $88.50. How many hands did she win?

15. Benedita has a jar containing 55 coins. Some are dimes, and the rest are nickels. If the value of the coins is $4.50, how many dimes does she have?

16. A movie theater charges $9.50 for an adult and $7.75 for a child to see a movie. If 92 people were at the 9:20 P.M. show and the total admission paid was $837.25, how many adults were at the movie?

17. Last week a softball team ordered three pizzas and two pitchers of soda after a game, and the bill was $41. This week they ordered five pizzas and three pitchers of soda, and the bill was $67. What is the price of a single pizza, and what is the price of a pitcher of soda?

18. Two business colleagues went shopping together. The first bought three shirts and one tie for a total of $182. The second bought two shirts and five ties for a total of $273. If the price for each shirt was the same and the price for each tie was the same, what is the cost of one shirt and what is the cost of one tie?

19. The perimeter of a rectangle is 92 inches. The width of the rectangle is 8 inches less than the length of the rectangle. Find the dimensions of the rectangle.

20. An oil painting is in the shape of a rectangle. The perimeter of the painting is 127.4 inches, and the length of the painting is 1.6 times the width of the painting. Find the dimensions of the painting.

21. The Woodbury vegetable garden is in the shape of a rectangle, surrounded by 330 feet of fence. The length

of the garden is 75 feet more than twice the width of the garden. Find the dimensions of this garden.

22. Tina needs 302 inches of fabric to use as binding around a rectangular quilt she is making. The width of the quilt is 10 inches less than the length of the quilt. Find the dimensions of the quilt.

23. The perimeter of a rectangular table is 500 centimeters. If the length of the table is 25 centimeters more than twice the width of the table, find the dimensions of the table.

24. Bjorn is building a tennis court in his backyard. He poured a rectangular concrete slab for the court and surrounded it with 360 feet of fencing. If the width of the slab is half the slab's length, find the dimensions of the concrete slab.

25. A swimming pool is in the shape of a rectangle. The length of the pool is 15 feet more than the width. A concrete deck 6 feet wide is added around the pool, and the outer perimeter of the deck is 178 feet.

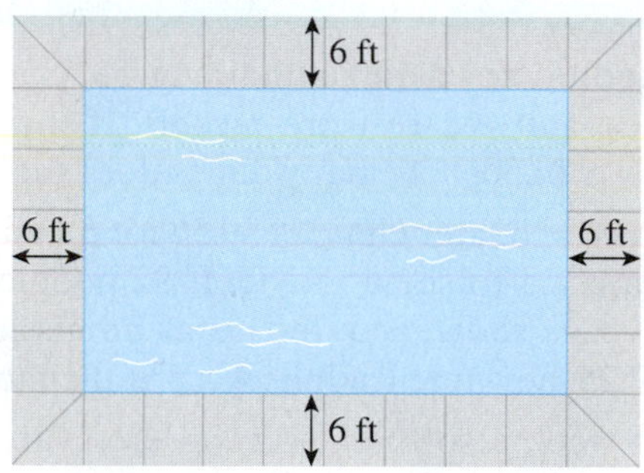

Find the dimensions of the swimming pool.

26. Next to a building, a farmer fences in a rectangular area as a pasture for her horses, using the existing building as the fourth side of the rectangle, as shown in the following diagram:

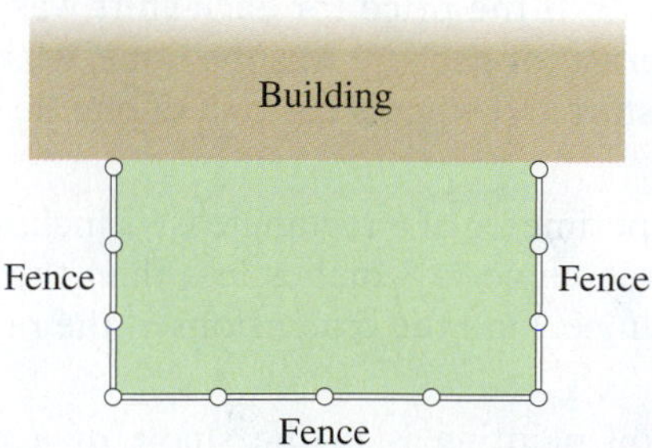

The side of the fence parallel to the building is 35 feet longer than either of the other two sides. If the farmer used 170 feet of fencing, find the dimensions of the pasture.

27. Corrine invested a total of $2000 in two mutual funds. One fund earned a 6% profit, while the other fund earned a 3% profit. If Corrine's total profit was $84, how much was invested in each mutual fund?

28. Angela deposited a total of $5200 in two different certificates of deposit (CDs). One CD paid 3% annual interest, and the other paid 4% annual interest. At the end of one year, she had earned $203 in interest from the two CDs. How much did Angela invest in each CD?

29. Mark deposited a total of $25,000 in two different CDs. One CD paid 4.25% annual interest, and the other paid 3.5% annual interest. At the end of one year, he had earned $1025 in interest from the two CDs. How much did Mark invest in each CD?

30. Maria invested $5500 in two mutual funds. One fund earned an 8% profit, while the other earned a 1.5% profit. Between the two accounts, Maria made a profit of $180. How much was invested in each account?

31. Fernando invested a total of $7500 in two mutual funds. While one fund earned a 4% profit, the other fund had a loss of 13%. Despite the loss, Fernando made a profit of $45 from the two funds. How much was invested in each fund?

32. Colin has had a bad run of luck in the stock market. He invested a total of $40,000 in two mutual funds. The first fund had a loss of 11%, and the second fund had a loss of 30%. Between the two funds, Colin lost $10,100. How much did he invest in each fund?

33. Jelani has two types of granola. One with dried fruit sells for $11.50 per pound, and one without dried fruit sells for $9.00 per pound. If Jelani wants to mix these two granolas to make 40 pounds of granola that can be sold for $10 per pound, how many pounds of each should she mix?

34. A feed store sells two types of cattle feed. One sells for $180 per ton, and another sells for $210 per ton. If the store manager wants to mix these two feeds to produce 60 tons of a mixture that can be sold for $200 per ton, how much of each feed should be mixed?

35. A coffee shop has pure Kona beans that sell for $45 per pound. The owner wants to create 80 pounds of a Kona blend mixture that can be sold for $20 per pound by mixing the Kona beans with some regular beans that sell for $5 per pound. How many pounds of each bean should the owner use?

36. Raheem can buy peanuts for $0.40 per pound and cashews for $1.05 per pound. How many pounds of each type should he buy to produce 120 pounds of a mixture that costs him $0.66 per pound?

37. A chemist has one solution that is 80% acid and a second solution that is 44% acid. She wants to combine the solutions to make a new solution that is 53% acid. If she needs to make 320 milliliters of this new solution, how many milliliters of the two solutions should be mixed?

38. A metallurgist has one brass alloy that is 65% copper and a second brass alloy that is 41% copper. He needs to melt and combine these alloys in such a way to make 60 grams of an alloy that is 47% copper. How much of each alloy should he use?

39. A bartender has two liquors—one that is 20% alcohol and another that is 50% alcohol. She needs to mix these liquors in such a way that she has 18 ounces of a mixture that is 30% alcohol. How much of each type of liquor does she need to mix together?

40. A farmer has two types of feed—one that is 12% protein and another that is 18% protein. How many pounds of each need to be combined to make a mixture of 36 pounds of feed that is 13.5% protein?

41. A bartender mixes a batch of vodka (40% alcohol) and tonic (no alcohol). How much of each needs to be mixed to make 4 liters of vodka and tonic that is 22% alcohol?

42. An auto mechanic has an antifreeze solution that is 10% antifreeze. He needs to mix this solution with pure antifreeze to produce 5 liters of a solution that is 19% antifreeze. How much of each does he need to mix?

43. George and Tina drove from their house to Palm Springs for a vacation. During the first part of the trip, George drove at a speed of 70 miles per hour. They switched drivers, and Tina drove the rest of the way at a speed of 85 miles per hour. It took them 6 hours to reach Palm Springs, which was 480 miles away. Find the length of time George drove as well as the length of time Tina drove.

44. Matt was driving to a conference in Salt Lake City, Utah. During the first day of driving, he averaged 72 miles per hour. On the second day, he drove at an average speed of 80 miles per hour in order to arrive before sundown. If his total driving time for the 2 days was 15 hours and the distance traveled was 1148 miles, what was Matt's driving time each day?

45. A racecar driver averaged 160 miles per hour over the first part of the Sequoia 500, but engine trouble dropped her average speed to 130 miles per hour for the rest of the race. If the 500-mile race took the driver 3.5 hours to complete, after what period did her engine trouble start?

46. A truck driver started for his destination at an average speed of 60 miles per hour. Once snow began to fall, he had to drop his speed to 40 miles per hour. It took the truck driver 9 hours to reach his destination, which was 425 miles away. How many hours after the driver started did the snow begin to fall?

47. A kayak can travel 18 miles downstream in 3 hours, while it would take 9 hours to make the same trip upstream. Find the speed of the kayak in still water as well as the speed of the current.

48. A rowboat would take 3 hours to travel 6 miles upstream, while it would take only $\frac{3}{4}$ hour to travel 6 miles downstream. Find the speed of the rowboat in still water as well as the speed of the current.

49. An airplane traveling with a tailwind can make a 3000-mile trip in 5 hours. However, traveling into the same wind would take 6 hours to fly 3000 miles. What is the speed of the plane in calm air, and what is the speed of the wind?

50. With a tailwind behind it, a small plane can fly 240 miles in 3 hours. Flying the same distance into the same wind would take 4 hours. Find the speed of the plane with no wind.

Writing in Mathematics

51. Write a word problem whose solution is "The length of the rectangle is 15 meters and the width is 8 meters."

52. Write a word problem whose solution is "$5000 at 6% and $3000 at 5%."

53. Write a word problem whose solution is "80 milliliters of 40% acid solution and 10 milliliters of 58% acid solution."

Quick Review Exercises

Section 4.4

Graph. Label any x- and y-intercepts.

1. $x - 4y = 6$

2. $y = 2x - 5$

3. $y = -\frac{1}{4}x - 2$

4. $y = 3$

Systems of Linear Inequalities

OBJECTIVES

1. **Solve a system of linear inequalities by graphing.**
2. **Solve a system of linear inequalities with more than two inequalities.**

In addition to systems of linear equations, we can have a **system of linear inequalities.** A system of linear inequalities is made up of two or more linear inequalities. As with systems of equations, an ordered pair is a solution to a system of linear inequalities if it is a solution to each linear inequality in the system.

Solving a System of Linear Inequalities by Graphing

Objective 1 Solve a system of linear inequalities by graphing. To find the solution set to a system of linear inequalities, we begin by graphing each inequality separately. If the solutions to the individual inequalities intersect, the region of intersection is the solution set to the system of inequalities.

EXAMPLE 1 Graph the system of inequalities.

$$y \geq 2x - 5$$
$$y < 3x - 7$$

SOLUTION

Begin by graphing the inequality $y \geq 2x - 5$. To do this, graph the line $y = 2x - 5$ as a solid line. Recall that we use a solid line when the points on the line are solutions to the inequality. Because this equation is in slope–intercept form, an efficient way to graph it is by plotting the y-intercept at $(0, -5)$ and then using the slope of 2 to find additional points on the line.

Because the origin is not on this line, we can use $(0, 0)$ as a test point. Substitute 0 for x and 0 for y in the inequality $y \geq 2x - 5$.

$$0 \geq 2(0) - 5$$
$$0 \geq 0 - 5$$
$$0 \geq -5$$

Substituting these coordinates into the inequality $y \geq 2x - 5$ produces a true statement; so we shade the half-plane containing $(0, 0)$.

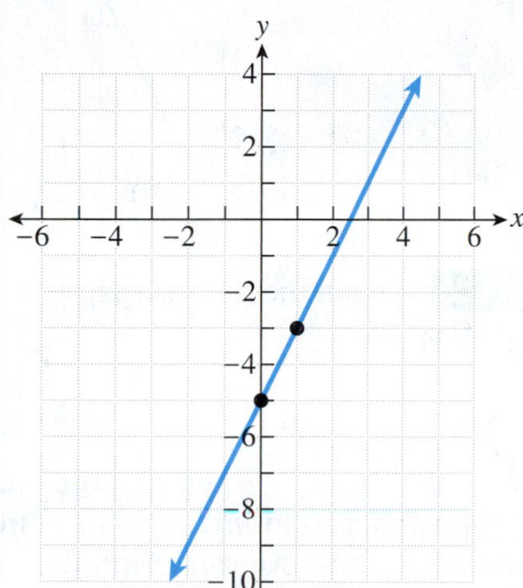

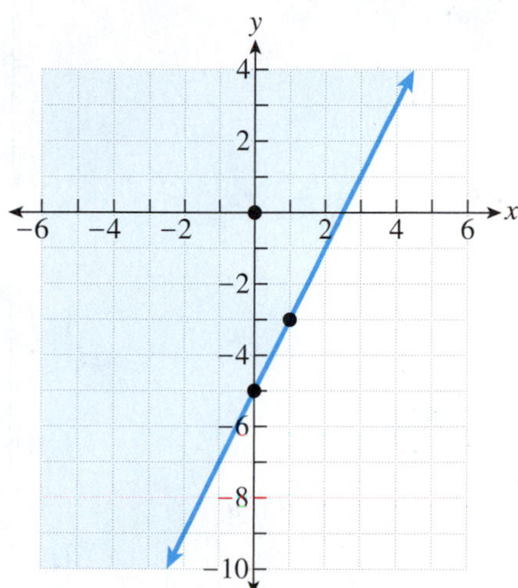

Now we turn our attention to the second inequality in the system. To graph the inequality $y < 3x - 7$, we use a dashed line to graph the equation $y = 3x - 7$. Recall that the dashed line is used to signify that no point on the line is a solution to the inequality. Again, this equation is in slope–intercept form, so we graph the line by plotting the y-intercept at $(0, -7)$ and using the slope of 3 to find additional points on the line.

We can use the origin as a test point for this inequality as well. Substituting 0 for x and 0 for y produces a false statement.

$$0 < 3(0) - 7$$
$$0 < 0 - 7$$
$$0 < -7$$

We shade the half-plane on the opposite side of the line from the origin.

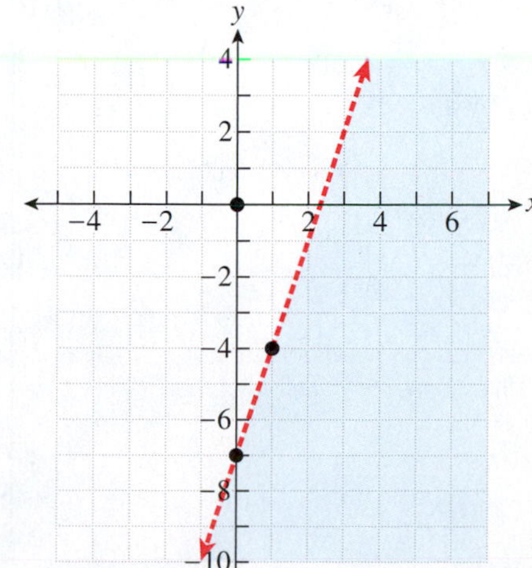

The solution of the system of linear inequalities is the region where the two solutions intersect.

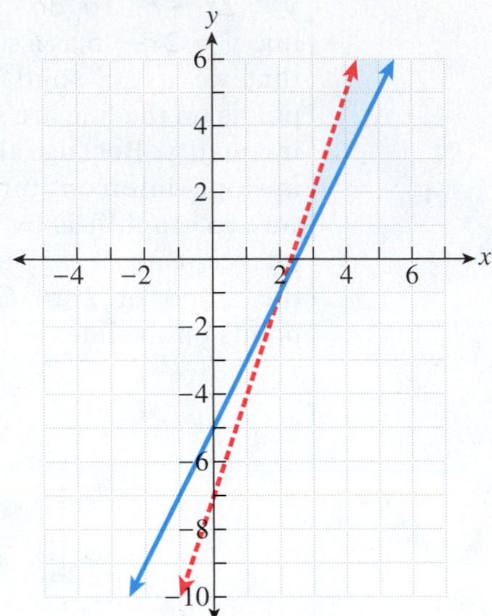

Quick Check 1

Graph the system of inequalities.

$$y > x + 3$$
$$2x + 3y < 6$$

EXAMPLE 2 Graph the following system of inequalities:

$$2x + y \leq 4$$
$$y > 3x$$

SOLUTION Begin by graphing the line $2x + y = 4$ using a solid line. This equation is in standard form, so we can graph the line by finding its x-intercept $(2, 0)$ and its y-intercept $(0, 4)$. Now we determine which half-plane to shade. Because the line does not pass through the origin, we can choose $(0, 0)$ as a test point.

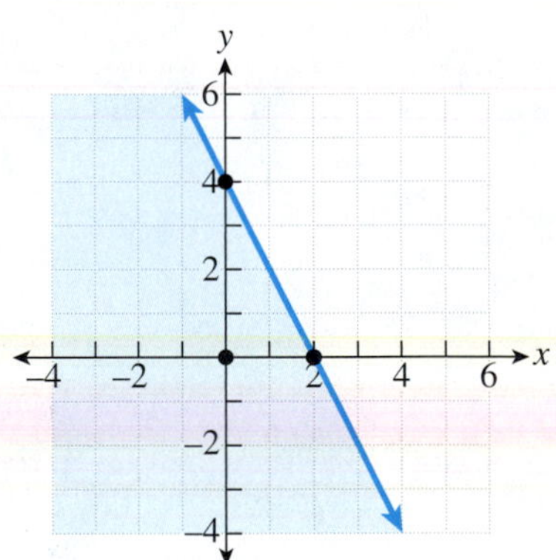

$$2(0) + (0) \leq 4$$
$$0 \leq 4$$

Because the test point produced a true statement, we shade the half-plane containing the origin.

Now we turn our attention to the inequality $y > 3x$. We begin by graphing the line $y = 3x$ as a dashed line. The y-intercept is at the origin, and the slope of the line is 3. We must choose a point that is clearly not on the line to use as a test point, such as $(5, 0)$.

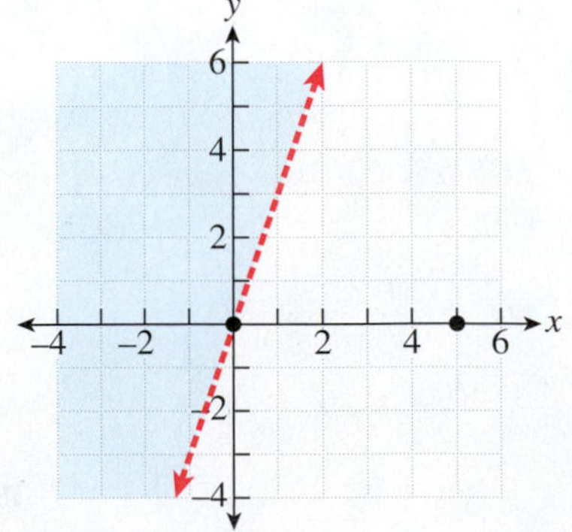

$$0 > 3(5)$$
$$0 > 15$$

The test point produced a false statement, so we shade the half-plane that does not contain the test point.

Quick Check 2

Graph the system of inequalities.

$$-3x + 4y \le 8$$
$$y \ge 0$$

The solution of the system of inequalities is the region where the two shaded regions intersect.

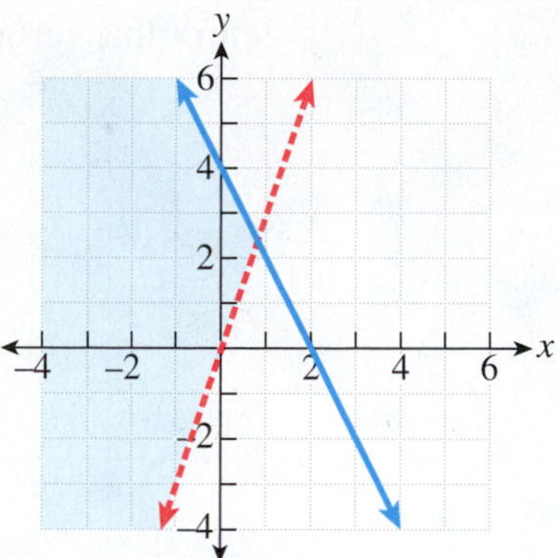

Solving a System of Linear Inequalities with More than Two Inequalities

Objective 2 Solve a system of linear inequalities with more than two inequalities. The next example involves a system of three linear inequalities. Regardless of the number of inequalities in a system, we still find the solution in the same manner. Graph each inequality individually and then find the region where the shaded regions intersect.

EXAMPLE 3 Graph the system of inequalities.

$$3x + 4y > 12$$
$$x < 6$$
$$y > 4$$

SOLUTION Begin by graphing the line $3x + 4y = 12$ with a dashed line. This is the line associated with the first inequality. Because the equation is in standard form, we can graph it by finding its x- and y-intercepts.

$3x + 4y = 12$

x-intercept ($y = 0$)	y-intercept ($x = 0$)
$3x + 4(0) = 12$	$3(0) + 4y = 12$
$3x = 12$	$4y = 12$
$x = 4$	$y = 3$
$(4, 0)$	$(0, 3)$

The x-intercept is at $(4, 0)$. The y-intercept of the line is at $(0, 3)$. Because the line does not pass through the origin, we can use $(0, 0)$ as a test point. When we substitute 0 for x and 0 for y in the inequality $3x + 4y > 12$, we obtain a false statement.

$$3(0) + 4(0) > 12$$
$$0 > 12$$

Because this statement is not true, we shade the half-plane on the side of the line that does not contain the point $(0, 0)$.

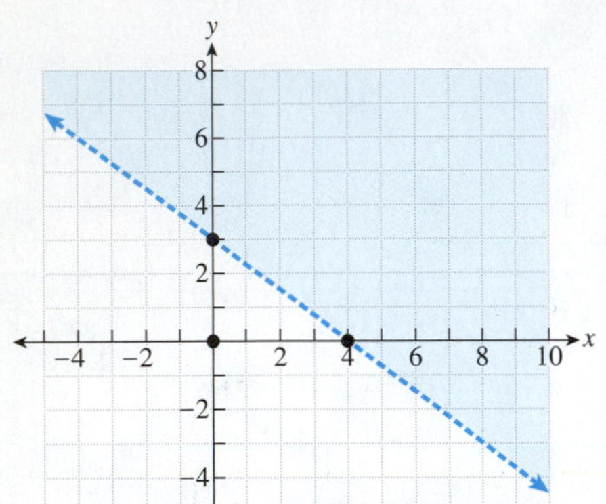

To graph the second inequality, begin with the graph of the vertical line $x = 6$. This line also is dashed. The solutions of this inequality are to the left of this vertical line.

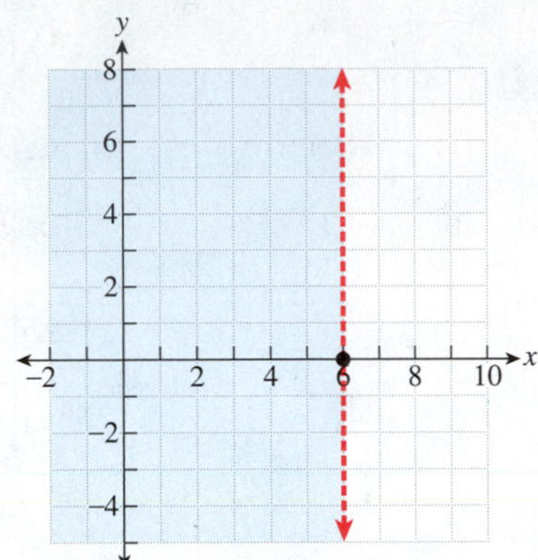

For the third inequality, graph the dashed horizontal line $y = 4$. The solutions of this inequality are above the horizontal line $y = 4$.

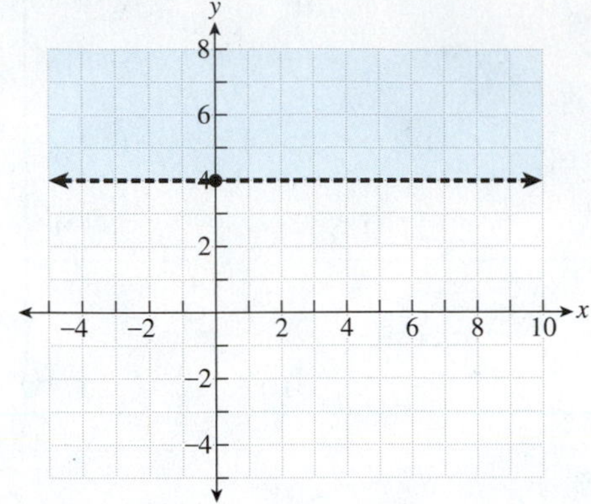

Now display the solutions to the system of inequalities by finding the region of intersection for these three inequalities, shown in the graph below.

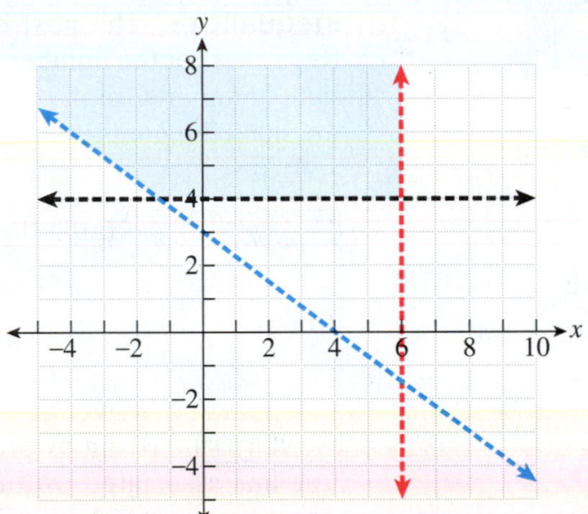

Quick Check 3

Graph the system of inequalities.

$$x + y \leq 8$$
$$x \geq 1$$
$$y \geq -1$$

BUILDING YOUR STUDY STRATEGY

Doing Your Homework, 5 **Homework Schedule** Try to establish a regular schedule for doing your homework. Some students find it advantageous to work on their homework as soon as possible after class, while the material is still fresh in their minds. You may be able to find other students in your class who can work with you in the library.

It is crucial that you complete an assignment before the next class. Because math skills build on previous concepts, if you do not understand the material one day, you will have difficulty understanding the next day's material.

Exercises 4.5

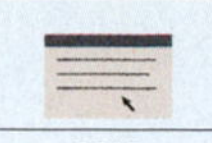

Vocabulary

1. A(n) ______________ is made up of two or more linear inequalities.
2. A linear inequality is graphed with a(n) _________ line if it involves either of the symbols $<$ or $>$.
3. A linear inequality is graphed with a(n) _________ line if it involves either of the symbols $\leq$ or $\geq$.
4. When solving a system of linear inequalities, shade the region where the solutions of the individual inequalities _________.

Graph the system of inequalities.

5. $y \geq 5x - 2$
   $y \leq -x + 1$
6. $y > 4x - 3$
   $y \leq 2x - 1$
7. $y \leq 3x - 1$
   $y > -\dfrac{3}{4}x + 6$
8. $y < 2x + 2$
   $y > 6x - 2$
9. $x + y \geq 5$
   $x - y \geq 3$
10. $x - y \geq 7$
    $x + y \leq 1$
11. $y < 5x + 5$
    $y > 5x - 5$
12. $y \leq 3x + 3$
    $y \geq 3x - 1$
13. $x + 4y < 8$
    $3x - 2y < 6$
14. $5x + 2y \leq 10$
    $-3x + y > 5$
15. $-7x + 3y > 21$
    $4x + 9y \leq 36$
16. $8x + y \geq -8$
    $3x + 4y < -12$
17. $y \geq 5$
    $5x + 2y < 0$
18. $-4x + y < 6$
    $4x - y < 2$

19. $y \geq 3x$
$y > 3$

20. $x < -4$
$2x + 6y > 12$

21. $y > -\frac{3}{2}x + 6$
$4x - 3y \geq -6$

22. $x < 4$
$y \geq -7$

23. $x > 1$
$8x - 4y > 12$

24. $3x + 10y > 15$
$y \leq -\frac{5}{2}x + 7$

25. $y < 2x$
$y < 6$
$x > 2$

26. $y > x - 4$
$y < -2x + 5$
$y > -2$

27. $3x - 2y \leq -10$
$y < -4x - 6$
$-x + 3y < 12$

28. $x \geq -5$
$x \leq 4$
$y < \frac{1}{5}x + 2$

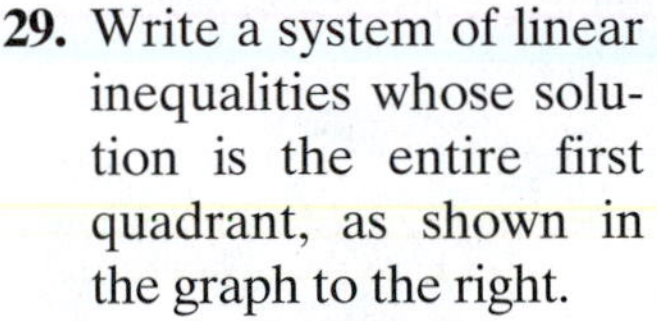

29. Write a system of linear inequalities whose solution is the entire first quadrant, as shown in the graph to the right.

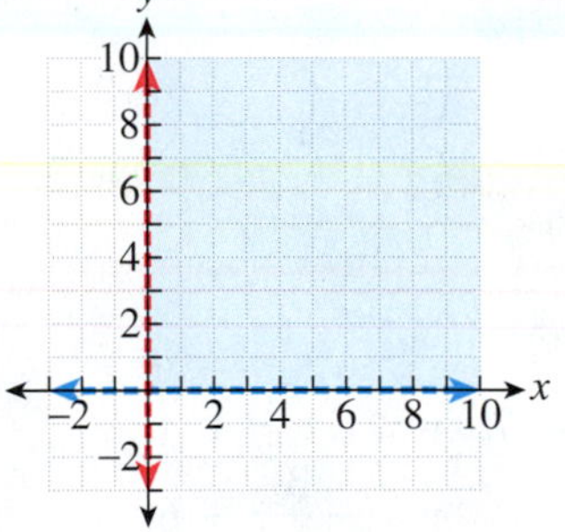

30. Write a system of linear inequalities whose solution is the entire third quadrant.

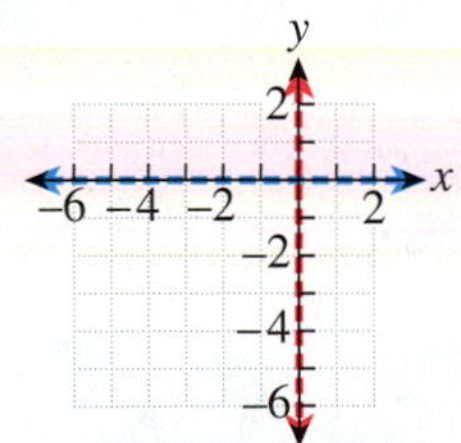

31. Find the area of the region defined by the following system of linear inequalities:

$x \geq 2$
$y \geq 1$
$y \leq -x + 9$

32. Write a system of linear inequalities whose solution set is shown on the following graph:

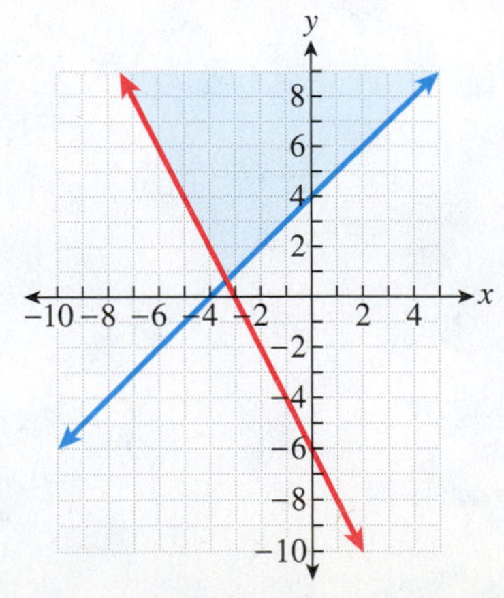

CHAPTER 4 SUMMARY

Section 4.1 Systems of Linear Equations; Solving Systems by Graphing

Systems of Linear Equations, p. 198

A system of linear equations consists of two or more linear equations.

$$\begin{aligned} 7x + 5y &= 62 \\ 3x - 2y &= 10 \end{aligned}$$

Solutions of a System of Linear Equations, p. 199

A solution of a system of linear equations is an ordered pair (x, y) that is a solution of each equation in the system.

Is $(5, -2)$ a solution of $\begin{aligned} 4x + 3y &= 14 \\ x + 2y &= 1 \end{aligned}$?

$$\begin{aligned} 4x + 3y &= 14 \\ 4(5) + 3(-2) &= 14 \\ 20 - 6 &= 14 \\ 14 &= 14 \end{aligned} \qquad \begin{aligned} x + 2y &= 1 \\ (5) + 2(-2) &= 1 \\ 5 - 4 &= 1 \\ 1 &= 1 \end{aligned}$$

$(5, -2)$ is a solution.

Solving a System of Linear Equations Graphically, pp. 199–202

To solve a system of linear equations graphically, begin by graphing each line. A solution of the system is an ordered pair that is on both lines.

Solve the system graphically. $\begin{aligned} y &= 3x - 7 \\ 2x + 3y &= 12 \end{aligned}$

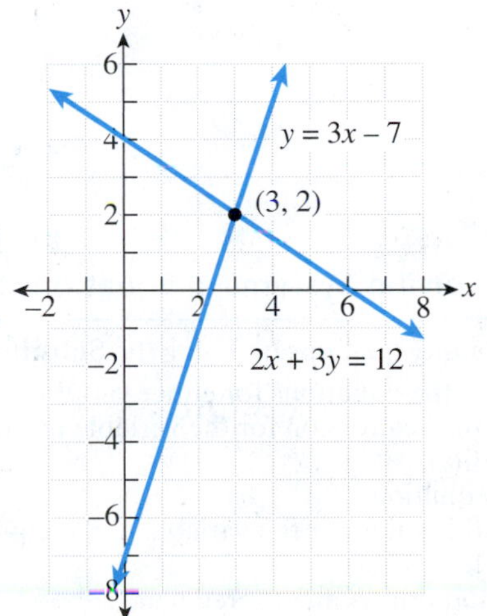

Independent Systems, p. 202

A system of equations that has a single ordered pair as its solution is called an independent system. Graphically, a system of linear equations is an independent system if the lines associated with the system intersect at a single point.

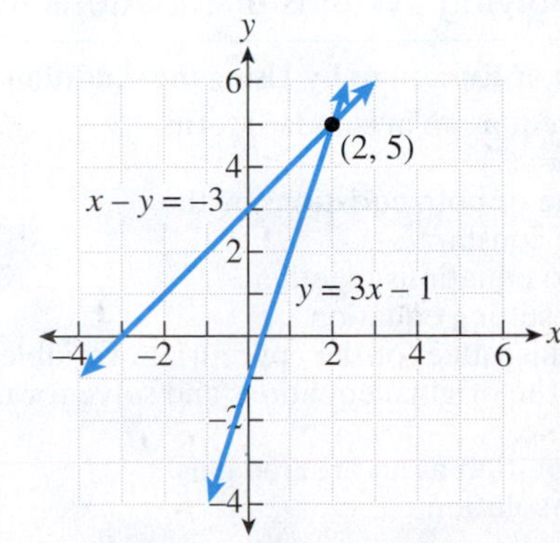

Inconsistent Systems, pp. 202–203

A system of equations that has no solution is called an inconsistent system.

Graphically, a system of linear equations is an inconsistent system if the lines associated with the system are distinct parallel lines.

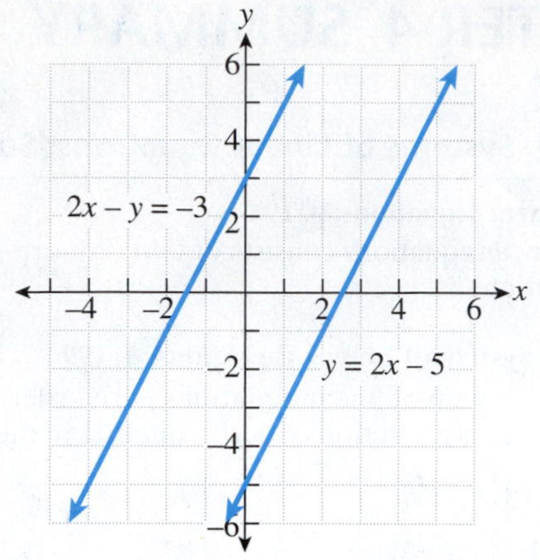

Dependent Systems, pp. 202–204

A system of equations that has infinitely many solutions is called a dependent system.

Graphically, a system of linear equations is a dependent system if the two lines associated with the system are the same line.

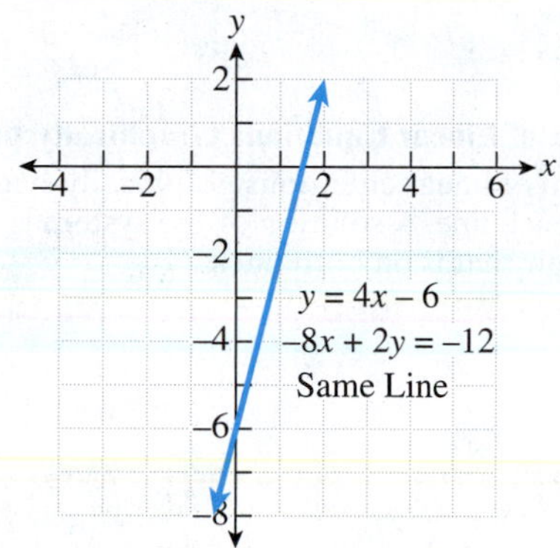

Section 4.2 Solving Systems of Equations by Using the Substitution Method

Solving a System of Equations by Using the Substitution Method, pp. 207–211

1. Solve one of the equations for either variable.
2. Substitute this expression for the variable in the other equation.
3. Solve this equation.
4. Substitute this value for the variable in the equation from Step 1.
5. Write the solution as an ordered pair.
6. Check your solution.

Solve: $\begin{aligned} y &= 2x + 7 \\ 3x - 4y &= -18 \end{aligned}$

$$\begin{aligned} 3x - 4(2x + 7) &= -18 \\ 3x - 8x - 28 &= -18 \\ -5x - 28 &= -18 \\ -5x &= 10 \\ x &= -2 \end{aligned} \qquad \begin{aligned} y &= 2(-2) + 7 \\ y &= 3 \end{aligned}$$

$(-2, 3)$

Section 4.3 Solving Systems of Equations by Using the Addition Method

Solving a System of Equations by Using the Addition Method, pp. 214–220

1. Write each equation in standard form ($Ax + By = C$).
2. Multiply one or both equations by the appropriate constant(s).
3. Add the two equations together.
4. Solve the resulting equation.
5. Substitute this value for the appropriate variable in either of the original equations and solve for the other variable.
6. Write the solution as an ordered pair.
7. Check your solution.

Solve: $\begin{aligned} 3x + 2y &= 4 \\ x - 6y &= 8 \end{aligned}$

$$\begin{aligned} 3x + 2y &= 4 \\ x - 6y &= 8 \end{aligned} \xrightarrow{\text{Multiply by 3}} \begin{aligned} 9x + 6y &= 12 \\ x - 6y &= 8 \\ \hline 10x \quad &= 20 \\ x \quad &= 2 \end{aligned}$$

$$\begin{aligned} 3(2) + 2y &= 4 \\ 6 + 2y &= 4 \\ 2y &= -2 \\ y &= -1 \end{aligned}$$

$(2, -1)$

Section 4.4 Applications of Systems of Equations

Solving Applied Problems by Using a System of Equations, pp. 224–230

1. Choose a variable for each unknown quantity.
2. Find two equations that relate the two unknown quantities.
3. Solve the system using substitution or addition.

Adults are charged \$6 and children are charged \$4 to see a play. Last night 103 people saw a play, paying a total of \$532. How many adults and how many children were at the play?

Unknowns:
Number of adults: A; number of children: C
System of Equations

$$\begin{aligned} A + C &= 103 \\ 6A + 4C &= 532 \end{aligned}$$

Solve by Addition

$$\begin{aligned} -4A - 4C &= -412 \\ 6A + 4C &= 532 \\ \hline 2A &= 120 \\ A &= 60 \end{aligned}$$

$$\begin{aligned} 60 + C &= 103 \\ C &= 43 \end{aligned}$$

There were 60 adults and 43 children at the play.

Perimeter of a Rectangle

The length of a rectangle is 3 inches more than twice its width. If the perimeter of the rectangle is 96 inches, find its length and width.

Unknowns:
Length: l; width: w
System of Equations

$$\begin{aligned} l &= 2w + 3 \\ 2l + 2w &= 96 \end{aligned}$$

Solve by Substitution

$$\begin{aligned} 2(2w + 3) + 2w &= 96 \\ 4w + 6 + 2w &= 96 \\ 6w + 6 &= 96 \\ 6w &= 90 \\ w &= 15 \end{aligned}$$

$$\begin{aligned} l &= 2(15) + 3 \\ l &= 33 \end{aligned}$$

The length is 33 inches, and the width is 15 inches.

Interest

Manny deposited a total of \$4000 in two bank accounts. One account paid 3% annual interest, and the other account paid 4% annual interest. If Manny earned \$145 interest in one year, how much was invested in each account?

Unknowns:
Principal at 3%: x; principal at 4%: y
System of Equations

$$\begin{aligned} x + y &= 4000 \\ 0.03x + 0.04y &= 145 \end{aligned}$$

Solve by Addition

$$\begin{aligned} x + y &= 4000 \\ 3x + 4y &= 14{,}500 \\ -3x - 3y &= -12{,}000 \\ 3x + 4y &= 14{,}500 \\ \hline y &= 2500 \\ x + 2500 &= 4000 \\ x &= 1500 \end{aligned}$$

Manny invested \$1500 at 3% and \$2500 at 4%.

Section 4.5 Systems of Linear Inequalities

Systems of Linear Inequalities, p. 234

A system of linear inequalities is made up of two or more linear inequalities. An ordered pair is a solution to a system of linear inequalities if it is a solution to each linear inequality in the system.

Is (6, 8) a solution of the system of inequalities?

$$2x + 5y > 40$$
$$y \le 2x + 3$$

$2(6) + 5(8) > 40$ $\quad$ $(8) \le 2(6) + 3$

$52 > 40$ $\quad$ $8 \le 15$

(6, 8) is a solution.

Solving a System of Linear Inequalities by Graphing, pp. 234–238

To find the solution set to a system of linear inequalities, begin by graphing each inequality separately. If the solutions to the individual inequalities intersect, the region of intersection is the solution set to the system of inequalities.

Solve: $\begin{aligned} y &\le x + 5 \\ 3x + 2y &> 18 \end{aligned}$

Graph $y \le x + 5$.

$y \le x + 5$

Test Point

Graph $3x + 2y > 18$.

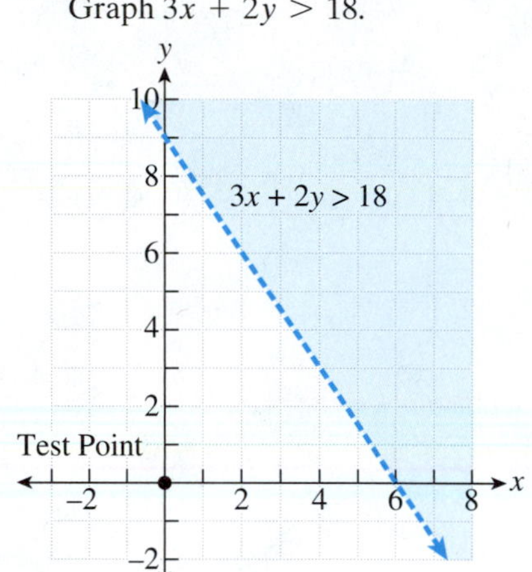

Graph the system.

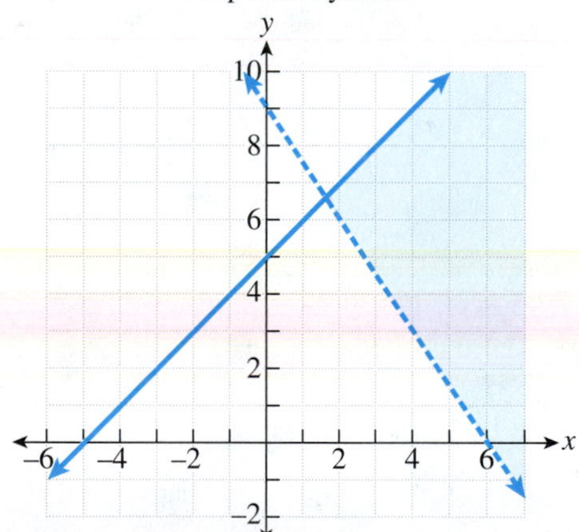

SUMMARY OF CHAPTER 4 STUDY STRATEGIES

You should not view a homework assignment as a chore to be completed; rather, you should consider it an opportunity to increase your understanding of mathematics. Here is a summary of the ideas presented in this chapter.

- Review your notes and the text before beginning a homework assignment.
- Be neat and complete when doing a homework assignment. *If your work is neat, you will be able to spot errors more easily. When you are reviewing for an exam, your homework will be easier to review if it is neat. Do not skip any steps, even if you think that showing them is not necessary.*
- Strategies for homework problems you cannot answer:
  - *Look in your notes and the text for similar problems or suggestions about this type of problem.*
  - *Call a classmate for help.*
  - *Get help from a tutor.*
  - *Ask your instructor for help at the first opportunity.*
- Complete a homework session by summarizing your work. *Prepare notes on difficult problems or concepts. These hints will be helpful when you begin to review for an exam.*
- Keep up to date with homework assignments. *Falling behind will make it difficult to learn the new material presented in class.*

CHAPTER 4 REVIEW

***Find the solution of the system of equations shown on each graph. If there is no solution, state this.* [4.1]**

1.

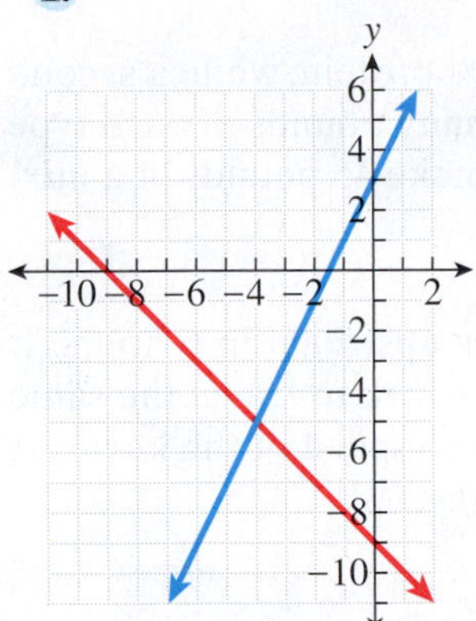

2.

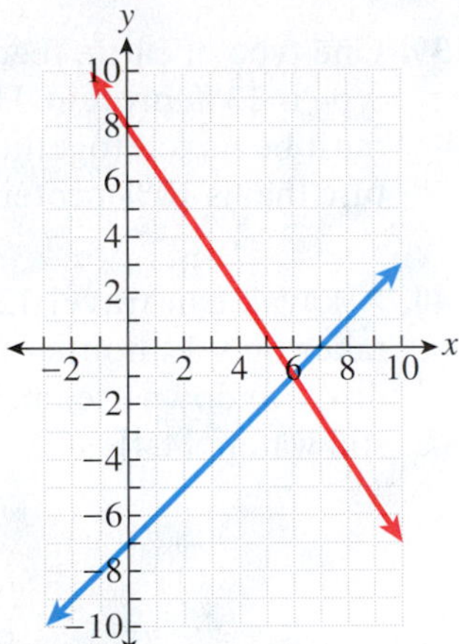

3.

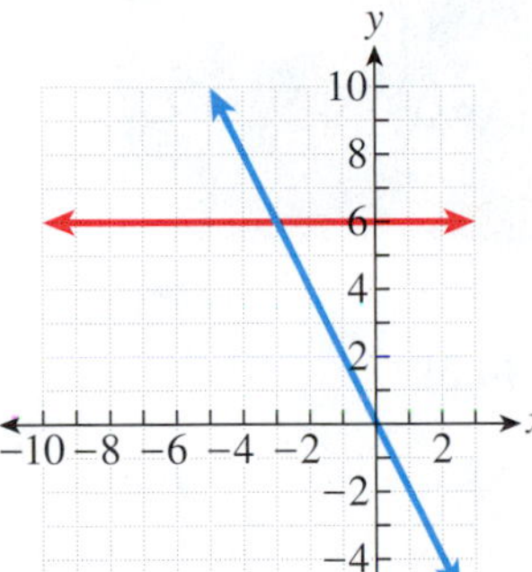

4.

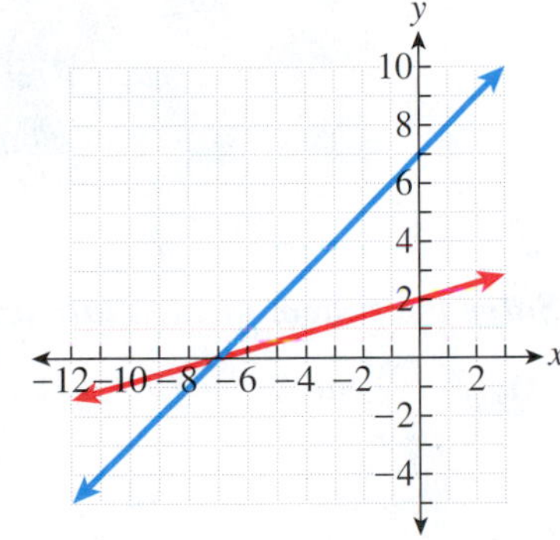

***Solve the system by graphing. If the system is inconsistent and has no solution, state this. If the system is dependent, write the form of the solution for any real number x.* [4.1]**

5. $y = x - 5$
$y = -\frac{1}{2}x + 7$

6. $y = \frac{1}{3}x - 4$
$y = -x$

7. $y = -2x + 8$
$y = \frac{1}{2}x + 3$

8. $2x - y = -8$
$-3x + 2y = 10$

***Solve the system by substitution. If the system is inconsistent and has no solution, state this. If the system is dependent, write the form of the solution for any real number x.* [4.2]**

9. $x = 2y + 1$
$2x - 5y = 4$

10. $y = -x - 3$
$3x + 4y = -5$

11. $4x + y = -5$
$5x - 3y = 15$

12. $-4x + 2y = 18$
$2x - y = 3$

13. $y = \frac{2}{3}x - 5$
$3x - 4y = 23$

***Solve the system by addition. If the system is inconsistent and has no solution, state this. If the system is dependent, write the form of the solution for any real number x.* [4.3]**

14. $3x + 2y = 15$
$5x - 2y = 17$

15. $6x + 7y = -66$
$-3x + 4y = -27$

16. $2x - y = 4$
$-4x + 2y = -8$

17. $y = 4x + 5$
$5x - 2y = -4$

18. $\frac{1}{2}x + \frac{2}{3}y = 2$
$3x - \frac{1}{2}y = -15$

***Solve the system using the most efficient method (substitution or addition). If the system is inconsistent and has no solution, state this. If the system is dependent, write the form of the solution for any real number x.* [4.2, 4.3]**

19. $4x - 3y = 20$
$-x + 6y = -26$

20. $x + y = 4$
$-5x + 4y = 43$

21. $y = 2x - 1$
$3x - 7y = 29$

Worked-out solutions to Review Exercises marked with can be found on page AN-15.

22. $8x - 9y = 67$
$x = 5$

23. $y = 3x - 2$
$-9x + 3y = -6$

24. $x + y = -4$
$4x - 6y = -1$

25. $3x - 2y = 12$
$y = \frac{4}{5}x - 6$

26. $x + 2y = -5$
$\frac{1}{3}x - \frac{1}{4}y = 13$

27. $-\frac{1}{2}x + 2y = 3$
$2x - 8y = -9$

28. $7x - 5y = -46$
$6x + 7y = 17$

29. $y = 0.3x + 2.4$
$3x - 2y = 7.2$

30. $x + y = 8000$
$0.04x + 0.05y = 340$

31. The sum of two numbers is 87. One number is 18 more than twice the other number. Find the two numbers. [4.4]

32. Hal has 284 more baseball cards in his collection than his brother Mac does. Together they have 1870 baseball cards. How many cards does Hal have? [4.4]

33. A hockey team charges $10 for tickets to its games. On Wednesday nights, the team has a Ladies' Night promotion where women of any age are admitted for half price. If total attendance on the last Ladies' Night was 1764 fans and the team collected $13,470, how many women were at the game? [4.4]

34. A movie theater charges $7.50 to see a movie, but senior citizens are admitted for only $4. If 206 people paid a total of $1384 to see a movie, how many were senior citizens? [4.4]

35. Jackie's home is on a rectangular lot. The fence that runs around the entire yard is 1320 feet long. If the length of the yard is 140 feet more than the width, find the dimensions of the yard. [4.4]

36. Jiana has a front door that lets in cold drafts. The height of the door is 4 inches more than twice its width. If she uses 224 inches of weather stripping around the entire door, how tall is her door? [4.4]

37. Juan deposits a total of $4000 in two different CDs. One CD pays 5% annual interest, while the other pays 4.25% annual interest. If Juan earned $179 interest during the first year, how much did he put in each CD? [4.4]

38. Marcia bought shares in two mutual funds, investing a total of $2500. One fund's shares went up by 20%, while the other fund went up in value by 8%. This was a total profit of $320 for Marcia. How much did she invest in each mutual fund? [4.4]

39. One type of cattle feed is 4% protein, while a second type is 13% protein. How many pounds of each type must be mixed together to make 45 pounds of a mixture that is 10% protein? [4.4]

40. A kayak can travel 12 miles upstream in 6 hours. It takes only $1\frac{1}{2}$ hours for the kayak to travel the same distance downstream. Find the speed of the kayak in still water. [4.4]

Solve the system of inequalities. **[4.5]**

41. $y \leq 2x + 7$
$y > x - 5$

42. $3x + 4y < 24$
$5x + y \geq 10$

43. $y \le -4x + 7$
$y \ge \frac{3}{2}x$

44. $y > \frac{1}{2}x - 3$
$y > -2$

CHAPTER 4 TEST

Step-by-step test solutions are found on the Chapter Test Prep Videos available via the Video Resources on DVD, in *MyMathLab*, and on YouTube (search "WoodburyElemIntAlg" and click on "Channels").

1. Find the solution of the system of equations shown on the graph. If there is no solution, state this.

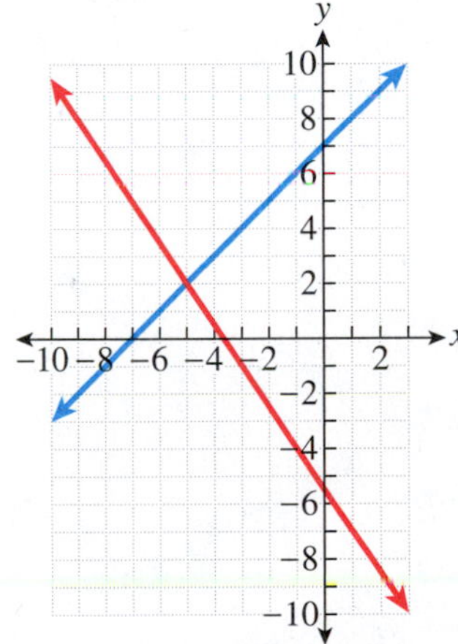

2. Solve the system that follows by graphing. If the system is inconsistent and has no solution, state this. If the system is dependent, write the form of the solution for any real number x.

$$y = -2x + 3$$
$$y = -3x + 7$$

3. Solve the system that follows by substitution. If the system is inconsistent and has no solution, state this. If the system is dependent, write the form of the solution for any real number x.

$$x + 4y = -9$$
$$2x - 5y = 21$$

4. Solve the system that follows by addition. If the system is inconsistent and has no solution, state this. If the system is dependent, write the form of the solution for any real number x.

$$5x + 2y = -22$$
$$2x - 5y = -3$$

Solve the given system by the method of your choice. If the system is inconsistent and has no solution, state this. If the system is dependent, write the form of the solution for any real number x.

5. $-x + 4y = 11$
$3x - 4y = -17$

6. $4x + 2y = 9$
$y = -2x + 18$

7. $x + y = 11$
$5x - 2y = 13$

8. $7x - 11y = -47$
$9x + 2y = 101$

9. $13x + 6y = 52$
$x = 7y + 4$

10. $4x - 6y = 18$
$-10x + 15y = -45$

11. Last season a baseball player had 123 more walks than he had home runs. If you add the number of times he walked to the number of times he hit a home run, the total is 237. How many walks did the player have?

12. Admission to an amusement park is \$20, but children under 8 years old are admitted for \$5. If 13,900 people were at the amusement park yesterday and they paid a total of \$197,000 for admission, how many children under 8 years old were at the park?

13. Steve's vegetable garden is in the shape of a rectangle, and it requires 66 feet of fencing to surround it. The length is 1 foot more than three times the width. Find the dimensions of Steve's garden.

14. Emma bought shares in two mutual funds, investing a total of \$32,000. One fund's shares went up by 7%, while the other fund went up in value by 3%. This was a total profit of \$1760 for Emma. How much did she invest in each mutual fund?

15. Solve the system of inequalities.

$$y < \frac{1}{2}x + 4$$
$$5x + 2y > -16$$

Mathematicians in History

Hypatia of Alexandria was one of the first women known to make major contributions to the field of mathematics. She also was active in the fields of astronomy and philosophy. One quote attributed to her is the following: "Reserve your right to think, for even to think wrongly is better than not to think at all."

Write a one-page summary (*or* make a poster) of the life of Hypatia of Alexandria and her accomplishments.

Interesting issues:

- Where and when was Hypatia born?
- Who was Hypatia's father? In what fields was he involved?
- Hypatia lectured on the philosophy of Neoplatonism. What are the principal ideas of Neoplatonism?
- What was Hypatia's relationship to Cyril and Orestes?
- In the year A.D. 415, Hypatia was murdered. Describe the circumstances of her death.

CUMULATIVE REVIEW CHAPTERS 1–4

***Write the appropriate symbol, < or >, between the following integers.* [1.1]**

1. -3 ____ -8

***Simplify.* [1.2]**

2. $-9 - 25$

3. $15(-11)$

4. $-234 \div (-18)$

***Write the prime factorization of the following number. (If the number is prime, state this.)* [1.3]**

5. 72

***Simplify. Your answer should be in lowest terms.* [1.4]**

6. $\frac{6}{35} \cdot \frac{5}{21}$

7. $\frac{3}{4} + \frac{9}{10}$

***Simplify the following decimal expressions.* [1.5]**

8. $9.8 - 3.72$

9. 4.7×3.8

10. A group of 32 baseball fans decided to attend a game together. If each ticket cost \$14, how much did the group spend on tickets? [1.2]

11. If one recipe calls for $1\frac{1}{3}$ cups of flour and a second recipe calls for $2\frac{3}{4}$ cups of flour, how much flour is needed to make both recipes? [1.4]

***Simplify the given expression.* [1.7]**

12. $9 + 3 \cdot 4 - 2^6$

13. $7 + 2(9 - 4 \cdot 5) - 6(-5)$

***Evaluate the following algebraic expression under the given conditions.* [1.8]**

14. $5x - 8$ for $x = -4$

***Simplify.* [1.8]**

15. $8x - 11 - 3x + 8$

16. $3(3x - 8) - 5(4x + 9)$

***Solve.* [2.1, 2.2]**

17. $3x - 4 = -31$

18. $-2x + 15 = 8$

19. $3x + 17 = 6x - 25$

20. $2(2x + 5) - 3(x - 8) = 20$

21. The width of a rectangle is 9 feet shorter than its length, and the perimeter is 70 feet. Find the length and the width of the rectangle. [2.3]

22. Celia has \$5 and \$10 bills in her purse, worth a total of \$205. She has 8 more \$5 bills than \$10 bills. How many \$5 bills does Celia have? [2.3]

23. Twenty percent of 125 is what number? [2.4]

24. What percent of 275 is 66? [2.4]

25. Monisha bought a new house for \$220,000. The house increased in value by 13% in one year. How much was the house worth after one year? [2.4]

***Solve the proportion.* [2.4]**

26. $\frac{n}{8} = \frac{27}{36}$

27. At a community college, 5 out of every 9 students plan to transfer to a four-year university. If the college has 8307 students, how many plan to transfer to a four-year university? [2.4]

***Solve. Graph your solution on a number line and express it in interval notation.* [2.5]**

28. $2x - 5 \geq 3x - 8$

29. $-7 \leq 3x + 8 \leq 11$

***Find the x- and y-intercepts and use them to graph the equation.* [3.2]**

30. $2x - y = 6$

31. $-3x + 4y = -24$

***Find the slope of the line that passes through the given points.* [3.3]**

32. $(-3, -7)$ and $(3, 5)$

***Find the slope and the y-intercept of the given line.* [3.3]**

33. $2x + 3y = 21$

***Graph using the slope and y-intercept. Find the x-intercept and label it on the graph.* [3.3]**

34. $y = 2x - 3$

35. $y = \frac{3}{4}x - 3$

***Are the two given lines parallel, perpendicular, or neither?* [3.5]**

36. $5x + 3y = 9$
$3x - 5y = -5$

37. Bruce is saving his money to buy a new guitar. So far he has saved \$220, and he plans to save \$40 each month. [3.4]

a) Create a function $f(x)$ for the amount of money Bruce has saved after x months.

b) Use the function from part a to determine how much money Bruce will have saved after 8 months.

c) Use the function from part a to determine how many months it will take until Bruce has saved enough to buy the guitar, which will cost \$800.

***Evaluate the given function.* [3.4]**

38. $f(x) = 24x + 55, f(14)$

39. $g(x) = -7x + 905, g(-32)$

***Find the slope–intercept form of the equation of a line with the given slope that passes through the given point.* [3.6]**

40. Slope -2 through $(3, -4)$

***Find the slope–intercept form of the equation of a line that passes through the two given points.* [3.6]**

41. $(2, 1)$ and $(4, -5)$

***Graph the inequality on a plane.* [3.7]**

42. $3x - 2y > 12$

***Solve the system by the method of your choice. If the system is inconsistent and has no solution, state this. If the system is dependent, write the form of the solution for any real number x.* [4.2, 4.3]**

43. $4x + y = 22$
$-3x + 8y = 1$

44. $y = 3x - 2$
$2x - 3y = -15$

45. $9x + 8y = 25$
$5x - 6y = -7$

46. $4x + y = 6$
$\frac{3}{2}x + \frac{3}{5}y = \frac{9}{2}$

47. A minor league baseball team charges admission of \$5 for adults and \$4 for children. If 1050 people paid a total of \$5000 to see last night's game, how many children were at the game? [4.4]

48. Bill's backyard is rectangular. The fence that runs around the backyard is 130 feet long. If the length of the backyard is 10 feet less than twice the width, find the dimensions of the backyard. [4.4]

49. Mya deposits a total of \$15,000 in two different CDs. One CD pays 3% annual interest, while the other pays 2.5% annual interest. If Mya earned \$430 in interest during the first year, how much did she put in each CD? [4.4]

***Solve the system of inequalities.* [4.5]**

50. $y < x + 4$
$y \geq 4x - 2$

CHAPTER 5

Exponents and Polynomials

The expressions and equations we have examined to this point in the text have been linear. In this chapter, we begin to examine expressions and functions that are nonlinear, including polynomials such as $-16t^2 + 32t + 128$ and $x^5 - 3x^4 + 8x^2 - 9x + 7$. After learning some basic definitions, we will learn to add, subtract, multiply, and divide polynomials. We will introduce properties for working with expressions containing exponents that make using these operations easier. We also will cover scientific notation, which is used to represent very large and very small numbers and to perform arithmetic calculations with them.

STUDY STRATEGY

Test Taking To be successful in a math class, understanding the material is important, but you also must be a good test taker. In this chapter, we will discuss the test-taking skills necessary for success in a math class.

5.1 Exponents

OBJECTIVES

1. Use the product rule for exponents.
2. Use the power rule for exponents.
3. Use the power of a product rule for exponents.
4. Use the quotient rule for exponents.
5. Use the zero exponent rule.
6. Use the power of a quotient rule for exponents.
7. Evaluate functions by using the rules for exponents.

We have used exponents to represent repeated multiplication of a particular factor. For example, the expression 3^7 is used to represent a product in which 3 is a factor 7 times.

$$3^7 = 3 \cdot 3 \cdot 3 \cdot 3 \cdot 3 \cdot 3 \cdot 3$$

Recall that in the expression 3^7, the number 3 is called the base and the number 7 is called the exponent.

In this section, we will introduce several properties of exponents and learn how to apply these properties to simplify expressions involving exponents.

Product Rule

Objective 1 Use the product rule for exponents.

Product Rule

For any base x, $x^m \cdot x^n = x^{m+n}$.

When multiplying two expressions with the same base, keep the base and add the exponents. Consider the example $x^3 \cdot x^7$.

$$x^3 = x \cdot x \cdot x \quad \text{and} \quad x^7 = x \cdot x \cdot x \cdot x \cdot x \cdot x \cdot x$$

So $x^3 \cdot x^7$ can be rewritten as $(x \cdot x \cdot x) \cdot (x \cdot x \cdot x \cdot x \cdot x \cdot x \cdot x)$.

Because x is being repeated as a factor 10 times, this expression can be rewritten as x^{10}.

$$x^3 \cdot x^7 = x^{3+7} = x^{10}$$

A WORD OF CAUTION When multiplying two expressions with the same base, keep the base and add the exponents. *Do not multiply the exponents.*

$$x^3 \cdot x^7 = x^{10}, \text{ not } x^{21}$$

EXAMPLE 1 Simplify $x^2 \cdot x^9$.

SOLUTION Because we are multiplying and the two bases are the same, we keep the base and add the exponents.

$$\begin{aligned} x^2 \cdot x^9 &= x^{2+9} && \text{Keep the base and add the exponents.} \\ &= x^{11} && \text{Add.} \end{aligned}$$

EXAMPLE 2 Simplify $2 \cdot 2^5 \cdot 2^4$.

SOLUTION In this example, we are multiplying three expressions that have the same base. To do this, we keep the base and add the three exponents. Keep in mind that if an exponent is not written for a factor, the exponent is a 1.

$$\begin{aligned} 2 \cdot 2^5 \cdot 2^4 &= 2^{1+5+4} && \text{Keep the base and add the exponents.} \\ &= 2^{10} && \text{Add.} \\ &= 1024 && \text{Raise 2 to the 10th power.} \end{aligned}$$

Quick Check 1

Simplify.

a) $x^5 \cdot x^4$ **b)** $3^3 \cdot 3^2 \cdot 3$

EXAMPLE 3 Simplify $(x + y)^4 \cdot (x + y)^9$.

Quick Check 2

Simplify $(a - 6)^{10} \cdot (a - 6)^{17}$.

SOLUTION The base for these expressions is the sum $x + y$.

$$\begin{aligned} (x + y)^4 \cdot (x + y)^9 &= (x + y)^{4+9} && \text{Keep the base and add the exponents.} \\ &= (x + y)^{13} && \text{Add.} \end{aligned}$$

EXAMPLE 4 Simplify $(x^2y^7)(x^3y)$.

SOLUTION When we have a product involving more than one base, we can use the associative and commutative properties of multiplication to simplify the product.

$$\begin{aligned}(x^2y^7)(x^3y) &= x^2y^7x^3y && \text{Write without parentheses.}\\ &= x^2x^3y^7y && \text{Group like bases together.}\\ &= x^5y^8 && \text{For each base, add the exponents and keep the base.}\end{aligned}$$

In future examples, we will not show the application of the associative and commutative properties of multiplication. Look for bases that are the same, add their exponents, and write each base to the calculated power.

Quick Check 3
Simplify $(a^6b^5)(a^9b^5)$.

Power Rule

Objective 2 Use the power rule for exponents. The second property of exponents involves raising an expression with an exponent to a power.

Power Rule

For any base x, $(x^m)^n = x^{m \cdot n}$.

In essence, this property says that when we raise a power to a power, we keep the base and multiply the exponents. To show why this is true, consider the expression $(x^3)^7$.

Whenever we raise an expression to the 7th power, the expression is repeated as a factor seven times. So $(x^3)^7 = x^3 \cdot x^3 \cdot x^3 \cdot x^3 \cdot x^3 \cdot x^3 \cdot x^3$. Applying the first property from this section, $x^3 \cdot x^3 \cdot x^3 \cdot x^3 \cdot x^3 \cdot x^3 \cdot x^3 = x^{3+3+3+3+3+3+3}$, or x^{21}. This exponent could be found by multiplying 3 by 7.

$$(x^3)^7 = x^{3 \cdot 7} = x^{21}$$

A WORD OF CAUTION When raising an expression with an exponent to another power, keep the base and multiply the exponents. *Do not add the exponents.*

$$(x^3)^7 = x^{21}, \text{ not } x^{10}$$

EXAMPLE 5 Simplify $(x^5)^4$.

SOLUTION In this example, we are raising a power to another power, so we keep the base and multiply the exponents.

$$\begin{aligned}(x^5)^4 &= x^{5 \cdot 4} && \text{Keep the base and multiply the exponents.}\\ &= x^{20} && \text{Multiply.}\end{aligned}$$

Quick Check 4
Simplify $(x^9)^7$.

EXAMPLE 6 Simplify $(x^3)^9(x^7)^8$.

SOLUTION To simplify this expression, we must use both properties introduced in this section.

$$\begin{aligned}(x^3)^9(x^7)^8 &= x^{3 \cdot 9}x^{7 \cdot 8} && \text{Apply the power rule.}\\ &= x^{27}x^{56} && \text{Multiply.}\\ &= x^{83} && \text{Keep the base and add the exponents.}\end{aligned}$$

Quick Check 5
Simplify $(x^4)^6(x^5)^2$.

Power of a Product Rule

Objective 3 Use the power of a product rule for exponents. The third property introduced in this section involves raising a product to a power.

Power of a Product Rule

For any bases x and y, $(xy)^n = x^n y^n$.

This property says that when we raise a product to a power, we may raise each factor to that power. Consider the expression $(xy)^5$. We know that this can be rewritten as $xy \cdot xy \cdot xy \cdot xy \cdot xy$. Using the associative and commutative properties of multiplication, we can rewrite this expression as $x \cdot x \cdot x \cdot x \cdot x \cdot y \cdot y \cdot y \cdot y \cdot y$, which simplifies to $x^5 y^5$.

EXAMPLE 7 Simplify $(x^2y^3)^6$.

SOLUTION This example combines two properties. When we raise each factor to the 6th power, we are raising a power to a power; therefore, we multiply the exponents.

$$\begin{aligned}(x^2y^3)^6 &= (x^2)^6(y^3)^6 && \text{Raise each factor to the 6th power.}\\ &= x^{12}y^{18} && \text{Keep the base and multiply the exponents.}\end{aligned}$$

Quick Check 6
Simplify $(a^5b^3)^8$.

EXAMPLE 8 Simplify $(x^2y^3)^7(3x^5)^2$.

SOLUTION In this example, begin by raising each factor to the appropriate power. Then multiply expressions with the same base.

$$\begin{aligned}(x^2y^3)^7(3x^5)^2 &= x^{14}y^{21} \cdot 3^2x^{10} && \text{Raise each base to the appropriate power.}\\ &= 9x^{24}y^{21} && \text{Simplify.}\end{aligned}$$

It is a good idea to write the variables in an expression in alphabetical order, as shown in this example, although it is not necessary.

Quick Check 7
Simplify $(4a^2b^3)^3(a^4b^2c^7)^5$.

A WORD OF CAUTION This property applies only when we raise a *product* to a power, not a sum or a difference. Although $(xy)^2 = x^2y^2$, a similar approach does not work for $(x + y)^2$ or $(x - y)^2$.

$$(x + y)^2 \neq x^2 + y^2$$
$$(x - y)^2 \neq x^2 - y^2$$

For example, let $x = 12$ and $y = 5$. In this case, $(x + y)^2$ is equal to 17^2, or 289, but $x^2 + y^2$ is equal to $12^2 + 5^2$, or 169. This shows that $(x + y)^2$ is not equal to $x^2 + y^2$ in general.

Quotient Rule

Objective 4 Use the quotient rule for exponents. The next property involves fractions containing the same base in their numerators and denominators. It also applies to dividing exponential expressions with the same base.

Quotient Rule

For any base x, $\dfrac{x^m}{x^n} = x^{m-n}$ $(x \neq 0)$.

Note that this property has a restriction; it does not apply when the base is 0. This is because division by 0 is undefined. This property tells us that when we are dividing two expressions with the same base, we subtract the exponent of the denominator from the exponent of the numerator. Consider the expression $\frac{x^7}{x^3}$. This can be rewritten as $\frac{x \cdot x \cdot x \cdot x \cdot x \cdot x \cdot x}{x \cdot x \cdot x}$. From our previous work, we know that this fraction can be simplified by dividing out common factors in the numerator and denominator.

$$\frac{\cancel{x} \cdot \cancel{x} \cdot \cancel{x} \cdot x \cdot x \cdot x \cdot x}{\cancel{x} \cdot \cancel{x} \cdot \cancel{x}} = x^4$$

One way to think of this is that the number of factors in the numerator has been reduced by three because there were three factors in the denominator. The property says that we can simplify this expression by subtracting the exponents, which produces the same result.

$$\frac{x^7}{x^3} = x^{7-3} = x^4$$

EXAMPLE 9 Simplify $\frac{y^9}{y^2}$. (Assume that $y \neq 0$.)

SOLUTION Because the bases are the same, keep the base and subtract the exponents.

$$\frac{y^9}{y^2} = y^{9-2} \quad \text{Keep the base and subtract the exponents.}$$
$$= y^7 \quad \text{Subtract.}$$

EXAMPLE 10 Simplify $a^{21} \div a^3$. (Assume that $a \neq 0$.)

SOLUTION Although in a different form than the previous example, $a^{21} \div a^3$ is equivalent to $\frac{a^{21}}{a^3}$. Keep the base and subtract the exponents.

$$a^{21} \div a^3 = a^{21-3} \quad \text{Keep the base and subtract the exponents.}$$
$$= a^{18} \quad \text{Subtract.}$$

Quick Check 8

Simplify. (Assume that $x \neq 0$.)

a) $\frac{x^{20}}{x^4}$ **b)** $x^{24} \div x^8$

A WORD OF CAUTION When dividing two expressions with the same base, keep the base and subtract the exponents. *Do not divide the exponents.*

$$a^{21} \div a^3 = a^{18}, \text{ not } a^7$$

EXAMPLE 11 Simplify $\frac{16x^{10}y^9}{2x^7y^4}$. (Assume that $x, y \neq 0$.)

SOLUTION Although we will subtract the exponents for the bases x and y, we will not subtract the numerical factors 16 and 2. We simplify numerical factors the same way we did in Chapter 1; so $\frac{16}{2} = 8$.

$$\frac{16x^{10}y^9}{2x^7y^4} = 8x^3y^5 \quad \text{Simplify } \tfrac{16}{2}\text{. Keep the variable bases and subtract the exponents.}$$

Quick Check 9

Simplify $\frac{20a^{10}b^{12}c^{15}}{5a^2b^6c^9}$. (Assume that $a, b, c \neq 0$.)

Zero Exponent Rule

Objective 5 Use the zero exponent rule. Consider the expression $\frac{2^7}{2^7}$. We know that any number (except 0) divided by itself is equal to 1. The previous property also tells us that we can subtract the exponents in this expression; so $\frac{2^7}{2^7} = 2^{7-7} = 2^0$.

Because we have shown that $\frac{2^7}{2^7}$ is equal to both 2^0 and 1, the expression 2^0 must be equal to 1. The following property involves raising a base to the power of zero.

Zero Exponent Rule

For any base x, $x^0 = 1$ $(x \neq 0)$.

EXAMPLE 12 Simplify 7^0.

SOLUTION By applying the zero exponent rule, we see that $7^0 = 1$.

A WORD OF CAUTION When we raise a nonzero base to the power of zero, the result is 1, not 0.

$$7^0 = 1, \text{ not } 0$$

Quick Check 10

Simplify. (Assume that $x \neq 0$.)

a) 124^0 **b)** $(3x^2)^0$ **c)** $9x^0$

The expressions $(4x)^0$ and $4x^0$ are different $(x \neq 0)$. In the first expression, the base being raised to the power of zero is $4x$. So $(4x)^0 = 1$ because any nonzero base raised to the power of zero is equal to 1. In the second expression, the base being raised to the power of zero is x, not $4x$. The expression $4x^0$ simplifies to be $4 \cdot 1$, or 4.

Power of a Quotient Rule

Objective 6 Use the power of a quotient rule for exponents. The power of a quotient rule is similar to the property that involves raising a product to a power.

Power of a Quotient Rule

For any bases x and y, $\left(\frac{x}{y}\right)^n = \frac{x^n}{y^n}$ $(y \neq 0)$.

To raise a quotient to a power, we raise both the numerator and denominator to that power. The restriction $y \neq 0$ is again due to the fact that division by 0 is undefined. Consider the expression $\left(\frac{r}{s}\right)^3$, where $s \neq 0$. The property tells us that this is equivalent to $\frac{r^3}{s^3}$, and here is why.

$$\left(\frac{r}{s}\right)^3 = \frac{r}{s} \cdot \frac{r}{s} \cdot \frac{r}{s} \quad \text{Repeat } \frac{r}{s} \text{ as a factor three times.}$$

$$= \frac{r \cdot r \cdot r}{s \cdot s \cdot s} \quad \text{Use the definition of multiplication for fractions.}$$

$$= \frac{r^3}{s^3} \quad \text{Rewrite the numerator and denominator using exponents.}$$

A WORD OF CAUTION When raising a quotient to a power, raise both the numerator and denominator to that power. Do not raise just the numerator to that power.

$$\left(\frac{r}{s}\right)^3 = \frac{r^3}{s^3}, \text{ not } \frac{r^3}{s} \qquad \text{(Assume that } s \neq 0.)$$

EXAMPLE 13 Simplify $\left(\frac{x^2y^3}{z^4}\right)^8$. (Assume that $z \neq 0$.)

SOLUTION Begin by raising each factor in the numerator and denominator to the 8th power; then simplify the resulting expression.

$$\left(\frac{x^2y^3}{z^4}\right)^8 = \frac{(x^2)^8(y^3)^8}{(z^4)^8} \quad \text{Raise each factor in the numerator and denominator to the 8th power.}$$

$$= \frac{x^{16}y^{24}}{z^{32}} \quad \text{Keep the bases and multiply the exponents.}$$

Quick Check 11

Simplify $\left(\frac{ab^7}{c^3d^4}\right)^5$. (Assume that $c, d \neq 0$.)

Here is a brief summary of the properties introduced in this section.

Properties of Exponents

1. **Product Rule** For any base x, $x^m \cdot x^n = x^{m+n}$.
2. **Power Rule** For any base x, $(x^m)^n = x^{m \cdot n}$.
3. **Power of a Product Rule** For any bases x and y, $(xy)^n = x^n y^n$.
4. **Quotient Rule** For any base x, $\frac{x^m}{x^n} = x^{m-n}$ $(x \neq 0)$.
5. **Zero Exponent Rule** For any base x, $x^0 = 1$ $(x \neq 0)$.
6. **Power of a Quotient Rule** For any bases x and y, $\left(\frac{x}{y}\right)^n = \frac{x^n}{y^n}$ $(y \neq 0)$.

Objective 7 Evaluate functions by using the rules for exponents. We conclude this section by investigating functions containing exponents.

EXAMPLE 14 For the function $f(x) = x^4$, evaluate $f(3)$.

SOLUTION Substitute 3 for x and simplify the resulting expression.

$$f(3) = (3)^4 \quad \text{Substitute 3 for } x.$$

$$= 81 \quad \text{Simplify.}$$

Quick Check 12

For the function $f(x) = x^6$, evaluate $f(-4)$.

EXAMPLE 15 For the function $f(x) = x^3$, evaluate $f(a^5)$.

SOLUTION Substitute a^5 for x and simplify.

$$\begin{aligned} f(a^5) &= (a^5)^3 && \text{Substitute } a^5 \text{ for } x. \\ &= a^{5 \cdot 3} && \text{Simplify using the power rule.} \\ &= a^{15} && \text{Multiply.} \end{aligned}$$

Quick Check 13

For the function $f(x) = x^7$, evaluate $f(a^8)$.

BUILDING YOUR STUDY STRATEGY

Test Taking, 1 Prepare Completely The first test-taking skill is preparing yourself completely. As legendary basketball coach John Wooden once said, "Failing to prepare is preparing to fail." Begin preparing for the exam well in advance; do not plan to study (or cram) the night before the exam.

Review your old homework assignments, notes, and note cards, spending more time on problems or concepts that were difficult for you. Work through the chapter review and chapter test in the text to identify any areas of weakness for you.

Create a practice test and take it under test conditions without using your text or notes. Allow yourself the same amount of time you will be given for the actual test so you know whether you are working fast enough.

Finally, get a good night's sleep the night before the exam. Tired students do not think as well as students who are rested.

Exercises 5.1

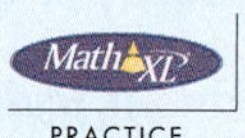

Vocabulary

1. The product rule states that for any base x, ________.
2. The power rule states that for any base x, ________.
3. The power of a product rule states that for any bases x and y, ________.
4. The quotient rule states that for any nonzero base x, ________.
5. The zero exponent rule states that for any nonzero base x, ______.
6. The power of a quotient rule states that for any bases x and y, where $y \neq 0$, ________.

Simplify.

7. $2^4 \cdot 2^2$
8. $5^2 \cdot 5^3$
9. $x^8 \cdot x^5$
10. $a^7 \cdot a^4$
11. $m^{21} \cdot m^{19}$
12. $n^{17} \cdot n^{45}$
13. $b^5 \cdot b^6 \cdot b^3$
14. $x^8 \cdot x^6 \cdot x$
15. $(2x - 3)^4 \cdot (2x - 3)^{10}$
16. $(4x + 1)^{12} \cdot (4x + 1)^9$
17. $(x^6y^5)(x^2y^8)$
18. $(5x^8y^8)(2x^5y^{11})$

Find the missing factor.

19. $x^4 \cdot ? = x^9$
20. $x^7 \cdot ? = x^{10}$
21. $? \cdot b^5 = b^6$
22. $? \cdot a^{11} = a^{20}$

Simplify.

23. $(x^4)^6$
24. $(x^5)^5$
25. $(a^7)^9$
26. $(b^{14})^4$
27. $(2^5)^2$
28. $(3^3)^3$
29. $(x^3)^7(x^4)^2$
30. $(x^2)^8(x^5)^3$

Find the missing exponent.

31. $(x^6)^? = x^{42}$
32. $(p^?)^8 = p^{112}$
33. $(x^3)^8(x^?)^4 = x^{44}$
34. $(x^9)^?(x^{11})^{12} = x^{195}$

Simplify.

35. $(3x)^4$
36. $(7m)^3$
37. $(-8x^3)^2$
38. $(-5n^6)^4$
39. $(m^8n^5)^7$
40. $(a^3b^{11})^5$
41. $(3x^4y)^4$
42. $(2x^5y^6)^8$
43. $(x^2y^5z^3)^7(y^4z^5)^8$
44. $(a^{10}b^{15}c^{20})^7(a^{11}b^7c^3)^8$

Find the missing exponent(s).

45. $(x^5y^4)^? = x^{35}y^{28}$
46. $(c^7d^{11})^? = c^{21}d^{33}$
47. $(2s^?t^?)^4 = 16s^{24}t^{52}$
48. $(m^?n^?)^9 = m^{54}n^{216}$

Simplify. (Assume that all variables are nonzero.)

49. $\frac{x^{12}}{x^3}$
50. $\frac{x^{32}}{x^4}$
51. $r^{25} \div r^{10}$
52. $z^{36} \div z^{15}$
53. $\frac{16x^{16}}{2x^2}$
54. $\frac{119x^{24}}{7x^8}$
55. $\frac{(a + 5b)^{13}}{(a + 5b)^{11}}$
56. $\frac{(2x - 17)^{11}}{(2x - 17)^5}$
57. $\frac{a^8b^6}{a^3b}$
58. $\frac{28r^6s^7t^9}{7r^3s^7t^6}$

Find the missing exponent(s). (Assume that all variables are nonzero.)

59. $\frac{x^{14}}{x^?} = x^2$
60. $\frac{y^?}{y^8} = y^9$
61. $m^? \div m^{11} = m^{19}$
62. $\frac{x^{15}y^?}{x^?y^{25}} = x^7y^6$

Simplify. (Assume that all variables are nonzero.)

63. 9^0
64. -8^0
65. $(-16)^0$
66. 5^0
67. $\frac{3}{7^0}$
68. $\left(\frac{4}{13}\right)^0$
69. $(13x)^0$
70. $(22x^5y^{12}z^9)^0$

Simplify. (Assume that all variables are nonzero.)

71. $\left(\frac{3}{4}\right)^3$
72. $\left(\frac{x}{2}\right)^7$
73. $\left(\frac{a}{b}\right)^8$
74. $\left(\frac{x}{y}\right)^{12}$
75. $\left(\frac{x^4}{y^3}\right)^5$
76. $\left(\frac{m^9}{n^6}\right)^8$
77. $\left(\frac{2a^6}{b^7}\right)^3$
78. $\left(\frac{x^{13}}{5y^9}\right)^5$
79. $\left(\frac{a^5b^4}{b^2c^7}\right)^9$
80. $\left(\frac{17x^5y^{19}z^{24}}{32a^{11}b^{70}c^{33}}\right)^0$

Find the missing exponent(s). (Assume that all variables are nonzero.)

81. $\left(\frac{3}{4}\right)^? = \frac{81}{256}$
82. $\left(\frac{x^4}{y^7}\right)^? = \frac{x^{12}}{y^{21}}$
83. $\left(\frac{a^3b^6}{c^8}\right)^? = \frac{a^{36}b^{72}}{c^{96}}$
84. $\left(\frac{ab^8}{c^5d^9}\right)^? = \frac{a^{15}b^{120}}{c^{75}d^{135}}$

Evaluate the given function.

85. $f(x) = x^2, f(4)$
86. $g(x) = x^3, g(9)$
87. $g(x) = x^4, g(-2)$
88. $f(x) = x^{10}, f(-1)$
89. $f(x) = x^6, f(a^4)$
90. $f(x) = x^5, f(b^7)$

Mixed Practice, 91–104

Simplify.

91. $(a^7)^6$
92. $(13a)^0$
93. $\left(\frac{5x^4}{2y^7}\right)^3$
94. $(x^6y^9z^2)^7$
95. $\frac{x^{13}}{x}$
96. $\frac{x^{100}}{x^{100}}$
97. $\left(\frac{a^8b^9}{2c^2}\right)^5$
98. $c^8 \cdot c^{12}$
99. $(b^{10})^2$
100. $\frac{x^8y^{13}}{xy^6}$
101. $(x^{15})^6$
102. $(2x^9yz^3)^4$
103. $\left(\frac{3a^2b^9}{7cd^4}\right)^4$
104. $(6x^7y^9z)^3$

The number of feet traveled by a free-falling object in t seconds is given by the function $f(t) = 16t^2$.

105. If a ball is dropped from the top of a building, how far will it fall in 3 seconds?
106. How far will a skydiver fall in 7 seconds?

The area of a square with side x is given by the function $A(x) = x^2$.

107. Use the function to find the area of a square if each side is 90 feet long.
108. Use the function to find the area of a square if each side is 17 meters long.

The area of a circle with radius r is given by the function $A(r) = \pi r^2$.

109. Use the function to find the area of the base of a circular storage tank if the radius is 14 feet long. Use $\pi \approx 3.14$.

110. Columbus Circle in New York City is a traffic circle. The radius of the inner circle is 107 feet. Use the function to find the area of the inner circle. Use $\pi \approx 3.14$.

Writing in Mathematics

Answer in complete sentences.

111. Which expression is equal to x^9: $(x^5)^4$ or $x^5 \cdot x^4$? Explain your answer.

112. Explain the difference between the expressions $\frac{a^3}{b}$ and $\left(\frac{a}{b}\right)^3$.

5.2 Negative Exponents; Scientific Notation

OBJECTIVES

1. Understand negative exponents.
2. Use the rules of exponents to simplify expressions containing negative exponents.
3. Convert numbers from standard notation to scientific notation.
4. Convert numbers from scientific notation to standard notation.
5. Perform arithmetic operations using numbers in scientific notation.
6. Use scientific notation to solve applied problems.

Negative Exponents

Objective 1 Understand negative exponents. In this section, we introduce the concept of a **negative exponent**.

Negative Exponents

For any nonzero base x, $x^{-n} = \frac{1}{x^n}$.

For example, $2^{-3} = \frac{1}{2^3}$, or $\frac{1}{8}$.

Let's examine this definition. Consider the expression $\frac{2^4}{2^7}$. If we apply the property $\frac{x^m}{x^n} = x^{m-n}$ from the previous section, we see that $\frac{2^4}{2^7}$ simplifies to 2^{-3}.

$$\frac{2^4}{2^7} = 2^{4-7}$$
$$= 2^{-3}$$

If we simplify $\frac{2^4}{2^7}$ by dividing out common factors, the result is $\frac{1}{2^3}$.

$$\frac{2^4}{2^7} = \frac{\overset{1}{\cancel{2}} \cdot \overset{1}{\cancel{2}} \cdot \overset{1}{\cancel{2}} \cdot \overset{1}{\cancel{2}}}{\underset{1}{\cancel{2}} \cdot \underset{1}{\cancel{2}} \cdot \underset{1}{\cancel{2}} \cdot \underset{1}{\cancel{2}} \cdot 2 \cdot 2 \cdot 2}$$
$$= \frac{1}{2^3}$$

Because $\frac{2^4}{2^7}$ is equal to 2^{-3} and $\frac{1}{2^3}$, $2^{-3} = \frac{1}{2^3}$.

A WORD OF CAUTION Raising a positive base to a negative exponent is not the same as raising the opposite of that base to a positive power.

$$2^{-3} = \frac{1}{2^3}, \text{ not } -2^3$$

EXAMPLE 1 Rewrite the expression 7^{-2} without using negative exponents and simplify.

SOLUTION When raising a number to a negative exponent, we begin by rewriting the expression without negative exponents. We finish by raising the base to the appropriate positive power.

$$7^{-2} = \frac{1}{7^2} \quad \text{Rewrite without negative exponents.}$$

$$= \frac{1}{49} \quad \text{Simplify.}$$

Quick Check 1

Rewrite 4^{-3} without using negative exponents and simplify.

EXAMPLE 2 Rewrite the expression $9x^{-6}$ without using negative exponents. (Assume that $x \neq 0$.)

SOLUTION The base in this example is x. The exponent does not apply to the 9 because there are no parentheses. When we rewrite the expression without a negative exponent, the number 9 is unaffected. Rather than writing x^{-6} as $\frac{1}{x^6}$ first, we can move the factor directly to the denominator of a fraction whose numerator is 9 by changing the sign of its exponent.

$$9x^{-6} = \frac{9}{x^6}$$

Note the difference between the expressions $9x^{-6}$ and $(9x)^{-6}$. In the first expression, the base x is being raised to the power of -6, leaving 9 in the numerator when the expression is rewritten without negative exponents. The base in the second expression is $9x$, so we rewrite $(9x)^{-6}$ as $\frac{1}{(9x)^6}$.

Quick Check 2

Rewrite the expression a^5b^{-4} without using negative exponents. (Assume that $b \neq 0$.)

EXAMPLE 3 Rewrite the expression $\frac{x^8}{z^{-5}}$ without using negative exponents. (Assume that $z \neq 0$.)

SOLUTION Begin by rewriting the denominator as $1/z^5$.

$$\frac{x^8}{z^{-5}} = \frac{x^8}{1/z^5} \quad \text{Rewrite the denominator without negative exponents.}$$

$$= x^8 \div \frac{1}{z^5} \quad \text{Rewrite as division.}$$

$$= x^8 \cdot \frac{z^5}{1} \quad \text{Invert the divisor and multiply.}$$

$$= x^8z^5 \quad \text{Simplify.}$$

When an expression has a factor in the denominator with a negative exponent, we can change its exponent to a positive number by moving the factor to the numerator.

EXAMPLE 4 Rewrite the expression $\frac{a^7b^{-3}}{c^{-5}d^{12}}$ without using negative exponents. (Assume that all variables are nonzero.)

SOLUTION Of the four factors in this example, two of them (b^{-3} and c^{-5}) need to be rewritten without using negative exponents. This can be done by changing the sign of their exponents and rewriting the factors on the other side of the fraction bar.

$$\frac{a^7b^{-3}}{c^{-5}d^{12}} = \frac{a^7c^5}{b^3d^{12}}$$

Quick Check 3

Rewrite without using negative exponents. (Assume that all variables are nonzero.)

a) $\frac{x^{12}}{y^{-7}}$ **b)** $\frac{x^{-2}y^4}{z^{-1}w^{-9}}$

Using the Rules of Exponents with Negative Exponents

Objective 2 Use the rules of exponents to simplify expressions containing negative exponents. All of the properties of exponents described in the previous section hold true for negative exponents. When we are simplifying expressions involving negative exponents, we can take two general routes. We can choose to apply the appropriate property first, then rewrite the expression without negative exponents. On the other hand, in some circumstances it will be more convenient to rewrite the expression without negative exponents before attempting to apply the appropriate property.

EXAMPLE 5 Simplify the expression $x^{11} \cdot x^{-6}$. Write the result without using negative exponents. (Assume that $x \neq 0$.)

SOLUTION This example uses the product rule $x^m \cdot x^n = x^{m+n}$.

$$x^{11} \cdot x^{-6} = x^{11+(-6)} \quad \text{Keep the base and add the exponents.}$$
$$= x^5 \quad \text{Simplify.}$$

An alternative approach is to rewrite the expression without using negative exponents.

$$x^{11} \cdot x^{-6} = \frac{x^{11}}{x^6} \quad \text{Rewrite without negative exponents.}$$
$$= x^5 \quad \text{Keep the base and subtract the exponents.}$$

Use the approach that seems clearer to you.

Quick Check 4

Simplify the expression $x^{-13} \cdot x^6$. Write the result without using negative exponents. (Assume that $x \neq 0$.)

EXAMPLE 6 Simplify the expression $(y^3)^{-2}$. Write the result without using negative exponents. (Assume that $y \neq 0$.)

SOLUTION This example uses the power rule $(x^m)^n = x^{m \cdot n}$.

$$(y^3)^{-2} = y^{-6} \quad \text{Keep the base and multiply the exponents.}$$
$$= \frac{1}{y^6} \quad \text{Rewrite without using negative exponents.}$$

Quick Check 5

Simplify the expression $(x^{-6})^{-7}$. Write the result without using negative exponents. (Assume that $x \neq 0$.)

EXAMPLE 7 Simplify the expression $(2a^{-7}b^{9})^{-4}$. Write the result without using negative exponents. (Assume that $a, b \neq 0$.)

SOLUTION This example uses the power of a product rule $(xy)^n = x^n y^n$.

$$(2a^{-7}b^{9})^{-4} = 2^{-4}a^{-7(-4)}b^{9(-4)} \quad \text{Raise each factor to the power of } -4.$$
$$= 2^{-4}a^{28}b^{-36} \quad \text{Simplify each exponent.}$$
$$= \frac{a^{28}}{2^4 b^{36}} \quad \text{Rewrite without using negative exponents.}$$
$$= \frac{a^{28}}{16b^{36}} \quad \text{Simplify } 2^4.$$

Quick Check 6

Simplify the expression $(9ab^{-3}c^{2})^{-2}$. Write the result without using negative exponents. (Assume that $a, b, c \neq 0$.)

EXAMPLE 8 Simplify the expression $\frac{x^3}{x^{15}}$. Write the result without using negative exponents. (Assume that $x \neq 0$.)

SOLUTION This example uses the quotient rule $\frac{x^m}{x^n} = x^{m-n}$.

$$\frac{x^3}{x^{15}} = x^{-12} \quad \text{Keep the base and subtract the exponents.}$$
$$= \frac{1}{x^{12}} \quad \text{Rewrite without using negative exponents.}$$

Quick Check 7

Simplify the expression $\frac{x^8}{x^{14}}$. Write the result without using negative exponents. (Assume that $x \neq 0$.)

EXAMPLE 9 Simplify the expression $\left(\frac{r^4}{s^2}\right)^{-3}$. Write the result without using negative exponents. (Assume that $r, s \neq 0$.)

SOLUTION This example uses the power of a quotient rule $\left(\frac{x}{y}\right)^n = \frac{x^n}{y^n}$.

$$\left(\frac{r^4}{s^2}\right)^{-3} = \frac{r^{4(-3)}}{s^{2(-3)}} \quad \text{Raise the numerator and the denominator to the power of } -3.$$
$$= \frac{r^{-12}}{s^{-6}} \quad \text{Multiply.}$$
$$= \frac{s^6}{r^{12}} \quad \text{Rewrite without using negative exponents.}$$

Quick Check 8

Simplify the expression $\left(\frac{x^5}{y^{-4}}\right)^{-4}$. Write the result without using negative exponents. (Assume that $x, y \neq 0$.)

Scientific Notation

Objective 3 Convert numbers from standard notation to scientific notation. **Scientific notation** is used to represent numbers that are very large, such as 93,000,000, or very small, such as 0.0000324.

Rewriting a Number in Scientific Notation

To convert a number to scientific notation, rewrite it in the form $a \times 10^b$, where $1 \leq a < 10$ and b is an integer.

First, consider the following table listing several powers of 10:

10^5	10^4	10^3	10^2	10^1	10^0	10^{-1}	10^{-2}	10^{-3}	10^{-4}	10^{-5}
100,000	10,000	1000	100	10	1	0.1	0.01	0.001	0.0001	0.00001

Notice that all of the positive powers of 10 are numbers that are 10 or higher. All of the negative powers of 10 are numbers between 0 and 1.

To convert a number to scientific notation, we first move the decimal point so that it immediately follows the first nonzero digit in the number. Count the number of decimal places the decimal point moves. This gives us the power of 10 when the number is written in scientific notation. If the original number is 10 or larger, the exponent is positive, but if the original number is between 0 and 1, the exponent is negative.

EXAMPLE 10 Convert 0.000321 to scientific notation.

SOLUTION Move the decimal point so that it follows the first nonzero digit, which is 3.

$$0.000321$$

To do this, move the decimal point four places to the right. Because 0.000321 is less than 1, this exponent must be negative.

$$0.000321 = 3.21 \times 10^{-4}$$

EXAMPLE 11 Convert 24,000,000,000 to scientific notation.

SOLUTION Move the decimal point so that it follows the digit 2. To do this, move the decimal point 10 places to the left. Because 24,000,000,000 is 10 or larger, the exponent will be positive.

$$24{,}000{,}000{,}000 = 2.4 \times 10^{10}$$

Quick Check 9

Convert to scientific notation.

a) 0.0046
b) 3,570,000

Objective 4 Convert numbers from scientific notation to standard notation.

EXAMPLE 12 Convert 5.28×10^4 from scientific notation to standard notation.

SOLUTION

$$\begin{aligned} 5.28 \times 10^4 &= 5.28 \times 10{,}000 && \text{Rewrite } 10^4 \text{ as } 10{,}000. \\ &= 52{,}800 && \text{Multiply.} \end{aligned}$$

To multiply a decimal number by 10^4 or 10,000, move the decimal point four places to the right. Note that the power of 10 is positive in this example.

EXAMPLE 13 Convert 2.0039×10^{-5} from scientific notation to standard notation.

SOLUTION

$$2.0039 \times 10^{-5} = 2.0039 \times 0.00001 \quad \text{Rewrite } 10^{-5} \text{ as } 0.00001.$$
$$= 0.000020039 \quad \text{Multiply.}$$

To multiply a decimal number by 10^{-5}, or 0.00001, move the decimal point five places to the left. Note that the power of 10 is negative in this example.

Quick Check 10

Convert from scientific notation to standard notation.

a) 3.2×10^6 **b)** 7.21×10^{-4}

If you are unsure about which direction to move the decimal point, think about whether you are making the number larger or smaller. Multiplying by positive powers of 10 makes the number larger; so move the decimal point to the right. Multiplying by negative powers of 10 makes the number smaller; so move the decimal point to the left.

Objective 5 Perform arithmetic operations using numbers in scientific notation. When we are performing calculations involving very large or very small numbers, using scientific notation can be convenient.

EXAMPLE 14 Multiply $(2.2 \times 10^7)(2.8 \times 10^{13})$. Express your answer using scientific notation.

SOLUTION When multiplying two numbers that are in scientific notation, we may multiply the two decimal numbers first. We then multiply the powers of 10 using the product rule, $x^m \cdot x^n = x^{m+n}$.

$$(2.2 \times 10^7)(2.8 \times 10^{13}) = (2.2)(2.8)(10^7)(10^{13}) \quad \text{Reorder the factors.}$$
$$= 6.16 \times 10^{20} \quad \text{Multiply decimal numbers. Add the exponents for the base 10.}$$

Quick Check 11

Multiply $(5.8 \times 10^4)(1.2 \times 10^9)$. Express your answer using scientific notation.

Using Your Calculator The TI-84 can help you perform calculations with numbers in scientific notation. To enter a number in scientific notation, use the second function **EE** above the key labeled [,]. For example, to enter the number 2.2×10^7, key 2.2 [2nd] [,] 7. Here is the screen shot showing how to multiply $(2.2 \times 10^7)(2.8 \times 10^{13})$.

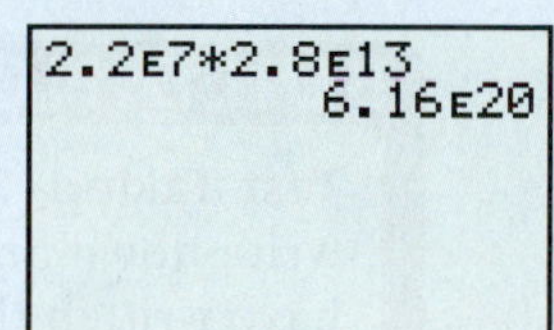

EXAMPLE 15 Divide (1.2×10^{-6}) by (2.4×10^{9}). Express your answer using scientific notation.

SOLUTION To divide numbers that are in scientific notation, begin by dividing the decimal numbers. Then divide the powers of 10 separately, using the property $\frac{x^m}{x^n} = x^{m-n}$.

$$(1.2 \times 10^{-6}) \div (2.4 \times 10^{9}) = \frac{1.2 \times 10^{-6}}{2.4 \times 10^{9}}$$ Rewrite as a fraction.

$$= \frac{1.2}{2.4} \times 10^{-6-9}$$ Divide the decimal numbers. Subtract the exponents for the powers of 10.

$$= 0.5 \times 10^{-15}$$ Divide 1.2 by 2.4. Simplify the exponent.

Although this is the correct quotient, the answer is not in scientific notation, as the number 0.5 does not have a nonzero digit to the left of the decimal point. To rewrite 0.5 as 5.0, move the decimal point one place to the right. This will decrease the exponent by one, from −15 to −16. $(1.2 \times 10^{-6}) \div (2.4 \times 10^{9}) = 5.0 \times 10^{-16}$

Quick Check 12

Divide (9.9×10^{5}) by (3.3×10^{-7}). Express your answer using scientific notation.

Applications

Objective 6 Use scientific notation to solve applied problems. We finish with an applied problem using scientific notation.

EXAMPLE 16 At its farthest point, the planet Mars is 155 million miles from the sun. How many seconds does it take light from the sun to reach Mars if the speed of light is 1.86×10^{5} miles per second?

SOLUTION To determine the length of time a trip takes, we divide the distance traveled by the rate of speed. To solve this problem, we need to divide the distance of 155 million (155,000,000) miles by the speed of light. We will convert the distance to scientific notation, which is 1.55×10^{8}.

$$\frac{1.55 \times 10^{8}}{1.86 \times 10^{5}} \approx 0.833 \times 10^{3}$$ Divide the decimal numbers; approximate with a calculator. Subtract the exponents for the powers of 10.

To write this result in scientific notation, we move the decimal point one place to the right. This decreases the exponent by one. The light will reach Mars in approximately 8.33×10^{2}, or 833, seconds, or 13 minutes and 53 seconds.

Quick Check 13

How many seconds does it take light to travel 1,488,000,000 miles? (The speed of light is 1.86×10^{5} miles per second.)

BUILDING YOUR STUDY STRATEGY

Test Taking, 2 Write Down Information As soon as you receive your test, write down any formulas, rules, or procedures that will help you. Once you have written these down, you can refer to them as you work through the test. By writing all of this information on your test, you will not have to worry about recalling everything you memorized for the test.

Exercises 5.2

MyMathLab

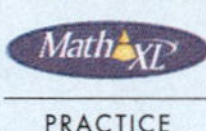
PRACTICE

WATCH

DOWNLOAD

READ

REVIEW

Vocabulary

1. For any *nonzero* base x, $x^{-n} =$ ____.
2. A number is in ____________ if it is in the form $a \times 10^b$, where $1 \le a < 10$ and b is an integer.

Rewrite the expression without using negative exponents. (Assume that all variables represent nonzero real numbers.)

3. 5^{-2}
4. 8^{-2}
5. 4^{-3}
6. 7^{-3}
7. -13^{-2}
8. -3^{-4}
9. b^{-12}
10. a^{-9}
11. $12x^{-6}$
12. $6x^{-8}$
13. $-5m^{-19}$
14. $-12m^{-6}$
15. $\dfrac{1}{x^{-5}}$
16. $\dfrac{1}{x^{-7}}$
17. $\dfrac{3}{y^{-4}}$
18. $\dfrac{x^7}{y^{-5}}$
19. $\dfrac{x^{-11}}{y^{-2}}$
20. $\dfrac{x^{-10}}{y^4}$
21. $\dfrac{-8a^{-6}b^5}{c^7}$
22. $\dfrac{-10x^3y^{-5}}{z^{-4}}$
23. $\dfrac{5a^{-6}b^{-9}}{c^{-4}d^5}$
24. $\dfrac{a^3b^{-3}}{c^{-4}d^{-9}}$

Simplify the expression. Write the result without using negative exponents. (Assume that all variables represent nonzero real numbers.)

25. $x^{-12} \cdot x^7$
26. $x^8 \cdot x^{-17}$
27. $a^{16} \cdot a^{-10}$
28. $b^{-13} \cdot b^{31}$
29. $m^{-9} \cdot m^{-12}$
30. $n^{-15} \cdot n^{-21}$
31. $(x^6)^{-3}$
32. $(x^{11})^{-4}$
33. $(x^{-6})^{-7}$
34. $(x^{-9})^{-2}$
35. $(x^5y^4z^{-6})^{-2}$
36. $(x^{-3}y^3z^{-2})^{-7}$
37. $(4a^{-5}b^{-8}z^2)^{-3}$
38. $(-2x^4b^{-6}c^{-7})^5$
39. $\dfrac{x^{-5}}{x^{10}}$
40. $\dfrac{x^{-3}}{x^{15}}$
41. $\dfrac{x^{11}}{x^{-6}}$
42. $\dfrac{x^8}{x^{-8}}$
43. $\dfrac{a^{-8}}{a^{-5}}$
44. $\dfrac{a^{-2}}{a^{-9}}$
45. $\dfrac{x^7}{x^{18}}$
46. $\dfrac{x^3}{x^{12}}$
47. $\dfrac{x^5 \cdot x^{-11}}{x^{-6}}$
48. $\dfrac{x^{-13}}{x^{-17} \cdot x^4}$
49. $\left(\dfrac{x^3}{y^4}\right)^{-5}$
50. $\left(\dfrac{2x^8}{y^5}\right)^{-3}$
51. $\left(\dfrac{a^4b^7}{3c^5d^{-8}}\right)^{-2}$
52. $\left(\dfrac{x^{-8}y^5}{wz^{-6}}\right)^{-8}$

Convert the given number to standard notation.

53. 3.07×10^{-7}
54. 2.3×10^5
55. 8.935×10^9
56. 6.001×10^{-8}
57. 9.021×10^4
58. 7.0×10^{10}

Convert the given number to scientific notation.

59. 0.00027
60. 0.00000621
61. 8,600,000
62. 92,000
63. 420,000,000,000
64. 0.0000000023

Perform the following calculations. Express your answer using scientific notation.

65. $(4.1 \times 10^8)(2.3 \times 10^{11})$
66. $(3.6 \times 10^{-13})(1.6 \times 10^6)$
67. $(1.598 \times 10^{-12}) \div (4.7 \times 10^9)$
68. $(3.286 \times 10^{14}) \div (6.2 \times 10^{-7})$
69. $(5.32 \times 10^{-15})(7.8 \times 10^3)$

70. $(6.0 \times 10^{13}) \div (7.5 \times 10^{-12})$

71. $(87{,}000{,}000{,}000)(0.000002)$

72. $0.0000000064 \div 160{,}000{,}000{,}000$

73. If a computer can perform a calculation in 0.0000000008 second, how long will it take to perform 42,000,000,000,000 calculations?

74. If a computer can perform a calculation in 0.0000000008 second, how many calculations can it perform in 2 minutes (120 seconds)?

75. The speed of light is 1.86×10^5 miles per second. How far can light from the sun travel in 10 minutes?

76. The speed of light is 1.86×10^5 miles per second. At its farthest point, Pluto is 4,555,000,000 miles from the sun. How many seconds does it take light from the sun to reach Pluto?

77. If the mass of a typical star is 2.2×10^{33} grams and there are approximately 500,000,000,000 stars in the Milky Way galaxy, what is the total mass of these stars?

78. The speed of light is 1.86×10^5 miles per second. How far can light from the sun travel in one year? (This distance is often referred to as a light-year.)

79. For the 12-month period ending March 31, 2009, Microsoft showed total revenues of $\$6.118 \times 10^{10}$ and Apple showed total revenues of $\$3.369 \times 10^{10}$. What was the combined revenue for the two companies? (*Source: Capital IQ/Yahoo! Finance*)

80. Each day Burger King restaurants serve 1.57×10^7 customers worldwide. How many customers do Burger King restaurants serve in one year? (*Source: Burger King Corporation*)

81. Approximately 3.3×10^6 high school students graduated in 2008. If 70% of these graduates attended college in the fall, how many attended college in the fall? (*Source: U.S. Department of Education*)

82. In the 2003–2004 school year, there were approximately 1.73×10^7 U.S. college students. If 63% of these students received some type of financial aid, how many received some type of financial aid? (*Source: U.S. Department of Education*)

Writing in Mathematics

Answer in complete sentences.

83. Explain the difference between the expressions 4^{-2} and $(-4)^2$.

84. Describe three real-world examples of numbers using scientific notation.

Quick Review Exercises

Section 5.2

Simplify.

1. $8x + 3x$
2. $9a - 16a$
3. $-9x^2 + 4x + 3x^2 + 7x$
4. $10y^3 - y^2 + 5y + 13 + 2y^3 + 6y^2 - 18y - 50$

5.3 Polynomials; Addition and Subtraction of Polynomials

OBJECTIVES

1. Identify polynomials and understand the vocabulary used to describe them.
2. Evaluate polynomials.
3. Add and subtract polynomials.
4. Understand polynomials in several variables.

Polynomials in a Single Variable

Objective 1 Identify polynomials and understand the vocabulary used to describe them. Many real-world phenomena cannot be described by linear expressions and linear functions. For example, if a car is traveling at x miles per hour on dry pavement, the distance in feet required for the car to come to a complete stop can be approximated by the expression $0.06x^2 + 1.1x + 0.02$. This nonlinear expression is an example of a **polynomial**.

Polynomials

A **polynomial in a single variable** x is a sum of **terms** of the form ax^n, where a is a real number and n is a whole number.

Here are some examples of a polynomial in a single variable x.

$$x^2 - 9x + 14 \qquad 4x^5 - 13 \qquad x^7 - 4x^5 + 8x^2 + 13x$$

An expression is *not* a polynomial if it contains a term with a variable that is raised to a power other than a whole number (such as $x^{3/4}$ or x^{-2}) or if it has a term with a variable in a denominator $\left(\text{such as } \frac{5}{x^3}\right)$.

Degree of a Term

The **degree of each term** in a polynomial in a single variable is equal to the variable's exponent.

For example, the degree of the term $8x^9$ is 9. A **constant term** does not contain a variable. The degree of a constant term is 0 because a constant term such as -15 can be rewritten as $-15x^0$. The polynomial $4x^5 - 9x^4 + 2x^3 - 7x - 11$ has five terms, and their degrees are 5, 4, 3, 1, and 0, respectively.

Term	$4x^5$	$-9x^4$	$2x^3$	$-7x$	-11
Degree	5	4	3	1	0

A polynomial with only one term is called a **monomial**. A **binomial** is a polynomial that has two terms, while a **trinomial** is a polynomial that has three terms. (We do not have special names to describe polynomials with four or more terms.) Here are some examples.

Monomial	Binomial	Trinomial
$2x$	$x^2 - 25$	$x^2 - 15x - 76$
$9x^2$	$7x + 3$	$4x^2 + 12x + 9$
$-4x^3$	$5x^3 - 135$	$x^6 + 11x^5 - 26x^4$

EXAMPLE 1 Classify the polynomial as a monomial, binomial, or trinomial. List the degree of each term.

a) $16x^4 - 81$ **b)** $-13x^7$ **c)** $x^6 - 7x^3 - 18x$

SOLUTION

a) $16x^4 - 81$ has two terms, so it is a binomial. The degrees of its two terms are 4 and 0.

b) $-13x^7$ has only one term, so it is a monomial. The degree of this term is 7.

c) $x^6 - 7x^3 - 18x$ is a trinomial as it has three terms. The degrees of those terms are 6, 3, and 1, respectively.

Quick Check 1

Classify as a monomial, binomial, or trinomial. List the degree of each term.

a) $x^2 - 9x + 20$ **b)** $x^2 - 81$
c) $-92x^7$

The **coefficient** of a term is the numerical part of a term. When determining the coefficient of a term, be sure to include its sign. For the polynomial $8x^3 - 6x^2 - 5x + 13$, the four terms have coefficients 8, −6, −5, and 13, respectively.

Term	$8x^3$	$-6x^2$	$-5x$	13
Coefficient	8	−6	−5	13

EXAMPLE 2 For the trinomial $x^2 - 9x + 18$, determine the coefficient of each term.

SOLUTION The first term of this trinomial, x^2, has a coefficient of 1. Even though we do not see a coefficient in front of the term, the coefficient is 1 because $x^2 = 1 \cdot x^2$. The second term, $-9x$, has a coefficient of −9. Finally, the coefficient of the constant term is 18.

Quick Check 2

For the polynomial $x^3 - 6x^2 - 11x + 32$, determine the coefficient of each term.

We write polynomials in **descending order**, writing the term of highest degree first, followed by the term of next highest degree, and so on. For example, the polynomial $3x^2 - 9 - 5x^5 + 7x$ is written in descending order as $-5x^5 + 3x^2 + 7x - 9$. The term of highest degree is called the **leading term**, and its coefficient is called the **leading coefficient**.

Degree of a Polynomial

The degree of the leading term also is called the **degree of the polynomial**.

EXAMPLE 3 Rewrite the polynomial $7x^2 - 9 + 3x^5 - x$ in descending order and identify the leading term, the leading coefficient, and the degree of the polynomial.

SOLUTION To write a polynomial in descending order, write the terms according to their degree from highest to lowest. In descending order, this polynomial is $3x^5 + 7x^2 - x - 9$. The leading term is $3x^5$, and the leading coefficient is 3. The leading term has degree 5, so this polynomial has degree 5.

Quick Check 3

Rewrite the polynomial $x + 9 + 4x^2 - x^3$ in descending order and identify the leading term, the leading coefficient, and the degree of the polynomial.

Evaluating Polynomials

Objective 2 Evaluate polynomials. To **evaluate a polynomial** for a particular value of a variable, we substitute the value for the variable in the polynomial and simplify the resulting expression.

EXAMPLE 4 Evaluate $4x^5 - 7x^3 + 12x^2 - 13x$ for $x = -2$.

SOLUTION

$4(-2)^5 - 7(-2)^3 + 12(-2)^2 - 13(-2)$ Substitute −2 for x.

$= 4(-32) - 7(-8) + 12(4) - 13(-2)$ Perform operations involving exponents.

$= -128 + 56 + 48 + 26$ Multiply.

$= 2$ Simplify.

Quick Check 4

Evaluate $x^4 - 5x^3 - 6x^2 + 10x - 21$ for $x = -3$.

A **polynomial function** is a function that is described by a polynomial, such as $f(x) = x^3 - 5x^2 - 5x + 7$. Linear functions of the form $f(x) = mx + b$ are first-degree polynomial functions.

EXAMPLE 5 For the polynomial function $f(x) = x^2 - 9x + 16$, find $f(3)$.

SOLUTION Recall that the notation $f(3)$ says to substitute 3 for x in the function. After substituting, simplify the resulting expression.

$f(3) = (3)^2 - 9(3) + 16$ Substitute 3 for x.
$= 9 - 9(3) + 16$ Perform operations involving exponents.
$= 9 - 27 + 16$ Multiply.
$= -2$ Simplify.

Quick Check 5

For the polynomial function $f(x) = x^3 + 12x^2 - 21x$, find $f(-5)$.

Adding and Subtracting Polynomials

Objective 3 Add and subtract polynomials. Just as we can perform arithmetic operations combining two numbers, we can perform operations combining two polynomials, such as adding and subtracting.

Adding Polynomials

To add two polynomials, combine their like terms.

Recall that two terms are like terms if they have the same variables with the same exponents.

EXAMPLE 6 Add $(7x^2 - 5x - 8) + (4x^2 - 9x - 17)$.

SOLUTION To add these polynomials, drop the parentheses and combine like terms.

$(7x^2 - 5x - 8) + (4x^2 - 9x - 17)$
$= 7x^2 - 5x - 8 + 4x^2 - 9x - 17$ Rewrite without using parentheses.
$= 11x^2 - 14x - 25$ Combine like terms. $7x^2 + 4x^2 = 11x^2$, $-5x - 9x = -14x$, $-8 - 17 = -25$

Quick Check 6

Add $(x^3 - x^2 - 9x + 25) + (x^2 + 12x + 144)$.

A WORD OF CAUTION The terms $11x^2$ and $-14x$ are not like terms, as the variables do not have the same exponents.

To subtract one polynomial from another, such as $(x^2 + 5x - 9) - (3x^2 - 2x + 15)$, change the sign of each term that is being subtracted and combine like terms. Changing the sign of each term in the parentheses being subtracted is equivalent to applying the distributive property with -1.

EXAMPLE 7 Subtract $(x^2 + 5x - 9) - (3x^2 - 2x + 15)$.

SOLUTION Remove the parentheses by changing the sign of each term in the polynomial that is being subtracted. Then combine like terms as follows:

$(x^2 + 5x - 9) - (3x^2 - 2x + 15)$
$= x^2 + 5x - 9 - 3x^2 + 2x - 15$ Distribute to remove parentheses.
$= -2x^2 + 7x - 24$ Combine like terms.

Quick Check 7

Subtract $(5x^2 + 6x + 27) - (3x^3 + 6x^2 - 9x + 22)$.

A WORD OF CAUTION When subtracting one polynomial from another polynomial, be sure to change the sign of each term in the polynomial that is being subtracted.

EXAMPLE 8 Given the polynomial functions $f(x) = x^2 + 5x + 7$ and $g(x) = 2x^2 - 7x + 11$, find $f(x) - g(x)$.

SOLUTION Substitute the appropriate expressions for $f(x)$ and $g(x)$ and simplify the resulting expression.

$$f(x) - g(x) = (x^2 + 5x + 7) - (2x^2 - 7x + 11)$$ Substitute.

$$= x^2 + 5x + 7 - 2x^2 + 7x - 11$$ Distribute to remove parentheses.

$$= -x^2 + 12x - 4$$ Combine like terms.

Quick Check 8

Given the polynomial functions $f(x) = 4x^2 + 30x - 45$ and $g(x) = -2x^2 + 17x + 52$, find $f(x) - g(x)$.

Polynomials in Several Variables

Objective 4 Understand polynomials in several variables. While the polynomials we have examined to this point contained a single variable, some polynomials contain two or more variables. Here are some examples.

$$x^2y^2 + 3xy - 10 \qquad 7a^3 - 8a^2b + 3ab^2 - 15b^3 \qquad x^2 + 2xh + h^2 + 5x + 5h$$

We evaluate polynomials in several variables by substituting values for each variable and simplifying the resulting expression.

EXAMPLE 9 Evaluate the polynomial $x^2 - 8xy + 2y^2$ for $x = 5$ and $y = -4$.

SOLUTION Substitute 5 for x and -4 for y. Be careful to substitute the correct value for each variable.

$$x^2 - 8xy + 2y^2$$

$$(5)^2 - 8(5)(-4) + 2(-4)^2$$ Substitute 5 for x and -4 for y.

$$= 25 - 8(5)(-4) + 2(16)$$ Perform operations involving exponents.

$$= 25 + 160 + 32$$ Multiply.

$$= 217$$ Add.

Quick Check 9

Evaluate the polynomial $b^2 - 4ac$ for $a = 1$, $b = -9$, and $c = -52$.

For a polynomial in several variables, the degree of a term is equal to the sum of the exponents of its variable factors. For example, the degree of the term $3x^2y^5$ is $2 + 5$, or 7.

Quick Check 10

Find the degree of each term in the polynomial $x^5 - 3x^2y^6 + 8x^9y$. In addition, find the degree of the polynomial.

EXAMPLE 10 Find the degree of each term in the polynomial $a^4b^3 - 3a^7b^5 + 9ab^6$. In addition, find the degree of the polynomial.

SOLUTION The first term, a^4b^3, has degree 7. The second term, $-3a^7b^5$, has degree 12. The third term, $9ab^6$, has degree 7. The degree of a polynomial is equal to the highest degree of any of its terms, so the degree of this polynomial is 12.

To add or subtract polynomials in several variables, we need to combine like terms. Two terms are like terms if they have the same variables with the same exponents.

EXAMPLE 11 Add $4xy^2 + 7x^3y^5$ and $9x^2y - 3x^3y^5$.

SOLUTION

$(4xy^2 + 7x^3y^5) + (9x^2y - 3x^3y^5)$ Write as a sum.
$= 4xy^2 + 7x^3y^5 + 9x^2y - 3x^3y^5$ Rewrite without parentheses.
$= 4xy^2 + 4x^3y^5 + 9x^2y$ Combine like terms ($7x^3y^5$ and $-3x^3y^5$).

Quick Check 11

Add $5x^4y^2 - 6x^3y^3$ and $8x^4y^2 + 2x^3y^3$.

BUILDING YOUR STUDY STRATEGY

Test Taking, 3 Read the Test In the same way you begin to solve a word problem, you should begin to take a test by briefly reading through it. This will give you an idea of how many problems you must solve, how many word problems there are, and roughly how much time you can devote to each problem. Keep an eye on the clock to be sure you are working fast enough.

Not all problems are assigned the same point value. Identify those problems that are worth more points. You do not want to have to rush through problems that have higher point values because you did not notice them until you got to the end of the test.

Exercises 5.3

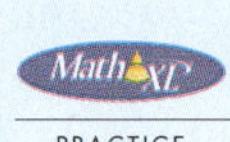

Vocabulary

1. A(n) ____________ in a single variable x is a sum of terms of the form ax^n, where a is a real number and n is a whole number.
2. For a polynomial in a single variable x, the __________ of a term is equal to its exponent.
3. A polynomial with one term is called a(n) __________.
4. A polynomial with two terms is called a(n) __________.
5. A polynomial with three terms is called a(n) __________.
6. The __________ of a term is the numerical part of a term.
7. When the terms of a polynomial are written from highest degree to lowest degree, the polynomial is said to be in ______________.
8. The ____________ of a polynomial is the term that has the greatest degree.

List the degree of each term in the given polynomial.

9. $9x^4 - 7x^2 + 3x - 8$
10. $-4x^3 + 10x^2 + 5x + 4$
11. $13x - 8x^7 - 11x^4$
12. $5x^5 - 3x^4 + 2x^3 + x^2 - 15$

List the coefficient of each term in the given polynomial.

13. $7x^2 + x - 15$
14. $3x^3 + x^2 - 8x - 13$
15. $10x^5 - 17x^4 + 6x^3 - x^2 + 2$
16. $-5x^3 - 12x^2 - 9x - 4$

Identify the given polynomial as a monomial, binomial, or trinomial.

17. $x^2 - 9x + 20$
18. $15x^9$
19. $4x - 7$
20. $2x^5 + 13x^4$
21. $-8x^3$
22. $5x^5 - 8x^3 + 17x$

Rewrite the polynomial in descending order. Identify the leading term, the leading coefficient, and the degree of the polynomial.

23. $8x - 7 + 3x^2$
24. $13 - 9x - x^2$
25. $6x^2 - 11x + 2x^4 + 10$
26. $2x^5 - x^6 + 24 - 3x^3 + x - 5x^2$

Evaluate the polynomial for the given value of the variable.

27. $x^2 + 3x - 10$ for $x = 5$

28. $2x^2 - 7x - 32$ for $x = 8$

29. $x^4 - 8x^3 - 48$ for $x = -2$

30. $x^3 + 16x^2 + 23$ for $x = -3$

31. $-x^2 + 7x + 22$ for $x = 10$

32. $-3x^2 - 8x + 16$ for $x = 6$

Evaluate the given polynomial function.

33. $f(x) = x^2 - 7x - 10, f(8)$

34. $f(x) = x^2 + 12x - 13, f(-9)$

35. $g(x) = 3x^3 - 8x^2 + 5x + 9, g(-5)$

36. $g(x) = -2x^3 - 5x^2 + 10x + 35, g(6)$

Add or subtract.

37. $(5x^2 + 8x - 11) + (3x^2 - 14x + 14)$

38. $(2x^2 - 9x - 35) + (8x^2 - 6x + 17)$

39. $(4x^2 - 7x + 30) - (2x^2 + 10x - 50)$

40. $(x^2 + 12x - 42) - (6x^2 - 19x - 23)$

41. $(2x^3 + 7x^2 - 19x) + (x^2 + 5x - 11)$

42. $(9x^4 + 7x^2 - 6x - 15) + (3x^3 - 7x^2 + 6x - 15)$

43. $(x^3 + 5x^2 - 16) - (6x^2 - 5x - 19)$

44. $(4x^3 + 9x^2 - 6x) - (x^4 - 9x^2 - 17)$

45. $(2x^9 - 5x^4 + 7x^2) - (-7x^6 + 4x^5 - 12)$

46. $(9x^3 + x^2 - x - 13) - (9x^3 + x^2 - x - 13)$

Find the missing polynomial.

47. $(3x^2 + 8x + 11) + ? = 7x^2 + 5x - 2$

48. $(2x^3 - 4x^2 - 7x + 19) + ? = x^2 + 3x + 24$

49. $(3x^4 - 5x^3 + 6x + 12) - ? =$
$x^4 - 2x^3 - x^2 - 5x + 1$

50. $? - (6x^3 - 11x^2 - 16x + 22) =$
$-x^3 + 19x^2 - 12x + 9$

For the given functions $f(x)$ and $g(x)$, find $f(x) + g(x)$ and $f(x) - g(x)$.

51. $f(x) = 7x^2 + 10x + 3, g(x) = 5x^2 - 9x + 6$

52. $f(x) = 2x^2 + 11x - 5, g(x) = -x^2 - 5x + 20$

53. $f(x) = x^3 - 8x - 31, g(x) = 4x^3 + x^2 + 3x + 25$

54. $f(x) = 2x^3 - 4x^2 + 8x - 16, \; g(x) = x^3 + 12x^2 - 15$

Evaluate the polynomial.

55. $x^2 - 7xy + 10y^2$ for $x = 2$ and $y = 5$

56. $3x^2 + 11xy + y^2$ for $x = 4$ and $y = 3$

57. $b^2 - 4ac$ for $a = -9, b = -2$, and $c = 4$

58. $b^2 - 4ac$ for $a = 5, b = -1$, and $c = 6$

59. $3x^2yz^3 - 4xy^2z - 6x^4y^2z$ for $x = 4, y = 5$, and $z = -2$

60. $7x^2yz + 9xy^2z - 10xyz^2$ for $x = 9, y = 3$, and $z = -2$

For each polynomial, list the degree of each term and the degree of the polynomial.

61. $6x^5y^5 - 7x^3y^3 + 9x^2y$

62. $-4x^8y^6 + 3x^3y^7 + x^5y^{11}$

63. $x^5y^2z - 5x^3y^3z^3 + 6x^2y^5z^4$

64. $a^{12}b^9c^{10} + 2a^{11}b^6c^{16} + 4a^{10}b^{10}c^3$

Add or subtract.

65. $(15x^2y + 8xy^2 - 7x^2y^3) + (-8x^2y + 3xy^2 + 4x^2y^3)$

66. $(x^4y - 5x^2y^3 - 14y^5) - (6x^4y + 11x^2y^3 - 7y^5)$

67. $(a^3b^2 + 11a^5b + 24a^2b^3) -$
$(19a^2b^3 - 2a^5b + 15a^3b^2)$

68. $(9m^4n^2 - 13m^3n^3 + 11m^2n^4) +$
$(-4m^2n^4 - 3m^4n^2 - m^3n^3)$

69. $(2x^3yz^2 - xy^4z^3 - 10x^2y^2z^5) -$
$(4x^2yz^3 + 14xy^4z^3 - 3x^2y^2z^5)$

70. $(22x^3y^9 - 21x^6y^4 - 7x^8y^2) -$
$(22x^3y^9 - 21x^6y^4 - 7x^8y^2)$

71. The manager of an amusement park is considering raising the price of admission to increase revenues.

She determines that if she raises the admission price by x dollars, the total daily revenues for the park can be approximated by the function $R(x) = -400x^2 + 13{,}600x + 1{,}230{,}000$. Find the daily revenues if she increases the cost of admission by \$6.

72. The average cost per shirt, in dollars, to produce x T-shirts is given by the function $f(x) = 0.00015x^2 - 0.06x + 10.125$. What is the average cost per shirt to produce 250 T-shirts?

73. The number of students earning master's degrees in mathematics or statistics in the United States in a particular year can be approximated by the function $g(x) = 6x^2 - 224x + 4941$, where x represents the number of years after 1971. Use the function to estimate the number of students who earned a master's degree in mathematics or statistics in the United States in 2010. (*Source: U.S. Department of Education*)

74. The number of births, in thousands, in the United States in a particular year can be approximated by the function $f(x) = 6x^2 - 72x + 4137$, where x represents the number of years after 1990. Use the function to estimate the number of births in the United States in 2010. *(Source: National Center for Health Statistics, U.S. Department of Health and Human Services)*

Writing in Mathematics

Answer in complete sentences.

75. Explain why it is a good idea to use parentheses when evaluating a polynomial for negative values of a variable. Do you believe that using parentheses is a good idea for evaluating any polynomial?

76. A classmate made the following error when subtracting two polynomials.

$$(3x^2 + 2x - 7) - (x^2 - x + 9)$$
$$= 3x^2 + 2x - 7 - x^2 - x + 9$$

Explain the error to your classmate and give advice on how to avoid making this error.

5.4 Multiplying Polynomials

OBJECTIVES

1. Multiply monomials.
2. Multiply a monomial by a polynomial.
3. Multiply polynomials.
4. Find special products.

Multiplying Monomials

Objective 1 Multiply monomials. Now that we have learned how to add and subtract polynomials, we will explore multiplication of polynomials. We begin by learning how to multiply monomials.

Multiplying Monomials

When multiplying monomials, begin by multiplying their coefficients. Then multiply the variable factors using the property of exponents that states that for any real number x, $x^m \cdot x^n = x^{m+n}$.

EXAMPLE 1 Multiply $3x \cdot 5x$.

SOLUTION Multiply the coefficients first; then multiply the variables.

$3x \cdot 5x = 3 \cdot 5 \cdot x \cdot x$ Multiply coefficients, then variables.

$= 15x^{1+1}$ Multiply variables using the property $x^m \cdot x^n = x^{m+n}$.

$= 15x^2$ Simplify the exponent.

Quick Check 1

Multiply $7x \cdot 8x$.

EXAMPLE 2 Multiply $2x^4y^3 \cdot 3x^2y^7z^4$.

SOLUTION The procedure for multiplying monomials containing more than one variable is the same as multiplying monomials containing a single variable. After multiplying the coefficients, multiply the variables one at a time.

$$2x^4y^3 \cdot 3x^2y^7z^4 = 6x^6y^{10}z^4 \quad \text{Multiply coefficients, then variables.}$$

Notice that the variable z was a factor of only one of the monomials. The exponent of z was not changed.

Quick Check 2

Multiply $5x^6yz^7 \cdot 12x^8z^6$.

Multiplying a Monomial by a Polynomial

Objective 2 Multiply a monomial by a polynomial. We now advance to multiplying a monomial by a polynomial containing two or more terms, such as $3x(2x^2 - 5x + 2)$. To do this, we use the distributive property $a(b + c) = ab + ac$. To find the product $3x(2x^2 - 5x + 2)$, we multiply the monomial $3x$ by each term of the polynomial $2x^2 - 5x + 2$.

Multiplying a Monomial by a Polynomial

To multiply a monomial by a polynomial containing two or more terms, multiply the monomial by each term of the polynomial.

EXAMPLE 3 Multiply $3x(2x^2 - 5x + 2)$.

SOLUTION Begin by distributing the monomial $3x$ to each term of the polynomial.

$$\begin{aligned} 3x(2x^2 - 5x + 2) &= 3x \cdot 2x^2 - 3x \cdot 5x + 3x \cdot 2 && \text{Distribute } 3x. \\ &= 6x^3 - 15x^2 + 6x && \text{Multiply.} \end{aligned}$$

Although we will continue to show the distribution of the monomial, your goal should be to perform this task mentally.

Quick Check 3

Multiply $4x(7x^3 - 6x^2 + 5x - 8)$.

EXAMPLE 4 Multiply $-8x^3(x^5 - 4x^4 - 9x^2)$.

SOLUTION Notice that the coefficient of the monomial being distributed is negative. We must distribute $-8x^3$, and multiplying by this negative term changes the sign of each term in the polynomial.

$$\begin{aligned} -8x^3(x^5 - 4x^4 - 9x^2) &= (-8x^3)(x^5) - (-8x^3)(4x^4) - (-8x^3)(9x^2) \\ & \qquad \text{Distribute } -8x^3. \\ &= -8x^8 - (-32x^7) - (-72x^5) && \text{Multiply.} \\ &= -8x^8 + 32x^7 + 72x^5 && \text{Simplify.} \end{aligned}$$

Quick Check 4

Multiply $-3x^5(-2x^4 + x^3 - x^2 - 15x + 21)$.

A WORD OF CAUTION When multiplying a polynomial by a term with a negative coefficient, be sure to change the sign of each term in the polynomial.

Multiplying Polynomials

Objective 3 Multiply polynomials.

Multiplying Two Polynomials

To multiply two polynomials when each contains two or more terms, multiply each term in the first polynomial by each term in the second polynomial.

Suppose we wanted to multiply $(x + 9)(x + 7)$. We could distribute the factor $(x + 9)$ to each term in the second polynomial as follows:

$$(x + 9)(x + 7) = (x + 9) \cdot x + (x + 9) \cdot 7$$

We could then perform the two multiplications by distributing x in the first product and distributing 7 in the second product.

$$\begin{aligned}(x + 9)(x + 7) &= (x + 9) \cdot x + (x + 9) \cdot 7 \\ &= x \cdot x + 9 \cdot x + x \cdot 7 + 9 \cdot 7 \\ &= x^2 + 9x + 7x + 63 \\ &= x^2 + 16x + 63\end{aligned}$$

We end up with each term in the first polynomial being multiplied by each term in the second polynomial.

EXAMPLE 5 Multiply $(x + 6)(x + 4)$.

SOLUTION Begin by taking the first term in the first polynomial, x, and multiplying it by each term in the second polynomial. Then repeat this for the second term in the first polynomial, 6.

$(x + 6)(x + 4) = x \cdot x + x \cdot 4 + 6 \cdot x + 6 \cdot 4$	Distribute the term x from the first polynomial; then distribute the 6 from the first polynomial.
$= x^2 + 4x + 6x + 24$	Multiply.
$= x^2 + 10x + 24$	Combine like terms.

Quick Check 5

Multiply $(x + 11)(x + 9)$.

We often refer to the process of multiplying a binomial by another binomial as "FOIL." FOIL is an acronym for **F**irst, **O**uter, **I**nner, **L**ast, which describes the four multiplications that occur when we multiply two binomials. Here are the four multiplications performed in the previous example.

First	Outer	Inner	Last
$x \cdot x$ $(x+6)(x+4)$	$x \cdot 4$ $(x+6)(x+4)$	$(x+6)(x+4)$ $6 \cdot x$	$(x+6)(x+4)$ $6 \cdot 4$

FOIL applies only when we multiply a binomial by another binomial. If we are multiplying a binomial by a trinomial, six multiplications must be performed and FOIL cannot be used. Keep in mind that each term in the first polynomial must be multiplied by each term in the second polynomial.

EXAMPLE 6 Multiply $(x - 6)(4x - 5)$.

SOLUTION When multiplying polynomials, we must be careful with the signs. When we distribute the second term of $x - 6$ to the second polynomial, we must distribute a negative 6.

$$(x - 6)(4x - 5) = x \cdot 4x - x \cdot 5 - 6 \cdot 4x + 6 \cdot 5$$ Distribute (FOIL). The product of two negative numbers is positive: $(-6)(-5) = 6 \cdot 5$.

$$= 4x^2 - 5x - 24x + 30$$ Multiply.

$$= 4x^2 - 29x + 30$$ Combine like terms.

Quick Check 6

Multiply $(5x - 2)(2x - 9)$.

EXAMPLE 7 Given the functions $f(x) = 3x - 2$ and $g(x) = x^2 + 4x - 7$, find $f(x) \cdot g(x)$.

SOLUTION We will substitute the appropriate expressions for $f(x)$ and $g(x)$ in parentheses and simplify the resulting expression. We need to multiply each term in the first polynomial by each term in the second polynomial.

$$f(x) \cdot g(x) = (3x - 2)(x^2 + 4x - 7)$$ Substitute for $f(x)$ and $g(x)$.

$$= 3x \cdot x^2 + 3x \cdot 4x - 3x \cdot 7 - 2 \cdot x^2 - 2 \cdot 4x + 2 \cdot 7$$ Distribute.

$$= 3x^3 + 12x^2 - 21x - 2x^2 - 8x + 14$$ Multiply.

$$= 3x^3 + 10x^2 - 29x + 14$$ Combine like terms.

Quick Check 7

Given the functions $f(x) = x + 8$ and $g(x) = 3x^2 + 6x - 2$, find $f(x) \cdot g(x)$.

Special Products

Objective 4 Find special products. We finish this section by examining some special products. The first special product is of the form $(a + b)(a - b)$. In words, this is the product of the sum and the difference of two terms. Here is the multiplication.

$$(a + b)(a - b) = a^2 - ab + ab - b^2$$ Distribute.

$$= a^2 - b^2$$ Combine like terms.

Notice that when we were combining like terms, two of the terms were opposites. This left only two terms. We can use this result whenever we multiply two binomials of the form $(a + b)(a - b)$.

$$(a + b)(a - b) = a^2 - b^2$$

EXAMPLE 8 Multiply $(x + 7)(x - 7)$.

SOLUTION

$$(x + 7)(x - 7) = x^2 - 7^2$$ Multiply using the pattern $(a + b)(a - b) = a^2 - b^2$.

$$= x^2 - 49$$ Simplify.

Quick Check 8

Multiply $(x + 10)(x - 10)$.

EXAMPLE 9 Multiply $(3x - 5)(3x + 5)$.

SOLUTION Although the difference is listed first, we can still multiply using the same pattern.

$$(3x - 5)(3x + 5) = (3x)^2 - 5^2 \quad \text{Multiply using the pattern } (a + b)(a - b) = a^2 - b^2.$$
$$= 9x^2 - 25 \quad \text{Simplify.}$$

Quick Check 9

Multiply $(2x + 7)(2x - 7)$.

The other special product we will examine is the square of a binomial, such as $(x + 3)^2$ and $(9x + 4)^2$. Here are the patterns for squaring binomials of the form $a + b$ and $a - b$.

$$(a + b)^2 = a^2 + 2ab + b^2$$
$$(a - b)^2 = a^2 - 2ab + b^2$$

Let's derive the first of these patterns.

$$(a + b)^2 = (a + b)(a + b) \quad \text{To square a binomial, multiply it by itself.}$$
$$= a^2 + ab + ab + b^2 \quad \text{Distribute (FOIL).}$$
$$= a^2 + 2ab + b^2 \quad \text{Combine like terms.}$$

The second pattern can be derived in the same fashion.

EXAMPLE 10 Multiply $(x - 3)^2$.

SOLUTION We can use the pattern $(a - b)^2 = a^2 - 2ab + b^2$, substituting x for a and 3 for b.

$$a^2 - 2ab + b^2$$
$$x^2 - 2 \cdot x \cdot 3 + 3^2 \quad \text{Substitute } x \text{ for } a \text{ and 3 for } b.$$
$$= x^2 - 6x + 9 \quad \text{Multiply.}$$

Quick Check 10

Multiply $(5x - 8)^2$.

EXAMPLE 11 Multiply $(2x + 5)^2$.

SOLUTION We can use the pattern $(a + b)^2 = a^2 + 2ab + b^2$, substituting $2x$ for a and 5 for b.

$$a^2 + 2ab + b^2$$
$$(2x)^2 + 2 \cdot (2x) \cdot 5 + 5^2 \quad \text{Substitute } 2x \text{ for } a \text{ and 5 for } b.$$
$$= 4x^2 + 20x + 25 \quad \text{Multiply.}$$

Quick Check 11

Multiply $(x + 6)^2$.

Although the patterns developed for these three special products may save us time when multiplying, keep in mind that we can find these types of products by multiplying as we did earlier in this section. When we square a binomial, we can start by rewriting the expression as the product of the binomial and itself. For example, we can rewrite $(8x - 7)^2$ as $(8x - 7)(8x - 7)$ and then multiply.

BUILDING YOUR STUDY STRATEGY

Test Taking, 4 Solve Easier Problems First When you take a test, work on easier problems first, saving more difficult problems for later. One benefit to this approach is that you will gain confidence as you progress through the test, making you confident when you attempt to solve a difficult problem. You also will save time to spend on the few difficult problems.

Exercises 5.4

MyMathLab

PRACTICE

WATCH

DOWNLOAD

READ

REVIEW

Vocabulary

1. To multiply a monomial by another monomial, multiply the coefficients and _______ the exponents of the variable factors.
2. To multiply a monomial by a polynomial, _______ the monomial to each term in the polynomial.
3. To multiply a polynomial by another polynomial, _______ each term in the first polynomial by each term in the second polynomial.
4. The acronym _______ can be used when you are multiplying a binomial by another binomial.

Multiply.

5. $4x \cdot 9x^4$
6. $8x^3 \cdot 7x^2$
7. $9m^7(-6m^{11})$
8. $-6n^5 \cdot 10n^3$
9. $13a^7b^{10} \cdot 7a^5b$
10. $20ab^9 \cdot 16a^{13}b^{15}$
11. $-5x^3y^2z^4 \cdot 14xz^5w^4$
12. $3a^5b^8c \cdot 9bc^7d^4$
13. $2x^5 \cdot 6x^3 \cdot 7x^2$
14. $-12x^8 \cdot 5x^{10} \cdot 6x^9$

Find the missing monomial.

15. $5x^5 \cdot ? = 30x^{30}$
16. $-7x^9 \cdot ? = -28x^{16}$
17. $2x^9 \cdot ? = -24x^{18}$
18. $-6x^{10} \cdot ? = 84x^{23}$

Multiply.

19. $5(3x - 4)$
20. $3(7x + 6)$
21. $-2(6x - 9)$
22. $-4(3x + 11)$
23. $6x(3x + 5)$
24. $9x(4x - 13)$
25. $x^3(3x^2 - 4x + 7)$
26. $-x^3(2x^5 + 9x^4 - 6x)$
27. $2xy^2(3x^2 - 6xy + 7y^2)$
28. $11x^4y^3(2xy - 5x^4y^2 - 7x)$

Multiply.

29. $(x + 7)(x - 9)$
30. $(x + 6)(x + 8)$
31. $(x - 3)(x - 9)$
32. $(x - 5)(x + 10)$
33. $(4x + 3)(x - 6)$
34. $(5x - 7)(2x + 3)$
35. $(x - 8)(x + 9)$
36. $(x + 6)(x - 3)$
37. $(2x + 3)(2x - 13)$
38. $(2x - 9)(3x - 4)$
39. $(x + 6)(3x - 5)$
40. $(5x + 8)(4x + 1)$
41. $(3x + 2)(x^2 - 5x - 9)$
42. $(7x - 4)(49x^2 + 28x + 16)$
43. $(x^2 - 7x + 10)(x^2 + 3x - 40)$
44. $(2x^2 - 5x - 8)(x^2 + 3x + 9)$
45. $(x + 2y)(x - 4y)$
46. $(x - 6y)(x - 7y)$
47. $(4xy + 3)(5xy + 6)$
48. $(2xy + 7)(3xy - 2)$

Find the missing factor or term.

49. $?(2x^2 - 7x - 10) = 6x^5 - 21x^4 - 30x^3$
50. $?(3x^4 + 9x^2 + 16) = 18x^9 + 54x^7 + 96x^5$
51. $4x^4(?) = 12x^7 - 20x^6 - 48x^4$
52. $8x^9(?) = 56x^{17} + 104x^{14} - 96x^{11}$
53. $(x + ?)(x + 5) = x^2 + 8x + 15$
54. $(x - 7)(x + ?) = x^2 + 3x - 70$
55. $(x + ?)(x + ?) = x^2 + 9x + 18$
56. $(? - 5)(? + 3) = 2x^2 + x - 15$

For the given functions $f(x)$ and $g(x)$, find $f(x) \cdot g(x)$.

57. $f(x) = x - 9, g(x) = x + 2$
58. $f(x) = 4x^3, g(x) = x^2 - 8x - 14$
59. $f(x) = 6x^5, g(x) = -3x^4$
60. $f(x) = 2x^7, g(x) = 11x^6$

61. $f(x) = x^3 - 5x^2 + 8x + 3, g(x) = -4x^5$

62. $f(x) = 7x + 4, g(x) = 6x - 13$

Find the special product using the appropriate formula.

63. $(x + 9)(x - 9)$
64. $(x - 6)(x + 6)$
65. $(3x - 7)(3x + 7)$
66. $(2x + 11)(2x - 11)$
67. $(x + 7)^2$
68. $(x - 10)^2$
69. $(4x - 3)^2$
70. $(9x + 5)^2$

Find the missing factor or term.

71. $(x + 9) \cdot (?) = x^2 - 81$
72. $(5x + 7) \cdot (?) = 25x^2 - 49$
73. $(?)^2 = x^2 - 12x + 36$
74. $(?)^2 = 4x^2 + 20x + 25$

Mixed Practice, 75–92

Multiply.

75. $(x + 7)(x - 1)$
76. $2x^5 \cdot 3x^2$
77. $5x(x^2 - 8x - 9)$
78. $(x + 2)(x - 2)$
79. $-7x^2y^3 \cdot 4x^4y^6$
80. $(x - 7)^2$
81. $-5x^6(-4x^8)$
82. $(2x + 15)(3x - 7)$
83. $(5x + 3)(5x - 3)$
84. $-3x^2(x^2 - 10x - 13)$
85. $(x + 13)^2$
86. $(x + 8)(3x^2 + 7x - 6)$
87. $2x^2y(3x^2 - x^4y^3 - 7y)$
88. $9x^3(-6x^3)$
89. $(4x - 9)^2$
90. $(6 + 5x)^2$
91. $(x^2 + 3x + 4)(x^2 - 5x + 4)$

92. $-6x^5(-x^5 + 3x^3 - 11)$

Writing in Mathematics

Answer in complete sentences.

93. Explain the difference between simplifying the expression $8x^2 + 3x^2$ and simplifying the expression $(8x^2)(3x^2)$. Discuss how the coefficients and exponents are handled differently.

94. To simplify the expression $(x + 8)^2$, you can multiply $(x + 8)(x + 8)$ or use the special product formula $(a + b)^2 = a^2 + 2ab + b^2$. Which method do you prefer? Explain your answer.

95. ***Newsletter*** Write a newsletter explaining how to multiply two binomials.

5.5 Dividing Polynomials

OBJECTIVES

1. Divide a monomial by a monomial.
2. Divide a polynomial by a monomial.
3. Divide a polynomial by a polynomial using long division.
4. Use placeholders when dividing a polynomial by a polynomial.

Dividing Monomials by Monomials

Objective 1 Divide a monomial by a monomial. In this section, we will learn to divide a polynomial by another polynomial. We will begin by reviewing how to divide a monomial by another monomial, such as $\frac{4x^5}{2x^2}$ or $\frac{30a^5b^4}{6ab^2}$.

Dividing a Monomial by a Monomial

To divide a monomial by another monomial, divide the coefficients first. Then divide the variables using the quotient rule $\frac{x^m}{x^n} = x^{m-n}$. Assume that no variable in the denominator is equal to 0.

EXAMPLE 1 Divide $\frac{18x^5}{3x^3}$. (Assume that $x \neq 0$.)

SOLUTION

$\frac{18x^5}{3x^3} = 6x^{5-3}$ Divide coefficients. Subtract exponents of x.

$= 6x^2$ Simplify the exponent.

We can check the quotient by using multiplication. If $\frac{18x^5}{3x^3} = 6x^2$, we know that $3x^3 \cdot 6x^2$ should equal $18x^5$.

Check

$3x^3 \cdot 6x^2 = 18x^{3+2}$ Multiply coefficients. Keep the base and add the exponents.

$= 18x^5$ Simplify the exponent.

The quotient of $6x^2$ checks.

Quick Check 1

Divide $\frac{32x^9}{4x^4}$.

(Assume that $x \neq 0$.)

EXAMPLE 2 Divide $\frac{24x^3y^2}{3xy^2}$. (Assume that $x, y \neq 0$.)

SOLUTION When there is more than one variable, as in this example, divide the coefficients and then divide the variables one at a time.

$\frac{24x^3y^2}{3xy^2} = 8x^2y^0$ Divide coefficients and subtract exponents.

$= 8x^2$ Rewrite without y as a factor. ($y^0 = 1$)

Quick Check 2

Divide $\frac{40x^3y^{11}}{8xy^2}$. (Assume that $x, y \neq 0$.)

Dividing Polynomials by Monomials

Objective 2 Divide a polynomial by a monomial. Now we move on to dividing a polynomial by a monomial, such as $\frac{3x^5 - 9x^3 - 18x^2}{3x}$.

Dividing a Polynomial by a Monomial

To divide a polynomial by a monomial, divide each term of the polynomial by the monomial.

EXAMPLE 3 Divide $\frac{15x^2 + 10x - 5}{5}$.

SOLUTION Divide each term in the numerator by 5.

$\frac{15x^2 + 10x - 5}{5} = \frac{15x^2}{5} + \frac{10x}{5} - \frac{5}{5}$ Divide each term in the numerator by 5.

$= 3x^2 + 2x - 1$ Divide.

If $\frac{15x^2 + 10x - 5}{5} = 3x^2 + 2x - 1$, $5(3x^2 + 2x - 1)$ should equal $15x^2 + 10x - 5$. We can use this to check our work.

Check

$$5(3x^2 + 2x - 1) = 5 \cdot 3x^2 + 5 \cdot 2x - 5 \cdot 1 \quad \text{Distribute.}$$
$$= 15x^2 + 10x - 5 \quad \text{Multiply.}$$

The quotient of $3x^2 + 2x - 1$ checks.

Quick Check 3

Divide $\frac{7x^2 - 21x - 49}{7}$.

EXAMPLE 4 Divide $\frac{3x^5 - 9x^3 - 18x^2}{3x}$. (Assume that $x \neq 0$.)

SOLUTION In this example, divide each term in the numerator by $3x$.

$$\frac{3x^5 - 9x^3 - 18x^2}{3x}$$
$$= \frac{3x^5}{3x} - \frac{9x^3}{3x} - \frac{18x^2}{3x} \quad \text{Divide each term in the numerator by } 3x.$$
$$= x^4 - 3x^2 - 6x \quad \text{Divide.}$$

Quick Check 4

Divide $\frac{48x^{10} + 12x^7 + 30x^5}{6x^2}$.

(Assume that $x \neq 0$.)

Dividing a Polynomial by a Polynomial (Long Division)

Objective 3 Divide a polynomial by a polynomial using long division. To divide a polynomial by another polynomial containing at least two terms, we use a procedure similar to long division. Before outlining this procedure, let's review some of the terms associated with long division.

$$2\overline{)6}^{\,3}$$

Quotient: 3 — Divisor: 2 — Dividend: 6

Suppose we were asked to divide $\frac{x^2 - 10x + 16}{x - 2}$. The polynomial in the numerator is the dividend, and the polynomial in the denominator is the divisor. We may rewrite this division as $x - 2\overline{)x^2 - 10x + 16}$. We must be sure to write both the divisor and the dividend in descending order. We perform the division using the following steps:

Division by a Polynomial

1. Divide the term in the dividend with the highest degree by the term in the divisor with the highest degree. Add this result to the quotient.
2. Multiply the monomial obtained in Step 1 by the divisor, writing the result underneath the dividend. (Align like terms vertically.)
3. Subtract the product obtained in Step 2 from the dividend. (Recall that to subtract a polynomial from another polynomial, we change the signs of its terms and then combine like terms.)
4. Repeat Steps 1–3 with the result of Step 3 as the new dividend. Keep repeating this procedure until the degree of the new dividend is less than the degree of the divisor.

EXAMPLE 5 Divide $\dfrac{x^2 - 10x + 16}{x - 2}$.

SOLUTION Begin by writing $x - 2\overline{)x^2 - 10x + 16}$. Divide the term in the dividend with the highest degree (x^2) by the term in the divisor with the highest degree (x). Because $\dfrac{x^2}{x} = x$, write x in the quotient and multiply x by the divisor $x - 2$, writing this product underneath the dividend.

$$\begin{array}{r} x \phantom{{}- 10x + 16} \\ x - 2\overline{)x^2 - 10x + 16} \\ x^2 - 2x \phantom{{}+ 16} \end{array} \quad \text{Multiply } x \text{ by } (x - 2).$$

To subtract, change the signs of the second polynomial and combine like terms.

$$\begin{array}{r} x \phantom{{}- 10x + 16} \\ x - 2\overline{)x^2 - 10x + 16} \\ \underline{\overset{-}{}x^2 \overset{+}{\cancel{-}} 2x \phantom{{}+ 16}\downarrow} \\ -8x + 16 \end{array} \quad \text{Change the signs and combine like terms.}$$

Begin the process again by dividing $-8x$ by x, which equals -8. Multiply -8 by $x - 2$ and subtract.

$$\begin{array}{r} x - 8 \\ x - 2\overline{)x^2 - 10x + 16} \\ \underline{\overset{-}{}x^2 \overset{+}{\cancel{-}} 2x \phantom{{}+ 16}\downarrow} \\ -8x + 16 \\ \underline{\overset{+}{\cancel{-}}8x \overset{-}{\cancel{+}} 16} \\ 0 \end{array} \quad \begin{array}{l} \text{Multiply } -8 \text{ by } (x - 2). \\ \text{Subtract the product.} \end{array}$$

The remainder of 0 says that we are finished because its degree is less than the degree of the divisor $x - 2$. The expression written above the division box $(x - 8)$ is the quotient.

$$\frac{x^2 - 10x + 16}{x - 2} = x - 8$$

We can check our work by multiplying the quotient $(x - 8)$ by the divisor $(x - 2)$, which should equal the dividend $(x^2 - 10x + 16)$.

Check

$$\begin{aligned} (x - 8)(x - 2) &= x^2 - 2x - 8x + 16 \\ &= x^2 - 10x + 16 \end{aligned}$$

The division checks; the quotient is $x - 8$.

Quick Check 5

Divide $\dfrac{x^2 + 13x + 36}{x + 4}$.

In the previous example, the remainder of 0 also says that $x - 2$ divides into $x^2 - 10x + 16$ evenly; so $x - 2$ is a **factor** of $x^2 - 10x + 16$. The quotient, $x - 8$, also is a factor of $x^2 - 10x + 16$.

In the next example, we will learn how to write a quotient when there is a nonzero remainder.

EXAMPLE 6 Divide $x^2 - 10x + 9$ by $x - 3$.

SOLUTION Because the divisor and dividend are already written in descending order, we may begin to divide.

$$\begin{array}{r} x - 7 \\ x - 3\overline{)x^2 - 10x + 9} \\ \underline{\overset{-}{x^2} \overset{+}{\cancel{-}} 3x \quad\downarrow} \\ -7x + 9 \\ \underline{\overset{+}{\cancel{-}} 7x \overset{-}{\cancel{+}} 21} \\ -12 \end{array}$$

Multiply x by $x - 3$ and subtract.

Multiply -7 by $x - 3$ and subtract.

The remainder is -12. After the quotient, we write a fraction with the remainder in the numerator and the divisor in the denominator. Because the remainder is negative, we subtract this fraction from the quotient. If the remainder had been positive, we would have added this fraction to the quotient.

$$\frac{x^2 - 10x + 9}{x - 3} = x - 7 - \frac{12}{x - 3}$$

Quick Check 6

Divide $\dfrac{x^2 - 3x - 8}{x + 7}$.

EXAMPLE 7 Divide $6x^2 + 17x + 17$ by $2x + 3$.

SOLUTION Because the divisor and dividend are already written in descending order, we may begin to divide.

$$\begin{array}{r} 3x + 4 \\ 2x + 3\overline{)6x^2 + 17x + 17} \\ \underline{\overset{-}{6x^2} \overset{-}{\cancel{+}} 9x \quad\downarrow} \\ 8x + 17 \\ \underline{\overset{-}{8x} \overset{-}{\cancel{+}} 12} \\ 5 \end{array}$$

Multiply $3x$ by $2x + 3$ and subtract.

Multiply 4 by $2x + 3$ and subtract.

The remainder is 5. We write the quotient and add a fraction with the remainder in the numerator and the divisor in the denominator.

$$\frac{6x^2 + 17x + 17}{2x + 3} = 3x + 4 + \frac{5}{2x + 3}$$

Quick Check 7

Divide $\dfrac{12x^2 - 7x + 4}{3x - 4}$.

Using Placeholders When Dividing a Polynomial by a Polynomial

Objective 4 Use placeholders when dividing a polynomial by a polynomial. Suppose we wanted to divide $x^3 - 12x - 11$ by $x - 3$. Notice that the dividend is missing an x^2 term. When this is the case, we add the term $0x^2$ as a **placeholder**. We add placeholders to dividends that are missing terms of a particular degree.

EXAMPLE 8 Divide $(x^3 - 12x - 11) \div (x - 3)$.

SOLUTION The degree of the dividend is 3, and each degree lower than 3 must be represented in the dividend. We will add the term $0x^2$ as a placeholder. The divisor does not have any missing terms.

$$\begin{array}{r|l} x^2 + 3x - 3 & \\ x - 3\overline{)x^3 + 0x^2 - 12x - 11} & \\ \underline{-x^3 + 3x^2} \qquad\qquad & \text{Multiply } x^2 \text{ by } x - 3 \text{ and subtract.} \\ 3x^2 - 12x - 11 & \\ \underline{-3x^2 + 9x} \qquad & \text{Multiply } 3x \text{ by } x - 3 \text{ and subtract.} \\ -3x - 11 & \\ \underline{+3x - 9} & \text{Multiply } -3 \text{ by } x - 3 \text{ and subtract.} \\ -20 & \end{array}$$

$$(x^3 - 12x - 11) \div (x - 3) = x^2 + 3x - 3 - \frac{20}{x - 3}.$$

Quick Check 8

Divide $\frac{x^3 - 3x^2 - 9}{x - 2}$.

BUILDING YOUR STUDY STRATEGY

Test Taking, 5 Review Your Test Try to leave yourself enough time to review the test at the end of the period. Check for careless errors, which can cost you a fair number of points. Also check that your answers make sense in the context of the problem. For example, if the question asks how tall a person is and your answer is 68 feet, chances are that something has gone astray.

Check for problems or parts of problems you may have skipped and left blank.

Take all of the allotted time to review the test. There is no reward for turning in a test early, and the more you work on a test, the more likely you are to find a mistake or a problem where you can gain points.

Exercises 5.5

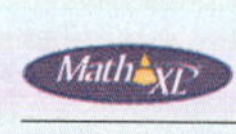

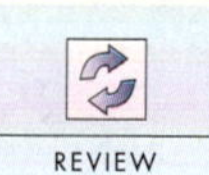

Vocabulary

1. To divide a polynomial by a monomial, divide each ________ by the monomial.
2. To divide a polynomial by another polynomial containing at least two terms, use ____________.
3. If one polynomial divides evenly into a second polynomial, the first polynomial is a(n) ________ of the second polynomial.
4. If a polynomial in the dividend is missing a term of a particular degree, use a(n) ____________ in the dividend.

Divide. Assume that all variables are nonzero.

5. $\frac{24x^{24}}{3x^3}$
6. $\frac{16x^{16}}{4x^4}$
7. $\frac{-30n^{11}}{6n^4}$
8. $\frac{-48n^9}{3n}$
9. $\frac{26a^7b^5}{2ab^3}$
10. $\frac{35x^9y^{10}}{5x^8y^3}$
11. $(15x^6) \div (3x^2)$
12. $(70x^{13}) \div (10x^5)$
13. $\frac{12x^6}{8x^4}$
14. $\frac{-35x^9}{25x^5}$
15. $\frac{7x^{12}}{21x^4}$
16. $\frac{3x^{15}}{18x^7}$

Find the missing monomial. Assume that $x \neq 0$.

17. $\frac{?}{5x^4} = 2x^7$
18. $\frac{?}{3x^6} = -15x^8$
19. $\frac{24x^9}{?} = -6x^4$
20. $\frac{20x^9}{?} = \frac{5x^4}{3}$

Divide. Assume that all variables are nonzero.

21. $\dfrac{15x^2 - 25x - 40}{5}$

22. $\dfrac{8x^5 - 2x^3 + 6x^2}{2}$

23. $\dfrac{24x^4 + 30x^2 - 27x}{3x}$

24. $\dfrac{32x^4 + 44x^3 + 60x^2}{4x}$

25. $(6x^7 + 9x^6 + 15x^5) \div (3x)$

26. $(-14x^8 + 21x^6 + 7x^4) \div (7x)$

27. $\dfrac{20x^6 - 30x^4}{-10x^2}$

28. $\dfrac{-8x^5 + 2x^4 - 6x^3 - 2x^2}{-2x^2}$

29. $\dfrac{x^6y^6 - x^4y^5 + x^2y^4}{xy^2}$

30. $\dfrac{2a^{10}b^7 - 8a^3b^6 - 10a^5b}{2a^3b}$

Find the missing dividend or divisor. Assume that x ≠ 0.

31. $\dfrac{24x^3 - 48x^2 + 36x}{?} = 4x^2 - 8x + 6$

32. $\dfrac{10x^7 - 50x^5 + 35x^3}{?} = 2x^5 - 10x^3 + 7x$

33. $\dfrac{?}{2x^3} = 3x^4 - 4x^2 - 9$

34. $\dfrac{?}{7x^5} = -x^2 - 9x + 1$

Divide using long division.

35. $\dfrac{x^2 + 13x + 40}{x + 8}$

36. $\dfrac{x^2 + 10x + 21}{x + 3}$

37. $\dfrac{x^2 - 8x - 84}{x - 14}$

38. $\dfrac{x^2 - x - 72}{x - 9}$

39. $(x^2 - 13x + 36) \div (x - 4)$

40. $(x^2 - 20x + 91) \div (x - 7)$

41. $\dfrac{x^2 - 4x - 29}{x - 8}$

42. $\dfrac{x^2 + 7x - 7}{x + 5}$

43. $\dfrac{x^3 - 11x^2 - 37x + 14}{x + 1}$

44. $\dfrac{x^3 - x^2 - 24x - 19}{x + 3}$

45. $\dfrac{2x^2 + 3x - 32}{x + 5}$

46. $\dfrac{3x^2 + 26x + 24}{x + 9}$

47. $\dfrac{6x^2 + 25x + 10}{2x + 7}$

48. $\dfrac{10x^2 - 53x + 28}{5x - 4}$

49. $\dfrac{x^2 - 169}{x + 13}$

50. $\dfrac{9x^2 - 25}{3x - 5}$

51. $\dfrac{x^4 + 2x^2 - 15x + 32}{x - 3}$

52. $\dfrac{x^5 - 9x^3 - 17x^2 + x - 15}{x + 8}$

53. $\dfrac{x^3 - 125}{x - 5}$

54. $\dfrac{x^3 + 343}{x + 7}$

Find the missing dividend or divisor.

55. $\dfrac{?}{x + 8} = x - 3$

56. $\dfrac{?}{x - 5} = x - 11$

57. $\dfrac{x^2 + 10x - 39}{?} = x - 3$

58. $\dfrac{6x^2 - 35x + 49}{?} = 3x - 7$

59. Is $x + 9$ a factor of $x^2 + 28x + 171$?

60. Is $x - 6$ a factor of $x^2 + 4x - 54$?

61. Is $2x - 5$ a factor of $4x^2 - 8x - 15$?

62. Is $3x + 4$ a factor of $9x^2 + 24x + 16$?

Mixed Practice, 63–80

Divide. Assume that all denominators are nonzero.

63. $\dfrac{x^3 + 12x^2 - 9}{x + 3}$

64. $\dfrac{x^2 + 3x - 15}{x + 7}$

65. $\dfrac{x^3 + 8x - 19}{x + 4}$

66. $\dfrac{8x^9}{2x^3}$

67. $\dfrac{12x^6 - 8x^4 + 20x^3 - 4x^2}{4x^2}$

68. $\dfrac{x^2 + 4x - 165}{x + 15}$

69. $\dfrac{8x^2 + 42x - 25}{x + 6}$

70. $\dfrac{4x^2 - 25x - 100}{x - 8}$

71. $\dfrac{20x^2 - 51x - 6}{4x - 3}$

72. $\dfrac{6x^2 - 19x + 30}{2x - 5}$

73. $\dfrac{27x^9 - 18x^7 - 6x^6 - 3x^3}{-3x^2}$

74. $\dfrac{8x^2 - 10x - 63}{2x - 7}$

75. $\dfrac{8x^3 - 38x - 39}{2x + 3}$

76. $\dfrac{16a^4b^8c^7}{4ab^7c^7}$

77. $\dfrac{6x^3 - 37x^2 + 11x + 153}{2x - 9}$

78. $\dfrac{6x^5 - 12x^4 + 18x^3}{4x^3}$

79. $\dfrac{21x^9y^6z^{17}}{-7x^7y^5z^{12}}$

80. $\dfrac{x^4 - 85}{x - 3}$

Writing in Mathematics

Answer in complete sentences.

81. ***Solutions Manual*** Write a solutions manual page for the following problem:

Divide $\dfrac{6x^2 - 17x - 19}{2x - 7}$.

CHAPTER 5 SUMMARY

Section 5.1 Exponents

Product Rule, pp. 252–253
For any base x, $x^m \cdot x^n = x^{m+n}$. | $7x^7 \cdot 4x^{12} = 28x^{7+12} = 28x^{19}$

Power Rule, p. 253
For any base x, $(x^m)^n = x^{m \cdot n}$. | $(x^9)^6 = x^{9 \cdot 6} = x^{54}$

Power of a Product Rule, p. 254
For any bases x and y, $(xy)^n = x^n y^n$. | $(4x^7)^3 = 4^3(x^7)^3 = 64x^{21}$

Quotient Rule, pp. 254–255
For any base x, $\frac{x^m}{x^n} = x^{m-n}$ $(x \neq 0)$. | $\frac{x^{16}}{x^6} = x^{16-6} = x^{10}$

Zero Exponent Rule, p. 256
For any base x, $x^0 = 1$ $(x \neq 0)$. | $9^0 = 1$

Power of a Quotient Rule, pp. 256–257
For any bases x and y, $\left(\frac{x}{y}\right)^n = \frac{x^n}{y^n}$ $(y \neq 0)$. | $\left(\frac{x^4}{y^5}\right)^3 = \frac{x^{4 \cdot 3}}{y^{5 \cdot 3}} = \frac{x^{12}}{y^{15}}$

Section 5.2 Negative Exponents; Scientific Notation

Negative Exponents, pp. 260–262
For any *nonzero* base x, $x^{-n} = \frac{1}{x^n}$ $(x \neq 0)$. | $5^{-2} = \frac{1}{5^2} = \frac{1}{25}$ $\quad x^{-7} = \frac{1}{x^7}$

Rules for Negative Exponents, pp. 262–263
The rules for positive exponents apply to negative exponents as well. | $m^{13} \cdot m^{-19} = m^{13+(-19)} = m^{-6} = \frac{1}{m^6}$

Converting to Scientific Notation, pp. 263–264
Move the decimal point so it follows the first nonzero digit. The number of places the decimal point moves is the power of 10.
If the original number is greater than 10, the exponent is positive.
If the original number is less than 1, the exponent is negative.

$320{,}000{,}000 = 3.2 \times 10^8$ (8 places)

$0.000294 = 2.94 \times 10^{-4}$ (4 places)

Converting from Scientific Notation, pp. 264–265
If the exponent for 10 is positive, move the decimal point that number of places to the right.
If the exponent for 10 is negative, move the decimal point that number of places to the left.

$5.2 \times 10^6 = 5{,}200{,}000$

$2.13 \times 10^{-4} = 0.000213$

Calculations with Scientific Notation, pp. 265–266
Perform arithmetic on decimal numbers first; then use the properties of exponents for the powers of 10.

$$(3.2 \times 10^{15})(1.4 \times 10^8) = (3.2 \cdot 1.4) \times 10^{15+8} = 4.48 \times 10^{23}$$

Section 5.3 Polynomials; Addition and Subtraction of Polynomials

Polynomials, p. 269
A polynomial in a single variable x is a sum of terms of the form ax^n, where a is a real number and n is a whole number. | $3x^5 - 4x^3 - 8x^2 + x + 5$

Degree of a Term, p. 269
For a polynomial in a single variable, the degree of a term is equal to the variable's exponent.

Polynomial: $4x^7 - 2x^3 + 9x + 12$

Term	$4x^7$	$-2x^3$	$9x$	12
Degree	7	3	1	0

Monomial, Binomial, Trinomial, p. 269

Monomial: A polynomial with only one term
Binomial: A polynomial with two terms
Trinomial: A polynomial with three terms

Monomial: $-9x^4$
Binomial: $x^2 - 49$
Trinomial: $3x^2 - 2x - 15$

Coefficients, p. 270

The coefficient of a term is the numerical part of a term, including its sign.

Polynomial: $4x^7 - 2x^3 + 9x + 12$

Term	$4x^7$	$-2x^3$	$9x$	12
Coefficient	4	−2	9	12

Degree of a Polynomial, p. 270

The degree of a polynomial is equal to the degree of the term with the highest degree.

$x^6 - 4x^4 + 9x^3 - x - 5$ Degree $= 6$

Evaluating a Polynomial, p. 270

Substitute the value for the variable in the polynomial and simplify the resulting expression.

Evaluate $x^2 + 5x + 22$ for $x = -8$.
$(-8)^2 + 5(-8) + 22 = 64 - 40 + 22 = 46$

Evaluating Polynomial Functions, p. 271

Substitute the value for the variable in the function and simplify the resulting expression.

For $f(x) = x^3 - 7x^2 + 40$, find $f(6)$.

$$\begin{aligned} f(6) &= (6)^3 - 7(6)^2 + 40 \\ &= 216 - 7(36) + 40 \\ &= 216 - 252 + 40 \\ &= 4 \end{aligned}$$

Adding Polynomials, p. 271

To add two polynomials together, combine like terms.

$$\begin{aligned} &(x^2 + 9x - 7) + (3x^2 - 4x - 11) \\ &= x^2 + 9x - 7 + 3x^2 - 4x - 11 \\ &= 4x^2 + 5x - 18 \end{aligned}$$

Subtracting Polynomials, pp. 271–272

To subtract one polynomial from another, change the sign of each term that is being subtracted and combine like terms.

$$\begin{aligned} &(4x^2 + x + 6) - (2x^2 + 7x - 24) \\ &= 4x^2 + x + 6 - 2x^2 - 7x + 24 \\ &= 2x^2 - 6x + 30 \end{aligned}$$

Section 5.4 Multiplying Polynomials

Multiplying Monomials, pp. 275–276

Multiply the coefficients; then multiply the variable factors by adding exponents.

$7x^4 \cdot 3x^9 = 21x^{4+9} = 21x^{13}$
$a^{12}b^9 \cdot 5a^7b^{13} = 5a^{12+7}b^{9+13} = 5a^{19}b^{22}$

Multiplying a Monomial by a Polynomial, p. 276

Apply the distributive property.

$$\begin{aligned} &3x^2(4x^2 - 5x - 10) \\ &= 3x^2 \cdot 4x^2 - 3x^2 \cdot 5x - 3x^2 \cdot 10 \\ &= 12x^4 - 15x^3 - 30x^2 \end{aligned}$$

Multiplying Polynomials, pp. 277–278

Multiply each term in the first polynomial by each term in the second polynomial. Combine any like terms.

$$\begin{aligned} &(2x - 7)(3x + 4) \\ &= 2x \cdot 3x + 2x \cdot 4 - 7 \cdot 3x - 7 \cdot 4 \\ &= 6x^2 + 8x - 21x - 28 \\ &= 6x^2 - 13x - 28 \end{aligned}$$

Special Products, pp. 278–279

$(a + b)(a - b) = a^2 - b^2$
$(a + b)^2 = a^2 + 2ab + b^2$
$(a - b)^2 = a^2 - 2ab + b^2$

$(4x + 5)(4x - 5) = (4x)^2 - 5^2 = 16x^2 - 25$
$(x + 3)^2 = x^2 + 2(x)(3) + 3^2 = x^2 + 6x + 9$

Section 5.5 Dividing Polynomials

Dividing Monomials, pp. 281–282

Divide the coefficients; then divide the variable factors by subtracting exponents.

$$\frac{20x^{12}}{4x^3} = 5x^{12-3} = 5x^9$$

Dividing a Polynomial by a Monomial, pp. 282–283

Divide each term of the polynomial by the monomial.

$$\frac{8x^5 - 12x^3 + 36x^2}{4x^2} = \frac{8x^5}{4x^2} - \frac{12x^3}{4x^2} + \frac{36x^2}{4x^2} = 2x^3 - 3x + 9$$

Dividing Polynomials, pp. 283–286

1. Divide the term in the dividend with the highest degree by the term in the divisor with the highest degree. Add this result to the quotient.
2. Multiply the monomial obtained in Step 1 by the divisor, writing the result underneath the dividend.
3. Subtract the product obtained in Step 2 from the dividend. (Change the signs of the product's terms and combine like terms.)

Repeat Steps 1–3 with the result of Step 3 as the new dividend. Keep repeating this procedure until the degree of the dividend is less than the degree of the divisor.

Divide: $\frac{6x^2 + 11x + 15}{2x - 5}$

$$\begin{array}{r} 3x + 13 \\ 2x - 5\overline{)6x^2 + 11x + 15} \\ \underline{-6x^2 \overset{+}{\not-} 15x \quad\downarrow} \\ 26x + 15 \\ \underline{-26x \overset{+}{\not-} 65} \\ 80 \end{array}$$

$$\frac{6x^2 + 11x + 15}{2x - 5} = 3x + 13 + \frac{80}{2x - 5}$$

SUMMARY OF CHAPTER 5 STUDY STRATEGIES

Attending class each day and doing all of your homework does not guarantee that you will earn a good grade when you take the exam. Through careful preparation and the adoption of the test-taking strategies introduced in this chapter, you can maximize your grade on a math exam. Here is a summary of the points that were introduced:

- Make the most of your time before the exam.
- Write down important facts as soon as you get your test.
- Briefly read through the test before you work any problems.
- Begin by solving easier problems first.
- Review your test as thoroughly as possible before turning it in.

CHAPTER 5 REVIEW

Simplify the expression. Write the result without using negative exponents. (Assume that all variables represent nonzero real numbers.) **[5.1–5.2]**

1. $\frac{x^9}{x^3}$

2. $(4x^7)^3$

3. $7x^0$

4. $\left(\frac{5x^9}{2y^2}\right)^4$

5. $7x^7 \cdot 4x^{12}$

6. $3x^8y^5 \cdot 8x^8y^3$

7. $(-6a^5b^3c^6)^2$

8. $15^0 - 3x^0$

9. $(x^{10}y^{13}z^4)^4$

10. $\frac{x^{16}}{x^6}$

11. 5^{-2}

12. 10^{-3}

13. $7x^{-4}$

14. $-6x^{-6}$

15. $\frac{-8}{y^{-5}}$

16. $(3x^{-4})^3$

17. $(2x^{-5})^{-2}$

18. $\frac{x^{10}}{x^{-5}}$

19. $\left(\frac{a^{-4}}{b^{-7}}\right)^3$

20. $m^{13} \cdot m^{-19}$

21. $(4x^{-5}y^4z^{-7})^{-3}$

22. $\left(\frac{x^{-11}}{y^{-6}}\right)^6$

23. $a^{-15} \cdot a^{-5}$

24. $\frac{x^{-20}}{x^{-9}}$

Rewrite in scientific notation. **[5.2]**

25. 1,400,000,000

26. 0.0000000000021

27. 0.000005002

Convert from scientific to standard notation. **[5.2]**

28. 1.23×10^{-4}

29. 4.075×10^7

30. 6.1275×10^{12}

Perform the following calculations. Express your answer using scientific notation. **[5.2]**

31. $(2.5 \times 10^7)(6.0 \times 10^6)$

32. $(3.0 \times 10^{-13}) \div (1.2 \times 10^5)$

33. The mass of a hydrogen atom is 1.66×10^{-24} grams. What is the mass of 1,500,000,000,000 hydrogen atoms? **[5.2]**

34. If a computer can perform a calculation in 0.000000000005 second, how many seconds would it take to perform 3,000,000,000,000,000 calculations? **[5.2]**

Evaluate the polynomial for the given value of the variable. **[5.3]**

35. $x^2 - 8x + 15$ for $x = -5$

36. $x^2 - 11x - 29$ for $x = 2$

37. $-x^2 + 16x - 30$ for $x = 6$

38. $-3x^2 - 4x + 7$ for $x = -7$

39. $5x^2 + 10x + 12$ for $x = -3$

40. $7x^2 - 14x + 35$ for $x = 8$

Add or subtract. **[5.3]**

41. $(x^2 + 3x - 15) + (x^2 - 9x + 6)$

42. $(3x^2 - 31) + (5x^2 - 19x - 19)$

43. $(x^2 - 6x - 13) - (x^2 - 14x + 8)$

44. $(2x^2 + 8x - 7) - (x^2 - x + 15)$

45. $(3x^3 + x^2 - 10) - (6x^2 - 13x + 20)$

46. $(10x^2 - 21x - 35) - (4x^3 + 6x^2 - 15x + 80)$

Evaluate the given polynomial function. **[5.3]**

47. $f(x) = x^2 - 25, f(-4)$

48. $f(x) = 2x^2 - 6x + 17, f(5)$

49. $f(x) = x^2 - 7x + 10, f(a^4)$

50. $f(x) = x^2 + 3x - 30, f(2a^3)$

Worked-out solutions to Review Exercises marked with can be found on page AN-18.

Given the functions $f(x)$ and $g(x)$, find $f(x) + g(x)$ and $f(x) - g(x)$. **[5.3]**

51. $f(x) = 3x^2 - 8x - 9, g(x) = 7x^2 + 6x + 40$

52. $f(x) = x^2 + 5x - 30, g(x) = -8x^2 - 15x + 6$

Multiply. **[5.4]**

53. $(x - 5)^2$

54. $(x - 4)(x + 13)$

55. $4x(3x^2 - 7x - 16)$

56. $(2x - 7)^2$

57. $5x^7 \cdot 3x^6$

58. $(x - 8)(x + 8)$

59. $(3x + 10)(2x - 13)$

60. $-9x^5 \cdot 2x$

61. $(6x + 5)(6x - 5)$

62. $(x + 10)^2$

63. $(4x + 7)^2$

64. $-8x^2(3x^5 - 7x^4 - 6x^3)$

65. $(x + 9)(x^2 - 6x + 12)$

66. $(x - 2)(3x^2 - 5x - 9)$

67. $(10x^5)(-3x^7)(4x^6)$

68. $-2x^4(-5x^2 - 11x + 12)$

Given the functions $f(x)$ and $g(x)$, find $f(x) \cdot g(x)$. **[5.4]**

69. $f(x) = x + 10, g(x) = x - 7$

70. $f(x) = -6x^6, g(x) = -4x^2 - 9x + 20$

Divide. Assume that all denominators are nonzero. **[5.5]**

71. $\dfrac{6x^4 - 8x^3 + 20x^2}{2x^2}$

72. $\dfrac{3x^2 - 8x - 25}{x - 4}$

73. $\dfrac{4x^5y^7}{-2x^5y}$

74. $\dfrac{x^3 + 3x^2 - 15}{x + 5}$

75. $\dfrac{x^4 + 81}{x + 3}$

76. $\dfrac{18x^3y^7z^6}{3x^3y^6z}$

77. $\dfrac{8x^2 + 87x + 171}{x + 9}$

78. $\dfrac{6x^2 - 30x - 84}{x - 7}$

79. $\dfrac{15x^2 - 43x - 238}{5x + 14}$

80. $\dfrac{16x^2 + 2x + 20}{8x - 3}$

CHAPTER 5 TEST

For Extra Help

Step-by-step test solutions are found on the Chapter Test Prep Videos available via the Video Resources on DVD, in *MyMathLab*, and on YouTube (search "WoodburyElemIntAlg" and click on "Channels").

Simplify the expression. Write the result without using negative exponents. (Assume that all variables represent nonzero real numbers.)

1. $\dfrac{x^{10}}{x^2}$

2. $\left(\dfrac{x^5}{3y^4}\right)^6$

3. $(x^9y^5z^8)^8$

4. 4^{-4}

5. $(2x^{-7})^5$

6. $\dfrac{x^4}{x^{-13}}$

7. $x^{22} \cdot x^{-15}$

8. $(3xy^{-6}z^5)^{-2}$

Rewrite in scientific notation.

9. 23,500,000

Convert to standard notation.

10. 4.7×10^{-8}

Perform the following calculations. Express your answer using scientific notation.

11. $(4.3 \times 10^{13})(1.8 \times 10^{-9})$

Evaluate the polynomial for the given value of the variable.

12. $x^2 + 12x - 38$ for $x = -8$

Add or subtract.

13. $(x^2 - 6x - 32) - (5x^2 + 8x - 33)$

14. $(4x^2 - 3x - 15) + (-2x^2 + 17x - 49)$

Evaluate the given polynomial function.

15. $f(x) = x^2 + 3x - 14, f(-7)$

Multiply.

16. $5x^3(6x^2 + 8x - 17)$

17. $(x + 6)(x - 6)$

18. $(5x - 9)(4x + 7)$

Divide. Assume that all denominators are nonzero.

19. $\dfrac{21x^7 + 33x^6 - 15x^5}{3x^2}$

20. $\dfrac{6x^2 - 11x + 38}{x + 3}$

Mathematicians in History

Carl Friedrich Gauss was a German mathematician who lived in the 18th and 19th centuries and was one of the most prominent and prolific mathematicians of his day. Gauss made significant contributions to the fields of analysis, probability, statistics, number theory, geometry, and astronomy, among others. The Prussian mathematician Leopold Kronecker once said of him, "Almost everything, which the mathematics of our century has brought forth in the way of original scientific ideas, attaches to the name of Gauss." Gauss once said, "It is not knowledge, but the act of learning, not possession but the act of getting there, which grants the greatest enjoyment."

Write a one-page summary (*or* make a poster) of the life of Carl Friedrich Gauss and his accomplishments. Also look up Gauss's most famous quotes and list your favorite.

Interesting issues:

- Where and when was Carl Friedrich Gauss born?
- Gauss displayed incredible genius at an early age. Relay the story of how Gauss found his father's accounting error at only 3 years of age.
- At what age did Gauss find a shortcut for summing the integers from 1 to 100?
- Gauss's motto was "Few, but ripe." In your own words, explain the meaning of this motto.
- In 1798, Gauss showed that it is possible to construct a regular 17-gon by ruler and compass. This was the first major advance in the area of regular polygons in approximately 2000 years. What is a regular 17-gon?
- Gauss earned his doctorate in 1799 with a proof of what major mathematical theorem?
- Gauss gained notoriety in the field of astronomy by accurately predicting the orbit of the newly discovered asteroid Ceres by inventing the method of least squares. Describe the method of least squares.
- In 1821, Gauss built the first heliotrope. What is a heliotrope?
- Describe the circumstances that led to Gauss's death.

CHAPTER 6

Factoring and Quadratic Equations

In this chapter, we will begin to learn how to solve quadratic equations. A *quadratic equation* is an equation of the form $ax^2 + bx + c = 0$, where a, b, and c are real numbers and $a \neq 0$. To solve a quadratic equation, we will attempt to rewrite the quadratic expression $ax^2 + bx + c$ as a product of two expressions. In other words, we will factor the quadratic expression. The first five sections of this chapter focus on factoring techniques.

Once we have learned how to factor polynomials, we will learn how to solve quadratic equations and to apply quadratic equations to real-world problems. We also will examine quadratic functions, which are functions of the form $f(x) = ax^2 + bx + c$, where $a \neq 0$.

STUDY STRATEGY

Overcoming Math Anxiety In this chapter, we will focus on how to overcome math anxiety, a condition that many students share. Math anxiety can prevent a student from learning mathematics, and it can seriously impact a student's performance on quizzes and exams.

6.1 An Introduction to Factoring; the Greatest Common Factor; Factoring by Grouping

OBJECTIVES

1. **Find the greatest common factor (GCF) of two or more integers.**
2. **Find the GCF of two or more variable terms.**
3. **Factor the GCF out of each term of a polynomial.**
4. **Factor a common binomial factor out of a polynomial.**
5. **Factor a polynomial by grouping.**

In the previous chapter, we learned about polynomials; in this section, we will begin to learn how to **factor** a polynomial. A polynomial has been **factored** when it is represented as the product of two or more polynomials. Being able to factor polynomials will help us solve quadratic equations such as $x^2 - 11x + 30 = 0$, as well as simplify rational expressions such as $\dfrac{x^2 + 3x - 10}{x^2 - 9x + 14}$.

Greatest Common Factor

Objective 1 Find the greatest common factor (GCF) of two or more integers. Before we begin to learn how to factor polynomials, we will go over the procedure for finding the **greatest common factor (GCF)** of two or more integers. The GCF of two or more integers is the largest whole number that is a factor of each integer.

Finding the GCF of Two or More Integers

Find the prime factorization of each integer.
Write the prime factors that are common to each integer; the GCF is the product of these prime factors.

EXAMPLE 1 Find the GCF of 30 and 42.

SOLUTION The prime factorization of 30 is $2 \cdot 3 \cdot 5$, and the prime factorization of 42 is $2 \cdot 3 \cdot 7$. (For a refresher on finding the prime factorization of a number, refer back to Section 1.3.) The prime factors they have in common are 2 and 3; so the GCF is $2 \cdot 3$, or 6. (This means that 6 is the greatest number that divides evenly into both 30 and 42.)

Quick Check 1

Find the GCF of 16 and 20.

EXAMPLE 2 Find the GCF of 108, 504, and 720.

SOLUTION Begin with the prime factorizations.

$$108 = 2^2 \cdot 3^3 \qquad 504 = 2^3 \cdot 3^2 \cdot 7 \qquad 720 = 2^4 \cdot 3^2 \cdot 5$$

The only primes that are factors of all three numbers are 2 and 3. The smallest power of 2 that is a factor of any of the three numbers is 2^2, and the smallest power of 3 that is a factor is 3^2. The GCF is $2^2 \cdot 3^2$, or 36.

Quick Check 2

Find the GCF of 48, 120, and 156.

Note that if two or more integers do not have any common prime factors, their GCF is 1. For example, the GCF of 8, 12, and 15 is 1 because no prime numbers are factors of all three numbers.

Objective 2 Find the GCF of two or more variable terms. We also can find the GCF of two or more variable terms. For a variable factor to be included in the GCF, it must be a factor of each term. As with prime factors, the exponent for a variable factor used in the GCF is the smallest exponent that can be found for that variable in any one term.

EXAMPLE 3 Find the GCF of $6x^2$ and $15x^4$.

SOLUTION The GCF of the two coefficients is 3. The variable x is a factor of each term, and its smallest exponent is 2. The GCF of these two terms is $3x^2$.

Quick Check 3

Find the GCF.

a) $12x^5$ and $28x^3$
b) $x^3y^4z^9$, $x^6y^2z^{10}$, and x^7z^4

EXAMPLE 4 Find the GCF of $a^5b^4c^2$, a^7b^3, $a^6b^9c^5$, and $a^3b^8c^3$.

SOLUTION Only the variables a and b are factors of all of the terms. (The variable c is not a factor of the second term.) The smallest power of a in any one term is 3, which is the case for the variable b as well. The GCF is a^3b^3.

Factoring Out the Greatest Common Factor

Objective 3 Factor the GCF out of each term of a polynomial. The first step for factoring a polynomial is to factor out the GCF of all of the terms. This process uses the distributive property, and you may think of it as "undistributing" the GCF from each term. Consider the polynomial $4x^2 + 8x + 20$. The GCF of these three terms is 4, and the polynomial can be rewritten as the product $4(x^2 + 2x + 5)$. Notice that the GCF has been factored out of each term and that the polynomial inside the parentheses is the polynomial we would multiply by 4 to equal $4x^2 + 8x + 20$.

$$\begin{aligned} 4(x^2 + 2x + 5) &= 4 \cdot x^2 + 4 \cdot 2x + 4 \cdot 5 && \text{Distribute.} \\ &= 4x^2 + 8x + 20 && \text{Multiply.} \end{aligned}$$

EXAMPLE 5 Factor $5x^3 - 30x^2 + 10x$ by factoring out the GCF.

SOLUTION Begin by finding the GCF, which is $5x$. The next task is to fill in the missing terms of the polynomial inside the parentheses so that the product of $5x$ and that polynomial is $5x^3 - 30x^2 + 10x$.

$$5x(? - ? + ?)$$

To find the missing terms, we can divide each term of the original polynomial by $5x$.

$$5x^3 - 30x^2 + 10x = 5x(x^2 - 6x + 2)$$

We can check the answer by multiplying $5x(x^2 - 6x + 2)$, which should equal $5x^3 - 30x^2 + 10x$. The check is left to the reader.

Quick Check 4

Factor $6x^4 - 42x^3 - 90x^2$ by factoring out the GCF.

EXAMPLE 6 Factor $6x^6 + 8x^4 + 2x^3$ by factoring out the GCF.

SOLUTION The GCF for these three terms is $2x^3$. Notice that this is also the third term.

$$6x^6 + 8x^4 + 2x^3 = 2x^3(3x^3 + 4x + 1) \quad \text{Factor out the GCF.}$$

When a term in a polynomial is the GCF of the polynomial, factoring out the GCF leaves a 1 in that term's place. Why? Because we need to determine what we multiply $2x^3$ by to equal $2x^3$, and $2x^3 \cdot 1 = 2x^3$.

Quick Check 5

Factor $15x^7 - 30x^5 + 3x^4$ by factoring out the GCF.

A WORD OF CAUTION When factoring out the GCF from a polynomial, be sure to write a 1 in the place of a term that was the GCF.

Objective 4 Factor a common binomial factor out of a polynomial. Occasionally, the GCF of two terms will be a binomial or some other polynomial with more than one term. Consider the expression $8x(x - 5) + 3(x - 5)$, which contains the two terms $8x(x - 5)$ and $3(x - 5)$. Each term has the binomial $x - 5$ as a factor. This common factor can be factored out of this expression.

EXAMPLE 7 Factor $x(x + 3) + 7(x + 3)$ by factoring out the GCF.

SOLUTION The binomial $x + 3$ is a common factor for these two terms. Begin by factoring out this common factor.

$$x(x + 3) + 7(x + 3)$$
$$= (x + 3)(? + ?)$$

After factoring out $x + 3$, what factors remain in the first term? The only factor that remains is x. Using the same method, we see that the only factor remaining in the second term is 7.

$$x(x + 3) + 7(x + 3) = (x + 3)(x + 7)$$

EXAMPLE 8 Factor $9x(2x + 3) - 8(2x + 3)$ by factoring out the GCF.

SOLUTION The GCF of these two terms is $2x + 3$, which can be factored out as follows:

$$9x(2x + 3) - 8(2x + 3) = (2x + 3)(9x - 8)$$ Factor out the common factor $2x + 3$.

Notice that the second term was negative, which led to $9x - 8$, rather than $9x + 8$, as the other factor.

Quick Check 6

Factor by factoring out the GCF.

a) $5x(x - 9) + 14(x - 9)$
b) $7x(x - 8) - 6(x - 8)$

Again, factoring out the GCF will be the first step in factoring a polynomial. Often this will make the factoring easier, and sometimes the expression cannot be factored without factoring out the GCF.

Factoring by Grouping

Objective 5 Factor a polynomial by grouping. We now turn our attention to a factoring technique known as **factoring by grouping**. Consider the polynomial $3x^3 + 33x^2 + 7x + 77$. The GCF for the four terms is 1, so we cannot factor the polynomial by factoring out the GCF. However, the first pair of terms has a common factor of $3x^2$ and the second pair of terms has a common factor of 7. If we factor $3x^2$ out of the first two terms and 7 out of the last two terms, we produce the following:

$$3x^2(x + 11) + 7(x + 11)$$

Notice that the two resulting terms have a common factor of $x + 11$. This common binomial factor can then be factored out.

$$3x^3 + 33x^2 + 7x + 77$$
$$= 3x^2(x + 11) + 7(x + 11)$$ Factor $3x^2$ from the first two terms and 7 from the last two terms.
$$= (x + 11)(3x^2 + 7)$$ Factor out the common binomial factor $x + 11$.

Factoring a Polynomial with Four Terms by Grouping

Factor a common factor out of the first two terms and another common factor out of the last two terms. If the two "groups" share a common binomial factor, this binomial can be factored out to complete the factoring of the polynomial.

If the two "groups" do not share a common binomial factor, we can try rearranging the terms of the polynomial in a different order. If we cannot find two "groups" that share a common factor, the polynomial cannot be factored by this method.

EXAMPLE 9 Factor $5x^3 - 30x^2 + 4x - 24$ by grouping.

SOLUTION First, check for a factor that is common to each of the four terms. Because there is no common factor other than 1, proceed to factoring by grouping.

$5x^3 - 30x^2 + 4x - 24 = 5x^2(x - 6) + 4x - 24$	Factor $5x^2$ out of the first two terms.
$= 5x^2(x - 6) + 4(x - 6)$	Factor 4 out of the last two terms.
$= (x - 6)(5x^2 + 4)$	Factor out the common factor $x - 6$.

As was the case earlier, we can check our factoring by multiplying the two factors. The check is left to the reader.

Quick Check 7

Factor $x^3 + 4x^2 + 7x + 28$ by grouping.

EXAMPLE 10 Factor $8x^2 + 32x - 7x - 28$ by grouping.

SOLUTION Because the four terms have no common factors other than 1, we factor by grouping. After factoring $8x$ out of the first two terms, we must factor a *negative* 7 out of the last two terms. If we factored a positive 7 rather than a negative 7 out of the last two terms, the two binomial factors would be different.

$8x^2 + 32x - 7x - 28 = 8x(x + 4) - 7(x + 4)$	Factor the common factor $8x$ out of the first two terms and the common factor -7 out of the last two terms.
$= (x + 4)(8x - 7)$	Factor out the common factor $x + 4$.

When the third of the four terms is negative, a negative common factor often needs to be factored from the last two terms.

Quick Check 8

Factor $2x^2 - 10x - 9x + 45$ by grouping.

A WORD OF CAUTION Factoring a negative common factor from two terms when factoring by grouping is often necessary.

EXAMPLE 11 Factor $10x^3 + 50x^2 + x + 5$ by grouping.

SOLUTION Again, the four terms have no common factor other than 1, so we may proceed to factor this polynomial by grouping. Notice that the last two terms have no common factor other than 1. When this is the case, we factor out the common factor of 1 from the two terms.

$10x^3 + 50x^2 + x + 5 = 10x^2(x + 5) + 1(x + 5)$	Factor the common factors of $10x^2$ and 1 from the first two terms and the last two terms, respectively.
$= (x + 5)(10x^2 + 1)$	Factor out the common factor $x + 5$.

Quick Check 9

Factor $4x^3 - 36x^2 + x - 9$ by grouping.

In the next example, the four terms have a common factor other than 1. We will factor out the GCF from the polynomial before attempting to use factoring by grouping.

EXAMPLE 12 Factor $2x^3 - 6x^2 - 20x + 60$ completely.

SOLUTION The four terms share a common factor of 2, and we will factor this out before proceeding with factoring by grouping.

$$2x^3 - 6x^2 - 20x + 60 = 2(x^3 - 3x^2 - 10x + 30)$$ Factor out the common factor 2.

$$= 2[x^2(x - 3) - 10(x - 3)]$$ Factor out the common factor x^2 from the first two terms. Factor out the common factor -10 from the last two terms.

$$= 2(x - 3)(x^2 - 10)$$ Factor out the common factor $x - 3$.

If we had not factored out the common factor 2 before factoring this polynomial by grouping, we would have factored $2x^3 - 6x^2 - 20x + 60$ to be $(x - 3)(2x^2 - 20)$. If we stop here, we have not factored this polynomial completely because $2x^2 - 20$ has a common factor of 2.

Quick Check 10

Factor $3x^2 + 24x - 12x - 96$ by grouping.

Occasionally, despite our best efforts, a polynomial cannot be factored. Here is an example of just such a polynomial. Consider the polynomial $x^2 - 5x + 3x + 15$. The first two terms have a common factor of x, and the last two terms have a common factor of 3.

$$x^2 - 5x + 3x + 15 = x(x - 5) + 3(x + 5)$$

Because the two binomials are not the same, we cannot factor a common factor out of the two terms. Reordering the terms $-5x$ and $3x$ leads to the same problem. The polynomial cannot be factored.

There are instances when factoring by grouping fails, yet the polynomial can be factored by other techniques. For example, the polynomial $x^3 + 2x^2 - 7x - 24$ cannot be factored by grouping. However, it can be shown that $x^3 + 2x^2 - 7x - 24$ is equal to $(x - 3)(x^2 + 5x + 8)$.

BUILDING YOUR STUDY STRATEGY

Overcoming Math Anxiety, 1 **Understanding Math Anxiety** If you have been avoiding taking a math class, if you panic when asked a mathematical question, or if you think you cannot learn mathematics, you may have a condition known as math anxiety. Like many other anxiety-related conditions, math anxiety may be traced back to an event that first triggered the negative feelings. If you are going to overcome your anxiety, the first step is to understand the cause of your anxiety.

Were you ridiculed as a child when you couldn't solve a math problem? Were you expected to be a mathematical genius like a parent or an older sibling but were not able to measure up? One negative event can trigger math anxiety, and the journey to overcome math anxiety can begin with a single success. If you are able to complete a homework assignment or show improvement on a quiz or an exam, celebrate your success. Let this be your vindication and consider your slate to have been wiped clean.

Exercises 6.1

MyMathLab

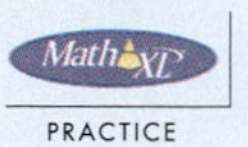
PRACTICE

WATCH

DOWNLOAD

READ

REVIEW

Vocabulary

1. A polynomial has been ________ when it is represented as the product of two or more polynomials.
2. The ______________________________ of two or more integers is the largest whole number that is a factor of each integer.
3. The exponent for a variable factor used in the GCF is the ________ exponent that can be found for that variable in any one term.
4. The process of factoring a polynomial by first breaking the polynomial into two sets of terms is called factoring by ________.

Find the GCF.

5. 6, 8
6. 15, 21
7. 30, 42
8. 5, 50
9. 16, 40, 60
10. 8, 24, 27
11. x^3, x^7
12. y^8, y^4
13. a^2b^3, a^5b^2
14. $a^8b^3c^4, a^5bc^2, a^3b^4c^5$
15. $4x^4, 6x^3$
16. $27x^5, 24x^{11}$
17. $15a^5b^2c, 25a^2c^3, 5a^3bc^2$
18. $30x^3y^4z^5, 12x^6y^{11}z^3, 24y^9z$

Factor the GCF out of the given expression.

19. $7x - 14$
20. $3a + 12$
21. $5x^2 + 4x$
22. $9x^2 - 20x$
23. $8x^3 + 20x$
24. $11x^5 - 33x^3$
25. $60x^2 + 36x + 6$
26. $15x^2 - 9x - 3$
27. $20x^6 - 35x^4 - 50x^3$
28. $36x^5 - 9x^4 - 45x^2$
29. $m^7n^3 - m^5n^4 + m^6n^6$
30. $a^7b^8 + a^5b^6 - a^3b^2$
31. $10x^5y^5 - 30x^7y^4 + 80x^2y^{10}$
32. $20x^6y^2 - 16xy^5 + 24x^3y^3$
33. $-8x^3 + 12x^2 - 16x$
34. $-18a^5 - 30a^3 + 24a^2$
35. $5x(2x - 7) + 8(2x - 7)$
36. $x(4x + 3) + 4(4x + 3)$
37. $x(3x - 4) - 9(3x - 4)$
38. $6x(x - 8) - 5(x - 8)$
39. $5x(4x + 7) - (4x + 7)$
40. $3x(7x - 2) + (7x - 2)$

Factor by grouping.

41. $x^2 + 10x + 3x + 30$
42. $x^2 - 7x + 5x - 35$
43. $x^3 - 9x^2 + 6x - 54$
44. $x^3 + 4x^2 + 11x + 44$
45. $x^2 - 5x - 12x + 60$
46. $x^2 + 3x - 9x - 27$
47. $3x^2 + 15x + 4x + 20$
48. $4x^2 - 24x + 7x - 42$
49. $7x^2 + 28x - 6x - 24$
50. $5x^2 - 25x - 12x + 60$
51. $3x^2 + 21x + x + 7$
52. $4x^2 - 48x + x - 12$
53. $2x^2 - 10x - x + 5$
54. $9x^2 - 81x - x + 9$
55. $x^3 + 8x^2 + 6x + 48$
56. $2x^3 - 12x^2 + 5x - 30$
57. $4x^3 + 12x + 3x^2 + 9$
58. $7x^3 + 49x + 4x^2 + 28$
59. $2x^2 + 10x + 6x + 30$
60. $3x^2 + 27x + 21x + 189$

Find a polynomial that has the given factor.

61. $x - 5$
62. $x + 3$
63. $2x + 9$
64. $3x - 4$

Writing in Mathematics

Answer in complete sentences.

65. When you are factoring a polynomial, how do you check your work?
66. Explain why it is necessary to factor a negative common factor from the last two terms of the polynomial $x^3 - 5x^2 - 7x + 35$ to factor it by grouping.

Quick Review Exercises

Section 6.1

Multiply.

1. $(x + 7)(x + 4)$
2. $(x - 9)(x - 11)$
3. $(x + 5)(x - 8)$
4. $(x - 10)(x + 6)$

6.2 Factoring Trinomials of the Form $x^2 + bx + c$

OBJECTIVES

1. Factor a trinomial of the form $x^2 + bx + c$ when c is positive.
2. Factor a trinomial of the form $x^2 + bx + c$ when c is negative.
3. Factor a perfect square trinomial.
4. Determine that a trinomial is prime.
5. Factor a trinomial by first factoring out a common factor.
6. Factor a trinomial in several variables.

In this section, we will learn how to factor trinomials of degree 2 with a leading coefficient of 1. Some examples of this type of polynomial are $x^2 + 12x + 32$, $x^2 - 9x + 14$, $x^2 + 8x - 20$, and $x^2 - x - 12$.

Factoring Trinomials of the Form $x^2 + bx + c$

If $x^2 + bx + c$, where b and c are integers, is factorable, it can be factored as the product of two binomials of the form $(x + m)(x + n)$, where m and n are integers.

We will begin by multiplying two binomials of the form $(x + m)(x + n)$ and using the result to help us learn to factor trinomials of the form $x^2 + bx + c$.

Multiply the two binomials $x + 4$ and $x + 8$.

$$\begin{aligned}(x + 4)(x + 8) &= x^2 + 8x + 4x + 32 && \text{Distribute.}\\ &= x^2 + 12x + 32 && \text{Combine like terms.}\end{aligned}$$

The two binomials have a product that is a trinomial of the form $x^2 + bx + c$ with $b = 12$ and $c = 32$. Notice that the two numbers 4 and 8 have a product of 32 and a sum of 12; in other words, their product is equal to c and their sum is equal to b.

$$4 \cdot 8 = 32$$
$$4 + 8 = 12$$

We will use this pattern to help us factor trinomials of the form $x^2 + bx + c$. We will look for two integers m and n with a product equal to c and a sum equal to b. If we can find two such integers, the trinomial will factor as $(x + m)(x + n)$.

Factoring $x^2 + bx + c$ When c Is Positive

Objective 1 **Factor a trinomial of the form $x^2 + bx + c$ when c is positive.**

EXAMPLE 1 Factor $x^2 + 9x + 18$.

SOLUTION We begin, as always, by looking for common factors. The terms in this trinomial have no common factors other than 1. We are looking for two integers m and n whose product is 18 and whose sum is 9. Here are the factors of 18.

Factors	$1 \cdot 18$	$2 \cdot 9$	$3 \cdot 6$
Sum	19	11	9

The pair of factors that have a sum of 9 are 3 and 6. The trinomial $x^2 + 9x + 18$ factors to be $(x + 3)(x + 6)$. Be aware that this also can be written as $(x + 6)(x + 3)$.

We can check our work by multiplying $x + 3$ by $x + 6$. If the product equals $x^2 + 9x + 18$, we have factored correctly. The check is left to the reader.

EXAMPLE 2 Factor $x^2 - 11x + 24$.

SOLUTION The major difference between this trinomial and the previous one is that the x term has a negative coefficient. We are looking for two integers m and n whose product is 24 and whose sum is -11. Because the product of these two integers is positive, they must have the same sign. The sum of these two integers is negative, so each integer must be negative. Here are the negative factors of 24.

Factors	$(-1)(-24)$	$(-2)(-12)$	$(-3)(-8)$	$(-4)(-6)$
Sum	-25	-14	-11	-10

The pair that has a sum of -11 is -3 and -8. $x^2 - 11x + 24 = (x - 3)(x - 8)$.

Quick Check 1

Factor.

a) $x^2 + 7x + 10$
b) $x^2 - 11x + 30$

From the previous examples, we see that when the constant term c is positive, such as in $x^2 + 9x + 18$ and $x^2 - 11x + 24$, the two numbers we are looking for (m and n) will have the same sign. Both numbers will be positive if b is positive, and both numbers will be negative if b is negative.

Factoring $x^2 + bx + c$ When c Is Negative

Objective 2 Factor a trinomial of the form $x^2 + bx + c$ when c is negative. If the product of two numbers is negative, one of the numbers must be negative and the other number must be positive. So if the constant term c is negative, m and n must have opposite signs.

EXAMPLE 3 Factor $x^2 + 8x - 33$.

SOLUTION These three terms do not have any common factors, so we look for two integers m and n with a product of -33 and a sum of 8. Because the product is negative, we are looking for one negative number and one positive number. Here are the factors of -33, along with their sums.

Factors	$-1 \cdot 33$	$-3 \cdot 11$	$-11 \cdot 3$	$-33 \cdot 1$
Sum	32	8	-8	-32

The pair of integers we are looking for are -3 and 11. $x^2 + 8x - 33 = (x - 3)(x + 11)$.

EXAMPLE 4 Factor $x^2 - 3x - 40$.

SOLUTION Because there are no common factors to factor out, we are looking for two integers m and n that have a product of -40 and a sum of -3. Here are the factors of -40.

Factors	$-1 \cdot 40$	$-2 \cdot 20$	$-4 \cdot 10$	$-5 \cdot 8$	$-8 \cdot 5$	$-10 \cdot 4$	$-20 \cdot 2$	$-40 \cdot 1$
Sum	39	18	6	3	-3	-6	-18	-39

The integers m and n are -8 and 5. $x^2 - 3x - 40 = (x - 8)(x + 5)$.

Quick Check 2

Factor.

a) $x^2 + 9x - 36$
b) $x^2 - x - 42$

It is not necessary to list each set of factors as in the previous examples. We can often find the two integers m and n quickly through trial and error.

EXAMPLE 5 Factor.

a) $x^2 - 10x - 24$

SOLUTION The two integers whose product is -24 and sum is -10 are -12 and 2.

$$x^2 - 10x - 24 = (x - 12)(x + 2)$$

b) $x^2 + 10x + 24$

SOLUTION This example is similar to the last example, but we are looking for two integers that have a product of *positive* 24 instead of -24. Also, the sum of these two integers is 10 instead of -10. The two integers whose product is 24 and whose sum is 10 are 4 and 6.

$$x^2 + 10x + 24 = (x + 4)(x + 6)$$

Quick Check 3

Factor.

a) $x^2 + 5x - 6$
b) $x^2 + 5x + 6$
c) $x^2 - 5x + 6$
d) $x^2 - 5x - 6$

c) $x^2 + 10x - 24$

SOLUTION The two integers that have a product of -24 and a sum of 10 are -2 and 12.

$$x^2 + 10x - 24 = (x - 2)(x + 12)$$

d) $x^2 - 10x + 24$

SOLUTION The integers -4 and -6 have a product of 24 and a sum of -10.

$$x^2 - 10x + 24 = (x - 4)(x - 6)$$

Factoring a Perfect Square Trinomial

Objective 3 Factor a perfect square trinomial.

EXAMPLE 6 Factor $x^2 - 6x + 9$.

SOLUTION We begin by looking for two integers that have a product of 9 and a sum of -6. Because the product of -3 and -3 is 9 and their sum is -6, $x^2 - 6x + 9 = (x - 3)(x - 3)$. Notice that the same factor is listed twice. We can rewrite this as $(x - 3)^2$.

Quick Check 4

Factor $x^2 + 10x + 25$.

When a trinomial factors to equal the square of a binomial, we call it a **perfect square trinomial**.

Trinomials That Are Prime

Objective 4 Determine that a trinomial is prime. Not every trinomial of the form $x^2 + bx + c$ can be factored. For example, there may not be a pair of integers that have a product of c and a sum of b. In this case, we say that the trinomial cannot be factored; it is **prime**. For example, $x^2 + 9x + 12$ is prime because we cannot find two numbers with a product of 12 and a sum of 9.

EXAMPLE 7 Factor $x^2 + 11x - 18$.

SOLUTION As there are no common factors to factor out, we are looking for two integers m and n that have a product of -18 and a sum of 11. Here are the factors of -18.

Factors	$-1 \cdot 18$	$-2 \cdot 9$	$-3 \cdot 6$	$-6 \cdot 3$	$-9 \cdot 2$	$-18 \cdot 1$
Sum	17	7	3	-3	-7	-17

No two integers have a product of -18 and a sum of 11, so the trinomial $x^2 + 11x - 18$ is prime.

Quick Check 5

Factor $x^2 + 9x - 20$.

Factoring a Trinomial Whose Terms Have a Common Factor

Objective 5 Factor a trinomial by first factoring out a common factor. We now turn our attention to factoring trinomials whose terms contain common factors other than 1. After we factor out a common factor, we will attempt to factor the remaining polynomial factor.

EXAMPLE 8 Factor $4x^2 + 48x + 80$.

SOLUTION The GCF of these three terms is 4.

A WORD OF CAUTION When factoring a polynomial, begin by factoring out the GCF of *all* of the terms.

After factoring out the GCF, we have $4(x^2 + 12x + 20)$. We now try to factor the trinomial $x^2 + 12x + 20$ by finding two integers that have a product of 20 and a sum of 12. The integers that satisfy these conditions are 2 and 10.

$$4x^2 + 48x + 80 = 4(x^2 + 12x + 20) \quad \text{Factor out the GCF 4.}$$
$$= 4(x + 2)(x + 10) \quad \text{Factor } x^2 + 12x + 20.$$

Quick Check 6

Factor $5x^2 + 40x + 60$.

A WORD OF CAUTION When a common factor is factored out of a polynomial, that common factor *must* be written in all following stages of factoring. *Do not "lose" the common factor!*

EXAMPLE 9 Factor $-x^2 - 7x + 60$.

SOLUTION The leading coefficient for this trinomial is -1, but to factor the trinomial using our technique, the leading coefficient must be a *positive* 1. We begin by factoring out a -1 from each term, which will change the sign of each term.

$$-x^2 - 7x + 60 = -(x^2 + 7x - 60) \quad \text{Factor out } -1 \text{ so that the leading coefficient is 1.}$$
$$= -(x - 5)(x + 12) \quad \text{Factor } x^2 + 7x - 60 \text{ by finding two integers whose product is } -60 \text{ and whose sum is 7.}$$

Quick Check 7

Factor $-x^2 + 14x - 48$.

If a trinomial's leading coefficient is not 1 or if the degree of the trinomial is not 2, we should factor out a common factor so that the trinomial factor is of the form $x^2 + bx + c$. At this point in the text, this is the only type of trinomial we know how to factor. We will learn techniques to factor other trinomials in the next section.

Factoring Trinomials in Several Variables

Objective 6 Factor a trinomial in several variables. Trinomials in several variables also can be factored. In the next example, we will explore the similarities and differences between factoring trinomials in two variables and trinomials in one variable.

EXAMPLE 10 Factor $x^2 + 13xy + 36y^2$.

SOLUTION These three terms have no common factors, and the leading coefficient is 1. Suppose the variable y were not included; in other words, suppose we were asked to factor $x^2 + 13x + 36$. After looking for two integers whose product is 36 and whose sum is 13, we would see that $x^2 + 13x + 36 = (x + 4)(x + 9)$. Now we can examine the changes that are necessary because of the second variable y. Note that $4 \cdot 9 = 36$, not $36y^2$. However, $4y \cdot 9y$ does equal $36y^2$ and $x \cdot 9y + x \cdot 4y$ does equal $13xy$. This trinomial factors to be $(x + 4y)(x + 9y)$.

Quick Check 8

Factor $x^2 - 4xy - 32y^2$.

We can factor a trinomial in two or more variables by ignoring all of the variables except the first variable. After determining how the trinomial factors for the first variable, we can determine where the rest of the variables appear in the factored form of the trinomial. We can check that our factoring is correct by multiplying, making sure the product is equal to the trinomial.

EXAMPLE 11 Factor $p^2r^2 - 9pr - 10$.

SOLUTION If the variable r did not appear in the trinomial and we were factoring $p^2 - 9p - 10$, we would look for two integers that have a product of -10 and a sum of -9. The two integers are -10 and 1, so the polynomial $p^2 - 9p - 10 = (p - 10)(p + 1)$. Now we work on the variable r. Because the first term in the trinomial is p^2r^2, the first term in each binomial must be pr. $p^2r^2 - 9pr - 10 = (pr - 10)(pr + 1)$.

Quick Check 9

Factor $x^2y^2 + 10xy - 24$.

BUILDING YOUR STUDY STRATEGY

Overcoming Math Anxiety, 2 Mathematical Autobiography One effective tool for understanding your past difficulties in mathematics and how they hinder your ability to learn mathematics today is to write a mathematical autobiography. Write down your successes and failures as well as the reasons you think you succeeded or failed. Go back to your mathematical past as far as you can remember. Write down how friends, relatives, and teachers affected you.

Let your autobiography sit for a few days and then read it over. Read it on an analytical level as if another person had written it, detaching yourself from it personally. Look for patterns, think about strategies to reverse the bad patterns, and continue any good patterns you can find.

Exercises 6.2

MyMathLab

PRACTICE

WATCH

DOWNLOAD

READ

REVIEW

Vocabulary

1. A polynomial of the form $x^2 + bx + c$ is a second-degree trinomial with a(n) ______________ of 1.
2. A(n) ____________ trinomial is a trinomial whose two binomial factors are identical.
3. A polynomial is ______ if it cannot be factored.
4. Before attempting to factor a second-degree trinomial, it is wise to factor out any ______________.

Factor completely. If the polynomial cannot be factored, write prime.

5. $x^2 - 8x - 20$
6. $x^2 - 2x - 15$
7. $x^2 + 5x - 36$
8. $x^2 - 10x - 39$
9. $x^2 + 12x + 24$
10. $x^2 - 12x + 36$
11. $x^2 + 14x + 48$
12. $x^2 + 14x - 33$
13. $x^2 + 13x - 30$
14. $x^2 - 13x + 30$
15. $x^2 - 5x + 36$
16. $x^2 - 15x + 54$
17. $x^2 - 17x - 60$
18. $x^2 - 17x + 60$
19. $x^2 + 11x - 12$
20. $x^2 + 15x + 56$
21. $x^2 + 14x + 49$
22. $x^2 - 4x - 12$
23. $x^2 + 20x + 91$
24. $x^2 + 14x + 13$
25. $x^2 - 24x + 144$
26. $x^2 + 5x - 14$
27. $x^2 - 13x + 40$
28. $x^2 - 5x - 24$
29. $x^2 - 13x - 30$
30. $x^2 + 18x + 81$
31. $x^2 - 9x + 20$
32. $x^2 - 11x - 28$
33. $x^2 - 16x + 60$
34. $x^2 + 16x + 48$
35. $6x^2 - 54x + 120$
36. $4x^2 - 28x - 72$
37. $5x^2 + 35x - 150$
38. $9x^2 + 126x + 432$
39. $-x^2 + 3x + 70$
40. $-x^2 + 8x - 15$
41. $x^3 + 8x^2 + 16x$
42. $-x^4 + x^3 + 30x^2$
43. $-3x^5 - 6x^4 + 240x^3$
44. $2x^3 - 42x^2 + 196x$
45. $10x^8 + 20x^7 - 990x^6$
46. $35x^4 - 70x^3 + 35x^2$
47. $x^2 + xy - 42y^2$
48. $3x^2 - 36xy - 84y^2$
49. $x^2 - 5xy - 36y^2$
50. $x^2 - xy - 30y^2$
51. $x^2y^2 + 17xy + 72$
52. $x^2y^2 - 9xy + 8$
53. $5x^5 + 50x^4y + 105x^3y^2$
54. $x^2y^2 + 11xy - 60$

Mixed Practice, 55–72

Factor completely using the appropriate technique. If the polynomial cannot be factored, write prime.

55. $x^4 + 9x^3 + 5x + 45$
56. $x^2 - 26x + 169$
57. $x^2 + x + 30$
58. $x^2 - 17x + 42$
59. $x^2 + 22x + 40$
60. $x^2 + 10x$
61. $x^2 + 40x + 400$
62. $x^2 + 4x - 60$
63. $4x^2 + 8x - 96$
64. $x^2 + 16x + 15$
65. $x^4 - 15x^3$
66. $x^2 + 2x - 120$
67. $x^2 - 16x + 63$
68. $x^2 - 7x - 78$
69. $x^2 + 10x - 25$
70. $x^3 - 12x^2 - 6x + 72$
71. $x^2 - 21x - 100$
72. $-x^2 + 17x - 60$

Find the missing value such that the given binomial is a factor of the given polynomial.

73. $x^2 + 10x + ?, x + 2$

74. $x^2 - 16x + ?, x - 7$

75. $x^2 + 11x - ?, x - 4$

76. $x^2 - 7x - ?, x + 17$

Quick Review Exercises

Section 6.2

Multiply.

1. $(2x + 5)(x + 4)$

2. $(3x - 8)(2x - 7)$

3. $(x + 6)(3x - 4)$

4. $(4x - 3)(2x + 9)$

6.3 Factoring Trinomials of the Form $ax^2 + bx + c$, Where $a \neq 1$

OBJECTIVES

1. **Factor a trinomial of the form $ax^2 + bx + c$, where $a \neq 1$, by grouping.**
2. **Factor a trinomial of the form $ax^2 + bx + c$, where $a \neq 1$, by trial and error.**

In this section, we continue to learn how to factor second-degree trinomials, focusing on trinomials with a leading coefficient that is not 1. Some examples of this type of trinomial are $2x^2 + 13x + 20$, $8x^2 - 22x - 21$, and $3x^2 - 11x + 8$.

We will examine two methods for factoring this type of trinomial: factoring by grouping and factoring by trial and error. Your instructor may prefer one of these methods to the other or an altogether different method and may ask you to use one method only. However, if you are allowed to use either method, use the one with which you are most comfortable.

Factoring by Grouping

Objective 1 Factor a trinomial of the form $ax^2 + bx + c$, where $a \neq 1$, by grouping. We begin with factoring by grouping. Consider the product $(3x + 2)(2x + 5)$.

$$\begin{aligned}(3x + 2)(2x + 5) &= 6x^2 + 15x + 4x + 10 && \text{Distribute.}\\ &= 6x^2 + 19x + 10 && \text{Combine like terms.}\end{aligned}$$

Because $(3x + 2)(2x + 5) = 6x^2 + 19x + 10$, we know that $6x^2 + 19x + 10$ can be factored as $(3x + 2)(2x + 5)$. By rewriting the middle term $19x$ as $15x + 4x$, we can factor by grouping. The important skill is being able to determine that $15x + 4x$ is the correct way to rewrite $19x$, instead of $16x + 3x$, $9x + 10x$, $22x - 3x$, or any other two terms whose sum is $19x$.

Factoring $ax^2 + bx + c$, Where $a \neq 1$, by Grouping

1. Multiply $a \cdot c$.
2. Find two integers with a product of $a \cdot c$ and a sum of b.
3. Rewrite the term bx as two terms, using the two integers found in Step 2.
4. Factor the resulting polynomial by grouping.

To factor the polynomial $6x^2 + 19x + 10$ by grouping, we begin by multiplying $6 \cdot 10$, which equals 60. Next, look for two integers with a product of 60 and a sum of 19; those two integers are 15 and 4. We can then rewrite $19x$ as $15x + 4x$. (These terms could be written in the opposite order and would still lead to the correct factoring.) We then factor the polynomial $6x^2 + 15x + 4x + 10$ by grouping.

$6x^2 + 19x + 10 = 6x^2 + 15x + 4x + 10$ — Rewrite $19x$ as $15x + 4x$.

$= 3x(2x + 5) + 2(2x + 5)$ — Factor the common factor $3x$ from the first two terms and the common factor 2 from the last two terms.

$= (2x + 5)(3x + 2)$ — Factor out the common factor $2x + 5$.

Now we will examine several examples using this technique.

EXAMPLE 1 Factor $2x^2 + 7x + 6$ by grouping.

SOLUTION We first check to see whether there are any common factors; in this case, there are not. We proceed with factoring by grouping. Because $2 \cdot 6 = 12$, we look for two integers with a product of 12 and a sum of 7, which is the coefficient of the middle term. The integers 3 and 4 meet those criteria, so we rewrite $7x$ as $3x + 4x$.

$2x^2 + 7x + 6 = 2x^2 + 3x + 4x + 6$ — Find two integers with a product of 12 and a sum of 7. Rewrite $7x$ as $3x + 4x$.

$= x(2x + 3) + 2(2x + 3)$ — Factor the common factor x from the first two terms and the common factor 2 from the last two terms.

$= (2x + 3)(x + 2)$ — Factor out the common factor $2x + 3$.

Quick Check 1

Factor $3x^2 + 13x + 12$ by grouping.

We check that this factoring is correct by multiplying $2x + 3$ by $x + 2$, which should equal $2x^2 + 7x + 6$. The check is left to the reader.

EXAMPLE 2 Factor $5x^2 - 7x + 2$ by grouping.

SOLUTION We begin by multiplying $5 \cdot 2$ as there are no common factors other than 1. We look for two integers with a product of 10 and a sum of -7. The integers are -5 and -2, so we rewrite $-7x$ as $-5x - 2x$ and then factor by grouping.

$5x^2 - 7x + 2 = 5x^2 - 5x - 2x + 2$ — Find two integers with a product of 10 and a sum of -7. Rewrite $-7x$ as $-5x - 2x$.

$= 5x(x - 1) - 2(x - 1)$ — Factor the common factor $5x$ from the first two terms and the common factor -2 from the last two terms.

$= (x - 1)(5x - 2)$ — Factor out the common factor $x - 1$.

Quick Check 2

Factor $4x^2 - 25x + 36$ by grouping.

EXAMPLE 3 Factor $12x^2 + 11x - 15$ by grouping.

SOLUTION Because the terms have no common factor other than 1, we begin by multiplying $12(-15)$, which equals -180. We need to find two integers with a product of -180 and a sum of 11. We can start by listing the different factors of 180. Because we know that one of the integers will be positive and one will be negative, we look for two factors in this list that have a difference of 11.

$1 \cdot 180 \quad 2 \cdot 90 \quad 3 \cdot 60 \quad 4 \cdot 45 \quad 5 \cdot 36 \quad 6 \cdot 30 \quad 9 \cdot 20 \quad 10 \cdot 18 \quad 12 \cdot 15$

Notice that the pair 9 and 20 has a difference of 11. The two integers are -9 and 20; their product is -180, and their sum is 11. We rewrite $11x$ as $-9x + 20x$ and then factor by grouping.

$12x^2 + 11x - 15 = 12x^2 - 9x + 20x - 15$	Find two integers with a product of -180 and a sum of 11. Rewrite $11x$ as $-9x + 20x$.
$= 3x(4x - 3) + 5(4x - 3)$	Factor the common factor $3x$ out of the first two terms and the common factor 5 out of the last two terms.
$= (4x - 3)(3x + 5)$	Factor out the common factor $4x - 3$.

Quick Check 3

Factor $16x^2 - 38x + 21$ by grouping.

Factoring by Trial and Error

Objective 2 Factor a trinomial of the form $ax^2 + bx + c$, where $a \neq 1$, by trial and error. We now turn our attention to the method of trial and error. We need to factor a trinomial $ax^2 + bx + c$ into the following form:

(Variable Term + Constant)(Variable Term + Constant)

The product of the variable terms will be ax^2, and the product of the constants will equal c. We will use these facts to get started by listing all of the factors of ax^2 and c.

Consider the trinomial $2x^2 + 7x + 6$, which has no common factors other than 1. We begin by listing all of the factors of $2x^2$ and 6.

Factors of $2x^2$: $x \cdot 2x$ Factors of 6: $1 \cdot 6$, $2 \cdot 3$

Because there is only one pair of factors with a product of $2x^2$, we know that if this trinomial is factorable, it will be of the form $(2x + ?)(x + ?)$. We will substitute the different factors of 6 in all possible orders in place of the question marks, satisfying the following products:

$2x^2$

$(2x + ?)(x + ?)$

6

We will keep substituting factors of 6 until the middle term of the product of the two binomials is $7x$. Rather than fully distributing for each trial, we need to look only at the two products shown in the following graphic:

$(2x + ?)(x + ?)$

When the sum of these two products is $7x$, we have found the correct factors.

We begin by using the factors 1 and 6, which leads to the binomial factors $(2x + 1)(x + 6)$. Because the middle term for these factors is $13x$, we have not found the correct factors. We then switch the positions of 1 and 6 and try again. However, when we multiply $(2x + 6)(x + 1)$, the middle term equals $8x$, not $7x$. We need to try another pairing, so we try the other factors of 6, which are 2 and 3. When we multiply $(2x + 3)(x + 2)$, the middle term is $7x$, so these are the correct factors.

$4x$

$(2x + 3)(x + 2)$

$3x$

Therefore, $2x^2 + 7x + 6 = (2x + 3)(x + 2)$.

EXAMPLE 4 Factor $10x^2 + 13x - 3$ by trial and error.

SOLUTION There are no common factors other than 1, so we begin by listing the factors of $10x^2$ and -3.

$$\text{Factors of } 10x^2\text{: } x \cdot 10x, \quad 2x \cdot 5x$$
$$\text{Factors of } -3\text{: } -1 \cdot 3, \quad -3 \cdot 1$$

We are looking for two factors that produce the middle term $13x$ when multiplied. We will begin by using x and $10x$ for the variable terms and -1 and 3 for the constants.

Factors	Middle Term
$(x - 1)(10x + 3)$	$-7x$
$(x + 3)(10x - 1)$	$29x$

Neither of these middle terms is equal to $13x$, so we have not found the correct factoring. Also, because neither middle term is the opposite of $13x$, switching the signs of the constants 1 and 3 will not lead to the correct factors either. So $(x + 1)(10x - 3)$ and $(x - 3)(10x + 1)$ are not correct. We switch to the pair $2x$ and $5x$.

Factors	Middle Term
$(2x - 1)(5x + 3)$	x
$(2x + 3)(5x - 1)$	$13x$

The second of these middle terms is the one we are looking for, so $10x^2 + 13x - 3 = (2x + 3)(5x - 1)$.

Quick Check 4

Factor $2x^2 + x - 28$ by trial and error.

Now we will try to factor another trinomial by trial and error using a trinomial with terms that have more factors.

EXAMPLE 5 Factor $12x^2 + 11x - 15$ by trial and error.

SOLUTION Because there are no common factors to factor out, we begin by listing the factors of $12x^2$ and -15.

$$\text{Factors of } 12x^2\text{: } x \cdot 12x, \quad 2x \cdot 6x, \quad 3x \cdot 4x$$
$$\text{Factors of } -15\text{: } -1 \cdot 15, \quad -15 \cdot 1, \quad -3 \cdot 5, \quad -5 \cdot 3$$

At first glance, this may seem like it will take a while, but we can shorten the process. For example, we may skip over any pairings resulting in a binomial factor in which both terms have a common factor other than 1, such as $3x - 3$. If the terms of a binomial factor had a common factor other than 1, the original polynomial also would have a common factor, but we know that $12x^2 + 11x - 15$ does not have a common factor that can be factored out.

If the pair of constants -1 and 15 does not produce the desired middle term and does not produce the opposite of the desired middle term, we do not need to use the pair of constants -15 and 1. Instead, we move on to the next pair of constants. Eventually, through trial and error, we find that $12x^2 + 11x - 15 = (3x + 5)(4x - 3)$.

Quick Check 5

Factor $20x^2 + 17x - 24$ by trial and error.

A WORD OF CAUTION If a trinomial has a leading coefficient other than 1, be sure to check whether the three terms have a common factor. If there is a common factor, we may be able to factor a trinomial using the techniques from Section 6.2.

EXAMPLE 6 Factor $8x^2 + 40x + 48$.

SOLUTION The three terms have a common factor of 8. After we factor out this common factor, the trinomial factor will be $x^2 + 5x + 6$, which is a trinomial with a leading coefficient of 1. We look for two integers with a product of 6 and a sum of 5. Those two integers are 2 and 3.

$$8x^2 + 40x + 48 = 8(x^2 + 5x + 6)$$ Factor out the common factor 8.

$$= 8(x + 2)(x + 3)$$ To factor $x^2 + 5x + 6$, we need to find two integers with a product of 6 and a sum of 5. The two integers are 2 and 3.

Quick Check 6

Factor $9x^2 + 27x - 162$.

Factoring out the common factor in the previous example makes factoring the trinomial more manageable. If we were to try factoring by grouping without factoring out the common factor, we would begin by multiplying 8 by 48, which equals 384. We would look for two integers with a product of 384 and a sum of 40, which would be challenging. If we tried to factor by trial and error, we would find that $8x^2$ has two pairs of factors $(x \cdot 8x, 2x \cdot 4x)$ but that 48 has several pairs of factors $(1 \cdot 48, 2 \cdot 24, 3 \cdot 16, 4 \cdot 12, 6 \cdot 8)$. Factoring by trial and error would be tedious at best.

BUILDING YOUR STUDY STRATEGY

Overcoming Math Anxiety, 3 Test Anxiety Many students confuse test anxiety with math anxiety. If you believe you can learn mathematics and you understand the material but you freeze up when taking tests, you may have test anxiety, especially if this is true in your other classes. Many colleges offer seminars or short-term classes about how to reduce test anxiety. Ask your counselor whether such a course would be appropriate for you.

Exercises 6.3

Vocabulary

1. A second-degree trinomial with a leading coefficient not equal to 1 can be written in the form ______________.
2. The two techniques for factoring trinomials of the form $ax^2 + bx + c$, where $a \neq 1$, are called ______________ and ______________.

Factor completely.

3. $5x^2 + 16x + 3$
4. $3x^2 + 16x + 5$
5. $4x^2 + 24x + 35$
6. $4x^2 + 4x - 3$
7. $6x^2 - 19x + 10$
8. $9x^2 + 3x - 20$
9. $14x^2 + 15x - 9$
10. $8x^2 - 34x + 21$
11. $12x^2 + 8x - 15$
12. $12x^2 + 28x - 5$
13. $12x^2 - 25x + 12$
14. $12x^2 + 13x - 35$
15. $2x^2 + 13x + 7$
16. $18x^2 + 93x - 16$
17. $3x^2 + 13x - 10$
18. $4x^2 + 18x - 15$
19. $12x^2 + 11x - 56$
20. $24x^2 + 55x - 24$
21. $16x^2 - 24x + 9$
22. $16x^2 + 66x - 27$
23. $15x^2 - 24x - 12$
24. $15x^2 + 32x - 60$
25. $16x^2 + 24x - 40$
26. $15x^2 + 45x - 150$
27. $16x^2 - 64x + 64$
28. $40x^2 - 100x - 60$
29. $15x^2 - 25x - 560$
30. $-12x^2 - 36x + 120$

Mixed Practice, 31–66

Factor completely using the appropriate technique. If the polynomial cannot be factored, write prime.

31. $x^2 + 14x + 13$
32. $x^2 - 10x - 11$
33. $18x - 9$
34. $3x^2 + 33x + 48$
35. $2x^2 - 7x + 6$
36. $x^2 + 30x + 225$
37. $3x^2 + 2x - 4$
38. $32x^2 - 20x - 25$
39. $x^5 - 6x^4$
40. $x^2 - 8x - 48$
41. $6x^2 - 48x$
42. $x^2 + 15x + 60$
43. $-x^2 - 4x + 77$
44. $2x^2 - 3x - 54$
45. $4x^2 - 7x - 10$
46. $3x^2 - 25x - 18$
47. $x^5 - 7x^3 + 4x^2 - 28$
48. $45x^3 + 30x^2 - 75x - 50$
49. $x^3 + 6x^2 - 15x - 90$
50. $x^2 + 25x + 156$
51. $x^2 - 24x + 144$
52. $x^2 - 3x + 2$
53. $9x^2 - 45x - 54$
54. $3x^3 - 30x^2 - 7x + 70$
55. $-5x^2 + 28x + 12$
56. $x^2 - 6x - 5$
57. $2x^2 - 15x - 27$
58. $x^2 + 8x + 16$
59. $x^2 + 12x - 45$
60. $3x^5 + 6x^4$
61. $-3x^2 - 12x + 96$
62. $x^2 - 17x + 60$
63. $15x^2 + 16x - 15$
64. $18x^2 + 51x + 8$
65. $2x^2 + 9x + 4$
66. $x^2 - 7x - 44$

Writing in Mathematics

Answer in complete sentences.

67. Compare the two factoring methods for trinomials of the form $ax^2 + bx + c$, where $a \neq 1$: factoring by grouping and factoring by trial and error. Which method do you prefer? Explain your reasoning.

68. Explain why it is a good idea to factor out a common factor from a trinomial of the form $ax^2 + bx + c$ before trying other factoring techniques.

Quick Review Exercises

Section 6.3

Multiply.

1. $(x + 8)(x - 8)$
2. $(4x + 3)(4x - 3)$
3. $(x + 8)(x^2 - 8x + 64)$
4. $(2x - 5)(4x^2 + 10x + 25)$

6.4 Factoring Special Binomials

OBJECTIVES

1. Factor a difference of squares.
2. Factor a difference of cubes.
3. Factor a sum of cubes.

Difference of Squares

Objective 1 Factor a difference of squares. Recall the special product $(a + b)(a - b) = a^2 - b^2$ from Section 5.4. The binomial $a^2 - b^2$ is a **difference of**

squares. Based on this special product, a difference of squares factors in the following manner:

Difference of Squares

$$a^2 - b^2 = (a + b)(a - b)$$

To identify a binomial as a difference of squares, we must verify that each term is a perfect square. Variable factors must have exponents that are multiples of 2 such as x^2, y^4, and a^6. Any constant also must be a perfect square. Here are the first ten perfect squares.

$$1^2 = 1 \quad 2^2 = 4 \quad 3^2 = 9 \quad 4^2 = 16 \quad 5^2 = 25$$
$$6^2 = 36 \quad 7^2 = 49 \quad 8^2 = 64 \quad 9^2 = 81 \quad 10^2 = 100$$

EXAMPLE 1 Factor $x^2 - 49$.

SOLUTION This binomial is a difference of squares, as it can be rewritten as $(x)^2 - (7)^2$. Rewriting the binomial in this form helps us use the formula $a^2 - b^2 = (a + b)(a - b)$.

$x^2 - 49 = (x)^2 - (7)^2$ Rewrite each term as a square.

$= (x + 7)(x - 7)$ Factor using the formula for a difference of squares, $a^2 - b^2 = (a + b)(a - b)$.

We can check our work by multiplying $x + 7$ by $x - 7$, which should equal $x^2 - 49$. The check is left to the reader.

Quick Check 1

Factor $x^2 - 25$.

Although we will continue to write each term as a perfect square, if you can factor a difference of squares without writing each term as a perfect square, do not think that you *must* rewrite each term as a perfect square before proceeding.

EXAMPLE 2 Factor $64a^{10} - 25b^4$.

SOLUTION The first term of this binomial is a square, as 64 is a square and the exponent for the variable factor a is even. The first term can be rewritten as $(8a^5)^2$. In a similar fashion, $25b^4$ can be rewritten as $(5b^2)^2$. This binomial is a difference of squares.

$64a^{10} - 25b^4 = (8a^5)^2 - (5b^2)^2$ Rewrite each term as a square.

$= (8a^5 + 5b^2)(8a^5 - 5b^2)$ Factor as a difference of squares.

Quick Check 2

Factor $81x^{12} - 100y^{16}$.

This binomial $8x^2 - 81$ is not a difference of squares because the coefficient of the first term (8) is not a square. Because there are no common factors other than 1, this binomial cannot be factored.

EXAMPLE 3 Factor $3x^2 - 108$.

SOLUTION Although this binomial is not a difference of squares (3 and 108 are not squares), the common factor of 3 can be factored out. After we factor out the common factor 3, we have $3x^2 - 108 = 3(x^2 - 36)$. The binomial in parentheses is a difference of squares and can be factored accordingly.

$3x^2 - 108 = 3(x^2 - 36)$ Factor out the common factor 3.

$= 3[(x)^2 - (6)^2]$ Rewrite each term in the binomial factor as a square.

$= 3(x + 6)(x - 6)$ Factor as a difference of squares.

Quick Check 3

Factor $7x^2 - 175$.

If a binomial is a **sum of squares**, it cannot be factored unless the two terms have a common factor. Some examples of a sum of squares are $x^2 + 49$, $4a^2 + 9b^2$, and $64a^{10} + 25b^4$.

Sum of Squares

A sum of squares, $a^2 + b^2$, cannot be factored.

Occasionally, after you have factored a difference of squares, one or more of the factors may still be factorable. For example, consider the binomial $x^4 - 81$, which factors to be $(x^2 + 9)(x^2 - 9)$. The first factor, $x^2 + 9$, is a sum of squares and cannot be factored. However, the second factor, $x^2 - 9$, is a difference of squares and can be factored further.

EXAMPLE 4 Factor $x^4 - 81$ completely.

SOLUTION This is a difference of squares, and we factor it accordingly. After we factor the binomial as a difference of squares, one of the binomial factors $(x^2 - 9)$ is a difference of squares and must be factored as well.

$$x^4 - 81 = (x^2)^2 - (9)^2 \quad \text{Rewrite each term as a perfect square.}$$
$$= (x^2 + 9)(x^2 - 9) \quad \text{Factor as a difference of squares.}$$
$$= (x^2 + 9)[(x)^2 - (3)^2] \quad \text{Rewrite each term in the binomial } x^2 - 9 \text{ as a square.}$$
$$= (x^2 + 9)(x + 3)(x - 3) \quad \text{Factor } x^2 - 9 \text{ as a difference of squares.}$$

Quick Check 4

Factor $x^4 - y^4$.

A WORD OF CAUTION When factoring a difference of squares, check the binomial factor containing a difference to see if it can be factored.

Difference of Cubes

Objective 2 Factor a difference of cubes. Another special binomial that can be factored is a **difference of cubes**. In this case, both terms are perfect cubes rather than squares. Here is the formula for factoring a difference of cubes.

Difference of Cubes

$$a^3 - b^3 = (a - b)(a^2 + ab + b^2)$$

Before proceeding, let's multiply $(a - b)(a^2 + ab + b^2)$ to show that it does equal $a^3 - b^3$.

$$(a - b)(a^2 + ab + b^2) = a \cdot a^2 + a \cdot ab + a \cdot b^2 - b \cdot a^2 - b \cdot ab - b \cdot b^2 \quad \text{Distribute.}$$
$$= a^3 + a^2b + ab^2 - a^2b - ab^2 - b^3 \quad \text{Multiply.}$$
$$= a^3 - b^3 \quad \text{Combine like terms.}$$

We see that the formula is correct.

For a term to be a cube, its variable factors must have exponents that are multiples of 3, such as x^3, y^6, and z^9. Also, each constant factor must be a perfect cube. Here are the first ten perfect cubes.

$1^3 = 1$ $\quad 2^3 = 8$ $\quad 3^3 = 27$ $\quad 4^3 = 64$ $\quad 5^3 = 125$

$6^3 = 216$ $\quad 7^3 = 343$ $\quad 8^3 = 512$ $\quad 9^3 = 729$ $\quad 10^3 = 1000$

We need to memorize the formula for factoring a difference of cubes. Some patterns can help us remember the formula. A difference of cubes, $a^3 - b^3$, has two factors: a binomial $(a - b)$ and a trinomial $(a^2 + ab + b^2)$. The binomial looks just like the difference of cubes without the cubes, including the sign between the terms. The signs between the three terms in the trinomial are addition signs. To find the actual terms in the trinomial factor, the following diagram may be helpful:

1st 2nd

$$a^3 - b^3 = (a - b)(a^2 + ab + b^2)$$

1st · 1st 2nd · 2nd

1st · 2nd

EXAMPLE 5 Factor $x^3 - 27$.

SOLUTION We begin by looking for common factors other than 1 that the two terms share, but there are none. This binomial is not a difference of squares, as the exponent of the variable term is not a multiple of 2 and the constant 27 is not a square. However, the binomial is a difference of cubes. We can rewrite the term x^3 as $(x)^3$, and we can rewrite 27 as $(3)^3$. Rewriting $x^3 - 27$ as $(x)^3 - (3)^3$ will help us identify the terms in the binomial and trinomial factors.

$x^3 - 27 = (x)^3 - (3)^3$ Rewrite each term as a perfect cube.

$= (x - 3)(x \cdot x + x \cdot 3 + 3 \cdot 3)$ Factor as a difference of cubes. The binomial factor is the same as $(x)^3 - (3)^3$ without the cubes. When x is treated as the 1st term in the binomial factor and 3 is treated as the 2nd term, the terms in the trinomial are 1st · 1st + 1st · 2nd + 2nd · 2nd.

$= (x - 3)(x^2 + 3x + 9)$ Simplify each term in the trinomial factor.

Quick Check 5

Factor $x^3 - 125$.

A WORD OF CAUTION A difference of cubes $x^3 - y^3$ cannot be factored as $(x - y)^3$.

EXAMPLE 6 Factor $y^{15} - 64$.

SOLUTION Although the number 64 is a square, this binomial is not a difference of squares because y^{15} is not a square. For a term to be a square, its variable factors must have exponents that are multiples of 2. This binomial is a difference of cubes. The exponent in the first term is a multiple of 3, and the number 64 is equal to 4^3.

$y^{15} - 64 = (y^5)^3 - (4)^3$ Rewrite each term as a perfect cube.

$= (y^5 - 4)(y^5 \cdot y^5 + y^5 \cdot 4 + 4 \cdot 4)$ Factor as a difference of cubes.

$= (y^5 - 4)(y^{10} + 4y^5 + 16)$ Simplify each term in the trinomial factor.

Quick Check 6

Factor $x^{12} - 8$.

A WORD OF CAUTION When factoring a difference of cubes $a^3 - b^3$, do not attempt to factor the trinomial factor $a^2 + ab + b^2$.

Sum of Cubes

Objective 3 Factor a sum of cubes. Unlike a sum of squares, a **sum of cubes** can be factored. Here is the formula.

Sum of Cubes

$$a^3 + b^3 = (a + b)(a^2 - ab + b^2)$$

Notice that the terms in the factors are the same as the factors in a *difference* of cubes, with the exception of some of their signs. When a sum of cubes is factored, the binomial factor is a sum rather than a difference. The sign of the middle term in the trinomial is negative rather than positive. The last term in the trinomial factor is positive, just as it was in the formula for a difference of cubes. The diagram shows the differences between the formula for a difference of cubes and a sum of cubes.

$$a^3 - b^3 = (a - b)(a^2 + ab + b^2)$$

$$\downarrow \qquad \downarrow \qquad \downarrow$$

$$a^3 + b^3 = (a + b)(a^2 - ab + b^2)$$

EXAMPLE 7 Factor $z^3 + 125$.

SOLUTION This binomial is a sum of cubes and can be rewritten as $(z)^3 + (5)^3$. Then we can factor the sum of cubes using the formula.

$$z^3 + 125 = (z)^3 + (5)^3 \quad \text{Rewrite each term as a perfect cube.}$$
$$= (z + 5)(z \cdot z - z \cdot 5 + 5 \cdot 5) \quad \text{Factor as a sum of cubes.}$$
$$= (z + 5)(z^2 - 5z + 25) \quad \text{Simplify each term in the trinomial factor.}$$

Quick Check 7

Factor $x^3 + 216$.

A WORD OF CAUTION A sum of cubes $x^3 + y^3$ cannot be factored as $(x + y)^3$.

We finish this section by summarizing the strategies for factoring binomials.

Factoring Binomials

- Factor out the GCF if there is a common factor other than 1.
- Determine whether both terms are perfect squares.
  If the binomial is a sum of squares, it cannot be factored.
  If the binomial is a difference of squares, factor it by using the formula $a^2 - b^2 = (a + b)(a - b)$. Keep in mind that some of the resulting factors can be differences of squares as well, which need to be factored further.
- If both terms are not perfect squares, determine whether both are perfect cubes.
  If the binomial is a difference of cubes, factor using the formula $a^3 - b^3 = (a - b)(a^2 + ab + b^2)$.
  If the binomial is a sum of cubes, factor using the formula $a^3 + b^3 = (a + b)(a^2 - ab + b^2)$.

BUILDING YOUR STUDY STRATEGY

Overcoming Math Anxiety, 4 Taking the Right Course Many students think they have math anxiety because they constantly feel as though they are in over their heads, when in fact they may not be prepared for their particular class. Taking a course without having the prerequisite skills is not a good idea, and success will be difficult to achieve. Your instructor will be able to advise you about whether you should be taking another course or whether you have the skills to be successful in the course you are currently taking.

Exercises 6.4

MyMathLab

PRACTICE

WATCH

DOWNLOAD

READ

REVIEW

Vocabulary

1. A binomial of the form $a^2 - b^2$ is a(n) ____________________.
2. A numerical factor is a(n) __________ if it can be written as n^2 for some integer n.
3. A variable factor is a perfect square if its exponents are __________.
4. A binomial of the form $a^2 + b^2$ is a(n) __________.
5. A binomial of the form $a^3 - b^3$ is a(n) __________.
6. A binomial of the form $a^3 + b^3$ is a(n) __________.

Factor completely. If the polynomial cannot be factored, write prime.

7. $x^2 - 64$
8. $x^2 - 81$
9. $x^2 - 100$
10. $x^2 - 1$
11. $25a^2 - 49$
12. $16a^2 - 9$
13. $x^2 - 20$
14. $x^2 - 27$
15. $a^2 - 16b^2$
16. $m^2 - 9n^2$
17. $25x^2 - 64y^2$
18. $36x^2 - 121y^2$
19. $16 - x^2$
20. $144 - x^2$
21. $4x^2 - 196$
22. $6x^2 - 486$
23. $x^2 + 25$
24. $x^2 + 1$
25. $5x^2 + 20$
26. $4x^2 + 100y^2$
27. $x^4 - 16$
28. $x^4 - 1$
29. $x^4 - 81y^2$
30. $x^8 - 64y^4$
31. $x^3 - 1$
32. $x^3 - y^3$
33. $x^3 - 8$
34. $a^3 - 216$
35. $1000 - y^3$
36. $343 - x^3$
37. $a^3 - 64b^3$
38. $m^3 - 343n^3$
39. $125x^3 - 8y^3$
40. $729x^3 - 1000y^3$
41. $x^3 + y^3$
42. $x^3 + 27$
43. $x^3 + 8$
44. $b^3 + 64$
45. $x^6 + 27$
46. $y^{12} + 125$
47. $64m^3 + n^3$
48. $x^3 + 343y^3$
49. $3x^3 - 375$
50. $7x^3 - 189$
51. $320a^3 + 135b^3$
52. $27a^7b^6 + a^4b^3$

Mixed Practice, 53–88

Factor completely using the appropriate technique. If the polynomial cannot be factored, write prime.

53. $x^2 - 16x + 64$
54. $3x^3 + 24x^2 + x + 8$
55. $x^2 + 16x + 28$
56. $4x^2 + 4x - 168$
57. $22x^4 - 55x^2$
58. $x^2 - 8x + 33$
59. $6x^2 - 5x - 25$
60. $-18x^2 - 69x + 12$
61. $x^5 - 16x^4 + 60x^3$
62. $x^2 + 18x + 77$
63. $-9x^2 - 225$
64. $25 - x^2$
65. $-6x^2 + 60x - 144$
66. $x^2 - 81x$
67. $x^2 - 10x - 39$
68. $x^2 + 4x - 140$
69. $25x^2 - 36y^8$
70. $x^2 + 3x - 40$
71. $2x^2 + x - 13$
72. $x^3 - 216y^3$
73. $x^2 - 64y^6$
74. $x^3 - 64y^6$
75. $8x^5 - 20x^4 - 52x^2$
76. $20x^2 - 61x + 36$
77. $5x^2 + 45$
78. $60x^2 - 470x + 900$
79. $x^2 + 12x + 35$
80. $343x^3 - 8y^6z^9$
81. $3x^2 - 13x - 38$
82. $2x^2 + 17x + 26$
83. $x^3 + 12x^2 - 6x - 72$
84. $x^2 + 16y^2$
85. $x^3 + 125y^3$
86. $125x^6 - 64y^9$
87. $4x^7 + 12x^6 + 9x^5$
88. $x^3 - 5x^2 + 2x - 10$

Determine which factoring technique is appropriate for the given polynomial.

89. ____ $8x^2 - 26x + 15$	**a.** factor out GCF
90. ____ $9a^2b^9 + 6a^4b^5 - 15a^6b^3$	**b.** factoring by grouping
91. ____ $x^2 + 121$	**c.** trinomial, leading coefficient of 1
92. ____ $x^2 - 3x - 54$	**d.** trinomial, leading coefficient other than 1
93. ____ $125x^3 - 216$	**e.** difference of squares
94. ____ $x^3 + 1331y^3$	**f.** sum of squares
95. ____ $x^3 - 3x^2 - 18x + 54$	**g.** difference of cubes
96. ____ $x^2 - 64$	**h.** sum of cubes

Writing in Mathematics

Answer in complete sentences.

97. Explain how to identify a binomial as a difference of squares.

98. Explain how to identify a binomial as a difference of cubes or as a sum of cubes.

6.5 Factoring Polynomials: A General Strategy

Objective 1 Understand the strategy for factoring a general polynomial. In this section, we will review the different factoring techniques introduced in this chapter and develop a general strategy for factoring any polynomial. Keep in mind that a polynomial must be factored completely, meaning each of its polynomial factors cannot be factored further. For example, if we factored the trinomial $4x^2 + 24x + 32$ to be $(2x + 4)(2x + 8)$, this would not be factored completely because the two terms in each binomial factor have a common factor of 2 that must be factored out as follows:

$$\begin{aligned} 4x^2 + 24x + 32 &= (2x + 4)(2x + 8) \\ &= 2(x + 2) \cdot 2(x + 4) \\ &= 4(x + 2)(x + 4) \end{aligned}$$

If we factor out all common factors as the first step when factoring a polynomial, we can avoid having to factor out common factors at the end of the process. We could have factored out the common factor of 4 from the polynomial $4x^2 + 24x + 32$ at the very beginning. In addition to ensuring that the polynomial has been factored completely, factoring out common factors often makes our work easier and occasionally allows us to factor polynomials that we would not have been able to factor otherwise.

Here is a general strategy for factoring any polynomial.

Factoring Polynomials

1. Factor out any common factors.
2. Determine the number of terms in the polynomial.
   - **(a)** If there are only *two terms*, check to see if the binomial is one of the special binomials discussed in Section 6.4.
     - Difference of Squares: $a^2 - b^2 = (a + b)(a - b)$
     - Sum of Squares: $a^2 + b^2$ is not factorable.
     - Difference of Cubes: $a^3 - b^3 = (a - b)(a^2 + ab + b^2)$
     - Sum of Cubes: $a^3 + b^3 = (a + b)(a^2 - ab + b^2)$
   - **(b)** If there are *three terms*, try to factor the trinomial using the techniques of Sections 6.2 and 6.3.
     - $x^2 + bx + c = (x + m)(x + n)$: find two integers m and n with a product of c and a sum of b.
     - $ax^2 + bx + c$, where $a \neq 1$: factor either by grouping or by trial and error.

       To use factoring by grouping, refer to objective 1 in Section 6.3.

       To factor by trial and error, refer to objective 2 in Section 6.3.
   - **(c)** If there are *four terms*, try factoring by grouping, discussed in Section 6.1.
3. After the polynomial has been factored, make sure that any factor with two or more terms does not have any common factors other than 1. If there are common factors, factor them out.
4. Check your factoring through multiplication.

We will now factor several polynomials of various forms. Some of the examples will have twists that we did not see in the previous sections. The focus will be on identifying the best technique.

EXAMPLE 1 Factor $-8x^2 - 80x + 192$ completely.

SOLUTION These three terms have a common factor of -8, and after we factor it out, we have $-8(x^2 + 10x - 24)$. This is a quadratic trinomial with a leading coefficient of 1, so we factor it using the method introduced in Section 6.2.

$$-8x^2 - 80x + 192 = -8(x^2 + 10x - 24) \quad \text{Factor out the GCF } -8.$$
$$= -8(x + 12)(x - 2) \quad \text{Find two integers with a product of } -24 \text{ and a sum of 10. The integers are 12 and } -2.$$

Quick Check 1

Factor $-6x^2 + 54x + 60$ completely.

EXAMPLE 2 Factor $343r^9 - 64s^6t^{12}$ completely.

SOLUTION There are no common factors other than 1, so we begin by noticing that this is a binomial. This is not a difference of squares, as $343r^9$ cannot be rewritten as a perfect square. It is, however, a difference of cubes. All of the variable factors have exponents that are multiples of 3, and the coefficients are perfect cubes.

(Recall the list of the first ten perfect cubes given in Section 6.4: $343 = 7^3$ and $64 = 4^3$.) We begin by rewriting each term as a perfect cube.

$$343r^9 - 64s^6t^{12} = (7r^3)^3 - (4s^2t^4)^3$$ Rewrite each term as a perfect cube.

$$= (7r^3 - 4s^2t^4)(7r^3 \cdot 7r^3 + 7r^3 \cdot 4s^2t^4 + 4s^2t^4 \cdot 4s^2t^4)$$ Factor as a difference of cubes. (Recall the pattern for the trinomial factor: 1st · 1st + 1st · 2nd + 2nd · 2nd.)

$$= (7r^3 - 4s^2t^4)(49r^6 + 28r^3s^2t^4 + 16s^4t^8)$$ Simplify each term in the trinomial factor.

For a difference or sum of cubes, check that the binomial factor cannot be factored further as one of the special binomials.

Quick Check 2

Factor $216x^{12}y^{21} + 125$ completely.

EXAMPLE 3 Factor $2x^3 - 22x^2 - 8x + 88$ completely.

SOLUTION The four terms have a common factor of 2, so we will factor out the GCF. Then we will try to factor by grouping. If we have trouble factoring the polynomial as written, we can rearrange the order of its terms.

$$2x^3 - 22x^2 - 8x + 88 = 2(x^3 - 11x^2 - 4x + 44)$$ Factor out the GCF.

$$= 2[x^2(x - 11) - 4(x - 11)]$$ Factor the common factor x^2 from the first two terms and factor the common factor -4 from the last two terms.

$$= 2(x - 11)(x^2 - 4)$$ Factor out the common factor $x - 11$.

Notice that the factor $x^2 - 4$ is a difference of squares and must be factored further.

$$2(x - 11)(x^2 - 4) = 2(x - 11)(x + 2)(x - 2)$$ Factor the difference of squares. $x^2 - 4 = (x + 2)(x - 2)$

Quick Check 3

Factor $x^4 + 5x^3 - 8x - 40$ completely.

If a binomial is both a difference of squares and a difference of cubes, such as $x^6 - 64$, to factor it completely, we must start by factoring it as a difference of squares. The next example illustrates this.

EXAMPLE 4 Factor $x^6 - 64$ completely.

SOLUTION There are no common factors to factor out, so we begin by factoring this binomial as a difference of squares.

$$x^6 - 64 = (x^3)^2 - (8)^2$$ Rewrite each term as a perfect square.

$$= (x^3 + 8)(x^3 - 8)$$ Factor as a difference of squares.

Notice that each factor can be factored further. The binomial $x^3 + 8$ is a sum of cubes, and the binomial $x^3 - 8$ is a difference of cubes. We use the fact that $x^3 + 8 = (x + 2)(x^2 - 2x + 4)$ and $x^3 - 8 = (x - 2)(x^2 + 2x + 4)$ to complete the factoring.

$$x^6 - 64 = (x + 2)(x^2 - 2x + 4)(x - 2)(x^2 + 2x + 4)$$

Note what would have occurred if we had first factored $x^6 - 64$ as a difference of cubes.

$$\begin{aligned} x^6 - 64 &= (x^2)^3 - (4)^3 && \text{Rewrite each term as a perfect cube.} \\ &= (x^2 - 4)(x^4 + 4x^2 + 16) && \text{Factor the difference of cubes.} \\ &= (x + 2)(x - 2)(x^4 + 4x^2 + 16) && \text{Factor the difference of squares.} \end{aligned}$$

At this point, we do not know how to factor $x^4 + 4x^2 + 16$, so the technique of Example 4 has factored the binomial $x^6 - 64$ more completely.

Quick Check 4

Factor $x^{12} - y^6$ completely.

EXAMPLE 5 Factor $x^{12} + 1$ completely.

SOLUTION The two terms do not have a common factor other than 1, so we need to determine whether this binomial matches one of the special forms. The binomial can be rewritten as a sum of cubes, $(x^4)^3 + (1)^3$.

$$\begin{aligned} x^{12} + 1 &= (x^4)^3 + (1)^3 && \text{Rewrite each term as a perfect cube.} \\ &= (x^4 + 1)(x^8 - x^4 + 1) && \text{Factor as a sum of cubes.} \end{aligned}$$

Quick Check 5

Factor $x^6 + 64$ completely.

EXAMPLE 6 Factor $15x^5 - 55x^4 + 40x^3$ completely.

SOLUTION The three terms have a common factor of $5x^3$, so we begin by factoring this out of the trinomial. After we have done this, we can factor the trinomial factor $3x^2 - 11x + 8$ using either of the techniques in Section 6.3. We will factor it by grouping, although you may prefer to use trial and error.

$$\begin{aligned} 15x^5 - 55x^4 + 40x^3 &= 5x^3(3x^2 - 11x + 8) && \text{Factor out the common factor } 5x^3. \\ &= 5x^3(3x^2 - 3x - 8x + 8) && \text{To factor by grouping, we need to find two integers whose product is equal to } 3 \cdot 8 = 24 \text{ and whose sum is } -11. \text{ The two integers are } -3 \text{ and } -8, \text{ so we rewrite the term } -11x \text{ as } -3x - 8x. \\ &= 5x^3[3x(x - 1) - 8(x - 1)] && \text{Factor the common factor } 3x \text{ from the first two terms and factor the common factor } -8 \text{ from the last two terms.} \\ &= 5x^3(x - 1)(3x - 8) && \text{Factor out the common factor } x - 1. \end{aligned}$$

Quick Check 6

Factor $8x^2 + 42x - 36$ completely.

BUILDING YOUR STUDY STRATEGY

Overcoming Math Anxiety, 5 Study Skills Many students who perform poorly in their math class attribute their performance to math anxiety, when, in reality, poor study skills are to blame. Make sure you are giving your best effort, including

- Working with a study group.
- Reading the text before the material is covered in class.
- Rereading the text after the material is covered in class.
- Completing each homework assignment.
- Making note cards for particularly difficult problems or procedures.
- Getting help if there is a problem you do not understand.
- Asking your instructor questions when you do not understand something.
- Seeing a tutor.

Exercises 6.5

MyMathLab  PRACTICE  WATCH  DOWNLOAD  READ  REVIEW

Vocabulary

1. The first step for factoring a trinomial is to factor out ______________, if there are any.
2. A binomial of the form $a^2 - b^2$ is a(n) ______________.
3. A binomial of the form $a^2 + b^2$ is a(n) ______________.
4. A binomial of the form $a^3 - b^3$ is a(n) ______________.
5. A binomial of the form $a^3 + b^3$ is a(n) ______________.
6. Factoring can be checked by __________ the factors.

Factor completely. If the polynomial cannot be factored, write prime.

7. $x^7 + 18x^5 - 36x$
8. $x^9 - 11x^7 + 37x^4$
9. $84x^2 + 4x - 20$
10. $12x + 20$
11. $a^5b^7 - 3a^4b^6 + 8a^2b^9$
12. $9m^7n^6 + 12m^{10}n^4 + 21m^3n^8$
13. $x^3 + 6x^2 + 5x + 30$
14. $x^3 - 9x^2 + 10x - 90$
15. $x^3 - 7x^2 - 3x + 21$
16. $x^3 + 13x^2 - 11x - 143$
17. $x^2 - 8x - 65$
18. $x^2 - 18x + 45$
19. $x^2 + 13x + 42$
20. $x^2 - 10x + 20$
21. $x^2 + 4x - 165$
22. $x^2 + 13x - 14$
23. $x^2 + 17x - 30$
24. $x^2 - 6x - 280$
25. $x^2 - 23x + 42$
26. $x^2 - 3x - 154$
27. $x^2 + 20x + 100$
28. $x^2 - 14x + 49$
29. $x^2 + 7xy - 60y^2$
30. $x^2 - 12xy + 32y^2$
31. $x^2y^2 + 17xy + 16$
32. $x^2y^2 + 10xy - 11$
33. $4x^2 + 56x + 192$
34. $9x^2 + 54x + 72$
35. $3x^2 - 42x + 135$
36. $4x^2 + 4x - 224$
37. $-10x^2 + 70x + 180$
38. $-6x^2 - 6x + 72$
39. $3x^2 + 10x - 48$
40. $6x^2 - 29x - 5$
41. $21x^2 + 31x + 4$
42. $6x^2 - 13x + 6$
43. $2x^2 + 7x + 14$
44. $6x^2 + 11x - 10$
45. $12x^2 - 41x + 22$
46. $2x^2 - 3x - 12$
47. $9x^2 - 30x + 25$
48. $4x^2 - 28x + 49$
49. $18x^2 - 60x + 42$
50. $6x^2 + 75x + 225$
51. $x^2 - 64$
52. $x^2 - 49$
53. $x^8 - 16y^6$
54. $49x^4 - 81y^{10}$
55. $x^4 - 36x^2$
56. $5x^2 - 125$
57. $x^4 + 81y^2$
58. $16x^{10} + 49y^6$
59. $x^3 - 1000$
60. $x^{21} - 1$
61. $27x^3 - 8y^3$
62. $x^6 - 64y^{15}$
63. $x^3 + 216$
64. $x^{15} + 8$
65. $x^6 + 729y^3$
66. $27x^9 + 125y^{15}$
67. $7x^3 + 189$
68. $3x^3 - 648$
69. $x^3 - 9x^2 - 4x + 36$
70. $x^3 + 5x^2 - 36x - 180$
71. $4x^3 - 28x^2 - x + 7$
72. $9x^3 + 18x^2 - 4x - 8$
73. $x^4 + 16x^3 + x + 16$
74. $x^4 + 7x^3 + 8x + 56$
75. $x^8 - 1$
76. $x^4 - y^4$

77. $x^6 - y^6$

78. $x^6 - 1$

79. $(x + 4)^2 - y^2$

80. $(x - 7)^2 - y^2$

81. $(x^2 - 10x + 25) - y^2$

82. $(x^2 + 16x + 64) - y^2$

83. $x^2 + 6x + 9 - y^2$

84. $x^2 - 20x + 100 - y^2$

Mixed Practice, 85–116

Factor completely. If the polynomial cannot be factored, write prime.

85. $4x^2 + 8x - 5$

86. $x^2 - 100$

87. $x^3 - 6x^2 + 8x - 48$

88. $x^3 + 7x^2 - 9x - 63$

89. $x^2 + 11x - 60$

90. $4x^2 + 28x + 48$

91. $x^2 + 13x - 36$

92. $2x^2 + 17x + 18$

93. $8x^3 + 343$

94. $25x^4 - 49z^2$

95. $x^2 + 21x + 98$

96. $6x^2 + 23x + 15$

97. $16x^4 - 81$

98. $x^2 - 3x - 180$

99. $6x^2 + 78x + 240$

100. $x^4 - 1296$

101. $9x^2 - 121y^6$

102. $x^2 + 15xy - 54y^2$

103. $x^3 + 4x^2 - 25x - 100$

104. $2x^3 - 8x^2 - 3x + 12$

105. $729x^3 - 512y^3$

106. $x^3 + 1331$

107. $x^2y^2 - 15xy + 54$

108. $x^2 - 2x + 48$

109. $3x^2 - 28x - 20$

110. $6x^2 - 35x + 49$

111. $2x^2 + 19x + 25$

112. $8x^2 + 24x + 32$

113. $5x^5 + 20x^3 - 35x^2$

114. $x^2 - 25x + 144$

115. $64x^2 - 169$

116. $27x^3 - 512$

Writing in Mathematics

Answer in complete sentences.

117. ***Newsletter*** Write a newsletter explaining the following strategies for factoring polynomials. Include a brief example of each.
- Factoring out the GCF
- Factoring by grouping
- Factoring trinomials of the form $x^2 + bx + c$
- Factoring polynomials of the form $ax^2 + bx + c$, where $a \neq 1$, by grouping
- Factoring polynomials of the form $ax^2 + bx + c$, where $a \neq 1$, by trial and error
- Factoring a difference of squares
- Factoring a difference of cubes
- Factoring a sum of cubes

Quick Review Exercises

Section 6.5

Solve.

1. $x - 7 = 0$
2. $x + 3 = 0$
3. $5x - 12 = 0$
4. $2x + 29 = 0$

6.6 Solving Quadratic Equations by Factoring

OBJECTIVES

1. Solve an equation by using the zero-factor property of real numbers.
2. Solve a quadratic equation by factoring.
3. Solve a quadratic equation that is not in standard form.
4. Solve a quadratic equation with coefficients that are fractions.
5. Find a quadratic equation, given its solutions.

Quadratic Equations

A **quadratic equation** is an equation that can be written as $ax^2 + bx + c = 0$, where a, b, and c are real numbers and $a \neq 0$. This form is the **standard form of a quadratic equation**.

We have already learned to solve linear equations ($ax + b = 0$). The difference between these two types of equations is that a quadratic equation has a second-degree term, ax^2. Because of the second-degree term, the techniques used to solve linear equations do not work for quadratic equations.

The Zero-Factor Property of Real Numbers

Objective 1 Solve an equation by using the zero-factor property of real numbers. To solve a quadratic equation, we will use the **zero-factor property of real numbers**.

Zero-Factor Property of Real Numbers

If $a \cdot b = 0$, then $a = 0$ or $b = 0$.

The principle behind this property is that if two or more unknown numbers have a product of zero, then at least one of the numbers must be zero. This property holds true only when the product is equal to 0, not for any other numbers.

EXAMPLE 1 Use the zero-factor property to solve the equation $(x + 3)(x - 7) = 0$.

SOLUTION In this example, the product of two unknown numbers, $x + 3$ and $x - 7$, is equal to 0. The zero-factor property says that either $x + 3 = 0$ or $x - 7 = 0$. Essentially, we have taken an equation that we do not know how to solve, $(x + 3)(x - 7) = 0$, and have rewritten it as two linear equations that we do know how to solve.

$$(x + 3)(x - 7) = 0$$

$x + 3 = 0$ or $x - 7 = 0$ Set each factor equal to 0.

$x = -3$ or $x = 7$ Solve each linear equation.

There are two solutions to this equation: -3 and 7. We write these solutions in a solution set as $\{-3, 7\}$.

EXAMPLE 2 Use the zero-factor property to solve the equation $x(3x - 5) = 0$.

SOLUTION Applying the zero-factor property gives us the equations $x = 0$ and $3x - 5 = 0$. The first equation, $x = 0$, is already solved, so we need to solve only the equation $3x - 5 = 0$ to find the second solution.

$$x(3x - 5) = 0$$

$x = 0$ or $3x - 5 = 0$ Set each factor equal to 0.

$x = 0$ or $3x = 5$ Solve each linear equation.

$x = 0$ or $x = \frac{5}{3}$

The solution set for this equation is $\{0, \frac{5}{3}\}$.

Quick Check 1

Use the zero-factor property to solve the equation.

a) $(x - 2)(x + 8) = 0$
b) $x(4x + 9) = 0$

If an equation has the product of more than two factors equal to 0, such as $(x + 4)(x + 1)(x - 2) = 0$, we set each factor equal to 0 and solve. The solution set to this equation is $\{-4, -1, 2\}$.

Solving Quadratic Equations by Factoring

Objective 2 Solve a quadratic equation by factoring. To solve a quadratic equation, we will use the following procedure:

Solving Quadratic Equations by Factoring

1. Write the equation in standard form: $ax^2 + bx + c = 0$. *We need to collect all of the terms on one side of the equation. It helps to collect all of the terms so that the coefficient of the squared term is positive.*
2. Factor the expression $ax^2 + bx + c$ completely. *If you are struggling with factoring, refer back to Sections 6.1–6.5.*
3. Set each factor equal to 0 and solve the resulting equations. *Each of these equations should be a linear equation.*
4. Finish by checking the solutions. *This is an excellent opportunity to catch mistakes in factoring.*

EXAMPLE 3 Solve $x^2 + 7x - 30 = 0$.

SOLUTION This equation is already in standard form, so we begin by factoring the expression $x^2 + 7x - 30$.

$$x^2 + 7x - 30 = 0$$

$$(x + 10)(x - 3) = 0 \qquad \text{Factor.}$$

$$x + 10 = 0 \quad \text{or} \quad x - 3 = 0 \qquad \text{Set each factor equal to 0.}$$

$$x = -10 \quad \text{or} \quad x = 3 \qquad \text{Solve each equation.}$$

The solution set is $\{-10, 3\}$. Now we will check the solutions.

Check ($x = -10$)

$$x^2 + 7x - 30 = 0$$
$$(-10)^2 + 7(-10) - 30 = 0$$
$$100 + 7(-10) - 30 = 0$$
$$100 - 70 - 30 = 0$$
$$0 = 0$$

Check ($x = 3$)

$$x^2 + 7x - 30 = 0$$
$$(3)^2 + 7(3) - 30 = 0$$
$$9 + 7(3) - 30 = 0$$
$$9 + 21 - 30 = 0$$
$$0 = 0$$

The check shows that the solutions are correct.

Quick Check 2

Solve $x^2 - x - 20 = 0$.

A WORD OF CAUTION Pay close attention to the instructions to a problem. If you are asked to solve a quadratic equation, do not just factor the quadratic expression and stop. If you are asked to factor a quadratic expression, do not set each factor equal to 0 and solve the resulting equations.

EXAMPLE 4 Solve $3x^2 - 18x - 48 = 0$.

SOLUTION The equation is in standard form. To factor $3x^2 - 18x - 48$, begin by factoring out the common factor 3.

$$3x^2 - 18x - 48 = 0$$

$$3(x^2 - 6x - 16) = 0 \quad \text{Factor out the GCF.}$$

$$3(x - 8)(x + 2) = 0 \quad \text{Factor } x^2 - 6x - 16.$$

$$x - 8 = 0 \quad \text{or} \quad x + 2 = 0 \quad \text{Set each factor containing a variable equal to 0. You can ignore the numerical factor 3.}$$

$$x = 8 \quad \text{or} \quad x = -2 \quad \text{Solve each equation.}$$

The solution set is $\{-2, 8\}$. The check is left to the reader.

We do not need to set a numerical factor equal to 0 because such an equation will not have a solution. For example, the equation $3 = 0$ has no solution. The zero-factor property says that at least one of the three factors must equal 0; we just know that it cannot be the factor 3 that is equal to 0.

Quick Check 3

Solve $4x^2 + 60x + 224 = 0$.

A WORD OF CAUTION A common numerical factor does not affect the solutions of a quadratic equation.

EXAMPLE 5 Solve $2x^2 - 7x - 15 = 0$.

SOLUTION When the expression $2x^2 - 7x - 15$ is factored, there is no common factor other than 1 that can be factored out. So the leading coefficient of this trinomial is not 1. We can factor by trial and error or by grouping. We will use factoring by grouping. We look for two integers whose product is equal to $(2)(-15)$, or -30, and whose sum is -7. The two integers are -10 and 3, so we can rewrite the term and then factor by grouping. (Refer to Section 6.3 to review this technique.)

$$2x^2 - 7x - 15 = 0$$

$$2x^2 - 10x + 3x - 15 = 0 \quad \text{Rewrite } -7x \text{ as } -10x + 3x.$$

$$2x(x - 5) + 3(x - 5) = 0 \quad \text{Factor the common factor } 2x \text{ from the first two terms and factor the common factor 3 from the last two terms.}$$

$$(x - 5)(2x + 3) = 0 \quad \text{Factor out the common factor } x - 5.$$

$$x - 5 = 0 \quad \text{or} \quad 2x + 3 = 0 \quad \text{Set each factor equal to 0.}$$

$$x = 5 \quad \text{or} \quad 2x = -3 \quad \text{Solve each equation.}$$

$$x = 5 \quad \text{or} \quad x = -\frac{3}{2}$$

The solution set is $\left\{-\frac{3}{2}, 5\right\}$. The check is left to the reader.

Quick Check 4

Solve $6x^2 - 23x + 7 = 0$.

Objective 3 Solve a quadratic equation that is not in standard form. We now turn our attention to equations that are not already in standard form. In each of the examples that follows, the check of the solutions is left to the reader.

EXAMPLE 6 Solve $x^2 = 49$.

SOLUTION Begin by rewriting this equation in standard form. This can be done by subtracting 49 from each side of the equation. Once this has been done, the expression to be factored is a difference of squares.

$$x^2 = 49$$
$$x^2 - 49 = 0 \quad \text{Subtract 49.}$$
$$(x + 7)(x - 7) = 0 \quad \text{Factor.}$$
$$x + 7 = 0 \quad \text{or} \quad x - 7 = 0 \quad \text{Set each factor equal to 0.}$$
$$x = -7 \quad \text{or} \quad x = 7 \quad \text{Solve each equation.}$$

The solution set is $\{-7, 7\}$.

EXAMPLE 7 Solve $x^2 + 25 = 10x$.

SOLUTION To rewrite this equation in standard form, we need to subtract $10x$ so that all terms will be on the left side of the equation. When we subtract $10x$, we must write the terms in descending order.

$$x^2 + 25 = 10x$$
$$x^2 - 10x + 25 = 0 \quad \text{Subtract } 10x.$$
$$(x - 5)(x - 5) = 0 \quad \text{Factor.}$$
$$x - 5 = 0 \quad \text{or} \quad x - 5 = 0 \quad \text{Set each factor equal to 0.}$$
$$x = 5 \quad \text{or} \quad x = 5 \quad \text{Solve each equation.}$$

Notice that both solutions are identical. In this case, we need to write the repeated solution only once. The solution set is $\{5\}$.

Quick Check 5

Solve $x^2 + 4x = 45$.

Occasionally, we need to simplify one or both sides of an equation to rewrite the equation in standard form. For instance, to solve the equation $x(x + 9) = 10$, we must first multiply x by $x + 9$. You may be wondering why we would want to multiply out the left side because it is already factored. Although it is factored, the product is equal to 10, not 0.

EXAMPLE 8 Solve $x(x + 9) = 10$.

SOLUTION As mentioned, first multiply x by $x + 9$. Then rewrite the equation in standard form.

$$x(x + 9) = 10$$
$$x^2 + 9x = 10 \quad \text{Multiply.}$$
$$x^2 + 9x - 10 = 0 \quad \text{Subtract 10.}$$
$$(x + 10)(x - 1) = 0 \quad \text{Factor.}$$
$$x + 10 = 0 \quad \text{or} \quad x - 1 = 0 \quad \text{Set each factor equal to 0.}$$
$$x = -10 \quad \text{or} \quad x = 1 \quad \text{Solve each equation.}$$

The solution set is $\{-10, 1\}$.

Quick Check 6

Solve $x(x - 3) = 70$.

A WORD OF CAUTION Make sure the equation you are solving is written as a product equal to 0 before you set each factor equal to 0 and solve.

Objective 4 Solve a quadratic equation with coefficients that are fractions. If an equation contains fractions, clearing those fractions makes it easier to factor the quadratic expression. We can clear the fractions by multiplying both sides of the equation by the LCM of the denominators.

EXAMPLE 9 Solve $\frac{1}{6}x^2 + x + \frac{4}{3} = 0$.

SOLUTION The LCM for these two denominators is 6, so we can clear the fractions by multiplying both sides of the equation by 6.

$$\frac{1}{6}x^2 + x + \frac{4}{3} = 0$$

$$6 \cdot \left(\frac{1}{6}x^2 + x + \frac{4}{3}\right) = 6 \cdot 0$$ Multiply both sides by 6, which is the LCM of the denominators.

$$\overset{1}{\cancel{6}} \cdot \frac{1}{\underset{1}{\cancel{6}}}x^2 + 6 \cdot x + \overset{2}{\cancel{6}} \cdot \frac{4}{\underset{1}{\cancel{3}}} = 6 \cdot 0$$ Distribute and divide out common factors.

$$x^2 + 6x + 8 = 0$$ Multiply.

$$(x + 2)(x + 4) = 0$$ Factor.

$$x + 2 = 0 \quad \text{or} \quad x + 4 = 0$$ Set each factor equal to 0.

$$x = -2 \quad \text{or} \quad x = -4$$ Solve each equation.

The solution set is $\{-4, -2\}$.

Quick Check 7

Solve $\frac{2}{45}x^2 + \frac{4}{15}x - \frac{6}{5} = 0$.

Finding a Quadratic Equation, Given Its Solutions

Objective 5 Find a quadratic equation, given its solutions. If we know the two solutions to a quadratic equation, we can determine an equation with these solutions. The next example illustrates this process.

EXAMPLE 10 Find a quadratic equation in standard form, with integer coefficients, that has the solution set $\{-9, 4\}$.

SOLUTION We know that $x = -9$ is a solution to the equation. This says that $x + 9$ is a factor of the quadratic expression. Similarly, knowing that $x = 4$ is a solution tells us that $x - 4$ is a factor of the quadratic expression. Multiplying these two factors will give us a quadratic equation with these two solutions.

$$x = -9 \quad \text{or} \quad x = 4$$ Begin with the solutions.

$$x + 9 = 0 \quad \text{or} \quad x - 4 = 0$$ Rewrite each equation so the right side is equal to 0.

$$(x + 9)(x - 4) = 0$$ Write an equation that has these two expressions as factors.

$$x^2 + 5x - 36 = 0$$ Multiply.

A quadratic equation that has the solution set $\{-9, 4\}$ is $x^2 + 5x - 36 = 0$.

Quick Check 8

Find a quadratic equation in standard form, with integer coefficients, that has the solution set $\{-6, 8\}$.

Notice that we say *a* quadratic equation rather than *the* quadratic equation. There are infinitely many quadratic equations with integer coefficients that have this solution set. For example, multiplying both sides of the equation by 2 gives the equation $2x^2 + 10x - 72 = 0$, which has the same solution set.

BUILDING YOUR STUDY STRATEGY

Overcoming Math Anxiety, 6 Relaxation Learning to relax in stressful situations will keep your anxiety under control. By relaxing, you will be able to avoid the physical symptoms associated with math anxiety, such as feeling tense and uncomfortable, sweating, feeling short of breath, and experiencing an accelerated heartbeat.

Many students use deep-breathing exercises. Close your eyes and take a deep breath. Try to clear your mind and relax your entire body. Slowly exhale; then repeat the process until you feel relaxed.

Exercises 6.6

MyMathLab

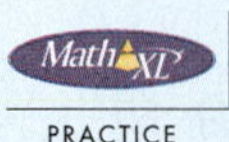
PRACTICE

WATCH

DOWNLOAD

READ

REVIEW

Vocabulary

1. A(n) ______________ is an equation that can be written as $ax^2 + bx + c = 0$, where a, b, and c are real numbers and $a \neq 0$.
2. A quadratic equation is in standard form if it is in the form ______________, where $a \neq 0$.
3. The __________ property of real numbers states that if $a \cdot b = 0$, then $a = 0$ or $b = 0$.
4. Once a quadratic expression has been set equal to 0 and factored you set each ______ equal to 0 and solve.

Solve.

5. $(x + 7)(x - 100) = 0$
6. $(x - 2)(x - 10) = 0$
7. $x(x - 12) = 0$
8. $x(x + 5) = 0$
9. $7(x + 8)(x - 6) = 0$
10. $-4(x + 5)(x - 12) = 0$
11. $(x + 7)(3x + 4) = 0$
12. $(x - 3)(5x - 22) = 0$
13. $x^2 - 11x + 18 = 0$
14. $x^2 - 13x + 42 = 0$
15. $x^2 + 15x + 44 = 0$
16. $x^2 + 14x + 48 = 0$
17. $x^2 + 6x - 40 = 0$
18. $x^2 - 3x - 88 = 0$
19. $x^2 - 12x - 45 = 0$
20. $x^2 + 7x - 44 = 0$
21. $x^2 - 16x + 64 = 0$
22. $x^2 + 12x + 36 = 0$
23. $x^2 - 81 = 0$
24. $x^2 - 4 = 0$
25. $x^2 + 7x = 0$
26. $x^2 - 10x = 0$
27. $5x^2 - 22x = 0$
28. $14x^2 + 13x = 0$
29. $3x^2 - 3x - 270 = 0$
30. $4x^2 + 16x - 308 = 0$
31. $-x^2 + 12x - 35 = 0$
32. $-x^2 - 4x + 12 = 0$
33. $3x^2 + 22x - 16 = 0$
34. $2x^2 - 17x + 35 = 0$
35. $27x^2 - 3x - 2 = 0$
36. $6x^2 + 7x - 68 = 0$
37. $x^2 - 2x = 35$
38. $x^2 + 8x = -15$
39. $x^2 = 8x - 16$
40. $x^2 - 13x = 48$
41. $x^2 + 12x = 3x - 18$
42. $x^2 + 11x + 20 = 10x + 76$
43. $x^2 - 3x - 7 = 4x + 11$
44. $2x^2 + 7x - 25 = x^2 + 23x - 40$
45. $x^2 = 9$
46. $x^2 = 25$
47. $x^2 + 11x = 11x + 36$
48. $x^2 - 3x = -3x + 100$
49. $x(x + 4) = 4(x + 16)$
50. $x(x + 7) = (4x + 3) + (3x + 13)$

51. $x(x - 7) = 30$

52. $x(x + 4) = 96$

53. $(x - 5)(x + 6) = -18$

54. $(x + 7)(x + 2) = 84$

55. $(x + 2)(x + 3) = (x + 7)(x - 4)$

56. $(x - 6)(x - 4) = (x + 12)(x - 10)$

57. $\frac{1}{6}x^2 - \frac{3}{2}x + 3 = 0$

58. $\frac{1}{8}x^2 + \frac{1}{4}x - 6 = 0$

59. $\frac{1}{6}x^2 - \frac{5}{4}x - \frac{9}{4} = 0$

60. $\frac{1}{2}x^2 + \frac{11}{12}x + \frac{1}{3} = 0$

61. $\frac{2}{15}x^2 - \frac{7}{10}x + \frac{2}{3} = 0$

62. $\frac{3}{4}x^2 - \frac{3}{4}x - \frac{5}{6} = 0$

Find a quadratic equation with integer coefficients that has the given solution set.

63. $\{4, 5\}$

64. $\{-10, 7\}$

65. $\{0, 6\}$

66. $\{4\}$

67. $\{-5, 5\}$

68. $\left\{\frac{3}{4}, 6\right\}$

69. $\left\{-\frac{2}{5}, \frac{1}{2}\right\}$

70. $\left\{-\frac{7}{3}, -\frac{3}{5}\right\}$

71. Use the fact that $x = \frac{7}{2}$ is a solution to the equation $6x^2 + 7x - 98 = 0$ to find the other solution to the equation.

72. Use the fact that $x = \frac{13}{3}$ is a solution to the equation $3x^2 - 28x + 65 = 0$ to find the other solution to the equation.

For the given quadratic equation, one of its solutions has been provided. Use it to find the other solution.

73. $x^2 + 6x - 667 = 0, x = 23$

74. $x^2 - 78x + 1517 = 0, x = 41$

75. $15x^2 + 11x - 532 = 0, x = -\frac{19}{3}$

76. $48x^2 - 154x - 735 = 0, x = \frac{35}{6}$

Writing in Mathematics

Answer in complete sentences.

77. Explain the zero-factor property of real numbers. Describe how this property is used when solving quadratic equations by factoring.

78. Explain why the common factor of 2 has no effect on the solutions of the equation $2(x - 7)(x + 5) = 0$.

79. ***Solutions Manual*** Write a solutions manual page for the following problem:

Solve $\frac{1}{6}x^2 - \frac{1}{3}x - 4 = 0$.

Quick Review Exercises

Section 6.6

Evaluate the given function.

1. $f(x) = 3x + 8, f(5)$

2. $f(x) = -10x + 21, f(-8)$

3. $f(x) = -7x - 19, f(2b + 3)$

4. $f(x) = x^2 + 7x - 20, f(6)$

6.7 Quadratic Functions

OBJECTIVES

1. Evaluate quadratic functions.
2. Solve equations involving quadratic functions.
3. Solve applied problems involving quadratic functions.

We first investigated linear functions in Chapter 3. A linear function is a function of the form $f(x) = mx + b$. In this section, we turn our attention to **quadratic functions**.

Quadratic Functions

A quadratic function is a function of the form $f(x) = ax^2 + bx + c$, where $a \neq 0$.

Evaluating Quadratic Functions

Objective 1 Evaluate quadratic functions. We begin our investigation of quadratic functions by learning to evaluate these functions for particular values of the variable. For example, if we are asked to find $f(3)$, we are being asked to evaluate the function $f(x)$ at $x = 3$. To do this, we substitute 3 for the variable x and simplify the resulting expression.

EXAMPLE 1 For the function $f(x) = x^2 - 5x - 8$, find $f(-4)$.

SOLUTION Substitute -4 for the variable x and then simplify.

$$
\begin{aligned}
f(-4) &= (-4)^2 - 5(-4) - 8 && \text{Substitute } -4 \text{ for } x.\\
&= 16 - 5(-4) - 8 && \text{Square } -4.\ (-4)(-4) = 16\\
&= 16 + 20 - 8 && \text{Multiply.}\\
&= 28 && \text{Simplify.}
\end{aligned}
$$

Quick Check 1

For the function $f(x) = x^2 - 9x + 405$, find $f(-6)$.

Solving Equations Involving Quadratic Functions

Objective 2 Solve equations involving quadratic functions. Now that we have learned to solve quadratic equations, we can solve equations involving quadratic functions. Suppose we were trying to find all values x for which some function $f(x)$ was equal to 0. We replace $f(x)$ with its formula and solve the resulting quadratic equation.

EXAMPLE 2 Let $f(x) = x^2 + 11x - 26$. Find all values x for which $f(x) = 0$.

SOLUTION Begin by replacing the function with its formula; then solve the resulting equation.

$$
\begin{aligned}
f(x) &= 0 &&\\
x^2 + 11x - 26 &= 0 && \text{Replace } f(x) \text{ with its formula } x^2 + 11x - 26.\\
(x + 13)(x - 2) &= 0 && \text{Factor.}\\
x + 13 = 0 \quad &\text{or} \quad x - 2 = 0 && \text{Set each factor equal to 0.}\\
x = -13 \quad &\text{or} \quad x = 2 && \text{Solve each equation.}
\end{aligned}
$$

The two values x for which $f(x) = 0$ are -13 and 2.

Quick Check 2

Let $f(x) = x^2 - 17x + 72$. Find all values x for which $f(x) = 0$.

EXAMPLE 3 Let $f(x) = x^2 - 24$. Find all values x for which $f(x) = 25$.

SOLUTION After setting the function equal to 25, we need to rewrite the equation in standard form in order to solve it.

$$f(x) = 25$$

$$x^2 - 24 = 25 \quad \text{Replace } f(x) \text{ with its formula.}$$

$$x^2 - 49 = 0 \quad \text{Subtract 25.}$$

$$(x + 7)(x - 7) = 0 \quad \text{Factor.}$$

$$x + 7 = 0 \quad \text{or} \quad x - 7 = 0 \quad \text{Set each factor equal to 0.}$$

$$x = -7 \quad \text{or} \quad x = 7 \quad \text{Solve each equation.}$$

The two values x for which $f(x) = 25$ are -7 and 7.

Quick Check 3

Let $f(x) = x^2 - 6x + 20$. Find all values x for which $f(x) = 92$.

Applications

Objective 3 Solve applied problems involving quadratic functions. We conclude the section with an applied problem that requires interpreting the graph of a quadratic function, a U-shaped curve called a **parabola**. Here are some examples. Note that these graphs are not linear.

$f(x) = x^2$

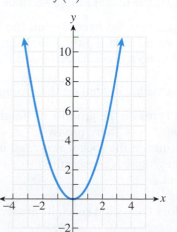

$f(x) = x^2 - 6x + 8$

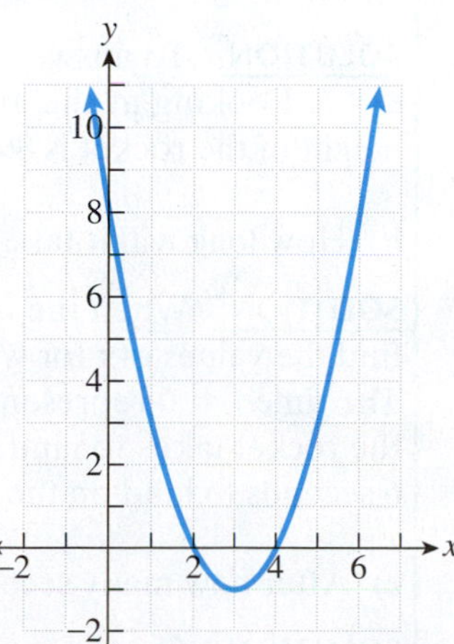

$f(x) = -x^2 + 4x$

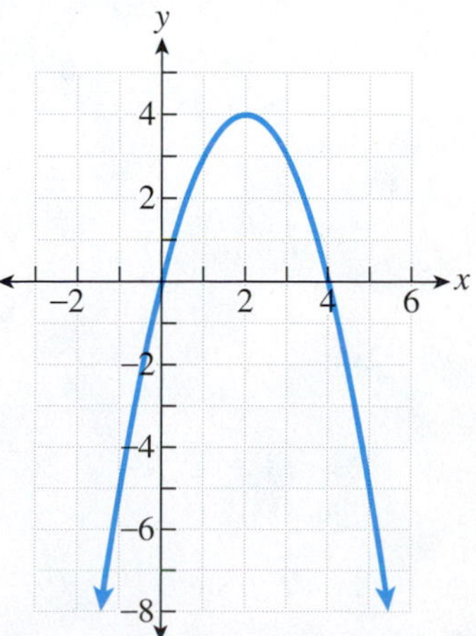

EXAMPLE 4 A rocket is fired into the air from the ground at a speed of 96 feet per second. Its height, in feet, after t seconds is given by the function $h(t) = -16t^2 + 96t$, which is graphed as follows:

Use the graph to answer the following questions:

a) How high above the ground is the rocket after 5 seconds?

SOLUTION To answer this question, we need to find the function's value when $t = 5$. Looking at the graph, we see that it passes through the point $(5, 80)$. The height of the rocket is 80 feet after 5 seconds have passed.

b) How long will it take for the rocket to land on the ground?

SOLUTION When the rocket lands on the ground, its height is 0 feet. We need to find the values of t for which $h(t) = 0$. There are two such values, $t = 0$ and $t = 6$. The time $t = 0$ represents the instant the rocket was fired into the air, so the time the rocket takes to land on the ground is represented by $t = 6$. It takes the rocket 6 seconds to land on the ground.

c) After how many seconds does the rocket reach its greatest height?

SOLUTION On the graph, we are looking for the point at which the function reaches its highest point. (This point where the parabola changes from rising to falling is called the **vertex** of the parabola.) We see that this point corresponds to a t value of 3, so it reaches its greatest height after 3 seconds.

d) What is the greatest height the rocket reaches?

SOLUTION From the graph, the best we can say is that the maximum height is between 140 feet and 150 feet. However, because we know the object reaches its greatest height after 3 seconds, we can evaluate the function $h(t) = -16t^2 + 96t$ at $t = 3$.

$$\begin{aligned} h(t) &= -16t^2 + 96t \\ h(3) &= -16(3)^2 + 96(3) && \text{Replace } t \text{ with 3.} \\ &= -16(9) + 96(3) && \text{Square 3.} \\ &= -144 + 288 && \text{Multiply.} \\ &= 144 && \text{Simplify.} \end{aligned}$$

The rocket reaches a height of 144 feet.

Quick Check 4

A ball is thrown into the air from the top of a building 48 feet above the ground at a speed of 32 feet per second. Its height, in feet, after t seconds is given by the function $h(t) = -16t^2 + 32t + 48$, which is graphed as follows.

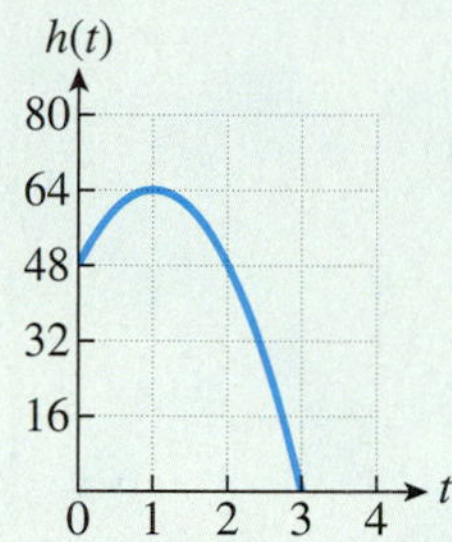

Use the graph to answer the following questions.

a) How high above the ground is the ball after 2 seconds?
b) How long will it take for the ball to land on the ground?
c) After how many seconds does the ball reach its greatest height?
d) What is the greatest height the ball reaches?

BUILDING YOUR STUDY STRATEGY

Overcoming Math Anxiety, 7 Positive Attitude Math anxiety causes many students to think negatively about math. A positive attitude will lead to more success than a negative attitude. Jot down some of the negative thoughts you have about math and your ability to learn and understand mathematics. Write each of these thoughts on a note card; then on the other side of the card, write the opposite statement.

A little confidence will go a long way toward improving your performance in class. Have the confidence to sit in the front row and ask questions during class, knowing that having your questions answered will increase your understanding and your chances for a better grade on the next exam. Finally, each time you experience success, no matter how small, reward yourself.

Exercises 6.7

MyMathLab

PRACTICE

WATCH

DOWNLOAD
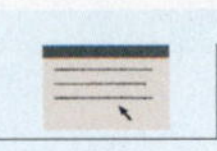
READ

REVIEW

Vocabulary

1. A(n) ______________ is a function that can be expressed as $f(x) = ax^2 + bx + c$, where $a \neq 1$.

2. The graph of a quadratic function is a U-shaped curve called a(n) ________.

Evaluate the given function.

3. $f(x) = x^2 + 8x + 20, f(6)$

4. $f(x) = x^2 + 5x + 40, f(10)$

5. $f(x) = x^2 + 7x - 33, f(-3)$

6. $f(x) = x^2 + 8x + 6, f(-5)$

7. $f(x) = x^2 - 7x + 10, f(5)$

8. $f(x) = x^2 - 10x - 39, f(8)$

9. $g(x) = x^2 + 12x - 45, g(-15)$

10. $g(x) = x^2 - 11x + 22, g(-10)$

11. $g(x) = 3x^2 + 2x - 14, g(7)$

12. $g(x) = 5x^2 - 4x + 8, g(-9)$

13. $f(x) = -2x^2 - 9x + 8, f(-3)$

14. $f(x) = -4x^2 + 16x + 13, f(4)$

15. $h(t) = -16t^2 + 96t + 32, h(4)$

16. $h(t) = -4.9t^2 + 26t + 10, h(2)$

17. $f(x) = x^2 + 4x - 20, f(4a)$

18. $f(x) = x^2 - 3x - 18, f(9a)$

19. $f(x) = x^2 - 7x - 25, f(a + 5)$

20. $f(x) = x^2 + 2x - 17, f(a - 4)$

21. Let $f(x) = x^2 + x - 42$. Find all values x for which $f(x) = 0$.

22. Let $f(x) = x^2 - 7x - 60$. Find all values x for which $f(x) = 0$.

23. Let $f(x) = x^2 - 9x + 20$. Find all values x for which $f(x) = 0$.

24. Let $f(x) = x^2 + 16x + 63$. Find all values x for which $f(x) = 0$.

25. Let $f(x) = x^2 - 23x + 144$. Find all values x for which $f(x) = 18$.

26. Let $f(x) = x^2 + 14x + 14$. Find all values x for which $f(x) = -31$.

27. Let $f(x) = x^2 + 11$. Find all values x for which $f(x) = 60$.

28. Let $f(x) = x^2 + 27$. Find all values x for which $f(x) = 108$.

A **fixed point** for a function $f(x)$ is a value a for which $f(a) = a$. For example, if $f(5) = 5$, $x = 5$ is a fixed point for the function $f(x)$.

Find all fixed points for the following functions by setting the function equal to x and solving the resulting equation for x.

29. $f(x) = x^2 + 7x - 16$

30. $f(x) = x^2 - 5x + 8$

31. $f(x) = x^2 + 14x + 40$

32. $f(x) = x^2 - 9x + 25$

33. A ball is dropped from an airplane flying at an altitude of 400 feet. The ball's height above the ground, in feet, after t seconds is given by the function $h(t) = 400 - 16t^2$, whose graph is shown.

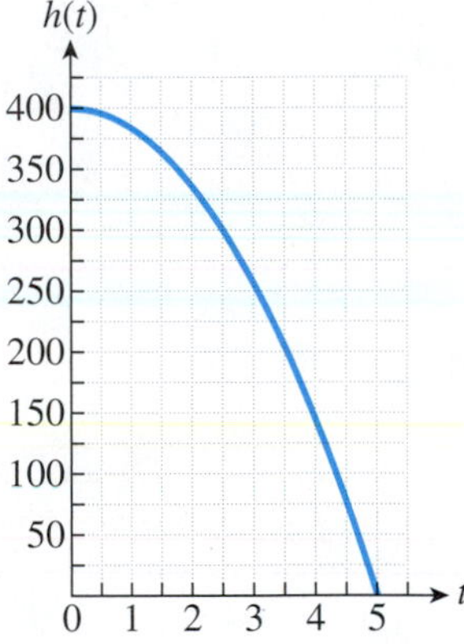

Use the function and its graph to answer the following questions.

a) How long will it take for the ball to land on the ground?

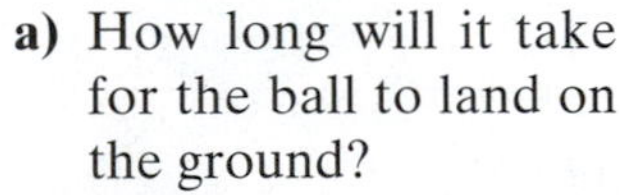

b) How high above the ground is the ball after 2 seconds?

34. A football is kicked up with an initial velocity of 64 feet/second. The football's height above the ground, in feet, after t seconds is given by the function $h(t) = -16t^2 + 64t$, whose graph is shown.

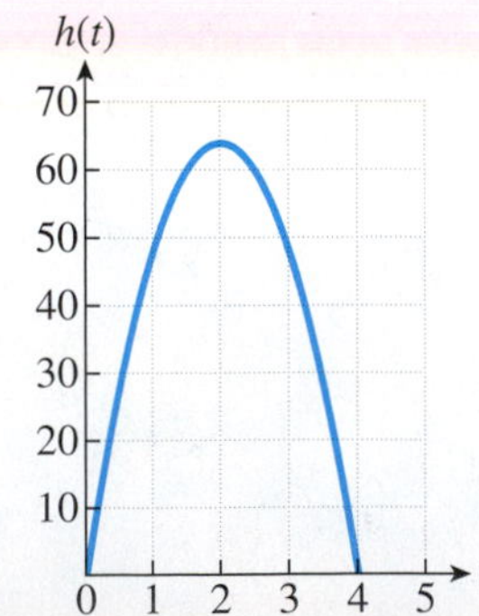

Use the function and its graph to answer the following questions.

a) Use the function to determine how high above the ground the football is after 1 second.

b) How long will it take for the football to land on the ground?

c) After how many seconds does the football reach its greatest height?

d) What is the greatest height the football reaches?

35. A man is standing on a cliff 240 feet above a beach. He throws a rock up off the cliff with an initial velocity of 32 feet/second. The rock's height above the beach, in feet, after t seconds is given by the function $h(t) = -16t^2 + 32t + 240$, whose graph is shown.

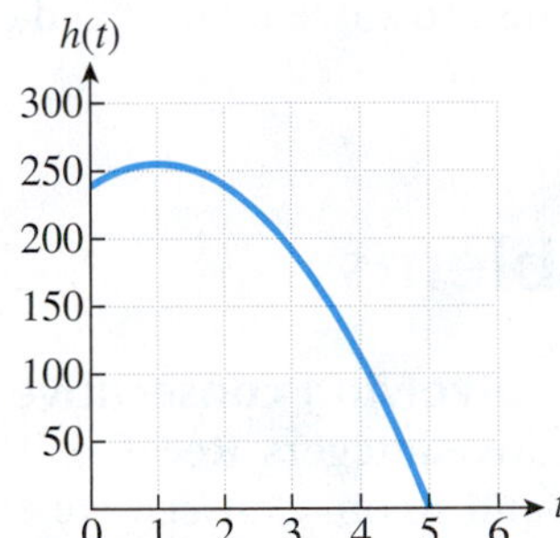

Use the function and its graph to answer the following questions.

a) Use the function to determine how high above the beach the rock is after 3 seconds.

b) How long will it take for the rock to land on the beach?

c) After how many seconds does the rock reach its greatest height above the beach?

d) What is the greatest height above the beach the rock reaches?

36. The average cost in dollars to produce x lawn chairs is given by the function $f(x) = 0.00000016x^2 - 0.0024x + 14.55$, whose graph is shown.

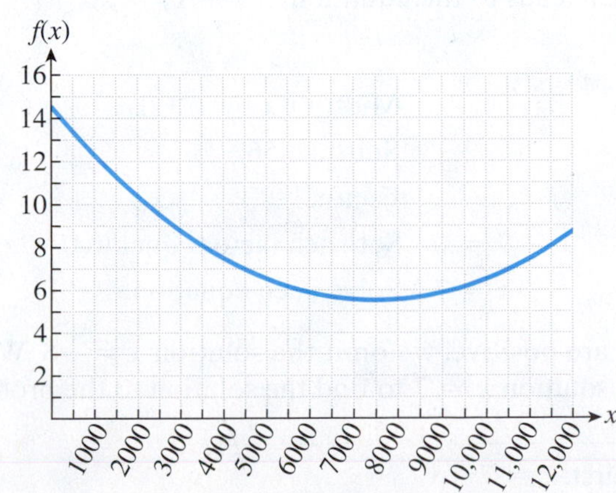

Use the function and its graph to answer the following questions.

a) Use the function to determine the average cost to produce 3000 lawn chairs.

b) What number of lawn chairs corresponds to the lowest average cost of production?

c) What is the lowest average cost that is possible?

37. A teenager starts a company selling personalized coffee mugs. The profit function, in dollars, for producing and selling x mugs is $f(x) = -0.4x^2 + 16x - 70$, whose graph is shown.

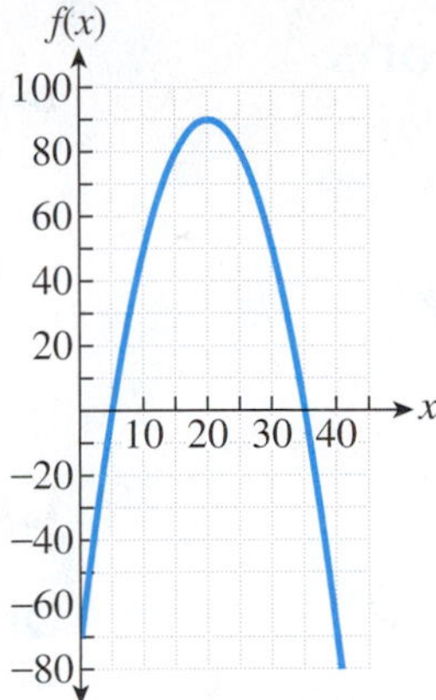

a) What are the start-up costs for the teenager's company?

b) How many mugs must the teenager sell before she breaks even?

c) How many mugs will give the maximum profit?

d) What will the profit be if she sells 25 mugs?

38. A peach farmer must determine how many peaches to thin from his trees. After he thins the peaches, the trees will produce fewer peaches, but they will be larger and of better quality. The expected profit, in dollars, if x percent of the peaches are removed during thinning is given by the function $f(x) = -38x^2 + 2280x + 244{,}625$, whose graph is shown.

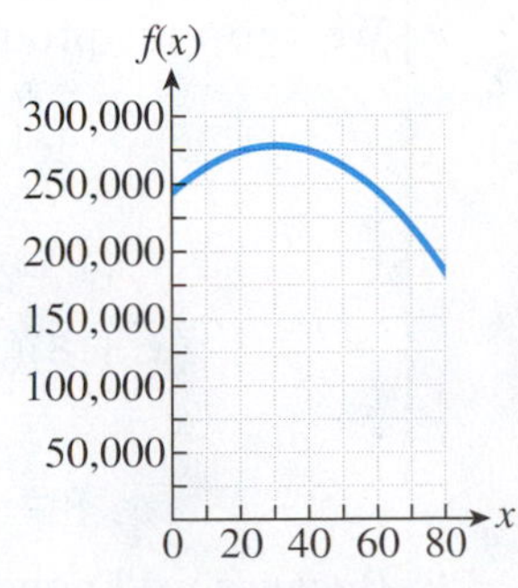

Use the function and its graph to answer the following questions.

a) What will the profit be if the farmer does not thin any peaches?

b) Use the function to determine the profit if 10% of the peaches are thinned.

c) What percent of the peaches must be thinned to produce the maximum profit?

d) What is the maximum potential profit?

6.8 Applications of Quadratic Equations and Quadratic Functions

OBJECTIVES

1. **Solve applied problems involving consecutive integers.**
2. **Solve applied problems involving the product of two unknown numbers.**
3. **Solve applied geometry problems.**
4. **Solve applied problems involving projectile motion.**
5. **Solve applied problems involving the sum of the first n natural numbers.**

In this section, we will learn to solve applied problems that involve quadratic equations. When problems become more complex, solving them by trial and error becomes difficult at best, so it is important to have a procedure to use. Even though some of the examples may be simple enough to solve by trial and error, they help us learn the new procedure.

Consecutive Integer Problems

Objective 1 Solve applied problems involving consecutive integers. We will begin with problems involving consecutive integers. Recall that consecutive integers follow the pattern $x, x + 1, x + 2$, and so on. Consecutive even integers, as well as consecutive odd integers, follow the pattern $x, x + 2, x + 4$, and so on.

EXAMPLE 1 The product of two consecutive positive integers is 56. Find the two integers.

SOLUTION We begin by creating a table of unknowns. It is important to represent each unknown in terms of the same variable. In this problem, we are looking for two consecutive positive integers. We can let x represent the first integer. Because the integers are consecutive, we can represent the second integer by $x + 1$.

Unknowns
First: x
Second: $x + 1$

We know the product is 56, which leads to the equation $x(x + 1) = 56$.

$$x(x + 1) = 56$$
$$x^2 + x = 56 \qquad \text{Multiply } x \text{ by } x + 1.$$
$$x^2 + x - 56 = 0 \qquad \text{Subtract 56.}$$
$$(x + 8)(x - 7) = 0 \qquad \text{Factor.}$$
$$x + 8 = 0 \quad \text{or} \quad x - 7 = 0 \qquad \text{Set each factor equal to 0.}$$
$$x = -8 \quad \text{or} \quad x = 7 \qquad \text{Solve each equation.}$$

Because we know the integers are positive, we omit the solution $x = -8$. We use the table of unknowns with the solution $x = 7$ to find the solution to the problem.

First: $x = 7$
Second: $x + 1 = 7 + 1 = 8$

The two integers are 7 and 8. The product of these two integers is 56.

Quick Check 1

The product of two consecutive positive integers is 132. Find the two integers.

A WORD OF CAUTION Make sure you check the practicality of your answers to an applied problem. In the previous example, the integers we were looking for were positive, so we omitted a solution that led to negative integers.

Problems Involving the Product of Two Unknown Numbers

Objective 2 Solve applied problems involving the product of two unknown numbers.

EXAMPLE 2 The sum of two numbers is 16, and their product is 55. Find the two numbers.

SOLUTION There are two unknowns in this problem: the two numbers. We let x represent the first number. If two numbers have a sum of 16 and we know the first number, we can find the second number by subtracting the first number from 16. Because the first number is x, the second number can be represented by $16 - x$.

Unknowns
First Number: x
Second Number: $16 - x$

Knowing that the product of these two numbers is 55 leads to the equation $x(16 - x) = 55$, which we now solve for x.

$x(16 - x) = 55$	
$16x - x^2 = 55$	Multiply.
$0 = x^2 - 16x + 55$	Collect all terms on the right side of the equation. This makes the leading coefficient positive.
$0 = (x - 5)(x - 11)$	Factor.
$x - 5 = 0$ or $x - 11 = 0$	Set each factor equal to 0.
$x = 5$ or $x = 11$	Solve each equation.

We return to the table of unknowns with the first solution, $x = 5$.

First Number: $x = 5$
Second Number: $16 - x = 16 - 5 = 11$

For this solution, the two numbers are 5 and 11. If we use the second solution, $x = 11$, we find the same two numbers.

First Number: $x = 11$
Second Number: $16 - x = 16 - 11 = 5$

The two numbers are 5 and 11, and their product is indeed 55.

Quick Check 2

The sum of two numbers is 21, and their product is 108. Find the two numbers.

Geometry Problems

Objective 3 Solve applied geometry problems. The next example involves the area of a rectangle. Recall that the area of a rectangle is equal to the product of its length and width.

$$\text{Area} = \text{Length} \cdot \text{Width}$$

EXAMPLE 3 The area of a rectangle is 108 square meters. If the length of the rectangle is 3 meters more than its width, find the dimensions of the rectangle.

SOLUTION For this problem, the unknowns are the length and the width of the rectangle. Because the length is defined in terms of the width, we let w represent the width. The length can be represented by the expression $w + 3$ as it is 3 meters longer than the width.

Unknowns
Length: $w + 3$
Width: w

Because the product of the length and the width is equal to the area of the rectangle, the equation we need to solve is $(w + 3) \cdot w = 108$.

$$(w + 3) \cdot w = 108$$
$$w^2 + 3w = 108 \quad \text{Multiply.}$$
$$w^2 + 3w - 108 = 0 \quad \text{Subtract 108.}$$
$$(w + 12)(w - 9) = 0 \quad \text{Factor.}$$
$$w + 12 = 0 \quad \text{or} \quad w - 9 = 0 \quad \text{Set each factor equal to 0.}$$
$$w = -12 \quad \text{or} \quad w = 9 \quad \text{Solve each equation.}$$

We omit the solution $w = -12$ because a rectangle cannot have a negative width or length. We can use the solution $w = 9$ and the table of unknowns to find the length and the width of the rectangle.

Length: $w + 3 = 9 + 3 = 12$
Width: $w = 9$

The length of the rectangle is 12 meters, and the width is 9 meters. We can verify that the area of a rectangle with these dimensions is 108 square meters.

Quick Check 3

The area of a rectangle is 105 square feet. If the width of the rectangle is 8 feet less than its length, find the dimensions of the rectangle.

EXAMPLE 4 A homeowner has installed an inground spa in the backyard. The spa is rectangular in shape, with a length that is 4 feet more than its width. The homeowner put a 1-foot-wide concrete border around the spa. If the area covered by the spa and the border is 96 square feet, find the dimensions of the spa.

SOLUTION

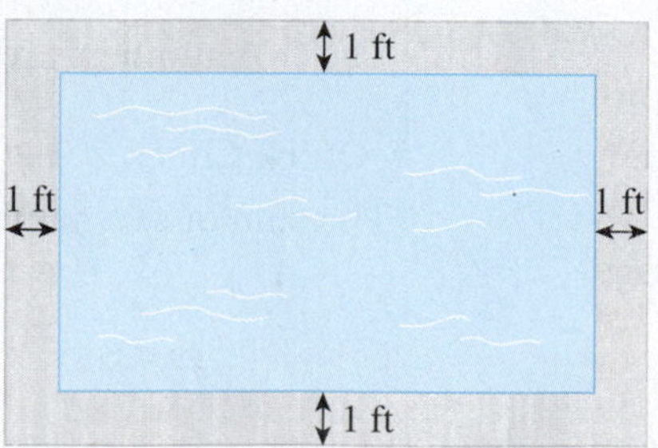

The unknowns are the length and width of the rectangular spa. We will let x represent the width of the spa.

Unknowns
Length: $x + 4$
Width: x

Because we know the area covered by the spa and the border, the equation must involve the outer rectangle in the picture. The width of the border can be represented by $x + 2$ because we need to add 2 feet to the width of the spa (x). The width of the spa is increased by 1 foot on both sides. In a similar fashion, the length of the border is $x + 6$ because we add a foot to the length of the spa on both sides.

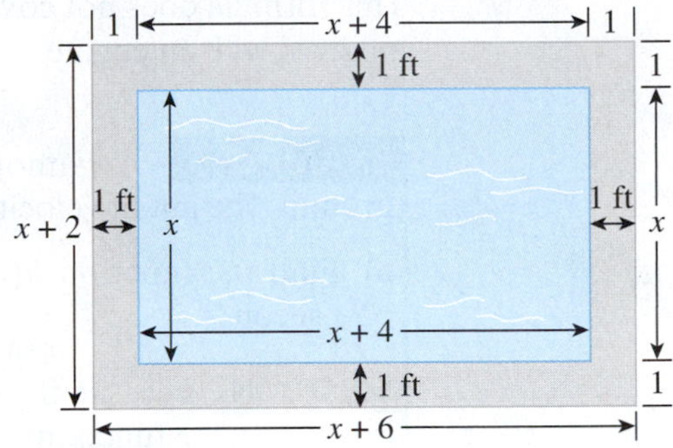

The equation we will solve is $(x + 6)(x + 2) = 96$ because the area covered is 96 square feet.

$$(x + 6)(x + 2) = 96$$
$$x^2 + 2x + 6x + 12 = 96 \quad \text{Multiply.}$$
$$x^2 + 8x + 12 = 96 \quad \text{Combine like terms.}$$
$$x^2 + 8x - 84 = 0 \quad \text{Subtract 96.}$$
$$(x + 14)(x - 6) = 0 \quad \text{Factor.}$$
$$x + 14 = 0 \quad \text{or} \quad x - 6 = 0 \quad \text{Set each factor equal to 0.}$$
$$x = -14 \quad \text{or} \quad x = 6 \quad \text{Solve each equation.}$$

We will omit the solution $x = -14$ because the dimensions of the spa cannot be negative. We use the solution $x = 6$ and the table of unknowns to find the length and the width of the rectangle.

Length: $x + 4 = 6 + 4 = 10$
Width: $x = 6$

The length of the spa is 10 feet, and the width of the spa is 6 feet. We can verify that the area covered by the spa and border is indeed 96 square feet.

Quick Check 4

A homeowner has poured a rectangular concrete slab in her backyard to use as a barbecue area. The length is 3 feet more than its width. There is a 2-foot-wide flower bed around the barbecue area. If the area covered by the barbecue area and the flower bed is 270 square feet, find the dimensions of the barbecue area.

Projectile Problems

Objective 4 Solve applied problems involving projectile motion. The height, in feet, of a projectile after t seconds can be found using the function $h(t) = -16t^2 + v_0t + s$, where v_0 is the initial velocity of the projectile and s is the initial height.

$$h(t) = -16t^2 + v_0t + s$$

t: Time (seconds)

v_0: Initial Velocity (feet/second)

s: Initial Height (feet)

This formula does not cover projectiles that continue to propel themselves, such as a rocket with an engine.

EXAMPLE 5 A cannonball is fired from a platform that is 256 feet above the ground. The initial velocity of the cannonball is 96 feet per second.

a) Find the function $h(t)$ that gives the height of the cannonball in feet after t seconds.

SOLUTION Because the initial velocity is 96 feet per second and the initial height is 256 feet, the function is $h(t) = -16t^2 + 96t + 256$.

b) How long will it take for the cannonball to land on the ground?

SOLUTION The cannonball's height when it lands on the ground is 0 feet; so we set the function equal to 0 and solve for the time t in seconds.

$-16t^2 + 96t + 256 = 0$	Set $h(t)$ equal to 0.
$0 = 16t^2 - 96t - 256$	Collect all terms on the right side of the equation so the leading coefficient is positive.
$0 = 16(t^2 - 6t - 16)$	Factor out the GCF (16).
$0 = 16(t - 8)(t + 2)$	Factor $t^2 - 6t - 16$.
$t - 8 = 0$ or $t + 2 = 0$	Set each variable factor equal to 0.
$t = 8$ or $t = -2$	Solve each equation.

We will omit the negative solution $t = -2$, as the time must be positive. It takes the cannonball 8 seconds to land on the ground.

c) What is the domain of the function $h(t)$?

SOLUTION Because the cannonball lands on the ground after 8 seconds, any length of time greater than 8 seconds does not apply. The domain of the function is $0 \le t \le 8$, which can be written in interval notation as $[0, 8]$.

Quick Check 5

A model rocket is launched from the ground with an initial velocity of 144 feet per second.

a) Find the function $h(t)$ that gives the height of the rocket in feet after t seconds.

b) How long will it take for the rocket to land on the ground?

c) What is the domain of the function $h(t)$?

Problems Involving the Sum of the First n Natural Numbers

Objective 5 Solve applied problems involving the sum of the first n natural numbers. The famous German mathematician Carl Gauss found a function for totaling the natural numbers from 1 to n, which is $f(n) = \frac{1}{2}n^2 + \frac{1}{2}n$. We use this function in the last example of the section.

EXAMPLE 6 **a)** Use Gauss's function to find the sum of the natural numbers from 1 to 46.

SOLUTION Evaluate Gauss's function at $n = 46$.

$$\begin{aligned} f(46) &= \frac{1}{2}(46)^2 + \frac{1}{2}(46) && \text{Substitute 46 for } n. \\ &= \frac{1}{2}(2116) + \frac{1}{2}(46) && \text{Square 46.} \\ &= 1058 + 23 && \text{Simplify each term.} \\ &= 1081 && \text{Add.} \end{aligned}$$

The sum of the first 46 natural numbers is 1081.

b) Find a natural number n such that the sum of the first n natural numbers is 78.

SOLUTION Begin by setting the function $f(n)$ equal to 78 and solving for n.

$$\begin{aligned} \frac{1}{2}n^2 + \frac{1}{2}n &= 78 && \text{Set the function equal to 78.} \\ 2\left(\frac{1}{2}n^2 + \frac{1}{2}n\right) &= 2 \cdot 78 && \text{Multiply both sides of the equation by 2 to clear the fractions.} \\ n^2 + n &= 156 && \text{Distribute and simplify each term.} \\ n^2 + n - 156 &= 0 && \text{Collect all terms on the left side of the equation.} \\ (n + 13)(n - 12) &= 0 && \text{Factor.} \end{aligned}$$

$$\begin{aligned} n + 13 = 0 \quad &\text{or} \quad n - 12 = 0 && \text{Set each factor equal to 0.} \\ n = -13 \quad &\text{or} \quad n = 12 && \text{Solve each equation.} \end{aligned}$$

We omit the solution $n = -13$ because -13 is not a natural number. The sum of the first 12 natural numbers is 78.

Quick Check 6

a) Use Gauss's function to find the sum of the natural numbers from 1 to 100.
b) Find a natural number n such that the sum of the first n natural numbers is 45.

BUILDING YOUR STUDY STRATEGY

Overcoming Math Anxiety, 8 Procrastination Some students put off doing their math homework or studying due to negative feelings for mathematics. Procrastination is your enemy. Convince yourself that you can do the task and get started. If you wait until you are tired, you will not be able to give your best effort. If you do not complete the homework assignment, you will have trouble following the next day's material. Try scheduling a time to devote to mathematics each day and stick to your schedule.

Exercises 6.8

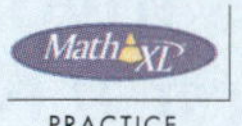
PRACTICE

WATCH

DOWNLOAD

READ

REVIEW

Vocabulary

1. Three consecutive integers can be represented by __________, __________, and __________.
2. Three consecutive odd integers can be represented by __________, __________, and __________.
3. State the formula for the area of a rectangle.
4. The function ____________________ gives the height of a projectile with initial velocity v_0 and initial height s in feet after t seconds.

5. Two consecutive positive integers have a product of 132. Find the integers.
6. Two consecutive positive integers have a product of 272. Find the integers.
7. Two consecutive positive even integers have a product of 528. Find the integers.
8. Two consecutive positive even integers have a product of 168. Find the integers.
9. Two consecutive positive odd integers have a product of 255. Find the integers.
10. Two consecutive positive odd integers have a product of 99. Find the integers.
11. The product of two consecutive positive integers is 27 more than five times the larger integer. Find the integers.
12. The product of two consecutive odd positive integers is 14 more than three times the larger integer. Find the integers.
13. The product of two consecutive even positive integers is 398 more than the sum of the two integers. Find the integers.
14. The product of two consecutive positive integers is 131 more than the sum of the two integers. Find the integers.
15. One positive number is 5 more than a second number, and their product is 66. Find the two numbers.
16. One positive number is 9 more than a second number, and their product is 112. Find the two numbers.
17. One positive number is 4 less than three times a second number, and their product is 84. Find the two numbers.
18. One positive number is 3 more than twice a second number, and their product is 189. Find the two numbers.
19. The sum of two numbers is 29, and their product is 210. Find the two numbers.
20. The sum of two numbers is 27, and their product is 140. Find the two numbers.
21. The sum of two numbers is 25, and their product is 66. Find the two numbers.
22. The sum of two numbers is 70, and their product is 1200. Find the two numbers.
23. The difference of two positive numbers is 3, and their product is 88. Find the two numbers.
24. The difference of two positive numbers is 9, and their product is 220. Find the two numbers.
25. Gabriela is 5 years older than her sister Zorayda. If the product of their ages is 126, how old is Gabriela?
26. Marco is 8 years younger than his brother Paolo. If the product of their ages is 105, how old is Marco?
27. Clyde's age is 5 years more than twice Bonnie's age. If the product of their ages is 168, how old is Clyde?
28. Edith's age is 1 year less than three times Archie's age. If the product of their ages is 70, how old is Edith?
29. The length of a rectangle is 7 inches more than its width. If the area of the rectangle is 120 square inches, find the length and width of the rectangle.
30. The length of a rectangular rug is twice its width. If the area of the rug is 450 square feet, find the length and width of the rug.
31. The length of a rectangular picture frame is 2 centimeters more than twice its width. If the area of the frame is 420 square centimeters, find the length and width of the frame.
32. The length of a rectangular room is 1 foot more than twice its width. If the area of the room is 300 square feet, find the length and width of the room.
33. Jose has a rectangular garden that covers 105 square meters. The length of the garden is 1 meter more than twice its width. Find the dimensions of the garden.
34. The height of a doorway is 7 feet less than five times its width. If the area of the doorway is 24 square feet, find the dimensions of the doorway.

35. The area of a rectangular lawn is 960 square feet. If the length of the lawn is 8 feet less than twice its width, find the dimensions of the lawn.

36. A rectangular soccer field has an area of 8800 square meters. If the length of the field is 30 meters more than its width, find the dimensions of the field.

37. The width of a photo is 3 inches less than its length. A border of 1 inch is placed around the photo, and the area covered by the photo and its border is 70 square inches. Find the dimensions of the photo itself.

38. The length of a rectangular quilt is 4 inches more than its width. After a 2-inch border is placed around the quilt, its area is 320 square inches. Find the original dimensions of the quilt.

39. A rectangular flower garden has a length that is 1 foot less than twice its width. A 2-foot brick border is added around the garden, and the area of the garden and brick border is a total of 117 square feet. Find the dimensions of the garden without the brick border.

40. Adjacent to his barn, Larry has a rectangular pen for his pet goats as shown in the following diagram:

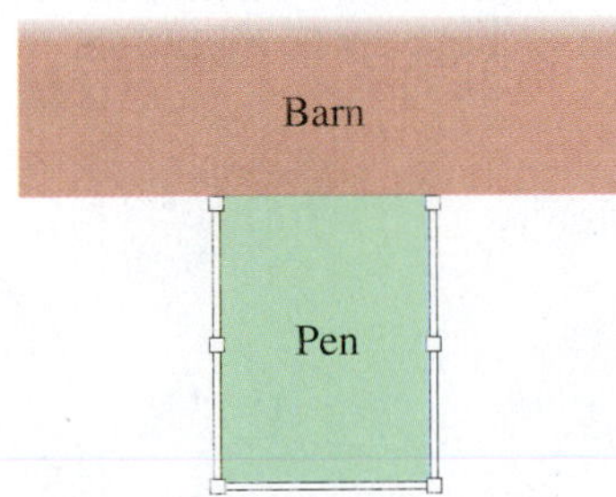

The side opposite the barn is 5 feet less than the side that projects out from the barn. He decides to expand the pen by 5 feet in all three directions, as shown in the following diagram:

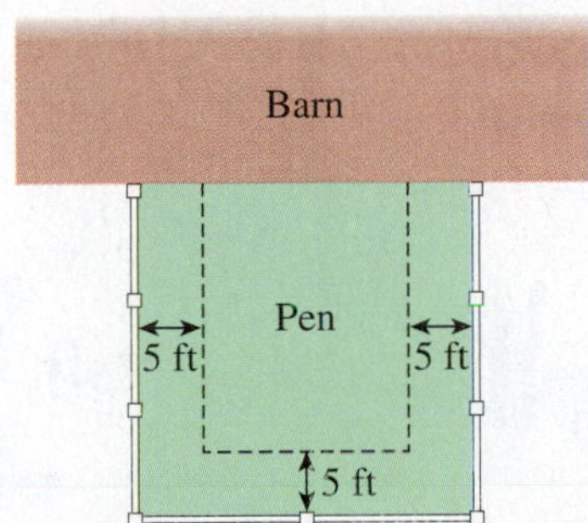

This expansion makes the area of the pen 625 square feet. Find the dimensions of the original pen.

For the following exercises, recall that the area of a square with side s is s^2 and that the area of a triangle with base b and height h is $\frac{1}{2}bh$.

41. The area of the floor in a square room is 64 square meters. Find the length of a wall of the room.

42. A dartboard is enclosed in a square that has an area of 1600 square centimeters. Find the length of a side of the square.

43. The height of a triangular sail is 5 feet more than twice its base. If the sail is made of 84 square feet of fabric, find the base and height of the triangle.

44. The three towns of Visalia, Hanford, and Fresno form a triangle. The base of the triangle extends from Visalia to Hanford, and the height of the triangle extends from Visalia to Fresno, as shown in the following diagram:

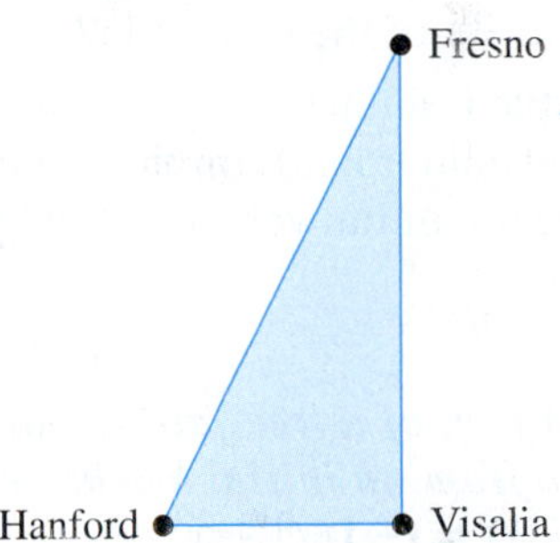

The distance from Visalia to Fresno is twice the distance from Visalia to Hanford. If the area of the triangle formed by these three cities is 400 square miles, find the distance from Visalia to Fresno.

For problems 45–54, use the following function:
$h(t) = -16t^2 + v_0t + s.$

45. A projectile is launched from a building 320 feet tall. If the initial velocity of the projectile was 128 feet per second, how long will it take the projectile to land on the ground?

46. Standing on a platform, Eesaeng throws a softball upward from a height 48 feet above the ground at a speed of 32 feet per second. How long will it take for the softball to land on the ground?

47. Standing on the edge of a cliff, Fernando throws a rock upward from a height 80 feet above the water at a speed of 64 feet per second. How long will it take for the rock to land in the water?

48. A projectile is launched from the top of a building 384 feet tall with an initial velocity of 160 feet per second. How long will it take for the projectile to land on the ground?

49. A projectile is launched from a platform 40 feet above the ground with an initial velocity of 64 feet per second. After how many seconds is the projectile at a height of 88 feet?

50. Standing atop a building, Angela throws a rock upward from a height 55 feet above the ground with an initial velocity of 32 feet per second. How long will it take before the rock is 7 feet above the ground?

51. A projectile is launched from the ground with an initial velocity of 128 feet per second. How long will it take for the projectile to land on the ground?

52. A projectile is launched from the ground with an initial velocity of 176 feet per second. How long will it take for the projectile to land on the ground?

53. A projectile is launched upward from the ground, and it lands on the ground after 5 seconds. What was the initial velocity of the projectile?

54. A projectile is launched upward from an initial height of 192 feet, and it lands on the ground after 6 seconds. What was the initial velocity of the projectile?

The height, in feet, of a free-falling object t seconds after being dropped from an initial height s can be found using the function $h(t) = -16t^2 + s$.

55. An object is dropped from a helicopter 144 feet above the ground. How long will it take for the object to land on the ground?

56. An object is dropped from an airplane that is 400 feet above the ground. How long will it take for the object to land on the ground?

Use Gauss's function to total the natural numbers from 1 to n, $f(n) = \frac{1}{2}n^2 + \frac{1}{2}n$, in the following exercises.

57. Find a natural number n such that the sum of the first n natural numbers is 36.

58. Find a natural number n such that the sum of the first n natural numbers is 55.

59. Find a natural number n such that the sum of the first n natural numbers is 210.

60. Find a natural number n such that the sum of the first n natural numbers is 325.

Writing in Mathematics

Answer in complete sentences.

61. Write a word problem involving a rectangle that leads to the equation $(x + 9)(x + 5) = 192$. Explain how you created your problem.

62. Write a word problem involving a rectangle with a length of 24 feet and width of 16 feet. Your problem must lead to a quadratic equation.

CHAPTER 6 SUMMARY

Section 6.1 An Introduction to Factoring; the Greatest Common Factor; Factoring by Grouping

Greatest Common Factor (GCF), p. 296

The greatest common factor of two or more integers is the largest whole number that is a factor of each integer.
For a variable factor to be included in the GCF, it must be a factor of each term.

Find the GCF: $35x^4, 14x^3, 49x^2$
GCF of 35, 14, and 49: 7
GCF: $7x^2$

Factoring Out the Greatest Common Factor, pp. 297–298

The first step for factoring any polynomial is to factor out the GCF of all of the terms.

Factor out the GCF: $8x^5 - 10x^4 + 14x^2$

$$8x^5 - 10x^4 + 14x^2 = 2x^2(4x^3 - 5x^2 + 7)$$

Factoring by Grouping, pp. 298–300

To factor a polynomial with four terms by grouping, split the polynomial into two groups.
Factor a common factor out of the first two terms and another common factor out of the last two terms. If the two "groups" share a common binomial factor, this binomial can be factored out to complete the factoring of the polynomial.

Factor by grouping: $2x^3 - 6x^2 + 7x - 21$

$$\begin{aligned} 2x^3 - 6x^2 + 7x - 21 &= 2x^2(x - 3) + 7(x - 3) \\ &= (x - 3)(2x^2 + 7) \end{aligned}$$

Section 6.2 Factoring Trinomials of the Form $x^2 + bx + c$

Factoring $x^2 + bx + c$, pp. 302–304

To factor a trinomial of this form, look for two integers m and n that have a product equal to c and a sum equal to b.
If you can find two such integers, the trinomial factors to be $(x + m)(x + n)$.

Factor: $x^2 - 3x - 54$
Multiply to equal -54; add to equal -3.

$$x^2 - 3x - 54 = (x - 9)(x + 6)$$

Prime Polynomials, p. 305

A polynomial that cannot be factored is said to be prime or unfactorable.

Factor: $x^2 + 7x - 10$
Multiply to equal -10; add to equal 7.
$x^2 + 7x - 10$ is prime.

Section 6.3 Factoring Trinomials of the Form $ax^2 + bx + c$, Where $a \neq 1$

$ax^2 + bx + c$ ($a \neq 1$): Factoring by Grouping, pp. 308–310

1. Multiply $a \cdot c$.
2. Find two integers whose product is $a \cdot c$ and whose sum is b.
3. Rewrite the term bx as two terms, using the two integers found in Step 2.
4. Factor the resulting polynomial by grouping.

Factor: $2x^2 + 15x + 28$
$2 \cdot 28 = 56$: Multiply to equal 56; add to equal 15.

$$\begin{aligned} 2x^2 + 15x + 28 &= 2x^2 + 7x + 8x + 28 \\ &= x(2x + 7) + 4(2x + 7) \\ &= (2x + 7)(x + 4) \end{aligned}$$

$ax^2 + bx + c$ ($a \neq 1$): Factoring by Trial and Error, pp. 310–312

If the trinomial $ax^2 + bx + c$ is factorable, it will be the product of two binomials that consist of a variable term and a constant.
The product of the variable terms will be ax^2, and the product of the constants will equal c. Use trial and error to find the pair of factors that has a product whose middle term is bx.

Factor: $3x^2 - 17x + 10$
$3x^2$: $x \cdot 3x$ 10: $1 \cdot 10, 2 \cdot 5$

$$\begin{aligned} (x - 1)(3x - 10) &= 3x^2 - 13x + 10 \\ (x - 10)(3x - 1) &= 3x^2 - 31x + 10 \\ (x - 2)(3x - 5) &= 3x^2 - 11x + 10 \\ (x - 5)(3x - 2) &= 3x^2 - 17x + 10 \leftarrow \end{aligned}$$

Section 6.4 Factoring Special Binomials

Difference of Squares, pp. 313–315

A difference of squares factors in the following manner:
$a^2 - b^2 = (a + b)(a - b)$

Factor: $x^2 - 49$

$$\begin{aligned} x^2 - 49 &= (x)^2 - (7)^2 \\ &= (x + 7)(x - 7) \end{aligned}$$

Sum of Squares, p. 315

A sum of squares, $a^2 + b^2$, is not factorable.

Factor: $x^2 + 64$
This is a sum of squares and is not factorable.

Difference of Cubes, pp. 315–316

A difference of cubes factors in the following manner:

$a^3 - b^3 = (a - b)(a^2 + ab + b^2)$

Factor: $x^3 - 125$

$$x^3 - 125 = (x)^3 - (5)^3 = (x - 5)(x^2 + 5x + 25)$$

Sum of Cubes, p. 317

A sum of cubes factors in the following manner:

$a^3 + b^3 = (a + b)(a^2 - ab + b^2)$

Factor: $x^3 + 216$

$$x^3 + 216 = (x)^3 + (6)^3 = (x + 6)(x^2 - 6x + 36)$$

Section 6.5 Factoring Polynomials: A General Strategy

Factoring Polynomials, pp. 319–322

1. Factor out any common factors.
2. Determine the number of terms in the polynomial.
   a) If there are only *two terms*, check to see if the binomial is one of the special binomials.
      - Difference of Squares: $a^2 - b^2 = (a + b)(a - b)$
      - Sum of Squares: $a^2 + b^2$ is not factorable.
      - Difference of Cubes: $a^3 - b^3 = (a - b)(a^2 + ab + b^2)$
      - Sum of Cubes: $a^3 + b^3 = (a + b)(a^2 - ab + b^2)$

   b) If there are *three terms*, try to use one of the following techniques:
      - $x^2 + bx + c = (x + m)(x + n)$: Find two integers m and n whose product is c and whose sum is b.
      - $ax^2 + bx + c$ $(a \neq 1)$: Factor by grouping or by trial and error.

   c) If there are *four terms*, try factoring by grouping.
3. After the polynomial has been factored, make sure that any factor with two or more terms does not have a common factor other than 1. If there are common factors, factor them out.
4. Check your factoring through multiplication.

Section 6.6 Solving Quadratic Equations by Factoring

Zero-Factor Property of Real Numbers, pp. 325–326

If $a \cdot b = 0$, then $a = 0$ or $b = 0$.

Solve: $(x - 4)(3x + 5) = 0$

$x - 4 = 0$, $x = 4$ $3x + 5 = 0$, $3x = -5$, $x = -\frac{5}{3}$

$\left\{-\frac{5}{3}, 4\right\}$

Solving Quadratic Equations by Factoring, pp. 326–329

1. Write the equation in standard form: $ax^2 + bx + c = 0$
2. Factor the expression $ax^2 + bx + c$ completely.
3. Set each factor equal to 0 and solve the resulting equations.
4. Finish by checking the solutions.

Solve: $x^2 + 9x - 36 = 0$

$$x^2 + 9x - 36 = 0$$
$$(x + 12)(x - 3) = 0$$

$x + 12 = 0$, $x = -12$ $x - 3 = 0$, $x = 3$

$\{-12, 3\}$

Finding a Quadratic Equation Given Two Solutions, p. 329

If $x = a$ and $x = b$ are solutions to a quadratic equation, the equation must be of the form $(x - a)(x - b) = 0$.

Find a quadratic equation with this solution set: $\{-4, 13\}$

$$[x - (-4)](x - 13) = 0$$
$$(x + 4)(x - 13) = 0$$
$$x^2 - 13x + 4x - 52 = 0$$
$$x^2 - 9x - 52 = 0$$

Section 6.7 Quadratic Functions

Quadratic Functions, p. 332

A quadratic function is a function of the form:

$$f(x) = ax^2 + bx + c\ (a \neq 0).$$

$$f(x) = x^2 + 37x - 205$$

Evaluating Quadratic Functions, p. 332

To evaluate a function for a particular value of the variable, substitute that value for the variable in the function's formula and simplify the resulting expression.

For $f(x) = 2x^2 + 7x - 19$, find $f(8)$.

$$\begin{aligned} f(8) &= 2(8)^2 + 7(8) - 19 \\ &= 2(64) + 7(8) - 19 \\ &= 128 + 56 - 19 \\ &= 165 \end{aligned}$$

Solving Equations Involving Quadratic Functions, pp. 332–333

To solve an equation involving a quadratic function, replace the function with its formula and solve the resulting equation.

For $f(x) = x^2 + 8x - 12$, solve $f(x) = 36$.

$$\begin{aligned} f(x) &= 36 \\ x^2 + 8x - 12 &= 36 \\ x^2 + 8x - 48 &= 0 \\ (x + 12)(x - 4) &= 0 \end{aligned}$$

$$x + 12 = 0 \qquad x - 4 = 0$$
$$x = -12 \qquad x = 4$$

Section 6.8 Applications of Quadratic Equations and Quadratic Functions

Problems Involving Consecutive Integers, p. 338

Let x represent the first integer.
If the integers are consecutive odd integers or consecutive even integers, let $x + 2$ represent the second integer.
If the integers are simply consecutive integers, let $x + 1$ represent the second integer.

The product of two consecutive positive integers is 72. Find the integers.

Unknowns
First: x; second: $x + 1$

$$\begin{aligned} x(x + 1) &= 72 \\ x^2 + x - 72 &= 0 \\ (x + 9)(x - 8) &= 0 \end{aligned}$$

$$x + 9 = 0 \qquad x - 8 = 0$$
$$x = -9 \qquad x = 8$$

Omit negative solution.

First: $x = 8$; second: $x + 1 = 8 + 1 = 9$
The consecutive integers are 8 and 9.

Area of a Rectangle, pp. 340–341

The unknowns are the length and the width.
The formula is

$$\text{Area} = \text{Length} \cdot \text{Width}.$$

The length of a rectangle is 7 inches more than its width. If the area is 60 square inches, find the dimensions of the rectangle.

Unknowns
Length: $x + 7$; width: x

$$\begin{aligned} x(x + 7) &= 60 \\ x^2 + 7x - 60 &= 0 \\ (x + 12)(x - 5) &= 0 \end{aligned}$$

$$x + 12 = 0 \qquad x - 5 = 0$$
$$x = -12 \qquad x = 5$$

Omit negative solution.

Length: $x + 7 = 5 + 7 = 12$; width: $x = 5$
The length is 12 inches, and the width is 5 inches.

Projectile Problems, p. 342

The height, in feet, of a projectile after t seconds can be found using the function

$$h(t) = -16t^2 + v_0 t + s$$

where v_0 is the initial velocity of the projectile and s is the initial height.

A projectile is launched upward at a speed of 96 feet per second from a building that is 432 feet tall. How long will it take for the projectile to land on the ground below?

$h(t) = -16t^2 + 96t + 432$

$$h(t) = 0$$
$$-16t^2 + 96t + 432 = 0$$
$$-16(t^2 - 6t - 27) = 0$$
$$-16(t + 3)(t - 9) = 0$$

$$t + 3 = 0 \qquad t - 9 = 0$$
$$t = -3 \qquad t = 9$$

Omit negative solution.

It will take 9 seconds.

SUMMARY OF CHAPTER 6 STUDY STRATEGIES

Math anxiety can hinder your success in a mathematics class, but you can overcome it.

- The first step is to understand what has caused your anxiety.
- Relaxation techniques can help you overcome the physical symptoms of math anxiety, allowing you to give your best effort.
- Developing a positive attitude and confidence in your abilities will help as well.
- By scheduling a set time for your mathematics studies and sticking to that schedule, you can avoid procrastinating.
- Many students believe they have math anxiety, but poor performance can be caused by other factors as well. Do you completely understand the material, only to freeze up when talking tests? Does this happen in other classes as well? If so, you may have test anxiety, not math anxiety.
- Some students do poorly because they are taking classes for which they are not prepared. Discuss placement with your instructor and your academic counselor if you believe this applies to you.
- Finally, make sure that poor study skills are not the cause of your difficulties. If you are not giving your best effort, you cannot expect to learn and understand mathematics.

CHAPTER 6 REVIEW

***Factor completely. If the polynomial cannot be factored, write prime.* [6.1–6.5]**

1. $3x - 21$

2. $x^2 - 20x$

3. $5x^7 + 15x^4 - 20x^3$

4. $6a^5b^2 - 10a^3b^3 + 15a^4b$

5. $x^3 + 6x^2 + 4x + 24$

6. $x^3 + 9x^2 - 7x - 63$

7. $x^3 - 3x^2 - 11x + 33$

8. $x^3 - 3x^2 - 9x + 27$

9. $x^2 - 8x + 16$

10. $x^2 + 3x - 28$

11. $x^2 - 11x + 24$

12. $-5x^2 - 20x + 160$

13. $x^2 + 14xy + 45y^2$

14. $x^2 + 9x - 23$

15. $x^2 + 17x + 30$

16. $x^2 - x - 42$

17. $10x^2 - 83x + 24$

18. $4x^2 - 3x - 10$

19. $18x^2 + 12x - 6$

20. $4x^2 - 12x + 9$

21. $3x^2 - 75$

22. $4x^2 - 25y^2$

23. $x^3 + 8y^3$

24. $x^2 + 121$

25. $x^3 - 216$

26. $7x^3 - 189$

***Solve.* [6.6]**

27. $(x - 5)(x + 7) = 0$

28. $(3x - 8)(2x + 1) = 0$

29. $x^2 - 25 = 0$

30. $x^2 - 14x + 40 = 0$

31. $x^2 + 16x + 64 = 0$

32. $x^2 + 3x - 108 = 0$

33. $x^2 - 9x - 70 = 0$

34. $x^2 - 16x = 0$

35. $x^2 + 19x + 60 = 0$

36. $x^2 - 18x + 80 = 0$

37. $x^2 - 15x - 16 = 0$

38. $2x^2 - 7x - 60 = 0$

39. $7x^2 - 252 = 0$

40. $4x^2 - 60x + 144 = 0$

41. $x(x + 10) = -21$

42. $(x + 4)(x + 5) = 72$

***Find a quadratic equation with integer coefficients that has the given solution set.* [6.6]**

43. $\{6, 7\}$

44. $\{-4, 4\}$

45. $\left\{-\frac{3}{5}, 2\right\}$

46. $\left\{-\frac{5}{4}, \frac{2}{7}\right\}$

***Evaluate the given quadratic function.* [6.7]**

47. $f(x) = x^2, f(-3)$

48. $f(x) = x^2 - 8x, f(4)$

49. $f(x) = x^2 + 10x + 24, f(-8)$

50. $f(x) = 3x^2 + 8x - 19, f(5)$

***For the given function f(x), find all values x for which f(x) = 0.* [6.7]**

51. $f(x) = x^2 - 36$

52. $f(x) = x^2 + 3x - 54$

53. $f(x) = x^2 + 12x + 20$

54. $f(x) = x^2 - 7x - 44$

55. A man is standing on a cliff above a beach. He throws a rock upward with an initial velocity of 32 feet/second from a height 128 feet above the beach. The rock's height above the beach, in feet, after t seconds is given by the function $h(t) = -16t^2 + 32t + 128$, whose graph is shown. [6.7]

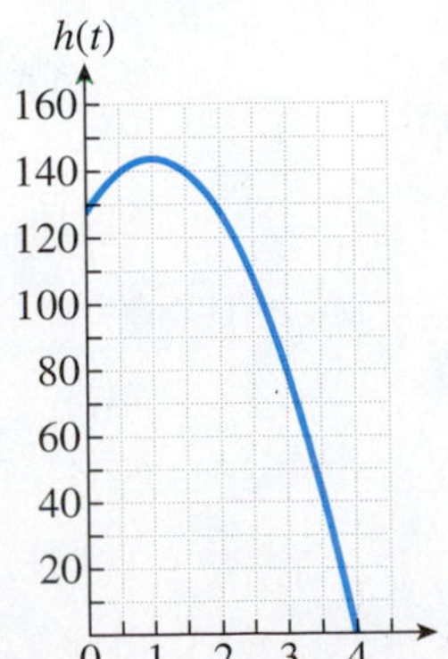

Use the function and its graph to answer the following questions.

a) Use the graph to determine how high above the beach the rock is after 3 seconds.

Worked-out solutions to Review Exercises marked with [icon] can be found on page AN-20.

b) How long will it take for the rock to land on the beach?

c) After how many seconds does the rock reach its greatest height above the beach?

d) Use the function to determine the greatest height above the beach that the rock reaches.

56. The average price in dollars for a manufacturer to produce x toys is given by the function $f(x) = 0.0001x^2 - 0.08x + 20.85$, whose graph is shown. [6.7]

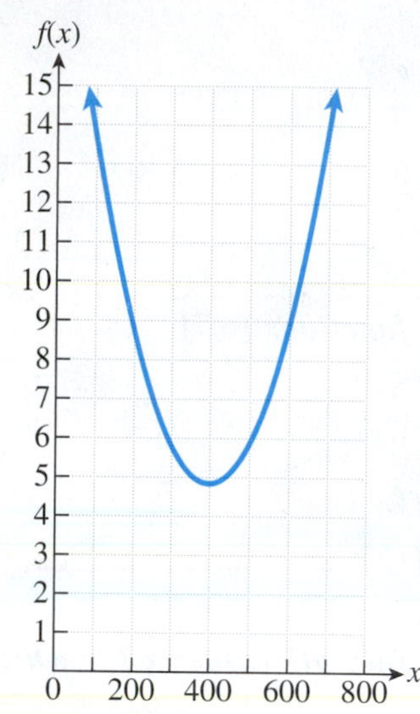

Use the function and its graph to answer the following questions.

a) Use the function to determine the average cost to produce 800 toys.

b) What number of toys corresponds to the lowest average cost of production?

c) Use the function to determine the lowest average cost that is possible.

57. Two consecutive even positive integers have a product of 288. Find the two integers. [6.8]

58. Rosa is 8 years younger than Dale. If the product of their ages is 105, how old is each person? [6.8]

59. The length of a rectangle is 3 meters less than twice the width. The area of the rectangle is 104 square meters. Find the length and the width of the rectangle. [6.8]

60. A ball is thrown upward with an initial velocity of 80 feet per second from the edge of a cliff that is 384 feet above a river.
Use the function $h(t) = -16t^2 + v_0t + s$. [6.8]

a) How long will it take for the ball to land in the river?

b) When is the ball 480 feet above the river?

CHAPTER 6 TEST

For Extra Help

Step-by-step test solutions are found on the Chapter Test Prep Videos available via the Video Resources on DVD, in MyMathLab, and on YouTube (search "WoodburyElemIntAlg" and click on "Channels").

Factor completely.

1. $x^2 - 25x$

2. $x^3 - 3x^2 - 5x + 15$

3. $x^2 - 9x + 20$

4. $x^2 - 14x - 72$

5. $3x^2 + 60x + 57$

6. $6x^2 - x - 5$

7. $x^2 - 25y^2$

8. $x^3 - 216y^3$

Solve.

9. $x^2 - 100 = 0$

10. $x^2 - 13x + 36 = 0$

11. $x^2 + 7x - 60 = 0$

12. $(x + 2)(x - 9) = 60$

Find a quadratic equation with integer coefficients that has the given solution set.

13. $\{-2, 2\}$

14. $\left\{-6, \frac{9}{5}\right\}$

Evaluate the given quadratic function.

15. $f(x) = x^2 - 12x + 35, f(-5)$

16. $f(x) = -x^2 + 2x + 17, f(7)$

For the given function $f(x)$, find all values x for which $f(x) = 0$.

17. $f(x) = x^2 + 14x + 48$

18. The average price in dollars for a manufacturer to produce x baseball hats is given by the function

$f(x) = 0.00000006x^2 - 0.00096x + 5.95$, whose graph is shown.

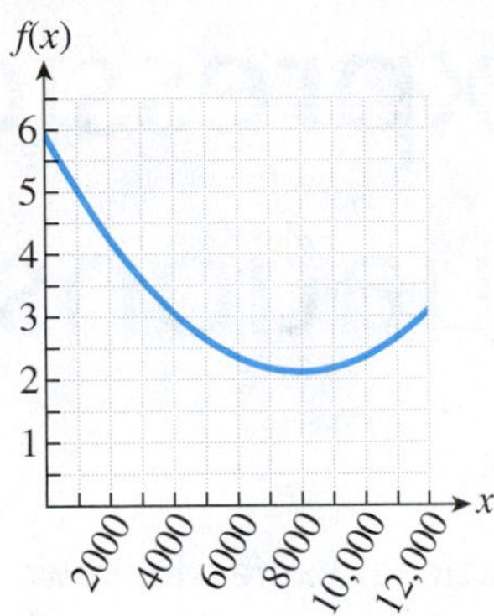

Use the function and its graph to answer the following questions.

a) Use the function to determine the average cost to produce 6500 baseball hats.

b) What number of baseball hats corresponds to the lowest average cost?

c) Use the function to determine the lowest average cost that is possible.

19. The length of a rectangle is 5 feet more than three times the width. The area of the rectangle is 100 square feet. Find the length and the width of the rectangle.

20. A ball is thrown upward with an initial velocity of 32 feet per second from the edge of a cliff 240 feet above a beach. Use the function $h(t) = -16t^2 + v_0t + s$.

a) How long will it take for the ball to land on the beach?

b) When is the ball 192 feet above the beach?

Mathematicians in History

Sir Isaac Newton was an English mathematician and scientist who lived in the 17th and 18th centuries. His work with motion and gravity helped us better understand the world as well as the solar system. Alexander Pope once said, "Nature and Nature's laws lay hid in night; God said, Let Newton be! And all was light."

Write a one-page summary (*or* make a poster) of the life of Sir Isaac Newton and his accomplishments.

Interesting issues:

- Where and when was Sir Isaac Newton born?
- Describe Newton's upbringing and his relationship with his mother and stepfather.
- It has been said that an apple was the inspiration for Newton's ideas about the force of gravity. Explain how the apple is believed to have inspired Newton's ideas.
- In a letter to Robert Hooke, Newton wrote, "If I have been able to see further, it was only because I stood on the shoulders of giants." Explain what this statement means.
- Newton is often referred to as the "Father of Calculus." What is calculus?
- What are Newton's three laws of motion? In your own words, explain what they mean.
- What did Newton invent for his pets?
- Where was Newton buried?

CHAPTER 7

Rational Expressions and Equations

In this chapter, we will examine rational expressions, which are fractions whose numerator and denominator are polynomials. We will learn to simplify, add, subtract, multiply, and divide rational expressions. Rational expressions are involved in many applied problems, such as determining the maximum load that a wooden beam can support, finding the illumination from a light source, and solving work-rate problems.

STUDY STRATEGY

Preparing for a Cumulative Exam In this chapter, we will focus on how to prepare for a cumulative exam, such as a final exam. Although some of the strategies are similar to those used to prepare for a chapter test or quiz, there are differences as well.

7.1 Rational Expressions and Functions

OBJECTIVES

1. Evaluate rational expressions.
2. Find the values for which a rational expression is undefined.
3. Evaluate rational functions.
4. Find the domain of a rational function.
5. Simplify rational expressions to lowest terms.
6. Identify factors that are opposites of each other.

A **rational expression** is a quotient of two polynomials, such as $\frac{x^2 + 15x + 44}{x^2 - 16}$. A major difference between rational expressions and linear or quadratic expressions is that a rational expression has one or more variables in the denominator. The denominator of a rational expression must not be zero, as division by zero is undefined.

Evaluating Rational Expressions

Objective 1 Evaluate rational expressions. We can evaluate a rational expression for a particular value of the variable just as we evaluated polynomials. We substitute the value for the variable in the expression and then simplify. When simplifying, we evaluate the numerator and denominator separately and then simplify the resulting fraction.

EXAMPLE 1 Evaluate the rational expression $\frac{x^2 + 2x - 24}{x^2 - 7x + 12}$ for $x = -5$.

SOLUTION Begin by substituting -5 for x.

$$\frac{(-5)^2 + 2(-5) - 24}{(-5)^2 - 7(-5) + 12} \quad \text{Substitute } -5 \text{ for } x.$$

$$= \frac{25 - 10 - 24}{25 + 35 + 12} \quad \text{Simplify the numerator and denominator separately.}$$

$$= \frac{-9}{72} \quad \text{Simplify the numerator and denominator.}$$

$$= -\frac{1}{8} \quad \text{Simplify.}$$

Quick Check 1

Evaluate the rational expression $\frac{x^2 - 3x - 18}{x^2 - 5x - 6}$ for $x = -4$.

Finding Values for Which a Rational Expression Is Undefined

Objective 2 Find the values for which a rational expression is undefined. Rational expressions are undefined for values of the variable that cause the denominator to equal 0, as division by 0 is undefined. In general, to find the values for which a rational expression is undefined, we set the denominator equal to 0, ignoring the numerator, and solve the resulting equation.

EXAMPLE 2 Find the values for which the rational expression $\frac{3}{5x + 4}$ is undefined.

SOLUTION Begin by setting the denominator, $5x + 4$, equal to 0. Then solve for x.

$$5x + 4 = 0 \quad \text{Set the denominator equal to 0.}$$

$$5x = -4 \quad \text{Subtract 4.}$$

$$x = -\frac{4}{5} \quad \text{Divide by 5.}$$

The expression $\frac{3}{5x + 4}$ is undefined for $x = -\frac{4}{5}$.

Quick Check 2

Find the values for which $\frac{3}{2x - 7}$ is undefined.

EXAMPLE 3 Find the values for which $\frac{x^2 - 7x + 12}{x^2 + 5x - 36}$ is undefined.

SOLUTION Begin by setting the denominator $x^2 + 5x - 36$ equal to 0, ignoring the numerator. Notice that the resulting equation is quadratic and can be solved by factoring.

$$x^2 + 5x - 36 = 0 \quad \text{Set the denominator equal to 0.}$$
$$(x + 9)(x - 4) = 0 \quad \text{Factor } x^2 + 5x - 36.$$
$$x + 9 = 0 \quad \text{or} \quad x - 4 = 0 \quad \text{Set each factor equal to 0.}$$
$$x = -9 \quad \text{or} \quad x = 4 \quad \text{Solve.}$$

The expression $\frac{x^2 - 7x + 12}{x^2 + 5x - 36}$ is undefined when $x = -9$ or $x = 4$.

Quick Check 3

Find the values for which $\frac{x - 4}{x^2 - 8x - 9}$ is undefined.

Rational Functions

A **rational function** $r(x)$ is a function of the form $r(x) = \frac{f(x)}{g(x)}$, where $f(x)$ and $g(x)$ are polynomials and $g(x) \neq 0$.

Evaluating Rational Functions

Objective 3 Evaluate rational functions. We begin our investigation of rational functions by learning to evaluate them.

EXAMPLE 4 For $r(x) = \frac{x^2 + 13x + 42}{x^2 + 9x + 20}$, find $r(-2)$.

SOLUTION Begin by substituting -2 for x in the function.

$$r(-2) = \frac{(-2)^2 + 13(-2) + 42}{(-2)^2 + 9(-2) + 20} \quad \text{Substitute } -2 \text{ for } x.$$
$$= \frac{4 - 26 + 42}{4 - 18 + 20} \quad \text{Simplify each term in the numerator and denominator.}$$
$$= \frac{20}{6} \quad \text{Simplify the numerator and denominator.}$$
$$= \frac{10}{3} \quad \text{Simplify.}$$

Quick Check 4

For $r(x) = \frac{x^2 - 3x - 12}{x^2 + 7x - 15}$, find $r(-6)$.

Finding the Domain of a Rational Function

Objective 4 Find the domain of a rational function. Rational functions differ from linear functions and quadratic functions in that the domain of a rational function is not always the set of real numbers. We have to exclude any value that causes the function to be undefined, namely, any value for which the denominator is equal to 0. Suppose the function $r(x)$ was undefined for $x = 6$. Then the domain of $r(x)$ is the set of all real numbers except 6. This can be expressed in interval notation as $(-\infty, 6) \cup (6, \infty)$, which is the union of the set of all real numbers that are less than 6 with the set of all real numbers that are greater than 6.

EXAMPLE 5 Find the domain of $r(x) = \dfrac{x^2 - 32x + 60}{x^2 - 9x}$.

SOLUTION Begin by setting the denominator equal to 0 and solving for x.

$$\begin{aligned} x^2 - 9x &= 0 && \text{Set the denominator equal to 0.} \\ x(x - 9) &= 0 && \text{Factor } x^2 - 9x. \\ x = 0 \quad \text{or} \quad x - 9 &= 0 && \text{Set each factor equal to 0.} \\ x = 0 \quad \text{or} \quad x &= 9 && \text{Solve each equation.} \end{aligned}$$

The domain of the function is the set of all real numbers except 0 and 9. In interval notation, this can be written as $(-\infty, 0) \cup (0, 9) \cup (9, \infty)$.

Quick Check 5

Find the domain of $r(x) = \dfrac{x^2 + 11x + 24}{x^2 - 4x - 45}$.

Simplifying Rational Expressions to Lowest Terms

Objective 5 Simplify rational expressions to lowest terms. Rational expressions are often referred to as *algebraic fractions*. As with numerical fractions, we will learn to simplify rational expressions to lowest terms. In later sections, we will learn to add, subtract, multiply, and divide rational expressions.

We simplified a numerical fraction to lowest terms by dividing out factors that were common to the numerator and denominator. For example, consider the fraction $\frac{30}{84}$. To simplify this fraction, we could begin by factoring the numerator and denominator.

$$\frac{30}{84} = \frac{2 \cdot 3 \cdot 5}{2 \cdot 2 \cdot 3 \cdot 7}$$

The numerator and denominator have common factors of 2 and 3, which are divided out to simplify the fraction to lowest terms.

$$\frac{\overset{1}{\cancel{2}} \cdot \overset{1}{\cancel{3}} \cdot 5}{\underset{1}{\cancel{2}} \cdot 2 \cdot \underset{1}{\cancel{3}} \cdot 7} = \frac{5}{2 \cdot 7}, \text{ or } \frac{5}{14}$$

Simplifying Rational Expressions

To simplify a rational expression to lowest terms, first factor the numerator and denominator completely. Then divide out any common factors in the numerator and denominator.

If P, Q, and R are polynomials, $Q \neq 0$, and $R \neq 0$, then $\dfrac{PR}{QR} = \dfrac{P}{Q}$.

EXAMPLE 6 Simplify the rational expression $\frac{10x^3}{8x^5}$. (Assume that $x \neq 0$.)

SOLUTION This rational expression has a numerator and denominator that are monomials. In this case, we can simplify the expression using the properties of exponents developed in Chapter 5.

$$\frac{10x^3}{8x^5} = \frac{\overset{5}{\cancel{10}}x^3}{\underset{4}{\cancel{8}}x^5} \quad \text{Divide out the common factor 2.}$$

$$= \frac{5x^3}{4x^5} \quad \text{Simplify.}$$

$$= \frac{5}{4x^2} \quad \text{Divide the numerator and denominator by } x^3.$$

Quick Check 6

Simplify the rational expression $\frac{15x^4}{21x^3}$. (Assume that $x \neq 0$.)

EXAMPLE 7 Simplify $\frac{x^2 + 8x - 20}{x^2 - 7x + 10}$. (Assume that the denominator is nonzero.)

SOLUTION The trinomials in the numerator and denominator must be factored before we can simplify this expression. (For a review of factoring techniques, you may refer to Sections 6.1–6.5.)

$$\frac{x^2 + 8x - 20}{x^2 - 7x + 10} = \frac{(x - 2)(x + 10)}{(x - 2)(x - 5)} \quad \text{Factor the numerator and denominator.}$$

$$= \frac{\overset{1}{\cancel{(x - 2)}}(x + 10)}{\underset{1}{\cancel{(x - 2)}}(x - 5)} \quad \text{Divide out common factors.}$$

$$= \frac{x + 10}{x - 5} \quad \text{Simplify.}$$

Quick Check 7

Simplify $\frac{x^2 + 10x + 24}{x^2 - 2x - 48}$. (Assume that the denominator is nonzero.)

A WORD OF CAUTION When we are simplifying a rational expression, we must be very careful that we divide out only expressions that are common factors of the numerator and denominator. We cannot *reduce* individual terms in the numerator and denominator as in the following examples.

$$\frac{x + 8}{x - 6} \neq \frac{x + \overset{4}{\cancel{8}}}{x - \underset{3}{\cancel{6}}} \qquad \frac{x^2 - 25}{x^2 - 36} \neq \frac{\overset{1}{\cancel{x^2}} - 25}{\underset{1}{\cancel{x^2}} - 36}$$

Factor the numerator and denominator completely before attempting to divide out common factors.

EXAMPLE 8 Simplify $\frac{x^2 - 8x}{2x^2 - 17x + 8}$. (Assume that the denominator is nonzero.)

SOLUTION The polynomials in the numerator and denominator must be factored before we can simplify this expression. The numerator $x^2 - 8x$ has a common factor of x that must be factored out first.

$$x^2 - 8x = x(x - 8)$$

The denominator $2x^2 - 17x + 8$ is a trinomial with a leading coefficient that is not equal to 1 and can be factored by grouping or by trial-and-error. (For a review of these factoring techniques, you may refer to Section 6.3.)

$$2x^2 - 17x + 8 = (2x - 1)(x - 8)$$

Now we can simplify the rational expression.

$$\frac{x^2 - 8x}{2x^2 - 17x + 8} = \frac{x(x - 8)}{(2x - 1)(x - 8)}$$ Factor the numerator and denominator.

$$= \frac{x\overset{1}{\cancel{(x - 8)}}}{(2x - 1)\underset{1}{\cancel{(x - 8)}}}$$ Divide out common factors.

$$= \frac{x}{2x - 1}$$ Simplify.

Quick Check 8

Simplify $\dfrac{3x^2 + 16x + 5}{x^2 + 8x + 15}$. (Assume that the denominator is nonzero.)

Identifying Factors in the Numerator and Denominator That Are Opposites

Objective 6 Identify factors that are opposites of each other. Two expressions of the form $a - b$ and $b - a$ are **opposites**. A difference written in the opposite order produces the opposite result. Consider the expressions $a - b$ and $b - a$ when $a = 10$ and $b = 4$. In this case, $a - b = 10 - 4$, or 6, and $b - a = 4 - 10$, or -6. We also can see that $a - b$ and $b - a$ are opposites by noting that their sum, $(a - b) + (b - a)$, is equal to 0.

This is useful to know when simplifying rational expressions. The rational expression $\dfrac{a - b}{b - a}$ simplifies to -1, as any fraction whose numerator is the opposite of its denominator is equal to -1. If a rational expression has a factor in the numerator that is the opposite of a factor in the denominator, these two factors can be divided out to equal -1, as in the next example. We write the -1 in the numerator.

EXAMPLE 9 Simplify $\dfrac{49 - x^2}{x^2 - 12x + 35}$. (Assume that the denominator is nonzero.)

SOLUTION Begin by factoring the numerator and denominator completely.

$$\frac{49 - x^2}{x^2 - 12x + 35} = \frac{(7 + x)(7 - x)}{(x - 5)(x - 7)}$$ Factor the numerator and denominator.

$$= \frac{(7 + x)\overset{-1}{\cancel{(7 - x)}}}{(x - 5)\underset{1}{\cancel{(x - 7)}}}$$ Divide out the opposite factors.

$$= -\frac{7 + x}{x - 5}$$ Simplify, writing the negative sign in front of the fraction.

Quick Check 9

Simplify $\dfrac{x^2 + 8x - 9}{1 - x^2}$. (Assume that the denominator is nonzero.)

A WORD OF CAUTION Two expressions of the form $a + b$ and $b + a$ are not opposites but are equal to each other. Addition in the opposite order produces the same result. When we divide two expressions of the form $a + b$ and $b + a$, the result is 1, not -1. For example, $\dfrac{x + 2}{2 + x} = 1$.

There is an alternative method for simplifying the expression $\frac{49 - x^2}{x^2 - 12x + 35}$ in Example 9. The numerator can be rewritten in standard form as $-x^2 + 49$; then we can factor -1 out of these terms.

$$\frac{49 - x^2}{x^2 - 12x + 35} = \frac{-x^2 + 49}{x^2 - 12x + 35}$$ Rewrite the numerator in standard form.

$$= \frac{-1(x^2 - 49)}{x^2 - 12x + 35}$$ Factor -1 from the terms in the numerator.

$$= \frac{-1(x + 7)(x - 7)}{(x - 5)(x - 7)}$$ Factor the numerator and denominator.

$$= \frac{-1(x + 7)\overset{1}{\cancel{(x - 7)}}}{(x - 5)\underset{1}{\cancel{(x - 7)}}}$$ Divide out common factors.

$$= -\frac{x + 7}{x - 5}$$ Simplify, writing the negative sign in front of the fraction.

BUILDING YOUR STUDY STRATEGY

Preparing for a Cumulative Review, 1 Study Plan To prepare for a cumulative exam effectively, develop a schedule and study plan. Your schedule should include study time every day, and you should increase the time you spend studying as you get closer to the exam date.

Before you begin to study, find out what material will be covered on the exam by visiting your instructor during office hours. You can find out the topics or chapters that will be emphasized and the format of the exam.

Exercises 7.1

PRACTICE

WATCH

DOWNLOAD

READ

REVIEW

Vocabulary

1. A(n) ______________ is a quotient of two polynomials.
2. Rational expressions are undefined for values of the variable that cause the ____________ to equal 0.
3. A(n) ______________ $r(x)$ is a function of the form $r(x) = \frac{f(x)}{g(x)}$, where $f(x)$ and $g(x)$ are polynomials and $g(x) \neq 0$.
4. The ________ of a rational function excludes all values for which the function is undefined.
5. A rational expression is said to be in __________ if its numerator and denominator do not have any common factors.
6. Two expressions of the form $a - b$ and $b - a$ are __________.

Evaluate the rational expression for the given value of the variable.

7. $\frac{6}{x + 4}$ for $x = 4$

8. $\frac{9}{x - 20}$ for $x = 5$

9. $\frac{x + 3}{x - 8}$ for $x = -25$

10. $\frac{x + 1}{x + 13}$ for $x = -7$

11. $\frac{x^2 - 13x - 48}{x^2 - 6x - 12}$ for $x = 3$

12. $\frac{x^2 + 5x - 23}{x^2 - 7x + 20}$ for $x = 2$

13. $\frac{x^2 - 6x + 15}{x^2 + 14x - 7}$ for $x = -9$

14. $\frac{x^2 + 10x - 56}{x - 4}$ for $x = -7$

Find all values of the variable for which the rational expression is undefined.

15. $\frac{6}{x-5}$

16. $\frac{x-2}{x+6}$

17. $\frac{x+3}{x^2+x-56}$

18. $\frac{x-8}{x^2+10x+9}$

19. $\frac{x^2+7x+10}{3x^2+13x-10}$

20. $\frac{x^2-7x+12}{2x^2-x-15}$

21. $\frac{6x}{x^2-36}$

22. $\frac{x-4}{x^2-16x}$

Evaluate the given rational function.

23. $r(x) = \frac{20}{x^2-5x+10}, r(-5)$

24. $r(x) = \frac{x}{x^2-13x-20}, r(8)$

25. $r(x) = \frac{x^2+10x+24}{x^2-5x-66}, r(4)$

26. $r(x) = \frac{x^2-100}{x^2+13x+30}, r(-3)$

27. $r(x) = \frac{x^3-7x^2-11x+20}{x^2+8x-20}, r(10)$

28. $r(x) = \frac{x^3+8x^2+17x+10}{x^3+3x^2-18x-40}, r(2)$

Find the domain of the given rational function.

29. $r(x) = \frac{x^2+18x+77}{x^2+10x}$

30. $r(x) = \frac{x^2-16x+60}{x^2-9x+8}$

31. $r(x) = \frac{x^2+2x-3}{x^2-2x-15}$

32. $r(x) = \frac{x^2+7x-8}{x^2-64}$

33. $r(x) = \frac{x^2+4x-60}{x^2+3x-18}$

34. $r(x) = \frac{x^2-13x+36}{x^2+14x+45}$

Identify the given function as a linear function, a quadratic function, or a rational function.

35. $f(x) = x^2 - 11x + 30$

36. $f(x) = 3x - 8$

37. $f(x) = \frac{x^2}{x^3-5x^2+11x-35}$

38. $f(x) = \frac{x^2+3x+8}{x^2-5x-5}$

39. $f(x) = \frac{1}{5}x + \frac{4}{9}$

40. $f(x) = -\frac{1}{2}x^2 + 3x - \frac{1}{7}$

Simplify the given rational expression. (Assume that all denominators are nonzero.)

41. $\frac{3x^5}{9x^8}$

42. $\frac{10x^7}{12x^{10}}$

43. $\frac{x+5}{x^2+11x+30}$

44. $\frac{x^2-64}{x-8}$

45. $\frac{x^2+11x+28}{x^2+4x-21}$

46. $\frac{x^2-7x-18}{x^2+8x+12}$

47. $\frac{x^2-4x}{x^2-14x+40}$

48. $\frac{x^2+10x+25}{x^2-25}$

49. $\frac{x^2-4x-45}{x^3+125}$

50. $\frac{x^2+6x-40}{x^3+1000}$

51. $\frac{2x^2+13x+6}{2x^2-7x-4}$

52. $\frac{4x^2-25}{2x^2-19x+35}$

53. $\frac{3x^2-25x-50}{x^2-3x-70}$

54. $\frac{x^2+7x+12}{4x^2+9x-9}$

55. $\frac{3x^2-6x-105}{x^2-11x+28}$

56. $\frac{2x^2+14x+20}{5x^2-25x-70}$

Determine whether the two given binomials are or are not opposites.

57. $x + 5$ and $5 + x$

58. $x - 11$ and $11 - x$

59. $14 - x$ and $x - 14$

60. $3x - 2$ and $2x - 3$

61. $2x + 13$ and $-2x - 13$

62. $6x - 7$ and $6x + 7$

Simplify the given rational expression. (Assume that all denominators are nonzero.)

63. $\dfrac{49 - x^2}{x^2 - 12x + 35}$

64. $\dfrac{18 - 2x}{x^2 - 3x - 54}$

65. $\dfrac{8x - x^2}{x^2 - 19x + 88}$

66. $\dfrac{36 - x^2}{x^2 - 12x + 36}$

67. $\dfrac{x^3 - 8}{4 - x^2}$

68. $\dfrac{x^3 - 27}{3 - x}$

Use the given graph of a rational function r(x) to solve the problems that follow.

69.

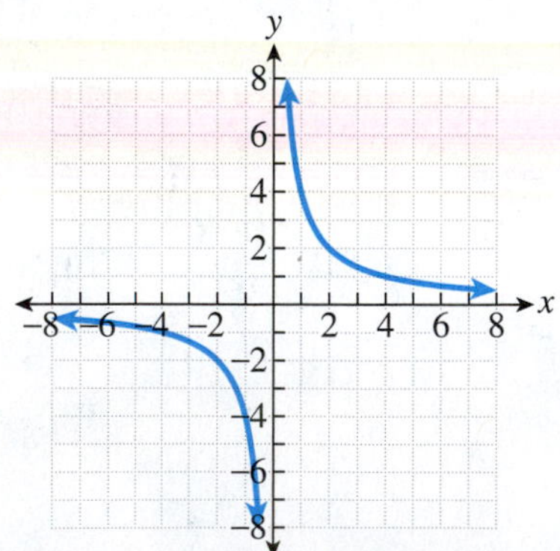

a) Find $r(1)$. **b)** Find $r(-4)$.
c) Find all values x such that $r(x) = 1$.

70.

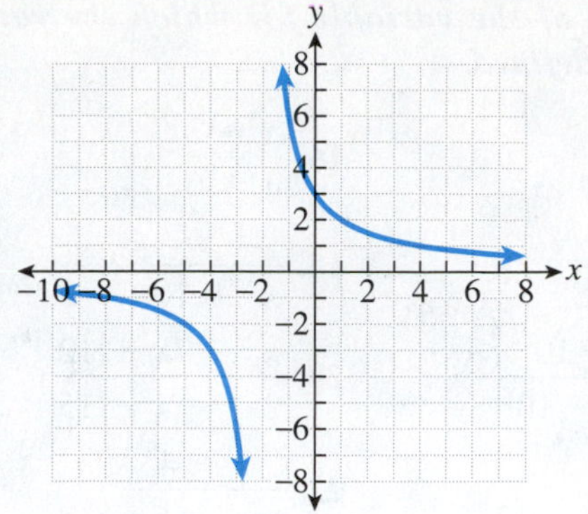

a) Find $r(0)$.
b) Find $r(4)$.
c) Find all values x such that $r(x) = -1$.

Writing in Mathematics

Answer in complete sentences.

71. Explain how to find the values for which a rational expression is undefined.

72. Is the rational expression $\dfrac{(x + 2)(x - 7)}{(x - 5)(x + 2)}$ undefined for the value $x = -2$? Explain your answer

73. Explain how to determine whether two factors are opposites. Use examples.

Quick Review Exercises

Section 7.1

Simplify.

1. $\dfrac{5}{12} + \dfrac{4}{15}$

2. $\dfrac{17}{24} - \dfrac{3}{10}$

3. $\dfrac{14}{45} \cdot \dfrac{165}{308}$

4. $\dfrac{20}{99} \div \dfrac{35}{51}$

7.2 Multiplication and Division of Rational Expressions

OBJECTIVES

1. Multiply two rational expressions.
2. Multiply two rational functions.
3. Divide a rational expression by another rational expression.
4. Divide a rational function by another rational function.

Multiplying Rational Expressions

Objective 1 Multiply two rational expressions. In this section, we will learn how to multiply and divide rational expressions. Multiplying rational expressions is similar to multiplying numerical fractions. Suppose we need to multiply $\frac{4}{9} \cdot \frac{21}{10}$. Before multiplying, we can divide out factors common to one of the numerators and one of the denominators. For example, the first numerator (4) and the second denominator (10) have a common factor of 2 that can be divided out of each. The second numerator (21) and the first denominator (9) have a common factor of 3 that can be divided out as well.

$$\frac{4}{9} \cdot \frac{21}{10} = \frac{2 \cdot 2}{3 \cdot 3} \cdot \frac{3 \cdot 7}{2 \cdot 5}$$ Factor each numerator and denominator.

$$= \frac{\overset{1}{\cancel{2}} \cdot 2}{\underset{1}{\cancel{3}} \cdot 3} \cdot \frac{\overset{1}{\cancel{3}} \cdot 7}{\underset{1}{\cancel{2}} \cdot 5}$$ Divide out common factors.

$$= \frac{14}{15}$$ Multiply remaining factors.

Multiplying Rational Expressions

$$\frac{A}{B} \cdot \frac{C}{D} = \frac{AC}{BD} \quad (B \neq 0 \text{ and } D \neq 0)$$

To multiply two rational expressions, we will begin by factoring each numerator and denominator completely. After dividing out factors common to a numerator and a denominator, we will express the product of the two rational expressions as a single rational expression, leaving the numerator and denominator in factored form.

EXAMPLE 1 Multiply: $\frac{x^2 - 11x + 30}{x^2 + 5x - 24} \cdot \frac{x^2 - 9}{x^2 - 2x - 15}$

SOLUTION Begin by factoring each numerator and denominator. Then divide out common factors.

$$\frac{x^2 - 11x + 30}{x^2 + 5x - 24} \cdot \frac{x^2 - 9}{x^2 - 2x - 15}$$

$$= \frac{(x - 5)(x - 6)}{(x + 8)(x - 3)} \cdot \frac{(x + 3)(x - 3)}{(x - 5)(x + 3)}$$ Factor the numerators and denominators completely.

$$= \frac{\overset{1}{\cancel{(x - 5)}}(x - 6)}{(x + 8)\underset{1}{\cancel{(x - 3)}}} \cdot \frac{\overset{1}{\cancel{(x + 3)}}\overset{1}{\cancel{(x - 3)}}}{\underset{1}{\cancel{(x - 5)}}\underset{1}{\cancel{(x + 3)}}}$$ Divide out common factors.

$$= \frac{x - 6}{x + 8}$$ Simplify.

Quick Check 1

Multiply:
$\frac{x^2 - 3x - 18}{x^2 - 9x + 20} \cdot \frac{x^2 - 3x - 10}{x^2 + 13x + 30}$

EXAMPLE 2 Multiply: $\dfrac{x^2 - 10x}{x^2 - 2x - 8} \cdot \dfrac{16 - x^2}{x^2 - 9x - 10}$

SOLUTION Again, begin by completely factoring both numerators and denominators.

$$\frac{x^2 - 10x}{x^2 - 2x - 8} \cdot \frac{16 - x^2}{x^2 - 9x - 10}$$

$$= \frac{x(x - 10)}{(x + 2)(x - 4)} \cdot \frac{(4 + x)(4 - x)}{(x + 1)(x - 10)}$$ Factor completely.

$$= \frac{x\overset{1}{\cancel{(x - 10)}}}{(x + 2)\underset{1}{\cancel{(x - 4)}}} \cdot \frac{(4 + x)\overset{-1}{\cancel{(4 - x)}}}{(x + 1)\underset{1}{\cancel{(x - 10)}}}$$ Divide out common factors. Notice that the factors $4 - x$ and $x - 4$ are opposites.

$$= \frac{-x(4 + x)}{(x + 2)(x + 1)}$$ Multiply remaining factors.

$$= -\frac{x(4 + x)}{(x + 2)(x + 1)}$$ Write the negative sign in the numerator in front of the fraction.

Quick Check 2

Multiply: $\dfrac{36 - x^2}{x^2 + 9x + 14} \cdot \dfrac{x^2 - 7x - 18}{x^2 - 15x + 54}$

Here is a summary of the procedure for multiplying rational expressions.

Multiplying Rational Expressions

- Completely factor each numerator and each denominator.
- Divide out factors that are common to a numerator and a denominator, as well as factors in a numerator and denominator that are opposites.
- Multiply the remaining factors, leaving the numerator and denominator in factored form.

Multiplying Rational Functions

Objective 2 Multiply two rational functions.

EXAMPLE 3 For $f(x) = \dfrac{x^2 - 3x - 18}{x^2 + 8x + 16}$ and $g(x) = \dfrac{x^2 + 3x - 4}{9 - x^2}$, find $f(x) \cdot g(x)$.

SOLUTION Replace $f(x)$ and $g(x)$ with their formulas and proceed to multiply.

$$f(x) \cdot g(x) = \frac{x^2 - 3x - 18}{x^2 + 8x + 16} \cdot \frac{x^2 + 3x - 4}{9 - x^2}$$ Replace $f(x)$ and $g(x)$ with their formulas.

$$= \frac{(x - 6)(x + 3)}{(x + 4)(x + 4)} \cdot \frac{(x + 4)(x - 1)}{(3 + x)(3 - x)}$$ Factor completely.

$$= \frac{(x - 6)\overset{1}{\cancel{(x + 3)}}}{\underset{1}{\cancel{(x + 4)}}(x + 4)} \cdot \frac{\overset{1}{\cancel{(x + 4)}}(x - 1)}{\underset{1}{\cancel{(3 + x)}}(3 - x)}$$ Divide out common factors.

$$= \frac{(x - 6)(x - 1)}{(x + 4)(3 - x)}$$ Multiply remaining factors.

Quick Check 3

For $f(x) = \dfrac{x + 4}{49 - x^2}$ and $g(x) = \dfrac{x^2 - 10x + 21}{x^2 + x - 12}$, find $f(x) \cdot g(x)$.

Dividing a Rational Expression by Another Rational Expression

Objective 3 Divide a rational expression by another rational expression. Dividing a rational expression by another rational expression is similar to dividing a numerical fraction by another numerical fraction. We replace the divisor, which is the rational expression by which we are dividing, with its reciprocal and then multiply.

Dividing Rational Expressions

$$\frac{A}{B} \div \frac{C}{D} = \frac{A}{B} \cdot \frac{D}{C} \qquad (B \neq 0, C \neq 0, \text{and } D \neq 0)$$

EXAMPLE 4 Divide: $\frac{x^2 + 13x + 42}{x^2 + x - 20} \div \frac{x^2 + 8x + 12}{x^2 - 4x}$

SOLUTION Begin by inverting the divisor and multiplying. Then factor each numerator and denominator completely.

$$\frac{x^2 + 13x + 42}{x^2 + x - 20} \div \frac{x^2 + 8x + 12}{x^2 - 4x}$$

$$= \frac{x^2 + 13x + 42}{x^2 + x - 20} \cdot \frac{x^2 - 4x}{x^2 + 8x + 12} \quad \text{Invert the divisor and multiply.}$$

$$= \frac{(x + 6)(x + 7)}{(x + 5)(x - 4)} \cdot \frac{x(x - 4)}{(x + 2)(x + 6)} \quad \text{Factor completely.}$$

$$= \frac{\overset{1}{\cancel{(x + 6)}}(x + 7)}{(x + 5)\underset{1}{\cancel{(x - 4)}}} \cdot \frac{x\overset{1}{\cancel{(x - 4)}}}{(x + 2)\underset{1}{\cancel{(x + 6)}}} \quad \text{Divide out common factors.}$$

$$= \frac{x(x + 7)}{(x + 5)(x + 2)} \quad \text{Multiply remaining factors.}$$

Quick Check 4

Divide:
$\frac{x^2 - 4}{x^2 - 5x + 4} \div \frac{x^2 - 6x - 16}{x^2 + 3x - 28}$

EXAMPLE 5 Divide: $\frac{x^2 - 14x - 15}{x^2 + 2x - 35} \div \frac{x^2 - 1}{5 - x}$

SOLUTION Rewrite the problem as a multiplication problem by inverting the divisor. Then factor each numerator and denominator completely before dividing out common factors. (You may want to factor at the same time you invert the divisor.)

$$\frac{x^2 - 14x - 15}{x^2 + 2x - 35} \div \frac{x^2 - 1}{5 - x}$$

$$= \frac{x^2 - 14x - 15}{x^2 + 2x - 35} \cdot \frac{5 - x}{x^2 - 1} \quad \text{Invert the divisor and multiply.}$$

$$= \frac{(x - 15)(x + 1)}{(x + 7)(x - 5)} \cdot \frac{5 - x}{(x + 1)(x - 1)} \quad \text{Factor completely.}$$

$$= \frac{(x - 15)\overset{1}{\cancel{(x + 1)}}}{(x + 7)\underset{1}{\cancel{(x - 5)}}} \cdot \frac{\overset{-1}{\cancel{5 - x}}}{\underset{1}{\cancel{(x + 1)}}(x - 1)} \quad \text{Divide out common factors. Note that } 5 - x \text{ and } x - 5 \text{ are opposites.}$$

$$= -\frac{x - 15}{(x + 7)(x - 1)} \quad \text{Multiply remaining factors.}$$

Quick Check 5

Divide:
$\frac{x^2 - 13x + 36}{x^2 - x - 12} \div \frac{9 - x}{x^2 + 8x + 15}$

Here is a summary of the procedure for dividing rational expressions.

Dividing Rational Expressions

- Invert the divisor and change the operation from division to multiplication.
- Completely factor each numerator and each denominator.
- Divide out factors that are common to a numerator and a denominator and factors in a numerator and denominator that are opposites.
- Multiply the remaining factors, leaving the numerator and denominator in factored form.

Dividing Rational Functions

Objective 4 Divide a rational function by another rational function.

EXAMPLE 6 For $f(x) = \frac{x^2 - x - 12}{x + 7}$ and $g(x) = x^2 + 3x - 28$, find $f(x) \div g(x)$.

SOLUTION Replace $f(x)$ and $g(x)$ with their formulas and divide. Treat $g(x)$ as a rational function with a denominator of 1. We can see that its reciprocal is $\frac{1}{x^2 + 3x - 28}$.

$f(x) \div g(x)$

$= \frac{x^2 - x - 12}{x + 7} \div x^2 + 3x - 28$ Replace $f(x)$ and $g(x)$ with their formulas.

$= \frac{x^2 - x - 12}{x + 7} \cdot \frac{1}{x^2 + 3x - 28}$ Invert the divisor and multiply.

$= \frac{(x - 4)(x + 3)}{x + 7} \cdot \frac{1}{(x + 7)(x - 4)}$ Factor completely.

$= \frac{\overset{1}{\cancel{(x - 4)}}(x + 3)}{x + 7} \cdot \frac{1}{(x + 7)\underset{1}{\cancel{(x - 4)}}}$ Divide out common factors.

$= \frac{x + 3}{(x + 7)^2}$ Multiply remaining factors.

Quick Check 6

For $f(x) = \frac{x^2 - 12x + 36}{x^2 - 11x + 10}$ and $g(x) = x^2 - 8x + 12$, find $f(x) \div g(x)$.

BUILDING YOUR STUDY STRATEGY

Preparing for a Cumulative Exam, 2 Old Quizzes and Tests To prepare for a cumulative exam, begin by reviewing old exams and quizzes. Review and make sure you understand any mistakes you made on them.

Try reworking all of the problems on the exam or quiz. So that you know which topics you need to review and which topics you have under control, you should rework the problems without referring to your notes or textbook. You need to focus on the topics you are struggling with and spend less time on the topics you understand.

Exercises 7.2

MyMathLab

PRACTICE

WATCH

DOWNLOAD

READ

REVIEW

Vocabulary

1. To multiply two rational expressions, begin by ________ each numerator and denominator completely.
2. When you are multiplying two rational expressions, ________ out factors common to a numerator and a denominator.
3. When you are multiplying two rational expressions, once the numerator and the denominator do not share any common factors, express the product as a single rational expression, leaving the numerator and denominator in ________ form.
4. When you are dividing by a rational expression, replace the divisor by its ________ and then multiply.

Multiply.

5. $\dfrac{x^2+15x+54}{x^2+2x-15}\cdot\dfrac{x^2-25}{x^2+4x-45}$
6. $\dfrac{x^2+14x+49}{x^2-10x+21}\cdot\dfrac{x^2+5x-24}{x^2-x-56}$
7. $\dfrac{121-x^2}{x^2-22x+120}\cdot\dfrac{x^2-24x+140}{x^2-18x+77}$
8. $\dfrac{x^2-6x}{x^2+2x-80}\cdot\dfrac{x^2-6x-16}{x^2-8x+12}$
9. $\dfrac{x^2-8x-33}{x^2-8x+16}\cdot\dfrac{2x^2-7x-4}{x^2+5x+6}$
10. $\dfrac{x^2+12x+20}{x^2-13x+40}\cdot\dfrac{x^2-x-56}{x^2+18x+80}$
11. $\dfrac{x^2-11x}{x^2+5x+4}\cdot\dfrac{x^2+10x+9}{x^2+6x}$
12. $\dfrac{x^2-6x-55}{36-x^2}\cdot\dfrac{x^2-4x-12}{2x^2+15x+25}$
13. $\dfrac{x^3-8}{x^2+x-42}\cdot\dfrac{x^2+9x+14}{x^2-4}$
14. $\dfrac{x^2+6x-55}{x^2+6x}\cdot\dfrac{x^3+216}{x^2-9x+20}$
15. $\dfrac{9-x^2}{x^2-x-90}\cdot\dfrac{x^2+13x+36}{x^2+10x+21}$
16. $\dfrac{x^2+3x-40}{4x^2+29x+7}\cdot\dfrac{x^2+5x-14}{x^2+14x+48}$
17. $\dfrac{x^2-11x+24}{x^2-7x+12}\cdot\dfrac{x^2+4x-32}{x^2-64}$
18. $\dfrac{2x^2+24x+22}{8x^2+104x+320}\cdot\dfrac{x^2+3x-10}{x^2+7x+6}$
19. $\dfrac{2x^2-x-36}{x^2+x-90}\cdot\dfrac{9x-x^2}{x^2-2x-24}$
20. $\dfrac{x^2-3x-10}{x^2+2x}\cdot\dfrac{x^2+8x}{x^2+3x-40}$

For the given functions $f(x)$ and $g(x)$, find $f(x)\cdot g(x)$.

21. $f(x)=\dfrac{2}{x-3},\ g(x)=\dfrac{x^2-12x+27}{x^2-5x-36}$
22. $f(x)=\dfrac{x^2-9x+20}{x^2+x-2},\ g(x)=\dfrac{x^2+6x-16}{x^2-25}$
23. $f(x)=\dfrac{x^2-13x+22}{x^2+11x-12},\ g(x)=\dfrac{x^2+4x-5}{x^2-6x+8}$
24. $f(x)=x+7,\ g(x)=\dfrac{x^2+3x-28}{x^2+14x+49}$

Divide.

25. $\dfrac{x^2+3x-28}{x^2-36}\div\dfrac{x^2-13x+36}{x^2+11x+30}$
26. $\dfrac{x^2-9x+8}{x^2+6x-16}\div\dfrac{x^2+x-72}{4-x^2}$
27. $\dfrac{x^2-6x+9}{x^2-8x-9}\div\dfrac{x^2-9}{x^2+13x+12}$
28. $\dfrac{x^2+5x-50}{x^2-9x+20}\div\dfrac{x^2-4x-60}{x^2+2x-24}$
29. $\dfrac{2x^2-13x+20}{x^2+x-6}\div\dfrac{5x-2x^2}{x^2+14x+33}$

30. $\frac{x^2 - 15x + 56}{x^2 + 11x + 18} \div \frac{2x^2 - 15x + 7}{4x^2 - 16x - 48}$

31. $\frac{x^2 + 6x - 7}{x^2 + 16x + 55} \div \frac{x^2 + 13x + 42}{x^2 + 7x - 44}$

32. $\frac{x^2 + 13x + 36}{x^2 - 12x + 32} \div \frac{x^2 + 6x - 27}{16 - x^2}$

33. $\frac{x^2 - 4x - 5}{x^2 + x - 30} \div \frac{x^2 - x - 2}{x^2 + 4x - 12}$

34. $\frac{x^2 - 12x + 20}{x^2 + 6x + 5} \div \frac{x^2 - 11x + 18}{x^2 - 3x - 40}$

35. $\frac{x^2 - 2x - 63}{x^2 - 4x - 21} \div \frac{81 - x^2}{x^2 - 7x - 30}$

36. $\frac{x^2 - 6x - 27}{x^2 + 3x - 4} \div \frac{x^2 - 18x + 81}{2x^2 - 9x + 7}$

37. $\frac{x^2 + 10x + 24}{4x^2 - 19x - 5} \div \frac{x^2 + 16x + 60}{x^2 + 5x - 50}$

38. $\frac{x^2 - 15x + 54}{x^2 - 16x + 48} \div \frac{x^2 - 14x + 48}{x^2 - 2x - 8}$

39. $\frac{x^2 + 16x + 63}{2x - x^2} \div \frac{x^2 + x - 72}{x^2 - 5x + 6}$

40. $\frac{x^2 - 12x + 36}{x^2 + 8x + 15} \div \frac{6x^2 - x^3}{x^2 + 5x}$

For the given functions $f(x)$ and $g(x)$, find $f(x) \div g(x)$.

41. $f(x) = \frac{x^2 + 8x - 9}{x^2 - 10x + 25}, g(x) = x + 9$

42. $f(x) = 2x + 14, g(x) = \frac{x^2 + 15x + 56}{x^2 - 10x + 21}$

43. $f(x) = \frac{2x^2 - 5x}{x^2 - 1}, g(x) = \frac{2x^2 + x - 15}{x^2 - 8x - 9}$

44. $f(x) = \frac{x^2 + 4x + 4}{x^2 + 14x + 40}, g(x) = \frac{3x^2 + 7x + 2}{5x^2 + 19x - 4}$

Find the missing numerator and denominator.

45. $\frac{x^2 - 10x + 16}{x^2 + 4x - 77} \cdot \frac{?}{?} = \frac{x^2 - 11x + 24}{x^2 - 5x - 14}$

46. $\frac{x^2 + 4x}{x^2 - 81} \cdot \frac{?}{?} = \frac{x^2 - 6x}{x^2 - 7x - 18}$

47. $\frac{x^2 - 10x + 9}{x^2 + 7x} \div \frac{?}{?} = \frac{x^2 - 13x + 12}{x^2 + 11x + 28}$

48. $\frac{x^2 + 12x + 36}{x^2 - 100} \div \frac{?}{?} = \frac{x^2 - 36}{x^2 + 10x}$

Mixed Practice 49–60

Simplify.

49. $\frac{x^2 - 9x - 10}{100 - x^2} \cdot \frac{x^2 + 4x - 60}{x^2 - 7x + 6}$

50. $\frac{x^2 - x - 12}{81 - x^2} \div \frac{x^2 - 3x - 4}{x^2 - 11x + 18}$

51. $\frac{3x^2 - 13x - 10}{2x^2 - 17x + 35}$

52. $\frac{x^2 - 6x - 27}{x^2 - 9x}$

53. $\frac{x^2 - 14x + 24}{x^2 + x - 6} \div \frac{x^2 + 2x - 80}{x^2 + 13x + 30}$

54. $\frac{x^2 + 9x + 14}{x^2 - 6x + 8} \cdot \frac{x^2 + x - 20}{x^2 - x - 6}$

55. $\frac{x^2 - 20x + 96}{144 - x^2}$

56. $\frac{2x^2 - 9x - 18}{3x^2 - 23x - 8} \div \frac{2x^2 + 13x + 15}{x^2 - 3x - 40}$

57. $\frac{x^2 + 5x - 84}{x^2 - x - 12} \cdot \frac{x^2 + 11x + 24}{2x^2 - 15x + 7}$

58. $\frac{x^3 + 27}{x^2 - 7x - 30}$

59. $\frac{x^2 + 3x - 54}{x^2 - 2x - 3} \div \frac{x^2 + 9x}{x^2 - 7x - 8}$

60. $\frac{x^2 + 6x - 16}{x^2 - x - 20} \cdot \frac{x^2 + 6x - 55}{x^2 + 17x + 72}$

Writing in Mathematics

Answer in complete sentences.

61. Explain the similarities between dividing numerical fractions and dividing rational expressions. Are there any differences?

Quick Review Exercises

Section 7.2

Simplify.

1. $(7x + 8) + (2x - 5)$
2. $(x^2 + 3x - 40) + (x^2 - 9x + 11)$
3. $(8x - 15) - (5x + 27)$
4. $(x^2 - 4x + 6) - (3x^2 - 2x + 35)$

7.3 Addition and Subtraction of Rational Expressions That Have the Same Denominator

OBJECTIVES

1. Add rational expressions that have the same denominator.
2. Subtract rational expressions that have the same denominator.
3. Add or subtract rational expressions that have opposite denominators.

Now that we have learned how to multiply and divide rational expressions, we move on to addition and subtraction. We know from our work with numerical fractions that two fractions must have the same denominator before we can add or subtract them. The same holds true for rational expressions. In this section, we will begin with rational expressions that already have the same denominator. In the next section, we will learn how to add and subtract rational expressions that have different denominators.

Adding Rational Expressions That Have the Same Denominator

Objective 1 Add rational expressions that have the same denominator. To add fractions that have the same denominator, we add the numerators and place the result over the common denominator. We will follow the same procedure when adding two rational expressions. Of course, we should check that the result is in simplest terms.

Adding Rational Expressions That Have the Same Denominator

$$\frac{A}{C} + \frac{B}{C} = \frac{A + B}{C} \qquad C \neq 0$$

EXAMPLE 1 Add: $\frac{7}{x + 3} + \frac{5}{x + 3}$

SOLUTION These two fractions have the same denominator, so we add the two numerators and place the result over the common denominator $x + 3$.

$$\frac{7}{x + 3} + \frac{5}{x + 3} = \frac{7 + 5}{x + 3} \quad \text{Add numerators, placing the sum over the common denominator.}$$

$$= \frac{12}{x + 3} \quad \text{Simplify the numerator.}$$

The numerator and denominator do not have any common factors, so this is the final result. A common error is to attempt to divide a common factor out of 12 in the numerator and 3 in the denominator, but the number 3 is a term of the denominator, not a factor.

Quick Check 1

Add: $\frac{9}{2x + 7} + \frac{12}{2x + 7}$

EXAMPLE 2 Add: $\dfrac{6x - 7}{x^2 + 5x + 6} + \dfrac{2x + 23}{x^2 + 5x + 6}$

SOLUTION The two denominators are the same, so we add.

$$\frac{6x - 7}{x^2 + 5x + 6} + \frac{2x + 23}{x^2 + 5x + 6}$$

$$= \frac{(6x - 7) + (2x + 23)}{x^2 + 5x + 6} \quad \text{Add the numerators.}$$

$$= \frac{8x + 16}{x^2 + 5x + 6} \quad \text{Combine like terms.}$$

$$= \frac{8\overset{1}{\cancel{(x + 2)}}}{\underset{1}{\cancel{(x + 2)}}(x + 3)} \quad \text{Factor the numerator and denominator and divide out the common factor.}$$

$$= \frac{8}{x + 3} \quad \text{Simplify.}$$

Quick Check 2

Add: $\dfrac{2x + 54}{x^2 + 4x - 32} + \dfrac{3x - 14}{x^2 + 4x - 32}$

EXAMPLE 3 Add: $\dfrac{x^2 - 10x}{x^2 - 5x - 36} + \dfrac{7x - 54}{x^2 - 5x - 36}$

SOLUTION The denominators are the same, so we add the numerators and then simplify.

$$\frac{x^2 - 10x}{x^2 - 5x - 36} + \frac{7x - 54}{x^2 - 5x - 36}$$

$$= \frac{(x^2 - 10x) + (7x - 54)}{x^2 - 5x - 36} \quad \text{Add numerators.}$$

$$= \frac{x^2 - 3x - 54}{x^2 - 5x - 36} \quad \text{Combine like terms.}$$

$$= \frac{(x + 6)\overset{1}{\cancel{(x - 9)}}}{(x + 4)\underset{1}{\cancel{(x - 9)}}} \quad \text{Factor the numerator and denominator and divide out the common factor.}$$

$$= \frac{x + 6}{x + 4} \quad \text{Simplify.}$$

Quick Check 3

Add: $\dfrac{x^2 + 6x + 4}{x^2 + 4x - 21} + \dfrac{5x + 24}{x^2 + 4x - 21}$

Subtracting Rational Expressions That Have the Same Denominator

Objective 2 Subtract rational expressions that have the same denominator. Subtracting two rational expressions that have the same denominator is just like adding them except that we subtract the two numerators rather than add them.

Subtracting Rational Expressions That Have the Same Denominator

$$\frac{A}{C} - \frac{B}{C} = \frac{A - B}{C} \qquad C \neq 0$$

EXAMPLE 4 Subtract: $\frac{9}{2x + 8} - \frac{5}{2x + 8}$

SOLUTION Because the two fractions have the same denominator, subtract the numerators and place the result over the common denominator.

$$\frac{9}{2x + 8} - \frac{5}{2x + 8} = \frac{4}{2x + 8}$$ Subtract the numerators and place the difference over the denominator.

$$= \frac{\overset{2}{\cancel{4}}}{\underset{1}{\cancel{2}}(x + 4)}$$ Factor the denominator and divide out common factors.

$$= \frac{2}{x + 4}$$ Simplify.

Quick Check 4

Subtract: $\frac{25}{6x + 36} - \frac{17}{6x + 36}$

EXAMPLE 5 Subtract: $\frac{2x^2 - 3x - 9}{7x - x^2} - \frac{x^2 + 9x - 44}{7x - x^2}$

SOLUTION The denominators are the same, so we can subtract these two rational expressions. When the numerator of the second fraction has more than one term, we must remember that we are subtracting the whole numerator, not just the first term. When we subtract the numerators and place the difference over the common denominator, it is a good idea to write each numerator within parentheses. This will remind us to subtract each term in the second numerator.

$$\frac{2x^2 - 3x - 9}{7x - x^2} - \frac{x^2 + 9x - 44}{7x - x^2}$$

$$= \frac{(2x^2 - 3x - 9) - (x^2 + 9x - 44)}{7x - x^2}$$ Subtract the numerators.

$$= \frac{2x^2 - 3x - 9 - x^2 - 9x + 44}{7x - x^2}$$ Change the sign of each term in the second set of parentheses by distributing -1.

$$= \frac{x^2 - 12x + 35}{7x - x^2}$$ Combine like terms.

$$= \frac{\overset{-1}{\cancel{(x - 7)}}(x - 5)}{x\underset{1}{\cancel{(7 - x)}}}$$ Factor the numerator and denominator and divide out the common factor. The -1 results from the fact that $x - 7$ and $7 - x$ are opposites.

$$= -\frac{x - 5}{x}$$ Simplify.

Quick Check 5

Subtract: $\frac{3x^2 + 6x - 37}{64 - x^2} - \frac{2x^2 + 17x - 61}{64 - x^2}$

A WORD OF CAUTION When subtracting a rational expression that has a numerator containing more than one term, be sure to subtract the *entire* numerator, not just the first term. One way to remember this is by placing the numerators inside parentheses.

Adding or Subtracting Rational Expressions That Have Opposite Denominators

Objective 3 Add or subtract rational expressions that have opposite denominators. Consider the expression $\frac{10}{x-2}+\frac{3}{2-x}$. Although the two denominators are not the same, they are opposites. We can rewrite the second denominator by factoring out a -1.

$$\frac{10}{x-2}+\frac{3}{2-x}=\frac{10}{x-2}+\frac{3}{-1(x-2)}$$ Factor -1 out of the denominator.

$$=\frac{10}{x-2}+(-1)\cdot\frac{3}{x-2}$$ Rewrite the second fraction.

$$=\frac{10}{x-2}-\frac{3}{x-2}$$ Rewrite as a difference.

$$=\frac{7}{x-2}$$ Subtract.

In general, when adding two rational expressions with opposite denominators, we can rewrite the second denominator as its opposite as long as we change the operation to subtraction.

Similarly, when subtracting two rational expressions with opposite denominators, we can rewrite the second denominator as its opposite by changing the operation to addition.

EXAMPLE 6 Add: $\frac{7x}{2x-16}+\frac{56}{16-2x}$

SOLUTION The two denominators are opposites, so we may change the second denominator to $2x-16$ by changing the operation from addition to subtraction.

$$\frac{7x}{2x-16}+\frac{56}{16-2x}$$

$$=\frac{7x}{2x-16}-\frac{56}{2x-16}$$ Rewrite the second denominator as $2x-16$ by changing the operation from addition to subtraction.

$$=\frac{7x-56}{2x-16}$$ Subtract the numerators.

$$=\frac{7\overset{1}{\cancel{(x-8)}}}{2\underset{1}{\cancel{(x-8)}}}$$ Factor the numerator and denominator and divide out the common factor.

$$=\frac{7}{2}$$ Simplify.

Quick Check 6

Add: $\frac{x}{5x-20}+\frac{4}{20-5x}$

EXAMPLE 7 Subtract: $\dfrac{x^2 - 4x}{x - 3} - \dfrac{11x - 30}{3 - x}$

SOLUTION These two denominators are opposites, so we begin by rewriting the second rational expression in such a way that the two rational expressions have the same denominator.

$$\frac{x^2 - 4x}{x - 3} - \frac{11x - 30}{3 - x}$$

$$= \frac{x^2 - 4x}{x - 3} + \frac{11x - 30}{x - 3}$$ Rewrite the second denominator as $x - 3$ by changing the operation from subtraction to addition.

$$= \frac{(x^2 - 4x) + (11x - 30)}{x - 3}$$ Add the numerators.

$$= \frac{x^2 + 7x - 30}{x - 3}$$ Combine like terms.

$$= \frac{(x + 10)\overset{1}{\cancel{(x - 3)}}}{\underset{1}{\cancel{x - 3}}}$$ Factor the numerator and divide out the common factor.

$$= x + 10$$ Simplify.

Quick Check 7

Subtract:

$\dfrac{x^2 - 2x - 11}{x^2 - 16} - \dfrac{7x - 25}{16 - x^2}$

BUILDING YOUR STUDY STRATEGY

Preparing for a Cumulative Exam, 3 Homework and Notes When you are studying a particular topic, look back at your old homework for that topic. If you struggled with a particular type of problem, your homework will reflect that. Your homework should contain some notes about how to do certain problems or how to avoid mistakes.

Your class notes should include examples of the problems in that section of the textbook, as well as pointers from your instructor.

If you have been creating note cards during the semester, they should focus on problems you considered difficult at the time and contain strategies for solving these types of problems.

Exercises 7.3

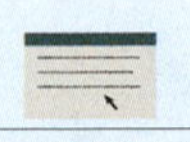

Vocabulary

1. To add fractions that have the same denominator, add the ____________ and place the result over the common denominator.
2. To subtract fractions that have the same denominator, ________ the numerators and place the result over the common denominator.
3. When subtracting a rational expression that has a numerator containing more than one term, subtract the entire numerator, not just the ________.
4. When adding two rational expressions that have opposite denominators, you can replace the second denominator with its opposite by changing the addition to ________.

Add.

5. $\dfrac{5}{x + 3} + \dfrac{8}{x + 3}$
6. $\dfrac{9}{x - 6} + \dfrac{7}{x - 6}$
7. $\dfrac{x}{x - 5} + \dfrac{10}{x - 5}$
8. $\dfrac{3x}{4x + 28} + \dfrac{21}{4x + 28}$
9. $\dfrac{x}{x^2 + 5x - 36} + \dfrac{9}{x^2 + 5x - 36}$
10. $\dfrac{2x + 7}{x^2 - 64} + \dfrac{x + 17}{x^2 - 64}$

11. $\frac{x^2 - 5x + 9}{x^2 - 8x + 15} + \frac{x - 6}{x^2 - 8x + 15}$

12. $\frac{x^2 - 13x + 20}{x^2 + 3x - 54} + \frac{2x + 10}{x^2 + 3x - 54}$

13. $\frac{x^2 + 3x + 11}{x^2 - 2x - 48} + \frac{12x + 43}{x^2 - 2x - 48}$

14. $\frac{x^2 + 5x + 20}{x^2 - 5x - 14} + \frac{5x - 4}{x^2 - 5x - 14}$

15. $\frac{x^2 - 16x - 45}{x^2 - 3x - 18} + \frac{x^2 - 2x - 27}{x^2 - 3x - 18}$

16. $\frac{x^2 - 11x - 34}{x^2 + 16x + 64} + \frac{x^2 + 17x - 46}{x^2 + 16x + 64}$

For the given rational functions $f(x)$ and $g(x)$, find $f(x) + g(x)$.

17. $f(x) = \frac{3}{x}, g(x) = \frac{5}{x}$

18. $f(x) = \frac{x}{5x + 30}, g(x) = \frac{6}{5x + 30}$

19. $f(x) = \frac{x^2 + 3x - 15}{x^2 - 4x - 45}, g(x) = \frac{4x + 25}{x^2 - 4x - 45}$

20. $f(x) = \frac{2x - 21}{x^2 + 20x + 99}, g(x) = \frac{x^2 + 2x - 56}{x^2 + 20x + 99}$

Subtract.

21. $\frac{13}{x + 1} - \frac{9}{x + 1}$

22. $\frac{6}{x - 4} - \frac{11}{x - 4}$

23. $\frac{x}{x + 7} - \frac{7}{x + 7}$

24. $\frac{2x}{3x - 15} - \frac{10}{3x - 15}$

25. $\frac{3x - 7}{x - 8} - \frac{x + 9}{x - 8}$

26. $\frac{5x - 23}{x - 9} - \frac{3x - 5}{x - 9}$

27. $\frac{x}{x^2 - 13x + 30} - \frac{3}{x^2 - 13x + 30}$

28. $\frac{x}{x^2 + 2x - 48} - \frac{6}{x^2 + 2x - 48}$

29. $\frac{4x - 7}{x^2 - 16} - \frac{2x - 15}{x^2 - 16}$

30. $\frac{6x + 1}{x^2 - 16x + 63} - \frac{3x + 22}{x^2 - 16x + 63}$

31. $\frac{x^2 - 2x - 3}{x^2 - x - 72} - \frac{5x + 15}{x^2 - x - 72}$

32. $\frac{x^2 + 3x + 17}{x^2 - 16} - \frac{11x + 1}{x^2 - 16}$

33. $\frac{x^2 - 3x + 67}{3x^2 + 17x + 10} - \frac{7 - 20x}{3x^2 + 17x + 10}$

34. $\frac{5x^2 + 4x}{x^2 - 7x + 6} - \frac{x^2 + 5x + 3}{x^2 - 7x + 6}$

35. $\frac{(x - 6)(x + 3)}{x^2 - 2x - 15} - \frac{9(x - 5)}{x^2 - 2x - 15}$

36. $\frac{(2x + 9)(x + 4)}{x^2 - 2x - 24} - \frac{(x + 6)(x - 6)}{x^2 - 2x - 24}$

For the given rational functions $f(x)$ and $g(x)$, find $f(x) - g(x)$.

37. $f(x) = \frac{12}{x + 8}, g(x) = \frac{23}{x + 8}$

38. $f(x) = \frac{21}{x - 7}, g(x) = \frac{3x}{x - 7}$

39. $f(x) = \frac{x^2 + 3x - 5}{x^2 - 11x + 18}, g(x) = \frac{8x + 9}{x^2 - 11x + 18}$

40. $f(x) = \frac{2x^2 + 10x - 13}{x^2 + 15x + 50}, g(x) = \frac{x^2 + 11x + 17}{x^2 + 15x + 50}$

Add or subtract.

41. $\frac{2x}{x - 5} + \frac{10}{5 - x}$

42. $\frac{5x}{2x - 14} + \frac{35}{14 - 2x}$

43. $\frac{x^2 + 8x}{x - 1} - \frac{2x - 11}{1 - x}$

44. $\frac{x^2 - 8x - 10}{x - 3} + \frac{x - 28}{3 - x}$

45. $\frac{x^2 - 3x - 9}{x - 9} + \frac{x^2 - 7x + 27}{9 - x}$

46. $\frac{3x^2 - 14x - 45}{x - 6} - \frac{-2x^2 + 14x + 9}{6 - x}$

Mixed Practice, 47–66

Simplify.

47. $\dfrac{x^2 - 3x + 5}{2x - 20} + \dfrac{4x + 35}{20 - 2x}$

48. $\dfrac{x^2 + 11x + 24}{x^2 - 12x + 32} \cdot \dfrac{x^2 - 4x}{x^3 + 27}$

49. $\dfrac{x^2 - 14x + 49}{x^2 + x - 30} \div \dfrac{3x^2 - 9x - 84}{x^2 - 6x + 5}$

50. $\dfrac{x^2 - 7x - 50}{x - 8} - \dfrac{x^2 + 3x - 46}{8 - x}$

51. $\dfrac{x^2 + 2x - 360}{x^2 + 30x + 200}$

52. $\dfrac{4x^2 - 6x + 3}{x^2 + 17x + 70} - \dfrac{3x^2 - 6x + 52}{x^2 + 17x + 70}$

53. $\dfrac{x^2 - 3x + 45}{x^2 - 81} + \dfrac{x^2 + 29x + 27}{x^2 - 81}$

54. $\dfrac{9x - x^2}{x^2 - 4x - 32} \div \dfrac{x^2 - 16x + 63}{x^2 + 11x + 28}$

55. $\dfrac{121 - x^2}{x^2 - 3x - 4} \cdot \dfrac{x^2 + 4x + 3}{x^2 - 3x - 88}$

56. $\dfrac{10}{5x - 25} + \dfrac{2x}{25 - 5x}$

57. $\dfrac{2x^2 - 15x + 39}{x^2 + 3x - 4} + \dfrac{x^2 - 18x - 9}{x^2 + 3x - 4}$

58. $\dfrac{2x^2 - 3x + 20}{x^2 + 6x - 55} - \dfrac{x^2 + 9x - 15}{x^2 + 6x - 55}$

59. $\dfrac{6x}{3x - 2} - \dfrac{4}{2 - 3x}$

60. $\dfrac{x^2 + 20x + 45}{x^2 + 15x + 56} - \dfrac{3x - 27}{x^2 + 15x + 56}$

61. $\dfrac{5x^2 + 3x - 13}{x^2 - 8x - 9} - \dfrac{4x^2 - 7x - 22}{x^2 - 8x - 9}$

62. $\dfrac{x^2 + 3x - 28}{2x^2 + 15x + 28} \cdot \dfrac{3x^2 + 16x + 16}{x^2 + 17x + 70}$

63. $\dfrac{x^2 + x - 90}{3x^2 + 2x - 16} \div \dfrac{x^2 - 9x}{3x^2 + 17x + 24}$

64. $\dfrac{5x^2 - 4x - 9}{x^2 - 6x - 7}$

65. $\dfrac{2x^2 + 5x + 20}{x^2 + 5x + 6} + \dfrac{3x^2 - 7x - 4}{x^2 + 5x + 6} - \dfrac{4x^2 - 2x + 25}{x^2 + 5x + 6}$

66. $\dfrac{x^2 + 3x - 7}{x^2 - 2x - 8} - \dfrac{x^2 + 5x - 9}{x^2 - 2x - 8} + \dfrac{x^2 - 6x + 14}{x^2 - 2x - 8}$

Find the missing numerator.

67. $\dfrac{?}{x + 6} + \dfrac{18}{x + 6} = 3$

68. $\dfrac{x^2 + 5x}{x - 2} - \dfrac{?}{x - 2} = x + 7$

69. $\dfrac{x^2 + 7x + 17}{(x + 8)(x + 5)} + \dfrac{?}{(x + 8)(x + 5)} = \dfrac{x + 3}{x + 5}$

70. $\dfrac{x^2 - 5x - 24}{(x + 2)(x - 4)} - \dfrac{?}{(x + 2)(x - 4)} = \dfrac{x + 2}{x - 4}$

Writing in Mathematics

Answer in complete sentences.

71. Explain how to determine whether two rational expressions have opposite denominators.

72. Explain why it is a good idea to use parentheses when subtracting a rational expression that has more than one term in the numerator.

7.4 Addition and Subtraction of Rational Expressions That Have Different Denominators

OBJECTIVES

1. Find the least common denominator (LCD) of two or more rational expressions.
2. Add or subtract rational expressions that have different denominators.

The Least Common Denominator of Two or More Rational Expressions

Objective 1 Find the least common denominator (LCD) of two or more rational expressions. If two numerical fractions do not have the same denominator, we cannot add or subtract the fractions until we rewrite them as equivalent fractions with a common denominator. The same holds true for rational expressions that have different denominators. We will begin this section by learning how to find the **least common denominator (LCD)** for two or more rational expressions. Then we will learn how to add or subtract rational expressions that have different denominators.

Finding the LCD of Two Rational Expressions

Begin by completely factoring each denominator; then identify each expression that is a factor of one or both denominators. The LCD is the product of these factors.

If an expression is a repeated factor in one or more of the denominators, we repeat it as a factor in the LCD as well. The exponent used for this factor is the greatest power to which the factor is raised in any one denominator.

EXAMPLE 1 Find the LCD of $\frac{7}{24a^3b}$ and $\frac{9}{16a}$.

SOLUTION We begin with the coefficients 24 and 16. The smallest number into which both divide evenly is 48. Moving on to variable factors in the denominator, we see that the variables a and b are factors of one or both denominators. Note that the variable a is raised to the third power in the first denominator; so the LCD must contain a factor of a^3. The LCD is $48a^3b$.

Quick Check 1

Find the LCD of $\frac{10}{9r^2s^3}$ and $\frac{1}{12rs^6}$.

EXAMPLE 2 Find the LCD of $\frac{8}{x^2 - 4x + 3}$ and $\frac{9}{x^2 - 9}$.

SOLUTION Begin by factoring each denominator.

$$\frac{8}{x^2 - 4x + 3} = \frac{8}{(x - 1)(x - 3)} \qquad \frac{9}{x^2 - 9} = \frac{9}{(x + 3)(x - 3)}$$

The factors in the denominators are $x - 1$, $x - 3$, and $x + 3$. Because no expression is repeated as a factor in any one denominator, the LCD is $(x - 1)(x - 3)(x + 3)$.

Quick Check 2

Find the LCD of $\frac{10}{x^2 - 13x + 40}$ and $\frac{3}{x^2 - 4x - 32}$.

EXAMPLE 3 Find the LCD of $\frac{2}{x^2 - 2x - 35}$ and $\frac{x}{x^2 + 10x + 25}$.

SOLUTION Again, begin by factoring each denominator.

$$\frac{2}{x^2 - 2x - 35} = \frac{2}{(x - 7)(x + 5)} \qquad \frac{x}{x^2 + 10x + 25} = \frac{x}{(x + 5)(x + 5)}$$

The two expressions that are factors are $x - 7$ and $x + 5$; the factor $x + 5$ is repeated twice in the second denominator. So the LCD also must have $x + 5$ as a factor twice. The LCD is $(x - 7)(x + 5)(x + 5)$, or $(x - 7)(x + 5)^2$.

Quick Check 3

Find the LCD of $\frac{x + 8}{x^2 - 81}$ and $\frac{x - 7}{x^2 + 18x + 81}$.

Adding or Subtracting Rational Expressions That Have Different Denominators

Objective 2 Add or subtract rational expressions that have different denominators. To add or subtract two rational expressions that do not have the same denominator, we begin by finding the LCD. We then convert each rational expression to an equivalent rational expression that has the LCD as its denominator. We can then add or subtract as we did in the previous section. As always, we should attempt to simplify the resulting rational expression.

EXAMPLE 4 Add: $\frac{2a}{3} + \frac{3a}{7}$

SOLUTION The LCD of these two fractions is 21. We will rewrite each fraction as an equivalent fraction whose denominator is 21. Because $\frac{7}{7} = 1$, multiplying the first fraction by $\frac{7}{7}$ will produce an equivalent fraction whose denominator is 21. In a similar fashion, we multiply the second fraction by $\frac{3}{3}$. Once both fractions have the same denominator, we add the numerators and write the sum above the common denominator.

$$\frac{2a}{3} + \frac{3a}{7} = \frac{2a}{3} \cdot \frac{7}{7} + \frac{3a}{7} \cdot \frac{3}{3}$$ Multiply the first fraction by $\frac{7}{7}$ and the second fraction by $\frac{3}{3}$ to rewrite each fraction with a denominator of 21.

$$= \frac{14a}{21} + \frac{9a}{21}$$ Multiply numerators and denominators.

$$= \frac{23a}{21}$$ Add the numerators and place the sum over the common denominator.

EXAMPLE 5 Add: $\frac{8}{x + 6} + \frac{2}{x - 7}$

SOLUTION The two denominators are not the same, so we begin by finding the LCD for these two rational expressions. Each denominator has a single factor, and the LCD is the product of these two denominators. The LCD is $(x + 6)(x - 7)$. We will multiply $\frac{8}{x + 6}$ by $\frac{x - 7}{x - 7}$ to write it as an equivalent fraction whose denominator is the LCD. We need to multiply $\frac{2}{x - 7}$ by $\frac{x + 6}{x + 6}$ to write it as an equivalent fraction whose denominator is the LCD.

$$\frac{8}{x + 6} + \frac{2}{x - 7}$$

$$= \frac{8}{x + 6} \cdot \frac{x - 7}{x - 7} + \frac{2}{x - 7} \cdot \frac{x + 6}{x + 6}$$ Multiply to rewrite each expression as an equivalent rational expression that has the LCD as its denominator.

$$= \frac{8x - 56}{(x + 6)(x - 7)} + \frac{2x + 12}{(x - 7)(x + 6)}$$ Distribute in each numerator, but do not distribute in the denominators.

$$= \frac{(8x - 56) + (2x + 12)}{(x + 6)(x - 7)}$$ Add the numerators, writing the sum over the common denominator.

$$= \frac{10x - 44}{(x + 6)(x - 7)}$$ Combine like terms.

$$= \frac{2(5x - 22)}{(x + 6)(x - 7)}$$ Factor the numerator.

Because the numerator and denominator do not have any common factors, this rational expression cannot be simplified any further.

Quick Check 4

Add: $\frac{5}{x - 2} + \frac{6}{x + 6}$

A WORD OF CAUTION When we add two rational expressions, we cannot simply add the two numerators and place their sum over the sum of the two denominators.

$$\frac{8}{x+6} + \frac{2}{x-7} \neq \frac{8+2}{(x+6)+(x-7)}$$

We must first find a common denominator and rewrite each rational expression as an equivalent expression whose denominator is equal to the common denominator.

When adding or subtracting rational expressions, we simplify the numerator, but leave the denominator in factored form. After we simplify the numerator, we factor, if possible, and check the denominator for common factors that can be divided out.

EXAMPLE 6 Subtract: $\frac{x}{x^2+8x+15} - \frac{5}{x^2+12x+35}$

SOLUTION In this example, we must factor each denominator to find the LCD.

$$\frac{x}{x^2+8x+15} - \frac{5}{x^2+12x+35}$$

$$= \frac{x}{(x+3)(x+5)} - \frac{5}{(x+5)(x+7)}$$

Factor each denominator. The LCD is $(x+3)(x+5)(x+7)$.

$$= \frac{x}{(x+3)(x+5)} \cdot \frac{x+7}{x+7} - \frac{5}{(x+5)(x+7)} \cdot \frac{x+3}{x+3}$$

Multiply to rewrite each expression as an equivalent rational expression that has the LCD as its denominator.

$$= \frac{x^2+7x}{(x+3)(x+5)(x+7)} - \frac{5x+15}{(x+5)(x+7)(x+3)}$$

Distribute in each numerator.

$$= \frac{(x^2+7x)-(5x+15)}{(x+3)(x+5)(x+7)}$$

Subtract the numerators, writing the difference over the LCD.

$$= \frac{x^2+7x-5x-15}{(x+3)(x+5)(x+7)}$$

Distribute.

$$= \frac{x^2+2x-15}{(x+3)(x+5)(x+7)}$$

Combine like terms.

$$= \frac{\overset{1}{\cancel{(x+5)}}(x-3)}{(x+3)\underset{1}{\cancel{(x+5)}}(x+7)}$$

Factor the numerator and divide out the common factor.

$$= \frac{x-3}{(x+3)(x+7)}$$

Simplify.

We must be careful to subtract the entire second numerator, not just the first term. In other words, the subtraction must change the sign of each term in the second numerator before we combine like terms. Using parentheses when subtracting the two numerators will help us remember to do this.

Quick Check 5

Add: $\frac{x}{x^2-6x+5} + \frac{1}{x^2+2x-3}$

EXAMPLE 7 Subtract: $\frac{x-6}{x^2-1} - \frac{5}{x^2-4x+3}$

SOLUTION Notice that the first numerator contains a binomial. We must be careful when multiplying to create equivalent rational expressions with the LCD as their denominators. We begin by factoring each denominator to find the LCD.

$$\frac{x-6}{x^2-1} - \frac{5}{x^2-4x+3}$$

$$= \frac{x-6}{(x+1)(x-1)} - \frac{5}{(x-3)(x-1)}$$

Factor each denominator. The LCD is $(x+1)(x-1)(x-3)$.

$$= \frac{x-6}{(x+1)(x-1)} \cdot \frac{x-3}{x-3} - \frac{5}{(x-3)(x-1)} \cdot \frac{x+1}{x+1}$$

Multiply to rewrite each expression as an equivalent rational expression with the LCD as its denominator.

$$= \frac{x^2-9x+18}{(x+1)(x-1)(x-3)} - \frac{5x+5}{(x-3)(x-1)(x+1)}$$

Distribute in each numerator.

$$= \frac{(x^2-9x+18)-(5x+5)}{(x+1)(x-1)(x-3)}$$

Subtract the numerators, writing the difference over the LCD.

$$= \frac{x^2-9x+18-5x-5}{(x+1)(x-1)(x-3)}$$

Distribute.

$$= \frac{x^2-14x+13}{(x+1)(x-1)(x-3)}$$

Combine like terms.

$$= \frac{\overset{1}{\cancel{(x-1)}}(x-13)}{(x+1)\underset{1}{\cancel{(x-1)}}(x-3)}$$

Factor the numerator and divide out the common factor.

$$= \frac{x-13}{(x+1)(x-3)}$$

Simplify.

Quick Check 6

Add: $\frac{x+5}{x^2+8x+12} + \frac{x+9}{x^2-36}$

BUILDING YOUR STUDY STRATEGY

Preparing for a Cumulative Exam, 4 **Problems from Your Instructor** Some of the most valuable materials for preparing for a cumulative exam are the materials provided by your instructor, such as a review sheet for the final, a practice final, or a list of problems to review from the text. If your instructor believes that these problems are important enough for you to review, they are important enough to appear on the cumulative exam.

Your performance on these problems will show you which topics you understand and which topics require further study.

If you are having difficulty with a certain problem, on a note card, write down the steps to solve that problem. Review your note cards for a short period each day before the exam. This will help you understand how to solve the problem, and it increases your chance of solving a similar problem on the exam.

Exercises 7.4

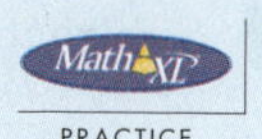
PRACTICE

WATCH

DOWNLOAD

READ

REVIEW

Vocabulary

1. The ______ of two rational expressions is an expression that is a product of each factor of the two denominators.
2. To add two rational expressions that have different denominators, begin by converting each rational expression to a(n) ________ rational expression that has the LCD as its denominator.

Find the LCD of the given rational expressions.

3. $\frac{5}{2a}, \frac{1}{3a}$
4. $\frac{9}{8x^2}, \frac{7}{6x^3}$
5. $\frac{6}{x-9}, \frac{8}{x+5}$
6. $\frac{x}{x-7}, \frac{x+2}{2x+3}$
7. $\frac{2}{x^2-9}, \frac{x+7}{x^2-9x+18}$
8. $\frac{x-6}{x^2-3x-28}, \frac{x+1}{x^2-x-20}$
9. $\frac{3}{x^2+7x+10}, \frac{x-1}{x^2+4x+4}$
10. $\frac{x-10}{x^2-12x+36}, \frac{8}{x^2-36}$

Add or subtract.

11. $\frac{3x}{8} + \frac{5x}{12}$
12. $\frac{8x}{15} - \frac{7x}{20}$
13. $\frac{9}{5n} - \frac{7}{4n}$
14. $\frac{7}{10a} + \frac{9}{14a}$
15. $\frac{10}{m^7n^2} + \frac{13}{m^5n^3}$
16. $\frac{7}{s^4t^3} - \frac{2}{t^8}$
17. $\frac{5}{x+2} + \frac{3}{x+4}$
18. $\frac{7}{x+5} + \frac{2}{x-8}$
19. $\frac{10}{x+3} - \frac{4}{x-3}$
20. $\frac{6}{x} - \frac{9}{x-4}$
21. $\frac{3}{x^2+7x+10} + \frac{8}{x^2-x-6}$
22. $\frac{10}{x^2-16x+60} + \frac{7}{x^2-100}$
23. $\frac{1}{x^2-5x-6} - \frac{5}{x^2+8x+7}$
24. $\frac{3}{x^2-6x-16} - \frac{2}{x^2-2x-48}$
25. $\frac{5}{x^2-15x+50} + \frac{9}{x^2-x-20}$
26. $\frac{4}{x^2+12x} - \frac{2}{x^2+18x+72}$
27. $\frac{3}{x^2-13x+40} - \frac{4}{x^2-12x+32}$
28. $\frac{2}{x^2+x-6} - \frac{4}{x^2-4x-21}$
29. $\frac{5}{2x^2+3x-2} + \frac{7}{2x^2-9x+4}$
30. $\frac{7}{2x^2-15x-27} - \frac{3}{2x^2-3x-9}$
31. $\frac{x+9}{x^2+8x+15} - \frac{2}{x^2+9x+20}$
32. $\frac{x-3}{x^2-12x+35} + \frac{7}{x^2-3x-10}$
33. $\frac{x-3}{x^2-x-20} + \frac{6}{x^2-2x-24}$

34. $\dfrac{x-10}{x^2-2x-3}-\dfrac{3}{x^2-9x+18}$

35. $\dfrac{x+14}{x^2+5x-50}-\dfrac{8}{x^2+15x+50}$

36. $\dfrac{x-6}{x^2-16}-\dfrac{2}{x^2+8x+16}$

37. $\dfrac{x-9}{x^2-49}+\dfrac{1}{x^2-7x}$

38. $\dfrac{x+7}{x^2-7x}+\dfrac{2}{x^2-15x+56}$

39. $\dfrac{3}{x^2+11x+28}+\dfrac{x-4}{x^2-x-56}$

40. $\dfrac{2}{x^2+4x-12}+\dfrac{x+3}{x^2-36}$

41. $\dfrac{x+1}{x^2+2x-8}+\dfrac{x+3}{x^2+10x+24}$

42. $\dfrac{x+5}{x^2-8x-9}+\dfrac{x-4}{x^2-4x-5}$

43. $\dfrac{x+9}{2x^2-9x-18}-\dfrac{x-4}{2x^2+17+21}$

44. $\dfrac{x+2}{x^2+9x+20}-\dfrac{x+8}{x^2+11x+30}$

Mixed Practice 45–60

45. $\dfrac{9}{x^2-11x+28}+\dfrac{3}{x^2-7x+12}$

46. $\dfrac{x-6}{x^2+9x+8}+\dfrac{2}{x^2+4x+3}$

47. $\dfrac{x+10}{x^2+10x+16}+\dfrac{x+7}{x^2+13x+40}$

48. $\dfrac{x^2+11x+24}{x^2-14x+40}\div\dfrac{x^2-5x-24}{100-x^2}$

49. $\dfrac{x^2-2x-24}{x^2+9x+20}\cdot\dfrac{2x^2+x-45}{x^2-7x+6}$

50. $\dfrac{4}{x^2-6x-27}-\dfrac{2}{x^2-12x+27}$

51. $\dfrac{x+2}{x^2+4x-32}-\dfrac{1}{x^2+18x+80}$

52. $\dfrac{x^2+17x+60}{x^2-8x-65}$

53. $\dfrac{3x+17}{x^2-5x}-\dfrac{x^2-15x+18}{5x-x^2}$

54. $\dfrac{x^2+10x+17}{x^2-13x+36}-\dfrac{7x+125}{x^2-13x+36}$

55. $\dfrac{2x^2+19x+42}{36-x^2}$

56. $\dfrac{x-2}{x^2-4x-32}-\dfrac{x+3}{x^2+6x+8}$

57. $\dfrac{3x^2+2x-8}{x^2-9x}\div\dfrac{x^3+8}{x^2-16x+63}$

58. $\dfrac{x^2-15x}{x^2+2x-48}+\dfrac{5x+24}{x^2+2x-48}$

59. $\dfrac{x^2-7x+23}{x^2-81}+\dfrac{12x-67}{81-x^2}$

60. $\dfrac{x^2-7x-18}{x^2-3x-40}\cdot\dfrac{8x-x^2}{x^2-15x+54}$

Writing in Mathematics

Answer in complete sentences.

61. Explain how to find the LCD of two rational expressions. Use an example to illustrate the process.

62. Here is a student's solution to a problem on an exam. Describe the student's error and provide the correct solution. Assuming that the problem was worth 10 points, how many points would you give to the student for this solution? Explain your reasoning.

$$\frac{5}{(x+3)(x+2)}+\frac{7}{(x+3)(x+6)}$$
$$=\frac{5}{(x+3)(x+2)}\cdot\frac{x+6}{x+6}+\frac{7}{(x+3)(x+6)}\cdot\frac{x+2}{x+2}$$
$$=\frac{5\overset{1}{\cancel{(x+6)}}+7\overset{1}{\cancel{(x+2)}}}{(x+3)\underset{1}{\cancel{(x+2)}}\underset{1}{\cancel{(x+6)}}}$$
$$=\frac{12}{x+3}$$

63. ***Solutions Manual*** Write a solutions manual page for the following problem:

Add: $\dfrac{x+6}{x^2+9x+20}+\dfrac{4}{x^2+6x+8}$

7.5

Complex Fractions

OBJECTIVES

1. Simplify complex numerical fractions.
2. Simplify complex fractions containing variables.

Complex Fractions

A **complex fraction** is a fraction or rational expression containing one or more fractions in its numerator or denominator. Here are some examples.

$$\frac{\frac{1}{2}+\frac{5}{3}}{\frac{10}{3}-\frac{7}{4}} \qquad \frac{\frac{x+5}{x-9}}{1-\frac{5}{x}} \qquad \frac{\frac{1}{2}+\frac{1}{x}}{\frac{1}{4}-\frac{1}{x^2}} \qquad \frac{1+\frac{3}{x}-\frac{28}{x^2}}{1+\frac{7}{x}}$$

Objective 1 Simplify complex numerical fractions. To simplify a complex fraction, we must rewrite it so that its numerator and denominator do not contain fractions. We can do this by finding the LCD of all fractions in the complex fraction and then multiplying the numerator and denominator by this LCD. This will clear the fractions in the complex fraction. We finish by simplifying the resulting rational expression, if possible.

We begin with a complex fraction made up of numerical fractions.

EXAMPLE 1 Simplify the complex fraction $\frac{1+\frac{4}{3}}{\frac{2}{3}+\frac{11}{4}}$.

SOLUTION The LCD of the three denominators (3, 3, and 4) is 12; so we begin by multiplying the complex fraction by $\frac{12}{12}$. Notice that when we multiply by $\frac{12}{12}$, we are really multiplying by 1, which does not change the value of the original expression.

$$\frac{1+\frac{4}{3}}{\frac{2}{3}+\frac{11}{4}} = \frac{12}{12}\cdot\frac{1+\frac{4}{3}}{\frac{2}{3}+\frac{11}{4}} \qquad \text{Multiply the numerator and denominator by the LCD.}$$

$$= \frac{12\cdot 1 + \overset{4}{\cancel{12}}\cdot\frac{4}{\underset{1}{\cancel{3}}}}{\overset{4}{\cancel{12}}\cdot\frac{2}{\underset{1}{\cancel{3}}} + \overset{3}{\cancel{12}}\cdot\frac{11}{\underset{1}{\cancel{4}}}} \qquad \text{Distribute and divide out common factors.}$$

$$= \frac{12+16}{8+33} \qquad \text{Multiply.}$$

$$= \frac{28}{41} \qquad \text{Simplify the numerator and denominator.}$$

There is another method for simplifying complex fractions. We can rewrite the numerator as a single fraction by adding $1+\frac{4}{3}$, which equals $\frac{7}{3}$.

$$1+\frac{4}{3} = \frac{3}{3}+\frac{4}{3} = \frac{7}{3}$$

We also can rewrite the denominator as a single fraction by adding $\frac{2}{3} + \frac{11}{4}$, which equals $\frac{41}{12}$.

$$\frac{2}{3} + \frac{11}{4} = \frac{8}{12} + \frac{33}{12} = \frac{41}{12}$$

Once the numerator and denominator are single fractions, we can rewrite $\frac{\frac{7}{3}}{\frac{41}{12}}$ as a division problem $\frac{7}{3} \div \frac{41}{12}$ and simplify from there. This method produces the same result of $\frac{28}{41}$.

$$\frac{7}{3} \div \frac{41}{12} = \frac{7}{\cancel{3}_1} \cdot \frac{\overset{4}{\cancel{12}}}{41} = \frac{28}{41}$$

Quick Check 1

Simplify the complex fraction $\dfrac{\frac{2}{5} + \frac{3}{8}}{\frac{1}{4} + \frac{7}{10}}$.

In most of the remaining examples, we will use the LCD method, as the technique is somewhat similar to the technique used to solve rational equations in the next section.

Simplifying Complex Fractions Containing Variables

Objective 2 Simplify complex fractions containing variables.

EXAMPLE 2 Simplify $\dfrac{1 - \frac{9}{x^2}}{1 + \frac{3}{x}}$.

SOLUTION The LCD for the two simple fractions with denominators x and x^2 is x^2; so we will begin by multiplying the complex fraction by $\frac{x^2}{x^2}$. Then once we have cleared the fractions, resulting in a rational expression, we simplify the rational expression by factoring and dividing out common factors.

$$\frac{1 - \frac{9}{x^2}}{1 + \frac{3}{x}} = \frac{x^2}{x^2} \cdot \frac{1 - \frac{9}{x^2}}{1 + \frac{3}{x}}$$ Multiply the numerator and denominator by the LCD.

$$= \frac{x^2 \cdot 1 - \overset{1}{\cancel{x^2}} \cdot \frac{9}{\cancel{x^2}_1}}{x^2 \cdot 1 + \overset{x}{\cancel{x^2}} \cdot \frac{3}{\cancel{x}_1}}$$ Distribute and divide out common factors, clearing the fractions.

$$= \frac{x^2 - 9}{x^2 + 3x}$$ Multiply.

$$= \frac{\overset{1}{\cancel{(x+3)}}(x-3)}{x\underset{1}{\cancel{(x+3)}}}$$ Factor the numerator and denominator and divide out the common factor.

$$= \frac{x - 3}{x}$$ Simplify.

Quick Check 2

Simplify $\dfrac{1 + \frac{3}{x} - \frac{10}{x^2}}{1 - \frac{2}{x}}$.

EXAMPLE 3 Simplify $\dfrac{\dfrac{1}{6}+\dfrac{1}{x}}{\dfrac{1}{36}-\dfrac{1}{x^2}}$.

SOLUTION Begin by multiplying the numerator and denominator by the LCD of all denominators. In this case, the LCD is $36x^2$.

$$\frac{\frac{1}{6}+\frac{1}{x}}{\frac{1}{36}-\frac{1}{x^2}} = \frac{36x^2}{36x^2}\cdot\frac{\frac{1}{6}+\frac{1}{x}}{\frac{1}{36}-\frac{1}{x^2}}$$ Multiply the numerator and denominator by the LCD.

$$= \frac{\overset{6}{\cancel{36}}x^2\cdot\frac{1}{\underset{1}{\cancel{6}}} + \overset{1}{\cancel{36}}\overset{x}{\cancel{x^2}}\cdot\frac{1}{\underset{1}{\cancel{x}}}}{\overset{1}{\cancel{36}}x^2\cdot\frac{1}{\underset{1}{\cancel{36}}} - 36\overset{1}{\cancel{x^2}}\cdot\frac{1}{\underset{1}{\cancel{x^2}}}}$$ Distribute and divide out common factors.

$$= \frac{6x^2+36x}{x^2-36}$$ Multiply.

$$= \frac{6x\overset{1}{\cancel{(x+6)}}}{\underset{1}{\cancel{(x+6)}}(x-6)}$$ Factor the numerator and denominator and divide out the common factor.

$$= \frac{6x}{x-6}$$ Simplify.

Quick Check 3

Simplify $\dfrac{\dfrac{1}{64}-\dfrac{1}{x^2}}{\dfrac{1}{8}-\dfrac{1}{x}}$.

EXAMPLE 4 Simplify $\dfrac{\dfrac{3}{x+1}-\dfrac{2}{x+2}}{\dfrac{x}{x+2}+\dfrac{6}{x+1}}$.

SOLUTION The LCD for the four simple fractions is $(x+1)(x+2)$.

$$\frac{\frac{3}{x+1}-\frac{2}{x+2}}{\frac{x}{x+2}+\frac{6}{x+1}}$$

$$= \frac{(x+1)(x+2)}{(x+1)(x+2)}\cdot\frac{\frac{3}{x+1}-\frac{2}{x+2}}{\frac{x}{x+2}+\frac{6}{x+1}}$$ Multiply the numerator and denominator by the LCD.

$$= \frac{\overset{1}{\cancel{(x+1)}}(x+2)\cdot\frac{3}{\underset{1}{\cancel{x+1}}} - (x+1)\overset{1}{\cancel{(x+2)}}\cdot\frac{2}{\underset{1}{\cancel{x+2}}}}{(x+1)\overset{1}{\cancel{(x+2)}}\cdot\frac{x}{\underset{1}{\cancel{x+2}}} + \overset{1}{\cancel{(x+1)}}(x+2)\cdot\frac{6}{\underset{1}{\cancel{x+1}}}}$$ Distribute and divide out common factors, clearing the fractions.

$$= \frac{3(x+2)-2(x+1)}{x(x+1)+6(x+2)}$$ Multiply.

$$= \frac{3x+6-2x-2}{x^2+x+6x+12}$$ Distribute.

$$= \frac{x + 4}{x^2 + 7x + 12}$$ Combine like terms.

$$= \frac{\overset{1}{\cancel{x + 4}}}{(x + 3)\underset{1}{\cancel{(x + 4)}}}$$ Factor the denominator and divide out the common factor.

$$= \frac{1}{x + 3}$$ Simplify.

Quick Check 4

Simplify $\dfrac{\dfrac{6}{x - 4} + \dfrac{5}{x + 7}}{\dfrac{x + 2}{x - 4}}$.

EXAMPLE 5 Simplify $\dfrac{\dfrac{x^2 + 5x - 14}{x^2 - 9}}{\dfrac{x^2 - 6x + 8}{x^2 + 8x + 15}}$.

SOLUTION In this case, it will be easier to rewrite the complex fraction as a division problem rather than multiply the numerator and denominator by the LCD. This is the best technique when we have a complex fraction with a single rational expression in its numerator and a single rational expression in its denominator. We have used this method when dividing rational expressions.

$$\frac{\dfrac{x^2 + 5x - 14}{x^2 - 9}}{\dfrac{x^2 - 6x + 8}{x^2 + 8x + 15}}$$

$$= \frac{x^2 + 5x - 14}{x^2 - 9} \div \frac{x^2 - 6x + 8}{x^2 + 8x + 15}$$ Rewrite as a division problem.

$$= \frac{x^2 + 5x - 14}{x^2 - 9} \cdot \frac{x^2 + 8x + 15}{x^2 - 6x + 8}$$ Invert the divisor and multiply.

$$= \frac{(x + 7)(x - 2)}{(x + 3)(x - 3)} \cdot \frac{(x + 3)(x + 5)}{(x - 2)(x - 4)}$$ Factor each numerator and denominator.

$$= \frac{(x + 7)\overset{1}{\cancel{(x - 2)}}}{\underset{1}{\cancel{(x + 3)}}(x - 3)} \cdot \frac{\overset{1}{\cancel{(x + 3)}}(x + 5)}{\underset{1}{\cancel{(x - 2)}}(x - 4)}$$ Divide out common factors.

$$= \frac{(x + 7)(x + 5)}{(x - 3)(x - 4)}$$ Simplify.

Quick Check 5

Simplify $\dfrac{\dfrac{x^2 - 2x - 48}{x^2 + 7x - 30}}{\dfrac{x^2 + 7x + 6}{x^2 - 5x + 6}}$.

BUILDING YOUR STUDY STRATEGY

Preparing for a Cumulative Exam, 5 Cumulative Review Exercises Use your cumulative review exercises to prepare for a cumulative exam. These exercises contain problems representative of the material covered in several previous chapters.

After you have been preparing for a while, try these exercises without referring to your notes, your note cards, or the text. In this way, you can determine which topics require more study. If you made a mistake while solving a problem, make note of the mistake and ways to avoid it in the future. If there are problems that you do not recognize or know how to begin, ask your instructor or a tutor for help.

Exercises 7.5

PRACTICE

WATCH

DOWNLOAD

READ

REVIEW

Vocabulary

1. A(n) ______________ is a fraction or rational expression containing one or more fractions in its numerator or denominator.

2. To simplify a complex fraction, multiply the numerator and denominator by the _____ of all fractions in the complex fraction.

Simplify the complex fraction.

3. $\dfrac{\frac{2}{5}-\frac{1}{4}}{\frac{9}{10}+\frac{5}{2}}$

4. $\dfrac{\frac{3}{7}+\frac{2}{3}}{\frac{16}{21}-\frac{2}{7}}$

5. $\dfrac{2-\frac{3}{8}}{\frac{5}{4}+\frac{1}{3}}$

6. $\dfrac{\frac{5}{6}-3}{\frac{3}{4}+\frac{11}{12}}$

7. $\dfrac{x+\frac{3}{5}}{x+\frac{4}{7}}$

8. $\dfrac{x-\frac{2}{9}}{x+\frac{5}{4}}$

9. $\dfrac{6+\frac{15}{x}}{x+\frac{5}{2}}$

10. $\dfrac{x+\frac{3}{7}}{14+\frac{6}{x}}$

11. $\dfrac{14-\frac{4}{x}}{21-\frac{6}{x}}$

12. $\dfrac{10+\frac{8}{x}}{25+\frac{20}{x}}$

13. $\dfrac{3+\frac{15}{x}}{1-\frac{25}{x^2}}$

14. $\dfrac{4-\frac{36}{x^2}}{x+2-\frac{15}{x}}$

15. $\dfrac{\frac{4}{x+3}+\frac{2}{x+6}}{\frac{x+5}{x+3}}$

16. $\dfrac{\frac{10}{x+4}-\frac{6}{x-6}}{\frac{3x-63}{x-6}}$

17. $\dfrac{\frac{12}{x}+\frac{3}{x-5}}{\frac{x+4}{x}+\frac{2}{x-5}}$

18. $\dfrac{\frac{8}{x+2}-\frac{2}{x-7}}{\frac{x+6}{x+2}-\frac{4}{x-7}}$

19. $\dfrac{1+\frac{4}{x}-\frac{32}{x^2}}{1+\frac{13}{x}+\frac{40}{x^2}}$

20. $\dfrac{1-\frac{9}{x}+\frac{18}{x^2}}{1+\frac{2}{x}-\frac{15}{x^2}}$

21. $\dfrac{\frac{8}{x^2}-\frac{8}{x}+2}{\frac{4}{x^2}-1}$

22. $\dfrac{\frac{18}{x}-3}{\frac{1}{x}-\frac{42}{x^2}+1}$

23. $\dfrac{\frac{x^2+14x+48}{x^2+3x-40}}{\frac{x^2-3x-54}{x^2-25}}$

24. $\dfrac{\frac{x^2+13x+30}{x^2+13x-30}}{\frac{x^2+5x+6}{x^2+10x-24}}$

25. $\dfrac{\frac{4x^2-16x-9}{8x-x^2}}{\frac{4x^2-1}{x^2-x-56}}$

26. $\dfrac{\frac{81-x^2}{x^2-36}}{\frac{x^2-15x+54}{3x^2+20x+12}}$

Mixed Practice, 27–52

Simplify the given rational expression using the techniques developed in Sections 7.1–7.5.

27. $\dfrac{x^2+4x}{x^2-x-42}+\dfrac{x-6}{x^2-x-42}$

28. $\dfrac{x^2-14x+40}{x^2+4x+3}\div\dfrac{x^2-6x+8}{x^2+10x+9}$

29. $\frac{x+3}{x^2-2x-24}+\frac{5}{x^2-8x+12}$

30. $\frac{6}{x^2-2x-35}-\frac{4}{x^2+2x-15}$

31. $\frac{x^2+7x}{3x^2-19x+20}\div\frac{x^2+12x+35}{x^2-3x-10}$

32. $\frac{36-x^2}{x^2-12x+36}$

33. $\frac{x+5}{x^2-49}-\frac{6}{x^2-7x}$

34. $\frac{x^2+3x-40}{x^2+x-6}\cdot\frac{x^2-7x-30}{2x^2+17x+8}$

35. $\frac{5x}{x-2}+\frac{10}{2-x}$

36. $\frac{1-\frac{11}{x}+\frac{18}{x^2}}{1-\frac{2}{x}-\frac{63}{x^2}}$

37. $\frac{\frac{7}{x+2}-\frac{2}{x-3}}{\frac{x-5}{x+2}}$

38. $\frac{5x^2-12x+4}{x^3-8}$

39. $\frac{x^2-13x+14}{x^2-6x}-\frac{x^2+4x-32}{6x-x^2}$

40. $\frac{x^2+6x-16}{x^2-4x-45}\cdot\frac{2x^2+13x+15}{x^2-4x+4}$

41. $\frac{\frac{1}{8}-\frac{1}{x}}{\frac{1}{x^2}-\frac{1}{64}}$

42. $\frac{x^2+9x}{x-5}-\frac{6x+50}{5-x}$

43. $\frac{8}{x^2+7x+12}+\frac{4}{x^2+10x+24}$

44. $\frac{8x}{2x-5}+\frac{20}{5-2x}$

45. $\frac{x^3+64}{x^2+14x+49}\cdot\frac{x^2+3x-28}{x^2-16}$

46. $\frac{10x^2+11x+1}{2x^2-x-15}\div\frac{x^2-x-2}{x^2-13x+30}$

47. $\frac{x^2+18x+77}{x^2-4x-32}\cdot\frac{x^2+6x+8}{x^2+9x+14}$

48. $\frac{x+3}{x^2-2x-3}+\frac{3}{x^2-8x+15}$

49. $\frac{x^2+7x}{x^2-2x-3}-\frac{4x+18}{x^2-2x-3}$

50. $\frac{x+\frac{5}{8}}{x-\frac{7}{2}}$

51. $\frac{x^3-1000}{x^2+10x-11}\div\frac{3x^2-37x+70}{x^2+4x-5}$

52. $\frac{x+5}{x^2-3x+2}-\frac{8}{x^2-8x+12}$

Writing in Mathematics

Answer in complete sentences.

53. Explain what a complex fraction is. Compare and contrast complex fractions and the rational expressions found in Section 7.1.

54. One method for simplifying complex fractions is to rewrite the complex fraction as one rational expression divided by another rational expression. When is this the most efficient way to simplify a complex fraction? Give an example.

7.6 Rational Equations

OBJECTIVES

1. Solve rational equations.
2. Solve literal equations containing rational expressions.

Solving Rational Equations

Objective 1 Solve rational equations. In this section, we will learn how to solve **rational equations**, which are equations containing at least one rational expression. The main goal is to rewrite the equation as an equivalent equation that does not contain a rational expression. We then solve the equation using methods developed in earlier chapters.

In Chapter 2, we learned how to solve an equation containing fractions, such as the equation $\frac{1}{4}x - \frac{3}{5} = \frac{9}{10}$. We began by finding the LCD of all fractions and then multiplied both sides of the equation by that LCD to clear the equation of fractions. We will use the same technique in this section. There is a major difference, though, when we solve equations containing a variable in a denominator. Occasionally, we will find a solution that causes one of the rational expressions in the equation to be undefined. If a denominator of a rational expression is equal to 0 when the value of a solution is substituted for the variable, the solution must be omitted and is called an **extraneous solution**. We must check each solution to make sure it is not an extraneous solution.

Solving Rational Equations

1. Find the LCD of all denominators in the equation.
2. Multiply both sides of the equation by the LCD to clear the equation of fractions.
3. Solve the resulting equation.
4. Check for extraneous solutions.

EXAMPLE 1 Solve $\frac{4}{x} + \frac{1}{3} = \frac{5}{6}$.

SOLUTION We begin by finding the LCD of these three fractions, which is $6x$. Then we multiply both sides of the equation by the LCD to clear the equation of fractions. Once we have done this, we can solve the resulting equation.

$$\frac{4}{x} + \frac{1}{3} = \frac{5}{6}$$

$$6x \cdot \left(\frac{4}{x} + \frac{1}{3}\right) = 6x \cdot \frac{5}{6} \quad \text{Multiply both sides of the equation by the LCD.}$$

$$\overset{1}{6\cancel{x}} \cdot \frac{4}{\underset{1}{\cancel{x}}} + \overset{2}{\cancel{6}}x \cdot \frac{1}{\underset{1}{\cancel{3}}} = \overset{1}{\cancel{6}}x \cdot \frac{5}{\underset{1}{\cancel{6}}} \quad \text{Distribute and divide out common factors.}$$

$$24 + 2x = 5x \quad \text{Multiply. The resulting equation is linear.}$$

$$24 = 3x \quad \text{Subtract } 2x \text{ to collect all variable terms on one side of the equation.}$$

$$8 = x \quad \text{Divide both sides by 3.}$$

Check

$$\frac{4}{(8)} + \frac{1}{3} = \frac{5}{6}$$ Substitute 8 for x.

$$\frac{1}{2} + \frac{1}{3} = \frac{5}{6}$$ Simplify the fraction $\frac{4}{8}$. The LCD of these fractions is 6.

$$\frac{3}{6} + \frac{2}{6} = \frac{5}{6}$$ Write each fraction with a common denominator of 6.

$$\frac{5}{6} = \frac{5}{6}$$ Add.

Because the solution $x = 8$ does not make any rational expression in the original equation undefined, this value is a valid solution. The solution set is $\{8\}$.

Quick Check 1

Solve $\frac{6}{x} - \frac{1}{8} = \frac{7}{40}$.

Checking for Extraneous Solutions

When checking whether a solution is an extraneous solution, we need only determine whether the solution causes the LCD to equal 0. If the LCD is equal to 0 for this solution, one or more rational expressions are undefined and the solution is an extraneous solution. Also, if the LCD is equal to 0, we have multiplied both sides of the equation by 0. The multiplication property of equality says that we can multiply both sides of an equation by any *nonzero* number without affecting the equality of both sides. In the previous example, the only solution that could possibly be an extraneous solution is $x = 0$ because it is the only value of x for which the LCD is equal to 0.

EXAMPLE 2 Solve $x - 5 - \frac{36}{x} = 0$.

SOLUTION The LCD in this example is x. The LCD is equal to 0 only if $x = 0$. If we find that $x = 0$ is a solution, we must omit that solution as an extraneous solution.

$$x - 5 - \frac{36}{x} = 0$$

$$x \cdot \left(x - 5 - \frac{36}{x}\right) = x \cdot 0$$ Multiply each side of the equation by the LCD, x.

$$x \cdot x - x \cdot 5 - \overset{1}{\cancel{x}} \cdot \frac{36}{\underset{1}{\cancel{x}}} = 0$$ Distribute and divide out common factors.

$$x^2 - 5x - 36 = 0$$ Multiply. The resulting equation is quadratic.

$$(x - 9)(x + 4) = 0$$ Factor.

$$x = 9 \quad \text{or} \quad x = -4$$ Set each factor equal to 0 and solve.

You may verify that neither solution causes the LCD to equal 0. The solution set is $\{-4, 9\}$.

Quick Check 2

Solve $1 = \frac{5}{x} + \frac{24}{x^2}$.

A WORD OF CAUTION When we are solving a rational equation, the use of the LCD is completely different than when we are adding or subtracting rational expressions. We use the LCD to clear the denominators of the rational expressions when we are solving a rational equation. When we are adding or subtracting rational expressions, we rewrite each expression as an equivalent expression whose denominator is the LCD.

EXAMPLE 3 Solve $\frac{x}{x+2} - 5 = \frac{3x+4}{x+2}$.

SOLUTION The LCD is $x + 2$, so we will begin to solve this equation by multiplying both sides of the equation by $x + 2$.

$$\frac{x}{x+2} - 5 = \frac{3x+4}{x+2}$$ The LCD is $x + 2$.

$$(x+2)\cdot\left(\frac{x}{x+2} - 5\right) = (x+2)\cdot\frac{3x+4}{x+2}$$ Multiply both sides by the LCD.

$$\overset{1}{\cancel{(x+2)}}\cdot\frac{x}{\underset{1}{\cancel{x+2}}} - (x+2)\cdot 5 = \overset{1}{\cancel{(x+2)}}\cdot\frac{3x+4}{\underset{1}{\cancel{x+2}}}$$ Distribute and divide out common factors.

$$x - 5x - 10 = 3x + 4$$ Multiply. The resulting equation is linear.

$$-4x - 10 = 3x + 4$$ Combine like terms.

$$-10 = 7x + 4$$ Add $4x$.

$$-14 = 7x$$ Subtract 4.

$$-2 = x$$ Divide both sides by 7.

The LCD is equal to 0 when $x = -2$, and two rational expressions in the original equation are undefined when $x = -2$. This solution is an extraneous solution, and because there are no other solutions, this equation has no solution. Recall that we write the solution set as $\varnothing$ when there is no solution.

Quick Check 3

Solve $\frac{7}{x-4} + 3 = \frac{2x-1}{x-4}$.

EXAMPLE 4 Solve $\frac{x+9}{x^2+9x+8} = \frac{2}{x^2+2x-48}$.

SOLUTION Begin by factoring the denominators to find the LCD.

$$\frac{x+9}{x^2+9x+8} = \frac{2}{x^2+2x-48}$$

$$\frac{x+9}{(x+1)(x+8)} = \frac{2}{(x+8)(x-6)}$$ The LCD is $(x + 1)(x + 8)(x - 6)$.

$$\overset{1}{\cancel{(x+1)}}\overset{1}{\cancel{(x+8)}}(x-6)\cdot\frac{x+9}{\underset{1}{\cancel{(x+1)}}\underset{1}{\cancel{(x+8)}}}$$

$$= (x+1)\underset{1}{\cancel{(x+8)}}\underset{1}{\cancel{(x-6)}}\cdot\frac{2}{\underset{1}{\cancel{(x+8)}}\underset{1}{\cancel{(x-6)}}}$$ Multiply by the LCD. Divide out common factors.

$$(x-6)(x+9) = 2(x+1)$$ Multiply remaining factors.

$$x^2 + 3x - 54 = 2x + 2$$ Multiply.

$$x^2 + x - 56 = 0$$ Collect all terms on the left side. The resulting equation is quadratic.

$$(x+8)(x-7) = 0$$ Factor.

$$x = -8 \quad \text{or} \quad x = 7$$ Set each factor equal to 0 and solve.

The solution $x = -8$ is an extraneous solution because it makes the LCD equal to 0. The reader may verify that the solution $x = 7$ checks. The solution set for this equation is $\{7\}$.

Quick Check 4

Solve $\frac{x+2}{x^2-3x-54} = \frac{2}{x^2-12x+27}$.

EXAMPLE 5 Solve $\frac{x+10}{x^2+4x-5} - \frac{1}{x-3} = \frac{x-6}{x^2-4x+3}$.

SOLUTION

$$\frac{x+10}{x^2+4x-5} - \frac{1}{x-3} = \frac{x-6}{x^2-4x+3}$$

$$\frac{x+10}{(x+5)(x-1)} - \frac{1}{x-3} = \frac{x-6}{(x-1)(x-3)}$$ The LCD is $(x+5)(x-1)(x-3)$.

$$(x+5)(x-1)(x-3)\cdot\left(\frac{x+10}{(x+5)(x-1)} - \frac{1}{x-3}\right)$$

$$= (x+5)(x-1)(x-3)\cdot\frac{x-6}{(x-1)(x-3)}$$ Multiply by the LCD.

$$\overset{1}{\cancel{(x+5)}}\overset{1}{\cancel{(x-1)}}(x-3)\cdot\frac{x+10}{\underset{1}{\cancel{(x+5)}}\underset{1}{\cancel{(x-1)}}} - (x+5)(x-1)\overset{1}{\cancel{(x-3)}}\cdot\frac{1}{\underset{1}{\cancel{(x-3)}}}$$

$$= (x+5)\overset{1}{\cancel{(x-1)}}\overset{1}{\cancel{(x-3)}}\cdot\frac{x-6}{\underset{1}{\cancel{(x-1)}}\underset{1}{\cancel{(x-3)}}}$$ Distribute and divide out common factors.

$(x-3)(x+10) - (x+5)(x-1) = (x+5)(x-6)$ Multiply remaining factors.

$(x^2+7x-30) - (x^2+4x-5) = x^2-x-30$ Multiply.

$x^2+7x-30-x^2-4x+5 = x^2-x-30$ Distribute.

$3x-25 = x^2-x-30$ Combine like terms.

$0 = x^2-4x-5$ Collect all terms on the right side. The resulting equation is quadratic.

$0 = (x+1)(x-5)$ Factor.

$x = -1$ or $x = 5$ Set each factor equal to 0 and solve.

Check to verify that neither solution is extraneous. The solution set is $\{-1, 5\}$.

Quick Check 5

Solve $\frac{x+3}{x^2+x-12} - \frac{3}{x^2-2x-3} = \frac{x+7}{x^2+5x+4}$.

Literal Equations

Objective 2 Solve literal equations containing rational expressions. Recall that a literal equation is an equation containing two or more variables and that we solve the equation for one of the variables by isolating that variable on one side of the equation. In this section, we will learn how to solve literal equations containing one or more rational expressions.

EXAMPLE 6 Solve the literal equation $\frac{1}{x} + \frac{1}{y} = \frac{2}{5}$ for x.

SOLUTION Begin by multiplying by the LCD $(5xy)$ to clear the equation of fractions.

$$\frac{1}{x} + \frac{1}{y} = \frac{2}{5}$$

$$5xy \cdot \left(\frac{1}{x} + \frac{1}{y}\right) = 5xy \cdot \frac{2}{5} \quad \text{Multiply by the LCD, } 5xy.$$

$$\overset{1}{5\cancel{x}y} \cdot \frac{1}{\underset{1}{\cancel{x}}} + \overset{1}{5x\cancel{y}} \cdot \frac{1}{\underset{1}{\cancel{y}}} = \overset{1}{\cancel{5}xy} \cdot \frac{2}{\underset{1}{\cancel{5}}} \quad \text{Distribute and divide out common factors.}$$

$$5y + 5x = 2xy \quad \text{Multiply remaining factors.}$$

Notice that two terms contain the variable for which we are solving. We need to collect both of these terms on the same side of the equation and then factor x out of those terms. This will allow us to divide and isolate x.

$$5y + 5x = 2xy$$

$$5y = 2xy - 5x \quad \text{Subtract } 5x \text{ to collect all terms with } x \text{ on the right side of the equation.}$$

$$5y = x(2y - 5) \quad \text{Factor out the common factor } x.$$

$$\frac{5y}{2y - 5} = \frac{x\overset{1}{\cancel{(2y - 5)}}}{\underset{1}{\cancel{(2y - 5)}}} \quad \text{Divide both sides by } 2y - 5 \text{ to isolate } x.$$

$$\frac{5y}{2y - 5} = x \quad \text{Simplify.}$$

$$x = \frac{5y}{2y - 5} \quad \text{Rewrite with the variable you are solving for on the left side of the equation.}$$

Quick Check 6

Solve the literal equation $\frac{2}{x} + \frac{3}{y} = \frac{4}{z}$ for x.

As in Chapter 2, we will rewrite the solution so that the variable for which we are solving is on the left side.

EXAMPLE 7 Solve the literal equation $y = \frac{2x}{3x - 5}$ for x.

SOLUTION Begin by multiplying both sides of the equation by $3x - 5$ to clear the equation of fractions.

$$y = \frac{2x}{3x - 5}$$

$$y(3x - 5) = \frac{2x}{\underset{1}{\cancel{3x - 5}}} \cdot \overset{1}{\cancel{(3x - 5)}} \quad \text{Multiply both sides by } 3x - 5.$$

$$3xy - 5y = 2x \quad \text{Simplify.}$$

$$3xy - 2x = 5y \quad \text{Collect all terms containing } x \text{ on the left side of the equation (subtract } 2x\text{) and all other terms on the right side of the equation (add } 5y\text{).}$$

$$x(3y - 2) = 5y \quad \text{Factor out the common factor } x \text{ on the left side of the equation.}$$

$$\frac{x\overset{1}{\cancel{(3y - 2)}}}{\underset{1}{\cancel{3y - 2}}} = \frac{5y}{3y - 2} \quad \text{Divide both sides by } 3y - 2 \text{ to isolate } x.$$

$$x = \frac{5y}{3y - 2} \quad \text{Simplify.}$$

Quick Check 7

Solve the literal equation $y = \frac{5x}{4x - 3}$ for x.

BUILDING YOUR STUDY STRATEGY

Preparing for a Cumulative Exam, 6 Applied Problems Many students have a difficult time with applied problems on a cumulative exam. Some students are unable to recognize what type of problem an applied problem is, and others cannot recall how to start to solve that type of problem. You will find the following strategy helpful:

- Make a list of the different applied problems you have covered this semester. Create a study sheet for each type of problem.
- Write down an example or two of each type of problem.

Review these study sheets frequently as the exam approaches. This should help you identify the applied problems on the exam as well as remember how to solve them.

Exercises 7.6

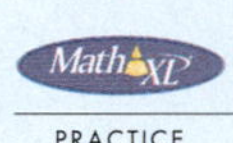

Vocabulary

1. A(n) ________________ is an equation containing at least one rational expression.
2. To solve a rational equation, begin by multiplying both sides of the equation by the ______ of the denominators in the equation.
3. A(n) __________________ of a rational equation is a solution that causes one or more of the rational expressions in the equation to be undefined.
4. A(n) ______ equation is an equation containing two or more variables.

Solve.

5. $\frac{x}{9} + \frac{11}{18} = \frac{7}{6}$
6. $\frac{x}{10} + \frac{5}{6} = \frac{26}{15}$
7. $\frac{5}{6} + \frac{11}{x} = \frac{7}{4}$
8. $\frac{1}{8} + \frac{13}{x} = \frac{2}{3}$
9. $\frac{25}{21} - \frac{17}{x} = \frac{7}{12}$
10. $\frac{21}{20} - \frac{19}{x} = \frac{5}{12}$
11. $x - 5 + \frac{12}{x} = 2$
12. $x - \frac{25}{x} = 2 + \frac{23}{x}$
13. $\frac{x}{2} + \frac{6}{x} = 4$
14. $\frac{x}{3} + 2 - \frac{9}{x} = 0$
15. $1 - \frac{7}{x} + \frac{10}{x^2} = 0$
16. $1 + \frac{11}{x} - \frac{42}{x^2} = 0$
17. $1 - \frac{25}{x^2} = 0$
18. $6 - \frac{13}{x} + \frac{6}{x^2} = 0$
19. $\frac{8x - 3}{x + 7} = \frac{2x + 15}{x + 7}$
20. $\frac{6x + 13}{x - 1} = \frac{4x + 15}{x - 1}$
21. $\frac{7x - 11}{5x - 2} = \frac{2x - 9}{5x - 2}$
22. $\frac{x^2 + 3x}{x + 8} = \frac{3x + 64}{x + 8}$
23. $8 + \frac{6}{x - 4} = \frac{x - 12}{x - 4}$
24. $1 + \frac{2x - 7}{x + 3} = \frac{4x - 1}{x + 3}$
25. $x + \frac{x + 11}{x - 6} = \frac{8x - 31}{x - 6}$
26. $x + \frac{3x - 10}{x + 1} = \frac{9x + 14}{x + 1}$

27. $\frac{3}{x-7} = \frac{7}{x+5}$

28. $\frac{6}{x+4} = \frac{9}{2x-3}$

29. $\frac{x-3}{x+9} = \frac{6}{x+2}$

30. $\frac{x-1}{3x-7} = \frac{x+1}{2x+7}$

31. $\frac{3}{x+2} - \frac{1}{x+1} = \frac{x+3}{x^2+3x+2}$

32. $\frac{2}{x} + \frac{3}{x+2} = \frac{7x-8}{x^2+2x}$

33. $\frac{4}{x+4} + \frac{3}{x-4} = \frac{24}{x^2-16}$

34. $\frac{x}{x+5} + \frac{3}{x-7} = \frac{36}{x^2-2x-35}$

35. $\frac{2}{x^2-x-2} + \frac{10}{x^2-2x-3} = \frac{x+12}{x^2-x-2}$

36. $\frac{20}{x^2+12x+27} - \frac{4}{x^2+14x+45} = \frac{x+10}{x^2+8x+15}$

37. $\frac{x-8}{x-5} + \frac{x-9}{x-4} = \frac{x+7}{x^2-9x+20}$

38. $\frac{x+4}{x^2+5x-14} + \frac{2}{x^2+3x-10} = \frac{38}{x^2+12x+35}$

39. $\frac{x+3}{x^2-4x-12} + \frac{x-11}{x^2-2x-24} = \frac{x+1}{x^2+6x+8}$

40. $\frac{x-2}{x^2+13x+40} + \frac{x+5}{x^2+7x-8} = \frac{x+3}{x^2+4x-5}$

41. $\frac{3x-4}{x^2-10x+21} - \frac{x-8}{x^2-18x+77} = \frac{x-5}{x^2-14x+33}$

42. $\frac{5x+4}{x^2+x-90} - \frac{1}{x-9} = \frac{3x-2}{x^2-100}$

Mixed Practice, 43–58

43. $1 + \frac{13}{x} + \frac{42}{x^2} = 0$

44. $3x - 8 = -29$

45. $x^2 - 14x - 120 = 0$

46. $\frac{2x}{x+9} = \frac{x-2}{x-1}$

47. $2(3x+4) - 19 = 4x - 1$

48. $\frac{x}{x+5} + \frac{2}{x-9} = \frac{28}{x^2-4x-45}$

49. $\frac{x}{x+3} + \frac{x-4}{x-3} = \frac{9x-5}{x^2-9}$

50. $\frac{1}{4}x^2 + \frac{1}{2}x - 2 = 0$

51. $6x^2 + 29x - 5 = 0$

52. $2 + \frac{9}{x} + \frac{4}{x^2} = 0$

53. $\frac{6}{x+2} = \frac{25}{3x+13}$

54. $7 - 5x = 52$

55. $\frac{x}{x+6} + \frac{3}{x+4} = \frac{8}{x^2+10x+24}$

56. $x^2 - 81 = 0$

57. $5x - 17 = 8x + 13$

58. $\frac{4}{x^2+4x-5} + \frac{x+9}{x^2-1} = \frac{41}{x^2+6x+5}$

Solve for the specified variable.

59. $L = \frac{A}{W}$ for W

60. $b = \frac{2A}{h}$ for A

61. $y = \frac{x}{2x+5}$ for x

62. $y = \frac{3x-7}{2x}$ for x

63. $y = \frac{2x-9}{3x-8}$ for x

64. $y = \frac{5x-4}{x-6}$ for x

65. $\frac{x}{r} + \frac{y}{2r} = 1$ for r

66. $r = \frac{d}{t}$ for t

67. $m = \frac{y-y_1}{x-x_1}$ for x

68. $\frac{1}{R} = \frac{1}{R_1} + \frac{1}{R_2}$ for R

69. If $x = 9$ is a solution to the equation

$$\frac{x-4}{x-1} + \frac{7}{x+3} = \frac{?}{x^2+2x-3}:$$

a) Find the constant in the missing numerator.
b) Find the other solution to the equation.

70. If $x = 12$ is a solution to the equation

$$\frac{x-8}{x-5} + \frac{?}{x-2} = \frac{7x+19}{x^2-7x+10}:$$

a) Find the constant in the missing numerator.
b) Find the other solution to the equation.

Writing in Mathematics

Answer in complete sentences.

71. Explain how to determine whether a solution is an extraneous solution.

72. You use the LCD when adding two rational expressions, as well as when solving a rational equation. Explain how the LCD is used differently for these two types of problems.

73. ***Newsletter*** Write a newsletter that explains how to solve rational equations.

7.7 Applications of Rational Equations

OBJECTIVES

1. **Solve applied problems involving the reciprocal of a number.**
2. **Solve applied work-rate problems.**
3. **Solve applied uniform motion problems.**
4. **Solve variation problems.**

In this section, we will look at applied problems requiring the use of rational equations to solve them. We begin with problems involving reciprocals.

Reciprocals

Objective 1 Solve applied problems involving the reciprocal of a number.

EXAMPLE 1 The sum of the reciprocal of a number and $\frac{1}{3}$ is $\frac{1}{2}$. Find the number.

SOLUTION There is only one unknown in this problem, and we will let x represent the unknown number.

Unknown
Number: x

The reciprocal of this number can be written as $\frac{1}{x}$. We are told that the sum of this reciprocal and $\frac{1}{3}$ is $\frac{1}{2}$, which leads to the equation $\frac{1}{x} + \frac{1}{3} = \frac{1}{2}$.

$$\frac{1}{x} + \frac{1}{3} = \frac{1}{2}$$ The LCD is $6x$.

$$6x \cdot \left(\frac{1}{x} + \frac{1}{3}\right) = 6x \cdot \frac{1}{2}$$ Multiply both sides by the LCD.

$$\overset{1}{6\cancel{x}} \cdot \frac{1}{\underset{1}{\cancel{x}}} + \overset{2}{\cancel{6}x} \cdot \frac{1}{\underset{1}{\cancel{3}}} = \overset{3}{\cancel{6}x} \cdot \frac{1}{\underset{1}{\cancel{2}}}$$ Distribute and divide out common factors.

$$6 + 2x = 3x$$ Multiply remaining factors. The resulting equation is linear.

$$6 = x$$ Subtract $2x$.

The unknown number is 6. The reader should verify that $\frac{1}{6} + \frac{1}{3} = \frac{1}{2}$.

Quick Check 1

The sum of the reciprocal of a number and $\frac{3}{8}$ is $\frac{19}{40}$. Find the number.

EXAMPLE 2 One positive number is 4 larger than another positive number. If the reciprocal of the smaller number is added to six times the reciprocal of the larger number, the sum is 1. Find the two numbers.

SOLUTION In this problem, there are two unknown numbers. If we let x represent the smaller number, we can write the larger number as $x + 4$.

Unknowns
Smaller Number: x
Larger Number: $x + 4$

The reciprocal of the smaller number is $\frac{1}{x}$, and six times the reciprocal of the larger number is $6 \cdot \frac{1}{x+4}$, or $\frac{6}{x+4}$. This leads to the equation $\frac{1}{x} + \frac{6}{x+4} = 1$.

$$\frac{1}{x} + \frac{6}{x+4} = 1$$ The LCD is $x(x + 4)$.

$$x(x+4) \cdot \left(\frac{1}{x} + \frac{6}{x+4}\right) = x(x+4) \cdot 1$$ Multiply both sides by the LCD.

$$\overset{1}{\cancel{x}}(x+4) \cdot \frac{1}{\underset{1}{\cancel{x}}} + x\overset{1}{\cancel{(x+4)}} \cdot \frac{6}{\underset{1}{\cancel{(x+4)}}} = x(x+4) \cdot 1$$ Distribute and divide out common factors.

$$x + 4 + 6x = x(x+4)$$ Multiply remaining factors.

$$x + 4 + 6x = x^2 + 4x$$ Distribute. The resulting equation is quadratic.

$$7x + 4 = x^2 + 4x$$ Combine like terms.

$$0 = x^2 - 3x - 4$$ Collect all terms on the right side of the equation.

$$0 = (x+1)(x-4)$$ Factor.

$$x = -1 \quad \text{or} \quad x = 4$$ Set each factor equal to 0 and solve.

Because we were told the numbers must be positive, we can omit the solution $x = -1$. We now return to the table of unknowns to find the two numbers.

Smaller Number: $x = 4$
Larger Number: $x + 4 = 4 + 4 = 8$

The two numbers are 4 and 8. The reader should verify that $\frac{1}{4} + 6 \cdot \frac{1}{8} = 1$.

Quick Check 2
One positive number is 9 less than another positive number. If two times the reciprocal of the smaller number is added to four times the reciprocal of the larger number, the sum is 1. Find the two numbers.

Work-Rate Problems

Objective 2 Solve applied work-rate problems. Now we turn our attention to **work-rate problems**. These problems usually involve two or more people or objects working together to perform a job, such as two people painting a room together or two copy machines processing an exam. In general, the equation we will be solving corresponds to the following:

Portion of the job completed by person 1	+	Portion of the job completed by person 2	=	1 (Completed job)

To determine the portion of the job completed by each person, we must know the **rate** at which the person works. If Kim takes 4 hours to paint a room, how much of the room could she paint in one hour? She could paint $\frac{1}{4}$ of the room in 1 hour, and this is her working rate. In general, the work rate for a person is equal to the reciprocal of the time it takes that person to complete the whole job. If we multiply a person's work rate by the time that person has been working, this tells us the portion of the job the person has completed.

EXAMPLE 3 Working alone, Kim can paint a room in 4 hours. Greg can paint the same room in only 3 hours. How long does it take the two of them to paint the room if they work together?

SOLUTION A good approach to any work-rate problem is to start with the following table and fill in the information:

Person	Time to Complete the Job Alone	Work Rate	Time Working	Portion of the Job Completed
Person 1				
Person 2				

- Because we know it takes Kim 4 hours to paint the room, her work rate is $\frac{1}{4}$ room per hour. Similarly, Greg's work rate is $\frac{1}{3}$ room per hour.
- The unknown in this problem is the amount of time they will work together, which we represent by the variable t.
- Finally, to determine the portion of the job completed by each person, we multiply the person's work rate by the time he or she has been working.

Person	Time to Complete the Job Alone	Work Rate	Time Working	Portion of the Job Completed
Kim	4 hours	$\frac{1}{4}$	t	$\frac{t}{4}$
Greg	3 hours	$\frac{1}{3}$	t	$\frac{t}{3}$

To find the equation we need to solve, we add the portion of the room that Kim painted to the portion of the room that Greg painted and set this sum equal to 1. The equation for this problem is $\frac{t}{4} + \frac{t}{3} = 1$.

$$\frac{t}{4} + \frac{t}{3} = 1 \qquad \text{The LCD is 12.}$$

$$12 \cdot \left(\frac{t}{4} + \frac{t}{3}\right) = 12 \cdot 1 \qquad \text{Multiply both sides by the LCD.}$$

$$\overset{3}{\cancel{12}} \cdot \frac{t}{\underset{1}{\cancel{4}}} + \overset{4}{\cancel{12}} \cdot \frac{t}{\underset{1}{\cancel{3}}} = 12 \cdot 1 \qquad \text{Distribute and divide out common factors.}$$

$$3t + 4t = 12 \qquad \text{Multiply remaining factors.}$$

$$7t = 12 \qquad \text{Combine like terms.}$$

$$t = \frac{12}{7} \text{ or } 1\frac{5}{7} \qquad \text{Divide both sides by 7.}$$

It takes them $1\frac{5}{7}$ hours to paint the room if they work together. To convert the answer to hours and minutes, we multiply $\frac{5}{7}$ of an hour by 60 minutes per hour, which is approximately 43 minutes. It takes them approximately one hour and 43 minutes. The check of this solution is left to the reader.

Quick Check 3

Javier's new printer can print a complete set of brochures in 20 minutes. His old printer can print the set of brochures in 35 minutes. How long does it take the two printers operating at the same time to print the set of brochures?

EXAMPLE 4 When used together, two drainpipes can drain a tank in 6 hours. Used alone, the smaller pipe takes 9 hours longer than the larger pipe to drain the tank. How long does it take the smaller pipe alone to drain the tank?

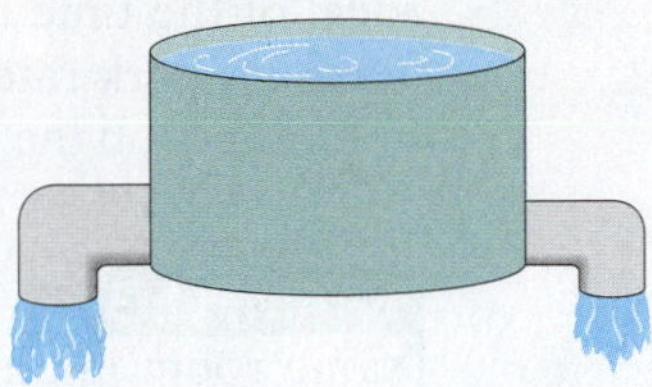

SOLUTION In this example, we know the amount of time it takes the pipes to drain the tank when they are used together, but we do not know how long it takes each pipe when used alone. If we let t represent the time, in hours, it takes the larger pipe to drain the tank, the smaller pipe takes $t + 9$ hours to drain the tank. We will begin with the following table:

Pipe	Time to Complete the Job Alone	Work Rate	Time Working	Portion of the Job Completed
Smaller	$t + 9$ hours	$\frac{1}{t+9}$	6 hours	$\frac{6}{t+9}$
Larger	t hours	$\frac{1}{t}$	6 hours	$\frac{6}{t}$

Adding the portion of the tank drained by the smaller pipe in 6 hours to the portion of the tank drained by the larger pipe, we find that the sum equals 1. The equation is $\frac{6}{t+9} + \frac{6}{t} = 1$.

$$\frac{6}{t+9} + \frac{6}{t} = 1$$ The LCD is $t(t + 9)$.

$$t(t+9) \cdot \left(\frac{6}{t+9} + \frac{6}{t}\right) = t(t+9) \cdot 1$$ Multiply both sides by the LCD.

$$\overset{1}{t(\cancel{t+9})} \cdot \frac{6}{\underset{1}{(\cancel{t+9})}} + \overset{1}{\cancel{t}}(t+9) \cdot \frac{6}{\underset{1}{\cancel{t}}} = t(t+9) \cdot 1$$ Distribute and divide out common factors.

$$6t + 6(t+9) = t(t+9)$$ Multiply remaining factors.

$$6t + 6t + 54 = t^2 + 9t$$ Multiply. The resulting equation is quadratic.

$$12t + 54 = t^2 + 9t$$ Combine like terms.

$$0 = t^2 - 3t - 54$$ Collect all terms on the right side of the equation.

$$0 = (t-9)(t+6)$$ Factor.

$$t = 9 \quad \text{or} \quad t = -6$$ Set each factor equal to 0 and solve.

Because the time spent by each pipe must be positive, we immediately omit the solution $t = -6$. The reader may check the solution $t = 9$ by verifying that $\frac{6}{9+9} + \frac{6}{9} = 1$. We now use $t = 9$ to find the amount of time it takes the smaller pipe to drain the tank. Because $t + 9$ represents the amount of time it takes the smaller pipe to drain the tank, it takes $9 + 9$, or 18, hours to drain the tank.

Quick Check 4

Used together, two drainpipes can drain a tank in 6 hours. The smaller pipe alone takes 16 hours longer than the larger pipe takes to drain the tank. How long does it take the smaller pipe alone to drain the tank?

Uniform Motion Problems

Objective 3 Solve applied uniform motion problems. We now turn our attention to uniform motion problems. Recall that if an object is moving at a constant rate of speed for a certain amount of time, the distance traveled by the object is equal to the product of its rate of speed and the length of time it traveled. The formula used earlier in the text was rate · time = distance, or $r \cdot t = d$. If we solve this formula for the time traveled, we have

$$\text{time} = \frac{\text{distance}}{\text{rate}}, \quad \text{or} \quad t = \frac{d}{r}.$$

EXAMPLE 5 Nick drove 24 miles, one way, to deliver a package. On the way home, he drove 20 miles per hour faster than he did on his way to deliver the package. If the total driving time was 1 hour for the entire trip, find Nick's driving speed on the way home.

SOLUTION

- We know that the distance traveled in each direction is 24 miles.
- Nick's speed on the way home was 20 miles per hour faster than it was on the way to deliver the package. We will let r represent his rate of speed on the way to deliver the package, and we can represent his rate of speed on the way home as $r + 20$.
- To find an expression for the time spent on each part of the trip, we divide the distance by the rate.

We can summarize this information in a table.

	Distance (d)	Rate (r)	Time (t)
To Deliver Package	24 miles	r	$\frac{24}{r}$
Return Trip	24 miles	$r + 20$	$\frac{24}{r+20}$

The equation we need to solve comes from the fact that the driving time on the way to deliver the package plus the driving time on the way home is equal to 1 hour. If we add the time spent on the way to deliver the package $\left(\frac{24}{r}\right)$ to the time spent on the way home $\left(\frac{24}{r+20}\right)$, it will be equal to 1. The equation we need to solve is $\frac{24}{r} + \frac{24}{r+20} = 1$.

$$\frac{24}{r} + \frac{24}{r+20} = 1 \qquad \text{The LCD is } r(r+20).$$

$$r(r+20) \cdot \left(\frac{24}{r} + \frac{24}{r+20}\right) = r(r+20) \cdot 1 \qquad \text{Multiply both sides by the LCD.}$$

$$\overset{1}{\cancel{r}}(r+20) \cdot \frac{24}{\underset{1}{\cancel{r}}} + r\overset{1}{\cancel{(r+20)}} \cdot \frac{24}{\underset{1}{\cancel{(r+20)}}} = r(r+20) \cdot 1 \qquad \text{Distribute and divide out common factors.}$$

$$24(r+20) + 24r = r(r+20) \cdot 1 \qquad \text{Multiply remaining factors.}$$

$$24r + 480 + 24r = r^2 + 20r \qquad \text{Multiply.}$$

$$48r + 480 = r^2 + 20r \qquad \text{Combine like terms.}$$

$$0 = r^2 - 28r - 480$$ Collect all terms on the right side of the equation by subtracting $48r$ and 480.

$$0 = (r - 40)(r + 12)$$ Factor.

$$r = 40 \quad \text{or} \quad r = -12$$ Set each factor equal to 0 and solve.

We omit the solution $r = -12$, as Nick's speed cannot be negative. The expression for Nick's driving speed on the way home is $r + 20$; so his speed on the way home was $40 + 20$, or 60, miles per hour. The reader can check the solution by verifying that $\frac{24}{40} + \frac{24}{60} = 1$.

Quick Check 5

Rehema drove 300 miles to pick up a friend and then returned home. On the way home, Rehema drove 15 miles per hour faster than she did on the way to pick up her friend. If the total driving time for the entire trip was 9 hours, find Rehema's driving speed on the way home.

EXAMPLE 6 Gretchen has taken her kayak to the Pawuxet River, which flows down-stream at a rate of 3 kilometers per hour. If the time it takes Gretchen to paddle 12 kilometers upstream is 1 hour longer than the time it takes her to paddle 12 kilometers downstream, find the speed that Gretchen can paddle in still water.

SOLUTION We will let r represent the speed that Gretchen can paddle in still water. Because the current of the river is 3 kilometers per hour, Gretchen's kayak travels at a speed of $r - 3$ kilometers per hour when she is paddling upstream. This is because the current is pushing against the kayak. Gretchen travels at a speed of $r + 3$ kilometers per hour when she is paddling downstream, as the current is flowing in the same direction as the kayak. The equation we will solve involves the time spent paddling upstream and downstream. To find expressions in terms of r for the time spent in each direction, we divide the distance (12 kilometers) by the rate of speed. Here is a table containing the relevant information.

	Distance	Rate	Time
Upstream	12 km	$r - 3$	$\frac{12}{r-3}$
Downstream	12 km	$r + 3$	$\frac{12}{r+3}$

We are told that the time needed to paddle 12 kilometers upstream is 1 hour more than the time needed to paddle 12 kilometers downstream. In other words, the time spent paddling upstream is equal to the time spent paddling downstream plus 1 hour, or $\frac{12}{r-3} = \frac{12}{r+3} + 1$.

$$\frac{12}{r-3} = \frac{12}{r+3} + 1$$ The LCD is $(r - 3)(r + 3)$.

$$(r-3)(r+3) \cdot \frac{12}{r-3} = (r-3)(r+3) \cdot \left(\frac{12}{r+3} + 1\right)$$ Multiply both sides by the LCD.

$$\overset{1}{\cancel{(r-3)}}(r+3) \cdot \frac{12}{\underset{1}{\cancel{(r-3)}}} = (r-3)\overset{1}{\cancel{(r+3)}} \cdot \frac{12}{\underset{1}{\cancel{(r+3)}}} + (r-3)(r+3) \cdot 1$$ Distribute and divide out common factors.

$$12(r+3) = 12(r-3) + (r-3)(r+3)$$ Multiply remaining factors.

$$12r + 36 = 12r - 36 + r^2 - 9$$ Multiply.

$$12r + 36 = r^2 + 12r - 45$$ Combine like terms, writing the right side in descending order.

$$0 = r^2 - 81$$ Collect all terms on the right side of the equation by subtracting $12r$ and 36.

$$0 = (r + 9)(r - 9)$$ Factor.

$$r = -9 \quad \text{or} \quad r = 9$$ Set each factor equal to 0 and solve.

We omit the negative solution, as the speed of the kayak in still water must be positive. Gretchen's kayak travels at a speed of 9 kilometers per hour in still water. The reader can check this solution by verifying that $\frac{12}{9 - 3} = \frac{12}{9 + 3} + 1$.

Quick Check 6

Lucy took her canoe to a river that flows downstream at a rate of 2 miles per hour. She paddled 8 miles downstream, then returned to the camp from which she started. If the round trip took her 3 hours, find the speed that Lucy can paddle in still water.

Variation

Objective 4 Solve variation problems. In the remaining examples, we will investigate the concept of **variation** between two or more quantities. Two quantities are said to **vary directly** if an increase in one quantity produces a proportional increase in the other quantity and a decrease in one quantity produces a proportional decrease in the other quantity. For example, suppose you have a part-time job that pays by the hour. The hours you work in a week and the amount of money you earn (before taxes) vary directly. As the hours you work increase, the amount of money you earn increases by the same factor. If there is a decrease in the number of hours you work, the amount of money you earn decreases by the same factor.

Direct Variation

If a quantity y varies directly as a quantity x, the two quantities are related by the equation

$$y = kx,$$

where k is called the **constant of variation.**

For example, if you are paid \$12 per hour at your part-time job, the amount of money you earn (y) and the number of hours you work (x) are related by the equation $y = 12x$. The value of k in this situation is 12, and it says that each time x increases by 1 hour, y increases by \$12.

EXAMPLE 7 y varies directly as x. If $y = 42$ when $x = 7$, find y when $x = 13$.

SOLUTION Because the variation is direct, we will use the equation $y = kx$. We begin by finding k. Using $y = 42$ when $x = 7$, we can use the equation $42 = k \cdot 7$ to find k.

$$y = kx$$

$$42 = k \cdot 7$$ Substitute 42 for y and 7 for x in $y = kx$.

$$6 = k$$ Divide both sides by 7.

Now we use this value of k to find y when $x = 13$.

$$y = 6 \cdot 13$$ Substitute 6 for k and 13 for x in $y = kx$.

$$y = 78$$ Multiply.

EXAMPLE 8 The distance a train travels varies directly as the time it is traveling. If a train can travel 424 miles in 8 hours, how far can it travel in 15 hours?

SOLUTION The first step in a variation problem is to find k. We will let y represent the distance traveled and x represent the time. To find the variation constant, we will substitute the related information (424 miles in 8 hours) into $y = kx$, the equation for direct variation.

$$y = kx$$

$$424 = k \cdot 8 \quad \text{Substitute 424 for } y \text{ and 8 for } x \text{ in } y = kx.$$

$$53 = k \quad \text{Divide both sides by 8.}$$

Now we will use this variation constant to find the distance traveled in 15 hours.

$$y = 53 \cdot 15 \quad \text{Substitute 53 for } k \text{ and 15 for } x \text{ in } y = kx.$$

$$y = 795 \quad \text{Multiply.}$$

The train travels 795 miles in 15 hours.

Quick Check 7

The tuition that a college student pays varies directly as the number of units the student is taking. If a student pays \$288 to take 12 units in a semester, how much does a student pay to take 15 units?

In some cases, an increase in one quantity produces a *decrease* in another quantity. In this case, we say that the quantities **vary inversely**. For example, suppose you and some of your friends are going to buy a birthday gift for your basketball coach, splitting the cost equally. As the number of people who are contributing increases, the cost for each person decreases. The cost for each person varies inversely as the number of people contributing.

Inverse Variation

If a quantity y varies inversely as a quantity x, the two quantities are related by the equation

$$y = \frac{k}{x},$$

where k is the constant of variation.

EXAMPLE 9 y varies inversely as x. If $y = 10$ when $x = 8$, find y when $x = 5$.

SOLUTION Because the quantities vary inversely, we will use the equation $y = \frac{k}{x}$. We begin by finding k.

$$y = \frac{k}{x}$$

$$10 = \frac{k}{8} \quad \text{Substitute 10 for } y \text{ and 8 for } x \text{ in } y = \tfrac{k}{x}.$$

$$80 = k \quad \text{Multiply both sides by 8.}$$

Now we use this value of k to find y when $x = 5$.

$$y = \frac{80}{5} \quad \text{Substitute 80 for } k \text{ and 5 for } x \text{ in } y = \tfrac{k}{x}.$$

$$y = 16 \quad \text{Divide.}$$

EXAMPLE 10 The time required to drive from Visalia to San Francisco varies inversely as the speed of the car. If it takes 3 hours to make the drive at 70 miles per hour, how long does it take to make the drive at 60 miles per hour?

SOLUTION Again, we begin by finding k. We will let y represent the time required to drive from Visalia to San Francisco and x represent the average speed of the car. To find k, we will substitute the related information (3 hours to make the trip at 70 miles per hour) into the equation for inverse variation.

$$y = \frac{k}{x}$$

$$3 = \frac{k}{70} \quad \text{Substitute 3 for } y \text{ and 70 for } x \text{ in } y = \frac{k}{x}.$$

$$210 = k \quad \text{Multiply both sides by 70.}$$

Now we will use this variation constant to find the time required to drive from Visalia to San Francisco at 60 miles per hour.

$$y = \frac{210}{60} \quad \text{Substitute 210 for } k \text{ and 60 for } x \text{ in } y = \frac{k}{x}.$$

$$y = 3.5 \quad \text{Divide.}$$

It takes 3.5 hours to drive from Visalia to San Francisco at 60 miles per hour.

Quick Check 8

The time required to complete the Boston Marathon varies inversely as the speed of the runner. If it takes 131 minutes to run the marathon at an average speed of 5 miles per hour, how long does it take to finish the marathon at an average speed of 4 miles per hour?

Often a quantity varies depending on two or more variables. A quantity y is said to **vary jointly** as two quantities x and z if it varies directly as the product of these two quantities. The equation in such a case is $y = kxz$.

EXAMPLE 11 y varies jointly as x and the square of z. If $y = 2400$ when $x = 8$ and $z = 10$, find y when $x = 12$ and $z = 20$.

SOLUTION Begin by finding k using the equation $y = kxz^2$.

$$y = kxz^2$$

$$2400 = k \cdot 8 \cdot 10^2 \quad \text{Substitute 2400 for } y\text{, 8 for } x\text{, and 10 for } z \text{ in } y = kxz^2.$$

$$2400 = k \cdot 800 \quad \text{Simplify.}$$

$$3 = k \quad \text{Divide both sides by 800.}$$

Now use this value of k to find y when $x = 12$ and $z = 20$.

$$y = 3 \cdot 12 \cdot 20^2 \quad \text{Substitute 3 for } k\text{, 12 for } x\text{, and 20 for } z \text{ in } y = kxz^2.$$

$$y = 14{,}400 \quad \text{Simplify.}$$

Quick Check 9

y varies jointly as x and the square root of z. If $y = 1000$ when $x = 50$ and $z = 25$, find y when $x = 72$ and $z = 9$.

BUILDING YOUR STUDY STRATEGY

Preparing for a Cumulative Exam, 7 Study Groups If you have participated in a study group throughout the semester, have each student in the group bring problems that he or she is struggling with or believes are important and work through them as a group.

At the end of your group study session, try to write a practice exam as a group. In determining which problems to include on the practice exam, you should focus on the types of problems that are most likely to appear on your cumulative exam.

Exercises 7.7

PRACTICE

WATCH

DOWNLOAD

READ
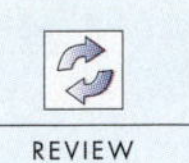
REVIEW

Vocabulary

1. The __________ of a number x is $\frac{1}{x}$.
2. The __________ of a person is the amount of work he or she does in one unit of time.
3. If an object is moving at a constant rate of speed for a certain amount of time, the length of time it travels is equal to the distance traveled by the object _______ by its rate of speed.
4. If a canoe can travel at a rate of r miles per hour in still water and a river's current is c miles per hour, the rate of speed the canoe travels while moving upstream is given by the expression ______.
5. Two quantities are said to ___________ if an increase in one quantity produces a proportional increase in the other quantity.
6. Two quantities are said to ___________ if an increase in one quantity produces a proportional decrease in the other quantity.

7. The sum of the reciprocal of a number and $\frac{23}{30}$ is $\frac{5}{6}$. Find the number.
8. The sum of the reciprocal of a number and $\frac{13}{15}$ is $\frac{19}{20}$. Find the number.
9. The sum of three times the reciprocal of a number and $\frac{13}{30}$ is $\frac{7}{12}$. Find the number.
10. The sum of five times the reciprocal of a number and $\frac{9}{40}$ is $\frac{13}{30}$. Find the number.
11. The difference of the reciprocal of a number and $\frac{5}{18}$ is $\frac{2}{9}$. Find the number.
12. The difference of the reciprocal of a number and $\frac{2}{11}$ is $\frac{5}{33}$. Find the number.
13. One positive number is three larger than another positive number. If four times the reciprocal of the smaller number is added to three times the reciprocal of the larger number, the sum is 1. Find the two numbers.
14. One positive number is six less than another positive number. If two times the reciprocal of the smaller number is added to five times the reciprocal of the larger number, the sum is 1. Find the two numbers.
15. One positive number is ten less than another positive number. If the reciprocal of the smaller number is added to three times the reciprocal of the larger number, the sum is $\frac{1}{4}$. Find the two numbers.
16. One positive number is nine larger than another positive number. If the reciprocal of the smaller number is added to four times the reciprocal of the larger number, the sum is $\frac{2}{3}$. Find the two numbers.
17. One copy machine can run off copies in 15 minutes. A newer machine can do the same job in 10 minutes. How long does it take the two machines, working together, to make all of the necessary copies?
18. Dylan can mow a lawn in 36 minutes, while Alycia takes 45 minutes to mow the same lawn. If Dylan and Alycia work together using two lawn mowers, how long does it take them to mow the lawn?
19. One hose can fill a 40,000-gallon swimming pool in 18 hours. A hose from the neighbor's house can fill a swimming pool of that size in 24 hours. If the two hoses run at the same time, how long does it take to fill the swimming pool?
20. One pipe can drain a tank in 10 minutes, while a smaller pipe takes 45 minutes to drain the same tank. How long does it take the two pipes to drain the tank if both are used at the same time?
21. Tina can weed her vegetable garden completely in 30 minutes. Her friend Marisa can do the same task in 40 minutes. If they work together, how long does it take Tina and Marisa to weed the vegetable garden?
22. A company has printed out 500 surveys to be put into preaddressed envelopes and mailed out. Dusty can fill all of the envelopes in 5 hours, and Grady can do it in 9 hours. If they work together, how long does it take them to stuff the 500 envelopes?
23. Two drainpipes used at the same time can drain a pool in 12 hours. Alone, the smaller pipe takes 7 hours longer than the larger pipe takes to drain the pool. How long does it take the smaller pipe alone to drain the pool?
24. Working together, Rosa and Dianne can plant 20 flats of pansies in 3 hours. If Rosa works alone, it takes her 8 hours longer than it takes Dianne to plant all 20 flats. How long does it take Rosa to plant the pansies by herself?
25. Steve and his assistant Ross can wire a house in 6 hours. If they work alone, Ross takes 9 hours longer than Steve to wire the house. How long does it take Ross to wire the house?
26. Earl and John can clean an entire building in 4 hours. Earl can clean the entire building by himself in 6 fewer hours than John can. How long does it take Earl to clean the building by himself?

27. Working together, Sarah and Jeff can feed all of the animals at the zoo in 2 hours. Jeff uses a motorized cart to carry the food, whereas Sarah uses a pushcart. If each person feeds all of the animals without the help of the other, it takes Sarah twice as long as it takes Jeff. How long does it take Sarah to feed all of the animals?

28. A small pipe takes three times longer to fill a tank than a larger pipe takes. If both pipes are used at the same time, it takes 12 minutes to fill the tank. How long does it take the smaller pipe alone to fill the tank?

29. When he works alone, Brent takes 2 hours longer to buff the gymnasium floor than Tracy takes when she works alone. After working for 3 hours, Brent quits and Tracy takes over. If it takes Tracy 2 hours to finish the rest of the floor, how long would it have taken Brent to buff the entire floor?

30. Frida can wallpaper a room in 3 fewer hours than Gerardo can. After they worked together for 2 hours, Frida had to leave. It took Gerardo an additional 4 hours to finish wallpapering the room. How long would it have taken Frida to wallpaper the room by herself?

31. Vang is training for a triathlon. He rides his bicycle at a speed that is 15 miles per hour faster than his running speed. If Vang can cycle 40 miles in the same amount of time it takes him to run 16 miles, what is his running speed?

32. Ariel drives 15 miles per hour faster than Sharon does. Ariel can drive 100 miles in the same amount of time it takes Sharon to drive 80 miles. Find Ariel's driving speed.

33. Jared is training to run a marathon. Today he ran 14 miles in 2 hours. After running the first 9 miles at a certain speed, he increased his speed by 4 miles per hour for the remaining 5 miles. Find the speed at which Jared was running for the first 9 miles.

34. Kyung ran 7 miles this morning. After running the first 3 miles, she increased her speed by 2 miles per hour. If it took her exactly 1 hour to finish her run, find the speed at which she was running for the last 4 miles.

35. A boat is traveling in a river that is flowing downstream at a speed of 10 kilometers per hour. The boat can travel 30 kilometers upstream in the same time it would take to travel 50 kilometers downstream. What is the speed of the boat in still water?

36. Denae is swimming in a river that flows downstream at a speed of 0.5 meter per second. It takes her the same amount of time to swim 500 meters upstream as it does to swim 1000 meters downstream. Find Denae's swimming speed in still water.

37. An airplane flies 72 miles with a 30-mph tailwind and then flies back into the 30-mph wind. If the time for the round trip was 1 hour, find the speed of the airplane in calm air.

38. Selma is kayaking in a river that flows downstream at a rate of 1 mile per hour. Selma paddles 5 miles downstream and then turns around and paddles 6 miles upstream. The trip takes 3 hours.

a) How fast can Selma paddle in still water?

b) Selma is now 1 mile upstream of her starting point. How many minutes will it take her to paddle back to her starting point?

39. y varies directly as x. If $y = 30$ when $x = 3$, find y when $x = 8$.

40. y varies directly as x. If $y = 78$ when $x = 6$, find y when $x = 15$.

41. y varies inversely as x. If $y = 8$ when $x = 7$, find y when $x = 4$.

42. y varies inversely as x. If $y = 6$ when $x = 12$, find y when $x = 9$.

43. y varies directly as the square of x. If $y = 150$ when $x = 5$, find y when $x = 7$.

44. y varies inversely as the square of x. If $y = 4$ when $x = 6$, find y when $x = 3$.

45. y varies jointly as x and z. If $y = 270$ when $x = 10$ and $z = 6$, find y when $x = 3$ and $z = 8$.

46. y varies directly as x and inversely as z. If $y = 78$ when $x = 26$ and $z = 7$, find y when $x = 30$ and $z = 18$.

47. Sam's gross pay varies directly as the number of hours he works. In a week that he worked 32 hours, his gross pay was \$304. What would Sam's gross pay be if he worked 20 hours?

48. The height that a ball bounces varies directly as the height from which it is dropped. If a ball dropped from a height of 55 inches bounces 25 inches, how high would the ball bounce if it were dropped from a height of 88 inches?

49. *Ohm's law.* In a circuit, the electric current (in amperes) varies directly as the voltage. If the current is 8 amperes when the voltage is 24 volts, find the current when the voltage is 9 volts.

50. *Hooke's law.* The distance that a hanging object stretches a spring varies directly as the mass of the object. If a 5-kilogram weight stretches a spring by 32 centimeters, how far would a 2-kilogram weight stretch the spring?

51. The amount of money that each person must contribute to buy a retirement gift for a coworker varies inversely as the number of people contributing. If 10 people must contribute $24 each, how much does each person have to contribute if there are 16 people?

52. The maximum load that a wooden beam can support varies inversely as its length. If a beam that is 8 feet long can support 700 pounds, what is the maximum load that can be supported by a beam that is 5 feet long?

53. In a circuit, the electric current (in amperes) varies inversely as the resistance (in ohms). If the current is 30 amperes when the resistance is 3 ohms, find the current when the resistance is 5 ohms.

54. *Boyle's law.* The volume of a gas varies inversely as the pressure upon it. The volume of a gas is 128 cubic centimeters when it is under a pressure of 50 kilograms per square centimeter. Find the volume when the pressure is reduced to 40 kilograms per square centimeter.

55. The illumination of an object varies inversely as the square of its distance from the source of light. If a light source provides an illumination of 30 foot-candles at a distance of 10 feet, find the illumination at a distance of 20 feet. (One foot-candle is the amount of illumination produced by a standard candle at a distance of 1 foot.)

56. The illumination of an object varies inversely as the square of its distance from the source of light. If a light source provides an illumination of 80 foot-candles at a distance of 5 feet, find the illumination at a distance of 40 feet.

Writing in Mathematics

Answer in complete sentences.

57. Write a work-rate word problem that can be solved by the equation $\frac{t}{8} + \frac{t}{10} = 1$. Explain how you created your problem.

58. Write a uniform motion word problem that can be solved by the equation $\frac{10}{r-4} + \frac{10}{r+4} = 6$. Explain how you created your problem.

CHAPTER 7 SUMMARY

Section 7.1 Rational Expressions and Functions

Rational Expressions, p. 354

A rational expression is a quotient of two polynomials.
The denominator of a rational expression must be nonzero, as division by zero is undefined.

$$\frac{x^2 - 7x + 10}{x^2 + 7x - 18}$$

Values for Which a Rational Expression Is Undefined, pp. 355–356

To find the values for which a rational expression is undefined, set the denominator equal to 0 and solve the resulting equation. These solutions are the values for which the expression is undefined.

Find the values for which the rational expression is undefined: $\frac{x^2 + 5x + 12}{x^2 - 9x + 18}$

$$x^2 - 9x + 18 = 0$$
$$(x - 3)(x - 6) = 0$$

$$x - 3 = 0 \qquad x - 6 = 0$$
$$x = 3 \qquad x = 6$$

The expression is undefined for $x = 3, 6$.

Rational Functions, pp. 356–357

A rational function $r(x)$ is a function of the form $r(x) = \frac{f(x)}{g(x)}$, where $f(x)$ and $g(x)$ are polynomials and $g(x) \neq 0$. You exclude from the domain of the function any value that causes the function to be undefined, namely, any value for which the denominator is equal to 0.

Find the domain of the rational function:

$$r(x) = \frac{x^2 + 24x + 29}{x^2 - 15x + 56}$$

$$x^2 - 15x + 56 = 0$$
$$(x - 7)(x - 8) = 0$$

$$x - 7 = 0 \qquad x - 8 = 0$$
$$x = 7 \qquad x = 8$$

Domain: All real numbers except 7 and 8; $(-\infty, 7) \cup (7, 8) \cup (8, \infty)$

Simplifying a Rational Expression to Lowest Terms, pp. 357–359

To simplify a rational expression to lowest terms, factor the numerator and denominator. Then divide out common factors in the numerator and denominator.

Simplify to lowest terms: $\frac{x^2 + 7x + 12}{x^2 - 5x - 36}$

$$\frac{x^2 + 7x + 12}{x^2 - 5x - 36} = \frac{(x + 3)(x + 4)}{(x + 4)(x - 9)}$$
$$= \frac{(x + 3)\overset{1}{\cancel{(x + 4)}}}{\underset{1}{\cancel{(x + 4)}}(x - 9)}$$
$$= \frac{x + 3}{x - 9}$$

Factors That Are Opposites, pp. 359–360

Two expressions of the form $a - b$ and $b - a$ are opposites. The rational expression $\frac{a - b}{b - a}$ simplifies to equal -1, as any fraction whose numerator is the opposite of its denominator is equal to -1.

Simplify to lowest terms: $\frac{x^2 - 13x + 40}{25 - x^2}$

$$\frac{x^2 - 13x + 40}{25 - x^2} = \frac{(x - 5)(x - 8)}{(5 + x)(5 - x)}$$
$$= \frac{\overset{-1}{\cancel{(x - 5)}}(x - 8)}{(5 + x)\underset{1}{\cancel{(5 - x)}}}$$
$$= -\frac{x - 8}{5 + x}$$

Section 7.2 Multiplication and Division of Rational Expressions

Multiplying Rational Expressions, pp. 363–364

To multiply two rational expressions, factor each numerator and denominator completely. Divide out factors that are common to a numerator and a denominator and multiply the remaining factors, leaving the numerator and denominator in factored form.

Multiply: $\dfrac{x^2 - 5x - 24}{x^2 - 7x - 18} \cdot \dfrac{x^2 + 13x + 22}{x^2 - 64}$

$$\frac{x^2 - 5x - 24}{x^2 - 7x - 18} \cdot \frac{x^2 + 13x + 22}{x^2 - 64}$$
$$= \frac{(x+3)(x-8)}{(x+2)(x-9)} \cdot \frac{(x+2)(x+11)}{(x+8)(x-8)}$$
$$= \frac{(x+3)\overset{1}{\cancel{(x-8)}}}{\underset{1}{\cancel{(x+2)}}(x-9)} \cdot \frac{\overset{1}{\cancel{(x+2)}}(x+11)}{(x+8)\underset{1}{\cancel{(x-8)}}}$$
$$= \frac{(x+3)(x+11)}{(x-9)(x+8)}$$

Dividing Rational Expressions, pp. 365–366

To divide a rational expression by another rational expression, replace the divisor by its reciprocal and then multiply.

Divide: $\dfrac{x^2 + 6x - 7}{x^2 + x - 30} \div \dfrac{x^2 - 49}{x + 6}$

$$\frac{x^2 + 6x - 7}{x^2 + x - 30} \div \frac{x^2 - 49}{x + 6}$$
$$= \frac{x^2 + 6x - 7}{x^2 + x - 30} \cdot \frac{x + 6}{x^2 - 49}$$
$$= \frac{(x-1)(x+7)}{(x-5)(x+6)} \cdot \frac{x+6}{(x+7)(x-7)}$$
$$= \frac{(x-1)\overset{1}{\cancel{(x+7)}}}{(x-5)\underset{1}{\cancel{(x+6)}}} \cdot \frac{\overset{1}{\cancel{x+6}}}{\underset{1}{\cancel{(x+7)}}(x-7)}$$
$$= \frac{x-1}{(x-5)(x-7)}$$

Section 7.3 Addition and Subtraction of Rational Expressions That Have the Same Denominator

Adding Rational Expressions That Have the Same Denominator, pp. 369–370

To add rational expressions that have the same denominator, add the numerators and place the result over the common denominator. Check to make sure the result is in simplest terms.

Add: $\dfrac{x^2 + 6x}{x^2 + 2x - 63} + \dfrac{7x + 36}{x^2 + 2x - 63}$

$$\frac{x^2 + 6x}{x^2 + 2x - 63} + \frac{7x + 36}{x^2 + 2x - 63}$$
$$= \frac{x^2 + 6x + 7x + 36}{x^2 + 2x - 63}$$
$$= \frac{x^2 + 13x + 36}{x^2 + 2x - 63}$$
$$= \frac{(x+4)(x+9)}{(x-7)(x+9)}$$
$$= \frac{x+4}{x-7}$$

Subtracting Rational Expressions That Have the Same Denominator, pp. 370–372

Subtracting two rational expressions that have the same denominator is just like adding them except that you subtract the two numerators rather than add them.

Subtract: $\dfrac{9x + 4}{x^2 + 15x + 50} - \dfrac{5x - 16}{x^2 + 15x + 50}$

$$\frac{9x + 4}{x^2 + 15x + 50} - \frac{5x - 16}{x^2 + 15x + 50}$$
$$= \frac{(9x + 4) - (5x - 16)}{x^2 + 15x + 50}$$
$$= \frac{9x + 4 - 5x + 16}{x^2 + 15x + 50}$$
$$= \frac{4x + 20}{x^2 + 15x + 50}$$
$$= \frac{4(x + 5)}{(x + 5)(x + 10)}$$
$$= \frac{4}{x + 10}$$

Adding or Subtracting Rational Expressions That Have Opposite Denominators, pp. 372–373

To add two fractions with opposite denominators, rewrite the denominator of the second fraction as its opposite by changing the operation to subtraction.

To subtract two fractions with opposite denominators, rewrite the denominator of the second fraction as its opposite by changing the operation to addition.

Add: $\dfrac{x^2 + x + 5}{x^2 - 4} + \dfrac{4x + 15}{4 - x^2}$

$$\frac{x^2 + x + 5}{x^2 - 4} + \frac{4x + 15}{4 - x^2}$$
$$= \frac{x^2 + x + 5}{x^2 - 4} - \frac{4x + 15}{x^2 - 4}$$
$$= \frac{x^2 + x + 5 - 4x - 15}{x^2 - 4}$$
$$= \frac{x^2 - 3x - 10}{x^2 - 4}$$
$$= \frac{(x + 2)(x - 5)}{(x + 2)(x - 2)}$$
$$= \frac{x - 5}{x - 2}$$

Section 7.4 Addition and Subtraction of Rational Expressions That Have Different Denominators

Finding the LCD of Two Rational Expressions, pp. 375–376

Completely factor each denominator; then identify each factor that is a factor of one or both denominators. The LCD is equal to the product of these factors. If a factor is repeated in one or both of the denominators, the exponent used for this factor is equal to the greatest power to which the factor is raised in any one denominator.

Find the LCD: $\dfrac{7}{x^2 + 5x + 6}, \dfrac{4}{x^2 - 4x - 21}$

$$x^2 + 5x + 6 = (x + 2)(x + 3)$$
$$x^2 - 4x - 21 = (x + 3)(x - 7)$$

LCD: $(x + 2)(x + 3)(x - 7)$

Adding or Subtracting Rational Expressions That Have Different Denominators, pp. 376–379

To add or subtract two rational expressions that do not have the same denominator, find the LCD of the two denominators and rewrite each rational expression as an equivalent rational expression that has the LCD as its denominator. Then add or subtract the numerators, placing the result over the common denominator. Simplify the resulting rational expression if possible.

Subtract: $\dfrac{7}{x^2 + 8x + 15} - \dfrac{4}{x^2 + 7x + 12}$

$$\frac{7}{x^2 + 8x + 15} - \frac{4}{x^2 + 7x + 12}$$
$$= \frac{7}{(x + 3)(x + 5)} - \frac{4}{(x + 3)(x + 4)}$$
$$= \frac{7}{(x + 3)(x + 5)} \cdot \frac{x + 4}{x + 4} - \frac{4}{(x + 3)(x + 4)} \cdot \frac{x + 5}{x + 5}$$
$$= \frac{7x + 28 - 4x - 20}{(x + 3)(x + 5)(x + 4)}$$
$$= \frac{3x + 8}{(x + 3)(x + 5)(x + 4)}$$

Section 7.5 Complex Fractions

Simplifying Complex Fractions, pp. 382–385

To simplify a complex fraction, find the LCD of all fractions in the complex fraction and then multiply the numerator and denominator by this LCD.

Simplify: $\dfrac{1 - \frac{7}{x} + \frac{12}{x^2}}{1 + \frac{4}{x} - \frac{32}{x^2}}$

$$\frac{1 - \frac{7}{x} + \frac{12}{x^2}}{1 + \frac{4}{x} - \frac{32}{x^2}} \cdot \frac{x^2}{x^2} = \frac{1 \cdot x^2 - \frac{7}{x} \cdot x^2 + \frac{12}{x^2} \cdot x^2}{1 \cdot x^2 + \frac{4}{x} \cdot x^2 - \frac{32}{x^2} \cdot x^2}$$

$$= \frac{1 \cdot x^2 - \frac{7}{\cancel{x}_1} \cdot \cancel{x^2}^{x} + \frac{12}{\cancel{x^2}_1} \cdot \cancel{x^2}^{1}}{1 \cdot x^2 + \frac{4}{\cancel{x}_1} \cdot \cancel{x^2}^{x} - \frac{32}{\cancel{x^2}_1} \cdot \cancel{x^2}^{1}}$$

$$= \frac{x^2 - 7x + 12}{x^2 + 4x - 32}$$

$$= \frac{(x - 3)(x - 4)}{(x - 4)(x + 8)}$$

$$= \frac{x - 3}{x + 8}$$

Section 7.6 Rational Equations

Solving Rational Equations, pp. 388–391

Find the LCD of all rational expressions in the equation and then multiply both sides of the equation by that LCD. Solve the resulting equation.

Solve: $1 - \dfrac{18}{x^2} = \dfrac{3}{x} + \dfrac{10}{x^2}$

LCD: x^2

$$x^2\left(1 - \frac{18}{x^2}\right) = x^2\left(\frac{3}{x} + \frac{10}{x^2}\right)$$

$$1 \cdot x^2 - \frac{18}{x^2} \cdot x^2 = \frac{3}{x} \cdot x^2 + \frac{10}{x^2} \cdot x^2$$

$$1 \cdot x^2 - \frac{18}{\cancel{x^2}_1} \cdot \cancel{x^2}^{1} = \frac{3}{\cancel{x}_1} \cdot \cancel{x^2}^{x} + \frac{10}{\cancel{x^2}_1} \cdot \cancel{x^2}^{1}$$

$$x^2 - 18 = 3x + 10$$

$$x^2 - 3x - 28 = 0$$

$$(x + 4)(x - 7) = 0$$

$$x + 4 = 0 \qquad x - 7 = 0$$

$$x = -4 \qquad x = 7$$

$\{-4, 7\}$

Checking for Extraneous Solutions, pp. 389–391

If a denominator of a rational expression is equal to 0 when the value of a solution is substituted for the variable, the solution must be omitted and is called an extraneous solution.

Solve: $\dfrac{x + 4}{x^2 - x - 2} = \dfrac{6}{x^2 - 4x - 5}$

$$\frac{x + 4}{(x + 1)(x - 2)} = \frac{6}{(x + 1)(x - 5)}$$

LCD: $(x + 1)(x - 2)(x - 5)$

$$\frac{x + 4}{(x + 1)(x - 2)} \cdot (x + 1)(x - 2)(x - 5)$$

$$= \frac{6}{(x + 1)(x - 5)} \cdot (x + 1)(x - 2)(x - 5)$$

$$(x + 4)(x - 5) = 6(x - 2)$$

$$x^2 - x - 20 = 6x - 12$$

$$x^2 - 7x - 8 = 0$$

$$(x + 1)(x - 8) = 0$$

$$x + 1 = 0 \qquad x - 8 = 0$$

$$x = -1 \qquad x = 8$$

Extraneous solution

$\{8\}$

Solving Literal Equations, pp. 391–392

A literal equation is an equation that contains two or more variables. Solve the equation for one of the variables by isolating that variable on one side of the equation. To solve a literal equation containing rational expressions, begin by multiplying both sides of the equation by the LCD.

Solve for x: $y = \frac{5x + 4}{2x}$

$$\begin{aligned} y \cdot 2x &= \frac{5x + 4}{2x} \cdot 2x \\ 2xy &= 5x + 4 \\ 2xy - 5x &= 4 \\ x(2y - 5) &= 4 \\ \frac{x(2y - 5)}{2y - 5} &= \frac{4}{2y - 5} \\ x &= \frac{4}{2y - 5} \end{aligned}$$

Section 7.7 Applications of Rational Equations

Reciprocal, pp. 395–396

If n represents a number, its reciprocal is $\frac{1}{n}$.

The sum of the reciprocal of a number and $\frac{7}{12}$ is $\frac{3}{4}$. Find the number.

Unknown number: n

$$\frac{1}{n} + \frac{7}{12} = \frac{3}{4}$$

LCD: $12n$

$$\begin{aligned} \frac{1}{n} \cdot 12n + \frac{7}{12} \cdot 12n &= \frac{3}{4} \cdot 12n \\ \frac{1}{\cancel{n}_1} \cdot 12\cancel{n}^1 + \frac{7}{\cancel{12}_1} \cdot \cancel{12}^1 n &= \frac{3}{\cancel{4}_1} \cdot \cancel{12}^3 n \\ 12 + 7n &= 9n \\ 12 &= 2n \\ 6 &= n \end{aligned}$$

The number is 6.

Work-Rate Problems, pp. 396–398

The work rate for a person or machine is equal to the reciprocal of the time required to complete the whole task.

The amount of work done by a person or machine is equal to the product of the work rate and the time spent working.

The total amount of work done by each person or machine must equal 1.

Jerry can mow a lawn in 40 minutes, while Bill takes 50 minutes to mow the same lawn. If Jerry and Bill work together using two lawn mowers, how long does it take them to mow the lawn?

Person	Time Alone	Work Rate	Time Working	Work Done
Jerry	40 min	$\frac{1}{40}$	t	$\frac{t}{40}$
Bill	50 min	$\frac{1}{50}$	t	$\frac{t}{50}$

$$\frac{t}{40} + \frac{t}{50} = 1$$

LCD: 200

$$\begin{aligned} \frac{t}{40} \cdot 200 + \frac{t}{50} \cdot 200 &= 1 \cdot 200 \\ 5t + 4t &= 200 \\ 9t &= 200 \\ t &= \frac{200}{9} \end{aligned}$$

It takes them $22\frac{2}{9}$ minutes working together.

Direct Variation, pp. 401–402

If a quantity y varies directly as a quantity x, the two quantities are related by the equation $y = kx$, where k is called the variation constant.

The number of calories in a glass of soda varies directly as the amount of soda. If a 12-ounce serving of soda has 180 calories, how many calories are in an 8-ounce glass of soda?

$$\begin{aligned} y &= kx \\ 180 &= k \cdot 12 \\ 15 &= k \end{aligned} \qquad \begin{aligned} y &= kx \\ y &= 15 \cdot 8 \\ y &= 120 \end{aligned}$$

There are 120 calories in an 8-ounce glass of soda.

Inverse Variation, pp. 402–403

If a quantity y varies inversely as a quantity x, the two quantities are related by the equation $y = \frac{k}{x}$, where k is the variation constant.

The maximum load that a wooden beam can support varies inversely as its length. If a beam that is 8 feet long can support 725 pounds, what is the maximum load that can be supported by a beam that is 10 feet long?

$$\begin{aligned} y &= \frac{k}{x} \\ 725 &= \frac{k}{8} \\ 5800 &= k \end{aligned} \qquad \begin{aligned} y &= \frac{k}{x} \\ y &= \frac{5800}{10} \\ y &= 580 \end{aligned}$$

A beam that is 10 feet long can support 580 pounds.

SUMMARY OF CHAPTER 7 STUDY STRATEGIES

- When preparing for a cumulative exam, begin approximately two weeks before the exam by determining what material will be on the exam.
- Begin studying by reviewing your old exams and quizzes. Although you should rework each problem, pay particular attention to the problems you got wrong on the exam or quiz. Also look over your old homework assignments for problems you struggled with and review these topics.
- Use review materials from your instructor and the cumulative review exercises in the text to check your progress. Try to work the exercises without referring to your notes, note cards, or text. This will give you a clear assessment as to which problems require further study.
- Create study sheets for each type of applied problem that may appear on the exam. These sheets should contain examples of the problems, the procedure for solving them, and the solutions.
- Continue to meet regularly with your study group. If your group has helped you succeed on previous exams, they will be able to help you succeed on this exam as well.

CHAPTER 7 REVIEW

***Evaluate the rational expression for the given value of the variable.* [7.1]**

1. $\frac{3}{x-9}$ for $x = -6$
2. $\frac{x-7}{x+13}$ for $x = 19$
3. $\frac{x^2+9x-11}{x^2+2x-5}$ for $x = 3$
4. $\frac{x^2-5x-15}{x^2+17x+66}$ for $x = -4$

***Find all values for which the rational expression is undefined.* [7.1]**

5. $\frac{-6}{x+9}$
6. $\frac{x^2+9x+18}{x^2-3x-54}$

***Simplify the given rational expression. (Assume that all denominators are nonzero.)* [7.1]**

7. $\frac{x+7}{x^2+4x-21}$
8. $\frac{x^2+6x-40}{x^2-9x+20}$
9. $\frac{36-x^2}{x^2-7x+6}$
10. $\frac{x^2+6x+9}{4x^2+15x+9}$

***Evaluate the given rational function.* [7.1]**

11. $r(x) = \frac{x+7}{x^2+8x-15}, r(3)$
12. $r(x) = \frac{x^2-15x}{x^2-11x+18}, r(-2)$

***Find the domain of the given rational function.* [7.1]**

13. $r(x) = \frac{x^2+3x-40}{x^2+6x}$
14. $r(x) = \frac{x^2+7x-8}{x^2-5x+4}$

***Multiply.* [7.2]**

15. $\frac{x+9}{x+5} \cdot \frac{x^2-7x}{x^2+2x-63}$
16. $\frac{x^2+14x+45}{x^2-6x-7} \cdot \frac{x^2+9x+8}{x^2+3x-10}$
17. $\frac{x^2-4x}{x^2-15x+54} \cdot \frac{x^2-x-72}{x^2+8x}$
18. $\frac{16-x^2}{x+2} \cdot \frac{7x^2+13x-2}{x^2+2x-24}$

***For the given functions $f(x)$ and $g(x)$, find $f(x) \cdot g(x)$.* [7.2]**

19. $f(x) = \frac{x-6}{x-1}, g(x) = \frac{x^2+6x+5}{x^2-3x-18}$
20. $f(x) = \frac{x^2+16x+64}{x^2+x-42}, g(x) = \frac{49-x^2}{x^2+11x+24}$

***Divide.* [7.2]**

21. $\frac{x^2+9x+18}{x^2-7x+12} \div \frac{x+3}{x-3}$
22. $\frac{x^2-x-6}{x^2-19x+88} \div \frac{x^2+10x+16}{x^2+4x-96}$
23. $\frac{x^2-3x-4}{x^2+x} \div \frac{3x^2-10x-8}{x^2-1}$
24. $\frac{9-x^2}{x-7} \div \frac{x^2-8x+15}{x^2-5x-14}$

***For the given functions $f(x)$ and $g(x)$, find $f(x) \div g(x)$.* [7.2]**

25. $f(x) = \frac{x^2+3x}{x^2+10x+25}, g(x) = \frac{x^2-6x-27}{x^2-x-30}$
26. $f(x) = \frac{x^2+8x+12}{x^2+5x+4}, g(x) = \frac{x^2-4x-60}{x^2+7x+12}$

Worked-out solutions to Review Exercises marked with ● can be found on page AN-23.

Add or subtract. **[7.3/7.4]**

27. $\frac{5}{x+6} + \frac{7}{x+6}$

28. $\frac{x^2+10x}{x^2+15x+56} - \frac{5x+24}{x^2+15x+56}$

29. $\frac{x^2-6x-13}{x^2-x-20} + \frac{10x-32}{x^2-x-20}$

30. $\frac{5x-34}{x-7} - \frac{x-8}{7-x}$

31. $\frac{2x^2+7x-3}{x-2} + \frac{x^2+3x+9}{2-x}$

32. $\frac{x^2-3x-12}{x^2-25} - \frac{2x-18}{25-x^2}$

33. $\frac{5}{x^2+3x-4} + \frac{2}{x^2-4x+3}$

34. $\frac{6}{x^2-10x+16} - \frac{1}{x^2-15x+56}$

35. $\frac{x}{x^2-5x-50} - \frac{4}{x^2-14x+40}$

36. $\frac{x-5}{x^2+14x+33} + \frac{4}{x^2+20x+99}$

37. $\frac{x+1}{x^2+2x-24} + \frac{x-5}{x^2-6x+8}$

For the given rational functions f(x) and g(x), find f(x) + g(x). **[7.4]**

38. $f(x) = \frac{x+3}{x^2-x-2}, g(x) = \frac{5}{x^2-7x+10}$

For the given rational functions f(x) and g(x), find f(x) − g(x). **[7.4]**

39. $f(x) = \frac{x+2}{x^2+4x-5}, g(x) = \frac{1}{x^2+12x+35}$

Simplify the complex fraction. **[7.5]**

40. $\frac{1-\frac{9}{x}}{1-\frac{81}{x^2}}$

41. $\frac{\frac{x-5}{x-3} - \frac{3}{x-7}}{\frac{x^2+7x-44}{x^2-10x+21}}$

42. $\frac{1+\frac{2}{x}-\frac{35}{x^2}}{1-\frac{7}{x}+\frac{10}{x^2}}$

43. $\frac{\frac{x^2-8x+15}{x^2-11x+18}}{\frac{x^2+x-30}{x^2-6x-27}}$

Solve. **[7.6]**

44. $\frac{9}{x} - \frac{3}{4} = \frac{7}{8}$

45. $x - 5 + \frac{12}{x} = 2$

46. $\frac{3}{x-7} = \frac{7}{x+5}$

47. $\frac{2x-11}{x-1} = \frac{x-7}{x+3}$

48. $\frac{3}{x+2} - \frac{1}{x+1} = \frac{x+3}{x^2+3x+2}$

49. $\frac{x}{x+5} + \frac{3}{x-7} = \frac{36}{x^2-2x-35}$

50. $\frac{2}{x^2-x-2} + \frac{10}{x^2-2x-3} = \frac{x+12}{x^2-x-2}$

51. $\frac{20}{x^2+12x+27} - \frac{4}{x^2+14x+45} = \frac{x+10}{x^2+8x+15}$

Solve for the specified variable. **[7.6]**

52. $\frac{1}{2} = \frac{A}{bh}$ for h

53. $y = \frac{3x}{4x-7}$ for x

54. $\frac{x}{2r} - \frac{y}{3r} = \frac{1}{5}$ for r

55. The sum of the reciprocal of a number and $\frac{5}{6}$ is $\frac{23}{24}$. Find the number. [7.7]

56. One positive number is five more than another positive number. If three times the reciprocal of the smaller number is added to four times the reciprocal of the larger number, the sum is 1. Find the two numbers. [7.7]

57. Rob can mow a lawn in 30 minutes, while Sean takes 60 minutes to mow the same lawn. If Rob and Sean work together using two lawn mowers, how long does it take them to mow the lawn? [7.7]

58. Two pipes, used together, can fill a tank in 28 minutes. Alone, the smaller pipe takes 42 minutes longer than the larger pipe takes to fill the tank. How long does it take the smaller pipe alone to fill the tank? [7.7]

59. Linda had to make a 275-mile trip to Omaha. After driving the first 100 miles, she increased her speed by 15 miles per hour. If the trip took her exactly 4 hours, find the speed at which she was driving for the first 100 miles. [7.7]

60. Dominique is kayaking in a river that is flowing downstream at a speed of 2 miles per hour. Dominique can paddle 10 miles downstream in the same amount of time she can paddle 5 miles upstream. What is the speed that Dominique can paddle in still water? [7.7]

61. The number of calories in a glass of milk varies directly as the amount of milk. If a 12-ounce serving of milk has 195 calories, how many calories are in an 8-ounce glass of milk? [7.7]

62. The distance required for a car to stop after the brakes are applied varies directly as the square of the speed of the car. If it takes 100 feet for a car traveling at 40 miles per hour to stop, how far does it take for a car traveling 60 miles per hour to come to a stop? [7.7]

63. The maximum load that a wooden beam can support varies inversely as its length. If a beam that is 6 feet long can support 900 pounds, what is the maximum load that can be supported by a beam that is 8 feet long? [7.7]

64. The illumination of an object varies inversely as the square of its distance from the source of light. If a light source provides an illumination of 20 foot-candles at a distance of 12 feet, find the illumination at a distance of 6 feet. [7.7]

CHAPTER 7 TEST

For Extra Help

Step-by-step test solutions are found on the Chapter Test Prep Videos available via the Video Resources on DVD, in MyMathLab, and on YouTube (search "WoodburyElemIntAlg" and click on "Channels").

Evaluate the rational expression for the given value of the variable.

1. $\dfrac{10}{x-5}$ for $x = -7$

Find all values for which the rational expression is undefined.

2. $\dfrac{x^2+5x-14}{x^2+10x+21}$

Simplify the given rational expression. (Assume that all denominators are nonzero.)

3. $\dfrac{x^2-8x+15}{x^2+7x-30}$

Evaluate the rational function.

4. $r(x) = \dfrac{x^2+3x-13}{x^2-4x-20}, r(-3)$

Multiply.

5. $\dfrac{x^2+x-42}{x^2+6x-7} \cdot \dfrac{x^2-2x+1}{x^2-9x+18}$

Divide.

6. $\dfrac{x^2-10x+25}{x^2+11x+24} \div \dfrac{25-x^2}{x^2+3x-40}$

Add or subtract.

7. $\dfrac{5x+3}{x-6} - \dfrac{2x+21}{x-6}$

8. $\dfrac{2x^2+6x-11}{x-4} + \dfrac{x^2+2x+21}{4-x}$

9. $\dfrac{6}{x^2+8x+7} + \dfrac{5}{x^2-3x-4}$

10. $\dfrac{x-4}{x^2-13x+40} - \dfrac{8}{x^2-10x+16}$

Simplify the complex fraction.

11. $\dfrac{1+\dfrac{4}{x}-\dfrac{45}{x^2}}{1+\dfrac{2}{x}-\dfrac{35}{x^2}}$

Simplify.

12. $\dfrac{x+5}{x^2+3x-54} \cdot \dfrac{x^2-11x+30}{x^2+16x+55}$

13. $\dfrac{x^2 - 3x - 40}{x^2 + 7x + 6} \div \dfrac{x^2 - 8x}{x^2 + 10x + 9}$

14. $\dfrac{x^2 + 7x - 14}{x^2 + 6x + 8} + \dfrac{3x + 38}{x^2 + 6x + 8}$

15. $\dfrac{x - 7}{x^2 + 7x - 8} - \dfrac{2}{x^2 + x - 2}$

Solve.

16. $\dfrac{15}{x} + \dfrac{3}{8} = \dfrac{19}{24}$

17. $\dfrac{x + 2}{x^2 + 5x + 4} - \dfrac{4}{x^2 + x - 12} = \dfrac{2}{x^2 - 2x - 3}$

Solve for the specified variable.

18. $x = \dfrac{2y}{7y - 5}$ for y

19. Two pipes together can fill a tank in 40 minutes. Alone, the smaller pipe takes 18 minutes longer than the larger pipe takes to fill the tank. How long does it take the smaller pipe alone to fill the tank?

20. Preparing for a race, Greg went for a 25-mile bicycle ride. After the first 10 miles, he increased his speed by 10 miles per hour. If the ride took him exactly 1 hour, find the speed at which he was riding for the first 10 miles.

Mathematicians in History

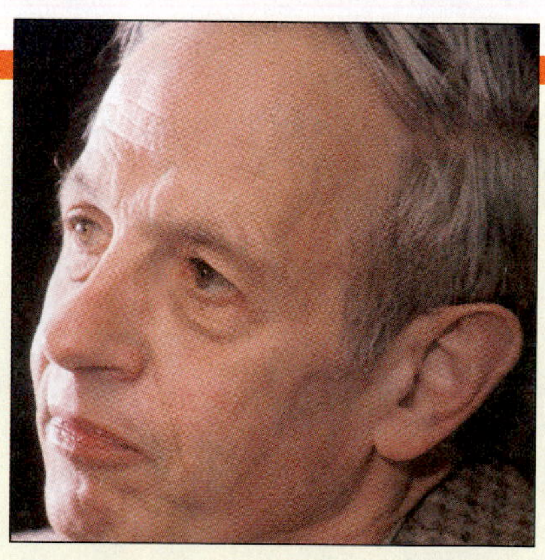

John Nash is an American mathematician whose research has greatly affected mathematics and economics as well as many other fields. When he was applying for graduate school at Princeton, one of his math professors said quite simply, "This man is a genius."

Write a one-page summary (*or* make a poster) of the life of John Nash and his accomplishments.

Interesting issues:

- Where and when was John Nash born?
- Describe Nash's childhood as well as his life as a college student.
- What mental illness struck Nash in the late 1950s?
- Nash was influential in the field of game theory. What is game theory?
- In 1994, Nash won the Nobel Prize for Economics. Exactly what did Nash win the prize for?
- One of Nash's nicknames is "The Phantom of Fine Hall." Why was this nickname chosen for him?
- What color sneakers did Nash wear?
- Sylvia Nasar wrote a biography of Nash's life, which was made into an Academy Award–winning movie. What was the title of the book and movie?
- What actor played John Nash in the movie?

CUMULATIVE REVIEW CHAPTERS 5–7

***Simplify the expression. Write the result without using negative exponents. (Assume that all variables represent nonzero real numbers.)* [5.1/5.2]**

1. $\dfrac{x^{11}}{x^4}$

2. $12m^{10}n^7 \cdot 9m^6n^{14}$

3. $\left(\dfrac{5a^6b^{11}}{c^7}\right)^4$

4. $\dfrac{t^{12}}{t^{-9}}$

5. $(a^{-6}b^7)^{-8}$

6. $x^{-16} \cdot x^{-19}$

***Rewrite in scientific notation.* [5.2]**

7. 330,000,000,000

8. 0.000000000000714

9. If a computer can perform a calculation in 0.000000000005 second, how long will it take to perform 200,000,000 calculations? [5.2]

***Evaluate the polynomial for the given value of the variable.* [5.3]**

10. $x^2 + 11x - 16$ for $x = -7$

***Add or subtract.* [5.3]**

11. $(x^2 + 6x - 20) + (x^2 - 13x - 39)$

12. $(x^2 - 15x - 9) - (2x^2 - 3x - 31)$

***Evaluate the given function.* [5.3]**

13. $f(x) = x^2 + 5x - 37, f(8)$

***Multiply.* [5.4]**

14. $3x(5x^2 - 8x + 12)$

15. $(4x - 7)(3x - 8)$

***Divide.* [5.5]**

16. $\dfrac{10x^{12} + 24x^9 - 18x^6}{2x^5}$

17. $\dfrac{x^2 + 13x + 39}{x + 6}$

***Factor completely.* [6.1–6.5]**

18. $x^3 - 4x^2 + 6x - 24$

19. $2x^2 - 26x + 80$

20. $x^2 + 5x - 36$

21. $2x^2 - x - 10$

22. $x^3 - 125$

23. $x^2 - 81$

***Solve.* [6.6]**

24. $x^2 - 49 = 0$

25. $x^2 - 11x + 24 = 0$

26. $x^2 + 14x + 49 = 0$

27. $x(x + 8) = 20$

***Find a quadratic equation with integer coefficients that has the given solution set.* [6.6]**

28. $\{-2, 7\}$

***For the given function $f(x)$, find all values x for which $f(x) = 0$.* [6.7]**

29. $f(x) = x^2 - 100$

30. $f(x) = x^2 + 2x - 35$

31. A man is standing on a cliff above a beach. He throws a rock upward from a height of 192 feet above the beach with an initial velocity of 64 feet/second. The rock's height above the beach, in feet, after t seconds is given by the function $h(t) = -16t^2 + 64t + 192$, which is graphed as follows:

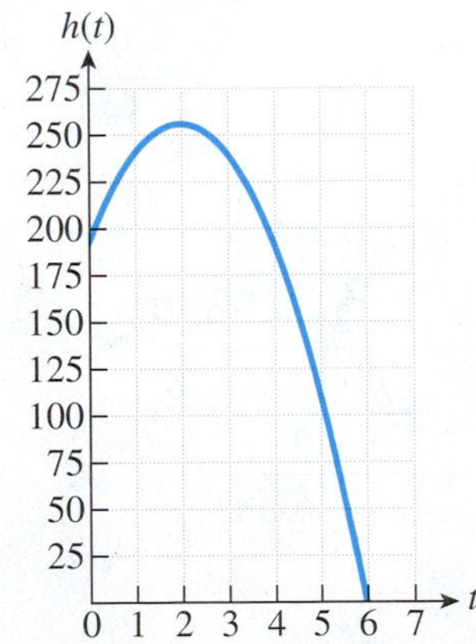

Use the function and its graph to answer the following questions: [6.7]

a) Use the function to determine how high above the beach the rock is after 3 seconds.

b) How long will it take for the rock to land on the beach?

c) After how many seconds does the rock reach its greatest height above the beach?

d) Use the function to determine the greatest height above the beach that the rock reaches.

32. Two consecutive even positive integers have a product of 168. Find the two integers. [6.8]

33. The length of a rectangle is 5 meters more than twice the width. The area of the rectangle is 133 square meters. Find the length and the width of the rectangle. [6.8]

Find all values for which the rational expression is undefined. **[7.1]**

34. $\frac{x^2 + 11x + 24}{x^2 - 14x + 45}$

Simplify the given rational expression. (Assume that all denominators are nonzero.) **[7.1]**

35. $\frac{x^2 + 3x - 54}{x^2 + 16x + 63}$

Evaluate the given rational function. **[7.1]**

36. $r(x) = \frac{x + 12}{x^2 + 7x + 5}, r(3)$

Multiply. **[7.2]**

37. $\frac{x^2 + 3x - 18}{x^2 + 15x + 44} \cdot \frac{x^2 + 2x - 8}{x^2 + 8x + 12}$

Divide. **[7.2]**

38. $\frac{x^2 + x - 90}{x^2 + 14x + 49} \div \frac{x^2 - 8x - 9}{x^2 + 5x - 14}$

Add or subtract. **[7.3, 7.4]**

39. $\frac{5x + 7}{x^2 - 2x - 8} - \frac{2x + 19}{x^2 - 2x - 8}$

40. $\frac{6}{x^2 + x - 2} + \frac{2}{x^2 - 6x + 5}$

41. $\frac{x - 5}{x^2 - 10x + 21} - \frac{3}{x^2 - 9}$

42. $\frac{x + 4}{x^2 + 13x + 42} + \frac{6}{x^2 + 12x + 35}$

Simplify the complex fraction. **[7.5]**

43. $\frac{\frac{x - 5}{x + 3} - \frac{1}{x - 6}}{\frac{x^2 + 5x - 24}{x^2 - 3x - 18}}$

44. $\frac{1 + \frac{1}{x} - \frac{20}{x^2}}{1 + \frac{13}{x} + \frac{40}{x^2}}$

Solve. **[7.6]**

45. $6 + \frac{4}{x - 3} = \frac{3x + 1}{x - 3}$

46. $\frac{2x - 3}{x + 4} = \frac{x + 3}{x - 4}$

47. $\frac{5}{x + 7} + \frac{4}{x - 3} = \frac{x - 11}{x^2 + 4x - 21}$

Solve for the specified variable.

48. $\frac{1}{x} + \frac{1}{y} = 2$ for x [7.6]

49. The sum of the reciprocal of a number and $\frac{4}{15}$ is $\frac{11}{30}$. Find the number. [7.7]

50. Together, two pipes can fill a tank in 3 hours. Alone, the smaller pipe takes 8 hours longer than the larger pipe takes to fill the tank. How long does it take the smaller pipe alone to fill the tank? [7.7]

CHAPTER 8

A Transition

This chapter provides a transition between beginning algebra and intermediate algebra. We will review concepts covered in the first half of the text and extend these ideas to new topics.

STUDY STRATEGY

Summary of Previous Study Strategies In this chapter, we will summarize the study strategies presented in the first half of the text.

8.1 Linear Equations and Absolute Value Equations

OBJECTIVES

1. Solve linear equations.
2. Solve absolute value equations.

Linear Equations

Objective 1 Solve linear equations. We begin this section by reviewing linear equations and their solutions. Recall the following guidelines for solving linear equations in Chapter 2:

- Simplify each side of the equation. This includes performing any distributive multiplications, clearing all fractions by multiplying each side of the equation by the least common multiple (LCM) of the denominators, and combining like terms.
- Collect all variable terms on one side of the equation and all constant terms on the other side. It is a good idea to collect the variable terms in such a way that the coefficient of the variable term will be positive.

- Divide both sides of the equation by the coefficient of the variable term to find the solution.
- Check your solution and write it in solution set notation.

We will proceed with some examples that are intended as a review.

EXAMPLE 1 Solve $2x - 6 = 10$.

SOLUTION Begin by isolating the variable term on the left side of the equation.

$$\begin{aligned} 2x - 6 &= 10 \\ 2x &= 16 && \text{Add 6 to both sides to isolate } 2x. \\ x &= 8 && \text{Divide both sides by 2.} \end{aligned}$$

Now check this solution.

Check

$$\begin{aligned} 2x - 6 &= 10 \\ 2(8) - 6 &= 10 && \text{Substitute 8 for } x. \\ 16 - 6 &= 10 && \text{Multiply.} \\ 10 &= 10 && \text{Subtract.} \end{aligned}$$

Because both sides simplify to be 10, the solution is valid. The solution set is $\{8\}$.

Quick Check 1

Solve $9x + 13 = -14$.

EXAMPLE 2 Solve $3(2x + 7) - 4x = 5x - 6$.

SOLUTION Begin by performing the distributive multiplication on the left side of the equation.

$$\begin{aligned} 3(2x + 7) - 4x &= 5x - 6 \\ 6x + 21 - 4x &= 5x - 6 && \text{Distribute 3.} \\ 2x + 21 &= 5x - 6 && \text{Combine like terms.} \\ 27 &= 3x && \text{Collect variable terms on the right side by subtracting } 2x \text{ from both sides. Collect constants on the left side by adding 6 to both sides.} \\ 9 &= x && \text{Divide both sides by 3.} \end{aligned}$$

The check is left to the reader. The solution set is $\{9\}$.

Quick Check 2

Solve
$3x - 2(4x - 5) = 2x - 11$.

EXAMPLE 3 Solve $\frac{4}{3}x + \frac{5}{6} = \frac{9}{4}x - 1$.

SOLUTION This equation includes fractions. We can clear these fractions by multiplying each side of the equation by the LCM of the denominators, which is 12.

$$\begin{aligned} \frac{4}{3}x + \frac{5}{6} &= \frac{9}{4}x - 1 \\ 12\left(\frac{4}{3}x + \frac{5}{6}\right) &= 12\left(\frac{9}{4}x - 1\right) && \text{Multiply both sides by 12.} \\ \overset{4}{\cancel{12}} \cdot \frac{4}{\underset{1}{\cancel{3}}}x + \overset{2}{\cancel{12}} \cdot \frac{5}{\underset{1}{\cancel{6}}} &= \overset{3}{\cancel{12}} \cdot \frac{9}{\underset{1}{\cancel{4}}}x - 12 \cdot 1 && \text{Distribute and divide out common factors.} \\ 16x + 10 &= 27x - 12 && \text{Multiply.} \\ 10 &= 11x - 12 && \text{Subtract } 16x \text{ from both sides.} \\ 22 &= 11x && \text{Add 12 to both sides.} \\ 2 &= x && \text{Divide both sides by 11.} \end{aligned}$$

The check is left to the reader. The solution set is $\{2\}$.

Quick Check 3

Solve $\frac{2}{3}x - \frac{1}{5} = \frac{1}{2}x + \frac{5}{6}$.

Equations That Are Identities

In the first three examples of this section, each linear equation had exactly one solution, but recall that this will not always be the case. An equation that is always true regardless of the value substituted for the variable is called an identity. The solution set for an identity is the set of all real numbers, denoted by $\mathbb{R}$.

EXAMPLE 4 Solve $2(2x - 3) + 1 = 4x - 5$.

SOLUTION

$$\begin{aligned} 2(2x - 3) + 1 &= 4x - 5 & \\ 4x - 6 + 1 &= 4x - 5 & \text{Distribute 2.} \\ 4x - 5 &= 4x - 5 & \text{Combine like terms.} \\ -5 &= -5 & \text{Subtract } 4x \text{ from both sides.} \end{aligned}$$

When we subtract $4x$ from each side in an attempt to collect all variable terms on one side of the equation, we eliminate the variable x from the resulting equation. Is the equation $-5 = -5$ a true statement? Yes, and it says that the equation is an identity and that the solution set is the set of all real numbers $\mathbb{R}$.

Quick Check 4

Solve $(2x + 1) - (5 - 3x) = 2(3x - 2) - x$.

Notice that at one point in the previous example, the equation was $4x - 5 = 4x - 5$. If you reach a line where both sides of the equation are identical, the equation is an identity.

Equations That Are Contradictions

Just as some equations are always true regardless of the value chosen for the variable, other equations are never true for any value of the variable. These equations are called contradictions and have no solution. The solution set for these equations is the empty set, or null set, and is denoted $\varnothing$. We can tell that an equation is a contradiction when we are solving it because the variable terms on each side of the equation are eliminated but in this case leave an equation that is false, such as $1 = 2$.

EXAMPLE 5 Solve $4x + 3 = 2(2x + 3) - 1$.

SOLUTION

$$\begin{aligned} 4x + 3 &= 2(2x + 3) - 1 & \\ 4x + 3 &= 4x + 6 - 1 & \text{Distribute 2.} \\ 4x + 3 &= 4x + 5 & \text{Combine like terms.} \\ 3 &= 5 & \text{Subtract } 4x \text{ from both sides.} \end{aligned}$$

After we subtract $4x$ from each side of the equation, the resulting equation $3 = 5$ is obviously false. This equation is a contradiction, and its solution set is the empty set $\varnothing$.

Quick Check 5

Solve $5x - 7 = 6(x + 2) - x$.

Absolute Value Equations

Objective 2 Solve absolute value equations. We now introduce equations involving absolute values of linear expressions.

EXAMPLE 6 Solve $|x| = 5$.

SOLUTION Recall that the absolute value of a number x, denoted $|x|$, is a measure of the distance between 0 and that number x on a real number line. In this equation, we are looking for a number x that is 5 units from 0.

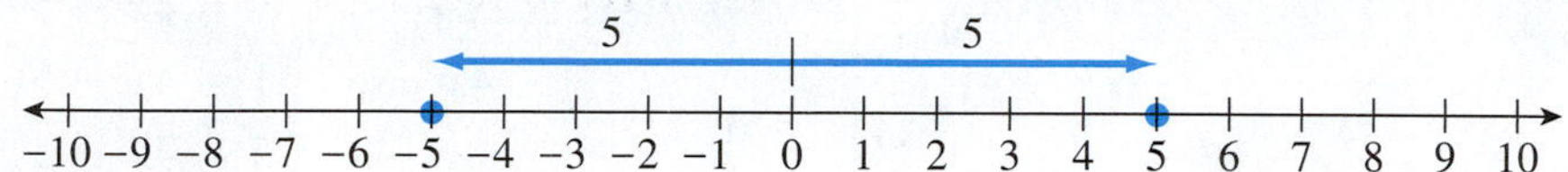

There are two such numbers: 5 and -5. The solution set is $\{-5, 5\}$.

Quick Check 6

Solve $|x| = 7$.

The equation in the previous example is an **absolute value equation**. An absolute value equation relates the absolute value of an expression to a constant, such as $|2x - 3| = 7$, or the absolute value of an expression to the absolute value of another expression, such as $|3x - 4| = |2x + 11|$.

Consider the absolute value equation $|2x - 3| = 7$. This equation involves a number, represented by $2x - 3$, whose absolute value is equal to 7. That number must be either 7 or -7. We begin to solve this equation by converting it into the following two equations:

$$2x - 3 = 7 \quad \text{and} \quad 2x - 3 = -7$$

In general, to solve an equation in which the absolute value of an expression is equal to a positive number, we set the expression (without the absolute value bars) equal to that number and to its opposite. Solving those two equations gives the solutions.

Solving Absolute Value Equations

For any expression X and any positive number a, the solutions to the equation

$$|X| = a$$

can be found by solving the two equations

$$X = a \quad \text{and} \quad X = -a.$$

EXAMPLE 7 Solve $|2x - 3| = 7$.

SOLUTION Begin by converting the absolute value equation to two equations that do not involve absolute values.

$$|2x - 3| = 7$$

$2x - 3 = 7$	or	$2x - 3 = -7$	Convert to two linear equations.
$2x = 10$	or	$2x = -4$	Add 3 to both sides of each equation.
$x = 5$	or	$x = -2$	Divide both sides of each equation by 2.

Check

$x = 5$	$x = -2$
$\lvert 2(5) - 3\rvert = 7$	$\lvert 2(-2) - 3\rvert = 7$
$\lvert 10 - 3\rvert = 7$	$\lvert -4 - 3\rvert = 7$
$\lvert 7\rvert = 7$	$\lvert -7\rvert = 7$
$7 = 7$	$7 = 7$

Because both values check, the solution set is $\{-2, 5\}$.

Quick Check 7

Solve $|3x + 8| = 5$.

A WORD OF CAUTION When solving an absolute value equation such as $|2x - 3| = 7$, be sure to rewrite the equation as two equations. Do not simply drop the absolute value bars and solve the resulting equation; if you do, you will miss one of the solutions.

EXAMPLE 8 Solve $3|2x + 7| - 4 = 20$.

SOLUTION We must isolate the absolute value before converting the equation to two linear equations.

$$3|2x + 7| - 4 = 20$$

$$3|2x + 7| = 24 \quad \text{Add 4 to both sides.}$$

$$|2x + 7| = 8 \quad \text{Divide both sides by 3.}$$

$$2x + 7 = 8 \quad \text{or} \quad 2x + 7 = -8 \quad \text{Convert to two linear equations.}$$

$$2x = 1 \quad \text{or} \quad 2x = -15 \quad \text{Subtract 7 from both sides of each equation.}$$

$$x = \frac{1}{2} \quad \text{or} \quad x = -\frac{15}{2} \quad \text{Divide both sides of each equation by 2.}$$

The solution set is $\left\{-\frac{15}{2}, \frac{1}{2}\right\}$. The check of these solutions is left as an exercise.

Quick Check 8

Solve
$4|2x - 5| - 9 = 7$.

A WORD OF CAUTION When solving an absolute value equation, isolate the absolute value before rewriting the equation as two equations without absolute values.

Because the absolute value of a number is a measure of its distance from 0 on the number line, the absolute value of a number cannot be negative. Therefore, an equation that has an absolute value equal to a negative number has no solution. For example, the equation $|x| = -3$ has no solution because there is no number whose absolute value is -3. We write $\varnothing$ for the solution set.

EXAMPLE 9 Solve $|3x - 4| + 8 = 6$.

SOLUTION Begin by isolating the absolute value.

$$|3x - 4| + 8 = 6$$

$$|3x - 4| = -2 \quad \text{Subtract 8 from both sides.}$$

Because an absolute value cannot equal -2, this equation has no solution. The solution set is $\varnothing$.

Quick Check 9

Solve $|2x + 3| = -6$.

A WORD OF CAUTION An equation in which an absolute value is equal to a negative number has no solution. Do not try to rewrite the equation as two equivalent equations.

If two unknown numbers have the same absolute values, either the two numbers are equal or they are opposites. We will use this idea to solve absolute value equations such as $|3x - 4| = |2x + 11|$. If these two absolute values are equal, either the two expressions inside the absolute value bars are equal $(3x - 4 = 2x + 11)$ or the first expression is the opposite of the second expression $(3x - 4 = -(2x + 11))$. To find the solution of the original equation, we will solve these two resulting equations.

Solving Absolute Value Equations Involving Two Absolute Values

For any expressions X and Y, the solutions of the equation

$$|X| = |Y|$$

can be found by solving the two equations

$$X = Y \quad \text{and} \quad X = -Y.$$

EXAMPLE 10 Solve $|3x - 4| = |2x + 11|$.

SOLUTION Begin by rewriting this equation as two equations that do not contain absolute values. For the first equation, set $3x - 4$ equal to $2x + 11$; for the second equation, set $3x - 4$ equal to the opposite of $2x + 11$. Finish by solving each equation.

$$3x - 4 = 2x + 11 \quad \text{or} \quad 3x - 4 = -(2x + 11)$$

Now solve each equation separately.

$3x - 4 = 2x + 11$

$x - 4 = 11$ Subtract $2x$ from both sides.

$x = 15$ Add 4 to both sides.

$3x - 4 = -(2x + 11)$

$3x - 4 = -2x - 11$ Distribute.

$5x - 4 = -11$ Add $2x$ to both sides.

$5x = -7$ Add 4 to both sides.

$x = -\frac{7}{5}$ Divide both sides by 5.

The solution set is $\{-\frac{7}{5}, 15\}$. The check of these solutions is left to the reader.

Quick Check 10

Solve $|x - 9| = |2x + 13|$.

BUILDING YOUR STUDY STRATEGY

Summary, 1 Study Groups Creating a study group is one of the best ways to learn mathematics and improve your performance on quizzes and exams. When a group of students works together, as long as one student understands the material being covered, there is a good chance that each student in the group will end up understanding the material. Sometimes a concept will be easier for you to understand if one of your peers explains it to you.

You also will find it helpful to have a support group: students who can lean on each other when times are bad. By being supportive of each other, you increase the chances that all of you will learn and be successful in your class.

Exercises 8.1

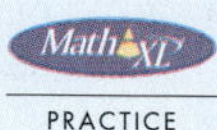

PRACTICE

WATCH

DOWNLOAD

READ

REVIEW

Vocabulary

1. A(n) __________ is a mathematical statement of equality between two expressions.
2. A(n) __________ of an equation is a value that, when substituted for the variable in the equation, produces a true statement.
3. To clear fractions from an equation, multiply both sides of the equation by the ______ of the denominators.
4. A contradiction is an equation that has ____ solutions.
5. The solution set to a contradiction can be denoted $\varnothing$, which represents ______________.
6. An equation that is always true is a(n) __________.
7. For any expression X and any positive number a, to solve the equation $|X| = a$, rewrite the equation as the two equations _______ and ________.
8. The equation $|X| = a$ has no solutions if a is a(n) __________ number.

Solve.

9. $m - 5 = -9$

10. $x + 7 = -3$

11. $5t = -8$

12. $-2b = -28$

13. $7n - 11 = -39$

14. $6n + 32 = 14$

15. $-2x + 9 = 31$

16. $13 - 4x = -43$

17. $9m - 30 = 39$

18. $8m + 35 = -51$

19. $4x + 17 = 2x - 19$

20. $7x + 22 = 10x + 40$

21. $3x - 7 = -x + 15$

22. $-6x + 40 = -3x - 11$

23. $-9r + 22 = 3r + 67$

24. $5r - 41 = 34 - 4r$

25. $2(2x - 4) - 7 = 6x - 11$

26. $5(3 - 2x) + 4x = 3(3x - 1) + 7$

27. $5x - 3(2x - 9) = 4(3x - 8) + 7x$

28. $10 - 6(2x + 1) = 13 - 9x$

29. $2(3x - 1) + 5(x + 4) = 3(x + 7) - (x + 6)$

30. $4(2x + 3) - 3(x + 3) = 2(3x - 2) + (x - 3)$

31. $5x - 3(4x + 1) = 3 + 4(3x + 8)$

32. $3(x + 4) - 4(3x - 10) = 4(2x - 3) + 6(3x + 1)$

33. $\frac{3}{4}t - 6 = \frac{1}{3}t - 1$

34. $\frac{1}{8}n - 7 = -\frac{1}{4}n - 13$

35. $\frac{1}{2}(2x - 7) - \frac{1}{3}x = \frac{4}{3}x - 5$

36. $\frac{2}{3}(x - 6) - \frac{1}{4}x = \frac{3}{8}(8 - x) + \frac{1}{2}x$

37. $2b - 9 = 2b - 9$

38. $4t + 11 = 4t - 11$

39. $3(2n - 1) + 5 = 2(3n + 2)$

40. $5(n - 3) - 3n = 2(n - 7) - 1$

41. $5x - 2(3 - 2x) = 3(3x - 2)$

42. $7x - (3x - 4) = 4x - 4$

43. $|x| = 2$

44. $|x| = 13$

45. $|x + 6| = 13$

46. $|x - 9| = 4$

47. $|3x + 2| = 20$

48. $|2x - 9| = 37$

49. $|x| + 7 = 3$

50. $|x + 7| = 3$

51. $|x + 4| - 12 = 6$

52. $|x - 2| - 8 = 2$

53. $|x - 3| + 7 = 12$

54. $|x + 12| - 10 = 10$

55. $|4x - 11| - 19 = -8$

56. $|7x - 52| + 18 = 14$

57. $2|3x + 2| - 9 = 17$

58. $2|5x - 3| + 7 = 21$

59. $-|x + 3| - 10 = -16$

60. $-3|2x + 1| + 13 = -29$

61. $|4x - 5| = |3x - 23|$

62. $|x - 14| = |3x + 2|$

63. $|x + 12| = |2x - 9|$

64. $|2x + 1| = |x - 19|$

Mixed Practice 65–80

65. $|x + 13| - 12 = -7$

66. $-9x + 10 = -23$

67. $|x - 6| + 14 = 9$

68. $|5x| = |2x - 21|$

69. $25 = 4 - 6x$

70. $3(6x - 1) + 4x = 16(x + 2) - (11 - 3x)$

71. $10x - 17 = 6x - 53$

72. $|4x - 3| + 7 = 16$

73. $\frac{3}{5}x - \frac{1}{2} = \frac{2}{3}x + \frac{3}{4}$

74. $9x - 11 = 13x + 23$

75. $|7x - 11| = |4x + 6|$

76. $\frac{3}{4}x - \frac{2}{3} = \frac{1}{9}x + 7$

77. $|3x + 16| - 4 = 9$

78. $|x + 10| + 7 = 6$

79. $3(4x - 5) + 7x = 5(2x - 9) + 12$

80. $|x - 9| - 10 = -2$

81. Find an absolute value equation whose solution is $\{-2, 2\}$.

82. Find an absolute value equation whose solution is $\{-6, 10\}$.

83. Find an absolute value equation whose solution is $\{-5, 3\}$.

84. Find an absolute value equation whose solution is $\{\frac{1}{2}, \frac{7}{2}\}$.

Writing in Mathematics

Answer in complete sentences.

85. Explain why the equation $|2x - 7| + 6 = 4$ has no solution.

86. Here is a student's work for solving the equation $|x - 3| = -5$.

$$|x - 3| = -5$$
$$x - 3 = -5 \quad \text{or} \quad x - 3 = 5$$
$$x = -2 \quad \text{or} \quad x = 8$$

$\{-2, 8\}$

Explain what the student's error is and how he or she can avoid making a similar error in the future.

Quick Review Exercises

Section 8.1

Evaluate.

1. $f(x) = 3x - 14, f(-6)$

2. $f(x) = x^2 + 17x - 60, f(-5)$

3. $f(x) = \dfrac{x^2 - 6x + 25}{3x + 7}, f(2)$

4. $f(x) = |x - 7| + 3, f(0)$

8.2 Linear Inequalities and Absolute Value Inequalities

OBJECTIVES

1. Graph the solutions of a linear inequality on a number line and express the solutions by using interval notation.
2. Solve linear inequalities.
3. Solve compound linear inequalities.
4. Solve absolute value inequalities.

Linear Inequalities and Their Solutions

Objective 1 Graph the solutions of a linear inequality on a number line and express the solutions by using interval notation. In this section, we will examine solving inequalities. We begin by reviewing linear inequalities and the presentation of their solutions. For example, the solutions of the inequality $x < 4$ are all real numbers that are less than 4. This can be represented on a number line as follows:

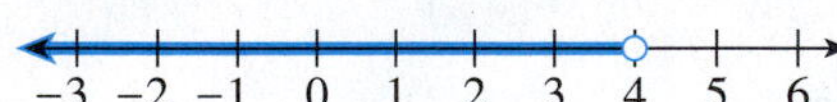

This can be expressed in interval notation as $(-\infty, 4)$.

The following table shows the number line and interval notation associated with several types of simple linear inequalities:

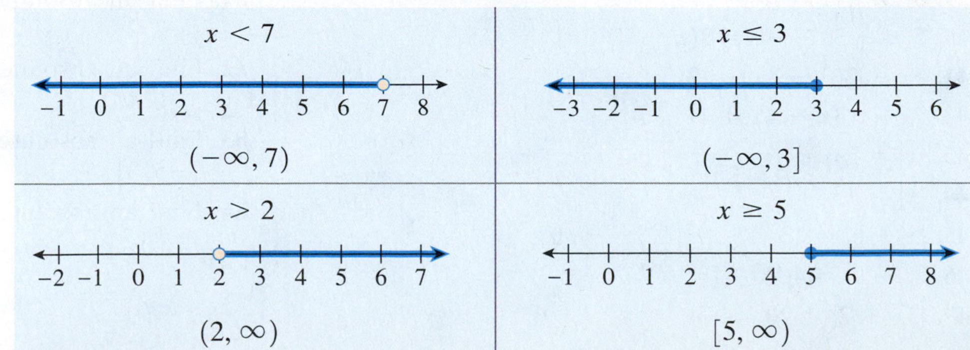

$x < 7$	$x \leq 3$
−1 0 1 2 3 4 5 6 7 8	−3 −2 −1 0 1 2 3 4 5 6
$(-\infty, 7)$	$(-\infty, 3]$
$x > 2$	$x \geq 5$
−2 −1 0 1 2 3 4 5 6 7	−1 0 1 2 3 4 5 6 7 8
$(2, \infty)$	$[5, \infty)$

Recall that we use a closed circle whenever the endpoint is included in the interval ($\leq$ or $\geq$) and an open circle when the endpoint is not included ($<$ or $>$). When expressing the interval in interval notation, we use square brackets when the endpoint is included and parentheses when the endpoint is not included.

Solving Linear Inequalities

Objective 2 Solve linear inequalities. Solving a linear inequality is similar to solving a linear equation. The exception is if we multiply or divide both sides of the inequality by a negative number, then the direction of the inequality changes.

EXAMPLE 1 Solve $3x - 5 < -14$.

SOLUTION

$$3x - 5 < -14$$

$3x < -9$ Add 5 to both sides.

$x < -3$ Divide both sides by 3.

Here is the number line showing the solution.

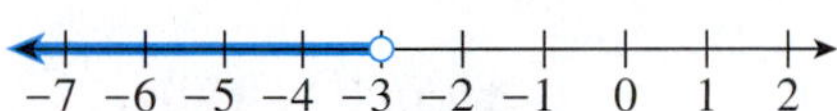

The solution expressed in interval notation is $(-\infty, -3)$.

Quick Check 1

Solve $2x + 5 \leq 9$.

EXAMPLE 2 Solve $5x + 11 \leq 2(2x + 4) - 3$.

SOLUTION

$5x + 11 \leq 2(2x + 4) - 3$

$5x + 11 \leq 4x + 8 - 3$ Distribute 2.

$5x + 11 \leq 4x + 5$ Combine like terms.

$x + 11 \leq 5$ Collect variables on the left side by subtracting $4x$ from both sides.

$x \leq -6$ Collect constant terms on the right side by subtracting 11 from both sides.

Here is the number line showing the solution.

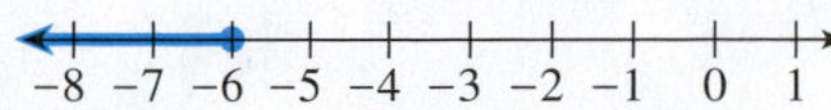

The solution expressed in interval notation is $(-\infty, -6]$.

Quick Check 2

Solve $3(2x - 1) + 4 \geq 2x - 7$.

EXAMPLE 3 Solve $-2x + 3 > 13$.

SOLUTION

$$-2x + 3 > 13$$

$-2x > 10$ Subtract 3 from both sides.

$\frac{-2x}{-2} < \frac{10}{-2}$ Divide both sides by -2. This changes the direction of the inequality from $>$ to $<$.

$x < -5$ Simplify.

Here is the number line showing the solution.

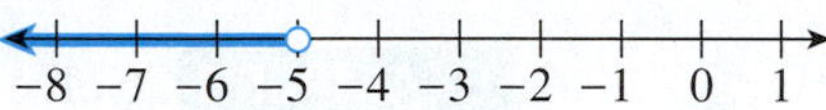

The solution expressed in interval notation is $(-\infty, -5)$.

Quick Check 3

Solve $5 - 2x > -5$.

A WORD OF CAUTION When we multiply or divide both sides of an inequality by a negative number, we must change the direction of the inequality.

Solving Compound Linear Inequalities

Objective 3 Solve compound linear inequalities. A compound inequality is made up of two simple inequalities. For instance, the compound inequality $3 \le x < 7$ represents numbers that are greater than or equal to 3 $(x \ge 3)$ and less than 7 $(x < 7)$. In set theory, this is referred to as the intersection of these two sets. The intersection of two sets contains the items that are in both sets. The inequality $3 \le x < 7$ can be represented graphically by the following number line:

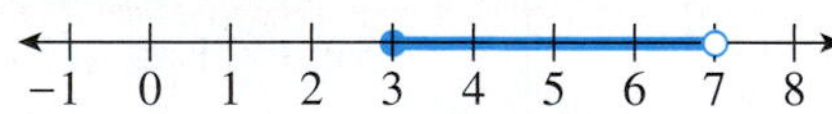

The interval notation for this interval is $[3, 7)$.

EXAMPLE 4 Solve $-3 < 2x - 1 < 5$.

SOLUTION The goal of solving a compound inequality such as this one is to isolate the variable x between two constants. We will work on all three parts of the inequality at the same time.

$$-3 < 2x - 1 < 5$$

$-2 < 2x < 6$ Add 1 to each part of the inequality.

$-1 < x < 3$ Divide each part of the inequality by 2.

The solution consists of all real numbers greater than -1 and less than 3. Here is the number line showing the solution.

−4 −3 −2 −1 0 1 2 3 4 5

The solution expressed in interval notation is $(-1, 3)$.

Quick Check 4

Solve $-8 \le 3x + 7 \le 1$.

Another type of compound inequality involves the word *or.* Consider the compound inequality $x < -2$ *or* $x > 1$. The solutions to this inequality are values that are solutions to either of the two inequalities. In other words, any number that is less than -2 ($x < -2$) *or* is greater than 1 ($x > 1$) is a solution of the compound inequality. Here are the solutions of this compound inequality graphed on a number line.

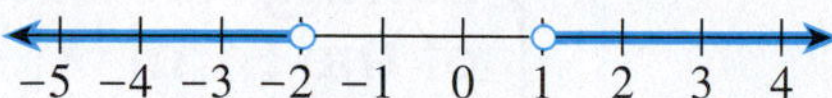

Notice that we have two different intervals that contain solutions. The solution expressed in interval notation is $(-\infty, -2) \cup (1, \infty)$. The symbol $\cup$ represents a union in set theory and combines the two intervals in one solution. The union of two sets is comprised of items that are elements of one set or the other.

EXAMPLE 5 Solve $x + 4 \leq 7$ or $2x - 1 > 11$.

SOLUTION To solve a compound inequality of this type, first solve each inequality separately. Begin by solving $x + 4 \leq 7$.

$$x + 4 \leq 7$$
$$x \leq 3 \quad \text{Subtract 4 from both sides.}$$

Now solve the other inequality.

$$2x - 1 > 11$$
$$2x > 12 \quad \text{Add 1 to both sides.}$$
$$x > 6 \quad \text{Divide both sides by 2.}$$

Here is a number line showing the solution.

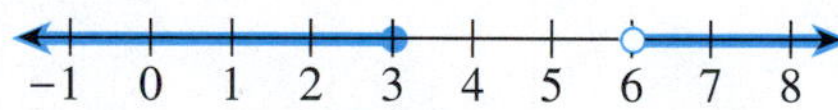

The solution expressed in interval notation is $(-\infty, 3] \cup (6, \infty)$.

Quick Check 5

Solve $x + 3 < -3$ or $2x + 5 > 14$.

Absolute Value Inequalities

Objective 4 Solve absolute value inequalities. As in the previous section, we will expand the topic to include absolute values. Consider the inequality $|x| < 4$. Any real number in the following interval has an absolute value less than 4:

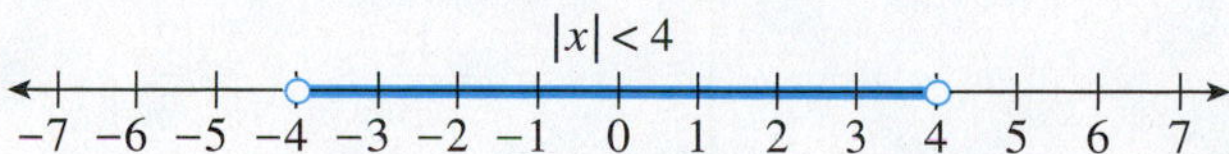

This can be expressed as the compound inequality $-4 < x < 4$. Using this idea, whenever we have an absolute value of an expression that is less than a positive number, we begin by "trapping" the expression between that positive number and its opposite, in this case, $-4 < x < 4$. We then proceed to solve the resulting compound inequality.

Solving Absolute Value Inequalities of the Form $|X| < a$ or $|X| \le a$

For any expression X and any nonnegative number a, the solutions of the inequality $|X| < a$ can be found by solving the compound inequality

$$-a < X < a.$$

(If $a = 0$, this inequality has no solutions.)

Similarly, for any expression X and any nonnegative number a, the solutions of the inequality $|X| \le a$ can be found by solving the compound inequality

$$-a \le X \le a.$$

(If $a = 0$, solve the equation $X = a$.)

EXAMPLE 6 Solve $|x - 4| < 2$.

SOLUTION The first step is to "trap" $x - 4$ between -2 and 2.

$|x - 4| < 2$

$-2 < x - 4 < 2$ "Trap" $x - 4$ between -2 and 2.

$2 < x < 6$ Add 4 to each part of the inequality.

Here is the number line showing the solution.

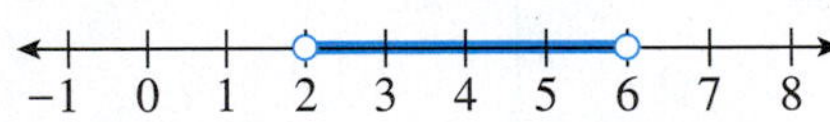

The solution expressed in interval notation is $(2, 6)$.

Quick Check 6

Solve $|x + 3| \le 1$.

As with equations involving absolute values, we must isolate the absolute value before we rewrite the inequality as a compound inequality.

EXAMPLE 7 Solve $|3x + 7| - 5 \le 4$.

SOLUTION

$|3x + 7| - 5 \le 4$

$|3x + 7| \le 9$ Add 5 to both sides.

$-9 \le 3x + 7 \le 9$ Rewrite as a compound inequality.

$-16 \le 3x \le 2$ Subtract 7 from each part of the inequality.

$-\frac{16}{3} \le x \le \frac{2}{3}$ Divide each part of the inequality by 3.

Here is the number line showing the solution.

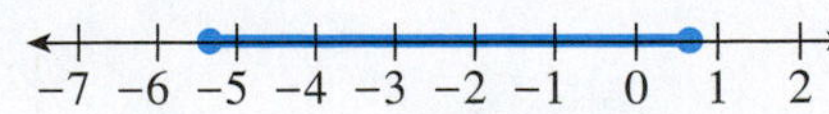

The solution expressed in interval notation is $\left[-\frac{16}{3}, \frac{2}{3}\right]$.

Quick Check 7

Solve $|2x - 1| + 4 < 9$.

A WORD OF CAUTION When solving an absolute value inequality, we must isolate the absolute value before we rewrite the inequality as a compound inequality.

If we have an inequality in which an absolute value is less than a negative number, such as $|x| < -2$, or less than 0, such as $|x| < 0$, this inequality has no solution. Because an absolute value is always 0 or greater, it can never be *less than* a negative number or zero.

EXAMPLE 8 Solve $|4x + 3| + 7 < 5$.

SOLUTION

$$|4x + 3| + 7 < 5$$
$$|4x + 3| < -2 \quad \text{Subtract 7 from both sides.}$$

Because an absolute value cannot be less than a negative number, this inequality has no solution: $\varnothing$.

Quick Check 8

Solve $|x + 1| - 3 < -8$.

A WORD OF CAUTION An inequality in which an absolute value is less than a negative number has no solution. Do not try to rewrite the inequality as a compound inequality.

Consider the inequality $|x| > 4$. Notice that we now have the absolute value of x *greater than* 4 rather than *less than* 4. Inequalities with an absolute value *greater* than a positive number must be approached in a different manner than absolute value inequalities with an absolute value *less* than a positive number. We cannot "trap" the expression between two constants. Solutions of the inequality $|x| > 4$ are in one of two categories. First, any positive number greater than 4 will have an absolute value greater than 4 as well. Second, any negative number less than -4 also will have an absolute value greater than 4. The following number line shows where the solutions of this inequality can be found:

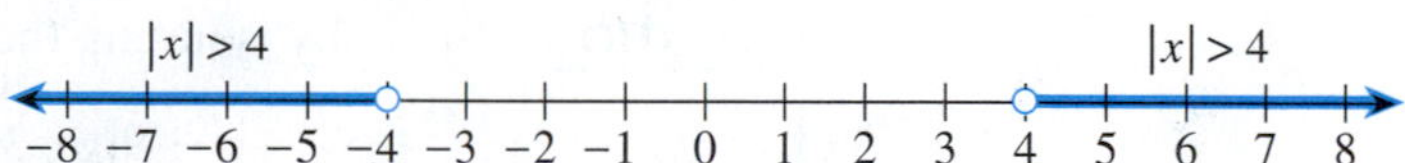

This can be expressed symbolically as $x < -4$ or $x > 4$.

Solving Absolute Value Inequalities of the Form $|X| > a$ or $|X| \geq a$

For any expression X and any nonnegative number a, the solutions of the inequality $|X| > a$ can be found by solving the compound inequality

$$X < -a \quad \text{or} \quad X > a.$$

Similarly, for any expression X and any nonnegative number a, the solutions of the inequality $|X| \geq a$ can be found by solving the compound inequality

$$X \leq -a \quad \text{or} \quad X \geq a.$$

EXAMPLE 9 Solve $|x + 3| > 2$.

SOLUTION Begin by converting this inequality to the compound inequality $x + 3 < -2$ or $x + 3 > 2$. Then solve each inequality.

$$|x + 3| > 2$$
$$x + 3 < -2 \quad \text{or} \quad x + 3 > 2 \quad \text{Rewrite as a compound inequality.}$$
$$x < -5 \quad \text{or} \quad x > -1 \quad \text{Subtract 3 from both sides to solve each inequality.}$$

Here is the number line showing the solution.

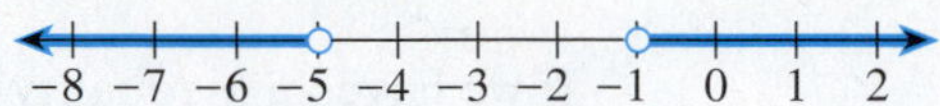

The solution expressed in interval notation is $(-\infty, -5) \cup (-1, \infty)$.

Quick Check 9

Solve $|x - 2| > 4$.

Whether we are solving an absolute value equation or an absolute value inequality, the first step is to isolate the absolute value. The following table shows the correct step to take when *comparing an absolute value with a nonnegative number*:

	Example	What to Do
"Equal To"	$\|2x + 9\| = 5$	$2x + 9 = 5$ or $2x + 9 = -5$
"Less Than"	$\|x - 4\| < 3$	$-3 < x - 4 < 3$
or	or	or
"Less Than or Equal To"	$\|x - 4\| \le 3$	$-3 \le x - 4 \le 3$
"Greater Than"	$\|3x + 5\| > 7$	$3x + 5 < -7$ or $3x + 5 > 7$
or	or	or
"Greater Than or Equal To"	$\|3x + 5\| \ge 7$	$3x + 5 \le -7$ or $3x + 5 \ge 7$

EXAMPLE 10 Solve $|4x - 9| - 3 \ge 5$.

SOLUTION Begin by isolating the absolute value.

$$|4x - 9| - 3 \ge 5$$

$$|4x - 9| \ge 8 \quad \text{Add 3 to both sides.}$$

Because this inequality involves an absolute value that is *greater than* or equal to 8, we will rewrite this absolute value inequality as two linear inequalities.

$4x - 9 \le -8$ or $4x - 9 \ge 8$ Convert to two linear inequalities.

$4x \le 1$ or $4x \ge 17$ Add 9 to both sides of each inequality.

$x \le \dfrac{1}{4}$ or $x \ge \dfrac{17}{4}$ Divide both sides of each inequality by 4.

Here is the number line showing the solution.

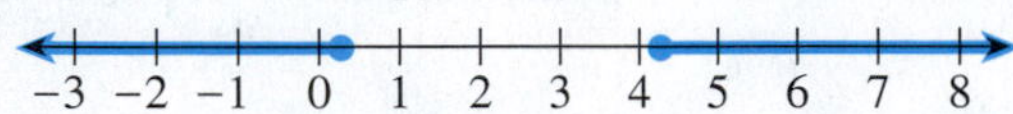

The solution expressed in interval notation is $\left(-\infty, \frac{1}{4}\right] \cup \left[\frac{17}{4}, \infty\right)$.

Quick Check 10

Solve $|2x + 7| + 5 > 10$.

Consider the inequality $|x| > -2$. Because the absolute value of any number must be 0 or greater, $|x|$ must be greater than -2 for any real number x. The solution set is the set of all real numbers $\mathbb{R}$.

EXAMPLE 11 Solve $|3x + 5| - 3 > -4$.

SOLUTION Begin by isolating the absolute value.

$$|3x + 5| - 3 > -4$$
$$|3x + 5| > -1 \quad \text{Add 3 to both sides to isolate the absolute value.}$$

This inequality must always be true, so the solution is the set of all real numbers $\mathbb{R}$. Here is the number line showing the solution.

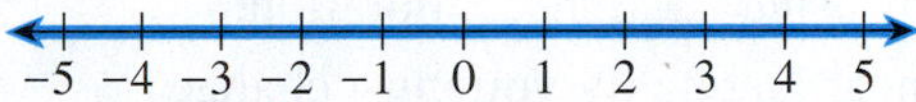

The interval notation associated with the set of real numbers is $(-\infty, \infty)$.

Quick Check 11

Solve $|7x - 13| \geq -1$.

The following table summarizes the solutions for an equation or inequality that *compares an absolute value with a negative number*:

Equation or Inequality	Example	Solution
Equal to a negative number	$\lvert x + 3\rvert = -2$	$\varnothing$
Less than a negative number	$\lvert 2x - 1\rvert < -4$	$\varnothing$
Greater than a negative number	$\lvert 3x + 4\rvert > -3$	All real numbers: $\mathbb{R}$

BUILDING YOUR STUDY STRATEGY

Summary, 2 Using Your Textbook and Other Resources Your textbook is a valuable resource when used properly. Read a section the night before your instructor covers it to familiarize yourself with the material. This way, you can prepare questions to ask your instructor in class. After a section is covered in class, reread the section.

As you reread a section, use the textbook to create note cards to help you memorize new terms or procedures. After you read through an example, attempt the Quick Check exercise that follows to reinforce the concept presented in the example. As you work through the homework exercises, refer back to the examples found in that section if you need assistance.

In each section, the feature titled A Word of Caution points out common mistakes that students make and offers advice for avoiding those mistakes.

Your instructor is a valuable resource for answering questions and providing advice. Take advantage of your instructor's office hours in addition to your time in class. The tutorial center is another valuable resource.

Finally, the online supplement to this textbook at MyMathLab.com can be helpful. At this site, you will find video clips, tutorial exercises, and more.

Exercises 8.2

PRACTICE

WATCH

DOWNLOAD
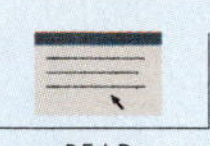
READ

REVIEW

Vocabulary

1. A(n) ______________ is an inequality containing linear expressions.
2. When graphing an inequality, use a(n) ________ circle to indicate an endpoint that is not included as a solution.
3. You can express the solutions of an inequality on a number line and by using ________ notation.
4. When solving a linear inequality, you must change the direction of the inequality whenever you multiply or divide both sides of the inequality by a(n) ________ number.
5. An inequality that is composed of two or more individual inequalities is called a(n) __________ inequality.
6. Before rewriting an absolute value inequality as a compound inequality, you must _________ the absolute value.

Solve the inequality. Graph your solution on a number line and write your solution in interval notation.

7. $x - 6 \le 2$
8. $x - 5 > 4$
9. $3x + 2 \ge -10$
10. $2x + 5 > 9$
11. $-3x + 2 < 17$
12. $-4x - 9 \ge 3$
13. $\frac{2}{3}x - \frac{5}{6} > -2$
14. $\frac{3}{5}x - \frac{1}{3} < \frac{2}{5}$
15. $5(2x + 1) - 7x \le 4x - 13$
16. $2x - 5(x - 3) < x + 9$
17. $5 \le x - 3 \le 11$
18. $-2 < x + 5 < 4$
19. $-10 \le 7x + 4 \le 53$
20. $7 < 6x - 5 < 49$
21. $-9 < 2x + 5 < -4$
22. $-8 \le 3x + 7 \le 28$
23. $-6 < \frac{3}{5}x + \frac{2}{5} < -2$
24. $\frac{7}{4} \le \frac{1}{2}x - \frac{3}{8} \le \frac{5}{2}$
25. $x + 6 < 8$ or $x - 3 > 5$
26. $x - 4 \le -7$ or $x - 8 \ge -1$
27. $3x + 4 \le -17$ or $2x - 9 \ge 9$
28. $\frac{2}{3}x - 2 < -4$ or $x + \frac{5}{3} > \frac{17}{2}$

29. $3x - 11 > 5x - 3$ or $2x + 31 < 6x + 7$

30. $2x + 17 \geq 7x - 13$ or $4x - 5 \leq 5x - 13$

31. $|x + 2| < 3$

32. $|2x - 9| < 7$

33. $|x - 5| + 7 \leq 13$

34. $|x + 2| - 6 \leq 1$

35. $|4x + 9| - 8 \leq -3$

36. $|7x + 14| + 9 < 6$

37. $|3x - 1| + 5 < 9$

38. $|6x - 5| - 2 \leq 11$

39. $2|2x + 10| + 19 \leq 11$

40. $3|x - 9| - 5 < 4$

41. $|x - 4| > 1$

42. $|2x + 6| > 4$

43. $|x + 6| + 5 \geq 8$

44. $|x - 2| - 3 \geq 2$

45. $|5x + 4| \geq -6$

46. $|8x - 21| + 7 \geq 10$

47. $|4x - 17| - 5 > 6$

48. $|x + 4| + 4 > 2$

49. $2|2x + 3| + 5 > 19$

50. $4|x + 5| - 11 > 17$

Mixed Practice, 51–64

Solve the inequality. Graph your solution on a number line and write your solution in interval notation

51. $|x + 4| \leq 6$

52. $|x + 10| > 4$

53. $-3x + 10 \geq 34$ or $6x - 9 \geq 33$

54. $-9 < 4x - 1 < 25$

55. $|x - 5| + 9 > 4$

56. $|x - 5| - 6 < -2$

57. $2x + 7 \geq -9$

58. $4x - 9 < 2x - 27$ or $5x + 11 > x - 13$

59. $|x - 9| + 8 \leq 3$

60. $-3x + 17 > -4$

61. $|2x + 5| - 9 \geq 4$

62. $|2x - 1| - 7 < 10$

63. $17 \leq 5x - 3 < 37$

64. $|4x - 3| + 6 > 11$

65. Find an absolute value inequality whose solution is $(-2, 2)$.

66. Find an absolute value inequality whose solution is $[-2, 6]$.

67. Find an absolute value inequality whose solution is $(-\infty, 1) \cup (9, \infty)$.

68. Find an absolute value inequality whose solution is $(-\infty, -4] \cup [-1, \infty)$.

Writing in Mathematics

Answer in complete sentences.

69. Explain why the inequality $|x + 5| < -3$ has no solutions, while every real number is a solution to the inequality $|x + 5| > -3$.

70. Explain why the inequality $|x| - 7 < -4$ has solutions, but the inequality $|x - 7| < -4$ does not.

71. ***Solutions Manual*** Write a solutions manual page for the following problem.
Solve. $|2x - 3| + 8 \leq 15$

8.3 Graphing Linear Equations and Linear Functions; Graphing Absolute Value Functions

OBJECTIVES

1. **Graph linear equations by using *x*- and *y*-intercepts.**
2. **Graph linear equations by using the slope and *y*-intercept.**
3. **Graph linear functions.**
4. **Graph absolute value functions by plotting points.**

In Chapter 3, we learned how to graph linear equations in two variables, such as $4x - y = 8$ and $y = -\frac{2}{3}x + 2$. We will begin this section with a brief review of how to graph this type of equation.

Graphing Linear Equations in Two Variables Using the *x*- and *y*-Intercepts

Objective 1 Graph linear equations by using *x*- and *y*-intercepts. We developed two techniques for graphing linear equations. The first involved finding the *x*- and *y*-intercepts of the line and then using these points to graph the line. Recall that the x-intercept of a line is the point where the line intersects the x-axis. To find the x-intercept, we substitute 0 for y in the equation and solve for x. The y-intercept is the point where the line crosses the y-axis, and we can find it by substituting 0 for x in the equation and solving for y.

EXAMPLE 1 Find any intercepts and graph $4x - y = 8$.

SOLUTION Find the x-intercept by substituting 0 for y and solving for x.

$$4x - 0 = 8 \quad \text{Substitute 0 for } y.$$
$$4x = 8 \quad \text{Simplify.}$$
$$x = 2 \quad \text{Divide both sides by 4.}$$

The x-intercept is at $(2, 0)$. To find the y-intercept, substitute 0 for x and solve for y.

$$4(0) - y = 8 \quad \text{Substitute 0 for } x.$$
$$-y = 8 \quad \text{Simplify.}$$
$$y = -8 \quad \text{Divide both sides by } -1.$$

The y-intercept is at $(0, -8)$.

It is helpful to find a third point before graphing the line to serve as a check for the other two points. If the three points do not lie on a straight line, at least one of the points is incorrect. To find a third point, we may select any value to substitute for x and solve for y. In this example, we will use 1 for x, but we could use any value we choose.

$$4(1) - y = 8 \quad \text{Substitute 1 for } x.$$
$$4 - y = 8 \quad \text{Multiply.}$$
$$-y = 4 \quad \text{Subtract 4 from both sides.}$$
$$y = -4 \quad \text{Divide both sides by } -1.$$

This tells us that the point $(1, -4)$ lies on the line. Here is a graph showing the line, as well as the three points we have found.

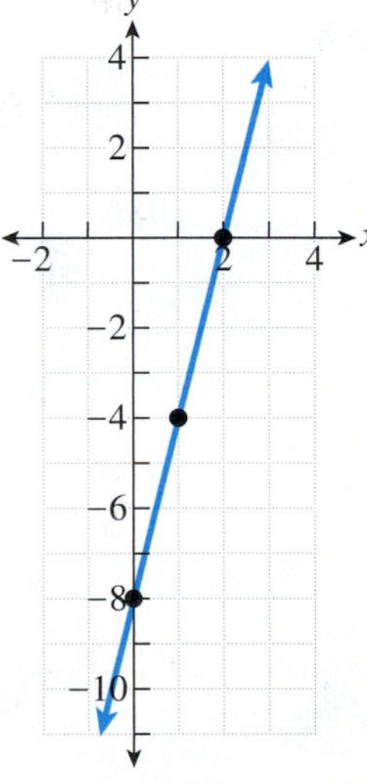

Quick Check 1

Find any intercepts and graph $5x + 2y = -10$.

Graphing a Linear Equation Using Its Slope and y-Intercept

Objective 2 Graph linear equations by using the slope and y-intercept. A linear equation is in slope–intercept form if it is in the form $y = mx + b$. The constant m that is multiplied by x is the slope of the line, and the constant b represents the y-coordinate of the y-intercept. For example, for the equation $y = -\frac{2}{3}x + 2$, the slope of the line is $-\frac{2}{3}$ and the y-intercept is $(0, 2)$. Recall that slope is a measure of how quickly the line rises or falls as it moves to the right. A line whose slope is 4 moves up four units for every one unit it moves to the right, while a line whose slope is $-\frac{2}{3}$ moves down two units for every three units it moves to the right.

If an equation is in slope–intercept form, it provides the information needed to graph the line. We begin by placing the y-intercept on the graph. From that point, we use the slope of the line to locate other points on the graph.

Quick Check 2

Graph $y = \frac{7}{2}x - 4$.

EXAMPLE 2 Graph $y = 2x + 4$.

SOLUTION Because this equation is in slope–intercept form, we know that the y-intercept is $(0, 4)$, and we plot this point on the graph. The slope of this line is 2; so we can find another point on the line by starting at the y-intercept, $(0, 4)$, and moving two units up and one unit to the right. This tells us that the point $(1, 6)$ also is on the line. We could continue to find more points on the line using the same procedure (up two and to the right one). A graph of the line is shown to the right.

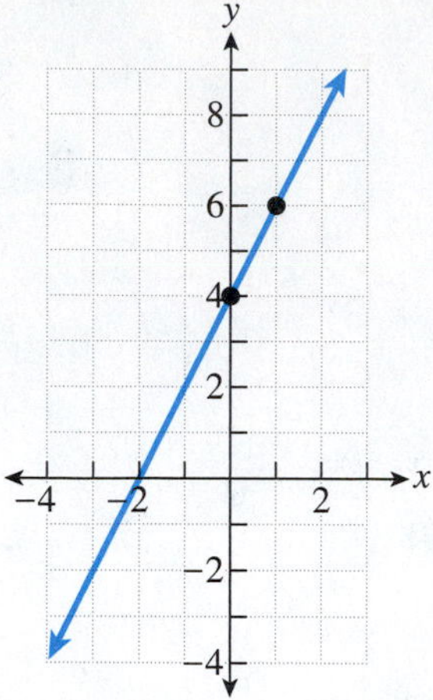

Graphing Linear Equations

- Equations in standard form $(Ax + By = C)$ are often easier to graph by finding and plotting the x- and y-intercepts.
- Equations in slope–intercept form $(y = mx + b)$ are often easier to graph by plotting the y-intercept and using the slope to find other points.

Linear Functions and Their Graphs

Objective 3 Graph linear functions. A linear function is a function of the form $f(x) = mx + b$. Recall that the notation $f(x)$, read "f of x," stands for the function f evaluated at x. For example, consider the linear function $f(x) = 3x - 8$. If we wanted to evaluate $f(6)$, we would substitute 6 for x in the function and simplify.

$$\begin{aligned} f(6) &= 3(6) - 8 && \text{Substitute 6 for } x. \\ &= 18 - 8 && \text{Multiply.} \\ &= 10 && \text{Subtract.} \end{aligned}$$

Graphing a linear function $f(x) = mx + b$ is identical to graphing a linear equation $y = mx + b$. We plot points of the form $(x, f(x))$, beginning with the y-intercept. We then use the slope of the line to find other points on the line.

EXAMPLE 3 Graph $f(x) = -\frac{3}{2}x + 6$.

SOLUTION The y-intercept of this graph is at $(0, 6)$, so we begin by placing that point on the graph. The slope of this line is $-\frac{3}{2}$; so beginning at the y-intercept, we move three units down and two units to the right. This tells us that the point $(2, 3)$ is on the graph.

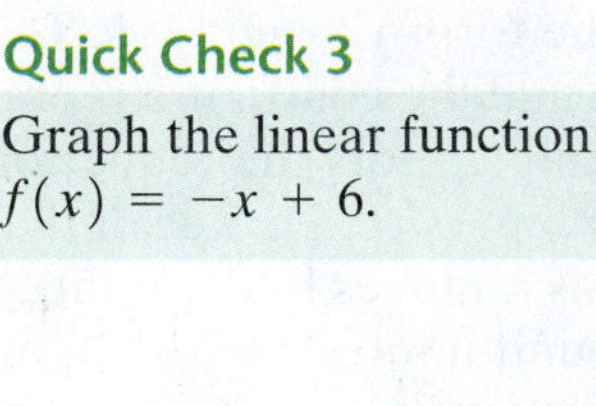

Quick Check 3

Graph the linear function $f(x) = -x + 6$.

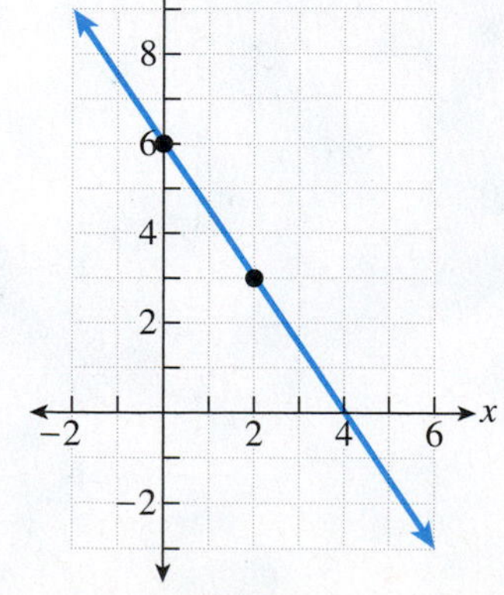

Absolute Value Functions

Objective 4 Graph absolute value functions by plotting points. We will now graph our first nonlinear function: the **absolute value function** $f(x) = |x|$. We will begin to graph this function by creating the following table of function values for various values of x:

x	$f(x) = \|x\|$	$(x, f(x))$
−2	$f(-2) = \|-2\| = 2$	(−2, 2)
−1	$f(-1) = \|-1\| = 1$	(−1, 1)
0	$f(0) = \|0\| = 0$	(0, 0)
1	$f(1) = \|1\| = 1$	(1, 1)
2	$f(2) = \|2\| = 2$	(2, 2)

Here is the graph. Notice that the graph is not a straight line; instead, it is V-shaped. *The most important point for graphing an absolute value function is the point of the V, where the graph changes from falling to rising.* The turning point is the origin for this particular function.

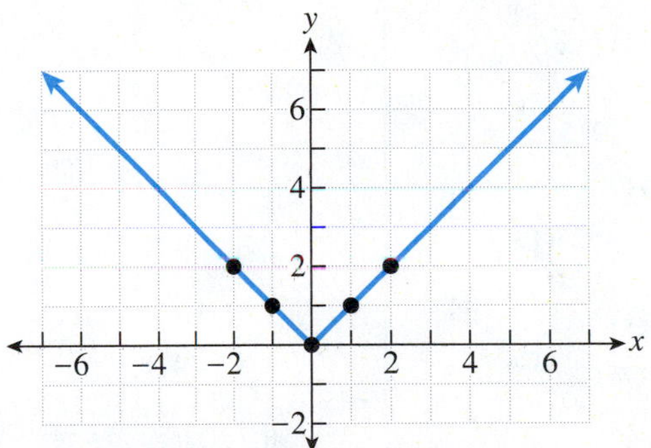

To graph an absolute value function, we find the x-coordinate of the turning point, select two values to the left and right of this value, and create a table of values. Choosing values for x that are to the left and to the right of the turning point is crucial. Suppose we had chosen only positive values for x when graphing $f(x) = |x|$. All of our points would have been on the straight line located to the right of the origin, and our graph would have been a straight line rather than a V shape.

The turning point occurs at the value of x that makes the expression inside the absolute value bars equal to 0. This is because the minimum output of an absolute value occurs when we take the absolute value of 0.

Graphing an Absolute Value Function

- Determine the value of x for which the expression inside the absolute value bars is equal to 0.
- In addition to this value, select two values that are less than this value and two that are greater.
- Create a table of function values for these values of x.
- Place the points $(x, f(x))$ on the graph and draw the V-shaped graph that passes through these points.

EXAMPLE 4 Graph $f(x) = |x + 3| - 2$.

SOLUTION For an absolute value function, begin by finding the value of x that makes the expression inside the absolute value bars equal to 0.

$$x + 3 = 0 \quad \text{Set } x + 3 \text{ equal to 0.}$$
$$x = -3 \quad \text{Subtract 3 from both sides.}$$

Now select two values that are less than −3, such as −5 and −4, and two values that are greater than −3, such as −2 and −1. Next, create a table of values.

x	$f(x) = \|x + 3\| - 2$	$(x, f(x))$
−5	$f(-5) = \|-5 + 3\| - 2 = 2 - 2 = 0$	$(-5, 0)$
−4	$f(-4) = \|-4 + 3\| - 2 = 1 - 2 = -1$	$(-4, -1)$
−3	$f(-3) = \|-3 + 3\| - 2 = 0 - 2 = -2$	$(-3, -2)$
−2	$f(-2) = \|-2 + 3\| - 2 = 1 - 2 = -1$	$(-2, -1)$
−1	$f(-1) = \|-1 + 3\| - 2 = 2 - 2 = 0$	$(-1, 0)$

Here is the graph, based on the points from the table.

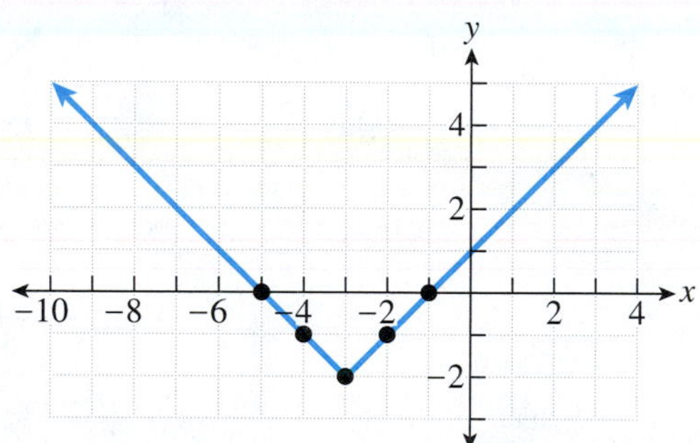

Quick Check 4

Graph the function $f(x) = |x + 1| + 3$.

Using Your Calculator You can graph absolute value functions using the TI-84. (Recall that to access the absolute value function on the TI-84, you press the MATH key; you will find the absolute value function under the **NUM** menu.) To enter the equation of the function from Example 4, $f(x) = |x + 3| - 2$, press the Y= key. Enter $|x + 3| - 2$ next to Y_1. Then press the key labeled GRAPH to graph the function.

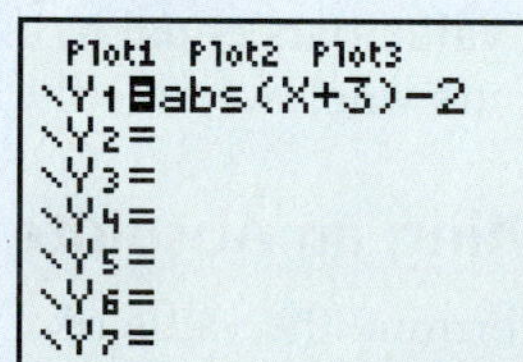

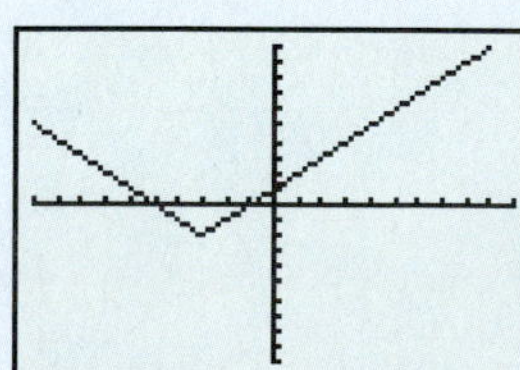

Finding the Domain and Range of an Absolute Value Function

Recall that the domain of a function is the set of all possible input values for the function and the range of a function is the set of all possible output values for the function. From the graph of a function, the domain can be read from left to right and the range can be read from bottom to top. Let's look at the graph of $f(x) = |x|$ again.

Notice that the graph extends to negative infinity on the left and to infinity on the right. Its domain is the set of real numbers $\mathbb{R}$, which is expressed in interval notation as $(-\infty, \infty)$. The lowest value of this function is 0, but it has no upper bound. The range of this function is $[0, \infty)$.

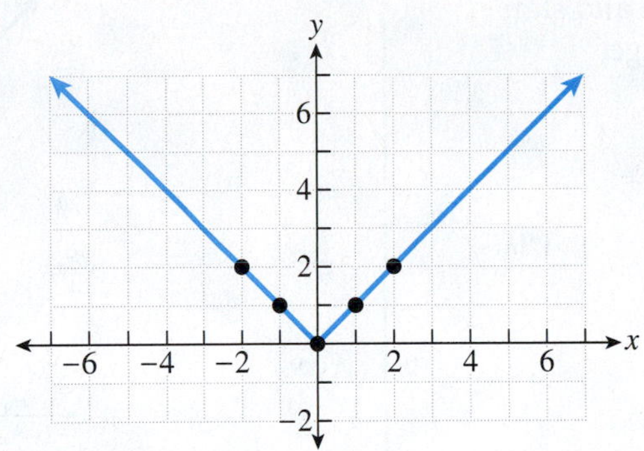

Let's reexamine the graph of $f(x) = |x + 3| - 2$. The domain of this function is also the set of real numbers. What is the range? The lowest function value for this function is -2, so the range is $[-2, \infty)$.

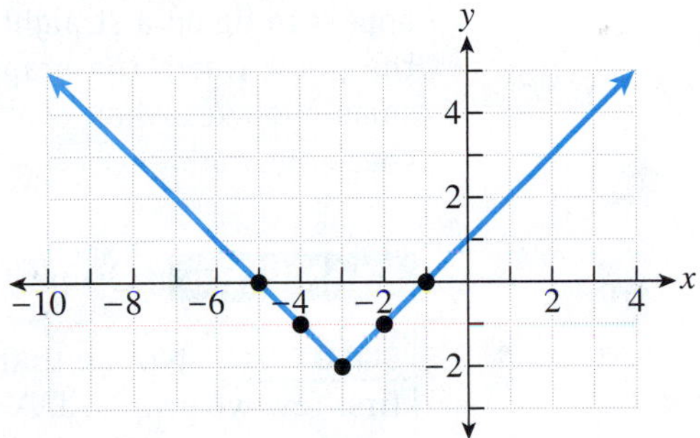

EXAMPLE 5 Graph the function $f(x) = |x - 1| + 4$ and find the domain and range of the function.

SOLUTION We begin by finding the x-coordinate of the turning point of the graph. To do this, we set the expression inside the absolute value bars equal to 0 and solve for x.

$$x - 1 = 0 \quad \text{Set } x - 1 \text{ equal to 0.}$$
$$x = 1 \quad \text{Add 1 to both sides.}$$

Now we can create a table of values using two values for x that are less than 1 and two that are greater than 1.

x	$f(x) = \|x - 1\| + 4$	$(x, f(x))$
-1	$f(-1) = \|-1 - 1\| + 4 = 6$	$(-1, 6)$
0	$f(0) = \|0 - 1\| + 4 = 5$	$(0, 5)$
1	$f(1) = \|1 - 1\| + 4 = 4$	$(1, 4)$
2	$f(2) = \|2 - 1\| + 4 = 5$	$(2, 5)$
3	$f(3) = \|3 - 1\| + 4 = 6$	$(3, 6)$

Below is the graph. From this graph, we can see that the domain is the set of all real numbers and the range is $[4, \infty)$.

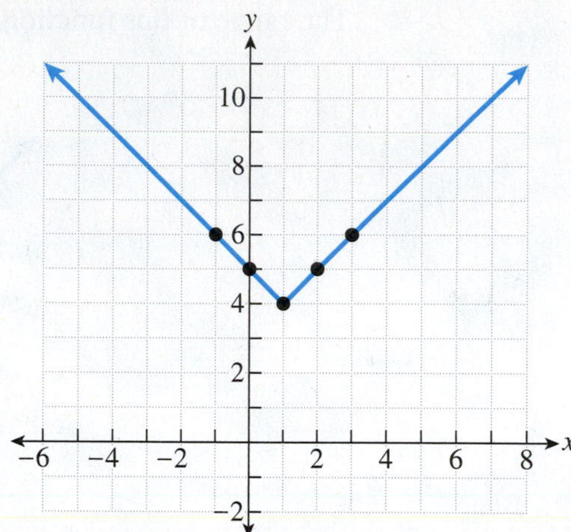

Quick Check 5

Graph the function $f(x) = |x - 5| - 4$ and state the domain and range.

A WORD OF CAUTION When we graph an absolute value function, if we choose values for x wisely, the graph will display a series of points that forms a V. If all of our points appear to lie on a straight line, we have chosen values for x that are all on one side of the point where the graph changes from decreasing to increasing or we have made some mistakes when evaluating the function. In either case, we should go back and check our work.

EXAMPLE 6 Graph $f(x) = -|x| + 2$ and state the domain and range.

SOLUTION Notice that this absolute value function is slightly different from the first few we graphed. A negative sign is in front of the absolute value bars, which will have a significant impact on how the graph looks. However, we will use the same approach to graphing this function as the previous absolute value functions. We begin by finding the x-coordinate of the turning point by setting the expression inside the absolute value bars equal to 0. In this case, $x = 0$ is the x-coordinate of the turning point. Now we can create a table of values using two values for x that are less than 0 and two that are greater than 0.

x	$f(x) = -\lvert x\rvert + 2$	$(x, f(x))$
-2	$f(-2) = -\lvert -2\rvert + 2 = -2 + 2 = 0$	$(-2, 0)$
-1	$f(-1) = -\lvert -1\rvert + 2 = -1 + 2 = 1$	$(-1, 1)$
0	$f(0) = -\lvert 0\rvert + 2 = 0 + 2 = 2$	$(0, 2)$
1	$f(1) = -\lvert 1\rvert + 2 = -1 + 2 = 1$	$(1, 1)$
2	$f(2) = -\lvert 2\rvert + 2 = -2 + 2 = 0$	$(2, 0)$

Quick Check 6

Graph the function $f(x) = -|x - 4| - 5$ and state the domain and range.

The graph is to the right. From this graph, we can see that the domain is the set of all real numbers. This time the function has a maximum value of 2 with no lower bound; so the range is $(-\infty, 2]$.

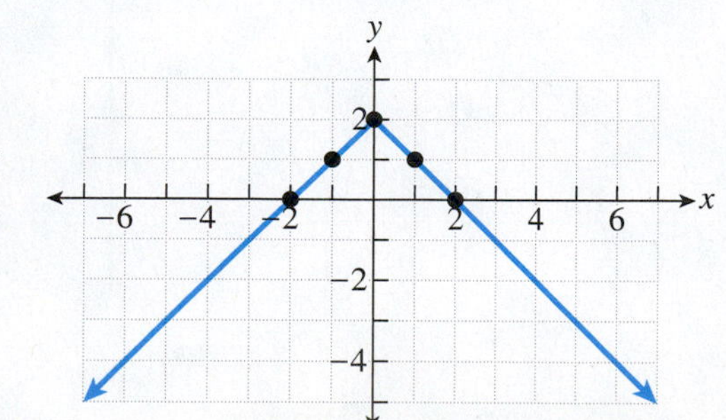

EXAMPLE 7 Find the absolute value function for the graph.

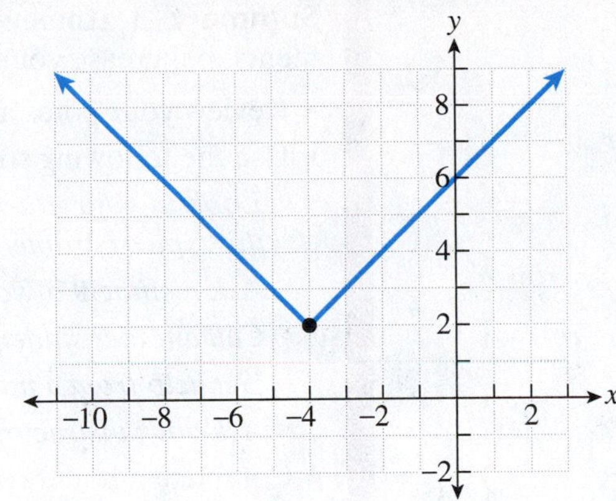

SOLUTION To find the function, let's focus on the turning point of this graph, which is $(-4, 2)$. Because the x-coordinate is -4, the expression $x + 4$ must be inside the absolute value bars. The y-coordinate of the turning point is 2, which tells us that 2 is added to the absolute value. The function is of the form $f(x) = a|x + 4| + 2$, where a is a real number. To determine the value of a, we need to use the coordinates of another point on the graph, such as the y-intercept $(0, 6)$. This point tells us that $f(0) = 6$. We will use this fact to find a.

$f(0) = 6$

$a|0 + 4| + 2 = 6$ Substitute 0 for x in the function $f(x)$.

$4a + 2 = 6$ Simplify the absolute value.

$4a = 4$ Subtract 2 from both sides.

$a = 1$ Divide both sides by 4.

Substituting 1 for a, we find that the function is $f(x) = |x + 4| + 2$.

Quick Check 7

Find the absolute value function for the graph.

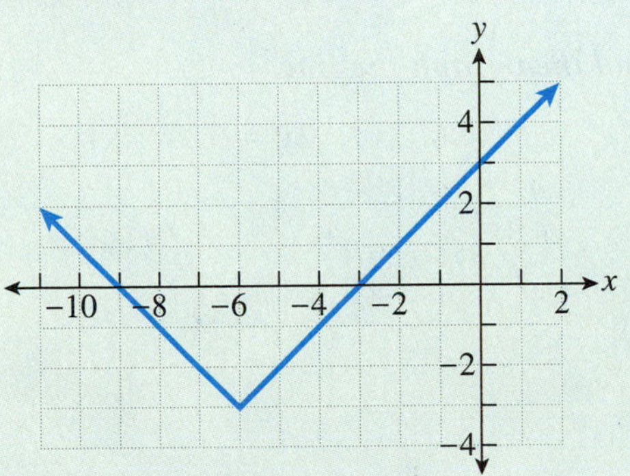

BUILDING YOUR STUDY STRATEGY

Summary, 3 Homework Here are some tips for using homework assignments to increase your understanding of mathematics.

- Review your notes and the text before beginning a homework assignment.
- Use the following strategies for homework problems you cannot do:
  - *Look in your notes and the text for similar problems or suggestions about this type of problem.*
  - *Ask members of your study group for help.*
  - *Call another student in the class for help.*
  - *Get help from a tutor at the tutorial center on your campus.*
  - *Ask your instructor for help at the first opportunity.*
- Complete a homework session by summarizing your work.
- Stay up to date with your homework assignments.

Exercises 8.3

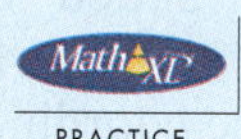
PRACTICE

WATCH

DOWNLOAD
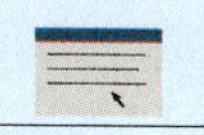
READ

REVIEW

Vocabulary

1. A(n) ______________ is a function of the form $f(x) = |x|$.

2. The graph of an absolute value function is ____-shaped.

3. Find $f(0)$ to find the __________ of a function.

4. Set $f(x) = 0$ to find the __________ of a function.

Find the intercepts and then graph the line.

5. $4x + 3y = -24$

6. $3x + 2y = 12$

7. $2x - 5y = 10$

8. $-4x + 6y = -12$

9. $\frac{2}{9}x + \frac{1}{3}y = 2$

10. $\frac{1}{6}x - \frac{1}{3}y = -\frac{2}{3}$

11. $8x - 3y = 12$

12. $7x + 10y = 35$

13. $y = -3x + 4$

14. $y = -\frac{1}{5}x + 2$

Graph the line using the slope and y-intercept.

15. $y = 3x + 2$

16. $y = 2x - 7$

17. $y = -4x + 8$

18. $y = -5x - 3$

19. $y = \frac{2}{3}x - 6$

20. $y = -\frac{5}{2}x + 5$

21. $y = 4x$

22. $y = -\frac{1}{3}x$

Graph the linear function.

23. $f(x) = x + 7$

24. $f(x) = x - 3$

25. $f(x) = 4x - 5$

26. $f(x) = 2x + 3$

27. $f(x) = -5x + 8$

28. $f(x) = -3x - 6$

29. $f(x) = \frac{3}{4}x - 6$

30. $f(x) = -\frac{5}{3}x + 2$

31. $f(x) = 2$

32. $f(x) = -7$

Graph the absolute value function. State the domain and range of the function.

33. $f(x) = |x - 7|$

34. $f(x) = |x + 6|$

35. $f(x) = |x| + 3$

36. $f(x) = |x| - 4$

37. $f(x) = |x + 3| - 4$

38. $f(x) = |x + 1| + 2$

39. $f(x) = |x - 6| + 3$

40. $f(x) = |x - 2| - 5$

41. $f(x) = |x + 4| - 2$

42. $f(x) = |x - 1| + 6$

43. $f(x) = |2x - 6|$

44. $f(x) = |3x + 12|$

45. $f(x) = |4x + 8| - 5$

46. $f(x) = |5x - 10| - 6$

47. $f(x) = 2|x - 1| + 3$

48. $f(x) = 3|x + 5| - 7$

49. $f(x) = -|x + 6|$

50. $f(x) = -|x| + 6$

51. $f(x) = -|x - 3| - 4$

52. $f(x) = -|x + 8| + 5$

53. $f(x) = -3|x + 2| + 9$

54. $f(x) = -4|x - 4| - 1$

Find the intercepts of the absolute value function.

55. $f(x) = |x - 3| - 7$
56. $f(x) = |x + 4| - 2$
57. $f(x) = |x| - 6$
58. $f(x) = |x + 5|$
59. $f(x) = |x + 2| + 4$
60. $f(x) = |x - 1| + 8$
61. $f(x) = -|x - 4| + 3$
62. $f(x) = -|x + 7| + 9$

Find the absolute value function on the graph.

63.

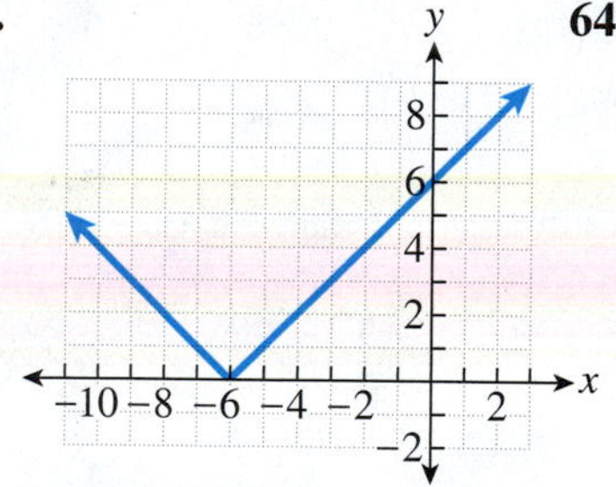

64.

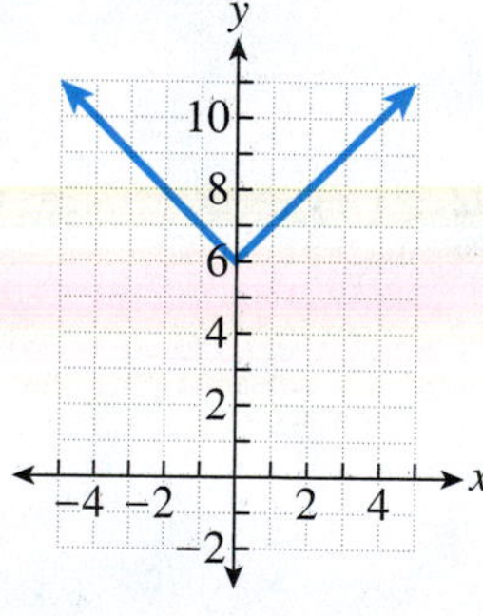

65.

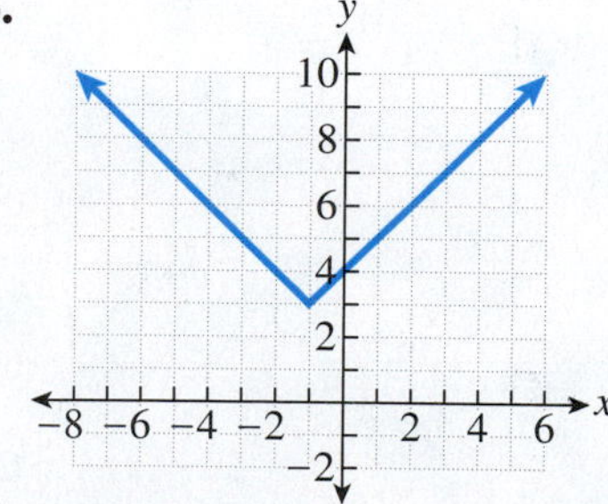

66.

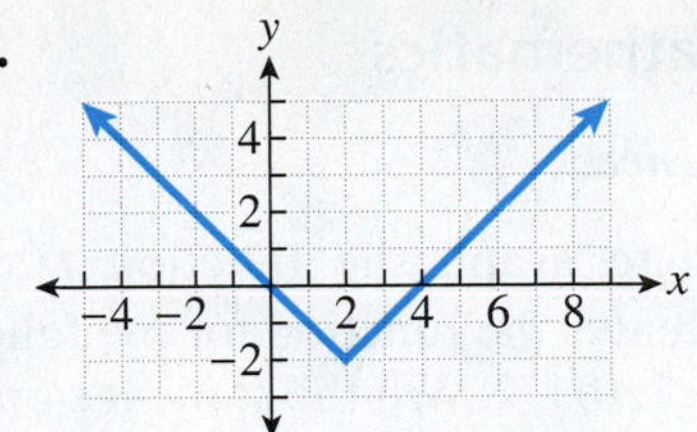

67.

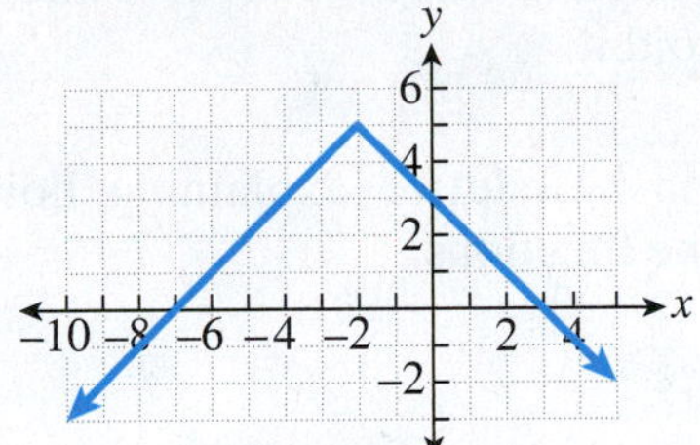

68.

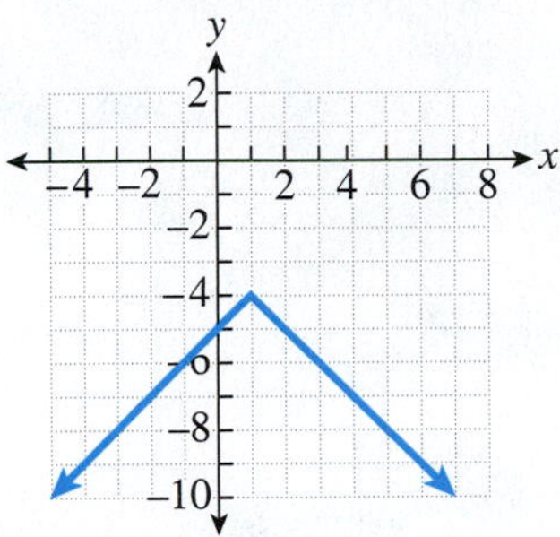

Mixed Practice, 69–80

Graph the function.

69. $f(x) = -\frac{3}{2}x + 7$

70. $f(x) = 2x - 5$

71. $f(x) = |x - 3| - 2$

72. $f(x) = |x + 5| + 4$

73. $f(x) = x + 3$

74. $f(x) = |x| + 3$

75. $f(x) = |x + 4|$

76. $f(x) = \frac{5}{2}x - 6$

77. $f(x) = |x + 2| - 7$

78. $f(x) = 7$

79. $f(x) = \frac{3}{7}x$

80. $f(x) = -|x - 2| + 4$

Writing in Mathematics

Answer in complete sentences.

81. A student is trying to graph the function $f(x) = |x - 5| + 1$ and evaluates the function for the following values of x: $-2, -1, 0, 1, 2$. Will the student's graph be correct? If not, explain what the error is and how the student can avoid it.

82. ***Newsletter*** Write a newsletter explaining how to graph absolute value functions.

8.4 Review of Factoring; Quadratic Equations and Rational Equations

OBJECTIVES

1. Factor polynomials.
2. Solve quadratic equations.
3. Solve rational equations.

Factoring Polynomials

Objective 1 Factor polynomials. A polynomial is factored when it has been written as a product of two or more expressions; these expressions are the factors.

Factoring Out the Greatest Common Factor (GCF)

Every time we attempt to factor a polynomial, we should begin by attempting to factor out the greatest common factor (GCF) of all of the terms.

EXAMPLE 1 Factor $12x^5 - 3x^4 + 15x^2$.

SOLUTION The largest number that divides into these three coefficients is 3. The variable x is a factor of all three terms. For the GCF, we use the largest power of x that is a factor of all of the terms, which is x^2. The GCF of these three terms is $3x^2$. We finish by factoring the GCF out of each term.

$$12x^5 - 3x^4 + 15x^2 = 3x^2(4x^3 - x^2 + 5)$$

Recall that we can check the factoring through multiplication. The reader can verify that $3x^2(4x^3 - x^2 + 5)$ equals $12x^5 - 3x^4 + 15x^2$.

Quick Check 1

Factor $36x^6 + 60x^5 - 12x^3$.

General Factoring Strategy

The first step in factoring any polynomial is to factor the GCF out of all terms. Once the GCF has been factored out, we must check to see if the resulting polynomial can be factored further. (In the previous example, the polynomial $4x^3 - x^2 + 5$ could not be factored further.)

Once we have dealt with the GCF, the factoring strategy is generally determined by the number of terms in the polynomial. Here is a summary from Chapter 6.

1. Factor out any common factors.
2. Determine the number of terms in the polynomial.
   a. If there are *four terms*, try factoring by grouping, discussed in Section 6.1.
   b. If there are *three terms*, try factoring the trinomial using the techniques of Sections 6.2 and 6.3.
      - $x^2 + bx + c = (x + m)(x + n)$: Find two integers m and n whose product is c and whose sum is b.
      - $ax^2 + bx + c$ $(a \neq 1)$: We have two methods for factoring this type of trinomial: by grouping or by trial and error. To use factoring by grouping, refer to Objective 1 in Section 6.3. To factor by trial and error, refer to Objective 2 in Section 6.3.
   c. If there are only *two terms*, check to see if the binomial is one of the special binomials discussed in Section 6.4.
      - Difference of Squares: $a^2 - b^2 = (a + b)(a - b)$
      - Sum of Squares: $a^2 + b^2$ is not factorable.
      - Difference of Cubes: $a^3 - b^3 = (a - b)(a^2 + ab + b^2)$
      - Sum of Cubes: $a^3 + b^3 = (a + b)(a^2 - ab + b^2)$
3. After the polynomial has been factored, make sure any factor with two or more terms does not have any common factors other than 1. If there are common factors, factor them out.
4. Check your factoring through multiplication.

Factoring by Grouping

An important factoring technique is factoring by grouping. This technique is often used when a polynomial has four or more terms. To factor a polynomial by grouping, we split the terms of the polynomial into two groups. We then factor out a common factor of each group. At this point, if the two groups share a common factor, this common factor can be factored out of each group, leaving the polynomial in factored form.

EXAMPLE 2 Factor $x^3 - 6x^2 - 10x + 60$.

SOLUTION There are no factors (other than 1) that are common to all four terms, so we proceed to factor by grouping. We will consider $x^3 - 6x^2$ to be the first group and $-10x + 60$ to be the second group.

$$x^3 - 6x^2 - 10x + 60 = x^2(x - 6) - 10(x - 6)$$ Factor x^2 out of the first two terms and -10 out of the last two terms.

$$= (x - 6)(x^2 - 10)$$ Factor out the common factor $x - 6$.

Recall that we can check the factoring by multiplying these two binomials. Their product should equal the original polynomial.

Quick Check 2

Factor $x^3 - 8x^2 + 4x - 32$.

Factoring Trinomials

We will now discuss factoring quadratic trinomials of the form $x^2 + bx + c$, where b and c are integers. To factor such a polynomial, we look for two integers m and n whose product is c (that is, $m \cdot n = c$) and whose sum is b (that is, $m + n = b$). If we can find two integers that satisfy these conditions, the polynomial factors to be $(x + m)(x + n)$.

When the constant term c of the trinomial is positive, this tells us that the two integers we are looking for have the same sign. The sign of b tells us what the sign of each integer will be. When the constant term is negative, we are looking for one positive integer and one negative integer. The sign of the integer with the largest absolute value is the same as the sign of b.

EXAMPLE 3 Factor $x^2 + 8x + 15$.

SOLUTION Because the three terms do not contain a common factor other than 1, begin by looking for two integers whose product is 15 and whose sum is 8. The integers that satisfy the conditions are 3 and 5, so $x^2 + 8x + 15$ can be factored as $(x + 3)(x + 5)$.

$$x^2 + 8x + 15 = (x + 3)(x + 5)$$

EXAMPLE 4 Factor $x^2 - 10x + 16$.

SOLUTION We cannot factor out a common factor, so we look for the integers with a product of 16 and a sum of -10. The two integers are -2 and -8.

$$x^2 - 10x + 16 = (x - 2)(x - 8)$$

Quick Check 3

Factor.

a) $x^2 + 11x + 28$
b) $x^2 - 13x + 30$

EXAMPLE 5 Factor $3x^2 - 15x - 42$.

SOLUTION We begin by factoring out the common factor 3. This gives us $3(x^2 - 5x - 14)$. We then factor the trinomial factor by finding two integers with a product of -14 and a sum of -5. Again, because the product must be negative, we are looking for one negative integer and one positive integer. In this case, the integers are -7 and 2.

$$\begin{aligned} 3x^2 - 15x - 42 &= 3(x^2 - 5x - 14) && \text{Factor out 3.} \\ &= 3(x - 7)(x + 2) && \text{Factor the trinomial.} \end{aligned}$$

Quick Check 4

Factor $2x^2 - 14x - 36$.

EXAMPLE 6 Factor $-x^2 - 6x + 55$.

SOLUTION When the coefficient of the leading term is negative, we begin by factoring a negative common factor from all of the terms. In this case, we will factor out -1 and then factor the resulting trinomial.

$$\begin{aligned} -x^2 - 6x + 55 &= -(x^2 + 6x - 55) && \text{Factor out } -1. \\ &= -(x + 11)(x - 5) && \text{Factor the trinomial.} \end{aligned}$$

Quick Check 5

Factor $-x^2 + 17x - 70$.

If the leading coefficient of a quadratic trinomial is not equal to 1 and that coefficient cannot be factored out as a common factor, we must use a different technique to factor the polynomial. We will use factoring by grouping to factor polynomials of the form $ax^2 + bx + c$ $(a \neq 1)$. (You also can use the method of trial and error, as presented in Section 6.3.)

We begin by multiplying a by c, then find two integers m and n whose product is equal to $a \cdot c$ and whose sum is b. Once we find these two integers, we rewrite the polynomial as $ax^2 + mx + nx + c$ and factor by grouping.

EXAMPLE 7 Factor $2x^2 - 3x - 20$.

SOLUTION We cannot factor out any common factors, so we begin by multiplying $2(-20)$, which equals -40. Next, we look for two integers whose product is -40 and whose sum is -3. The two integers are -8 and 5; so we will replace the term $-3x$ in the original polynomial by $-8x + 5x$ and then factor by grouping.

$2x^2 - 3x - 20 = 2x^2 - 8x + 5x - 20$ — Rewrite $-3x$ as $-8x + 5x$.

$= 2x(x - 4) + 5(x - 4)$ — Factor the common factor $2x$ from the first two terms and the common factor 5 from the last two terms.

$= (x - 4)(2x + 5)$ — Factor out the common factor $x - 4$.

Quick Check 6

Factor $6x^2 - 7x - 20$.

Factoring a Difference of Squares

A binomial of the form $a^2 - b^2$ is called a difference of squares, and it can be factored using the following formula:

$$a^2 - b^2 = (a + b)(a - b)$$

EXAMPLE 8 Factor $4x^2 - 25$.

SOLUTION We cannot factor any common factors out of this binomial, so we check to see whether it is a difference of squares. The binomial can be rewritten as $(2x)^2 - (5)^2$, so this is a difference of squares and can be factored using the formula.

$4x^2 - 25 = (2x)^2 - (5)^2$ — Rewrite as the difference of two squares.

$= (2x + 5)(2x - 5)$ — Factor using the formula $a^2 - b^2 = (a + b)(a - b)$.

Quick Check 7

Factor $9x^2 - 100$.

Sum of Squares

A binomial that is a sum of squares $(a^2 + b^2)$ cannot be factored.

Factoring a Difference or a Sum of Cubes

A binomial that is a difference of cubes $(a^3 - b^3)$ or a sum of cubes $(a^3 + b^3)$ can be factored using the following formulas:

Difference of Cubes: $a^3 - b^3 = (a - b)(a^2 + ab + b^2)$

Sum of Cubes: $a^3 + b^3 = (a + b)(a^2 - ab + b^2)$

EXAMPLE 9 Factor $8x^3 - 125$.

SOLUTION This binomial is a difference of cubes: $8x^3 = (2x)^3$ and $125 = 5^3$.

$8x^3 - 125 = (2x)^3 - (5)^3$ — Rewrite each term as a perfect cube.

$= (2x - 5)(2x \cdot 2x + 2x \cdot 5 + 5 \cdot 5)$ — Factor as a difference of cubes.

$= (2x - 5)(4x^2 + 10x + 25)$ — Simplify each term in the trinomial factor.

Quick Check 8

Factor $27x^3 - 64y^3$.

Recall from Chapter 6 that the major difference between factoring a difference of cubes and factoring a sum of cubes is keeping track of the signs.

Solving Quadratic Equations

Objective 2 Solve quadratic equations. We now review the solution of quadratic equations by factoring. A quadratic equation is an equation that can be written in the form $ax^2 + bx + c = 0$, where a, b, and c are real numbers and $a \neq 0$. Recall the following guidelines for solving quadratic equations:

- Simplify each side of the equation. This includes performing any distributive multiplications, clearing all fractions by multiplying each side of the equation by the LCD, and combining like terms.
- Collect all terms on one side of the equation, leaving 0 on the other side. It is a good idea to move all terms to one side in such a way that the coefficient of the second-degree term is positive.
- Factor the quadratic expression.
- Set each factor equal to 0 and solve each equation.

The principle behind this procedure is the zero-factor property, which says that if two unknown numbers have a product of zero, at least one of these factors must be equal to zero.

EXAMPLE 10 Solve $x^2 + 7x = 30 - 6x$.

SOLUTION We begin to solve this equation by collecting all terms on the left side of the equation.

$$x^2 + 7x = 30 - 6x$$

$$x^2 + 13x - 30 = 0 \quad \text{Add } 6x \text{ and subtract 30 (on both sides).}$$

$$(x + 15)(x - 2) = 0 \quad \text{Factor.}$$

$$x + 15 = 0 \quad \text{or} \quad x - 2 = 0 \quad \text{Set each factor equal to 0.}$$

$$x = -15 \quad \text{or} \quad x = 2 \quad \text{Solve.}$$

The solution set is $\{-15, 2\}$.

Quick Check 9

Solve $x(x + 3) = 7x + 32$.

EXAMPLE 11 Solve $\frac{1}{2}x^2 - \frac{1}{4}x = 2x^2 - 3$.

SOLUTION We begin to solve this equation by clearing all fractions. We do this by multiplying each side of the equation by 4, which is the least common denominator (LCD).

$$\frac{1}{2}x^2 - \frac{1}{4}x = 2x^2 - 3$$

$$4 \cdot \left(\frac{1}{2}x^2 - \frac{1}{4}x\right) = 4 \cdot (2x^2 - 3) \quad \text{Multiply both sides by 4.}$$

$$\overset{2}{\cancel{4}} \cdot \frac{1}{\underset{1}{\cancel{2}}}x^2 - \overset{1}{\cancel{4}} \cdot \frac{1}{\underset{1}{\cancel{4}}}x = 4 \cdot 2x^2 - 4 \cdot 3 \quad \text{Distribute and divide out common factors.}$$

$$2x^2 - x = 8x^2 - 12 \quad \text{Simplify.}$$

$$0 = 6x^2 + x - 12 \quad \text{Collect all terms on the right side. Subtract } 2x^2 \text{ and add } x \text{ (on both sides).}$$

$$0 = (3x - 4)(2x + 3) \quad \text{Factor.}$$

$$3x - 4 = 0 \quad \text{or} \quad 2x + 3 = 0 \quad \text{Set each factor equal to 0.}$$

$$x = \frac{4}{3} \quad \text{or} \quad x = -\frac{3}{2} \quad \text{Solve.}$$

The solution set is $\left\{-\frac{3}{2}, \frac{4}{3}\right\}$.

Quick Check 10

Solve $\frac{1}{5}x^2 - \frac{1}{10}x - 1 = 0$.

Solving Rational Equations

Objective 3 Solve rational equations. We have learned to solve rational equations, which are equations that contain one or more rational expressions. An example of such an equation is $\frac{2x}{x-2} - \frac{5}{x+1} = 3$. A first step in solving a rational equation is to multiply each side of the equation by the LCD. Then we proceed to solve the resulting equation. We must check that a solution does not cause a denominator to be 0. Such solutions are called extraneous solutions and are omitted from the solution set.

To find the LCD for two or more rational expressions:

- Completely factor each denominator.
- Identify each factor that is a factor of at least one denominator. The LCD is equal to the product of these factors.
- If an expression is a repeated factor of one or more of the denominators, repeat it as a factor in the LCD as well. The exponent used for this factor is equal to the greatest power to which the factor is raised in any one denominator.

EXAMPLE 12 Solve $\frac{1}{x+3} + \frac{1}{x} = \frac{9}{x(x+3)}$.

SOLUTION The LCD for these rational expressions is $x(x+3)$. We see that the values 0 and -3 for x cause the LCD to be equal to 0 and therefore cannot be solutions of this equation.

$$\frac{1}{x+3} + \frac{1}{x} = \frac{9}{x(x+3)}$$

$$x(x+3)\left(\frac{1}{x+3} + \frac{1}{x}\right) = x(x+3)\left(\frac{9}{x(x+3)}\right) \quad \text{Multiply both sides by the LCD.}$$

$$x\cancel{(x+3)}^{1} \cdot \frac{1}{\cancel{x+3}_{1}} + \cancel{x}^{1}(x+3) \cdot \frac{1}{\cancel{x}_{1}} = \cancel{x(x+3)}^{1} \cdot \frac{9}{\cancel{x(x+3)}_{1}} \quad \text{Distribute and divide out common factors.}$$

$$x + x + 3 = 9 \quad \text{Simplify.}$$

$$2x + 3 = 9 \quad \text{Combine like terms.}$$

$$2x = 6 \quad \text{Subtract 3 from both sides.}$$

$$x = 3 \quad \text{Divide both sides by 2.}$$

This value does not cause a denominator to equal 0. The solution set is {3}.

Quick Check 11

Solve $\frac{x}{x-4} - \frac{8}{x-3} = 1$.

EXAMPLE 13 Solve $\frac{1}{x-5} + \frac{1}{x+5} = \frac{10}{x^2-25}$.

SOLUTION Factor each denominator to find the LCD.

$$\frac{1}{x-5} + \frac{1}{x+5} = \frac{10}{x^2-25}$$

$$\frac{1}{x-5} + \frac{1}{x+5} = \frac{10}{(x+5)(x-5)} \quad \text{Factor. The LCD is } (x+5)(x-5).$$

$$(x+5)(x-5)\left(\frac{1}{x-5} + \frac{1}{x+5}\right) = \frac{10}{(x+5)(x-5)} \cdot (x+5)(x-5) \quad \text{Multiply both sides by the LCD.}$$

$$(x + 5)\overset{1}{\cancel{(x - 5)}} \cdot \frac{1}{\underset{1}{\cancel{x - 5}}} + \overset{1}{\cancel{(x + 5)}}(x - 5) \cdot \frac{1}{\underset{1}{\cancel{x + 5}}}$$

$$= \frac{10}{\underset{1}{\cancel{(x + 5)(x - 5)}}} \cdot \overset{1}{\cancel{(x + 5)(x - 5)}} \quad \text{Distribute and divide out common factors.}$$

$$x + 5 + x - 5 = 10 \quad \text{Simplify.}$$
$$2x = 10 \quad \text{Combine like terms.}$$
$$x = 5 \quad \text{Divide both sides by 2.}$$

This solution must be omitted, as $x = 5$ causes a denominator to be equal to 0. Therefore, this equation has no solution and its solution set is $\varnothing$.

Quick Check 12

Solve $\frac{x}{x + 3} + \frac{1}{x + 2} = \frac{7x}{x^2 + 5x + 6}$.

BUILDING YOUR STUDY STRATEGY

Summary, 4 Test-Taking Strategies: Preparation Through careful preparation and effective test-taking strategies, you can maximize your grade on a math exam. Here is a summary of some test-taking strategies.

- Write down important facts as soon as you get your test.
- Quickly read through the test.
- Begin by solving easier problems first.
- Review your test as thoroughly as possible before turning it in.

When preparing for a cumulative exam,

- Begin preparing early.
- Review your old exams, quizzes, and homework.
- Use review materials from your instructor and the cumulative review exercises in the text.
- Continue to meet regularly with your study group.

Exercises 8.4

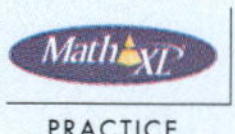

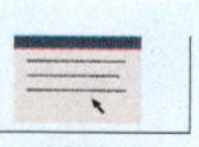

Vocabulary

1. A polynomial has been ________ when it is represented as the product of two or more polynomials.
2. The process of factoring a polynomial by first breaking the polynomial into two sets of terms is called factoring by ________.
3. Before attempting to factor a second-degree trinomial, it is wise to factor out any ________.
4. A binomial of the form $a^2 - b^2$ is a(n) ________.
5. A binomial of the form $a^3 - b^3$ is a(n) ________.
6. A binomial of the form $a^3 + b^3$ is a(n) ________.
7. The ________ property of real numbers states that if $a \cdot b = 0$, then $a = 0$ or $b = 0$.
8. Once a quadratic expression has been set equal to 0 and factored, you set each ________ equal to 0 and solve.

Factor completely.

9. $36x^8 - 16x^4 + 50x^3$

10. $9a^4 + 24a^3 - 3$

11. $20a^3b - 28a^2b^4 - 16a^5b^2$

12. $15x^3y^2z^4 - 5xy^6z^{10} + 40x^6y - 25x^4y^4z^4$

13. $n^3 + 4n^2 + 5n + 20$

14. $x^3 - 9x^2 + 2x - 18$

15. $x^3 + 7x^2 - 3x - 21$

16. $2t^3 - 16t^2 - 10t + 80$

17. $x^2 - 13x + 40$

18. $n^2 + 3n - 54$

19. $2x^2 - 8x - 64$

20. $x^2 - 6x - 16$

21. $a^2 + 13ab + 36b^2$

22. $m^2n^2 - 11mn + 28$

23. $4x^2 - 20x - 56$

24. $3x^2 - 33x + 90$

25. $2x^2 + x - 36$

26. $3x^2 + 11x - 20$

27. $6x^2 + 25x + 24$

28. $12x^2 - 23x + 10$

29. $x^2 - 36$

30. $x^2 - 81$

31. $16b^2 - 9$

32. $2a^2 - 98$

33. $x^2 + 16$

34. $x^2 + 81$

35. $x^3 - 8$

36. $x^3 + 64$

37. $27x^3 + 125$

38. $8x^3 - 343y^3$

39. $x^4 - 81$

40. $625n^4 - 1$

41. $x^3 - 7x^2 - 9x + 63$

42. $2x^3 + 5x^2 - 98x - 245$

43. $n^4 + 4n^3 + 8n + 32$

44. $x^4 - 9x^3 - x + 9$

Solve.

45. $x^2 + 3x - 4 = 0$

46. $x^2 - 7x - 30 = 0$

47. $t^2 + 17t + 72 = 0$

48. $3a^2 - 18a + 24 = 0$

49. $x^2 - 2x - 48 = 0$

50. $x^2 + 12x + 36 = 0$

51. $4x^2 - 44x + 72 = 0$

52. $x^2 + 6x - 55 = 0$

53. $2x^2 + 7x + 6 = 0$

54. $4x^2 + 4x - 35 = 0$

55. $x^2 - 25 = 0$

56. $x^2 - 121 = 0$

57. $9x^2 - 4 = 0$

58. $4x^2 - 81 = 0$

59. $x^2 - 25x = 0$

60. $x^2 - 16x = 0$

61. $x^2 + 4x = 21$

62. $x^2 = 11x - 18$

63. $x^2 + 11x = 4x - 6$

64. $x^2 + x + 5 = 9x - 7$

65. $\frac{4}{3}x^2 + \frac{14}{3}x + 2 = 0$

66. $\frac{1}{15}x^2 + \frac{3}{5}x + \frac{4}{3} = 0$

Solve.

67. $|x^2 - 13x| = 30$

68. $|x^2 + 5x| = 6$

69. $|x^2 + 10x| = 24$

70. $|x^2 + 8x + 6| = 6$

Find a quadratic equation with integer coefficients whose solution set is given.

71. $\{-9, 4\}$

72. $\{-3, 0\}$

73. $\{\frac{1}{3}, 5\}$

74. $\{-\frac{2}{3}, \frac{5}{2}\}$

75. $\{-12, 12\}$

76. $\{-7\}$

Solve.

77. $\frac{8}{x} = \frac{7}{x - 2}$

78. $\frac{5}{x - 6} = \frac{3}{x + 12}$

79. $x - 3 = \frac{10}{x}$

80. $x + \frac{4}{x} = 5$

81. $\frac{x^2 - 23}{x - 3} + 7 = 0$

82. $x + \frac{6x - 17}{x - 5} = -3$

83. $\frac{x + 1}{x + 2} = \frac{6}{x + 6}$

84. $\frac{x - 1}{x + 9} = \frac{4}{x - 3}$

85. $\frac{x}{x - 3} + \frac{4}{x + 5} = \frac{8x}{(x - 3)(x + 5)}$

86. $\frac{x + 2}{x + 4} + \frac{3}{x + 7} = \frac{6}{(x + 4)(x + 7)}$

87. $\frac{x-4}{x-1}+\frac{4}{x+1}=\frac{x+17}{x^2-1}$

88. $\frac{x}{x^2+13x+40}+\frac{3}{x^2+2x-15}=\frac{4}{x^2+5x-24}$

89. $\frac{3}{x-1}+\frac{7}{x+2}=\frac{9}{x^2+x-2}$

90. $\frac{3}{x+1}+\frac{1}{x-5}=\frac{2x-4}{x^2-4x-5}$

Solve.

91. $\left|\frac{2x-15}{x}\right|=3$

92. $\left|\frac{3x+1}{2x-5}\right|=4$

93. $\left|\frac{x^2+5x-48}{x}\right|=3$

94. $\left|\frac{x^2+9x-20}{x}\right|=10$

Writing in Mathematics

Answer in complete sentences.

95. What is an extraneous solution of a rational equation? Why do you omit them from the solution set? Explain your answer.

8.5 Systems of Equations (Two Equations in Two Unknowns and Three Equations in Three Unknowns)

OBJECTIVES

1. **Solve systems of two linear equations in two unknowns.**
2. **Determine whether an ordered triple is a solution of a system of three equations in three unknowns.**
3. **Solve systems of three linear equations in three unknowns.**
4. **Solve applied problems by using a system of three linear equations in three unknowns.**

Systems of Two Linear Equations in Two Unknowns

Objective 1 Solve systems of two linear equations in two unknowns. We begin this section by reviewing how to solve a system of two linear equations in two variables. A solution of a system of two linear equations is an ordered pair that satisfies both equations simultaneously. In Chapter 4, we learned two techniques for solving a system of linear equations: by substitution and by addition.

The graph of each equation in a system of linear equations is a line, and if these two lines cross at exactly one point, this ordered pair is the only solution of the system. In this case, the system is said to be independent.

Independent System

The figure shows the graphs of $y = 2x + 5$ and $y = -x - 1$. These two lines intersect at the ordered pair $(-2, 1)$, and this point of intersection is the solution to this independent system of equations.

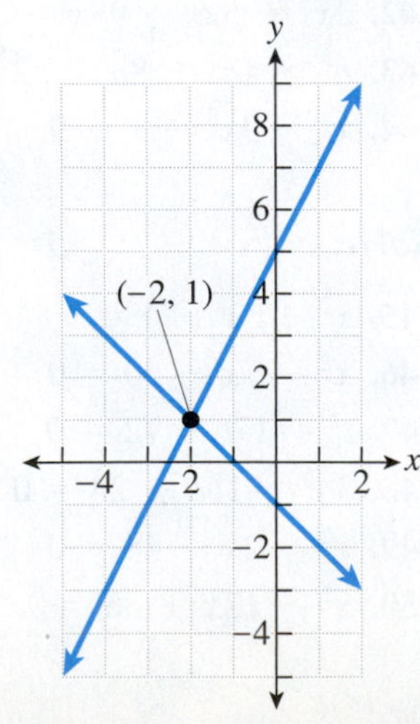

Inconsistent System

If the two lines associated with a system of equations are parallel lines, the system has no solution and is said to be inconsistent.

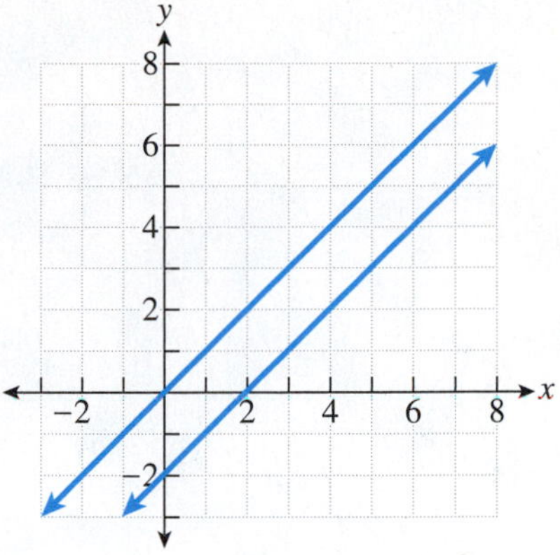

Dependent System

If the two lines associated with a system of equations are the same line, the system is said to be a dependent system. For example, consider the system $\begin{aligned} x + y &= 5 \\ 2x + 2y &= 10 \end{aligned}$.

The figure shows the graph associated with this system. Each ordered pair that is on the line is a solution. We write the solution in the form $(x, 5 - x)$ by solving one of the equations for y in terms of x.

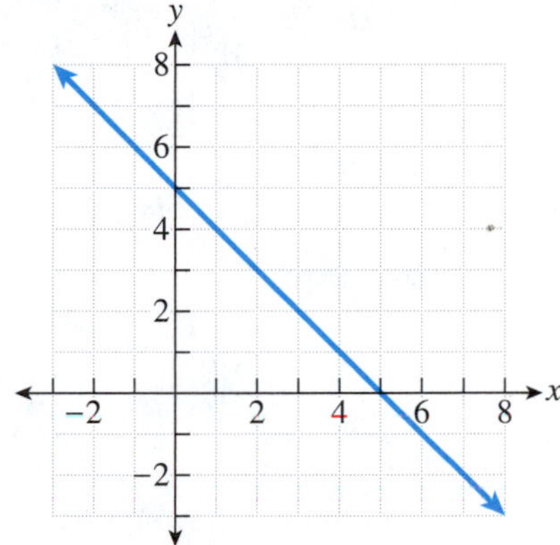

The Substitution Method

Now we turn our attention to solving a system of equations by substitution. This technique requires us to solve one of the equations for one of the variables in terms of the other variable. The expression that we find is then *substituted* for that variable in the other equation, giving us one equation with one variable. After solving that equation, we substitute this value into the expression we originally substituted to solve for the other variable.

EXAMPLE 1 Solve by substitution. $\begin{aligned} x + 3y &= 1 \\ 3x - 2y &= 14 \end{aligned}$

SOLUTION If there is a variable with a coefficient of 1, it is a good idea to solve the equation for that variable. (Using this strategy will help us avoid the use of fractions.) Solving the first equation for x gives us $x = 1 - 3y$. The expression $1 - 3y$ is then substituted into the equation $3x - 2y = 14$ for x.

$$\begin{aligned} 3x - 2y &= 14 \\ 3(1 - 3y) - 2y &= 14 && \text{Substitute } 1 - 3y \text{ for } x. \\ 3 - 9y - 2y &= 14 && \text{Distribute.} \\ 3 - 11y &= 14 && \text{Combine like terms.} \\ -11y &= 11 && \text{Subtract 3 from both sides.} \\ y &= -1 && \text{Divide both sides by } -11. \end{aligned}$$

This value is then substituted for y into the equation $x = 1 - 3y$.

$$x = 1 - 3(-1) \quad \text{Substitute } -1 \text{ for } y.$$
$$x = 1 + 3 \quad \text{Multiply.}$$
$$x = 4 \quad \text{Add.}$$

The ordered pair solution is $(4, -1)$.

Quick Check 1

Solve by substitution.

$2x - 5y = -24$

$4x + y = -4$

The Addition Method

Another technique for solving a system of equations is solving by addition. We multiply one or both equations by a constant in such a way that one of the variables has opposite coefficients in the two equations. When we add the two equations, this variable is eliminated, resulting in an equation that contains only one variable. We finish finding the solution just as we did when solving by substitution.

EXAMPLE 2 Solve. $\begin{aligned} 4x + 5y &= 13 \\ 2x + 7y &= 11 \end{aligned}$

SOLUTION If we multiply the second equation by -2, the coefficient for the x term will be -4, which is the opposite of the coefficient of the x term in the first equation. (Recall that there will be times when we must multiply each equation by a different constant to eliminate one variable.)

$$\begin{aligned} 4x + 5y &= 13 & & & 4x + 5y &= 13 \\ 2x + 7y &= 11 & &\xrightarrow{\text{Multiply by } -2.} & -4x - 14y &= -22 \end{aligned}$$

$$\begin{aligned} 4x + 5y &= 13 & & \\ -4x - 14y &= -22 & & \text{Add.} \\ \hline -9y &= -9 & & \\ y &= 1 & & \text{Divide both sides by } -9. \end{aligned}$$

We now substitute 1 for y into either original equation to find x.

$$\begin{aligned} 4x + 5y &= 13 & & \\ 4x + 5(1) &= 13 & & \text{Substitute 1 for } y. \\ 4x + 5 &= 13 & & \text{Multiply.} \\ 4x &= 8 & & \text{Subtract 5 from both sides.} \\ x &= 2 & & \text{Divide both sides by 4.} \end{aligned}$$

The solution is $(2, 1)$.

Quick Check 2

Solve by addition.

$3x + 2y = 13$

$5x - 3y = 9$

Applications of Systems of Two Linear Equations in Two Unknowns

We can use a system of two linear equations in two unknowns to solve many applied problems, including perimeter problems, mixture problems, and interest problems. Refer to Section 4.4 to review these topics in more detail. Here is an example of a mixture problem.

EXAMPLE 3 A chemist has one solution that is 16% sodium and a second solution that is 28% sodium. She wants to combine these solutions to make 90 milliliters of a solution that is 20% sodium. How many milliliters of each original solution does she need to use?

SOLUTION: The two unknowns are the volume of the 16% sodium solution and the volume of the 28% sodium solution that need to be mixed together. We will use x and y to represent those quantities. To determine the volume of sodium contributed by each solution as well as the volume of sodium in the mixture, we multiply the volume of the solution by the concentration of sodium in the solution. The following table summarizes the information:

Solution	Volume of Solution	% Concentration of Sodium	Volume of Sodium
16%	x	0.16	$0.16x$
28%	y	0.28	$0.28y$
Mixture (20%)	90	0.20	18

The first equation in the system comes from the fact that the volume of the two solutions must equal 90 milliliters. As an equation, this can be represented by $x + y = 90$. (This equation can be found in the table in the column labeled "Volume of Solution.") The second equation comes from the total volume of sodium, in the column labeled "Volume of Sodium." We know that the total volume of sodium is 18 milliliters (20% of the 90 milliliters in the mixture is sodium); so the second equation in the system is $0.16x + 0.28y = 18$. Here is the system we must solve.

$$\begin{aligned} x + \quad y &= 90 \\ 0.16x + 0.28y &= 18 \end{aligned}$$

We begin by clearing the second equation of decimals. This can be done by multiplying both sides of the second equation by 100.

$$\begin{aligned} x + \quad y &= 90 & & x + \quad y = \quad 90 \\ 0.16x + 0.28y &= 18 & \xrightarrow{\text{Multiply by 100.}} & \ 16x + 28y = 1800 \end{aligned}$$

To solve this system of equations, we can use the addition method. If we multiply the first equation by -16, the variable terms containing x are opposites.

$$\begin{aligned} x + \quad y &= \quad 90 & \xrightarrow{\text{Multiply by } -16.} & \ -16x - 16y = -1440 \\ 16x + 28y &= 1800 & & \ 16x + 28y = \quad 1800 \end{aligned}$$

$$\begin{aligned} -16x - 16y &= -1440 & \\ 16x + 28y &= \quad 1800 & \text{Add to eliminate } x. \\ \hline 12y &= \quad 360 & \\ y &= 30 & \text{Divide both sides by 12.} \end{aligned}$$

We know that the chemist must use 30 milliliters of the solution that is 28% sodium. To determine how much of the 16% sodium solution she must use, we substitute 30 for y in the equation $x + y = 90$.

$$x + (30) = 90 \quad \text{Substitute 30 for } y.$$
$$x = 60 \quad \text{Subtract 30.}$$

She must use 60 milliliters of the 16% sodium solution and 30 milliliters of the 28% sodium solution.

Quick Check 3

A chemist has one solution that is 60% acid and a second solution that is 50% acid. He wants to combine the solutions to make a new solution that is 52% acid. If the chemist needs to make 400 milliliters of this new solution, how many milliliters of the two solutions should he mix?

Solutions of a System of Three Equations in Three Unknowns

Objective 2 Determine whether an ordered triple is a solution of a system of three equations in three unknowns. We now turn our attention to systems of three equations in three unknowns. Each of the three equations will be of the form $ax + by + cz = d$, where a, b, c, and d are real numbers. If there is a unique solution of such a system, it will be an **ordered triple** (x, y, z).

EXAMPLE 4 Is (3, 2, 4) a solution of the following system of equations?

$$\begin{aligned} 2x - 3y + z &= 4 \\ x + 2y - z &= 3 \\ 2x - y - 2z &= -4 \end{aligned}$$

SOLUTION To determine whether the ordered triple is a solution of the system of equations, we will substitute 3 for x, 2 for y, and 4 for z in each equation. If each equation is true for these three values, the ordered triple is a solution of the system of equations.

$2x - 3y + z = 4$	$x + 2y - z = 3$	$2x - y - 2z = -4$
$2(3) - 3(2) + (4) = 4$	$(3) + 2(2) - (4) = 3$	$2(3) - (2) - 2(4) = -4$
$6 - 6 + 4 = 4$	$3 + 4 - 4 = 3$	$6 - 2 - 8 = -4$
$4 = 4$	$3 = 3$	$-4 = -4$

Because the ordered triple makes each equation in the system true, the ordered triple (3, 2, 4) is a solution of the system of equations.

Quick Check 4

Is (4, −2, 5) a solution of the following system of equations?

$$\begin{aligned} x - 2y - z &= 3 \\ 3x + 2y + 2z &= 26 \\ 5x - 3y - 4z &= 6 \end{aligned}$$

Solving a System of Three Linear Equations in Three Unknowns

Objective 3 Solve systems of three linear equations in three unknowns. The graph of a linear equation with three variables is a two-dimensional **plane**.

For a system of three linear equations in three unknowns, if the planes associated with the equations intersect at a single point, this ordered triple (x, y, z) is the solution of the system of equations.

Some systems of three equations in three unknowns have no solution. Such a system is called an inconsistent system. When we try to solve an inconsistent system, we obtain an equation that is a contradiction, such as $0 = 1$. Graphically, the following systems are inconsistent, as the three planes do not have a point in common:

Some systems of three equations in three unknowns have infinitely many solutions. Such a system is called a dependent system of equations. In solving a dependent system of equations, we will obtain an equation that is an identity, such as $0 = 0$. Graphically, if at least two of the planes are identical or if the intersection of the three planes is a line, the system is a dependent system.

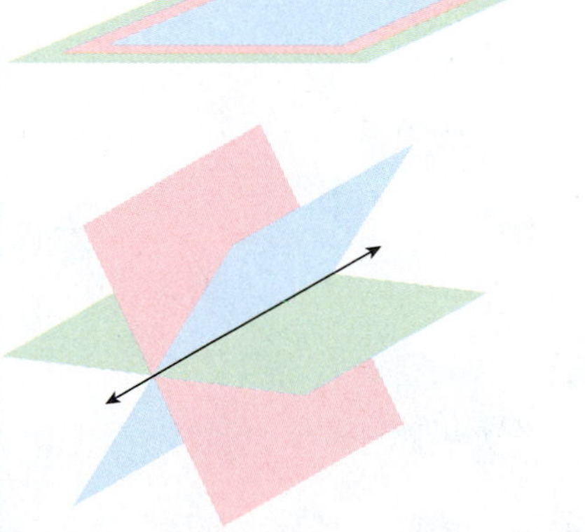

Because graphing equations in three dimensions is beyond the scope of this text, we will focus on solving systems of three linear equations in three unknowns algebraically.

Solving a System of Three Linear Equations in Three Unknowns

- Select two of the equations and use the addition method to eliminate one of the variables.
- Select a different pair of equations and use the addition method to eliminate the same variable, leaving two equations in two unknowns.
- Solve the system of two equations for the two unknowns.
- Substitute these two values in any of the original three equations and solve for the unknown variable.

We will now apply this technique to the system of equations from the previous example.

EXAMPLE 5 Solve the system.

$$\begin{aligned} 2x - 3y + z &= 4 \quad \text{(Eq. 1)} \\ x + 2y - z &= 3 \quad \text{(Eq. 2)} \\ 2x - y - 2z &= -4 \quad \text{(Eq. 3)} \end{aligned}$$

SOLUTION Notice that the three equations have been labeled. This will help us keep track of the equations as we solve the system. We must choose a variable to eliminate first, and in this example, it will be z. We may begin by adding Equations 1 and 2 together, as the coefficients of the z-terms are opposites.

$$\begin{aligned} 2x - 3y + z &= 4 \quad \text{(Eq. 1)} \\ x + 2y - z &= 3 \quad \text{(Eq. 2)} \\ \hline 3x - y \quad &= 7 \quad \text{(Eq. 4)} \end{aligned}$$

We need to select a different pair of equations and eliminate z. By multiplying Equation 1 by 2 and adding the product to Equation 3, we can eliminate z once again.

$$\begin{aligned} 4x - 6y + 2z &= 8 \quad (2 \cdot \text{Eq. 1}) \\ 2x - y - 2z &= -4 \quad \text{(Eq. 3)} \\ \hline 6x - 7y \quad &= 4 \quad \text{(Eq. 5)} \end{aligned}$$

We now work to solve the system of equations formed by Equations 4 and 5, which is a system of two equations in two unknowns. We will solve this system using the addition method as well, although we could choose the substitution method if it was convenient. We will eliminate the variable y by multiplying Equation 4 by -7.

$$\begin{aligned} 3x - y &= 7 \quad \text{(Eq. 4)} \xrightarrow{\text{Multiply by } -7.} & -21x + 7y &= -49 \\ 6x - 7y &= 4 \quad \text{(Eq. 5)} & 6x - 7y &= 4 \\ & & -15x \quad &= -45 \quad \text{Add.} \\ & & x \quad &= 3 \quad \text{Divide both sides by } -15. \end{aligned}$$

We now substitute 3 for x in either of the equations that contained only two variables. We will use Equation 4.

$$\begin{aligned} 3x - y &= 7 \\ 3(3) - y &= 7 \quad \text{Substitute 3 for } x \text{ in Equation 4.} \\ 9 - y &= 7 \quad \text{Multiply.} \\ -y &= -2 \quad \text{Subtract 9 from both sides.} \\ y &= 2 \quad \text{Divide both sides by } -1. \end{aligned}$$

To find z, we will substitute 3 for x and 2 for y in any of the original equations that contained three variables. We will use Equation 1.

$$\begin{aligned} 2x - 3y + z &= 4 \\ 2(3) - 3(2) + z &= 4 \quad \text{Substitute 3 for } x \text{ and 2 for } y \text{ in Equation 1.} \\ 6 - 6 + z &= 4 \quad \text{Multiply.} \\ 0 + z &= 4 \quad \text{Combine like terms.} \\ z &= 4 \end{aligned}$$

The solution of this system is the ordered triple $(3, 2, 4)$.

Quick Check 5

Solve the system.

$$\begin{aligned} x + y + z &= 0 \\ 2x - y + 3z &= -9 \\ -3x + 2y - 4z &= 13 \end{aligned}$$

EXAMPLE 6 Solve the system.

$$\begin{aligned} 3x + 2y + 2z &= 7 && \text{(Eq. 1)} \\ 4x + 4y - 3z &= -17 && \text{(Eq. 2)} \\ -x - 6y + 4z &= 13 && \text{(Eq. 3)} \end{aligned}$$

SOLUTION In this example, we will eliminate x first. It is a good idea to choose x because the coefficient of the x-term in Equation 3 can easily be multiplied to become the opposite of the coefficients of the x-terms in Equations 1 and 2. We begin by multiplying Equation 3 by 3 and adding it to Equation 1.

$$\begin{array}{rrrrl} 3x + & 2y + & 2z = & 7 & \text{(Eq. 1)} \\ -3x - & 18y + & 12z = & 39 & (3 \cdot \text{Eq. 3}) \\ \hline & -16y + & 14z = & 46 & \text{(Eq. 4)} \end{array}$$

We need to select a different pair of equations and eliminate x. By multiplying Equation 3 by 4 and adding it to Equation 2, we can eliminate x once again.

$$\begin{array}{rrrrl} 4x + & 4y - & 3z = & -17 & \text{(Eq. 2)} \\ -4x - & 24y + & 16z = & 52 & (4 \cdot \text{Eq. 3}) \\ \hline & -20y + & 13z = & 35 & \text{(Eq. 5)} \end{array}$$

We now solve the system of equations formed by Equations 4 and 5. We will use the addition method to eliminate the variable y. To determine what to multiply each equation by, we should find the LCM of 16 and 20, which is 80. If we multiply Equation 4 by 5, the coefficient of the y-term will be -80. If we multiply Equation 5 by -4, the coefficient of the y-term will be 80. This will make the coefficients of the y-terms opposites, allowing us to add the equations together and eliminate y.

$$\begin{aligned} -16y + 14z &= 46 \quad \text{(Eq. 4)} \xrightarrow{\text{Multiply by 5.}} -80y + 70z = 230 \\ -20y + 13z &= 35 \quad \text{(Eq. 5)} \xrightarrow{\text{Multiply by } -4.} 80y - 52z = -140 \end{aligned}$$

$$\begin{array}{rl} -80y + 70z = 230 & \\ 80y - 52z = -140 & \\ \hline 18z = 90 & \text{Add.} \\ z = 5 & \text{Divide both sides by 18.} \end{array}$$

We now substitute 5 for z in either of the equations that contained only two variables. We will use Equation 4.

$$\begin{aligned} -16y + 14z &= 46 \\ -16y + 14(5) &= 46 && \text{Substitute 5 for } z \text{ in Equation 4.} \\ -16y + 70 &= 46 && \text{Multiply.} \\ -16y &= -24 && \text{Subtract 70 from both sides.} \\ y &= \frac{3}{2} && \text{Divide both sides by } -16 \text{ and simplify.} \end{aligned}$$

To find x, we will substitute 5 for z and $\frac{3}{2}$ for y in any of the original equations that contained three variables. We will use Equation 1.

$$\begin{aligned} 3x + 2y + 2z &= 7 \\ 3x + 2\left(\frac{3}{2}\right) + 2(5) &= 7 && \text{Substitute } \tfrac{3}{2} \text{ for } y \text{ and 5 for } z \text{ in Equation 1.} \\ 3x + 3 + 10 &= 7 && \text{Multiply.} \end{aligned}$$

$$3x + 13 = 7 \quad \text{Combine like terms.}$$
$$3x = -6 \quad \text{Subtract 13 from both sides.}$$
$$x = -2 \quad \text{Divide both sides by 3.}$$

The solution of this system is the ordered triple $\left(-2, \frac{3}{2}, 5\right)$.

Quick Check 6

Solve the system.

$$\begin{aligned} 2x + y - z &= -8 \\ -4x - 3y + 3z &= 25 \\ 6x + 5y + 4z &= 3 \end{aligned}$$

Applications of Systems of Three Linear Equations in Three Unknowns

Objective 4 Solve applied problems by using a system of three linear equations in three unknowns. We finish the section with an application involving a system of three equations in three unknowns.

EXAMPLE 7 A baseball team has three prices for admission to their games. Adults are charged \$7, senior citizens are charged \$5, and children are charged \$4. At last night's game, 1000 people were in attendance, generating \$5600 in receipts. If we know that 300 more children than senior citizens were at the game, find how many adult tickets, senior citizen tickets, and child tickets were sold.

SOLUTION There are three unknowns in this problem: the number of adult tickets sold, the number of senior citizen tickets sold, and the number of child tickets sold. We will let a, s, and c represent these three quantities, respectively. Here is a table of the unknowns.

Unknowns

Number of adult tickets sold: a

Number of senior citizen tickets sold: s

Number of child tickets sold: c

Solving a problem with three unknowns requires that we establish a system of three equations. Because we know that 1000 people were at last night's game, the first equation in the system is $a + s + c = 1000$.

We also know that the revenue from all of the tickets was \$5600, which will lead to the second equation. The amount received from the adult tickets was \$$7a$ (a tickets at \$7 each). In the same way, the amount received from senior citizen tickets was \$$5s$ and the amount received from child tickets was \$$4c$. The second equation in the system is $7a + 5s + 4c = 5600$.

The third equation comes from knowing that 300 more children than senior citizens were at the game. In other words, the number of children at the game was equal to the number of senior citizens plus 300. The third equation is $c = s + 300$, which can be rewritten as $-s + c = 300$.

Here is the system with which we will be working.

$$\begin{aligned} a + s + c &= 1000 \quad \text{(Eq. 1)} \\ 7a + 5s + 4c &= 5600 \quad \text{(Eq. 2)} \\ -s + c &= 300 \quad \text{(Eq. 3)} \end{aligned}$$

Because Equation 3 has only two variables, we will use the addition method to combine Equations 1 and 2, eliminating the variable a. We begin by multiplying Equation 1 by -7 and adding it to Equation 2.

$$\begin{aligned} -7a - 7s - 7c &= -7000 \quad (-7 \cdot \text{Eq. 1}) \\ 7a + 5s + 4c &= 5600 \quad \text{(Eq. 2)} \\ \hline -2s - 3c &= -1400 \quad \text{(Eq. 4)} \end{aligned}$$

We now have two equations that contain only the two variables s and c. We can solve the system of equations formed by Equations 3 and 4. We will use the addition method to eliminate the variable c by multiplying Equation 3 by 3 and adding it to Equation 4.

$$\begin{array}{rcl} -s + c = 300 & \xrightarrow{\text{Multiply by 3.}} & -3s + 3c = 900 \\ -2s - 3c = -1400 & & -2s - 3c = -1400 \end{array}$$

$$\begin{aligned} -3s + 3c &= 900 \\ -2s - 3c &= -1400 \\ \hline -5s &= -500 && \text{Add.} \\ s &= 100 && \text{Divide both sides by } -5. \end{aligned}$$

There were 100 senior citizens at the game. To find the number of children at the game, we will substitute 100 for s in Equation 3.

$$\begin{aligned} -s + c &= 300 \\ -100 + c &= 300 && \text{Substitute 100 for } s \text{ in Equation 3.} \\ c &= 400 && \text{Add 100 to both sides.} \end{aligned}$$

There were 400 children at the game. (Because we knew that 300 more children than senior citizens were at the game, we could have just added 300 to 100.) To find the number of adults at the game, we will substitute 100 for s and 400 for c in Equation 1.

$$\begin{aligned} a + s + c &= 1000 \\ a + (100) + (400) &= 1000 && \text{Substitute 100 for } s \text{ and 400 for } c \text{ in Equation 1.} \\ a &= 500 && \text{Solve for } a. \end{aligned}$$

There were 500 adults, 100 senior citizens, and 400 children at last night's game. The reader can verify that the total revenue for 500 adults, 100 senior citizens, and 400 children is \$5600.

Quick Check 7

Gilbert looks in his wallet and finds \$1, \$5, and \$10 bills totaling \$111. There are 30 bills in all, and Gilbert has 4 more \$5 bills than he has \$10 bills. How many bills of each type does Gilbert have in his wallet?

BUILDING YOUR STUDY STRATEGY

Summary, 5 Math Anxiety Math anxiety can hinder your success in a mathematics class, but you can overcome it. The first step is to understand what has caused your anxiety. Relaxation techniques can help you overcome the physical symptoms of math anxiety. Developing a positive attitude and confidence in your abilities will help as well.

Poor performance on math exams can be caused by other factors, such as test anxiety, improper placement, and poor study skills. If you freeze when taking exams in other classes and do poorly on them even when you understand the material, you may have test anxiety. An academic counselor or learning resource specialist can help.

Exercises 8.5

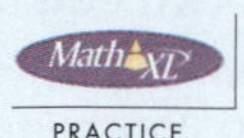
PRACTICE

WATCH

DOWNLOAD
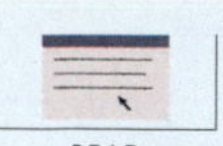
READ

REVIEW

Vocabulary

1. A(n) ____________ consists of two or more linear equations.
2. An ordered pair (x, y) is a solution to a system of two linear equations if ____________ ____________.
3. A system of two linear equations is a(n) ____________ system if it has exactly one solution.
4. A system of two linear equations is a(n) ____________ system if it has no solution.
5. A system of two linear equations is a(n) ____________ system if it has infinitely many solutions.
6. To solve a system of equations by substitution, begin by solving one of the equations for one of the ____________.
7. The solution of a system of three equations in three unknowns is a(n) ____________.
8. To solve a system of three equations in three unknowns, begin by eliminating one of the variables to create a system of ______ equations in ______ unknowns.

Solve by addition or substitution.

9. $y = 2x - 16$
$3x + 2y = 17$

10. $y = 3x - 7$
$4x + 5y = 136$

11. $2x - 7y = 41$
$x = 4y + 23$

12. $-5x + 3y = -54$
$x = 5y + 2$

13. $y = \frac{2}{3}x + 11$
$4x - 3y = -45$

14. $x = \frac{5}{2}y - 1$
$4x + 9y = 72$

15. $y = 3x + 7$
$6x - 2y = -14$

16. $x = 5 - 2y$
$4x + 8y = -20$

17. $2x - 3y = 10$
$4x - 2y = 28$

18. $x + 2y = -12$
$3x - 4y = 26$

19. $x - 4y = 7$
$3x + 2y = 14$

20. $5x + y = -2$
$3x - 7y = -43$

21. $4x + 3y = -28$
$3x - 2y = -4$

22. $-5x + 4y = 22$
$2x + 5y = 11$

23. $6x - 4y = 11$
$9x - 6y = 17$

24. $4x + y = 13$
$12x + 3y = 39$

25. $\frac{3}{4}x + \frac{1}{3}y = 1$
$\frac{2}{5}x - \frac{1}{4}y = \frac{31}{10}$

26. $\frac{1}{6}x + \frac{1}{2}y = -3$
$\frac{2}{3}x - \frac{3}{4}y = \frac{7}{4}$

Solve using a system of two equations in two unknowns.

27. The length of a rectangle is 3 inches more than twice its width. If the perimeter of the rectangle is 54 inches, find its length and width.

28. The length of a rectangle is 4 feet less than 3 times its width. If the perimeter of the rectangle is 96 feet, find its length and width.

29. A jar contains 65 coins; some are dimes, and the rest are quarters. If the coins have a value of $13.10, how many of each type of coin are in the jar?

30. A bouquet with 12 carnations and 5 roses costs $54, while a bouquet with 6 carnations and 4 roses costs $36. What is the cost of a single carnation? What is the cost of a single rose?

31. Faustino deposited a total of $7500 in two bank accounts. One of the accounts paid 6% annual interest, while the other paid 3% annual interest. If Faustino earned $387 in interest during the first year, how much did he deposit in each account?

32. Hasinah invested $6000 in two mutual funds. During the first year, one fund increased in value by 13% and the other increased in value by 5%. If Hasinah earned a profit of $660 during the first year, how much did he invest in each fund?

33. A chemist has two solutions: one is 15% acid, and the other is 30% acid. How much of each solution should she mix to create 60 milliliters of a solution that is 25% acid?

34. A chemist has two solutions: one is 20% acid, and the other is 60% acid. How much of each solution should he mix to produce 120 milliliters of a solution that is 28% acid?

Is the ordered triple a solution to the given system of equations?

35. (3, 2, 6)

$$\begin{aligned} x + y + z &= 11 \\ x - y + 2z &= 17 \\ 4x - 3y - 2z &= -6 \end{aligned}$$

36. (2, −4, −1)

$$\begin{aligned} x + y - z &= -1 \\ 5x + 2y + 2z &= 0 \\ \frac{1}{2}x + 2y - 6z &= -1 \end{aligned}$$

37. (−5, 2, 0)

$$\begin{aligned} x + 2y + 3z &= -1 \\ 3x - y - 6z &= -17 \\ 4x - 3y + z &= -26 \end{aligned}$$

38. (1, −1, −6)

$$\begin{aligned} x - y - z &= 8 \\ 5x + 7y + z &= -8 \\ -2x + 5y - 4z &= 21 \end{aligned}$$

Solve.

39.
$$\begin{aligned} x + 2y + 3z &= 23 \\ -2x + 3y + z &= 17 \\ -3x + 5y - 4z &= 12 \end{aligned}$$

40.
$$\begin{aligned} 4x - y + 6z &= 25 \\ 2x + 3y + 4z &= 49 \\ 5x + 2y - 3z &= 8 \end{aligned}$$

41.
$$\begin{aligned} x + y + z &= 2 \\ 2x + 3y - z &= 5 \\ 7x + 4y + 2z &= 25 \end{aligned}$$

42.
$$\begin{aligned} x + y - z &= -13 \\ 3x + 2y + 3z &= 6 \\ 4x + 9y + 12z &= 45 \end{aligned}$$

43.
$$\begin{aligned} 3x + 2y + 5z &= -8 \\ 9x - 4y + 6z &= -38 \\ -6x - 6y + 7z &= -62 \end{aligned}$$

44.
$$\begin{aligned} 4x + 3y + 2z &= 19 \\ 2x + 5y - 4z &= 18 \\ -8x + y + 6z &= 29 \end{aligned}$$

45.
$$\begin{aligned} 2x + 3y + z &= 5 \\ -2x - y + 4z &= 60 \\ x + 2y - 8z &= -30 \end{aligned}$$

46.
$$\begin{aligned} x - y - z &= 16 \\ 2x + 3y + 2z &= -5 \\ 4x + 5y + 4z &= -5 \end{aligned}$$

47.
$$\begin{aligned} 2x - 3y - z &= -6 \\ -3x + 2y + 4z &= 24 \\ 5x - 4y + 3z &= -18 \end{aligned}$$

48.
$$\begin{aligned} x + 4y + 2z &= -3 \\ 2x - 3y - z &= 19 \\ 4x + 9y + 5z &= 3 \end{aligned}$$

49.
$$\begin{aligned} 4x + 2y + z &= -5 \\ 2x - y - 6z &= 0 \\ -8x + 6y + 3z &= -30 \end{aligned}$$

50.
$$\begin{aligned} x + 3y - 6z &= 8 \\ 3x + 4y - 2z &= 9 \\ 5x + 6y - 12z &= 22 \end{aligned}$$

51.
$$\begin{aligned} \frac{1}{2}x + \frac{1}{3}y + z &= 3 \\ \frac{2}{3}x - 4y - 3z &= 13 \\ 3x + 4y + 5z &= 11 \end{aligned}$$

52.
$$\begin{aligned} \frac{3}{4}x - \frac{5}{2}y - \frac{1}{3}z &= -14 \\ x + \frac{3}{4}y + \frac{7}{2}z &= -26 \\ 2x - 3y - 4z &= -4 \end{aligned}$$

53.
$$\begin{aligned} x + 2y + z &= 8 \\ 3x + 4y + 2z &= 9 \\ 3x - 4y &= -37 \end{aligned}$$

54.
$$\begin{aligned} 2x + 3y - z &= 10 \\ -x + 4y + 2z &= 200 \\ 5x + 2y &= 88 \end{aligned}$$

Solve using a system of three equations in three unknowns.

55. Jonah opens his wallet and finds that all of the bills are \$1 bills, \$5 bills, or \$10 bills. There are 35 bills in his wallet. There are four more \$5 bills than \$1 bills. If the total amount of bills is \$246, how many \$10 bills does Jonah have?

56. A cash register has no \$1 bills in it. All of the bills are \$5 bills, \$10 bills, or \$20 bills. There are a total of 37 bills in the cash register, and the total value of these bills is \$310. If the number of \$10 bills in the register is twice the number of \$20 bills, how many \$5 bills are in the register?

57. A campus club held a bake sale as a fund-raiser, selling coffee, muffins, and bacon-and-egg sandwiches. The club members charged \$1 for a cup of coffee, \$2 for a muffin, and \$3 for a bacon-and-egg sandwich. They sold a total of 50 items, raising \$85 dollars. If the club members sold five more muffins than cups of coffee, how many bacon-and-egg sandwiches did they sell?

58. A grocery store was supporting a local charity by asking shoppers to donate \$1, \$5, or \$20 at the checkout. A total of 110 shoppers donated \$430. The amount of money raised from \$5 donations was \$50 higher than the amount of money raised from \$20 donations. How many shoppers donated \$20?

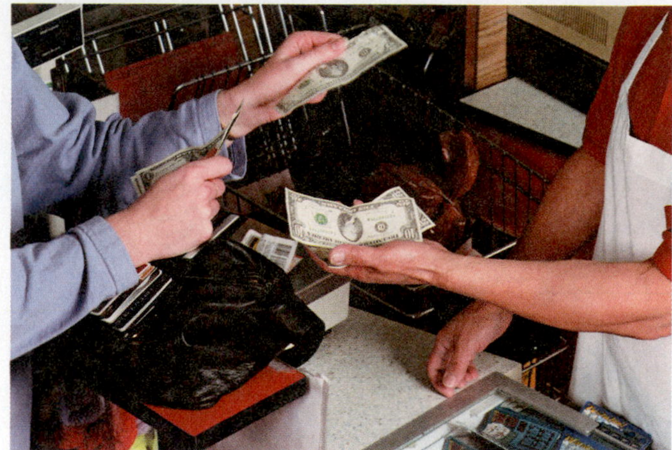

59. Abraham is 8 years older than Belen. The sum of Belen's age and Celeste's age is three years more than twice Abraham's age. If the sum of their three ages is 138, how old are Abraham, Belen, and Celeste?

60. The total of Dharma's, Eugenio's, and Fern's final exam scores is 271. If Dharma's score is added to Eugenio's score, the sum is 11 points less than twice Fern's score. If Eugenio's score is 5 points higher than Dharma's score, what is each student's final exam score?

61. The measures of the three angles in a triangle must total 180°. The measure of angle A is 20° less than the measure of angle C. The measure of angle C is twice the measure of angle B. Find the measure of each angle.

62. The measures of the three angles in a triangle must total 180°. The measure of angle A is 15° more than the measure of angle B. The measure of angle B is 15° greater than four times the measure of angle C. Find the measure of each angle.

63. Recall the introduction to problems about interest in Section 4.4. Mary invested a total of \$40,000 in three different bank accounts. One account pays an annual interest rate of 3%, the second account pays 5% annual interest, and the third account pays 6% annual interest. In one year, Mary earned a total of \$1960 in interest from these three accounts. If Mary invested \$8000 more in the account that pays 5% interest than she did in the account that pays 6% interest, find the amount she invested in each account.

64. Hua-Ling invested a total of \$10,000 in three bank accounts. The accounts paid 4%, 5%, and 6% annual interest. In the first year, Hua-Ling earned \$545 in interest. If she invested \$1000 more at 5% than she invested at 4%, how much did she invest in each account?

65. Pilar invested a total of \$8000 in three mutual funds. During the first year, the first fund increased in value by 2%, the second fund increased by 10%, and the third fund increased by 25%. In the first year, she earned a profit of \$550. If Pilar invested \$1000 less in the fund that increased by 25% than she invested in the fund that increased by 10%, how much did she invest in each fund?

66. Lev invested a total of \$5000 in three stocks. During the first year, the first stock increased in value by 8%, the second stock increased by 6%, and the third stock increased by 1%. In the first year, he earned a profit of \$310. If Lev invested the same amount in the stock that increased by 6% as he invested in the stock that increased by 1%, how much did he invest in each stock?

67. Salim invested a total of \$25,000 in three stocks. During the first year, the first stock increased in value by 5% and the second stock increased by 7%, but the third stock decreased by 40%. In the first year, he lost \$1330. If Salim invested the same amount in the stock that increased by 7% as he did in the stock that decreased by 40%, how much did he invest in each stock?

68. Dennis invested a total of \$100,000 in three different mutual funds. After one year, the first fund showed a profit of 8%, the second fund showed a loss of 5%, and the third fund showed a profit of 2%. Dennis invested \$10,000 more in the fund that lost 5% than he invested in the fund that made a 2% profit. In the first year, Dennis made a profit of \$2900 from the three funds. How much did he invest in each fund?

69. A chemist wants to mix three different solutions to create 100 milliliters of a solution that is 13.5% alcohol. Solution A is 10% alcohol, solution B is 15% alcohol, and solution C is 30% alcohol. The amount of solution A that is used must be twice the amount of solution B that is used. How many milliliters of each solution should the chemist combine?

70. A chemist wants to mix three different solutions to create 50 milliliters of a solution that is 28% alcohol. Solution A is 20% alcohol, solution B is 30% alcohol, and solution C is 50% alcohol. The amount of solution A that is used must be 20 milliliters more than the amount of solution B that is used. How many milliliters of each solution should the chemist combine?

Writing in Mathematics

Answer in complete sentences.

71. Write a word problem that has the following solution: There were 40 children, 200 adults, and 60 senior citizens in attendance. Your problem must lead to a system of three equations in three unknowns. Explain how you created your problem.

72. Write a word problem for the following system of equations.

$$x + y + z = 350{,}000$$
$$0.05x + 0.07y - 0.04z = 17{,}000$$
$$y = 2x$$

Explain how you created your problem.

CHAPTER 8 SUMMARY

Section 8.1 Linear Equations and Absolute Value Equations

Solving Linear Equations, pp. 419–420

1. Simplify each side of the equation completely.
   - Use the distributive property to clear any parentheses.
   - If there are fractions in the equation, multiply both sides of the equation by the LCM of the denominators to clear the fractions from the equation.
   - Combine any like terms that are on the same side of the equation.
2. Collect all variable terms on one side of the equation.
3. Collect all constant terms on the other side of the equation.
4. Divide both sides of the equation by the coefficient of the variable term.
5. Check your solution.

Solve $5x - 7 = 3x + 11$.

$$\begin{aligned} 2x - 7 &= 11 \\ 2x &= 18 \\ x &= 9 \end{aligned}$$

$\{9\}$

Identities, p. 421

An identity is an equation that is always true. The solution set for an identity is the set of all real numbers, denoted $\mathbb{R}$. An identity has infinitely many solutions.

Solve $4x + 18 - 11 = 2(x + 7) + 2x - 7$.

$$\begin{aligned} 4x + 7 &= 2x + 14 + 2x - 7 \\ 4x + 7 &= 4x + 7 \\ 7 &= 7 \end{aligned}$$

$\mathbb{R}$

Contradictions, p. 421

A contradiction is an equation that has no solution, so its solution set is the empty set { }.

The empty set also is known as the null set and is denoted by the symbol $\varnothing$.

Solve $3(x - 4) + 5 = 3x + 7$.

$$\begin{aligned} 3x - 12 + 5 &= 3x + 7 \\ 3x - 7 &= 3x + 7 \\ -7 &= 7 \quad \text{(False)} \end{aligned}$$

$\varnothing$

Solving Absolute Value Equations, pp. 422–423

For any expression X and any positive number a, the solutions to the equation $|X| = a$ can be found by solving the two equations $X = a$ and $X = -a$.

Solve $|2x - 7| - 5 = 14$.

$$|2x - 7| = 19$$

$$\begin{aligned} 2x - 7 &= 19 \quad \text{or} \quad & 2x - 7 &= -19 \\ 2x &= 26 & 2x &= -12 \\ x &= 13 & x &= -6 \end{aligned}$$

$\{-6, 13\}$

Solving Absolute Value Equations Involving Two Absolute Values, p. 424

For any expressions X and Y, the solutions to the equation $|X| = |Y|$ can be found by solving the two equations $X = Y$ and $X = -Y$.

Solve $|2x + 5| = |x - 7|$.

$$\begin{aligned} 2x + 5 &= x - 7 \quad \text{or} \quad & 2x + 5 &= -(x - 7) \\ x + 5 &= -7 & 2x + 5 &= -x + 7 \\ x &= -12 & 3x &= 2 \\ & & x &= \frac{2}{3} \end{aligned}$$

$\left\{-12, \frac{2}{3}\right\}$

Section 8.2 Linear Inequalities and Absolute Value Inequalities

Solving Linear Inequalities, pp. 427–428

Solving a linear inequality is similar to solving a linear equation except that multiplying or dividing both sides of an inequality by a negative number changes the direction of the inequality.

Solve $5x - 4 \geq 26$.

$$\begin{aligned} 5x - 4 &\geq 26 \\ 5x &\geq 30 \\ x &\geq 6 \end{aligned}$$

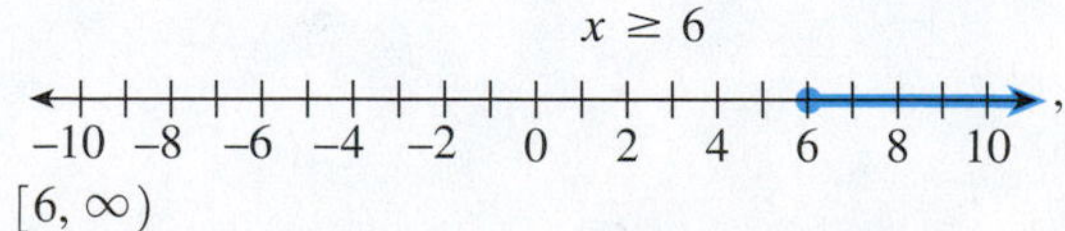

$[6, \infty)$

Compound Inequalities, pp. 428–429

A compound inequality is made up of two or more individual inequalities.

Solve $2x + 11 < -3$ or $3x - 10 > 14$.

$$2x + 11 < -3 \qquad 3x - 10 > 14$$
$$2x < -14 \qquad 3x > 24$$
$$x < -7 \qquad x > 8$$

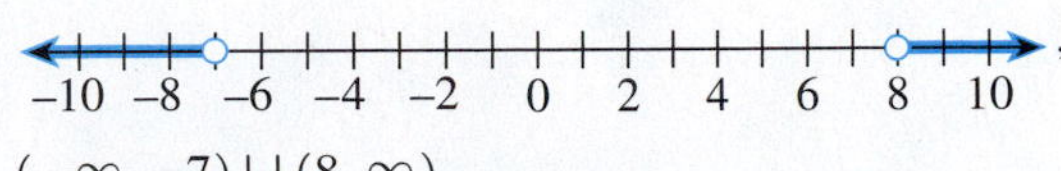

$(-\infty, -7) \cup (8, \infty)$

Solve $9 \le 8x + 17 \le 57$.

$$9 \le 8x + 17 \le 57$$
$$-8 \le 8x \le 40$$
$$-1 \le x \le 5$$

–10 –8 –6 –4 –2 0 2 4 6 8 10

$[-1, 5]$

Solving Absolute Value Inequalities, pp. 429–431

- For any expression X and any nonnegative number a, the solutions to the inequality $|X| < a$ can be found by solving the compound inequality $-a < X < a$.
- For any expression X and any nonnegative number a, the solutions to the inequality $|X| \le a$ can be found by solving the compound inequality $-a \le X \le a$.

Solve $|x - 4| < 5$.

$$-5 < x - 4 < 5$$
$$-1 < x < 9$$

–10 –8 –6 –4 –2 0 2 4 6 8 10

$(-1, 9)$

Solving Absolute Value Inequalities, pp. 431–433

- For any expression X and any nonnegative number a, the solutions to the inequality $|X| > a$ can be found by solving the compound inequality $X < -a$ or $X > a$.
- For any expression X and any nonnegative number a, the solutions to the inequality $|X| \ge a$ can be found by solving the compound inequality $X \le -a$ or $X \ge a$.

Solve $|x + 3| + 2 \ge 7$.

$$|x + 3| + 2 \ge 7$$
$$|x + 3| \ge 5$$

$$x + 3 \le -5 \quad \text{or} \quad x + 3 \ge 5$$
$$x \le -8 \qquad x \ge 2$$

–10 –8 –6 –4 –2 0 2 4 6 8 10

$(-\infty, -8] \cup [2, \infty)$

Section 8.3 Graphing Linear Equations and Linear Functions; Graphing Absolute Value Functions

Graphing a Linear Equation Using Its Intercepts, pp. 436–437

Find the x-intercept and the y-intercept and plot them on the plane. A third point with an arbitrarily chosen value of x can be used as a check point.

Graph $3x - 2y = -12$.

x-intercept	y-intercept
$3x - 2(0) = -12$ $3x = -12$ $x = -4$	$3(0) - 2y = -12$ $-2y = -12$ $y = 6$
$(-4, 0)$	$(0, 6)$

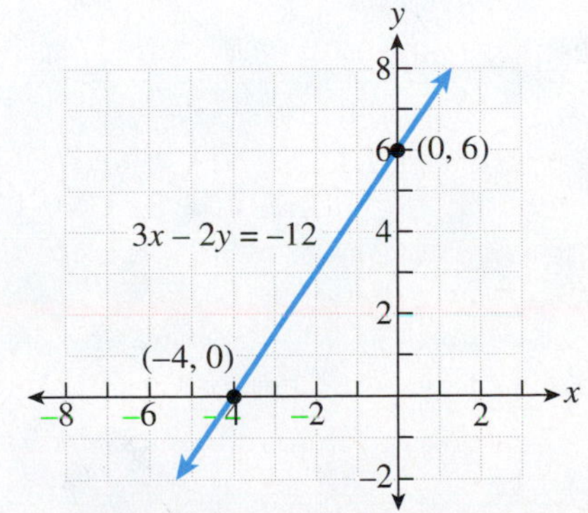

Graphing a Line Using Its Slope and y-Intercept, pp. 437–438

1. Plot the y-intercept.
2. Use the slope of the line to find a second point.
3. Draw a line that passes through these two points.

Graph $y = -3x + 7$.

y-intercept: $(0, 7)$
Slope: -3
(down 3 units,
1 to the right)

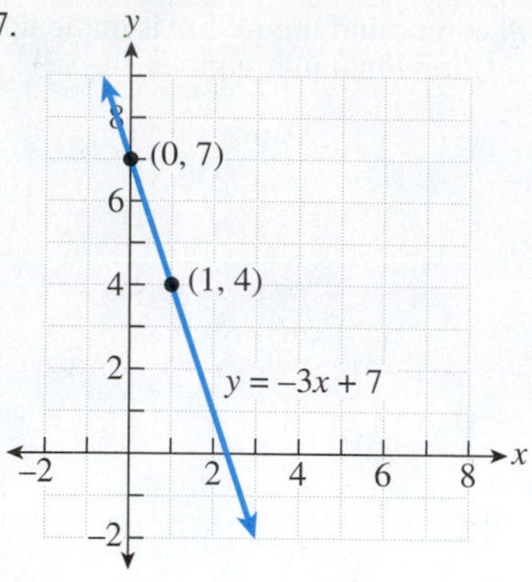

Graphing Linear Functions, p. 438

A linear function is a function of the form $f(x) = mx + b$, where m and b are real numbers.
To graph a linear function, begin by plotting its y-intercept at $(0, b)$. Then use the slope m to find other points.

Graph $f(x) = \frac{3}{2}x - 6$.

y-intercept:
$(0, -6)$
Slope: $\frac{3}{2}$
(up 3 units,
2 to the right)

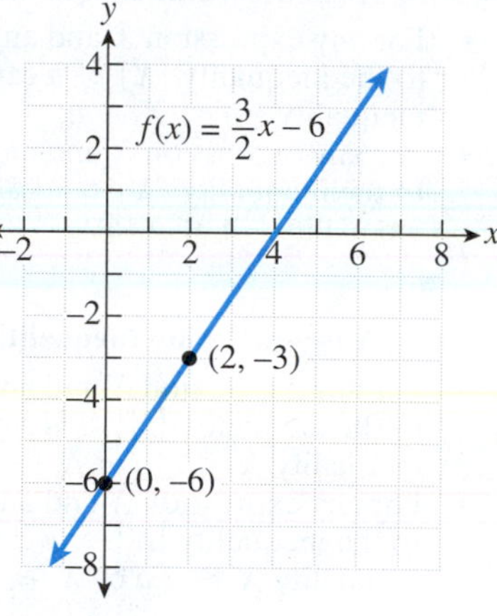

Graphing an Absolute Value Function, pp. 439–442

1. Determine the value of x for which the expression inside the absolute value bars is equal to 0.
2. In addition to this value, select two values that are less than this value and two that are higher.
3. Create a table of function values for these values of x.
4. Place the points $(x, f(x))$ on the graph and draw the V-shaped graph that passes through these points.

Graph $f(x) = |x - 2| + 3$.

$$x - 2 = 0$$
$$x = 2$$

x	$\lvert x - 2\rvert + 3$
0	$\lvert 0 - 2\rvert + 3 = \lvert -2\rvert + 3 = 2 + 3 = 5$
1	$\lvert 1 - 2\rvert + 3 = \lvert -1\rvert + 3 = 1 + 3 = 4$
2	$\lvert 2 - 2\rvert + 3 = \lvert 0\rvert + 3 = 0 + 3 = 3$
3	$\lvert 3 - 2\rvert + 3 = \lvert 1\rvert + 3 = 1 + 3 = 4$
4	$\lvert 4 - 2\rvert + 3 = \lvert 2\rvert + 3 = 2 + 3 = 5$

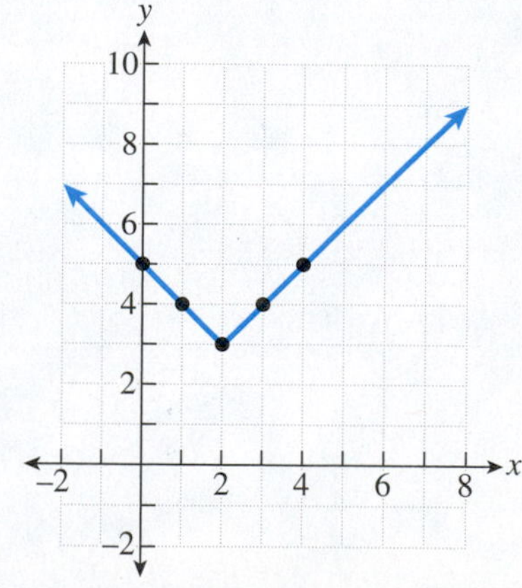

Domain: $(-\infty, \infty)$; range: $[3, \infty)$

Section 8.4 Review of Factoring; Quadratic Equations and Rational Equations

Factoring Polynomials, pp. 450–453

1. Factor out any common factors.
2. Determine the number of terms in the polynomial.
   a. If there are *four terms*, try factoring by grouping.
   b. If there are *three terms*, try to use one of the following techniques:
      - $x^2 + bx + c = (x + m)(x + n)$: Find two integers m and n whose product is c and whose sum is b.
      - $ax^2 + bx + c\ (a \neq 1)$: Factor by grouping or by trial and error.
   c. If there are only *two terms*, check to see if the binomial is one of the special binomials.
      - Difference of Squares: $a^2 - b^2 = (a + b)(a - b)$
      - Sum of Squares: $a^2 + b^2$ is not factorable.
      - Difference of Cubes: $a^3 - b^3 = (a - b)(a^2 + ab + b^2)$
      - Sum of Cubes: $a^3 + b^3 = (a + b)(a^2 - ab + b^2)$
3. After the polynomial has been factored, make sure any factor with two or more terms does not have a common factor other than 1. If there are common factors, factor them out.
4. Check your factoring through multiplication.

Factor: $x^2 - 81$ (Difference of Squares)

$$\begin{aligned} x^2 - 81 &= x^2 - 9^2 \\ &= (x + 9)(x - 9) \end{aligned}$$

Factor: $x^3 + 125$ (Sum of Cubes)

$$\begin{aligned} x^3 + 125 &= x^3 + 5^3 \\ &= (x + 5)(x^2 - 5x + 25) \end{aligned}$$

Factor: $x^2 - 17x + 72$
(Trinomial, Leading Coefficient of 1)

$$x^2 - 17x + 72 = (x - 8)(x - 9)$$

Factor: $3x^2 - 2x - 21$
(Trinomial, Leading Coefficient Other Than 1)

$$\begin{aligned} 3x^2 - 2x - 21 &= 3x^2 - 9x + 7x - 21 \\ &= 3x(x - 3) + 7(x - 3) \\ &= (x - 3)(3x + 7) \end{aligned}$$

Solving Quadratic Equations by Factoring, p. 454

1. Write the equation in standard form: $ax^2 + bx + c = 0$
2. Factor the expression $ax^2 + bx + c$ completely.
3. Set each factor equal to 0 and solve the resulting equations.
4. Check your solutions.

Solve $x^2 + 7x = 2x + 36$.

$$\begin{aligned} x^2 + 7x &= 2x + 36 \\ x^2 + 5x - 36 &= 0 \\ (x + 9)(x - 4) &= 0 \end{aligned}$$

$$x + 9 = 0 \quad \text{or} \quad x - 4 = 0$$
$$x = -9 \qquad\qquad x = 4$$

$\{-9, 4\}$

Solving Rational Equations, pp. 455–456

Find the LCD of all rational expressions in the equation and multiply both sides of the equation by that LCD. Solve the resulting equation. Check for extraneous solutions.

Solve $\dfrac{x + 8}{x^2 - 10x + 24} = \dfrac{6}{x^2 - 9x + 20}$.

$$\frac{x + 8}{(x - 4)(x - 6)} = \frac{6}{(x - 4)(x - 5)}$$

LCD: $(x - 4)(x - 6)(x - 5)$

$$\frac{x + 8}{(x - 4)(x - 6)} \cdot (x - 4)(x - 6)(x - 5)$$
$$= \frac{6}{(x - 4)(x - 5)} \cdot (x - 4)(x - 6)(x - 5)$$

$$\begin{aligned} (x + 8)(x - 5) &= 6(x - 6) \\ x^2 + 3x - 40 &= 6x - 36 \\ x^2 - 3x - 4 &= 0 \\ (x + 1)(x - 4) &= 0 \end{aligned}$$

$$x + 1 = 0 \quad \text{or} \quad x - 4 = 0$$
$$x = -1 \qquad\qquad x = 4$$

Extraneous Solution

$\{-1\}$

Section 8.5 Systems of Equations (Two Equations in Two Unknowns and Three Equations in Three Unknowns)

Solving a System of Equations Using the Substitution Method, pp. 459–460

1. Solve one of the equations for either variable.
2. Substitute this expression for the variable in the other equation.
3. Solve this equation.
4. Substitute this value for the variable into the equation from Step 1.
5. Write the solution as an ordered pair.
6. Check your solution.

Solve: $\begin{aligned} y &= 2x - 9 \\ 3x - 4y &= 11 \end{aligned}$

$$\begin{aligned} 3x - 4(2x - 9) &= 11 \\ 3x - 8x + 36 &= 11 \\ -5x + 36 &= 11 \\ -5x &= -25 \\ x &= 5 \end{aligned} \qquad \begin{aligned} y &= 2(5) - 9 \\ y &= 1 \end{aligned}$$

$(5, 1)$

Solving a System of Equations Using the Addition Method, p. 460

1. Write each equation in standard form $(Ax + By = C)$.
2. Multiply one or both equations by the appropriate constant(s).
3. Add the two equations.
4. Solve the resulting equation.
5. Substitute this value for the appropriate variable in either of the original equations and solve for the other variable.
6. Write the solution as an ordered pair.
7. Check your solution.

Solve:
$$\begin{aligned} -3x + 4y &= 2 \\ 5x + 3y &= 45 \end{aligned}$$

$$-3x + 4y = 2 \xrightarrow{\text{Multiply by 5.}} -15x + 20y = 10$$
$$5x + 3y = 45 \xrightarrow{\text{Multiply by 3.}} \underline{15x + 9y = 135}$$
$$29y = 145$$

$$\begin{aligned} 29y &= 145 \\ y &= 5 \end{aligned} \qquad \begin{aligned} 5x + 3(5) &= 45 \\ 5x + 15 &= 45 \\ 5x &= 30 \\ x &= 6 \end{aligned}$$

$(6, 5)$

Solving Systems of Three Equations in Three Unknowns, pp. 462–466

1. Select two of the equations and use the addition method to eliminate one of the variables.
2. Select a different pair of equations and use the addition method to eliminate the same variable, leaving two equations in two unknowns.
3. Solve the system of two equations for the two unknowns.
4. Substitute these two values in any of the original three equations and solve for the unknown variable.

The solution to the system of equations will be an ordered triple in the form (x, y, z).

Solve:
$$\begin{aligned} x + y + z &= 6 \\ 3x - y + 4z &= -1 \\ 2x - 3y - 6z &= 16 \end{aligned}$$

Two equations in terms of x and z:

$$\begin{aligned} x + y + z &= 6 \\ \underline{3x - y + 4z} &\underline{= -1} \\ 4x + 5z &= 5 \end{aligned}$$

$$x + y + z = 6 \xrightarrow{\text{Multiply by 3.}} 3x + 3y + 3z = 18$$
$$2x - 3y - 6z = 16 \qquad \underline{2x - 3y - 6z = 16}$$
$$5x - 3z = 34$$

Solve for x:

$$4x + 5z = 5 \xrightarrow{\text{Multiply by 3.}} 12x + 15z = 15$$
$$5x - 3z = 34 \xrightarrow{\text{Multiply by 5.}} \underline{25x - 15z = 170}$$
$$37x = 185$$
$$x = 5$$

Back-substitute:

Find z:
$$x = 5 \qquad \begin{aligned} 4(5) + 5z &= 5 \\ 20 + 5z &= 5 \\ 5z &= -15 \\ z &= -3 \end{aligned}$$

Find y:
$$\begin{aligned} 5 + y + (-3) &= 6 \\ y + 2 &= 6 \\ y &= 4 \end{aligned}$$

$(5, 4, -3)$

SUMMARY OF CHAPTER 8 STUDY STRATEGIES

The study tips in this chapter summarized the ideas presented in the first seven chapters of the text. For further information on a particular topic, refer back to the appropriate chapter.

Chapter 1—Study Groups
Chapter 2—Using Your Textbook
Chapter 3—Making the Best Use of Your Resources
Chapter 4—Doing Your Homework
Chapter 5—Test Taking
Chapter 6—Overcoming Math Anxiety
Chapter 7—Preparing for a Cumulative Exam
Chapter 8—Review

CHAPTER 8 REVIEW

Solve. **[8.1]**

1. $2x - 17 = -5$

2. $3x - 9 = 17 - 5x$

3. $\frac{1}{3}x - \frac{1}{6} = \frac{2}{15}x + \frac{1}{3}$

4. $4x + 2(3x - 1) = 17x - 3(4x - 11)$

Solve. **[8.1]**

5. $|x| = 6$

6. $|x| = -6$

7. $|x + 7| - 14 = -6$

8. $|3x + 13| + 20 = 28$

9. $|6x + 25| = |x + 10|$

10. $|4x - 3| = |2x - 15|$

Solve the inequality. Graph your solution on a number line and write your solution in interval notation. **[8.2]**

11. $3x + 16 < 10$

12. $-5x + 19 \geq -16$

13. $2x - 13 \leq 4x - 31$

14. $13 \leq 2x - 21 \leq 18$

15. $-14 < 3x + 10 < 37$

16. $6x + 13 < 25$ or $2x - 19 \geq -7$

Solve the inequality. Graph your solution on a number line and write your solution in interval notation. **[8.2]**

17. $|x| \leq 10$

18. $|3x - 11| < 7$

19. $|2x + 1| - 9 < 6$

20. $|x| + 4 > 13$

21. $|6x + 21| - 10 \geq 5$

22. $|x - 8| + 13 \geq 8$

Find the intercepts and graph the line. **[8.3]**

23. $5x + 6y = 30$

24. $8x - 3y = -12$

Graph the line using the slope and y-intercept. **[8.3]**

25. $y = -2x + 9$

26. $y = \frac{2}{3}x + 6$

Worked-out solutions to Review Exercises marked with ● can be found on page AN-28.

Graph the linear function. **[8.3]**

27. $f(x) = -x + 5$

28. $f(x) = 3x$

29. $f(x) = \frac{3}{5}x - 6$

30. $f(x) = -\frac{1}{6}x + 1$

Graph the absolute value function. State the domain and range of the function. **[8.3]**

31. $f(x) = |x| + 3$

32. $f(x) = |x - 5|$

33. $f(x) = |x + 2| - 4$

34. $f(x) = -|x - 1| + 2$

Factor completely. **[8.4]**

35. $6x^9 - 4x^7 + 10x^3 - 18x^2 + 2x$

36. $5a^4b^3 - 10a^2b^5 - 25a^3b$

37. $x^3 - 8x^2 + 11x - 88$

38. $x^3 + 9x^2 - 4x - 36$

39. $x^2 - 5x - 14$

40. $3x^2 - 33x + 72$

41. $x^2 + 18xy + 80y^2$

42. $x^2 + 13x - 30$

43. $4x^2 + 20x - 24$

44. $x^2 - 3x - 54$

45. $6x^2 - 29x + 9$

46. $4x^2 - 8x - 21$

47. $x^2 - 36$

48. $9x^2 - 1$

49. $4x^2 - 49$

50. $25x^2 - 64y^2$

51. $x^3 + 27$

52. $x^3 - 729$

Solve. **[8.4]**

53. $x^2 - x - 6 = 0$

54. $x^2 + 10x + 24 = 0$

55. $x^2 + 13x + 25 = 6x + 43$

56. $x^2 = 144$

Solve. **[8.4]**

57. $\frac{9}{x + 2} = \frac{3}{x - 8}$

58. $x - 5 - \frac{24}{x} = 0$

59. $\frac{4}{x + 5} + \frac{1}{x - 4} = \frac{9}{x^2 + x - 20}$

60. $\frac{x - 2}{x + 3} - \frac{3}{x + 6} = \frac{9}{x^2 + 9x + 18}$

Solve. [8.5]

61. $2x + y = 7$
$3x + 4y = 8$

62. $x = 7 - 4y$
$6x - 5y = -45$

63. $2x + 3y = 23$
$-6x + 2y = 8$

64. $3x + 5y = 19$
$5x - 4y = -67$

Solve. [8.5]

65. $x - 4y + 2z = -15$
$2x + 3y = 22$
$y = 4$

66. $x + y - z = -9$
$3x + 5y + 6z = -7$
$4y + 7z = -4$

67. $x + y + z = 0$
$2x + 3y - z = 24$
$3x - 2y + 4z = -36$

68. $x + 3y - 4z = -13$
$-3x + 9y + 8z = -5$
$-5x + 6y + 6z = -19$

69. A golf course charges \$8 for children under age 13 to play a round of golf. Senior citizens are charged \$10 to play a round. Everyone else pays \$15 per round. Yesterday 200 people played a round of golf at the course, paying a total of \$2540. If there were 20 more senior citizens who played than there were children under 13 years old who played, find the number of senior citizens who played at the course. [8.5]

70. The measures of the three angles in a triangle must total 180°. The measure of angle A is 24° more than the measure of angle B. The sum of the measures of angles A and B is twice the measure of angle C. Find the measure of each angle. [8.5]

CHAPTER 8 TEST

For Extra Help
CHAPTER Test Prep VIDEOS

Step-by-step test solutions are found on the Chapter Test Prep Videos available via the Video Resources on DVD, in *MyMathLab*, and on YouTube (search "WoodburyElemIntAlg" and click on "Channels").

Solve.

1. $6(2x - 5) + 8 = 5(3x + 7) - 9$

Solve.

2. $|x + 9| = 7$

3. $|x - 5| - 11 = -3$

Solve the inequality. Graph your solution on a number line and write your solution in interval notation.

4. $-5x < 3x + 32$

5. $-47 \leq 6x - 11 \leq 19$

6. $2x + 15 < 21$ or $3x - 5 > 10$

Solve the inequality. Graph your solution on a number line and write your solution in interval notation.

7. $|x + 4| \leq 5$

8. $|2x - 5| + 8 > 13$

Graph the line.

9. $4x - 3y = 6$

Graph the linear function.

10. $f(x) = x - 2$

Graph the absolute value function. State the domain and range of the function.

11. $f(x) = |x - 6| + 3$

Factor completely.

12. $x^2 + 3x - 40$

13. $4x^2 - 28x + 40$

14. $6x^2 + 19x + 10$

15. $9x^2 - 100$

Solve.

16. $x^2 + 6x - 27 = 0$

Solve.

17. $\dfrac{x + 7}{x^2 + 5x + 4} + \dfrac{6}{x^2 - 3x - 4} = \dfrac{3}{x^2 - 16}$

Solve.

18. $\begin{aligned} 5x + y &= -11 \\ 4x + 3y &= 0 \end{aligned}$

Solve.

19. $\begin{aligned} x + y + z &= 11 \\ 3x - 2y + 4z &= 24 \\ -2x + 3y - 5z &= -15 \end{aligned}$

20. The measures of the three angles in a triangle must total 180°. The measure of angle B is 25° more than the measure of angle C. The measure of angle A is 25° more than twice the measure of angle B. Find the measure of each angle.

Mathematicians in History

Pierre de Fermat was a French mathematician who lived in the 17th century. His work focused on number theory and probability. Although he did little publishing in his lifetime, he did pose many problems as challenges to the mathematical community of Europe.

Write a one-page summary (*or* make a poster) of the life of Pierre de Fermat and his accomplishments.

Interesting issues:

- Where and when was Pierre de Fermat born?
- What is number theory?
- Although Fermat lived until 1665, his health was so bad that it was believed that he had died in 1653. What was the cause of his near death?
- The Fronde greatly disrupted Fermat's communication with the math community in 1648. What was the Fronde?
- Fermat is best known today for Fermat's Last Theorem. What is Fermat's Last Theorem? Who finally proved this theorem in November 1994?
- Fermat had a public feud with which well-known mathematician? Describe what the feud was about and how it was resolved.

CHAPTER 9

Radical Expressions and Equations

In this chapter, we will investigate radical expressions and equations and their applications. Among the applications is the method for finding the distance between any two objects and a way to determine the speed a car was traveling by measuring the skid marks left by its tires.

STUDY STRATEGY

Note Taking If you have poor note-taking skills, you will not learn as much during the class period as you might if your skills were better. In this chapter, we will focus on how to take notes in a math class. We will discuss how to be more efficient when taking notes, what material to include in your notes, and how to rework your notes.

9.1 Square Roots; Radical Notation

OBJECTIVES

1. Find the square root of a number.
2. Simplify the square root of a variable expression.
3. Approximate the square root of a number by using a calculator.
4. Find *n*th roots.
5. Multiply radical expressions.
6. Divide radical expressions.
7. Evaluate radical functions.
8. Find the domain of a radical function.

Square Roots

Consider the equation $x^2 = 36$. There are two solutions to this equation: $x = 6$ and $x = -6$.

Square Root

A number a is a **square root** of a number b if $a^2 = b$.

The numbers 6 and -6 are square roots of 36 because $6^2 = 36$ and $(-6)^2 = 36$. The number 6 is the positive square root of 36, while the number -6 is the negative square root of 36.

Objective 1 Find the square root of a number.

Principal Square Root

The **principal square root** of b, denoted $\sqrt{b}$, for $b > 0$, is the positive number a such that $a^2 = b$.

The expression $\sqrt{b}$ is called a **radical expression**. The sign $\sqrt{}$ is called a **radical sign**, while the expression contained inside the radical sign is called the **radicand**.

EXAMPLE 1 Simplify $\sqrt{25}$.

SOLUTION We are looking for a positive number a such that $a^2 = 25$. The number is 5, so $\sqrt{25} = 5$.

EXAMPLE 2 Simplify $\sqrt{\dfrac{1}{9}}$.

SOLUTION Because $\left(\dfrac{1}{3}\right)^2 = \dfrac{1}{3} \cdot \dfrac{1}{3} = \dfrac{1}{9}$, $\sqrt{\dfrac{1}{9}} = \dfrac{1}{3}$.

EXAMPLE 3 Simplify $-\sqrt{49}$.

SOLUTION In this example, we are looking for the negative square root of 49, which is -7. So $-\sqrt{49} = -7$. We can find the principal square root of 49 first and then make it negative.

Quick Check 1

Simplify.

a) $\sqrt{49}$ **b)** $\sqrt{\dfrac{4}{25}}$

c) $-\sqrt{36}$

The principal square root of a negative number, such as $\sqrt{-49}$, is not a real number because there is no real number a such that $a^2 = -49$. We will learn in Section 9.6 that $\sqrt{-49}$ is called an *imaginary* number.

Square Roots of Variable Expressions

Objective 2 Simplify the square root of a variable expression. Now we turn our attention to simplifying radical expressions containing variables, such as $\sqrt{x^6}$.

Simplifying $\sqrt{a^2}$

For any real number a, $\sqrt{a^2} = |a|$.

You may be wondering why the absolute value bars are necessary. For any nonnegative number a, $\sqrt{a^2} = a$. For example, $\sqrt{8^2} = 8$. The absolute value bars are necessary in order to include negative values of a. Suppose $a = -10$. Then

$\sqrt{a^2} = \sqrt{(-10)^2}$, which is equal to $\sqrt{100}$, or 10. So $\sqrt{a^2}$ equals the opposite of a. In either case, the principal square root of a^2 will be a nonnegative number.

$$\sqrt{a^2} = \begin{cases} a & \text{if } a \geq 0 \\ -a & \text{if } a < 0 \end{cases}$$

The absolute value bars address both cases. We must use absolute values when dealing with variables because we do not know whether the variable is negative.

EXAMPLE 4 Simplify $\sqrt{x^6}$.

SOLUTION We begin by rewriting the radicand as a square. Note that $x^6 = (x^3)^2$.

$$\begin{aligned} \sqrt{x^6} &= \sqrt{(x^3)^2} && \text{Rewrite the radicand as a square.} \\ &= |x^3| && \text{Simplify.} \end{aligned}$$

Because we do not know whether x is negative, the absolute value bars are necessary.

EXAMPLE 5 Simplify $\sqrt{64y^{10}}$.

SOLUTION Again, we rewrite the radicand as a square.

$$\begin{aligned} \sqrt{64y^{10}} &= \sqrt{(8y^5)^2} && \text{Rewrite the radicand as a square.} \\ &= |8y^5| && \text{Simplify.} \end{aligned}$$

Because we know that 8 is a positive number, we can remove it from the absolute value bars. This allows us to write the expression as $8|y^5|$.

Quick Check 2

Simplify.

a) $\sqrt{a^{14}}$ **b)** $\sqrt{25b^{22}}$

EXAMPLE 6 Simplify $\sqrt{36x^4}$.

SOLUTION We rewrite the radicand as a square.

$$\begin{aligned} \sqrt{36x^4} &= \sqrt{(6x^2)^2} && \text{Rewrite the radicand as a square.} \\ &= |6x^2| && \text{Simplify.} \end{aligned}$$

Because we know that 6 is a positive number, it can be removed from the absolute value bars. Because x^2 cannot be negative either, it may be removed from the absolute value bars as well.

$$\sqrt{36x^4} = 6x^2$$

Quick Check 3

Simplify $\sqrt{64x^{24}}$.

From this point on, we will assume that all variable factors in a radicand represent nonnegative real numbers. This eliminates the need to use absolute value bars when simplifying radical expressions whose radicand contains variables.

Approximating Square Roots Using a Calculator

Objective 3 Approximate the square root of a number by using a calculator. Consider the expression $\sqrt{12}$. There is no positive integer that is the square root of 12. In such a case, we can use a calculator to approximate the radical expression. All calculators have a function for calculating square roots. Rounding to the nearest thousandth, we see that $\sqrt{12} \approx 3.464$. The symbol $\approx$ is read as *is approximately equal to;* so the principal square root of 12 is approximately equal to 3.464. If we square 3.464, it is equal to 11.999296, which is very close to 12.

EXAMPLE 7 Approximate $\sqrt{42}$ to the nearest thousandth using a calculator.

SOLUTION Because $6^2 = 36$ and $7^2 = 49$, we know that $\sqrt{42}$ must be a number between 6 and 7.

$$\sqrt{42} \approx 6.481$$

Quick Check 4

Approximate $\sqrt{109}$ to the nearest thousandth using a calculator.

Using Your Calculator We can approximate square roots by using the TI-84.

```
√(42)
       6.480740698
```

*n*th Roots

Objective 4 Find *n*th roots. Now we move on to discuss roots other than square roots.

Principal *n*th Root

For any positive integer $n > 1$ and any number b, if $a^n = b$ and both a and b have the same sign, a is the **principal *n*th root** of b, denoted $a = \sqrt[n]{b}$. The number n is called the **index** of the radical.

A square root has an index of 2, and the radical is written without the index. If the index is 3, this is called a **cube root**. If n is even, the principal nth root is a nonnegative number, but if n is odd, the principal nth root has the same sign as the radicand. As with square roots, an even root of a negative number is not a real number. For example, $\sqrt[6]{-64}$ is not a real number. However, an odd root of a negative number, such as $\sqrt[3]{-64}$, is a negative real number.

EXAMPLE 8 Simplify $\sqrt[3]{125}$.

SOLUTION Look for a number that, when cubed, is equal to 125. Because $125 = 5^3$, $\sqrt[3]{125} = \sqrt[3]{(5)^3} = 5$.

EXAMPLE 9 Simplify $\sqrt[4]{81}$.

SOLUTION In this example, we are looking for a number that when raised to the fourth power is equal to 81. (We can use a factor tree to factor 81.) Because $81 = 3^4$, $\sqrt[4]{81} = \sqrt[4]{(3)^4} = 3$.

Using Your Calculator Find the function for calculating nth roots on the TI-84 by pressing the MATH key and selecting option 5 under the MATH menu.

```
MATH NUM CPX PRB
1:▸Frac
2:▸Dec
3:³
4:³√(
5:ˣ√
6:fMin(
7↓fMax(
```

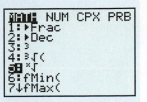

Press the index for the radical, the nth root function, and then the radicand inside a set of parentheses. Here is the screen shot of the calculation of $\sqrt[4]{81}$.

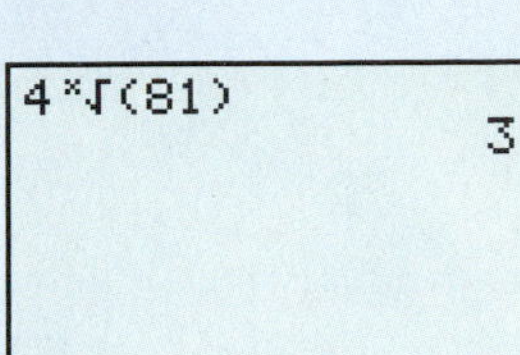

Quick Check 5

Simplify.

a) $\sqrt[3]{27}$ **b)** $\sqrt[4]{256}$
c) $\sqrt[3]{-343}$

EXAMPLE 10 Simplify $\sqrt[5]{-32}$.

SOLUTION Notice that the radicand is negative. So we are looking for a negative number that when raised to the fifth power is equal to -32. (If the index were even, this expression would not be a real number.) Because $(-2)^5 = -32$, $\sqrt[5]{-32} = \sqrt[5]{(-2)^5} = -2$.

For any nonnegative number x, $\sqrt[n]{x^n} = x$.

EXAMPLE 11 Simplify $\sqrt[4]{x^{20}}$. (Assume that x is nonnegative.)

SOLUTION We begin by rewriting the radicand as an expression raised to the fourth power. We can rewrite x^{20} as $(x^5)^4$.

$$\sqrt[4]{x^{20}} = \sqrt[4]{(x^5)^4} \quad \text{Rewrite the radicand as an expression raised to the fourth power.}$$
$$= x^5 \quad \text{Simplify.}$$

Quick Check 6

Simplify $\sqrt[6]{x^{42}}$. (Assume that x is nonnegative.)

EXAMPLE 12 Simplify $\sqrt[3]{64a^3b^9c^{21}}$. (Assume that a, b, and c are nonnegative.)

SOLUTION We begin by rewriting the radicand as a cube. We can rewrite $64a^3b^9c^{21}$ as $(4ab^3c^7)^3$.

$$\sqrt[3]{64a^3b^9c^{21}} = \sqrt[3]{(4ab^3c^7)^3} \quad \text{Rewrite the radicand as a cube.}$$
$$= 4ab^3c^7 \quad \text{Simplify.}$$

Quick Check 7

Simplify $\sqrt[5]{-243x^{35}y^{40}}$. (Assume that x and y are nonnegative.)

EXAMPLE 13 Simplify $\sqrt{x^2 - 8x + 16}$. (Assume that $x \geq 4$.)

SOLUTION This is a square root, so we must begin by rewriting the radicand as a square. If we factor the radicand, we see that it can be expressed as $(x - 4)(x - 4)$, or $(x - 4)^2$.

$$\sqrt{x^2 - 8x + 16} = \sqrt{(x - 4)^2} \quad \text{Rewrite the radicand as a square.}$$
$$= x - 4 \quad \text{Simplify. (Because } x \geq 4,\ x - 4 \text{ is nonnegative.)}$$

Quick Check 8

Simplify $\sqrt{x^2 + 10x + 25}$. (Assume that $x \geq -5$.)

Multiplying Radical Expressions

Objective 5 Multiply radical expressions. We know that $\sqrt{9} \cdot \sqrt{100} = 3 \cdot 10$, or 30. We also know that $\sqrt{9 \cdot 100} = \sqrt{900}$, or 30. In this case, we see that $\sqrt{9} \cdot \sqrt{100} = \sqrt{9 \cdot 100}$. If two radical expressions with nonnegative radicands have the same index, we can multiply the two expressions by multiplying the two radicands and writing the product inside the same radical.

Product Rule for Radicals

For any root n, if $\sqrt[n]{a}$ and $\sqrt[n]{b}$ are real numbers, then $\sqrt[n]{a} \cdot \sqrt[n]{b} = \sqrt[n]{ab}$.

For example, the product of $\sqrt{2}$ and $\sqrt{8}$ is $\sqrt{16}$, which simplifies to equal 4.

Quick Check 9

Multiply $\sqrt{6b^5} \cdot \sqrt{150b^3}$. (Assume that b is nonnegative.)

EXAMPLE 14 Multiply $\sqrt{45n} \cdot \sqrt{5n}$. (Assume that n is nonnegative.)

SOLUTION Because both radicals are square roots, we can multiply the radicands.

$\sqrt{45n} \cdot \sqrt{5n} = \sqrt{225n^2}$ Multiply the radicands.

$= 15n$ Simplify the square root.

EXAMPLE 15 Multiply $7\sqrt{2} \cdot 9\sqrt{2}$.

SOLUTION We multiply the factors in front of the radicals by each other and multiply the radicands by each other.

$7\sqrt{2} \cdot 9\sqrt{2} = 63\sqrt{4}$ Multiply the factors in front of the radicals $(7 \cdot 9)$. Multiply the radicands.

$= 63 \cdot 2$ Simplify the square root.

$= 126$ Multiply.

Quick Check 10

Multiply $3\sqrt{6} \cdot 8\sqrt{6}$.

Dividing Radical Expressions

Objective 6 Divide radical expressions. We can rewrite the quotient of two radical expressions that have the same index as the quotient of the two radicands inside the same radical.

Quotient Rule for Radicals

For any root n, if $\sqrt[n]{a}$ and $\sqrt[n]{b}$ are real numbers and $b \neq 0$, then $\frac{\sqrt[n]{a}}{\sqrt[n]{b}} = \sqrt[n]{\frac{a}{b}}$.

EXAMPLE 16 Simplify $\frac{\sqrt{108}}{\sqrt{3}}$.

SOLUTION Because both radicals are square roots, we begin by dividing the radicands. We write the quotient of the radicands under a single square root.

$\frac{\sqrt{108}}{\sqrt{3}} = \sqrt{\frac{108}{3}}$ Rewrite as the square root of the quotient of the radicands.

$= \sqrt{36}$ Divide.

$= 6$ Simplify the square root.

EXAMPLE 17 Simplify $\frac{\sqrt[3]{40b^7}}{\sqrt[3]{5b^4}}$. (Assume that b is nonnegative.)

SOLUTION Because the index of both radicals is the same, begin by dividing the radicands.

$$\frac{\sqrt[3]{40b^7}}{\sqrt[3]{5b^4}} = \sqrt[3]{\frac{40b^7}{5b^4}} \quad \text{Divide the radicands.}$$
$$= \sqrt[3]{8b^3} \quad \text{Simplify the radicand.}$$
$$= 2b \quad \text{Simplify the radical.}$$

Quick Check 11

Simplify.

a) $\frac{\sqrt{350}}{\sqrt{14}}$ **b)** $\frac{\sqrt[5]{2916a^{18}}}{\sqrt[5]{12a^3}}$

(Assume that a is nonnegative.)

Radical Functions

Objective 7 Evaluate radical functions. A **radical function** is a function that involves radicals, such as $f(x) = \sqrt{x-4} + 3$.

EXAMPLE 18 For the radical function $f(x) = \sqrt{x+5} - 2$, find $f(-1)$.

SOLUTION To evaluate this function, substitute -1 for x and simplify.

$$f(-1) = \sqrt{(-1)+5} - 2 \quad \text{Substitute } -1 \text{ for } x.$$
$$= \sqrt{4} - 2 \quad \text{Simplify the radicand.}$$
$$= 2 - 2 \quad \text{Take the square root of 4.}$$
$$= 0 \quad \text{Subtract.}$$

Quick Check 12

For the radical function $f(x) = \sqrt{3x+13} + 33$, find $f(-3)$.

Finding the Domain of Radical Functions

Objective 8 Find the domain of a radical function. Radical functions involving even roots are different from many of the functions we have seen to this point in that their domain is restricted. To find the domain of a radical function involving an even root, we need to find the values of the variable that make the radicand nonnegative. In other words, set the radicand greater than or equal to zero and solve. The domain of a radical function involving odd roots is the set of all real numbers. Remember that square roots are considered to be even roots with an index of 2.

EXAMPLE 19 Find the domain of the radical function $f(x) = \sqrt{x-9} + 7$. Express your answer in interval notation.

SOLUTION Because the radical has an even index, we begin by setting the radicand $(x - 9)$ greater than or equal to zero. We solve this inequality to find the domain.

$$x - 9 \geq 0 \quad \text{Set the radicand greater than or equal to 0.}$$
$$x \geq 9 \quad \text{Add 9.}$$

The domain of the function is $[9, \infty)$.

Quick Check 13

Find the domain of the radical function $f(x) = \sqrt[6]{x+18} - 30$. Express your answer in interval notation.

EXAMPLE 20 Find the domain of the radical function $f(x) = \sqrt[5]{14x - 9} + 21$. Express your answer in interval notation.

SOLUTION Because this radical function involves an odd root, its domain is the set of all real numbers $\mathbb{R}$. This can be expressed in interval notation as $(-\infty, \infty)$.

Quick Check 14

Find the domain of the radical function $f(x) = \sqrt[3]{16x - 409} + 38$. Express your answer in interval notation.

BUILDING YOUR STUDY STRATEGY

Note Taking, 1 Choosing a Seat One important yet frequently overlooked aspect of effective note taking is choosing an appropriate seat in class. You must sit where you can clearly see the board, preferably in the center of the room. Try not to sit behind anyone who will obstruct your vision.

Location also impacts your ability to hear your instructor clearly. You will find that your instructor speaks toward the middle of the classroom; so finding a seat toward the center of the classroom should ensure that you hear everything your instructor says. Try not to sit close to students who talk to each other during class; their discussions may distract you from what your instructor is saying.

In summary, try to choose your classroom seat the same way you choose a seat at a movie theater. Make sure you can see and hear everything.

Exercises 9.1

Vocabulary

1. A number a is a(n) __________ of a number b if $a^2 = b$.
2. The __________ square root of b, denoted $\sqrt{b}$, for $b > 0$, is the positive number a such that $a^2 = b$.
3. For any positive integer $n > 1$ and any number b, if $a^n = b$ and both a and b have the same sign, a is the principal __________ of b, denoted $a = \sqrt[n]{b}$.
4. For the expression $a = \sqrt[n]{b}$, n is called the __________ of the radical.
5. The expression contained inside a radical is called the __________.
6. A radical with an index of 3 is also known as a(n) __________ root.
7. A(n) __________ is a function that involves radicals.
8. The domain of a radical function involving an even root consists of values of the variable for which the radicand is __________.

Simplify the radical expression. Indicate if the expression is not a real number.

9. $\sqrt{36}$
10. $\sqrt{64}$
11. $\sqrt{4}$
12. $\sqrt{100}$
13. $\sqrt{\frac{1}{81}}$
14. $\sqrt{\frac{1}{49}}$
15. $\sqrt{\frac{25}{36}}$
16. $\sqrt{\frac{121}{4}}$
17. $\sqrt{-16}$
18. $\sqrt{-36}$
19. $-\sqrt{49}$
20. $-\sqrt{81}$

Simplify the radical expression. Where appropriate, include absolute values.

21. $\sqrt{a^{14}}$
22. $\sqrt{b^{10}}$
23. $\sqrt{x^{34}}$
24. $\sqrt{x^{42}}$
25. $\sqrt{9x^6}$
26. $\sqrt{16x^2}$
27. $\sqrt{\frac{1}{49}x^4}$
28. $\sqrt{\frac{1}{100}x^{24}}$
29. $\sqrt{a^{18}b^{22}}$
30. $\sqrt{m^6n^{22}}$
31. $\sqrt{x^8y^{10}z^{14}}$
32. $\sqrt{x^{20}y^{16}z^{26}}$

Find the missing number or expression. Assume that all variables represent nonnegative real numbers.

33. $\sqrt{?} = 9$

34. $\sqrt{?} = 16$

35. $\sqrt{?} = 3x$

36. $\sqrt{?} = 5a^3$

Approximate to the nearest thousandth using a calculator.

37. $\sqrt{55}$

38. $\sqrt{98}$

39. $\sqrt{326}$

40. $\sqrt{409}$

41. $\sqrt{0.53}$

42. $\sqrt{0.06}$

Simplify the radical expression. Assume that all variables represent nonnegative real numbers.

43. $\sqrt[3]{125}$

44. $\sqrt[3]{-64}$

45. $\sqrt[5]{-243}$

46. $\sqrt[4]{1296}$

47. $\sqrt[4]{x^{20}}$

48. $\sqrt[7]{x^{56}}$

49. $\sqrt[6]{a^{42}b^{24}}$

50. $\sqrt[3]{m^{33}n^{27}}$

51. $\sqrt[3]{343m^{21}}$

52. $\sqrt[3]{216n^{42}}$

53. $\sqrt[4]{81x^{12}y^{20}}$

54. $\sqrt[6]{64s^{36}t^{54}}$

55. $\sqrt{x^2 + 6x + 9}, x \geq -3$

56. $\sqrt{x^2 + 12x + 36}, x \geq -6$

Simplify. Assume that all variables represent nonnegative real numbers.

57. $\sqrt{27} \cdot \sqrt{3}$

58. $\sqrt{8} \cdot \sqrt{8}$

59. $\sqrt{10} \cdot \sqrt{90}$

60. $\sqrt{2} \cdot \sqrt{72}$

61. $\sqrt{a^9} \cdot \sqrt{a^{17}}$

62. $\sqrt{b^{15}} \cdot \sqrt{b}$

63. $\sqrt{x^{19}} \cdot \sqrt{x^{19}}$

64. $\sqrt{x^{13}} \cdot \sqrt{x^{13}}$

65. $\dfrac{\sqrt{180}}{\sqrt{5}}$

66. $\dfrac{\sqrt{63}}{\sqrt{7}}$

67. $\dfrac{\sqrt{800}}{\sqrt{8}}$

68. $\dfrac{\sqrt{1872}}{\sqrt{13}}$

69. $\dfrac{\sqrt{x^{23}}}{\sqrt{x^5}}$

70. $\dfrac{\sqrt{x^{31}}}{\sqrt{x^{27}}}$

71. $\dfrac{\sqrt{a^{13}}}{\sqrt{a}}$

72. $\dfrac{\sqrt{a^{21}}}{\sqrt{a^7}}$

73. $\sqrt[3]{6} \cdot \sqrt[3]{36}$

74. $\sqrt[3]{12} \cdot \sqrt[3]{18}$

75. $\sqrt[4]{48} \cdot \sqrt[4]{27}$

76. $\sqrt[5]{16} \cdot \sqrt[5]{64}$

77. $\sqrt[4]{x^7} \cdot \sqrt[4]{x^{13}}$

78. $\sqrt[3]{x^{10}} \cdot \sqrt[3]{x^{41}}$

79. $\sqrt[5]{a^{11}b^8} \cdot \sqrt[5]{a^4b^2}$

80. $\sqrt[6]{m^{15}n^8} \cdot \sqrt[6]{m^9n^{28}}$

81. $\dfrac{\sqrt[4]{324}}{\sqrt[4]{4}}$

82. $\dfrac{\sqrt[3]{875}}{\sqrt[3]{7}}$

83. $\dfrac{\sqrt[5]{x^{32}}}{\sqrt[5]{x^7}}$

84. $\dfrac{\sqrt[4]{x^{29}}}{\sqrt[4]{x^{17}}}$

85. $7\sqrt{6} \cdot 8\sqrt{6}$

86. $9\sqrt{5} \cdot 4\sqrt{5}$

87. $5\sqrt{12} \cdot 10\sqrt{27}$

88. $11\sqrt{8} \cdot 2\sqrt{50}$

89. $9\sqrt{x} \cdot 12\sqrt{x}$

90. $15\sqrt{x} \cdot 8\sqrt{x}$

Find the missing number.

91. $\sqrt{20} \cdot \sqrt{?} = 10$

92. $\sqrt{21} \cdot \sqrt{?} = 42$

93. $\dfrac{\sqrt{?}}{\sqrt{8}} = 7$

94. $\dfrac{\sqrt{?}}{\sqrt{6}} = 6$

Evaluate the radical function. Round to the nearest thousandth if necessary.

95. $f(x) = \sqrt{x - 5}$; find $f(21)$.

96. $f(x) = \sqrt{3x + 4}$; find $f(20)$.

97. $f(x) = \sqrt{2x + 19} - 2$; find $f(3)$.

98. $f(x) = \sqrt{4x - 11} - 10$; find $f(5)$.

99. $f(x) = \sqrt{2x + 9} + 7$; find $f(10)$.

100. $f(x) = \sqrt{5x - 4} - 13$; find $f(16)$.

101. $f(x) = \sqrt[3]{x^2 + 7x + 81}$; find $f(4)$.

102. $f(x) = \sqrt[4]{x^2 + 4x - 36}$; find $f(9)$.

Find the domain of the radical function. Express your answer in interval notation.

103. $f(x) = \sqrt{x - 4} + 11$

104. $f(x) = \sqrt{2x + 9} - 20$

105. $f(x) = \sqrt[4]{3x - 8} - 17$

106. $f(x) = \sqrt[6]{4x} - 34$

107. $f(x) = \sqrt{13 - 2x} + 5$

108. $f(x) = \sqrt{8 - x} - 2$

109. $f(x) = \sqrt[3]{15x - 42} + 6$

110. $f(x) = \sqrt[5]{10x + 115} - 45$

Writing in Mathematics

Answer in complete sentences.

111. Which of the following are real numbers, and which are not real numbers: $-\sqrt{64}$, $\sqrt{-64}$, $-\sqrt[3]{64}$, $\sqrt[3]{-64}$? Explain your reasoning.

112. Explain how to find the domain of a radical function. Use examples.

Rational Exponents

OBJECTIVES

1. Simplify expressions containing exponents of the form $1/n$.
2. Simplify expressions containing exponents of the form m/n.
3. Simplify expressions containing rational exponents.
4. Simplify expressions containing negative rational exponents.
5. Use rational exponents to simplify radical expressions.

Rational Exponents of the Form $1/n$

Objective 1 Simplify expressions containing exponents of the form $1/n$. We have used exponents to represent repeated multiplication. For example, x^n says that the base x is a factor n times.

$$x^3 = x \cdot x \cdot x$$
$$x^6 = x \cdot x \cdot x \cdot x \cdot x \cdot x$$

We run into a problem with this definition when we encounter an expression with a fractional exponent, such as $x^{1/2}$. Saying that the base x is a factor $\frac{1}{2}$ times does not make any sense. Using the properties of exponents introduced in Chapter 5, we know that $x^{1/2} \cdot x^{1/2} = x^{1/2+1/2}$, or x. If we multiply $x^{1/2}$ by itself, the result is x. The same is true when we multiply $\sqrt{x}$ by itself, suggesting that $x^{1/2} = \sqrt{x}$.

> For any integer $n > 1$, we define $a^{1/n}$ to be the nth root of a, or $\sqrt[n]{a}$.

EXAMPLE 1 Rewrite $81^{1/4}$ as a radical expression and simplify if possible.

SOLUTION

$81^{1/4} = \sqrt[4]{81}$ Rewrite as a radical expression. An exponent of 1/4 is equivalent to a fourth root.

$= \sqrt[4]{3^4}$ Rewrite the radicand as a number to the fourth power.

$= 3$ Simplify.

EXAMPLE 2 Rewrite $(-125x^6)^{1/3}$ as a radical expression and simplify if possible.

SOLUTION

$(-125x^6)^{1/3} = \sqrt[3]{(-5x^2)^3}$ Rewrite as a radical expression. Rewrite the radicand as a cube.

$= -5x^2$ Simplify. Keep in mind that an odd root of a negative number is negative.

Quick Check 1

Rewrite as a radical expression and simplify if possible.

a) $36^{1/2}$ **b)** $(-32x^{15})^{1/5}$

Quick Check 2

Rewrite $(x^{32}y^4z^{24})^{1/4}$ as a radical expression and simplify if possible. (Assume that all variables represent nonnegative values.)

EXAMPLE 3 Rewrite $(a^{10}b^5c^{20})^{1/5}$ as a radical expression and simplify if possible.

SOLUTION

$(a^{10}b^5c^{20})^{1/5} = \sqrt[5]{a^{10}b^5c^{20}}$ Rewrite as a radical expression.

$= \sqrt[5]{(a^2bc^4)^5}$ Rewrite the radicand.

$= a^2bc^4$ Simplify.

For any negative number x and even integer n, $x^{1/n}$ is not a real number. For example, $(-64)^{1/6}$ is not a real number because $(-64)^{1/6} = \sqrt[6]{-64}$ and an even root of a negative number is not a real number.

EXAMPLE 4 Rewrite $\sqrt[7]{x}$ by using rational exponents.

SOLUTION

$\sqrt[7]{x} = x^{1/7}$ Rewrite using the definition $\sqrt[n]{x} = x^{1/n}$.

Quick Check 3

Rewrite $\sqrt[9]{a}$ using rational exponents.

Rational Exponents of the Form m/n

Objective 2 Simplify expressions containing exponents of the form m/n.

We now turn our attention to rational exponents of the form m/n for any integers m and $n > 1$. The expression $a^{m/n}$ can be rewritten as $(a^{1/n})^m$, which is equivalent to $(\sqrt[n]{a})^m$.

For any integers m and n, $n > 1$, we define $a^{m/n}$ to be $(\sqrt[n]{a})^m$. This is also equivalent to $\sqrt[n]{a^m}$. The denominator in the exponent, n, is the root we are taking. The numerator in the exponent, m, is the power to which we raise this radical.

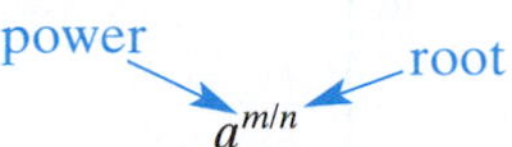

When simplifying an expression of the form $a^{m/n}$, take the nth root of a first if $\sqrt[n]{a}$ can be simplified. Otherwise, raise a to the m power and then attempt to simplify $\sqrt[n]{a^m}$.

EXAMPLE 5 Rewrite $(243x^{10})^{3/5}$ as a radical expression and simplify if possible.

SOLUTION

$$
\begin{aligned}
(243x^{10})^{3/5} &= (\sqrt[5]{243x^{10}})^3 && \text{Rewrite as a radical expression.} \\
&= (\sqrt[5]{(3x^2)^5})^3 && \text{Rewrite the radicand.} \\
&= (3x^2)^3 && \text{Simplify the radical.} \\
&= 27x^6 && \text{Raise } 3x^2 \text{ to the third power.}
\end{aligned}
$$

Quick Check 4

Rewrite $(4096x^{18})^{5/6}$ as a radical expression and simplify if possible. (Assume that x is nonnegative.)

EXAMPLE 6 Rewrite $\sqrt[5]{x^3}$ by using rational exponents.

SOLUTION

$\sqrt[5]{x^3} = x^{3/5}$ Rewrite using the definition $\sqrt[n]{x^m} = x^{m/n}$.

Quick Check 5

Rewrite $\sqrt[12]{x^5}$ by using rational exponents. (Assume that x is nonnegative.)

Simplifying Expressions Containing Rational Exponents

Objective 3 Simplify expressions containing rational exponents. The properties of exponents developed in Chapter 5 for integer exponents are true for fractional exponents as well. Here is a summary of those properties.

Properties of Exponents

For any bases x and y:

1. $x^m \cdot x^n = x^{m+n}$
2. $(x^m)^n = x^{m \cdot n}$
3. $(xy)^n = x^n y^n$
4. $\frac{x^m}{x^n} = x^{m-n}\ (x \neq 0)$
5. $x^0 = 1\ (x \neq 0)$
6. $\left(\frac{x}{y}\right)^n = \frac{x^n}{y^n}\ (y \neq 0)$
7. $x^{-n} = \frac{1}{x^n}\ (x \neq 0)$

EXAMPLE 7 Simplify the expression $x^{4/3} \cdot x^{5/6}$. (Assume that x is nonnegative.)

SOLUTION When multiplying two expressions with the same base, add the exponents and keep the base.

$$\begin{aligned} x^{4/3} \cdot x^{5/6} &= x^{\frac{4}{3}+\frac{5}{6}} && \text{Add the exponents; keep the base.} \\ &= x^{\frac{8}{6}+\frac{5}{6}} && \text{Rewrite the fractions with a common denominator.} \\ &= x^{13/6} && \text{Add.} \end{aligned}$$

Quick Check 6

Simplify the expression $x^{5/8} \cdot x^{7/12}$. (Assume that x is nonnegative.)

EXAMPLE 8 Simplify the expression $(x^{3/4})^{2/5}$. (Assume that x is nonnegative.)

SOLUTION When raising an exponential expression to another power, multiply the exponents and keep the base.

$$\begin{aligned} (x^{3/4})^{2/5} &= x^{\frac{3}{4}\cdot\frac{2}{5}} && \text{Multiply the exponents; keep the base.} \\ &= x^{3/10} && \text{Multiply.} \end{aligned}$$

Quick Check 7

Simplify the expression $(x^{7/10})^{4/9}$. (Assume that x is nonnegative.)

EXAMPLE 9 Simplify the expression $\left(\frac{b^4}{c^2d^6}\right)^{3/2}$, where $c \neq 0, d \neq 0$. (Assume that all variables represent nonnegative values.)

SOLUTION Begin by raising each factor to the $\frac{3}{2}$ power.

$$\begin{aligned} \left(\frac{b^4}{c^2d^6}\right)^{3/2} &= \frac{b^{4\cdot\frac{3}{2}}}{c^{2\cdot\frac{3}{2}}d^{6\cdot\frac{3}{2}}} && \text{Raise each factor to the } \tfrac{3}{2} \text{ power.} \\ &= \frac{b^6}{c^3d^9} && \text{Multiply exponents.} \end{aligned}$$

Quick Check 8

Simplify the expression $\left(\frac{x^9}{y^3z^{12}}\right)^{5/3}$, where $y \neq 0, z \neq 0$.

Simplifying Expressions Containing Negative Rational Exponents

Objective 4 Simplify expressions containing negative rational exponents.

EXAMPLE 10 Simplify the expression $64^{-5/6}$.

SOLUTION Begin by rewriting the expression with a positive exponent.

$$64^{-5/6} = \frac{1}{64^{5/6}} \quad \text{Rewrite the expression with a positive exponent.}$$

$$= \frac{1}{(\sqrt[6]{64})^5} \quad \text{Rewrite in radical notation.}$$

$$= \frac{1}{2^5} \quad \text{Simplify the radical.}$$

$$= \frac{1}{32} \quad \text{Raise 2 to the fifth power.}$$

Quick Check 9

Simplify the expression $81^{-3/2}$.

Using Rational Exponents to Simplify Radical Expressions

Objective 5 Use rational exponents to simplify radical expressions. In the previous section, we learned that we may multiply two radicals if they have the same index. In other words, $\sqrt[n]{a} \cdot \sqrt[n]{b} = \sqrt[n]{ab}$ if $\sqrt[n]{a}$ and $\sqrt[n]{b}$ are real numbers. If the two indices are not the same, we can use fractional exponents to multiply the radicals.

EXAMPLE 11 Simplify the expression $\sqrt{a} \cdot \sqrt[5]{a}$. (Assume that a is nonnegative.) Express your answer in radical notation.

SOLUTION Begin by rewriting the radicals using fractional exponents.

$$\sqrt{a} \cdot \sqrt[5]{a} = a^{1/2} \cdot a^{1/5} \quad \text{Rewrite both radicals using fractional exponents.}$$

$$= a^{\frac{1}{2}+\frac{1}{5}} \quad \text{Add the exponents, keeping the base.}$$

$$= a^{\frac{5}{10}+\frac{2}{10}} \quad \text{Rewrite each fraction with a common denominator.}$$

$$= a^{7/10} \quad \text{Add.}$$

$$= \sqrt[10]{a^7} \quad \text{Rewrite in radical notation.}$$

Quick Check 10

Simplify the expression $\sqrt[3]{x} \cdot \sqrt[4]{x}$. (Assume that x is nonnegative.) Express your answer in radical notation.

BUILDING YOUR STUDY STRATEGY

Note Taking, 2 Note-Taking Speed "I'm a slow writer. There's no way I can take down all of the notes in time." This is a common concern for students. Try to focus on writing down enough information so that you will understand the material later, rather than thinking you must copy every word your instructor writes or speaks. Then rewrite your notes, filling in any blanks or adding explanations, as soon as possible after class.

During class, try to use abbreviations whenever possible. Write your notes using phrases rather than complete sentences.

With your instructor's approval, you may want to record a lecture. After class, you can use the recording to help fill in holes in your notes.

If you are falling behind in your note taking, leave space in your notes. Afterward ask a classmate if you can borrow his or her notes to get the information you missed.

Exercises 9.2

MyMathLab

PRACTICE

WATCH

DOWNLOAD

READ
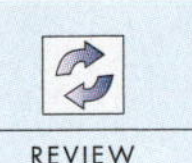
REVIEW

Vocabulary

1. For any integer $n > 1$, $a^{1/n} =$ ______.
2. For any integers m and n, $n > 1$, $a^{m/n} =$ ______ ______.
3. An exponent of the form m/n is said to be a(n) ______ exponent.
4. For the expression $x^{m/n}$, m represents the ______ to which x is raised and n represents the ______ that is being taken.

Rewrite each radical expression using rational exponents.

5. $\sqrt[4]{x}$
6. $\sqrt[5]{a}$
7. $\sqrt{7}$
8. $\sqrt[3]{10}$
9. $9\sqrt[4]{d}$
10. $\sqrt[4]{9d}$

Rewrite as a radical expression and simplify if possible. Assume that all variables represent nonnegative real numbers.

11. $64^{1/2}$
12. $25^{1/2}$
13. $(-343)^{1/3}$
14. $(-1024)^{1/5}$
15. $(x^{30})^{1/5}$
16. $(a^{28})^{1/4}$
17. $(64a^{33}b^{21})^{1/3}$
18. $(32x^{35}y^{40})^{1/5}$
19. $(-243x^{55}y^{25}z^{5})^{1/5}$
20. $(-125a^{24}b^{3}c^{27})^{1/3}$

Rewrite each radical expression using rational exponents. Assume that all variables represent nonnegative real numbers.

21. $(\sqrt[5]{x})^3$
22. $(\sqrt[4]{x})^7$
23. $(\sqrt[9]{y})^8$
24. $(\sqrt[12]{a})^{17}$
25. $(\sqrt[3]{3x^2})^7$
26. $(\sqrt[9]{2x^4})^2$
27. $(\sqrt[8]{10x^4y^5})^3$
28. $(\sqrt[5]{a^6b^7c^8})^4$

Rewrite as a radical expression and simplify if possible. Assume that all variables represent nonnegative real numbers.

29. $25^{3/2}$
30. $16^{3/4}$
31. $32^{2/5}$
32. $1000^{7/3}$
33. $(x^{12})^{2/3}$
34. $(x^{20})^{6/5}$
35. $(256a^8b^{24})^{3/4}$
36. $(4x^{16}y^{32}z^{64})^{5/2}$
37. $(-125x^9y^{15}z^3)^{4/3}$
38. $(256a^4b^8c^{20})^{7/4}$

Simplify the expression. Assume that all variables represent nonnegative real numbers.

39. $x^{7/10} \cdot x^{1/10}$
40. $x^{3/4} \cdot x^{11/4}$
41. $x^{4/3} \cdot x^{1/2}$
42. $x^{10/7} \cdot x^{1/14}$
43. $a^{5/3} \cdot a^{3/4} \cdot a^{1/6}$
44. $m^{4/5} \cdot m^{5/2} \cdot m^{7/10}$
45. $x^{1/3}y^{3/4} \cdot x^{1/6}y^{1/6}$
46. $x^{3/8}y^{1/3} \cdot x^{7/10}y^{1/4}$
47. $(x^{3/4})^{2/7}$
48. $(x^{5/6})^{9/20}$
49. $(a^{3/5})^{3/5}$
50. $(b^{2/7})^{7/2}$
51. $(x^{5/12})^4$
52. $(x^{12})^{7/6}$
53. $\dfrac{x^{4/5}}{x^{3/10}}$ $(x \neq 0)$
54. $\dfrac{x^{7/6}}{x^{11/12}}$ $(x \neq 0)$
55. $\dfrac{x^{5/8}}{x^{1/3}}$ $(x \neq 0)$
56. $\dfrac{x^{9/10}}{x^{21/40}}$ $(x \neq 0)$
57. $(x^{5/4})^0$ $(x \neq 0)$
58. $(x^{2/9})^0$ $(x \neq 0)$
59. $32^{-1/5}$
60. $125^{-1/3}$
61. $27^{-4/3}$
62. $100^{-3/2}$
63. $216^{-4/3} \cdot 216^{-1/3}$
64. $16^{-3/4} \cdot 16^{-7/4}$
65. $\dfrac{125^{2/3}}{125^{7/3}}$
66. $\dfrac{16^{3/4}}{16^{9/4}}$
67. $\dfrac{216^{7/3}}{216^{8/3}}$
68. $\dfrac{4^{3/2}}{4^5}$

Simplify each expression. Assume that all variables represent nonnegative real numbers. Express your answer in radical notation.

69. $\sqrt[4]{x} \cdot \sqrt[5]{x}$
70. $\sqrt[9]{x} \cdot \sqrt[18]{x}$
71. $\sqrt[12]{a} \cdot \sqrt[4]{a}$
72. $\sqrt[3]{b} \cdot \sqrt[8]{b}$
73. $\dfrac{\sqrt[4]{m}}{\sqrt[12]{m}}$ $(m \neq 0)$
74. $\dfrac{\sqrt{n}}{\sqrt[3]{n}}$ $(n \neq 0)$
75. $\dfrac{\sqrt[30]{x}}{\sqrt[5]{x}}$ $(x \neq 0)$
76. $\dfrac{\sqrt[10]{x}}{\sqrt[5]{x}}$ $(x \neq 0)$

Writing in Mathematics

Answer in complete sentences.

77. Is $-16^{1/2}$ a real number? Explain your answer.

78. Is $16^{-1/2}$ a real number? Explain your answer.

9.3 Simplifying, Adding, and Subtracting Radical Expressions

OBJECTIVES

1. **Simplify radical expressions by using the product property.**
2. **Add or subtract radical expressions containing like radicals.**
3. **Simplify radical expressions before adding or subtracting.**

Simplifying Radical Expressions Using the Product Property

Objective 1 Simplify radical expressions by using the product property. A radical expression is considered simplified if the radicand contains no factors with exponents greater than or equal to the index of the radical. For example, $\sqrt[3]{x^5}$ is not simplified because an exponent inside the radical is greater than the index of the radical. The goal for simplifying radical expressions is to remove as many factors as possible from the radicand. We will use the product property for radical expressions to help us with this.

We could rewrite $\sqrt[3]{x^5}$ as $\sqrt[3]{x^3 \cdot x^2}$, then rewrite this radical expression as the product of two radicals. Using the product property for radicals, $\sqrt[n]{a} \cdot \sqrt[n]{b} = \sqrt[n]{ab}$, we know that $\sqrt[3]{x^3 \cdot x^2} = \sqrt[3]{x^3} \cdot \sqrt[3]{x^2}$. The reason for rewriting $\sqrt[3]{x^5}$ as $\sqrt[3]{x^3} \cdot \sqrt[3]{x^2}$ is that the radical $\sqrt[3]{x^3}$ equals x.

$$\begin{aligned}\sqrt[3]{x^5} &= \sqrt[3]{x^3 \cdot x^2} \\ &= \sqrt[3]{x^3} \cdot \sqrt[3]{x^2} \\ &= x\sqrt[3]{x^2}\end{aligned}$$

Now the radicand contains no factors with an exponent that is greater than or equal to the index 3 and is simplified.

Simplifying Radical Expressions

- Completely factor any numerical factors in the radicand.
- Rewrite each factor as a product of two factors. The exponent for the first factor should be the largest multiple of the radical's index that is less than or equal to the factor's original exponent.
- Use the product property to remove factors from the radicand.

EXAMPLE 1 Simplify $\sqrt[4]{a^{23}}$. (Assume that a is nonnegative.)

SOLUTION We begin by rewriting a^{23} as a product of two factors. The largest multiple of the index (4) that is less than or equal to the exponent for this factor (23) is 20; so we will rewrite a^{23} as $a^{20} \cdot a^3$.

$$\begin{aligned}\sqrt[4]{a^{23}} &= \sqrt[4]{a^{20} \cdot a^3} && \text{Rewrite } a^{23} \text{ as the product of two factors.} \\ &= \sqrt[4]{a^{20}} \cdot \sqrt[4]{a^3} && \text{Use the product property of radicals to rewrite the radical as the product of two radicals.} \\ &= a^5\sqrt[4]{a^3} && \text{Simplify the radical.}\end{aligned}$$

Quick Check 1

Simplify $\sqrt[5]{x^{17}}$. (Assume that x is nonnegative.)

EXAMPLE 2 Simplify $\sqrt{x^{11}y^{10}z^5}$. (Assume that all variables represent nonnegative values.)

SOLUTION Again, we begin by rewriting factors as a product of two factors. In this example, the exponent of the factor y is a multiple of the index 2. We do not need to rewrite this factor as the product of two factors.

$$\begin{aligned}\sqrt{x^{11}y^{10}z^5} &= \sqrt{(x^{10}\cdot x)y^{10}(z^4\cdot z)} && \text{Rewrite factors.}\\ &= \sqrt{x^{10}y^{10}z^4}\cdot\sqrt{xz} && \text{Rewrite as the product of two radicals.}\\ &= x^5y^5z^2\sqrt{xz} && \text{Simplify the radical.}\end{aligned}$$

Quick Check 2

Simplify $\sqrt{a^8b^{15}c^7d}$. (Assume that all variables represent nonnegative values.)

EXAMPLE 3 Simplify $\sqrt{24}$.

SOLUTION Begin by rewriting 24 using its prime factorization $(2^3\cdot 3)$.

$$\begin{aligned}\sqrt{24} &= \sqrt{2^3\cdot 3} && \text{Factor 24.}\\ &= \sqrt{(2^2\cdot 2)\cdot 3} && \text{Rewrite } 2^3 \text{ as } 2^2\cdot 2.\\ &= \sqrt{2^2}\cdot\sqrt{2\cdot 3} && \text{Rewrite as the product of two radicals.}\\ &= 2\sqrt{6} && \text{Simplify.}\end{aligned}$$

We could have used a different tactic to simplify this square root. The largest factor of 24 that is a perfect square is 4; so we could begin by rewriting 24 as $4\cdot 6$. Because we know that the square root of 4 is 2, we could factor 4 out of the radicand and write it as 2 in front of the radical.

$$\begin{aligned}\sqrt{24} &= \sqrt{4\cdot 6}\\ &= 2\sqrt{6}\end{aligned}$$

Quick Check 3

Simplify $\sqrt{90}$.

EXAMPLE 4 Simplify $\sqrt[3]{324}$.

SOLUTION Begin by factoring 324 to be $2^2\cdot 3^4$.

$$\begin{aligned}\sqrt[3]{324} &= \sqrt[3]{2^2\cdot 3^4} && \text{Factor 324.}\\ &= \sqrt[3]{2^2\cdot(3^3\cdot 3)} && \text{Rewrite } 3^4 \text{ as } 3^3\cdot 3.\\ &= \sqrt[3]{3^3}\cdot\sqrt[3]{2^2\cdot 3} && \text{Rewrite as the product of two radicals.}\\ &= 3\sqrt[3]{12} && \text{Simplify.}\end{aligned}$$

Quick Check 4

Simplify $\sqrt[3]{280}$.

There is an alternative approach for simplifying radical expressions. Suppose we were trying to simplify $\sqrt[5]{a^{48}}$. Using the previous method, we would arrive at the answer $a^9\sqrt[5]{a^3}$.

$$\begin{aligned}\sqrt[5]{a^{48}} &= \sqrt[5]{a^{45}\cdot a^3} && \text{Rewrite } a^{48} \text{ as } a^{45}\cdot a^3 \text{ because 45 is the highest multiple of 5 that is less than or equal to 48.}\\ &= \sqrt[5]{a^{45}}\cdot\sqrt[5]{a^3} && \text{Rewrite as the product of two radicals.}\\ &= a^9\sqrt[5]{a^3} && \text{Simplify.}\end{aligned}$$

We know that for every five times a is repeated as a factor in the radicand, we can take a^5 out of the radicand and write it as a in front of the radical. We need to determine how many groups of five can be removed from the radicand, using division. If we divide the exponent 48 by the index 5, the quotient is 9 with a remainder of 3. When we divide the exponent of a factor in the radicand by the index of the radical, the quotient tells us the exponent of the factor removed from the radicand and the remainder tells us the exponent of the factor remaining in the radicand.

An Alternative Approach for Simplifying $\sqrt[n]{x^p}$

- Divide p by n: $\frac{p}{n} = q + \frac{r}{n}$
- The quotient q tells us how many times x will be a factor in front of the radical.
- The remainder r tells us how many times x will remain as a factor in the radicand.

$$\sqrt[n]{x^p} = x^q\sqrt[n]{x^r}$$

EXAMPLE 5 Simplify $\sqrt[6]{a^{31}b^{18}c^5d^{53}}$. (Assume that all variables represent nonnegative values.)

SOLUTION We will work with one factor at a time, beginning with a. The index, 6, divides into the exponent, 31, five times with a remainder of one. This tells us we can write a^5 as a factor in front of the radical and can write a^1, or a, in the radicand.

$$\sqrt[6]{a^{31}b^{18}c^5d^{53}} = a^5\sqrt[6]{ab^{18}c^5d^{53}}$$

For the factor b, $18 \div 6 = 3$ with a remainder of 0. We will write b^3 as a factor in front of the radical, and because the remainder is 0, we will not write b as a factor in the radicand. For the factor c, the index does not divide into 5, so c^5 remains as a factor in the radicand. Finally, for the factor d, $53 \div 6 = 8$ with a remainder of 5. We will write d^8 as a factor in front of the radical and d^5 as a factor in the radicand.

$$\sqrt[6]{a^{31}b^{18}c^5d^{53}} = a^5b^3d^8\sqrt[6]{ac^5d^5}$$

Quick Check 5

Simplify $\sqrt[5]{x^{33}y^6z^{50}w^{18}}$.

Adding and Subtracting Radical Expressions Containing Like Radicals

Objective 2 **Add and subtract radical expressions containing like radicals.**

Like Radicals

Two radical expressions are called **like radicals** if they have the same index and the same radicand.

The radical expressions $5\sqrt[3]{4x}$ and $9\sqrt[3]{4x}$ are like radicals because they have the same index (3) and the same radicand ($4x$). Here are some examples of radical expressions that are not like radicals.

$\sqrt{5}$ and $\sqrt[3]{5}$ — The two radicals have different indices.

$\sqrt[4]{7x^2y^3}$ and $\sqrt[4]{7x^3y^2}$ — The two radicands are different.

We can add and subtract radical expressions by combining like radicals similar to the way we combine like terms. We add or subtract the coefficients in front of the like radicals.

$$6\sqrt{2} + 3\sqrt{2} = 9\sqrt{2}$$

EXAMPLE 6 Simplify $12\sqrt{13} + \sqrt{3} + \sqrt{3} - 6\sqrt{13}$.

SOLUTION There are two pairs of like radicals in this example. There are two radical expressions containing $\sqrt{13}$ and two radical expressions containing $\sqrt{3}$.

$$\begin{aligned} &12\sqrt{13} + \sqrt{3} + \sqrt{3} - 6\sqrt{13} \\ &= 6\sqrt{13} + \sqrt{3} + \sqrt{3} && \text{Subtract } 12\sqrt{13} - 6\sqrt{13}. \\ &= 6\sqrt{13} + 2\sqrt{3} && \text{Add } \sqrt{3} + \sqrt{3}. \end{aligned}$$

Quick Check 6

Simplify $9\sqrt{10} - 13\sqrt{5} + 6\sqrt{10} + 8\sqrt{5}$.

EXAMPLE 7 Simplify $18\sqrt[5]{x^3y^2} - 2\sqrt[5]{x^2y^3} - 8\sqrt[5]{x^3y^2}$.

SOLUTION Of the three terms, only the first and third contain like radicals.

$$18\sqrt[5]{x^3y^2} - 2\sqrt[5]{x^2y^3} - 8\sqrt[5]{x^3y^2} = 10\sqrt[5]{x^3y^2} - 2\sqrt[5]{x^2y^3}$$

Subtract $18\sqrt[5]{x^3y^2} - 8\sqrt[5]{x^3y^2}$.

Quick Check 7

Simplify $16\sqrt[5]{a^4b^3} - 11\sqrt[7]{a^4b^3} - 5\sqrt[5]{a^4b^3}$.

Objective 3 Simplify radical expressions before adding or subtracting.

Are the expressions $\sqrt{24}$ and $\sqrt{54}$ like radicals? We must simplify each radical completely before we can determine whether the two expressions are like radicals. In this case, $\sqrt{24} = 2\sqrt{6}$ and $\sqrt{54} = 3\sqrt{6}$; so $\sqrt{24}$ and $\sqrt{54}$ are like radicals.

EXAMPLE 8 Simplify $\sqrt{12} + \sqrt{3}$.

SOLUTION We begin by simplifying each radical completely. Because 12 can be written as $4 \cdot 3$, $\sqrt{12}$ can be simplified to $2\sqrt{3}$. We also could use the prime factorization of 12 $(2^2 \cdot 3)$ to simplify $\sqrt{12}$.

$$\begin{aligned} \sqrt{12} + \sqrt{3} &= \sqrt{4 \cdot 3} + \sqrt{3} && \text{Factor 12.} \\ &= 2\sqrt{3} + \sqrt{3} && \text{Simplify } \sqrt{4 \cdot 3}. \\ &= 3\sqrt{3} && \text{Add.} \end{aligned}$$

Quick Check 8

Simplify $\sqrt{63} + \sqrt{7}$.

EXAMPLE 9 Simplify $\sqrt{45} - \sqrt{80} - \sqrt{20}$.

SOLUTION In this example, we must simplify all three radicals before proceeding.

$$\begin{aligned} \sqrt{45} - \sqrt{80} - \sqrt{20} &= \sqrt{9 \cdot 5} - \sqrt{16 \cdot 5} - \sqrt{4 \cdot 5} && \text{Factor each radicand.} \\ &= 3\sqrt{5} - 4\sqrt{5} - 2\sqrt{5} && \text{Simplify each radical.} \\ &= -3\sqrt{5} && \text{Combine like radicals.} \end{aligned}$$

Quick Check 9

Simplify $\sqrt{18} - \sqrt{32} + \sqrt{98}$.

BUILDING YOUR STUDY STRATEGY

Note Taking, 3 Things to Include What belongs in your notes? Begin by including anything your instructor writes on the board. Instructors give verbal cues when they want you to include something in your notes. Your instructor may pause to give you enough time to finish writing in your notebook or may repeat the phrase to make sure you accurately write down the statement in your notes.

Some instructors warn the class that a particular topic, problem, or step is difficult. When you hear this, your instructor is telling you to take the best notes you can because you will need them later.

Exercises 9.3

MyMathLab

PRACTICE

WATCH

DOWNLOAD

READ

REVIEW

Vocabulary

1. Two radical expressions are called ____________ if they have the same index and the same radicand.
2. To add radical expressions with like radicals, add the ____________ of the radicals and place the sum in front of the like radical.

Simplify.

3. $\sqrt{12}$
4. $\sqrt{18}$
5. $\sqrt{45}$
6. $\sqrt{252}$
7. $\sqrt{432}$
8. $\sqrt{448}$
9. $\sqrt[3]{192}$
10. $\sqrt[3]{800}$
11. $\sqrt[3]{1296}$
12. $\sqrt[3]{2187}$
13. $\sqrt[4]{1200}$
14. $\sqrt[5]{448}$

Simplify the radical expression. Assume that all variables represent nonnegative real numbers.

15. $\sqrt{x^9}$
16. $\sqrt{a^{15}}$
17. $\sqrt[3]{m^{13}}$
18. $\sqrt[3]{x^{40}}$
19. $\sqrt[5]{x^{74}}$
20. $\sqrt[4]{b^{82}}$
21. $\sqrt{x^{15}y^{12}}$
22. $\sqrt{a^{23}b^3}$
23. $\sqrt{xy^7}$
24. $\sqrt{x^{20}y^{11}}$
25. $\sqrt{x^{33}y^{17}z^{16}}$
26. $\sqrt{r^{19}s^{18}t^{28}}$
27. $\sqrt[3]{x^{13}y^{11}z}$
28. $\sqrt[3]{x^{27}y^{24}z^{32}}$
29. $\sqrt[4]{a^{14}b^{35}c^9}$
30. $\sqrt[5]{a^{33}b^5c^{24}}$
31. $\sqrt{27x^{11}y^{12}}$
32. $\sqrt{150x^{14}y^5}$
33. $\sqrt[4]{162x^{23}y^6z^{16}}$
34. $\sqrt[3]{250x^{11}y^{25}z^3}$

Add or subtract. Assume that all variables represent nonnegative real numbers.

35. $10\sqrt{5} + 17\sqrt{5}$
36. $8\sqrt{11} - 19\sqrt{11}$
37. $9\sqrt[3]{4} - 15\sqrt[3]{4}$
38. $14\sqrt[3]{20} + 6\sqrt[3]{20}$
39. $10\sqrt{15} - 3\sqrt{15} + 8\sqrt{15}$
40. $7\sqrt{6} - 19\sqrt{6} - 32\sqrt{6}$
41. $10\sqrt{7} - 3\sqrt{3} - 18\sqrt{3} - 5\sqrt{7}$
42. $9\sqrt{10} + \sqrt{17} - 3\sqrt{17} + 8\sqrt{10}$
43. $(8\sqrt{2} - 4\sqrt{3}) - (6\sqrt{2} + 9\sqrt{3})$
44. $(2\sqrt{5} + 7\sqrt{6}) - (11\sqrt{6} - 13\sqrt{5})$
45. $5\sqrt{x} - 7\sqrt{x}$
46. $13\sqrt{y} + 12\sqrt{y}$
47. $3\sqrt[5]{a} + 8\sqrt[5]{a}$
48. $11\sqrt[4]{x} - 2\sqrt[4]{x}$
49. $16\sqrt{x} - 9\sqrt{x} + 15\sqrt{x}$
50. $-24\sqrt{x} - 17\sqrt{x} + 3\sqrt{x}$
51. $4\sqrt{x} - 11\sqrt{y} - 6\sqrt{y} + \sqrt{x}$
52. $-2\sqrt{a} + 9\sqrt{b} - 18\sqrt{a} - 5\sqrt{b}$
53. $8\sqrt[3]{x} + 7\sqrt[4]{x} - 10\sqrt[4]{x} + 13\sqrt[3]{x}$
54. $\sqrt[4]{ab^3} + 12\sqrt[4]{a^2b^3} + 23\sqrt[4]{ab^3} - 25\sqrt[4]{a^2b^3}$
55. $\sqrt{50} + \sqrt{128}$
56. $\sqrt{48} - \sqrt{300}$
57. $7\sqrt{24} - 9\sqrt{150}$
58. $8\sqrt{320} + 13\sqrt{125}$
59. $10\sqrt{3} + 2\sqrt{27} + 5\sqrt{108}$
60. $3\sqrt{98} - 6\sqrt{200} - 14\sqrt{2}$

61. $6\sqrt[4]{2} + 7\sqrt[4]{1250} - 2\sqrt[4]{162}$

62. $3\sqrt[3]{3} - 5\sqrt[3]{24} - 8\sqrt[3]{192}$

63. $20\sqrt{12} - 9\sqrt{18} - 13\sqrt{147} - 8\sqrt{72}$

64. $6\sqrt{50} + 7\sqrt{180} - 10\sqrt{162} + 15\sqrt{405}$

Find the missing radical expression.

65. $(8\sqrt{3} + 7\sqrt{2}) + (?) = 15\sqrt{3} - 4\sqrt{2}$

66. $(3\sqrt{200} - 6\sqrt{108}) + (?) = 19\sqrt{2} - 50\sqrt{3}$

67. $(3\sqrt{50} - \sqrt{405}) - (?) = 3\sqrt{98} - 13\sqrt{20}$

68. $(4\sqrt{224} - 2\sqrt{360}) - (?) = \sqrt{640} + 4\sqrt{350}$

Writing in Mathematics

Answer in complete sentences.

69. Explain how to simplify $\sqrt{360}$.

70. Explain how to determine whether radical expressions are like radicals.

9.4 Multiplying and Dividing Radical Expressions

OBJECTIVES

1. Multiply radical expressions.
2. Use the distributive property to multiply radical expressions.
3. Multiply radical expressions that have two or more terms.
4. Multiply radical expressions that are conjugates.
5. Rationalize a denominator that has one term.
6. Rationalize a denominator that has two terms.

Multiplying Two Radical Expressions

Objective 1 Multiply radical expressions. In Section 9.1, we learned how to multiply one radical by another as well as how to divide one radical by another.

Multiplying Radicals with the Same Index

For any positive integer $n > 1$, if $\sqrt[n]{x}$ and $\sqrt[n]{y}$ are real numbers, then

$$\sqrt[n]{x} \cdot \sqrt[n]{y} = \sqrt[n]{xy} \quad \text{and} \quad \frac{\sqrt[n]{x}}{\sqrt[n]{y}} = \sqrt[n]{\frac{x}{y}}.$$

In this section, we will build on that knowledge and learn how to multiply and divide expressions containing two or more radicals. We will begin with a review of multiplying radicals.

EXAMPLE 1 Multiply $\sqrt{18} \cdot \sqrt{8}$.

SOLUTION Because the index of each radical is the same and neither radicand is a perfect square, we begin by multiplying the two radicands.

$$\begin{aligned} \sqrt{18} \cdot \sqrt{8} &= \sqrt{144} && \text{Multiply the radicands.} \\ &= 12 && \text{Simplify the radical.} \end{aligned}$$

EXAMPLE 2 Multiply $\sqrt[3]{a^{13}b^7c^2} \cdot \sqrt[3]{a^{10}b^8c^{17}}$.

SOLUTION Because powers in each radical are greater than the index of that radical, we could simplify each radical first. However, then we would have to multiply and simplify the radical again. A more efficient approach is to multiply first and then simplify only once.

$$\sqrt[3]{a^{13}b^7c^2} \cdot \sqrt[3]{a^{10}b^8c^{17}} = \sqrt[3]{a^{23}b^{15}c^{19}}$$ Multiply the radicands by adding the exponents for each factor.

$$= a^7b^5c^6\sqrt[3]{a^2c}$$ Simplify the radical. For each factor, take out as many groups of 3 as possible.

EXAMPLE 3 Multiply $9\sqrt{6} \cdot 7\sqrt{10}$.

SOLUTION Because the index is the same for each radical, we can multiply the radicands together. The factors in front of each radical, 9 and 7, will be multiplied by each other as well. After multiplying, we finish by simplifying the radical completely.

$$9\sqrt{6} \cdot 7\sqrt{10} = 63\sqrt{60}$$ Multiply factors in front of the radicals and multiply the radicands.

$$= 63\sqrt{2^2 \cdot 3 \cdot 5}$$ Factor the radicand.

$$= 63 \cdot 2\sqrt{3 \cdot 5}$$ Simplify the radical.

$$= 126\sqrt{15}$$ Multiply.

Quick Check 1

Multiply.

a) $\sqrt{45} \cdot \sqrt{80}$
b) $\sqrt[3]{x^4y^2z} \cdot \sqrt[3]{x^8yz^4}$
c) $4\sqrt{8} \cdot 9\sqrt{6}$

It is important to note that whenever we multiply the square root of an expression by the square root of the same expression, the product is equal to the expression itself as long as the expression is nonnegative.

Multiplying a Square Root by Itself

For any nonnegative x,

$$\sqrt{x} \cdot \sqrt{x} = x.$$

Using the Distributive Property with Radical Expressions

Objective 2 Use the distributive property to multiply radical expressions. Now we will use the distributive property to multiply radical expressions.

EXAMPLE 4 Multiply $\sqrt{5}(\sqrt{10} - \sqrt{5})$.

SOLUTION Begin by distributing $\sqrt{5}$ to each term in the parentheses; then multiply the radicals as in the previous examples.

$$\sqrt{5}(\sqrt{10} - \sqrt{5}) = \sqrt{5} \cdot \sqrt{10} - \sqrt{5} \cdot \sqrt{5}$$ Distribute $\sqrt{5}$.

$$= \sqrt{50} - 5$$ Multiply. Recall that $\sqrt{5} \cdot \sqrt{5} = 5$.

$$= 5\sqrt{2} - 5$$ Simplify the radical.

Quick Check 2

Multiply $\sqrt{12}(\sqrt{3} + \sqrt{15})$.

EXAMPLE 5 Multiply $\sqrt[4]{x^{11}y^6}(\sqrt[4]{x^5y^{19}} + \sqrt[4]{x^{11}y^2})$. (Assume that x and y are nonnegative.)

SOLUTION We begin by using the distributive property. Because each radical has an index of 4, we can then multiply the radicals.

$$\sqrt[4]{x^{11}y^6}(\sqrt[4]{x^5y^{19}} + \sqrt[4]{x^{11}y^2}) = \sqrt[4]{x^{11}y^6} \cdot \sqrt[4]{x^5y^{19}} + \sqrt[4]{x^{11}y^6} \cdot \sqrt[4]{x^{11}y^2}$$ Distribute $\sqrt[4]{x^{11}y^6}$.

$$= \sqrt[4]{x^{16}y^{25}} + \sqrt[4]{x^{22}y^8}$$ Multiply by adding exponents for each factor.

$$= x^4y^6\sqrt[4]{y} + x^5y^2\sqrt[4]{x^2}$$ Simplify each radical.

Because the radicals are not like radicals, we cannot simplify this expression any further.

Quick Check 3

Multiply $\sqrt{x^9y^6}(\sqrt{x^3y^6} - \sqrt{x^8y^7})$. (Assume that x and y are nonnegative.)

Multiplying Radical Expressions That Have at Least Two Terms

Objective 3 Multiply radical expressions that have two or more terms.

EXAMPLE 6 Multiply $(8\sqrt{6} + \sqrt{2})(3\sqrt{12} - 4\sqrt{3})$.

SOLUTION We begin by multiplying each term in the first set of parentheses by each term in the second set of parentheses using the distributive property. Because there are two terms in each set of parentheses we can use the FOIL technique. Multiply factors outside a radical by factors outside a radical and multiply radicands by radicands.

$$(8\sqrt{6} + \sqrt{2})(3\sqrt{12} - 4\sqrt{3})$$

$$= 8 \cdot 3\sqrt{6 \cdot 12} - 8 \cdot 4\sqrt{6 \cdot 3} + 3\sqrt{2 \cdot 12} - 4\sqrt{2 \cdot 3}$$ Distribute.

$$= 24\sqrt{72} - 32\sqrt{18} + 3\sqrt{24} - 4\sqrt{6}$$ Multiply.

$$= 24 \cdot 6\sqrt{2} - 32 \cdot 3\sqrt{2} + 3 \cdot 2\sqrt{6} - 4\sqrt{6}$$ Simplify each radical.

$$= 144\sqrt{2} - 96\sqrt{2} + 6\sqrt{6} - 4\sqrt{6}$$ Multiply.

$$= 48\sqrt{2} + 2\sqrt{6}$$ Combine like radicals.

EXAMPLE 7 Multiply $(\sqrt{5} + \sqrt{6})^2$.

SOLUTION To square any binomial, multiply it by itself.

$$(\sqrt{5} + \sqrt{6})^2 = (\sqrt{5} + \sqrt{6})(\sqrt{5} + \sqrt{6})$$ Rewrite as $(5 + \sqrt{6})(5 + \sqrt{6})$.

$$= \sqrt{5} \cdot \sqrt{5} + \sqrt{5} \cdot \sqrt{6} + \sqrt{6} \cdot \sqrt{5} + \sqrt{6} \cdot \sqrt{6}$$ Distribute.

$$= 5 + \sqrt{30} + \sqrt{30} + 6$$ Multiply.

$$= 11 + 2\sqrt{30}$$ Combine like terms.

Quick Check 4

Multiply. **a)** $(5\sqrt{3} + 4\sqrt{2})(7\sqrt{3} - 6\sqrt{2})$ **b)** $(\sqrt{7} - \sqrt{10})^2$

A WORD OF CAUTION Whenever we square a binomial, such as $(\sqrt{5} + \sqrt{6})^2$, we must multiply the binomial by itself. We cannot simply square each term.

$$(a + b)^2 \neq a^2 + b^2$$

Multiplying Conjugates

Objective 4 Multiply radical expressions that are conjugates. The expressions $\sqrt{13} + \sqrt{5}$ and $\sqrt{13} - \sqrt{5}$ are called **conjugates**. Two expressions are conjugates if they are of the form $x + y$ and $x - y$. Notice that the two terms are the same, with the exception of the sign of the second term.

The multiplication of two conjugates follows a pattern. Let's look at the product $(\sqrt{x} + \sqrt{y})(\sqrt{x} - \sqrt{y})$.

$$(\sqrt{x} + \sqrt{y})(\sqrt{x} - \sqrt{y}) = x - \sqrt{xy} + \sqrt{xy} - y$$ Distribute. Note that $\sqrt{x} \cdot \sqrt{x} = x$ and $\sqrt{y} \cdot \sqrt{y} = y$.

$$= x - y$$ Combine the two opposite terms $-\sqrt{xy}$ and $\sqrt{xy}$.

Whenever we multiply conjugates, the two middle terms will be opposites of each other; therefore, their sum is 0. We can multiply the first term in the first set of parentheses by the first term in the second set of parentheses, multiply the second term in the first set of parentheses by the second term in the second set of parentheses, and then place a minus sign between the two products.

Multiplication of Two Conjugates

$$\overbrace{(a + b)(a - b)}^{a^2} = a^2 - b^2$$

(the first terms multiply to a^2; the second terms multiply to b^2)

EXAMPLE 8 Multiply $(\sqrt{17} + \sqrt{23})(\sqrt{17} - \sqrt{23})$.

SOLUTION These two expressions are conjugates, so we multiply them accordingly.

$$(\sqrt{17} + \sqrt{23})(\sqrt{17} - \sqrt{23}) = (\sqrt{17})^2 - (\sqrt{23})^2$$ Multiply using the rule for multiplying conjugates.

$$= 17 - 23$$ Square each square root.

$$= -6$$ Subtract.

EXAMPLE 9 Multiply $(2\sqrt{6} - 8\sqrt{5})(2\sqrt{6} + 8\sqrt{5})$.

SOLUTION When we multiply two conjugates, we must remember to multiply the factors in front of the radicals by each other and to multiply the radicands by each other.

$$(2\sqrt{6} - 8\sqrt{5})(2\sqrt{6} + 8\sqrt{5}) = 2\sqrt{6} \cdot 2\sqrt{6} - 8\sqrt{5} \cdot 8\sqrt{5}$$ Multiply the conjugates.

$$= 4 \cdot 6 - 64 \cdot 5$$ Multiply.

$$= 24 - 320$$ Multiply.

$$= -296$$ Subtract.

Quick Check 5

Multiply. **a)** $(\sqrt{38} - \sqrt{29})(\sqrt{38} + \sqrt{29})$ **b)** $(8\sqrt{11} - 5\sqrt{7})(8\sqrt{11} + 5\sqrt{7})$

Rationalizing the Denominator

Objective 5 Rationalize a denominator that has one term. Earlier in this chapter, we introduced a criterion for determining whether a radical was simplified. We stated that for a radical to be simplified, its index must be greater than any power in the radical. We now add two other rules.

- There can be no fractions in a radicand.
- There can be no radicals in the denominator of a fraction.

For example, we would not consider the following expressions to be simplified: $\sqrt{\frac{3}{10}}, \frac{9}{\sqrt{2}}, \frac{6}{\sqrt{4}-\sqrt{3}}$, and $\frac{\sqrt{6}+\sqrt{12}}{\sqrt{3}-8}$. The process of rewriting an expression without a radical in its denominator is called **rationalizing the denominator**.

The rational expression $\frac{\sqrt{16}}{\sqrt{49}}$ is not simplified, as there is a radical in the denominator. However, we know that $\sqrt{49} = 7$; so we can simplify the denominator in such a way that it no longer contains a radical.

$$\frac{\sqrt{16}}{\sqrt{49}} = \frac{4}{7} \quad \text{Simplify the numerator and denominator.}$$

The radical expression $\sqrt{\frac{75}{3}}$ is not simplified, as there is a fraction inside the radical. We can simplify $\frac{75}{3}$ to be 25, rewriting the radical without a fraction inside.

$$\sqrt{\frac{75}{3}} = \sqrt{25} = 5 \quad \text{Simplify the fraction, then } \sqrt{25}.$$

Suppose we needed to simplify $\frac{\sqrt{15}}{\sqrt{2}}$. We cannot simplify $\sqrt{2}$, and the fraction itself cannot be simplified. In such a case, we will multiply both the numerator and denominator by an expression that will allow us to rewrite the denominator without a radical. Then we simplify.

EXAMPLE 10 Rationalize the denominator: $\frac{\sqrt{15}}{\sqrt{2}}$

SOLUTION If we multiply the denominator by $\sqrt{2}$, the denominator will equal 2 and will be rationalized.

$$\frac{\sqrt{15}}{\sqrt{2}} = \frac{\sqrt{15}}{\sqrt{2}} \cdot \frac{\sqrt{2}}{\sqrt{2}}$$ Multiply by $\frac{\sqrt{2}}{\sqrt{2}}$, which makes the denominator equal to 2. Multiplying by $\frac{\sqrt{2}}{\sqrt{2}}$ is equivalent to multiplying by 1.

$$= \frac{\sqrt{30}}{2}$$ Multiply.

Because $\sqrt{30}$ cannot be simplified, this expression cannot be simplified further.

Quick Check 6

Rationalize the denominator. $\frac{\sqrt{70}}{\sqrt{3}}$

EXAMPLE 11 Rationalize the denominator: $\frac{11}{\sqrt{12}}$

SOLUTION At first glance, we might think that multiplying the numerator and denominator by $\sqrt{12}$ is the correct way to proceed. However, if we multiply the numerator and denominator by $\sqrt{3}$, the radicand in the denominator will be 36, which is a perfect square.

$$\frac{11}{\sqrt{12}} = \frac{11}{\sqrt{12}} \cdot \frac{\sqrt{3}}{\sqrt{3}}$$ Multiply by $\frac{\sqrt{3}}{\sqrt{3}}$ to make the radicand in the denominator a perfect square.

$$= \frac{11\sqrt{3}}{\sqrt{36}}$$ Multiply.

$$= \frac{11\sqrt{3}}{6}$$ Simplify the radical in the denominator.

Multiplying by $\frac{\sqrt{12}}{\sqrt{12}}$ also would be valid, but the subsequent process of simplifying would be difficult. One way to determine the best expression by which to multiply is to completely factor the radicand in the denominator. In this example, $12 = 2^2 \cdot 3$. The factor 2 is already a perfect square, but the factor 3 is not. Multiplying by $\sqrt{3}$ makes the factor 3 a perfect square as well.

Quick Check 7

Rationalize the denominator: $\frac{2}{\sqrt{20}}$

EXAMPLE 12 Rationalize the denominator. $\sqrt{\frac{5a^3b^{14}c^6}{80a^9b^7c^{11}}}$ (Assume that all variables represent nonnegative values.)

SOLUTION Notice that the numerator and denominator have common factors. We begin by simplifying the fraction to lowest terms.

$$\sqrt{\frac{5a^3b^{14}c^6}{80a^9b^7c^{11}}} = \sqrt{\frac{b^7}{16a^6c^5}}$$ Divide out common factors and simplify.

$$= \frac{\sqrt{b^7}}{\sqrt{16a^6c^5}}$$ Rewrite as the quotient of two square roots. Notice that the factors 16 and a^6 are already perfect squares, but c^5 is not.

$$= \frac{\sqrt{b^7}}{\sqrt{16a^6c^5}} \cdot \frac{\sqrt{c}}{\sqrt{c}}$$ Multiply by $\frac{\sqrt{c}}{\sqrt{c}}$.

$$= \frac{\sqrt{b^7c}}{\sqrt{16a^6c^6}}$$ Multiply.

$$= \frac{b^3\sqrt{bc}}{4a^3c^3}$$ Simplify both radicals.

Quick Check 8

Rationalize the denominator. $\sqrt{\frac{12x^2y^8z^5}{75x^5y^3z^{15}}}$ (Assume that all variables represent nonnegative values.)

Rationalizing a Denominator That Has Two Terms

Objective 6 Rationalize a denominator that has two terms. In the previous examples, each denominator had only one term. If a denominator is a binomial that contains one or two square roots, we rationalize the denominator by multiplying the numerator and denominator by the conjugate of the denominator. For example, consider the expression $\frac{6}{\sqrt{11} + \sqrt{7}}$. We know from earlier in this section that multiplying $\sqrt{11} + \sqrt{7}$ by its conjugate $\sqrt{11} - \sqrt{7}$ produces a product that does not contain a radical.

EXAMPLE 13 Rationalize the denominator: $\frac{6}{\sqrt{11} + \sqrt{7}}$

SOLUTION Because this denominator is a binomial, we multiply the numerator and denominator by the conjugate of the denominator.

$$\frac{6}{\sqrt{11} + \sqrt{7}} = \frac{6}{\sqrt{11} + \sqrt{7}} \cdot \frac{\sqrt{11} - \sqrt{7}}{\sqrt{11} - \sqrt{7}}$$ Multiply the numerator and denominator by the conjugate of the denominator $(\sqrt{11} - \sqrt{7})$.

$$= \frac{6\sqrt{11} - 6\sqrt{7}}{\sqrt{11} \cdot \sqrt{11} - \sqrt{7} \cdot \sqrt{7}}$$ Distribute in the numerator. Multiply conjugates in the denominator.

$$= \frac{6\sqrt{11} - 6\sqrt{7}}{11 - 7}$$ $\sqrt{11} \cdot \sqrt{11} = 11$, $\sqrt{7} \cdot \sqrt{7} = 7$

$$= \frac{6\sqrt{11} - 6\sqrt{7}}{4}$$ Subtract.

$$= \frac{6(\sqrt{11} - \sqrt{7})}{4}$$ Factor the numerator.

$$= \frac{\overset{3}{\cancel{6}}(\sqrt{11} - \sqrt{7})}{\underset{2}{\cancel{4}}}$$ Divide out the common factor 2.

$$= \frac{3(\sqrt{11} - \sqrt{7})}{2}$$ Simplify.

Quick Check 9

Rationalize the denominator: $\frac{\sqrt{15}}{\sqrt{5} + \sqrt{3}}$

EXAMPLE 14 Rationalize the denominator: $\frac{4\sqrt{3} - 3\sqrt{5}}{2\sqrt{3} - \sqrt{5}}$

SOLUTION Because the denominator is a binomial, we begin by multiplying the numerator and denominator by the conjugate of the denominator, which is $2\sqrt{3} + \sqrt{5}$.

$$\frac{4\sqrt{3} - 3\sqrt{5}}{2\sqrt{3} - \sqrt{5}} = \frac{4\sqrt{3} - 3\sqrt{5}}{2\sqrt{3} - \sqrt{5}} \cdot \frac{2\sqrt{3} + \sqrt{5}}{2\sqrt{3} + \sqrt{5}}$$ Multiply the numerator and denominator by the conjugate of the denominator.

$$= \frac{4\sqrt{3} \cdot 2\sqrt{3} + 4\sqrt{3} \cdot \sqrt{5} - 3\sqrt{5} \cdot 2\sqrt{3} - 3\sqrt{5} \cdot \sqrt{5}}{2\sqrt{3} \cdot 2\sqrt{3} - \sqrt{5} \cdot \sqrt{5}}$$ Multiply numerators and denominators.

$$= \frac{8 \cdot 3 + 4\sqrt{15} - 6\sqrt{15} - 3 \cdot 5}{4 \cdot 3 - 5}$$ Simplify each product.

$$= \frac{24 + 4\sqrt{15} - 6\sqrt{15} - 15}{12 - 5}$$ Multiply.

$$= \frac{9 - 2\sqrt{15}}{7}$$ Combine like terms and like radicals.

Quick Check 10

Rationalize the denominator: $\frac{2\sqrt{6} - 7\sqrt{2}}{2\sqrt{6} + 3\sqrt{2}}$

BUILDING YOUR STUDY STRATEGY

Note Taking, 4 Being an Active Learner Some students fall into the trap of taking notes mechanically without thinking about the material. In a math class, there can be no learning without thinking. Try to be an active learner while taking notes. Do your best to understand each statement you write in your notes.

When your instructor writes a problem on the board, try to solve it yourself in your notes. When you have finished solving the problem, compare your solution to your instructor's solution. If you make a mistake or your solution is different from your instructor's solution, you can make editor's notes on your solution.

Exercises 9.4

MyMathLab

PRACTICE

WATCH

DOWNLOAD

READ

REVIEW

Vocabulary

1. For any nonnegative x, $\sqrt{x} \cdot \sqrt{x} =$ ______.
2. Two expressions of the form $x + y$ and ______ are called conjugates.
3. The process of rewriting an expression without a radical in its denominator is called ______ the denominator.
4. To rationalize a denominator containing two terms and at least one square root, multiply the numerator and denominator by the ______ of the denominator.

Multiply. Assume that all variables represent nonnegative real numbers.

5. $\sqrt{12} \cdot \sqrt{75}$
6. $\sqrt{98} \cdot \sqrt{8}$
7. $4\sqrt{27} \cdot 6\sqrt{3}$
8. $8\sqrt{125} \cdot 2\sqrt{20}$
9. $7\sqrt{32} \cdot 5\sqrt{6}$
10. $9\sqrt{30} \cdot 4\sqrt{15}$
11. $6\sqrt[3]{24} \cdot 2\sqrt[3]{18}$
12. $5\sqrt[3]{36} \cdot 9\sqrt[3]{30}$
13. $\sqrt{x^{17}} \cdot \sqrt{x^5}$
14. $\sqrt{x^{33}} \cdot \sqrt{x^{13}}$
15. $\sqrt{6a^7b^6} \cdot \sqrt{10a^5b^9}$
16. $\sqrt{21a^{10}b^3} \cdot \sqrt{14a^9b^9}$
17. $\sqrt[4]{x^{15}y^{10}z^5} \cdot \sqrt[4]{x^9y^2z^{18}}$
18. $\sqrt[3]{x^{10}y^{14}z^{17}} \cdot \sqrt[3]{x^6y^{19}z^8}$

Multiply. Assume that all variables represent nonnegative real numbers.

19. $\sqrt{2}(\sqrt{8} - \sqrt{6})$
20. $\sqrt{6}(\sqrt{2} + \sqrt{3})$
21. $\sqrt{98}(\sqrt{18} + \sqrt{10})$
22. $\sqrt{15}(\sqrt{30} - \sqrt{35})$
23. $6\sqrt{3}(9\sqrt{6} - 4\sqrt{15})$
24. $17\sqrt{10}(8\sqrt{2} + 3\sqrt{5})$
25. $\sqrt{m^5n^9}(\sqrt{m^3n^4} + \sqrt{m^8n^7})$
26. $\sqrt{xy^{11}}(\sqrt{x^7y^3} - \sqrt{x^{39}y^{15}})$
27. $\sqrt[3]{a^7b^{10}c^{13}}(\sqrt[3]{a^{11}b^{35}} + \sqrt[3]{a^{23}c^8})$
28. $\sqrt[3]{x^4y^7z^{10}}(\sqrt[3]{x^8y^{16}z^{24}} - \sqrt[3]{x^{16}y^{12}z^2})$

Multiply.

29. $(\sqrt{2} + \sqrt{5})(\sqrt{2} - \sqrt{3})$
30. $(\sqrt{3} + \sqrt{8})(\sqrt{6} - \sqrt{2})$
31. $(7\sqrt{2} + 4\sqrt{3})(5\sqrt{2} - \sqrt{3})$
32. $(8\sqrt{5} - 3\sqrt{2})(2\sqrt{5} - 4\sqrt{2})$
33. $(2\sqrt{7} - 7\sqrt{3})(2\sqrt{7} - 6\sqrt{3})$
34. $(\sqrt{6} + 2\sqrt{11})(9\sqrt{6} + \sqrt{11})$
35. $(6\sqrt{3} + 4\sqrt{5})(3\sqrt{5} - 2\sqrt{3})$
36. $(2\sqrt{2} - 9\sqrt{7})(8\sqrt{7} - 5\sqrt{2})$
37. $(4\sqrt{5} - 3\sqrt{2})^2$
38. $(6 + 7\sqrt{3})^2$

Multiply the conjugates.

39. $(\sqrt{7} - \sqrt{10})(\sqrt{7} + \sqrt{10})$
40. $(\sqrt{26} + \sqrt{3})(\sqrt{26} - \sqrt{3})$
41. $(9\sqrt{6} + 3\sqrt{8})(9\sqrt{6} - 3\sqrt{8})$
42. $(5\sqrt{18} - 2\sqrt{24})(5\sqrt{18} + 2\sqrt{24})$
43. $(8 - 11\sqrt{2})(8 + 11\sqrt{2})$
44. $(4\sqrt{3} + 15)(4\sqrt{3} - 15)$

Simplify. Assume that all variables represent nonnegative real numbers.

45. $\sqrt{\dfrac{12}{x^4y^6}}$

46. $\sqrt{\dfrac{b^7}{a^{10}c^2}}$

47. $\sqrt{\dfrac{126}{7}}$

48. $\sqrt{\dfrac{100}{5}}$

49. $\sqrt{\dfrac{135}{20}}$

50. $\sqrt{\dfrac{14}{18}}$

Rationalize the denominator and simplify. Assume that all variables represent nonnegative real numbers.

51. $\dfrac{\sqrt{8}}{\sqrt{3}}$

52. $\dfrac{\sqrt{24}}{\sqrt{7}}$

53. $\dfrac{1}{\sqrt{2}}$

54. $\dfrac{5}{\sqrt{5}}$

55. $\sqrt{\dfrac{27}{11a^7}}$

56. $\sqrt{\dfrac{80}{3n^{10}}}$

57. $\sqrt[3]{\dfrac{s^8}{2r^2t}}$

58. $\sqrt[3]{\dfrac{2x^{10}}{7y^{16}z^{22}}}$

59. $\dfrac{14}{\sqrt{12}}$

60. $\dfrac{15}{\sqrt{50}}$

61. $\dfrac{8}{\sqrt{18a}}$

62. $\dfrac{6}{\sqrt{20b^5}}$

63. $\dfrac{x^4z^8}{\sqrt{x^3y^5z^{10}}}$

64. $\dfrac{xy^2z^9}{\sqrt{x^5y^6z^7}}$

Rationalize the denominator and simplify.

65. $\dfrac{9}{\sqrt{13}-\sqrt{7}}$

66. $\dfrac{10}{\sqrt{20}+\sqrt{2}}$

67. $\dfrac{4\sqrt{3}}{\sqrt{3}+\sqrt{11}}$

68. $\dfrac{5\sqrt{5}}{\sqrt{15}-\sqrt{5}}$

69. $\dfrac{6\sqrt{3}}{\sqrt{13}-3}$

70. $\dfrac{8\sqrt{2}}{4-\sqrt{6}}$

71. $\dfrac{5\sqrt{5}-3\sqrt{3}}{4\sqrt{5}+2\sqrt{3}}$

72. $\dfrac{9\sqrt{3}+4\sqrt{2}}{5\sqrt{3}-4\sqrt{2}}$

73. $\dfrac{7\sqrt{11}+2\sqrt{2}}{4\sqrt{11}-5\sqrt{2}}$

74. $\dfrac{2\sqrt{5}+\sqrt{7}}{4\sqrt{5}-\sqrt{7}}$

75. $\dfrac{4\sqrt{2}-\sqrt{3}}{2\sqrt{2}-3\sqrt{3}}$

76. $\dfrac{2\sqrt{3}-9\sqrt{5}}{6\sqrt{3}-4\sqrt{5}}$

Mixed Practice, 77–98

Simplify. Assume that all variables represent nonnegative real numbers.

77. $\sqrt{\dfrac{90}{98}}$

78. $\sqrt{10x^5}\cdot\sqrt{18x^9}$

79. $(8\sqrt{7}+5\sqrt{5})^2$

80. $\dfrac{28}{\sqrt{32}}$

81. $\sqrt[3]{25a^8bc^7}\cdot\sqrt[3]{25ab^5c^7}$

82. $\sqrt[3]{\dfrac{2x^4y^{13}}{54x^{10}y^2}}$

83. $\dfrac{10\sqrt{5}-7\sqrt{7}}{\sqrt{5}+\sqrt{7}}$

84. $\sqrt{162x^9y^8z^{15}}$

85. $\sqrt[3]{1080}$

86. $\sqrt{60}(7\sqrt{3}-2\sqrt{15})$

87. $(3\sqrt{10}-7\sqrt{2})(4\sqrt{10}+9\sqrt{5})$

88. $\dfrac{8\sqrt{2}}{\sqrt{10}}$

89. $-\sqrt{7x^9}(2\sqrt{14x}-\sqrt{7x^{13}})$

90. $(9\sqrt{7}-8\sqrt{8})(9\sqrt{7}+8\sqrt{8})$

91. $\sqrt[4]{a^{17}b^{34}c^{44}}$

92. $\sqrt{1008}$

93. $\sqrt{\dfrac{a^5b^8}{12c^9}}$

94. $\dfrac{13\sqrt{5}+6}{10-7\sqrt{5}}$

95. $\dfrac{10x^2}{\sqrt[3]{36x^8}}$

96. $(20\sqrt{3}-7\sqrt{10})(8\sqrt{3}+15\sqrt{2})$

97. $(18\sqrt{7}-\sqrt{11})(18\sqrt{7}+\sqrt{11})$

98. $(4\sqrt{14}-3\sqrt{7})^2$

99. Develop a general formula for the product $(\sqrt{a}+\sqrt{b})^2$.

100. Develop a general formula for the product $(a\sqrt{b}+c\sqrt{d})^2$.

Writing in Mathematics

Answer in complete sentences.

101. Explain how to determine whether two radical expressions are conjugates.

102. ***Solutions Manual*** Write a solutions manual page for the following problem.
Simplify $(5\sqrt{2} + 4\sqrt{3})(8\sqrt{2} - 3\sqrt{3})$.

Quick Review Exercises

Section 9.4

Find the prime factorization of the given number.

1. 144

2. 1400

3. 1024

4. 29,106

9.5 Radical Equations and Applications of Radical Equations

OBJECTIVES

1. Solve radical equations.
2. Solve equations containing radical functions.
3. Solve equations containing rational exponents.
4. Solve equations in which a radical is equal to a variable expression.
5. Solve equations containing two radicals.
6. Solve applied problems involving a pendulum and its period.
7. Solve other applied problems involving radicals.

Solving Radical Equations

Objective 1 Solve radical equations. A **radical equation** is an equation containing one or more radicals. Here are some examples of radical equations.

$$\sqrt{x} = 9 \qquad \sqrt[3]{2x-5} = 3 \qquad x + \sqrt{x} = 20$$

$$\sqrt[4]{3x-8} = \sqrt[4]{2x+11} \qquad \sqrt{x-4} + \sqrt{x+8} = 6$$

In this section, we will learn how to solve radical equations. We will find a way to convert a radical equation to an equivalent equation that we already know how to solve.

Raising Equal Numbers to the Same Power

If two numbers a and b are equal, then for any n, $a^n = b^n$.

If we raise two equal numbers to the same power, they remain equal to each other. We will use this fact to solve radical equations.

Solving Radical Equations

- Isolate a radical term containing the variable on one side of the equation.
- Raise both sides of the equation to the nth power, where n is the index of the radical. *For any nonnegative number x and any integer n* > 1, $(\sqrt[n]{x})^n = x$.
- If the resulting equation does not contain a radical, solve the equation. If the resulting equation does contain a radical, begin the process again by isolating the radical on one side of the equation.
- Check the solution(s).

It is crucial that we check all solutions when solving a radical equation, as raising both sides of an equation to an even power can introduce **extraneous solutions**. When raising both sides of an equation to an nth power, it is possible to arrive at an extraneous solution—a solution that does not satisfy the original equation. In solving a radical equation, all solutions must be checked.

EXAMPLE 1 Solve $\sqrt{x-5} = 3$.

SOLUTION Because the radical $\sqrt{x-5}$ is already isolated on the left side of the equation, we begin by squaring both sides of the equation.

$$\sqrt{x-5} = 3$$
$$(\sqrt{x-5})^2 = 3^2 \quad \text{Square both sides.}$$
$$x - 5 = 9 \quad \text{Simplify.}$$
$$x = 14 \quad \text{Add 5 to both sides.}$$

We need to check this solution using the original equation.

$$\sqrt{14-5} = 3 \quad \text{Substitute 14 for } x.$$
$$\sqrt{9} = 3 \quad \text{Subtract.}$$
$$3 = 3 \quad \text{Simplify the square root.}$$

The solution $x = 14$ checks, so the solution set is {14}.

EXAMPLE 2 Solve $\sqrt[3]{6x+4} + 7 = 11$.

SOLUTION In this example, begin by isolating the radical.

$$\sqrt[3]{6x+4} + 7 = 11$$
$$\sqrt[3]{6x+4} = 4 \quad \text{Subtract 7 from both sides to isolate the radical.}$$
$$(\sqrt[3]{6x+4})^3 = 4^3 \quad \text{Raise both sides of the equation to the third power.}$$
$$6x + 4 = 64 \quad \text{Simplify.}$$
$$6x = 60 \quad \text{Subtract 4 from both sides.}$$
$$x = 10 \quad \text{Divide both sides by 6.}$$

Now check the solution using the original equation.

$$\sqrt[3]{6(10)+4} + 7 = 11 \quad \text{Substitute 10 for } x.$$
$$\sqrt[3]{64} + 7 = 11 \quad \text{Simplify the radicand.}$$
$$4 + 7 = 11 \quad \text{Simplify the cube root.}$$
$$11 = 11 \quad \text{Add. The solution checks.}$$

The solution set is {10}. Because we raised both sides of the equation to an odd power, we did not introduce any extraneous solutions.

Quick Check 1

Solve.

a) $\sqrt{x+2} = 7$
b) $\sqrt[3]{x+9} + 10 = 6$

EXAMPLE 3 Solve $\sqrt{x} + 8 = 5$.

SOLUTION Begin by isolating $\sqrt{x}$ on the left side of the equation.

$$\sqrt{x} + 8 = 5$$
$$\sqrt{x} = -3 \quad \text{Subtract 8 from both sides to isolate the square root.}$$
$$(\sqrt{x})^2 = (-3)^2 \quad \text{Square both sides of the equation.}$$
$$x = 9 \quad \text{Simplify.}$$

Now check this solution using the original equation.

$\sqrt{9} + 8 = 5$	Substitute 9 for x.
$3 + 8 = 5$	Simplify the square root. The principal square root of 9 is 3, not -3.
$11 = 5$	Add.

This solution does not check, so it is an extraneous solution. The equation has no solution; the solution set is $\varnothing$.

Quick Check 2

Solve $\sqrt{2x - 9} - 8 = -11$.

> If we obtain an equation in which an even root is equal to a negative number, such as $\sqrt{x} = -3$, this equation will not have any solutions. This is because the principal even root of a number, if it exists, cannot be negative.

Solving Equations Involving Radical Functions

Objective 2 Solve equations containing radical functions.

EXAMPLE 4 For $f(x) = \sqrt{x^2 + 5x + 11} - 2$, find all values for which $f(x) = 3$.

SOLUTION Begin by setting the function equal to 3.

$f(x) = 3$	
$\sqrt{x^2 + 5x + 11} - 2 = 3$	Replace $f(x)$ with its formula.
$\sqrt{x^2 + 5x + 11} = 5$	Add 2 to both sides to isolate the radical.
$(\sqrt{x^2 + 5x + 11})^2 = 5^2$	Square both sides.
$x^2 + 5x + 11 = 25$	Simplify.
$x^2 + 5x - 14 = 0$	The equation is quadratic, so collect all terms on the left side of the equation.
$(x + 7)(x - 2) = 0$	Factor.
$x + 7 = 0$ or $x - 2 = 0$	Set each factor equal to 0.
$x = -7$ or $x = 2$	Solve.

It is left to the reader to verify that neither solution is an extraneous solution: $f(-7) = 3$ and $f(2) = 3$.

Quick Check 3

For $f(x) = \sqrt{3x - 8} + 4$, find all values x for which $f(x) = 9$.

Solving Equations with Rational Exponents

Objective 3 Solve equations containing rational exponents. The next examples involve equations containing fractional exponents. Recall that $x^{1/n} = \sqrt[n]{x}$.

EXAMPLE 5 Solve $x^{1/3} + 7 = -3$.

SOLUTION Begin by rewriting $x^{1/3}$ as $\sqrt[3]{x}$.

$x^{1/3} + 7 = -3$	
$\sqrt[3]{x} + 7 = -3$	Rewrite $x^{1/3}$ as $\sqrt[3]{x}$ using radical notation.
$\sqrt[3]{x} = -10$	Subtract 7 to isolate the radical.
$(\sqrt[3]{x})^3 = (-10)^3$	Raise both sides to the third power.
$x = -1000$	Simplify.

It is left to the reader to verify that the solution is not an extraneous solution. The solution set is $\{-1000\}$.

EXAMPLE 6 Solve $(1 - 5x)^{1/2} - 3 = 3$.

SOLUTION Begin by rewriting the equation using a radical.

$(1 - 5x)^{1/2} - 3 = 3$	
$\sqrt{1 - 5x} - 3 = 3$	Rewrite using radical notation.
$\sqrt{1 - 5x} = 6$	Add 3 to isolate the radical.
$(\sqrt{1 - 5x})^2 = 6^2$	Square both sides.
$1 - 5x = 36$	Simplify.
$-5x = 35$	Subtract 1 from both sides.
$x = -7$	Divide both sides by -5.

It is left to the reader to verify that the solution is not an extraneous solution. The solution set is $\{-7\}$.

Quick Check 4

Solve.

a) $x^{1/2} - 10 = -7$
b) $(x + 4)^{1/3} + 8 = 2$

Solving Equations in Which a Radical Is Equal to a Variable Expression

Objective 4 Solve equations in which a radical is equal to a variable expression. After we isolated the radical in all of the previous examples, the resulting equation had a radical expression equal to a constant. In the next example, we will learn how to solve equations that result in a radical equal to a variable expression.

EXAMPLE 7 Solve $\sqrt{6x + 16} = x$.

SOLUTION Because the radical is already isolated, we begin by squaring both sides. This will result in a quadratic equation, which we solve by collecting all terms on one side of the equation and factoring.

$\sqrt{6x + 16} = x$	
$(\sqrt{6x + 16})^2 = x^2$	Square both sides.
$6x + 16 = x^2$	Simplify.
$0 = x^2 - 6x - 16$	Collect all terms on the right side of the equation by subtracting $6x$ and 16 from both sides.
$0 = (x - 8)(x + 2)$	Factor.
$x = 8$ or $x = -2$	Set each factor equal to 0 and solve.

Now we check both solutions.

$x = 8$	$x = -2$
$\sqrt{6(8) + 16} = 8$	$\sqrt{6(-2) + 16} = -2$
$\sqrt{48 + 16} = 8$	$\sqrt{-12 + 16} = -2$
$\sqrt{64} = 8$	$\sqrt{4} = -2$
$8 = 8$	$2 = -2$
True	False

The solution $x = -2$ is an extraneous solution. The solution set is $\{8\}$.

Quick Check 5

Solve $\sqrt{12x - 20} = x$.

EXAMPLE 8 Solve $\sqrt{x} + 6 = x$.

SOLUTION Begin by isolating the radical.

$$\sqrt{x} + 6 = x$$

$$\sqrt{x} = x - 6 \quad \text{Subtract 6 from both sides to isolate the radical.}$$

$$(\sqrt{x})^2 = (x - 6)^2 \quad \text{Square both sides.}$$

$$x = (x - 6)(x - 6) \quad \text{Square the binomial by multiplying it by itself.}$$

$$x = x^2 - 12x + 36 \quad \text{Multiply. The resulting equation is quadratic.}$$

$$0 = x^2 - 13x + 36 \quad \text{Subtract } x \text{ to collect all terms on the right side of the equation.}$$

$$0 = (x - 4)(x - 9) \quad \text{Factor.}$$

$$x = 4 \quad \text{or} \quad x = 9 \quad \text{Set each factor equal to 0 and solve.}$$

Now check both solutions.

$x = 4$	$x = 9$
$\sqrt{4} + 6 = 4$	$\sqrt{9} + 6 = 9$
$2 + 6 = 4$	$3 + 6 = 9$
$8 = 4$	$9 = 9$
False	True

The solution $x = 4$ is an extraneous solution. The solution set is $\{9\}$.

Quick Check 6

Solve $\sqrt{2x} + 4 = x$.

Solving Radical Equations Containing Two Radicals

Objective 5 Solve equations containing two radicals.

EXAMPLE 9 Solve $\sqrt[5]{6x + 5} = \sqrt[5]{4x - 3}$.

SOLUTION We raise both sides to the fifth power. Because both radicals have the same index, this will result in an equation that does not contain a radical.

$$\sqrt[5]{6x + 5} = \sqrt[5]{4x - 3}$$

$$(\sqrt[5]{6x + 5})^5 = (\sqrt[5]{4x - 3})^5 \quad \text{Raise both sides to the fifth power.}$$

$$6x + 5 = 4x - 3 \quad \text{Simplify.}$$

$$2x + 5 = -3 \quad \text{Subtract } 4x \text{ from both sides.}$$
$$2x = -8 \quad \text{Subtract 5 from both sides.}$$
$$x = -4 \quad \text{Divide both sides by 2.}$$

It is left to the reader to verify that the solution is not an extraneous solution. This solution checks, and the solution set is $\{-4\}$.

Quick Check 7

Solve $\sqrt[3]{5x - 11} = \sqrt[3]{7x + 33}$.

Occasionally, equations containing two square roots will still contain a square root after we have squared both sides. This will require us to square both sides a second time.

EXAMPLE 10 Solve $\sqrt{x+6} - \sqrt{x-1} = 1$.

SOLUTION We must begin by isolating one of the two radicals on the left side of the equation. We will isolate $\sqrt{x+6}$, as it is positive.

$$\sqrt{x+6} - \sqrt{x-1} = 1$$
$$\sqrt{x+6} = 1 + \sqrt{x-1} \quad \text{Add } \sqrt{x-1} \text{ to isolate the radical } \sqrt{x+6} \text{ on the left side.}$$
$$(\sqrt{x+6})^2 = (1 + \sqrt{x-1})^2 \quad \text{Square both sides.}$$
$$x + 6 = (1 + \sqrt{x-1})(1 + \sqrt{x-1}) \quad \text{Square the binomial on the right side by multiplying it by itself.}$$
$$x + 6 = 1 \cdot 1 + 1 \cdot \sqrt{x-1} + 1 \cdot \sqrt{x-1} + \sqrt{x-1} \cdot \sqrt{x-1} \quad \text{Distribute.}$$
$$x + 6 = 1 + 2\sqrt{x-1} + x - 1 \quad \text{Simplify.}$$
$$x + 6 = 2\sqrt{x-1} + x \quad \text{Combine like terms.}$$
$$6 = 2\sqrt{x-1} \quad \text{Subtract } x \text{ from both sides isolate the to radical.}$$
$$3 = \sqrt{x-1} \quad \text{Divide both sides by 2.}$$
$$3^2 = (\sqrt{x-1})^2 \quad \text{Square both sides.}$$
$$9 = x - 1 \quad \text{Simplify.}$$
$$10 = x \quad \text{Add 1 to both sides.}$$

It is left to the reader to verify that the solution is not an extraneous solution. The solution set is $\{10\}$.

Quick Check 8

Solve: $\sqrt{x+3} - \sqrt{x-2} = 1$.

A Pendulum and Its Period

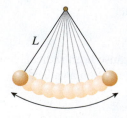

Objective 6 Solve applied problems involving a pendulum and its period. The **period** of a pendulum is the amount of time it takes to swing from one extreme to the other and back again. The period T of a pendulum in seconds can be found by the formula $T = 2\pi\sqrt{\frac{L}{32}}$, where L is the length of the pendulum in feet.

EXAMPLE 11 A pendulum has a length of 3 feet. Find its period, rounded to the nearest hundredth of a second.

SOLUTION Substitute 3 for L in the formula and simplify to find the period T.

$$T = 2\pi\sqrt{\frac{L}{32}}$$

$$T = 2\pi\sqrt{\frac{3}{32}} \quad \text{Substitute 3 for } L.$$

$$T \approx 1.92 \quad \text{Approximate using a calculator.}$$

The period of a pendulum that is 3 feet long is approximately 1.92 seconds.

Quick Check 9

A pendulum has a length of 6 feet. Find its period, rounded to the nearest hundredth of a second.

EXAMPLE 12 If a pendulum has a period of 1 second, find its length in feet. Round to the nearest hundredth of a foot.

SOLUTION In this example, substitute 1 for T and solve for L. To solve this equation for L, isolate the radical and square both sides.

$$T = 2\pi\sqrt{\frac{L}{32}}$$

$$1 = 2\pi\sqrt{\frac{L}{32}} \quad \text{Substitute 1 for } T.$$

$$\frac{1}{2\pi} = \sqrt{\frac{L}{32}} \quad \text{Divide both sides by } 2\pi \text{ to isolate the radical.}$$

$$\left(\frac{1}{2\pi}\right)^2 = \left(\sqrt{\frac{L}{32}}\right)^2 \quad \text{Square both sides.}$$

$$\frac{1}{4\pi^2} = \frac{L}{32} \quad \text{Simplify.}$$

$$\overset{8}{\cancel{32}} \cdot \frac{1}{\underset{1}{\cancel{4}}\pi^2} = \overset{1}{\cancel{32}} \cdot \frac{L}{\underset{1}{\cancel{32}}} \quad \text{Multiply both sides by 32 and simplify to isolate } L.$$

$$\frac{8}{\pi^2} = L \quad \text{Simplify.}$$

$$L \approx 0.81 \quad \text{Approximate using a calculator.}$$

The length of the pendulum is approximately 0.81 foot.

Quick Check 10

If a pendulum has a period of 3 seconds, find its length in feet. Round to the nearest hundredth of a foot.

Other Applications Involving Radicals

Objective 7 **Solve other applied problems involving radicals.**

EXAMPLE 13 A vehicle made 150 feet of skid marks on the asphalt before crashing. The speed, s, in miles per hour, the vehicle was traveling when it started skidding can be approximated by the formula $s = \sqrt{30df}$, where d represents the length of the skid marks in feet and f represents the drag factor of the road. If the drag factor for asphalt is 0.75, find the speed the car was traveling. Round to the nearest mile per hour.

SOLUTION Begin by substituting 150 for d and 0.75 for f.

$$s = \sqrt{30df}$$

$$s = \sqrt{30(150)(0.75)} \quad \text{Substitute 150 for } d \text{ and 0.75 for } f.$$

$$s = \sqrt{3375} \quad \text{Simplify the radicand.}$$

$$s \approx 58 \quad \text{Approximate using a calculator.}$$

The car was traveling approximately 58 miles per hour when it started skidding.

Quick Check 11

A vehicle made 215 feet of skid marks on the asphalt before crashing. The speed, s, the vehicle was traveling in miles per hour when it started skidding can be approximated by the formula $s = \sqrt{30df}$, where d represents the length of the skid marks in feet and f represents the drag factor of the road. If the drag factor for asphalt is 0.75, find the speed the car was traveling when it started skidding. Round to the nearest mile per hour.

BUILDING YOUR STUDY STRATEGY

Note Taking, 5 Formatting Your Notes A good note-taking system takes advantage of the margins for specific tasks. You can use the left margin as a place to write down key words or to denote important material. You can then use the right margin to clarify steps in problems or to take down advice from your instructor.

Exercises 9.5

PRACTICE

WATCH

DOWNLOAD

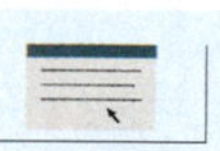
READ

REVIEW

Vocabulary

1. A(n) __________ equation is an equation containing one or more radicals.
2. If two numbers a and b are equal, then for any n, $a^n =$ ______.
3. To solve a radical equation, first __________ one radical containing the variable on one side of the equation.
4. A(n) __________ solution is a solution to the equation that results when we raise a given equation to a power but is not a solution to the original equation.
5. An object suspended from a support so that it swings freely back and forth under the influence of gravity is called a(n) __________.
6. The __________ of a pendulum is the amount of time it takes the pendulum to swing from one extreme to the other and back again.

Solve. Check for extraneous solutions.

7. $\sqrt{x+3} = 10$
8. $\sqrt{x-6} = 8$
9. $\sqrt[3]{2x+9} = 5$
10. $\sqrt[4]{4x-3} = 3$
11. $\sqrt{5x+17} = -6$
12. $\sqrt[3]{x-8} = -4$
13. $\sqrt{x+5} - 7 = -2$
14. $\sqrt{3x-8} + 11 = 4$
15. $\sqrt[3]{2x+7} + 3 = 6$
16. $\sqrt[4]{5x-4} + 12 = 16$
17. $\sqrt{x^2+5x-1} - 2 = 5$
18. $\sqrt{x^2+x-4} + 6 = 10$
19. For the function $f(x) = \sqrt{3x+9}$, find all values x for which $f(x) = 12$.
20. For the function $f(x) = \sqrt{x-8} + 7$, find all values x for which $f(x) = 13$.
21. For the function $f(x) = \sqrt[3]{5x-1} + 5$, find all values x for which $f(x) = 9$.
22. For the function $f(x) = \sqrt[4]{2x-3} - 7$, find all values x for which $f(x) = -4$.

23. For the function $f(x) = \sqrt{x^2 - 5x + 2} + 10$, find all values x for which $f(x) = 14$.

24. For the function $f(x) = \sqrt{x^2 + 8x + 40} - 3$, find all values x for which $f(x) = 2$.

Solve. Check for extraneous solutions.

25. $x^{1/2} + 8 = 10$
26. $x^{1/3} - 11 = -20$
27. $(x + 10)^{1/2} - 4 = 3$
28. $(2x - 35)^{1/2} + 6 = 17$
29. $(5x + 6)^{1/3} - 8 = -2$
30. $(x^2 - 10x + 49)^{1/2} + 2 = 7$
31. $x = \sqrt{2x + 48}$
32. $\sqrt{3x + 10} = x$
33. $\sqrt{4x + 13} = x - 2$
34. $x + 9 = \sqrt{6x + 46}$
35. $\sqrt{2x - 5} + 4 = x$
36. $\sqrt{3x + 13} - 3 = x$
37. $x = \sqrt{49 - 8x} + 7$
38. $x = \sqrt{2x + 9} - 5$
39. $2x - 3 = \sqrt{30 - 7x}$
40. $3x + 5 = \sqrt{27x + 27}$
41. $3x = 1 + \sqrt{4x^2 + x + 7}$
42. $x = \sqrt{54 + 5x - x^2} - 3$
43. $\sqrt{4x - 15} = \sqrt{3x + 11}$
44. $\sqrt[3]{6x + 7} = \sqrt[3]{x - 5}$
45. $\sqrt[4]{x^2 - 8x + 4} = \sqrt[4]{3x - 14}$
46. $\sqrt{5x^2 + 3x - 11} = \sqrt{4x^2 - 6x - 25}$
47. $\sqrt{x + 4} = \sqrt{x - 1} + 1$
48. $\sqrt{x + 14} - \sqrt{x - 10} = 2$
49. $\sqrt{2x + 3} = 1 + \sqrt{x + 1}$
50. $\sqrt{2x + 11} = 2 + \sqrt{x + 2}$
51. $\sqrt{2x + 12} = 1 + \sqrt{x + 5}$
52. $\sqrt{3x - 2} + \sqrt{x + 3} = 3$
53. $\sqrt{3x + 1} - \sqrt{x + 4} = 1$
54. $\sqrt{3x + 3} - \sqrt{2x - 3} = 2$

For Exercises 55–60, use the formula $T = 2\pi\sqrt{\frac{L}{32}}$.

55. A pendulum has a length of 5 feet. Find its period, rounded to the nearest hundredth of a second.

56. A pendulum has a length of 2.2 feet. Find its period, rounded to the nearest hundredth of a second.

57. A pendulum has a length of 1.8 feet. Find its period, rounded to the nearest hundredth of a second.

58. A pendulum has a length of 3.5 feet. Find its period, rounded to the nearest hundredth of a second.

59. If a pendulum has a period of 1.8 seconds, find its length in feet. Round to the nearest tenth of a foot.

60. If a pendulum has a period of 3.5 seconds, find its length in feet. Round to the nearest tenth of a foot.

Skid-mark analysis is one way to estimate the speed a car was traveling prior to an accident. The speed, s, the vehicle was traveling in miles per hour can be approximated by the formula $s = \sqrt{30df}$, where d represents the length of the skid marks in feet and f represents the drag factor of the road.

61. A vehicle involved in an accident made 70 feet of skid marks on the asphalt before crashing. If the drag factor for asphalt is 0.75, find the speed the car was traveling when it started skidding. Round to the nearest mile per hour.

62. A vehicle involved in an accident made 240 feet of skid marks on the asphalt before crashing. If the drag factor for asphalt is 0.75, find the speed the car was traveling when it started skidding. Round to the nearest mile per hour.

63. A vehicle involved in an accident made 185 feet of skid marks on a concrete road before crashing. If the drag factor for concrete is 0.95, find the speed the car was traveling when it started skidding. Round to the nearest mile per hour.

64. A vehicle involved in an accident made 60 feet of skid marks on a concrete road before crashing. If the drag factor for concrete is 0.95, find the speed the car was traveling when it started skidding. Round to the nearest mile per hour.

Body Surface Area (BSA), a measure of the surface area of a human body, is involved in many medical applications including chemotherapy dosing. The BSA, in square meters, can be approximated using the Mosteller formula $BSA = \sqrt{\frac{hw}{3600}}$, where h is the person's height in centimeters and w is the person's weight in kilograms. Find the BSA of a person with the given measurements. Round to the nearest hundredth of a square meter.

65. Adult male—height: 177 centimeters; weight: 89 kilograms

66. Adult female—height: 163 centimeters; weight: 75 kilograms

67. 12-year-old female—height: 151 centimeters; weight: 50 kilograms

68. 16-year-old male—height: 171 centimeters; weight: 70 kilograms

69. If a person whose height is 190 centimeters has a BSA of 1.9 square meters, find the person's weight to the nearest kilogram.

70. If a person whose height is 155 centimeters has a BSA of 1.6 square meters, find the person's weight to the nearest kilogram.

A water tank has a hole at the bottom, and the rate r at which water flows out of the hole in gallons per minute can be found by the formula $r = 19.8\sqrt{d}$, where d represents the depth of the water in the tank, in feet.

71. Find the rate of water flow if the depth of water in the tank is 49 feet.

72. Find the rate of water flow if the depth of water in the tank is 9 feet.

73. Find the rate of water flow if the depth of water in the tank is 13 feet. Round to the nearest tenth of a gallon per minute.

74. Find the rate of water flow if the depth of water in the tank is 22 feet. Round to the nearest tenth of a gallon per minute.

75. If water is flowing out of the tank at a rate of 30 gallons per minute, find the depth of water in the tank. Round to the nearest tenth of a foot.

76. If water is flowing out of the tank at a rate of 100 gallons per minute, find the depth of water in the tank. Round to the nearest tenth of a foot.

The sight distance d, in miles, to the horizon can be approximated using the formula $d = \sqrt{1.5h}$, where h is the eyelevel of the person, in feet. Round all answers to the nearest tenth of a mile.

77. A kayaker is paddling in the ocean when he sees the shore on the horizon. If the kayaker's eyelevel is 2 feet above the water, how far does he have to paddle to reach the shore?

78. Jon is climbing a rock wall on a cruise ship when he sees an island on the horizon. If his eyelevel is 150 feet above the water, how far from the island is the cruise ship?

79. The crow's nest of a tall ship is a perch on the main mast that puts an observer at an eyelevel 100 feet above the water. If the observer sees the base of another ship on the horizon, how far away is the other ship?

80. While on a cruise, Jean watches the sun set on the horizon. If Jean's eyelevel is 35 feet above the water, what is the distance to the horizon?

Writing in Mathematics

Answer in complete sentences.

81. Write a real-world word problem that involves estimating the speed a car was traveling when it started skidding, based on the length of its skid marks. Solve your problem, explaining each step of the process.

82. What is an extraneous solution to an equation? Explain how to determine that a solution to a radical equation is actually an extraneous solution.

83. ***Newsletter*** Write a newsletter that explains how to solve a radical equation.

9.6 The Complex Numbers

OBJECTIVES

1. Rewrite square roots of negative numbers as imaginary numbers.
2. Add and subtract complex numbers.
3. Multiply imaginary numbers.
4. Multiply complex numbers.
5. Divide by a complex number.
6. Divide by an imaginary number.
7. Simplify expressions containing powers of *i*.

Imaginary Numbers

Objective 1 Rewrite square roots of negative numbers as imaginary numbers. Section 9.1 stated that the square root of a negative number, such as $\sqrt{-1}$ and $\sqrt{-25}$, is not a real number. This is because no real number equals a negative number when it is squared. The square root of a negative number is an **imaginary number**.

We define the **imaginary unit** i to be a number that is equal to $\sqrt{-1}$. The number i has the property that $i^2 = -1$.

Imaginary Unit i

$$i = \sqrt{-1}$$
$$i^2 = -1$$

All imaginary numbers can be expressed in terms of i because $\sqrt{-1}$ is a factor of every imaginary number.

EXAMPLE 1 Express $\sqrt{-25}$ in terms of i.

SOLUTION Whenever we have a square root with a negative radicand, we begin by factoring out i. Then we simplify the resulting square root.

$\sqrt{-25} = \sqrt{25(-1)}$ Rewrite -25 as $25(-1)$.
$= \sqrt{25} \cdot \sqrt{-1}$ Rewrite as the product of two square roots.
$= \sqrt{25}i$ Rewrite $\sqrt{-1}$ as i.
$= 5i$ Simplify the square root.

As you become more experienced at working with imaginary numbers, you may want to combine a few of the previous steps into one step.

EXAMPLE 2 Express $\sqrt{-40}$ in terms of i.

SOLUTION

$\sqrt{-40} = \sqrt{40} \cdot \sqrt{-1}$ Rewrite as the product of two square roots.
$= 2\sqrt{10} \cdot i$ Simplify the square root. Rewrite $\sqrt{-1}$ as i.
$= 2i\sqrt{10}$ Rewrite with i in front of the radical.

A WORD OF CAUTION After rewriting the square root of a negative number, such as $\sqrt{-40}$, as an imaginary number, be sure that i does not appear in the radicand. Instead, i should be written in front of the radical.

EXAMPLE 3 Express $-\sqrt{-12}$ in terms of i.

SOLUTION Notice that in this example, a negative sign is in front of the square root as well as in the radicand. We simplify the square root first, making the result negative.

$-\sqrt{-12} = -\sqrt{12} \cdot \sqrt{-1}$ Rewrite as the product of two square roots.
$= -2i\sqrt{3}$ Simplify the square root. Rewrite $\sqrt{-1}$ as i.

Quick Check 1

Express in terms of i.

a) $\sqrt{-36}$
b) $\sqrt{-63}$
c) $-\sqrt{-54}$

Complex Numbers

The set of imaginary numbers and the set of real numbers are subsets of the set of **complex numbers**.

Complex Numbers

A **complex number** is a number of the form $a + bi$, where a and b are real numbers.

Real numbers are complex numbers for which $b = 0$: the real number 7 can be written as $7 + 0i$. Imaginary numbers are complex numbers for which $a = 0$ but $b \neq 0$: the imaginary number $3i$ can be written as $0 + 3i$. We often associate the word *complex* with something that is difficult, but here *complex* refers to the fact that these numbers are made up of two parts.

Real and Imaginary Parts of a Complex Number

For the complex number $a + bi$, the number a is the **real part** and the number b is the **imaginary part**.

Addition and Subtraction of Complex Numbers

Objective 2 Add and subtract complex numbers. We now focus on operations involving complex numbers. We add two complex numbers by adding the two real parts and adding the two imaginary parts. We can follow a similar technique for subtraction.

EXAMPLE 4 Simplify $(6 + 5i) + (7 - 2i)$.

SOLUTION Begin by removing the parentheses; then combine the two real parts and the two imaginary parts of these complex numbers.

$$(6 + 5i) + (7 - 2i) = 6 + 5i + 7 - 2i \quad \text{Remove parentheses.}$$
$$= 13 + 3i \quad \text{Combine the two real parts. Combine the two imaginary parts.}$$

Notice that the process of adding these two complex numbers is similar to simplifying the expression $(6 + 5x) + (7 - 2x)$.

EXAMPLE 5 Simplify $(-2 + 9i) - (8 - 5i)$.

SOLUTION We must distribute the negative sign to both parts of the second complex number, just as we distributed negative signs when we subtracted variable expressions.

$$(-2 + 9i) - (8 - 5i) = -2 + 9i - 8 + 5i \quad \text{Distribute.}$$
$$= -10 + 14i \quad \text{Combine the two real parts and combine the two imaginary parts.}$$

Quick Check 2

Simplify.

a) $(3 + 12i) + (-8 + 15i)$
b) $(14 - 6i) - (3 + 22i)$

Multiplying Imaginary Numbers

Objective 3 Multiply imaginary numbers. Before learning to multiply two complex numbers, we will discuss the multiplication of two imaginary numbers. Suppose we wanted to multiply $6i$ by $9i$. Just as $6x \cdot 9x = 54x^2$, the product $6i \cdot 9i$ is equal to $54i^2$. However, recall that $i^2 = -1$. So this product is $54(-1)$, or -54. When multiplying two imaginary numbers, we substitute -1 for i^2.

EXAMPLE 6 Multiply $8i \cdot 13i$.

SOLUTION

$$\begin{aligned} 8i \cdot 13i &= 104i^2 && \text{Multiply.} \\ &= 104(-1) && \text{Rewrite } i^2 \text{ as } -1. \\ &= -104 && \text{Multiply.} \end{aligned}$$

Quick Check 3

Multiply $4i \cdot 15i$.

EXAMPLE 7 Multiply $\sqrt{-24} \cdot \sqrt{-45}$.

SOLUTION Although it may be tempting to multiply -24 by -45 and combine the two square roots, we cannot do this. The property $\sqrt{x} \cdot \sqrt{y} = \sqrt{xy}$ holds true only if x or y is nonnegative. We must rewrite each square root as an imaginary number before multiplying.

$$\begin{aligned} \sqrt{-24} \cdot \sqrt{-45} &= i\sqrt{24} \cdot i\sqrt{45} && \text{Rewrite each radical as an imaginary number.} \\ &= i\sqrt{2^3 \cdot 3} \cdot i\sqrt{3^2 \cdot 5} && \text{Factor each radicand.} \\ &= i^2\sqrt{2^3 \cdot 3^3 \cdot 5} && \text{Multiply the two radicands.} \\ &= 2 \cdot 3 \cdot i^2\sqrt{2 \cdot 3 \cdot 5} && \text{Simplify the square root.} \\ &= -6\sqrt{30} && \text{Rewrite } i^2 \text{ as } -1 \text{ and simplify.} \end{aligned}$$

When we multiply $\sqrt{24}$ by $\sqrt{45}$, our work will be easier if we factor 24 and 45 before multiplying, rather than trying to simplify $\sqrt{1080}$.

Quick Check 4

Multiply $\sqrt{-5} \cdot \sqrt{-120}$.

Multiplying Complex Numbers

Objective 4 Multiply complex numbers. We multiply two complex numbers by using the distributive property. Often, a product of two complex numbers will contain a term with i^2, and we will rewrite i^2 as -1.

EXAMPLE 8 Multiply $3i(4 - 5i)$.

SOLUTION Begin by multiplying $3i$ by both terms in the parentheses.

$$\begin{aligned} 3i(4 - 5i) &= 3i \cdot 4 - 3i \cdot 5i && \text{Distribute.} \\ &= 12i - 15i^2 && \text{Multiply.} \\ &= 12i + 15 && \text{Rewrite } i^2 \text{ as } -1 \text{ and simplify.} \\ &= 15 + 12i && \text{Rewrite in the form } a + bi. \end{aligned}$$

Quick Check 5

Multiply $-6i(7 + 8i)$.

EXAMPLE 9 Multiply $(9 + i)(2 + 7i)$.

SOLUTION In this example, we must multiply each term in the first set of parentheses by each term in the second set.

$$\begin{aligned}(9 + i)(2 + 7i) &= 9 \cdot 2 + 9 \cdot 7i + i \cdot 2 + i \cdot 7i && \text{Distribute (FOIL).}\\ &= 18 + 63i + 2i + 7i^2 && \text{Multiply.}\\ &= 18 + 65i - 7 && \text{Add } 63i + 2i. \text{ Rewrite } i^2 \text{ as } -1 \text{ and simplify.}\\ &= 11 + 65i && \text{Combine like terms.}\end{aligned}$$

Quick Check 6

Multiply $(3 - 2i)(8 + i)$.

EXAMPLE 10 Multiply $(7 + 2i)^2$.

SOLUTION Recall that we square a binomial by multiplying it by itself.

$$\begin{aligned}(7 + 2i)^2 &= (7 + 2i)(7 + 2i) && \text{Multiply } 7 + 2i \text{ by itself.}\\ &= 7 \cdot 7 + 7 \cdot 2i + 2i \cdot 7 + 2i \cdot 2i && \text{Distribute.}\\ &= 49 + 14i + 14i + 4i^2 && \text{Multiply.}\\ &= 49 + 28i - 4 && \text{Add } 14i + 14i. \text{ Rewrite } i^2 \text{ as } -1 \text{ and simplify.}\\ &= 45 + 28i && \text{Combine like terms.}\end{aligned}$$

Quick Check 7

Multiply $(9 - 4i)^2$.

EXAMPLE 11 Multiply $(5 + 6i)(5 - 6i)$.

SOLUTION

$$\begin{aligned}(5 + 6i)(5 - 6i) &= 5 \cdot 5 - 5 \cdot 6i + 6i \cdot 5 - 6i \cdot 6i && \text{Distribute.}\\ &= 25 - 30i + 30i - 36i^2 && \text{Multiply.}\\ &= 25 - 36i^2 && \text{Combine like terms.}\\ &= 25 + 36 && \text{Rewrite } i^2 \text{ as } -1 \text{ and simplify.}\\ &= 61 && \text{Add.}\end{aligned}$$

Quick Check 8

Multiply $(7 + 4i)(7 - 4i)$.

In the previous example, the two complex numbers that were multiplied were conjugates. Their product was a real number that did not have an imaginary part. Two complex numbers of the form $a + bi$ and $a - bi$ are conjugates, and their product will always be equal to $a^2 + b^2$.

$$\begin{aligned}(a + bi)(a - bi) &= a^2 - abi + abi - b^2i^2 && \text{Distribute.}\\ &= a^2 - b^2i^2 && \text{Combine like terms.}\\ &= a^2 + b^2 && \text{Rewrite } i^2 \text{ as } -1 \text{ and simplify.}\end{aligned}$$

EXAMPLE 12 Multiply $(10 - 3i)(10 + 3i)$.

SOLUTION We will use the fact that $(a + bi)(a - bi) = a^2 + b^2$ for two complex numbers that are conjugates.

$$\begin{aligned}(10 - 3i)(10 + 3i) &= 10^2 + 3^2 && \text{The product equals } a^2 + b^2.\\ &= 100 + 9 && \text{Square 10 and 3.}\\ &= 109 && \text{Add.}\end{aligned}$$

Quick Check 9

Multiply $(11 - 4i)(11 + 4i)$.

Dividing by a Complex Number

Objective 5 Divide by a complex number. Because the imaginary number i is a square root ($\sqrt{-1}$), a simplified expression cannot contain i in its denominator. In a procedure similar to rationalizing a denominator (Section 9.4), we will use conjugates to rewrite the expression without i in the denominator.

EXAMPLE 13 Simplify $\frac{2}{3+i}$.

SOLUTION Begin by multiplying the numerator and denominator by the conjugate of the denominator, which is $3 - i$.

$$\frac{2}{3+i} = \frac{2}{3+i}\cdot\frac{3-i}{3-i}$$ Multiply the numerator and denominator by the conjugate of the denominator.

$$= \frac{6-2i}{3^2+1^2}$$ Multiply the numerators and denominators.

$$= \frac{6-2i}{10}$$ Simplify the denominator.

$$= \frac{\overset{1}{\cancel{2}}(3-i)}{\underset{5}{\cancel{10}}}$$ Factor the numerator. Divide out factors common to the numerator and denominator.

$$= \frac{3-i}{5}$$ Simplify.

$$= \frac{3}{5} - \frac{1}{5}i$$ Rewrite in the form $a + bi$.

Quick Check 10

Simplify $\frac{15}{8+6i}$.

EXAMPLE 14 Simplify $\frac{1+6i}{2+5i}$.

SOLUTION In this example, the numerator has two terms and we must multiply using the distributive property.

$$\frac{1+6i}{2+5i} = \frac{1+6i}{2+5i}\cdot\frac{2-5i}{2-5i}$$ Multiply the numerator and denominator by the conjugate of the denominator.

$$= \frac{2-5i+12i-30i^2}{2^2+5^2}$$ Multiply.

$$= \frac{2+7i-30i^2}{4+25}$$ Combine like terms in the numerator. Square 2 and 5 in the denominator.

$$= \frac{2+7i+30}{29}$$ Rewrite i^2 as -1 and simplify. Simplify the denominator.

$$= \frac{32+7i}{29}$$ Combine like terms.

$$= \frac{32}{29} + \frac{7}{29}i$$ Rewrite in the form $a + bi$.

Quick Check 11

Simplify $\frac{9+2i}{7-3i}$.

Note that the expression $\frac{1+6i}{2+5i}$ is equivalent to $(1 + 6i) \div (2 + 5i)$. If we are asked to divide a complex number by another complex number, we begin by rewriting the expression as a fraction and then proceed as in the previous example.

Dividing by an Imaginary Number

Objective 6 Divide by an imaginary number. When a denominator is an imaginary number, we multiply the numerator and denominator by i to rewrite the fraction without i in the denominator.

EXAMPLE 15 Simplify $\frac{5 + 2i}{3i}$.

SOLUTION Because the denominator is an imaginary number, we begin by multiplying the fraction by $\frac{i}{i}$.

$$\frac{5 + 2i}{3i} = \frac{5 + 2i}{3i} \cdot \frac{i}{i} \quad \text{Multiply the numerator and denominator by } i.$$

$$= \frac{5i + 2i^2}{3i^2} \quad \text{Multiply.}$$

$$= \frac{5i - 2}{-3} \quad \text{Rewrite } i^2 \text{ as } -1 \text{ and simplify.}$$

$$= -\frac{5i - 2}{3} \quad \text{Factor } -1 \text{ out of the denominator.}$$

$$= \frac{-5i + 2}{3} \quad \text{Distribute.}$$

$$= \frac{2}{3} - \frac{5}{3}i \quad \text{Rewrite in } a - bi \text{ form.}$$

Quick Check 12

Simplify $\frac{15 - 8i}{12i}$.

Powers of i

Objective 7 Simplify expressions containing powers of i. Occasionally, i may be raised to a power greater than 2. We finish this section by learning to simplify such expressions. The expression i^3 can be rewritten as $i^2 \cdot i$, which is equivalent to $-1 \cdot i$, or $-i$. The expression i^4 can be rewritten as $i^2 \cdot i^2$, which is equivalent to $-1(-1)$, or 1. The following table shows the first four positive powers of i:

i	i^2	i^3	i^4
↓	↓	↓	↓
i	-1	$-i$	1

We can use the fact that $i^4 = 1$ to simplify greater powers of i.

$$i^5 = i^4 \cdot i = 1 \cdot i = i \qquad i^6 = i^4 \cdot i^2 = 1(-1) = -1$$

$$i^7 = i^4 \cdot i^3 = 1 \cdot i^3 = 1(-i) = -i \qquad i^8 = i^4 \cdot i^4 = 1 \cdot 1 = 1$$

We can now see a pattern.

i	i^2	i^3	i^4	i^5	i^6	i^7	i^8
↓	↓	↓	↓	↓	↓	↓	↓
i	-1	$-i$	1	i	-1	$-i$	1

In general, to simplify i^n, where n is a whole number, we divide n by 4. If the remainder is equal to r, $i^n = i^r$.

Remainder	0	1	2	3
i^n	1	i	-1	$-i$

EXAMPLE 16 Simplify i^{27}.

Quick Check 13
Simplify i^{30}.

SOLUTION If we divide the exponent 27 by 4, the remainder is 3. So $i^{27} = i^3 = -i$.

BUILDING YOUR STUDY STRATEGY

Note Taking, 6 **Rewriting Your Notes** A good strategy is to rewrite your notes as soon as possible after class. The simple task of rewriting your notes serves as a review of the material that was covered in class. If you rewrite your notes while the material is still fresh in your mind, you have a better chance of being able to read what you have written.

As you rework your notes, take the time to supplement them. Replace a brief definition with the full definition from the text. Add notes to clarify what you wrote down in class.

Finally, consider creating a page in your notes that contains the problems (without solutions) your instructor solved during class. When you begin to review for the exam, this list of problems will be a good review sheet.

Exercises 9.6

PRACTICE
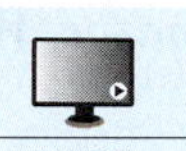
WATCH

DOWNLOAD

READ

REVIEW

Vocabulary

1. The square root of a negative number is a(n) ___________ number.
2. The imaginary unit i is a number that is equal to ________.
3. The number i has the property that $i^2 =$ ______.
4. A(n) _________ number is a number of the form $a + bi$, where a and b are real numbers.
5. For the complex number $a + bi$, the number a is the _______ part.
6. For the complex number $a + bi$, the number b is the ___________ part.

Express in terms of i.

7. $\sqrt{-4}$
8. $\sqrt{-16}$
9. $\sqrt{-81}$
10. $\sqrt{-100}$
11. $-\sqrt{-169}$
12. $-\sqrt{-225}$
13. $\sqrt{-50}$
14. $\sqrt{-18}$
15. $-\sqrt{-252}$
16. $\sqrt{-243}$

Add or subtract the complex numbers.

17. $(8 + 9i) + (3 + 5i)$
18. $(13 + 11i) - (6 + 2i)$
19. $(6 - 7i) - (2 + 10i)$
20. $(1 + 6i) + (5 - 12i)$
21. $(-3 + 8i) + (3 - 4i)$
22. $(5 - 9i) + (5 + 9i)$
23. $(12 + 13i) - (12 - 13i)$
24. $(14 - i) - (21 + 11i)$
25. $(1 - 12i) - (8 + 8i) + (5 - 4i)$
26. $(-9 + 2i) - (5 + 6i) - (10 - 20i)$

Find the missing complex number.

27. $(4 - 7i) + ? = 10 + 2i$
28. $(11 + 10i) + ? = 20 - 3i$
29. $(3 + 4i) - ? = 1 - 5i$
30. $? - (-9 + 7i) = -15 - 3i$

Multiply.

31. $7i \cdot 14i$
32. $3i \cdot 8i$
33. $-4i \cdot 5i$
34. $i(-9i)$
35. $\sqrt{-49} \cdot \sqrt{-64}$
36. $\sqrt{-25} \cdot \sqrt{-25}$
37. $\sqrt{-12} \cdot \sqrt{-18}$
38. $\sqrt{-28} \cdot \sqrt{-7}$
39. $\sqrt{-20} \cdot \sqrt{-45}$
40. $\sqrt{-27} \cdot \sqrt{-50}$

Multiply. Write your answer in a + bi form.

41. $7i(8 + 5i)$
42. $3i(10 - 9i)$
43. $-6i(9 + 4i)$
44. $-5i(11 - 14i)$
45. $(7 + 2i)(5 + i)$
46. $(9 + 8i)(1 + 4i)$
47. $(3 - 5i)(4 + 10i)$
48. $(5 - 7i)(4 - i)$
49. $(6 + 7i)^2$
50. $(10 + 4i)^2$
51. $(1 - 5i)^2$
52. $(3 - 8i)^2$

Multiply the conjugates using the fact that $(a + bi)(a - bi) = a^2 + b^2$.

53. $(5 + 2i)(5 - 2i)$
54. $(7 + 4i)(7 - 4i)$
55. $(12 - 8i)(12 + 8i)$
56. $(6 - i)(6 + i)$
57. $(5 + 6i)(5 - 6i)$
58. $(11 + 3i)(11 - 3i)$

Simplify by rationalizing the denominator. Write your answer in a + bi form.

59. $\frac{6}{5 + i}$
60. $\frac{20}{4 - 3i}$
61. $\frac{25i}{9 - 7i}$
62. $\frac{10i}{6 + 7i}$
63. $\frac{4 + 7i}{7 - 2i}$
64. $\frac{9 - 5i}{2 - 3i}$
65. $\frac{1 + 6i}{7 + 6i}$
66. $\frac{5 - 4i}{5 + 4i}$

Simplify by rationalizing the denominator. Write your answer in a + bi form.

67. $\frac{8}{i}$
68. $\frac{5}{6i}$
69. $\frac{5 - 7i}{3i}$
70. $\frac{3 + 10i}{5i}$
71. $\frac{6 + i}{-4i}$
72. $\frac{2 - 9i}{-8i}$

Divide. Write your answer in a + bi form.

73. $(7 + 3i) \div 2i$
74. $(2 - 13i) \div 6i$
75. $(8 - 7i) \div (1 + 3i)$
76. $(5 + 8i) \div (5 - 11i)$

Find the missing complex number.

77. $(4 + i)(?) = 4 + 35i$
78. $(7 - 5i)(?) = 31 - i$
79. $\frac{?}{1 + 9i} = 4 - 3i$
80. $\frac{?}{10 - 3i} = 6 - 4i$

Simplify.

81. i^{45}
82. i^{72}
83. i^{42}
84. i^{99}
85. $\frac{i^{37}}{i^{15}}$
86. $i^{26} \cdot i^{31}$
87. $i^{21} + i^{53}$
88. $i^{66} - i^{80}$

Mixed Practice, 89–114

Simplify. Write your answer in a + bi form when appropriate.

89. $(7 + 2i)(6 - 3i)$
90. $(4 - 6i)(4 + 6i)$
91. $\frac{9 + 4i}{5i}$
92. $8i(17 - 3i)$
93. $-9i(12 + 5i)$
94. $(18 - 13i) - (8 - 5i)$
95. $12i \div (4 - 2i)$
96. $\frac{10 - 9i}{5 - i}$
97. $(2 - 5i)(2 + 5i)$
98. $(14 - 11i) + (-23 + 6i)$
99. $\sqrt{-24} \cdot \sqrt{-27}$
100. $(6 - 4i) \div (8i)$
101. $\frac{15i}{9 - 2i}$
102. $(16 - 9i)^2$
103. $\sqrt{-252}$
104. $(11 + 4i)(6 - 13i)$
105. $(9 - 3i)^2$
106. $7i \cdot 12i$
107. $(8 - 11i) + (15 + 21i)$
108. i^{207}

109. $-14i \cdot 9i$

110. $\sqrt{-20} \cdot \sqrt{-15}$

111. i^{323}

112. $\dfrac{7 - 6i}{-2i}$

113. $(30 - 17i) - (54 - 33i)$

114. $\sqrt{-2673}$

Writing in Mathematics

Answer in complete sentences.

115. Explain how adding two complex numbers is similar to adding two variable expressions.

CHAPTER 9 SUMMARY

Section 9.1 Square Roots; Radical Notation

Radical Expressions, Square Roots, pp. 481–483

The principal square root of b, denoted $\sqrt{b}$, is the positive number a such that $a^2 = b$.
The expression $\sqrt{b}$ is called a radical expression.
The sign $\sqrt{\ }$ is called a radical sign.
The expression contained inside the radical sign is called the radicand.

Simplify assuming that x is nonnegative: $\sqrt{81x^{10}}$
$\sqrt{81x^{10}} = 9x^5$ because $(9x^5)^2 = 81x^{10}$.

Radical Expressions, *n*th Roots, pp. 484–485

For any positive integer $n > 1$ and any nonnegative number b, the nth root of b, denoted $\sqrt[n]{b}$, is the number that when raised to the nth power equals b: $\sqrt[n]{b} = a$ if $a^n = b$ and a and b have the same sign.
The third root of b, $\sqrt[3]{b}$, is called the cube root of b.
The number n is called the index of the radical.

Simplify: $\sqrt[5]{32a^{15}b^{20}}$
$\sqrt[5]{32a^{15}b^{20}} = 2a^3b^4$ because $(2a^3b^4)^5 = 32a^{15}b^{20}$.

Product Property of Radicals, p. 486

For any index $n > 1$ and any nonnegative numbers a and b, $\sqrt[n]{a} \cdot \sqrt[n]{b} = \sqrt[n]{ab}$.

Simplify: $\sqrt{50} \cdot \sqrt{18}$

$$\begin{aligned} \sqrt{50} \cdot \sqrt{18} &= \sqrt{900} \\ &= 30 \end{aligned}$$

Quotient Property of Radicals, pp. 486–487

For any index $n > 1$ and any nonnegative numbers a and positive number b, $\dfrac{\sqrt[n]{a}}{\sqrt[n]{b}} = \sqrt[n]{\dfrac{a}{b}}$.

Simplify: $\sqrt{\dfrac{396}{11}}$

$$\sqrt{\frac{396}{11}} = \sqrt{36} = 6$$

Radical Functions, p. 487

A radical function $f(x)$ is a function that contains radicals.

For $f(x) = \sqrt{7x + 65} + 8$, find $f(8)$.

$$\begin{aligned} f(8) &= \sqrt{7(8) + 65} + 8 \\ &= \sqrt{121} + 8 \\ &= 11 + 8 \\ &= 19 \end{aligned}$$

Domain of Radical Functions, pp. 487–488

For an even root, the domain of a radical function is the set of all values for which the radicand is greater than or equal to 0.
For an odd root, the domain of a radical function is the set of all real numbers.

Find the domain of the radical function:
$f(x) = \sqrt{2x - 9} + 6$

$$\begin{aligned} 2x - 9 &\geq 0 \\ 2x &\geq 9 \\ x &\geq \frac{9}{2} \end{aligned}$$

Domain: $\left[\dfrac{9}{2}, \infty\right)$

Section 9.2 Rational Exponents

Rational Exponents of the form $1/n$, pp. 490–491

For any integer $n > 1$, $a^{1/n}$ is defined to be the nth root of a, or $\sqrt[n]{a}$.

Simplify: $100^{1/2}$

$$100^{1/2} = \sqrt{100} = 10$$

Rational Exponents of the form m/n, p. 491

For any integers m and n, $n > 1$, $a^{m/n}$ is defined to be $(\sqrt[n]{a})^m$. This is also equivalent to $\sqrt[n]{a^m}$.
The denominator in the exponent, n, is the root you are taking.
The numerator in the exponent, m, is the power to which you raise this radical.

Simplify: $216^{2/3}$

$$\begin{aligned} 216^{2/3} &= (\sqrt[3]{216})^2 \\ &= (6)^2 \\ &= 36 \end{aligned}$$

Properties of Exponents, p. 492

The properties of exponents also hold for rational exponents.

1. For any base x, $x^m \cdot x^n = x^{m+n}$.
2. For any base x, $(x^m)^n = x^{m \cdot n}$.
3. For any bases x and y, $(xy)^n = x^n y^n$.
4. For any base x, $\dfrac{x^m}{x^n} = x^{m-n} \quad (x \neq 0)$.
5. For any base x, $x^0 = 1 \quad (x \neq 0)$.
6. For any bases x and y, $\left(\dfrac{x}{y}\right)^n = \dfrac{x^n}{y^n} \quad (y \neq 0)$.
7. For any *nonzero* base x, $x^{-n} = \dfrac{1}{x^n} \quad (x \neq 0)$.

Simplify: $\left(x^{3/4}\right)^{2/7}$

$$\left(x^{3/4}\right)^{2/7} = x^{\frac{3}{4}\cdot\frac{2}{7}} = x^{3/14}$$

Simplify: $x^{5/6} \cdot x^{2/3}$

$$x^{5/6} \cdot x^{2/3} = x^{\frac{5}{6}+\frac{2}{3}} = x^{\frac{5}{6}+\frac{4}{6}} = x^{9/6} = x^{3/2}$$

Section 9.3 Simplifying, Adding, and Subtracting Radical Expressions

Simplifying Radicals, pp. 495–497

1. Rewrite the radicand as a product of two factors. The exponent for the first factor should be the largest multiple of the radical's index that is less than or equal to the factor's original exponent.
2. Use the product property to remove factors from the radicand.

Simplify: $\sqrt[3]{a^{19}b^5c^{12}}$

$$\sqrt[3]{a^{19}b^5c^{12}} = \sqrt[3]{a^{18}b^3c^{12} \cdot a^1b^2} = \sqrt[3]{(a^6bc^4)^3 \cdot ab^2} = a^6bc^4\sqrt[3]{ab^2}$$

Like Radicals, p. 497

Two radical expressions are called like radicals if they have the same index and the same radicand.

$3\sqrt{2}$ and $7\sqrt{2}$

Adding and Subtracting Radical Expressions, pp. 497–498

1. Simplify each radical completely.
2. Combine like radicals by adding/subtracting the factors in front of the radicals.

Add: $7\sqrt{50} + 6\sqrt{32}$

$$7\sqrt{50} + 6\sqrt{32} = 7\sqrt{25 \cdot 2} + 6\sqrt{16 \cdot 2} = 7 \cdot 5\sqrt{2} + 6 \cdot 4\sqrt{2} = 35\sqrt{2} + 24\sqrt{2} = 59\sqrt{2}$$

Section 9.4 Multiplying and Dividing Radical Expressions

Multiplying Radical Expressions, pp. 500–501

If the index of each radical is the same, multiply factors in front of the radical by each other and multiply the radicands by each other. Simplify the resulting radical if possible.

Multiply: $5\sqrt{6} \cdot 9\sqrt{14}$

$$5\sqrt{6} \cdot 9\sqrt{14} = 45\sqrt{84} = 45\sqrt{4 \cdot 21} = 45 \cdot 2\sqrt{21} = 90\sqrt{21}$$

Multiplying Radical Expressions with Two or More Terms, pp. 501–503

If one or more of the expressions contains at least two terms, multiply by using the distributive property.

Multiply: $(7\sqrt{2} + 3\sqrt{3})(3\sqrt{2} + 4\sqrt{3})$

$$(7\sqrt{2} + 3\sqrt{3})(3\sqrt{2} + 4\sqrt{3}) = 21\sqrt{4} + 28\sqrt{6} + 9\sqrt{6} + 12\sqrt{9} = 21 \cdot 2 + 28\sqrt{6} + 9\sqrt{6} + 12 \cdot 3 = 42 + 28\sqrt{6} + 9\sqrt{6} + 36 = 78 + 37\sqrt{6}$$

Squaring a Radical Expression with Two or More Terms, pp. 502–503

To square a radical expression containing two or more terms, multiply the expression by itself.

Multiply: $(6\sqrt{5} + \sqrt{7})^2$

$$(6\sqrt{5} + \sqrt{7})^2 = (6\sqrt{5} + \sqrt{7})(6\sqrt{5} + \sqrt{7}) = 36\sqrt{25} + 6\sqrt{35} + 6\sqrt{35} + \sqrt{49} = 36 \cdot 5 + 6\sqrt{35} + 6\sqrt{35} + 7 = 180 + 6\sqrt{35} + 6\sqrt{35} + 7 = 187 + 12\sqrt{35}$$

Conjugates, p. 503

Two radical expressions of the form $a + b$ and $a - b$ are conjugates. The product of two conjugates $a + b$ and $a - b$ is $a^2 - b^2$.

Multiply: $(8\sqrt{2} + 7\sqrt{11})(8\sqrt{2} - 7\sqrt{11})$

$$\begin{aligned}&(8\sqrt{2} + 7\sqrt{11})(8\sqrt{2} - 7\sqrt{11})\\ &= (8\sqrt{2})^2 - (7\sqrt{11})^2\\ &= 64\sqrt{4} - 49\sqrt{121}\\ &= 64\cdot 2 - 49\cdot 11\\ &= 128 - 539\\ &= -411\end{aligned}$$

Rationalizing a Denominator with Only One Term, pp. 504–505

1. Multiply both the numerator and denominator by a radical expression that results in the radicand being a perfect nth power.
2. Simplify the radical in the denominator.
3. If possible, simplify the fraction by dividing out factors that are common to the numerator and denominator.

Rationalize the denominator: $\dfrac{4\sqrt{7}}{\sqrt{12}}$

$$\begin{aligned}\frac{4\sqrt{7}}{\sqrt{12}}\cdot\frac{\sqrt{3}}{\sqrt{3}} &= \frac{4\sqrt{21}}{\sqrt{36}}\\ &= \frac{4\sqrt{21}}{6}\\ &= \frac{2\sqrt{21}}{3}\end{aligned}$$

Rationalizing a Denominator with Two Terms Containing at Least One Square Root, pp. 505–506

1. Multiply both the numerator and denominator by the conjugate of the denominator.
2. Simplify the numerator and denominator.
3. If possible, simplify the fraction by dividing out factors that are common to the numerator and denominator.

Rationalize the denominator: $\dfrac{6}{8 - 3\sqrt{2}}$

$$\begin{aligned}\frac{6}{8 - 3\sqrt{2}}\cdot\frac{8 + 3\sqrt{2}}{8 + 3\sqrt{2}} &= \frac{48 + 18\sqrt{2}}{64 - 9\cdot 2}\\ &= \frac{48 + 18\sqrt{2}}{46}\\ &= \frac{2(24 + 9\sqrt{2})}{46}\\ &= \frac{24 + 9\sqrt{2}}{23}\end{aligned}$$

Section 9.5 Radical Equations and Applications of Radical Equations

Solving Rational Equations, pp. 509–514

1. Isolate one radical containing the variable on one side of the equation.
2. Raise both sides of the equation to the nth power, where n is the index of the radical.
   For any nonnegative number x and any integer $n > 1$, $(\sqrt[n]{x})^n = x$.
3. If the resulting equation does not contain a radical, solve the equation. If the resulting equation does contain a radical, begin the process again by isolating the radical on one side of the equation.
4. Check the solution(s).

Solve: $\sqrt{15x - 29} - 1 = x$

$$\begin{aligned}\sqrt{15x - 29} - 1 &= x\\ \sqrt{15x - 29} &= x + 1\\ (\sqrt{15x - 29})^2 &= (x + 1)^2\\ 15x - 29 &= x^2 + 2x + 1\\ 0 &= x^2 - 13x + 30\\ 0 &= (x - 3)(x - 10)\end{aligned}$$

$$x - 3 = 0 \qquad x - 10 = 0$$
$$x = 3 \qquad x = 10$$

$\{3, 10\}$

Extraneous Solutions, pp. 510–511

A solution obtained in solving an equation that is not a solution to the original equation is called an extraneous solution. Extraneous solutions are excluded from the solution set.

Solve: $\sqrt{x + 9} = x + 7$

$$\begin{aligned}(\sqrt{x + 9})^2 &= (x + 7)^2\\ x + 9 &= (x + 7)(x + 7)\\ x + 9 &= x^2 + 14x + 49\\ 0 &= x^2 + 13x + 40\\ 0 &= (x + 5)(x + 8)\end{aligned}$$

$$x + 5 = 0 \qquad x + 8 = 0$$
$$x = -5 \qquad x = -8$$

$$\sqrt{(-8) + 9} = (-8) + 7$$
$$\sqrt{1} = -1 \text{ (False)}$$

Extraneous solution

$\{-5\}$

Pendulum, pp. 514–515

The **period** of a pendulum is the amount of time it takes to swing from one extreme to the other and back again.

The period T of a pendulum in seconds can be found using the formula

$$T = 2\pi\sqrt{\frac{L}{32}},$$

where L is the length of the pendulum in feet.

A pendulum has a length of 6 feet. Find its period, rounded to the nearest hundredth of a second.

$$T = 2\pi\sqrt{\frac{L}{32}}$$
$$T = 2\pi\sqrt{\frac{6}{32}}$$
$$T \approx 2.72$$

The period is approximately 2.72 seconds.

Other Applications, pp. 515–516

Use the formula provided to solve the given problem.

A water tank has a hole at the bottom, and the rate r at which water flows out of the hole in gallons per minute can be found by the formula

$$r = 19.8\sqrt{d},$$

where d represents the depth of the water in the tank, in feet.

Find the rate of water flow if the depth of water in the tank is 13 feet. Round to the nearest tenth of a gallon per minute.

$$r = 19.8\sqrt{d}$$
$$r = 19.8\sqrt{13}$$
$$r \approx 71.4$$

The water flow is approximately 71.4 gallons per minute.

Section 9.6 The Complex Numbers

Imaginary Numbers, p. 519

The imaginary number i is defined to be equal to $\sqrt{-1}$.

$$i = \sqrt{-1}$$
$$i^2 = -1$$

Simplify: $\sqrt{-81}$

$$\sqrt{-81} = 9i$$

Complex Numbers, p. 520

A complex number is a number of the form $a + bi$, where a and b are real numbers.

$7 + 3i$

Adding/Subtracting Complex Numbers, p. 520

To add two complex numbers, add the two real parts and add the two imaginary parts.

To subtract two complex numbers, subtract the two real parts and subtract the two imaginary parts.

Subtract: $(8 + 7i) - (13 - 4i)$

$$(8 + 7i) - (13 - 4i) = 8 + 7i - 13 + 4i$$
$$= -5 + 11i$$

Multiplying Imaginary and Complex Numbers, pp. 521–522

When multiplying complex numbers, use the distributive property. For any occurrence of i^2 in the product, replace i^2 with -1.

Multiply: $(11 + 8i)(9 - 5i)$

$$(11 + 8i)(9 - 5i) = 99 - 55i + 72i - 40i^2$$
$$= 99 - 55i + 72i + 40$$
$$= 139 + 17i$$

Complex Conjugates, p. 522

Two complex numbers of the form $a + bi$ and $a - bi$ are conjugates, and their product is equal to $a^2 + b^2$.

Multiply: $(2 - 9i)(2 + 9i)$

$$(2 - 9i)(2 + 9i) = 2^2 + 9^2$$
$$= 4 + 81$$
$$= 85$$

Rationalizing Denominators—Complex Denominator, p. 523

If a denominator contains a complex number, rationalize the denominator by multiplying the numerator and denominator by the conjugate of the denominator.

Simplify by rationalizing the denominator: $\frac{3 + 4i}{2 - 3i}$

$$\frac{3 + 4i}{2 - 3i} \cdot \frac{2 + 3i}{2 + 3i} = \frac{6 + 9i + 8i + 12i^2}{2^2 + 3^2}$$
$$= \frac{6 + 9i + 8i - 12}{13}$$
$$= \frac{-6 + 17i}{13}$$
$$= -\frac{6}{13} + \frac{17}{13}i$$

Rationalizing Denominators—Imaginary Denominator, p. 524

If a denominator contains an imaginary number, rationalize the denominator by multiplying the numerator and denominator by i.

Simplify by rationalizing the denominator: $\frac{8}{13i}$

$$\frac{8}{13i} = \frac{8}{13i} \cdot \frac{i}{i}$$
$$= \frac{8i}{13i^2}$$
$$= \frac{8i}{-13}$$
$$= -\frac{8}{13}i$$

SUMMARY OF CHAPTER 9 STUDY STRATEGIES

- A good set of class notes is a valuable study resource. To ensure that you create the best set of notes you can, begin by choosing your seat wisely. Make sure you can see the entire board and can hear the instructor from your location.
- Write down each step of the solution to a problem, whether or not you believe the step is necessary. Often, solutions seem clear while you are sitting in class but will not be so clear when you sit down to work the homework exercises. If you find yourself completely lost in class, do your best to write down every step so that you can make sense of the steps when reworking your notes after class.
- If you find that you are having trouble keeping up with your instructor, write down enough information so that you will be able to understand the material after class has ended. Try using abbreviations when appropriate and write down phrases rather than complete sentences. The use of a recorder, with your instructor's permission, can help you fill in missing sections in your notes. A classmate can help you fill in missing notes as well. If your instructor gives you a cue that material is important, write that material in your notes immediately. You can catch up on missing material after class.
- Reworking your notes after class serves as an excellent review of what was covered in class and gives you an opportunity to make your notes more legible and user friendly. During this time, you can fill in any holes in your notes and supplement your notes with extra material.

CHAPTER 9 REVIEW

***Simplify the radical expression. Assume that all variables represent nonnegative real numbers.* [9.1]**

1. $\sqrt{25}$

2. $\sqrt{169}$

3. $\sqrt[3]{-64}$

4. $\sqrt[3]{x^9}$

5. $\sqrt{81x^{12}}$

6. $\sqrt[5]{243x^{15}y^{10}z^{25}}$

***Approximate to the nearest thousandth using a calculator.* [9.1]**

7. $\sqrt{17}$

8. $\sqrt{41}$

***Evaluate the radical function. Round to the nearest thousandth if necessary.* [9.1]**

9. $f(x) = \sqrt{x - 10}, f(91)$

10. $f(x) = \sqrt{x^2 + 6x - 19}, f(3)$

***Find the domain of the radical function. Express your answer in interval notation.* [9.1]**

11. $f(x) = \sqrt{2x + 6}$

12. $f(x) = \sqrt[4]{3x + 25}$

***Rewrite each radical expression by using rational exponents. Assume that all variables represent nonnegative real numbers.* [9.2]**

13. $\sqrt[5]{a}$

14. $(\sqrt[4]{x})^3$

15. $(\sqrt{x})^{13}$

16. $\sqrt[5]{n^8}$

***Rewrite as a radical expression and simplify if possible. Assume that all variables represent nonnegative real numbers.* [9.2]**

17. $4^{1/2}$

18. $(343x^{21})^{2/3}$

***Simplify the expression. Assume that all variables represent nonnegative real numbers.* [9.2]**

19. $x^{3/2} \cdot x^{1/6}$

20. $(x^{3/8})^{7/6}$

21. $\dfrac{x^{2/3}}{x^{5/12}}$

22. $49^{-1/2}$

23. $64^{-5/3}$

24. $\dfrac{27^{4/3}}{27^{8/3}}$

***Simplify the radical expression. Assume that all variables represent nonnegative real numbers.* [9.3]**

25. $\sqrt{28}$

26. $\sqrt[3]{x^{10}}$

27. $\sqrt[5]{x^{11}y^8z^{15}}$

28. $\sqrt[3]{405a^{25}b^{12}c^{16}}$

29. $\sqrt{180r^3s^{11}t^{12}}$

30. $\sqrt{175r^6s^8t^4}$

***Add or subtract. Assume that all variables represent nonnegative real numbers.* [9.3]**

31. $4\sqrt{2} + 8\sqrt{2}$

32. $7\sqrt[3]{x} - \sqrt[3]{x}$

33. $5\sqrt{20} + 3\sqrt{125}$

34. $6\sqrt{12} - 8\sqrt{50} + 9\sqrt{75}$

***Multiply. Assume that all variables represent nonnegative real numbers.* [9.4]**

35. $\sqrt[3]{4x^2} \cdot \sqrt[3]{18x^4}$

36. $5\sqrt{3}(2\sqrt{6} - 4\sqrt{3})$

37. $(\sqrt{3} + \sqrt{8})(\sqrt{6} - \sqrt{2})$

38. $(10\sqrt{5} - 7\sqrt{11})(9\sqrt{5} + 2\sqrt{11})$

39. $(2\sqrt{15} - 9\sqrt{3})^2$

40. $(2\sqrt{13} - 10\sqrt{7})(2\sqrt{13} + 10\sqrt{7})$

***Simplify. Assume that all variables represent nonnegative real numbers.* [9.4]**

41. $\dfrac{\sqrt{224x^5}}{\sqrt{2x}}$

42. $\sqrt{\dfrac{50a^9b^8c^3}{8a^3bc^{13}}}$

***Rationalize the denominator and simplify. Assume that all variables represent nonnegative real numbers.* [9.4]**

43. $\sqrt[3]{\dfrac{16}{5}}$

44. $\dfrac{8}{\sqrt{14}}$

Worked-out solutions to Review Exercises marked with (highlighted) can be found on page AN-32.

45. $\sqrt{\dfrac{11}{18}}$

46. $\dfrac{a^5bc^3}{\sqrt{a^4b^5c}}$

47. $\dfrac{8}{\sqrt{11}-\sqrt{5}}$

48. $\dfrac{2\sqrt{3}}{10\sqrt{15}-3\sqrt{12}}$

49. $\dfrac{\sqrt{7}+5\sqrt{2}}{2\sqrt{7}-3\sqrt{2}}$

50. $\dfrac{3-\sqrt{5}}{6+7\sqrt{5}}$

Solve. **[9.5]**

51. $\sqrt{4x+1}=7$
52. $\sqrt[3]{7x-15}=5$
53. $\sqrt{3x+22}+2=x$
54. $\sqrt{17-4x}-x=1$
55. $\sqrt{x+5}=\sqrt{3x-13}$
56. $\sqrt{x+8}=2-\sqrt{x-12}$
57. For the function $f(x)=\sqrt{3x+4}$, find all values of x for which $f(x)=4$. [9.5]
58. For the function $f(x)=\sqrt[3]{x^2-6x+11}+8$, find all values of x for which $f(x)=11$. [9.5]
59. A pendulum has a length of 5 feet. Find its period, rounded to the nearest hundredth of a second. Use the formula $T=2\pi\sqrt{\dfrac{L}{32}}$. [9.5]
60. A vehicle involved in an accident made 200 feet of skid marks on the asphalt. The speed, s, the vehicle was traveling in miles per hour when it started skidding can be approximated by the formula $s=\sqrt{30df}$, where d represents the length of the skid marks in feet and f represents the drag factor of the road. If the drag factor for asphalt is 0.75, find the speed the car was traveling. Round to the nearest mile per hour. [9.5]

Express in terms of i. **[9.6]**

61. $\sqrt{-49}$
62. $\sqrt{-40}$
63. $\sqrt{-252}$
64. $-\sqrt{-675}$

Add or subtract the complex numbers. **[9.6]**

65. $(4+2i)+(7+3i)$
66. $(9-4i)+(7+2i)$
67. $(11-8i)-(10-13i)$
68. $(6+12i)-(5+5i)$

Multiply. **[9.6]**

69. $2i\cdot 8i$
70. $\sqrt{-6}\cdot\sqrt{-50}$
71. $7i(3-4i)$
72. $(10+2i)(13+8i)$
73. $(9-3i)^2$
74. $(3+14i)(3-14i)$

Simplify by rationalizing the denominator. Write your answer in a + bi from. **[9.6]**

75. $\dfrac{6}{7-3i}$

76. $\dfrac{15i}{6+8i}$

77. $\dfrac{2+i}{9-5i}$

78. $\dfrac{16}{i}$

Simplify. **[9.6]**

79. i^{53}
80. i^{22}

CHAPTER 9 TEST

For Extra Help

Step-by-step test solutions are found on the Chapter Test Prep Videos available via the Video Resources on DVD, in MyMathLab, and on YouTube (search "WoodburyElemIntAlg" and click on "Channels").

Simplify the radical expression. Assume that all variables represent nonnegative real numbers.

1. $\sqrt{25x^{10}y^{14}}$
2. For $f(x)=\sqrt{x^2-9x+6}$, evaluate $f(-5)$. Round to the nearest thousandth.

Simplify the expression. Assume that all variables represent nonnegative real numbers.

3. $a^{3/10}\cdot a^{1/5}$
4. $\dfrac{b^{7/8}}{b^{5/12}}$
5. $125^{-2/3}$

Simplify the radical expression. Assume that all variables represent nonnegative real numbers.

6. $\sqrt{72}$

7. $\sqrt[4]{a^{21}b^{42}c^4}$

Add or subtract.

8. $8\sqrt{18} + 11\sqrt{2} - 3\sqrt{200}$

Multiply.

9. $4\sqrt{5}(7\sqrt{10} - 2\sqrt{35})$

10. $(3\sqrt{6} - 4\sqrt{8})(9\sqrt{6} - 2\sqrt{8})$

Rationalize the denominator and simplify. Assume that all variables represent nonnegative real numbers.

11. $\dfrac{x^3y^2}{\sqrt{x^3y^{11}}}$

12. $\dfrac{15\sqrt{7} - 8\sqrt{2}}{4\sqrt{7} + 9\sqrt{2}}$

Solve.

13. $\sqrt[3]{6x + 5} = 5$

14. $\sqrt{x + 28} + 2 = x$

15. If a pendulum has a period of 4 seconds, find its length, rounded to the nearest tenth of a foot. Use the formula $T = 2\pi\sqrt{\dfrac{L}{32}}$.

16. Express $\sqrt{-99}$ in terms of i.

Add or subtract the complex numbers.

17. $(2 - 3i) - (20 + 8i)$

Multiply.

18. $(6 + 7i)(6 - 7i)$

Simplify by rationalizing the denominator. Write your answer in a + bi form.

19. $\dfrac{4 + 9i}{3 - 5i}$

Mathematicians in History

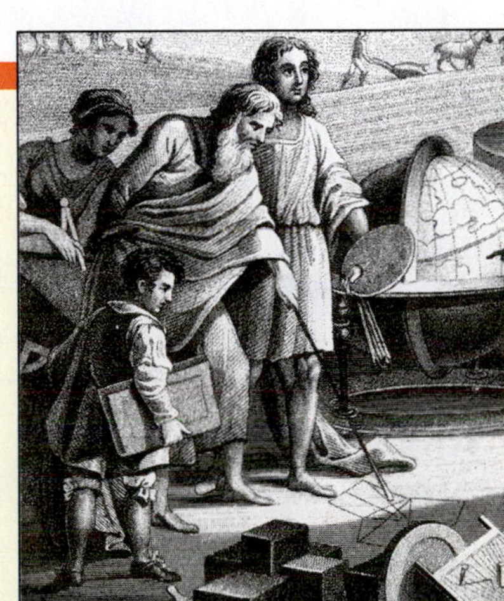

Pythagoras of Samos, often referred to as simply Pythagoras, was the leader of a society known as the Pythagoreans. This society was partially religious and partially scientific in nature. One of its mathematical achievements was the first proof of a theorem that related the lengths of the sides of a right triangle. This theorem is known as the Pythagorean theorem.

Write a one-page summary (*or* make a poster) of the life of Pythagoras, the mathematical achievements of Pythagoras and his society, and the beliefs of the Pythagoreans.

Interesting issues:

- What was the "semicircle"?
- Who were the *mathematikoi*? By what rules did they lead their lives?
- What number did the Pythagoreans consider to be the "best" number? Why?
- Which society was the first to know of the Pythagorean theorem?
- The Pythagoreans believed that all relationships could be expressed numerically as a ratio of two integers. They discovered, however, that the diagonal of a square whose side has length 1 is an irrational number. This number is $\sqrt{2}$. This discovery rocked the foundation of their system of beliefs, and they swore each other to secrecy regarding this discovery. A Pythagorean named Hippasus told others outside the society of this irrational number. What was the fate of Hippasus?

CHAPTER 10

Quadratic Equations

In this chapter, we will learn how to apply techniques other than factoring to solve quadratic equations. *Quadratic equations* are equations that can be written in the form $ax^2 + bx + c = 0$, where $a \neq 0$. One important goal of this chapter is to determine which technique will provide the most efficient solution to a particular equation. We also will apply these techniques to new application problems that previously could not be solved because they produce equations that are not factorable. In this chapter, we also will examine the graphs of quadratic equations and functions. The graph of a quadratic equation is called a parabola and is U-shaped. The techniques used to graph linear equations also will be applied in graphing quadratic equations, along with some new techniques. The chapter ends by examining inequalities that involve quadratic and rational expressions.

STUDY STRATEGY

Time Management When asked why they are having difficulties in a particular class, some students claim that they do not have enough time. However, many of these students have enough time but lack the time management skills to make the most of their time. In this chapter, we will discuss effective time management strategies, focusing on how to make more efficient use of available time.

10.1 Solving Quadratic Equations by Extracting Square Roots; Completing the Square

OBJECTIVES

1. Solve quadratic equations by factoring.
2. Solve quadratic equations by extracting square roots.
3. Solve quadratic equations by extracting square roots involving a linear expression that is squared.
4. Solve applied problems by extracting square roots.
5. Solve quadratic equations by completing the square.
6. Find a quadratic equation, given its solutions.

Solving Quadratic Equations by Factoring

Objective 1 Solve quadratic equations by factoring. In Chapter 6, we learned how to solve quadratic equations through the use of factoring. We begin this section with a review of this technique. We start by simplifying both sides of the equation completely. This includes any distributing that must be done and combining like terms when possible. After that, we need to collect all of the terms on one side of the equation with 0 on the other side of the equation. Recall that it is a good idea to move the terms to the side of the equation that will produce a positive second-degree term. After the equation contains an expression equal to 0, we need to factor the expression. Finally, we set each factor equal to 0 and solve the resulting equation.

EXAMPLE 1 Solve $x^2 - 7x = -10$.

SOLUTION

$$x^2 - 7x = -10$$

$$x^2 - 7x + 10 = 0 \quad \text{Add 10 to collect all terms on the left side.}$$

$$(x-2)(x-5) = 0 \quad \text{Factor.}$$

$$x - 2 = 0 \quad \text{or} \quad x - 5 = 0 \quad \text{Set each factor equal to 0.}$$

$$x = 2 \quad \text{or} \quad x = 5 \quad \text{Solve the resulting equations.}$$

The solution set is $\{2, 5\}$.

EXAMPLE 2 Solve $x^2 = 49$.

SOLUTION

$$x^2 = 49$$

$$x^2 - 49 = 0 \quad \text{Collect all terms on the left side.}$$

$$(x+7)(x-7) = 0 \quad \text{Factor (difference of squares).}$$

$$x + 7 = 0 \quad \text{or} \quad x - 7 = 0 \quad \text{Set each factor equal to 0.}$$

$$x = -7 \quad \text{or} \quad x = 7 \quad \text{Solve the resulting equations.}$$

The solution set is $\{-7, 7\}$.

Quick Check 1

Solve.

a) $x^2 = 2x + 15$
b) $x^2 = 121$

Solving Quadratic Equations by Extracting Square Roots

Objective 2 Solve quadratic equations by extracting square roots. In the preceding example, we tried to find a number x that, when squared, equals 49. The principal square root of 49 is 7, which gives us one of the two solutions. In an equation where we have a squared term equal to a constant, we can take the square root of both sides of the equation to find the solution as long as we take both the positive and negative square root of the constant. The symbol $\pm$ is used to represent the positive and negative square root and is read as *plus or minus.* For example, $x = \pm 7$ means $x = 7$ or $x = -7$. This technique for solving quadratic equations is called **extracting square roots**.

Extracting Square Roots

1. Isolate the squared term.
2. Take the square root of each side. (Remember to take both the *positive* and *negative* ($\pm$) square root of the constant.)
3. Simplify the square root.
4. Solve by isolating the variable.

Now we will use extracting square roots to solve an equation that we would not have been able to solve by factoring.

EXAMPLE 3 Solve $x^2 - 50 = 0$.

SOLUTION The expression $x^2 - 50$ cannot be factored because 50 is not a perfect square. We proceed to solve this equation by extracting square roots.

$$x^2 - 50 = 0$$

$$x^2 = 50 \quad \text{Add 50 to both sides to isolate } x^2.$$

$$\sqrt{x^2} = \pm\sqrt{50} \quad \text{Take the square root of each side.}$$

$$x = \pm 5\sqrt{2} \quad \text{Simplify the square root. } \sqrt{50} = \sqrt{25 \cdot 2} = 5\sqrt{2}$$

The solution set is $\{-5\sqrt{2}, 5\sqrt{2}\}$. Using a calculator, we find that the solutions are approximately ± 7.07.

A WORD OF CAUTION When taking the square root of both sides of an equation, do not forget to use the symbol $\pm$ on the side of the equation containing the constant.

This technique also can be used to find complex solutions to equations, as well as real solutions.

EXAMPLE 4 Solve $x^2 = -16$.

SOLUTION If we had added the 16 to the left side of the equation, the resulting equation would have been $x^2 + 16 = 0$. The expression $x^2 + 16$ is not factorable. (Recall that the sum of 2 squares is not factorable.) The only way to find a solution would be to take the square roots of both sides of the original equation. We proceed to solve this equation by extracting square roots.

$$x^2 = -16$$

$$\sqrt{x^2} = \pm\sqrt{-16} \quad \text{Take the square root of each side.}$$

$$x = \pm 4i \quad \text{Simplify the square root.}$$

The solution set is $\{-4i, 4i\}$. Note that these two solutions are not real numbers.

Quick Check 2

Solve.

a) $x^2 = 28$
b) $x^2 + 32 = 0$

Objective 3 Solve quadratic equations by extracting square roots involving a linear expression that is squared. Extracting square roots is an excellent technique for solving quadratic equations whenever the equation is made up of a squared term and a constant term. Consider the equation $(2x - 7)^2 = 25$. To use factoring to solve this equation, we would have to square $2x - 7$, collect all terms on the left side of the equation by subtracting 25, and hope that the resulting expression could be factored. Extracting square roots is a more efficient way to solve this equation.

EXAMPLE 5 Solve $(2x - 7)^2 = 25$.

SOLUTION

$$(2x - 7)^2 = 25$$

$$\sqrt{(2x - 7)^2} = \pm\sqrt{25}$$ Take the square root of each side.

$$2x - 7 = \pm 5$$ Simplify the square root. Note that the square root of $(2x - 7)^2$ is $2x - 7$.

$$2x = 7 \pm 5$$ Add 7.

$$x = \frac{7 \pm 5}{2}$$ Divide by 2.

$\frac{7 + 5}{2} = 6$ and $\frac{7 - 5}{2} = 1$, so the solution set is $\{1, 6\}$.

EXAMPLE 6 Solve $(3x + 4)^2 + 35 = 15$.

SOLUTION

$$(3x + 4)^2 + 35 = 15$$

$$(3x + 4)^2 = -20$$ Subtract 35 to isolate the squared term.

$$\sqrt{(3x + 4)^2} = \pm\sqrt{-20}$$ Take the square root of each side.

$$3x + 4 = \pm 2i\sqrt{5}$$ Simplify the square root.

$$3x = -4 \pm 2i\sqrt{5}$$ Subtract 4.

$$x = \frac{-4 \pm 2i\sqrt{5}}{3}$$ Divide both sides by 3.

Because we cannot simplify the numerator in this case, these are the solutions. The solution set is $\left\{\frac{-4 - 2i\sqrt{5}}{3}, \frac{-4 + 2i\sqrt{5}}{3}\right\}$.

Quick Check 3

Solve.

a) $(3x + 1)^2 = 100$ **b)** $2(2x - 3)^2 + 17 = -1$

Solving Applied Problems by Extracting Square Roots

Objective 4 Solve applied problems by extracting square roots. Problems involving the area of a geometric figure often lead to quadratic equations, as area is measured in square units. Here are two useful area formulas that can lead to quadratic equations that can be solved by extracting square roots.

Figure	Square	Circle
	s	r
Area	$A = s^2$	$A = \pi r^2$

EXAMPLE 7 A square has an area of 40 square centimeters. Find the length of its side, rounded to the nearest tenth of a centimeter.

SOLUTION In this problem, the unknown quantity is the length of the side of the square, while we know that the area is 40 cm². We will let x represent the length of the side. Because the area of a square is equal to its side squared, the equation we need to solve is $x^2 = 40$.

$$x^2 = 40$$
$$\sqrt{x^2} = \pm\sqrt{40} \quad \text{Take the square root of each side.}$$
$$x = \pm 2\sqrt{10} \quad \text{Simplify the square roots.}$$

Because x represents the length of the side of the square, it must be positive. We may omit the solution $x = -2\sqrt{10}$. The solution is $x = 2\sqrt{10}$ cm. Rounding this to the nearest tenth, we know that the length of the side of the square is 6.3 cm.

Quick Check 4

A square has an area of 150 square inches. Find the length of its side, rounded to the nearest tenth of an inch.

EXAMPLE 8 A circle has an area of 100 square feet. Find the radius of the circle, rounded to the nearest tenth of a foot.

SOLUTION In this problem, the unknown quantity is the radius of the circle and we will let r represent the radius. Because the area of a circle is given by the formula $A = \pi r^2$, the equation we need to solve is $\pi r^2 = 100$.

$$\pi r^2 = 100$$
$$\frac{\overset{1}{\cancel{\pi}} r^2}{\underset{1}{\cancel{\pi}}} = \frac{100}{\pi} \quad \text{Divide both sides by } \pi.$$
$$\sqrt{r^2} = \pm\sqrt{\frac{100}{\pi}} \quad \text{Take the square root of each side.}$$
$$r = \pm\sqrt{\frac{100}{\pi}} \quad \text{Simplify.}$$

Again, we may omit the negative solution. The solution is $r = \sqrt{\frac{100}{\pi}}$ ft. Rounding this to the nearest tenth, the radius is 5.6 feet.

Quick Check 5

A circle has an area of 20 square centimeters. Find the radius of the circle, rounded to the nearest tenth of a centimeter.

Solving Quadratic Equations by Completing the Square

Objective 5 Solve quadratic equations by completing the square. We can solve any quadratic equation by converting it to an equation with a squared term equal to a constant. Rewriting the equation in this form allows us to solve it by extracting square roots. We will now examine a procedure for doing this called **completing the square**. This procedure is used for an equation such as $x^2 - 6x - 16 = 0$, which has both a second-degree term (x^2) and a first-degree term ($-6x$).

The first step is to isolate the variable terms on one side of the equation with the constant term on the other side. This can be done by adding 16 to both sides of the equation. The resulting equation is $x^2 - 6x = 16$.

The next step is to add a number to both sides of the equation that makes the side of the equation containing the variable terms a perfect square trinomial. To do this, we take half of the coefficient of the first-degree term, which in this case is -6, and square it.

$$\left(\frac{-6}{2}\right)^2 = (-3)^2 = 9$$

After adding 9 to both sides of the equation, we obtain the equation $x^2 - 6x + 9 = 25$. The expression on the left side of the equation, $x^2 - 6x + 9$, is a perfect square trinomial and can be factored as $(x - 3)^2$. The resulting equation, $(x - 3)^2 = 25$, is in the correct form for extracting square roots. (This equation will be solved completely in the next example.)

Here is the procedure for solving a quadratic equation by completing the square, provided the coefficient of the squared term is 1.

Completing the Square

1. Isolate all variable terms on one side of the equation, with the constant term on the other side of the equation.
2. Identify the coefficient of the first-degree term. Take half of that number, square it, and add that to both sides of the equation.
3. Factor the resulting perfect square trinomial.
4. Take the square root of each side of the equation. Be sure to include $\pm$ on the side where the constant is.
5. Solve the resulting equation.

In the following example, we will complete the solution of $x^2 - 6x - 16 = 0$ by completing the square.

EXAMPLE 9 Solve $x^2 - 6x - 16 = 0$ by completing the square.

SOLUTION

$$x^2 - 6x - 16 = 0$$

$$x^2 - 6x = 16 \quad \text{Add 16.}$$

$$x^2 - 6x + 9 = 16 + 9 \quad \left(\frac{-6}{2}\right)^2 = (-3)^2 = 9$$

Add 9 to complete the square.

$$(x - 3)^2 = 25 \quad \text{Simplify. Factor the left side.}$$

$$\sqrt{(x - 3)^2} = \pm\sqrt{25} \quad \text{Take the square root of each side.}$$

$$x - 3 = \pm 5 \quad \text{Simplify the square root.}$$

$$x = 3 \pm 5 \quad \text{Add 3.}$$

$3 + 5 = 8$ and $3 - 5 = -2$, so the solution set is $\{-2, 8\}$.

Note that the equation in the previous example could have been solved with less work by factoring $x^2 - 6x - 16$. Use factoring whenever possible because it is often the quickest and most direct way to find the solutions.

EXAMPLE 10 Solve $x^2 + 8x + 20 = 0$ by completing the square.

SOLUTION

$$x^2 + 8x + 20 = 0$$

$$x^2 + 8x = -20 \quad \text{Subtract 20.}$$

$$x^2 + 8x + 16 = -20 + 16$$

Half of 8 is 4, which equals 16 when squared: $\left(\frac{8}{2}\right)^2 = (4)^2 = 16$. Add 16.

$$x^2 + 8x + 16 = -4 \quad \text{Simplify.}$$

$$(x + 4)^2 = -4 \quad \text{Factor the left side.}$$

$$\sqrt{(x+4)^2} = \pm\sqrt{-4}$$ Take the square root of each side.

$$x + 4 = \pm 2i$$ Simplify the square root.

$$x = -4 \pm 2i$$ Subtract 4.

The solution set is $\{-4 - 2i, -4 + 2i\}$.

Quick Check 6

Solve by completing the square.

a) $x^2 + 8x + 12 = 0$
b) $x^2 - 6x + 10 = 0$

Note that the expression $x^2 + 8x + 20$ does not factor; so completing the square is the only technique we have for solving this equation. In the first two examples of completing the square, the coefficient of the first-degree term was an even integer. If this is not the case, we must use fractions to complete the square.

EXAMPLE 11 Solve $x^2 - 5x - 5 = 0$ by completing the square.

SOLUTION The expression $x^2 - 5x - 5$ does not factor, so we proceed with completing the square.

$$x^2 - 5x - 5 = 0$$

$$x^2 - 5x = 5$$ Add 5.

$$x^2 - 5x + \frac{25}{4} = 5 + \frac{25}{4}$$ Half of -5 is $-\frac{5}{2}$, which equals $\frac{25}{4}$ when squared: $\left(-\frac{5}{2}\right)^2 = \frac{25}{4}$. Add $\frac{25}{4}$.

$$x^2 - 5x + \frac{25}{4} = \frac{45}{4}$$ Add 5 and $\frac{25}{4}$ by rewriting 5 as a fraction whose denominator is 4: $5 + \frac{25}{4} = \frac{20}{4} + \frac{25}{4}$.

$$\left(x - \frac{5}{2}\right)^2 = \frac{45}{4}$$ Factor.

$$\sqrt{\left(x - \frac{5}{2}\right)^2} = \pm\sqrt{\frac{45}{4}}$$ Take the square root of each side.

$$x - \frac{5}{2} = \pm\frac{3\sqrt{5}}{2}$$ Simplify the square root.

$$x = \frac{5}{2} \pm \frac{3\sqrt{5}}{2}$$ Add $\frac{5}{2}$.

The solution set is $\left\{\frac{5 - 3\sqrt{5}}{2}, \frac{5 + 3\sqrt{5}}{2}\right\}$.

Quick Check 7

Solve $x^2 + 7x - 18 = 0$ by completing the square.

In the previous example, half of b was a fraction, causing us to add a fraction to both sides of the equation. In this case, factoring the resulting trinomial can be difficult. Keep this fact in mind: when adding $\left(\frac{b}{2}\right)^2$, or $\frac{b^2}{4}$, to both sides while completing the square, the trinomial will factor to be of the form $\left(x + \frac{b}{2}\right)^2$. This fact can be verified by simplifying $\left(x + \frac{b}{2}\right)^2$ and seeing that it is equal to $x^2 + bx + \frac{b^2}{4}$.

$$\left(x + \frac{b}{2}\right)^2 = \left(x + \frac{b}{2}\right)\left(x + \frac{b}{2}\right)$$ Rewrite as a product.

$$= x^2 + \frac{b}{2}x + \frac{b}{2}x + \frac{b^2}{4}$$ Multiply.

$$= x^2 + bx + \frac{b^2}{4}$$ Combine like terms.

A WORD OF CAUTION To solve a quadratic equation by completing the square, we must be sure that the leading coefficient is positive 1. If the leading coefficient is not equal to 1, we divide both sides of the equation by the leading coefficient.

EXAMPLE 12 Solve $2x^2 - 9x + 5 = 0$ by completing the square.

SOLUTION The leading coefficient is 2, so we begin by dividing each term on both sides of the equation by 2. We can then solve this equation by completing the square.

$$2x^2 - 9x + 5 = 0$$

$$x^2 - \frac{9}{2}x + \frac{5}{2} = 0 \quad \text{Divide both sides of the equation by 2 so that the leading coefficient is equal to 1.}$$

$$x^2 - \frac{9}{2}x = -\frac{5}{2} \quad \text{Subtract } \frac{5}{2}.$$

$$x^2 - \frac{9}{2}x + \frac{81}{16} = -\frac{5}{2} + \frac{81}{16} \quad \text{Half of } -\frac{9}{2} \text{ is } -\frac{9}{4}. \left(-\frac{9}{4}\right)^2 = \frac{81}{16}. \text{ Add } \frac{81}{16} \text{ to both sides of the equation.}$$

$$x^2 - \frac{9}{2}x + \frac{81}{16} = \frac{41}{16} \quad \text{Simplify: } -\frac{5}{2} + \frac{81}{16} = -\frac{40}{16} + \frac{81}{16}.$$

$$\left(x - \frac{9}{4}\right)^2 = \frac{41}{16} \quad \text{Factor. In completing the square with } x^2 - bx, \text{ the expression on the left side will factor to be of the form } \left(x - \frac{b}{2}\right)^2.$$

$$\sqrt{\left(x - \frac{9}{4}\right)^2} = \pm\sqrt{\frac{41}{16}} \quad \text{Take the square root of each side.}$$

$$x - \frac{9}{4} = \pm\frac{\sqrt{41}}{4} \quad \text{Simplify each square root.}$$

$$x = \frac{9 \pm \sqrt{41}}{4} \quad \text{Add } \frac{9}{4}.$$

The solution set is $\left\{\frac{9 - \sqrt{41}}{4}, \frac{9 + \sqrt{41}}{4}\right\}$.

Quick Check 8

Solve $2x^2 - 32x + 92 = 0$ by completing the square.

Finding a Quadratic Equation, Given Its Solutions

Objective 6 Find a quadratic equation, given its solutions.

EXAMPLE 13 Find a quadratic equation in standard form, with integer coefficients, that has the solution set $\{-6, \frac{1}{2}\}$.

SOLUTION We know that $x = -6$ is a solution to the equation. This tells us that $x + 6$ is a factor of the quadratic expression. Similarly, knowing that $x = \frac{1}{2}$ is a solution tells us that $2x - 1$ is a factor of the quadratic expression.

$$x = \frac{1}{2}$$

$$2 \cdot x = \frac{1}{\cancel{2}_1} \cdot \overset{1}{\cancel{2}} \quad \text{Multiply both sides by 2 to clear the fractions.}$$

$$2x = 1 \quad \text{Simplify.}$$

$$2x - 1 = 0 \quad \text{Subtract 1.}$$

Multiplying these two factors will give us a quadratic equation with these two solutions.

$x = -6$ or $x = \frac{1}{2}$	Begin with the solutions.
$x + 6 = 0$ or $2x - 1 = 0$	Rewrite each equation so that the right side is equal to 0.
$(x + 6)(2x - 1) = 0$	Write an equation that has these two expressions as factors.
$2x^2 - x + 12x - 6 = 0$	Multiply.
$2x^2 + 11x - 6 = 0$	Simplify.

A quadratic equation that has the solution set $\left\{-6, \frac{1}{2}\right\}$ is $2x^2 + 11x - 6 = 0$.

Quick Check 9

Find a quadratic equation in standard form, with integer coefficients, that has the solution set $\left\{\frac{5}{3}, -2\right\}$.

Notice that we say *a* quadratic equation rather than *the* quadratic equation. There are infinitely many quadratic equations with integer coefficients that have this solution set. For example, multiplying both sides of the equation by 3 gives us the equation $6x^2 + 33x - 18 = 0$, which has the same solution set.

EXAMPLE 14 Find a quadratic equation in standard form, with integer coefficients, that has the solution set $\{-6i, 6i\}$.

SOLUTION We know that $x = -6i$ is a solution to the equation. This tells us that $x + 6i$ is a factor of the quadratic expression. Similarly, knowing that $x = 6i$ is a solution tells us that $x - 6i$ is a factor of the quadratic expression. Multiplying these two factors will give us a quadratic equation with these two solutions.

$x = -6i$ or $x = 6i$	Begin with the solutions.
$x + 6i = 0$ or $x - 6i = 0$	Rewrite each equation so that the right side is equal to 0.
$(x + 6i)(x - 6i) = 0$	Write an equation with one side having these two expressions as factors.
$x^2 - 6i\,x + 6i\,x - 36i^2 = 0$	Multiply.
$x^2 + 36 = 0$	Simplify. Recall that $i^2 = -1$.

A quadratic equation that has the solution set $\{-6i, 6i\}$ is $x^2 + 36 = 0$.

Quick Check 10

Find a quadratic equation in standard form, with integer coefficients, that has the solution set $\{-10i, 10i\}$.

BUILDING YOUR STUDY STRATEGY

Time Management, 1 Keeping Track The first step in achieving effective time management is keeping track of your time over a one-week period. Write down the times you spend on school, work, and leisure activities. This will give you a good idea of how much extra time you have to devote to studying, as well as how much time you devote to other activities.

Once you have made needed changes to your current schedule, try to set aside more time for studying math. You should be studying between two and four hours per week for each hour you spend in class. Try to schedule some time each day rather than cramming all of your studying into the weekend.

Exercises 10.1

MyMathLab

PRACTICE

WATCH

DOWNLOAD
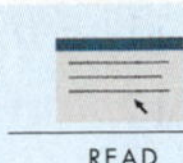
READ

REVIEW

Vocabulary

1. The method of solving an equation by taking the square root of both sides is called solving by ____________________.
2. To solve an equation by extracting square roots, first isolate the expression that is ________.
3. The method of solving an equation by rewriting the variable expression as a square of a binomial is called solving by ____________________.
4. To solve the equation $x^2 + bx = c$ by completing the square, first add ______ to both sides of the equation.

Solve by factoring.

5. $x^2 - 3x - 10 = 0$
6. $x^2 + 9x + 14 = 0$
7. $x^2 + 7x = 0$
8. $x^2 - 4x = 0$
9. $3x^2 - 13x + 4 = 0$
10. $2x^2 - 21x + 27 = 0$
11. $x^2 - 2x = 24$
12. $x^2 - 45 = -4x$
13. $x^2 - 9 = 0$
14. $x^2 - 25 = 0$
15. $x^2 - 4x = 7x + 26$
16. $x^2 + 8x + 13 = 5x + 41$

Solve by extracting square roots.

17. $x^2 = 64$
18. $x^2 = 81$
19. $x^2 - 24 = 0$
20. $x^2 - 27 = 0$
21. $x^2 = -72$
22. $x^2 = -25$
23. $x^2 + 52 = 32$
24. $x^2 - 17 = 63$
25. $(x - 7)^2 = 18$
26. $(x + 3)^2 = 100$
27. $(x + 6)^2 = -49$
28. $(x - 1)^2 = -60$
29. $(x + 5)^2 + 33 = 15$
30. $(x + 3)^2 - 11 = 38$
31. $2(x - 9)^2 - 17 = 83$
32. $3(x + 1)^2 + 11 = 83$
33. $\left(x - \frac{2}{5}\right)^2 = \frac{49}{25}$
34. $\left(x + \frac{3}{4}\right)^2 = -\frac{25}{16}$

Fill in the missing term that makes the expression a perfect square trinomial. Factor the resulting expression.

35. $x^2 + 12x +$ ___
36. $x^2 - 2x +$ ___
37. $x^2 - 5x +$ ___
38. $x^2 + 13x +$ ___
39. $x^2 - 8x +$ ___
40. $x^2 + 18x +$ ___
41. $x^2 + \frac{3}{4}x +$ ___
42. $x^2 - \frac{7}{3}x +$ ___

For Exercises 43–50, approximate to the nearest tenth when necessary.

43. The area of a square is 81 square meters. Find the length of a side of the square.
44. The area of a square is 144 square feet. Find the length of a side of the square.
45. The area of a square is 128 square inches. Find the length of a side of the square.
46. The area of a square is 200 square centimeters. Find the length of a side of the square.
47. The area of a circle is 24 square inches. Find the radius of the circle.
48. The area of a circle is 60 square centimeters. Find the radius of the circle.
49. The area of a circle is 250 square meters. Find the radius of the circle.
50. The area of a circle is 48 square feet. Find the radius of the circle.

Solve by completing the square.

51. $x^2 - 2x - 63 = 0$
52. $x^2 + 10x - 39 = 0$

53. $x^2 + 6x - 11 = 0$

54. $x^2 - 8x - 17 = 0$

55. $x^2 - 4x = -12$

56. $x^2 + 12x = -48$

57. $x^2 - 14x + 30 = 0$

58. $x^2 + 20x + 125 = 0$

59. $x^2 + 8x + 44 = 0$

60. $x^2 + 26x - 84 = 0$

61. $x^2 - 16 = -6x$

62. $x^2 - 48 = 2x$

63. $x^2 - 3x - 54 = 0$

64. $x^2 + 9x + 20 = 0$

65. $x^2 + 7x - 10 = 0$

66. $x^2 - 5x - 8 = 0$

67. $x^2 - 9x + 36 = 0$

68. $x^2 + 11x + 37 = 0$

69. $x^2 + \frac{5}{2}x + 1 = 0$

70. $x^2 - \frac{23}{6}x + \frac{7}{2} = 0$

71. $2x^2 - 2x - 144 = 0$

72. $4x^2 - 56x + 172 = 0$

73. $2x^2 - 9x + 4 = 0$

74. $3x^2 - 13x - 30 = 0$

75. $2x^2 - 6x + 5 = 0$

76. $4x^2 - 20x + 29 = 0$

Find a quadratic equation with integer coefficients that has the following solution set.

77. $\{-5, 2\}$

78. $\{-4, -3\}$

79. $\{6, 8\}$

80. $\{0, 5\}$

81. $\left\{-2, \frac{3}{4}\right\}$

82. $\left\{\frac{1}{2}, \frac{11}{4}\right\}$

83. $\{-3, 3\}$

84. $\{-6, 6\}$

85. $\{-5i, 5i\}$

86. $\{-i, i\}$

Mixed Practice, 87–114

Solve by any method (factoring, extracting square roots, or completing the square).

87. $x^2 + 13x + 30 = 0$

88. $x^2 + 12 = 0$

89. $x^2 + 6x - 17 = 0$

90. $(x - 5)^2 = 1$

91. $x^2 - 6x - 7 = 0$

92. $x^2 - 8x = 0$

93. $2x^2 - 15x + 25 = 0$

94. $x^2 - 6x = -10$

95. $(x - 4)^2 - 11 = 16$

96. $x^2 - 5x - 36 = 0$

97. $x^2 - 4x = 20$

98. $x^2 + 85 = 18x$

99. $x^2 - 5x - 6 = 0$

100. $(2x + 8)^2 - 5 = 11$

101. $x^2 - 8x + 19 = 0$

102. $x^2 - 20x + 91 = 0$

103. $3(x + 3)^2 + 13 = -11$

104. $x^2 - 9 = 0$

105. $x^2 + 3x + 9 = 0$

106. $x^2 - 2x + 50 = 0$

107. $x^2 + 16x = 0$

108. $6x^2 + 13x + 6 = 0$

109. $x(x + 5) - 7(x + 5) = 0$

110. $(4x - 3)(4x - 3) = 25$

111. $-16x^2 + 64x + 80 = 0$

112. $\left(x + \frac{b}{2a}\right)^2 = \frac{b^2 - 4ac}{4a^2}$, where a, b, and c are constants and $a \neq 0$.

113. $\frac{2}{3}x^2 + \frac{8}{3}x - \frac{10}{3} = 0$

114. $x^2 - 10x + 11 = 0$

Writing in Mathematics

Answer in complete sentences.

115. Explain why you use the symbol ± when taking the square root of each side of an equation.

10.2 The Quadratic Formula

OBJECTIVES

1. **Derive the quadratic formula.**
2. **Identify the coefficients *a*, *b*, and *c* of a quadratic equation.**
3. **Solve quadratic equations by using the quadratic formula.**
4. **Use the discriminant to determine the number and type of solutions of a quadratic equation.**
5. **Use the discriminant to determine whether a quadratic expression is factorable.**
6. **Solve projectile motion problems.**

The Quadratic Formula

Objective 1 Derive the quadratic formula. Completing the square to solve a quadratic equation can be tedious. As an alternative, in this section, we will develop and use the **quadratic formula**, which is a numerical formula that gives the solution(s) of *any* quadratic equation.

We derive the quadratic formula by solving the general equation $ax^2 + bx + c = 0$ (where a, b, and c are real numbers and $a \neq 0$) by completing the square. Recall that the coefficient of the second-degree term must be equal to 1; so the first step is to divide both sides of the equation $ax^2 + bx + c = 0$ by a.

$$ax^2 + bx + c = 0$$

$$x^2 + \frac{b}{a}x + \frac{c}{a} = 0$$ Divide both sides by a.

$$x^2 + \frac{b}{a}x = -\frac{c}{a}$$ Subtract the constant term $\frac{c}{a}$.

$$x^2 + \frac{b}{a}x + \frac{b^2}{4a^2} = -\frac{c}{a} + \frac{b^2}{4a^2}$$ Half of $\frac{b}{a}$ is $\frac{b}{2a}$. $\left(\frac{b}{2a}\right)^2 = \frac{b^2}{4a^2}$. Add $\frac{b^2}{4a^2}$ to both sides of the equation.

$$x^2 + \frac{b}{a}x + \frac{b^2}{4a^2} = \frac{b^2 - 4ac}{4a^2}$$ Simplify the right side of the equation as a single fraction by rewriting $-\frac{c}{a}$ as $-\frac{4ac}{4a^2}$.

$$\left(x + \frac{b}{2a}\right)^2 = \frac{b^2 - 4ac}{4a^2}$$ Factor the left side of the equation.

$$\sqrt{\left(x + \frac{b}{2a}\right)^2} = \pm\sqrt{\frac{b^2 - 4ac}{4a^2}}$$ Take the square root of both sides.

$$x + \frac{b}{2a} = \pm\frac{\sqrt{b^2 - 4ac}}{\sqrt{4a^2}}$$ Simplify the square root on the left side. Rewrite the right side as the quotient of two square roots.

$$x + \frac{b}{2a} = \pm\frac{\sqrt{b^2 - 4ac}}{2a}$$ Simplify the square root in the denominator.

$$x = -\frac{b}{2a} \pm \frac{\sqrt{b^2 - 4ac}}{2a}$$ Subtract $\frac{b}{2a}$.

$$x = \frac{-b \pm \sqrt{b^2 - 4ac}}{2a}$$ Rewrite as a single fraction.

The Quadratic Formula

For a general quadratic equation $ax^2 + bx + c = 0$ (where a, b, and c are real numbers and $a \neq 0$), the quadratic formula tells us that the solutions of this equation are given by

$$x = \frac{-b \pm \sqrt{b^2 - 4ac}}{2a}.$$

If we can identify the coefficients a, b, and c in a quadratic equation, we can find the solutions of the equation by substituting these values for a, b, and c in the quadratic formula.

Identifying the Coefficients to Be Used in the Quadratic Formula

Objective 2 Identify the coefficients *a, b,* and *c* of a quadratic equation. The first step in using the quadratic formula is to identify the coefficients a, b, and c. We can do this only after the equation is in standard form: $ax^2 + bx + c = 0$.

EXAMPLE 1 For the quadratic equation, identify a, b, and c.

a) $x^2 - 6x + 8 = 0$

SOLUTION $a = 1$, $b = -6$, and $c = 8$. Be sure to include the negative sign when identifying coefficients that are negative.

b) $3x^2 - 5x = 7$

SOLUTION Before identifying the coefficients, we must convert this equation to standard form: $3x^2 - 5x - 7 = 0$. So $a = 3$, $b = -5$, and $c = -7$.

c) $x^2 + 8 = 0$

SOLUTION In this example, we do not see a first-degree term. In this case, $b = 0$. The coefficients that we do see tell us that $a = 1$ and $c = 8$.

Quick Check 1

For the given quadratic equation, identify a, b, and c.

a) $x^2 + 11x - 13 = 0$
b) $5x^2 - 11 = 9x$
c) $x^2 = 30$

Solving Quadratic Equations by the Quadratic Formula

Objective 3 Solve quadratic equations by using the quadratic formula. Now we will solve several equations using the quadratic formula.

EXAMPLE 2 Solve $x^2 - 6x + 8 = 0$.

SOLUTION We can use the quadratic formula with $a = 1$, $b = -6$, and $c = 8$.

$$x = \frac{-b \pm \sqrt{b^2 - 4ac}}{2a}$$

$$x = \frac{6 \pm \sqrt{(-6)^2 - 4(1)(8)}}{2(1)}$$ Substitute 1 for a, -6 for b, and 8 for c. When you use the formula, think of the $-b$ in the numerator as the "opposite of b." The opposite of -6 is 6.

$$x = \frac{6 \pm \sqrt{36 - 32}}{2}$$ Simplify each term in the radicand.

$$x = \frac{6 \pm \sqrt{4}}{2}$$ Subtract.

$$x = \frac{6 \pm 2}{2}$$ Simplify the square root.

$\frac{6 + 2}{2} = 4$ and $\frac{6 - 2}{2} = 2$; so the solution set is $\{2, 4\}$.

Note that the equation in the previous example could have been solved more quickly by factoring $x^2 - 6x + 8$ to be $(x - 4)(x - 2)$ and then solving. Use factoring whenever possible, treating the quadratic formula as an alternative.

EXAMPLE 3 Solve $3x^2 + 8x = -3$.

SOLUTION To solve this equation, we must rewrite the equation in standard form by collecting all terms on the left side of the equation. In other words, we will rewrite the equation as $3x^2 + 8x + 3 = 0$.

If an equation does not quickly factor, we should proceed directly to the quadratic formula rather than try to factor an expression that may not be factorable. Here $a = 3$, $b = 8$, and $c = 3$.

$$x = \frac{-8 \pm \sqrt{(8)^2 - 4(3)(3)}}{2(3)}$$ Substitute 3 for a, 8 for b, and 3 for c in the quadratic formula.

$$x = \frac{-8 \pm \sqrt{28}}{6}$$ Simplify the radicand.

$$x = \frac{-8 \pm 2\sqrt{7}}{6}$$ Simplify the square root.

$$x = \frac{\overset{1}{\cancel{2}}(-4 \pm \sqrt{7})}{\underset{3}{\cancel{6}}}$$ Factor the numerator and divide out the common factor.

$$x = \frac{-4 \pm \sqrt{7}}{3}$$ Simplify.

The solution set is $\left\{\frac{-4 - \sqrt{7}}{3}, \frac{-4 + \sqrt{7}}{3}\right\}$. These solutions are approximately -2.22 and -0.45, respectively.

Using Your Calculator Here is the screen that shows how to approximate $\frac{-4 + \sqrt{7}}{3}$ and $\frac{-4 - \sqrt{7}}{3}$.

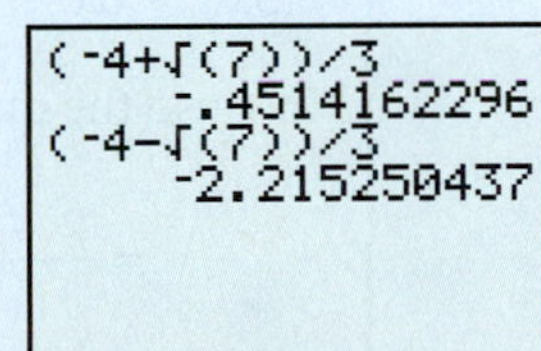

Quick Check 2

Solve by using the quadratic formula.

a) $x^2 + 7x - 30 = 0$

b) $7x^2 + 21x = -9$

EXAMPLE 4 Solve $x^2 - 4x = -5$.

SOLUTION We begin by rewriting the equation in standard form: $x^2 - 4x + 5 = 0$. The quadratic expression in this equation does not factor, so we use the quadratic formula with $a = 1$, $b = -4$, and $c = 5$.

$x = \frac{4 \pm \sqrt{(-4)^2 - 4(1)(5)}}{2(1)}$ Substitute 1 for a, -4 for b, and 5 for c in the quadratic formula.

$x = \frac{4 \pm \sqrt{-4}}{2}$ Simplify the radicand.

$x = \frac{4 \pm 2i}{2}$ Simplify the square root. Be sure to include i, as you are taking the square root of a negative number.

$x = \frac{\overset{1}{\cancel{2}}(2 \pm i)}{\underset{1}{\cancel{2}}}$ Factor the numerator and divide out the common factor.

$x = 2 \pm i$ Simplify.

The solution set is $\{2 - i, 2 + i\}$.

Quick Check 3

Solve $x^2 - 7x = -19$.

If an equation has coefficients that are fractions, multiply each side of the equation by the LCD to clear the equation of fractions. If we can solve the equation by factoring, it will be easier to factor without the fractions. If we cannot solve by factoring, the quadratic formula will be easier to simplify using integers rather than fractions.

EXAMPLE 5 Solve $\frac{1}{2}x^2 - x + \frac{1}{3} = 0$.

SOLUTION The first step is to clear the fractions by multiplying each side of the equation by the LCD, which in this example is 6.

$6 \cdot \left(\frac{1}{2}x^2 - x + \frac{1}{3}\right) = 6 \cdot 0$ Multiply both sides of the equation by the LCD.

$3x^2 - 6x + 2 = 0$ Distribute and simplify.

Now use the quadratic formula with $a = 3$, $b = -6$, and $c = 2$.

$x = \frac{6 \pm \sqrt{(-6)^2 - 4(3)(2)}}{2(3)}$ Substitute 3 for a, -6 for b, and 2 for c in the quadratic formula.

$x = \frac{6 \pm \sqrt{12}}{6}$ Simplify the radicand.

$x = \frac{6 \pm 2\sqrt{3}}{6}$ Simplify the square root.

$x = \frac{\overset{1}{\cancel{2}}(3 \pm \sqrt{3})}{\underset{3}{\cancel{6}}}$ Factor the numerator and divide out common factors.

$x = \frac{3 \pm \sqrt{3}}{3}$ Simplify.

The solution set is $\left\{\frac{3 - \sqrt{3}}{3}, \frac{3 + \sqrt{3}}{3}\right\}$. These solutions are approximately 0.42 and 1.58, respectively.

Quick Check 4

Solve $\frac{1}{4}x^2 + \frac{1}{3}x + \frac{1}{2} = 0$.

In addition to clearing any fractions, make sure the coefficient a of the second-degree term is positive. If this term is negative, you can collect all terms on the other side of the equation or multiply each side of the equation by negative 1.

EXAMPLE 6 Solve $-2x^2 + 11x + 6 = 0$.

SOLUTION We begin by rewriting the equation so that a is positive because factoring a quadratic expression is more convenient when the leading coefficient is positive. Also, we may find that simplifying the quadratic formula is easier when the leading coefficient a is positive.

$0 = 2x^2 - 11x - 6$ Collect all terms on the right side of the equation.

When the leading coefficient is not 1, factoring often is difficult or time-consuming—if the expression is factorable at all. In such a situation, go directly to the quadratic formula. In this example, we can use the quadratic formula with $a = 2$, $b = -11$, and $c = -6$.

$$x = \frac{11 \pm \sqrt{(-11)^2 - 4(2)(-6)}}{2(2)}$$ Substitute 2 for a, -11 for b, and -6 for c in the quadratic formula.

$$x = \frac{11 \pm \sqrt{169}}{4}$$ Simplify the radicand.

$$x = \frac{11 \pm 13}{4}$$ Simplify the square root.

$\frac{11 + 13}{4} = 6$ and $\frac{11 - 13}{4} = -\frac{1}{2}$, so the solution set is $\left\{-\frac{1}{2}, 6\right\}$.

Quick Check 5

Solve $-6x^2 - 7x + 20 = 0$.

Although the quadratic formula can be used to solve *any* quadratic equation, it does not always provide the most efficient way to solve a particular equation. We should check to see whether factoring or extracting square roots can be used before we use the quadratic formula.

General Strategy for Solving Quadratic Equations

- If the equation has one squared term that contains a variable and all other terms are constant terms, solve the equation by extracting square roots.
  Examples: $x^2 - 52 = 0$ $(x - 7)^2 - 15 = 33$
- If the equation is of the form $x^2 + bx + c = 0$ or $ax^2 + bx + c = 0$ and you can quickly factor the trinomial, solve the equation by factoring.
  Examples: $x^2 - 7x + 12 = 0$ $2x^2 - 5x + 2 = 0$
- If you have an equation of the form $x^2 + bx + c = 0$ in which the trinomial cannot be factored and the coefficient b is even, consider solving the equation by completing the square. If b is odd, as in $x^2 + 5x + 7 = 0$, or if the leading coefficient is not equal to 1, as in $2x^2 - 9x + 11 = 0$, using the quadratic formula is often more efficient than completing the square.
  Examples: $x^2 - 4x + 8 = 0$ $x^2 + 16x - 3 = 0$
- In all other cases, use the quadratic formula.

Using the Discriminant to Determine the Number and Type of Solutions of a Quadratic Equation

Objective 4 Use the discriminant to determine the number and type of solutions of a quadratic equation. In the quadratic formula, the expression $b^2 - 4ac$ is called the **discriminant**. The discriminant can provide some information about the solutions. If the discriminant is negative $(b^2 - 4ac < 0)$, the equation has two nonreal complex solutions. This is because we take the square root of a negative number in the quadratic formula. If the discriminant is zero, $b^2 - 4ac = 0$, the equation has one real solution. In this case, the quadratic formula simplifies to be $x = \frac{-b \pm \sqrt{0}}{2a}$, or simply $x = \frac{-b}{2a}$. If the discriminant is positive, $b^2 - 4ac > 0$, the equation has two real solutions. The square root of a positive discriminant is a real number; so the quadratic formula produces two real solutions. The following chart summarizes what the discriminant tells us about the number and type of solutions of an equation:

Solutions of a Quadratic Equation Based on the Discriminant

$b^2 - 4ac$	Number and Type of Solutions
Negative	Two Nonreal Complex Solutions
Zero	One Real Solution
Positive	Two Real Solutions

When working on an applied problem, we need to know if the corresponding equation has no real solutions. (It also will be important to know this when we are graphing quadratic equations, which is covered later in this chapter.)

EXAMPLE 7 For the quadratic equation, use the discriminant to determine the number and type of solutions.

a) $x^2 - 6x - 16 = 0$

SOLUTION

$$(-6)^2 - 4(1)(-16) = 100 \quad \text{Substitute 1 for } a, -6 \text{ for } b, \text{ and } -16 \text{ for } c.$$

Because the discriminant is positive, this equation has two real solutions.

b) $x^2 + 36 = 0$

SOLUTION

$$(0)^2 - 4(1)(36) = -144 \quad \text{Substitute 1 for } a, 0 \text{ for } b, \text{ and } 36 \text{ for } c.$$

The discriminant is negative. This equation has two nonreal complex solutions.

c) $x^2 + 10x + 25 = 0$

SOLUTION

$$(10)^2 - 4(1)(25) = 0 \quad \text{Substitute 1 for } a, 10 \text{ for } b, \text{ and } 25 \text{ for } c.$$

Because the discriminant equals 0, this equation has one real solution.

Quick Check 6

For each given quadratic equation, use the discriminant to determine the number and type of solutions.

a) $x^2 + 18x - 63 = 0$
b) $x^2 + 5x + 42 = 0$
c) $x^2 - 20x + 100 = 0$

Using the Discriminant to Determine Whether a Quadratic Expression Is Factorable

Objective 5 Use the discriminant to determine whether a quadratic expression is factorable. The discriminant also can be used to tell us whether a quadratic expression is factorable. If the discriminant is equal to 0 or a positive number that is a perfect square (1, 4, 9, and so on), the expression is factorable.

EXAMPLE 8 For the quadratic expression, use the discriminant to determine whether the expression can be factored.

a) $x^2 - 6x - 27$

SOLUTION

$$(-6)^2 - 4(1)(-27) = 144 \quad \text{Substitute 1 for } a, -6 \text{ for } b, \text{ and } -27 \text{ for } c.$$

The discriminant is a perfect square ($\sqrt{144} = 12$), so the expression can be factored.

$$x^2 - 6x - 27 = (x - 9)(x + 3)$$

b) $2x^2 + 7x + 4$

SOLUTION

$$(7)^2 - 4(2)(4) = 17 \quad \text{Substitute 2 for } a, 7 \text{ for } b, \text{ and 4 for } c.$$

The discriminant is not a perfect square, so the expression is not factorable.

Quick Check 7

For each given quadratic expression, use the discriminant to determine whether the expression can be factored.

a) $x^2 + 14x + 12$
b) $5x^2 - 36x - 32$

Projectile Motion Problems

Objective 6 Solve projectile motion problems. An application that leads to a quadratic equation involves the height, in feet, of an object propelled into the air after t seconds. Recall that the height, in feet, of a projectile after t seconds can be found by the function $h(t) = -16t^2 + v_0t + s$, where v_0 is the initial velocity of the projectile and s is the initial height.

Height of a Projectile

$$h(t) = -16t^2 + v_0t + s$$

t: Time (seconds)
v_0: Initial velocity (feet/second)
s: Initial height (feet)

EXAMPLE 9 A rock is thrown at a speed of 48 feet/second from ground level. How long will it take until the rock lands on the ground?

SOLUTION The initial velocity of the rock is 48 feet per second, so $v_0 = 48$. Because the rock is thrown from ground level, the initial height is 0 feet. The function for the height of the rock after t seconds is $h(t) = -16t^2 + 48t$.

The rock's height when it lands on the ground is 0 feet, so we set the function equal to 0 and solve for the time t in seconds.

$$-16t^2 + 48t = 0$$ Set the function equal to 0.

$$0 = 16t^2 - 48t$$ Collect all terms on the right side of the equation so that the leading coefficient is positive.

$$0 = 16t(t - 3)$$ Factor out the GCF ($16t$).

$$16t = 0 \quad \text{or} \quad t - 3 = 0$$ Set each variable factor equal to 0.

$$t = 0 \quad \text{or} \quad t = 3$$ Solve each equation.

The time of 0 seconds corresponds to the precise moment the rock was thrown and does not represent the time required to land on the ground. The solution is 3 seconds.

Quick Check 8

A rock is thrown upward at an initial velocity of 80 feet per second from ground level. After how many seconds will the rock land on the ground?

EXAMPLE 10 A golf ball is launched by a slingshot at an initial velocity of 88 feet per second from a platform that is 95 feet high. When will the golf ball be at a height of 175 feet? (Round to the nearest hundredth of a second.)

SOLUTION The initial velocity of the golf ball is 88 feet per second, so $v_0 = 88$. Because the golf ball was launched from a platform 95 feet high, the initial height $s = 95$. The function for the height of the golf ball after t seconds is $h(t) = -16t^2 + 88t + 95$.

To find when the golf ball is at a height of 175 feet, we set the function equal to 175 and solve for the time t in seconds.

$$-16t^2 + 88t + 95 = 175$$ Set the function equal to 175.

$$0 = 16t^2 - 88t + 80$$ Collect all terms on the right side of the equation.

$$0 = 8(2t^2 - 11t + 10)$$ Factor out the GCF (8).

$$0 = 2t^2 - 11t + 10$$ Divide both sides by 8.

This trinomial does not factor, so we will use the quadratic formula with $a = 2$, $b = -11$, and $c = 10$ to solve the equation.

$$t = \frac{11 \pm \sqrt{(-11)^2 - 4(2)(10)}}{2(2)}$$ Substitute 2 for a, -11 for b, and 10 for c.

$$t = \frac{11 \pm \sqrt{41}}{4}$$ Simplify the radicand and the denominator.

We use a calculator to approximate these solutions.

$$\frac{11 + \sqrt{41}}{4} \approx 4.35 \qquad \frac{11 - \sqrt{41}}{4} \approx 1.15$$

The golf ball is at a height of 175 feet after approximately 1.15 seconds and again after approximately 4.35 seconds.

Quick Check 9

A projectile is launched at an initial velocity of 36 feet per second from the top of a building 40 feet high. When will the projectile be at a height of 50 feet? (Round to the nearest hundredth of a second.)

EXAMPLE 11 If a projectile is launched upward at a speed of 96 feet per second from a platform 7 feet above the ground, will it ever reach a height of 160 feet? If so, when will it be at this height?

SOLUTION The function for the height of the projectile after t seconds is $h(t) = -16t^2 + 96t + 7$. We begin by setting the function equal to 160 and solving for t.

$$-16t^2 + 96t + 7 = 160 \quad \text{Set the function equal to 160.}$$
$$0 = 16t^2 - 96t + 153 \quad \text{Collect all terms on the right side.}$$

The trinomial does not have any common factors other than 1, so we will use the quadratic formula with $a = 16$, $b = -96$, and $c = 153$ to solve this equation.

$$t = \frac{96 \pm \sqrt{(-96)^2 - 4(16)(153)}}{2(16)} \quad \text{Substitute 16 for } a, -96 \text{ for } b, \text{ and 153 for } c.$$
$$t = \frac{96 \pm \sqrt{-576}}{32} \quad \text{Simplify the radicand and the denominator.}$$

Because the discriminant is negative, this equation has no real-number solutions. The projectile does not reach a height of 160 feet.

Quick Check 10

A projectile is launched at an initial velocity of 36 feet per second from the top of a building 40 feet high. Will the projectile ever reach a height of 80 feet? If so, when will it be at this height?

A WORD OF CAUTION When the quadratic formula is used to solve an applied problem, a negative discriminant indicates that there are no real solutions to this problem.

BUILDING YOUR STUDY STRATEGY

Time Management 2, Study after Class The best time to study new material is as soon as possible after class. Look for a block of time close to your class period. You can study the new material on campus if necessary. Try to study each day at the same time. Begin each study session by reworking your notes; then move on to attempting the homework exercises.

If possible, establish a second study period during the day to use for review purposes. This second study period should take place later in the day and can be used to review homework or notes or to read ahead for the next class period.

Exercises 10.2

Vocabulary

1. The ________ formula is a formula for calculating the solutions of a quadratic equation.
2. In the quadratic formula, the expression $b^2 - 4ac$ is called the ________.
3. If the discriminant is negative, the quadratic equation has ______ real solutions.
4. If the discriminant is zero, the quadratic equation has ______ unique real solution.
5. If the discriminant is positive, the quadratic equation has ______ real solutions.
6. If the discriminant is 0 or a positive perfect square, the quadratic expression in the equation is ________.

Solve by using the quadratic formula.

7. $x^2 - 5x - 36 = 0$
8. $x^2 + 4x - 45 = 0$
9. $x^2 - 4x + 2 = 0$
10. $x^2 + 10x + 13 = 0$
11. $x^2 + x + 7 = 0$
12. $x^2 - 3x + 18 = 0$
13. $x^2 + 12 = 0$
14. $x^2 + 12x = 0$
15. $x^2 + 7x + 11 = 0$
16. $x^2 + 8x + 21 = 0$
17. $x^2 - 3x - 88 = 0$

18. $x^2 - 6x - 10 = 0$

19. $2x^2 - 5x - 12 = 0$

20. $3x^2 + 4x + 8 = 0$

21. $x^2 - 4x = -32$

22. $x^2 + 6x = 8$

23. $x^2 - 3x = 9$

24. $x^2 = 7x$

25. $-4 = 19x - 5x^2$

26. $4x - x^2 = 3$

27. $x^2 - 6x + 9 = 0$

28. $x^2 + 10x + 25 = 0$

29. $x^2 - 24 = 0$

30. $x^2 + 49 = 0$

31. $x(x - 4) + 3x = 20$

32. $(2x + 1)(x - 3) = -9$

33. $x^2 - \frac{1}{5}x + \frac{3}{4} = 0$

34. $\frac{2}{3}x^2 - \frac{3}{5}x + \frac{1}{4} = 0$

35. $-x^2 + 7x - 12 = 0$

36. $-2x^2 + 15x = 8$

For each of the following quadratic equations, use the discriminant to determine the number and type of solutions.

37. $x^2 + 12x - 30 = 0$

38. $x^2 - 9x + 21 = 0$

39. $2x^2 - 3x + 5 = 0$

40. $25x^2 - 20x + 4 = 0$

41. $x^2 + \frac{2}{5}x + \frac{5}{6} = 0$

42. $x^2 - 5x - 9 = 0$

43. $9x^2 - 12x + 4 = 0$

44. $x^2 - \frac{2}{3}x + \frac{1}{9} = 0$

Use the discriminant to determine whether each of the given quadratic expressions is factorable. If the expression can be factored, write factorable. Otherwise, write prime.

45. $x^2 + 12x - 35$

46. $x^2 + 30x + 221$

47. $x^2 + 52x + 667$

48. $x^2 - 88x + 1886$

49. $35x^2 - 116x + 65$

50. $2x^2 + 13x - 25$

51. $5x^2 - 16x - 18$

52. $32x^2 + 76x - 33$

Mixed Practice, 53–88

Solve each of the following quadratic equations using the most efficient technique (factoring, extracting square roots, completing the square, or using the quadratic formula).

53. $x^2 - 5x - 15 = 0$

54. $x^2 - 68 = 0$

55. $3x^2 + 2x - 1 = 0$

56. $x^2 - 20x + 91 = 0$

57. $(5x - 4)^2 = 36$

58. $7x(8x - 3) + 4(8x - 3) = 0$

59. $2x^2 + 8x = -9$

60. $x^2 + 179x = 0$

61. $x^2 + 324 = 0$

62. $x^2 + \frac{3}{5}x - \frac{1}{12} = 0$

63. $6x^2 - 29x + 28 = 0$

64. $x^2 + 8x - 9 = 0$

65. $3(2x + 1)^2 - 7 = 23$

66. $5x^2 + 11x - 9 = 0$

67. $x^2 - 9x - 21 = 0$

68. $4x^2 - 25 = 0$

69. $16x^2 - 24x + 9 = 0$

70. $x^2 - 15x + 50 = 0$

71. $x^2 + x + 20 = 0$

72. $x^2 + 13x + 36 = 0$

73. $x^2 - 4x - 2 = 0$

74. $3x^2 - 2x - 16 = 0$

75. $x^2 - 6x + 10 = 0$

76. $(x - 8)^2 + 13 = 134$

77. $\frac{3}{4}x^2 + \frac{2}{3}x - \frac{1}{2} = 0$

78. $x^2 + 3x - 18 = 0$

79. $2x(3x + 7) - 5(3x + 7) = 0$

80. $(2x + 3)^2 - 10 = 71$

81. $2(2x - 9)^2 + 13 = 77$

82. $x^2 + x - 72 = 0$

83. $x^2 + 3x - 4 = 0$

84. $x^2 + 10x + 21 = 0$

85. $x^2 - 10x + 18 = 0$

86. $x^2 + 2x + 4 = 0$

87. $x^2 - 16x + 63 = 0$

88. $4x^2 - 12x - 11 = 0$

For Exercises 89–102, use the function $h(t) = -16t^2 + v_0t + s$.

89. An object is launched upward from the ground at an initial speed of 128 feet per second. How long will it take until the object lands on the ground?

90. An object is launched upward at an initial speed of 48 feet per second from a platform 160 feet above the ground. How long will it take until it lands on the ground?

91. Ubaldo Jimenez is standing on a cliff above a beach. He throws a rock upward at a speed of 70 feet per second from a height 90 feet above the beach. How long will it take until the rock lands on the beach? Round to the nearest tenth of a second.

92. Jan is standing on the roof of a building. She launches a water balloon upward at an initial speed of 44 feet per second from a height of 20 feet. How long will it take until the balloon lands on the ground? Round to the nearest tenth of a second.

93. An object is launched upward at an initial velocity of 80 feet per second from ground level. At what time(s) is the object 64 feet above the ground?

94. An object is launched upward at an initial speed of 112 feet per second from ground level. At what time(s) is the object 160 feet above the ground?

95. An object is launched upward at an initial velocity of 120 feet per second from the top of a building 50 feet high. At what time(s) is the object 195 feet above the ground? Round to the nearest tenth of a second.

96. An object is launched upward at an initial velocity of 116 feet per second from a cliff 300 feet above a beach. At what time(s) is the object 350 feet above the ground? Round to the nearest tenth of a second.

A rock is thrown upward at the given velocity from a height of 5 feet above the ground. Does the rock reach a height of 30 feet above the ground? If it does, state how long the rock takes to reach the height of 30 feet. (Round to the nearest tenth of a second.) If it does not, explain.

97. 16 feet per second

98. 48 feet per second

99. 42 feet per second

100. 30 feet per second

101. A projectile is launched straight up from the ground. If it takes 6 seconds for the projectile to land on the ground, find the original velocity of the projectile.

102. A projectile is launched straight up from the ground. If it takes 7.5 seconds for the projectile to land on the ground, find the original velocity of the projectile.

Writing in Mathematics

Answer in complete sentences.

103. Explain how the discriminant tells whether an equation has two real solutions, one real solution, or two nonreal complex solutions.

10.3 Equations That Are Quadratic in Form

OBJECTIVES

1. Solve equations by making a u-substitution.
2. Solve radical equations.
3. Solve rational equations.
4. Solve work-rate problems.

In this section, we will learn how to solve several types of equations that are **quadratic in form**. For example, $x^4 - 13x^2 + 36 = 0$ is not a quadratic equation, but if we rewrite it as $(x^2)^2 - 13(x^2) + 36 = 0$, we can see that it looks like a quadratic equation.

Solving Equations by Making a *u*-Substitution

Objective 1 Solve equations by making a *u*-substitution. One approach to solving equations that are quadratic in form is to use a ***u*-substitution**. We substitute the variable u for an expression such as x^2 so that the resulting equation is a quadratic equation in u. In other words, the equation can be rewritten in the form $au^2 + bu + c = 0$. We can then solve this quadratic equation applying the methods presented in Sections 10.1 and 10.2 (factoring, extracting square roots, completing the square, and using the quadratic formula). After solving this equation for u, we replace u with the expression it previously substituted for and then solve the resulting equations for the original variable.

EXAMPLE 1 Solve $x^4 - 13x^2 + 36 = 0$.

SOLUTION Let $u = x^2$. We can then replace x^2 in the original equation with u, and we can replace x^4 with u^2. The resulting equation will be $u^2 - 13u + 36 = 0$, which is quadratic.

$$x^4 - 13x^2 + 36 = 0$$

$$u^2 - 13u + 36 = 0 \quad \text{Substitute } u \text{ for } x^2.$$

$$(u - 4)(u - 9) = 0 \quad \text{Factor.}$$

$$u - 4 = 0 \quad \text{or} \quad u - 9 = 0 \quad \text{Set each factor equal to 0.}$$

$$u = 4 \quad \text{or} \quad u = 9 \quad \text{Solve.}$$

Now we replace u with x^2 and solve the resulting equations for x.

$$u = 4 \quad \text{or} \quad u = 9$$

$$x^2 = 4 \quad \text{or} \quad x^2 = 9 \quad \text{Substitute } x^2 \text{ for } u.$$

$$\sqrt{x^2} = \pm\sqrt{4} \quad \text{or} \quad \sqrt{x^2} = \pm\sqrt{9} \quad \text{Solve by taking the square roots of both sides of the equation.}$$

$$x = \pm 2 \quad \text{or} \quad x = \pm 3 \quad \text{Simplify the square root.}$$

The solution set is $\{-2, 2, -3, 3\}$.

Quick Check 1

Solve $x^4 - x^2 - 12 = 0$.

A WORD OF CAUTION When solving an equation by using a u-substitution, *do not stop after solving for u.* You must solve for the variable in the original equation.

The challenge is determining when a u-substitution will be helpful and determining what to let u represent. Look for an equation in which the variable part of the first term is the square of the variable part of a second term; in other words, its exponent is twice the exponent of the second term. Then let u represent the variable part with the smaller exponent.

EXAMPLE 2 Find the u-substitution that will convert the equation to a quadratic equation.

a) $x - 7\sqrt{x} - 30 = 0$

SOLUTION Let $u = \sqrt{x}$. This allows us to replace $\sqrt{x}$ with u and x with u^2, because $u^2 = (\sqrt{x})^2 = x$. The resulting equation is $u^2 - 7u - 30 = 0$, which is quadratic.

b) $x^{2/3} + 9x^{1/3} + 8 = 0$

SOLUTION Let $u = x^{1/3}$. We can then replace $x^{2/3}$ with u^2, because $(x^{1/3})^2 = x^{2/3}$. The resulting equation is $u^2 + 9u + 8 = 0$.

c) $(x^2 + 6x)^2 + 13(x^2 + 6x) + 40 = 0$

SOLUTION Let $u = x^2 + 6x$ as this is the expression that is being squared. The resulting equation is $u^2 + 13u + 40 = 0$.

Quick Check 2

Find the u-substitution that will convert the given equation to a quadratic equation.

a) $x + 11\sqrt{x} - 26 = 0$
b) $2x^{2/3} - 17x^{1/3} + 8 = 0$
c) $(x^2 - 4x)^2 - 9(x^2 - 4x) - 36 = 0$

EXAMPLE 3 Solve $x + 3\sqrt{x} - 10 = 0$.

SOLUTION Let $u = \sqrt{x}$. The resulting equation is $u^2 + 3u - 10 = 0$, which is a quadratic equation solvable by factoring. If we could not use factoring, we would have to use the quadratic formula to solve for u.

$x + 3\sqrt{x} - 10 = 0$		
$u^2 + 3u - 10 = 0$		Replace $\sqrt{x}$ with u.
$(u - 2)(u + 5) = 0$		Factor.
$u - 2 = 0$ or	$u + 5 = 0$	Set each factor equal to 0.
$u = 2$ or	$u = -5$	Solve for u.
$\sqrt{x} = 2$ or	$\sqrt{x} = -5$	Replace u with $\sqrt{x}$.
$(\sqrt{x})^2 = 2^2$ or	$(\sqrt{x})^2 = (-5)^2$	Square both sides of the equation.
$x = 4$ or	$x = 25$	Simplify.

Recall that any time we square each side of an equation, we must check for extraneous roots.

Check $(x = 25)$

$(25) + 3\sqrt{(25)} - 10 = 0$	Substitute 25 for x in the original equation.
$25 + 3 \cdot 5 - 10 = 0$	Simplify the square root.
$30 = 0$	Simplify.

So $x = 25$ is not a solution of the equation. We could have seen this before squaring each side of the equation $\sqrt{x} = -5$. The square root of x cannot be negative, so the equation $\sqrt{x} = -5$ cannot have a solution. The check whether $x = 4$ is actually a solution is left to the reader. The solution set is $\{4\}$.

A WORD OF CAUTION Whenever we square both sides of an equation, such as in the previous example, we must check the solutions for extraneous roots.

EXAMPLE 4 Solve $x^{2/3} - 6x^{1/3} - 7 = 0$.

SOLUTION Let $u = x^{1/3}$. The resulting equation is $u^2 - 6u - 7 = 0$, which can be solved by factoring.

$x^{2/3} - 6x^{1/3} - 7 = 0$	
$u^2 - 6u - 7 = 0$	Replace $x^{1/3}$ with u.
$(u - 7)(u + 1) = 0$	Factor.

$u - 7 = 0$ or $u + 1 = 0$		Set each factor equal to 0.
$u = 7$ or $u = -1$		Solve for u.
$x^{1/3} = 7$ or $x^{1/3} = -1$		Replace u with $x^{1/3}$.
$(x^{1/3})^3 = 7^3$ or $(x^{1/3})^3 = (-1)^3$		Raise each side to the third power.
$x = 343$ or $x = -1$		Simplify.

The solution set is $\{-1, 343\}$.

Quick Check 3

Solve.

a) $x - 6x^{1/2} + 5 = 0$
b) $x^{2/3} - 4x^{1/3} + 3 = 0$

Solving Radical Equations

Objective 2 Solve radical equations. Some equations that contain square roots cannot be solved by a u-substitution. When this happens, we will use the techniques developed in Section 9.5. We begin by isolating the radical; then we proceed to square both sides of the equation. This can lead to an equation that is quadratic.

EXAMPLE 5 Solve $\sqrt{x + 7} + 5 = x$.

SOLUTION We begin by isolating the radical so that we may square each side of the equation.

$$\sqrt{x+7} + 5 = x$$

$\sqrt{x+7} = x - 5$	Subtract 5 to isolate the radical.
$(\sqrt{x+7})^2 = (x-5)^2$	Square both sides.
$x + 7 = (x-5)(x-5)$	Square the binomial by multiplying it by itself.
$x + 7 = x^2 - 10x + 25$	Multiply.
$0 = x^2 - 11x + 18$	Subtract x and 7 to collect all terms on the right side of the equation.
$0 = (x-2)(x-9)$	Factor.
$x - 2 = 0$ or $x - 9 = 0$	Set each factor equal to 0.
$x = 2$ or $x = 9$	Solve.

Because we have squared each side of the equation, we must check for extraneous roots.

$x = 2$	$x = 9$
$\sqrt{(2)+7} + 5 = (2)$	$\sqrt{(9)+7} + 5 = (9)$
$\sqrt{9} + 5 = 2$	$\sqrt{16} + 5 = 9$
$3 + 5 = 2$	$4 + 5 = 9$
$8 = 2$	$9 = 9$
False	True

The solution $x = 2$ is an extraneous solution and must be omitted. The solution set is $\{9\}$.

Quick Check 4

Solve $x + 7 = \sqrt{x + 9}$.

Solving Rational Equations

Objective 3 Solve rational equations. Solving rational equations, which were covered in Chapter 7, often requires that we solve a quadratic equation. We begin to solve a rational equation by finding the LCD and multiplying each side of the equation by it to clear the equation of fractions. The resulting equation could be quadratic, as demonstrated in the next example. Once we solve the resulting equation, any solution that causes a denominator in the original equation to be equal to 0 must be omitted.

EXAMPLE 6 Solve $\frac{x}{x-4} + \frac{2}{x+3} = \frac{6}{x^2 - x - 12}$.

SOLUTION We begin by factoring the denominators to find the LCD. The LCD is $(x-4)(x+3)$, and solutions of $x = 4$ and $x = -3$ must be omitted because either would result in a denominator of 0.

$$\frac{x}{x-4} + \frac{2}{x+3} = \frac{6}{x^2 - x - 12}$$

$$\frac{x}{x-4} + \frac{2}{x+3} = \frac{6}{(x-4)(x+3)} \quad \text{The LCD is } (x-4)(x+3).$$

$$(x-4)(x+3) \cdot \left(\frac{x}{x-4} + \frac{2}{x+3}\right) = (x-4)(x+3) \cdot \frac{6}{(x-4)(x+3)} \quad \text{Multiply by the LCD.}$$

$$\overset{1}{\cancel{(x-4)}}(x+3) \cdot \frac{x}{\underset{1}{\cancel{(x-4)}}} + (x-4)\overset{1}{\cancel{(x+3)}} \cdot \frac{2}{\underset{1}{\cancel{(x+3)}}}$$

$$= \overset{1}{\cancel{(x-4)}}\overset{1}{\cancel{(x+3)}} \cdot \frac{6}{\underset{1}{\cancel{(x-4)}}\underset{1}{\cancel{(x+3)}}} \quad \text{Distribute and divide out common factors.}$$

$$x(x+3) + 2(x-4) = 6 \quad \text{Multiply remaining factors.}$$

$$x^2 + 3x + 2x - 8 = 6 \quad \text{Multiply.}$$

$$x^2 + 5x - 8 = 6 \quad \text{Combine like terms.}$$

$$x^2 + 5x - 14 = 0 \quad \text{Collect all terms on the left side by subtracting 6.}$$

$$(x+7)(x-2) = 0 \quad \text{Factor.}$$

$$x + 7 = 0 \quad \text{or} \quad x - 2 = 0 \quad \text{Set each factor equal to 0.}$$

$$x = -7 \quad \text{or} \quad x = 2 \quad \text{Solve.}$$

Because neither solution causes a denominator to equal 0, we do not need to omit either solution. The solution set is $\{-7, 2\}$.

Quick Check 5

Solve $\frac{2}{x+1} + \frac{1}{x-1} = 1$.

Solving Work-Rate Problems

Objective 4 Solve work-rate problems. The last example of the section is a work-rate problem. Work-rate problems involve rational equations and were introduced in Chapter 7.

EXAMPLE 7 A water tower has two drainpipes attached to it. Alone, the smaller pipe takes 15 minutes longer than the larger pipe to empty the tower. If both drainpipes are used together, the tower can be drained in 30 minutes. How long does it take the small pipe alone to drain the tower? (Round your answer to the nearest tenth of a minute.)

SOLUTION If we let t represent the amount of time it takes for the larger pipe to drain the tower, then the time required for the small pipe to drain the tower can be represented by $t + 15$. Recall that the work-rate is the reciprocal of the time required to complete the entire job. So the work-rate for the smaller pipe is $\frac{1}{t+15}$ and the work-rate for the large pipe is $\frac{1}{t}$. To determine the portion of the job completed by each pipe when both are working, we multiply the work-rate for each pipe by the amount of time it takes for the two pipes together to drain the tower. Here is a table showing the important information.

Pipe	Time to Complete the Job Alone	Work-Rate	Time Working	Portion of the Job Completed
Smaller	$t + 15$ minutes	$\frac{1}{t+15}$	30	$\frac{30}{t+15}$
Larger	t minutes	$\frac{1}{t}$	30	$\frac{30}{t}$

After adding the portion of the tower drained by the smaller pipe in 30 minutes to the portion of the tower drained by the larger pipe, the sum will equal 1, which represents finishing the entire job. The equation is $\frac{30}{t+15} + \frac{30}{t} = 1$.

$$\frac{30}{t+15} + \frac{30}{t} = 1$$ The LCD is $t(t + 15)$.

$$t(t+15) \cdot \left(\frac{30}{t+15} + \frac{30}{t}\right) = t(t+15) \cdot 1$$ Multiply both sides by the LCD.

$$\overset{1}{t\cancel{(t+15)}} \cdot \frac{30}{\underset{1}{\cancel{(t+15)}}} + \overset{1}{\cancel{t}}(t+15) \cdot \frac{30}{\underset{1}{\cancel{t}}} = t(t+15) \cdot 1$$ Distribute and divide out common factors.

$$30t + 30(t + 15) = t(t + 15)$$ Multiply remaining factors.

$$30t + 30t + 450 = t^2 + 15t$$ Multiply. The resulting equation is quadratic.

$$60t + 450 = t^2 + 15t$$ Combine like terms.

$$0 = t^2 - 45t - 450$$ Collect all terms on the right side of the equation.

The quadratic expression does not factor, so we will use the quadratic formula.

$$t = \frac{45 \pm \sqrt{(-45)^2 - 4(1)(-450)}}{2(1)}$$ Substitute 1 for a, -45 for b, and -450 for c.

$$t = \frac{45 \pm \sqrt{3825}}{2}$$ Simplify the radicand.

At this point, we must use a calculator to approximate the solutions for t.

$$\frac{45 + \sqrt{3825}}{2} \approx 53.4 \qquad \frac{45 - \sqrt{3825}}{2} \approx -8.4$$

We omit the negative solution, so $t \approx 53.4$. The amount of time required by the small pipe is represented by $t + 15$; so it takes the small pipe approximately $53.4 + 15$, or 68.4, minutes to drain the tank.

Quick Check 6

Working alone, Gabe can clean the gymnasium floor in 50 minutes less time than it takes Rob. If both janitors work together, it takes them 45 minutes to clean the gymnasium floor. How long does it take Gabe, working alone, to clean the gymnasium floor? (Round your answer to the nearest tenth of a minute.)

BUILDING YOUR STUDY STRATEGY

Time Management, 3 When to Study When should you study math? Try to schedule your math study sessions for the time of day you are at your sharpest. For example, if you feel most alert in the mornings, reserve as much time as possible in the mornings to study math. If you get tired while studying late at night, change your study schedule so that you can study math when you are more alert and focused.

Exercises 10.3

MyMathLab

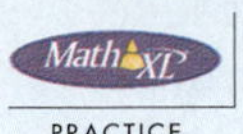
PRACTICE

WATCH

DOWNLOAD
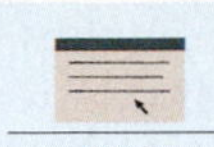
READ

REVIEW

Vocabulary

1. Replacing a variable expression with the variable u to rewrite an equation as a quadratic equation is called solving by __________.
2. Whenever both sides of an equation are squared, it is necessary to check for __________ solutions.

Solve by making a u-substitution.

3. $x^4 - 5x^2 + 4 = 0$
4. $x^4 + 5x^2 - 36 = 0$
5. $x^4 - 6x^2 + 9 = 0$
6. $x^4 + 12x^2 + 32 = 0$
7. $x^4 + 7x^2 - 18 = 0$
8. $x^4 + x^2 - 2 = 0$
9. $x^4 - 13x^2 + 36 = 0$
10. $x^4 + 8x^2 + 16 = 0$
11. $x - 13\sqrt{x} + 36 = 0$
12. $x - 10\sqrt{x} + 21 = 0$
13. $x + 2\sqrt{x} - 48 = 0$
14. $x - 5\sqrt{x} - 24 = 0$
15. $x + 9\sqrt{x} + 20 = 0$
16. $x + 11\sqrt{x} + 18 = 0$
17. $x - 8x^{1/2} + 7 = 0$
18. $x + 4x^{1/2} - 21 = 0$
19. $x + 9x^{1/2} + 18 = 0$
20. $x - 11x^{1/2} + 30 = 0$
21. $x - 2x^{1/2} - 80 = 0$
22. $x + 10x^{1/2} + 25 = 0$
23. $x^{2/3} - 5x^{1/3} + 6 = 0$
24. $x^{2/3} + 3x^{1/3} - 4 = 0$
25. $x^{2/3} + 5x^{1/3} - 6 = 0$
26. $x^{2/3} - 12x^{1/3} + 20 = 0$
27. $x^{2/3} + 9x^{1/3} + 20 = 0$
28. $x^{2/3} - 4x^{1/3} - 45 = 0$
29. $(x - 3)^2 + 5(x - 3) + 4 = 0$
30. $(x + 7)^2 + 2(x + 7) - 24 = 0$
31. $(2x - 9)^2 - 6(2x - 9) - 27 = 0$
32. $(4x + 2)^2 - 4(4x + 2) - 60 = 0$

Solve.

33. $\sqrt{x^2 + 6x} = 4$
34. $\sqrt{x^2 - 8x} = 3$
35. $\sqrt{3x - 6} = x - 2$
36. $\sqrt{4x + 52} = x + 5$
37. $\sqrt{x + 15} - x = 3$
38. $\sqrt{3x^2 + 8x + 5} - 5 = 2x$
39. $\sqrt{x - 1} + 2 = \sqrt{2x + 5}$
40. $\sqrt{2x + 3} - \sqrt{x - 2} = 2$

Solve.

41. $x = \dfrac{40}{x + 6}$
42. $x - 9 = \dfrac{52}{x}$
43. $1 + \dfrac{7}{x} - \dfrac{60}{x^2} = 0$
44. $1 - \dfrac{19}{x} + \dfrac{90}{x^2} = 0$
45. $\dfrac{1}{x} + \dfrac{7}{x + 2} = \dfrac{10}{x(x + 2)}$
46. $\dfrac{4}{x + 3} + \dfrac{1}{x - 5} = \dfrac{3}{x^2 - 2x - 15}$
47. $\dfrac{2}{x + 3} + \dfrac{x + 7}{x + 1} = \dfrac{9}{4}$
48. $\dfrac{x}{x + 7} - \dfrac{4}{x + 2} = \dfrac{7}{x^2 + 9x + 14}$
49. One small pipe takes twice as long to fill a tank as a larger pipe does. If the two pipes together take 40 minutes to fill the tank, how long does it take each pipe individually to fill the tank?
50. Rob takes 5 hours longer than Genevieve to paint a room. If they work together, they can paint a room in 6 hours. How long does it take Rob to paint a room by himself?

51. A small hose takes 1 hour longer than a large hose to fill a tank. If the two hoses are used at the same time, they can fill the tank in 3 hours. How long does it take to fill the tank using the smaller hose alone? Round to the nearest tenth of an hour.

52. A water tank has two drainpipes attached to it. The larger pipe can drain the tank in 2 hours less than the smaller pipe can. If both pipes are being used, they can drain the tank in 5 hours. How long does it take the smaller pipe alone to drain the tank? Round to the nearest tenth of an hour.

53. A new printer can print a set of newsletters in 15 minutes less than an older printer can. If both printers work simultaneously, they can print the set of newsletters in 40 minutes. How long does it take the older printer alone to print the set of newsletters? Round to the nearest tenth of a minute.

54. It takes Alison 10 minutes more than it takes Elaine to stain a cedar fence. If they work together, the two of them can stain a cedar fence in 25 minutes. How long does it take Alison to stain a cedar fence? Round to the nearest tenth of a minute.

Mixed Practice, 55–78

Solve using the technique of your choice.

55. $x^2 - 8x - 13 = 0$

56. $(7x - 11)^2 - 6(7x - 11) = 0$

57. $x - 2\sqrt{x} - 48 = 0$

58. $x^2 + 15x + 54 = 0$

59. $(3x - 2)^2 = 32$

60. $x^{2/3} + 3x^{1/3} - 4 = 0$

61. $x^2 - 5x + 14 = 0$

62. $x^4 - 5x^2 - 36 = 0$

63. $x^2 - 6x - 91 = 0$

64. $(2x + 15)^2 = 18$

65. $\dfrac{x+1}{x+3} + \dfrac{7}{x+4} = \dfrac{5}{x^2 + 7x + 12}$

66. $3x^2 + x - 8 = 0$

67. $\sqrt{3x - 2} = x - 2$

68. $x + \sqrt{x} - 12 = 0$

69. $(5x + 3)^2 + 2(5x + 3) - 15 = 0$

70. $x = \sqrt{6x - 27} + 3$

71. $(x + 8)^2 = -144$

72. $\dfrac{x+3}{x-4} + \dfrac{9}{x+2} = \dfrac{2}{x^2 - 2x - 8}$

73. $x^4 - 3x^2 - 4 = 0$

74. $x^2 + 13x + 50 = 0$

75. $x^{2/3} + 12x^{1/3} + 35 = 0$

76. $x^2 - 15x + 56 = 0$

77. $x^2 + 10x + 25 = 0$

78. $(x + 4)^2 = -108$

Writing in Mathematics

Answer in complete sentences.

79. Suppose you are solving an equation that is quadratic in form by using the substitution $u = x^2$. Once you have solved the equation for u, explain how you would solve for x.

80. Suppose you are solving an equation that is quadratic in form by using the substitution $u = \sqrt{x}$. Once you have solved the equation for u, explain how you would solve for x.

Quick Review Exercises

Section 10.3

Solve.

1. $(x - 3)^2 = 49$

2. $x^2 - 9x - 36 = 0$

3. $x^2 - 6x - 13 = 0$

4. $x^2 + 11x + 40 = 0$

10.4 Graphing Quadratic Equations

OBJECTIVES

1. Graph quadratic equations in standard form.
2. Graph parabolas that open downward.
3. Graph quadratic equations of the form $y = a(x - h)^2 + k$.

Graphing Quadratic Equations in Standard Form

Objective 1 Graph quadratic equations in standard form. The graphs of quadratic equations are not lines like the graphs of linear equations or V-shaped like the graphs of absolute value equations. The graphs of quadratic equations are U-shaped and are called **parabolas**. Let's consider the graph of the most basic quadratic equation: $y = x^2$. We will first create a table of ordered pairs to represent points on the graph.

x	$y = x^2$
−2	4
−1	1
0	0
1	1
2	4

The figure at the right shows these points and the graph of $y = x^2$. Notice that the shape, called a parabola, is not a straight line, but is U-shaped.

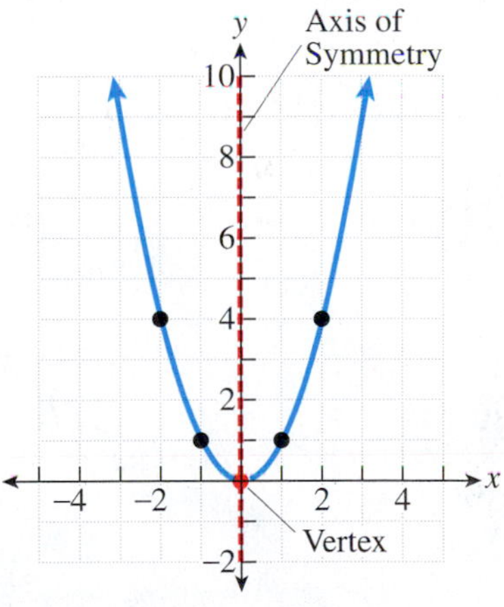

The point where the graph changes from decreasing to increasing is called the **vertex.** In the figure, the vertex is located at the bottom of the parabola. Notice that if we draw a vertical line through the vertex of the parabola, the left and right sides become mirror images. The graph of an equation is said to be **symmetric** if we can fold the graph along a line and the two sides of the graph coincide; in other words, a graph is symmetric if one side of the graph is a mirror image of the other side. A parabola is always symmetric, and we call the vertical line through the vertex the **axis of symmetry**. For this parabola, the equation of the axis of symmetry is $x = 0$.

We graph parabolas by plotting points, and the choice of our points is very important. We look for the y-intercept, the x-intercept(s) if there are any, and the vertex. We also use the axis of symmetry to help us find "mirror" points that are symmetric to points we have already graphed.

As before, we find the y-intercept by substituting 0 for x and solving for y. We will see that the y-intercept of a quadratic equation in standard form ($y = ax^2 + bx + c$) is always the point $(0, c)$. We find the x-intercepts, if there are any, by substituting 0 for y and solving for x. This equation will be quadratic, and we solve it using previous techniques.

A parabola opens upward if $a > 0$, and the vertex will be at the lowest point of the parabola. A parabola opens downward if $a < 0$, and the vertex will be at the highest point of the parabola. Parabolas that open downward will be covered later in this section. To learn how to find the coordinates of the vertex, we begin by completing the square for the equation $y = ax^2 + bx + c$.

$$y = ax^2 + bx + c$$

$$y - c = a\left(x^2 + \frac{b}{a}x\right)$$ Subtract c from both sides. Factor a from the two terms containing x.

$$y - c + \frac{b^2}{4a} = a\left(x^2 + \frac{b}{a}x + \frac{b^2}{4a^2}\right)$$ Half of $\frac{b}{a}$ is $\frac{b}{2a}$. Add $\left(\frac{b}{2a}\right)^2$, or $\frac{b^2}{4a^2}$, to the terms inside the parentheses. Because there is a factor in front of the parentheses, add $a \cdot \frac{b^2}{4a^2}$, or $\frac{b^2}{4a}$, to the left side.

$$y = a\left(x + \frac{b}{2a}\right)^2 + \frac{4ac - b^2}{4a}$$ Factor the trinomial inside the parentheses. Solve for y and collect all terms on the right side of the equation.

Note: $c - \frac{b^2}{4a} = \frac{4ac - b^2}{4a}$

Because a squared expression cannot be negative, the minimum value of y occurs when $x + \frac{b}{2a} = 0$, or in other words, when $x = \frac{-b}{2a}$.

If the equation is in standard form, $y = ax^2 + bx + c$, we can find the x-coordinate of the vertex using the formula $x = \frac{-b}{2a}$. We then find the y-coordinate of the vertex by substituting this value for x in the original equation.

Strategy for Graphing Quadratic Equations in Standard Form ($y = ax^2 + bx + c$)

- ***Find the vertex.***
  Use the formula $x = \frac{-b}{2a}$ to find the x-coordinate of the vertex. Substitute this value for x in the original equation to find the y-coordinate.
- ***Find the y-intercept.***
  Substitute 0 for x to find the y-coordinate of the y-intercept. The y-intercept is $(0, c)$.
- ***Find the x-intercept(s) if there are any.***
  Substitute 0 for y in the original equation and solve for x if possible. If you cannot solve the equation by factoring, use the quadratic formula.
- ***Use the axis of symmetry to add additional points to the graph.***
  The axis of symmetry is a vertical line extending upward from the vertex. It can be used to find the point on the parabola that is symmetric to the y-intercept.

EXAMPLE 1 Graph $y = x^2 + 6x + 8$.

SOLUTION We begin by finding the vertex using the formula $x = \frac{-b}{2a}$ to find its x-coordinate.

$$x = \frac{-6}{2(1)} = -3$$ Substitute 1 for a and 6 for b in $x = \frac{-b}{2a}$.

Now we substitute this value for x in the original equation and solve for y.

$$y = (-3)^2 + 6(-3) + 8$$ Substitute -3 for x.

$$y = -1$$ Simplify.

The vertex is at $(-3, -1)$.

Because the equation is in standard form, the y-intercept is the point $(0, c)$. In this case, the y-intercept is $(0, 8)$. (Alternatively, we could substitute 0 for x and solve for y.)

To find the x-intercepts, we substitute 0 for y and attempt to solve for x.

$0 = x^2 + 6x + 8$ Substitute 0 for y in the original equation.

$0 = (x + 2)(x + 4)$ Factor the trinomial.

$x = -2$ or $x = -4$ Set each factor equal to 0 and solve.

The x-intercepts are $(-2, 0)$ and $(-4, 0)$.

On the left is a sketch showing the vertex, y-intercept, x-intercepts, and axis of symmetry. The y-intercept is three units to the right of the axis of symmetry; so there is a mirror point with the same y-coordinate located three units to the left of this axis. The completed graph of the parabola is shown below.

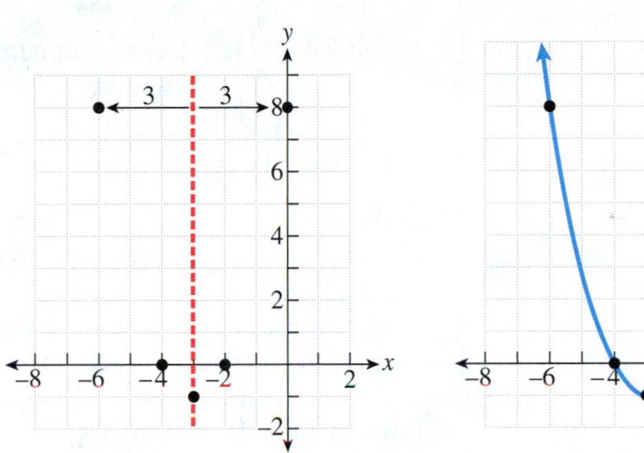

Quick Check 1

Graph $y = x^2 - 8x + 7$.

Using Your Calculator You can use the TI–84 to graph parabolas. To graph $y = x^2 + 6x + 8$, begin by tapping the [Y=] key and keying $x^2 + 6x + 8$ next to Y_1 as shown in the screen shot on the left.

To graph the parabola, tap the [GRAPH] key. On the right is the screen you should see in the standard viewing window.

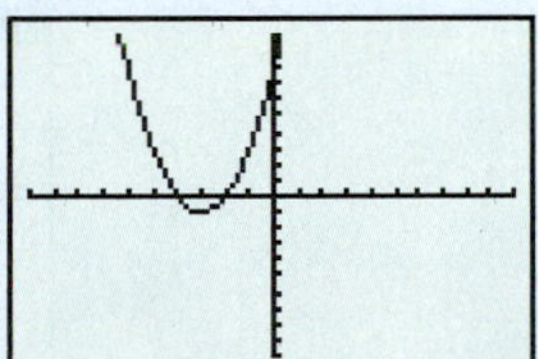

EXAMPLE 2 Graph $y = x^2 - 4x - 7$.

SOLUTION Again, we begin by finding the vertex using the formula $x = \frac{-b}{2a}$.

$$x = \frac{-(-4)}{2(1)} = 2 \quad \text{Substitute 1 for } a \text{ and } -4 \text{ for } b.$$

We substitute 2 for x in the original equation to find the y-coordinate of the vertex.

$y = (2)^2 - 4(2) - 7$ Substitute 2 for x.

$y = -11$ Simplify.

The vertex is at $(2, -11)$.

Because this equation is in standard form, we see that the y-intercept is $(0, -7)$. Now we find the x-intercepts by substituting 0 for y and solving the equation for x.

$$0 = x^2 - 4x - 7 \quad \text{Substitute 0 for } y.$$

Because we cannot factor this expression, we use the quadratic formula to find the x-intercepts.

$$x = \frac{-(-4) \pm \sqrt{(-4)^2 - 4(1)(-7)}}{2(1)} \quad \text{Substitute 1 for } a, -4 \text{ for } b, \text{ and } -7 \text{ for } c.$$

$$x = \frac{4 \pm \sqrt{44}}{2} \quad \text{Simplify the radicand and denominator.}$$

$$x = \frac{4 \pm 2\sqrt{11}}{2} \quad \text{Simplify the square root.}$$

$$x = \frac{\overset{1}{\cancel{2}}(2 \pm \sqrt{11})}{\underset{1}{\cancel{2}}} \quad \text{Divide out common factors.}$$

$$x = 2 \pm \sqrt{11} \quad \text{Simplify.}$$

Because $2 + \sqrt{11} \approx 5.3$ and $2 - \sqrt{11} \approx -1.3$, the x-intercepts are approximately $(5.3, 0)$ and $(-1.3, 0)$.

At the right is a sketch of the parabola showing the vertex, y-intercept, x-intercepts, and axis of symmetry ($x = 2$). Also shown is the point $(4, -7)$, which is symmetric to the y-intercept.

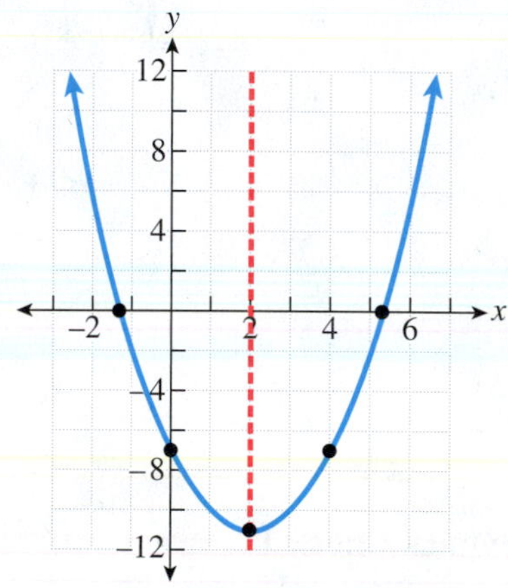

Quick Check 2

Graph $y = x^2 + 6x - 9$.

Occasionally, a parabola will not have any x-intercepts. In this case, we will have a negative discriminant ($b^2 - 4ac$) when using the quadratic formula. If the vertex is above the x-axis, the parabola does not have x-intercepts. If the vertex is on the x-axis, the vertex is the only x-intercept of the parabola.

EXAMPLE 3 Graph $y = x^2 - 2x + 2$.

SOLUTION The x-coordinate of the vertex is found using the formula $x = \frac{-b}{2a}$.

$$x = \frac{-(-2)}{2(1)} = 1 \quad \text{Substitute 1 for } a \text{ and } -2 \text{ for } b.$$

Now we substitute 1 for x in the original equation and solve for y.

$$y = (1)^2 - 2(1) + 2 \quad \text{Substitute 1 for } x.$$
$$y = 1 \quad \text{Simplify.}$$

The vertex is at $(1, 1)$.

Because the equation is in standard form, we see that the y-intercept is $(0, 2)$.

If we plot these two points on the graph, we see that there cannot be any x-intercepts. The vertex is above the x-axis, and the parabola moves only in an upward direction from there. Therefore, this parabola does not have any x-intercepts.

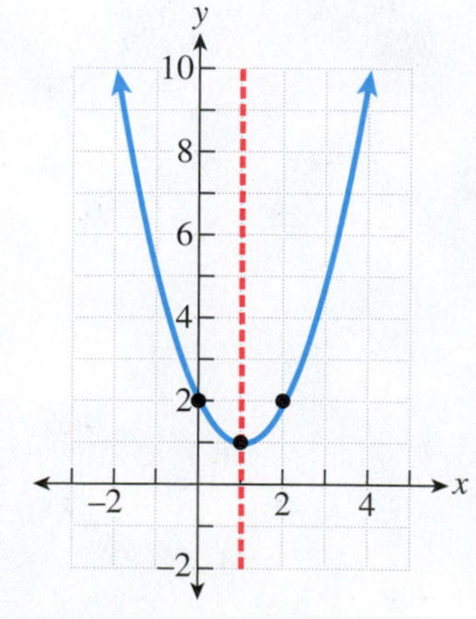

If we overlooked the visual evidence and chose to use the quadratic formula to find the x-intercepts, we would have ended up with $x = \frac{2 \pm \sqrt{-4}}{2}$. Because the discriminant is negative, the equation does not have any real solutions and the parabola does not have any x-intercepts.

The graph shows the axis of symmetry ($x = 1$) and a third point, $(2, 2)$, that is symmetric to the y-intercept.

Quick Check 3

Graph $y = x^2 + 6x + 12$.

A WORD OF CAUTION If a parabola that opens upward has its vertex above the x-axis, there are no x-intercepts.

Graphing Parabolas That Open Downward

Objective 2 Graph parabolas that open downward. Some parabolas open downward rather than upward. The way to determine which way a parabola will open is by writing the equation in standard form: $y = ax^2 + bx + c$. If a is positive, as it was in the previous examples, the parabola will open upward. If a is negative, the parabola will open downward. For instance, the graph of $y = -3x^2 + 5x - 7$ would open downward because the coefficient of the second-degree term is negative. The following example shows how to graph a parabola that opens downward.

EXAMPLE 4 Graph $y = -x^2 + 4x + 12$.

SOLUTION Begin by finding the x-coordinate of the vertex.

$$x = \frac{-4}{2(-1)} = 2 \quad \text{Substitute } -1 \text{ for } a \text{ and } 4 \text{ for } b.$$

Now substitute 2 for x in the original equation to find the y-coordinate.

$$y = -(2)^2 + 4(2) + 12 \quad \text{Substitute 2 for } x.$$
$$y = 16 \quad \text{Simplify.}$$

The vertex is at $(2, 16)$.

Because this equation is already in standard form, the y-intercept is $(0, 12)$.

If we plot the vertex and the y-intercept on the graph, we will see that there must be two x-intercepts. The vertex is above the x-axis, and the parabola opens downward; so the graph must cross the x-axis. To find the x-intercepts, we substitute 0 for y in the original equation and solve for x.

$$0 = -x^2 + 4x + 12 \quad \text{Substitute 0 for } y.$$
$$x^2 - 4x - 12 = 0 \quad \text{Collect all terms on the left side of the equation so that the coefficient of the squared term is positive.}$$
$$(x - 6)(x + 2) = 0 \quad \text{Factor.}$$
$$x = 6 \quad \text{or} \quad x = -2 \quad \text{Set each factor equal to 0 and solve.}$$

Quick Check 4

Graph $y = -x^2 + 9x + 10$.

The x-intercepts are $(6, 0)$ and $(-2, 0)$.

At the right is the graph, showing the axis of symmetry ($x = 2$) and the point $(4, 12)$ that is symmetric to the y-intercept.

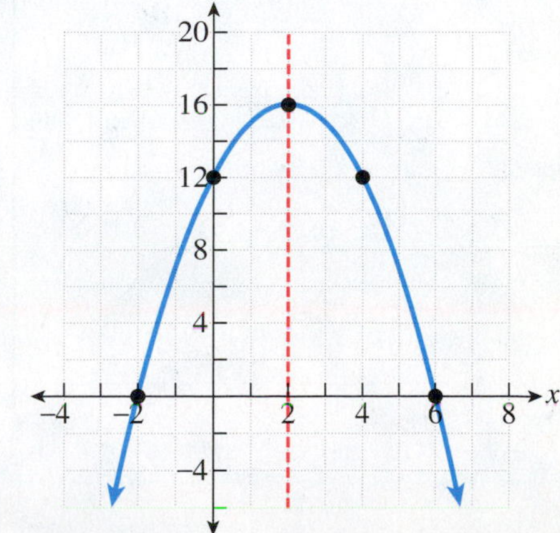

EXAMPLE 5 Graph $y = -\frac{1}{2}x^2 - 2x + 4$.

SOLUTION This parabola will open downward because the second-degree term has a negative coefficient. Begin by finding the x-coordinate of the vertex.

$$x = \frac{-(-2)}{2\left(-\frac{1}{2}\right)} = -2 \quad \text{Substitute } -\tfrac{1}{2} \text{ for } a \text{ and } -2 \text{ for } b.$$

Now substitute -2 for x in the original equation to find the y-coordinate.

$$y = -\frac{1}{2}(-2)^2 - 2(-2) + 4 \quad \text{Substitute } -2 \text{ for } x.$$
$$y = -2 + 4 + 4 \quad \text{Simplify each term.}$$
$$y = 6 \quad \text{Simplify.}$$

The vertex is at $(-2, 6)$.

The equation is in general form, so the y-intercept is $(0, 4)$.

Because the vertex is above the x-axis and the parabola opens downward, the graph must have two x-intercepts. To find the x-intercepts, substitute 0 for y in the original equation and solve for x.

$$0 = -\frac{1}{2}x^2 - 2x + 4 \quad \text{Substitute 0 for } y.$$
$$\frac{1}{2}x^2 + 2x - 4 = 0 \quad \text{Collect all terms on the left side of the equation so that the coefficient of the squared term is positive.}$$
$$2\left(\frac{1}{2}x^2 + 2x - 4\right) = 2 \cdot 0 \quad \text{Multiply both sides of the equation by the LCD (2) to clear the equation of fractions.}$$
$$x^2 + 4x - 8 = 0 \quad \text{Distribute and simplify.}$$

Because we cannot factor this expression, we must use the quadratic formula to find the x-intercepts.

$$x = \frac{-4 \pm \sqrt{(4)^2 - 4(1)(-8)}}{2(1)} \quad \text{Substitute 1 for } a, \text{ 4 for } b, \text{ and } -8 \text{ for } c.$$
$$x = \frac{-4 \pm \sqrt{48}}{2} \quad \text{Simplify the radicand and denominator.}$$
$$x = \frac{-4 \pm 4\sqrt{3}}{2} \quad \text{Simplify the square root.}$$
$$x = -2 \pm 2\sqrt{3} \quad \text{Simplify.}$$

Because $-2 + 2\sqrt{3} \approx 1.5$ and $-2 - 2\sqrt{3} \approx -5.5$, the x-intercepts are approximately $(1.5, 0)$ and $(-5.5, 0)$.

To the right is the graph, showing the axis of symmetry ($x = -2$) and the point $(-4, 4)$ that is symmetric to the y-intercept.

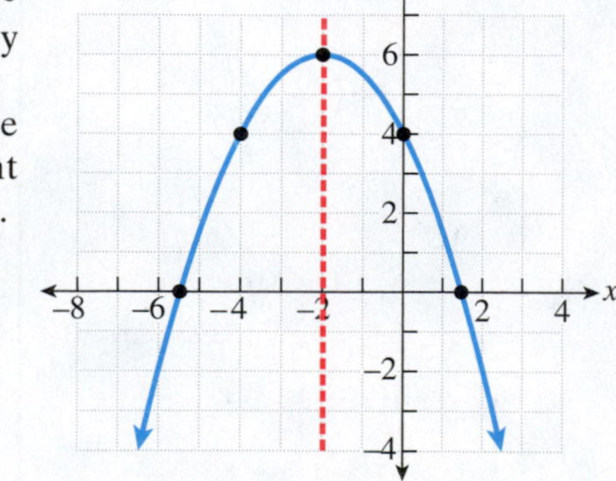

Quick Check 5

Graph $y = -\frac{1}{4}x^2 - \frac{5}{2}x - 4$.

A WORD OF CAUTION If a parabola that opens downward has its vertex below the x-axis, it has no x-intercepts.

Graphing Quadratic Equations of the Form $y = a(x - h)^2 + k$

Objective 3 Graph quadratic equations of the form $y = a(x - h)^2 + k$. Now that we have covered graphing quadratic equations in standard form, we turn our attention to graphing equations of the form $y = a(x - h)^2 + k$. One of the differences between graphing equations of this form is in the way we find the vertex.

Graphing Equations of the Form $y = a(x - h)^2 + k$

The graph of the quadratic equation $y = a(x - h)^2 + k$ is a parabola with vertex (h, k) and axis of symmetry $x = h$. The parabola opens upward if a is positive and opens downward if a is negative.

If $a > 0$, the expression $a(x - h)^2 + k$ achieves its minimum value k when $x = h$; so the vertex of $y = a(x - h)^2 + k$ is the point (h, k). Recall that $(x - h)^2$ is nonnegative and has its minimum value when $x - h = 0$, or $x = h$. In this case, $a(x - h)^2$ is equal to 0 and $a(x - h)^2 + k = k$. A similar argument holds true for (h, k) being the vertex of $y = a(x - h)^2 + k$ in the case of $a < 0$.

EXAMPLE 6 Find the vertex and axis of symmetry for the parabola.

a) $y = (x - 4)^2 + 3$

SOLUTION This equation is in the form $y = a(x - h)^2 + k$. The axis of symmetry is $x = 4$ and the vertex is $(4, 3)$.

b) $y = -(x + 1)^2 - 4$

SOLUTION This parabola opens downward, but that does not affect how we find the axis of symmetry or the vertex. The axis of symmetry is $x = -1$, and the vertex is $(-1, -4)$.

Quick Check 6

Find the vertex and axis of symmetry for the parabola.

a) $y = (x + 2)^2 - 8$
b) $y = -2(x - 8)^2 + 7$

EXAMPLE 7 Graph $y = (x + 4)^2 - 9$.

SOLUTION This parabola opens upward, and the vertex is $(-4, -9)$. Next, we find the y-intercept.

$y = (0 + 4)^2 - 9$ Substitute 0 for x.
$y = 7$ Simplify.

The y-intercept is $(0, 7)$. Because the parabola opens upward and the vertex is below the x-axis, the parabola has two x-intercepts. We find the coordinates of the x-intercepts by substituting 0 for y and solving for x by extracting square roots.

$0 = (x + 4)^2 - 9$ Substitute 0 for y.
$9 = (x + 4)^2$ Add 9 to isolate $(x + 4)^2$.
$\pm\sqrt{9} = \sqrt{(x + 4)^2}$ Take the square root of each side.
$\pm 3 = x + 4$ Simplify each square root.
$-4 \pm 3 = x$ Subtract 4 to isolate x.

The x-coordinates of the x-intercepts are $-4 + 3 = -1$ and $-4 - 3 = -7$. The x-intercepts are $(-1, 0)$ and $(-7, 0)$. The axis of symmetry is $x = -4$, and the point $(-8, 7)$ is symmetric to the y-intercept.

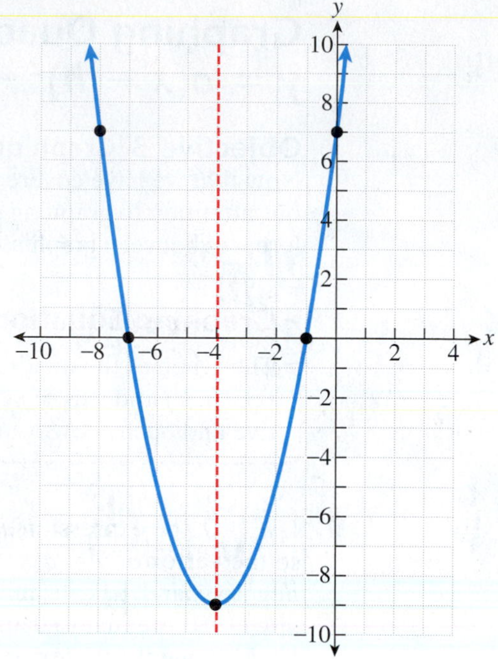

Quick Check 7

Graph $y = (x + 2)^2 + 3$.

EXAMPLE 8 Graph $y = -(x - 2)^2 - 1$.

SOLUTION This parabola opens downward, and the vertex is $(2, -1)$. Next, we find the y-intercept by substituting 0 for x.

$$y = -(0 - 2)^2 - 1 \quad \text{Substitute 0 for } x.$$
$$y = -4 - 1 \quad \text{Simplify.}$$
$$y = -5 \quad \text{Simplify.}$$

The y-intercept is $(0, -5)$.

Because the parabola opens downward and the vertex is below the x-axis, the parabola does not have any x-intercepts.

The axis of symmetry is $x = 2$, and the point $(4, -5)$ is symmetric to the y-intercept.

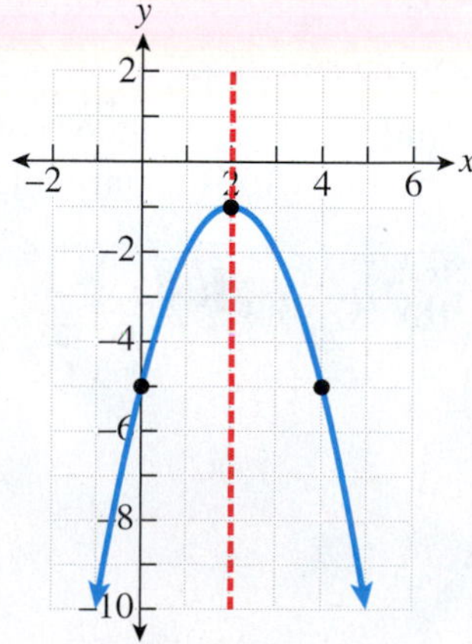

Quick Check 8

Graph $y = -(x + 1)^2 + 4$.

BUILDING YOUR STUDY STRATEGY

Time Management, 4 Setting Goals One way to get the most from a study session is to establish a set of goals to accomplish for each session. Setting goals will encourage you to work quickly and efficiently. Many students set a goal of studying for a certain amount of time, but time alone is not a worthy goal. Create a to-do list each time you start a study session, and you will find that you have a greater chance of reaching your goals.

Exercises 10.4

MyMathLab

PRACTICE

WATCH

DOWNLOAD

READ

REVIEW

Vocabulary

1. The graph of a quadratic equation is a U-shaped graph called a(n) ________.
2. A parabola opens upward if the leading coefficient is ________.
3. A parabola opens downward if the leading coefficient is ________.
4. The turning point of a parabola is called its ________.
5. To find the ____________ of a parabola, substitute 0 for x and solve for y.
6. To find the ____________ of a parabola, substitute 0 for y and solve for x.

Find the vertex of the parabola as well as the equation of the axis of symmetry associated with each of the following quadratic equations.

7. $y = x^2 + 8x - 22$
8. $y = x^2 - 6x - 45$
9. $y = x^2 - 7x + 10$
10. $y = x^2 + 9x + 40$
11. $y = -x^2 + 12x + 62$
12. $y = -x^2 - 4x + 17$
13. $y = 2x^2 + 8x - 19$
14. $y = 3x^2 + 6x + 16$
15. $y = -4x^2 + 24x - 11$
16. $y = -2x^2 - 20x + 175$
17. $y = x^2 + \frac{3}{2}x + 8$
18. $y = \frac{1}{4}x^2 + 6x - 32$
19. $y = x^2 + 10x$
20. $y = x^2 + 10$
21. $y = (x - 7)^2 + 12$
22. $y = (x + 2)^2 + 9$
23. $y = (x + 1)^2 - 5$
24. $y = (x - 6)^2 - 13$
25. $y = -(x - 9)^2 - 7$
26. $y = -(x + 4)^2 - 6$

Find the x- and y-intercepts of the parabola associated with the given quadratic equations. If necessary, round to the nearest tenth. If the parabola does not have any x-intercepts, state no x-intercepts.

27. $y = x^2 + 3x - 40$
28. $y = x^2 - 11x + 28$
29. $y = x^2 - 5x + 3$
30. $y = x^2 + 6x - 9$
31. $y = x^2 + 5x + 15$
32. $y = x^2 - 2x + 6$
33. $y = 2x^2 + 6x - 15$
34. $y = 2x^2 + 5x - 42$
35. $y = -x^2 - 5x + 6$
36. $y = -x^2 + 8x - 12$
37. $y = -x^2 + 6x + 20$
38. $y = -x^2 - 10x + 32$
39. $y = -x^2 + 7x - 15$
40. $y = -x^2 + 8x - 22$
41. $y = x^2 - 8x + 16$
42. $y = x^2 + 12x$
43. $y = (x - 7)^2 - 9$
44. $y = (x + 2)^2 - 6$
45. $y = (x + 5)^2 + 4$
46. $y = (x + 9)^2 - 4$
47. $y = -(x - 1)^2 + 12$
48. $y = -(x - 8)^2 - 1$

Graph the given parabolas. Label the vertex and all intercepts.

49. $y = x^2 - 2x - 3$
50. $y = x^2 + x - 6$

51. $y = x^2 - 5x$

52. $y = x^2 + 6x + 9$

53. $y = x^2 - 2x - 1$

54. $y = \frac{1}{2}x^2 - 2x + \frac{3}{2}$

55. $y = x^2 - 6x + 10$

56. $y = x^2 + 6x + 10$

57. $y = -x^2 + 4x - 4$

58. $y = -x^2 - 4x + 5$

59. $y = -\frac{1}{2}x^2 + 3x - 3$

60. $y = -x^2 + 4x - 1$

61. $y = -x^2 + 3x - 4$

62. $y = -x^2 - 2x - 3$

63. $y = (x - 3)^2 - 4$

64. $y = (x + 1)^2 - 9$

65. $y = -(x-2)^2 + 1$

66. $y = -(x+3)^2 + 9$

67. $y = (x-4)^2 - 2$

68. $y = -(x+1)^2 + 3$

69. $y = -(x-2)^2 - 4$

70. $y = (x-2)^2 + 3$

Writing in Mathematics

Answer in complete sentences.

71. Explain how to determine that a parabola does not have any x-intercepts.

72. ***Newsletter*** Write a newsletter explaining how to graph an equation of the form $y = ax^2 + bx + c$.

10.5 Applications Using Quadratic Equations

OBJECTIVES

1. Solve applied geometric problems.
2. Solve problems by using the Pythagorean theorem.
3. Solve applied problems by using the Pythagorean theorem.

In this section, we will learn how to solve applied problems resulting in quadratic equations.

Solving Applied Geometric Problems

Objective 1 Solve applied geometric problems. Problems involving the area of a geometric figure often lead to quadratic equations, as area is measured in square units. Here are some useful area formulas.

	Rectangle	Triangle
Figure	w, l	h, b
Dimensions	Length l, Width w	Base b, Height h
Area	$A = l \cdot w$	$A = \frac{1}{2}bh$

EXAMPLE 1 The height of a triangle is 5 centimeters less than its base. The area is 18 square centimeters. Find the base and the height of the triangle.

SOLUTION In this problem, the unknown quantities are the base and the height, while we know that the area is 18 square centimeters. Because the height is given in terms of the base, a wise choice is to represent the base of the triangle as x. Because the height is 5 centimeters less than the base, it can be represented by $x - 5$. This information is summarized in the following table:

Unknowns	***Known***	
Base: x Height: $x - 5$	Area: 18 cm^2	$x - 5$, x

Because the area of a triangle is given by the formula $A = \frac{1}{2}bh$, the equation we need to solve is $\frac{1}{2}x(x - 5) = 18$.

$$\frac{1}{2}x(x - 5) = 18$$

$$2 \cdot \frac{1}{2}x(x - 5) = 2 \cdot 18 \quad \text{Multiply both sides by 2.}$$

$$x^2 - 5x = 36 \quad \text{Simplify each side of the equation.}$$

$$x^2 - 5x - 36 = 0 \quad \text{Rewrite in standard form.}$$

$$(x - 9)(x + 4) = 0 \quad \text{Factor.}$$

$$x - 9 = 0 \quad \text{or} \quad x + 4 = 0 \quad \text{Set each factor equal to 0.}$$

$$x = 9 \quad \text{or} \quad x = -4 \quad \text{Solve.}$$

Look back at the table of unknowns. If $x = -4$, the base is -4 centimeters and the height is -9 centimeters, which is not possible. The solutions derived from $x = -4$ are omitted. If $x = 9$, the base is 9 centimeters, while the height is $9 - 5 = 4$ centimeters.

$$\text{Base: } x = 9$$
$$\text{Height: } x - 5 = 9 - 5 = 4$$

Now put the answer in a complete sentence with the proper units. The base of the triangle is 9 centimeters, and the height is 4 centimeters.

Quick Check 1

The base of a triangle is 2 inches longer than three times its height. If the area of the triangle is 60 square inches, find the base and height of the triangle.

EXAMPLE 2 The length of a rectangle is 7 inches less than twice its width. The area of the rectangle is 240 square inches. Find the length and the width of the rectangle, rounded to the nearest tenth of an inch.

SOLUTION In this problem, the unknown quantities are the length and the width, while we know that the area is 240 square inches. Because the length is given in terms of the width, a wise choice is to represent the width of the rectangle as x. Because the length is 7 inches less than twice the width, it can be represented by $2x - 7$. This information is summarized in the following table:

Unknowns	***Known***	
Length: $2x - 7$	Area: 240 in.^{2}	x
Width: x		$2x - 7$

Because the area of a rectangle is equal to its length times its width, the equation we need to solve is $x(2x - 7) = 240$.

$$x(2x - 7) = 240$$

$$2x^2 - 7x = 240 \quad \text{Distribute.}$$

$$2x^2 - 7x - 240 = 0 \quad \text{Rewrite in standard form.}$$

This trinomial does not factor, so we will use the quadratic formula with $a = 2$, $b = -7$, and $c = -240$ to solve the equation.

$$x = \frac{-(-7) \pm \sqrt{(-7)^2 - 4(2)(-240)}}{2(2)} \quad \text{Substitute 2 for } a, -7 \text{ for } b, \text{ and } -240 \text{ for } c.$$

$$x = \frac{7 \pm \sqrt{1969}}{4} \quad \text{Simplify the radicand and the denominator.}$$

We use a calculator to approximate these solutions.

$$\frac{7 + \sqrt{1969}}{4} \approx 12.8 \qquad \frac{7 - \sqrt{1969}}{4} \approx -9.3$$

We may omit the negative solution, as both the length and width of the rectangle would be negative. If $x \approx 12.8$, the width is approximately 12.8 inches, while the length is approximately $2(12.8) - 7 = 18.6$ inches.

$$\text{Width: } x \approx 12.8$$

$$\text{Length: } 2x - 7 \approx 2(12.8) - 7 = 18.6$$

The width of the rectangle is approximately 12.8 inches, and the length is approximately 18.6 inches.

Quick Check 2

The length of a rectangle is 8 inches more than its width. The area of the rectangle is 350 square inches. Find the length and the width of the rectangle, rounded to the nearest tenth of an inch.

Solving Problems by Using the Pythagorean Theorem

Objective 2 Solve problems by using the Pythagorean theorem. Other applications of geometry that lead to quadratic equations involve the Pythagorean theorem, which is an equation that relates the lengths of the three sides of a right triangle.

The side opposing the right angle is called the **hypotenuse** and is labeled c in the figure at the top of the next page. The other two sides that form the right angle are called the **legs** of the triangle and are labeled a and b. (It makes no difference which is a and which is b.)

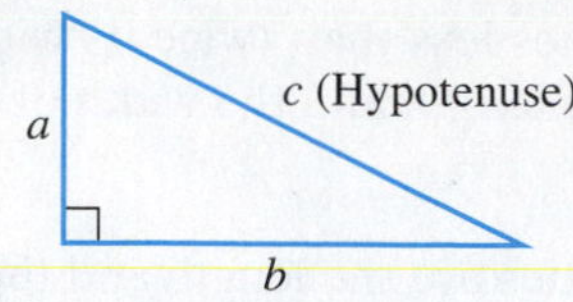

Pythagorean Theorem

For any right triangle whose hypotenuse has length c and whose legs have lengths a and b, respectively,

$$a^2 + b^2 = c^2.$$

EXAMPLE 3 A right triangle has a hypotenuse that measures 18 inches, and one of its legs is 6 inches long. Find the length of the other leg, to the nearest tenth of an inch.

SOLUTION In this problem, the length of one of the legs is unknown. We can label the unknown leg as either a or b.

Unknowns	***Known***
a	b: 6 in. Hypotenuse (c): 18 in.

$a^2 + 6^2 = 18^2$ Substitute 6 for b and 18 for c in $a^2 + b^2 = c^2$.

$a^2 + 36 = 324$ Square 6 and 18.

$a^2 = 288$ Subtract 36.

$\sqrt{a^2} = \pm\sqrt{288}$ Solve by extracting square roots.

$a = \pm 12\sqrt{2}$ Simplify the square root.

Because the length of a leg must be a positive number, we are concerned only with $12\sqrt{2}$, which rounds to 17.0 inches. The length of the other leg is approximately 17.0 inches.

Quick Check 3

One leg of a right triangle measures 5 inches, while the hypotenuse measures 11 inches. Find, to the nearest hundredth of an inch, the length of the other leg of the triangle.

Applications of the Pythagorean Theorem

Objective 3 Solve applied problems by using the Pythagorean theorem. Now we turn our attention to solving applied problems using the Pythagorean theorem. In these problems, we begin by drawing a picture of the situation. We must be able to identify a right triangle in the figure in order to apply the Pythagorean theorem.

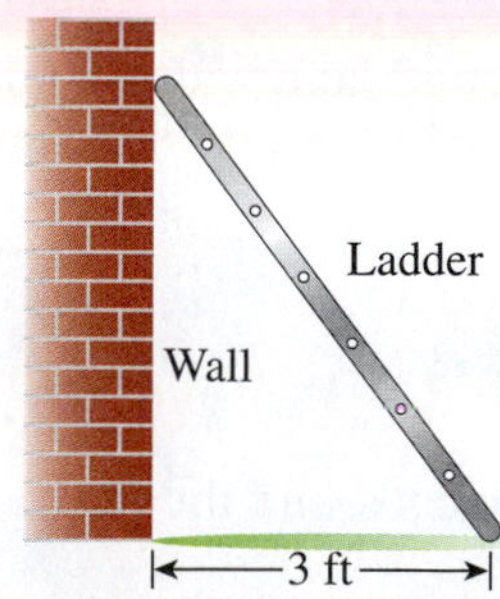

EXAMPLE 4 A 5-foot ladder is leaning against a wall. If the bottom of the ladder is 3 feet from the base of the wall, how high up the wall is the top of the ladder?

SOLUTION The ladder, the wall, and the ground form a right triangle, with the ladder being the hypotenuse. In this problem, the height of the wall, which is the length of one of the legs in the right triangle, is unknown.

Unknowns	***Known***
a	b: 3 ft Hypotenuse (c): 5 ft

$a^2 + 3^2 = 5^2$ Substitute 3 for b and 5 for c in $a^2 + b^2 = c^2$.

$a^2 + 9 = 25$ Square 3 and 5.

$a^2 = 16$ Subtract 9.

$\sqrt{a^2} = \pm\sqrt{16}$ Solve by extracting square roots.

$a = \pm 4$ Simplify the square root.

Again, the negative solution does not make sense in this problem. The ladder is resting at a point on the wall that is 4 feet above the ground.

Quick Check 4

An 8-foot ladder is leaning against a wall. If the bottom of the ladder is 2 feet from the base of the wall, how high up the wall is the top of the ladder? (Round to the nearest tenth of a foot.)

EXAMPLE 5 The Modesto airport is located 120 miles north and 50 miles west of the Visalia airport. If a plane flies directly from Visalia to Modesto, how many miles is the flight?

SOLUTION The directions of north and west form a 90-degree angle, so the picture shows a right triangle whose hypotenuse (the direct distance from Visalia to Modesto) is unknown.

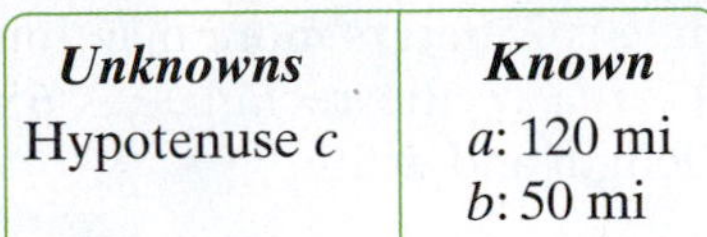

Unknowns	*Known*
Hypotenuse c	a: 120 mi b: 50 mi

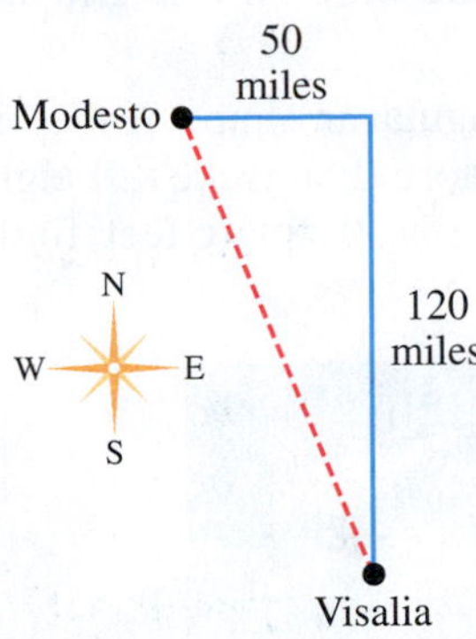

$$
\begin{aligned}
120^2 + 50^2 &= c^2 && \text{Substitute 120 for } a \text{ and 50 for } b \text{ in } a^2 + b^2 = c^2.\\
14{,}400 + 2500 &= c^2 && \text{Square 120 and 50.}\\
16{,}900 &= c^2 && \text{Simplify.}\\
\pm\sqrt{16{,}900} &= \sqrt{c^2} && \text{Solve by extracting square roots.}\\
\pm 130 &= c && \text{Simplify the square root.}
\end{aligned}
$$

The negative solution does not make sense in this problem. The direct distance from the Visalia airport to the Modesto airport is 130 miles.

Quick Check 5

Cassie's backyard is in the shape of a rectangle whose dimensions are 70 feet by 240 feet. She needs a hose that will extend from one corner of her yard to the corner that is diagonally opposite to it. How long does the hose have to be?

BUILDING YOUR STUDY STRATEGY

Time Management, 5 Studying Difficult Subjects First and Taking Brief Breaks If you have more than one subject to study, as most students do, the order in which you study the subjects is important. A good idea is to arrange your study schedule so that you study the most difficult subject first, while you are most alert.

Another suggestion to keep your mental energy at its highest while you are studying is to take brief ten-minute study breaks. One study break per hour will help keep you from feeling fatigued.

Exercises 10.5

PRACTICE

WATCH

DOWNLOAD

READ

REVIEW

Vocabulary

1. State the formula for the area of a triangle. ________
2. In a right triangle, the side opposite the right angle is called the ________.
3. In a right triangle, the sides adjacent to the right angle are called ________.
4. The ________ theorem states that for any right triangle whose hypotenuse has length c and whose two legs have lengths a and b, respectively, $a^2 + b^2 = c^2$.

For all problems, approximate to the nearest tenth when necessary.

5. The length of a rectangle is 5 inches more than its width. If the area of the rectangle is 66 square inches, find its length and width.
6. The width of a rectangle is 6 feet less than its length. If the area of the rectangle is 112 square feet, find its length and width.
7. The length of a rectangle is twice its width. If the area of the rectangle is 120 square meters, find its length and width.

8. The length of a rectangle is 3 inches more than twice its width. If the area of the rectangle is 75 square inches, find its length and width.

9. The width of a rectangle is 13 feet less than twice its length. If the area of the rectangle is 68 square feet, find its length and width.

10. The length of a rectangle is 7 meters more than four times its width. If the area of the rectangle is 650 square meters, find its length and width.

11. The base of a triangle is 5 inches more than its height. If the area of the triangle is 42 square inches, find the base and height of the triangle.

12. The height of a triangle is 1 foot less than three times its base. If the area of the triangle is 22 square feet, find the base and height of the triangle.

13. The height of a triangle is 1 inch more than twice its base. If the area of the triangle is 15 square inches, find the base and height of the triangle.

14. The height of a triangle is 7 inches less than its base. If the area of the triangle is 32 square inches, find the base and height of the triangle.

15. A rectangular photograph has an area of 80 square inches. If the width of the photograph is 2 inches less than its height, find the dimensions of the photograph.

16. The area of a rectangular patio is 700 square feet. If the length of the patio is 5 feet less than twice its width, find the dimensions of the patio.

17. The length of a rectangular rug is 10 inches less than twice its width. If the area of the rug is 2160 square inches, find its dimensions.

18. The width of a rectangular room is 14 meters less than twice its length. If the area of the room is 125 square meters, find its dimensions.

19. The width of a rectangular table is 16 inches less than its length. If the area of the table is 540 square inches, find its dimensions.

20. Steve has a rectangular lawn, and the length of the lawn is 25 feet more than its width. If the area of the lawn is 7000 square feet, find its dimensions.

21. Tina made a quilt that was 8 inches taller than it was wide. She then sewed a 4-inch border around the entire quilt. If the area of the quilt and its border is 1920 square inches, find the dimensions of the quilt without the border.

22. The length of a rectangular swimming pool is 2 meters more than twice its width. The pool is surrounded by a concrete deck that is 2 meters wide. If the area of the surface of the pool and deck is 180 square meters, find the dimensions of the pool.

23. A kite is in the shape of a triangle. The base of the kite is twice its height. If the area of the kite is 256 square inches, find the base and height of the kite.

24. A hang glider is triangular in shape. If the base of the hang glider is 2 feet more than twice its height and the area of the hang glider is 30 square feet, find the hang glider's base and height.

25. The sail on a sailboat is shaped like a triangle, with an area of 46 square feet. The height of the sail is 8 feet more than the base of the sail. Find the base and height of the sail.

26. A kite is in the shape of a triangle and is made with 250 square inches of material. The base of the kite is 20 inches longer than the height of the kite. Find the base and the height of the kite.

27. The two legs of a right triangle are 10 inches and 16 inches. Find the hypotenuse of the triangle.

28. The two legs of a right triangle are 7 centimeters and 12 centimeters. Find the hypotenuse of the triangle.

29. A right triangle with a hypotenuse of 15 feet has a leg that measures 9 feet. Find the length of the other leg.

30. A right triangle with a hypotenuse of 25 inches has a leg that measures 24 inches. Find the length of the other leg.

31. A right triangle with a leg that measures 12 meters has a hypotenuse of 24 meters. Find the length of the other leg.

32. A right triangle with a leg that measures 8 feet has a hypotenuse of 23 feet. Find the length of the other leg.

33. Patti drove 80 miles to the west and then drove 60 miles south. How far is she from her starting location?

34. Victor flew to a city that was 700 miles north and 2400 miles east of his starting point. How far did he fly to reach this city?

35. George is casting a shadow on the ground. If George is 2 feet shorter than the length of the shadow on the ground and the tip of the shadow is 10 feet from the top of George's head, how tall is George?

36. A 15-foot ladder is leaning against the wall. If the distance between the base of the ladder and the wall is 3 feet less than the height of the top of the ladder on the wall, how high up the wall does the ladder reach?

37. A guy wire 40 feet long runs from the top of a pole to a spot on the ground. If the height of the pole is 5 feet more than the distance from the base of the pole to the spot where the guy wire is anchored, how tall is the pole?

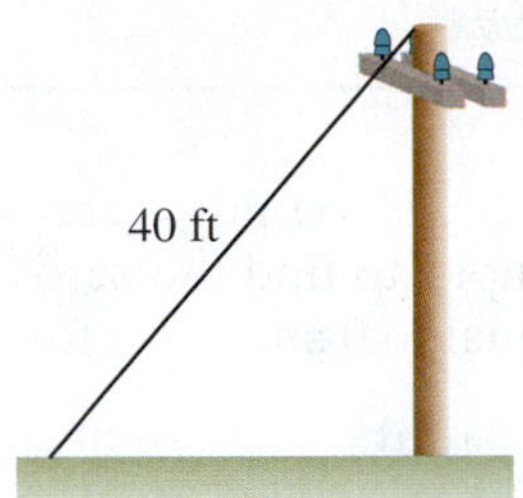

38. A 12-foot ramp leads to a doorway. If the distance between the ground and the doorway is 8 feet less than the horizontal distance covered by the ramp, how high above the ground is the doorway?

39. A rectangular computer screen is 13 inches wide and 10 inches high. Find the length of its diagonal.

40. The bases on a baseball diamond form a square whose side is 90 feet. How far is it from home plate to second base?

41. The length of a rectangular quilt is 1 foot more than its width. If the diagonal of the quilt is 6.5 feet, find the length and width of the quilt.

42. The diagonal of a rectangular table is 9 feet, and the length of the table is 5 feet more than its width. Find the length and width of the table.

Use the following fact about right triangles for Exercises 43–46: The legs of a right triangle represent the base and height of that triangle.

43. The height of a right triangle is 3 feet less than the base of the triangle. The area of the triangle is 54 square feet.

a) Use the information to find the base and height of the triangle.

b) Find the hypotenuse of the triangle.

44. The base of a right triangle is 4 feet less than twice the height of the triangle. The area of the triangle is 24 square feet.

a) Use the information to find the base and height of the triangle.

b) Use the information to find the hypotenuse of the triangle.

45. A road sign indicating falling rocks is in the shape of an equilateral triangle, with each side measuring 80 centimeters.

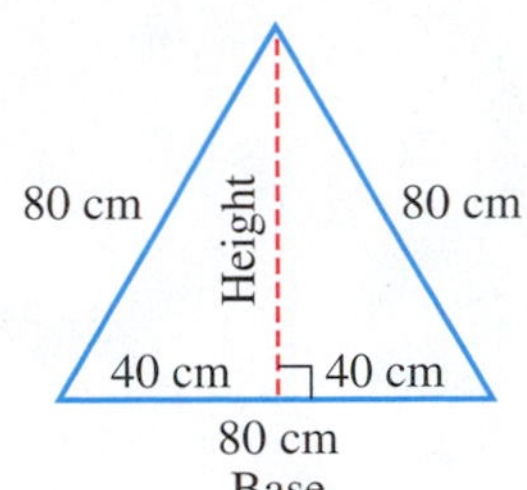

a) Use the Pythagorean theorem to find the height of the triangle. Round to the nearest tenth of a centimeter.

b) Find the area of the sign.

46. A farmer fenced in a corral in the shape of an equilateral triangle, with each side measuring 30 feet.

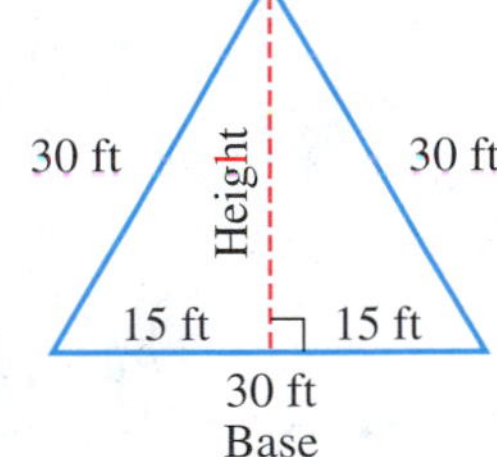

a) Use the Pythagorean theorem to find the height of the triangle. Round to the nearest tenth of a foot.

b) Find the area of the corral.

Writing in Mathematics

Answer in complete sentences.

47. Write a word problem whose solution is *The length of the rectangle is 14 feet, and the width is 9 feet.* The problem must lead to a quadratic equation.

48. Write a word problem associated with the equation $60^2 + b^2 = 90^2$. Explain how you created the problem. Solve your problem, explaining each step.

10.6 Quadratic and Rational Inequalities

OBJECTIVES

1. Solve quadratic inequalities.
2. Solve rational inequalities.
3. Solve inequalities involving functions.
4. Solve applied problems involving inequalities.

Quadratic Inequalities

Objective 1 Solve quadratic inequalities. In this section, we will expand our knowledge of inequalities to include two different types of inequalities: quadratic inequalities and rational inequalities.

Quadratic Inequalities

A **quadratic inequality** is an inequality that can be rewritten as $ax^2 + bx + c < 0$, $ax^2 + bx + c \leq 0$, $ax^2 + bx + c > 0$, or $ax^2 + bx + c \geq 0$.

The first step to solving a quadratic inequality is to find its **zeros**, which are values of x for which $ax^2 + bx + c = 0$. For example, to find the zeros for the inequality $x^2 + 9x - 22 > 0$, we set $x^2 + 9x - 22$ equal to 0 and solve for x.

$$x^2 + 9x - 22 = 0 \quad \text{Set the quadratic expression equal to 0.}$$
$$(x + 11)(x - 2) = 0 \quad \text{Factor.}$$
$$x = -11 \quad \text{or} \quad x = 2 \quad \text{Set each factor equal to 0 and solve.}$$

The zeros for this inequality are -11 and 2.

The zeros of an inequality divide the number line into intervals and often act as boundary points. In any particular interval created by the zeros of an inequality, each real number in the interval is a solution of the inequality or each real number in the interval is not a solution of the inequality. By picking one value in each interval and testing it, we can determine which intervals contain solutions and which do not.

When solving a strict inequality involving the symbol $<$ or $>$, the zeros are not included in the solutions of the inequality and we place an open circle on each zero on the number line. When solving a weak inequality involving the symbol $\leq$ or $\geq$, the zeros are included in the solutions of the inequality and we place a closed circle on each zero.

EXAMPLE 1 Solve $x^2 + x - 20 \leq 0$.

SOLUTION We begin by finding the zeros.

$$x^2 + x - 20 = 0 \quad \text{Set the quadratic expression equal to 0.}$$
$$(x + 5)(x - 4) = 0 \quad \text{Factor.}$$
$$x = -5 \quad \text{or} \quad x = 4 \quad \text{Set each factor equal to 0 and solve.}$$

The zeros for this inequality are -5 and 4. We plot these zeros on a number line and create boundaries between the three intervals, which have been labeled I, II, and III in the figure that follows. The zeros are included as solutions, and we place closed circles on the number line at $x = -5$ and $x = 4$.

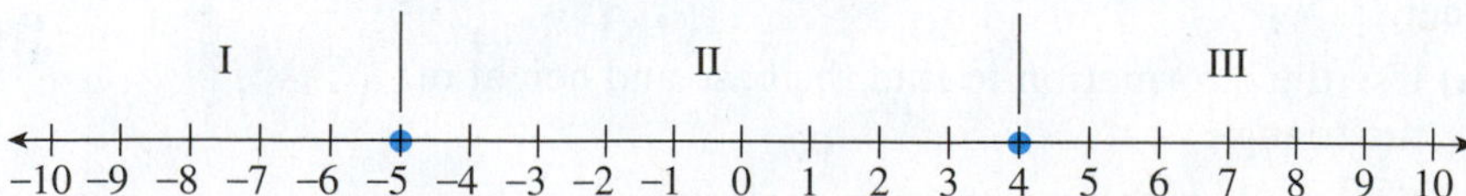

Now we choose a value from each of the three intervals we have created to be the test points. A **test point** is a value of the variable x that we use to evaluate the expression in the inequality, allowing us to determine which intervals contain solutions of the inequality. We will use $x = -6$, $x = 0$, and $x = 5$. Because we are looking for intervals where $x^2 + x - 20 \le 0$, the solution will be made up of the intervals whose test points produce a negative result when substituted into the expression $x^2 + x - 20$.

Test Point (x)	-6	0	5
$x^2 + x - 20$	$(-6)^2 + (-6) - 20$ $= 36 - 6 - 20$ $= 10$	$(0)^2 + (0) - 20$ $= 0 + 0 - 20$ $= -20$	$(5)^2 + (5) - 20$ $= 25 + 5 - 20$ $= 10$

The only test point for which the expression $x^2 + x - 20$ is negative is $x = 0$, which is in interval II; so the interval $[-5, 4]$ is the solution of this inequality.

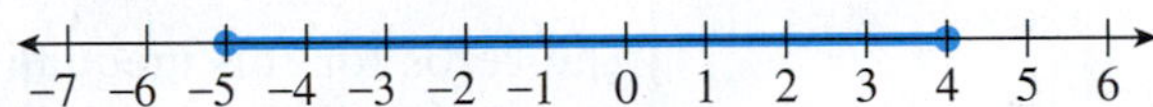

When the expression with which we are working can be factored, we can use a sign chart to determine which intervals are solutions to the inequality. A **sign chart** is a chart that can be used to determine whether an expression is positive or negative for certain intervals of real numbers. A sign chart focuses on whether a factor is positive or negative for each interval, by the use of a test point. If we know the sign of each factor, we can easily find the sign of the product, allowing us to solve the inequality. The initial sign chart should look like the following:

Test Point	$x + 5$	$x - 4$	$(x + 5)(x - 4)$
-6			
0			
5			

We fill in the first row by determining the sign of each factor when $x = -6$. Because $-6 + 5$ is negative, we put a negative sign in the first row underneath $x + 5$. In the same way, $x - 4$ is negative when $x = -6$; so we put another negative sign in the first row underneath $x - 4$. The product of two negative factors is positive; so we put a positive sign in the first row under $(x + 5)(x - 4)$. Here is the sign chart after it has been completed.

Test Point	$x + 5$	$x - 4$	$(x + 5)(x - 4)$
-6	$-$	$-$	$+$
0	$+$	$-$	$-$
5	$+$	$+$	$+$

Notice that the product is negative in the interval containing the test point $x = 0$. Therefore, the solution is the interval $[-5, 4]$.

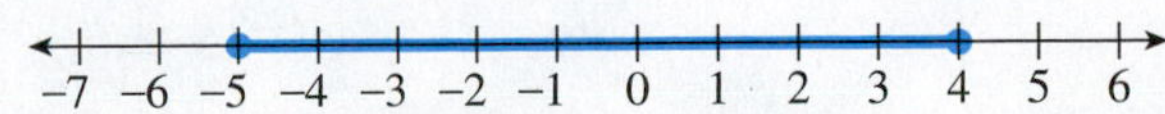

Quick Check 1

Solve $x^2 - 14x + 48 \le 0$.

If the quadratic expression does not factor, we cannot use a sign chart. In that case, we will find the zeros of the expression by using the quadratic formula. We will then substitute the values for the test points directly into the quadratic expression.

EXAMPLE 2 Solve $x^2 + 6x + 4 > 0$.

SOLUTION We find the zeros by solving the equation $x^2 + 6x + 4 = 0$. Because $x^2 + 6x + 4$ does not factor, we will use the quadratic formula to solve the equation.

$$x = \frac{-6 \pm \sqrt{6^2 - 4(1)(4)}}{2(1)}$$ Substitute 1 for a, 6 for b, and 4 for c.

$$x = \frac{-6 \pm \sqrt{20}}{2}$$ Simplify the radicand.

$$x = \frac{-6 \pm 2\sqrt{5}}{2}$$ Simplify the square root.

$$x = \frac{\overset{1}{\cancel{2}}(-3 \pm \sqrt{5})}{\underset{1}{\cancel{2}}}$$ Divide out common factors.

$$x = -3 \pm \sqrt{5}$$ Simplify.

The zeros for this inequality are $-3 - \sqrt{5}$ and $-3 + \sqrt{5}$. We need an approximate value for each zero to determine where each zero belongs on the number line. These two zeros are approximately equal to -0.8 and -5.2. The zeros are not included as solutions, so we place open circles at the zeros.

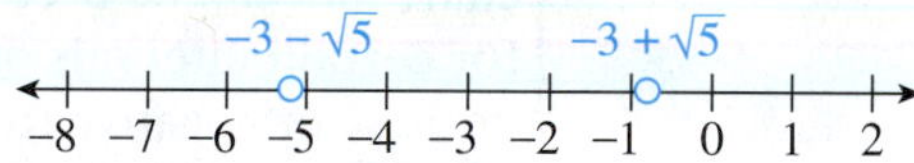

We will use $x = -6$, $x = -1$, and $x = 0$ as test points. We are looking for intervals where the expression $x^2 + 6x + 4$ is greater than 0, and our solution will be made up of the intervals whose test points produce a positive result when substituted into $x^2 + 6x + 4$.

Test Point	$x^2 + 6x + 4$	Sign
-6	$(-6)^2 + 6(-6) + 4 = 4$	+
-1	$(-1)^2 + 6(-1) + 4 = -1$	−
0	$(0)^2 + 6(0) + 4 = 4$	+

Notice that the product is positive in the intervals containing the test points $x = -6$ and $x = 0$. Here is the solution.

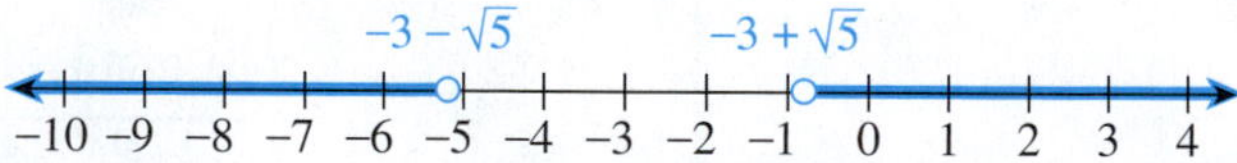

In interval notation, we can express this solution as $(-\infty, -3 - \sqrt{5}) \cup (-3 + \sqrt{5}, \infty)$. (Notice that we used the exact values in the solution, not the approximate values.)

Quick Check 2

Solve $x^2 + 3x - 15 \geq 0$.

If the expression in the inequality has no zeros, every real number is a solution of the inequality or the inequality has no solutions at all.

EXAMPLE 3 Solve $x^2 - 5x + 7 \le 0$.

SOLUTION We begin by looking for the zeros of $x^2 - 5x + 7$. Because this expression does not factor, we will use the quadratic formula.

$$x = \frac{-(-5) \pm \sqrt{(-5)^2 - 4(1)(7)}}{2(1)}$$ Substitute 1 for a, -5 for b, and 7 for c.

$$x = \frac{5 \pm \sqrt{-3}}{2}$$ Simplify the radicand and denominator.

The discriminant is negative, so this expression has no real zeros. In this case, we can pick any real number and use it as the only test point. When $x = 0$, we can see that $x^2 - 5x + 7$ is equal to 7, which is not less than or equal to 0. This inequality has no solution.

Quick Check 3

Solve $x^2 - 5x + 20 > 0$.

Rational Inequalities

Objective 2 Solve rational inequalities.

Rational Inequalities

A **rational inequality** is an inequality that involves a rational expression, such as

$$\frac{x + 2}{x - 4} < 0.$$

There are two differences between solving a rational inequality and solving a quadratic inequality. The first difference is that the zeros of the numerator and the denominator are used to divide the real number line into intervals. The second difference is that the zeros from the denominator are never included in the solution, even when the inequality is a weak inequality involving $\le$ or $\ge$. This is because if a value causes a denominator to equal 0, the rational expression is undefined for that value.

EXAMPLE 4 Solve $\frac{(x + 3)(x + 5)}{x - 1} \ge 0$.

SOLUTION Begin by finding the zeros of the numerator.

$(x + 3)(x + 5) = 0$ Set the numerator equal to 0.

$x = -3$ or $x = -5$ Set each factor equal to 0 and solve.

The zeros of the numerator are -3 and -5. Now find the zeros of the denominator.

$x - 1 = 0$ Set the denominator equal to 0.

$x = 1$ Solve.

The zero of the denominator is 1. Here are all of the zeros on a single number line, dividing the number line into four intervals.

−7 −6 −5 −4 −3 −2 −1 0 1 2 3

Keep in mind that the value $x = 1$ cannot be included in any solution, as it is a zero of the denominator. We will use the values $x = -6$, $x = -4$, $x = 0$, and $x = 2$ as test points. Because the numerator factors and the denominator is a linear expression, we can use a sign chart. We are looking for intervals where the expression is positive.

Test Point	$x + 3$	$x + 5$	$x - 1$	$\frac{(x+3)(x+5)}{x-1}$
−6	−	−	−	−
−4	−	+	−	+
0	+	+	−	−
2	+	+	+	+

The intervals containing $x = -4$ and $x = 2$ are solutions of this inequality, as represented on the following number line:

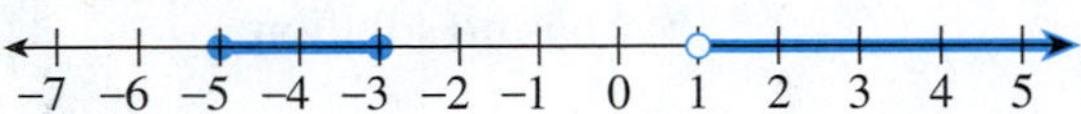

This can be represented in interval notation as $[-5, -3] \cup (1, \infty)$.

Quick Check 4

Solve $\frac{(x+2)(x-6)}{(x-8)(x+7)} \leq 0$.

A WORD OF CAUTION When we solve a rational inequality, the zeros of the denominator are excluded as solutions because the rational expression is undefined for those values.

EXAMPLE 5 Solve $\frac{x^2 - 4x - 12}{x^2 - 9} < 0$.

SOLUTION Begin by finding the zeros of the numerator.

$$x^2 - 4x - 12 = 0 \quad \text{Set the numerator equal to 0.}$$
$$(x - 6)(x + 2) = 0 \quad \text{Factor.}$$
$$x - 6 = 0 \quad \text{or} \quad x + 2 = 0 \quad \text{Set each factor equal to 0.}$$
$$x = 6 \quad \text{or} \quad x = -2 \quad \text{Solve.}$$

The zeros of the numerator are 6 and −2. Now find the zeros of the denominator.

$$x^2 - 9 = 0 \quad \text{Set the denominator equal to 0.}$$
$$(x + 3)(x - 3) = 0 \quad \text{Factor.}$$
$$x + 3 = 0 \quad \text{or} \quad x - 3 = 0 \quad \text{Set each factor equal to 0.}$$
$$x = -3 \quad \text{or} \quad x = 3 \quad \text{Solve.}$$

The zeros of the denominator are −3 and 3. Here are all of the zeros on a single number line, dividing the number line into five intervals.

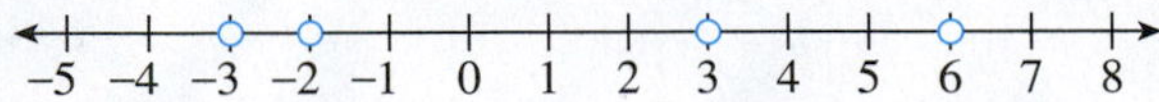

In this example, none of the zeros can be included in any solution, as the inequality was strictly less than 0 and not less than or equal to 0. We will use the values $x = -4$, $x = -2.5$, $x = 0$, $x = 4$, and $x = 7$ as test points. Here is the sign chart; keep in mind that we are looking for intervals in which this expression is negative.

Test Point	$x - 6$	$x + 2$	$x + 3$	$x - 3$	$\frac{(x-6)(x+2)}{(x+3)(x-3)}$
-4	$-$	$-$	$-$	$-$	$+$
-2.5	$-$	$-$	$+$	$-$	$-$
0	$-$	$+$	$+$	$-$	$+$
4	$-$	$+$	$+$	$+$	$-$
7	$+$	$+$	$+$	$+$	$+$

The intervals containing $x = -2.5$ and $x = 4$ are solutions of this inequality, as represented on the following number line:

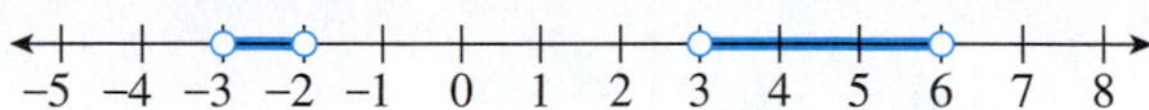

This can be represented in interval notation as $(-3, -2) \cup (3, 6)$.

Quick Check 5

Solve $\frac{x^2 - 6x}{x^2 + 5x + 4} > 0$.

When solving rational inequalities, if the inequality contains expressions on both sides, we must rewrite the inequality in such a way that there is a single rational expression on one side of the inequality and 0 on the other side. We then solve the inequality the same way we solved the previous examples.

EXAMPLE 6 Solve $\frac{x^2 + 4x + 9}{x + 6} \geq 2$.

SOLUTION Begin by subtracting 2 from both sides of the inequality; then combine the left side of the inequality as a single rational expression.

$$\frac{x^2 + 4x + 9}{x + 6} \geq 2$$

$$\frac{x^2 + 4x + 9}{x + 6} - 2 \geq 0 \quad \text{Subtract 2 from both sides.}$$

$$\frac{x^2 + 4x + 9}{x + 6} - 2 \cdot \frac{x + 6}{x + 6} \geq 0 \quad \text{Multiply 2 by } \frac{x+6}{x+6} \text{ so that both expressions have the same denominator.}$$

$$\frac{x^2 + 4x + 9 - 2x - 12}{x + 6} \geq 0 \quad \text{Distribute and combine numerators.}$$

$$\frac{x^2 + 2x - 3}{x + 6} \geq 0 \quad \text{Simplify the numerator.}$$

Now find the zeros of the numerator.

$$x^2 + 2x - 3 = 0 \quad \text{Set the numerator equal to 0.}$$

$$(x - 1)(x + 3) = 0 \quad \text{Factor.}$$

$$x - 1 = 0 \quad \text{or} \quad x + 3 = 0 \quad \text{Set each factor equal to 0.}$$

$$x = 1 \quad \text{or} \quad x = -3 \quad \text{Solve.}$$

The zeros of the numerator are 1 and -3. Now find the zeros of the denominator.

$$x + 6 = 0 \quad \text{Set the denominator equal to 0.}$$

$$x = -6 \quad \text{Solve.}$$

The only zero of the denominator is −6. Here are all of the zeros on a single number line, dividing the number line into four intervals.

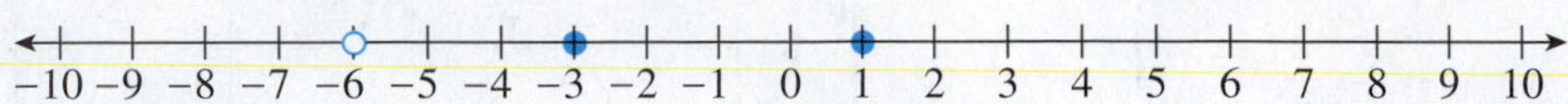

In this example, the zeros from the numerator are included as solutions, while the zeros from the denominator are excluded as solutions. We will use the values $x = -7$, $x = -4$, $x = 0$, and $x = 2$ as test points. Here is the sign chart; keep in mind that we are looking for intervals in which the expression $\frac{(x+3)(x-1)}{x+6}$ is positive.

Test Point	$x+3$	$x-1$	$x+6$	$\frac{(x+3)(x-1)}{x+6}$
−7	−	−	−	−
−4	−	−	+	+
0	+	−	+	−
2	+	+	+	+

The intervals containing $x = -4$ and $x = 2$ are solutions to this inequality, as represented on the following number line:

$-10\ -9\ -8\ -7\ -6\ -5\ -4\ -3\ -2\ -1\ 0\ 1\ 2\ 3\ 4\ 5\ 6\ 7\ 8\ 9\ 10$

This can be represented in interval notation as $(-6, -3] \cup [1, \infty)$.

Quick Check 6

Solve $\frac{x^2 - 17}{x - 3} < 4$.

Solving Inequalities Involving Functions

Objective 3 Solve inequalities involving functions. We move on to examining inequalities involving functions.

EXAMPLE 7 Given $f(x) = x^2 - 2x - 16$, find all values x for which $f(x) \le 8$.

SOLUTION We begin by setting the function less than or equal to 8.

$$f(x) \le 8$$

$$x^2 - 2x - 16 \le 8 \quad \text{Set the function less than or equal to 8.}$$

$$x^2 - 2x - 24 \le 0 \quad \text{Subtract 8 to write the inequality in standard form.}$$

We now find the zeros for this inequality.

$$x^2 - 2x - 24 = 0 \quad \text{Set the quadratic expression equal to 0.}$$

$$(x-6)(x+4) = 0 \quad \text{Factor.}$$

$$x - 6 = 0 \quad \text{or} \quad x + 4 = 0 \quad \text{Set each factor equal to 0.}$$

$$x = 6 \quad \text{or} \quad x = -4 \quad \text{Solve.}$$

The zeros for this inequality are 6 and −4.

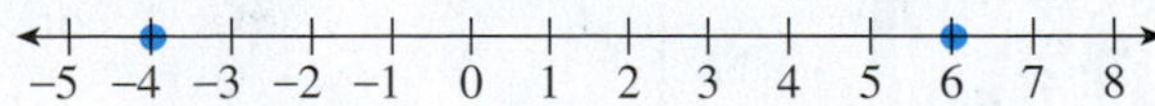

We will use $x = -5$, $x = 0$, and $x = 7$ as our test points. Our solution will be made up of the intervals whose test points produce a negative result when substituted into the expression $x^2 - 2x - 24$. Here is the sign chart after it has been completed.

Test Point	$x - 6$	$x + 4$	$(x - 6)(x + 4)$
-5	$-$	$-$	$+$
0	$-$	$+$	$-$
7	$+$	$+$	$+$

The product is negative in the interval containing the test point $x = 0$. Here is our solution represented on a number line.

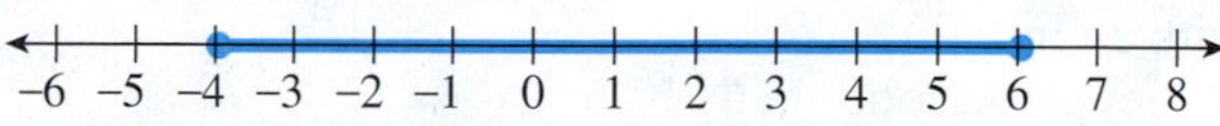

In interval notation, the solution is expressed as $[-4, 6]$.

Quick Check 7

Given $f(x) = x^2 + 10x + 35$, find all values x for which $f(x) < 14$.

Solving Applied Problems Involving Inequalities

Objective 4 Solve applied problems involving inequalities.

EXAMPLE 8 A projectile with an initial velocity of 96 feet per second is fired from the roof of a building 72 feet tall. The height of the projectile, in feet, after t seconds is given by the function $h(t) = -16t^2 + 96t + 72$. During what period of time is the projectile at least 200 feet above the ground?

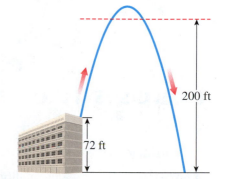

SOLUTION To find the time interval that the projectile is at least 200 feet above the ground, we must solve the inequality $h(t) \geq 200$.

$$h(t) \geq 200$$

$$-16t^2 + 96t + 72 \geq 200 \quad \text{Replace } h(t) \text{ with } -16t^2 + 96t + 72.$$

$$-16t^2 + 96t - 128 \geq 0 \quad \text{Subtract 200 to rewrite the inequality in standard form.}$$

$$16t^2 - 96t + 128 \leq 0 \quad \text{Multiply both sides of the inequality by } -1. \text{ Change the direction of the inequality.}$$

Now we find the zeros of the inequality.

$$16t^2 - 96t + 128 = 0$$

$$16(t^2 - 6t + 8) = 0 \quad \text{Factor out the common factor 16.}$$

$$16(t - 2)(t - 4) = 0 \quad \text{Factor the trinomial.}$$

$$t = 2 \quad \text{or} \quad t = 4 \quad \text{Set each variable factor equal to 0 and solve.}$$

The zeros of the inequality are $t = 2$ and $t = 4$.

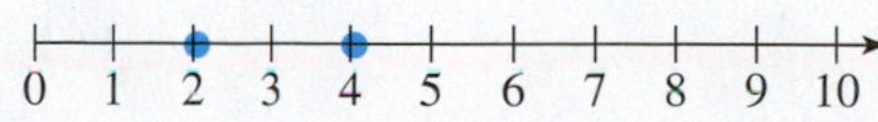

We will use $t = 1$, $t = 3$, and $t = 5$ as test points. (Note that t must be greater than or equal to 0 because it represents the amount of time since the projectile was fired.) In this case, we will substitute the values for the test points in the function

$h(t)$. If the function's output for one of the test points is 200 or higher, each point in the interval containing the test point is a solution to the inequality $h(t) \geq 200$.

$t = 1$	$t = 3$	$t = 5$
$h(1) = -16(1)^2 + 96(1) + 72$	$h(3) = -16(3)^2 + 96(3) + 72$	$h(5) = -16(5)^2 + 96(5) + 72$
$= -16(1) + 96 + 72$	$= -16(9) + 288 + 72$	$= -16(25) + 480 + 72$
$= -16 + 96 + 72$	$= -144 + 288 + 72$	$= -400 + 480 + 72$
$= 152$	$= 216$	$= 152$

The inequality $h(t) \geq 200$ is true only in the interval containing the test point $t = 3$; so the projectile's height is at least 200 feet from 2 seconds after launch until 4 seconds after launch.

Quick Check 8

A projectile with an initial velocity of 272 feet per second is fired from ground level. The height of the projectile, in feet, after t seconds is given by the function $h(t) = -16t^2 + 272t$. During what period of time is the projectile at least 960 feet above the ground?

BUILDING YOUR STUDY STRATEGY

Time Management, 6 Studying at Work Many students have jobs, and this commitment takes away a great deal of time from their studies. If possible, find a job that allows you opportunities to study while at work. Be sure to take full advantage of any breaks you receive. A ten-minute work break may not seem like much time, but you can read through a series of note cards or review your class notes during the break.

Exercises 10.6

PRACTICE

WATCH

DOWNLOAD

READ

REVIEW

Vocabulary

1. A(n) ____________ inequality is an inequality involving a quadratic expression.
2. The ____________ of a quadratic inequality in standard form are values of the variable for which the quadratic expression is equal to 0.
3. The zeros of a(n) ____________ inequality are not included in the solutions of that inequality.
4. The zeros from the ____________ of a rational inequality are never included as a solution.

Solve each quadratic inequality. Express your solution on a number line and using interval notation.

5. $(x - 5)(x - 1) < 0$
6. $(x + 3)(x - 2) > 0$
7. $-2(x - 6)(x + 3) \leq 0$
8. $x(x - 4) > 0$
9. $x^2 - 4x - 21 < 0$
10. $x^2 + 9x + 18 \leq 0$
11. $x^2 - 10x + 16 \geq 0$
12. $x^2 + 3x - 54 > 0$
13. $x^2 - 25 \leq 0$
14. $x^2 - 4x \geq 0$
15. $x^2 + 6x + 3 > 0$

16. $x^2 - 5x - 11 < 0$

17. $3x^2 - 5x - 1 \geq 0$

18. $2x^2 + 7x + 4 \leq 0$

19. $x^2 + 3x + 5 > 0$

20. $x^2 + 7x + 14 > 0$

21. $x^2 - 2x + 24 < 0$

22. $x^2 + x + 8 < 0$

23. $x^2 - x \geq 56$

24. $x^2 - 8x \leq 20$

25. $-x^2 + 10x - 5 > 0$

26. $-x^2 + 6x + 1 < 0$

Solve each polynomial inequality. Express your solution on a number line and using interval notation.

27. $(x - 1)(x + 2)(x - 4) \geq 0$

28. $(x + 8)(x - 5)(x - 2) \leq 0$

29. $(x - 7)^2(x + 5)(x + 9) \leq 0$

30. $(x - 10)^3(x + 4)^2(x - 4) \geq 0$

Solve each rational inequality. Express your solution on a number line and using interval notation.

31. $\dfrac{x - 3}{x + 4} < 0$

32. $\dfrac{x + 8}{x - 6} > 0$

33. $\dfrac{(x + 2)(x - 2)}{x + 1} \leq 0$

34. $\dfrac{(x + 7)(x - 4)}{(x - 5)(x + 10)} < 0$

35. $\dfrac{x - 4}{x^2 + 9x + 18} < 0$

36. $\dfrac{x^2 + 11x + 24}{x + 5} > 0$

37. $\dfrac{x^2 + x - 20}{x^2 - 4} \geq 0$

38. $\dfrac{x^2 - 7x - 8}{x^2 - 4x - 21} \leq 0$

39. $\dfrac{x^2 - 2x - 48}{x^2 + 13x + 36} \leq 0$

40. $\dfrac{x^2 + 11x + 30}{x^2 + 3x - 28} \geq 0$

41. $\dfrac{x^2 - 4x + 3}{x^2 + 4x - 32} < 0$

42. $\dfrac{x^2 - 81}{x^2 - 15x + 54} \leq 0$

43. $\dfrac{4x - 7}{x - 4} > 3$

44. $\dfrac{6x + 18}{x + 2} \le 5$

45. $\dfrac{x^2 + 9x + 12}{x + 3} \ge 2$

46. $\dfrac{x^2 - 3x - 46}{x - 7} < 4$

47. $\dfrac{x^2 + 3x + 10}{x - 2} \le x$

48. $\dfrac{x^2 + 7x + 21}{x + 4} \ge x$

49. A projectile with an initial velocity of 64 feet per second is fired upward from the roof of a building 27 feet tall. The height of the projectile, in feet, after t seconds is given by the function $h(t) = -16t^2 + 64t + 27$. For what length of time is the projectile at least 75 feet above the ground?

50. A projectile with an initial velocity of 208 feet per second is fired upward from the roof of a building 98 feet tall. The height of the projectile, in feet, after t seconds is given by the function $h(t) = -16t^2 + 208t + 98$. For what length of time is the projectile at least 450 feet above the ground?

51. A projectile with an initial velocity of 32 feet per second is fired upward from ground level. The height of the projectile, in feet, after t seconds is given by the function $h(t) = -16t^2 + 32t$. For what length of time is the projectile at least 12 feet above the ground?

52. A projectile with an initial velocity of 48 feet per second is fired upward from ground level. The height of the projectile, in feet, after t seconds is given by the function $h(t) = -16t^2 + 48t$. For what length of time is the projectile at least 20 feet above the ground?

53. A projectile with an initial velocity of 125 feet per second is fired upward from the roof of a building 48 feet tall.

a) List the function $h(t)$ that gives the height of the projectile in feet after t seconds.

b) For what time interval is the projectile at least 200 feet above the ground?

54. A projectile with an initial velocity of 150 feet per second is fired upward from the roof of a building 80 feet tall.

a) List the function $h(t)$ that gives the height of the projectile in feet after t seconds.

b) For what time interval is the projectile at least 300 feet above the ground?

55. Given $f(x) = x^2 - 8x + 12$, find all values x for which $f(x) \le 0$.

56. Given $f(x) = x^2 + 13x - 5$, find all values x for which $f(x) \ge 0$.

57. Given $f(x) = x^2 - 5x + 13$, find all values x for which $f(x) \ge 7$.

58. Given $f(x) = x^2 + 4x - 31$, find all values x for which $f(x) < 14$.

59. Given $f(x) = \dfrac{x + 9}{x - 6}$, find all values x for which $f(x) \le 0$.

60. Given $f(x) = \dfrac{x^2 - 17x + 70}{x^2 + 2x - 48}$, find all values x for which $f(x) < 0$.

61. Given $f(x) = \dfrac{x^2 - 8x - 33}{x^2 + 13x + 42}$, find all values x for which $f(x) > 0$.

62. Given $f(x) = \dfrac{x^2 + 14x + 24}{x^2 + 6x + 5}$, find all values x for which $f(x) \ge 0$.

Find each quadratic inequality whose solution is given.

63.

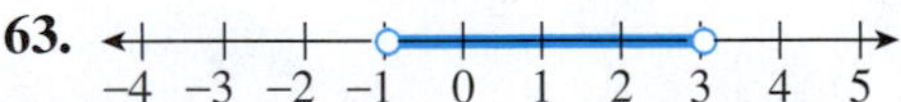

64. −9 −8 −7 −6 −5 −4 −3 −2 −1 0 1 2 3

65. −5 −4 −3 −2 −1 0 1 2 3 4 5

66.

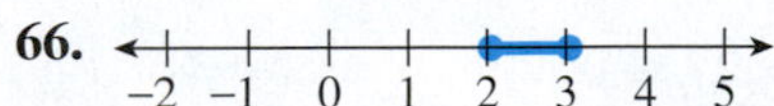

Find each rational inequality whose solution is given.

67. −6 −5 −4 −3 −2 −1 0 1 2 3 4 5 6 7

68. 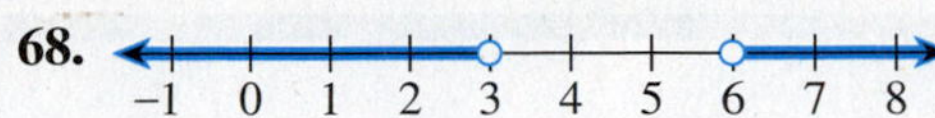

69. 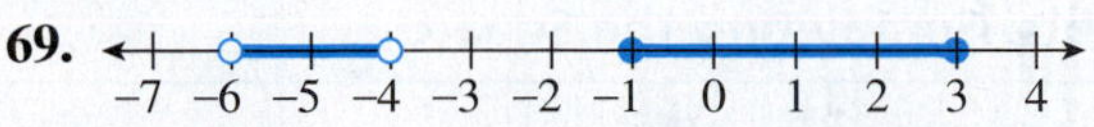

70. –4 –3 –2 –1 0 1 2 3 4 5 6

Writing in Mathematics

Answer in complete sentences.

71. ***Solutions Manual*** Write a solutions manual page for the following problem.

Solve $\dfrac{x^2 - 6x - 16}{x^2 - 16} \geq 0.$

CHAPTER 10 SUMMARY

Section 10.1 Solving Quadratic Equations by Extracting Square Roots; Completing the Square

Solving a Quadratic Equation by Factoring, p. 537

1. Write the equation in standard form:

$$ax^2 + bx + c = 0$$

2. Completely factor the quadratic expression.
3. Set each factor equal to 0 and solve.

Solve: $x^2 + 10x - 35 = 9x - 5$

$$x^2 + 10x - 35 = 9x - 5$$
$$x^2 + x - 30 = 0$$
$$(x + 6)(x - 5) = 0$$

$$x + 6 = 0 \qquad x - 5 = 0$$
$$x = -6 \qquad x = 5$$

$\{-6, 5\}$

Solving a Quadratic Equation by Extracting Square Roots, pp. 537–540

1. Isolate the squared term.
2. Take the square root of each side. (Remember $\pm$ when taking the square root of the constant.)
3. Simplify the square root.
4. Solve the resulting equation.

Solve: $(x + 4)^2 - 7 = 42$

$$(x + 4)^2 - 7 = 42$$
$$(x + 4)^2 = 49$$
$$\sqrt{(x + 4)^2} = \pm\sqrt{49}$$
$$x + 4 = \pm 7$$
$$x = -4 \pm 7$$

$$-4 + 7 = 3 \qquad -4 - 7 = -11$$

$\{-11, 3\}$

Solving a Quadratic Equation by Completing the Square, pp. 540–543

1. Isolate all variable terms on one side of the equation, with the constant term on the other side of the equation.
2. If the coefficient of the second-degree term is not equal to 1, divide both sides of the equation by that coefficient.
3. Identify the coefficient of the first-degree term. Take half of that number, square it, and add that to both sides of the equation.
4. Factor the resulting perfect square trinomial.
5. Take the square root of each side of the equation. Be sure to include $\pm$ on the side where the constant is.
6. Solve the resulting equation.

Solve by completing the square: $x^2 + 8x + 4 = 0$

$$x^2 + 8x + 4 = 0$$
$$x^2 + 8x = -4 \qquad \left(\frac{8}{2}\right)^2 = 4^2 = 16$$
$$x^2 + 8x + 16 = -4 + 16$$
$$(x + 4)^2 = 12$$
$$\sqrt{(x + 4)^2} = \pm\sqrt{12}$$
$$x + 4 = \pm 2\sqrt{3}$$
$$x = -4 \pm 2\sqrt{3}$$

$\{-4 - 2\sqrt{3}, -4 + 2\sqrt{3}\}$

Section 10.2 The Quadratic Formula

Quadratic Formula, pp. 547–551

The quadratic formula can be used to solve any quadratic equation that is written in standard form:

$$ax^2 + bx + c = 0$$

The quadratic formula is

$$x = \frac{-b \pm \sqrt{b^2 - 4ac}}{2a}.$$

Solve using the quadratic formula: $x^2 + 6x - 20 = 0$.

$$a = 1, b = 6, c = -20$$
$$x = \frac{-6 \pm \sqrt{6^2 - 4(1)(-20)}}{2(1)}$$
$$x = \frac{-6 \pm \sqrt{116}}{2}$$
$$x = \frac{-6 \pm 2\sqrt{29}}{2}$$
$$x = \frac{2(-3 \pm \sqrt{29})}{2}$$
$$x = -3 \pm \sqrt{29}$$

$\{-3 - \sqrt{29}, -3 + \sqrt{29}\}$

Discriminant, p. 552

The radicand, $b^2 - 4ac$, is called the discriminant.
If the discriminant is positive, the quadratic equation has two real solutions.
If the discriminant is equal to 0, the quadratic equation has one real solution.
If the discriminant is negative, the quadratic equation has two nonreal complex solutions.

Use the discriminant to determine the number and type of solutions.

$2x^2 + 9x + 11 = 0$

$b^2 - 4ac = 9^2 - 4(2)(11) = 81 - 88 = -7$

Because the discriminant is negative, the equation has two nonreal complex solutions.

Projectile Problems, pp. 553–555

The height, in feet, of a projectile after t seconds can be found using the function

$$h(t) = -16t^2 + v_0 t + s,$$

where v_0 is the initial velocity of the projectile and s is the initial height.

An object with an initial velocity of 80 feet per second is launched upward from a cliff 240 feet above a beach. When will the object land on the ground? Round to the nearest tenth of a second.

$$v_0 = 80, s = 240$$
$$h(t) = -16t^2 + 80t + 240$$
$$h(t) = 0$$
$$-16t^2 + 80t + 240 = 0$$
$$-16(t^2 - 5t - 15) = 0$$
$$t = \frac{-(-5) \pm \sqrt{(-5)^2 - 4(1)(-15)}}{2(1)}$$
$$t = \frac{5 \pm \sqrt{85}}{2}$$

$\frac{5 - \sqrt{85}}{2} \approx -2.1$ Omit negative solution. $\qquad \frac{5 + \sqrt{85}}{2} \approx 7.1$

The object will land on the ground in approximately 7.1 seconds.

Section 10.3 Equations That Are Quadratic in Form

Solving Equations Using a *u*-Substitution, pp. 558–560

Concept

1. Substitute the variable u for an expression in such a way that the resulting equation is a quadratic equation in terms of u.

Equation	Substitution
$ax^4 + bx^2 + c = 0$	$u = x^2$
$ax + b\sqrt{x} + c = 0$	$u = \sqrt{x}$
$ax + bx^{1/2} + c = 0$	$u = x^{1/2}$
$ax^{2/3} + bx^{1/3} + c = 0$	$u = x^{1/3}$

2. Solve for u.
3. Replace u with the expression for which u was substituted and solve for the original variable.
4. Check for extraneous roots.

Example

Solve: $x^4 - 5x^2 - 36 = 0$

$$x^4 - 5x^2 - 36 = 0$$

Let $u = x^2$.

$$u^2 - 5u - 36 = 0$$
$$(u + 4)(u - 9) = 0$$

$u + 4 = 0$, $u = -4$, $x^2 = -4$, $\sqrt{x^2} = \pm\sqrt{-4}$, $x = \pm 2i$

$u - 9 = 0$, $u = 9$, $x^2 = 9$, $\sqrt{x^2} = \pm\sqrt{9}$, $x = \pm 3$

$\{-2i, 2i, -3, 3\}$

Solving Radical Equations, p. 560

1. Isolate one radical containing the variable on one side of the equation.
2. Raise both sides of the equation to the nth power, where n is the index of the radical.
   For any nonnegative number x and any integer $n > 1$, $\left(\sqrt[n]{x}\right)^n = x$.
3. If the resulting equation does not contain a radical, solve the equation. If the resulting equation does contain a radical, begin the process again by isolating the radical on one side of the equation.
4. Check the solution(s).

Solve: $\sqrt{4x + 33} + 3 = x$

$$\sqrt{4x + 33} + 3 = x$$
$$\sqrt{4x + 33} = x - 3$$
$$(\sqrt{4x + 33})^2 = (x - 3)^2$$
$$4x + 33 = x^2 - 6x + 9$$
$$0 = x^2 - 10x - 24$$
$$0 = (x + 2)(x - 12)$$

$x + 2 = 0$, $x = -2$ Omit extraneous root. $\qquad x - 12 = 0$, $x = 12$

$\{12\}$

Solving Rational Equations, pp. 560–561

Find the LCD of all rational expressions in the equation and then multiply both sides of the equation by that LCD. Solve the resulting equation. Any solution that causes a denominator in the original equation to be zero must be omitted.

Solve: $1 + \frac{5}{x} - \frac{14}{x^2} = 0$

LCD: x^2

$$x^2\left(1 + \frac{5}{x} - \frac{14}{x^2}\right) = x^2 \cdot 0$$

$$1 \cdot x^2 + \frac{5}{x} \cdot x^2 - \frac{14}{x^2} \cdot x^2 = 0$$

$$1 \cdot x^2 + \frac{5}{\cancel{x}_1} \cdot \cancel{x^2}^{x} - \frac{14}{\cancel{x^2}_1} \cdot \cancel{x^2}^{1} = 0$$

$$x^2 + 5x - 14 = 0$$

$$(x + 7)(x - 2) = 0$$

$x + 7 = 0$ $\quad$ $x - 2 = 0$

$x = -7$ $\quad$ $x = 2$

$\{-7, 2\}$

Work-Rate Problems, pp. 561–562

The work-rate for a person or machine is equal to the reciprocal of the time required to complete the whole task.

The amount of work done by a person or machine is equal to the product of the work-rate and the time spent working.

The amount of work done by each person or machine must add up to equal 1.

Jerry takes 30 minutes longer than George takes to mow the same lawn. If Jerry and George work together using two lawn mowers, they can mow the lawn in 20 minutes. How long would it take each of them to mow the lawn?

Person	Time Alone	Work Rate	Time Working	Work Done
Jerry	$t + 30$	$\frac{1}{t + 30}$	20	$\frac{20}{t + 30}$
George	t	$\frac{1}{t}$	20	$\frac{20}{t}$

$$\frac{20}{t + 30} + \frac{20}{t} = 1$$

LCD: $t(t + 30)$

$$t(t + 30) \cdot \frac{20}{t + 30} + t(t + 30) \cdot \frac{20}{t} = 1 \cdot t(t + 30)$$

$$20t + 20(t + 30) = t(t + 30)$$

$$20t + 20t + 600 = t^2 + 30t$$

$$0 = t^2 - 10t - 600$$

$$0 = (t + 20)(t - 30)$$

$t + 20 = 0$ $\quad$ $t - 30 = 0$

$t = -20$ $\quad$ $t = 30$

Omit negative solution.

It would take Jerry 60 minutes and George 30 minutes.

Section 10.4 Graphing Quadratic Equations

Graphing Quadratic Equations of the Form $y = ax^2 + bx + c$, pp. 565–571

1. Find the vertex.
   The x-coordinate is found using the formula $x = \dfrac{-b}{2a}$.
   Substitute this value for x and solve for y to find the y-coordinate.
2. Find the y-intercept by substituting 0 for x and solving for y.
3. Determine whether the parabola opens upward or downward.
   If a is positive, the parabola opens upward.
   If a is negative, the parabola opens downward.
4. Find the x-intercept(s) if any exist.
   If the parabola has x-intercepts, find them by substituting 0 for y and solving the resulting quadratic equation for x.
5. Find the axis of symmetry $\left(x = \dfrac{-b}{2a}\right)$ and use symmetry to find other points on the parabola.

Graph: $y = x^2 - 6x - 7$

$$x = \frac{-(-6)}{2(1)} = \frac{6}{2} = 3$$

$$y = 3^2 - 6(3) - 7 = 9 - 18 - 7 = -16$$

Vertex: $(3, -16)$
y-intercept: $(0, -7)$

$$x^2 - 6x - 7 = 0$$
$$(x + 1)(x - 7) = 0$$

$$x + 1 = 0 \qquad x - 7 = 0$$
$$x = -1 \qquad x = 7$$

x-intercepts: $(-1, 0), (7, 0)$

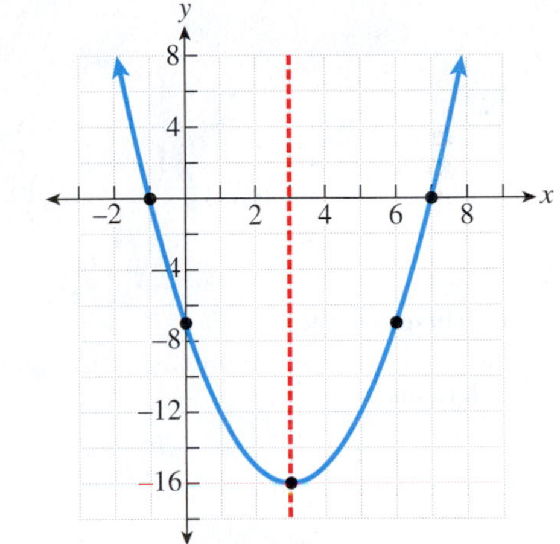

Graphing Quadratic Equations of the Form $y = a(x - h)^2 + k$, pp. 571–572

1. Find the vertex, which is (h, k).
2. Find the y-intercept by substituting 0 for x and solving for y.
3. Determine whether the parabola opens upward or downward.
   If a is positive, the parabola opens upward.
   If a is negative, the parabola opens downward.
4. Find the x-intercept(s) if any exist.
   If the parabola has x-intercepts, find them by substituting 0 for y and solving the resulting quadratic equation for x.
5. Find the axis of symmetry $(x = h)$ and use symmetry to find other points on the parabola.

Graph: $y = (x + 2)^2 - 9$
Vertex: $(-2, -9)$

$$y = (0 + 2)^2 - 9 = 4 - 9 = -5$$

y-intercept: $(0, -5)$

$$(x + 2)^2 - 9 = 0$$
$$(x + 2)^2 = 9$$
$$\sqrt{(x + 2)^2} = \pm\sqrt{9}$$
$$x + 2 = \pm 3$$
$$x = -2 \pm 3$$

$$-2 - 3 = -5 \qquad -2 + 3 = 1$$

x-intercepts: $(-5, 0), (1, 0)$

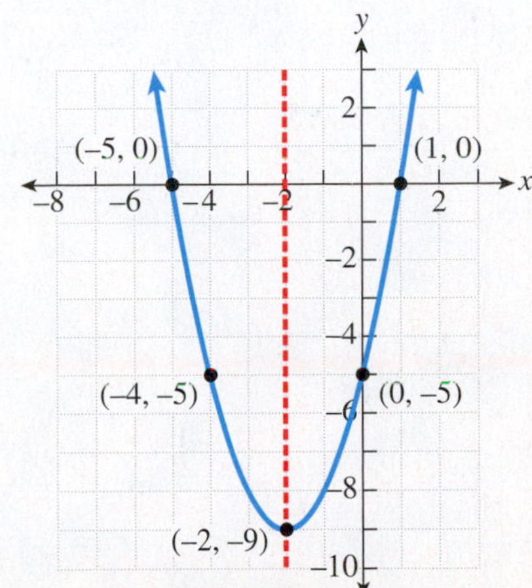

Section 10.5 Applications Using Quadratic Equations

Area of a Rectangle, p. 577

The unknowns will be the length and the width.
The formula is

$$\text{Area} = \text{Length} \cdot \text{Width}.$$

The length of a rectangle is 3 inches more than its width. If the area of the rectangle is 100 inches, find its dimensions. Round to the nearest tenth of an inch.

Unknowns
Length: $x + 3$; width: x

$$x(x+3) = 100$$
$$x^2 + 3x - 100 = 0$$
$$x = \frac{-3 \pm \sqrt{3^2 - 4(1)(-100)}}{2(1)}$$
$$x = \frac{-3 \pm \sqrt{409}}{2}$$

$\frac{-3 - \sqrt{409}}{2} \approx -11.6$ Omit negative solution. $\qquad \frac{-3 + \sqrt{409}}{2} \approx 8.6$

Length: $x + 3 \approx 8.6 + 3 = 11.6$
Width: $x \approx 8.6$
The length is approximately 11.6 inches, and the width is approximately 8.6 inches.

Pythagorean Theorem, pp. 577–578

For any right triangle whose hypotenuse has length c and whose other two legs have lengths a and b,

$$a^2 + b^2 = c^2.$$

A right triangle has a hypotenuse that measures 25 centimeters, and one of the legs measures 7 centimeters. Find the length of the other leg.

Unknown
Leg: a

$$a^2 + b^2 = c^2$$
$$a^2 + 7^2 = 25^2$$
$$a^2 + 49 = 625$$
$$a^2 = 576$$
$$a = \sqrt{576}$$
$$a = 24$$

The length of the other leg is 24 centimeters.

Applications of the Pythagorean Theorem, pp. 578–579

To solve an applied problem using the Pythagorean theorem, find a right triangle.
The Pythagorean theorem also can be used to solve problems involving the diagonal of a rectangle.

The diagonal of a rectangular table is 10 feet, and the length of the table is 5 feet more than its width. Find the length and width of the table. Round to the nearest tenth of a foot.

Unknowns
Length: $x + 5$; width: x

$$a^2 + b^2 = c^2$$
$$(x+5)^2 + x^2 = 10^2$$
$$x^2 + 10x + 25 + x^2 = 100$$
$$2x^2 + 10x - 75 = 0$$
$$x = \frac{-10 \pm \sqrt{10^2 - 4(2)(-75)}}{2(2)}$$
$$x = \frac{-10 \pm \sqrt{700}}{4}$$

$\frac{-10 - \sqrt{700}}{4} \approx -9.1$ Omit negative solution. $\qquad \frac{-10 + \sqrt{700}}{4} \approx 4.1$

Length: $x + 5 \approx 4.1 + 5 = 9.1$
Width: $x \approx 4.1$
The length of the table is approximately 9.1 feet and the width is approximately 4.1 feet.

Section 10.6 Quadratic and Rational Inequalities

Solving Quadratic Inequalities, pp. 582–585

1. Rewrite the inequality in standard form by collecting all terms on one side of the inequality.
2. Find the zeros of the inequality by setting $ax^2 + bx + c$ equal to 0 and solving for x.
3. Use the zeros of the inequality to break a number line into intervals.
4. Select a test point from each interval and use it to determine whether each point in that interval is a solution. This can be done by using a sign chart or by substituting the test point into the original inequality.
5. Express the solution on a number line and using interval notation.
   - If the inequality is a strict inequality ($<$ or $>$), the zeros cannot be included as solutions.
   - If the inequality is a weak inequality ($\leq$ or $\geq$), the zeros are included as solutions.

Solve: $x^2 - 6x - 27 > 0$

Zeros

$$x^2 - 6x - 27 = 0$$
$$(x + 3)(x - 9) = 0$$
$$x = -3 \quad \text{or} \quad x = 9$$

Test Point	$x^2 - 6x - 27$	> 0?
-4	$(-4)^2 - 6(-4) - 27 = 13$	True
0	$(0)^2 - 6(0) - 27 = -27$	False
10	$(10)^2 - 6(10) - 27 = 13$	True

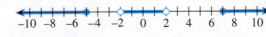

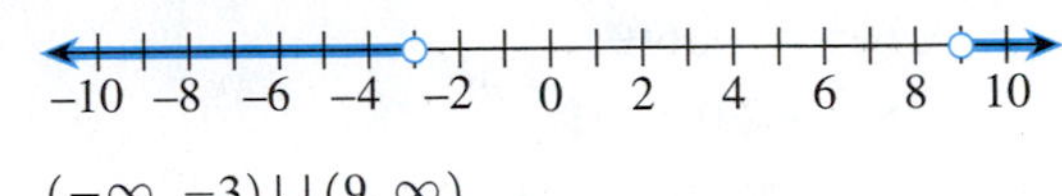

$(-\infty, -3) \cup (9, \infty)$

Solving Rational Inequalities, pp. 585–588

1. Rewrite the inequality in standard form by collecting all terms on one side of the inequality.
2. Find the zeros of the inequality by setting the numerator equal to 0 and the denominator equal to 0, solving each equation for x.
3. Use the zeros of the inequality to break a number line into intervals.
4. Select a test point from each interval and use it to determine whether each point in that interval is a solution. This can be done by using a sign chart or by substituting the test point into the original inequality.
5. Express the solution on a number line and using interval notation.
   - If the inequality is a strict inequality ($<$ or $>$), the zeros from the numerator cannot be included as solutions.
   - If the inequality is a weak inequality ($\leq$ or $\geq$), the zeros from the numerator are included as solutions.
   - Regardless of the type of inequality, the zeros from the denominator cannot be included as solutions.

Solve: $\dfrac{x^2 - 2x - 35}{x^2 - 4} \geq 0$

Zeros

$$x^2 - 2x - 35 = 0 \qquad x^2 - 4 = 0$$
$$(x + 5)(x - 7) = 0 \qquad (x + 2)(x - 2) = 0$$
$$x = -5 \text{ or } x = 7 \qquad x = -2 \text{ or } x = 2$$

Test Point	$x + 5$	$x - 7$	$x + 2$	$x - 2$	Sign
-6	$-$	$-$	$-$	$-$	$+$
-4	$+$	$-$	$-$	$-$	$-$
0	$+$	$-$	$+$	$-$	$+$
3	$+$	$-$	$+$	$+$	$-$
8	$+$	$+$	$+$	$+$	$+$

$(-\infty, -5] \cup (-2, 2) \cup [7, \infty)$

REVIEW OF CHAPTER 10 STUDY STRATEGIES

Poor time management can cause a student to do poorly in a math class. To maximize the time you have available to study mathematics and to use your time effectively, consider the following:

- Record how you spend your time over the period of one week. Look for ways you might be wasting time. Create an efficient schedule to follow.
- Study new material as soon as possible after class. Save review work for later in the day.
- Arrange your schedule so that you can study mathematics when your mental energy is at its highest.
- Set goals for each study session and record the goals in a to-do list.
- Study your most difficult subject first, before you become fatigued.
- If you must work, try to find a job that will allow you to do some studying while at work. At the very least, make efficient use of your work breaks to study.

CHAPTER 10 REVIEW

Solve by factoring. **[10.1]**

1. $x^2 + 11x + 28 = 0$
2. $x^2 + 7x - 30 = 0$
3. $x^2 + 4x = 0$
4. $2x^2 - 13x + 15 = 0$
5. $x^2 + 7x - 12 = 3x - 7$
6. $x^2 - 9x + 23 = 9(x - 6)$

Solve by extracting square roots. **[10.1]**

7. $x^2 = 49$
8. $x^2 - 72 = 0$
9. $x^2 + 4 = 0$
10. $(x - 6)^2 = -36$
11. $(x - 8)^2 + 30 = 21$
12. $(2x + 7)^2 + 20 = 141$

Solve by completing the square. **[10.1]**

13. $x^2 - 2x - 8 = 0$
14. $x^2 - 6x + 3 = 0$
15. $x^2 + 14x + 47 = 0$
16. $x^2 + 5x + 10 = 0$

Solve by using the quadratic formula. **[10.2]**

17. $x^2 + 7x + 1 = 0$
18. $x^2 + 4x - 2 = 0$
19. $x^2 - 3x + 9 = 0$
20. $4x^2 - 20x + 29 = 0$
21. $2x^2 - 5x + 9 = x^2 - 9x + 30$
22. $x(x + 4) = 16x - 37$

Solve by using a u-substitution. **[10.3]**

23. $x^4 - 5x^2 + 4 = 0$
24. $x - 4\sqrt{x} - 32 = 0$
25. $x + 7x^{1/2} - 8 = 0$
26. $x^{2/3} - 4x^{1/3} - 21 = 0$

Solve. **[10.3]**

27. $\sqrt{x^2 + x - 2} = 2$
28. $\sqrt{x + 15} - x = 9$

Solve. **[10.3]**

29. $1 + \dfrac{11}{x} + \dfrac{30}{x^2} = 0$
30. $\dfrac{x}{x - 3} - \dfrac{7}{x - 2} = \dfrac{3}{x^2 - 5x + 6}$

Solve using the most efficient method. **[10.1, 10.2, 10.3]**

31. $x^2 - 19x + 90 = 0$
32. $x^4 + 5x^2 - 36 = 0$
33. $(x + 6)^2 + 11 = 3$
34. $x^2 + 4x + 16 = 0$
35. $x^2 + 10x = 0$
36. $x^2 - 6x - 20 = 0$
37. $\sqrt{3x + 49} = x + 3$
38. $\sqrt{2x + 9} = \sqrt{4x + 1} - 2$
39. $x - 1 - \dfrac{20}{x} = 0$
40. $3x^2 - 8x + 2 = 0$
41. $x + 5x^{1/2} - 24 = 0$
42. $(x + 9)^2 + 64 = 0$

For each function f(x), find all values x for which f(x) = 0. **[10.1, 10.2, 10.3]**

43. $f(x) = (x + 12)^2 - 16$
44. $f(x) = x^2 - 7x - 20$

Find a quadratic equation with integer coefficients that has the given solution(s). **[10.1]**

45. $\{-9, 4\}$
46. $\{4\}$
47. $\left\{\frac{4}{3}, \frac{13}{2}\right\}$
48. $\{-2\sqrt{2}, 2\sqrt{2}\}$
49. $\{-5i, 5i\}$
50. $\{-8i, 8i\}$

Find the vertex and axis of symmetry of the parabola associated with the given quadratic equation. **[10.4]**

51. $y = x^2 - 8x - 20$
52. $y = x^2 + 5x - 16$
53. $y = -x^2 - 2x + 35$
54. $y = (x - 3)^2 - 4$

Worked-out solutions to Review Exercises marked with ● can be found on page AN-35.

***Find the x- and y-intercepts of the parabola associated with the given equation. If the parabola does not have any x-intercepts, state no x-intercepts.* [10.4]**

55. $y = x^2 + 6x + 11$

56. $y = x^2 - 4x - 3$

57. $y = -x^2 + 10x - 25$

58. $y = (x + 4)^2 - 8$

***Graph each quadratic equation. Label the vertex, the y-intercept, and any x-intercepts.* [10.4]**

59. $y = x^2 + 8x + 15$

60. $y = -x^2 + 2x + 3$

61. $y = x^2 + 4x - 2$

62. $y = x^2 + 4x + 8$

63. $y = -(x + 3)^2 + 4$

64. $y = (x - 2)^2 - 9$

***Solve each quadratic inequality. Express your solution on a number line and using interval notation.* [10.6]**

65. $x^2 - 9x + 14 \leq 0$

66. $x^2 + 9x + 8 > 0$

67. $x^2 + 4x - 11 > 0$

68. $x^2 + 3x + 40 \geq 0$

***Solve each rational inequality. Express your solution on a number line and using interval notation.* [10.6]**

69. $\dfrac{x^2 - 15x + 54}{x - 7} \leq 0$

70. $\dfrac{x^2 - 4}{x^2 + 2x - 80} \geq 0$

***Given each function f(x), solve the inequality f(x) ≥ 0. Express your solution on a number line and using interval notation.* [10.6]**

71. $f(x) = x^2 + 15x + 50$

72. $f(x) = \dfrac{x^2 - 16}{x^2 - 10x + 9}$

For Exercises 73 and 74, use the function $h(t) = -16t^2 + v_0t + s$. [10.2]

73. A projectile is launched upward from the ground at a speed of 104 feet per second. How long will it take until the projectile lands on the ground?

74. A projectile with an initial velocity of 66 feet per second is launched upward from the roof of a building 30 feet high. How long will it take until the projectile lands on the ground? Round to the nearest hundredth of a second.

75. A water tank is connected to two pipes. When both pipes are open, they can fill the tank in 4 hours. The larger of the pipes can fill the tank in 2 hours less time than the smaller pipe can. How long will it take the smaller pipe alone to fill the tank? Round to the nearest tenth of an hour. [10.3]

76. A tree-trimming company has two crews. If both crews work together, they can trim all of the trees on campus in 32 hours. The faster crew can trim all of the trees in 6 hours less than the slower crew can. How long will it take the faster crew, working alone, to trim all of the trees on campus? Round to the nearest tenth of an hour. [10.3]

77. A rectangular photograph has an area of 288 square centimeters. If the length of the photo is 2 centimeters longer that its width, find the dimensions of the photo. [10.5]

78. A rectangular swimming pool covers an area of 500 square feet. If the length of the pool is 13 feet more than its width, find the dimensions of the pool. Round to the nearest tenth of a foot. [10.5]

79. Ann needs to fly to a city that is 1000 miles north and 800 miles east of where she lives. If the airplane flies a direct route, how far will the flight be? Round to the nearest tenth of a mile. [10.5]

80. The diagonal of a rectangular workbench is 12 feet long. If the width of the workbench is 3 feet less than the length of the workbench, find the dimensions of the workbench. Round to the nearest tenth of a foot. [10.5]

CHAPTER 10 TEST

For Extra Help

Step-by-step test solutions are found on the Chapter Test Prep Videos available via the Video Resources on DVD, in MyMathLab, and on YouTube (search "WoodburyElemIntAlg" and click on "Channels").

1. Solve by factoring. $x^2 + 9x + 18 = 0$

2. Solve by extracting square roots. $(3x - 8)^2 - 6 = 58$

3. Solve by completing the square. $x^2 - 14x - 3 = 0$

4. Solve by using the quadratic formula.
$x^2 - 6x + 12 = 0$

5. Solve by using a u-substitution. $x^4 - 8x^2 - 48 = 0$

6. Solve by using a u-substitution. $x + 8x^{1/2} - 9 = 0$

Solve.

7. $(4x - 1)^2 - 13 = 12$

8. $2x^2 - 15x - 27 = 0$

9. $x^2 - 14x + 40 = 0$

10. $\dfrac{x}{x + 8} - \dfrac{6}{x - 4} = \dfrac{8}{x^2 + 4x - 32}$

11. $x^2 - 10x + 34 = 0$

Find a quadratic equation with integer coefficients that has the given solutions.

12. $\{\frac{4}{7}, 5\}$

13. $\{-4i, 4i\}$

Find the vertex of the parabola associated with the given quadratic equation.

14. $y = x^2 + 8x - 39$

15. $y = (x - 7)^2 + 20$

16. Find the x- and y-intercepts of the parabola associated with $y = x^2 - 10x + 20$. Round to the nearest tenth if necessary.

17. Graph $y = -(x + 1)^2 + 4$.
Label the vertex, the y-intercept, and any x-intercepts.

18. Graph $y = x^2 - 10x + 13$. Label the vertex, the y-intercept, and any x-intercepts.

19. Solve. $x^2 + 3x - 28 > 0$

20. Solve. $\dfrac{x^2 + 11x + 24}{x^2 - 5x - 14} \leq 0$

21. The area of a rectangular playing field is 3600 square yards. If the width of the field is 60 yards less than the length of the field, find the dimensions of the field. Round to the nearest tenth of a yard.

22. Daniel flies to a resort town 1500 miles south and 800 miles west of where he lives. How far away is the resort town from where he lives?

Mathematicians in History

Evariste Galois was a brilliant mathematician who was interested in the algebraic solutions of equations. The amount and depth of work accomplished by Galois, who died before his 21st birthday, is legendary. Galois showed that there is no general solution of equations that are fifth degree or higher:

$$ax^5 + bx^4 + cx^3 + dx^2 + ex + f = 0$$

Write a one-page summary (*or* make a poster) of the mathematical achievements of Galois, his fascinating (but short) life, and the details surrounding his untimely death.

Interesting issues:

- When and where was Galois born?
- What was the fate of Galois's father?
- Galois twice failed the admittance exam to École Polytechnique. On his second attempt, what did he do to one of the examiners?
- Galois was expelled from École Normale in December 1830. Why was he expelled?
- Galois was arrested in 1831 after raising a toast to King Louis-Philippe. What was the toast? How did Galois make it?
- In 1831, Galois was arrested on Bastille Day. What was the charge?
- On the night before his death, Galois wrote a letter to his friend Auguste Chevalier. What were the contents of the letter?
- What were the circumstances that led to Galois's death? How old was Galois at the time?
- Which mathematician published Galois's papers in 1846?

CHAPTER 11

Functions

In this chapter, we will continue to explore functions. In addition to linear and quadratic functions, we will examine new functions such as the square root function and the cubic function. We also will cover the algebra of functions, learning ways to create a new function from two or more existing functions. We will finish the chapter by discussing inverse functions.

STUDY STRATEGY

Study Environment It is very important to study in the proper environment. This chapter will focus on how to create an environment that allows you to get the most out of your study sessions. We will discuss where to study and what the conditions should be in that location.

11.1 Review of Functions

OBJECTIVES

1. Review of functions, domain, and range.
2. Represent a function as a set of ordered pairs.
3. Use the vertical-line test to determine whether a graph represents a function.
4. Use function notation.
5. Evaluate functions.
6. Interpret graphs of functions.

Functions; Domain and Range

Objective 1 Review of functions, domain, and range. Suppose a college charges its students a \$50 registration fee in addition to an enrollment fee of \$26 per unit. A student who signs up for 12 units would be charged \$312 for the classes (12 units times \$26 per unit) plus the \$50 registration fee, or a total of \$362. The amount a student pays depends on the number of units the student takes. We say that the amount a student pays is a function of the number of units in which the student is enrolled. In general, if a student is taking x units, the student's fees can be found using the formula $50 + 26x$.

Recall from Chapter 3 that a **function** is a relation that takes an input value and assigns a particular output value to it. In the previous example, the input value is the number of units the student is taking and the output value is the total fees for that student. For a relation to define a function, each input value must be assigned one and only one output value. In terms of the previous example, this means that students who are taking the same number of units pay the same amount of money. It is not possible for two students to each take 12 units but pay a different amount.

As we first saw in Chapter 3, the set of input values for a function is called the **domain** of the function. The domain for the previous example is the set of units in which a student can be enrolled. The possible values in the domain would be the set $\{1, 2, 3, \ldots\}$. The set of output values for a function is called the **range** of the function. Here is a table showing possible values in the range of this function.

Domain (Units)	1	2	3	...	x
Range (Fees, $)	$50 + 26(1) = 76$	$50 + 26(2) = 102$	$50 + 26(3) = 128$	...	$50 + 26x$

If there is no maximum number of units that a student can take, this table would continue indefinitely.

Although many of our functions will be given by a formula, many functions will not. For instance, consider the set of students enrolled in your math class. If we asked each student in your class to tell us the month of his or her birthday, a function would exist in which the input value is a student and the output value is the month of the student's birthday. The domain of this function is the set of students in your math class, and the range is the set of months from January through December.

EXAMPLE 1 The World Cup is a soccer tournament that is held every four years. Here are the winners of the World Cup for the years 1930–2010. (The tournament was not held in 1942 or 1946 due to World War II.)

Year	Winner	Year	Winner
1930	Uruguay	1978	Argentina
1934	Italy	1982	Italy
1938	Italy	1986	Argentina
1950	Uruguay	1990	West Germany
1954	West Germany	1994	Brazil
1958	Brazil	1998	France
1962	Brazil	2002	Brazil
1966	England	2006	Italy
1970	Brazil	2010	Spain
1974	West Germany		

a) Does a function exist for which the input value is the year of the tournament and the output value is the winner of that tournament?

SOLUTION Because each year listed has only one winner, this correspondence is a function. This is true even though some teams, such as Italy, have won more than once. By the definition of a function, each input value must be associated with one and only one output value. It is possible for more than one input value to be associated with the same output value.

The domain of this function is the set of years in which the World Cup has been held, and the range of this function is the set of teams that have won the World Cup.

b) Does a function exist for which the input value is the winner of a tournament and the output value is the year the country won the tournament?

SOLUTION No, this correspondence is not a function, as one input value could be associated with more than one output value. For instance, the input value Italy is associated with the years 1934, 1938, 1982, and 2006.

Quick Check 1

Here are the names and ages of the five members of a study group.

Name	Tiffany	Tyler	Teresa	Tim	Tina
Age	19	22	19	20	24

a) Does a function exist for which the input value is the name of a member of the study group and the output value is the age of that person?

b) Does a function exist for which the input value is the age of a member of the study group and the output value is the name of the person?

Ordered-Pair Notation

Objective 2 **Represent a function as a set of ordered pairs.** Functions can be expressed as a set of ordered pairs. The first coordinate of each ordered pair will be the input value of the function, and the second coordinate will be the corresponding output value. A set of ordered pairs represents a function if none of the first coordinates are repeated.

EXAMPLE 2 Determine whether the set of ordered pairs is a function. If it is a function, state its domain and range.

$$\{(-3, 9), (-2, 4), (-1, 1), (0, 0), (1, 1), (2, 4), (3, 9)\}$$

SOLUTION This set of ordered pairs is a function because no first coordinate is repeated. Notice that three pairs of ordered pairs share the same second coordinate. This does not violate the definition of a function. Although each possible input value of a function can be associated with only one output value, a particular output value of a function can be associated with several different input values.

The domain of this function is the set $\{-3, -2, -1, 0, 1, 2, 3\}$.

The range is $\{0, 1, 4, 9\}$.

Quick Check 2

Determine whether the set of ordered pairs is a function. If it is a function, state its domain and range.

$$\{(16, -3), (13, -2), (10, -1), (7, 0), (10, 1)\}$$

Vertical-Line Test

Objective 3 **Use the vertical-line test to determine whether a graph represents a function.** Although we have examined only sets of ordered pairs that have a **finite,** or limited, number of ordered pairs, the graph of an equation may represent a function as well. The graph of an equation represents an **infinite,** or unlimited, number of ordered pairs. If there are two ordered pairs on the graph of an equation that share the same x-coordinate but they have different y-coordinates, the graph does not represent a function.

EXAMPLE 3 Determine whether the graph below represents a function.

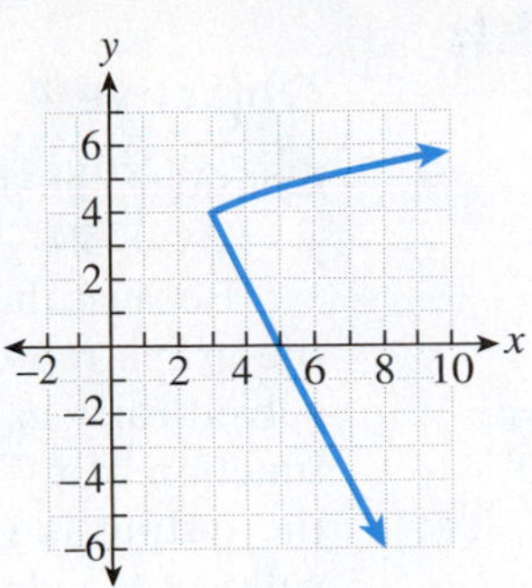

SOLUTION This graph does not represent a function because there are points on the graph that have the same x-coordinates but different y-coordinates. For example, the points $(6, 5)$ and $(6, -2)$ have the same x-coordinate but different y-coordinates. This means that one input value is associated with more than one output value, violating the definition of a function.

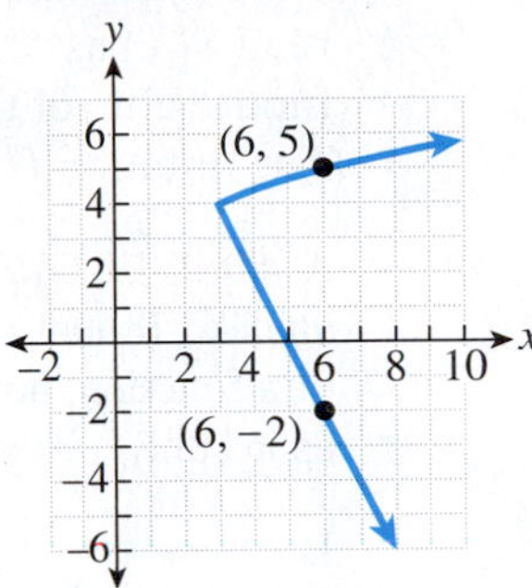

To determine whether a graph represents a function, we will apply the **vertical-line test.**

The Vertical-Line Test

If a vertical line can be drawn that intersects a graph at more than one point, the graph does not represent a function.

The graph from the previous example fails the vertical-line test because a vertical line can be drawn that falls through both $(6, 5)$ and $(6, -2)$; so the graph does not represent a function.

EXAMPLE 4 Use the vertical-line test to determine whether the graph below represents a function.

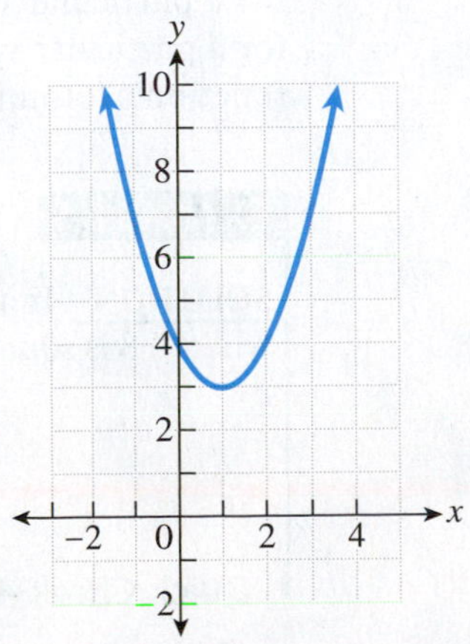

SOLUTION We cannot draw a vertical line that intersects this graph at more than one point. This graph passes the vertical-line test, so it does represent a function.

Quick Check 3

Use the vertical-line test to determine whether the following graph represents a function.

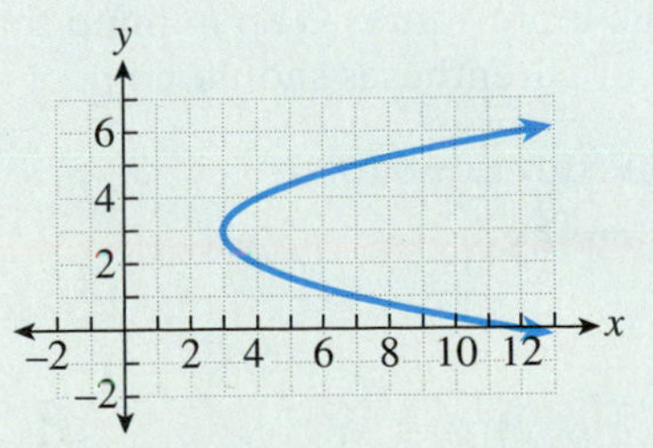

Function Notation

Objective 4 Use function notation.

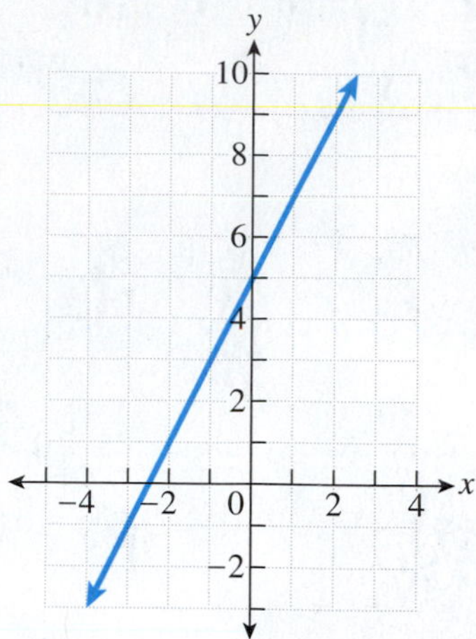

Consider the following graph of the equation $y = 2x + 5$.

Because the graph passes the vertical-line test, the graph represents a function. The equation is solved for y in terms of x, and we say that y is a function of x. The input of this function is x, and the output is y. The value of y depends on the value of x, and we say that y is the **dependent variable** in the equation. Typically, the equation will be solved for the dependent variable in terms of the other variable. The other variable in an equation is called the **independent variable.** In this equation, the variable x is the independent variable.

Saying that y is a function of x can be expressed symbolically as $y = f(x)$. We read $f(x)$ as "f of x." The letter f is the name of the function, and x represents the input value of the function. The input variable is always the independent variable. The notation $f(x)$ represents the output value of a function f for the input value x.

A WORD OF CAUTION The notation $f(x)$ represents the *output* of function f for the input value x. It does not indicate the product of f and x. We often use the letter f for the name of a function, but we may choose any letter, such as g, h, or F or even a Greek letter such as ϕ (phi).

Because $y = 2x + 5$ is a function, we also can write it as $f(x) = 2x + 5$. When an equation contains $f(x)$ instead of y, the equation is expressed in **function notation.**

$f(x) = 2x + 5$ ⇑ The variable inside the parentheses is the input variable for the function.	$f(x) = 2x + 5$ ⇑ $f(x)$ is the output value of the function f when the input value is x.	$f(x) = 2x + 5$ ⇑ The expression on the right side of the equation is the formula for this function.

Evaluating Functions

Objective 5 Evaluate functions. Finding the output value of a function $f(x)$ for a particular value of x is called **evaluating** the function. To evaluate a function for a particular value of the variable, we substitute that value for the variable in the function's formula, then simplify the resulting expression.

EXAMPLE 5 Let $h(x) = x^2 + 7x - 30$. Find $h(-8)$.

SOLUTION In this example, we need to square the input value. Keep in mind that when -8 is squared, the result is positive. The use of parentheses should help.

$$h(-8) = (-8)^2 + 7(-8) - 30 \quad \text{Substitute } -8 \text{ for } x.$$
$$= -22 \quad \text{Simplify.}$$

Quick Check 4

Let $f(x) = x^2 - 3x - 20$. Find $f(-5)$.

EXAMPLE 6 Let $f(x) = 7x - 5$. Find $f(3a + 2)$.

SOLUTION In this example, we are substituting a variable expression, $3a + 2$, for the input variable x rather than substituting a constant as in previous examples.

$$\begin{aligned} f(3a + 2) &= 7(3a + 2) - 5 && \text{Substitute } 3a + 2 \text{ for } x. \\ &= 21a + 14 - 5 && \text{Multiply.} \\ &= 21a + 9 && \text{Combine like terms.} \end{aligned}$$

Quick Check 5

Let $f(x) = -4x + 21$. Find $f(3m - 9)$.

Interpreting Graphs

Objective 6 Interpret graphs of functions. The ability to read and interpret a graph is a valuable skill because the graph of a function reveals important information about the function itself. To begin, we can determine the domain and range of a function from its graph. Recall that the domain of a function is the set of all possible input values for that function. This can be read from left to right along the x-axis on a graph. The range of a function is the set of all possible output values for the function, and this can be read vertically from the bottom to the top of the graph along the y-axis. The following graph of a function $f(x)$ will help us understand these concepts.

The graph begins at the point $(5, 2)$ and extends upward to the right. We can express the domain of the function in interval notation as $[5, \infty)$, as the set of possible input values begins at $x = 5$ and continues without bound. The range of the function is $[2, \infty)$, as the lowest function value is 2, and the values of the function increase without bound from there.

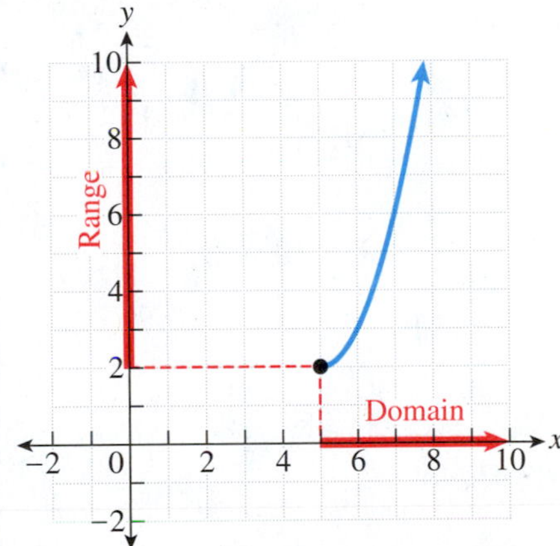

Consider the graph of $f(x)$ shown here. This graph extends indefinitely to the left and to the right, so the domain of the function $f(x)$ is the set of all real numbers $\mathbb{R}$. For the range, this function has a minimum value of -6, as the point $(-2, -6)$ is the lowest point on the graph. The function increases without limit, so the range of the function $f(x)$ is $[-6, \infty)$.

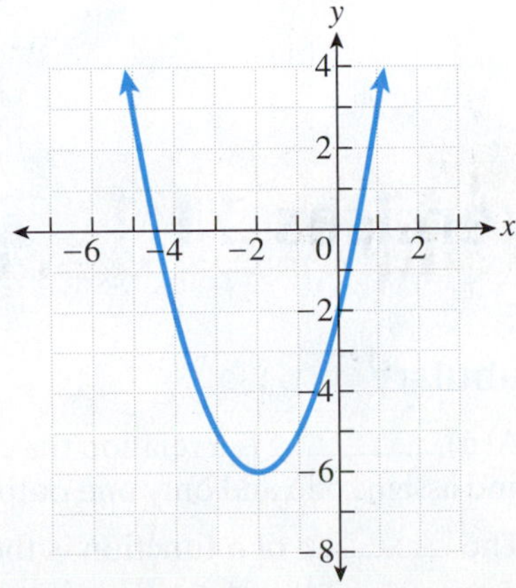

We can use the graph of a function to evaluate the function for a particular input value even when we do not know the formula for the function.

Suppose we wanted to use the graph of a function $f(x)$ to evaluate the function when $x = 6$, or in other words, find $f(6)$. The point on the graph that has an x-coordinate of 6 is the point $(6, 5)$. Because the y-coordinate is 5, we know that $f(6) = 5$.

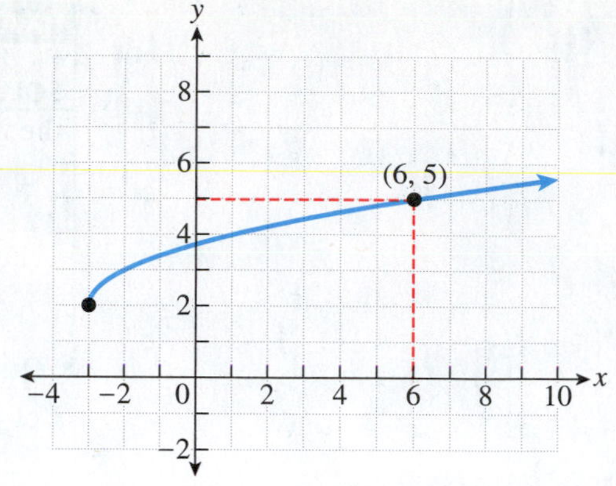

We also can use the graph of a function to determine which input value(s) produce a particular output value.

Suppose we wanted to use the graph of a function $f(x)$ to determine the values of x for which the function $f(x) = -7$. We are looking for a point or points on the graph that have a y-coordinate of -7. There are two such points on this graph: $(-4, -7)$ and $(-6, -7)$. $f(x) = -7$ when $x = -4$ and when $x = -6$.

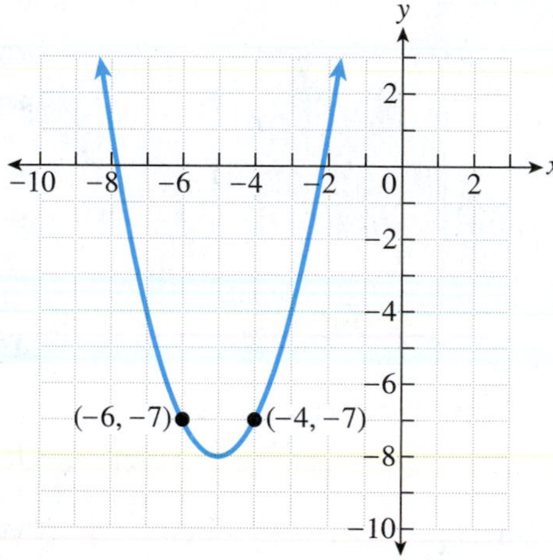

BUILDING YOUR STUDY STRATEGY

Study Environment, 1 Location When you are choosing a place to study, select one that is free from distractions. For example, if you study in a room that has a television on, you will find yourself repeatedly glancing up at the television. This steals time and attention from the task at hand. Working in a room that has people entering and exiting can be distracting as well.

Exercises 11.1

Vocabulary

1. A(n) ________ is a relation that takes one input value and assigns one and only one output value to it.

2. The ________ of a function is the set of all possible input values, and the ________ of a function is the set of all possible output values.

3. Substituting a value for the function's variable and simplifying the resulting expression is called ________ the function.

4. The ________________ is a method that determines whether a graph represents a function.

5. Does a function exist for which the input value is a college graduate and the output value is the starting salary for that graduate's first job? Explain.

6. Does a function exist for which the input value is a woman and the output value is the number of children the woman has? Explain.

7. Does a function exist for which the input value is a person and the output value is the number of jobs the person has held in his or her lifetime? Explain.

8. Does a function exist for which the input value is an orchestra and the output value is the number of players in that orchestra? Explain.

9. Does a function exist for which the input value is a temperature in degrees Fahrenheit and the output value is a city that had that temperature as its high temperature last Tuesday? Explain.

10. Does a function exist for which the input value is the price per share of a stock and the output value is the name of the stock? Explain.

Determine whether the set of ordered pairs is a function. If it is a function, state its domain and range.

11. {(1960, Kennedy), (1964, Johnson), (1968, Nixon), (1972, Nixon), (1976, Carter), (1980, Reagan), (1984, Reagan)}

12. {(2003, Marlins), (2004, Red Sox), (2005, White Sox), (2006, Cardinals), (2007, Red Sox)}

13. $\{(5, -5), (3, -3), (1, -1), (-1, 1), (-3, 3), (-5, 5)\}$

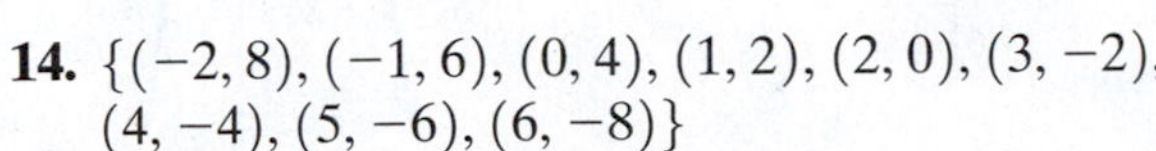

14. $\{(-2, 8), (-1, 6), (0, 4), (1, 2), (2, 0), (3, -2), (4, -4), (5, -6), (6, -8)\}$

15. $\{(-6, 5), (-3, 5), (0, 5), (3, 5), (6, 5)\}$

16. $\{(9, -3), (4, -2), (1, -1), (0, 0), (1, 1), (4, 2), (9, 3)\}$

Use the vertical-line test to determine whether the graph represents a function.

17.

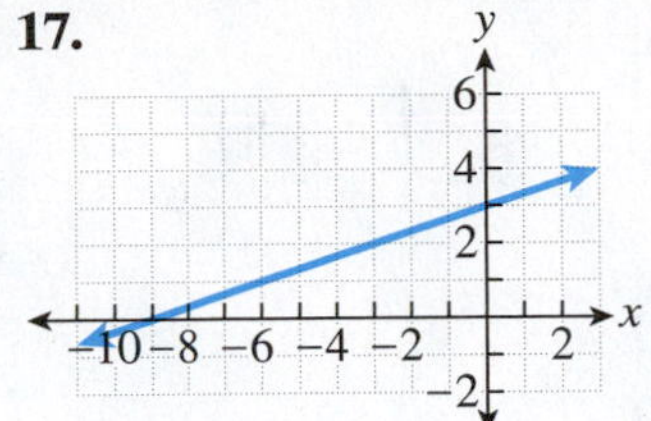

18.

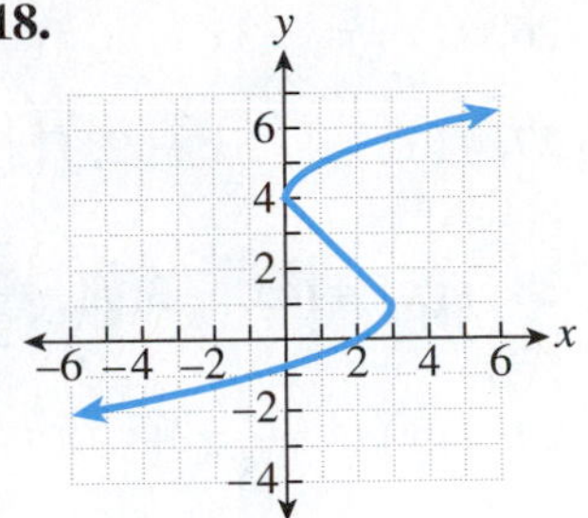

19.

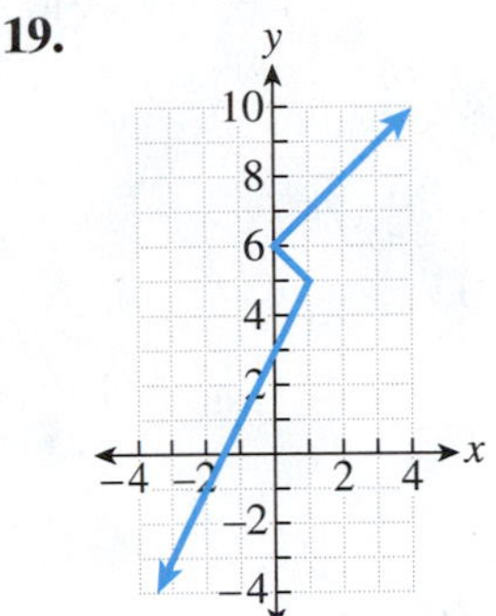

20.

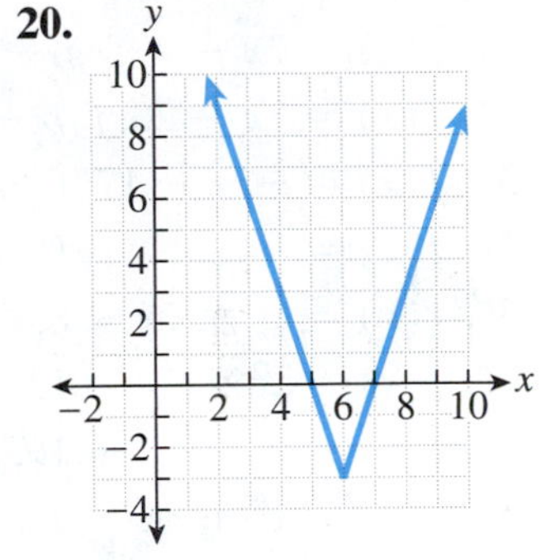

21.

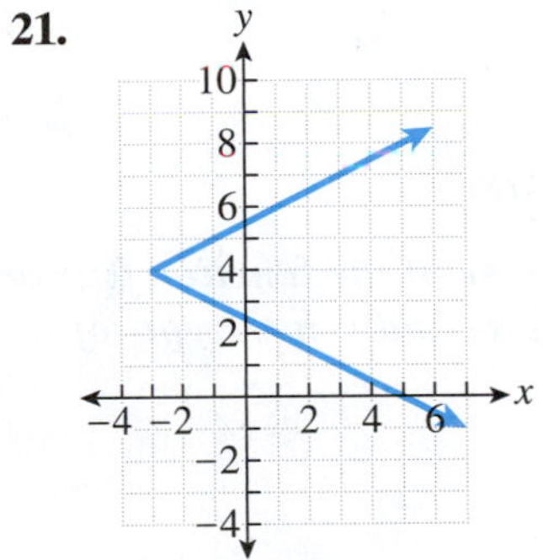

22.

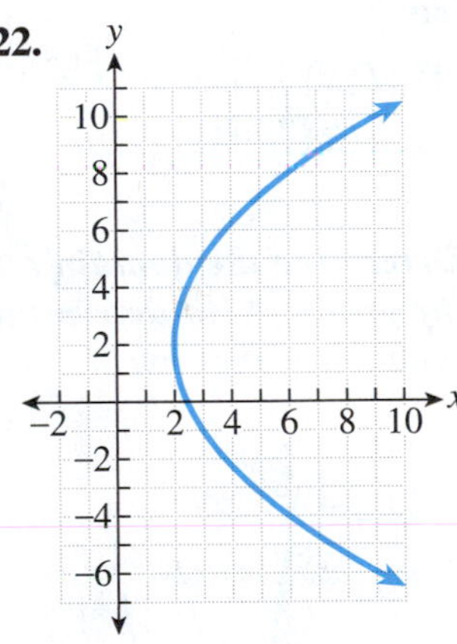

23.

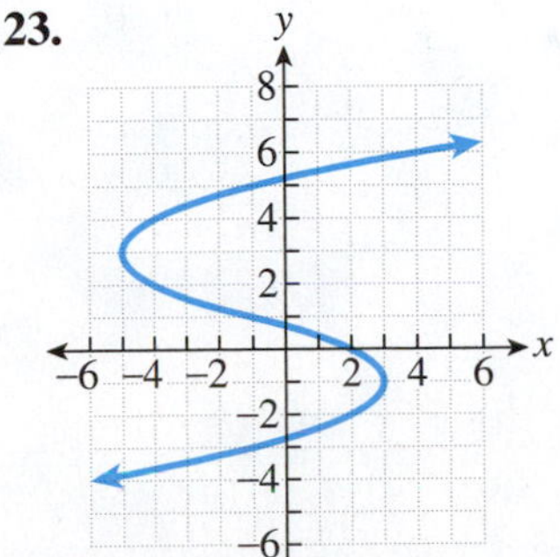

24.

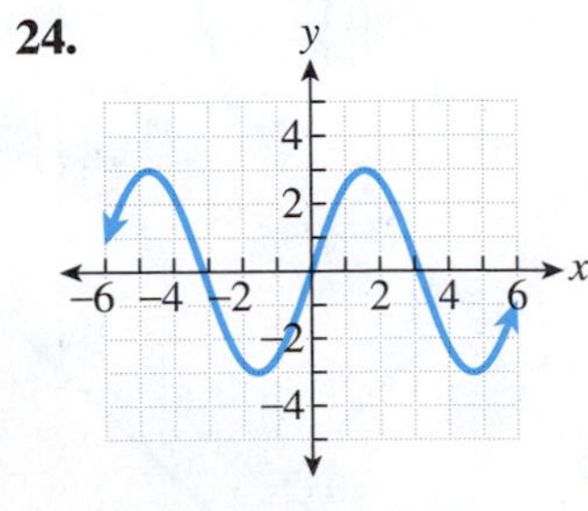

Evaluate the given function.

25. $f(x) = 2x - 9, f(-6)$

26. $f(x) = -3x + 8, f(-17)$

27. $f(x) = -5x - 2, f\left(\frac{3}{5}\right)$

28. $f(x) = 4x - 3, f\left(-\frac{7}{2}\right)$

29. $f(x) = -\frac{2}{3}x + 10, f(9)$

30. $f(x) = \frac{9}{4}x - 8, f(-16)$

31. $f(x) = 7x - 4, f(a + 6)$

32. $f(x) = 3x - 20, f(b - 8)$

33. $f(x) = -2x - 17, f(5n + 2)$

34. $f(x) = -4x + 11, f(3m - 10)$

35. $f(x) = x^2 + 7x - 8, f(5)$

36. $f(x) = x^2 - 3x - 12, f(6)$

37. $f(x) = x^2 - 9x + 16, f(-7)$

38. $f(x) = x^2 + 8x - 23, f(-4)$

39. $f(x) = (x + 6)^2 + 37, f(3)$

40. $f(x) = (x - 4)^2 - 60, f(-9)$

41. $f(x) = |3x - 2| + 17, f(-4)$

42. $f(x) = |4x + 19| - 21, f(-8)$

Determine the domain and range of the function f(x) on the graph. Also determine the x- and y-intercepts of the function if they exist.

43.

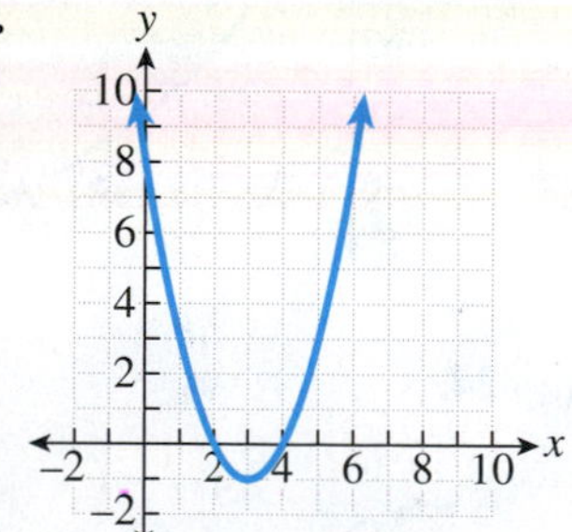

44.

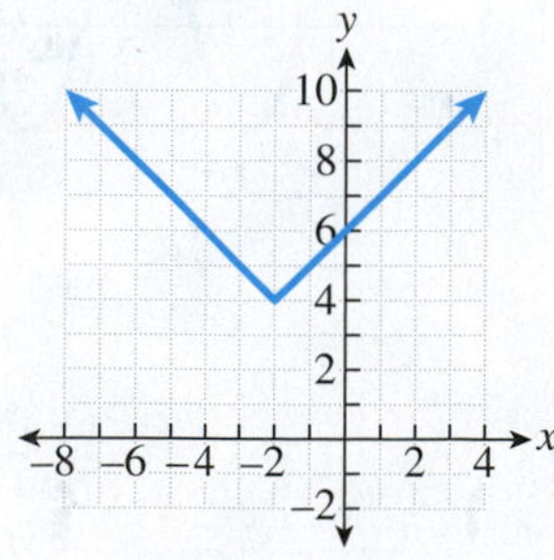

45.

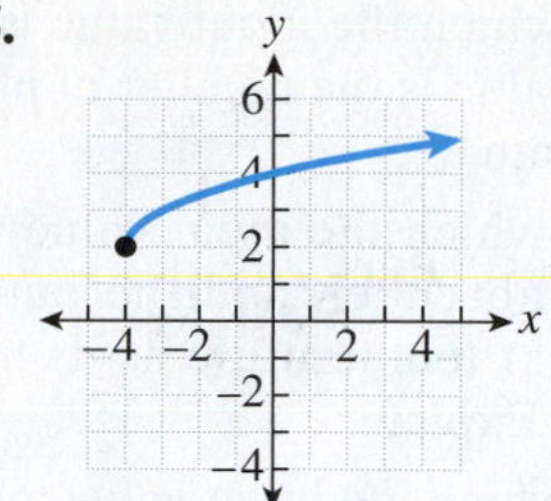

46.

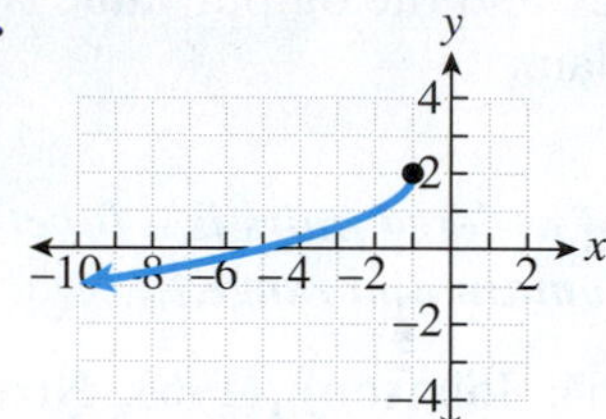

Use the graph of the function f(x) to find the indicated function value.

47. $f(-3)$

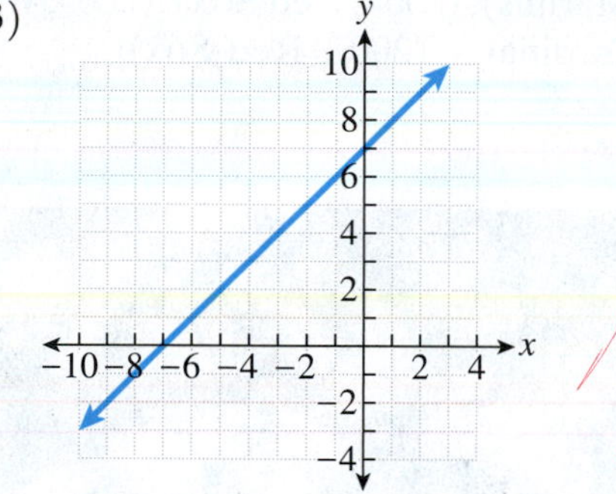

48. $f(4)$

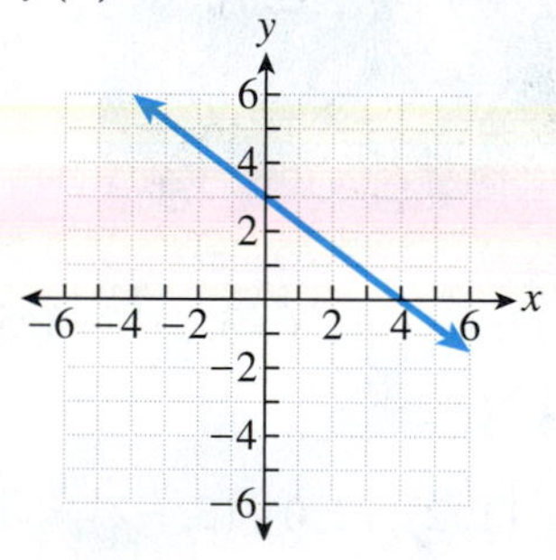

49. $f(-2)$

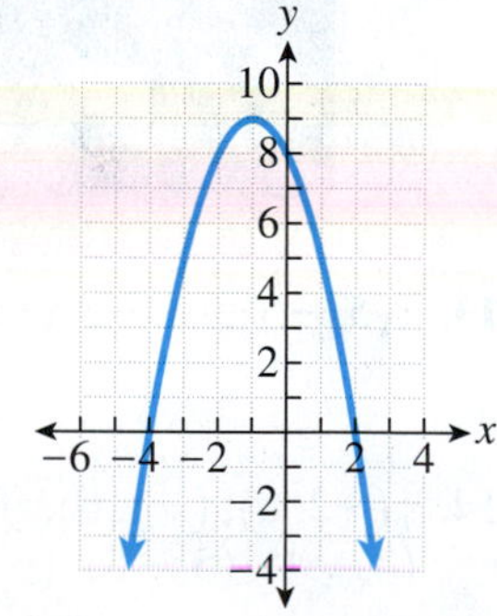

50. $f(4)$

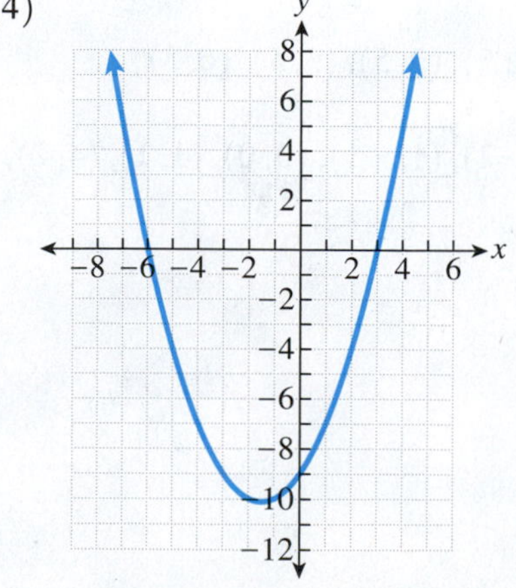

Use the graph of the function f(x) to determine which values of x satisfy the given equation.

51. $f(x) = -4$

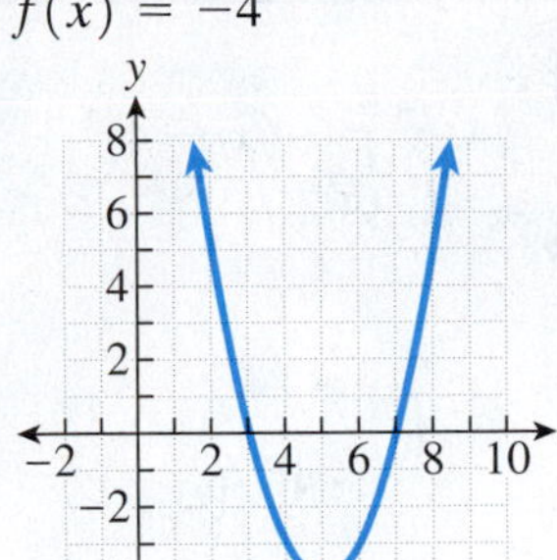

52. $f(x) = -2$

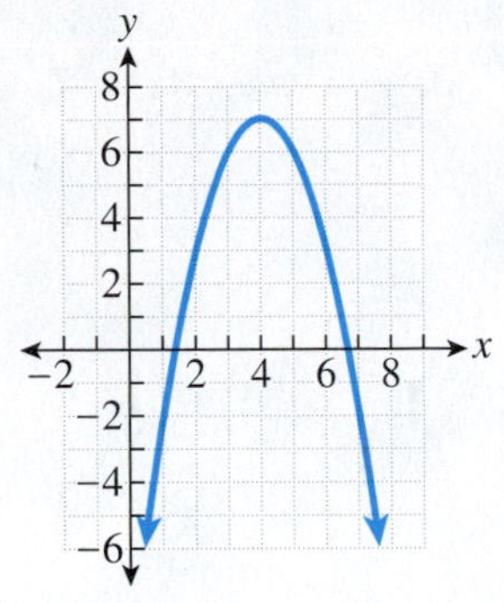

53. $f(x) = -9$

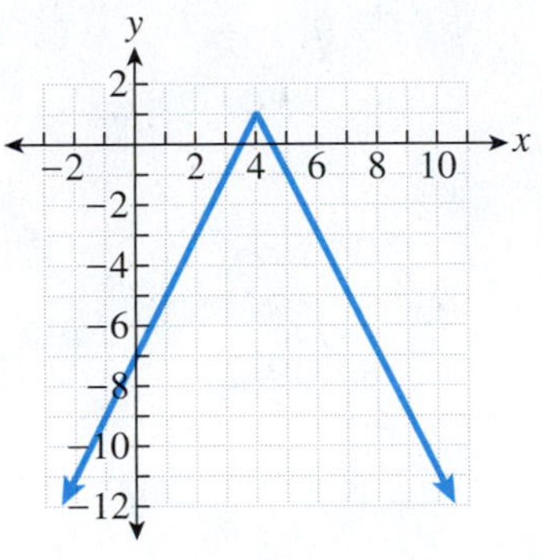

54. $f(x) = -4$

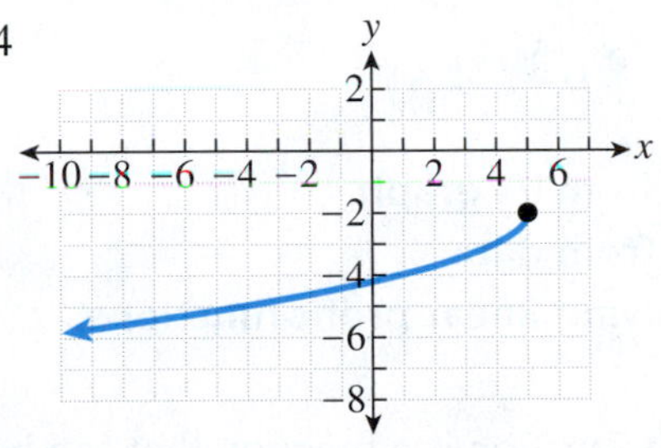

Draw a graph of a function f(x) that meets the given conditions.

55. $f(1) = 2, f(3) = 7, f(5) = 4$, and $f(8) = 6$

56. $f(-5) = -4, f(-2) = 1, f(0) = 3, f(2) = -4$, and $f(5) = -1$

57. Domain: $(-\infty, \infty)$; range: $[-4, \infty)$, $f(1) = -3$

58. Domain: $[-2, \infty)$, range: $[3, \infty)$, $f(2) = 5$

Writing in Mathematics

Answer in complete sentences.

59. If a vertical line can be drawn that intersects a graph at two points, explain why this graph does not represent a function.

60. Explain how to find the domain and range of a function from its graph.

61. Given the graph of a function $f(x)$, explain how to find all values x for which $f(x) = 2$. Use an example to illustrate the process.

62. Given the graph of a function $f(x)$, explain how to find $f(2)$. Use an example to illustrate the process.

Quick Review Exercises

Section 11.1

Graph. Label all intercepts.

1. $y = 3x - 9$

2. $4x - 3y = -24$

3. $y = x^2 - 8x + 7$

4. $y = -(x - 4)^2 + 1$

11.2 Linear Functions

OBJECTIVES

1. **Graph linear functions.**
2. **Determine a linear function from its graph.**
3. **Determine a linear function from data.**
4. **Solve applied problems involving linear profit functions.**

As discussed in Chapter 3, a linear function is a function that can be written in the form $f(x) = mx + b$, where m and b are real numbers.

Graphing Linear Functions

Objective 1 Graph linear functions. Recall that the graph of a linear function is a line and that the two characteristics that determine the graph of a line are its y-intercept and its slope. To graph a linear function, $f(x) = mx + b$, we begin by putting the y-intercept, which is the point $(0, b)$, on the graph. Next, we use the slope m of the line to find additional points on the graph.

EXAMPLE 1 Graph $f(x) = -\frac{2}{3}x + 4$.

SOLUTION We begin by plotting the y-intercept $(0, 4)$ on the graph.

The slope of this line is $m = -\frac{2}{3}$. Beginning at the y-intercept, we can find a second point on the line by moving two units down and three units to the right, which is at the point $(3, 2)$.

Quick Check 1

Graph $f(x) = \frac{3}{4}x - 6$.

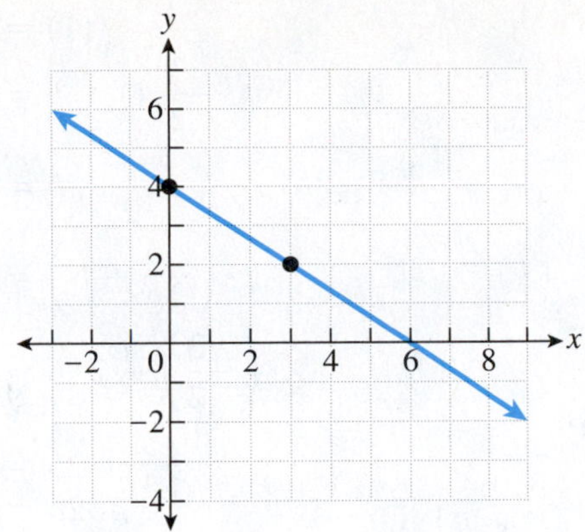

Some functions do not have an x-intercept. One example is a **constant function** of the form $f(x) = b$. The graph of this function is a horizontal line that has a y-intercept at the point $(0, b)$. The reason this type of function is called a constant function is that its output value remains the same, or stays constant, regardless of the input value x.

EXAMPLE 2 Graph $f(x) = 2$.

SOLUTION The graph of this constant function is a horizontal line with a y-intercept at $(0, 2)$.

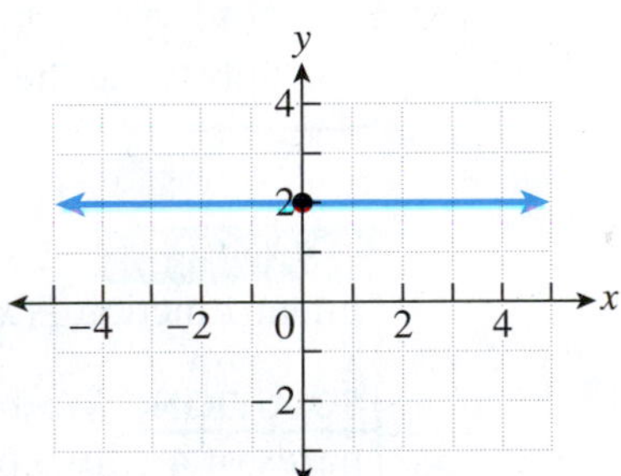

Quick Check 2

Graph $f(x) = -5$.

Finding a Linear Function for a Given Situation

Objective 2 Determine a linear function from its graph. We can determine the linear function $f(x) = mx + b$ from its graph if we can find the coordinates of at least two points that lie on its graph. We begin by finding m. If the line passes through the points (x_1, y_1) and (x_2, y_2), we can find the slope of the line using the formula $m = \frac{y_2 - y_1}{x_2 - x_1}$.

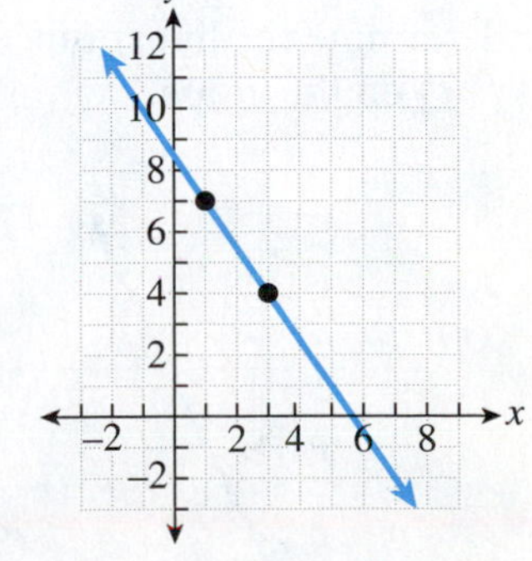

Consider the line passing through the points $(1, 7)$ and $(3, 4)$, as shown in the graph at the right. We can substitute 1 for x_1, 7 for y_1, 3 for x_2, and 4 for y_2 in the formula $m = \frac{y_2 - y_1}{x_2 - x_1}$ to find m.

$m = \frac{4 - 7}{3 - 1}$ Substitute into $m = \frac{y_2 - y_1}{x_2 - x_1}$.

$m = -\frac{3}{2}$ Simplify the numerator and denominator.

Now that we have found the slope, we know that the function is of the form $f(x) = -\frac{3}{2}x + b$. We turn our attention to finding b. We choose one of the points on the line, substitute its coordinates in the function, and solve for b. Because the point $(1, 7)$ is on the graph of the function, we know that $f(1) = 7$.

$$f(1) = 7$$

$$-\frac{3}{2}(1) + b = 7 \quad \text{Substitute 1 for } x \text{ in the function } f(x).$$

$$-\frac{3}{2} + b = 7 \quad \text{Multiply.}$$

$$2\cdot\left(-\frac{3}{2} + b\right) = 2\cdot 7 \quad \text{Multiply both sides of the equation by 2 to clear the equation of fractions.}$$

$$\overset{1}{\cancel{2}}\cdot\left(-\frac{3}{\underset{1}{\cancel{2}}}\right) + 2\cdot b = 2\cdot 7 \quad \text{Distribute and divide out common factors.}$$

$$-3 + 2b = 14 \quad \text{Simplify.}$$

$$2b = 17 \quad \text{Add 3 to both sides to isolate the term containing } b.$$

$$b = \frac{17}{2} \quad \text{Divide both sides by 2.}$$

Replacing b with $\frac{17}{2}$, we find that the function is $f(x) = -\frac{3}{2}x + \frac{17}{2}$.

To Determine a Linear Function from Its Graph

1. Determine the coordinates of two points (x_1, y_1) and (x_2, y_2) on the line.
2. Find the slope m of the line using the formula $m = \frac{y_2 - y_1}{x_2 - x_1}$.
3. Substitute the value found for m in the formula for a linear function, $f(x) = mx + b$. Then set $f(x_1) = y_1$, or $f(x_2) = y_2$, and solve for b.
4. Substitute the value for b in the formula $f(x) = mx + b$.

EXAMPLE 3 Use the given graph to determine the linear function $f(x)$ on the graph.

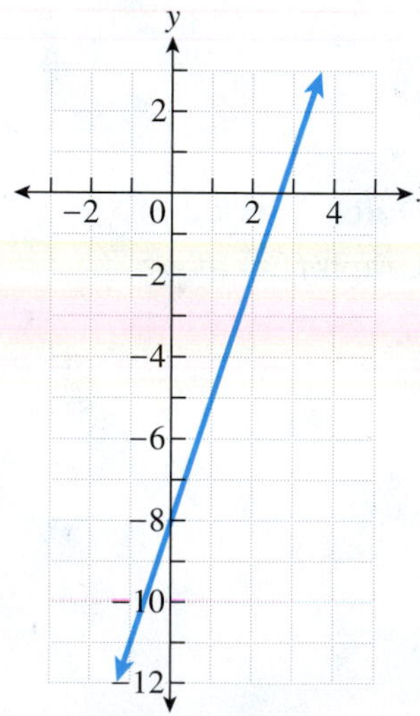

SOLUTION We begin by determining the coordinates of any two points on the line. This line passes through the points $(2, -2)$ and $(3, 1)$. Now we find the slope m of the line.

$$m = \frac{1 - (-2)}{3 - 2} \quad \text{Substitute into } m = \frac{y_2 - y_1}{x_2 - x_1}.$$

$$m = 3 \quad \text{Simplify.}$$

The function is of the form $f(x) = 3x + b$. Because the point $(3, 1)$ is on the graph of the function, we know that $f(3) = 1$. We can use this fact to find b.

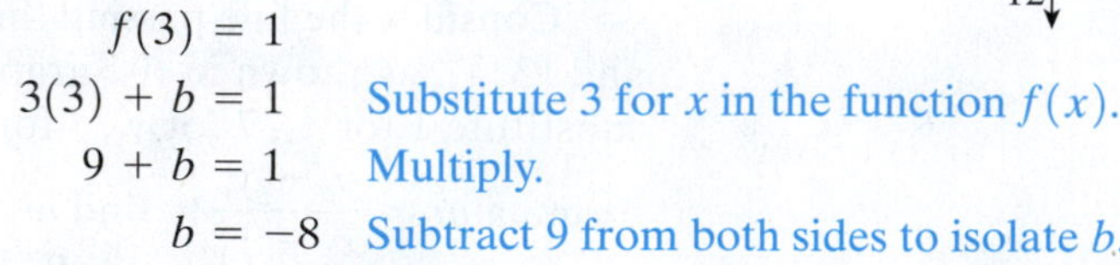

$$f(3) = 1$$

$$3(3) + b = 1 \quad \text{Substitute 3 for } x \text{ in the function } f(x).$$

$$9 + b = 1 \quad \text{Multiply.}$$

$$b = -8 \quad \text{Subtract 9 from both sides to isolate } b.$$

Replacing b with -8, we find that the function is $f(x) = 3x - 8$.

Quick Check 3

Use the given graph to determine the linear function $f(x)$ on the graph.

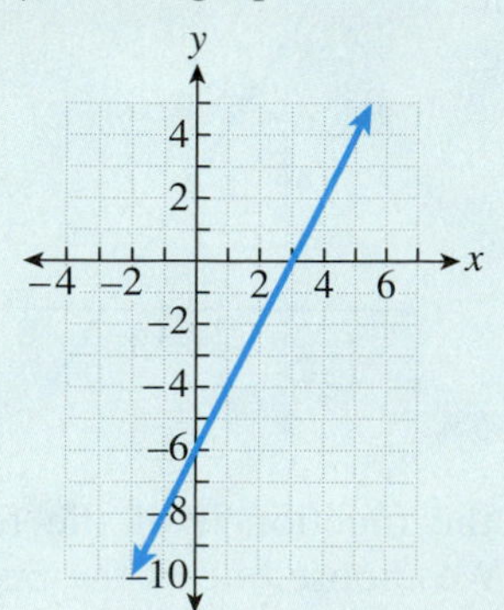

Finding a Linear Function from Data

Objective 3 **Determine a linear function from data.**

EXAMPLE 4 The freezing point of water is 0°C on the Celsius scale and 32°F on the Fahrenheit scale. The boiling point of water is 100°C on the Celsius scale and 212°F on the Fahrenheit scale. Find a linear function $F(x) = mx + b$ whose input is a Celsius temperature and whose output is the corresponding Fahrenheit temperature.

SOLUTION Because a Celsius temperature of 0°C corresponds to a Fahrenheit temperature of 32°F, we know that $F(0) = 32$. In a similar fashion, we know that $F(100) = 212$. We can find the slope of this linear function by using $(0, 32)$ for (x_1, y_1) and $(100, 212)$ for (x_2, y_2).

$$m = \frac{212 - 32}{100 - 0} \quad \text{Substitute.}$$

$$m = \frac{9}{5} \quad \text{Simplify the fraction to lowest terms.}$$

The function is of the form $F(x) = \frac{9}{5}x + b$. We will now use the fact that $F(0) = 32$ to find b.

$$F(0) = 32$$

$$\frac{9}{5}(0) + b = 32 \quad \text{Substitute 0 for } x \text{ in the function } F(x).$$

$$b = 32 \quad \text{Simplify.}$$

The function that converts a Celsius temperature x to its corresponding Fahrenheit temperature is $F(x) = \frac{9}{5}x + 32$.

Quick Check 4

Mary held her wedding reception at a country club. There were 150 guests at the reception, and Mary was charged $3500. Juliet held her wedding reception at the same country club and was charged $2100 for 80 guests. Find a linear function $f(x) = mx + b$ whose input is the number of guests at a reception and whose output is the corresponding charge by the country club.

Applications of Linear Functions

Objective 4 **Solve applied problems involving linear profit functions.** A specific business application that can involve linear functions is determining the cost to manufacture x units of a product. Functions also can be used to determine the revenue that will be generated by selling those x units as well as the profit that will result from their sale.

Suppose a company is manufacturing items. Such a company has fixed costs (rent, utilities, and so on.) as well as variable costs that depend on the number of items manufactured (materials, payroll, and so on.). The **cost function** $C(x)$ is equal to the fixed costs plus the variable costs associated with producing x items. The **revenue function** $R(x)$ tells us how much money will be earned by selling these x items. The revenue function is equal to the product of the selling price of each item and the number of those particular items sold. Finally, the **profit function** $P(x)$ is equal to the difference of the revenue function and the cost function; that is, $P(x) = R(x) - C(x)$ because profit is how much revenue is left over after costs are paid.

EXAMPLE 5 A student club decides to sell burgers on campus to raise money. The club must pay the college a $25 fee to reserve space for a booth. In addition, it must spend another $40 for condiments and paper goods. The club determines that it will cost $0.55 for ingredients to make each burger, and it plans to sell the burgers for $3 each.

a) Find the cost function $C(x)$, the revenue function $R(x)$, and the profit function $P(x)$.

SOLUTION The cost function is equal to the fixed costs plus the cost to make x burgers. The fixed costs are \$65 (\$25 for booth space and \$40 for condiments and paper goods). Each burger costs \$0.55 for ingredients, so it costs \$0.55x for the ingredients in x burgers. The cost function is $C(x) = 65 + 0.55x$.

Because each burger is sold for \$3, the revenue function for selling x burgers is $R(x) = 3x$. The profit function is equal to the difference of the revenue function and the cost function, $P(x) = R(x) - C(x)$. In this example, $P(x) = 3x - (65 + 0.55x)$, which simplifies to $P(x) = 2.45x - 65$.

b) How much profit will be generated if the club makes and sells 50 burgers?

SOLUTION To determine the profit, evaluate the profit function $P(x) = 2.45x - 65$ when $x = 50$.

$$\begin{aligned} P(50) &= 2.45(50) - 65 && \text{Substitute 50 for } x. \\ &= 57.5 && \text{Simplify.} \end{aligned}$$

Making and selling 50 burgers will generate a profit of \$57.50.

c) How many burgers does the club need to make and sell to break even?

SOLUTION The club breaks even when the profit is 0.

$$\begin{aligned} P(x) &= 0 && \text{Set } P(x) \text{ equal to 0.} \\ 2.45x - 65 &= 0 && \text{Replace } P(x) \text{ with its formula.} \\ 2.45x &= 65 && \text{Add 65 to both sides.} \\ x &\approx 26.5 && \text{Divide both sides by 2.45.} \end{aligned}$$

Because the club cannot break even until it sells 26.5 burgers, it needs to make and sell 27 burgers to reach this point.

d) If the goal of the club is to raise \$500, how many burgers does it need to make and sell?

SOLUTION Begin by setting the profit function equal to \$500 and solving for x.

$$\begin{aligned} P(x) &= 500 && \text{Set } P(x) \text{ equal to 500.} \\ 2.45x - 65 &= 500 && \text{Replace } P(x) \text{ with its formula.} \\ 2.45x &= 565 && \text{Add 65 to both sides.} \\ x &\approx 230.6 && \text{Divide both sides by 2.45.} \end{aligned}$$

This solution needs to be a whole number, as the club cannot make 230.6 burgers. Making and selling only 230 burgers would result in a profit of only $2.45(230) - 65$, or \$498.50. The club needs to make and sell 231 burgers to raise at least \$500.

Quick Check 5

Ross decides to sell calendars that feature the pictures of 12 renowned mathematicians. To do this, he must pay \$50 for a business license. Each calendar costs Ross \$1.25 to produce, and he plans to sell them for \$9.95 each.

a) Find the cost function $C(x)$, the revenue function $R(x)$, and the profit function $P(x)$.

b) How many calendars will Ross have to produce and sell to make a \$1000 profit?

BUILDING YOUR STUDY STRATEGY

Study Environment, 2 Noise Noise can be a major distraction while you are studying. Some students need total silence while they are working; if you are one of those students, make sure you select a location that is free of noise. This is not easy to do in a public location such as a library, but you may be able to find a room or cubicle that is devoted to silent studying.

The amount of noise you can handle may depend on the type of studying you are doing. Music and other noises may provide a pleasant background while you are reviewing, but can be distracting while you are learning new material.

Exercises 11.2

Vocabulary

1. A(n) ______ function is a function of the form $f(x) = mx + b$.
2. The ______ function describes how much must be spent to produce x items.
3. The ________ function describes how much money is earned from selling x items.
4. The ______ function is equal to the difference between the revenue function and the cost function.

Graph each function.

5. $f(x) = x - 7$
6. $f(x) = x + 4$
7. $f(x) = -x + 3$
8. $f(x) = -x - 5$
9. $f(x) = 2x + 4$
10. $f(x) = 3x - 9$
11. $f(x) = -3x + 3$
12. $f(x) = -4x - 8$
13. $g(x) = 4x - 6$
14. $g(x) = -6x + 4$

15. $f(x) = \frac{2}{3}x - 4$

16. $f(x) = -\frac{1}{5}x - 1$

17. $f(x) = -\frac{3}{4}x + 6$

18. $f(x) = \frac{2}{5}x - 2$

19. $f(x) = \frac{3}{7}x$

20. $f(x) = -\frac{5}{8}x$

21. $f(x) = 7$

22. $f(x) = -2$

Use the graph to determine the linear function f(x) on the graph.

23.

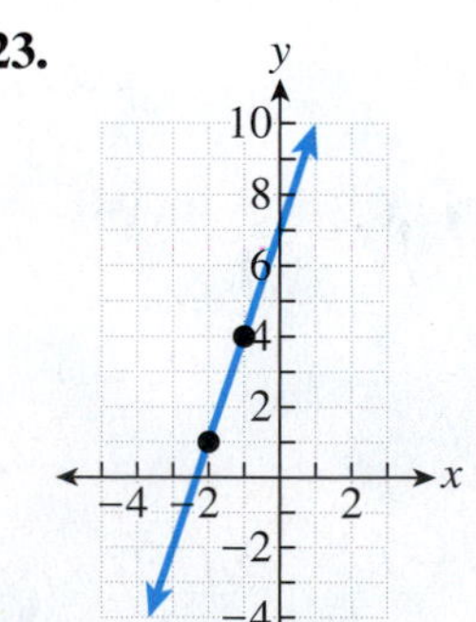

24.

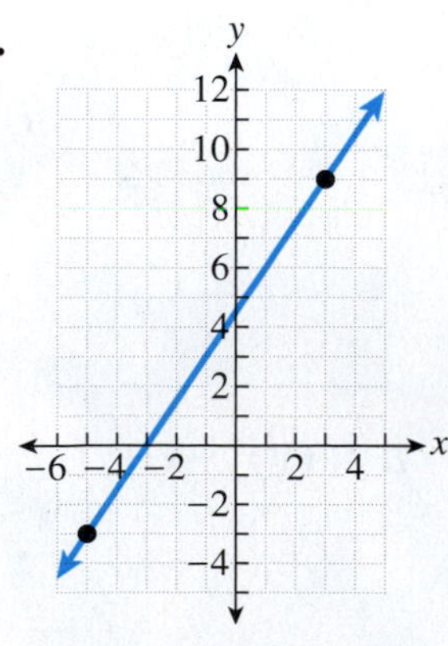

25.

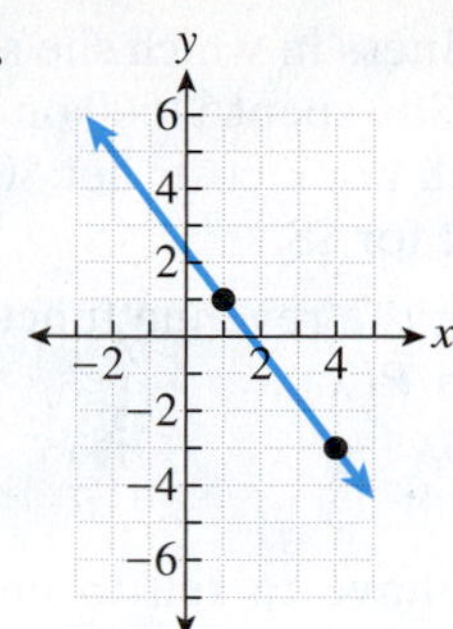

26.

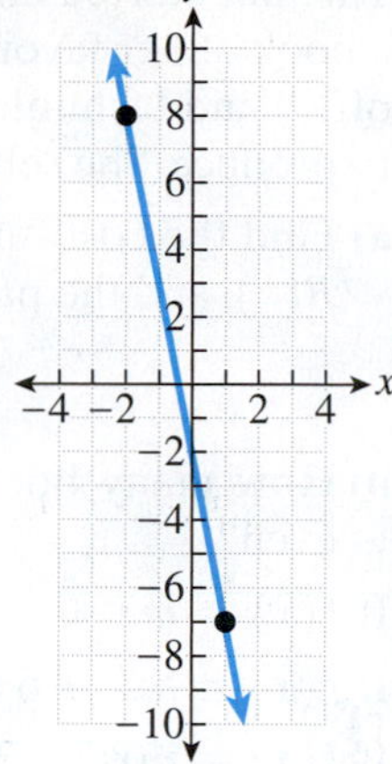

27.

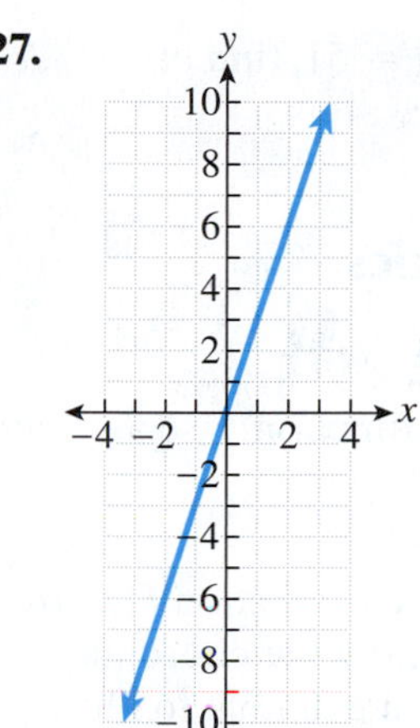

28.

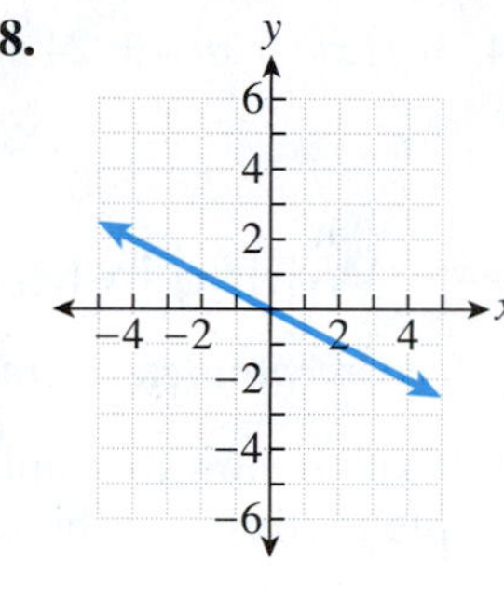

29.

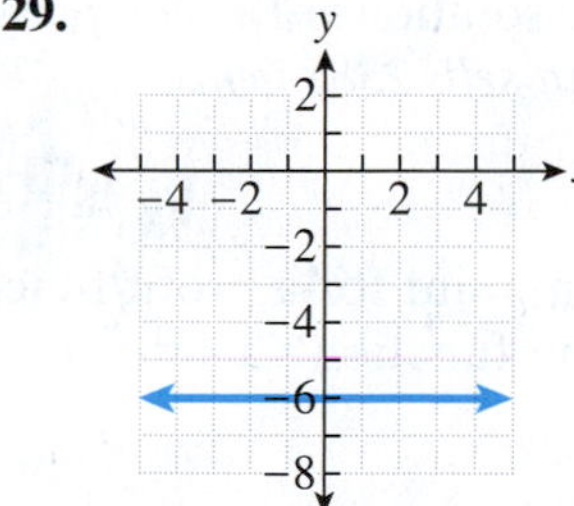

30.

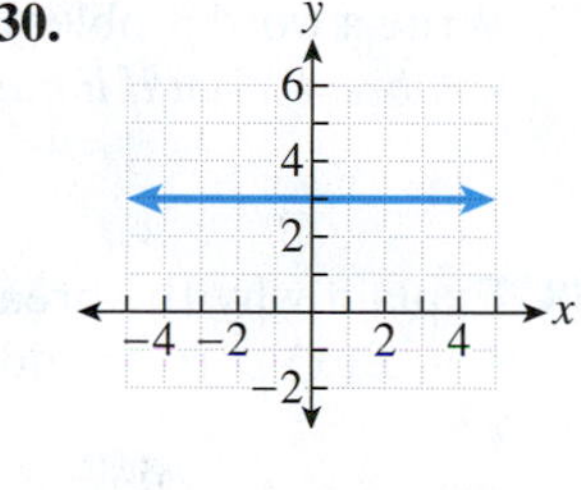

31. John's telephone plan charges him a connection fee for each call, in addition to charging him per minute. A 10-minute phone call costs a total of $0.69, while a 30-minute phone call costs $1.29.

a) Find a linear function $f(x) = mx + b$ whose input is the length of a call in minutes and whose output is the corresponding charge for that call.

b) In your own words, explain what the slope and y-intercept of this function represent.

c) Use the function from part a to determine the cost of a phone call that lasts 44 minutes.

32. A community college charges each of its students a registration fee. Students also pay a tuition fee for each academic unit they take. Last semester Amadou took 12 units and paid a total of $855, while Jia Li took 15 units and paid a total of $1050.

a) Find a linear function $f(x) = mx + b$ whose input is the number of units taken by a student and whose output is the total fees paid.

b) In your own words, explain what the slope and y-intercept of this function represent.

c) Use the function from part a to determine the total fees paid by a student taking 18 units.

33. A car rental company charges a base fee for each car rented, in addition to a fee for each mile driven. If a person drives a car for 100 miles, the total charge is $35. If a person drives a car for 150 miles, the total charge is $42.50.

a) Find a linear function $f(x) = mx + b$ whose input is the number of miles driven and whose output is the corresponding total charge for renting the car.

b) In your own words, explain what the slope and y-intercept of this function represent.

c) Use the function from part a to determine the cost of renting a car and driving it 260 miles.

34. A certain county charges a flat fine for speeding, in addition to a fee for each mile per hour a driver goes over the speed limit. A driver who was going 63 mph in a 55 mph zone was fined $159, while another driver who was traveling 58 mph in a 40 mph zone was fined $249.

a) Find a linear function $f(x) = mx + b$ whose input is the number of miles per hour over the speed limit that a person was driving and whose output is the fine that must be paid.

b) Use the function from part a to determine the fine for a person who was driving 56 mph in a 45 mph zone.

35. An elementary school booster club is wrapping gifts at a local mall as a fund-raiser. In addition to paying $50 to rent space for the day, the booster club also had to spend $75 on materials. It decides to charge $4 to wrap each gift.

a) Find the cost function $C(x)$, the revenue function $R(x)$, and the profit function $P(x)$.

b) How much profit will be generated if the booster club wraps 75 gifts?

36. Members of a sorority have set up a car wash to raise money for a scholarship fund. They paid a gas station $20 to use the parking lot and water. They also paid $10 for soap and buckets. They are charging $5 per car.

a) Find the cost function $C(x)$, the revenue function $R(x)$, and the profit function $P(x)$.

b) How much profit will be generated if they wash 185 cars?

37. The art club at a community college is selling burgers at lunchtime as a fund-raiser. The club spent \$12 for condiments and supplies. The cost for the meat and bun for each burger is \$0.40, and the club is charging \$3 for each burger.

a) Find the cost function $C(x)$, the revenue function $R(x)$, and the profit function $P(x)$.

b) How much profit will be generated if the art club sells 60 burgers?

38. A student has started a business selling her homemade fruit jams at a local farmer's market. She spent \$25 to rent a space at the market. Each jar of jam costs her \$3.25 to make, and she sells each jar for \$6.

a) Find the cost function $C(x)$, the revenue function $R(x)$, and the profit function $P(x)$.

b) How much profit will be generated if she sells 30 jars of jam?

39. Marge has started a business selling gift baskets out of her home. She spent \$1200 remodeling her home to accommodate the business. Each basket costs her \$7.50 to make, and she sells them for \$29.95 each. Customers pay all shipping fees.

a) Find the cost function $C(x)$, the revenue function $R(x)$, and the profit function $P(x)$.

b) How many gift baskets will she have to make and sell to generate a profit of at least \$10,000?

40. Tina has started an online business in which she sells a book of her favorite recipes. She spent \$1500 on her office and technology, and each book costs her \$0.50 to produce. She sells each book for \$8.

a) Find the cost function $C(x)$, the revenue function $R(x)$, and the profit function $P(x)$.

b) How many books will she have to sell to break even?

41. If $f(x) = mx + 7$ and $f(5) = 22$, find m.

42. If $f(x) = mx - 6$ and $f(8) = 34$, find m.

43. If $f(x) = mx - 13$ and $f(-6) = -25$, find m.

44. If $f(x) = mx + 24$ and $f(-9) = 51$, find m.

Writing in Mathematics

Answer in complete sentences.

45. Explain how to graph a linear function. Use an example to illustrate the process.

46. Explain how to determine a linear function $f(x)$ from its graph, provided that you know the coordinates of two points on the graph. Use an example to illustrate the process.

47. Write a word problem whose solution is *The company will break even if it makes and sells 2385 items.*

48. Explain why the break-even point for a profit function $P(x)$ corresponds to the function's x-intercept.

11.3 Quadratic Functions

OBJECTIVES

1. **Graph quadratic functions of the form $f(x) = a(x - h)^2 + k$ by shifting.**
2. **Find the maximum or minimum value of a quadratic function.**
3. **Solve applied maximum–minimum problems.**
4. **Simplify difference quotients for quadratic functions.**

Objective 1 Graph quadratic functions of the form $f(x) = a(x - h)^2 + k$ by shifting. Before we learn to graph a quadratic function of the form $f(x) = a(x - h)^2 + k$ by shifting, let's recall a few facts from Chapter 10 about this type of function.

Graph of $f(x) = a(x - h)^2 + k$

- The graph is a U-shaped parabola.
- The parabola opens upward if $a > 0$, and it opens downward if $a < 0$.
- The vertex, or turning point, of the parabola is the point (h, k).
- The axis of symmetry is $x = h$.
- The y-coordinate of the y-intercept of the parabola is equal to $f(0)$.
- If the parabola has any x-intercepts, they can be found by setting $f(x) = 0$ and solving for x.
- The domain of a quadratic function is the set of all real numbers: $(-\infty, \infty)$. The range can be determined from the graph.

The basic quadratic function $f(x) = x^2$ has its vertex at the origin. Here is its graph.

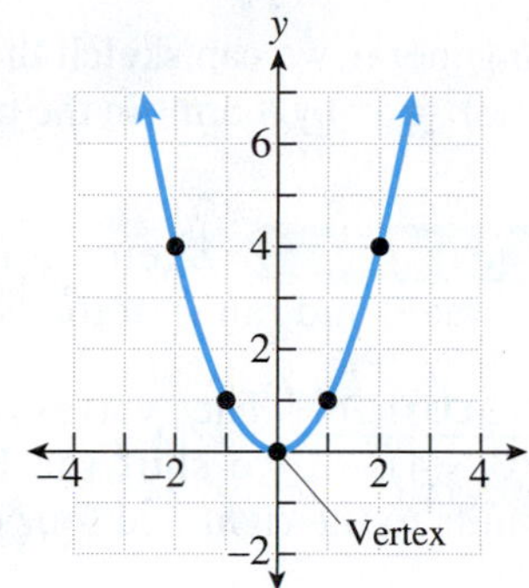

x	$f(x)$
−2	4
−1	1
0	0
1	1
2	4

Other quadratic functions can be graphed by **shifting** or **translating** this basic graph.

Vertical Translations

The function $f(x) = x^2 + k$ can be graphed by shifting the parabola $f(x) = x^2$ vertically by k units. The vertex moves from $(0, 0)$ to $(0, k)$.

$f(x) = x^2 + 3$
Shift the basic parabola three units upward.

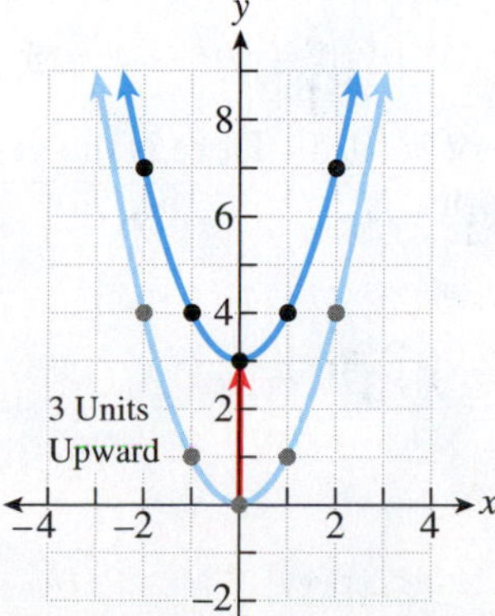

$f(x) = x^2 - 2$
Shift the basic parabola two units downward.

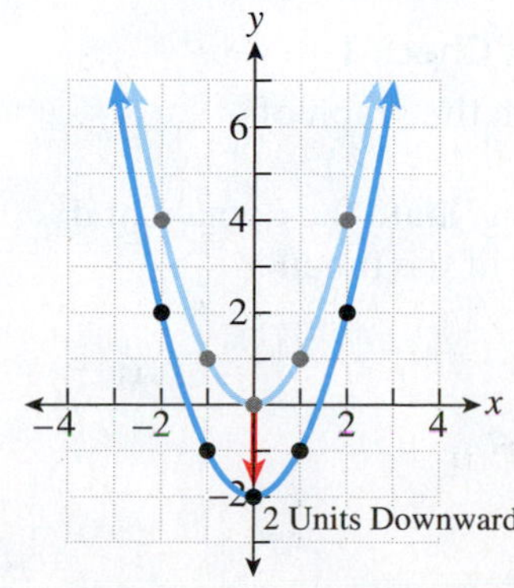

Horizontal Translations

The function $f(x) = (x - h)^2$ can be graphed by shifting the parabola $f(x) = x^2$ horizontally by h units. The vertex moves from $(0, 0)$ to $(h, 0)$.

$f(x) = (x - 1)^2$
Shift the basic parabola one unit to the right.

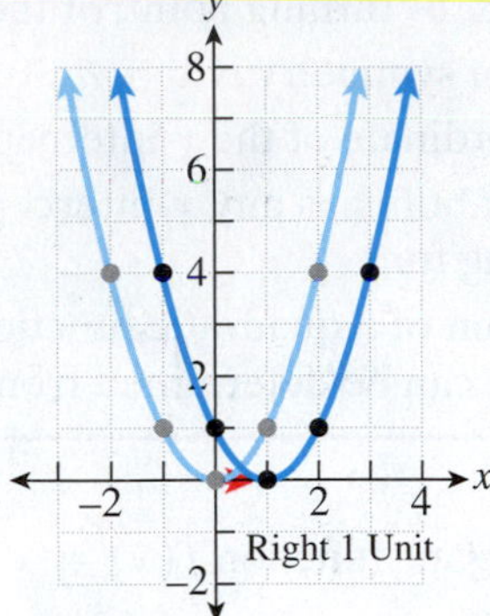

$f(x) = (x + 4)^2$
Shift the basic parabola four units to the left.

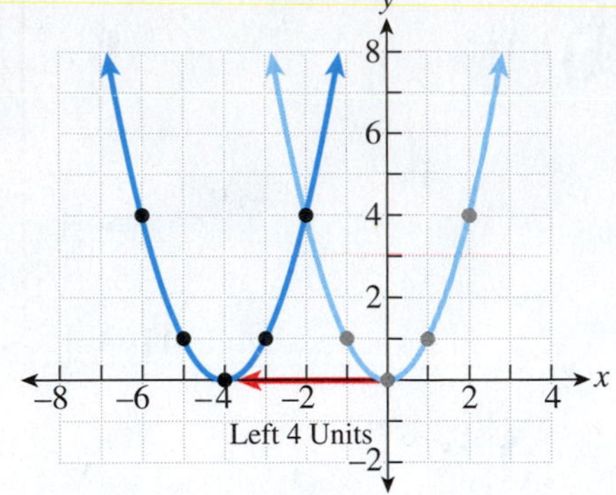

In general, we can sketch the graph of $f(x) = (x - h)^2 + k$ by shifting the graph of $f(x) = x^2$ by h units in the horizontal direction and by k units in the vertical direction.

EXAMPLE 1 Sketch the graph of $f(x) = (x - 3)^2 - 4$ by shifting. State the domain and range of the function.

SOLUTION The vertex of this parabola is $(3, -4)$, so we shift the basic parabola three units to the right and four units down.

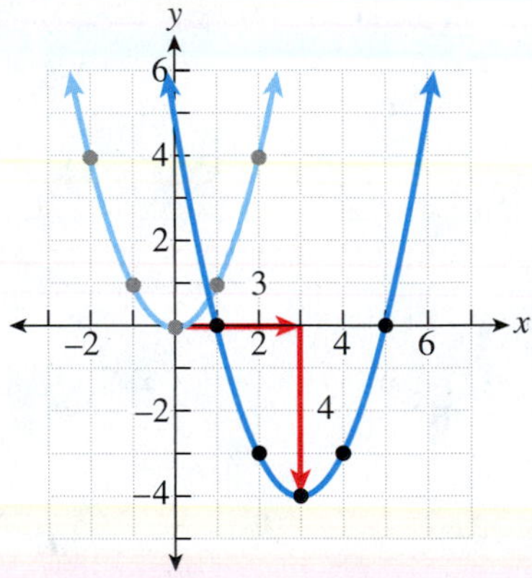

We can add more detail to the graph by adding its intercepts. Notice that by shifting, we already know that the x-intercepts are $(1, 0)$ and $(5, 0)$. We can determine the y-intercept by finding $f(0)$.

$$f(0) = (0 - 3)^2 - 4 = 5 \quad \text{Substitute 0 for } x \text{ and simplify.}$$

The y-intercept is $(0, 5)$. Here is the graph, including all intercepts.

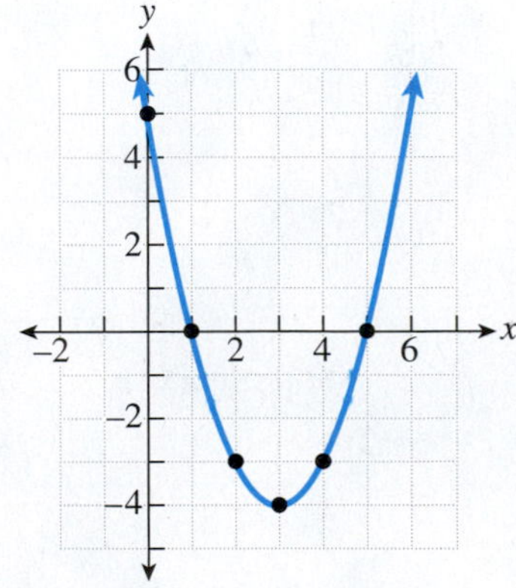

The domain of the function is $(-\infty, \infty)$. By looking at the graph, we see that the lowest value of this function is -4. So the range of the function is $[-4, \infty)$.

Quick Check 1

Sketch the graph of $f(x) = (x + 2)^2 - 1$ by shifting. State the domain and range of the function.

EXAMPLE 2 Sketch the graph of $f(x) = (x + 2)^2 + 5$ by shifting. State the domain and range of the function.

SOLUTION The vertex of this parabola is $(-2, 5)$, so we shift the basic parabola two units to the left and five units up.

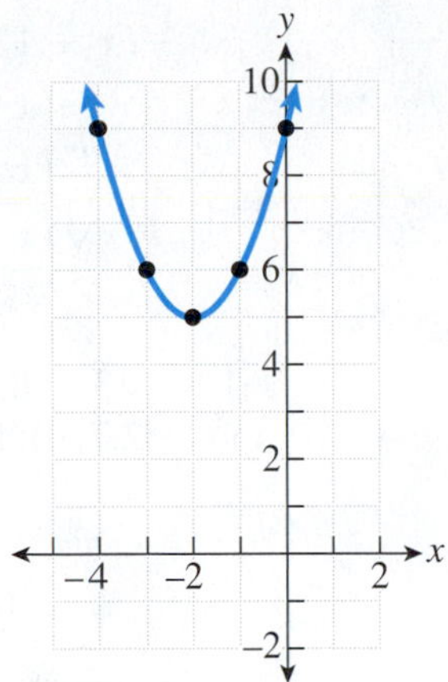

We can determine that the y-intercept is $(0, 9)$ by finding $f(0)$.

$$f(0) = (0 + 2)^2 + 5 = 4 + 5 = 9$$

Also, because the vertex is above the x-axis and the parabola opens upward, there are no x-intercepts. The domain of the function is $(-\infty, \infty)$. The lowest value of this function is 5, so the range of the function is $[5, \infty)$.

Quick Check 2

Sketch the graph of $f(x) = (x - 1)^2 + 4$ by shifting. State the domain and range of the function.

We know that the graph of $f(x) = a(x - h)^2 + k$ opens downward when $a < 0$. To sketch the graph of a parabola in this situation, we begin by rotating the graph of $f(x) = x^2$ about the x-axis as shown. Following this rotation, we can shift the graph horizontally and/or vertically, depending on the values of h and k.

$f(x) = x^2$

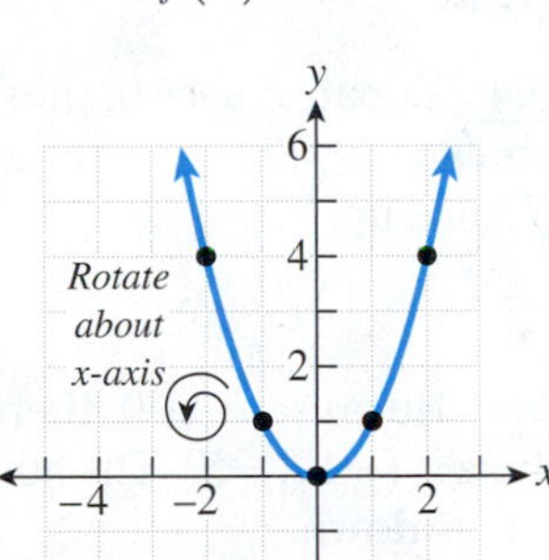

$f(x) = -x^2$

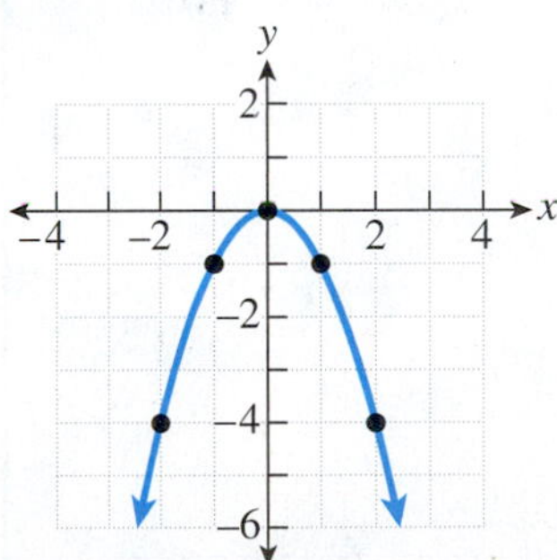

EXAMPLE 3 Sketch the graph of $f(x) = -(x + 1)^2 + 3$ by shifting. State the domain and range of the function.

SOLUTION Because $a < 0$, we know that this parabola opens downward. We begin by rotating the parabola $f(x) = x^2$ about the x-axis. The vertex of this parabola is $(-1, 3)$, so we now shift the parabola one unit to the left and three units up.

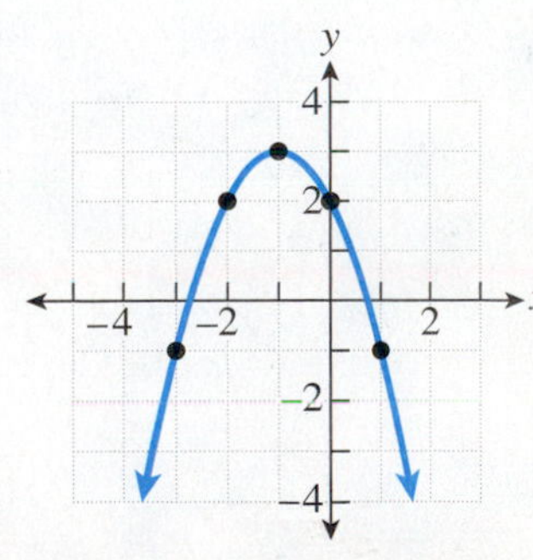

We can add more detail to the graph by adding the x- and y-intercepts. To find the y-intercept, find $f(0)$.

$$f(0) = -(0 + 1)^2 + 3 = -1 + 3 = 2$$

We can find the x-intercepts by setting the function equal to 0 and solving for x.

$-(x + 1)^2 + 3 = 0$	Set the function equal to 0.
$-(x + 1)^2 = -3$	Subtract 3 from both sides.
$(x + 1)^2 = 3$	Divide both sides by -1.
$\sqrt{(x + 1)^2} = \pm\sqrt{3}$	Take the square root of each side.
$x = -1 \pm \sqrt{3}$	Simplify and solve for x by subtracting 1.

$-1 + \sqrt{3} \approx 0.7$ and $-1-\sqrt{3} \approx -2.7$, so the x-intercepts can be plotted at $(0.7, 0)$ and $(-2.7, 0)$. Here is the graph, including all intercepts.

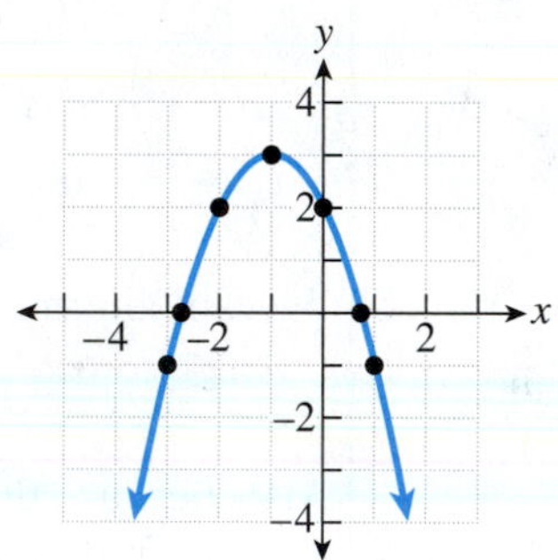

The domain of the function is $(-\infty, \infty)$. The highest value of this function is 3, so the range of the function is $(-\infty, 3]$.

Quick Check 3

Sketch the graph of $f(x) = -(x - 2)^2 + 4$ by shifting. State the domain and range of the function.

EXAMPLE 4 State the rotation, horizontal shift, and vertical shift used to graph the given function.

a) $f(x) = (x - 9)^2 + 17$

SOLUTION Because $a > 0$, there is no rotation about the x-axis. The vertex for this parabola is $(9, 17)$. The horizontal shift is 9 units to the right, and the vertical shift is 17 units up.

b) $f(x) = (x + 12)^2 - 5$

SOLUTION Because $a > 0$, there is no rotation about the x-axis. The vertex for this parabola is $(-12, -5)$. The horizontal shift is 12 units to the left, and the vertical shift is 5 units down.

c) $f(x) = -(x + 8)^2 + 21$

SOLUTION Because $a < 0$, there is a rotation about the x-axis. The vertex for the rotated parabola is $(-8, 21)$. The horizontal shift is 8 units to the left, and the vertical shift is 21 units up.

Quick Check 4

State the rotation, horizontal shift, and vertical shift used to graph the given function.

a) $f(x) = (x + 13)^2 - 7$

b) $f(x) = -(x - 4)^2 + 30$

Minimum and Maximum Values of Quadratic Functions

Objective 2 **Find the maximum or minimum value of a quadratic function.**

Maximum and Minimum Values of a Function

The smallest possible output of a function is called the **minimum value** of the function. The greatest possible output of a function is called the **maximum value** of the function.

Each parabola that opens upward has a minimum value at its vertex. It has no maximum value.

To find the minimum value of a quadratic function whose graph is a parabola that opens upward, we need to find the y-coordinate of its vertex. For example, we can see that the following parabola does not go below the line $y = -4$, which is the y-coordinate of the vertex. The minimum value of this parabola is -4.

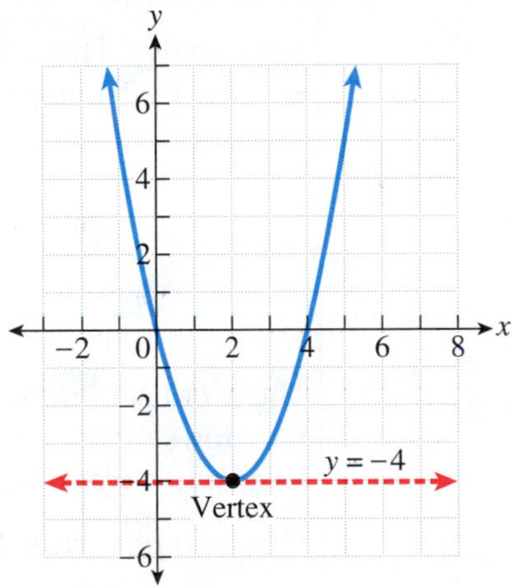

Recall that for quadratic functions of the form $f(x) = ax^2 + bx + c$, we can find the x-coordinate of the vertex using the formula $x = -\frac{b}{2a}$. To find the y-coordinate of the vertex, we evaluate the function for this value of x. If $a > 0$, this y-coordinate will be the minimum value of the function.

For quadratic functions of the form $f(x) = a(x - h)^2 + k$, the vertex is (h, k). If $a > 0$, the minimum value of the function is k.

EXAMPLE 5 Find the minimum value of the function $f(x) = 3x^2 - 12x + 17$.

SOLUTION Because the graph of this function is a parabola that opens upward ($a > 0$), the function has a minimum value at its vertex. To find the vertex, we begin by finding the x-coordinate using the formula $x = \frac{-b}{2a}$.

$$x = \frac{-(-12)}{2(3)} = 2 \quad \text{Substitute 3 for } a \text{ and } -12 \text{ for } b \text{ and simplify.}$$

The minimum value of this function occurs when $x = 2$. To find the minimum value, we evaluate the function when $x = 2$.

$$\begin{aligned} f(2) &= 3(2)^2 - 12(2) + 17 && \text{Substitute 2 for } x. \\ &= 5 && \text{Simplify.} \end{aligned}$$

The minimum value for this function is 5.

Do all quadratic functions have a minimum value? No, if the graph of a quadratic function is a parabola that opens downward, the function does not have a minimum value. Such a function does have a maximum value, which can be found at the vertex.

EXAMPLE 6 Find the maximum or minimum value of the function $f(x) = -3(x + 2)^2 - 5$.

SOLUTION Because this parabola will open downward, the function has a maximum value at its vertex. The vertex of this function is $(-2, -5)$. The maximum value of the function is -5, which occurs when $x = -2$.

Quick Check 5

Find the maximum or minimum value of a quadratic function.

a) $f(x) = -x^2 + 10x - 35$
b) $f(x) = 5(x - 2)^2 + 47$

Applied Maximum–Minimum Problems

Objective 3 Solve applied maximum–minimum problems.

EXAMPLE 7 What is the maximum product of two numbers whose sum is 40?

SOLUTION There are two unknown numbers in this problem. If we let x represent the first number, we can represent the second number with $40 - x$.

Unknowns
First number: x
Second number: $40 - x$

The product of these two numbers is given by the function $f(x) = x(40 - x)$. This function simplifies to be $f(x) = -x^2 + 40x$. Because this function is quadratic and its graph is a parabola that opens downward $(a < 0)$, its maximum value can be found at the vertex. The maximum product occurs when $x = \frac{-b}{2a}$.

$$x = \frac{-40}{2(-1)} = 20$$ Substitute -1 for a and 40 for b and simplify.

The maximum product occurs when $x = 20$.

First number: $x = 20$
Second number: $40 - x = 40 - 20 = 20$

The two numbers whose sum is 40 that have the greatest product are 20 and 20. The maximum product is 400.

Quick Check 6

What is the maximum product of two numbers whose sum is 62?

EXAMPLE 8 A farmer wants to fence off a rectangular pen for some pigs. If he has 120 feet of fencing, what dimensions would give the pen the largest possible area?

SOLUTION There are two unknowns in this problem: the length and the width of the rectangle.

Suppose we let x represent the length of the rectangle. Because the perimeter is equal to 120 feet, the total length of the remaining two sides is $120 - 2x$. Therefore, each remaining side is $\frac{120 - 2x}{2}$, or $60 - x$.

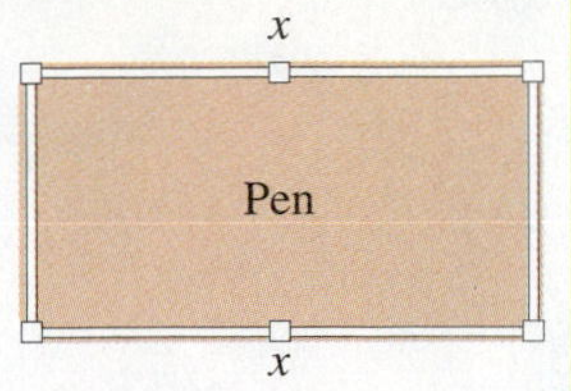

The width of a rectangle is equal to the difference of half the perimeter and the length, $W = \frac{P}{2} - L$. To verify this, solve the formula for the perimeter of a rectangle, $P = 2L + 2W$, for W.

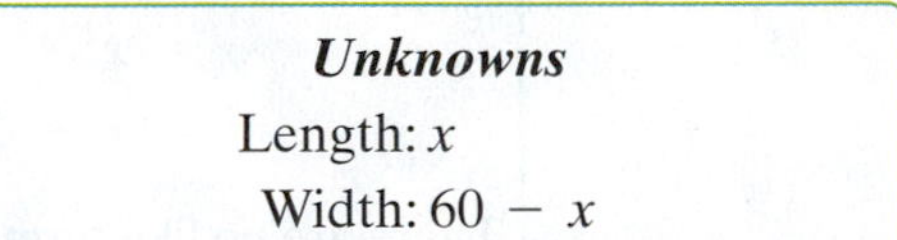
Unknowns
Length: x
Width: $60 - x$

The area of a rectangle is equal to its length times its width. The area of this pen is given by the function $A(x) = x(60 - x) = -x^2 + 60x$. Because this function is quadratic and its graph is a parabola that opens downward $(a < 0)$, its maximum value can be found at the vertex. The maximum area occurs when $x = \frac{-b}{2a}$.

$$x = \frac{-60}{2(-1)} = 30 \quad \text{Substitute } -1 \text{ for } a \text{ and } 60 \text{ for } b \text{ and simplify.}$$

The maximum area occurs when $x = 30$.

Length: $x = 30$
Width: $60 - x = 60 - 30 = 30$

The maximum area occurs when both the length and the width are 30 feet. The maximum area that can be enclosed is 900 square feet.

Quick Check 7

What is the largest possible rectangular area that can be fenced in with 264 meters of fencing?

Difference Quotients

Objective 4 Simplify difference quotients for quadratic functions.

The **difference quotient** $\frac{f(x + h) - f(x)}{h}$ is used to determine the rate of change for a function $f(x)$ at a particular value of x.

We will now learn how to simplify this difference quotient for quadratic functions.

EXAMPLE 9 For the function $f(x) = x^2 - 8x + 14$, simplify the difference quotient $\dfrac{f(x+h) - f(x)}{h}$.

SOLUTION We begin to simplify the difference quotient by simplifying $f(x + h)$.

$f(x+h) = (x+h)^2 - 8(x+h) + 14$	Substitute $x + h$ for x.
$= (x+h)(x+h) - 8(x+h) + 14$	Square $x + h$ by multiplying it by itself.
$= x^2 + xh + xh + h^2 - 8(x+h) + 14$	Multiply using the distributive property.
$= x^2 + 2xh + h^2 - 8(x+h) + 14$	Combine like terms.
$= x^2 + 2xh + h^2 - 8x - 8h + 14$	Distribute.

There are no like terms to combine, so we can substitute into the difference quotient.

$$\frac{f(x+h) - f(x)}{h} = \frac{(x^2 + 2xh + h^2 - 8x - 8h + 14) - (x^2 - 8x + 14)}{h}$$

Substitute for $f(x + h)$ and $f(x)$.

$$= \frac{x^2 + 2xh + h^2 - 8x - 8h + 14 - x^2 + 8x - 14}{h}$$

Simplify the numerator by removing parentheses.

$$= \frac{2xh + h^2 - 8h}{h}$$

Combine like terms.

$$= \frac{\overset{1}{\cancel{h}}(2x + h - 8)}{\underset{1}{\cancel{h}}}$$

Divide out common factors.

$$= 2x + h - 8$$

Simplify.

$$\frac{f(x+h) - f(x)}{h} = 2x + h - 8 \quad \text{for} \quad f(x) = x^2 - 8x + 14$$

Quick Check 8

For the function $f(x) = x^2 + 3x + 317$, simplify the difference quotient $\dfrac{f(x+h) - f(x)}{h}$.

BUILDING YOUR STUDY STRATEGY

Study Environment, 3 Work space When you sit down to study, you need a space large enough to accommodate all of your materials. Make sure you can fit your textbook, notes, homework notebook, study cards, calculator, ruler, and other supplies on your desk or table. Anything you might use should be within easy reach. If your work space is disorganized and cluttered, you will waste time searching for something you need and shuffling through your materials.

Exercises 11.3

MyMathLab

PRACTICE

WATCH

DOWNLOAD

READ

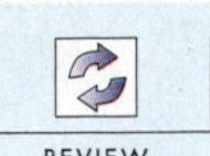
REVIEW

Vocabulary

1. The graph of a quadratic equation is a U-shaped graph called a(n) ________.
2. The turning point of a parabola is called its ________.
3. To find the __________ of a parabola, substitute 0 for x and solve for y.
4. To find the __________ of a parabola, substitute 0 for y and solve for x.
5. To graph the function $f(x) = (x - h)^2$, apply a(n) __________ shift of h units.
6. To graph the function $f(x) = x^2 + k$, apply a(n) ________ shift of k units.
7. A quadratic function whose graph opens upward has a(n) __________ value at its vertex.
8. A quadratic function whose graph opens downward has a(n) __________ value at its vertex.

Without actually graphing the function, state the shift(s) that are applied to the graph of $f(x) = x^2$ to graph the given function. If the graph of $f(x) = x^2$ must be rotated about the x-axis, state this first.

9. $f(x) = (x + 6)^2$
10. $f(x) = (x - 8)^2$
11. $f(x) = x^2 + 9$
12. $f(x) = x^2 - 11$
13. $f(x) = (x - 5)^2 - 9$
14. $f(x) = (x + 12)^2 + 30$
15. $f(x) = (x + 10)^2 - 40$
16. $f(x) = (x - 4)^2 + 49$
17. $f(x) = -(x - 2)^2 + 13$
18. $f(x) = -(x + 7)^2 - 16$

Graph the function by shifting. Label the vertex, y-intercept, and any x-intercepts. (Round x-intercepts to the nearest tenth.) State the domain and range of the function.

19. $f(x) = x^2 + 7$
20. $f(x) = x^2 - 4$
21. $f(x) = (x - 3)^2$
22. $f(x) = (x + 2)^2$
23. $f(x) = (x - 4)^2 - 1$
24. $f(x) = (x - 3)^2 - 9$
25. $f(x) = (x - 1)^2 - 4$
26. $f(x) = (x - 2)^2 - 1$

27. $f(x) = (x + 3)^2 - 5$

28. $f(x) = (x + 4)^2 - 3$

29. $f(x) = (x + 1)^2 - 8$

30. $f(x) = (x + 2)^2 - 6$

31. $f(x) = (x - 2)^2 + 4$

32. $f(x) = (x - 3)^2 + 2$

33. $f(x) = -(x + 4)^2 + 9$

34. $f(x) = -(x - 1)^2 + 4$

35. $f(x) = -(x - 3)^2 + 7$

36. $f(x) = -(x + 1)^2 + 8$

37. $f(x) = -(x + 2)^2 - 3$

38. $f(x) = -(x + 3)^2 - 5$

Determine whether the given quadratic function has a maximum value or a minimum value. Then find that maximum or minimum value.

39. $f(x) = x^2 - 8x + 20$
40. $f(x) = x^2 - 6x - 35$
41. $f(x) = -x^2 + 4x - 30$
42. $f(x) = -x^2 + 14x - 76$
43. $f(x) = (x - 7)^2 + 18$
44. $f(x) = (x + 3)^2 - 19$
45. $f(x) = -(x + 16)^2 - 33$
46. $f(x) = -(x - 14)^2 + 47$
47. $f(x) = 7(x - 4)^2 + 46$
48. $f(x) = -8(x + 11)^2 - 50$
49. $f(x) = 3x^2 - 20x + 121$
50. $f(x) = -2x^2 - 14x + 207$
51. A girl throws a rock upward with an initial velocity of 32 feet/second. The height of the rock (in feet) after t seconds is given by the function $h(t) = -16t^2 + 32t + 5$. What is the maximum height the rock reaches?
52. A pilot flying at a height of 5000 feet determines that she must eject from the plane. The ejection seat launches with an initial velocity of 144 feet/second. The height of the pilot (in feet) t seconds after ejection is given by the function $h(t) = -16t^2 + 144t + 5000$. What is the maximum height the pilot reaches?
53. At the start of a college football game, a referee tosses a coin to determine which team will receive the opening kickoff. The height of the coin (in meters) after t seconds is given by the function $h(t) = -4.9t^2 + 2.1t + 1.5$. What is the maximum height the coin reaches?
54. If an astronaut on the moon throws a rock upward with an initial velocity of 78 feet per second, the height of the rock (in feet) after t seconds is given by the function $h(t) = -2.6t^2 + 78t + 5$. What is the maximum height the rock reaches?
55. What is the maximum product of two numbers whose sum is 24?
56. What is the maximum product of two numbers whose sum is 100?
57. A farmer has 300 feet of fencing to make a rectangular corral. What dimensions will result in the maximum area? What is the maximum area possible?
58. Emeril wants to fence off a rectangular herb garden adjacent to his house, using the house to form the fourth side of the rectangle as shown.

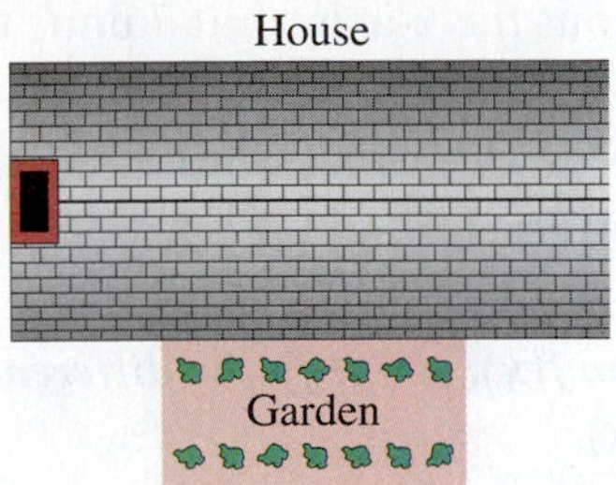

If Emeril has 80 feet of fencing, what dimensions of the herb garden will produce the maximum area? What is the maximum possible area?

59. A company produces small frying pans. The average cost to produce each pan in dollars is given by the function $f(x) = 0.000625x^2 - 0.0875x + 6.5$, where x is the number of pans produced (in thousands). For what number of pans is the average cost per pan minimized? What is the minimum average cost per pan that is possible?
60. A company produces radios. The average cost to produce each radio in dollars is given by the function $f(x) = 0.009x^2 - 0.432x + 9.3$, where x is the number of radios produced (in thousands). For what number of radios is the average cost per radio minimized? What is the minimum cost per radio that is possible?
61. A minor league baseball team is considering raising ticket prices for the upcoming season. If it raises ticket prices by x dollars, the expected revenue function per game is $r(x) = -250(x - 3)^2 + 37{,}750$. (This takes into account lower attendance that will be caused by higher prices.) What increase in ticket prices will produce the maximum revenue for the baseball team? What is the maximum revenue that is possible?
62. A movie theater is considering lowering ticket prices. If it lowers ticket prices by x dollars, the expected revenue function per week is $r(x) = -500(x - 2.5)^2 + 28{,}025$. What decrease in ticket prices will produce the maximum revenue for the theater? What is the maximum revenue that is possible?

Determine a quadratic function that results when applying the given shifts to the graph of $f(x) = x^2$.

63. Shift six units to the right.

64. Shift four units up.

65. Shift 3 units to the left and 11 units down.

66. Shift two units to the right and five units down.

67. Rotate about the x-axis, shift 7 units to the right and 13 units up.

68. Rotate about the x-axis, shift 9 units to the left and 22 units down.

For the function $f(x)$, simplify the difference quotient $\frac{f(x+h) - f(x)}{h}$.

69. $f(x) = x^2 - 4$

70. $f(x) = x^2 - 10x$

71. $f(x) = x^2 - 9x + 17$

72. $f(x) = x^2 + 8x + 45$

73. $f(x) = 3x^2 - 11x + 49$

74. $f(x) = -x^2 - 16x + 105$

Writing in Mathematics

Answer in complete sentences.

75. Explain how to determine that a parabola does not have any x-intercepts.

76. Explain how to determine whether a quadratic function has a minimum value or a maximum value. Explain how to find the minimum or maximum value of a quadratic function. Use an example to illustrate the process.

77. ***Newsletter*** Write a newsletter explaining how to graph a function of the form $f(x) = (x - h)^2 + k$.

Other Functions and Their Graphs

OBJECTIVES

1. Graph absolute value functions by shifting.
2. Graph square-root functions.
3. Graph cubic functions.
4. Determine a function from its graph.

In this section, we will learn how to graph absolute value functions by shifting. We first learned how to graph these functions by plotting points in Section 8.3. We will then move on to learn how to graph square root functions and cubic functions.

Objective 1 Graph absolute value functions by shifting. Recall from Section 8.3 that the graph of an absolute value function $f(x) = a|x - h| + k$ is V-shaped. Here is the graph of the basic absolute function $f(x) = |x|$.

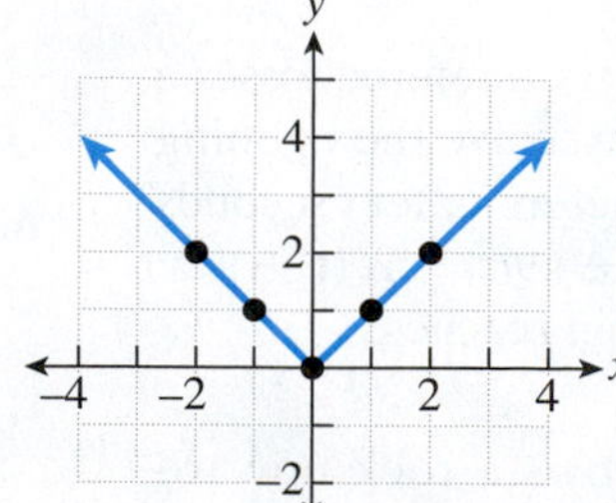

x	$f(x)$
−2	2
−1	1
0	0
1	1
2	2

To sketch the graph of the function $f(x) = |x - h| + k$, we shift the graph of $f(x) = |x|$ by h units horizontally and by k units vertically. The tip of the V, the point where the function changes from decreasing to increasing, should be located at (h, k).

We can add details such as x- and y-intercepts to the graph. To determine the y-coordinate of the y-intercept, we find $f(0)$. To find the x-intercepts, if any exist, we set the function equal to 0 and solve for x.

The domain of an absolute value function is the set of all real numbers: $(-\infty, \infty)$. The range of the function can be determined by examining the graph.

EXAMPLE 1 Sketch the graph of $f(x) = |x - 3| + 4$ by shifting. State the domain and range of the function.

SOLUTION In this example, $h = 3$ and $k = 4$. We need to shift the graph of $f(x) = |x|$ by three units to the right and by four units up.

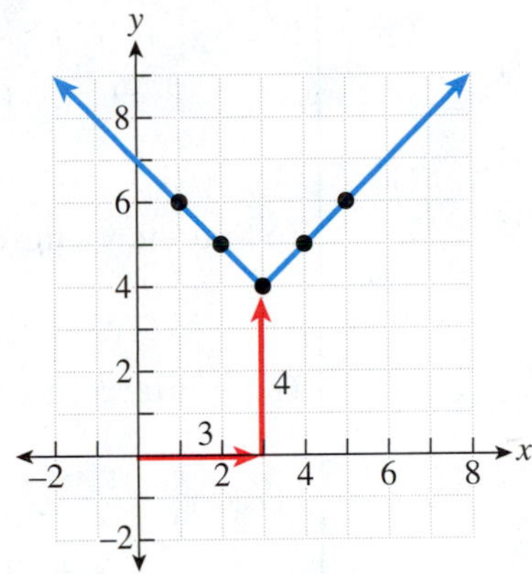

We can add more detail to the graph by adding its intercepts. Notice that by shifting, we already know that there are no x-intercepts. We can determine the y-intercept by finding $f(0)$.

$$f(0) = |0 - 3| + 4 = 7 \quad \text{Substitute 0 for } x \text{ and simplify.}$$

The y-intercept is $(0, 7)$. Here is the graph, including all intercepts.

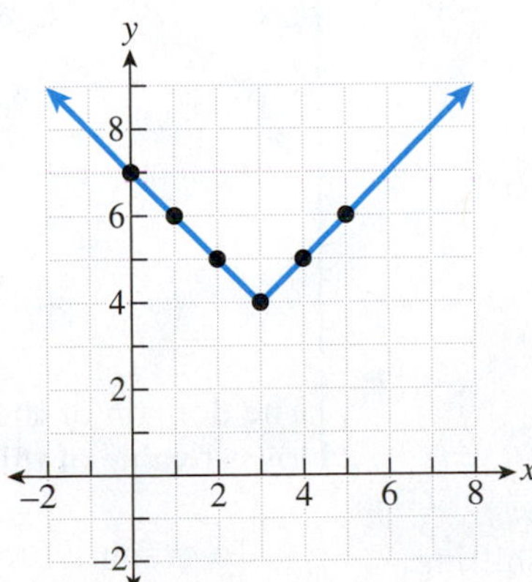

The domain of the function is $(-\infty, \infty)$. By looking at the graph, we see that the lowest value of this function is 4; so the range of the function is $[4, \infty)$.

Quick Check 1

Sketch the graph of $f(x) = |x + 1| + 5$ by shifting. State the domain and range of the function.

EXAMPLE 2 Sketch the graph of $f(x) = |x + 2| - 5$ by shifting. State the domain and range of the function.

SOLUTION In this example, $h = -2$ and $k = -5$. We need to shift the graph of $f(x) = |x|$ by two units to the left and by five units down.

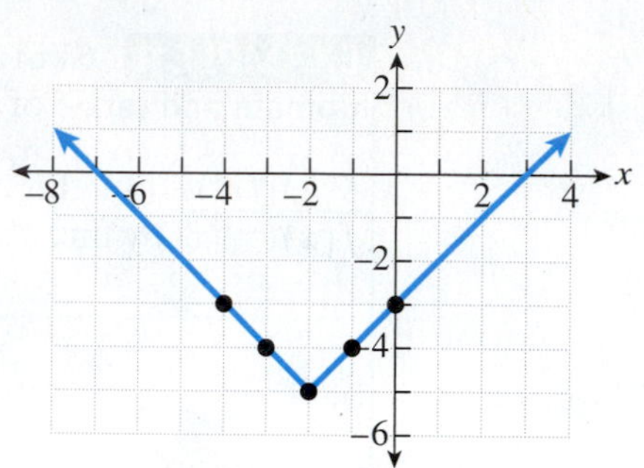

To find the y-intercept, we need to find $f(0)$.

$$f(0) = |0 + 2| - 5 = 2 - 5 = -3$$

The y-intercept is $(0, -3)$. To find the x-intercepts, we set the function equal to 0 and solve.

$$|x + 2| - 5 = 0 \quad \text{Set the function equal to 0.}$$
$$|x + 2| = 5 \quad \text{Isolate the absolute value.}$$
$$x + 2 = 5 \quad \text{or} \quad x + 2 = -5 \quad \text{Rewrite as two linear equations.}$$
$$x = 3 \quad \text{or} \quad x = -7 \quad \text{Solve each equation by subtracting 2.}$$

The x-intercepts are $(3, 0)$ and $(-7, 0)$. Here is the graph, including all intercepts.

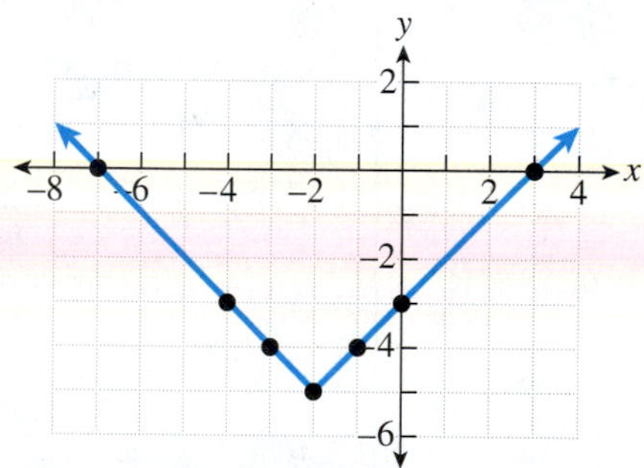

The domain of the function is $(-\infty, \infty)$. By looking at the graph, we see that the lowest value of this function is -5; so the range of the function is $[-5, \infty)$.

Quick Check 2

Sketch the graph of $f(x) = |x - 4| - 2$ by shifting. State the domain and range of the function.

To graph a function of the form $f(x) = -|x - h| + k$, we begin by rotating the graph of $f(x) = |x|$ about the x-axis. (Recall that the negative sign in front of the absolute value indicates that the graph opens downward.) We finish by applying any horizontal or vertical shifts that are necessary.

EXAMPLE 3 Sketch the graph of $f(x) = -|x - 4| - 1$ by shifting. State the domain and range of the function.

SOLUTION In this example, $h = 4$ and $k = -1$. After rotating the graph of $f(x) = |x|$ about the x-axis, we need to shift this rotated graph by four units to the right and by one unit down.

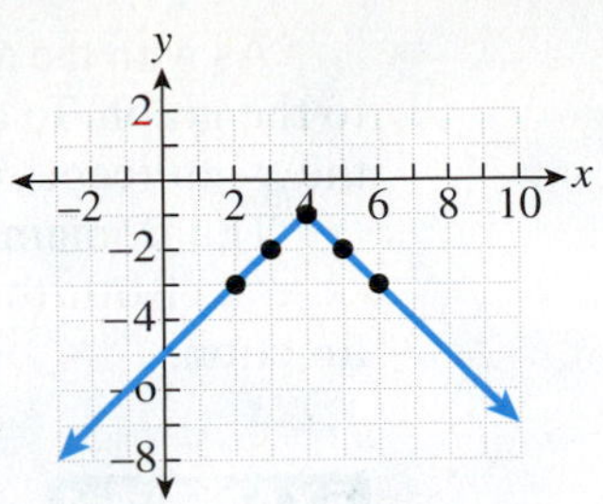

Notice that by shifting, we already know that there are no x-intercepts. We can determine the y-intercept by finding $f(0)$.

$f(0) = -|0 - 4| - 1 = -4 - 1 = -5$ Substitute 0 for x and simplify.

The y-intercept is $(0, -5)$. Here is the graph, including all intercepts.

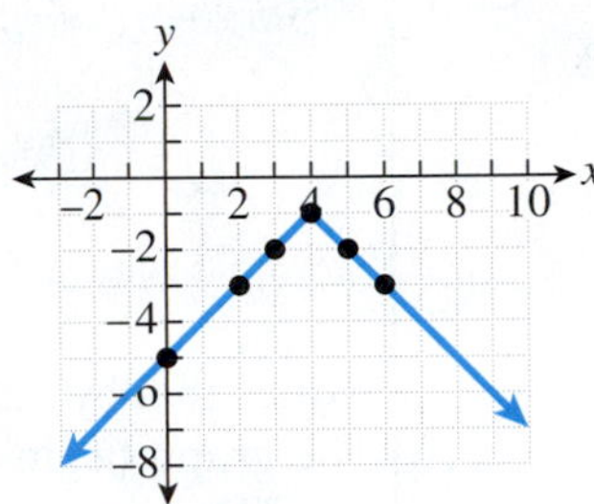

The domain of the function is $(-\infty, \infty)$. By looking at the graph, we see that the greatest value of this function is -1; so the range of the function is $(-\infty, -1]$.

Quick Check 3

Sketch the graph of $f(x) = -|x - 2| - 3$ by shifting. State the domain and range of the function.

Graphing Square-Root Functions

Objective 2 Graph square-root functions. Now we will learn to graph the **square-root function.** The basic square-root function is the function $f(x) = \sqrt{x}$. Recall that the square root of a negative number is an imaginary number, so the radicand must be nonnegative. We begin to graph this function by selecting values for x and evaluating the function for these values. It is a good idea to choose values of x that are perfect squares, such as 0, 1, and 4.

x	$f(x) = \sqrt{x}$	$(x, f(x))$
0	$f(0) = \sqrt{0} = 0$	$(0, 0)$
1	$f(1) = \sqrt{1} = 1$	$(1, 1)$
4	$f(4) = \sqrt{4} = 2$	$(4, 2)$

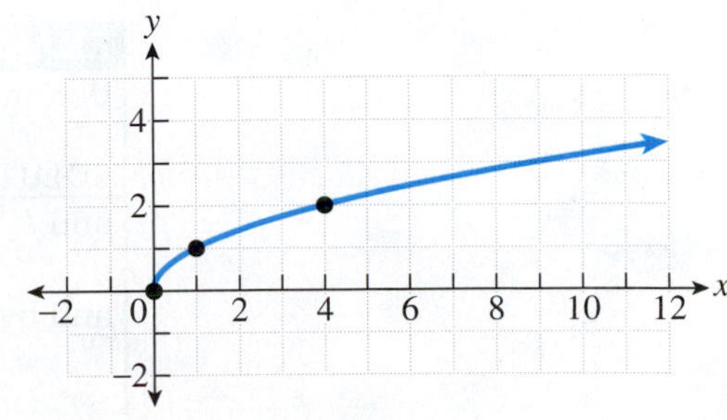

The domain of this function, which is read from left to right along the x-axis on this graph, is $[0, \infty)$. The range of this function, which is read from bottom to top along the y-axis, is also $[0, \infty)$.

We now turn our attention to graphing square-root functions of the form $f(x) = a\sqrt{x - h} + k$.

To sketch the graph of the function $f(x) = \sqrt{x - h} + k$, we shift the graph of $f(x) = \sqrt{x}$ by h units horizontally and by k units vertically. The graph of the function should begin at the point (h, k).

As with the absolute value function, we can add details such as x- and y-intercepts to the graph. To determine the y-coordinate of the y-intercept, we find $f(0)$. To find the x-intercepts, if any exist, we set the function equal to 0 and solve for x.

The domain of the function $f(x) = \sqrt{x - h} + k$ is $[h, \infty)$, and the range is $[k, \infty)$. Both the domain and range can be determined from the graph of the function.

EXAMPLE 4 Sketch the graph of $f(x) = \sqrt{x - 5} + 3$ by shifting. State the domain and range of the function.

SOLUTION In this example, $h = 5$ and $k = 3$. We need to shift the graph of $f(x) = \sqrt{x}$ by five units to the right and by three units up.

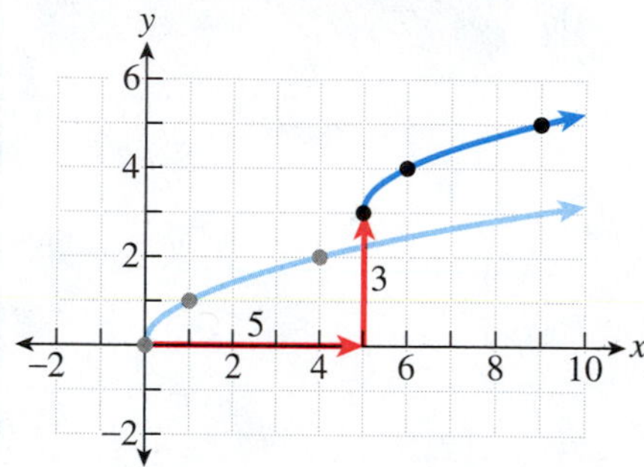

Notice that by shifting, we already know that there are no x-intercepts or y-intercepts. The graph begins at $(5, 3)$ and moves only up and to the right from there.

The domain of the function is $[5, \infty)$, and the range is $[3, \infty)$.

Quick Check 4

Sketch the graph of $f(x) = \sqrt{x - 2} + 6$ by shifting. State the domain and range of the function.

Using Your Calculator You can graph square-root functions using the TI-84. To enter the function from Example 4, $f(x) = \sqrt{x - 5} + 3$, tap the [Y=] key. Enter $\sqrt{x - 5} + 3$ next to Y_1. Use parentheses to separate the radicand from the rest of the expression. Tap the key labeled [GRAPH] to graph the function in the standard window.

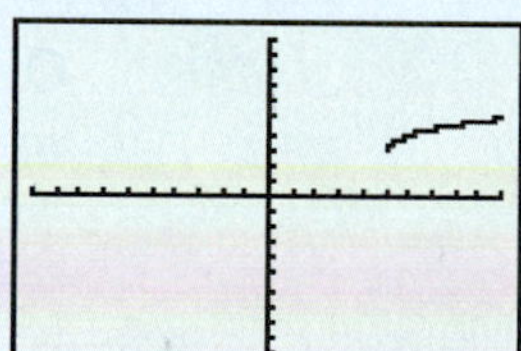

EXAMPLE 5 Sketch the graph of $f(x) = \sqrt{x + 2} - 3$ by shifting. State the domain and range of the function.

SOLUTION In this example, $h = -2$ and $k = -3$. We need to shift the graph of $f(x) = \sqrt{x}$ by two units to the left and by three units down.

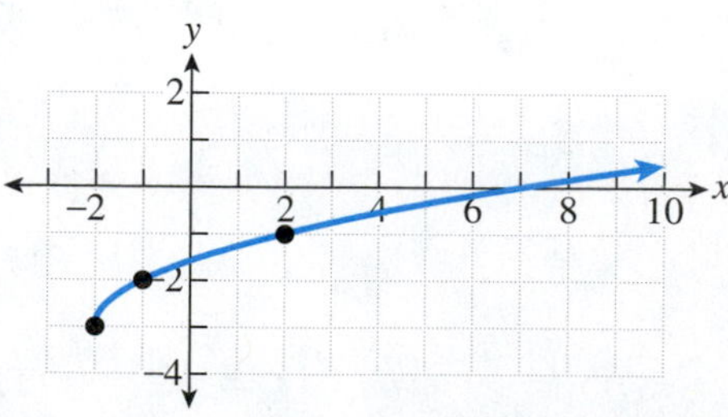

We can add more detail to the graph by plotting the x- and y-intercepts. To find the y-intercept of this function, we evaluate $f(0)$.

$$\begin{aligned} f(0) &= \sqrt{0 + 2} - 3 && \text{Substitute 0 for } x. \\ &= \sqrt{2} - 3 && \text{Simplify the radicand.} \\ &\approx -1.6 && \text{Approximate using a calculator.} \end{aligned}$$

The y-intercept is located approximately at $(0, -1.6)$. To find the x-intercept, we set the function equal to 0 and solve for x.

$$\sqrt{x+2} - 3 = 0 \quad \text{Set the function equal to 0.}$$
$$\sqrt{x+2} = 3 \quad \text{Add 3 to isolate the square root.}$$
$$(\sqrt{x+2})^2 = 3^2 \quad \text{Square both sides.}$$
$$x + 2 = 9 \quad \text{Simplify each side.}$$
$$x = 7 \quad \text{Subtract 2.}$$

The x-intercept is at $(7, 0)$. Here is the graph, including all intercepts.

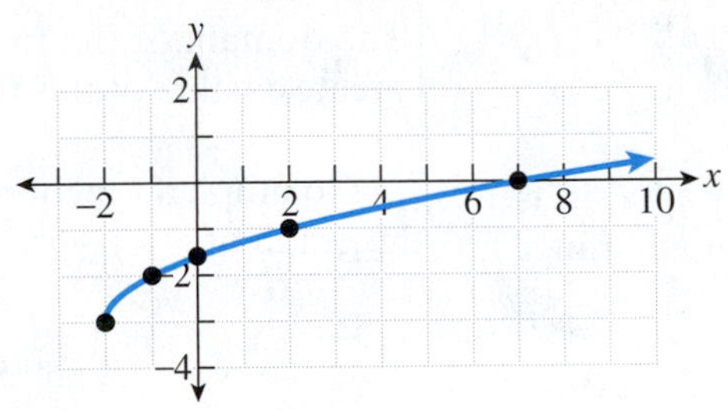

The domain of this function is $[-2, \infty)$, and the range is $[-3, \infty)$.

Quick Check 5

Sketch the graph of $f(x) = \sqrt{x+1} - 2$ by shifting. State the domain and range of the function.

To graph a function of the form $f(x) = -\sqrt{x-h} + k$, we begin by rotating the graph of $f(x) = \sqrt{x}$ about the x-axis.

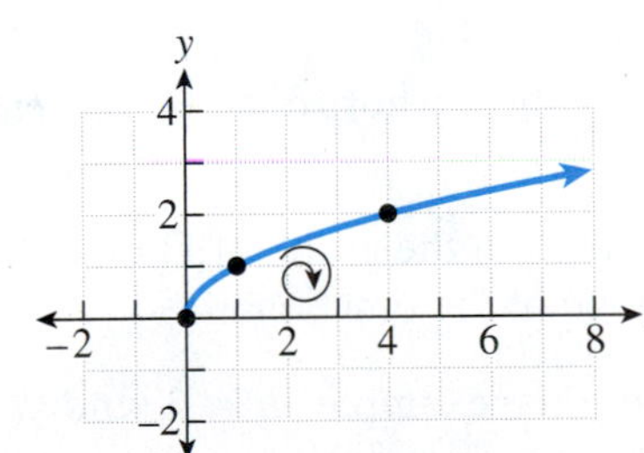

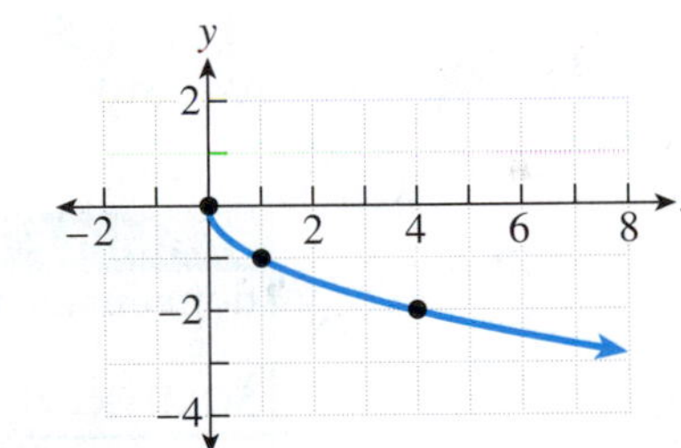

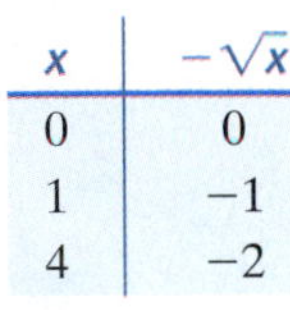

x	$-\sqrt{x}$
0	0
1	−1
4	−2

We finish by applying any horizontal or vertical shifts that are necessary.

EXAMPLE 6 Sketch the graph of $f(x) = -\sqrt{x+7} + 1$ by shifting. State the domain and range of the function.

SOLUTION In this example, $h = -7$ and $k = 1$. After rotating the graph of $f(x) = \sqrt{x}$ about the x-axis, we need to shift the rotated graph by seven units to the left and by one unit up.

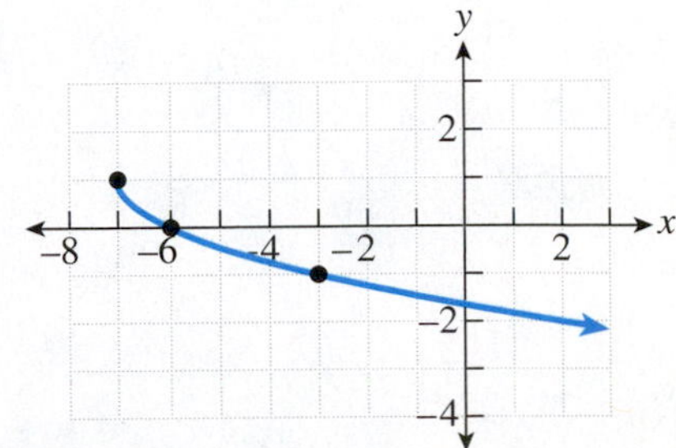

Notice that by shifting, we already know that the x-intercept is $(-6, 0)$. We can determine the y-intercept by finding $f(0)$.

$$f(0) = -\sqrt{0+7} + 1 = -\sqrt{7} + 1 \approx -1.6 \quad \text{Substitute 0 for } x \text{ and simplify.}$$

The y-intercept is approximately $(0, -1.6)$. Here is the graph, including all intercepts.

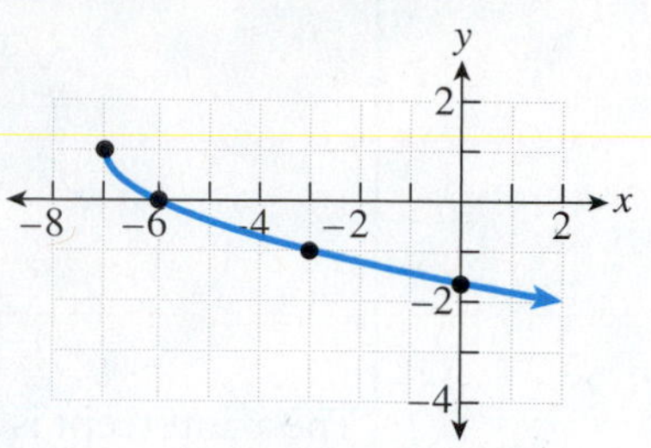

The domain of the function is $[-7, \infty)$. By looking at the graph, we see that the greatest value of this function is 1; so the range of the function is $(-\infty, 1]$.

Quick Check 6

Sketch the graph of $f(x) = -\sqrt{x+1} - 5$ by shifting. State the domain and range of the function.

To graph a function of the form $f(x) = \sqrt{-(x-h)} + k$, we begin by rotating the graph of $f(x) = \sqrt{x}$ about the y-axis.

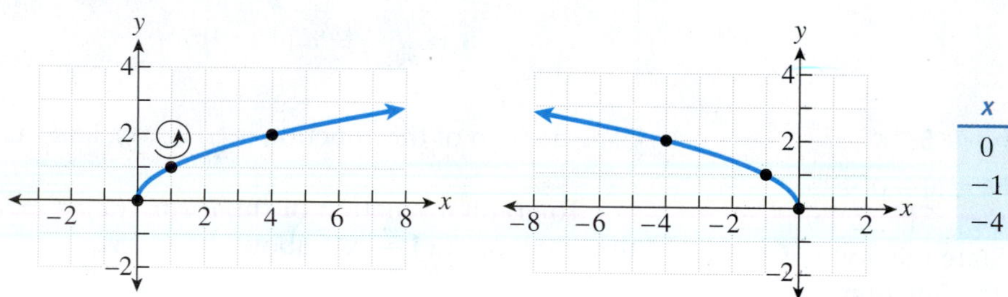

x	$\sqrt{-x}$
0	0
−1	1
−4	2

We finish by applying any horizontal or vertical shifts that are necessary.

EXAMPLE 7 Sketch the graph of $f(x) = \sqrt{-(x-3)} + 6$ by shifting. State the domain and range of the function.

SOLUTION In this example, $h = 3$ and $k = 6$. After rotating the graph of $f(x) = \sqrt{x}$ about the y-axis, we need to shift the rotated graph by three units to the right and by six units up.

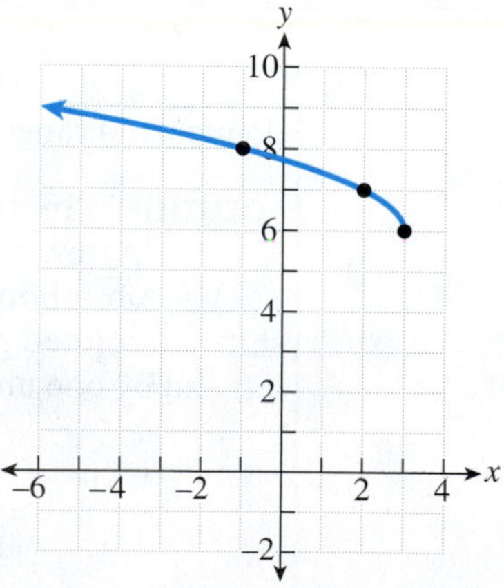

Notice that by shifting, we already know that there are no x-intercepts. We can determine the y-intercept by finding $f(0)$.

$$f(0) = \sqrt{-(0-3)} + 6 = \sqrt{3} + 6 \approx 7.7 \quad \text{Substitute 0 for } x \text{ and simplify.}$$

The y-intercept is approximately $(0, 7.7)$. Here is the graph, including the y-intercept.

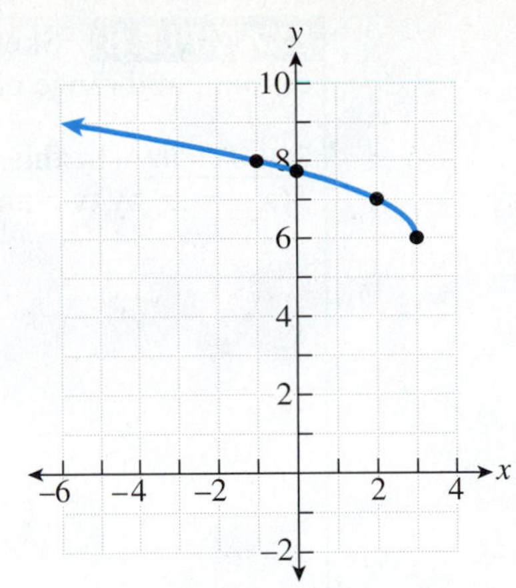

From the graph, we see that the domain of the function is $(-\infty, 3]$ and the range is $[6, \infty)$.

Quick Check 7

Sketch the graph of $f(x) = \sqrt{-(x + 4)} - 2$ by shifting. State the domain and range of the function.

Graphing Cubic Functions

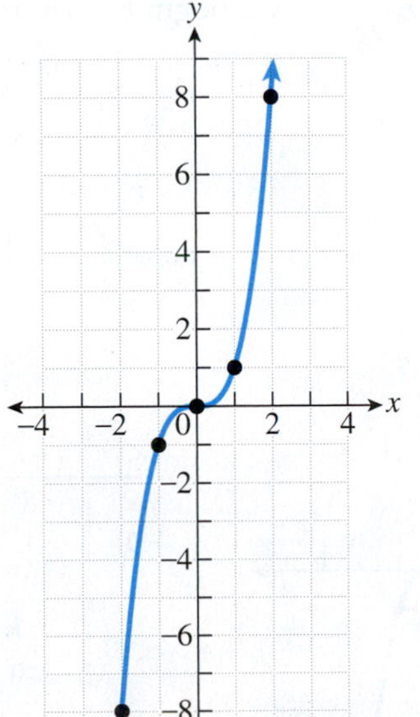

Objective 3 Graph cubic functions. The other function we will investigate in this section is the **cubic function.** The basic cubic function is the function $f(x) = x^3$. Here is a table of function values for values of x from -2 to 2.

x	$f(x) = x^3$	$(x, f(x))$
-2	$f(-2) = (-2)^3 = -8$	$(-2, -8)$
-1	$f(-1) = (-1)^3 = -1$	$(-1, -1)$
0	$f(0) = 0^3 = 0$	$(0, 0)$
1	$f(1) = 1^3 = 1$	$(1, 1)$
2	$f(2) = 2^3 = 8$	$(2, 8)$

We plot these points and draw a smooth curve through them. The graph at the left extends forever to the left and to the right; so the domain is the set of all real numbers $\mathbb{R}$, which can be expressed as $(-\infty, \infty)$. The graph also extends forever upward and downward; so the range of this function is also $(-\infty, \infty)$. This is true for all cubic functions.

To sketch the graph of the function $f(x) = (x - h)^3 + k$, we shift the graph of $f(x) = x^3$ by h units horizontally and by k units vertically.

EXAMPLE 8 Sketch the graph of $f(x) = (x + 2)^3 - 1$ by shifting. State the domain and range of the function.

SOLUTION In this example, $h = -2$ and $k = -1$. We need to shift the graph of $f(x) = x^3$ by two units to the left and by one unit down.

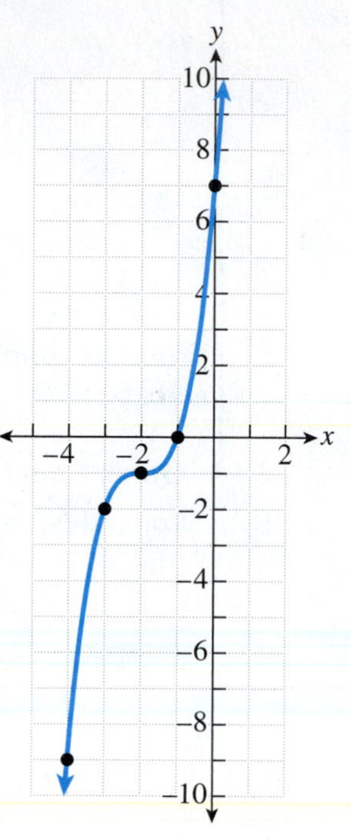

The domain of the function is $(-\infty, \infty)$, and the range is $(-\infty, \infty)$.

Quick Check 8

Sketch the graph of $f(x) = (x - 1)^3 + 1$ by shifting. State the domain and range of the function.

To graph a function of the form $f(x) = -(x - h)^3 + k$, we begin by rotating the graph of $f(x) = x^3$ about the x-axis.

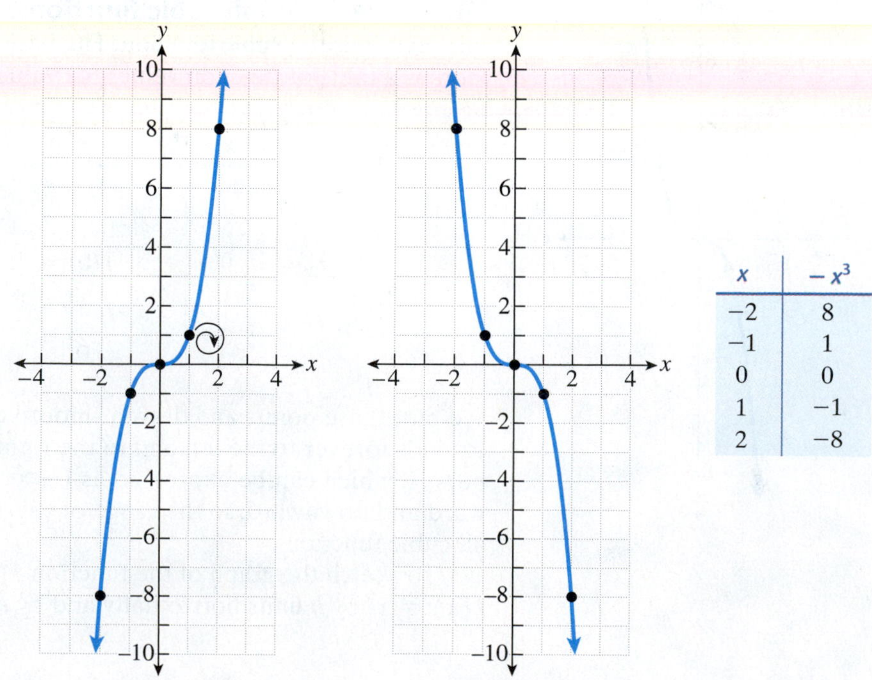

x	$-x^3$
−2	8
−1	1
0	0
1	−1
2	−8

We finish by applying any horizontal or vertical shifts that are necessary.

EXAMPLE 9 Sketch the graph of $f(x) = -(x - 1)^3 + 4$ by shifting. State the domain and range of the function.

SOLUTION In this example, $h = 1$ and $k = 4$. After rotating the graph of $f(x) = x^3$ about the x-axis, we need to shift the rotated graph by one unit to the right and by four units up.

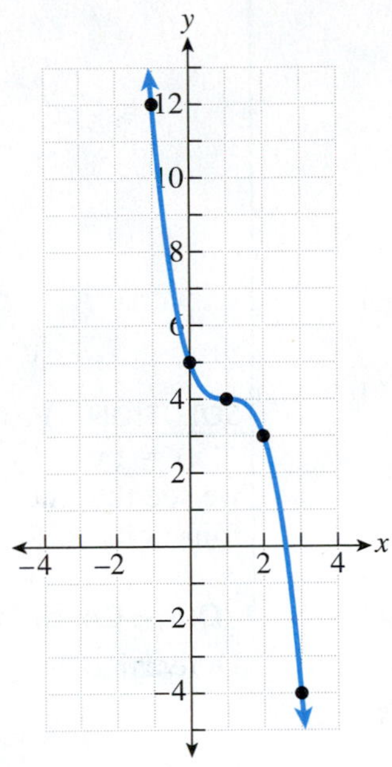

The domain of the function is $(-\infty, \infty)$, and the range is $(-\infty, \infty)$.

Quick Check 9

Sketch the graph of $f(x) = -(x + 3)^3$ by shifting. State the domain and range of the function.

Determining a Function from Its Graph

Objective 4 **Determine a function from its graph.**

EXAMPLE 10 Determine the function $f(x) = \sqrt{x - h} + k$ that has been graphed.

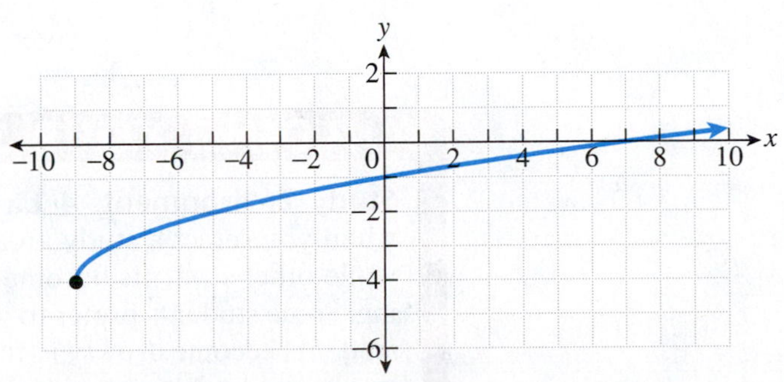

SOLUTION The graph of this function tells us that $f(x)$ is a square-root function. To find the function, let's focus on the point where this graph begins, which is at $(-9, -4)$. The function $f(x) = \sqrt{x}$ has been shifted by nine units to the left and by four units down, so $h = -9$ and $k = -4$. The function is $f(x) = \sqrt{x + 9} - 4$.

EXAMPLE 11 Determine the function $f(x) = (x - h)^3 + k$ that has been graphed.

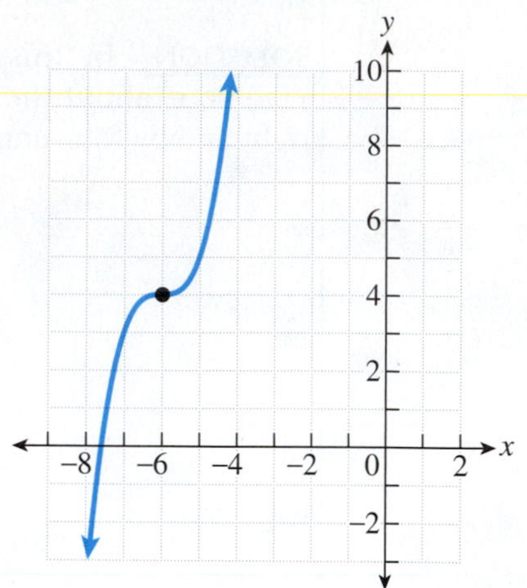

SOLUTION From the graph, we see that $f(x)$ is a function of the form $f(x) = (x - h)^3 + k$. Focusing on the point where the graph flattens out, $(-6, 4)$, we see that the function $f(x) = x^3$ has been shifted by six units to the left and by four units up. So $h = -6$ and $k = -4$, and the function is $f(x) = (x + 6)^3 + 4$.

Quick Check 10

Determine the function $f(x)$ that has been graphed.

a)

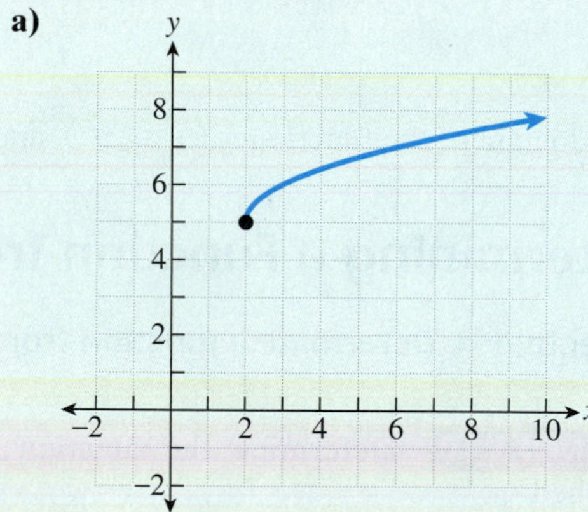

b)

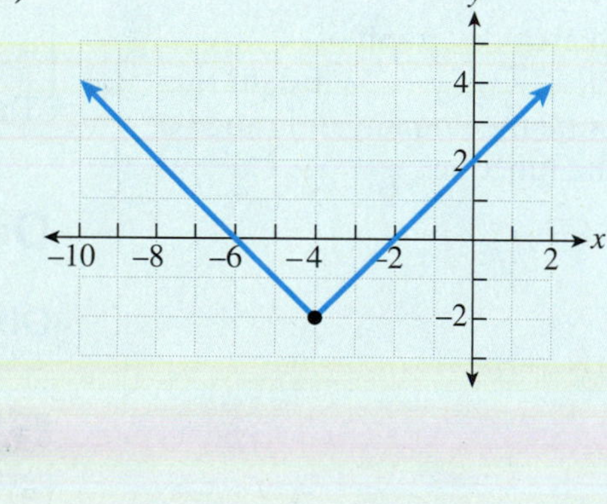

BUILDING YOUR STUDY STRATEGY

Study Environment, 4 Lighting Lighting is an important consideration when you select a study area. Some students prefer a bright, well-lit room, while other students become fidgety and uncomfortable in this type of lighting. Some students prefer to work in a room with soft, warm light, while other students become drowsy in this type of lighting. Some students prefer to work in a room with ample natural light. Whatever your preference, make sure there is enough light so that you can see clearly without straining your eyes. Working with insufficient light will make your eyes tired, which will make you tired as well.

Exercises 11.4

MyMathLab

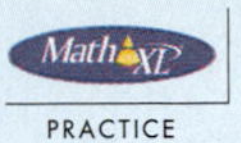
PRACTICE

WATCH

DOWNLOAD

READ

REVIEW

Vocabulary

1. To find the domain of a square root function, set the __________ greater than or equal to 0 and solve.

2. The domain and range of any cubic function are the set of ______________.

Graph the given absolute value function and state its domain and range.

3. $f(x) = |x - 2| + 6$

4. $f(x) = |x - 4| - 3$

5. $f(x) = |x + 1| - 5$

6. $f(x) = |x + 3| - 2$

7. $f(x) = -|x + 6| + 3$

8. $f(x) = -|x + 2| - 1$

9. $f(x) = -|x - 7| - 2$

10. $f(x) = -|x - 5| + 4$

Graph each given square-root function and state its domain and range.

11. $f(x) = \sqrt{x + 2}$

12. $f(x) = \sqrt{x} - 3$

13. $f(x) = \sqrt{x + 4} + 6$

14. $f(x) = \sqrt{x + 1} - 7$

15. $f(x) = \sqrt{x - 5} - 2$

16. $f(x) = \sqrt{x - 9} - 10$

17. $f(x) = -\sqrt{x - 2} + 3$

18. $f(x) = -\sqrt{x + 5} - 4$

19. $f(x) = \sqrt{-(x - 3)} + 1$

20. $f(x) = \sqrt{-(x + 2)} + 4$

Graph the given cubic function and state its domain and range.

21. $f(x) = x^3 - 1$

22. $f(x) = (x + 1)^3$

23. $f(x) = (x + 4)^3 + 1$

24. $f(x) = (x - 2)^3 - 4$

25. $f(x) = (x + 2)^3 + 7$

26. $(x + 1)^3 - 10$

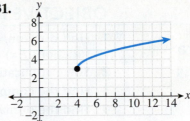

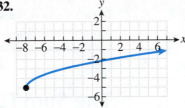

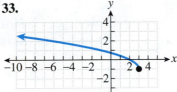

27. $f(x) = -(x - 1)^3 + 4$

28. $f(x) = -(x + 2)^3 - 5$

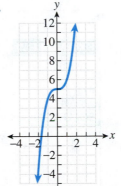

Determine the function $f(x)$ that has been graphed.

29.

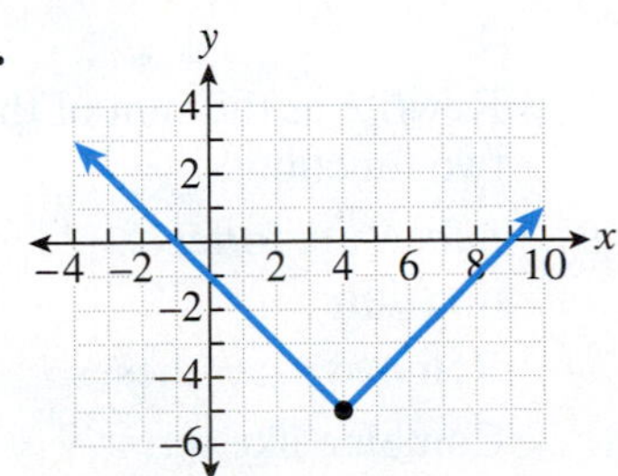

30.

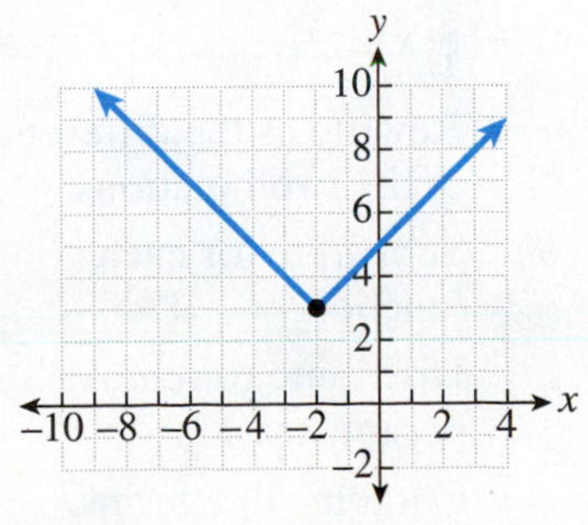

31.

32.

33.

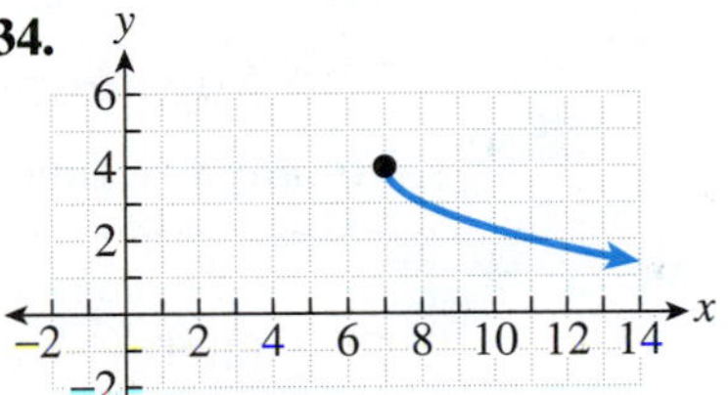

34.

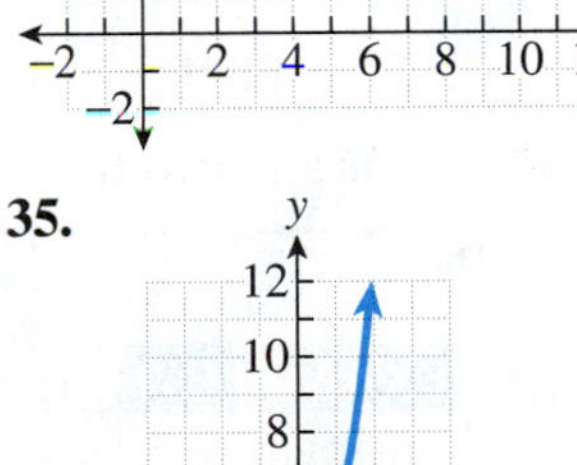

35.

36.

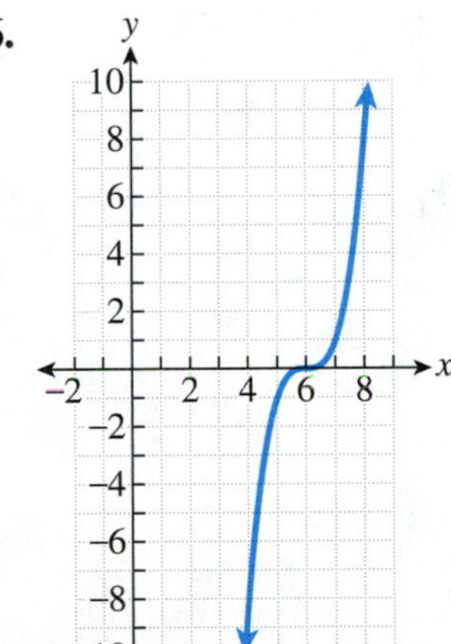

Writing in Mathematics

Answer in complete sentences.

37. Explain the similarities in and differences between graphing a quadratic function and graphing a square-root function.

38. Explain why both the domain and the range of a cubic function are the set of real numbers.

11.5 The Algebra of Functions

OBJECTIVES

1. Find the sum, difference, product, and quotient functions for two functions $f(x)$ and $g(x)$.
2. Solve applications involving the sum function or the difference function.
3. Find the composite function of two functions $f(x)$ and $g(x)$.
4. Find the domain of a composite function $(f \circ g)(x)$.

In this section, we will examine several ways to combine two or more functions into a single function. Just as we use addition, subtraction, multiplication, and division to combine two numbers, we can use these operations to combine functions as well.

The Sum Function and the Difference Function

Objective 1 Find the sum, difference, product, and quotient functions for two functions $f(x)$ and $g(x)$. The function that results when two functions $f(x)$ and $g(x)$ are added together is called the **sum function** and is denoted by $(f + g)(x)$. The function $(f - g)(x)$ is the difference of the two functions $f(x)$ and $g(x)$.

The Sum Function

For any two functions $f(x)$ and $g(x)$, $(f + g)(x) = f(x) + g(x)$.

The Difference Function

For any two functions $f(x)$ and $g(x)$, $(f - g)(x) = f(x) - g(x)$.

EXAMPLE 1 If $f(x) = x^2 + 3x - 10$ and $g(x) = 3x^2 - 5x - 9$, find the following:

a) $(f + g)(x)$

SOLUTION Begin by rewriting $(f + g)(x)$ as $f(x) + g(x)$.

$(f + g)(x) = f(x) + g(x)$	Rewrite as the sum of the two functions.
$= (x^2 + 3x - 10) + (3x^2 - 5x - 9)$	Substitute for each function.
$= x^2 + 3x - 10 + 3x^2 - 5x - 9$	Remove parentheses.
$= 4x^2 - 2x - 19$	Combine like terms.

b) $(f - g)(x)$

SOLUTION Begin by rewriting $(f - g)(x)$ as $f(x) - g(x)$.

$(f - g)(x) = f(x) - g(x)$	Rewrite as the difference of the two functions.
$= (x^2 + 3x - 10) - (3x^2 - 5x - 9)$	Substitute for each function.
$= x^2 + 3x - 10 - 3x^2 + 5x + 9$	Distribute to remove parentheses.
$= -2x^2 + 8x - 1$	Combine like terms.

c) $(f + g)(-3)$

SOLUTION Because we have already found that $(f + g)(x) = 4x^2 - 2x - 19$, we can substitute -3 for x in this function.

$$(f + g)(x) = 4x^2 - 2x - 19$$
$$(f + g)(-3) = 4(-3)^2 - 2(-3) - 19 \quad \text{Substitute } -3 \text{ for } x.$$
$$= 23 \quad \text{Simplify.}$$

$(f + g)(-3) = 23$. An alternative method to find $(f + g)(-3)$ is to substitute -3 for x in the functions $f(x)$ and $g(x)$ and then add the results. The reader can verify that $f(-3) = -10$ and $g(-3) = 33$; so $(f + g)(-3) = -10 + 33 = 23$.

Quick Check 1

If $f(x) = x^2 - 9x + 20$ and $g(x) = 3x - 44$, find the given function.

a) $(f + g)(x)$
b) $(f - g)(x)$
c) $(f + g)(7)$
d) $(f - g)(-2)$

The Product Function and the Quotient Function

The function $(f \cdot g)(x)$ is the product of the two functions $f(x)$ and $g(x)$. The function $\left(\frac{f}{g}\right)(x)$ is the quotient of the two functions $f(x)$ and $g(x)$.

The Product Function

For any two functions $f(x)$ and $g(x)$, $(f \cdot g)(x) = f(x) \cdot g(x)$.

The Quotient Function

For any two functions $f(x)$ and $g(x)$, $\left(\frac{f}{g}\right)(x) = \frac{f(x)}{g(x)}$, $g(x) \neq 0$.

EXAMPLE 2 If $f(x) = x - 5$ and $g(x) = x^2 + 6x + 8$, find $(f \cdot g)(x)$.

SOLUTION Begin by writing $(f \cdot g)(x)$ as $f(x) \cdot g(x)$.

$$(f \cdot g)(x) = f(x) \cdot g(x) \quad \text{Rewrite as the product of the two functions.}$$
$$= (x - 5)(x^2 + 6x + 8) \quad \text{Substitute for each function.}$$
$$= x^3 + 6x^2 + 8x - 5x^2 - 30x - 40 \quad \text{Multiply using the distributive property.}$$
$$= x^3 + x^2 - 22x - 40 \quad \text{Combine like terms.}$$

Quick Check 2

If $f(x) = 3x + 4$ and $g(x) = 4x - 7$, find $(f \cdot g)(x)$.

EXAMPLE 3 If $f(x) = x^2 - 3x - 40$ and $g(x) = x^2 - 25$, find $\left(\frac{f}{g}\right)(x)$ and list any restrictions on its domain.

SOLUTION Begin by writing $\left(\frac{f}{g}\right)(x)$ as $\frac{f(x)}{g(x)}$. Then simplify the resulting rational expression, if possible, using the techniques in Section 7.1.

$$\left(\frac{f}{g}\right)(x) = \frac{f(x)}{g(x)}$$ Rewrite as the quotient of the two functions.

$$= \frac{x^2 - 3x - 40}{x^2 - 25}$$ Substitute for each function.

$$= \frac{(x-8)(x+5)}{(x+5)(x-5)}$$ Factor the numerator and denominator.

$$= \frac{(x-8)\overset{1}{\cancel{(x+5)}}}{\underset{1}{\cancel{(x+5)}}(x-5)}$$ Divide out common factors.

$$= \frac{x-8}{x-5}$$ Simplify.

Recall that any value x for which the denominator of a rational expression, such as $\left(\frac{f}{g}\right)(x)$, is equal to 0 is not in the domain of the function. In this case, $x \neq -5, 5$.

Quick Check 3

If $f(x) = x^2 + 3x - 18$ and $g(x) = x^2 + 13x + 42$, find $\left(\frac{f}{g}\right)(x)$ and list any restrictions on its domain.

Applications

Objective 2 Solve applications involving the sum function or the difference function.

EXAMPLE 4 The number of public elementary and secondary schools in the United States in a particular year can be approximated by the function $f(x) = 950x + 82{,}977$, where x represents the number of years after 1990. The number of private elementary and secondary schools in the United States in a particular year can be approximated by the function $g(x) = 188x + 25{,}941$, where again x represents the number of years after 1990. (*Source:* U.S. Department of Education, National Center for Education Statistics)

a) Find $(f + g)(x)$. In your own words, explain what this function represents.

SOLUTION

$(f + g)(x) = f(x) + g(x)$ Rewrite as the sum of the two functions.

$= (950x + 82{,}977) + (188x + 25{,}941)$ Substitute for each function.

$= 950x + 82{,}977 + 188x + 25{,}941$ Remove parentheses.

$= 1138x + 108{,}918$ Combine like terms.

$(f + g)(x) = 1138x + 108{,}918$. This function gives the combined number of public and private elementary and secondary schools x years after 1990.

b) Find $(f + g)(19)$. In your own words, explain what this number represents.

SOLUTION

$$\begin{aligned}(f + g)(x) &= 1138x + 108{,}918 \\ (f + g)(19) &= 1138(19) + 108{,}918 \quad \text{Substitute 19 for } x. \\ &= 130{,}540 \quad \text{Simplify.}\end{aligned}$$

This number tells us that there will be a total of approximately 130,540 public and private elementary and secondary schools in 2009, which is 19 years after 1990.

Quick Check 4

The number of students enrolled at a public college in the United States in a particular year can be approximated by the function $f(x) = 111{,}916x + 9{,}528{,}000$, where x represents the number of years after 1990. The number of students enrolled at a private college in the United States in a particular year can be approximated by the function $g(x) = 43{,}973x + 2{,}591{,}000$, where again x represents the number of years after 1990. (*Source:* U.S. Department of Education, National Center for Education Statistics)

a) Find $(f + g)(x)$. In your own words, explain what this function represents.
b) Find $(f - g)(x)$. In your own words, explain what this function represents.

Composition of Functions

Objective 3 Find the composite function of two functions $f(x)$ and $g(x)$.
Another way to combine two functions is to use the output of one function as the input for the other function. When this is done, it is called the **composition** of the two functions. For example, consider the functions $f(x) = x + 5$ and $g(x) = 2x + 1$. If we evaluated the function $g(x)$ when $x = 7$, the output will be 15.

$$\begin{aligned}g(7) &= 2(7) + 1 \quad \text{Substitute 7 for } x. \\ &= 15 \quad \text{Simplify.}\end{aligned}$$

We can visualize this with a picture of a function "machine." The machine takes an input of 7 and creates an output of 15.

$$7 \longrightarrow \boxed{g(x) = 2x + 1} \longrightarrow 15$$

Now if we evaluate the function $f(x)$ when $x = 15$, this is a composition of the two functions. The output of function $g(x)$ is the input of the function $f(x)$.

$$\begin{aligned}f(15) &= 15 + 5 \quad \text{Substitute 15 for } x. \\ &= 20 \quad \text{Add.}\end{aligned}$$

The function $f(x)$ takes an input of 15 and creates an output of 20.

$$7 \longrightarrow \boxed{g(x) = 2x + 1} \longrightarrow 15 \longrightarrow \boxed{f(x) = x + 5} \longrightarrow 20$$

Symbolically, this can be represented as $f(g(7)) = 20$.

Composite Function

For any two functions $f(x)$ and $g(x)$, the **composite function** $(f \circ g)(x)$ is defined as $(f \circ g)(x) = f(g(x))$ and read "f of g of x."

The symbol $\circ$ is used to denote the **composition of two functions.**

Suppose that $f(x) = 3x - 5$ and $g(x) = 4x + 9$ and that we want to find $(f \circ g)(5)$. We rewrite $(f \circ g)(5)$ as $f(g(5))$ and begin by evaluating $g(5)$.

$$\begin{aligned} g(5) &= 4(5) + 9 && \text{Substitute 5 for } x. \\ &= 29 && \text{Simplify.} \end{aligned}$$

Now we evaluate $f(29)$.

$$\begin{aligned} (f \circ g)(5) &= f(g(5)) \\ &= f(29) && \text{Substitute 29 for } g(5). \\ &= 3(29) - 5 && \text{Substitute 29 for } x \text{ in the function } f(x). \\ &= 82 && \text{Simplify.} \end{aligned}$$

$$5 \longrightarrow \boxed{g(x) = 4x + 9} \longrightarrow 29 \longrightarrow \boxed{f(x) = 3x - 5} \longrightarrow 82$$

Again, suppose that $f(x) = 3x - 5$ and $g(x) = 4x + 9$ and that we want to find the composite function $(f \circ g)(x)$. We begin by rewriting $(f \circ g)(x)$ as $f(g(x))$; then we replace $g(x)$ with the expression $4x + 9$.

$$\begin{aligned} (f \circ g)(x) &= f(g(x)) \\ &= f(4x + 9) && \text{Replace } g(x) \text{ with } 4x + 9. \\ &= 3(4x + 9) - 5 && \text{Substitute } 4x + 9 \text{ for } x \text{ in the function } f(x). \\ &= 12x + 27 - 5 && \text{Distribute.} \\ &= 12x + 22 && \text{Combine like terms.} \end{aligned}$$

Notice that if we evaluate this composite function for $x = 5$, the output will be $12(5) + 22$, or 82. This is the same result we obtained by first evaluating $g(5)$ and then evaluating $f(x)$ for this value.

EXAMPLE 5 Given that $f(x) = 2x + 11$ and $g(x) = 3x - 10$, find $(f \circ g)(x)$.

SOLUTION When finding a composite function, it is crucial to substitute the correct expression into the correct function. A safe way to ensure this is by rewriting $(f \circ g)(x)$ as $f(g(x))$ and then replacing the "inner" function with the appropriate expression. In this example, we will replace $g(x)$ with $3x - 10$. Then we will evaluate the "outer" function for this expression.

$$\begin{aligned} (f \circ g)(x) &= f(g(x)) \\ &= f(3x - 10) && \text{Replace } g(x) \text{ with } 3x - 10. \\ &= 2(3x - 10) + 11 && \text{Substitute } 3x - 10 \text{ for } x \text{ in the function } f(x). \\ &= 6x - 20 + 11 && \text{Distribute.} \\ &= 6x - 9 && \text{Combine like terms.} \end{aligned}$$

Quick Check 5

Given that $f(x) = 4x - 9$ and $g(x) = 2x + 7$, find $(f \circ g)(x)$.

EXAMPLE 6 Given that $f(x) = 5x - 6$ and $g(x) = x^2 + 2x - 8$, find the given composite functions.

a) $(f \circ g)(x)$

SOLUTION

$$\begin{aligned} (f \circ g)(x) &= f(g(x)) \\ &= f(x^2 + 2x - 8) && \text{Replace } g(x) \text{ with } x^2 + 2x - 8. \end{aligned}$$

$= 5(x^2 + 2x - 8) - 6$ Substitute $x^2 + 2x - 8$ for x in the function $f(x)$.

$= 5x^2 + 10x - 40 - 6$ Distribute.

$= 5x^2 + 10x - 46$ Combine like terms.

b) $(g \circ f)(x)$

SOLUTION In this example, the outer function $g(x)$ is a quadratic function.

$(g \circ f)(x) = g(f(x))$

$= g(5x - 6)$ Replace $f(x)$ with $5x - 6$.

$= (5x - 6)^2 + 2(5x - 6) - 8$ Substitute $5x - 6$ for x in the function $g(x)$.

$= (5x - 6)(5x - 6) + 2(5x - 6) - 8$ Square the binomial $5x - 6$ by multiplying it by itself.

$= 25x^2 - 30x - 30x + 36 + 10x - 12 - 8$ Multiply $(5x - 6)(5x - 6)$ and $2(5x - 6)$.

$= 25x^2 - 50x + 16$ Combine like terms.

Quick Check 6

Given that $f(x) = x + 3$ and $g(x) = x^2 - 3x - 28$, find the given composite functions.

a) $(f \circ g)(x)$
b) $(g \circ f)(x)$

The Domain of a Composite Function

Objective 4 Find the domain of a composite function $(f \circ g)(x)$. The domain of a composite function $(f \circ g)(x)$ is the set of all input values x in the domain of $g(x)$ whose output from $g(x)$ is in the domain of $f(x)$. In all of the previous examples, the domain was the set of all real numbers $\mathbb{R}$, as none of the functions had any restrictions on their domains.

EXAMPLE 7 Given that $f(x) = \dfrac{x + 9}{2x - 1}$ and $g(x) = 6x + 15$, find $(f \circ g)(x)$ and state its domain.

SOLUTION Because $g(x)$ is a linear function, there are no restrictions on its domain. Once we find the composite function $(f \circ g)(x)$, we will find the domain of that function.

$(f \circ g)(x) = f(g(x))$

$= f(6x + 15)$ Replace $g(x)$ with $6x + 15$.

$= \dfrac{(6x + 15) + 9}{2(6x + 15) - 1}$ Substitute $6x + 15$ for x in the function $f(x)$.

$= \dfrac{6x + 15 + 9}{12x + 30 - 1}$ Distribute.

$= \dfrac{6x + 24}{12x + 29}$ Combine like terms.

So $(f \circ g)(x) = \frac{6x + 24}{12x + 29}$. The composite function is a rational function, and we find the restrictions on the domain by setting the denominator equal to 0 and solving.

$$12x + 29 = 0 \quad \text{Set the denominator equal to 0.}$$
$$12x = -29 \quad \text{Subtract 29.}$$
$$x = -\frac{29}{12} \quad \text{Divide by 12.}$$

The domain of this composite function $(f \circ g)(x)$ is the set of all real numbers except $-\frac{29}{12}$.

Quick Check 7

Given that $f(x) = \dfrac{2x + 5}{x + 7}$ and $g(x) = x - 4$, find $(f \circ g)(x)$ and state its domain.

BUILDING YOUR STUDY STRATEGY

Study Environment, 5 Being Comfortable It is important that you are comfortable while studying, without being too comfortable. If the room is too warm, you may get tired. If the room is too cold, you may be distracted by thinking about how cold you are. If your chair is not comfortable, you may fidget and lose your concentration. If your chair is too comfortable, you may get sleepy. You may have the same problem if you eat a large meal before sitting down to study. In conclusion, choose a situation where you will be comfortable but not too comfortable.

Exercises 11.5

PRACTICE

WATCH

DOWNLOAD

READ

REVIEW

Vocabulary

1. For two functions $f(x)$ and $g(x)$, the function denoted by $(f + g)(x)$ is called the ______ function.
2. For two functions $f(x)$ and $g(x)$, the difference function is denoted by ________.
3. For two functions $f(x)$ and $g(x)$, the product function is denoted by ________.
4. For two functions $f(x)$ and $g(x)$, the function denoted by $\left(\frac{f}{g}\right)(x)$ is called the ______ function.
5. When the output of one function is used as the input for another function, this is called the ________ of the two functions.
6. For any two functions $f(x)$ and $g(x)$, the ________ function $(f \circ g)(x)$ is defined as $(f \circ g)(x) = f(g(x))$.

For the given functions f(x) and g(x), find
a) $(f + g)(x)$ ***b)*** $(f + g)(8)$ ***c)*** $(f + g)(-3)$

7. $f(x) = 4x + 11, g(x) = 3x - 17$
8. $f(x) = 2x - 13, g(x) = -7x + 6$
9. $f(x) = 5x + 8, g(x) = x^2 - x - 19$
10. $f(x) = x^2 - 7x + 12, g(x) = x^2 + 4x - 32$

For the given functions f(x) and g(x), find
a) $(f - g)(x)$ ***b)*** $(f - g)(6)$ ***c)*** $(f - g)(-5)$

11. $f(x) = 6x - 11, g(x) = -2x - 5$
12. $f(x) = 8x - 13, g(x) = 8x + 13$
13. $f(x) = 10x + 37, g(x) = x^2 - 5x - 12$
14. $f(x) = x^2 + 7x + 100, g(x) = -x^2 + 12x - 25$

For the given functions f(x) and g(x), find
a) $(f \cdot g)(x)$ ***b)*** $(f \cdot g)(5)$ ***c)*** $(f \cdot g)(-3)$

15. $f(x) = 2x - 3, g(x) = x + 4$
16. $f(x) = 3x + 7, g(x) = 4x - 1$
17. $f(x) = x - 6, g(x) = x^2 + 3x - 4$

18. $f(x) = 2x - 5$, $g(x) = x^2 - 5x + 8$

For the given functions $f(x)$ and $g(x)$, find
a) $\left(\frac{f}{g}\right)(x)$ *b)* $\left(\frac{f}{g}\right)(7)$ *c)* $\left(\frac{f}{g}\right)(-2)$

19. $f(x) = 3x + 1$, $g(x) = x + 7$

20. $f(x) = 6x - 22$, $g(x) = x^2 + 11x + 10$

21. $f(x) = x^2 - 9$, $g(x) = x^2 + 7x + 12$

22. $f(x) = x^2 + 10x + 16$, $g(x) = x^2 - x - 72$

Let $f(x) = 2x + 9$ and $g(x) = 5x - 1$. Find the following.

23. $(f + g)(8)$
24. $(f + g)(-2)$
25. $(f - g)(-4)$
26. $(f - g)(5)$
27. $(f \cdot g)(10)$
28. $(f \cdot g)(-3)$
29. $\left(\frac{f}{g}\right)(-2)$
30. $\left(\frac{f}{g}\right)(0)$

Let $f(x) = 5x - 9$ and $g(x) = 2x - 17$. Find the following and simplify completely.

31. $(f + g)(x)$
32. $(f - g)(x)$
33. $(f \cdot g)(x)$
34. $\left(\frac{f}{g}\right)(x)$

Let $f(x) = x - 10$ and $g(x) = x^2 - 8x - 20$. Find the following and simplify completely.

35. $(f - g)(x)$
36. $(f + g)(x)$
37. $\left(\frac{f}{g}\right)(x)$
38. $(f \cdot g)(x)$
39. The number of male doctors, in thousands, in the United States in a particular year can be approximated by the function $f(x) = 10.4x + 398$, where x represents the number of years after 1980. The number of female doctors, in thousands, in the United States in a particular year can be approximated by the function $g(x) = 9.1x + 49$, where again x represents the number of years after 1980. (*Source:* American Medical Association)

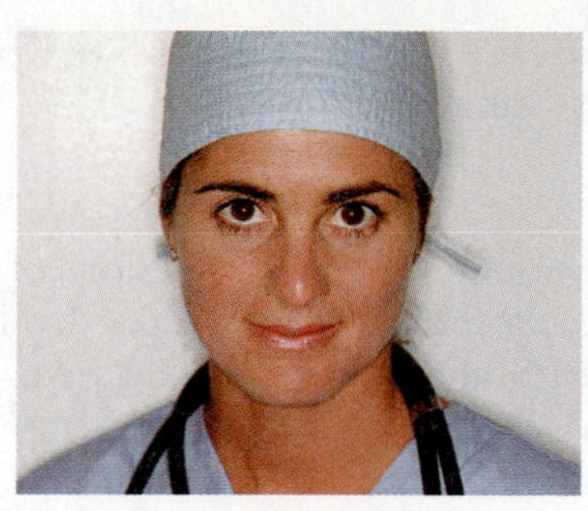

a) Find $(f + g)(x)$. In your own words, explain what this function represents.

b) Find $(f + g)(40)$. In your own words, explain what this number represents.

40. The number of male inmates who are HIV positive in the United States in a particular year can be approximated by the function $f(x) = 223x + 20{,}969$, where x represents the number of years after 1995. The number of female inmates who are HIV positive in the United States in a particular year can be approximated by the function $g(x) = 83x + 2102$, where again x represents the number of years after 1995. (*Source:* U.S. Department of Justice, Bureau of Justice Statistics)

a) Find $(f + g)(x)$. In your own words, explain what this function represents.

b) Find $(f + g)(27)$. In your own words, explain what this number represents.

41. The number of master's degrees, in thousands, earned by females in the United States in a particular year can be approximated by the function $f(x) = 9.2x + 181.8$, where x represents the number of years after 1990. The number of master's degrees, in thousands, earned by males in the United States in a particular year can be approximated by the function $g(x) = 3.4x + 160.9$, where again x represents the number of years after 1990. (*Source:* U.S. Department of Education, National Center for Education Statistics)

a) Find $(f - g)(x)$. In your own words, explain what this function represents.

b) Find $(f - g)(30)$. In your own words, explain what this number represents.

42. The amount of money, in billions of dollars, spent on health care that was covered by insurance in the United States in a particular year can be approximated by the function $f(x) = x^2 + 11x + 244$, where x represents the number of years after 1990. The amount of money, in billions of dollars, spent on health care that was paid out of pocket in the United States in a particular year can be approximated by the function $g(x) = 6x + 129$, where again x represents the number of years after 1990. (*Source:* U.S. Centers for Medicare and Medicaid Services)

a) Find $(f - g)(x)$. In your own words, explain what this function represents.

b) Find $(f - g)(40)$. In your own words, explain what this number represents.

Let $f(x) = 4x + 7$ and $g(x) = x - 3$. Find the following.

43. $(f \circ g)(5)$

44. $(f \circ g)(-2)$

45. $(g \circ f)(-1)$

46. $(g \circ f)(10)$

For the given functions $f(x)$ and $g(x)$, find a) $(f \circ g)(x)$ b) $(f \circ g)(5)$

47. $f(x) = 3x + 2, g(x) = x^2 + 5x - 6$

48. $f(x) = 2x - 7, g(x) = x^2 - 8x + 10$

49. $f(x) = x^2 - 7x + 12, g(x) = x + 8$

50. $f(x) = x^2 + 9x - 22, g(x) = 4x - 3$

For the given functions $f(x)$ and $g(x)$, find a) $(f \circ g)(x)$ b) $(g \circ f)(x)$

51. $f(x) = 3x + 5, g(x) = 2x + 4$

52. $f(x) = 5x - 9, g(x) = -x + 6$

53. $f(x) = x + 4, g(x) = x^2 + 3x - 40$

54. $f(x) = x^2 - 7x - 18, g(x) = x - 9$

For the given functions $f(x)$ and $g(x)$, find $(f \circ g)(x)$ and state its domain.

55. $f(x) = \frac{5x}{x + 7}, g(x) = 2x + 13$

56. $f(x) = \frac{2x - 3}{3x + 5}, g(x) = x + 8$

57. $f(x) = \sqrt{2x - 10}, g(x) = x + 11$

58. $f(x) = \sqrt{x - 2}, g(x) = 3x + 14$

For the given function $f(x)$, find $(f \circ f)(x)$.

59. $f(x) = 3x - 10$

60. $f(x) = -5x + 18$

61. $f(x) = x^2 + 6x + 12$

62. $f(x) = x^2 - 2x - 15$

For the given functions $f(x)$ and $g(x)$, find a value of x for which $(f \circ g)(x) = (g \circ f)(x)$.

63. $f(x) = x^2 + 4x + 3, g(x) = x + 5$

64. $f(x) = x^2 + 6x - 9, g(x) = x - 3$

65. $f(x) = x + 7, g(x) = x^2 - 25$

66. $f(x) = x - 4, g(x) = 2x^2 + 15x + 18$

For the given function $g(x)$, find a function $f(x)$ such that $(f \circ g)(x) = x$.

67. $g(x) = x + 8$

68. $g(x) = -5x$

69. $g(x) = 3x + 8$

70. $g(x) = 4x - 17$

Writing in Mathematics

Answer in complete sentences.

71. In general, is $(f + g)(x)$ equal to $(g + f)(x)$? If your answer is yes, explain why. If your answer is no, give an example that shows why, in general, they are not equal.

72. In general, is $(f - g)(x)$ equal to $(g - f)(x)$? If your answer is yes, explain why. If your answer is no, give an example that shows why, in general, they are not equal.

11.6 Inverse Functions

OBJECTIVES

1. Determine whether a function is one-to-one.
2. Use the horizontal-line test to determine whether a function is one-to-one.
3. Understand inverse functions.
4. Determine whether two functions are inverse functions.
5. Find the inverse of a one-to-one function.
6. Find the inverse of a function from its graph.

One-to-One Functions

Objective 1 Determine whether a function is one-to-one. Recall that for any function $f(x)$, each value in the domain corresponds to only one value in the range.

One-to-One Function

If every element in the range of a function $f(x)$ has at most one value in the domain that corresponds to it, the function $f(x)$ is called a **one-to-one function.**

In other words, if $f(x)$ is different for each input value x in the domain, $f(x)$ is a one-to-one function.

Suppose for some function $f(x)$, both $f(-1)$ and $f(3)$ are equal to 2. This function is not a one-to-one function because one output value is associated with two different input values. However, the function shown in the following table is a one-to-one function, because each function value corresponds to only one input value.

x	0	1	4	9	16
$f(x)$	0	1	2	3	4

A function whose input is a person's name and whose output is that person's birthday is not a one-to-one function because several people (input) can have the same birthday (output). A function whose input is a person's name and whose output is that person's Social Security number is a one-to-one function because no two people can have the same Social Security number.

EXAMPLE 1 Determine whether the function represented by the set of ordered pairs $\{(-7, -5), (-5, -2), (-2, 4), (3, 3), (6, -7)\}$ is a one-to-one function.

SOLUTION Because each x-value corresponds to only one y-value and each y-value corresponds to only one x-value, the function is one-to-one.

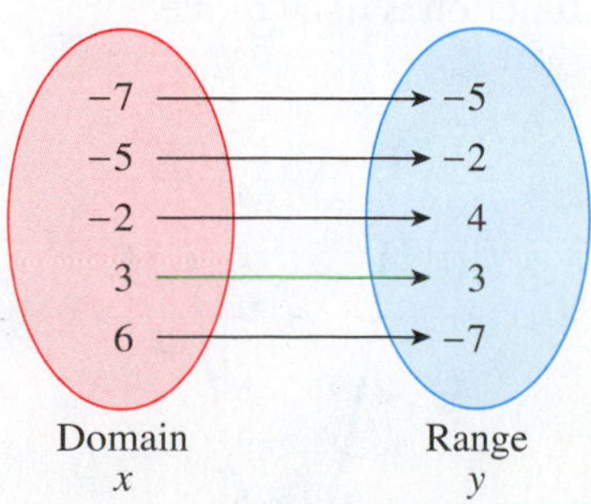

Quick Check 1

Determine whether the function represented by the set of ordered pairs $\{(-4, -9), (-2, -7), (1, -6), (2, -7), (6, -11)\}$ is a one-to-one function.

Horizontal-Line Test

Objective 2 Use the horizontal-line test to determine whether a function is one-to-one. We can determine whether a function is one-to-one by applying the **horizontal-line test** to its graph.

Horizontal-Line Test

If a horizontal line can intersect the graph of a function at more than one point, the function is not one-to-one.

Consider the graph of a quadratic function shown. We can draw a horizontal line that intersects the function at two points, as indicated. The coordinates of these two points are $(-2, 6)$ and $(2, 6)$. This shows that two different input values, -2 and 2, correspond to the same output value (6). Therefore, the function is not one-to-one.

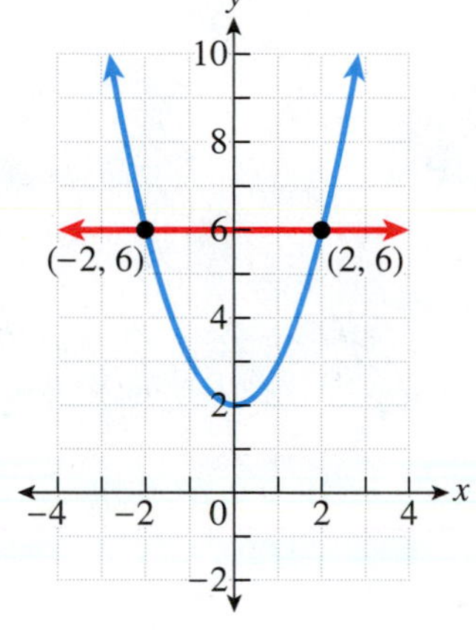

The linear function graphed is a one-to-one function. No horizontal line crosses the graph of this function at more than one point, so the function passes the horizontal-line test.

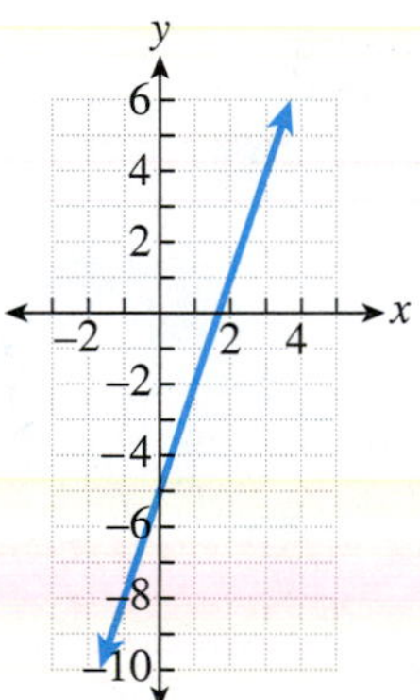

Quick Check 2

Determine whether the function is one-to-one.

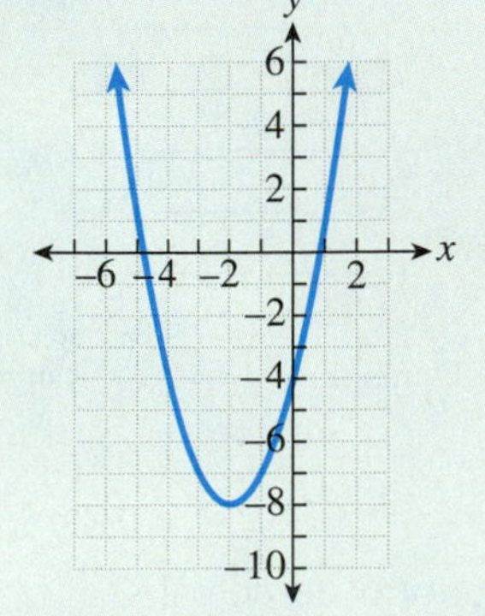

EXAMPLE 2 Use the horizontal-line test to determine whether the function is one-to-one.

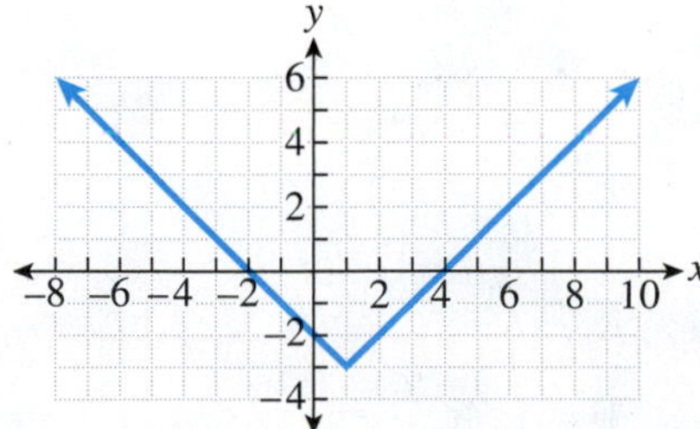

SOLUTION We can draw a horizontal line that crosses the graph of this absolute value function at more than one point, as shown here. This function fails the horizontal-line test and is not one-to-one.

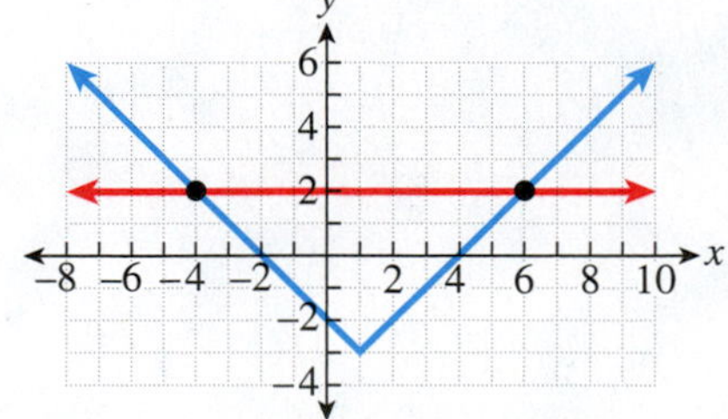

Inverse Functions

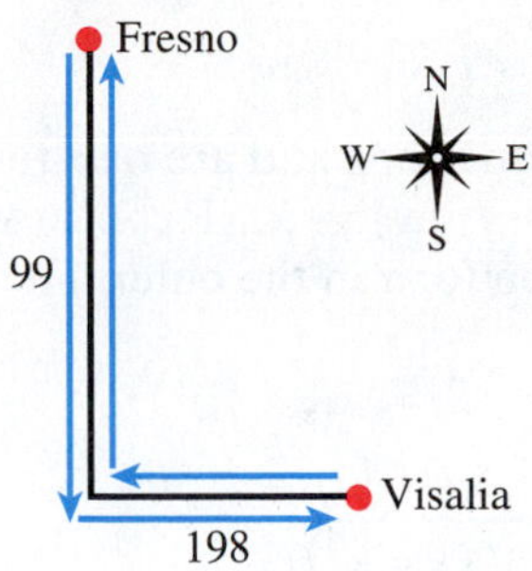

Objective 3 Understand inverse functions. Anne commutes to work from Visalia to Fresno each day. She starts by taking Highway 198 west and then takes Highway 99 north. How does she return home each day? She begins by taking Highway 99 south and then takes Highway 198 east. Her trips are depicted in the map at the left.

The second route takes Anne back to the starting point. The two routes have an **inverse** relationship. In this section, we will be examining **inverse functions.**

The inverse of a one-to-one function "undoes" what the function does. For example, if a function $f(x)$ takes an input value of 3 and produces an output value of 7, its inverse function takes an input value of 7 and produces an output value of 3.

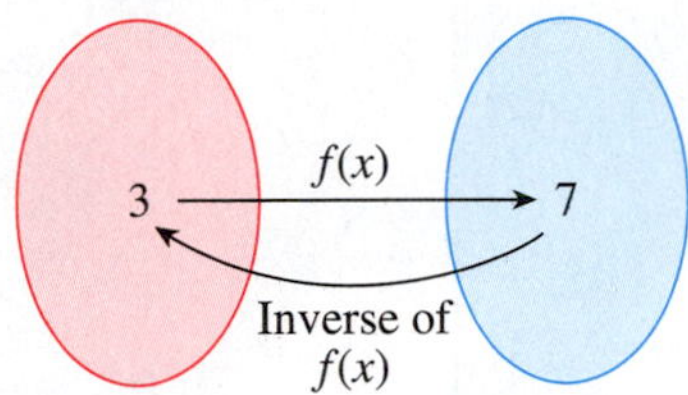

We see that the ordered pair $(3, 7)$ is in the function $f(x)$ and that the ordered pair $(7, 3)$ is in the inverse function of $f(x)$. The inputs of a function are the outputs of its inverse function, and the outputs of a function are the inputs of its inverse function. In general, if (a, b) is in a function, (b, a) is in the inverse of that function.

Verifying That Two Functions Are Inverses

Objective 4 Determine whether two functions are inverse functions. We can determine whether two functions are inverse functions algebraically by finding their composite functions.

Inverse Functions

Two one-to-one functions $f(x)$ and $g(x)$ are inverse functions if $(f \circ g)(x) = x$ and $(g \circ f)(x) = x$.

If $f(x)$ and $g(x)$ are inverse functions, each function "undoes" the actions of the other function.

The function g takes an input value x and produces an output value $g(x)$.

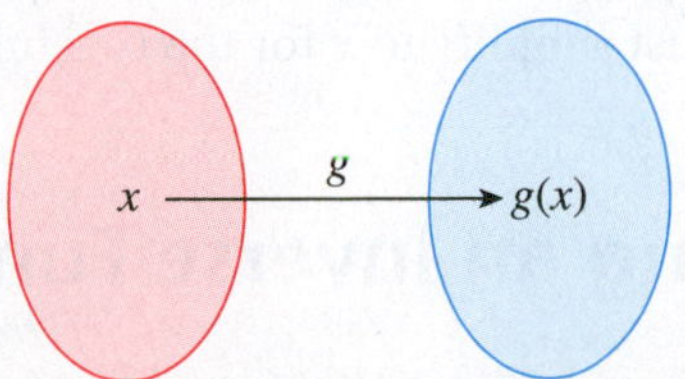

The function f takes this value and produces an output value of x.

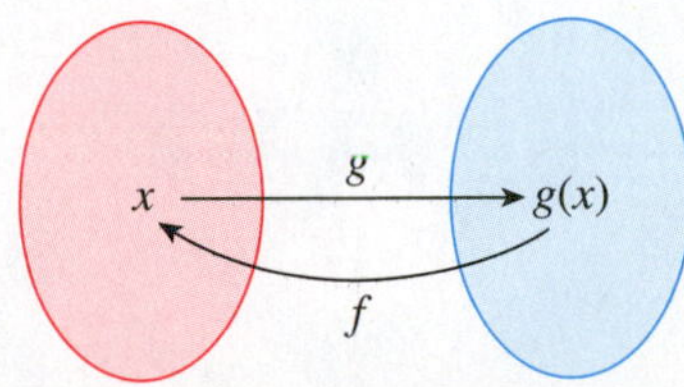

So $f(g(x)) = x$, or in other words, $(f \circ g)(x) = x$. In a similar fashion, we could show that $(g \circ f)(x) = x$ as well.

EXAMPLE 3 Determine whether the following two functions are inverse functions: $f(x) = 5x - 2$, $g(x) = \frac{x+2}{5}$

SOLUTION Both of these functions are linear functions and are one-to-one. We will determine whether $(f \circ g)(x) = x$ and $(g \circ f)(x) = x$. (Replace the inner function with its formula, substitute that expression for x in the outer function, and simplify.)

$(f \circ g)(x) = x$	$(g \circ f)(x) = x$
$(f \circ g)(x) = f(g(x))$	$(g \circ f)(x) = g(f(x))$
$= f\left(\frac{x+2}{5}\right)$	$= g(5x - 2)$
$= 5\left(\frac{x+2}{5}\right) - 2$	$= \frac{(5x-2)+2}{5}$
$= \overset{1}{\cancel{5}}\left(\frac{x+2}{\underset{1}{\cancel{5}}}\right) - 2$	$= \frac{5x-2+2}{5}$
$= x + 2 - 2$	$= \frac{5x}{5}$
$= x$	$= \frac{\overset{1}{\cancel{5}}x}{\underset{1}{\cancel{5}}}$
	$= x$

Because $(f \circ g)(x) = x$ and $(g \circ f)(x) = x$, these two functions are inverses.

Quick Check 3

Determine whether the following two functions are inverse functions:

$f(x) = \frac{x-7}{4}$, $g(x) = 4x + 7$

EXAMPLE 4 Determine whether the following two functions are inverse functions: $f(x) = \frac{1}{4}x - \frac{3}{4}$, $g(x) = 3x + 4$

SOLUTION These two functions are linear and one-to-one. We begin by determining whether $(f \circ g)(x) = x$.

$$\begin{aligned}(f \circ g)(x) &= f(g(x)) \\ &= f(3x+4) && \text{Replace } g(x) \text{ with } 3x+4. \\ &= \frac{1}{4}(3x+4) - \frac{3}{4} && \text{Substitute } 3x+4 \text{ for } x \text{ in the function } f(x). \\ &= \frac{3}{4}x + 1 - \frac{3}{4} && \text{Distribute.} \\ &= \frac{3}{4}x + \frac{1}{4} && \text{Combine like terms.}\end{aligned}$$

$(f \circ g)(x) \neq x$. These two functions are not inverses.

Note that it is not necessary to check $(g \circ f)(x)$ because both composite functions must simplify to x for the two functions to be inverse functions.

Quick Check 4

Determine whether the following two functions are inverse functions:

$f(x) = \frac{1}{3}x + 4$,

$g(x) = 3x - 12$

Finding an Inverse Function

Objective 5 Find the inverse of a one-to-one function. We use the notation $f^{-1}(x)$ to denote the inverse of a function $f(x)$. We read $f^{-1}(x)$ as *f inverse of x*.

A WORD OF CAUTION When a superscript of −1 is written after the name of a function, the −1 is not an exponent; it is used to denote the inverse of a function. In other words, $f^{-1}(x)$ is not the same as $\frac{1}{f(x)}$.

Here is a procedure that can be used to find the inverse of a one-to-one function $f(x)$.

Finding $f^{-1}(x)$

1. **Determine whether $f(x)$ is one-to-one.** We can use the horizontal-line test to determine this. If $f(x)$ is not one-to-one, it does not have an inverse function.
2. **Replace $f(x)$ with y.**
3. **Interchange x and y.** We interchange these two variables because we know that the input of $f(x)$ must be the output of $f^{-1}(x)$ and the output of $f(x)$ must be the input of $f^{-1}(x)$.
4. **Solve the resulting equation for y.** This allows us to express the inverse function as a function of x.
5. **Replace y with $f^{-1}(x)$.**

EXAMPLE 5 Find $f^{-1}(x)$ for the function $f(x) = 5x - 3$.

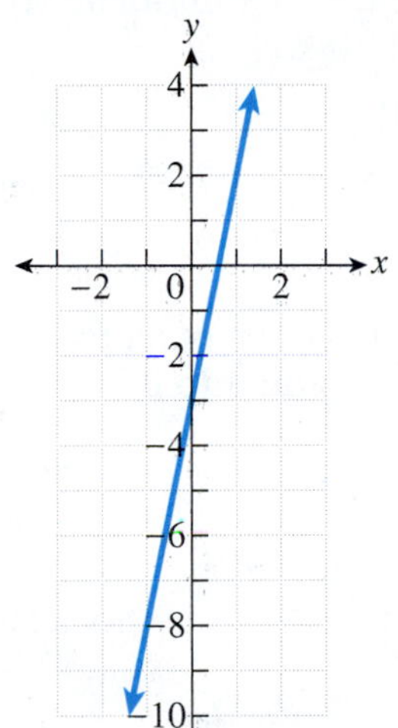

SOLUTION We begin by determining whether the function is one-to-one. Because the function is a linear function, it is one-to-one. Shown at the left is the graph of $f(x)$. The function clearly passes the horizontal-line test.

We begin to find $f^{-1}(x)$ by replacing $f(x)$ with y.

$$f(x) = 5x - 3$$

$$y = 5x - 3 \quad \text{Replace } f(x) \text{ with } y.$$

$$x = 5y - 3 \quad \text{Interchange } x \text{ and } y.$$

$$x + 3 = 5y \quad \text{Add 3 to isolate the term containing } y.$$

$$\frac{x+3}{5} = y \quad \text{Divide both sides by 5 to isolate } y.$$

$$f^{-1}(x) = \frac{x+3}{5} \quad \text{Replace } y \text{ with } f^{-1}(x). \text{ It is customary to write } f^{-1}(x) \text{ on the left side.}$$

The inverse of $f(x) = 5x - 3$ is $f^{-1}(x) = \frac{x+3}{5}$, which could be written as $f^{-1}(x) = \frac{1}{5}x + \frac{3}{5}$.

Quick Check 5

Find $f^{-1}(x)$ for the function $f(x) = \frac{1}{2}x + 10$.

EXAMPLE 6 Find $f^{-1}(x)$ for the function $f(x) = \frac{8}{x+1}$ and state the restrictions on its domain.

SOLUTION In this example, $f(x)$ is a rational function. We have not yet learned to graph rational functions, but the graph of $f(x)$, which has been generated by technology, is shown at the right. (We could use a graphing calculator or software to generate this graph.) The function passes the horizontal-line test.

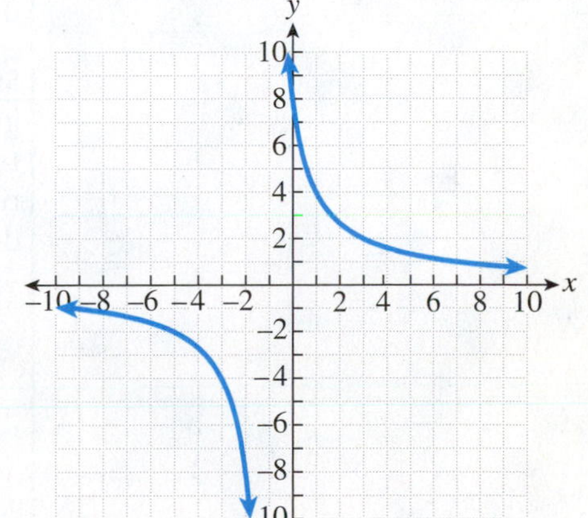

Begin to find $f^{-1}(x)$ by replacing $f(x)$ with y.

$$f(x) = \frac{8}{x+1}$$

$$y = \frac{8}{x+1}$$ Replace $f(x)$ with y.

$$x = \frac{8}{y+1}$$ Interchange x and y.

$$x(y+1) = \frac{8}{\cancel{y+1}} \cdot (\cancel{y+1})$$ Multiply both sides by $y + 1$ to clear the equation of fractions.

$$x(y+1) = 8$$ Simplify.

$$xy + x = 8$$ Multiply.

$$xy = 8 - x$$ Subtract x from both sides to isolate the term containing y.

$$y = \frac{8-x}{x}$$ Divide both sides by x to isolate y.

$$f^{-1}(x) = \frac{8-x}{x}$$ Replace y with $f^{-1}(x)$.

The inverse of $f(x) = \frac{8}{x+1}$ is $f^{-1}(x) = \frac{8-x}{x}$.

Note that the domain of this inverse function is the set of all real numbers except 0, because the denominator of the function is equal to 0 when $x = 0$. This makes the inverse function undefined when $x = 0$.

Quick Check 6

Find $f^{-1}(x)$ for the function $f(x) = \frac{3+5x}{x}$ and state the restrictions on its domain.

Recall that the output values of a one-to-one function are the input values of its inverse function. So the range of a one-to-one function $f(x)$ is the domain of its inverse function $f^{-1}(x)$. Occasionally, if the range of a one-to-one function $f(x)$ is not the set of real numbers, we must restrict the domain of $f^{-1}(x)$ accordingly. This is illustrated in the next example.

EXAMPLE 7 Find $f^{-1}(x)$ for the function $f(x) = \sqrt{x+4} - 2$. State the domain of $f^{-1}(x)$.

SOLUTION Begin by showing that the function is one-to-one. Following is the graph of $f(x)$. (For help on graphing square-root functions, refer back to Section 11.4.) The function passes the horizontal-line test, so it is a one-to-one function that has an inverse function. The range of $f(x)$ is $[-2, \infty)$, and this is the domain of its inverse function $f^{-1}(x)$.

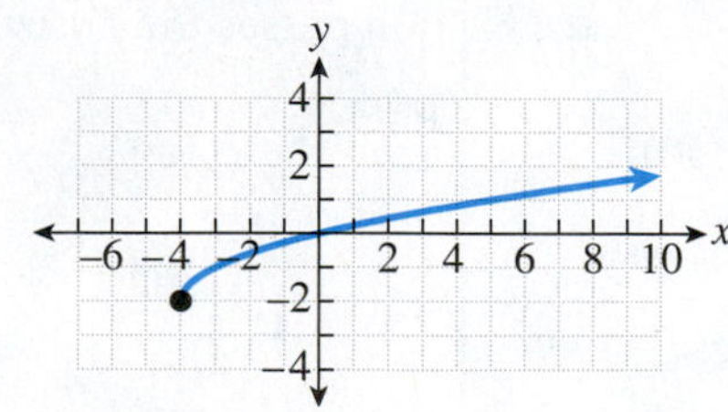

Begin to find $f^{-1}(x)$ by replacing $f(x)$ with y.

$$f(x) = \sqrt{x+4} - 2$$
$$y = \sqrt{x+4} - 2 \quad \text{Replace } f(x) \text{ with } y.$$
$$x = \sqrt{y+4} - 2 \quad \text{Interchange } x \text{ and } y.$$
$$x + 2 = \sqrt{y+4} \quad \text{Add 2 to both sides to isolate the radical.}$$
$$(x+2)^2 = \left(\sqrt{y+4}\right)^2 \quad \text{Square both sides.}$$
$$(x+2)(x+2) = y + 4 \quad \text{Square } x + 2 \text{ by multiplying it by itself.}$$
$$x^2 + 4x + 4 = y + 4 \quad \text{Multiply } (x+2)(x+2).$$
$$x^2 + 4x = y \quad \text{Subtract 4 from both sides to isolate } y.$$
$$f^{-1}(x) = x^2 + 4x \quad \text{Replace } y \text{ with } f^{-1}(x).$$

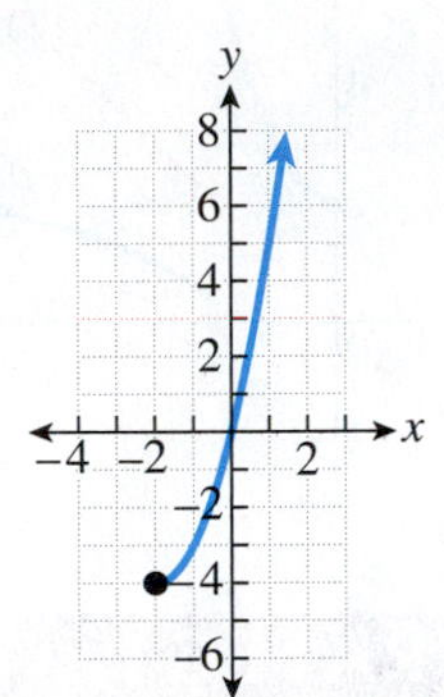

The inverse of $f(x) = \sqrt{x+4} - 2$ is $f^{-1}(x) = x^2 + 4x$. The domain of $f^{-1}(x)$ is $[-2, \infty)$.

At the left is the graph of $f^{-1}(x)$. Notice that the graph does pass the horizontal-line test with this restricted domain. If we did not restrict this domain, the graph of $f^{-1}(x)$ would be a parabola and the function would not be one-to-one.

Quick Check 7

Find $f^{-1}(x)$ for the function $f(x) = \sqrt{x-9} + 8$. State the domain of $f^{-1}(x)$.

Finding an Inverse Function from a Graph

Objective 6 Find the inverse of a function from its graph. We know that if $f(x)$ is a one-to-one function and $f(a) = b$, $f^{-1}(b) = a$. This tells us that if the point (a, b) is on the graph of $f(x)$, the point (b, a) is on the graph of $f^{-1}(x)$.

EXAMPLE 8 If the function is one-to-one, find its inverse.

$$\{(-2, -3), (-1, -2), (0, 1), (1, 5), (2, 13)\}$$

SOLUTION Because each value of x corresponds to only one value of y and each value of y corresponds to only one value of x, this is a one-to-one function.

To find the inverse function, we interchange each x-coordinate and y-coordinate. The inverse function is $\{(-3, -2), (-2, -1), (1, 0), (5, 1), (13, 2)\}$.

EXAMPLE 9 For the given graph of a one-to-one function $f(x)$, graph its inverse function $f^{-1}(x)$.

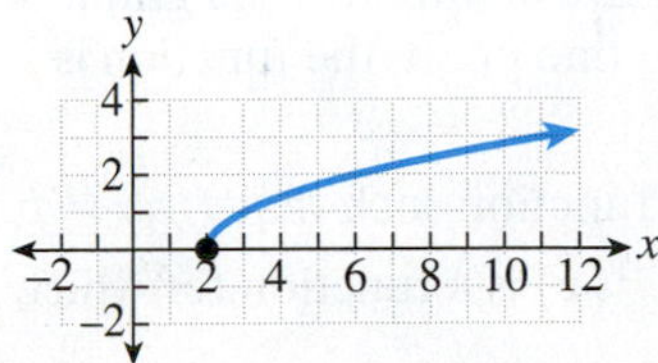

SOLUTION We begin by identifying the coordinates of some points that are on the graph of this function. We will use the points $(2, 0)$, $(3, 1)$, and $(6, 2)$.

By interchanging each x-coordinate with its corresponding y-coordinate, we know that the points $(0, 2)$, $(1, 3)$, and $(2, 6)$ are on the graph of $f^{-1}(x)$. We finish by drawing a graph that passes through these points. Shown at the right is the graph of $f^{-1}(x)$.

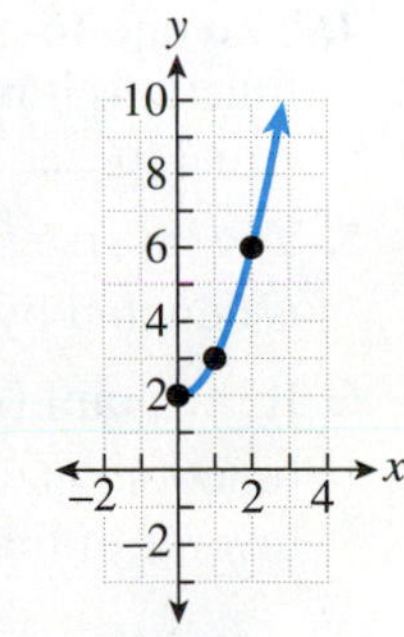

Quick Check 8

For the given graph of a one-to-one function $f(x)$, graph its inverse function $f^{-1}(x)$.

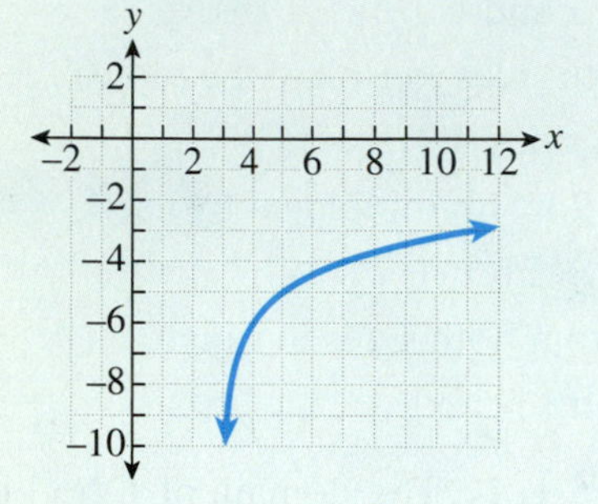

Here are the graphs of $f(x)$ and $f^{-1}(x)$ from the previous example on the same set of axes, along with the line $y = x$. Notice that the two functions look somewhat similar; in fact, they are mirror images of each other. For any one-to-one function $f(x)$ and its inverse function $f^{-1}(x)$, their graphs are symmetric to each other about the line $y = x$. If we fold the graph along the line $y = x$, the graphs of $f(x)$ and $f^{-1}(x)$ will lie on top of each other.

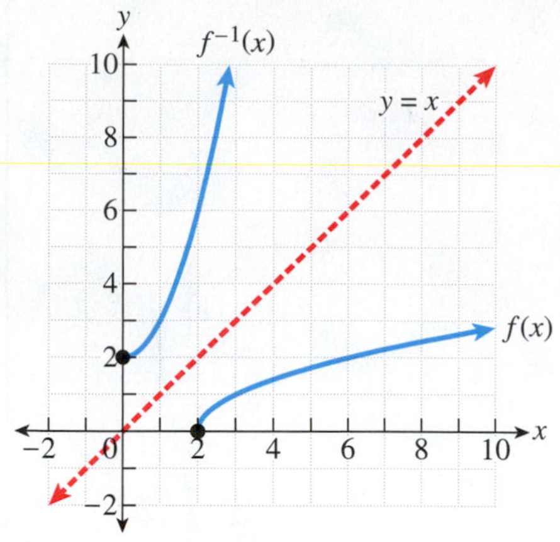

BUILDING YOUR STUDY STRATEGY

Study Environment, 6 Distractions Under ideal circumstances, you should do some of your studying on campus and some at home. The best time for studying new material is as soon as possible after class. Because studying new material takes more concentration, find a location with few distractions. Most campus libraries have an area designated as a silent study area.

While studying at home, stay away from the computer, phone, and television. These devices will steal time away from you, without your even realizing it.

Exercises 11.6

PRACTICE

WATCH

DOWNLOAD

READ

REVIEW

Vocabulary

1. If $f(x)$ is different for each input value x in its domain, $f(x)$ is a(n) __________ function.
2. If a(n) ______________ can intersect the graph of a function at more than one point, the function is not one-to-one.
3. If $f(x)$ is a one-to-one function such that $f(a) = b$, its ________ function $f^{-1}(x)$ is a function for which $f^{-1}(b) = a$.
4. Two one-to-one functions $f(x)$ and $g(x)$ are inverse functions if both $(f \circ g)(x)$ and $(g \circ f)(x)$ are equal to ____.
5. The inverse function of a one-to-one function $f(x)$ is denoted by ________.
6. If the point (a, b) is on the graph of a one-to-one function, the point ______ is on the graph of the inverse of that function.

Determine whether the function f(x) is a one-to-one function.

7.

x	−3	0	3	6	9
$f(x)$	−8	−2	4	10	16

8.

x	−5	−1	1	4	10
$f(x)$	9	5	3	0	−6

9.

x	−4	−2	0	2	4
$f(x)$	16	4	0	4	16

10.

x	−3	−2	−1	0	1
$f(x)$	3	2	1	0	1

Determine whether the function represented by the set of ordered pairs is a one-to-one function.

11. $\{(-9, -3), (-4, 0), (1, 1), (5, -2), (10, -3)\}$
12. $\{(1, 4), (3, 8), (5, 12), (7, 8), (9, 4)\}$
13. $\{(-8, -4), (-5, 3), (-1, 0), (4, 3), (9, 10)\}$

14. $\{(0, 1), (1, 2), (2, 4), (3, 8), (4, 16)\}$

15. Is the function whose input is a person and whose output is that person's mother a one-to-one function? Explain your answer in your own words.

16. Is the function whose input is a student and whose output is that student's favorite math teacher a one-to-one function? Explain your answer in your own words.

17. Is the function whose input is a state and whose output is that state's governor a one-to-one function? Explain your answer in your own words.

18. Is the function whose input is a licensed driver and whose output is that driver's license number a one-to-one function? Explain your answer in your own words.

Use the horizontal-line test to determine whether the function is one-to-one.

19.

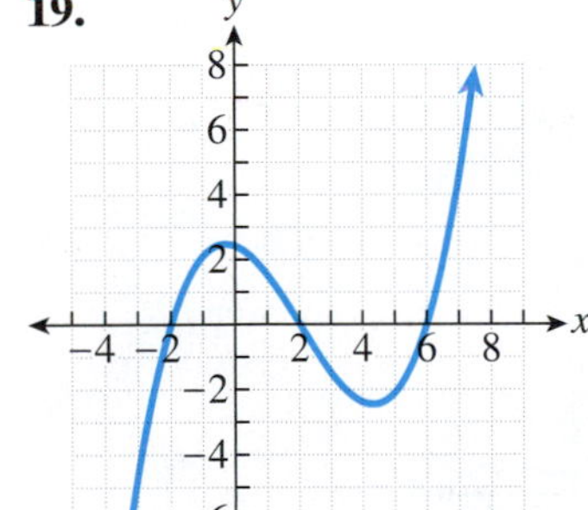

20.

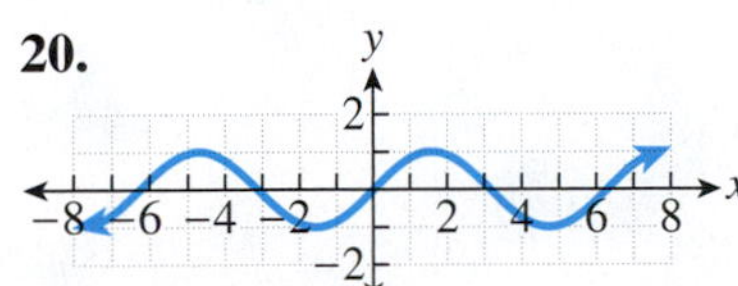

21.

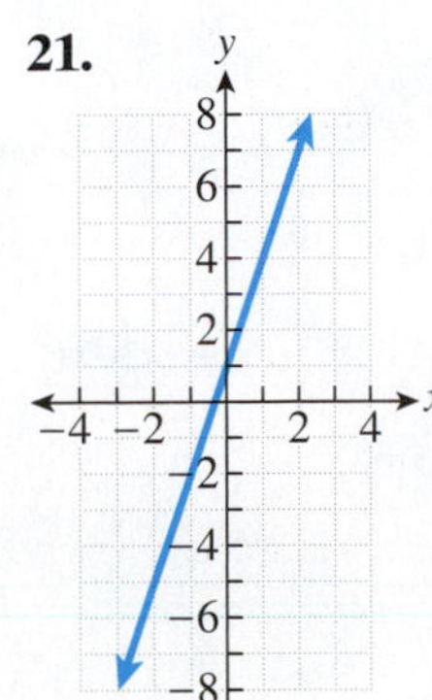

22.

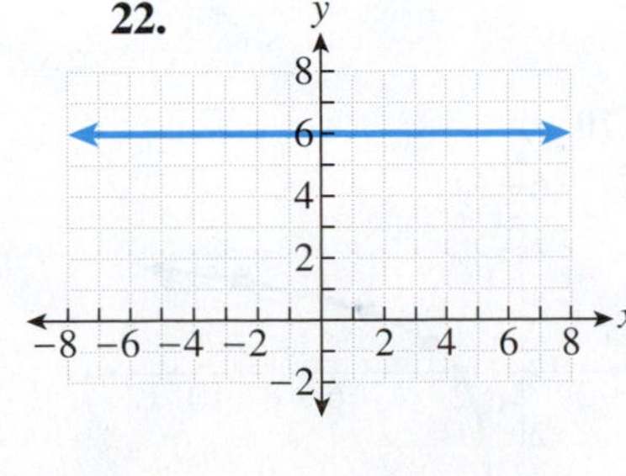

23.

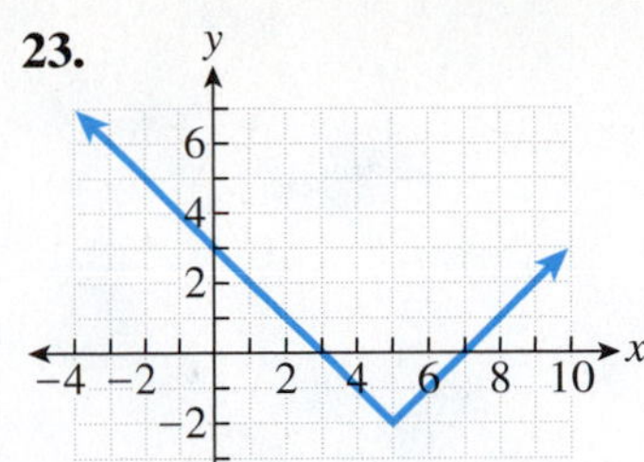

24.

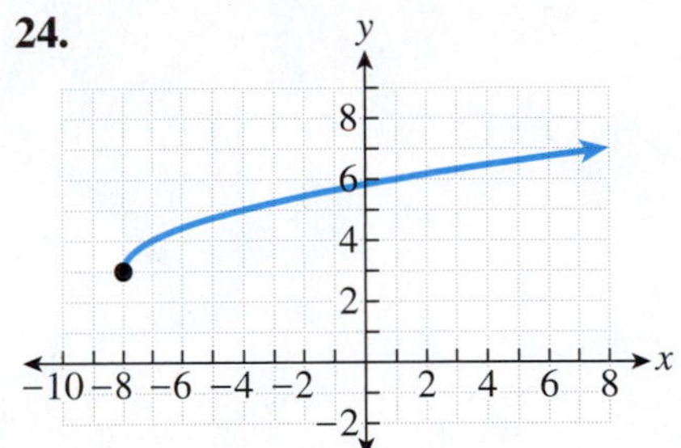

Determine whether the functions $f(x)$ and $g(x)$ are inverse functions by showing that $(f \circ g)(x) = x$ and $(g \circ f)(x) = x$.

25. $f(x) = x + 5, g(x) = 5 + x$

26. $f(x) = x + 8, g(x) = x - 8$

27. $f(x) = 4x, g(x) = \frac{x}{4}$

28. $f(x) = 7x, g(x) = -7x$

29. $f(x) = 3x - 4, g(x) = \frac{x + 4}{3}$

30. $f(x) = 2x + 5, g(x) = \frac{1}{2}x - 5$

31. $f(x) = -x + 13, g(x) = x - 13$

32. $f(x) = 8x + 32, g(x) = \frac{1}{8}x - 4$

For the given function $f(x)$, find $f^{-1}(x)$.

33. $f(x) = x + 9$

34. $f(x) = x - 4$

35. $f(x) = \frac{x}{3}$

36. $f(x) = -8x$

37. $f(x) = 3x + 14$

38. $f(x) = 2x - 11$

39. $f(x) = 7x - 6$

40. $f(x) = -5x + 16$

41. $f(x) = -\frac{3}{4}x + 6$

42. $f(x) = \frac{2}{3}x - 10$

43. $f(x) = mx$

44. $f(x) = mx + b$

45. $f(x) = \frac{1}{x} + 3$

46. $f(x) = \frac{1}{x} - 5$

47. $f(x) = \frac{1}{x + 3}$

48. $f(x) = \frac{1}{x - 5}$

49. $f(x) = \frac{5 - 6x}{x}$

50. $f(x) = \frac{1 + 10x}{x}$

51. $f(x) = \frac{8 + 4x}{3x}$

52. $f(x) = \frac{2x}{7x - 10}$

For the given function $f(x)$, find $f^{-1}(x)$. State the domain of $f^{-1}(x)$.

53. $f(x) = \sqrt{x}$

54. $f(x) = \sqrt{x - 4}$

55. $f(x) = \sqrt{x} - 7$

56. $f(x) = \sqrt{x - 6} - 2$

57. $f(x) = \sqrt{x - 8} + 3$

58. $f(x) = \sqrt{x + 11} + 5$

59. $f(x) = x^2 - 9\ (x \geq 0)$

60. $f(x) = (x + 6)^2 + 5\ (x \geq -6)$

61. $f(x) = x^2 - 6x + 8\ (x \geq 3)$ (*Hint:* Try completing the square.)

62. $f(x) = x^2 + 4x - 12\ (x \geq -2)$ (*Hint:* Try completing the square.)

If the function represented by the set of ordered pairs is one-to-one, find its inverse.

63. $\{(-5, -17), (-1, -9), (1, -5), (4, 1), (7, 7)\}$

64. $\{(-2, -1), (0, 5), (1, 8), (4, 17), (7, 26)\}$

65. $\{(-2, 21), (-1, 17), (0, 13), (1, 9), (2, 13)\}$

66. $\{(-10, 4), (-9, 5), (-6, 6), (-1, 7), (6, 8)\}$

For the given graph of a one-to-one function $f(x)$, graph its inverse function $f^{-1}(x)$ without finding a formula for $f(x)$ or $f^{-1}(x)$.

67.

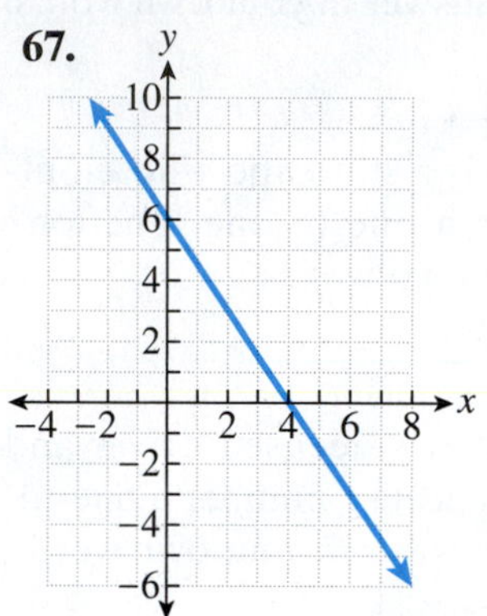

68.

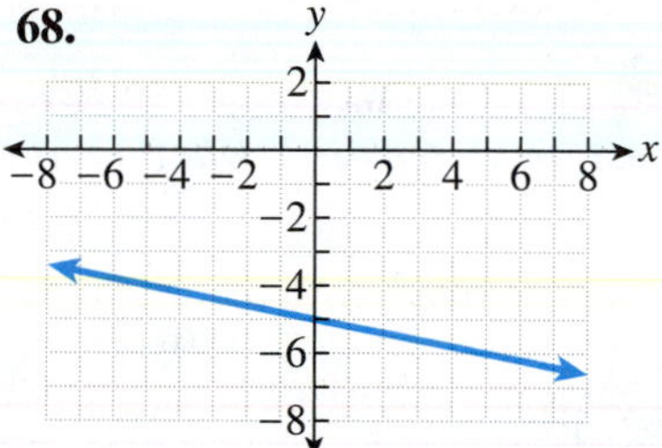

69.

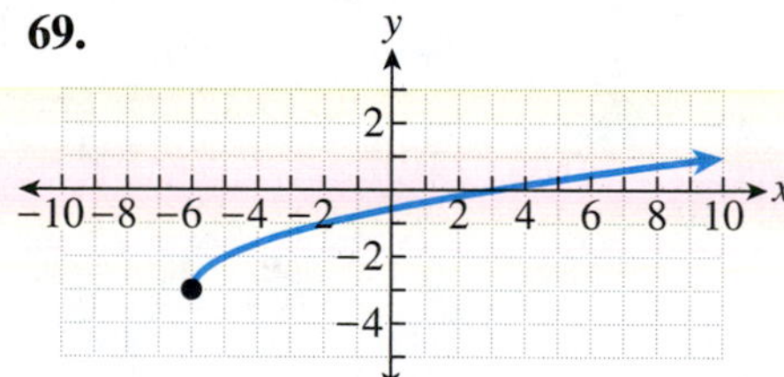

70.

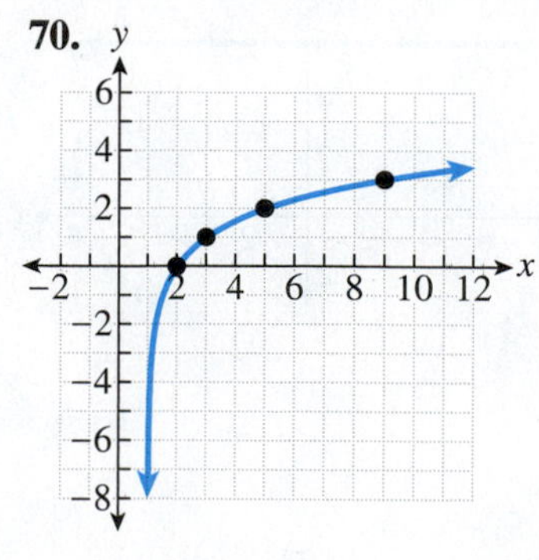

Graph a one-to-one function f(x) that meets the given criteria.

71. $f(x)$ is a linear function, $f(4) = 9$, and $f^{-1}(5) = -2$.

72. $f(x)$ is a linear function, $f^{-1}(5) = -7$, and $f^{-1}(-6) = 8$.

73. $f(x)$ is a quadratic function, the domain of $f(x)$ is restricted to $[3, \infty)$, $f(3) = -1$, $f(4) = 0$, $f^{-1}(3) = 5$, and $f^{-1}(8) = 6$.

74. $f(x)$ is a quadratic function, the domain of $f(x)$ is restricted to $[1, \infty)$, $f^{-1}(-8) = 1$, $f^{-1}(-2) = 3$, and $f^{-1}(8) = 5$.

Writing in Mathematics

Answer in complete sentences.

75. Explain how the horizontal-line test shows whether a function is or is not one-to-one.

76. Explain how to find the inverse function of a one-to-one function $f(x)$. Use an example to illustrate the process.

77. ***Solutions Manual*** Write a solutions manual page for the following problem.

For the one-to-one function $f(x) = \dfrac{3 + 2x}{4x}$, *find* $f^{-1}(x)$.

CHAPTER 11 SUMMARY

Section 11.1 Review of Functions

Functions, pp. 604–606

A function is a rule that takes an input value and assigns a particular output value to it. For a rule to define a function, each input value must be assigned one and only one output value. The set of input values for a function is called the domain of the function.

The set of output values for a function is called the range of the function.

Is a relation whose input value is a student at your college and whose output value is the month the student was born a function?

Yes, because each student has only one birth month.

The Vertical-Line Test, pp. 606–607

If a vertical line can be drawn that intersects a graph at more than one point, the graph does not represent a function.

Not a function

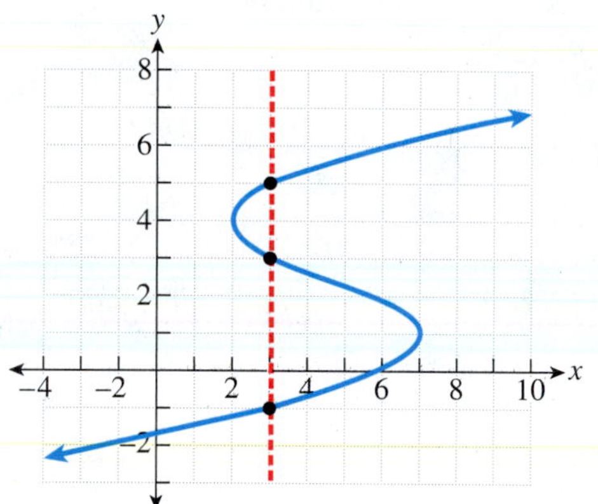

Function Notation, p. 608

Function notation is a way to present the formula for the output value of a function for the input x.

$f(x) = 3x + 14$

Function name: f

Input variable: x

Formula for output: $3x + 14$

Evaluating Functions, pp. 608–609

To evaluate a function for a particular value of the variable, substitute that value for the variable in the function's formula, then simplify the resulting expression.

For $f(x) = x^2 + 7x - 24$, find $f(-5)$.

$$\begin{aligned} f(-5) &= (-5)^2 + 7(-5) - 24 \\ &= 25 - 35 - 24 \\ &= -34 \end{aligned}$$

Interpreting Graphs, pp. 609–610

The domain of a function can be read from left to right along the x-axis on a graph.

The range of a function can be read vertically from the bottom to the top of the graph along the y-axis.

An x-intercept is a point at which a graph intersects the x-axis. The y-coordinate of an x-intercept is 0.

A y-intercept is a point at which a graph intersects the y-axis. The x-coordinate of a y-intercept is 0.

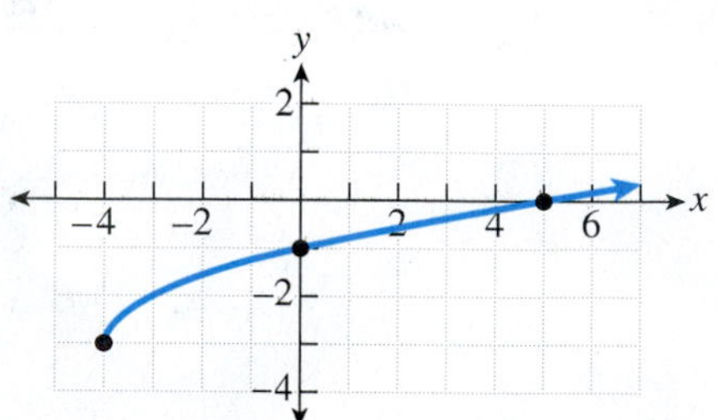

Domain: $[-4, \infty)$; range: $[-3, \infty)$

x-intercept: $(5, 0)$; y-intercept: $(0, -1)$

Section 11.2 Linear Functions

Linear Functions, p. 614

A linear function is a function that can be written in the form $f(x) = mx + b$, where m and b are real numbers.

$f(x) = 7x - 18$

Graphing Linear Functions, pp. 614–615

A linear function is a function of the form $f(x) = mx + b$, where m and b are real numbers.

To graph a linear function, begin by plotting its y-intercept at $(0, b)$. Then use the slope m to find other points.

Graph: $f(x) = -\frac{2}{3}x + 5$

y-intercept: $(0, 5)$; slope: $-\frac{2}{3}$

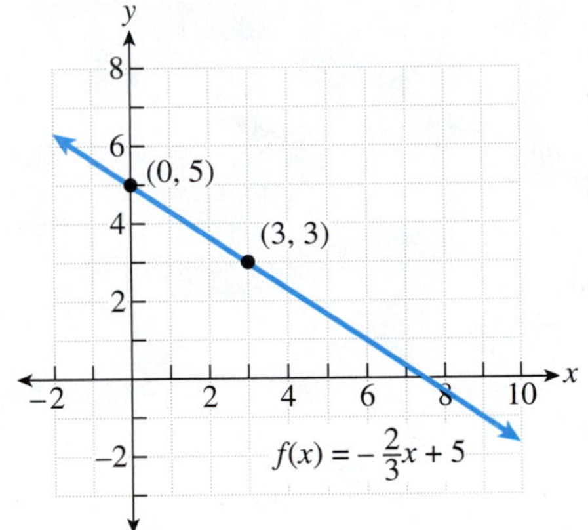

Constant Functions, p. 615

A function of the form $f(x) = b$ is called a constant function. The graph of a constant function is a horizontal line whose y-intercept is the point $(0, b)$.

Graph: $f(x) = -6$

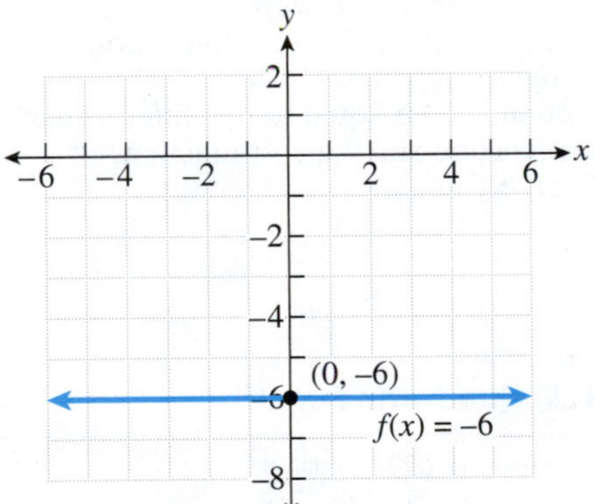

Determining a Linear Function from Its Graph, pp. 615–616

1. Determine the coordinates of two points (x_1, y_1) and (x_2, y_2) on the line.
2. Find the slope m of the line using the formula

$$m = \frac{y_2 - y_1}{x_2 - x_1}.$$

3. Set $f(x_1) = y_1$ or $f(x_2) = y_2$, and solve for b.
4. Write the formula for the function using $f(x) = mx + b$.

Find the linear function on the graph.

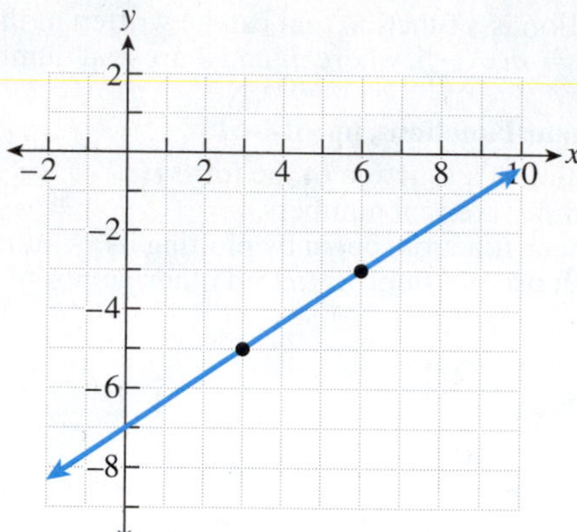

The graph passes through $(3, -5)$ and $(6, -3)$.

$$m = \frac{-3 - (-5)}{6 - 3} = \frac{2}{3} \qquad f(3) = -5$$

$$f(x) = \frac{2}{3}x + b \qquad \frac{2}{3}(3) + b = -5$$

$$2 + b = -5$$
$$b = -7$$

$$f(x) = \frac{2}{3}x - 7$$

Business Applications, pp. 617–618

The cost function $C(x)$ is equal to the fixed costs plus the variable costs associated with producing x items.
The revenue function $R(x)$ tells how much money will be taken in by selling x items.
The profit function $P(x)$ is equal to the difference of the revenue function and the cost function, $P(x) = R(x) - C(x)$.

Emma decided to start selling her homemade jam. She spent $250 on new kitchen equipment, and each jar of jam costs her $2 to make. She will sell each jar for $5. How many jars does she need to make and sell to earn a profit of $800?

$$C(x) = 2x + 250,\ R(x) = 5x$$
$$P(x) = 5x - (2x + 250) = 3x - 250$$
$$3x - 250 = 800$$
$$3x = 1050$$
$$x = 350$$

Emma needs to make and sell 350 jars of jam.

Section 11.3 Quadratic Functions

Quadratic Functions, p. 622

A quadratic function is a function that can be written in the form $f(x) = ax^2 + bx + c$, where a, b, and c are real numbers and $a \neq 0$.

$$f(x) = x^2 - 7x - 18$$

Graphing Functions of the Form $f(x) = a(x - h)^2 + k$, pp. 622–626

- Determine whether the parabola opens upward or downward.
- Shift the graph of $f(x) = x^2$ by h units horizontally and by k units vertically.
- Find the y-intercept of the parabola by finding $f(0)$.
- If there are any x-intercepts, set the function equal to 0 and solve for x.

Graph: $f(x) = (x + 3)^2 - 5$

Opens upward

$h = -3$; $k = -5$; vertex: $(-3, -5)$

Shift basic graph of $f(x) = x^2$ left by 3 units and down by 5 units.

y-intercept: $(0, 4)$

$$f(x) = (0 + 3)^2 - 5 = 9 - 5 = 4$$

x-intercepts: $(-5.2, 0), (-0.8, 0)$

$$\begin{aligned}(x + 3)^2 - 5 &= 0\\ (x + 3)^2 &= 5\\ x + 3 &= \pm\sqrt{5}\\ x &= -3 \pm \sqrt{5}\end{aligned}$$

$$-3 - \sqrt{5} \approx -5.2, -3 + \sqrt{5} \approx -0.8$$

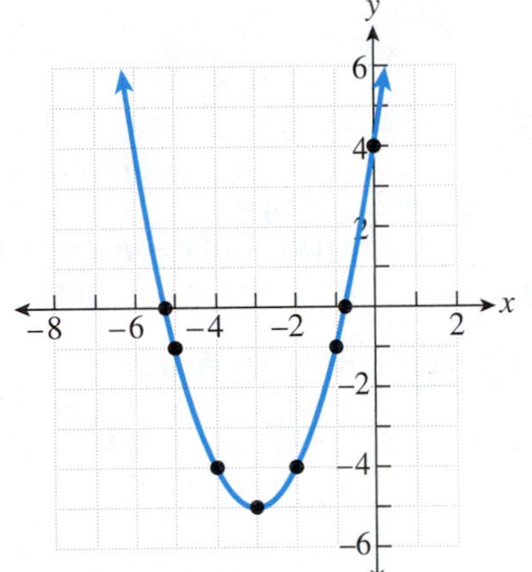

Minimum and Maximum Values of Quadratic Functions, pp. 627–629

If the graph of a quadratic function is a parabola that opens upward, the function has a minimum value. The minimum value of the function is the y-coordinate of the vertex.

If the graph of a quadratic function is a parabola that opens downward, the function has a maximum value. The maximum value of the function is the y-coordinate of the vertex.

Find the maximum or minimum value: $f(x) = x^2 - 14x + 90$

The function has a minimum value.

$$x = \frac{-(-14)}{2(1)} = 7$$

$$\begin{aligned}f(7) &= (7)^2 - 14(7) + 90\\ &= 49 - 98 + 90\\ &= 41\end{aligned}$$

The minimum value is 41.

Section 11.4 Other Functions and Their Graphs

Graphing an Absolute Value Function $f(x) = a|x - h| + k$, pp. 634–637

Shift the graph of $f(x) = |x|$ by h units horizontally and by k units vertically.

The domain of the function is the set of all real numbers, while the range is restricted and can be read from the graph.

Graph: $f(x) = |x - 2| + 3$

$h = 2, k = 3$

Shift basic graph of $f(x) = |x|$ right by 2 units and up by 3 units.

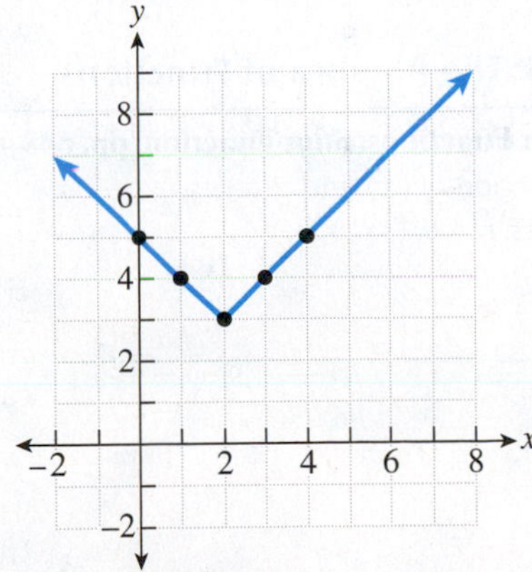

Domain: $(-\infty, \infty)$; range: $[3, \infty)$

Graphing a Square Root Function $f(x) = a\sqrt{x-h} + k$, pp. 637–641

Shift the graph of $f(x) = \sqrt{x}$ by h units horizontally and by k units vertically.

The domain and range of the function are restricted and can be read from the graph.

Graph: $f(x) = \sqrt{x+2} - 7$

Shift basic graph of $f(x) = \sqrt{x}$ left by 2 units and down by 7 units.

$h = -2, k = -7$

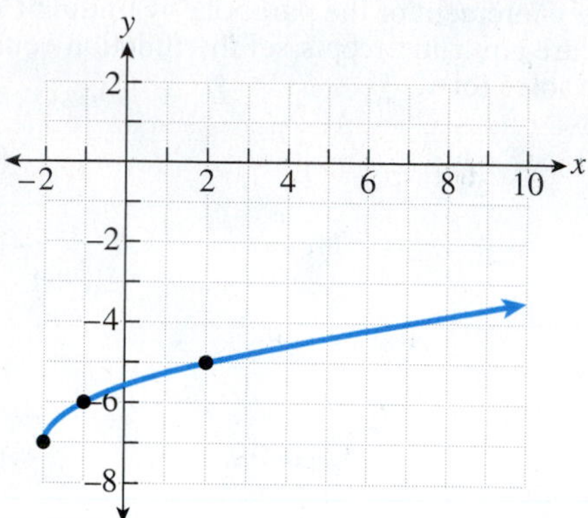

Domain: $[-2, \infty)$; range: $[-7, \infty)$

Graphing a Cubic Function $f(x) = a(x-h)^3 + k$, pp. 641–643

Shift the graph of $f(x) = x^3$ by h units horizontally and by k units vertically.

The domain and range of the function are both unrestricted.

Graph: $f(x) = (x+1)^3 - 4$

Shift basic graph of $f(x) = x^3$ left by 1 unit and down by 4 units.

$h = -1, k = -4$

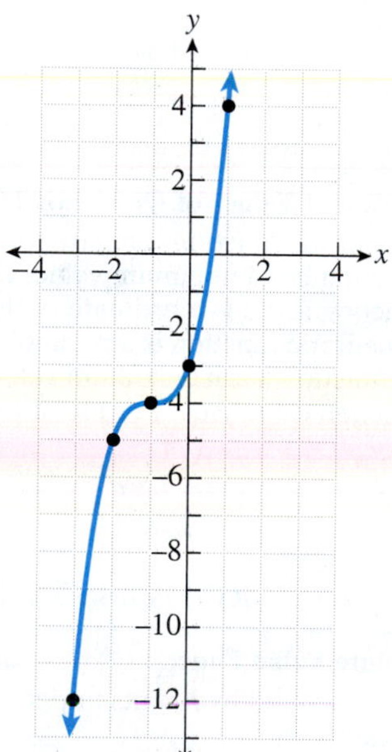

Domain: $(-\infty, \infty)$; range: $(-\infty, \infty)$

Section 11.5 The Algebra of Functions

Operations with Functions: Sum Function, pp. 648–649

For any two functions $f(x)$ and $g(x)$,
$(f+g)(x) = f(x) + g(x)$.

For $f(x) = x^2 - 9x + 22$ and $g(x) = x^2 + 5x + 17$, find $(f+g)(x)$.

$$\begin{aligned}(f+g)(x) &= (x^2 - 9x + 22) + (x^2 + 5x + 17)\\ &= x^2 - 9x + 22 + x^2 + 5x + 17\\ &= 2x^2 - 4x + 39\end{aligned}$$

Operations with Functions: Difference Function, pp. 648–649

For any two functions $f(x)$ and $g(x)$, $(f - g)(x) = f(x) - g(x)$.

For $f(x) = 3x^2 + 4x - 19$ and $g(x) = 2x^2 - 8x + 35$, find $(f - g)(x)$.

$$\begin{aligned}(f - g)(x) &= (3x^2 + 4x - 19) - (2x^2 - 8x + 35)\\ &= 3x^2 + 4x - 19 - 2x^2 + 8x - 35\\ &= x^2 + 12x - 54\end{aligned}$$

Operations with Functions: Product Function, pp. 649–650

For any two functions $f(x)$ and $g(x)$, $(f \cdot g)(x) = f(x) \cdot g(x)$.

For $f(x) = 5x - 4$ and $g(x) = 2x + 7$, find $(f \cdot g)(x)$.

$$\begin{aligned}(f \cdot g)(x) &= (5x - 4)(2x + 7)\\ &= 10x^2 + 35x - 8x - 28\\ &= 10x^2 + 27x - 28\end{aligned}$$

Operations with Functions: Quotient Function, pp. 649–650

For any two functions $f(x)$ and $g(x)$,

$$\left(\frac{f}{g}\right)(x) = \frac{f(x)}{g(x)},\ g(x) \neq 0.$$

For $f(x) = x^2 + 15x + 56$ and $g(x) = x^2 - 64$, find $\left(\frac{f}{g}\right)(x)$.

$$\begin{aligned}\left(\frac{f}{g}\right)(x) &= \frac{x^2 + 15x + 56}{x^2 - 64}\\ &= \frac{(x + 7)(x + 8)}{(x + 8)(x - 8)}\\ &= \frac{x + 7}{x - 8},\ x \neq -8, 8\end{aligned}$$

Composition of Functions, pp. 651–654

For any two functions $f(x)$ and $g(x)$, the composite function $(f \circ g)(x)$ is defined as $(f \circ g)(x) = f(g(x))$.

For $f(x) = 3x + 2$ and $g(x) = 4x - 9$, find $(f \circ g)(8)$.

$$\begin{aligned}(f \circ g)(8) &= f(g(8))\\ &= f(4(8) - 9)\\ &= f(23)\\ &= 3(23) + 2\\ &= 71\end{aligned}$$

Section 11.6 Inverse Functions

One-to-One Functions, pp. 657–658

If each value in the range of a function $f(x)$ corresponds to only one value in the domain, the function is called a one-to-one function.

If two different input values for $f(x)$ are associated with the same output value, $f(x)$ is not a one-to-one function.

Horizontal-Line Test

If a horizontal-line can intersect the graph of a function at more than one point, the function is not one-to-one.

Not one-to-one

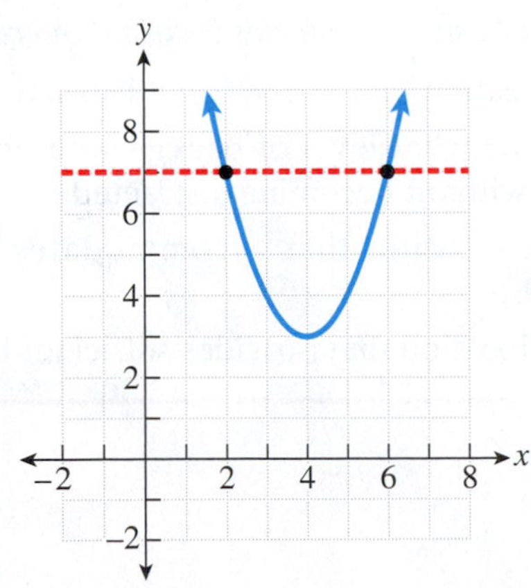

Inverse of a Function, pp. 659–660

If a one-to-one function assigns an input value x to an output value y, that function's inverse assigns the input value y to an output value x.

The notation $f^{-1}(x)$ is used to denote the inverse of a one-to-one function $f(x)$.

If $f(x)$ is a one-to-one function and $f(3) = 10$, then $f^{-1}(10) = 3$.

Verifying That Two Functions Are Inverses, pp. 659–660

Two one-to-one functions $f(x)$ and $g(x)$ are inverse functions if $(f \circ g)(x) = x$ and $(g \circ f)(x) = x$.

Verify that $f(x) = 2x - 3$ and $g(x) = \dfrac{x+3}{2}$ are inverse functions.

$$\begin{aligned}(f \circ g)(x) &\\ &= f\left(\frac{x+3}{2}\right)\\ &= 2\left(\frac{x+3}{2}\right) - 3\\ &= x + 3 - 3\\ &= x\end{aligned}$$

$$\begin{aligned}(g \circ f)(x) &\\ &= g(2x - 3)\\ &= \frac{(2x-3)+3}{2}\\ &= \frac{2x}{2}\\ &= x\end{aligned}$$

The functions are inverses.

Finding an Inverse Function for a One-To-One Function $f(x)$, pp. 660–663

1. Replace $f(x)$ with y.
2. Interchange x and y.
3. Solve the resulting equation for y.
4. Replace y with $f^{-1}(x)$.

For $f(x) = 7x + 19$, find $f^{-1}(x)$.

$$\begin{aligned}y &= 7x + 19\\ x &= 7y + 19\\ x - 19 &= 7y\\ \frac{x-19}{7} &= y\end{aligned}$$

$$f^{-1}(x) = \frac{x-19}{7}$$

SUMMARY OF CHAPTER 11 STUDY STRATEGIES

This chapter's study strategies focused on creating the proper study environment.

- Find a location that is free from distractions.
- Determine what level of background noise you can tolerate without becoming distracted.
- Select a location that accommodates all of your materials.
- Select a location that provides sufficient lighting.
- Select a location that is comfortable, including a temperature that is neither too hot nor too cold.
- Find locations at home and on campus that are conducive to studying and stay away from distractions such as TV while you are studying.

CHAPTER 11 REVIEW

1. Does a function exist for which the input value is a mathematics instructor and the output value is the state in which the instructor was born? [11.1]
2. Does a function exist for which the input value is the political party of a state's governor and the output value is the state? [11.1]

***Use the vertical-line test to determine whether the graph represents a function.* [11.1]**

3.

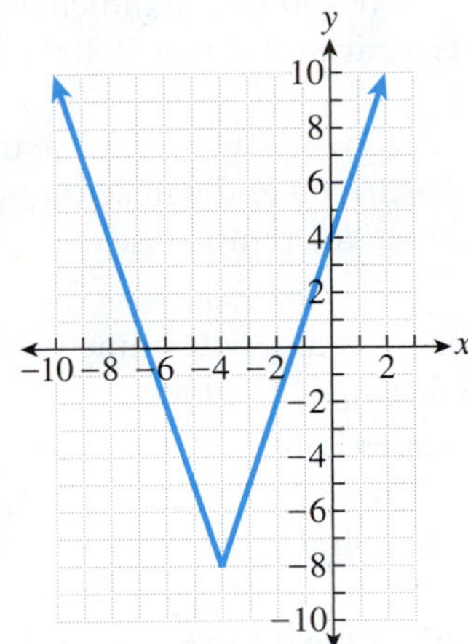

4.

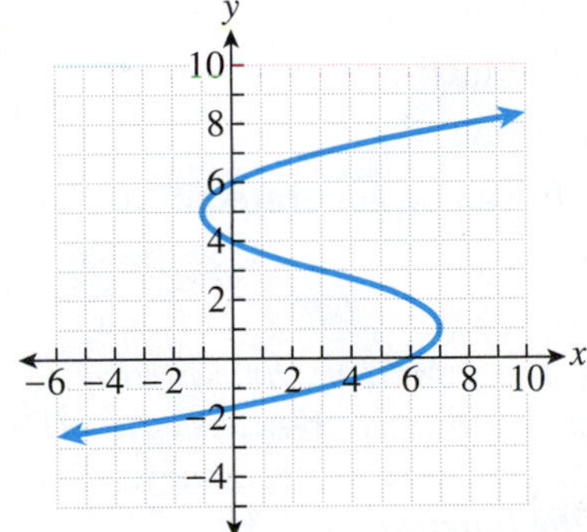

***Evaluate the given function.* [11.1]**

5. $f(x) = 2x, f(-5)$
6. $f(x) = 3x - 7, f(4)$
7. $f(x) = 7x - 11, f(3b - 8)$
8. $f(x) = x^2 - 7x + 32, f(7)$
9. $f(x) = x^2 - 9x - 36, f(-3)$
10. $f(x) = x^2 + 10x - 25, f(2n + 5)$
11. $f(x) = |x + 2| - 17, f(-9)$
12. $f(x) = \sqrt{2x + 16} + 27, f(10)$

***Determine the domain and range of the function graphed.* [11.1]**

13.

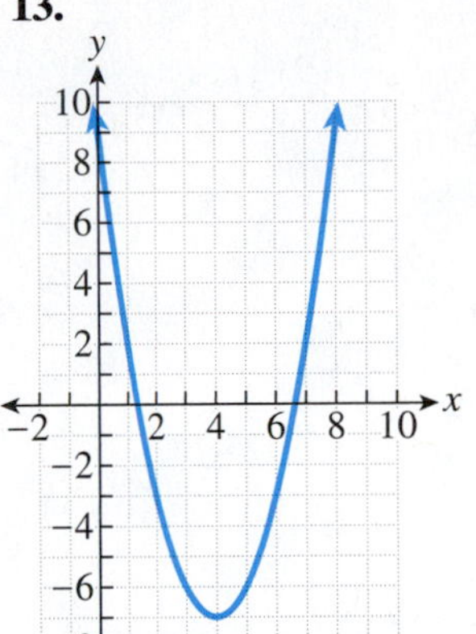

14.

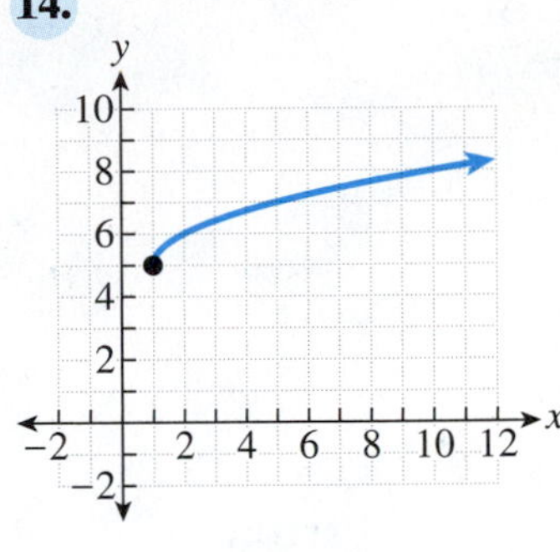

***Use the graph of the function $f(x)$ to find the indicated function value.* [11.1]**

15. $f(4)$
16. $f(-1)$

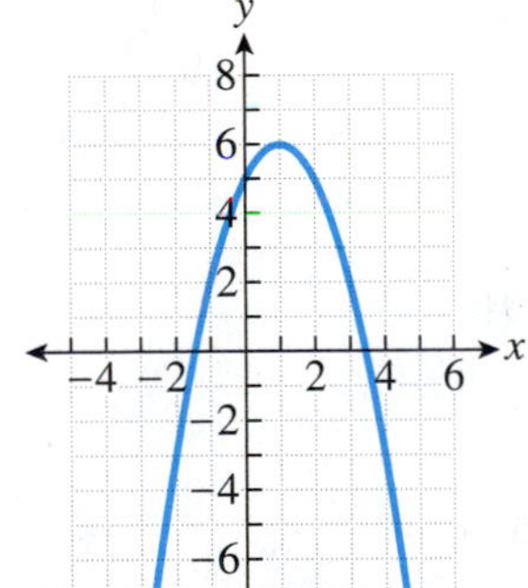

***Use the graph of the function $f(x)$ to determine which values of x satisfy the given equation.* [11.1]**

17. $f(x) = 3$
18. $f(x) = -5$

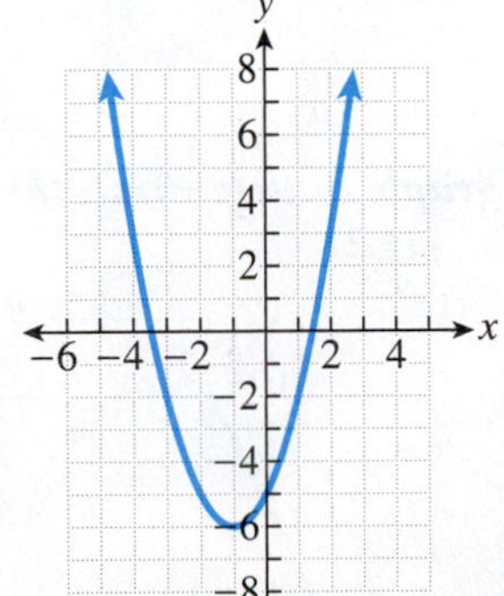

Worked-out solutions to Review Exercises marked with ◯ can be found on page AN-43.

***Graph each function.* [11.2]**

19. $f(x) = -3x - 6$

20. $f(x) = 2x + 4$

21. $f(x) = \frac{2}{5}x - 5$

22. $f(x) = -\frac{3}{4}x + 6$

***Use the graph to determine the linear function f(x) on the graph.* [11.2]**

23.

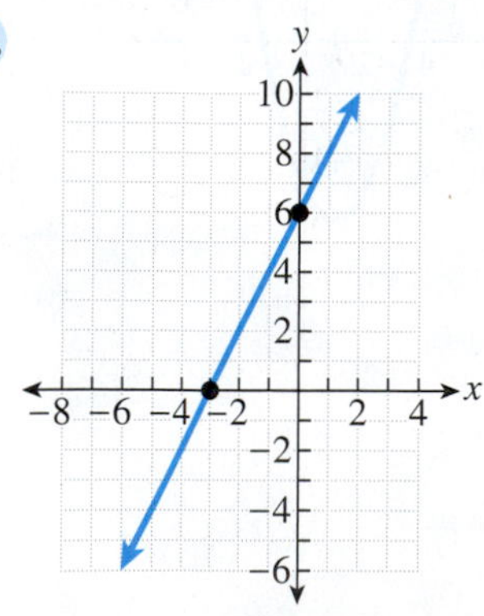

24.

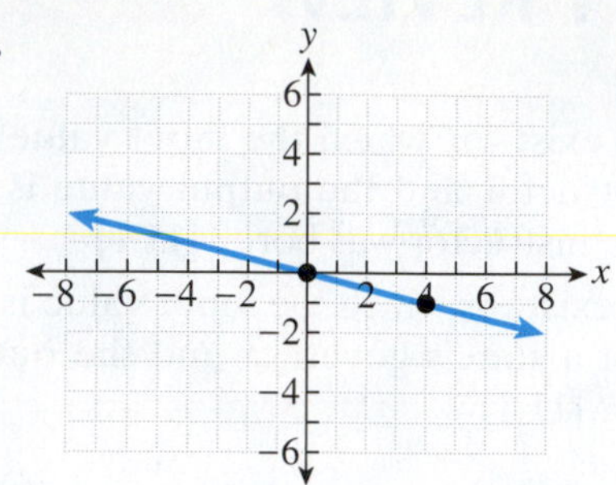

25. A rental car company charges a fixed fee for renting a car, in addition to charging per mile driven. If a person rents a car and drives it 250 miles, the charge is \$42.50. If a person rents a car and drives it 400 miles, the charge is \$50. [11.2]

a) Find a linear function $f(x) = mx + b$ whose input is the number of miles driven and whose output is the corresponding charge for renting the car.

b) Use the function from part a to determine the cost of renting a car and driving it 75 miles.

26. A college club is selling snow cones to raise money. In addition to paying \$40 to rent space for the day, the club also had to spend \$60 on a snow cone machine. The supplies to make each snow cone cost \$0.10, and the club is charging \$1.50 for each snow cone. [11.2]

a) Find the cost function $C(x)$.

b) Find the revenue function $R(x)$.

c) Find the profit function $P(x)$.

d) How much profit will be generated if the club sells 400 snow cones?

***Sketch the graph of the function by shifting. Label the vertex, y-intercept, and any x-intercepts. (Round x-intercepts to the nearest tenth if necessary.) State the domain and range of the function.* [11.3]**

27. $f(x) = (x - 2)^2 - 9$

28. $f(x) = (x + 1)^2 + 3$

29. $f(x) = (x - 3)^2 - 7$

30. $f(x) = -(x + 2)^2 + 4$

***Find the maximum or minimum value for each of the given quadratic functions.* [11.3]**

31. $f(x) = x^2 - 12x + 60$
32. $f(x) = -x^2 + 16x - 70$
33. $f(x) = (x + 9)^2 - 25$
34. $f(x) = -(x + 8)^2 - 39$
35. A projectile is launched upward with an initial velocity of 176 feet/second from the roof of a building. The height of the projectile (in feet) after t seconds is given by the function $h(t) = -16t^2 + 176t + 42$. What is the maximum height the projectile reaches? [11.3]
36. A farmer has 144 feet of fencing to make a rectangular corral. What dimensions will result in the maximum area? What is the maximum area possible? [11.3]

***For the given function f(x), simplify the difference quotient* $\frac{f(x + h) - f(x)}{h}$. [11.3]**

37. $f(x) = x^2 + 13$
38. $f(x) = x^2 - 6x - 33$

***Sketch the graph of the given function by shifting and state its domain and range.* [11.4]**

39. $f(x) = |x - 2| - 4$

40. $f(x) = -|x + 3| - 2$

***Graph the given square-root function and state its domain and range.* [11.4]**

41. $f(x) = \sqrt{x - 2} + 7$

42. $f(x) = \sqrt{x + 5} - 3$

43. $f(x) = -\sqrt{x+1} + 2$

44. $f(x) = \sqrt{-(x-2)} + 5$

Graph the given cubic function and state its domain and range. **[11.4]**

45. $f(x) = (x-2)^3$

46. $f(x) = x^3 - 1$

Determine the function f(x) that has been graphed. **[11.4]**

47.

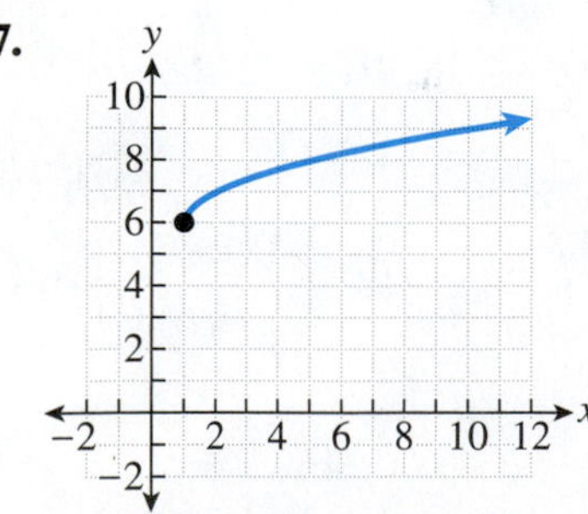

48.

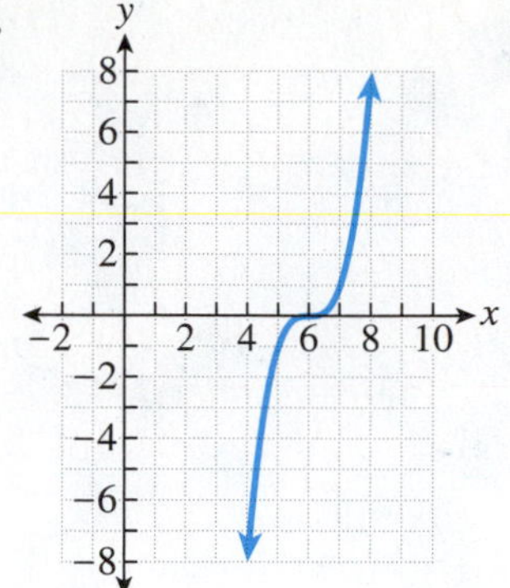

49.

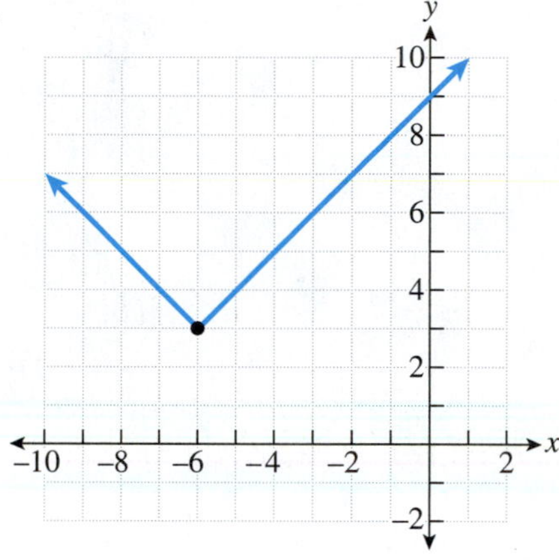

Let $f(x) = x^2 + 9x - 22$ and $g(x) = x + 11$. Find the following. **[11.5]**

50. $(f+g)(-5)$

51. $(f-g)(15)$

52. $(f \cdot g)(8)$

53. $\left(\frac{f}{g}\right)(-1)$

For the given functions f(x) and g(x), find $(f+g)(x)$. **[11.5]**

54. $f(x) = 6x - 11$, $g(x) = 4x + 35$

For the given functions f(x) and g(x), find $(f-g)(x)$. **[11.5]**

55. $f(x) = 8x - 33$, $g(x) = -8x + 55$

For the given functions f(x) and g(x), find $(f \cdot g)(x)$. **[11.5]**

56. $f(x) = x - 9$, $g(x) = x + 4$

For the given functions f(x) and g(x), find $\left(\frac{f}{g}\right)(x)$. **[11.5]**

57. $f(x) = x^2 - x - 6$, $g(x) = x^2 - 10x + 21$

Let $f(x) = x + 4$ and $g(x) = x^2 - 14x + 48$. Find the following. **[11.5]**

58. $(f \circ g)(8)$

59. $(g \circ f)(9)$

***For the given functions f(x) and g(x), find $(f \circ g)(x)$. and $(g \circ f)(x)$.* [11.5]**

60. $f(x) = 6x - 17, g(x) = 3x + 20$

61. $f(x) = 2x + 15, g(x) = -4x + 9$

62. $f(x) = x - 8, g(x) = x^2 + 7x - 56$

63. $f(x) = 3x + 7, g(x) = x^2 - 9x + 41$

64. Is the function whose input is a person and whose output is that person's favorite fast-food restaurant a one-to-one function? Explain your answer in your own words. [11.6]

65. Is the function whose input is a player on your school's basketball team and whose output is that player's uniform number a one-to-one function? Explain your answer in your own words. [11.6]

***Use the horizontal-line test to determine whether the function is one-to-one.* [11.6]**

66.

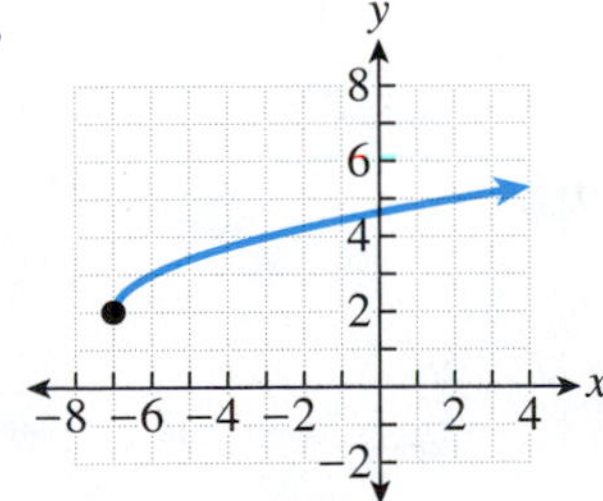

67.

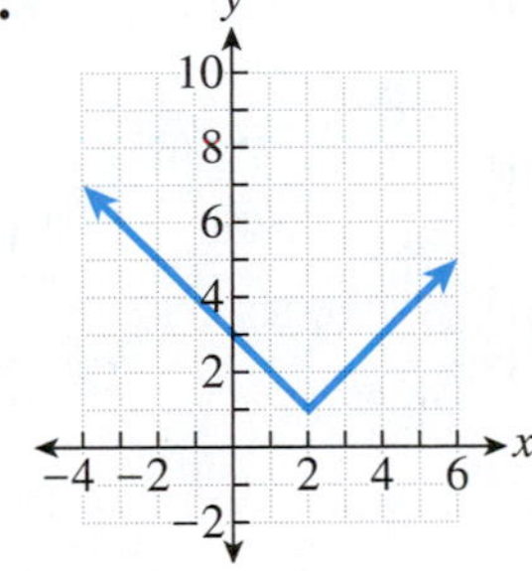

***Determine whether the functions f(x) and g(x) are inverse functions by evaluating whether $(f \circ g)(x) = x$ and $(g \circ f)(x) = x$.* [11.6]**

68. $f(x) = 3x + 4, g(x) = 4x - 3$

69. $f(x) = 2x - 10, g(x) = \dfrac{x + 10}{2}$

70. $f(x) = 4x + 18, g(x) = \dfrac{1}{4}x + \dfrac{9}{2}$

71. $f(x) = -5x + 9, g(x) = \dfrac{9 - x}{5}$

***For the given function f(x), find $f^{-1}(x)$. State the domain of $f^{-1}(x)$.* [11.6]**

72. $f(x) = x - 10$

73. $f(x) = 2x - 15$

74. $f(x) = -x + 19$

75. $f(x) = -8x + 27$

76. $f(x) = \dfrac{9x + 2}{3x}$

77. $f(x) = \dfrac{7x}{4x + 21}$

***For the given graph of a one-to-one function f(x), graph its inverse function $f^{-1}(x)$.* [11.6]**

78.

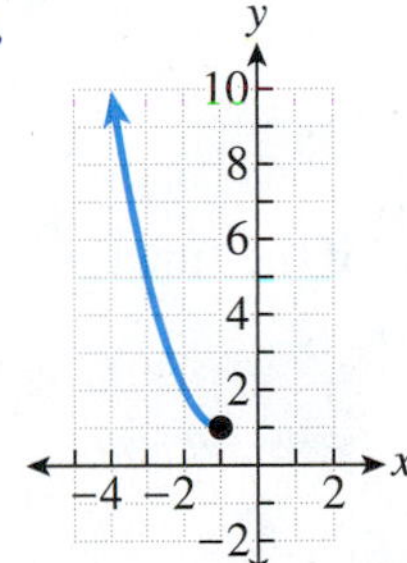

79.

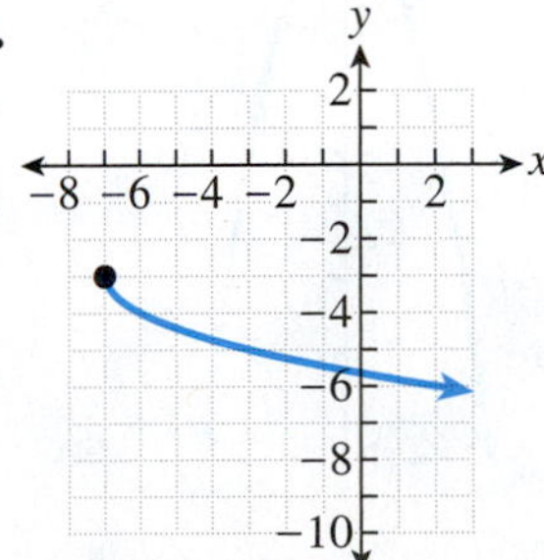

CHAPTER 11 TEST

For Extra Help

CHAPTER Test Prep VIDEOS

Step-by-step test solutions are found on the Chapter Test Prep Videos available via the Video Resources on DVD, in MyMathLab, and on YouTube (search "WoodburyElemIntAlg" and click on "Channels").

1. Does a function exist for which the input value is a state and the output value is the state's capital city?

2. Use the vertical-line test to determine whether the graph represents a function.

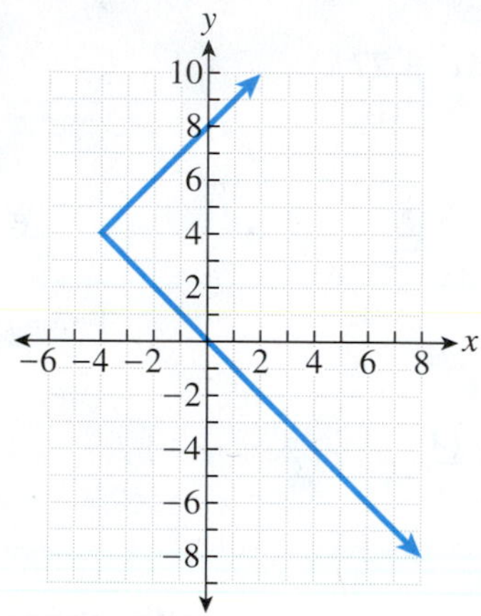

Evaluate the given function.

3. $f(x) = -2x + 35, f(-8)$

4. $f(x) = 5x + 16, f(6a - 5)$

5. $f(x) = x^2 - 15x + 54, f(-9)$

6. Determine the domain and range of the function $f(x)$ that has been graphed.

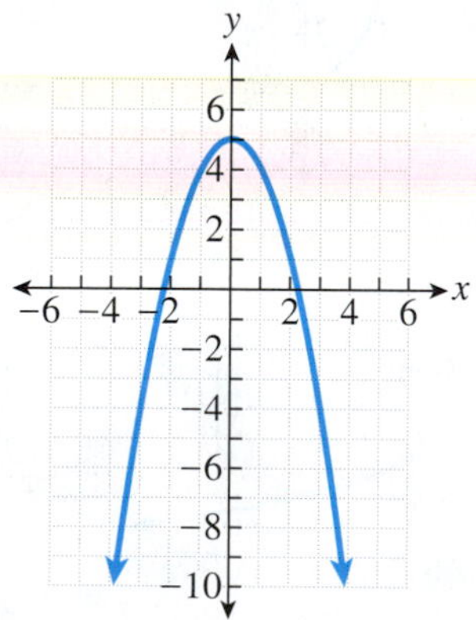

Graph each function.

7. $f(x) = -4x + 6$

8. $f(x) = \frac{2}{3}x - 4$

9. A ceramics shop holds birthday parties for children. They charge a party fee in addition to a fee for each guest attending. A party with 10 guests costs \$105, while a party with 25 guests costs \$210. Find a linear function $f(x) = mx + b$ whose input is the number of guests and whose output is the cost of the party.

Graph the function by shifting. Label the vertex, y-intercept, and any x-intercepts. Round to the nearest tenth if necessary.

10. $f(x) = (x + 4)^2 - 6$

11. $f(x) = -(x + 3)^2 + 4$

12. A young child throws a ball upward with an initial velocity of 20 feet/second. The height of the ball (in feet) after t seconds is given by the function $h(t) = -16t^2 + 20t + 3$. What is the maximum height the ball reaches?

Graph the given function and state its domain and range.

13. $f(x) = |x + 4| - 5$

14. $f(x) = \sqrt{x + 1} - 3$

15. $f(x) = (x + 1)^3 + 8$

16. Determine the function $f(x)$ on the graph.

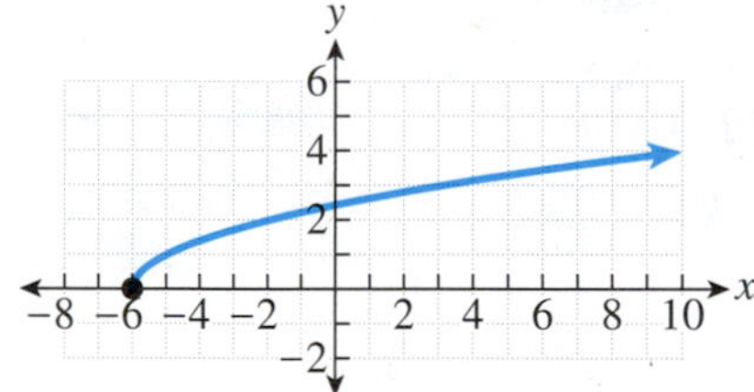

17. If $f(x) = 5x - 2$ and $g(x) = x^2 - 4x - 13$, find $(f + g)(-2)$.

18. If $f(x) = x - 6$ and $g(x) = x^2 + 6x + 36$, find $(f \cdot g)(x)$.

Let $f(x) = x + 7$ and $g(x) = 2x - 8$. Find the following.

19. $(f \circ g)(6)$

20. $(g \circ f)(6)$

For the given functions $f(x)$ and $g(x)$, find $(f \circ g)(x)$.

21. $f(x) = 4x + 13, g(x) = -2x + 23$

22. $f(x) = x^2 - 9x - 30, g(x) = x + 12$

23. Use the horizontal-line test to determine whether the function is one-to-one.

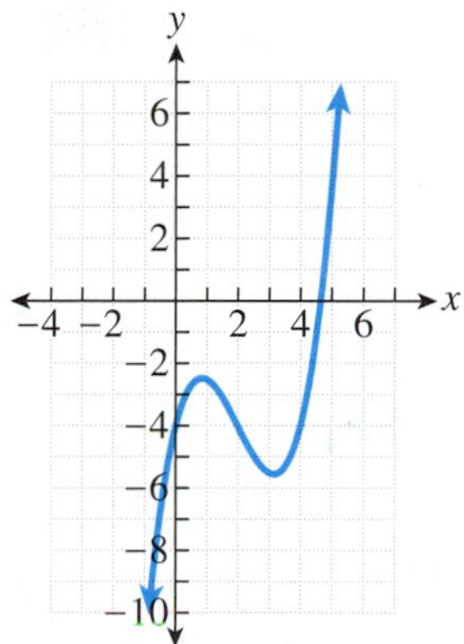

24. Determine whether $f(x) = 2x - 16$ and $g(x) = \frac{1}{2}x + 8$ are inverse functions. Recall that $f(x)$ and $g(x)$ are inverse functions if $(f \circ g)(x) = x$ and $(g \circ f)(x) = x$.

For the given function $f(x)$, find $f^{-1}(x)$.

25. $f(x) = -4x + 20$

26. $f(x) = \dfrac{x - 9}{4x}$

Mathematicians in History

Leonhard Euler (pronounced "oiler") is believed by many to have been the greatest mathematician of the 18th century. He was clearly one of the most prolific, authoring more than 800 papers and books. He is credited with introducing the function notation $f(x)$ that is still used today.

Write a one-page summary (*or* make a poster) of the life of Euler and his mathematical achievements.

Interesting issues:

- Where and when was Leonhard Euler born?
- On what currency did Euler's face appear?
- What are the Seven Bridges of Königsberg?
- Who invited Euler to Berlin in 1741?
- On the day he died, Euler was lecturing to his grandchildren. What was the subject? What were his last words?
- Besides function notation, what other notation did Euler introduce?
- What physical condition did Euler have for the last 20 years of his life?

CUMULATIVE REVIEW CHAPTERS 8–11

***Solve.* [8.1]**

1. $|x + 3| - 8 = -2$

***Solve the inequality. Graph your solution on a number line and write your solution in interval notation.* [8.2]**

2. $-2x - 9 \geq -17$

3. $-11 \leq 4x - 13 \leq 39$

***Solve the inequality. Graph your solution on a number line and write your solution in interval notation.* [8.2]**

4. $|2x + 1| > 5$

5. $|x| - 8 < -3$

***Factor completely.* [8.4]**

6. $x^2 - 3x - 54$
7. $4x^2 - 36x + 32$
8. $x^2 + 11x + 30$
9. $49x^2 - 64$

***Solve.* [8.5]**

10. $3x - 2y = 22$
    $7x + 4y = 34$

***Solve.* [8.5]**

11. $x + y + z = 6$
    $x + 3y - z = 20$
    $2x - 4y + z = -16$

***Evaluate the radical function. (Round to the nearest thousandth if necessary.)* [9.1]**

12. $f(x) = \sqrt{6x - 58}, f(19)$

***Rewrite as a radical expression and simplify if possible. Assume that all variables represent nonnegative values.* [9.2]**

13. $(16x^{10})^{3/2}$

***Simplify the expression. Assume that all variables represent nonnegative values.* [9.2]**

14. $(x^{6/7})^{3/4}$

***Simplify the radical expression. Assume that all variables represent nonnegative values.* [9.3]**

15. $\sqrt{75r^{12}s^{10}t^9}$

***Subtract. Assume that all variables represent nonnegative values.* [9.3]**

16. $7\sqrt{108} - 2\sqrt{48}$

***Multiply. Assume that all variables represent nonnegative values.* [9.4]**

17. $(4\sqrt{3} - 2\sqrt{5})(5\sqrt{3} + 4\sqrt{5})$

***Rationalize the denominator and simplify. Assume that all variables represent nonnegative values.* [9.4]**

18. $\dfrac{ab^3c^2}{\sqrt{a^3b^8c}}$

19. $\dfrac{5\sqrt{5} + 6\sqrt{2}}{4\sqrt{5} - 3\sqrt{2}}$

***Solve. Check for extraneous solutions.* [9.5]**

20. $\sqrt{2x + 7} - 3 = 4$
21. $\sqrt{x + 6} = x$
22. For the function $f(x) = \sqrt{2x - 9}$, find all values x for which $f(x) = 5$. [9.5]

***Express in terms of i.* [9.6]**

23. $\sqrt{-360}$

***Multiply.* [9.6]**

24. $(2 + 5i)(9 - 2i)$

***Rationalize the denominator.* [9.6]**

25. $\dfrac{5i}{3 + 4i}$

***Solve by extracting square roots.* [10.1]**

26. $(x + 2)^2 + 41 = 23$

***Solve by using the quadratic formula.* [10.2]**

27. $3x^2 - 10x - 21 = 0$

***Solve by using a u-substitution.* [10.3]**

28. $x^4 - 11x^2 + 18 = 0$
29. $x + 3\sqrt{x} - 10 = 0$

Solve. **[10.3]**

30. $\sqrt{x + 23} - x = 3$

Solve. **[10.1–10.2]**

31. $x^2 - 12x + 27 = 0$

Graph the quadratic equation. Label the vertex, the y-intercept, and any x-intercepts. **[10.4]**

32. $y = -x^2 + 8x + 6$

33. $y = x^2 - 4x + 7$

Solve the quadratic inequality. Express your solution on a number line and in interval notation. **[10.6]**

34. $x^2 - 10x + 24 \leq 0$

Solve the rational inequality. Express your solution on a number line and in interval notation. **[10.6]**

35. $\dfrac{x^2 - 2x - 48}{x + 5} \leq 0$

36. The length of a rectangular lawn is 5 feet more than its width. If the area of the lawn is 980 square feet, find the dimensions of the lawn. Round to the nearest tenth of a foot. [10.5]

37. A projectile is launched upward with an initial velocity of 80 feet per second from the roof of a building 12 feet high. How long will it take until the projectile lands on the ground? Round to the nearest hundredth of a second. (Use the function $h(t) = -16t^2 + v_0t + s$.) [10.2]

38. A water tank is connected to two pipes. When both pipes are open, they can fill the tank in 5 hours. The larger of the pipes can fill the tank 3 hours sooner than the smaller pipe can. How long will it take the smaller pipe, working alone, to fill the tank? Round to the nearest tenth of an hour. [10.3]

Evaluate the given function. **[11.1]**

39. $f(x) = 3x - 10, f(4n - 11)$

40. $f(x) = x^2 - 10x + 51, f(-5)$

Graph the function. **[11.2]**

41. $f(x) = -\dfrac{2}{3}x - 4$

42. A company charges a fixed fee for renting a moving van, in addition to charging per mile driven. If a person rents a moving van and drives it 100 miles, the charge is \$35. If a person rents a moving van and drives it 160 miles, the charge is \$44. [11.2]

a) Find a linear function $f(x) = mx + b$ whose input is the number of miles driven and whose output is the corresponding charge for renting the moving van.

b) Use the function from part a to determine the cost of renting a moving van and driving it 300 miles.

Find the maximum or minimum value for the given quadratic function. **[11.3]**

43. $f(x) = x^2 - 30x + 129$

44. A projectile is launched upward with an initial velocity of 144 feet/second from the roof of a building. The height of the projectile (in feet) after t seconds is given by the function $h(t) = -16t^2 + 144t + 50$. What is the maximum height the projectile reaches? [11.3]

Graph the given function by shifting and state its domain and range. **[8.3, 11.3, 11.4]**

45. $f(x) = (x - 3)^2 - 4$

46. $f(x) = |x - 2| - 5$

47. $f(x) = \sqrt{x + 4} - 1$

48. $f(x) = (x + 1)^3$

Find $(f + g)(x)$. **[11.5]**

49. $f(x) = 7x + 11, g(x) = 3x - 42$

Find $(f \circ g)(x)$ and $(g \circ f)(x)$. **[11.5]**

50. $f(x) = x + 6, g(x) = x^2 - 2x + 12$

Determine whether the functions $f(x)$ and $g(x)$ are inverse functions. Recall that $f(x)$ and $g(x)$ are inverse functions if $(f \circ g)(x) = x$ and $(g \circ f)(x) = x$. **[11.6]**

51. $f(x) = 3x + 8, g(x) = \dfrac{x - 8}{3}$

For the given function $f(x)$, find $f^{-1}(x)$. **[11.6]**

52. $f(x) = 3x - 19$

53. $f(x) = \dfrac{3}{x - 9}$

CHAPTER 12

Logarithmic and Exponential Functions

In this chapter, we will investigate exponential functions and logarithmic functions. These types of functions have many practical applications, including calculating compound interest and population growth, using radiocarbon dating, and determining the intensity of an earthquake or the pH of a substance.

STUDY STRATEGY

Practice Quizzes Creating practice quizzes is an excellent way to prepare for exams. It is important to put yourself in a test environment before taking a test, without facing the consequences of a test. Practice quizzes will tell you which topics you understand and on which topics you need further work. Throughout this chapter, we will revisit this study strategy and help you incorporate it into your study habits.

12.1 Exponential Functions

OBJECTIVES

1. Define exponential functions.
2. Evaluate exponential functions.
3. Graph exponential functions.
4. Define the natural exponential function.
5. Solve exponential equations.
6. Use exponential functions in applications.

Definition of Exponential Functions

Objective 1 Define exponential functions. Would you rather have $1 million or $1 that gets doubled every day for 30 days? You may be surprised by just how much that \$1 is worth at the end of 30 days. The following table shows you:

Day	Amount (\$)	Day	Amount (\$)	Day	Amount (\$)
1	\$1	11	\$1024	21	\$1,048,576
2	\$2	12	\$2048	22	\$2,097,152
3	\$4	13	\$4096	23	\$4,194,304
4	\$8	14	\$8192	24	\$8,388,608
5	\$16	15	\$16,384	25	\$16,777,216
6	\$32	16	\$32,768	26	\$33,554,432
7	\$64	17	\$65,536	27	\$67,108,864
8	\$128	18	\$131,072	28	\$134,217,728
9	\$256	19	\$262,144	29	\$268,435,456
10	\$512	20	\$524,288	30	\$536,870,912

After 30 days, the \$1 would turn into over \$500 million. Notice that although the total grew quite slowly at first, it increased quickly toward the end of the 30 days. The amount of money after x days can be expressed as $f(x) = 2^{x-1}$. This is an example of exponential growth, which is based on exponential functions.

Exponential Function

A function that can be written in the form $f(x) = b^x$, where $b > 0$ and $b \neq 1$, is called an **exponential function** with base b.

[If $b = 1$, then $f(x)$ is simply the constant function $f(x) = 1$.]

This type of function is called an exponential function because the variable is in an exponent. Some examples of exponential functions are as follows:

$$f(x) = 2^x \quad f(x) = 10^x \quad f(x) = \left(\frac{1}{3}\right)^x \quad f(x) = 4^{-x} \quad f(x) = 3^{2x+1} + 7$$

Evaluating Exponential Functions

Objective 2 Evaluate exponential functions. We evaluate exponential functions the same way we evaluate other functions: substitute the specified value for the function's variable and then simplify the resulting expression.

EXAMPLE 1 Let $f(x) = 3^x$. Find the following:

a) $f(4)$

SOLUTION Begin by substituting 4 for x and then simplify the resulting expression.

$f(4) = 3^4$ Substitute 4 for x.

$= 81$ Simplify.

b) $f(0)$

SOLUTION Substitute 0 for x.

$f(0) = 3^0$ Substitute 0 for x.

$= 1$ Simplify. Recall that raising a nonzero base to the exponent 0 is equal to 1.

c) $f(-2)$

SOLUTION Substitute -2 for x.

$$f(-2) = 3^{-2} \quad \text{Substitute } -2 \text{ for } x.$$
$$= \frac{1}{3^2} \quad \text{Rewrite the expression without a negative exponent. Recall that } b^{-n} = \frac{1}{b^n}.$$
$$= \frac{1}{9} \quad \text{Simplify.}$$

Quick Check 1

Let $f(x) = 4^x$. Find $f(5)$ and $f(-3)$.

EXAMPLE 2 Let $f(x) = 5 \cdot 3^{x-4} + 16$. Find $f(6)$.

SOLUTION This function, although more complex than those in previous examples, also is an exponential function. Substitute 6 for x and then simplify.

$$f(6) = 5 \cdot 3^{6-4} + 16 \quad \text{Substitute 6 for } x.$$
$$= 5 \cdot 3^2 + 16 \quad \text{Simplify the exponent.}$$
$$= 61 \quad \text{Simplify.}$$

Quick Check 2

Let $f(x) = 2 \cdot 7^{x+3} - 19$. Find $f(-1)$.

Using Your Calculator To simplify the expression $5 \cdot 3^2 + 16$ with the TI-84, you use the button labeled [^]. Here is how the calculator screen should look.

```
5*3^2+16
              61
```

Graphs of Exponential Functions

Objective 3 Graph exponential functions. To understand the behavior of exponential functions, we must examine the graphs of these functions. In this section, we will learn to create a quick sketch of the graph of an exponential function. A more thorough presentation of graphing exponential functions, including finding x-intercepts, will be given in Section 12.6. For now, we will plot points for selected values of x.

Consider the function $f(x) = 2^x$. Here is a table of values of $f(x)$ for values of x ranging from -3 to 3.

x	-3	-2	-1	0	1	2	3
$f(x)$	$\frac{1}{8}$	$\frac{1}{4}$	$\frac{1}{2}$	1	2	4	8

We plot the ordered pairs and then draw a smooth curve passing through those points. The domain of this function is the set of all real numbers $(-\infty, \infty)$. As we look to the left of the graph, the function gets closer to the x-axis without actually touching it. In this case, we say that the x-axis is a horizontal asymptote for the graph of the function $f(x) = 2^x$.

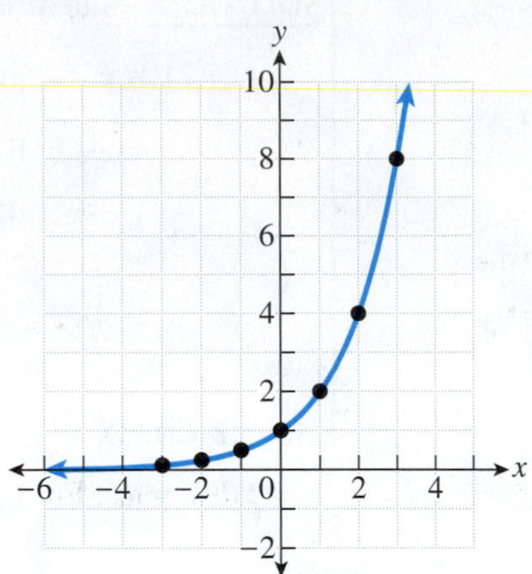

A **horizontal asymptote** is a horizontal line that the graph of a function approaches as it moves to the left or to the right. The horizontal asymptote is placed on the graph as a dashed line because the points on the asymptote are not part of the graph of the exponential function.

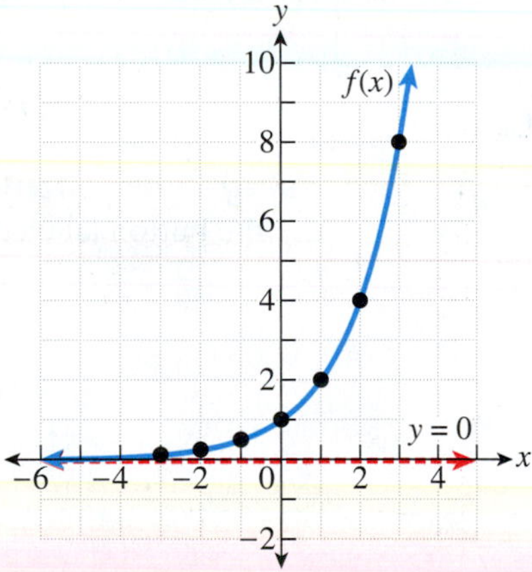

The horizontal asymptote also helps us determine the range of an exponential function. As the values of x decrease, the values of $f(x)$ get closer and closer to 0 without ever reaching it. As the values of x increase, the function values increase without limit. The range of this function is $(0, \infty)$.

An exponential function of the form $f(x) = b^x$ $(b > 1)$ is an **increasing function.** This means that as x increases, so does $f(x)$. The graph of an increasing function moves upward as it moves from left to right. Notice that the function increases at a slow rate for values of x that are less than 0, but increases more rapidly as the values of x increase.

To graph an exponential function of the form $f(x) = b^x$ where $b > 1$, we begin by graphing the horizontal asymptote on the x-axis, which is the line $y = 0$. Next, we plot a point on the y-axis at $(0, 1)$, as well as two other points at $(1, b)$ and $\left(-1, \frac{1}{b}\right)$. Then we draw the graph of the function through these points, increasing toward the right and approaching the horizontal asymptote toward the left.

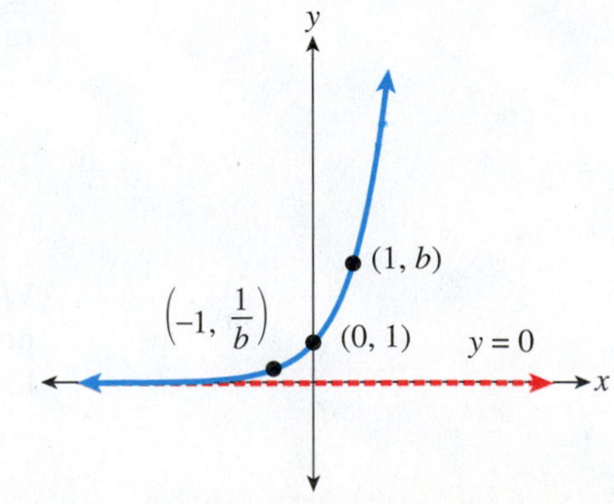

Quick Check 3

Graph $f(x) = 5^x$ and state the domain and range of $f(x)$.

EXAMPLE 3 Graph $f(x) = 3^x$ and state the domain and range of $f(x)$.

SOLUTION We begin by plotting the horizontal asymptote on the x-axis. Next, we plot points at $(0, 1)$, $(1, 3)$, and $\left(-1, \frac{1}{3}\right)$. We finish by drawing the smooth curve passing through these points.

The domain of any exponential function is $(-\infty, \infty)$. From the graph, we see that the range is $(0, \infty)$.

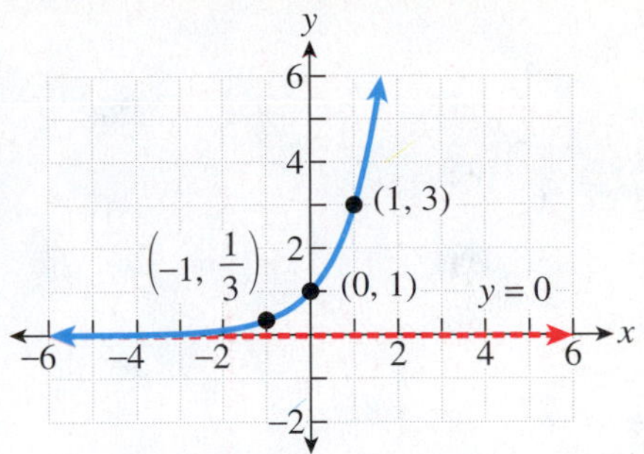

Here are the graphs of $f(x) = 2^x$ and $g(x) = 3^x$ on the same set of axes. Notice that for positive values of x, the function $g(x) = 3^x$ increases more quickly than the function $f(x) = 2^x$ does. For exponential functions of the form $f(x) = b^x, b > 1$, the function increases more quickly for larger values of the base b.

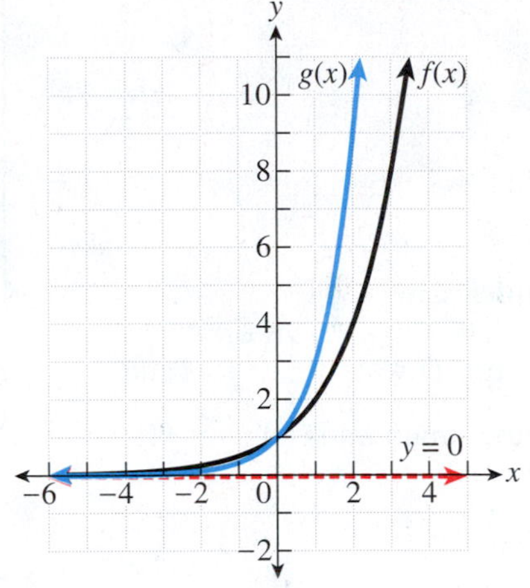

The graph of an exponential function of the form $f(x) = b^x$ where $0 < b < 1$, such as $f(x) = \left(\frac{1}{4}\right)^x$, *decreases* from left to right. To graph an exponential function of this form, we begin by graphing the horizontal asymptote on the x-axis, which is the line $y = 0$. Next, we plot a point on the y-axis at $(0, 1)$, as well as two other points at $(1, b)$ and $\left(-1, \frac{1}{b}\right)$. Then we draw the graph of the function through these points, decreasing toward the right and approaching the horizontal asymptote.

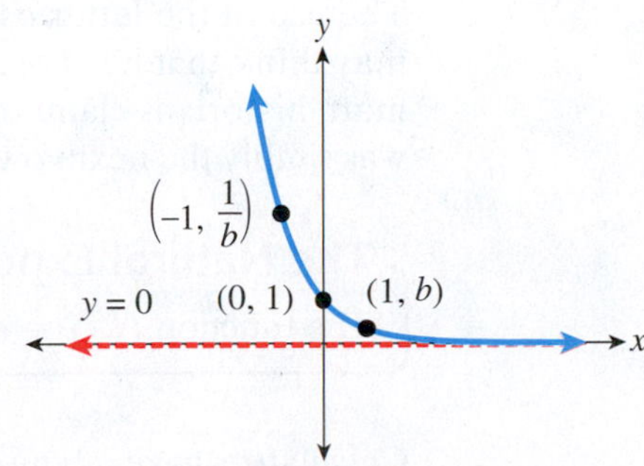

Note that a function such as $f(x) = 5^{-x}$ is also a function of this form because it can be rewritten as $f(x) = \left(\frac{1}{5}\right)^x$.

EXAMPLE 4 Graph $f(x) = \left(\frac{1}{3}\right)^x$ and state the domain and range of $f(x)$.

SOLUTION Begin by plotting the horizontal asymptote on the x-axis. Next, plot points at $(0, 1)$, $\left(1, \frac{1}{3}\right)$, and $(-1, 3)$. Finish by drawing the smooth curve passing through these points.

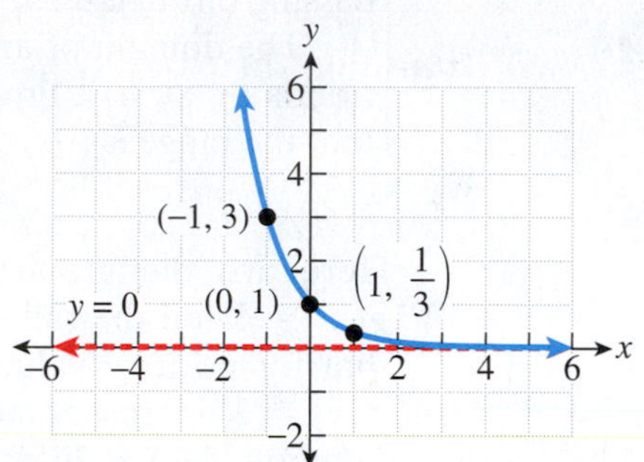

The domain of any exponential function is $(-\infty, \infty)$. From the graph, we see that the range is $(0, \infty)$.

Quick Check 4

Graph $f(x) = \left(\frac{1}{2}\right)^x$ and state the domain and range of $f(x)$.

One last fact about exponential functions is that they are one-to-one functions. We can determine that exponential functions are one-to-one functions by applying the horizontal-line test.

The Natural Exponential Function

Objective 4 Define the natural exponential function. One number that will be used as the base of an exponential function is an irrational number that is denoted by the letter e and is called the natural base.

The Natural Base e

$$e \approx 2.7182818284\ldots$$

The use of the letter e to denote this number is credited to Leonhard Euler. You may think that he chose e because it was the first letter of his last name, but many math historians claim that he had been using the letter a in his writings and that e was simply the next vowel available.

The Natural Exponential Function

The function $f(x) = e^x$ is called the **natural exponential function.**

Calculators have a built-in function for evaluating the natural exponential function. It usually is listed above the key labeled "ln."

EXAMPLE 5 Let $f(x) = e^x$. Find the following and round to the nearest thousandth:

a) $f(5)$

SOLUTION

$f(5) = e^5$	Substitute 5 for x.
≈ 148.413	Approximate using a calculator.

b) $f(-1.7)$

SOLUTION

$$f(-1.7) = e^{-1.7} \quad \text{Substitute } -1.7 \text{ for } x.$$
$$\approx 0.183 \quad \text{Approximate using a calculator.}$$

Quick Check 5

Let $f(x) = e^x$. Find $f(-1)$ and $f(4.3)$. Round to the nearest thousandth.

Using Your Calculator The TI-84 has a built-in function e^x, which is a 2nd function above the key labeled [LN]. To calculate e^5, begin by tapping the [2nd] key, followed by the [LN] key. Notice that the calculator gives you a left-hand parenthesis. After you key the exponent 5, close the parentheses by tapping [)]. After you tap [ENTER], your calculator screen should look as follows:

```
e^(5)
        148.4131591
```

To graph $f(x) = e^x$, we will begin by creating a table of values for this function.

x	-2	-1	0	1	2
$f(x) = e^x$	$e^{-2} \approx 0.14$	$e^{-1} \approx 0.37$	$e^0 = 1$	$e^1 \approx 2.72$	$e^2 \approx 7.39$

As with any exponential function of the form $f(x) = b^x$, the horizontal asymptote is the line $y = 0$. The domain of this function is the set of all real numbers $(-\infty, \infty)$, while the range is $(0, \infty)$.

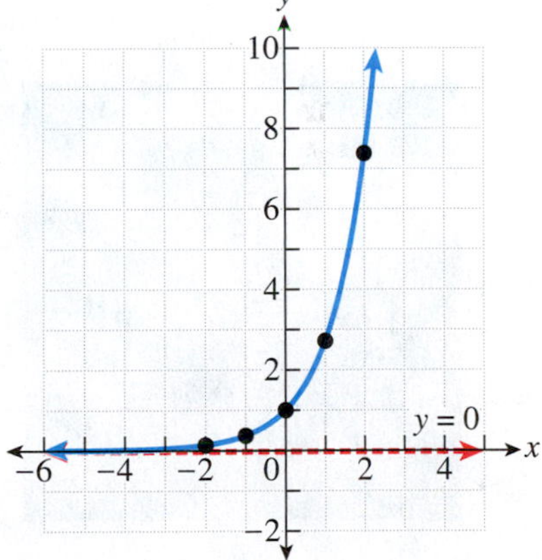

Solving Exponential Equations

Objective 5 Solve exponential equations. An **exponential equation** is an equation containing a variable in an exponent, such as $2^x = 16$. One technique for solving exponential equations is to write both sides of the equation in terms of the same base. If this is possible, we can then apply the one-to-one property of exponential functions to solve the equation.

One-to-One Property of Exponential Functions

For any positive real number b ($b \neq 1$) and any real numbers r and s, if $b^r = b^s$, then $r = s$.

EXAMPLE 6 Solve $2^x = 16$.

SOLUTION Begin by expressing 16 as a power of 2.

$$2^x = 16$$
$$2^x = 2^4 \quad \text{Rewrite 16 as } 2^4.$$
$$x = 4 \quad \text{Use the one-to-one property of exponential functions.}$$

The solution set is $\{4\}$.

EXAMPLE 7 Solve $3^{x+2} = \frac{1}{27}$.

SOLUTION We can rewrite $\frac{1}{27}$ as 3^{-3}.

$$3^{x+2} = \frac{1}{27}$$
$$3^{x+2} = 3^{-3} \quad \text{Rewrite } \tfrac{1}{27} \text{ as } 3^{-3}.$$
$$x + 2 = -3 \quad \text{If } b^r = b^s, \text{ then } r = s.$$
$$x = -5 \quad \text{Subtract 2.}$$

The solution set is $\{-5\}$.

Quick Check 6

Solve.

a) $4^x = 256$ **b)** $9^{2x-17} = 9$

Applications

Objective 6 Use exponential functions in applications. We conclude this section with an application of exponential functions.

EXAMPLE 8 The population of a particular small town can be approximated by the function $f(x) = 75{,}000 \cdot e^{0.018x}$, where x represents the number of years after 1990. What will the population of the town be in 2020?

SOLUTION To predict the town's population in 2020, we evaluate the function for $x = 30$, which is $2020 - 1990$.

$$f(30) = 75{,}000 \cdot e^{0.018(30)} \quad \text{Substitute 30 for } x.$$
$$= 75{,}000 \cdot e^{0.54} \quad \text{Simplify the exponent.}$$
$$\approx 128{,}701 \quad \text{Approximate using a calculator. Round to the nearest person.}$$

The town's population in 2020 will be approximately 128,701.

Quick Check 7

The total revenues generated in a given year by McDonald's (in billions of dollars) can be approximated by the function $f(x) = 7.4 \cdot 1.08^x$, where x represents the number of years after 1992. Use this function to predict the total revenues of McDonald's in 2018. (*Source:* McDonald's Corporation)

BUILDING YOUR STUDY STRATEGY

Practice Quizzes, 1 Assess Your Knowledge Many students have an easy time following along in class or doing their homework but struggle when attempting to do problems on an exam or a quiz. While you work through your exercises, all of your resources are available—textbooks, solutions manual, notes, tutor, and study group members. However, when you are taking an exam or a quiz, you do not have access to these resources. A student who is relying too heavily on his or her resources may not realize this fact until it is too late.

Creating and taking your own practice quiz is an excellent way to assess your knowledge. You can put yourself in a test situation without facing the consequences of taking an actual test. Attempt the problems without using any of your resources, and you will determine which topics require more study.

Exercises 12.1

Vocabulary

1. A function that can be written in the form $f(x) = b^x, b > 0$ and $b \neq 1$, is called a(n) ________ function with base b.
2. A(n) ______________ is a horizontal line that the graph of a function approaches as it moves to the left or to the right.
3. If $f(x)$ increases as x increases, $f(x)$ is said to be a(n) ________ function.
4. If $f(x)$ decreases as x increases, $f(x)$ is said to be a(n) ________ function.
5. The one-to-one property of exponential functions states that for any positive real number b ($b \neq 1$) and any real numbers r and s, if $b^r = b^s$, then ______.
6. The function $f(x) = e^x$ is called the ______ exponential function.

Let $f(x) = 2^x$. Find the following.

7. $f(4)$
8. $f(7)$
9. $f(0)$
10. $f(-1)$

Let $f(x) = \left(\frac{2}{5}\right)^x$. Find the following.

11. $f(2)$
12. $f(3)$
13. $f(-4)$
14. $f(-2)$

Evaluate the given function.

15. $f(x) = 4^{x-3}, f(5)$
16. $f(x) = 4^x - 3, f(5)$
17. $f(x) = 3^{x+2} - 19, f(4)$
18. $f(x) = 3^{x-4} + 46, f(8)$
19. $f(x) = 5^{2x+1} + 400, f(2)$
20. $f(x) = 2^{3x-4} - 225, f(3)$
21. $f(x) = \left(\frac{1}{5}\right)^{x+4}, f(2)$
22. $f(x) = \left(\frac{1}{3}\right)^{x-5}, f(10)$
23. $f(x) = 4^{-x} + 18, f(-2)$
24. $f(x) = 8^{-x} - 52, f(-3)$

Graph $f(x)$. Label the horizontal asymptote. State the domain and range of the function.

25. $f(x) = 4^x$
26. $f(x) = 10^x$

27. $f(x) = \left(\frac{1}{4}\right)^x$

28. $f(x) = \left(\frac{1}{5}\right)^x$

29. $f(x) = 7^x$

30. $f(x) = 2^x$

31. $f(x) = 6^{-x}$

32. $f(x) = 8^{-x}$

Let $f(x) = e^x$. Find the following, rounded to the nearest thousandth.

33. $f(2)$

34. $f(-1)$

35. $f(-5.2)$

36. $f(4.13)$

Evaluate the given function. Round to the nearest thousandth.

37. $f(x) = e^{x+4}, f(-1)$

38. $f(x) = e^{x+2}, f(3)$

39. $f(x) = e^{3x-1}, f(-2)$

40. $f(x) = e^{2x-5}, f(2.7)$

Solve.

41. $3^x = 81$

42. $2^x = 32$

43. $5^x = 125$

44. $10^x = 100{,}000$

45. $6^x = \frac{1}{6}$

46. $4^x = \frac{1}{64}$

47. $2^{x-6} = 32$

48. $7^{x-3} = 343$

49. $5^{x+4} = \frac{1}{25}$

50. $10^{x+5} = \frac{1}{1000}$

51. $\left(\frac{1}{3}\right)^{x+1} = \frac{1}{729}$

52. $\left(\frac{1}{4}\right)^{x-2} = \frac{1}{256}$

53. $6^{2x-13} = 216$

54. $8^{3x+14} = 64$

55. $3^{4x+3} = \frac{1}{243}$

56. $2^{5x-22} = \frac{1}{128}$

57. $9^{3x-12} = 1$

58. $12^{4x+20} = 1$

59. $2^{x^2+3x-6} = 16$

60. $3^{x^2-8x+15} = 27$

61. Wendy deposited \$5000 in an account that pays 13% annual interest, compounded annually. The balance after t years is given by the function $f(t) = 5000(1.13)^t$. What will the balance of this account be after 40 years?

62. Mary deposited \$350 in an account that pays 5% annual interest, compounded annually. The balance after t years is given by the function $f(t) = 350(1.05)^t$. What will the balance of this account be after 6 years?

63. Dylan deposited \$10,000 in an account that pays 9% annual interest, compounded quarterly. The balance after t years is given by the function $f(t) = 10{,}000(1.0225)^{4t}$. What will the balance of this account be after 5 years?

64. Adam deposited \$50,000 in an account that pays 6% annual interest, compounded quarterly. The balance after t years is given by the function $f(t) = 50{,}000(1.015)^{4t}$. What will the balance of this account be after ten years?

65. In 1995, a particular Cal Ripken baseball card was valued at \$8. By 2003, the card had increased in value to \$20. The value of the card can be approximated by the function $f(t) = 8 \cdot e^{0.1145t}$, where t represents the number of years after 1995. Assuming that the value of the card continues to grow exponentially, predict the value of the card in 2017.

66. The average ticket price for attending a movie has been increasing exponentially. The average ticket price in a particular year can be approximated by the function $f(t) = 4.69 \cdot e^{0.0627t}$, where t represents the number of years after 1998. (This average price includes both first run and subsequent runs as well as all special pricing.) Use this function to predict the average ticket price in 2015. (*Source:* Motion Picture Association of America (MPAA))

67. The net revenues of Amazon.com in a particular year, in billions of dollars, can be approximated by the function $f(t) = 2.8 \cdot e^{0.1767t}$, where t represents the

number of years after 2000. If net revenues continue to grow at this rate, predict the net revenues of Amazon.com in 2016. (*Source:* Amazon.com, Inc.)

68. The number of Walmart stores in a particular year can be approximated by the function $f(t) = 2440 \cdot e^{0.0726t}$, where t represents the number of years after 1993. Assuming that this rate of growth continues, use this function to predict the number of Walmart stores in 2018. (*Source:* Walmart Stores, Inc.)

69. If a container of water whose temperature is 80°C is placed in a refrigerator whose temperature remains a constant 4°C, the temperature of the water, in degrees Celsius, after t minutes is given by the function $f(t) = 4 + 76 \cdot e^{-0.014t}$. What will the temperature of the water be after 60 minutes?

70. If a person dies in a room that is 70°F, the body's temperature in degrees Fahrenheit after t hours is given by the function $f(t) = 70 + 28.6 \cdot e^{-0.44t}$. What will the body's temperature be after 3 hours?

71. The average cost of a loaf of white bread in a particular year can be approximated by the function $f(t) = 0.74 \cdot 1.033^t$, where t represents the number of years after 1993. Use this function to approximate the cost of a loaf of white bread in 2020. (*Source:* U.S. Bureau of Labor Statistics)

72. The number of U.S. wineries in a particular year can be approximated by the function $f(t) = 669 \cdot 1.055^t$, where t represents the number of years after 1975. Assuming that this rate of growth continues, use this function to predict the number of U.S. wineries in 2025. (*Source:* Alcohol and Tobacco Tax and Trade Bureau)

Writing in Mathematics

Answer in complete sentences.

73. Explain the difference between the graph of $f(x) = b^x$ for $b > 1$ and $0 < b < 1$.

74. Explain the difference between the graph of $f(x) = 2^x$ and $g(x) = x^2$.

75. Create an exponential equation whose solution is $x = 4$ and explain how to solve the equation.

12.2 Logarithmic Functions

OBJECTIVES

1. Define logarithms and logarithmic functions.
2. Evaluate logarithms and logarithmic functions.
3. Define the common logarithm and natural logarithm.
4. Convert back and forth between exponential form and logarithmic form.
5. Solve logarithmic equations.
6. Graph logarithmic functions.
7. Use logarithmic functions in applications.

Logarithms

Objective 1 Define logarithms and logarithmic functions. In this section, we will examine numbers called **logarithms.** Prior to the development of modern technology such as computers and handheld calculators, logarithms were used to help perform difficult numeric calculations. Although this is no longer a practice, logarithms are still important in today's society and are used in a variety of applications, such as the study of earthquakes and the pH of a chemical substance.

Suppose we are trying to determine what power of 2 is equal to 10; in other words, we are trying to solve the equation $2^a = 10$ for a. This unknown exponent a is a logarithm that can be expressed as $\log_2 10$. We read $\log_2 10$ as *the logarithm, base 2, of 10.*

Logarithm

For any positive real number b $(b \neq 1)$ and any positive real number x, we define $\log_b x$ to be the exponent to which b must be raised to equal x. The number x is called the **argument** of the logarithm. The argument must be a positive number, as it is the result obtained from raising a positive number to a power.

So $\log_5 25$ is the exponent to which 5 must be raised to equal 25. In other words, $\log_5 25$ is the value of y that is a solution to the equation $5^y = 25$. Because we know that $5^2 = 25$, we know that $\log_5 25 = 2$.

Logarithmic Functions

For any positive real number b such that $b \neq 1$, a function of the form $f(x) = \log_b x$ is called a **logarithmic function.** This function is defined for values of x that are greater than 0; so the domain of this function is the set of positive real numbers: $(0, \infty)$.

Later in this section, we will see that logarithmic functions are the inverses of exponential functions.

Evaluating Logarithms and Logarithmic Functions

Objective 2 Evaluate logarithms and logarithmic functions. Now we will shift our focus to evaluating logarithms.

EXAMPLE 1 Evaluate $\log_2 8$.

SOLUTION To evaluate this logarithm, we need to determine what power of 2 is equal to 8. Because $2^3 = 8$, $\log_2 8 = 3$.

EXAMPLE 2 Evaluate $\log_4 \frac{1}{16}$.

SOLUTION We need to determine what power of 4 is equal to the fraction $\frac{1}{16}$. The only way a number greater than 1 can be raised to a power and produce a result less than 1 is if the exponent is negative. Because $4^{-2} = \frac{1}{4^2}$, or $\frac{1}{16}$, $\log_4 \frac{1}{16} = -2$.

Quick Check 1

Evaluate.

a) $\log_4 16$ **b)** $\log_3 \frac{1}{27}$
c) $\log_{12} 12$ **d)** $\log_6 1$

One important property of logarithms is that for any positive number b $(b \neq 1)$, $\log_b b = 1$. For example, $\log_7 7 = 1$. Another important property of logarithms is that for any positive number b $(b \neq 1)$, $\log_b 1 = 0$. For example, $\log_9 1 = 0$.

EXAMPLE 3 Given the function $f(x) = \log_2 x$, find $f(16)$.

SOLUTION We evaluate a logarithmic function the same way we evaluate any other function. We substitute 16 for x and then simplify the resulting expression.

$$f(16) = \log_2 16 \quad \text{Substitute 16 for } x.$$
$$= 4 \quad \text{Because } 2^4 = 16, \log_2 16 = 4.$$

Quick Check 2

Given the function $f(x) = \log_4 x$, find $f\left(\frac{1}{256}\right)$.

Common Logarithms, Natural Logarithms

Objective 3 Define the common logarithm and natural logarithm. We will use two logarithms more often than others. The first is called the common logarithm.

Common Logarithm

The **common logarithm** is a logarithm whose base is 10.

In this case, we will write $\log x$ rather than $\log_{10} x$, and the base is understood to be 10.

A WORD OF CAUTION When working with a logarithm whose base is not 10, we must be sure to write the base because $\log x$ is understood to have a base of 10.

Scientific calculators have built-in functions to calculate common logarithms. This function is typically denoted by a button labeled "log."

EXAMPLE 4 Evaluate the given common logarithm. Round to the nearest thousandth if necessary.

a) $\log 100$

SOLUTION To evaluate $\log 100$, we must determine what power of the base 10 is equal to 100. Because $10^2 = 100$, $\log 100 = 2$. A calculator would produce the same result.

b) $\log 321$

SOLUTION To evaluate $\log 321$, we use a calculator because we do not know what power of 10 is 321. Rounding to the nearest thousandth, we find that $\log 321 \approx 2.507$. To check this result, we evaluate $10^{2.507}$ on a calculator, which should be approximately equal to 321.

Quick Check 3

Evaluate. Round to the nearest thousandth if necessary.

a) $\log 10{,}000$

b) $\log 75$

Using Your Calculator The TI-84 has a built-in function for calculating the common logarithm of a number. To evaluate log 321, tap the button labeled LOG. Notice that the calculator gives you a left-hand parenthesis. The argument of the logarithm needs to be enclosed in parentheses. Then key 321 and close the parentheses by tapping). After you tap ENTER, your calculator screen should look as follows:

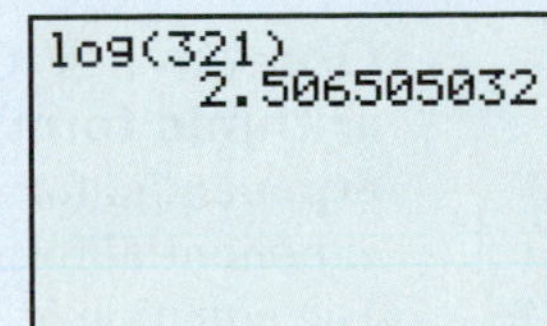

The other logarithm that we will use frequently is called the **natural logarithm.**

Natural Logarithm

The natural logarithm, denoted $\ln x$, is the logarithm whose base is the natural base e.

Recall from the previous section that e is a real number that is approximately equal to 2.718281828. Scientific calculators have built-in functions for calculating natural logarithms by using a button labeled "ln."

EXAMPLE 5 Evaluate the given natural logarithm. Round to the nearest thousandth if necessary.

a) $\ln e^7$

SOLUTION To evaluate $\ln e^7$, we must find an exponent a such that $e^a = e^7$. In this case, $a = 7$; so $\ln e^7 = 7$.

Example 5a illustrates another important property of logarithms.

$\log_b b^r$

For any positive base b $(b \neq 1)$, $\log_b b^r = r$.

b) $\ln 45$

SOLUTION This logarithm must be evaluated by a calculator. Rounded to the nearest thousandth, $\ln 45 \approx 3.807$.

Quick Check 4

Evaluate the following and round to the nearest thousandth if necessary.

a) $\ln e^{-3}$ **b)** $\ln 605$

Using Your Calculator The TI-84 also has a built-in function for calculating the natural logarithm of a number. To evaluate ln 45, begin by tapping the [LN] key. After keying the argument 45, close the parentheses by tapping [)] and then tap [ENTER]. Your calculator screen should look as follows:

```
ln(45)
           3.80666249
```

Exponential Form, Logarithmic Form

Objective 4 Convert back and forth between exponential form and logarithmic form. Consider the equation $5^3 = 125$. This equation is said to be in **exponential form**, as it has a base (5) being raised to a power (3). Any equation in exponential form can be rewritten as an equivalent equation in **logarithmic form**. The equation $5^3 = 125$ also can be written in logarithmic form as $\log_5 125 = 3$.

Exponential Form	Logarithmic Form
$b^y = n$	$\log_b n = y$

The following images will help when we are rewriting equations from one form to the other:

$b^y = n$ $\log_b n = y$ Base(b) Exponent(y) $b^y = n$ $\log_b n = y$ $b^y = n$ $\log_b n = y$ Number(n)

EXAMPLE 6 Write the exponential equation $2^{-8} = \frac{1}{256}$ in logarithmic form.

SOLUTION This equation can be written in logarithmic form as $\log_2\left(\frac{1}{256}\right) = -8$.

Quick Check 5

Write the exponential equation $7^x = 20$ in logarithmic form.

EXAMPLE 7 Write the logarithmic equation $\ln x = 7$ in exponential form.

SOLUTION Recall that the base is e. In exponential form, this equation can be written as $e^7 = x$.

Quick Check 6

Write the logarithmic equation $\log_3 x = 6$ in exponential form.

Consider the function $f(x) = b^x$, where $b > 0$ and $b \neq 1$. To find the inverse of this function, we replace $f(x)$ with y and interchange x and y, which produces the equation $x = b^y$. If we rewrite this equation in logarithmic form, we obtain the equation $y = \log_b x$. Because this equation is solved for y, we have shown that $f^{-1}(x) = \log_b x$. The inverse of an exponential function is a logarithmic function.

Solving Logarithmic Equations

Objective 5 Solve logarithmic equations. When we are solving an equation that is in logarithmic form, it will often be helpful to rewrite the equation in its equivalent exponential form. We can then attempt to solve the resulting equation. For example, suppose we were trying to solve the equation $\log_3 x = 2$. The exponential form of this equation is $3^2 = x$, which simplifies to $x = 9$.

EXAMPLE 8 Solve $\log_5 x = 4$.

SOLUTION Begin by rewriting the equation in exponential form.

$$\log_5 x = 4$$
$$5^4 = x \quad \text{Rewrite in exponential form.}$$
$$625 = x \quad \text{Simplify.}$$

The solution set is $\{625\}$.

Quick Check 7

Solve $\log_4 x = 5$.

EXAMPLE 9 Let $f(x) = \ln(x - 8)$. Solve $f(x) = 3$. Round to the nearest thousandth.

SOLUTION Begin by rewriting the equation in exponential form, keeping in mind that the base of this logarithm is e.

$$\ln(x - 8) = 3 \quad \text{Set the function equal to 3.}$$
$$e^3 = x - 8 \quad \text{Rewrite in exponential form.}$$
$$e^3 + 8 = x \quad \text{Add 8.}$$

The exact solution to the equation is $x = e^3 + 8$, so the solution set is $\{e^3 + 8\}$. Approximating this solution to the nearest thousandth by calculator, we get $x \approx 28.086$.

Quick Check 8

Let $f(x) = \log_3(x + 6)$. Solve $f(x) = 2$.

Graphing Logarithmic Functions

Objective 6 Graph logarithmic functions. We will graph logarithmic functions by plotting points. As with many other functions, the choice of which points to plot is crucial. We find this choice made easier by rewriting a logarithmic function in exponential form. Suppose we wanted to graph the function $f(x) = \log_2 x$. Replacing $f(x)$ with y, we can rewrite the equation $y = \log_2 x$ in exponential form as $2^y = x$. We can now choose values for y and find the corresponding values of x. We will use the values $-2, -1, 0, 1$, and 2 for y. At this point, we notice that this is similar to the way we graphed exponential functions in the previous section, except that we are substituting values for y rather than x.

$x = 2^y$	$2^{-2} = \frac{1}{4}$	$2^{-1} = \frac{1}{2}$	$2^0 = 1$	$2^1 = 2$	$2^2 = 4$
y	-2	-1	0	1	2

Shown next is the graph of $f(x) = \log_2 x$, which passes through the preceding points. Notice that the graph includes only positive values of x. Recall that the domain of a logarithmic function of the form $f(x) = \log_b x$ is the set of all positive real numbers, written in interval notation as $(0, \infty)$. Also notice that as the values of x get closer and closer to 0, the graph of the function decreases without bound. The y-axis, whose equation is $x = 0$, is the **vertical asymptote** for the graph of this function.

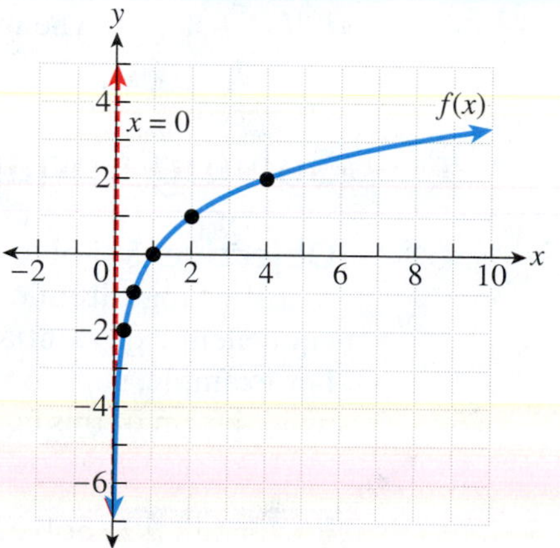

In general, to graph a logarithmic function of the form $f(x) = \log_b x$, where $b > 1$, we begin by graphing the vertical asymptote on the y-axis, which is the line $x = 0$. Next, we plot a point on the x-axis at $(1, 0)$, as well as two other points at $(b, 1)$ and $\left(\frac{1}{b}, -1\right)$. Then we draw the graph of the function, which increases to the right, through these points.

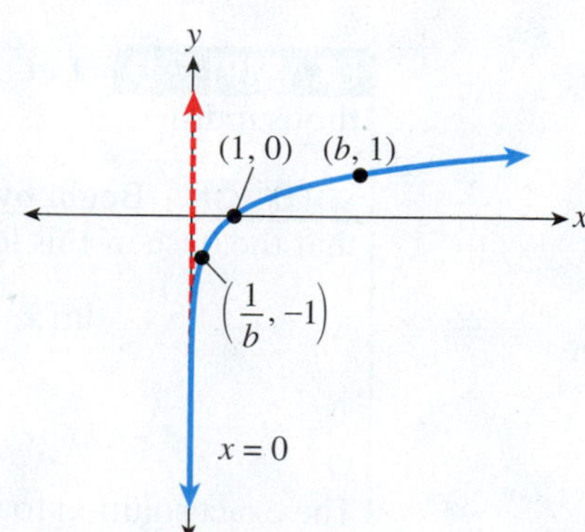

EXAMPLE 10 Graph $f(x) = \log_4 x$ and state the domain and range of $f(x)$.

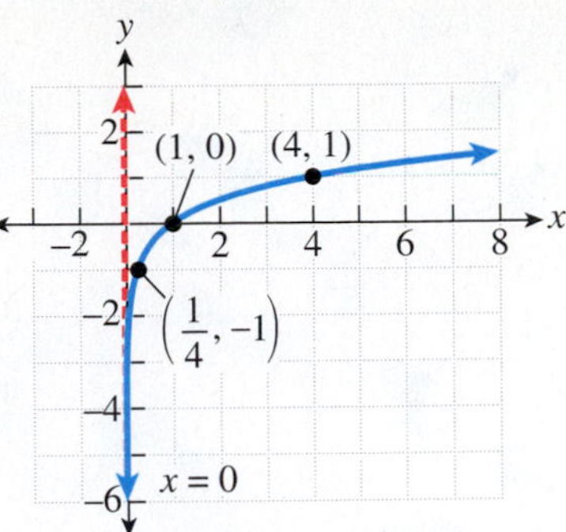

SOLUTION For this logarithmic function, $b = 4$. We begin by plotting the vertical asymptote on the y-axis. Next, we plot points at $(1, 0)$, $(4, 1)$, and $\left(\frac{1}{4}, -1\right)$. We finish by drawing the smooth curve passing through these points.

The domain of this logarithmic function is $(0, \infty)$, and the range is $(-\infty, \infty)$.

Note that as the values of x increase, the function $f(x) = \log_4 x$ does not increase nearly as quickly as the function $f(x) = \log_2 x$ does. What would we expect of the graph of the function $f(x) = \log x$ as x increases? It increases at a much slower rate than either of the two previous functions. In general, the larger the base of a logarithmic function, the slower the function increases as x increases. Here is a graph showing all three functions.

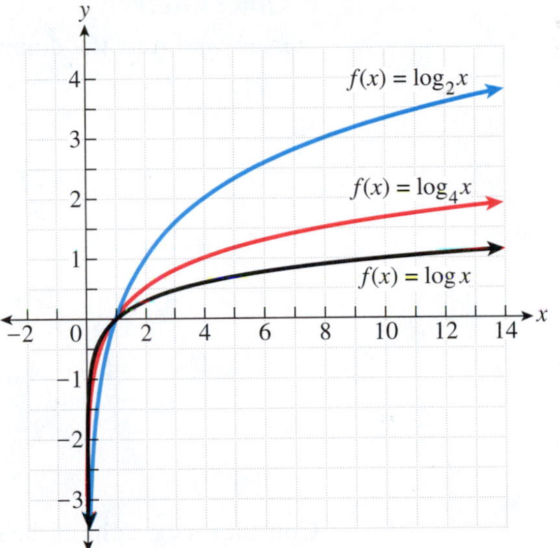

EXAMPLE 11 Graph $f(x) = \ln x$ and state the domain and range of $f(x)$.

SOLUTION We begin by plotting the vertical asymptote on the y-axis. Recall that the base of $\ln x$ is e, so we plot points at $(1, 0)$, $(e, 1)$, and $\left(\frac{1}{e}, -1\right)$. Note that $e \approx 2.7$ and $\frac{1}{e} \approx 0.4$. We finish by drawing the smooth curve passing through these points.

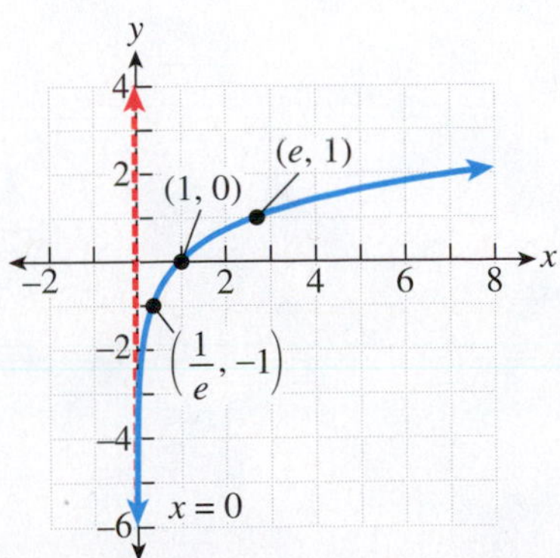

The domain of this logarithmic function is $(0, \infty)$, and the range is $(-\infty, \infty)$.

If $0 < b < 1$, the graph of the function $f(x) = \log_b x$ decreases from left to right. After placing the vertical asymptote on the graph ($x = 0$), we plot points at $(1, 0)$, $(b, 1)$, and $\left(\frac{1}{b}, -1\right)$. We finish by drawing the graph passing through these points.

EXAMPLE 12 Graph $f(x) = \log_{1/3} x$ and state the domain and range of $f(x)$.

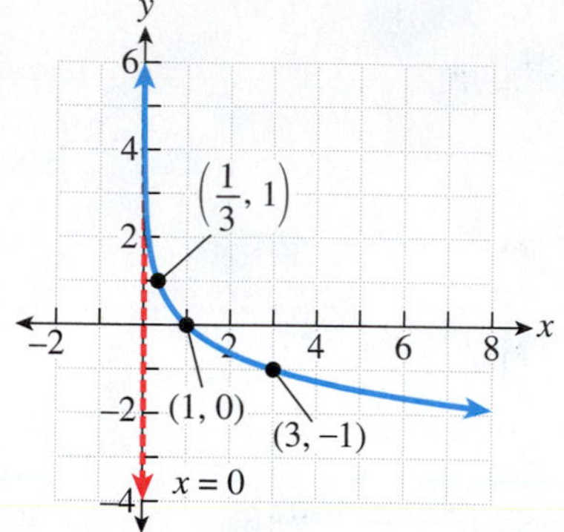

SOLUTION For this logarithmic function, $b = \frac{1}{3}$, we begin by plotting the vertical asymptote on the y-axis, as well as points at $(1, 0)$, $\left(\frac{1}{3}, 1\right)$, and $(3, -1)$. We finish by drawing the smooth curve passing through these points.

The domain of this logarithmic function is $(0, \infty)$, and the range is $(-\infty, \infty)$.

Quick Check 9

Graph and state the domain and range of the function.

a) $f(x) = \log_5 x$ **b)** $f(x) = \log_{1/4} x$

Consider the graphs of the functions $f(x) = 2^x$ and $g(x) = \log_2 x$, which have been graphed together on the same set of axes along with the line $y = x$.

We can see that the two functions are symmetric to each other about the line $y = x$ and therefore are inverses of each other.

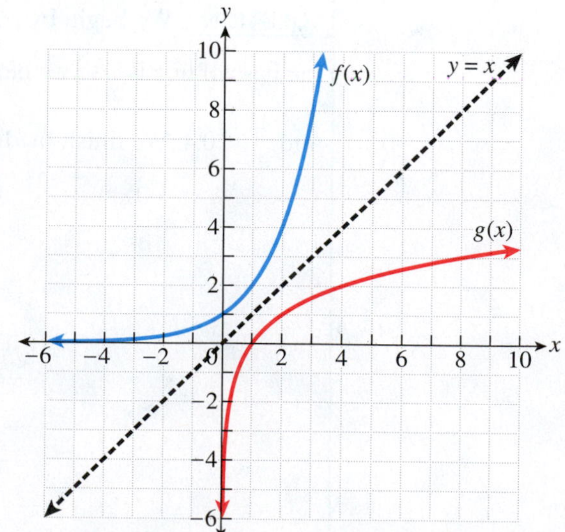

In general, for any positive real number b ($b \neq 1$), the functions $f(x) = b^x$ and $g(x) = \log_b x$ are inverse functions. We will explore this idea more thoroughly in Section 12.4.

Applications of Logarithmic Functions

Objective 7 Use logarithmic functions in applications.

EXAMPLE 13 The average height of boys, in inches, during their first year of life can be approximated by the function $f(x) = 1.95 \ln x + 23.53$, where x is the age in months. Use this function to estimate the average height of boys who are 6 months old. (*Source:* Centers for Disease Control and Prevention (CDC))

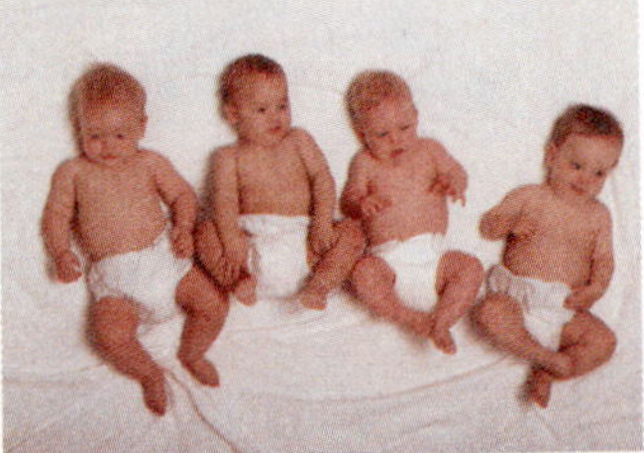

SOLUTION To estimate the average height of boys who are 6 months old, we need to evaluate the function when $x = 6$.

$$f(6) = 1.95 \ln 6 + 23.53 \quad \text{Substitute 6 for } x.$$
$$\approx 27.02 \quad \text{Approximate by calculator.}$$

The average height of 6-month-old boys is approximately 27.02 inches.

Quick Check 10

The annual sales of a company, in billions of dollars, can be approximated by the function $f(x) = 7.95 \ln x + 20.95$, where x represents the number of years after 1995. Use the function to predict the annual sales of this company in 2020.

The **Richter scale** is used to measure the magnitude, or strength, of an earthquake. The magnitude of an earthquake is a function of the size of the shock wave that it creates on a seismograph. If an earthquake creates a shock wave that is I times larger than the smallest measurable shock wave that is recordable by a seismograph, its magnitude R on the Richter scale is given by the formula $R = \log I$.

EXAMPLE 14 If an earthquake creates a shock wave that is 500 times the smallest measurable shock wave recordable by a seismograph, find the magnitude of the earthquake on the Richter scale.

SOLUTION In this problem, we are given the fact that $I = 500$; so we substitute 500 for I in the formula $R = \log I$ and solve for R.

$$R = \log I$$
$$R = \log 500 \quad \text{Substitute 500 for } I.$$
$$R \approx 2.7 \quad \text{Approximate on a calculator.}$$

The magnitude of this earthquake on the Richter scale is 2.7.

Quick Check 11

If an earthquake creates a shock wave that is 2500 times the smallest measurable shock wave recordable by a seismograph, find the magnitude of the earthquake on the Richter scale.

BUILDING YOUR STUDY STRATEGY

Practice Quizzes, 2 Outline Make your own study quiz for this section by selecting odd-numbered problems from the exercise set. Here is a suggested outline.

Number of Problems	Topic
4	Evaluating logarithms and logarithmic functions (including common logarithms and natural logarithms)
2	Rewriting an equation in exponential form
2	Rewriting an equation in logarithmic form
4	Solving logarithmic equations
2	Graphing logarithmic functions
1	Solving applied problems

Choose problems of varying levels of difficulty. By choosing odd-numbered problems from the exercise set, you will have an answer key for your practice quiz in the back of the text.

Exercises 12.2

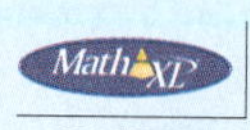

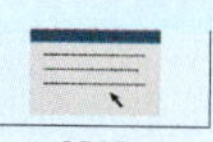

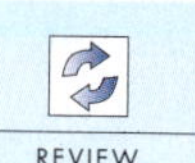

Vocabulary

1. For any positive real number b $(b \neq 1)$ and any positive real number x, ______ is the exponent to which b must be raised to equal x.
2. For any positive real number b such that $b \neq 1$, a function of the form $f(x) = \log_b x$ is called a(n) ________ function.
3. The ________ logarithm is a logarithm whose base is 10 and is denoted by $\log x$.
4. The ________ logarithm is a logarithm whose base is e and is denoted by $\ln x$.
5. An equation of the form $b^y = n$ is said to be in ________ form.
6. An equation of the form $\log_b n = y$ is said to be in ________ form.
7. The graph of a logarithmic function has a(n) ______ asymptote.
8. The __________ is used to measure the magnitude, or strength, of an earthquake.

Evaluate.

9. $\log_2 4$
10. $\log_2 32$
11. $\log_3 27$
12. $\log_3 81$
13. $\log_4 64$
14. $\log_5 625$
15. $\log_8 1$
16. $\log_5 1$
17. $\log_3\left(\frac{1}{9}\right)$
18. $\log_2\left(\frac{1}{16}\right)$
19. $\log_8\left(\frac{1}{512}\right)$
20. $\log_6\left(\frac{1}{6}\right)$

Evaluate the given function.

21. $f(x) = \log_3 x$ **a)** $f(9)$ **b)** $f(243)$
22. $f(x) = \log_5 x$ **a)** $f(25)$ **b)** $f(1)$
23. $f(x) = \log_2(x - 7)$ **a)** $f(8)$ **b)** $f(135)$
24. $f(x) = \log_3(2x + 11)$ **a)** $f(35)$ **b)** $f(116)$

Evaluate. Round to the nearest thousandth if necessary.

25. $\log 1000$
26. $\log 10$
27. $\log 0.01$
28. $\log\left(\frac{1}{10{,}000}\right)$
29. $\log 112$
30. $\log 47$
31. $\log 7.42$
32. $\log 0.0018$
33. $\ln e$
34. $\ln(e^3)$
35. $\ln\left(\frac{1}{e^6}\right)$
36. $\ln(e^{-2})$
37. $\ln 32$
38. $\ln 105$
39. $\ln 16.23$
40. $\ln 0.5$

Rewrite in logarithmic form.

41. $4^3 = 64$

42. $3^5 = 243$

43. $8^{-2} = \frac{1}{64}$

44. $7^{-3} = \frac{1}{343}$

45. $5^x = 42$

46. $2^x = 95$

47. $10^x = 321$

48. $e^x = 6$

Rewrite in exponential form.

49. $\log_2 128 = 7$

50. $\log_5 125 = 3$

51. $\log_9 \frac{1}{6561} = -4$

52. $\log_4 \frac{1}{1024} = -5$

53. $\ln x = 6$

54. $\log x = 2$

55. $\log x + 5 = 12$

56. $\ln x - 8 = -4$

Solve. Round to the nearest thousandth if necessary.

57. $\log_2 x = 5$

58. $\log_2 x = -2$

59. $\log_3 x = -4$

60. $\log_3 x = 5$

61. $\log x = 4$

62. $\log x = 7$

63. $\ln x = 2$

64. $\ln x = -1$

65. $\log_2 x + 5 = 8$

66. $\log_2(x + 5) = 8$

67. $\ln(x - 3) + 6 = 10$

68. $\log(5x + 20) + 3 = 6$

69. Let $f(x) = \log_3(2x - 1)$. Solve $f(x) = 4$.

70. Let $f(x) = \log_5(3x + 32) - 5$. Solve $f(x) = -2$.

71. Let $f(x) = \log(x - 8) + 5$. Solve $f(x) = 7$.

72. Let $f(x) = \ln(x + 4) + 6$. Solve $f(x) = 9$.

Solve.

73. $\log_5 125 = x$

74. $\log_2 256 = x$

75. $\log 100{,}000 = x$

76. $\log 0.001 = x$

77. $\log_9 27 = x$

78. $\log_4 32 = x$

Graph the given function $f(x)$. Label the vertical asymptote. State the domain and range of the function.

79. $f(x) = \log_3 x$

80. $f(x) = \log_6 x$

81. $f(x) = \log x$

82. $f(x) = \log_8 x$

83. $f(x) = \log_{1/5} x$

84. $f(x) = \log_{1/2} x$

85. The number of U.S. airports in a particular year can be approximated by the function $f(x) = 14650 + 1344 \ln x$, where x represents the number of years after 1979. Use this function to predict the number of U.S. airports in 2030. (*Source:* Bureau of Transportation Statistics)

86. The percent of U.S. workers who drive themselves to work can be approximated by the function $f(x) = 72.5 + 2.01 \ln x$, where x represents the number of years after 1984. Use this function to predict what percent of U.S. workers will drive themselves to work in 2031. (*Source:* Bureau of Transportation Statistics)

87. The average height, in inches, of girls who are 3 years old or younger can be approximated by the function $f(x) = 18.6 + 4.78 \ln x$, where x represents the age in months. Use the function to determine the average height of girls who are 24 months old. (*Source:* CDC)

88. The average weight, in pounds, of boys who are 3 years old or younger can be approximated by the function $f(x) = 7.6 + 6.36 \ln x$, where x represents the age in months. Use the function to determine the average weight of boys who are 18 months old. (*Source:* CDC)

For Exercises 89–92, use the formula R = log I, where R is the magnitude of an earthquake on the Richter scale whose shock wave is I times larger than the smallest measurable shock wave that is recordable on a seismograph.

89. If an earthquake creates a shock wave that is 16,000 times the smallest measurable shock wave recordable by a seismograph, find the magnitude of the earthquake on the Richter scale.

90. A 1950 earthquake along the India–China border created a shock wave that was 400,000,000 times the smallest measurable shock wave recordable by a seismograph. Find the magnitude of this earthquake on the Richter scale.

91. The 1906 earthquake off the coast of Ecuador had a magnitude of 8.8 on the Richter scale. How many times larger was the shock wave created by this earthquake than the smallest measurable shock wave?

92. The 1992 earthquake centered in Landers, California, had a magnitude of 7.3 on the Richter scale. How many times larger was the shock wave created by this earthquake than the smallest measurable shock wave?

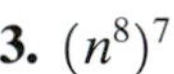

Writing in Mathematics

Answer in complete sentences.

93. What is a logarithm? Explain clearly, using your own words.

Quick Review Exercises

Section 12.2

Simplify. Assume that all variables are nonzero real numbers.

1. $x^7 \cdot x^9$

2. $\dfrac{b^{16}}{b^4}$

3. $(n^8)^7$

4. z^0

12.3 Properties of Logarithms

OBJECTIVES

1. Use the product rule for logarithms.
2. Use the quotient rule for logarithms.
3. Use the power rule for logarithms.
4. Use additional properties of logarithms.
5. Use properties to rewrite two or more logarithmic expressions as a single logarithmic expression.
6. Use properties to rewrite a single logarithmic expression as a sum or difference of logarithmic expressions whose arguments have an exponent of 1.
7. Use the change-of-base formula for logarithms.

In this section, we will learn several properties of logarithms. These properties, in addition to being useful for simplifying logarithmic expressions, also will be helpful when we solve logarithmic equations.

Product Rule for Logarithms

Objective 1 Use the product rule for logarithms. We know that $\log_2 8 = 3$ because $2^3 = 8$. We also know that $\log_2 16 = 4$ because $2^4 = 16$. So $\log_2 8 + \log_2 16 = 3 + 4$, or 7. To find this sum, we had to evaluate two logarithms. Consider $\log_2(8 \cdot 16)$, which is equal to $\log_2 128$. This logarithm also is equal to 7 because $2^7 = 128$. We see that the sum $\log_2 8 + \log_2 16$ is equal to the logarithm whose argument is the product of their two arguments, $\log_2(8 \cdot 16)$. This leads to the **product rule for logarithms**.

Product Rule for Logarithms

For any positive real number b ($b \neq 1$) and any positive real numbers x and y, $\log_b x + \log_b y = \log_b(xy)$.

This rule says that we can rewrite the sum of two logarithms as a single logarithm, provided the two logarithms have the same base. It also says that the logarithm of a product can be rewritten as a sum of logarithms. Here is a proof of the product rule for logarithms.

Let $M = \log_b x$ and $N = \log_b y$.
Rewriting $M = \log_b x$ in exponential form, we see that $b^M = x$. Similarly, $b^N = y$. So $xy = b^M \cdot b^N = b^{M+N}$.

Rewriting $xy = b^{M+N}$ in logarithmic form, we see that $\log_b(xy) = M + N$.
But $M + N = \log_b x + \log_b y$, so $\log_b(xy) = \log_b x + \log_b y$.

EXAMPLE 1 Rewrite $\ln 7 + \ln x$ as a single logarithm. Assume that x is a positive real number.

SOLUTION Both logarithms have the same base, so we will use the product rule to simplify this expression.

$$\ln 7 + \ln x = \ln(7x) \quad \text{Rewrite as the logarithm of a product.}$$

Quick Check 1

Rewrite $\log 5 + \log 7 + \log x$ as a single logarithm. Assume that x is a positive real number.

EXAMPLE 2 Rewrite $\log(10xy)$ as the sum of two or more logarithms. Simplify if possible. Assume that all variables are positive real numbers.

SOLUTION This expression is the logarithm of three factors, and we may use the product rule to rewrite it as the sum of three logarithms.

$$\begin{aligned} \log(10xy) &= \log 10 + \log x + \log y && \text{Rewrite as the sum of three logarithms.} \\ &= 1 + \log x + \log y && \text{Simplify log 10.} \end{aligned}$$

Quick Check 2

Rewrite $\log_2(5b)$ as the sum of two or more logarithms. Assume that b is a positive real number.

A WORD OF CAUTION The product rule for logarithms applies only to expressions of the form $\log_b x + \log_b y$, not to expressions such as $\log_b(x + y)$. In general,

$$\log_b(x + y) \neq \log_b x + \log_b y.$$

Quotient Rule for Logarithms

Objective 2 Use the quotient rule for logarithms. Just as the sum of two logarithms can be expressed as a single logarithm, so can the difference of two logarithms.

Quotient Rule for Logarithms

For any positive real number b ($b \neq 1$) and any positive real numbers x and y, $\log_b x - \log_b y = \log_b\left(\frac{x}{y}\right)$.

This rule says that we can rewrite the difference of two logarithms as a single logarithm of a quotient, provided the two logarithms have the same base. It also says that the logarithm of a quotient can be rewritten as a difference of logarithms. Using the fact that $\frac{b^M}{b^N} = b^{M-N}$, we find that the proof of the quotient rule for logarithms is similar to the proof of the product rule for logarithms. The proof is left as an exercise for the reader.

EXAMPLE 3 Rewrite $\log_3 36 - \log_3 4$ as a single logarithm. Simplify if possible.

SOLUTION Both logarithms have the same base, so we use the quotient rule to simplify this expression.

$$\log_3 36 - \log_3 4 = \log_3\left(\frac{36}{4}\right) \quad \text{Rewrite as the logarithm of a quotient.}$$
$$= \log_3 9 \quad \text{Divide.}$$
$$= 2 \quad \text{Simplify the logarithm.}$$

Quick Check 3

Rewrite $\log_2 60 - \log_2 15$ as a single logarithm. Simplify if possible.

EXAMPLE 4 Rewrite $\ln\left(\frac{12x}{y}\right)$ in terms of two or more logarithms. Assume that all variables are positive real numbers.

SOLUTION Begin by applying the quotient rule.

$$\ln\left(\frac{12x}{y}\right) = \ln(12x) - \ln y \quad \text{Rewrite as the logarithm of the numerator minus the logarithm of the denominator.}$$
$$= \ln 12 + \ln x - \ln y \quad \text{Apply the product rule to } \ln(12x).$$

Quick Check 4

Rewrite $\log_b\left(\frac{x}{b}\right)$ in terms of two logarithms. Simplify if possible. Assume that x and b are positive real numbers.

A WORD OF CAUTION The quotient rule for logarithms applies only to expressions of the form $\log_b x - \log_b y$, not to expressions such as $\log_b(x - y)$. In general,

$$\log_b(x - y) \neq \log_b x - \log_b y.$$

Power Rule for Logarithms

Objective 3 Use the power rule for logarithms. The properties of logarithms discussed in this section will help us solve exponential equations, and one of the most important properties is the power rule for logarithms.

Power Rule for Logarithms

For any positive real number b ($b \neq 1$), any positive real number x, and any real number r,

$$\log_b x^r = r \cdot \log_b x.$$

This rule says that we can rewrite the logarithm of a positive number that is raised to a power by multiplying the exponent by the logarithm of the number. Here is a proof of the power rule, which uses the product rule for logarithms.

$$\log_b x^r = \log_b\left(\underbrace{x \cdot x \cdot \cdots \cdot x}_{r \text{ times}}\right)$$ Rewrite x^r by listing x as a factor r times.

$$= \underbrace{\log_b x + \log_b x + \cdots + \log_b x}_{r \text{ times}}$$ Use the product rule for logarithms.

$$= r \cdot \log_b x$$ Rewrite the repeated addition of $\log_b x$ as $r \cdot \log_b x$.

EXAMPLE 5 Rewrite $\log_2 x^7$ as a logarithm of a single factor without exponents. Assume that x is a positive real number.

SOLUTION The power rule says that we may rewrite this expression by removing 7 as an exponent and multiplying it by $\log_2 x$. In other words, we are moving the number 7 from the exponent of x and placing it as a factor in front of $\log_2 x$.

$$\log_2 x^{⑦}$$

$$\log_2 x^7 = 7 \log_2 x$$ Use the power rule.

Quick Check 5

Rewrite $\log_3(9x^2y^4)$ using logarithms of single factors that do not have exponents. Simplify if possible. Assume that x and y are positive real numbers.

Quick Check 6

Rewrite $\ln \sqrt[7]{x}$ as a logarithm of a single factor without any radicals. Assume that x is a positive real number.

Recall that $\sqrt[n]{x} = x^{1/n}$. So we can apply the power rule when the argument of a logarithm is an expression involving radicals. For example, $\log_9 \sqrt[4]{x} = \log_9 x^{1/4} = \frac{1}{4} \log_9 x$.

EXAMPLE 6 Rewrite $15 \ln x$ in such a way that the logarithm is not being multiplied by a number. Assume that x is a positive real number.

SOLUTION We will use the power rule to rewrite 15 as an exponent of x.

$$15 \ln x = \ln x^{15}$$ Use the power rule.

Quick Check 7

Rewrite $8 \log_8 x$ in such a way that the logarithm is not being multiplied by a number. Assume that x is a positive real number.

A WORD OF CAUTION Be careful when working with logarithmic expressions involving exponents. In the expression $\log_b x^n$, only the number x is being raised to the *nth* power. However, in the expression $(\log_b x)^n$, it is the logarithm $\log_b x$ that is being raised to the *nth* power. We cannot apply the power rule for logarithms to the expression $(\log_b x)^n$. In general,

$$\log_b x^n \neq (\log_b x)^n.$$

For example, $\log_2 4^3 \neq (\log_2 4)^3$.

Objective 4 Use additional properties of logarithms. The power rule leads us to another property of logarithms. Consider the expression $\log_5 5^7$. The power rule says that $\log_5 5^7 = 7 \log_5 5$. Because $\log_5 5$ is equal to 1, this expression simplifies to equal $7 \cdot 1$, or 7. This result can be generalized as follows:

$\log_b b^r$

For any positive real number b $(b \neq 1)$ and any real number r, $\log_b b^r = r$.

This property will be particularly helpful when we take the natural logarithm of e raised to a power.

EXAMPLE 7 Simplify $\ln e^x$.

SOLUTION The base of this logarithm is e, so the property $\log_b b^r = r$ may be applied.

$$\ln e^x = x$$

Quick Check 8
Simplify $\ln e^{-4}$.

Another property of logarithms involves raising a positive number to a power containing a logarithm whose base is that number. This property follows from the fact that exponential and logarithmic functions are inverses.

$b^{\log_b x}$

For any positive real number b $(b \neq 1)$ and any positive real number x,

$$b^{\log_b x} = x.$$

EXAMPLE 8 Simplify $8^{\log_8 15}$.

SOLUTION By the previous property, $8^{\log_8 15} = 15$.

Quick Check 9
Simplify $e^{\ln 16}$.

Rewriting Logarithmic Expressions as a Single Logarithm

Objective 5 Use properties to rewrite two or more logarithmic expressions as a single logarithmic expression. We will often need to rewrite an expression containing two or more logarithms in terms of a single logarithm. To do this, we will be using the power rule first if possible; then we will use the quotient rule and the product rule.

EXAMPLE 9 Rewrite $\log a + \log b - \log c - \log d$ as a single logarithm. Assume that $a, b, c,$ and d are positive real numbers.

SOLUTION The first two logarithms can be combined by the product rule, as can the last two logarithms. Then we can combine these logarithms using the quotient rule.

$\log a + \log b - \log c - \log d$

$= \log a + \log b - (\log c + \log d)$ Factor -1 from the logarithms being subtracted.

$= \log(ab) - \log(cd)$ Use the product rule.

$= \log\left(\frac{ab}{cd}\right)$ Use the quotient rule.

Quick Check 10

Rewrite $\log_2 a - \log_2 b + \log_2 c + \log_2 d$ as a single logarithm. Assume that $a, b, c,$ and d are positive real numbers.

EXAMPLE 10 Rewrite $\frac{1}{4}\ln a - 10\ln b - 2\ln c$ as a single logarithm. Assume that $a, b,$ and c are positive real numbers.

SOLUTION Begin by using the power rule to rewrite each logarithm. Then use the quotient rule.

$\frac{1}{4}\ln a - 10\ln b - 2\ln c = \ln a^{1/4} - \ln b^{10} - \ln c^2$ Use the power rule.

$= \ln a^{1/4} - (\ln b^{10} + \ln c^2)$ Factor -1 from the last two terms.

$= \ln\left(\frac{a^{1/4}}{b^{10}c^2}\right)$ Use the quotient rule. Because both $\ln b^{10}$ and $\ln c^2$ are being subtracted, b^{10} and c^2 are written in the denominator.

$= \ln\left(\frac{\sqrt[4]{a}}{b^{10}c^2}\right)$ Rewrite the numerator using radical notation.

Quick Check 11

Rewrite $6\log a + \frac{1}{5}\log b - 9\log c - 12\log d$ as a single logarithm. Assume that $a, b, c,$ and d are positive real numbers.

Rewriting a Logarithmic Expression as a Sum or Difference of Logarithms Whose Arguments Have an Exponent of 1

Objective 6 Use properties to rewrite a single logarithmic expression as a sum or difference of logarithmic expressions whose arguments have an exponent of 1. We now turn our attention to expanding a single logarithm into the sum or difference of separate logarithms. This will be done in such a way that each logarithm is of a single factor that does not have an exponent. To do this, we will be using the product rule, the quotient rule, and the power rule.

EXAMPLE 11 Rewrite $\log\left(\frac{a^5b^8}{c^7}\right)$ in terms of two or more logarithms. Each argument should contain a single factor, and each exponent should be written as a factor. Assume that a, b, and c represent positive real numbers.

SOLUTION Begin by applying the product and quotient rules to rewrite this logarithm in terms of three logarithms. The power rule can then be applied to each of these logarithms.

$$\log\left(\frac{a^5b^8}{c^7}\right) = \log a^5 + \log b^8 - \log c^7 \quad \text{Use the product and quotient rules.}$$

$$= 5\log a + 8\log b - 7\log c \quad \text{Use the power rule.}$$

Quick Check 12

Rewrite $\log_5\left(\frac{z^3}{x^9y^4}\right)$ in terms of two or more logarithms. Each argument should contain a single factor, and each exponent should be written as a factor. Assume that x, y, and z represent positive real numbers.

Change-of-Base Formula

Objective 7 Use the change-of-base formula for logarithms. The final property of logarithms in this section is the change-of-base formula.

Change-of-Base Formula

For any positive numbers a, b, and x $(a, b \neq 1)$, $\log_b x = \frac{\log_a x}{\log_a b}$.

This formula allows us to rewrite any logarithm in terms of two logarithms that have a different base than the original logarithm. The advantage of this is that we can rewrite logarithms in terms of common logarithms or natural logarithms, allowing us to approximate these logarithms using a calculator. For instance, suppose we wanted to evaluate $\log_2 7$. Because we do not know what power of 2 is equal to 7, we are not able to evaluate this logarithm. We can rewrite this logarithm as $\frac{\log 7}{\log 2}$, or as $\frac{\ln 7}{\ln 2}$, using the change-of-base formula. Either of these two expressions can be evaluated on a calculator. Rounding to three decimal places, we see that $\log_2 7 \approx 2.807$.

EXAMPLE 12 Evaluate $\log_5 13$ using the change-of-base formula.

SOLUTION Using common logarithms, we can rewrite this logarithm as $\frac{\log 13}{\log 5}$.

$$\log_5 13 = \frac{\log 13}{\log 5} \quad \text{Use the change-of-base formula.}$$

$$\approx 1.594 \quad \text{Approximate by calculator.}$$

We can check this result by raising 5 to the 1.594 power. Notice that the result is approximately equal to 13. Also notice that if we use natural logarithms instead of common logarithms, $\frac{\ln 13}{\ln 5} \approx 1.594$ as well.

Quick Check 13

Evaluate $\log_3 125$ using the change-of-base formula.

Using Your Calculator You can evaluate logarithms such as $\log_5 13$ on the TI-84 using the change-of-base formula, which says that $\log_5 13 = \frac{\log 13}{\log 5}$. This can be entered as follows:

```
log(13)/log(5)
        1.593692641
```

BUILDING YOUR STUDY STRATEGY

Practice Quizzes, 3 Make Your Own Make your own practice quiz for the first three sections of this chapter by selecting odd-numbered problems from the exercise sets of the first three sections. Select four or five odd-numbered problems from each section that you think are representative of the types of problems you would be expected to solve on an exam. Choose problems of varying levels of difficulty. By choosing odd-numbered problems from the exercise sets, you will have an answer key for your practice quiz in the back of the text.

Exercises 12.3

PRACTICE

WATCH

DOWNLOAD

READ

REVIEW

Vocabulary

1. The product rule for logarithms states that for any positive real number b $(b \neq 1)$ and any positive real numbers x and y, $\log_b x + \log_b y =$ ________.
2. The quotient rule for logarithms states that for any positive real number b $(b \neq 1)$ and any positive real numbers x and y, $\log_b x - \log_b y =$ ________.
3. The power rule for logarithms states that for any positive real number b $(b \neq 1)$, any positive real number x, and any real number r, $\log_b x^r =$ ________.
4. State the change-of-base formula.

Rewrite as a single logarithm using the product rule for logarithms. Simplify if possible. Assume that all variables represent positive real numbers.

5. $\log_4 3 + \log_4 a$
6. $\log_8 7 + \log_8 b$
7. $\log x + \log y$
8. $\ln m + \ln n$
9. $\log_9 3 + \log_9 27$
10. $\log_6 12 + \log_6 18$

Rewrite as the sum of two or more logarithms using the product rule for logarithms. Simplify if possible. Assume that all variables represent positive real numbers.

11. $\log_3(10x)$
12. $\log_6(8b)$
13. $\ln(3e)$
14. $\log_5(25z)$
15. $\log_4(abc)$
16. $\log_2(16mn)$

Rewrite as a single logarithm using the quotient rule for logarithms. Simplify if possible. Assume that all variables represent positive real numbers.

17. $\log_3 56 - \log_3 8$
18. $\log_4 24 - \log_4 144$
19. $\ln x - \ln 3$

20. $\log 20 - \log x$

21. $\log_4 2 - \log_4 128$

22. $\log_5 150 - \log_5 6$

Rewrite in terms of two or more logarithms using the quotient and product rules for logarithms. Simplify if possible. Assume that all variables represent positive real numbers.

23. $\log_8\left(\frac{10a}{b}\right)$

24. $\log_5\left(\frac{7x}{3y}\right)$

25. $\log_4\left(\frac{6xyz}{w}\right)$

26. $\log_7\left(\frac{1}{30}\right)$

27. $\log_5\left(\frac{xy}{25z}\right)$

28. $\log_2\left(\frac{8a}{bc}\right)$

Rewrite using the power and product rules. Simplify if possible. Assume that all variables represent positive real numbers.

29. $\log_2 x^7$

30. $\log b^{10}$

31. $\log_5(x^4y^7)$

32. $\ln(x^3y^8)$

33. $\log_3 \sqrt{x}$

34. $\log_4 \sqrt[6]{b}$

35. $\log_2(16x^5y^7)$

36. $\log_5(25a^6 \sqrt[3]{b})$

Rewrite using the power rule. Assume that all variables represent positive real numbers.

37. $8 \log_5 x$

38. $3 \log_4 b$

39. $\frac{1}{3} \ln x$

40. $\frac{1}{2} \log_2 x$

41. $5 \log_3 x^2$

42. $4 \log b^6$

Simplify.

43. $\log_5 5^{12}$

44. $\log_2 2^{21}$

45. $\ln e^9$

46. $\log 10^{25}$

47. $2^{\log_2 7}$

48. $5^{\log_5 36}$

49. $3^{\log_3 b}$

50. $e^{\ln 0.026}$

Rewrite as a single logarithm. Assume that all variables represent positive real numbers.

51. $\log_7 4 + \log_7 5 - \log_7 3$

52. $\log_3 5 - \log_3 7 - \log_3 8$

53. $4 \log_9 2 + 3 \log_9 5$

54. $3 \log 7 - 5 \log 3$

55. $\frac{1}{2} \ln 16 - \frac{1}{3} \ln 125$

56. $\frac{1}{2} \log_6 144 + 4 \log_6 5$

57. $\log_2 x - \log_2 y - \log_2 z - \log_2 w$

58. $5 \log_3 a - 7 \log_3 b + 2 \log_3 c$

59. $4 \ln x + 9 \ln y - 13 \ln z$

60. $10 \log_5 x - 3 \log_5 y - 6 \log_5 z$

61. $\log_b(x + 6) + \log_b(x - 8)$

62. $\log_5(x + 7) + \log_5(x - 7)$

63. $\log(3x - 4) + \log(x + 10)$

64. $\ln(x^2 - 3x - 108) - \ln(x + 9)$

65. $3 \ln a + \frac{1}{3} \ln b - 10 \ln c$

66. $\frac{1}{5} \ln x - 7 \ln y - 6 \ln z$

Suppose for some base $b > 0$ ($b \neq 1$) that $\log_b 2 = A$, $\log_b 3 = B$, $\log_b 5 = C$, and $\log_b 7 = D$. Express the given logarithms in terms of A, B, C, or D.

(Hint: Rewrite the logarithm in terms of $\log_b 2$, $\log_b 3$, $\log_b 5$, or $\log_b 7$.)

67. $\log_b 6$

68. $\log_b 15$

69. $\log_b 35$

70. $\log_b 14$

71. $\log_b 8$

72. $\log_b 9$

73. $\log_b 125$

74. $\log_b\left(\frac{1}{49}\right)$

Expand. Simplify if possible. Assume that all variables represent positive real numbers.

75. $\log_3\left(\frac{ab}{cd}\right)$

76. $\log_5\left(\frac{25xy}{z}\right)$

77. $\log_2\left(\frac{a^3}{8b^4}\right)$

78. $\log_6(1296x^7y^{10})$

79. $\ln\left(\frac{a^6\sqrt{b}}{c^7}\right)$

80. $\ln\left(\frac{\sqrt[7]{x}}{y^{10}z^3}\right)$

81. $-\ln\left(\frac{a^3d^2}{bc^7}\right)$

82. $-2\log\left(\frac{xy^2}{z^3}\right)$

Evaluate using the change-of-base formula. Round to the nearest thousandth.

83. $\log_2 11$

84. $\log_9 320$

85. $\log_{15} 30$

86. $\log_5 789$

87. $\log_3 12{,}695$

88. $\log_2 325{,}000$

89. $\log_6 4$

90. $\log_{16} 7$

91. $\log_8 0.13$

92. $\log_4 0.065$

Find the missing number. Round to the nearest thousandth. (Hint: Rewrite in logarithmic form and use the change-of-base formula.)

93. $5^? = 19$

94. $3^? = 30$

95. $15^? = 100$

96. $2^? = 725$

Writing in Mathematics

Answer in complete sentences.

97. Give examples showing that the following statements are false.

- $\log_b(x + y) = \log_b x + \log_b y$
- $\log_b(x - y) = \log_b x - \log_b y$
- $\log_b x^n = (\log_b x)^n$

12.4 Exponential and Logarithmic Equations

OBJECTIVES

1. Solve an exponential equation in which both sides have the same base.
2. Solve an exponential equation by using logarithms.
3. Solve a logarithmic equation in which both sides have logarithms with the same base.
4. Solve a logarithmic equation by converting it to exponential form.
5. Use properties of logarithms to solve a logarithmic equation.
6. Find the inverse function for exponential and logarithmic functions.

In this section, we will solve exponential and logarithmic equations. An exponential equation is an equation containing a variable in an exponent. A logarithmic equation is an equation involving logarithms of variable expressions.

Solving Exponential Equations

Objective 1 Solve an exponential equation in which both sides have the same base. Given an exponential equation to solve, we will first try to express both sides of the equation in terms of the same base. If we can do this, we can solve the equation with the techniques developed in Section 12.1. Here are two examples reviewing this process.

EXAMPLE 1 Solve $2^{x+5} = 8$.

SOLUTION We begin by rewriting 8 as 2^3, as both sides of the equation will then have the same base. We can then apply the one-to-one property of exponential functions.

$$2^{x+5} = 8$$

$2^{x+5} = 2^3$ Rewrite 8 as 2^3.

$x + 5 = 3$ Apply the one-to-one property of exponential functions. If $b^r = b^s$, then $r = s$.

$x = -2$ Subtract 5.

The solution of the equation is $x = -2$. We could check this solution by substituting -2 for x in the original equation. This check is left to the reader. The solution set is $\{-2\}$.

EXAMPLE 2 Solve $9^{4x-1} = 27^{2x}$.

SOLUTION Although we cannot write 27 as a power of 9, we can rewrite both bases as powers of 3.

$$9^{4x-1} = 27^{2x}$$

$(3^2)^{4x-1} = (3^3)^{2x}$ Rewrite 9 as 3^2 and 27 as 3^3.

$3^{8x-2} = 3^{6x}$ Simplify the exponents.

$8x - 2 = 6x$ If $b^r = b^s$, then $r = s$.

$-2 = -2x$ Subtract $8x$.

$1 = x$ Divide both sides by -2.

The solution set is $\{1\}$.

Quick Check 1

Solve.

a) $3^{3x-2} = 81$
b) $4^{3x-1} = 32^{x+2}$

Solving Exponential Equations by Using Logarithms

Objective 2 Solve an exponential equation by using logarithms. Often it is not possible to rewrite both sides of an exponential equation in terms of the same base, such as in the equation $3^x = 12$. We cannot rewrite 12 as a power of 3. In such a case, we will apply the one-to-one property of logarithmic functions. We can see that logarithmic functions are one-to-one functions by applying the horizontal line test to the graph of any logarithmic function.

One-to-One Property of Logarithmic Functions

For any positive real number b ($b \neq 1$) and any real numbers r and s, if $r = s$, then $\log_b r = \log_b s$. In addition, if $\log_b r = \log_b s$, then $r = s$.

This property says that if two positive numbers are equal, the logarithms of those numbers are equal as well as long as both logarithms have the same base. When solving an exponential equation such as $3^x = 12$, we first take the common logarithm (base 10) of both sides of the equation. Then we use the power rule for logarithms to help us solve the resulting equation. We will use the common logarithm because of its availability on calculators, although using the natural logarithm (base e) would be a good choice as well.

EXAMPLE 3 Solve $3^x = 12$. Find the exact solution and then approximate the solution to the nearest thousandth.

SOLUTION We begin by taking the common logarithm of each side of the equation.

$$3^x = 12$$

$$\log 3^x = \log 12 \quad \text{Take the common logarithm of both sides.}$$

$$x \cdot \log 3 = \log 12 \quad \text{Use the power rule for logarithms. This moves the variable from the exponent.}$$

$$\frac{x \cdot \cancel{\log 3}^{1}}{\cancel{\log 3}_{1}} = \frac{\log 12}{\log 3} \quad \text{Divide both sides by log 3.}$$

$$x = \frac{\log 12}{\log 3} \quad \text{This is the exact solution.}$$

$$x \approx 2.262 \quad \text{Approximate using a calculator.}$$

The solution set is $\left\{\frac{\log 12}{\log 3}\right\}$. The exact solution is $x = \frac{\log 12}{\log 3}$, which is approximately equal to 2.262. Substituting this value back into the original equation, we see that $3^{2.262} \approx 12.00185$, which is approximately equal to 12.

Quick Check 2

Solve $4^{x-5} - 9 = 3$.

EXAMPLE 4 Let $f(x) = 13^{x+8} - 22$. Solve the equation $f(x) = 108$. Find the exact solution and then approximate the solution to the nearest thousandth.

SOLUTION Begin by setting the function equal to 108.

$$f(x) = 108 \quad \text{Set } f(x) \text{ equal to 108.}$$

$$13^{x+8} - 22 = 108 \quad \text{Replace } f(x) \text{ with } 13^{x+8} - 22.$$

$$13^{x+8} = 130 \quad \text{Add 22.}$$

$$\log 13^{x+8} = \log 130 \quad \text{Take the common logarithm of both sides.}$$

$$(x + 8) \cdot \log 13 = \log 130 \quad \text{Use the power rule for logarithms.}$$

$$\frac{(x + 8) \cdot \cancel{\log 13}^{1}}{\cancel{\log 13}_{1}} = \frac{\log 130}{\log 13} \quad \text{Divide both sides by log 13.}$$

$$x + 8 = \frac{\log 130}{\log 13} \quad \text{Simplify.}$$

$$x = \frac{\log 130}{\log 13} - 8 \quad \text{Subtract 8. This is the exact solution.}$$

$$x \approx -6.102 \quad \text{Approximate using a calculator.}$$

The solution set is $\left\{\frac{\log 130}{\log 13} - 8\right\}$. The exact solution is $x = \frac{\log 130}{\log 13} - 8$, which is approximately equal to -6.102.

Quick Check 3

Let $f(x) = 9^{x+3} - 17$. Solve the equation $f(x) = -6$. Find the exact solution and then approximate the solution to the nearest thousandth.

If the exponential expression in an equation has e as its base, it is a good idea to use the natural logarithm rather than the common logarithm. This is because $\ln e^x = x$ for any real number x.

EXAMPLE 5 Solve $2 \cdot e^{x+4} + 19 = 107$. Find the exact solution and then approximate the solution to the nearest thousandth.

SOLUTION Because the base is e, we use the natural logarithm rather than the common logarithm. Begin by isolating the exponential expression.

$2 \cdot e^{x+4} + 19 = 107$

$2 \cdot e^{x+4} = 88$ Subtract 19.

$e^{x+4} = 44$ Divide both sides by 2, isolating the exponential expression.

$\ln e^{x+4} = \ln 44$ Take the natural logarithm of both sides.

$x + 4 = \ln 44$ Simplify the left side of the equation.

$x = \ln 44 - 4$ Subtract 4. This is the exact solution. Do not subtract 4 from 44—you must take the natural logarithm of 44 before subtracting.

$x \approx -0.216$ Approximate using a calculator.

The solution set is $\{\ln 44 - 4\}$. The exact solution is $x = \ln 44 - 4$, which is approximately equal to -0.216.

Quick Check 4

Solve $e^{x+2} - 10 = 18$. Round to the nearest thousandth.

Solving Logarithmic Equations

Objective 3 Solve a logarithmic equation in which both sides have logarithms with the same base. Suppose we have a logarithmic equation in which two logarithms with the same base are equal to each other, such as $\log_5 x = \log_5 2$. The one-to-one property of logarithmic functions can be applied to this situation, which states that for any positive real number b ($b \neq 1$) and any real numbers r and s, if $\log_b r = \log_b s$, then $r = s$. This means that for the equation $\log_5 x = \log_5 2$, x must be equal to 2.

When solving logarithmic equations, we must check the solutions for extraneous solutions. Recall that the domain for a logarithmic function is the set of positive real numbers. Any tentative solution that would require us to take the logarithm of a number that is not positive must be omitted from the solution set.

EXAMPLE 6 Solve $\ln(x^2 - 2x - 19) = \ln 5$.

SOLUTION Because both logarithms have the same base, we use the one-to-one property of logarithmic functions to solve the equation.

$\ln(x^2 - 2x - 19) = \ln 5$

$x^2 - 2x - 19 = 5$ Use the one-to-one property of logarithmic functions. The resulting equation is quadratic.

$x^2 - 2x - 24 = 0$ Subtract 5.

$(x - 6)(x + 4) = 0$ Factor.

$x = 6 \quad \text{or} \quad x = -4$ Set each factor equal to 0 and solve.

We must check each solution in the original equation. Although one of these solutions is negative, it does not mean this solution is an extraneous solution. We must determine whether a solution requires us to take a logarithm of a nonpositive number in the original equation.

Check

$x = 6$ | $x = -4$
$\ln((6)^2 - 2(6) - 19) = \ln 5$ | $\ln((-4)^2 - 2(-4) - 19) = \ln 5$
$\ln 5 = \ln 5$ | $\ln 5 = \ln 5$

Both solutions check. The solution set is $\{6, -4\}$.

Quick Check 5

Solve $\log(x^2 + 15x - 30) = \log 2x$.

EXAMPLE 7 Solve $\log(4x - 20) = \log(x - 14)$.

SOLUTION Because both logarithms have the same base, we use the one-to-one property of logarithmic functions to solve the equation.

$$\log(4x - 20) = \log(x - 14)$$

$4x - 20 = x - 14$ — Use the one-to-one property of logarithmic functions.

$3x - 20 = -14$ — Subtract x.

$3x = 6$ — Add 20.

$x = 2$ — Divide both sides by 3.

We must check this tentative solution in the original equation.

Check ($x = 2$)

$\log(4(2) - 20) = \log((2) - 14)$ — Substitute 2 for x.

$\log(-12) = \log(-12)$ — Simplify.

Because we cannot take the logarithm of a negative number, we must omit the solution $x = 2$. Because this was the only tentative solution, the equation has no solutions. The solution set is $\varnothing$.

Quick Check 6

Solve $\log_4(2x - 7) = \log_4(3x + 2)$.

Solving Logarithmic Equations by Rewriting the Equation in Exponential Form

Objective 4 Solve a logarithmic equation by converting it to exponential form. If we are solving an equation in which the logarithm of a variable expression is equal to a number, such as $\log_2(x - 3) = 4$, we begin by rewriting the equation in exponential form and then solve the resulting equation. As with the previous logarithmic equations, we must check the solutions.

EXAMPLE 8 Solve $\log_2(x - 3) = 4$.

SOLUTION Begin by rewriting the equation $\log_2(x - 3) = 4$ in exponential form, which is $2^4 = x - 3$. Then solve the resulting equation.

$$\log_2(x - 3) = 4$$

$2^4 = x - 3$ — Rewrite in exponential form.

$16 = x - 3$ — Raise 2 to the fourth power.

$19 = x$ — Add 3.

The solution set is $\{19\}$. The check is left to the reader.

Quick Check 7

Solve $\log_7(x^2 - 3x - 3) = 1$.

When solving an equation involving a natural logarithm, keep in mind that the base of this logarithm is e when you are rewriting the equation in exponential form.

EXAMPLE 9 Let $f(x) = \ln(x - 3) + 15$. Solve the equation $f(x) = 21$. Find the exact solution and then approximate the solution to the nearest thousandth.

SOLUTION We begin by setting the function equal to 21. We then rewrite the equation in exponential form and solve the resulting equation.

$$f(x) = 21 \quad \text{Set } f(x) \text{ equal to 21.}$$
$$\ln(x - 3) + 15 = 21 \quad \text{Replace } f(x) \text{ with } \ln(x - 3) + 15.$$
$$\ln(x - 3) = 6 \quad \text{Subtract 15.}$$
$$e^6 = x - 3 \quad \text{Convert to exponential form.}$$
$$e^6 + 3 = x \quad \text{Add 3.}$$
$$x \approx 406.429 \quad \text{Approximate the solution using a calculator.}$$

We now must check the tentative solution in the original equation. We use the approximate solution.

$$f(x) = \ln(x - 3) + 15$$
$$f(406.429) = \ln(406.429 - 3) + 15 \quad \text{Substitute 406.429 for } x.$$
$$= \ln 403.429 + 15 \quad \text{Subtract.}$$
$$\approx 21 \quad \text{Approximate using a calculator.}$$

This solution checks, and the solution set is $\{e^6 + 3\}$.

Quick Check 8

Let $f(x) = \log(3x - 14) + 9$. Solve the equation $f(x) = 15$.

Using the Properties of Logarithms When Solving Logarithmic Equations

Objective 5 Use properties of logarithms to solve a logarithmic equation. We now know how to solve a logarithmic equation in which two logarithms are equal to each other ($\log_b x = \log_b y$) or a logarithm is equal to a number ($\log_b x = a$). However, we will often solve a logarithmic equation that is not in one of these forms, such as $\log_3 x + \log_3(x - 6) = 3$. In such a case, we will use the properties of logarithms to rewrite the equation in the form of an equation that we already know how to solve. In other words, we need to rewrite the equation as an equation that has one logarithm equal to another logarithm with the same base or as an equation that has a logarithm equal to a number. The product and quotient rules will be particularly helpful.

EXAMPLE 10 Solve $\log_3 x + \log_3(x - 6) = 3$.

SOLUTION We begin to solve this equation by using the product rule for logarithms to rewrite the left-hand side of this equation as a single logarithm. We can then rewrite the resulting equation in exponential form and solve it.

$$\log_3 x + \log_3(x - 6) = 3$$
$$\log_3(x(x - 6)) = 3 \quad \text{Use the product rule for logarithms.}$$
$$\log_3(x^2 - 6x) = 3 \quad \text{Multiply.}$$
$$3^3 = x^2 - 6x \quad \text{Convert to exponential form. This equation is quadratic.}$$
$$27 = x^2 - 6x \quad \text{Raise 3 to the third power.}$$
$$0 = x^2 - 6x - 27 \quad \text{Subtract 27.}$$
$$0 = (x - 9)(x + 3) \quad \text{Factor.}$$
$$x = 9 \quad \text{or} \quad x = -3 \quad \text{Set each factor equal to 0 and solve.}$$

We must check each tentative solution in the original equation. We must determine whether a solution requires us to take a logarithm of a nonpositive number.

Check

$x = 9$	$x = -3$
$\log_3(9) + \log_3((9) - 6) = 3$	$\log_3(-3) + \log_3((-3) - 6) = 3$
$\log_3 9 + \log_3 3 = 3$	$\log_3(-3) + \log_3(-9) = 3$
$2 + 1 = 3$	
$3 = 3$	

The solution $x = 9$ checks. Because the tentative solution $x = -3$ produces logarithms of negative numbers in the original equation, it must be omitted as a solution. The solution set is $\{9\}$.

EXAMPLE 11 Solve $\log(9x + 2) - \log(x - 7) = 1$.

SOLUTION We begin to solve this equation by using the quotient rule for logarithms to rewrite the left side of this equation as a single logarithm. We can then rewrite the resulting equation in exponential form and solve it.

$$\log(9x + 2) - \log(x - 7) = 1$$

$$\log\left(\frac{9x + 2}{x - 7}\right) = 1 \quad \text{Use the quotient rule for logarithms.}$$

$$10^1 = \frac{9x + 2}{x - 7} \quad \text{Rewrite in exponential form. This is a rational equation.}$$

$$10(x - 7) = \frac{9x + 2}{\cancel{x - 7}_1} \cdot (\cancel{x - 7}^1) \quad \text{Multiply both sides by } x - 7.$$

$$10x - 70 = 9x + 2 \quad \text{Multiply.}$$

$$x - 70 = 2 \quad \text{Subtract } 9x.$$

$$x = 72 \quad \text{Add 70.}$$

The check of this solution is left to the reader. The solution set is $\{72\}$.

Quick Check 9

Solve.

a) $\log_2 x + \log_2(x + 4) = 5$
b) $\log_3(41x - 42) - \log_3(x - 2) = 4$

EXAMPLE 12 Solve $\ln(x - 1) + \ln(x + 5) = \ln(3x + 7)$.

SOLUTION We begin to solve this equation by using the product rule for logarithms to rewrite the left-hand side of this equation as a single logarithm. We can then solve the resulting equation by using the one-to-one property of logarithms.

$$\ln(x - 1) + \ln(x + 5) = \ln(3x + 7)$$

$$\ln((x - 1)(x + 5)) = \ln(3x + 7) \quad \text{Use the product rule for logarithms.}$$

$$\ln(x^2 + 4x - 5) = \ln(3x + 7) \quad \text{Multiply.}$$

$$x^2 + 4x - 5 = 3x + 7 \quad \text{Use the one-to-one property of logarithmic functions. This equation is quadratic.}$$

$$x^2 + x - 12 = 0 \quad \text{Subtract } 3x \text{ and } 7.$$

$$(x + 4)(x - 3) = 0 \quad \text{Factor.}$$

$$x = -4 \quad \text{or} \quad x = 3 \quad \text{Set each factor equal to 0 and solve.}$$

When we substitute the tentative solution -4 for x in the original equation, it produces the following equation:

$$\ln(-5) + \ln(1) = \ln(-5)$$

Because we can find the logarithm of only positive numbers, the tentative solution $x = -4$ must be omitted. When checking the solution $x = 3$, we obtain the equation $\ln(2) + \ln(8) = \ln(16)$. This solution checks, so the solution set is $\{3\}$.

Quick Check 10

Solve $\log_2(x - 3) + \log_2(x - 7) = \log_2(3x - 1)$.

Finding the Inverse Function of Exponential and Logarithmic Functions

Objective 6 Find the inverse function for exponential and logarithmic functions. One important characteristic of the exponential function $f(x) = b^x$ and the logarithmic function $f(x) = \log_b x$ $(b > 0, b \neq 1)$ is that they are inverse functions. Consider the function $f(x) = e^x$. We will find its inverse by using the technique developed in Section 11.6.

$f(x) = e^x$	
$y = e^x$	Replace $f(x)$ with y.
$x = e^y$	Exchange x and y.
$\ln x = \ln e^y$	To solve for y, take the natural logarithm of each side.
$\ln x = y$	Simplify $\ln e^y$.
$f^{-1}(x) = \ln x$	Replace y with $f^{-1}(x)$.

The inverse function of $f(x) = e^x$ is $f^{-1}(x) = \ln x$.

In the examples that follow, we will learn to find the inverse function of exponential functions as well as logarithmic functions.

EXAMPLE 13 Find the inverse function of $f(x) = e^{x-2} - 5$.

SOLUTION

$f(x) = e^{x-2} - 5$	
$y = e^{x-2} - 5$	Replace $f(x)$ with y.
$x = e^{y-2} - 5$	Exchange x and y.
$x + 5 = e^{y-2}$	Add 5.
$\ln(x + 5) = \ln e^{y-2}$	Take the natural logarithm of each side.
$\ln(x + 5) = y - 2$	Simplify $\ln e^{y-2}$.
$\ln(x + 5) + 2 = y$	Add 2.
$f^{-1}(x) = \ln(x + 5) + 2$	Replace y with $f^{-1}(x)$.

Quick Check 11

Find the inverse function of $f(x) = e^{x+6} + 9$.

The process for finding the inverse function of a logarithmic function is similar to finding the inverse function of an exponential function. However, rather than taking the logarithm of both sides of an equation, we will rewrite the equation in exponential form to solve for y.

EXAMPLE 14 Find the inverse function of $f(x) = \ln(x - 8) + 3$.

SOLUTION

$f(x) = \ln(x - 8) + 3$	
$y = \ln(x - 8) + 3$	Replace $f(x)$ with y.
$x = \ln(y - 8) + 3$	Exchange x and y.
$x - 3 = \ln(y - 8)$	Subtract 3.
$e^{x-3} = y - 8$	Rewrite in exponential form.
$e^{x-3} + 8 = y$	Add 8.
$f^{-1}(x) = e^{x-3} + 8$	Replace y with $f^{-1}(x)$.

Quick Check 12

Find the inverse function of $f(x) = \ln(x + 10) - 18$.

BUILDING YOUR STUDY STRATEGY

Practice Quizzes, 4 Trading with a Classmate Practice quizzes can be effective when you are working with classmates. Make a practice quiz for a classmate by selecting odd-numbered problems associated with each objective in this section. Select problems for each objective that you think are representative of the types of problems you would be expected to solve on an exam. Make sure you understand how to do each of the problems that you select. Have this classmate prepare a practice quiz for you as well.

After the two of you have traded practice quizzes and completed them, return the quiz to the student who created it so that student can grade the quiz. When grading the quiz, do not simply mark the problems as correct or incorrect, but provide feedback on each problem. After you receive your practice quiz back, review any mistakes you made. When you understand what you did wrong, select a similar problem from the exercise set and try it.

Exercises 12.4

PRACTICE
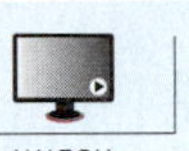
WATCH

DOWNLOAD

READ

REVIEW

Vocabulary

1. When solving a logarithmic equation, it is necessary to check for ____________ solutions.
2. The inverse of a logarithmic function is a(n) ____________ function.

Solve.

3. $3^x = 27$
4. $8^x = 64$
5. $2^{x-4} = 32$
6. $5^{x+2} = 625$
7. $7^{3x-25} = 49$
8. $10^{5x+19} = 10{,}000$

Solve. (Rewrite both sides of the equation in terms of the same base.)

9. $9^x = 27$
10. $4^x = 128$
11. $36^{x-3} = 216$
12. $16^{x+4} = 8$
13. $100^{x-6} = 1000^x$
14. $4^{x+9} = 512^x$

Solve. Round to the nearest thousandth.

15. $4^x = 17$
16. $9^x = 40$
17. $10^x = 129$
18. $e^x = 65$
19. $2^{x-5} = 27$
20. $2^{x+10} = 4295$
21. $7^{x+4} - 8 = 192$
22. $e^{x-3} + 32 = 97$
23. $6^{2x} = 45$
24. $4^{5x} = 165$
25. $e^{2x-7} = 987$
26. $3^{4x+5} = 209$
27. $8^{2x+9} - 17 = 1233$
28. $10^{6x-50} + 250 = 37{,}500$

Solve.

29. $\ln(3x + 1) = \ln(5x - 7)$
30. $\log_3 10x = \log_3(4x + 12)$
31. $\log_8(3x + 13) = \log_8(7x - 23)$
32. $\ln(2x + 5) = \ln(6x - 1)$
33. $\log_4(4x - 1) = \log_4(5x + 3)$
34. $\log_{14}(x - 6) = \log_{14}(3x - 8)$
35. $\log_4(x^2 - 5x + 30) = \log_4 44$
36. $\log_3(x^2 + 13x + 90) = \log_3 50$
37. $\log_7(x^2 + 7x - 29) = \log_7(3x + 3)$
38. $\log_2(x^2 - 3x - 31) = \log_2(4x + 13)$

Solve. Round to the nearest thousandth.

39. $\log_3 x = 5$
40. $\ln x = 2$
41. $\log_5 x = -4$
42. $\log_3 x = 0$
43. $\ln(x - 4) = 6$
44. $\log(x + 11) = 4$
45. $\log_2(x + 8) - 7 = -2$
46. $\ln(x - 3) + 5 = 9$
47. $\log_5(x^2 + 5x - 11) = 2$
48. $\log_4(x^2 - 36) = 3$
49. $\log(x^2 - 9x + 30) = 1$
50. $\log_2(x^2 + 2x - 16) = 6$

51. $\log_4\left(\frac{3x+8}{x-6}\right) = 2$

52. $\log_2\left(\frac{x-1}{9x-4}\right) = -3$

Solve.

53. $\log_4(x+5) + \log_4(x+7) = \log_4 3$

54. $\ln(x+4) + \ln(x-9) = \ln 14$

55. $\log(3x+8) - \log(x-7) = \log 4$

56. $\log_5(7x+10) - \log_5(2x+1) = \log_5 4$

57. $\log_2(x+2) + \log_2(x+6) = 5$

58. $\log_3(x-1) + \log_3(x-7) = 3$

59. $\log_2 x + \log_2(x-4) = 5$

60. $\log(x+15) + \log x = 3$

61. $\log_4(3x-1) + \log_4(x+5) = 3$

62. $\log_3 3x + \log_3(x+6) = 4$

63. $\log_6(7x+11) - \log_6(2x-9) = 1$

64. $\log_2(9x+50) - \log_2(2x+5) = 5$

65. Let $f(x) = 3^{2x-15} + 4$. Solve $f(x) = 31$.

66. Let $f(x) = \ln(x-6) + 8$. Solve $f(x) = 13$. Round to the nearest thousandth.

67. Let $f(x) = \log_5(x^2 + 3x - 45) + 6$. Solve $f(x) = 8$.

68. Let $f(x) = 5^{x+2} - 7$. Solve $f(x) = 29$. Round to the nearest thousandth.

Mixed Practice, 69–98

Solve. Round to the nearest thousandth.

69. $\log_8(x^2 - 10x - 20) = \log_8(3x+10)$

70. $4^{6x-15} = \frac{1}{64}$

71. $5^{x+9} = 407$

72. $\log_6(7x+8) = 2$

73. $9^{x-4} = 32$

74. $9^{x+6} = 27$

75. $2^{7x+30} = \frac{1}{32}$

76. $\log_5(x^2 - 4x - 20) = 2$

77. $\log(3x-41) = 3$

78. $\log_3(x^2 + 8x - 3) = 4$

79. $2^{6x+25} = 128$

80. $3^{x^2-x-15} = 243$

81. $7^{x-8} = 56$

82. $\log(7x-185) = 4$

83. $\log_{12}(9x-11) = \log_{12} 4$

84. $4^{5x-2} = 32^{x+3}$

85. $\log_2 x + \log_2(x+4) = 5$

86. $\log_6\left(\frac{5x+13}{x-8}\right) = 1$

87. $\log_7(4x-19) - \log_7(x+6) = \log_7 3$

88. $\log_3(x+1) + \log_3(x-7) = 2$

89. $\log_8(5x-43) + 4 = 7$

90. $5^{5x-3} = 25^{2x+11}$

91. $\log_4(x^2 + 4x + 16) = \log_4(x^2 + x - 5)$

92. $\ln(7x-13) - 8 = -3$

93. $\log_5(x-7) + \log_5(x+3) = \log_5(5x+15)$

94. $3\log_2(x-10) - 4 = 14$

95. $\ln(2x+5) = -2$

96. $3^{2x-11} = 49$

97. $2^{x^2+12x+42} = 64$

98. $\log(15x-27) = \log(11x-17)$

Solve.

99. $\log_2(\log_3(2x-1)) = 2$

100. $\log_8(\log_2(3x+10)) = 1$

101. $\log_3(\log_3(x^2 - 4x - 5)) = 1$

102. $\log_2(\log_3(x^2 + x + 25)) = 2$

103. $(\log_2 x)^2 - 3\log_2 x - 10 = 0$

104. $(\log_5 x)^2 + 6\log_5 x - 16 = 0$

105. $(\log_3 x)^2 + \log_3 x^2 = 8$

106. $(\log_4 x)^2 + \log_4 x^5 = 14$

For the given function f(x), find its inverse function $f^{-1}(x)$.

107. $f(x) = e^{x-2}$

108. $f(x) = e^x - 2$

109. $f(x) = e^{x+5} + 4$

110. $f(x) = e^{x-8} + 10$

111. $f(x) = 2^{x-6} - 9$

112. $f(x) = 3^{x+1} + 7$

113. $f(x) = \log(x+6)$

114. $f(x) = \log x + 6$

115. $f(x) = \ln(x-4) - 9$

116. $f(x) = \ln(x+2) - 5$

117. $f(x) = \log_3(x+1) + 8$

118. $f(x) = \log_2(x-7) + 3$

Writing in Mathematics

Answer in complete sentences.

119. Explain why you must check logarithmic equations, not exponential equations, for extraneous roots.

120. Solve the exponential equation $25^{x-7} = 125^{2x+9}$ in two ways: by rewriting each base in terms of 5 and by taking the common logarithm of both sides. Which method do you prefer? Clearly explain your reasoning.

121. Explain how to find the inverse of an exponential function. Explain how to find the inverse of a logarithmic function. Use examples to illustrate the process.

122. ***Solutions Manual*** Write a solutions manual page for the following problem.

Solve $\log_3 x + \log_3(x + 6) = 3$.

12.5 Applications of Exponential and Logarithmic Functions

OBJECTIVES

1. Solve compound interest problems.
2. Solve exponential growth problems.
3. Solve exponential decay problems.
4. Solve applications involving exponential functions.
5. Solve applications involving logarithmic functions.

In this section, we will solve applied problems that involve exponential and logarithmic equations and functions.

Compound Interest

Objective 1 Solve compound interest problems. Suppose you deposit \$1000 in a savings account at a bank that pays interest monthly. At the end of the first month, the bank calculates the interest earned and places it in your account. During the second month, besides the original \$1000 earning interest, the interest earned during the first month earns interest as well. This type of interest is called **compound interest**.

Compound Interest Formula

If P dollars are deposited in an account that pays an annual interest rate r that is compounded n times per year, the amount A in the account after t years is given by the formula

$$A = P\left(1 + \frac{r}{n}\right)^{nt}.$$

The amount P is referred to as the principal. The annual interest rate r is often reported as a percent, but it must be converted to a decimal number for this formula.

Interest that is compounded monthly is compounded 12 times per year. Here are some other common types of compound interest.

	Quarterly	Semiannually	Annually
n	4	2	1

EXAMPLE 1 Mona is planning on taking a trip in 4 years and deposits \$3000 in an account that pays 6% annual interest, compounded monthly. If she does not withdraw any money from this account, how much money will be in the account after 4 years?

SOLUTION It is often helpful to collect pertinent information in a table before attempting to use the equation.

Amount in the Account	A	Unknown
Principal	P	$3000
Interest Rate	r	0.06 (6%)
Number of Times Compounded/Year	n	12 (monthly)
Time (Years)	t	4

Now use the equation for compound interest.

$$A = P\left(1 + \frac{r}{n}\right)^{nt}$$

$$A = 3000\left(1 + \frac{0.06}{12}\right)^{12 \cdot 4}$$ Substitute for $P, r, n,$ and t in the equation.

$$A = 3000(1.005)^{48}$$ Simplify the expression inside parentheses.

$$A = 3811.47$$ Approximate using a calculator. Round to the nearest cent.

After 4 years, Mona's account balance will be $3811.47.

Quick Check 1

Sergio deposited $400 in an account that pays 3% annual interest, compounded monthly. What will the balance of this account be in 5 years?

EXAMPLE 2 Pablo deposited $20,000 in an account that pays 8% annual interest, compounded quarterly. How long will it take his account to reach $40,000?

SOLUTION Here is a table containing the pertinent information.

A	P	r	n	t
$40,000	$20,000	0.08 (8%)	4 (quarterly)	Unknown

Now we can use the equation for compound interest. In this problem, the unknown is the variable t. Because this variable is an exponent, we need to use logarithms to solve this equation.

$$40{,}000 = 20{,}000\left(1 + \frac{0.08}{4}\right)^{4t}$$ Substitute for $A, P, r,$ and n in the equation.

$$40{,}000 = 20{,}000(1.02)^{4t}$$ Simplify the expression inside parentheses.

$$2 = 1.02^{4t}$$ Divide both sides by 20,000.

$$\log 2 = \log 1.02^{4t}$$ Take the common logarithm of both sides.

$$\log 2 = 4t \log 1.02$$ Use the power rule for logarithms.

$$\frac{\log 2}{4 \log 1.02} = \frac{4t \log 1.02}{4 \log 1.02}$$ Divide by 4 log 1.02 to isolate t.

$$\frac{\log 2}{4 \log 1.02} = t$$ Simplify.

$$t \approx 8.8$$ Approximate using a calculator. On your calculator, enter the denominator inside a set of parentheses.

Pablo's account will grow to $40,000 in approximately 8.8 years.

Quick Check 2

If $25,000 is deposited in an account that pays 10% annual interest, compounded quarterly, how long will it take until there is $100,000 in the account?

Using Your Calculator To approximate the expression $\frac{\log 2}{4 \log 1.02}$ using the TI-84, it is necessary to enter the denominator in a pair of parentheses, as shown in the following screen shot:

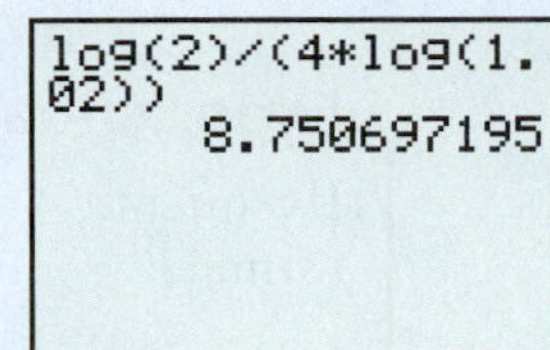

In the previous example, we were trying to determine the length of time for the balance of a bank account to double. The length of time for any quantity to double in size is called the **doubling time**.

Some banks use a method for calculating compound interest known as **continuous compounding**.

Continuous Compound Interest Formula

If P dollars are deposited in an account that pays an annual interest rate r that is compounded continuously, the amount A in the account after t years is given by the formula

$$A = Pe^{rt}.$$

EXAMPLE 3 If an account pays 12% annual interest, compounded continuously, how long will it take a deposit of \$50,000 to produce an account balance of \$1,000,000?

SOLUTION Here is a table containing the pertinent information.

Amount in the Account	A	\$1,000,000
Principal	P	\$50,000
Interest Rate	r	0.12 (12%)
Time (Years)	t	Unknown

Now we can use the equation for continuous compound interest: $A = Pe^{rt}$.

$$1{,}000{,}000 = 50{,}000e^{0.12t} \quad \text{Substitute for } A, P, \text{ and } r \text{ in the equation.}$$
$$20 = e^{0.12t} \quad \text{Divide both sides by 50,000.}$$
$$\ln 20 = \ln e^{0.12t} \quad \text{Take the natural logarithm of both sides.}$$
$$\ln 20 = 0.12t \quad \text{Simplify the right side of the equation.}$$
$$\frac{\ln 20}{0.12} = t \quad \text{Divide both sides by 0.12 to isolate } t.$$
$$t \approx 24.96 \quad \text{Approximate using a calculator.}$$

The account balance will reach \$1 million in approximately 25 years.

Quick Check 3

If \$50,000 is invested in an account that pays 15% annual interest, compounded continuously, how long will it take until the balance of the account is \$500,000?

Exponential Growth

Objective 2 Solve exponential growth problems. The previous example involving interest that is compounded continuously is an example of a quantity that grows exponentially.

Exponential Growth Formula

If a quantity increases at an exponential rate, its size P at time t is given by the formula

$$P = P_0e^{kt}.$$

P_0 represents the initial size of the quantity, while k is the exponential growth rate.

Often we do not know the exponential growth rate k, but we can find it if we know the size of the quantity at two different times.

EXAMPLE 4 The average annual tuition and fees paid to attend a public two-year college increased from \$641 in 1985 to \$1735 in 2002. If tuition and fees are growing exponentially, what will it cost to attend a public two-year college in 2018? (*Source:* The College Board)

SOLUTION We begin by determining the exponential growth rate k. We treat \$641 as the initial cost P_0. Because it took 17 years for the costs to reach \$1735, we can substitute these values for t and P in the exponential growth formula. Here is a table containing the pertinent information.

P	P_0	k	t
1735	641	Unknown	17

We now substitute into the equation $P = P_0e^{kt}$ and solve for k.

$$1735 = 641e^{k \cdot 17} \quad \text{Substitute for } P, P_0, \text{ and } t.$$

$$\frac{1735}{641} = e^{k \cdot 17} \quad \text{Divide both sides by 641.}$$

$$\ln\left(\frac{1735}{641}\right) = \ln e^{k \cdot 17} \quad \text{Take the natural logarithm of both sides.}$$

$$\ln\left(\frac{1735}{641}\right) = k \cdot 17 \quad \text{Simplify the right side of the equation.}$$

$$\frac{\ln\left(\frac{1735}{641}\right)}{17} = k \quad \text{Divide both sides by 17.}$$

$$k \approx 0.058573 \quad \text{Approximate using a calculator.}$$

Now that we have found the exponential growth rate k, we can turn our attention to finding the cost to attend a public two-year college in 2018. If we continue to let 641 be the initial cost (P_0), then $t = 2018 - 1985$, or 33. Here is a table containing information that is pertinent to this part of the problem.

P	P_0	k	t
Unknown	641	0.058573	33

Now we substitute the appropriate values into the formula for exponential growth to determine the cost to attend a public two-year college in 2018.

$P = P_0e^{kt}$

$P = 641e^{0.058573 \cdot 33}$ Substitute for P_0, k, and t.

$P = 641e^{1.932909}$ Simplify the exponent.

$P \approx 4429.04$ Approximate using a calculator.

The cost to attend a public two-year college in 2018 will be approximately $4429.04.

Quick Check 4

In 1974, a first-class U.S. postage stamp cost 10 cents. By 1994, the cost of a stamp had risen to 32 cents. If the cost of a first-class U.S. postage stamp continues to rise exponentially, how much will it cost to buy a stamp in 2020?

Exponential Decay

Objective 3 Solve exponential decay problems. Just as some quantities grow exponentially, other quantities decay exponentially. The equation for exponential decay is the same as the formula for exponential growth, with the exception that k is the exponential *decay* rate. The exponential decay rate will be a negative number.

Exponential Decay Formula

If a quantity decreases at an exponential rate, its size P at time t is given by the formula

$$P = P_0e^{kt}.$$

P_0 represents the initial size of the quantity, while k is the exponential decay rate.

The **half-life** of a radioactive element is the amount of time it takes for half of the substance to decay into another element. For example, the half-life of carbon-14 is 5730 years. This means that after 5730 years, 50%, or half, of the carbon-14 that was present in a sample will have decayed into another element.

EXAMPLE 5 All living things contain the radioactive element carbon-14. When a living thing dies, the radioactive carbon-14 begins to change to the stable nitrogen-14. By measuring the percentage of carbon-14 that remains, scientists can determine how much time has passed since death.

If an old table is made of wood that still contains 95% of its original carbon-14, how much time has passed since the tree it was made of died? (The half-life of carbon-14 is 5730 years.)

SOLUTION We use the half-life of carbon-14 to determine the value of the exponential decay rate k. We know that after 5730 years, the wood from the tree will contain 50% of its original carbon-14.

P	P_0	k	t
0.5 (50%)	1 (100%)	Unknown	5730

We now substitute into the equation and solve for k.

$0.5 = 1e^{k \cdot 5730}$ Substitute for P, P_0, and t.

$\ln 0.5 = \ln e^{k \cdot 5730}$ Take the natural logarithm of both sides.

$\ln 0.5 = k \cdot 5730$ Simplify the right side of the equation.

$\frac{\ln 0.5}{5730} = k$ Divide both sides by 5730.

$k \approx -0.000121$ Approximate using a calculator.

Now we can determine how long it has been since the death of the tree. Here is a table containing information that is pertinent to this part of the problem.

P	P_0	k	t
0.95 (95%)	1 (100%)	−0.000121	Unknown

Now we substitute the appropriate values into the formula for exponential growth.

$$0.95 = 1e^{-0.000121t} \quad \text{Substitute for } P, P_0, \text{ and } k.$$

$$\ln 0.95 = \ln e^{-0.000121t} \quad \text{Take the natural logarithm of both sides.}$$

$$\ln 0.95 = -0.000121t \quad \text{Simplify the right side of the equation.}$$

$$\frac{\ln 0.95}{-0.000121} = t \quad \text{Divide both sides by } -0.000121.$$

$$t \approx 423.9 \quad \text{Approximate using a calculator.}$$

The tree that the table was made of died approximately 423.9 years ago.

Quick Check 5

The radioactive element strontium-89 has a half-life of 50.6 days. How long will it take for 50 grams of strontium-89 to decay to 10 grams?

Applications of Other Exponential Functions

Objective 4 Solve applications involving exponential functions.

EXAMPLE 6 The number of Subway restaurants in business in a particular year can be described by the function $f(x) = 1828.65 \cdot 1.06^x$, where x represents the number of years after 1964. If this pattern of exponential growth continues, when will there be 50,000 Subway restaurants? (*Source:* Doctor's Associates, Inc.)

SOLUTION Begin by setting the function equal to 50,000 and solving for x.

$$1828.65 \cdot 1.06^x = 50{,}000 \quad \text{Set the function equal to 50,000.}$$

$$1.06^x = \frac{50{,}000}{1828.65} \quad \text{Divide both sides by 1828.65.}$$

$$\log 1.06^x = \log\left(\frac{50{,}000}{1828.65}\right) \quad \text{Take the common logarithm of both sides.}$$

$$x \cdot \log 1.06 = \log\left(\frac{50{,}000}{1828.65}\right) \quad \text{Apply the power rule for logarithms.}$$

$$x = \frac{\log\left(\frac{50{,}000}{1828.65}\right)}{\log 1.06} \quad \text{Divide both sides by log 1.06.}$$

$$x \approx 57 \quad \text{Approximate using a calculator. Round to the nearest year.}$$

Recall that x represents the number of years after 1964; so Subway will have 50,000 restaurants in 2021 (1964 + 57).

Quick Check 6

The average annual income of a person with a bachelor's degree can be approximated by the function $f(x) = 13400 \cdot 1.055^x$, where x represents the number of years after 1975. If this exponential growth in income continues, in what year will the average annual income of a person with a bachelor's degree be $125,000? (*Source:* U.S. Census Bureau)

Applications of Logarithmic Functions

Objective 5 Solve applications involving logarithmic functions.

pH

Every liquid can be classified as acid, base, or neutral. Acids contain an excess of hydrogen (H^+) ions, while bases contain an excess of hydroxide (OH^-) ions. In a neutral liquid such as pure water, the concentration of these two ions is equal. The pH scale, based on the concentration of hydrogen ions in the solution, is used to determine how acidic or basic a solution is.

pH of a Liquid

The pH of a liquid is calculated by the formula $\text{pH} = -\log[H^+]$, where $[H^+]$ is the concentration of hydrogen ions in moles per liter.

Solutions that are neutral have a pH of 7. The pH of an acid is less than 7, and the pH of a base is greater than 7. The pH of a liquid is generally between 0 and 14. The stronger an acid is, the closer its pH is to 0. The stronger a base is, the closer its pH is to 14.

EXAMPLE 7 Determine the pH of ammonia, whose concentration of hydrogen ions is 3.16×10^{-12} moles/liter. Determine whether ammonia is an acid or a base.

SOLUTION Substitute 3.16×10^{-12} for $[H^+]$ in the formula $\text{pH} = -\log[H^+]$ to determine the pH of ammonia.

$$\begin{aligned} \text{pH} &= -\log(3.16 \times 10^{-12}) && \text{Substitute } 3.16 \times 10^{-12} \text{ for } [H^+]. \\ &= -(-11.5) && \text{Approximate } \log(3.16 \times 10^{-12}) \text{ using a calculator.} \\ &= 11.5 && \text{Simplify.} \end{aligned}$$

The pH of ammonia is 11.5. Ammonia is a base, as it has a pH greater than 7.

Quick Check 7

The concentration of hydrogen ions $[H^+]$ in vinegar is 1.58×10^{-3} moles/liter. Find the pH of vinegar. Is vinegar an acid or a base?

EXAMPLE 8 The pH of the water in a swimming pool is 7.8. Find the concentration of hydrogen ions in moles per liter.

SOLUTION We substitute 7.8 for pH in the formula for pH and solve for the concentration of hydrogen ions, $[H^+]$.

$$\begin{aligned} 7.8 &= -\log[H^+] && \text{Substitute 7.8 for pH.} \\ -7.8 &= \log[H^+] && \text{Divide by } -1. \\ 10^{-7.8} &= [H^+] && \text{Convert to exponential form.} \\ [H^+] &\approx 1.58 \times 10^{-8} && \text{Approximate using a calculator.} \end{aligned}$$

The concentration of hydrogen ions in the water of the swimming pool is approximately 1.58×10^{-8} moles per liter.

Quick Check 8

The pH of battery acid is 0.3. Find the concentration of hydrogen ions in moles per liter.

Volume of a Sound

The volume of a sound is measured in decibels (db) and depends on the intensity of the sound, which is measured in watts per square meter. 85 db is considered the threshold level for safety, and prolonged exposure to noise at 85 db can cause hearing loss.

Volume of a Sound

The volume (L) of a sound in decibels is given by the formula $L = 10\log\left(\frac{I}{10^{-12}}\right)$, where I is the intensity of the sound in watts per square meter. The constant 10^{-12} is the intensity of the softest sound that humans can hear.

EXAMPLE 9 The intensity of a sound generated by personal stereo headphones is 10^{-2} watts per square meter. Find the volume of this sound.

SOLUTION Begin by substituting 10^{-2} for I in the formula for the volume of a sound, $L = 10\log\left(\frac{I}{10^{-12}}\right)$; then find L.

$$L = 10\log\left(\frac{10^{-2}}{10^{-12}}\right) \quad \text{Substitute } 10^{-2} \text{ for } I.$$

$$L = 10\log(10^{-2-(-12)}) \quad \text{Simplify the fraction using the rule } \frac{x^m}{x^n} = x^{m-n}.$$

$$L = 10\log(10^{10}) \quad \text{Simplify the exponent.}$$

$$L = 10 \cdot 10 \quad \text{Simplify the logarithm.}$$

$$L = 100 \quad \text{Multiply.}$$

The volume generated by personal stereo headphones is 100 db.

Quick Check 9

The intensity of a sound generated by normal conversation is 10^{-6} watts per square meter. Find the volume of this sound.

EXAMPLE 10 The threshold for pain is defined to be at a volume of 130 db. Find the intensity of a sound with this volume.

SOLUTION Begin by substituting 130 for L in the formula for the volume of a sound; then solve for I.

$$130 = 10\log\left(\frac{I}{10^{-12}}\right) \quad \text{Substitute 130 for } L.$$

$$13 = \log\left(\frac{I}{10^{-12}}\right) \quad \text{Divide both sides by 10.}$$

$$\frac{I}{10^{-12}} = 10^{13} \quad \text{Convert to exponential form.}$$

$$10^{-12} \cdot \frac{I}{10^{-12}} = 10^{-12} \cdot 10^{13} \quad \text{Multiply by } 10^{-12} \text{ to isolate } I.$$

$$I = 10^1 \quad \text{Simplify.}$$

A sound that meets the threshold for pain has an intensity of 10 watts per square meter.

Quick Check 10

The volume of a vacuum cleaner is 80 db. Find the intensity of a sound with this volume.

BUILDING YOUR STUDY STRATEGY

Practice Quizzes, 5 Study Groups Practice quizzes can be effective when you are working with a group of classmates. Here is a plan for incorporating a practice quiz into a group study session. Have each student bring one question to the session. Randomly select a student to do each problem for the group. The student should explain each step of the solution as he or she works through it. The rest of the group should act as coaches, using resources such as their notes or text as needed. If you are not the student solving the problem, ask any questions you may have. At the end of the session, the group should create a practice quiz based on these problems.

Exercises 12.5

MyMathLab

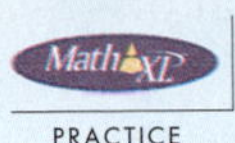
PRACTICE

WATCH

DOWNLOAD

READ

REVIEW

Vocabulary

1. When a bank pays ______ interest, the amount of interest earned in one period is added to the principal before the interest earned in the next period is calculated.
2. The length of time for any quantity to double in size is called the ______.
3. The formula $A = Pe^{rt}$ is used to calculate interest that is compounded ______.
4. If a quantity increases at an exponential rate, its size P at time t is given by the formula ______.
5. If a quantity decreases at an exponential rate, the exponential decay rate k in the formula $P = P_0e^{kt}$ is a(n) ______ number.
6. The ______ scale is used to determine how acidic or basic a solution is.

7. Kasumi deposited \$7500 in a certificate of deposit (CD) account that pays 3% annual interest, compounded quarterly. How much will the balance of the account be in 4 years?
8. Shoshanna deposited \$900 in an account that pays 5% annual interest, compounded monthly. What will the balance of the account be in 2 years?
9. If Tina deposits \$12,413 in an account that pays 8% annual interest, compounded semiannually, how much money will be in the account in 10 years?
10. Vladimir sold his baseball card collection for \$3325 and deposited the money in a bank account that pays 7% interest, compounded annually. How much money will be in the account in 20 years?

11. Dustin deposited \$4000 in an account that pays 9% annual interest, compounded monthly. How long will it take until Dustin's balance is \$6000?
12. Vanessa deposited \$5600 in an account that pays 8% annual interest, compounded quarterly. How long will it take until Vanessa's balance is \$10,000?
13. Nomar started a bank account with an initial deposit of \$240. If the bank account pays 3% annual interest, compounded monthly, how long will it take until the account's balance doubles?
14. If a bank account pays 10% annual interest, compounded annually, find the amount of time it would take a deposit to double in value.
15. The interest on a bank account is compounded continuously, with an annual interest rate of 4.3%. If a deposit of \$2000 is made, what will the balance be in 6 years?
16. Vineta has a bank account that pays 5% annual interest, compounded continuously. If Vineta made an initial deposit of \$10,000, what will the balance be in 12 years?
17. How long will it take an investment of \$4500 to grow to \$6750 if it is invested at 7% annual interest, compounded continuously?
18. An investor has deposited \$80,000 in an account that pays 5.3% annual interest, compounded continually. How long will it take until the account's balance is \$100,000?
19. In 2000, Home Depot had 1134 stores nationwide. By 2002, this total had grown to 1532. If the number of stores continues to grow exponentially at the same rate, how many stores will there be in 2016? (*Source:* The Home Depot, Inc.)
20. In 1997, Lowe's had 427 stores nationwide. By 1999, this total had grown to 520. If the number of stores continues to grow exponentially at this rate, how many stores will there be in 2017? (*Source:* Lowe's Companies, Inc.)
21. In 1974, the total world population reached 4 billion people. The total world population reached 6 billion people in 1999. If the total world population continues to grow exponentially at this rate, what will the population be in 2050? (*Source:* U.S. Census Bureau)
22. During its exponential growth phase, a colony of the bacteria *Bacillus megaterium* grew from 20,000 cells to 40,000 cells in 25 minutes. How many cells will be present after 60 minutes?
23. The average monthly bill for U.S. cell phone users rose from \$39.43 in 1998 to \$48.40 in 2002. If the average monthly bill continues to increase exponentially at this rate, when will it reach \$100? (*Source:* Cellular Telecommunications and Internet Associates)

24. The average tuition and fees paid by students at private four-year universities grew from \$9340 in 1990 to \$16,233 in 2000. If the average tuition and fees continue to grow exponentially at this rate, when will the average tuition and fees reach \$40,000? (*Source:* The College Board)

25. During its exponential growth phase, a colony of the bacterium *E. coli* grew from 5000 cells to 40,000 cells in 60 minutes. How long will it take the colony of 5000 cells to grow to 100,000 cells?

26. During its exponential growth phase, a colony of the bacterium *E. coli* grew from 15,000 cells to 60,000 cells in 40 minutes. How long will it take the colony of 15,000 cells to grow to 250,000 cells?

27. In 1993, there were 593 drive-in theaters in the United States. By 2001, the number of drive-in theaters had dropped to 474. If the number of drive-in theaters continues to decline exponentially, how many drive-in theaters will there be in 2020? (*Source:* MPAA)

28. The circulation of all daily U.S. newspapers in 1992 was 60.1 million. Total circulation had dropped to 55.1 million by 2002. If circulation continues to decline exponentially, what will the total circulation of all daily U.S. newspapers be in 2025? (*Source: USA Today*)

29. Fluorine-18 is a radioactive version of glucose and is used in PET scanning. The half-life of fluorine-18 is 112 minutes. If 100 milligrams of fluorine-18 is injected into a patient, how much will remain after 240 minutes?

30. The half-life of bismuth-210 is 5 days. If 10 grams of bismuth-210 are present initially, how much will remain after 2 weeks?

31. If a bone contains 98% of its original carbon-14, how old is the bone? (Carbon-14 has a half-life of 5730 years.)

32. If a wooden bowl contains 85% of its original carbon-14, how old is the bowl? (Carbon-14 has a half-life of 5730 years.)

33. Calcium-41 is being used in studies testing the effectiveness of drugs for preventing osteoporosis. The half-life of calcium-41 is 100,000 years. If 5 grams of calcium-14 are present initially, how long will it take until only 1 gram remains?

34. Radon-222 has a half-life of 3.8 days. If 200 milligrams of radon-222 are present initially, how long will it take until only 10 milligrams remain?

35. In 1940, there were 1878 daily newspapers in the United States. In 2002, this number had dropped to 1457 daily newspapers. If the number of daily newspapers in the United States continues to decline exponentially at this rate, when will there be only 1000 daily newspapers in the United States? (*Source:* USA Today)

36. The number of drive-in screens in the United States dropped from 3561 in 1980 to only 717 in 2000. If the number of screens continues to decline exponentially at this rate, when will there be only 100 drive-in screens in the United States? (*Source:* MPAA)

37. The value of a car purchased in 2008 for \$30,000 had dropped to \$24,000 in 2009. If the value of the car continues to decrease exponentially at this rate, when will it be worth \$5000?

38. The value of a car purchased in 2001 for \$12,000 had dropped in value to \$9720 in 2003. If the value of the car continues to decrease exponentially at this rate, when will it be worth \$2000?

39. The number of Starbucks stores in business in a particular year can be described by the function $f(x) = 4794 \cdot 1.21^x$, where x represents the number of years after 2001. Use this function to predict the number of Starbucks stores in 2016. (*Source:* Starbucks Corporation)

40. The number of McDonald's restaurants in business in a particular year can be described by the function $f(x) = 12{,}436 \cdot 1.095^x$, where x represents the number of years after 1991. Use this function to predict the number of McDonald's restaurants in 2016. (*Source:* McDonald's Corporation)

41. The average premium, in dollars, of homeowner's insurance in a particular year can be described by the function $f(x) = 410.57 \cdot 1.035^x$, where x represents the number of years after 1994. Use this function to determine when the average homeowner's premium is \$1000. (*Source:* National Association of Insurance Commissioners and Insurance Information Institute)

42. The annual domestic gross box office receipts in the United States, in billions of dollars, can be approximated by the function $f(x) = 4.8 \cdot 1.057^x$, where x represents the number of years after 1992. Use this function to determine when the annual domestic box office receipts reach \$25 billion. (*Source*: www.boxofficemojo.com)

43. The average annual income of a person without a high school diploma can be approximated by the function $f(x) = 6830 \cdot 1.038^x$, where x represents the number of years after 1975. Use this function to predict when the average annual income of a person without a high school diploma will be \$100,000. (*Source:* U.S. Census Bureau)

44. The average annual income of a person with an advanced college degree (master's degree or higher) can be approximated by the function $f(x) = 17{,}775 \cdot 1.059^x$, where x represents the number of years after 1975. Use this function to predict when the average annual income of a person with an advanced college degree will be \$200,000. (*Source:* U.S. Census Bureau)

45. The value of a new car, in dollars, x years after its purchase can be described by the function $f(x) = 24{,}000 \cdot 0.8^x$. Use this function to determine when the value of the car will be \$5000.

46. The value of a new copy machine, in dollars, x years after its purchase can be described by the function $f(x) = 3500 \cdot 0.85^x$. Use this function to determine when the value of the copier will be \$2000.

47. Forensic investigators often use body temperature to determine time of death. For a person who was found dead in a walk-in refrigerator that maintained a constant temperature of 35° F, the body's temperature, in degrees Fahrenheit, can be approximated by the function $f(t) = 35 + 61e^{-0.0165t}$, where $t = 0$ corresponds to the time the body was found.

a) What will the body's temperature be 6 hours after it was found?

b) How many hours before being found did the person die? Assume that the person had a normal temperature of 98.6° F when alive.

48. The body temperature, in degrees Fahrenheit, of a person who was found dead in an apartment that maintained a constant temperature of 70° F can be approximated by the function $f(t) = 70 + 12e^{-0.182t}$, where $t = 0$ corresponds to the time the body was found.

a) What will the body's temperature be 12 hours after it was found?

b) How many hours before being found did the person die?

49. The average annual salary, in dollars, of teachers in public elementary and secondary schools in a particular year can be described by the function $f(x) = 32178 \cdot 1.027^x$, where x represents the number of years after 1989. Use this function to determine when the average annual salary will reach \$70,000. (*Source:* National Education Association)

50. The expenditure, in dollars, per pupil in public elementary and secondary schools in a particular year can be described by the function $f(x) = 4454 \cdot 1.038^x$, where x represents the number of years after 1988. Use this function to determine when the expenditure per pupil will reach \$16,000. (*Source:* U.S. Department of Education)

51. The concentration of hydrogen ions $[H^+]$ in bleach is 2.51×10^{-13} moles/liter. Find the pH of bleach. Is bleach an acid or a base?

52. The concentration of hydrogen ions $[H^+]$ in orange juice is 6.31×10^{-5} moles/liter. Find the pH of orange juice. Is orange juice an acid or a base?

53. The concentration of hydrogen ions $[H^+]$ in lemon juice is 5.01×10^{-3} moles/liter. Find the pH of lemon juice. Is lemon juice an acid or a base?

54. The concentration of hydrogen ions $[H^+]$ in seawater is 1×10^{-8} moles/liter. Find the pH of seawater. Is seawater an acid or a base?

55. The pH of boric acid is 5.0. Find the concentration of hydrogen ions in moles per liter.

56. The pH of borax is 9.3. Find the concentration of hydrogen ions in moles per liter.

57. The pH of milk of magnesia is 10.2. Find the concentration of hydrogen ions in moles per liter.

58. The pH of corn is 6.2. Find the concentration of hydrogen ions in moles per liter.

59. The intensity of a sound generated by a vacuum cleaner is 10^{-4} watts per square meter. Find the volume of this sound.

60. The intensity of a sound generated by an alarm clock is 10^{-5} watts per square meter. Find the volume of this sound.

61. The intensity of a sound generated by a leaf blower is 10^{-1} watts per square meter. Find the volume of this sound.

62. The intensity of a sound generated by an airplane taking off is 10^2 watts per square meter. Find the volume of this sound.

63. The volume of a crying baby is 110 db. Find the intensity of a sound with this volume.

64. The volume of a refrigerator is 50 db. Find the intensity of a sound with this volume.

65. The volume of an electric drill is 95 db. Find the intensity of a sound with this volume.

66. The volume of a chainsaw is 125 db. Find the intensity of a sound with this volume.

67. The life expectancy at birth for Americans of both sexes in a particular year can be described by the function $f(x) = 11.03 + 14.29 \ln x$, where x represents the number of years after 1900. Use this function to determine the life expectancy for Americans born in 2025. (*Source:* National Center for Health Statistics)

68. The number of U.S. cell phone subscribers (in millions) in a particular year can be described by the function $f(x) = 63.1 + 45.43 \ln x$, where x represents the number of years after 1997. Use this function to predict how many subscribers there will be in 2015. (*Source:* Cellular Telecommunications & Internet Association)

69. The number of public elementary and secondary school teachers (in millions) in a particular year can be described by the function $f(x) = 2.648 + 0.181 \ln x$, where x represents the number of years after 1995. Use this function to predict when there will be 3.2 million teachers. (*Source:* U.S. Department of Education)

70. The percent of U.S. full-time college students who receive some financial aid in a particular year can be described by the function $f(x) = 58.8 + 6.7 \ln x$, where x represents the number of years after 1991. Use this function to predict when 80% of all students will receive some sort of financial aid. (*Source:* U.S. Department of Education)

Writing in Mathematics

Answer in complete sentences.

71. Write a compound interest word problem whose solution is *Gretchen's balance will reach the correct amount in 17 years.*

72. Write a compound interest word problem whose solution is *Lucy needs to invest $10,000.*

73. Explain what the half-life of a radioactive element is.

74. ***Newsletter*** Write a newsletter explaining how to solve applied problems involving compound interest.

12.6 Graphing Exponential and Logarithmic Functions

OBJECTIVES

1. Graph exponential functions.
2. Graph logarithmic functions.

In this section, we will learn how to graph exponential and logarithmic functions by shifting basic exponential and logarithmic functions. In addition, now that we know how to solve exponential equations and logarithmic equations, we will be able to find the x-intercept of the function if one exists.

Graphing Exponential Functions

Objective 1 Graph exponential functions. Recall the following facts about the graph of an exponential function of the form $f(x) = b^x$, where $b > 0$ and $b \neq 1$:

- The graph has a horizontal asymptote at the line $y = 0$.
- The y-intercept is at $(0, 1)$.
- The graph also passes through the points $\left(-1, \frac{1}{b}\right)$ and $(1, b)$.
- If $b > 1$, the function increases from left to right, and if $0 < b < 1$, the function decreases from left to right.

To graph a function of the form $f(x) = b^{x-h} + k$, we will shift the graph of $f(x) = b^x$ horizontally by h units and vertically by k units. A good strategy is to begin by shifting the horizontal asymptote by k units vertically. We will follow up by applying the shifts to the points $\left(-1, \frac{1}{b}\right)$, $(0, 1)$, and $(1, b)$. After drawing the smooth curve passing through these points, we can add additional detail to the graph by finding any intercepts.

EXAMPLE 1 Graph $f(x) = 2^{x+1} - 4$. Label any intercepts and state the domain and range of $f(x)$.

SOLUTION In this example, $h = -1$ and $k = -4$; so we shift the graph of $f(x) = 2^x$ by one unit to the left and by four units down. This places the horizontal asymptote at $y = -4$. Here are three points on the graph of $f(x) = 2^x$ as well as their location after the shifts have been applied to the graph. (Subtract 1 from each x-coordinate and subtract 4 from each y-coordinate.)

$f(x) = 2^x$	$\left(-1, \frac{1}{2}\right)$	$(0, 1)$	$(1, 2)$
$f(x) = 2^{x+1} - 4$	$\left(-2, -3\frac{1}{2}\right)$	$(-1, -3)$	$(0, -2)$

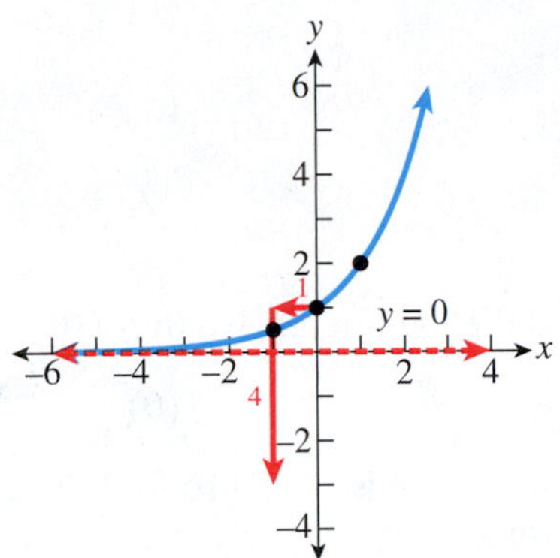

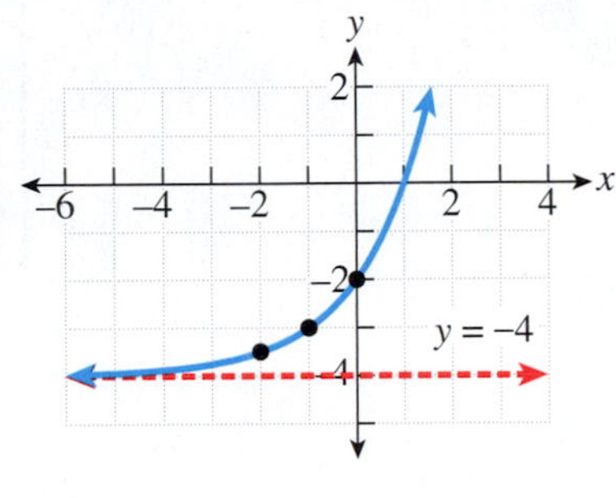

By shifting, we see that the y-intercept is $(0, -2)$. To find the x-intercept, we set the function equal to 0 and solve for x.

$2^{x+1} - 4 = 0$ Set $f(x) = 0$.

$2^{x+1} = 4$ Add 4 to both sides to isolate the exponential expression.

$2^{x+1} = 2^2$ Rewrite 4 as 2^2.

$x + 1 = 2$ Apply the one-to-one property of exponential functions. If $b^r = b^s$, then $r = s$.

$x = 1$ Subtract 1.

The x-intercept is $(1, 0)$.

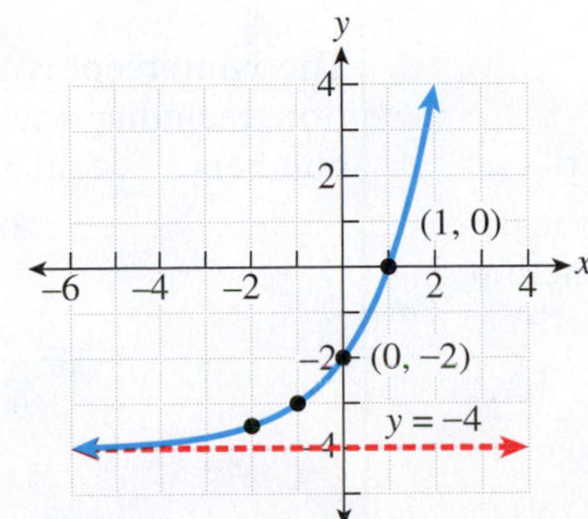

The domain of the function is $(-\infty, \infty)$. From the graph, we see that the range is $(-4, \infty)$.

Quick Check 1

Graph $f(x) = 3^{x-1} - 9$. Label any intercepts and state the domain and range of $f(x)$.

In general, for a function of the form $f(x) = b^{x-h} + k$, the equation of the horizontal asymptote will be $y = k$ and the range of the function will be (k, ∞).

EXAMPLE 2 Graph $f(x) = 3^{x+2} - 7$. Label any intercepts and state the domain and range of the function.

SOLUTION In this example, $h = -2$ and $k = -7$; so we shift the graph of $f(x) = 3^x$ by two units to the left and by seven units down. This places the horizontal asymptote at $y = -7$.

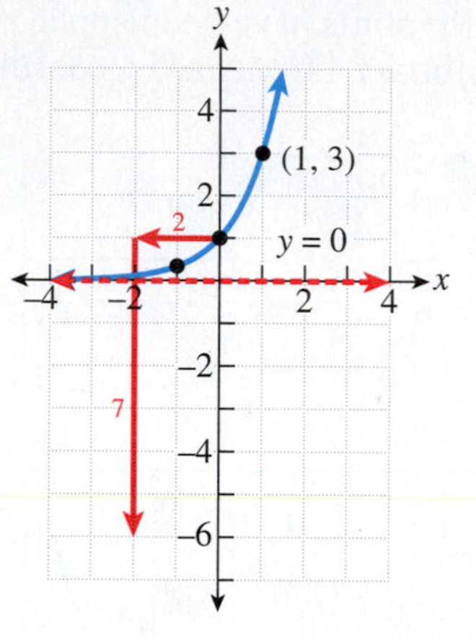

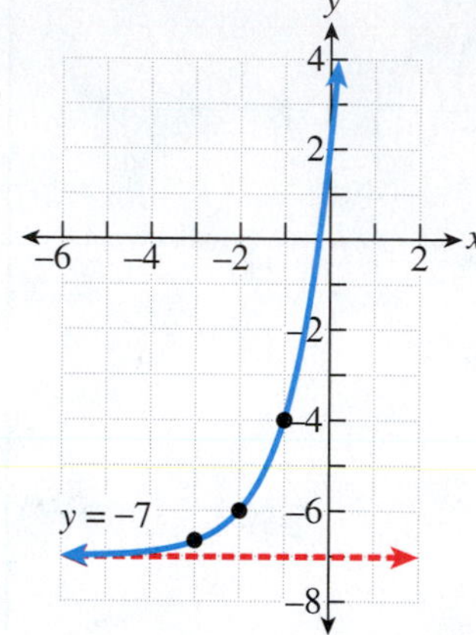

To find the y-intercept, we find $f(0)$.

$$f(0) = 3^{0+2} - 7 = 9 - 7 = 2$$

The y-intercept is $(0, 2)$. To find the x-intercept, we set the function equal to 0 and solve for x.

$3^{x+2} - 7 = 0$	Set $f(x) = 0$.
$3^{x+2} = 7$	Add 7 to both sides to isolate the exponential expression.
$\log 3^{x+2} = \log 7$	Take the common logarithm of each side.
$(x+2) \cdot \log 3 = \log 7$	Apply the power rule for logarithms.
$\dfrac{(x+2) \cdot \overset{1}{\cancel{\log 3}}}{\underset{1}{\cancel{\log 3}}} = \dfrac{\log 7}{\log 3}$	Divide both sides by log 3.
$x = \dfrac{\log 7}{\log 3} - 2$	Subtract 2.
$x \approx -0.2$	Approximate using a calculator.

The x-intercept is approximately at $(-0.2, 0)$. Following is the graph of the function, including the x-intercept. The domain of this function is the set of all real numbers $(-\infty, \infty)$. The range is $(-7, \infty)$.

Quick Check 2

Graph $f(x) = 4^{x-2} - 6$. Label any intercepts and state the domain and range of the function.

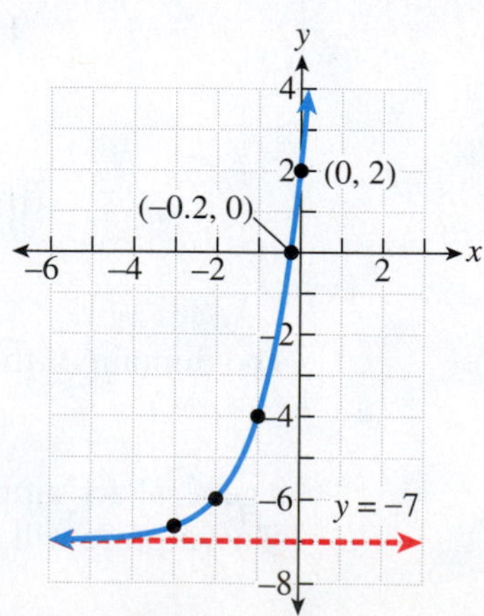

Using Your Calculator You can use the TI-84 to graph $f(x) = 3^{x+2} - 7$ as well as its horizontal asymptote, $y = -7$. Begin by tapping the Y= key. Next to Y_1, key $3^{x+2} - 7$. Key the number -7 next to Y_2. Tap the GRAPH key to display the graph.

Enter Expressions

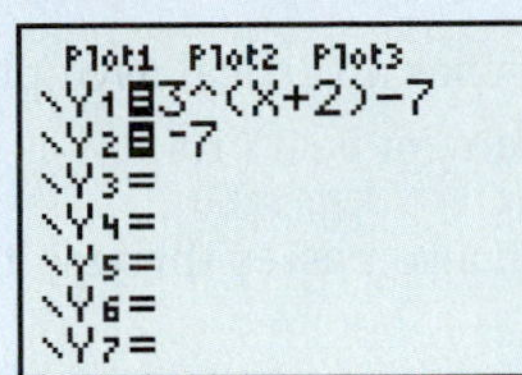

Graph

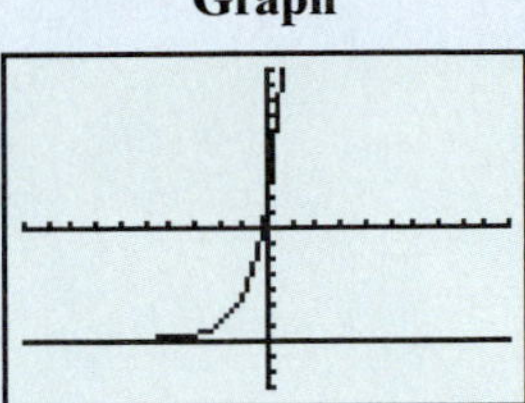

EXAMPLE 3 Graph $f(x) = e^{x-1} - 5$. Label any intercepts and state the domain and range of $f(x)$.

SOLUTION In this example, $h = 1$ and $k = -5$; so we shift the graph of $f(x) = e^x$ by one unit to the right and by five units down. This places the horizontal asymptote at $y = -5$.

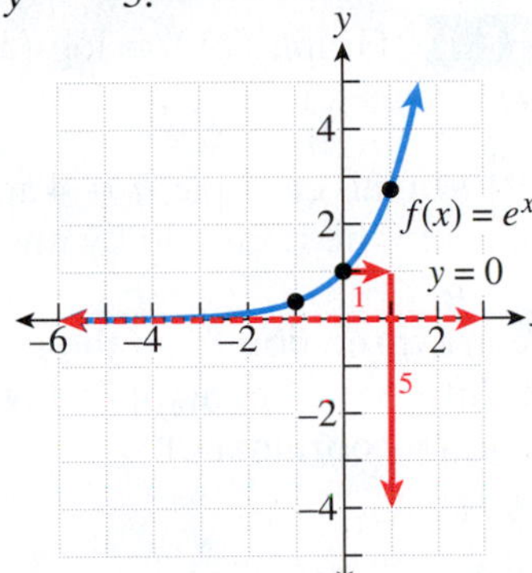

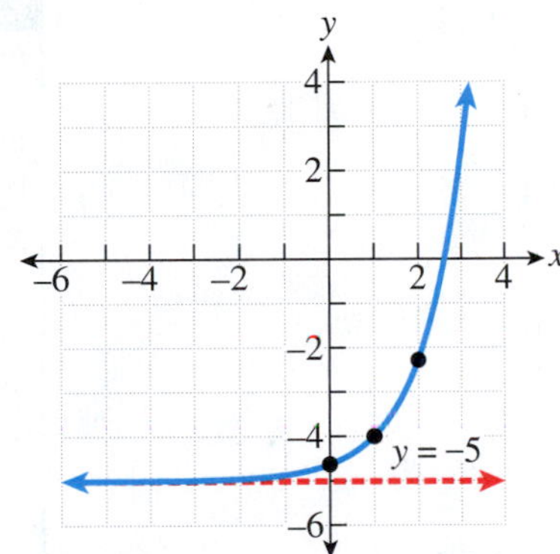

To find the y-intercept, we find $f(0)$.

$$f(0) = e^{0-1} - 5 = e^{-1} - 5 \approx -4.6$$

The y-intercept can be plotted at $(0, -4.6)$.

To find the x-intercept, we set the function equal to 0 and solve for x.

$e^{x-1} - 5 = 0$	Set $f(x) = 0$.
$e^{x-1} = 5$	Add 5 to both sides of the equation.
$\ln e^{x-1} = \ln 5$	Take the natural logarithm of each side.
$x - 1 = \ln 5$	Simplify the left side of the equation.
$x = \ln 5 + 1$	Add 1 to each side.
$x \approx 2.6$	Approximate using a calculator.

Quick Check 3

Graph $f(x) = e^{x-3} - 9$. Label any intercepts and state the domain and range of the function.

The x-intercept is approximately located at $(2.6, 0)$.

The domain of the function is $(-\infty, \infty)$, and the range is $(-5, \infty)$.

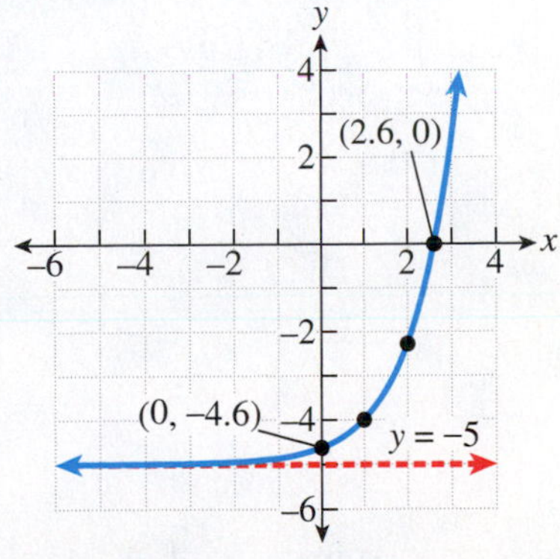

Graphing Logarithmic Functions

Objective 2 Graph logarithmic functions. Recall the following facts about the graph of a logarithmic function of the form $f(x) = \log_b x$, where $b > 0$ and $b \neq 1$:

- The graph has a vertical asymptote at the line $x = 0$.
- The x-intercept is at $(1, 0)$.
- The graph also passes through the points $\left(\frac{1}{b}, -1\right)$ and $(b, 1)$.

To graph a function of the form $f(x) = \log_b(x - h) + k$, we will shift the graph of $f(x) = \log_b x$ horizontally by h units and vertically by k units. A good strategy is to begin by shifting the vertical asymptote by h units horizontally. We will follow up by applying the shifts to the points $\left(\frac{1}{b}, -1\right)$, $(1, 0)$, and $(b, 1)$. After drawing the smooth curve passing through these points, we can add additional detail to the graph by finding any intercepts.

EXAMPLE 4 Graph $f(x) = \log_2(x - 4) + 1$. Label any intercepts and state the domain and range of $f(x)$.

SOLUTION In this example, $h = 4$ and $k = 1$; so we shift the graph of $f(x) = \log_2 x$ by four units to the right and by one unit up. This places the vertical asymptote at $x = 4$.

Here are three points on the graph of $f(x) = \log_2 x$ as well as their location after the shifts have been applied to the graph. (Add 4 to each x-coordinate and add 1 to each y-coordinate.)

$f(x) = \log_2 x$	$\left(\frac{1}{2}, -1\right)$	$(1, 0)$	$(2, 1)$
$f(x) = \log_2(x - 4) + 1$	$\left(4\frac{1}{2}, 0\right)$	$(5, 1)$	$(6, 2)$

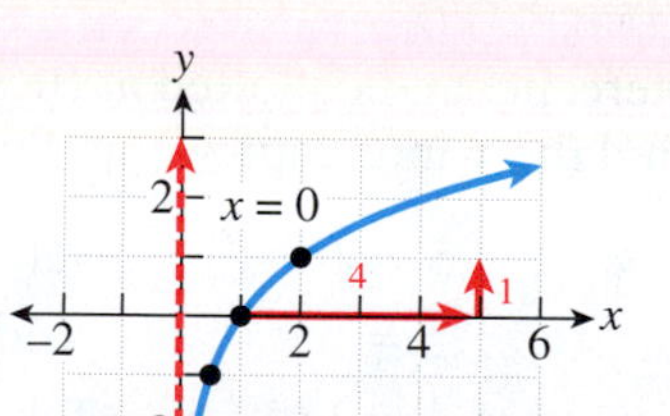

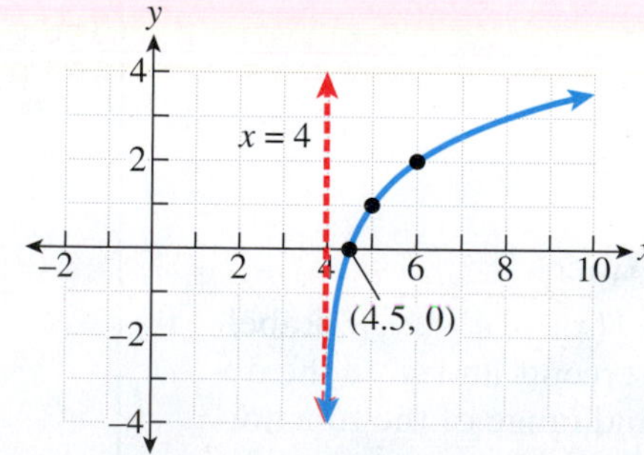

The x-intercept is already on the graph. From the graph, we can see that there is no y-intercept, as the vertical asymptote is to the right of the y-axis. The domain of the function is $(4, \infty)$, and the range is $(-\infty, \infty)$.

Quick Check 4

Graph $f(x) = \log_3(x - 2) - 1$. Label any intercepts and state the domain and range of the function.

Using Your Calculator You can use the TI-84 to graph the logarithmic function $f(x) = \log_2(x - 4) + 1$. Begin by tapping the Y= key. Next to Y_1, enter the function. You must use the change-of-base formula to rewrite $\log_2(x - 4) + 1$ as $\frac{\log(x - 4)}{\log 2} + 1$ to graph this function on the calculator. Tap the GRAPH key to display the graph. Using the standard window, your calculator screen should look like this. (Note that this function has a vertical asymptote at $x = 4$.)

Enter Expressions

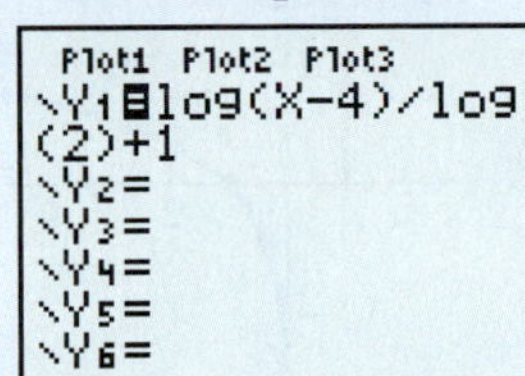

Graph

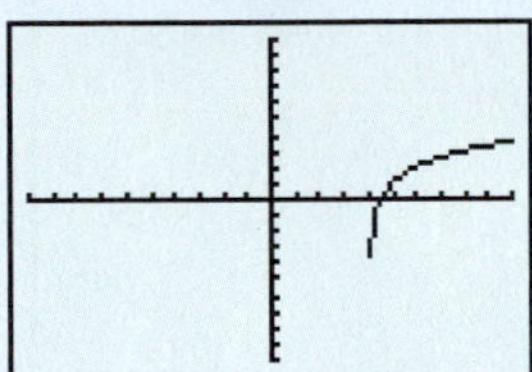

In general, for a function of the form $f(x) = \log_b(x - h) + k$, the equation of the vertical asymptote will be $x = h$ and the domain of the function will be (h, ∞).

EXAMPLE 5 Graph $f(x) = \log_6(x + 3) - 1$ and state the domain and range of $f(x)$.

SOLUTION In this example, $h = -3$ and $k = -1$; so we shift the graph of $f(x) = \log_6 x$ by three units to the left and by one unit down. This places the vertical asymptote at $x = -3$.

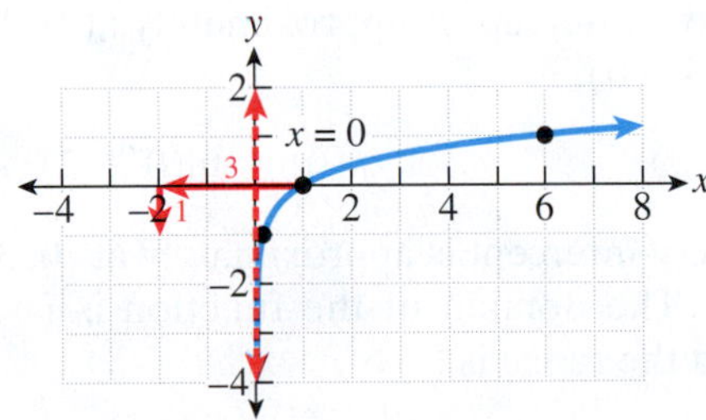

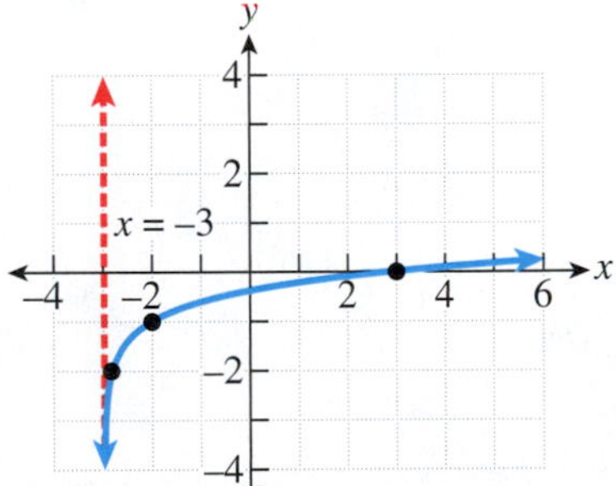

After shifting the graph of $f(x) = \log_6 x$, we see that the x-intercept is at $(3, 0)$. To find the y-intercept, we need to find $f(0)$.

$$\begin{aligned} f(0) &= \log_6(0 + 3) - 1 && \text{Substitute 0 for } x. \\ &= \log_6 3 - 1 && \text{Simplify.} \\ &= \frac{\log 3}{\log 6} - 1 && \text{Apply the change-of-base formula.} \\ &\approx -0.4 && \text{Approximate using a calculator.} \end{aligned}$$

The y-intercept is approximately at $(0, -0.4)$.

The domain of the function is $(-3, \infty)$, and the range is $(-\infty, \infty)$.

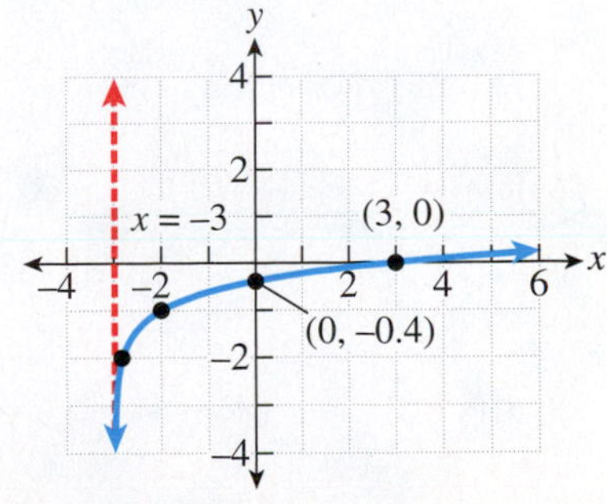

Quick Check 5

Graph $f(x) = \log_3(x + 7) - 2$. Label any intercepts and state the domain and range of $f(x)$.

EXAMPLE 6 Graph $f(x) = \ln(x + 2) + 3$. Label any intercepts and state the domain and range of $f(x)$.

SOLUTION In this example, $h = -2$ and $k = 3$; so we shift the graph of $f(x) = \ln x$ two units to the left and three units up. This places the vertical asymptote at $x = -2$.

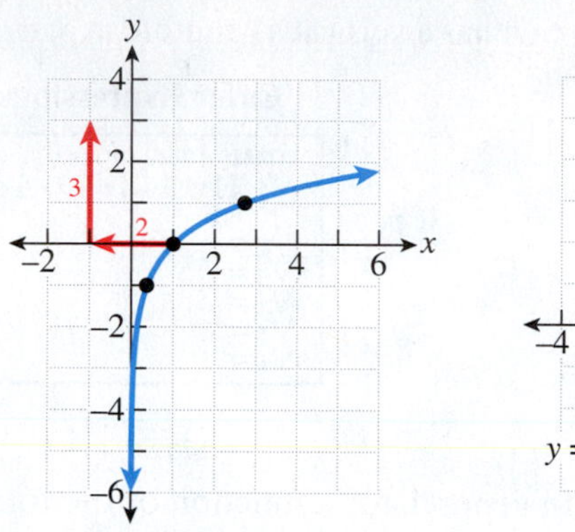

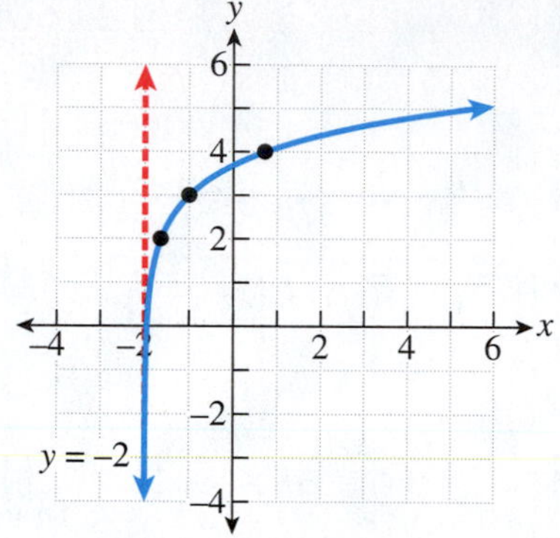

To find the x-intercept, we set the function equal to 0 and solve for x.

$$\ln(x + 2) + 3 = 0 \quad \text{Set } f(x) = 0.$$
$$\ln(x + 2) = -3 \quad \text{Subtract 3 from each side.}$$
$$e^{-3} = x + 2 \quad \text{Rewrite in exponential form.}$$
$$e^{-3} - 2 = x \quad \text{Subtract 2 from each side.}$$
$$x \approx -1.95 \quad \text{Approximate using a calculator.}$$

The x-intercept is approximately at $(-1.95, 0)$. To find the y-intercept, we need to find $f(0)$.

$$f(0) = \ln(0 + 2) + 3 = \ln(2) + 3 \approx 3.7$$

The y-intercept is approximately at $(0, 3.7)$.

The domain of the function is $(-2, \infty)$, and the range is $(-\infty, \infty)$.

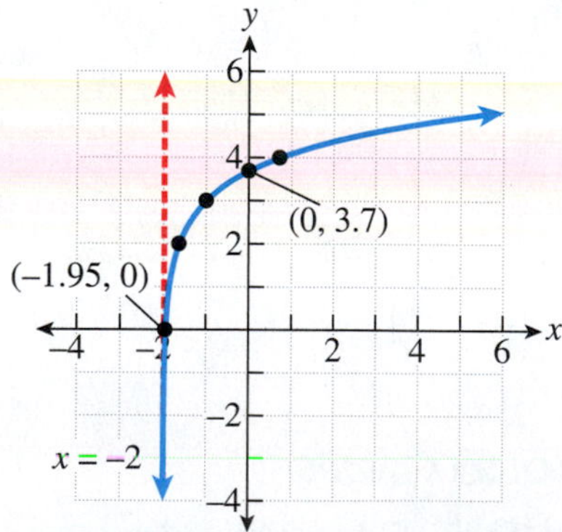

Quick Check 6

Graph $f(x) = \ln(x - 1) - 2$. Label any intercepts and state the domain and range of the function.

BUILDING YOUR STUDY STRATEGY

Practice Quizzes, 6 **Your Own Chapter Quiz** Create your own practice quiz for this chapter by selecting odd-numbered problems from the exercise sets of each section. Select four or five odd-numbered problems from each section that you think are representative of the types of problems you would be expected to solve on an exam. Choose problems of varying levels of difficulty.

After you have finished writing a practice chapter test, compare your problems with the chapter test exercises in the textbook. If you are missing certain types of problems, go back and add them to your practice test.

Exercises 12.6

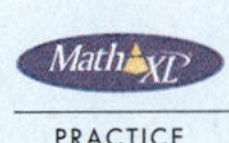

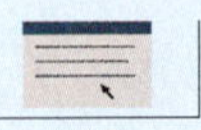

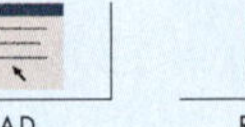

Vocabulary

1. The graph of an exponential function has a(n) __________ asymptote.
2. The graph of a logarithmic function has a(n) __________ asymptote.
3. The domain of a(n) __________ function is the set of all real numbers.
4. The domain of a(n) __________ function is restricted.
5. The graph of every exponential function has a(n) ______ intercept.
6. The graph of every logarithmic function has a(n) ______ intercept.

Find all intercepts for the given function. Round to the nearest tenth if necessary.

7. $f(x) = 2^{x+4} + 6$
8. $f(x) = 3^{x+4} - 9$
9. $f(x) = 6^{x+2} - 21$
10. $f(x) = e^{x+4} + 1$
11. $f(x) = e^{x+3} - 10$
12. $f(x) = e^{x-2} - 76$
13. $f(x) = \log_2(x - 5) - 3$
14. $f(x) = \log_3(x + 1) - 3$
15. $f(x) = \ln(x + 3) - 4$
16. $f(x) = \ln(x + 8) - 10$

Determine the equation of the horizontal asymptote for the graph of this function and state the domain and range of this function.

17. $f(x) = 3^{x-5} + 8$
18. $f(x) = 6^{x+2} - 10$
19. $f(x) = 4^{x-12} - 9$
20. $f(x) = 5^{x-6}$

Determine the equation of the vertical asymptote for the graph of this function and state the domain and range of this function.

21. $f(x) = \log_7(x - 5) - 9$
22. $f(x) = \log_8(x + 4) + 4$
23. $f(x) = \log(2x - 1) + 16$
24. $f(x) = \log_6 x - 12$

Graph. Label any intercepts and the horizontal asymptote. If necessary, round to the nearest tenth. State the domain and range of the function.

25. $f(x) = 2^{x+2} + 4$

26. $f(x) = 2^{x-3} + 5$

27. $f(x) = 3^{x+1} - 9$

28. $f(x) = 4^{x-2} - 4$

29. $f(x) = \left(\frac{1}{2}\right)^{x+4} - 4$

30. $f(x) = \left(\frac{1}{3}\right)^{x+1} - 9$

31. $f(x) = 5^{x+1} - 6$

32. $f(x) = 4^{x+2} - 7$

33. $f(x) = e^{x-2} + 3$

34. $f(x) = e^{x+2} + 1$

35. $f(x) = e^{x+1} - 7$

36. $f(x) = e^{x+2} - 5$

Graph. Label any intercepts and the vertical asymptote. If necessary, round to the nearest tenth. State the domain and range of the function.

37. $f(x) = \log_4(x - 2) - 1$

38. $f(x) = \log_3(x - 1) + 1$

39. $f(x) = \log_3(x + 8) + 2$ **40.** $f(x) = \log_5(x + 2) - 1$

41. $f(x) = \log_2(x + 5) - 3$ **42.** $f(x) = \log_3(x + 4) - 2$

43. $f(x) = \ln(x - 3) + 1$ **44.** $f(x) = \ln(x - 2) + 2$

45. $f(x) = \ln(x + 1) - 2$ **46.** $f(x) = \ln(x + 4) + 2$

Mixed Practice, 47–56

Graph. Label any intercepts and asymptotes. If necessary, round to the nearest tenth. State the domain and range of the function.

47. $f(x) = e^{x-1} - 4$ **48.** $f(x) = \ln(x - 2) - 1$

49. $f(x) = \log_3(x + 4) - 2$ **50.** $f(x) = 3^{x-3} - 6$

51. $f(x) = \ln(x + 6)$ **52.** $f(x) = 4^{x+1} + 4$

53. $f(x) = 2^{x-1} - 8$ **54.** $f(x) = \log_2(x + 7) + 1$

55. $f(x) = \left(\frac{1}{3}\right)^x - 3$

56. $f(x) = e^{x-2} + 5$

Given the graph of a function f(x), graph the function $f^{-1}(x)$.

57.

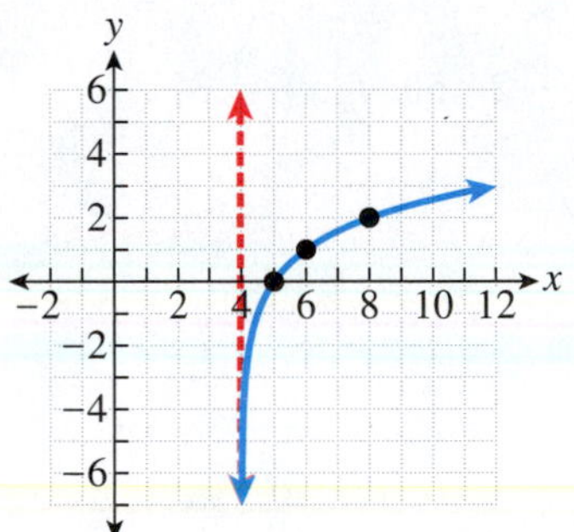

58.

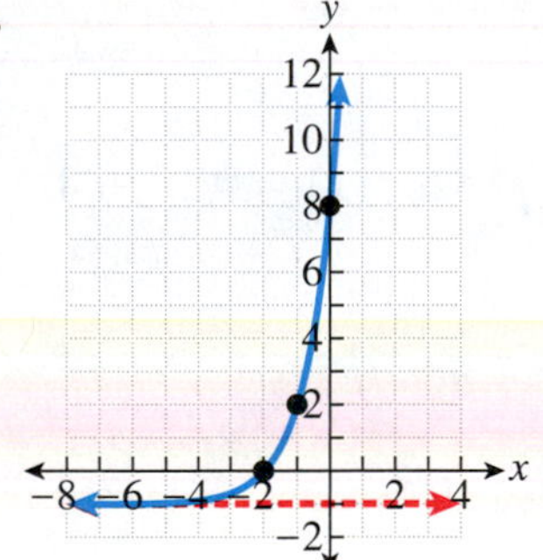

59.

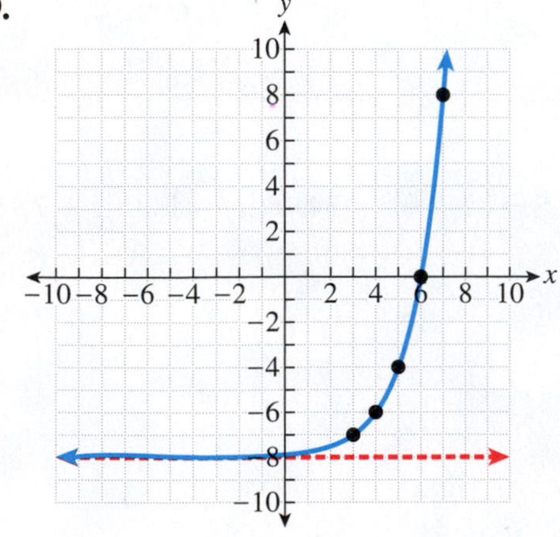

60.

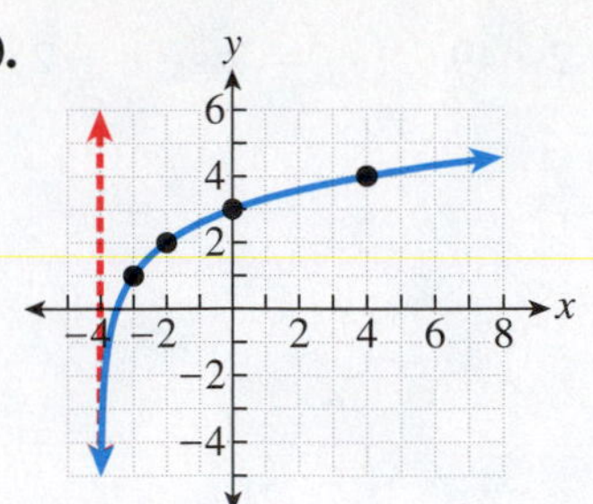

Determine which equation is associated with the given graph.

a) $y = \log_2(x - 3) + 4$

b) $y = \log_2(x - 4) + 3$

c) $y = \log_4 x$

d) $y = \ln x$

e) $y = 2^{x-4} + 3$

f) $y = 2^{x-3} + 4$

g) $y = e^x$

h) $y = 4^x$

61.

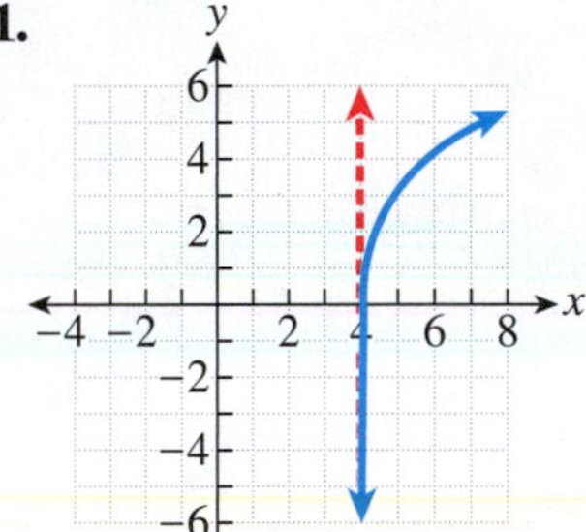

62.

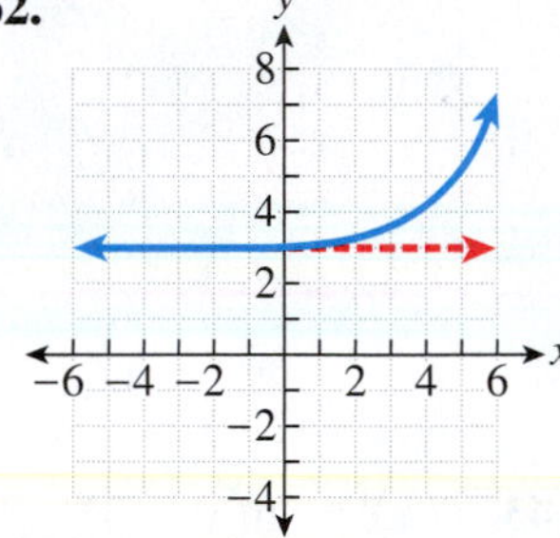

63.

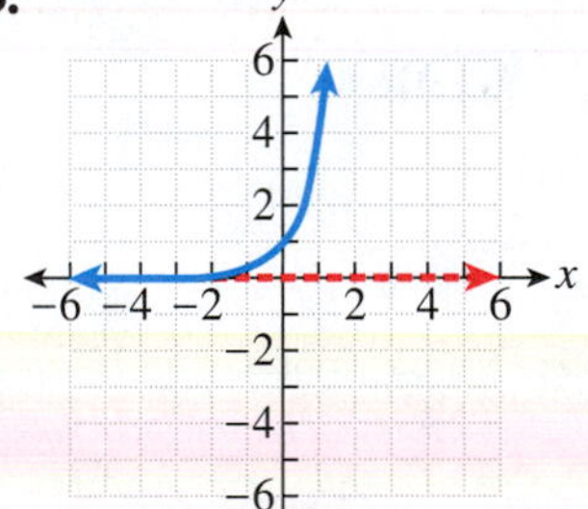

64.

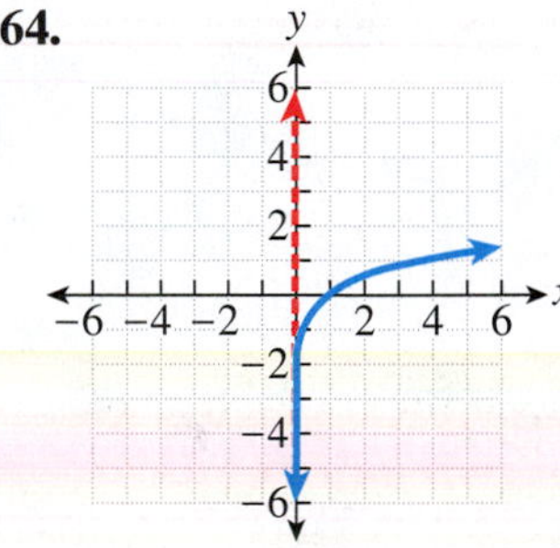

65.

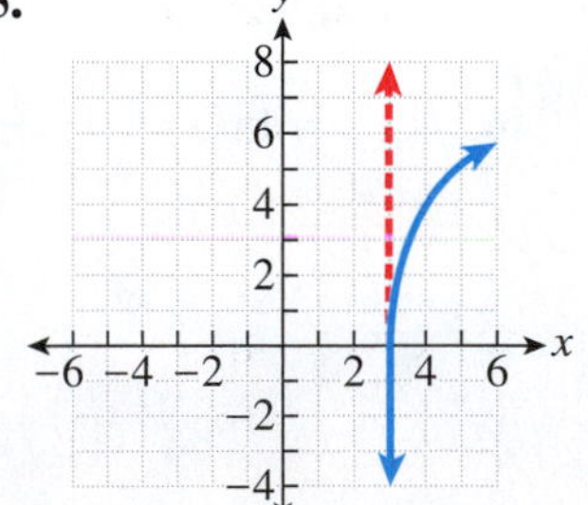

66.

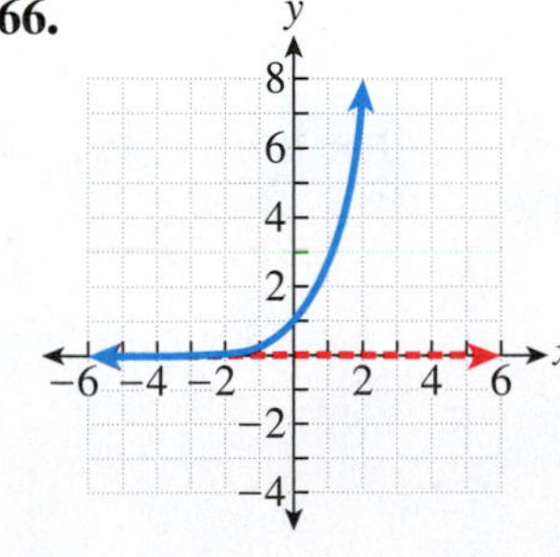

67.

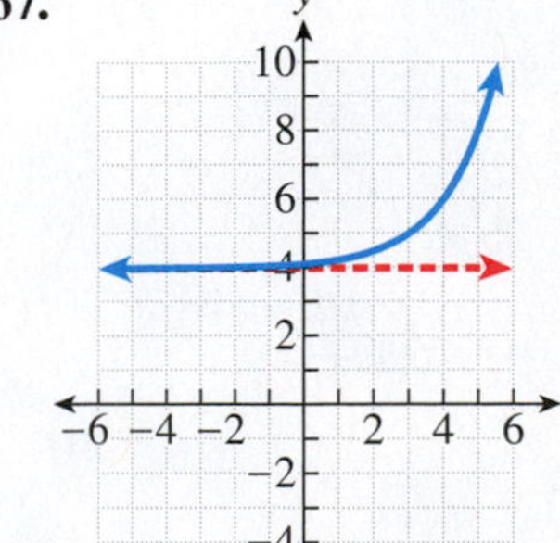

68.

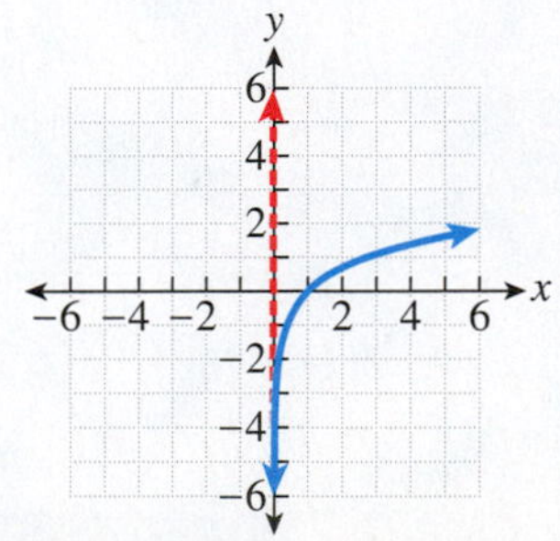

Writing in Mathematics

Answer in complete sentences.

69. Explain how to graph an exponential function, including how to find any intercepts and the horizontal asymptote. Use an example to illustrate the process.

70. Explain how to graph a logarithmic function, including how to find any intercepts and the vertical asymptote. Use an example to illustrate the process.

CHAPTER 12 SUMMARY

Section 12.1 Exponential Functions

Exponential Function, pp. 685–687

A function of the form $f(x) = b^x, b > 0$ and $b \neq 1$, is called an **exponential function**.

For $f(x) = 2^{x-4} + 3$, find $f(9)$.

$$\begin{aligned} f(9) &= 2^{9-4} + 3 \\ &= 2^5 + 3 \\ &= 32 + 3 \\ &= 35 \end{aligned}$$

Graphing Exponential Functions, pp. 687–690

For an exponential function $f(x) = b^x$:

- Graph the horizontal asymptote $y = 0$ using a dashed line.
- Plot the points $(1, b)$, $(0, 1)$, and $\left(-1, \frac{1}{b}\right)$.
- Draw the graph passing through the points.

The domain of the function is the set of all real numbers, but the range is restricted.

Graph: $f(x) = 3^x$

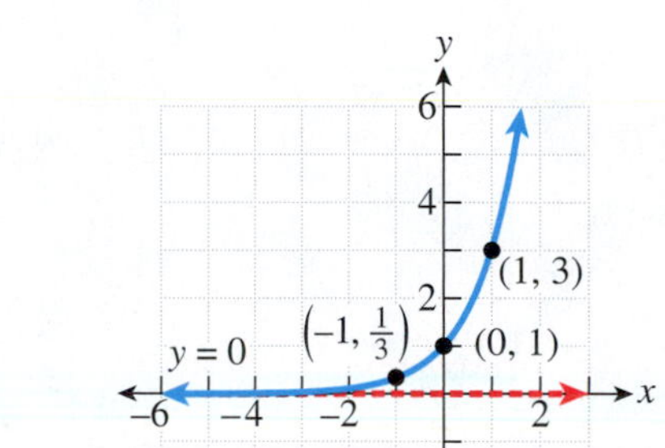

Domain: $(-\infty, \infty)$; range: $(0, \infty)$

The Natural Exponential Function, pp. 690–691

The function $f(x) = e^x$ is the natural exponential function, where $e \approx 2.7182818284\ldots$.

For $f(x) = e^{x+1} - 7$, find $f(3)$. Round to the nearest thousandth.

$$\begin{aligned} f(3) &= e^{3+1} - 7 \\ &= e^4 - 7 \\ &\approx 47.598 \end{aligned}$$

One-to-One Property of Exponential Functions, pp. 691–692

For any positive real number b $(b \neq 1)$ and any real numbers x and y, if $b^x = b^y$, then $x = y$.

Solve: $3^{x-4} = 243$

$$\begin{aligned} 3^{x-4} &= 243 \\ 3^{x-4} &= 3^5 \\ x - 4 &= 5 \\ x &= 9 \end{aligned}$$

$\{9\}$

Section 12.2 Logarithmic Functions

Logarithm, pp. 695–696

For any positive real number b $(b \neq 1)$ and any positive real number x, $\log_b x$ is defined as the exponent to which b must be raised to equal x.

$\log_5 125 = 3$ because $5^3 = 125$.

Logarithmic Function, pp. 696–697

For any positive real number b $(b \neq 1)$, a function of the form $f(x) = \log_b x$ is a **logarithmic function**.

This function is defined for values of x that are greater than 0; so the domain of this function is the set of positive real numbers: $(0, \infty)$.

For $f(x) = \log_2 x - 5$, find $f\left(\frac{1}{8}\right)$.

$$\begin{aligned} f\left(\frac{1}{8}\right) &= \log_2\left(\frac{1}{8}\right) - 5 \\ &= -3 - 5 \\ &= -8 \end{aligned}$$

Common Logarithm, p. 697

A logarithm whose base is 10 is called a common logarithm; it is written without its base. The common logarithm of x is written simply as $\log x$. The common logarithm can be approximated with a calculator.

For $f(x) = \log x$, find $f(57)$.

$$\begin{aligned} f(57) &= \log 57 \\ &\approx 1.756 \end{aligned}$$

Natural Logarithm, pp. 697–698

A logarithm whose base is e is called a natural logarithm. The natural logarithm of x is written as $\ln x$.

For $f(x) = \ln(x + 17)$, find $f(30)$. Round to the nearest thousandth.

$$\begin{aligned} f(30) &= \ln(30 + 17) \\ &= \ln 47 \\ &\approx 3.850 \end{aligned}$$

Exponential and Logarithmic Form of an Equation, pp. 698–699

An equation written as $b^y = n$ is said to be in exponential form. The equivalent logarithmic form of this equation is $\log_b n = y$.

Rewrite in logarithmic form: $2^x + 7 = 25$

$$\begin{aligned} 2^x + 7 &= 25 \\ 2^x &= 18 \\ \log_2 18 &= x \end{aligned}$$

Solving Logarithmic Equations, p. 699

Some logarithmic equations can be solved by rewriting the equation in exponential form and solving the resulting equation.

Solve: $\log_4 x = 3$

$$\begin{aligned} \log_4 x &= 3 \\ 4^3 &= x \\ 64 &= x \end{aligned}$$

$\{64\}$

Graphing Logarithmic Functions, pp. 700–702

For a logarithmic function $f(x) = \log_b x$:

- Graph the vertical asymptote $x = 0$ using a dashed line.
- Plot the points $(b, 1)$, $(1, 0)$, and $\left(\frac{1}{b}, -1\right)$.
- Draw the graph passing through the points.

The domain of the function is restricted, but the range is the set of all real numbers.

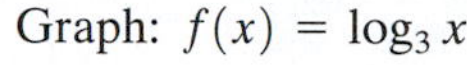

Graph: $f(x) = \log_3 x$

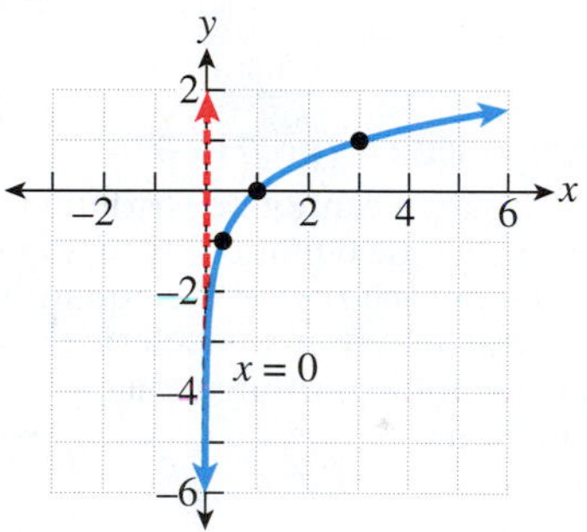

Domain: $(0, \infty)$; range: $(-\infty, \infty)$

Section 12.3 Properties of Logarithms

Product Rule for Logarithms, pp. 707–708

For any positive real number b $(b \neq 1)$ and any positive real numbers x and y,

$$\log_b x + \log_b y = \log_b(xy).$$

Rewrite as a single logarithm: $\log_2 5 + \log_2 x$

$$\log_2 5 + \log_2 x = \log_2(5x)$$

Quotient Rule for Logarithms, p. 708

For any positive real number b $(b \neq 1)$ and any positive real numbers x and y,

$$\log_b x - \log_b y = \log_b\left(\frac{x}{y}\right).$$

Rewrite as a single logarithm: $\log_4 x - \log_4 9$

$$\log_4 x - \log_4 9 = \log_4\left(\frac{x}{9}\right)$$

Power Rule for Logarithms, pp. 709–710

For any positive real number b $(b \neq 1)$, any positive real number x, and any real number r,

$$\log_b x^r = r \cdot \log_b x.$$

Rewrite without an exponent: $\log_7 x^8$

$$\log_7 x^8 = 8 \cdot \log_7 x$$

Rewriting Logarithmic Expressions as a Single Logarithm, pp. 710–711

To rewrite two or more logarithmic expressions in terms of a single logarithm that is not being multiplied by a constant, use the properties of logarithms. This is referred to as condensing logarithms.

Rewrite as a single logarithm: $8 \log_b x + 5 \log_b y - 2 \log_b z$

$$\begin{aligned} &8 \log_b x + 5 \log_b y - 2 \log_b z \\ &= \log_b x^8 + \log_b y^5 - \log_b z^2 \\ &= \log_b\left(\frac{x^8 y^5}{z^2}\right) \end{aligned}$$

Rewriting a Logarithmic Expression as a Sum or Difference of Logarithms, pp. 711–712

To rewrite a single logarithm as a sum or difference of two logarithms without using exponents is referred to as expanding logarithms.

Rewrite as a sum or difference of logarithms without exponents: $\log\left(\frac{a^7}{b^6c^5}\right)$

$$\log\left(\frac{a^7}{b^6c^5}\right) = \log a^7 - \log b^6 - \log c^5$$
$$= 7\log a - 6\log b - 5\log c$$

Change-of-Base Formula, pp. 712–713

For any positive numbers $a, b,$ and x $(a, b \neq 1)$,

$$\log_b x = \frac{\log_a x}{\log_a b}.$$

Evaluate $\log_7 60$ using the change of base formula. Round to the nearest thousandth.

$$\log_7 60 = \frac{\ln 60}{\ln 7}$$
$$\approx 2.104$$

Section 12.4 Exponential and Logarithmic Equations

Solving Exponential Equations, pp. 715–716

If both sides of an exponential equation can be rewritten in terms of the same base, the equation can be solved using the one-to-one property of exponential functions.

Solve: $5^{2x-9} = 125$

$$5^{2x-9} = 5^3$$
$$2x - 9 = 3$$
$$2x = 12$$
$$x = 6$$

$\{6\}$

Solving Exponential Equations, pp. 716–718

If both sides of the equation cannot be written in terms of the same base, solve the equation by taking the common logarithm of each side of the equation or by taking the natural logarithm of each side. After using the power rule for logarithms, you can solve the resulting equation.

Solve $3^{x+4} = 20$. Round to the nearest thousandth.

$$3^{x+4} = 20$$
$$\log 3^{x+4} = \log 20$$
$$(x + 4)\log 3 = \log 20$$
$$x + 4 = \frac{\log 20}{\log 3}$$
$$x = \frac{\log 20}{\log 3} - 4$$
$$x \approx -1.273$$

Solving Logarithmic Equations, pp. 718–721

1. If one side of an equation contains two or more logarithms, rewrite these logarithms as a single logarithmic expression.
2. To solve an equation in which two logarithms with the same base are equal, use the one-to-one property of logarithmic functions. This allows you to set the two arguments equal to each other and solve the resulting equation.
3. To solve an equation in which a logarithm is equal to a number, convert the equation to its exponential form and solve the resulting equation.
4. Check the solution to make sure it does not cause an argument of a logarithm to be negative or 0 because the domain is $(0, \infty)$. If it does, this solution must be omitted from the solution set.

Solve: $\log(x^2 - 5x - 9) = \log(2x + 9)$

$$\log(x^2 - 5x - 9) = \log(2x + 9)$$
$$x^2 - 5x - 9 = 2x + 9$$
$$x^2 - 7x - 18 = 0$$
$$(x + 2)(x - 9) = 0$$

$\{-2, 9\}$

Solve: $\log_3 x + \log_3(x + 6) = 3$

$$\log_3 x + \log_3(x + 6) = 3$$
$$\log_3(x^2 + 6x) = 3$$
$$x^2 + 6x = 3^3$$
$$x^2 + 6x - 27 = 0$$
$$(x + 9)(x - 3) = 0$$

Extraneous solution: $x = -9$

$\{3\}$

Finding the Inverse Function of an Exponential or Logarithmic Function, p. 722

1. Replace $f(x)$ with y.
2. Interchange x and y.
3. Solve the resulting equation for y using the techniques for solving exponential and logarithmic equations.
4. Replace y with $f^{-1}(x)$.

Find $f^{-1}(x)$: $f(x) = e^{x+4} + 9$

$$y = e^{x+4} + 9$$
$$x = e^{y+4} + 9$$
$$x - 9 = e^{y+4}$$
$$\ln(x - 9) = \ln e^{y+4}$$
$$\ln(x - 9) = y + 4$$
$$\ln(x - 9) - 4 = y$$
$$f^{-1}(x) = \ln(x - 9) - 4$$

Section 12.5 Applications of Exponential and Logarithmic Functions

Compound Interest, pp. 725–727

If P dollars are deposited in an account that pays an annual interest rate r that is compounded n times per year, the amount A in the account after t years is given by the formula

$$A = P\left(1 + \frac{r}{n}\right)^{nt}.$$

If \$6000 is invested in an account that pays 6% annual interest, compounded quarterly, what will the balance be in 10 years?

$$A = 6000\left(1 + \frac{0.06}{4}\right)^{4 \cdot 10}$$
$$A = 6000(1.015)^{40}$$
$$A = 10{,}884.11$$

The balance will be \$10,884.11.

Continuous Compound Interest, p. 727

If P dollars are deposited in an account that pays an annual interest rate r that is compounded continuously, the amount A in the account after t years is given by the formula

$$A = Pe^{rt}.$$

If \$10,000 is invested in an account that pays 7% annual interest, compounded continuously, how long will it take for the investment to double?

$$20{,}000 = 10{,}000e^{0.07t}$$
$$2 = e^{0.07t}$$
$$\ln 2 = \ln e^{0.07t}$$
$$\ln 2 = 0.07t$$
$$\frac{\ln 2}{0.07} = t$$
$$t \approx 9.9$$

It will take approximately 9.9 years.

Exponential Growth, pp. 728–729

If a quantity increases at an exponential rate, its size P at time t is given by the formula

$$P = P_0e^{kt}.$$

P_0 represents the initial size of the quantity, while k is the exponential growth rate.

In 1993, Visalia's population was 80,000. In 2008, the population reached 120,000. When will the population reach 200,000?

Find k.

$$120{,}000 = 80{,}000e^{15k}$$
$$1.5 = e^{15k}$$
$$\ln 1.5 = \ln e^{15k}$$
$$\frac{\ln 1.5}{15} = k$$
$$k \approx 0.027031$$

$$P = P_0e^{kt}$$
$$200{,}000 = 80{,}000e^{0.027031t}$$
$$2.5 = e^{0.027031t}$$
$$\ln 2.5 = \ln e^{0.027031t}$$
$$\frac{\ln 2.5}{0.027031} = t$$
$$t \approx 34$$

In 2027.

Exponential Decay, pp. 729–730

If a quantity decreases at an exponential rate, its size P at time t is given by the formula

$$P = P_0e^{kt}.$$

P_0 represents the initial size of the quantity, while k is the exponential decay rate.

Half-Life

The half-life of a radioactive element is the amount of time it takes for half of the substance to decay.

The half-life of a particular radioactive substance is 8 days. If 20 grams of the substance is initially present, how many grams will remain in 30 days?

$$10 = 20e^{8k}$$
$$0.5 = e^{8k}$$
$$\ln 0.5 = \ln e^{8k}$$
$$\frac{\ln 0.5}{8} = k$$
$$k \approx -0.086643$$

$$P = 20e^{-0.086643(30)}$$
$$P = 20e^{-2.59929}$$
$$P \approx 1.5$$

Approximately 1.5 grams will remain.

pH, p. 731

The pH of a liquid is calculated using the formula $\text{pH} = -\log[\text{H}^+]$, where $[\text{H}^+]$ is the concentration of hydrogen ions in moles per liter.

If the concentration of hydrogen ions for a certain liquid is 5.0×10^{-11} moles per liter, what is its pH?

$$\text{pH} = -\log(5.0 \times 10^{-11})$$
$$\text{pH} = 10.3$$

The pH is 10.3.

Volume of a Sound, pp. 731–732

The volume (L) of a sound in decibels is given by the formula $L = 10\log\left(\frac{I}{10^{-12}}\right)$, where I is the intensity of the sound in watts per square meter.

Find the intensity of a sound whose volume is 115 db.

$$115 = 10\log\left(\frac{I}{10^{-12}}\right)$$
$$11.5 = \log\left(\frac{I}{10^{-12}}\right)$$
$$10^{11.5} = \frac{I}{10^{-12}}$$
$$10^{11.5} \cdot 10^{-12} = I$$
$$10^{-0.5} = I$$

The intensity is $10^{-0.5}$ watts per square meter.

Section 12.6 Graphing Exponential and Logarithmic Functions

Graphing Exponential Functions of the Form $f(x) = b^{x-h} + k$, pp. 736–739

1. Graph the horizontal asymptote.
   Using a dashed line, graph the line $y = k$.
2. Shift the graph of $f(x) = b^x$ by h units horizontally and k units vertically.
3. Find the y-intercept.
   Evaluate $f(0)$.
4. Find the x-intercept if one exists.
   Set the function $f(x)$ equal to 0 and solve for x.

Graph: $f(x) = 2^{x-3} - 5$

Horizontal asymptote: $y = -5$

$h = 3, k = -5$

y-intercept: $(0, -4.9)$

$$f(0) = 2^{-3} - 5 \approx -4.9$$

x-intercept: $(5.3, 0)$

$$2^{x-3} - 5 = 0$$
$$2^{x-3} = 5$$
$$\ln 2^{x-3} = \ln 5$$
$$(x - 3)\ln 2 = \ln 5$$
$$x = \frac{\ln 5}{\ln 2} + 3 \approx 5.3$$

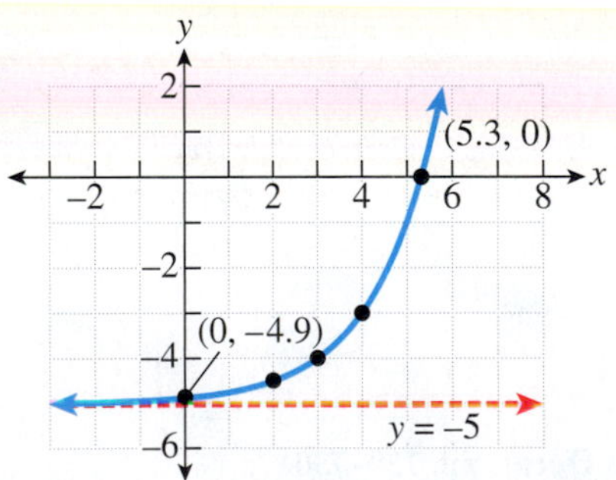

Domain: $(-\infty, \infty)$; range: $(-5, \infty)$

Graphing Logarithmic Functions of the Form $f(x) = \log_b(x - h) + k$, pp. 740–742

1. Graph the vertical asymptote.
   Using a dashed line, graph the line $x = h$.
2. Shift the graph of $f(x) = \log_b x$ by h units horizontally and k units vertically.
3. Find the y-intercept if one exists.
   Evaluate the original function when $x = 0$. This value of x may not be in the domain of the function, and in this case, there is no y-intercept.
4. Find the x-intercept.
   Set the function equal to 0 and solve this equation for x.

Graph: $f(x) = \log_3(x + 2) - 1$

Vertical asymptote: $x = -2$
$h = -2, k = -1$
y-intercept: $(0, -0.4)$

$$\begin{aligned} f(0) &= \log_3(0 + 2) - 1 \\ &= \log_3 2 - 1 \\ &= \frac{\ln 2}{\ln 3} - 1 \approx -0.4 \end{aligned}$$

x-intercept: $(1, 0)$

$$\begin{aligned} \log_3(x + 2) - 1 &= 0 \\ \log_3(x + 2) &= 1 \\ 3^1 &= x + 2 \\ 1 &= x \end{aligned}$$

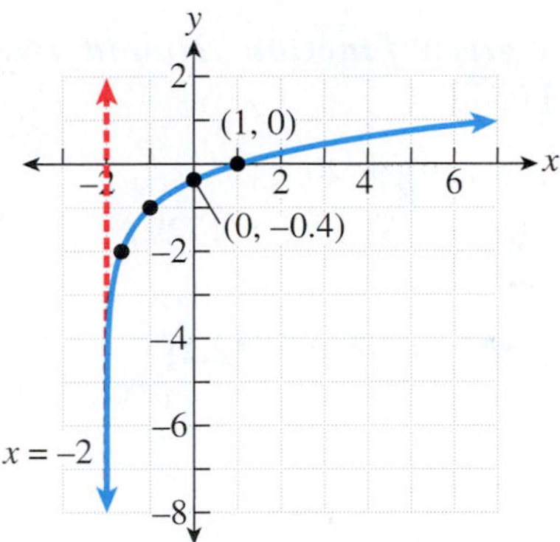

Domain: $(-2, \infty)$; range: $(-\infty, \infty)$

SUMMARY OF CHAPTER 12 STUDY STRATEGIES

The chapter test at the end of this chapter can be treated as a practice quiz. Take this chapter test as if it were an actual exam, without using your notes, homework, or text. Do this at least two days before the actual exam if possible. After you check your answers, you will have a good idea as to which types of problems you understand and which types of problems you need to practice. At this point, review your notes and homework regarding these problems and read the appropriate sections of the text again. Now you can create focused practice quizzes on these topics by choosing problems from the chapter review. If you are still having trouble at this point, review the material again and visit your instructor or a tutorial center to get your questions answered. When you believe you understand these topics, create another practice quiz by choosing problems from the appropriate sections to confirm that you are ready for the test.

CHAPTER 12 REVIEW

Evaluate the given function. Round to the nearest thousandth. [12.1]

1. $f(x) = 4^x - 9, f(5)$
2. $f(x) = e^{x-8} - 13, f(11)$

Simplify. [12.2]

3. $\log_6 \frac{1}{36}$
4. $\log_5 3125$
5. $\log_9 1$
6. $\ln e^{12}$

Evaluate the given function. Round to the nearest thousandth. [12.2]

7. $f(x) = \ln(x - 9) + 20, f(52)$
8. $f(x) = \log(x + 23) - 6, f(1729)$

Rewrite in logarithmic form. [12.2]

9. $4^x = 1024$
10. $e^x = 20$

Rewrite in exponential form. [12.2]

11. $\log x = 3.2$
12. $\ln x - 3 = 8$

Rewrite as a single logarithmic expression. Simplify if possible. Assume that all variables represent positive real numbers. [12.3]

13. $\log_2 13 + \log_2 8$
14. $2 \log a - 5 \log b$
15. $4 \log_2 x - 3 \log_2 y + 9 \log_2 z$
16. $-5 \log_b 3 - 2 \log_b 2$

Rewrite as the sum or difference of logarithmic expressions whose arguments have an exponent of 1. Simplify if possible. Assume that all variables represent positive real numbers. [12.3]

17. $\log_8 64x$
18. $\log_3\left(\frac{n}{27}\right)$
19. $\log_b\left(\frac{a^6c^3}{b^2}\right)$
20. $\log_2\left(\frac{8x^9}{yz^7}\right)$

Evaluate using the change-of-base formula. Round to the nearest thousandth. [12.3]

21. $\log_4 25$
22. $\log_{1/2} 295$
23. $\log_2 1597$
24. $\log_{37} 6{,}403{,}200$

Solve. Round to the nearest thousandth. [12.4]

25. $10^x = 1000$
26. $8^{x-5} = 64$
27. $16^x = 32$
28. $9^{x+6} = 27^x$
29. $7^x = 11$
30. $6^{x+4} = 3000$
31. $e^{x-8} = 62$
32. $5^{2x-9} = 67$

Solve. Round to the nearest thousandth. [12.4]

33. $\log_4(x + 3) = \log_4 17$
34. $\log(x^2 + 12x + 6) = \log(3x + 16)$
35. $\ln x = 5$
36. $\log_3(x^2 - 13x + 57) - 1 = 2$
37. $\log_9 x + \log_9(x + 8) = \log_9 20$
38. $\ln(x + 10) - \ln(x - 6) = \ln 5$
39. $\log_3(x + 5) + \log_3(x - 1) = 3$
40. $\log x + \log(x + 3) + 5 = 6$
41. Let $f(x) = 2^{3x-7} + 8$. Solve $f(x) = 13$. Round to the nearest thousandth.
42. Let $f(x) = \ln x + 4$. Solve $f(x) = 8$. Round to the nearest thousandth.

For the given function $f(x)$, find its inverse function $f^{-1}(x)$. [12.4]

43. $f(x) = e^x + 4$
44. $f(x) = 2^{x-5} - 1$
45. $f(x) = \ln(x + 8)$
46. $f(x) = \log_3(x + 6) - 10$
47. Grady has deposited \$3500 in an account that pays 8% annual interest, compounded quarterly. How long will it take until the balance of the account is \$10,000? [12.5]
48. The population of a city increased from 200,000 in 1992 to 250,000 in 2002. If the population continues to increase exponentially at the same rate, what will the population be in 2015? [12.5]
49. The half-life of bismuth-210 is 5 days. If 25 grams of bismuth-210 are present initially, how much will remain after 10 days? [12.5]
50. If researchers find a wooden arrow that contains 98% of its original carbon-14, how old is the arrow? (Carbon-14 has a half-life of 5730 years.) [12.5]

Worked-out solutions to Review Exercises marked with ● can be found on page AN-50.

51. During the 1990s, the percentage of American adults who were obese increased exponentially. For any particular year, the percentage of adult Americans who were obese can be approximated by the function $f(x) = 11.5 \cdot 1.057^x$, where x represents the number of years after 1990. Use this function to determine when the percentage of Americans who were obese reached 30%. (*Source:* CDC) [12.5]

52. The body temperature, in degrees Fahrenheit, of a person who was found dead in an apartment that maintained a constant temperature of 75° F can be approximated by the function $f(t) = 75 + 12e^{-0.206t}$, where $t = 0$ corresponds to the time the body was found. What was the body's temperature 6 hours after it was found? [12.5]

53. If an earthquake creates a shock wave that is 20,000 times the smallest measurable shock wave that is recordable by a seismograph, find the magnitude of the earthquake on the Richter scale. [12.5]

54. The pH of a particular bottle of wine is 3.5. Find the concentration of hydrogen ions in moles per liter. [12.5]

***Graph. Label any intercepts and asymptotes. State the domain and range of the function.* [12.6]**

55. $f(x) = 3^x + 2$

56. $f(x) = 2^{x+2} - 8$

57. $f(x) = e^{x+1} - 6$

58. $f(x) = \log_2(x - 5)$

59. $f(x) = \log_3(x + 9) - 2$

60. $f(x) = \ln(x + 8) + 1$

CHAPTER 12 TEST

For Extra Help

Step-by-step test solutions are found on the Chapter Test Prep Videos available via the Video Resources on DVD, in MyMathLab, and on YouTube (search "WoodburyElemIntAlg" and click on "Channels").

Evaluate the given function. Round to the nearest thousandth.

1. $f(x) = e^{x-4} - 10, f(6)$

2. $f(x) = \ln(x + 7) + 4, f(11)$

Simplify.

3. $\log_{12} 144$

4. $\ln e^{-3}$

Rewrite as a single logarithmic expression. Simplify if possible. Assume that all variables represent positive real numbers.

5. $\log_4 3136 - \log_4 49$

6. $2 \log_b x + 5 \log_b y - 3 \log_b z$

Rewrite as the sum or difference of logarithmic expressions whose arguments have an exponent of 1. Simplify if possible. Assume that all variables represent positive real numbers.

7. $\ln\left(\frac{a^4b^6}{c^8d}\right)$

Evaluate using the change-of-base formula. Round to the nearest thousandth.

8. $\log_9 62$

Solve. Round to the nearest thousandth.

9. $6^{x+4} = \frac{1}{36}$

10. $4^{x-9} = 76$

11. $e^{x+2} - 6 = 490$

Solve. Round to the nearest thousandth.

12. $\log_3(2x - 11) = \log_3 5$

13. $\log_5(x + 65) + 5 = 8$

14. $\log_2 x + \log_2(x - 4) = 5$

15. For the function $f(x) = e^{x+9} - 17$, find its inverse function $f^{-1}(x)$.

16. Sonia started a bank account with an initial deposit of \$4200. The bank account pays 6% interest, compounded monthly. How long will it take for the deposit to grow to \$6000?

17. A painting that was purchased for \$15,000 in 1992 was valued at \$25,000 in 2002. If the value of the painting keeps rising exponentially at the same rate, how much will it be worth in 2025?

18. If researchers find a wooden chalice that contains 90% of its original carbon-14, how old is the chalice? (Carbon-14 has a half-life of 5730 years.)

Graph. Label any intercepts and asymptotes. State the domain and range of the function.

19. $f(x) = e^{x-2} - 9$

20. $f(x) = \log_3(x + 4) + 1$

Mathematicians in History

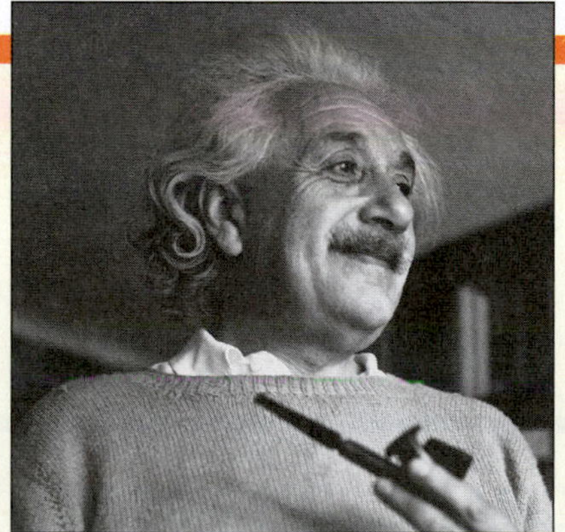

Albert Einstein was named *Time* magazine's "Man of the Century" for the twentieth century. That is quite an accomplishment for a man who once failed an exam that would have allowed him to study to be an electrical engineer. In 1901, working as a temporary high school math teacher, Einstein had given up the ambition to go to a university. However, while working in Switzerland's patent office, Einstein earned a doctorate in physics from the University of Zurich in 1908. Einstein introduced the theory of relativity and became one of the most popular scientists ever. Einstein once said that compound interest was the most powerful force in the universe and that it is the greatest mathematical discovery of all time.

Write a one-page summary (*or* make a poster) of the life of Albert Einstein and his accomplishments. Also look up Einstein's most famous quotes and list your five favorites.

Interesting issues:

- Where and when was Albert Einstein born?
- What famous formula is credited to Einstein?
- Einstein married Mileva Maric in 1903. What became of their sons Hans Albert and Eduard?
- After divorcing Mileva in 1919, Einstein married his second wife, Elsa. How did he know Elsa?
- When did Einstein win the Nobel prize? For what did he win it?
- Einstein came to Princeton University in 1932, planning to teach part of the year at Princeton and the remainder of the year in Berlin. What event prohibited Einstein from returning to Berlin in 1933?
- In 1939, Einstein sent a letter to Franklin Roosevelt. What was the subject of that letter?
- What job was Einstein offered in 1952?
- When did Einstein die? What were the circumstances of his death?

CHAPTER 13

Conic Sections

In this chapter, we will learn to graph *conic sections:* parabolas, circles, ellipses, and hyperbolas. The equations of the different conic sections are all nonlinear equations. These graphs are called conic sections because they can be constructed by intersecting a plane with a cone, as shown in the following figure:

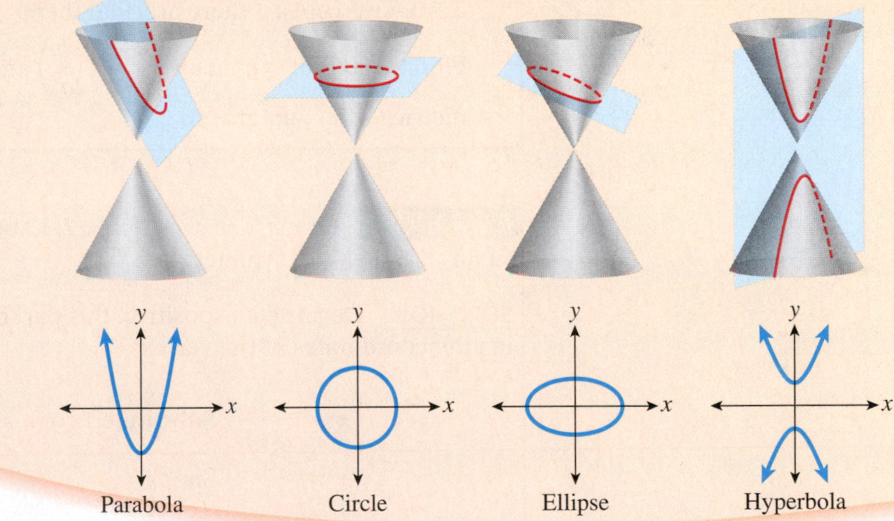

The chapter concludes with a section on systems of nonlinear equations.

STUDY STRATEGY

Using the Previous Study Strategies to Prepare for a Final Exam (Part 1)
The study strategies in this chapter will focus on how to incorporate the study strategies from previous chapters into your preparation for a final exam. Although some of the strategies are similar to those used to prepare for a chapter test or quiz, there are some differences as well. These strategies will be continued in Chapter 14.

13.1 Parabolas

OBJECTIVES

1. Graph equations of the form $y = ax^2 + bx + c$.
2. Graph equations of the form $y = a(x - h)^2 + k$.
3. Graph equations of the form $x = ay^2 + by + c$.
4. Graph equations of the form $x = a(y - k)^2 + h$.
5. Find a vertex by completing the square.
6. Find the equation of a parabola that meets the given conditions.

In Chapter 10, we graphed parabolas that opened upward or downward. These parabolas were associated with equations that could be expressed as functions of x. After reviewing these graphs, we will learn to graph parabolas that open to the right or to the left.

Graphing Equations of the Form $y = ax^2 + bx + c$

Objective 1 Graph equations of the form $y = ax^2 + bx + c$. Here is a brief summary for graphing equations of the form $y = ax^2 + bx + c$.

Graphing an Equation of the Form $y = ax^2 + bx + c$

- Determine whether the parabola opens upward or downward.
- Find the vertex of the parabola.
- Find the y-intercept of the parabola.
- After plotting the vertex and the y-intercept, determine whether there are any x-intercepts; if there are, find them.
- Find the axis of symmetry $\left(x = \frac{-b}{2a}\right)$ and use it to find the point that is symmetric to the y-intercept.

EXAMPLE 1 Graph $y = x^2 + 6x - 2$. Label the vertex, y-intercept, x-intercept(s) (if any), and axis of symmetry.

SOLUTION Because a is positive, this parabola opens upward. We begin by finding the coordinates of the vertex.

$$x = \frac{-6}{2(1)} \quad \text{Substitute 1 for } a \text{ and 6 for } b \text{ into } x = \frac{-b}{2a}.$$
$$x = -3 \quad \text{Simplify.}$$

To find the y-coordinate, we substitute -3 for x.

$$y = (-3)^2 + 6(-3) - 2 \quad \text{Substitute } -3 \text{ for } x.$$
$$y = -11 \quad \text{Simplify.}$$

The vertex is $(-3, -11)$.

Now we turn our attention to the y-intercept. We substitute 0 for x in the original equation.

$$y = (0)^2 + 6(0) - 2 \quad \text{Substitute 0 for } x.$$
$$y = -2 \quad \text{Simplify.}$$

The y-intercept is $(0, -2)$.

Because the vertex is located below the x-axis and the parabola opens upward, the graph must cross the x-axis. To find the x-intercepts, we substitute 0 for y and solve for x.

$$0 = x^2 + 6x - 2 \quad \text{Substitute 0 for } y. \text{ Because } x^2 + 6x - 2 \text{ does not factor, use the quadratic formula.}$$
$$x = \frac{-6 \pm \sqrt{(6)^2 - 4(1)(-2)}}{2(1)} \quad \text{Substitute 1 for } a, \text{ 6 for } b, \text{ and } -2 \text{ for } c.$$
$$x = \frac{-6 \pm \sqrt{44}}{2} \quad \text{Simplify the radicand.}$$

$$x = \frac{-6 \pm 2\sqrt{11}}{2}$$ Simplify the radical.

$$x = \frac{\overset{1}{\cancel{2}}(-3 \pm \sqrt{11})}{\underset{1}{\cancel{2}}}$$ Factor the numerator and divide out common factors.

$$x = -3 \pm \sqrt{11}$$ Simplify.

Using a calculator, we find that $-3 + \sqrt{11} \approx 0.3$ and $-3 - \sqrt{11} \approx -6.3$. The x-intercepts are approximately at $(0.3, 0)$ and $(-6.3, 0)$.

Here is the graph of the given equation. The axis of symmetry, $x = -3$, has been used to find the coordinates of the point symmetric to the y-intercept, $(-6, -2)$.

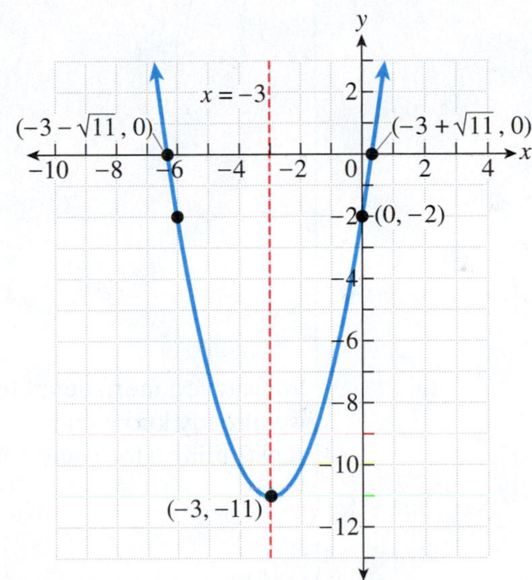

Quick Check 1

Graph $y = -x^2 + 2x + 15$. Label the vertex, y-intercept, x-intercept(s) (if any), and axis of symmetry.

Graphing Equations of the Form $y = a(x - h)^2 + k$

Objective 2 Graph equations of the form $y = a(x - h)^2 + k$. Now we will review graphing equations of the form $y = a(x - h)^2 + k$ by shifting.

Graphing an Equation of the Form $y = a(x - h)^2 + k$

- Begin with the graph of the basic parabola: $y = x^2$.
- If $a < 0$, the graph opens downward. Rotate the graph of $y = x^2$ about the x-axis.
- The vertex of the parabola is (h, k). Shift the graph h units horizontally and k units vertically.
- To find the y-intercept, substitute 0 for x and solve for y.
- To find the x-intercepts, if there are any, substitute 0 for y and solve for x.

EXAMPLE 2 Sketch the graph of $y = -(x - 1)^2 + 9$ by shifting. Label the vertex and y-intercept as well as any x-intercept(s).

SOLUTION Because $a < 0$, we know that this parabola opens downward. We begin by rotating the parabola $y = x^2$ about the x-axis. The vertex of this parabola is $(1, 9)$, so we shift this parabola one unit to the right and nine units up.

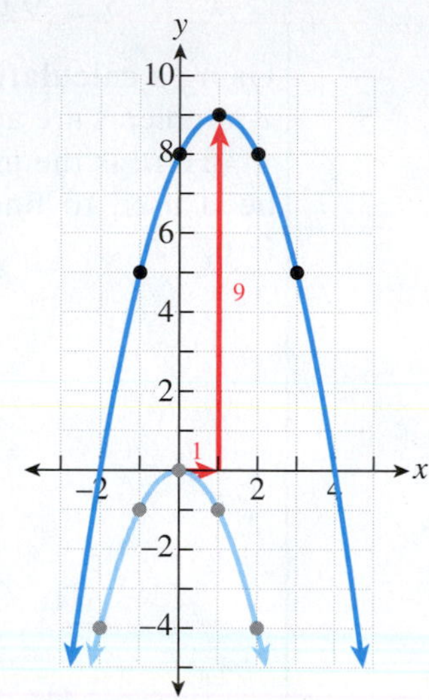

We can add more detail to the graph by adding its intercepts. Notice that by shifting, we already know that the y-intercept is $(0, 8)$. We can determine the x-intercepts by substituting 0 for y and solving for x.

$$0 = -(x - 1)^2 + 9 \quad \text{Substitute 0 for } y.$$
$$(x - 1)^2 = 9 \quad \text{Add } (x - 1)^2 \text{ to both sides.}$$
$$\sqrt{(x - 1)^2} = \pm\sqrt{9} \quad \text{Take the square root of each side.}$$
$$x - 1 = \pm 3 \quad \text{Simplify.}$$
$$x = 1 \pm 3 \quad \text{Add 1 to both sides.}$$
$$x = 1 + 3 = 4 \quad \text{or} \quad x = 1 - 3 = -2 \quad \text{Simplify.}$$

The x-intercepts are $(4, 0)$ and $(-2, 0)$. Here is the graph, including all intercepts.

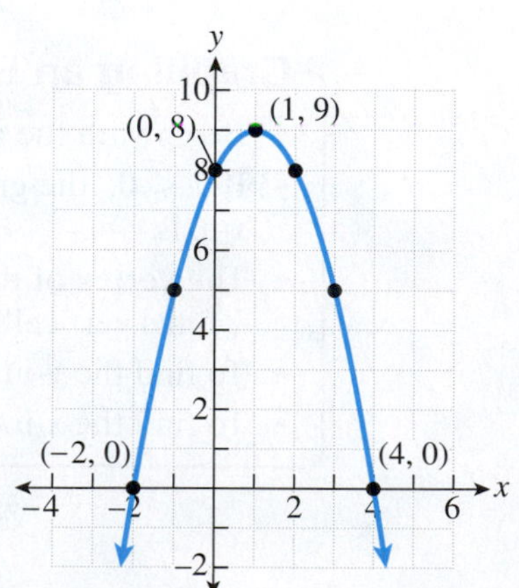

Quick Check 2

Sketch the graph of $y = (x + 4)^2 + 3$ by shifting. Label the vertex and y-intercept as well as any x-intercept(s).

Graphing Equations of the Form $x = ay^2 + by + c$

Objective 3 Graph equations of the form $x = ay^2 + by + c$. If the variables x and y are interchanged in the equation $y = ax^2 + bx + c$, the graph of the equation will be a parabola that opens to the right or to the left. The process for graphing this type of parabola is similar to graphing a parabola that opens up or down.

Graphing an Equation of the Form $x = ay^2 + by + c$

- Determine whether the parabola opens to the right or to the left. The parabola will open to the right if a is positive and to the left if a is negative.
- Find the vertex of the parabola by using the formula $y = \frac{-b}{2a}$. Substitute this result in the original equation for y to find the x-coordinate.
- Find the x-intercept of the parabola by substituting 0 for y in the original equation.
- After plotting the vertex and the x-intercept, find the y-intercepts, if there are any, by letting $x = 0$ in the original equation and solving the resulting quadratic equation for y.
- Find the axis of symmetry $\left(y = \frac{-b}{2a}\right)$ and use it to find the point on the parabola that is symmetric to the x-intercept.

EXAMPLE 3 Graph $x = y^2 + 2y - 3$. Label the vertex, any intercepts, and the axis of symmetry.

SOLUTION Notice that the squared variable in the equation is y, not x; so the parabola will open to the right or to the left. This parabola opens to the right because a is positive. We begin by finding the y-coordinate of the vertex using the formula $y = \frac{-b}{2a}$.

$$y = \frac{-2}{2(1)} \quad \text{Substitute 1 for } a \text{ and 2 for } b \text{ in } y = \frac{-b}{2a}.$$

$$y = -1 \quad \text{Simplify.}$$

To find the x-coordinate of the vertex, we substitute -1 for y in the original equation.

$$x = (-1)^2 + 2(-1) - 3 \quad \text{Substitute } -1 \text{ for } y.$$

$$x = -4 \quad \text{Simplify.}$$

The vertex of the parabola is $(-4, -1)$.

A WORD OF CAUTION Keep in mind that even though we find the y-coordinate first and the x-coordinate second, we must write the ordered pair as (x, y).

Next, we find the x-intercept by substituting 0 for y in the equation.

$$x = (0)^2 + 2(0) - 3 \quad \text{Substitute 0 for } y.$$

$$x = -3 \quad \text{Simplify.}$$

The x-intercept is $(-3, 0)$.

Because the vertex is located to the left of the y-axis and the parabola opens to the right, the parabola has two y-intercepts. We can find these intercepts by substituting 0 for x and solving the resulting equation for y.

$$0 = y^2 + 2y - 3 \quad \text{Substitute 0 for } x.$$
$$0 = (y + 3)(y - 1) \quad \text{Factor.}$$
$$y = -3 \quad \text{or} \quad y = 1 \quad \text{Set each factor equal to 0 and solve.}$$

The y-intercepts are $(0, -3)$ and $(0, 1)$.

Quick Check 3

Graph $x = y^2 + 4y - 2$. Label the vertex, any intercepts, and the axis of symmetry.

We can find another point on the graph by finding the point on the parabola that is symmetric to the x-intercept $(-3, 0)$. We can do this using the axis of symmetry, which in this case is the horizontal line $y = -1$.

Because the x-intercept is one unit above the axis of symmetry, the point symmetric to it will be one unit below the axis of symmetry at $(-3, -2)$. Here is the graph.

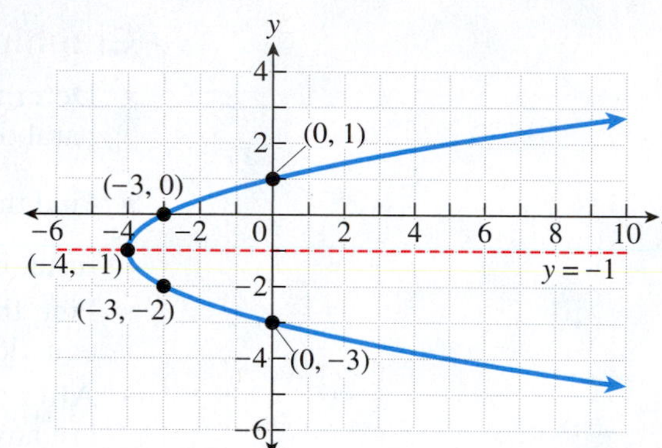

If a parabola opens to the right and the vertex is to the right of the y-axis, the parabola does not have any y-intercepts. The same holds true for parabolas that open to the left whose vertex is to the left of the y-axis.

EXAMPLE 4 Graph $x = -y^2 + 2y - 5$. Label the vertex, any intercepts, and the axis of symmetry.

SOLUTION Because a is negative, the parabola opens to the left. We begin by finding the vertex.

$$y = \frac{-2}{2(-1)} = 1 \quad \text{Substitute } -1 \text{ for } a \text{ and 2 for } b \text{ in } y = \frac{-b}{2a}.$$
$$x = -(1)^2 + 2(1) - 5 = -4 \quad \text{Substitute 1 for } y \text{ in } x = -y^2 + 2y - 5.$$

The vertex of the parabola is $(-4, 1)$. Now we find the x-intercept.

$$x = -(0)^2 + 2(0) - 5 = -5 \quad \text{Substitute 0 for } y \text{ in } x = -y^2 + 2y - 5.$$

The x-intercept is $(-5, 0)$.

The vertex is located to the left of the y-axis. Because the parabola opens to the left, the parabola does not have any y-intercepts. The point $(-5, 2)$ shown on the graph is symmetric to the x-intercept and can be found using the axis of symmetry $y = 1$.

Quick Check 4

Graph $x = -y^2 + 8y - 12$. Label the vertex, any intercepts, and the axis of symmetry.

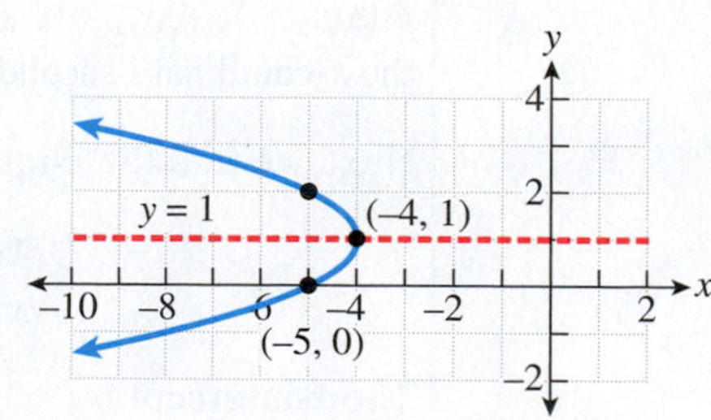

Graphing Equations of the Form $x = a(y - k)^2 + h$

Objective 4 Graph equations of the form $x = a(y - k)^2 + h$. Now we turn our attention to the graphs of equations of the form $x = a(y - k)^2 + h$. We will graph equations of this form by transforming the graph of the basic parabola $x = y^2$.

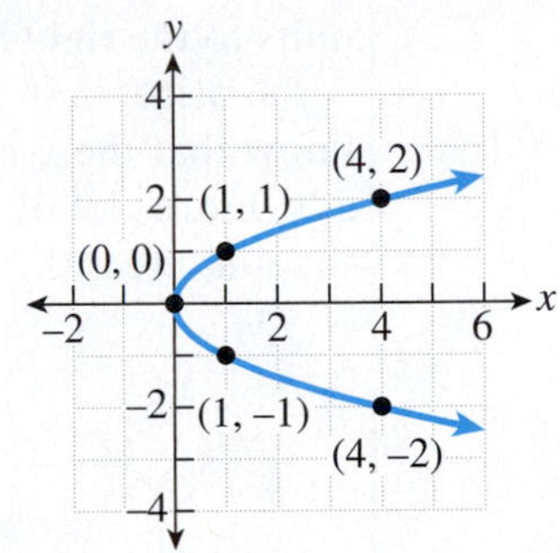

Graphing an Equation of the Form $x = a(y - k)^2 + h$

- Begin with the graph of the basic parabola: $x = y^2$.
- If $a < 0$, the graph opens to the left. Rotate the graph of $x = y^2$ about the y-axis.
- The vertex of the parabola is (h, k). Shift the graph h units horizontally and k units vertically.
- To find the x-intercept, substitute 0 for y and solve for x.
- To find the y-intercepts, if there are any, substitute 0 for x and solve for y.

EXAMPLE 5 Sketch the graph of $x = (y - 2)^2 + 3$ by shifting. Label the vertex and x-intercept as well as any y-intercept(s).

SOLUTION Because a is positive, we know that this parabola opens to the right. The vertex of this parabola is $(3, 2)$, so we shift the basic parabola $x = y^2$ by three units to the right and by two units up.

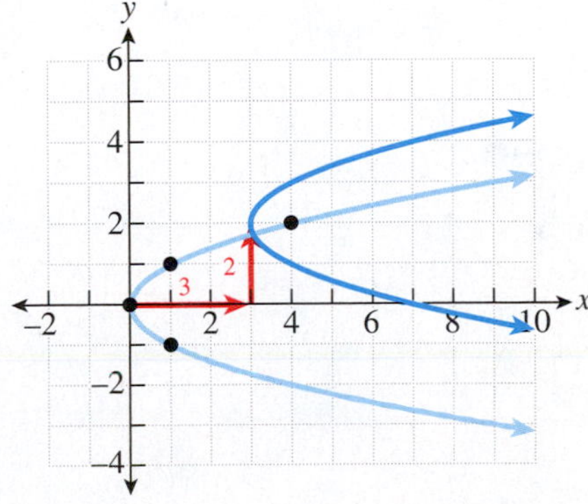

Quick Check 5

Sketch the graph of $x = (y - 3)^2 - 1$ by shifting. Label the vertex and x-intercept as well as any y-intercept(s).

Notice that by shifting, we already know that the x-intercept is $(7, 0)$. We also can see that this graph does not have any y-intercepts because the vertex is located to the right of the y-axis and opens to the right.

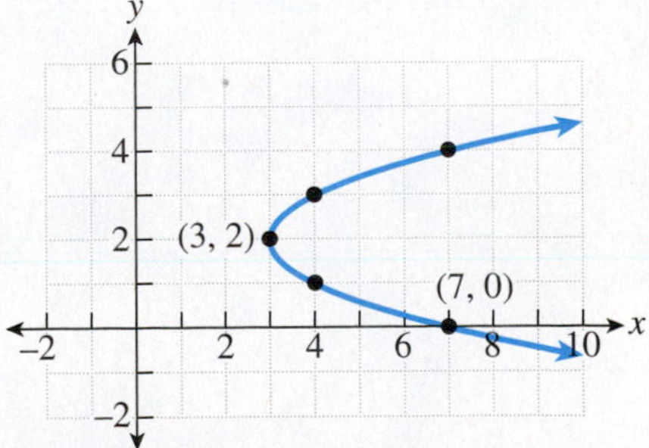

EXAMPLE 6 Sketch the graph of $x = -(y-1)^2 + 5$ by shifting. Label the vertex and x-intercept as well as any y-intercept(s).

SOLUTION Because a is negative, we know that this parabola opens to the left. We start by rotating the graph of the basic parabola $x = y^2$ about the y-axis. The vertex of the parabola we are graphing is $(5, 1)$, so we shift the parabola by five units to the right and by one unit up.

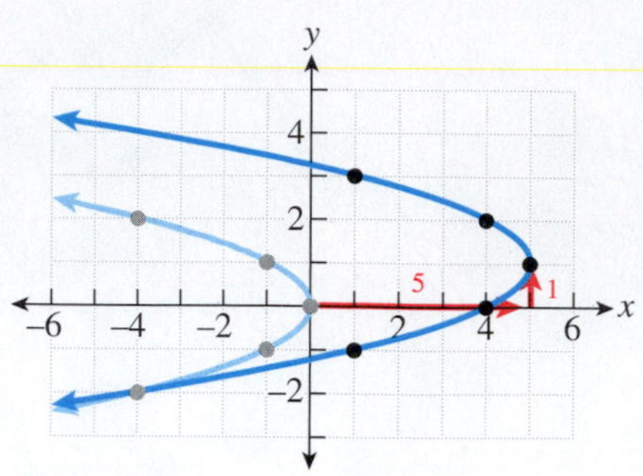

Notice that by shifting, we already know that the x-intercept is $(4, 0)$. We can add more detail to the graph by finding the y-intercepts.

$$0 = -(y-1)^2 + 5 \quad \text{Substitute 0 for } y.$$
$$(y-1)^2 = 5 \quad \text{Add } (y-1)^2 \text{ to both sides.}$$
$$\sqrt{(y-1)^2} = \pm\sqrt{5} \quad \text{Take the square root of each side.}$$
$$y - 1 = \pm\sqrt{5} \quad \text{Simplify.}$$
$$y = 1 \pm \sqrt{5} \quad \text{Add 1 to both sides.}$$

Because $y = 1 + \sqrt{5} \approx 3.2$ and $y = 1 - \sqrt{5} \approx -1.2$, the y-intercepts are approximately located at $(3.2, 0)$ and $(-1.2, 0)$.

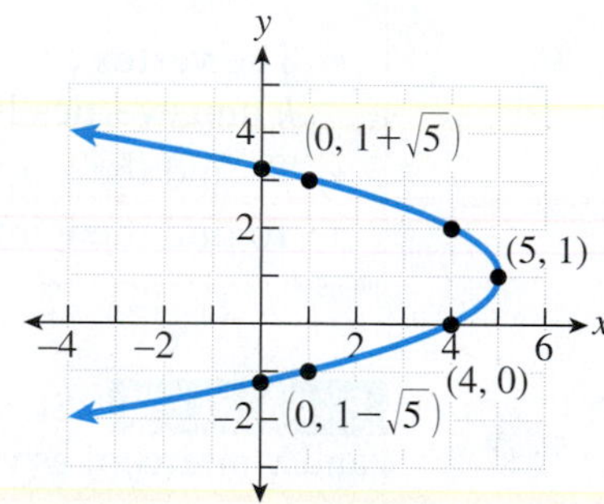

Quick Check 6

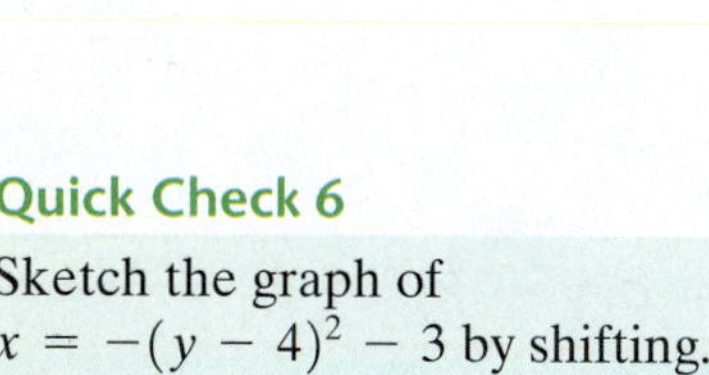
Sketch the graph of $x = -(y-4)^2 - 3$ by shifting. Label the vertex and x-intercept as well as any y-intercept(s).

Finding the Vertex by Completing the Square

Objective 5 Find a vertex by completing the square. Consider the equation $y = x^2 - 8x + 33$. Although we could find the vertex by using the formula $x = \frac{-b}{2a}$, we also can find the vertex by completing the square to write the equation in the form $y = a(x-h)^2 + k$, where (h, k) is the vertex.

EXAMPLE 7 Find the vertex of the parabola defined by the equation $y = x^2 - 8x + 33$ by completing the square.

SOLUTION

$$y = x^2 - 8x + 33$$
$$y - 33 = x^2 - 8x \quad \text{Subtract 33 from both sides.}$$
$$y - 33 + 16 = x^2 - 8x + 16 \quad \left(\tfrac{-8}{2}\right)^2 = 16. \text{ Add 16 to both sides.}$$
$$y - 17 = (x-4)^2 \quad \text{Combine like terms on the left side. Factor the quadratic expression on the right side.}$$
$$y = (x-4)^2 + 17 \quad \text{Add 17 to both sides.}$$

The vertex is $(4, 17)$.

Although it may seem easier to use the formula $x = \frac{-b}{2a}$ to find the vertex in this case, the ability to complete the square will be an important skill in the sections that follow.

EXAMPLE 8 Find the vertex of the parabola defined by the equation $x = -y^2 + 4y - 50$ by completing the square.

SOLUTION

$$x = -y^2 + 4y - 50$$

$$x + 50 = -y^2 + 4y$$ Add 50 to both sides.

$$x + 50 = -(y^2 - 4y)$$ Factor out a negative 1 on the right side.

$$x + 50 - 4 = -(y^2 - 4y + 4)$$ $\left(\frac{-4}{2}\right)^2 = 4$. Add 4 to the expression in parentheses on the right side. Subtract 4 from the left side because there is a factor of -1 in front of the parentheses.

$$x + 46 = -(y - 2)^2$$ Combine like terms on the left side. Factor the quadratic expression on the right side.

$$x = -(y - 2)^2 - 46$$ Subtract 46 from both sides.

The vertex is $(-46, 2)$.

Quick Check 7

Find the vertex of the parabola defined by the given equation by completing the square.

a) $y = x^2 + 4x + 20$
b) $x = -y^2 - 6y + 17$

Finding the Equation of a Parabola

Objective 6 Find the equation of a parabola that meets the given conditions. Several techniques can be used to find the equation of a parabola. Here is one that uses the intercepts of the parabola.

EXAMPLE 9 Find the equation of the parabola whose graph is shown in the accompanying figure.

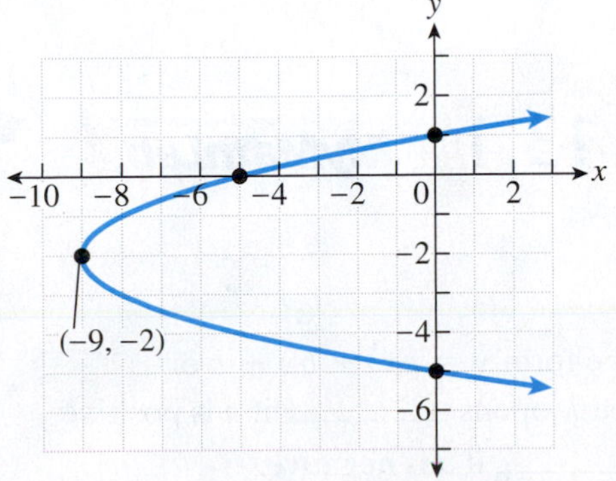

SOLUTION We begin by noting that the parabola opens to the right; so its equation is of the form $x = ay^2 + by + c$, where $a > 0$. We can observe that the y-intercepts are $(0, 1)$ and $(0, -5)$ and the x-intercept is $(-5, 0)$.

Because the y-coordinates of the y-intercept are $y = 1$ and $y = -5$, we know that $y - 1$ and $y + 5$ are factors of the expression $ay^2 + by + c$. Multiplying $(y - 1)(y + 5)$ yields the expression $y^2 + 4y - 5$, so the equation is of the form $x = a(y^2 + 4y - 5)$. To find the value of a, we use the coordinates of the x-intercept $(-5, 0)$. We also could use the coordinates of any other point on the graph of the parabola, including the vertex. After substituting -5 for x and 0 for y in the equation $x = a(y^2 + 4y - 5)$, we solve for a.

$$x = a(y^2 + 4y - 5)$$
$$-5 = a((0)^2 + 4(0) - 5) \quad \text{Substitute } -5 \text{ for } x \text{ and } 0 \text{ for } y.$$
$$-5 = a(-5) \quad \text{Simplify.}$$
$$1 = a \quad \text{Divide both sides by } -5.$$

Because $a = 1$, the equation of the parabola is $x = 1(y^2 + 4y - 5)$, or simply $x = y^2 + 4y - 5$.

Note that we could have used the vertex to find the equation of the parabola. Because the vertex is $(-9, -2)$ and the parabola opens to the right, we know that the equation is of the form $x = a(y + 2)^2 - 9$. Substituting the coordinates of any other point on the parabola, such as the x-intercept, for x and y will lead to the equation $x = 1(y + 2)^2 - 9$, or $x = (y + 2)^2 - 9$, which is equivalent to $x = y^2 + 4y - 5$.

Quick Check 8

Find the equation of the parabola whose graph is shown in the accompanying figure.

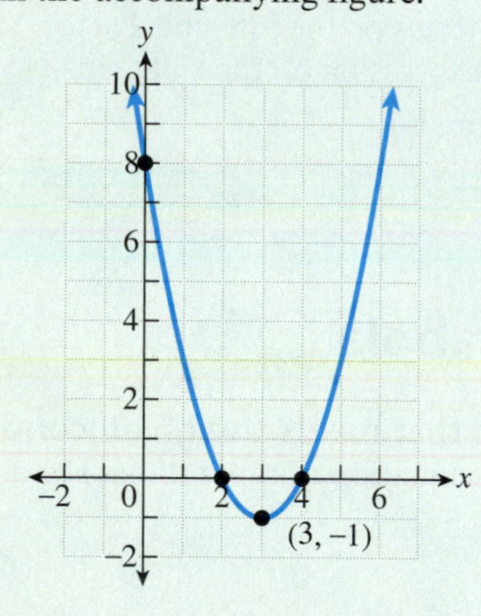

BUILDING YOUR STUDY STRATEGY

Preparing for a Final Exam (Part 1), 1 Schedule and Study Plan To prepare for a final exam effectively, develop a schedule and study plan. Ideally, you should begin studying approximately two weeks before the exam. Your schedule should include study time every day, and you should increase the time spent studying as to the exam date gets closer.

For further information on preparing for a cumulative exam, see the Study Strategies in Chapter 7.

Exercises 13.1

PRACTICE

WATCH

DOWNLOAD

READ

REVIEW

Vocabulary

1. A parabola of the form $y = ax^2 + bx + c$ or $y = a(x - h)^2 + k$ opens ________ if a is positive and opens ________ if a is negative.
2. A parabola of the form $y = a(x - h)^2 + k$ has its vertex at the point ______.
3. A parabola of the form $x = ay^2 + by + c$ or $x = a(y - k)^2 + h$ opens to the _____ if a is positive and opens to the _____ if a is negative.
4. The _____ coordinate of the vertex of a parabola of the form $x = ay^2 + by + c$ is $\frac{-b}{2a}$.
5. The vertex of a parabola of the form $x = a(y - k)^2 + h$ is at the point ______.
6. For a parabola that opens to the right or to the left and has a vertex at the point (h, k), the axis of symmetry is a horizontal line whose equation is ______.

Graph the parabola. Label the vertex and any intercepts.

7. $y = x^2 + 8x + 8$

8. $y = x^2 + 6x + 9$

9. $y = -x^2 + 2x + 15$

10. $y = -x^2 + 4x - 7$

11. $y = (x + 5)^2 + 5$

12. $y = (x - 1)^2 - 15$

13. $y = -(x + 2)^2 + 9$

14. $y = -(x - 3)^2 - 1$

15. $x = y^2 + 4y - 5$

16. $x = y^2 + 6y + 2$

17. $x = y^2 - 4y + 6$

18. $x = y^2 - 8y + 16$

19. $x = -y^2 + 6y - 8$

20. $x = -y^2 - 5y - 4$

21. $x = -y^2 + 5y + 1$

22. $x = -y^2 + 2y - 6$

23. $x = (y + 3)^2 - 1$

24. $x = (y - 2)^2 - 5$

25. $x = (y + 2)^2 + 2$

26. $x = (y - 5)^2 - 4$

27. $x = -(y - 2)^2 + 4$

28. $x = -(y - 3)^2 + 7$

Mixed Practice 29–40

29. $x = -y^2 - 6y - 15$

30. $x = -y^2 + 4y - 7$

31. $x = (y - 1)^2 - 7$

32. $x = y^2 - 8y - 20$

33. $y = -x^2 - 4x - 7$

34. $x = -(y - 1)^2 + 9$

35. $y = -(x - 1)^2 + 9$

36. $y = x^2 + 5x - 6$

37. $x = (y + 2)^2 - 4$

38. $y = (x - 4)^2 - 9$

39. $x = -(y + 1)^2 - 3$

40. $x = y^2 - 5y + 7$

Find the vertex of the parabola by completing the square.

41. $y = x^2 - 2x + 15$

42. $y = x^2 + 10x + 62$

43. $y = -x^2 + 14x + 76$

44. $y = -x^2 - 8x + 21$

45. $x = y^2 + 6y - 40$

46. $x = y^2 - 12y - 33$

47. $x = -y^2 + 4y - 51$

48. $x = -y^2 - 16y + 109$

Find the equation of the parabola that has been graphed, using the intercepts.

49.

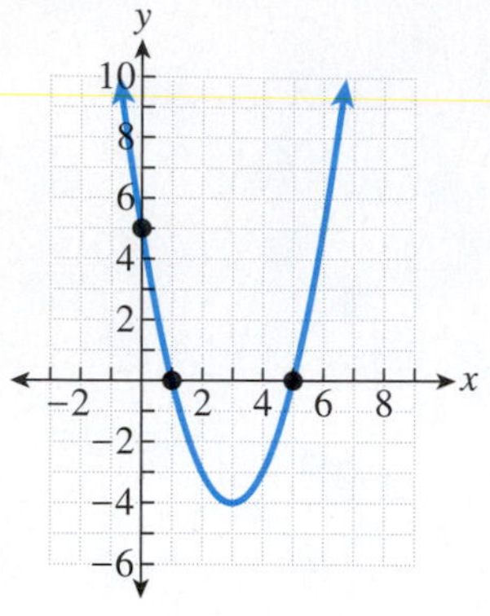

50.

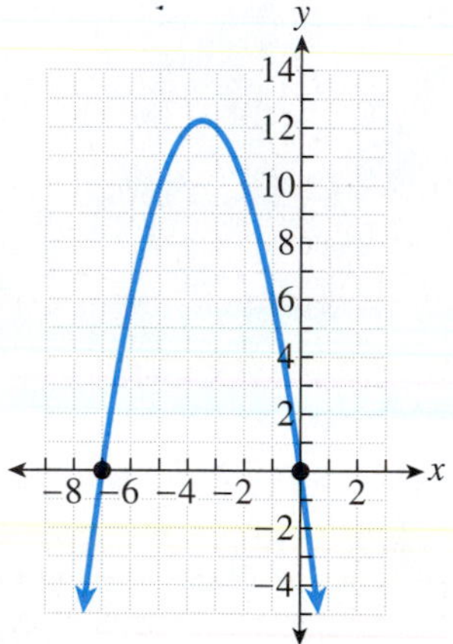

51.

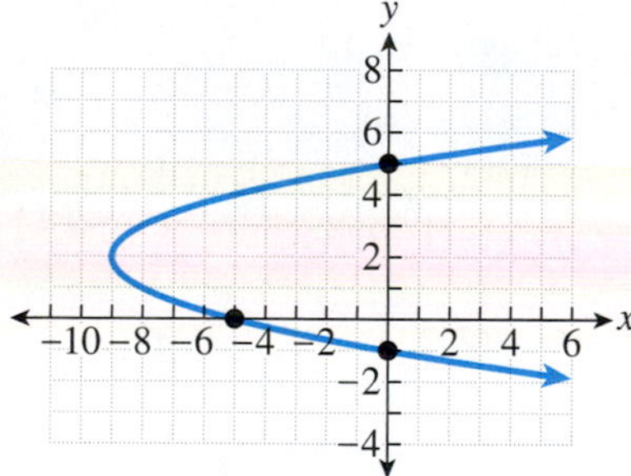

52.

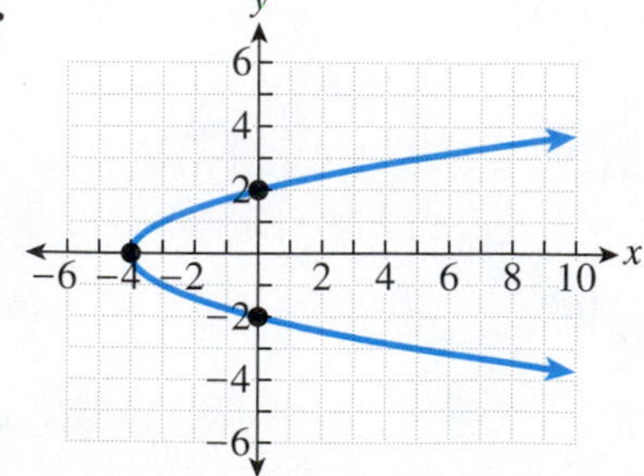

53. A footbridge in the Sequoia National Park is in the shape of a parabola. It spans a distance of 100 feet, and at its lowest point, it drops 5 feet in elevation.

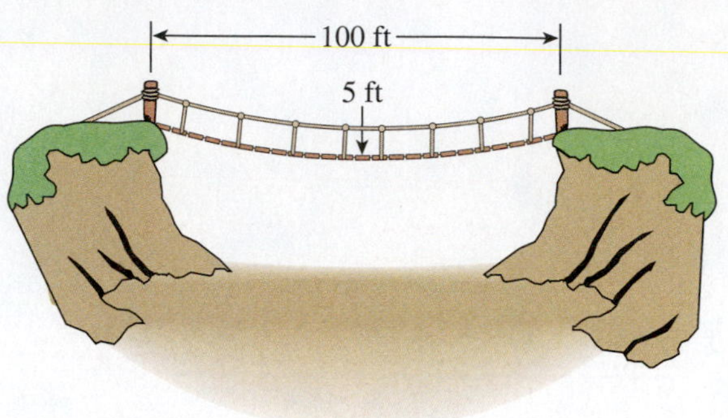

The bridge is displayed as follows on a rectangular coordinate plane:

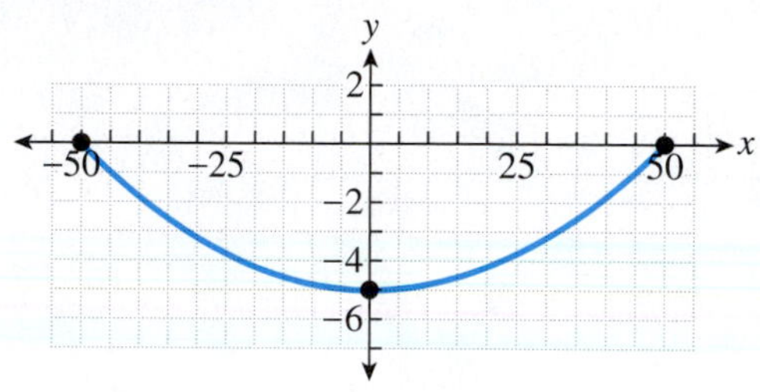

a) Find the equation for the parabola.

b) After a person walks 30 feet from one side of the bridge, how far has the person dropped in elevation?

54. A freeway overpass is built atop a parabolic arch. The width of the arch at the base is 120 feet, and its height at its peak is 30 feet.

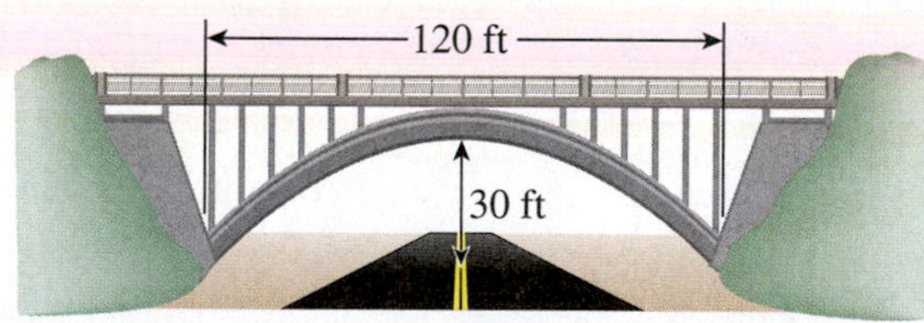

The overpass is displayed as follows on a rectangular coordinate plane:

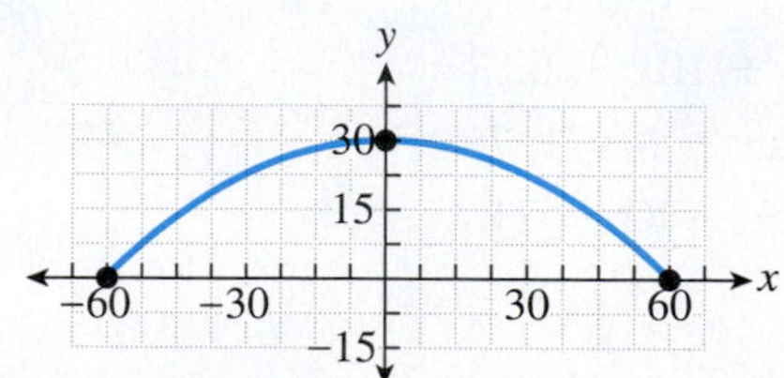

a) Find the equation for the parabola.

b) How far from the center of the arch does the height drop to below 16 feet? Round to the nearest tenth of a foot.

For 55–62, determine which equation is associated with each given graph.

a)

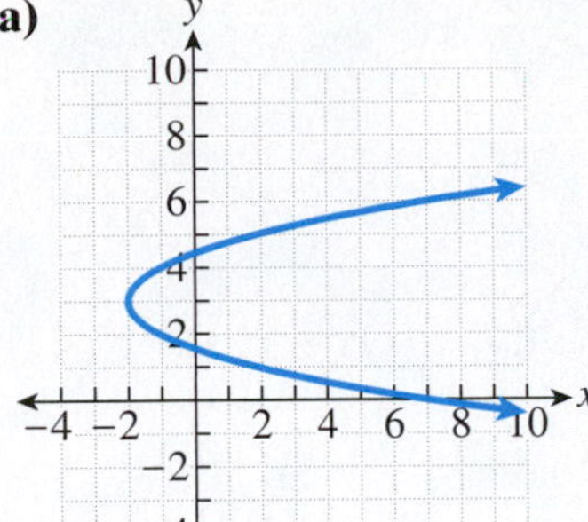

b)

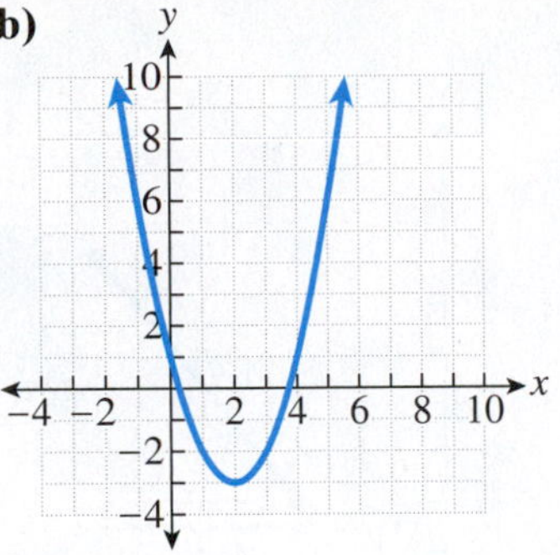

c)

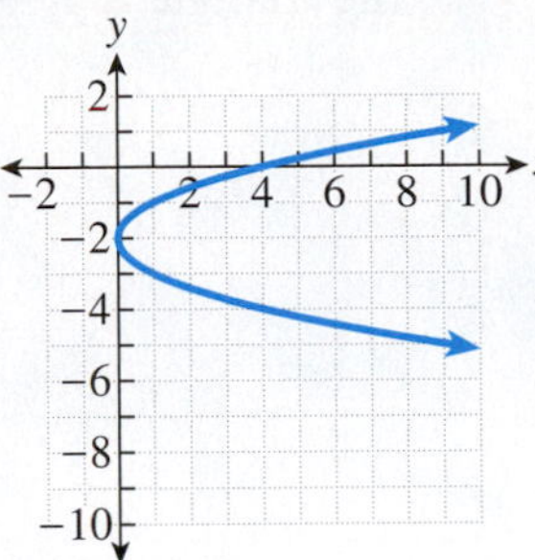

d)

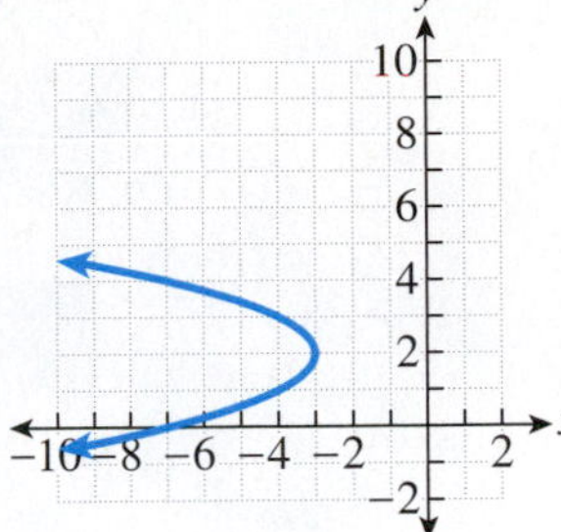

e)

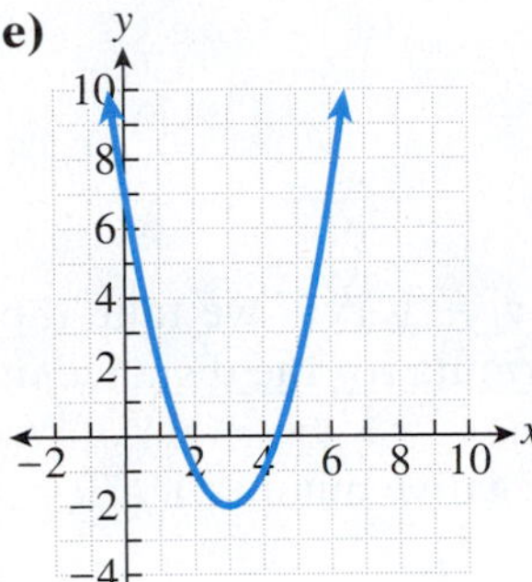

f)

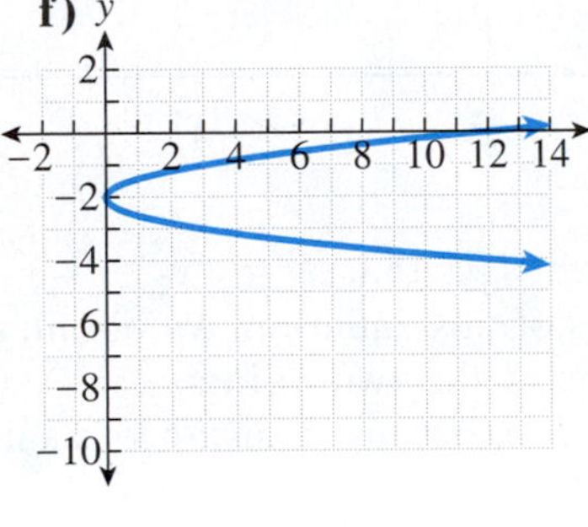

g)

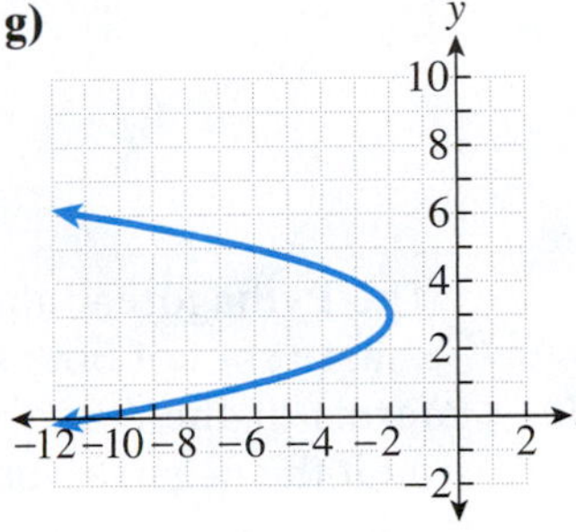

h)

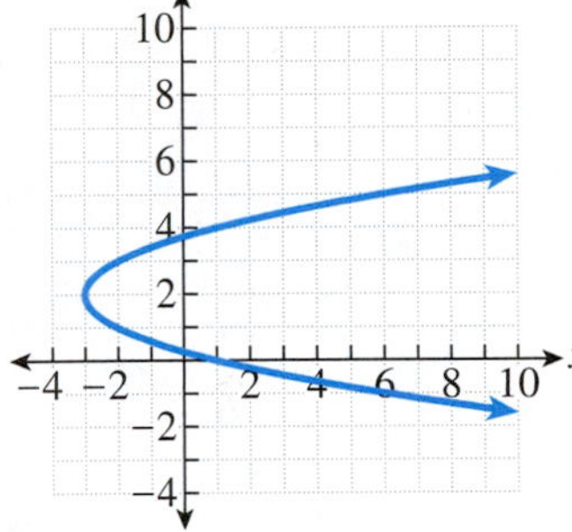

55. $y = (x - 2)^2 - 3$

56. $y = x^2 - 6x + 7$

57. $x = (y - 2)^2 - 3$

58. $x = y^2 - 6y + 7$

59. $x = -y^2 + 6y - 11$

60. $x = -(y - 2)^2 - 3$

61. $x = (y + 2)^2$

62. $x = 3(y + 2)^2$

Writing in Mathematics

Answer in complete sentences.

63. Explain how to determine whether a parabola opens to the left or to the right, as opposed to opening up or down.

64. Suppose a parabola opens to the right. Explain how you would determine that the parabola has no y-intercepts.

13.2 Circles

OBJECTIVES

1. Use the distance and midpoint formulas.
2. Graph circles centered at the origin.
3. Graph circles centered at a point (h, k).
4. Find the center of a circle and its radius by completing the square.
5. Find the equation of a circle that meets the given conditions.

The Distance Formula

Objective 1 Use the distance and midpoint formulas. Suppose we are given two points, (x_1, y_1) and (x_2, y_2), and are asked to find the distance between these two points. If we plot the points on a graph, we see that we can construct a right triangle whose hypotenuse has these points as its endpoints. The distance between the two points, d, is the length of the hypotenuse. The length of the leg on the bottom of the triangle is $x_2 - x_1$, and the length of the leg on the right side of the triangle is $y_2 - y_1$.

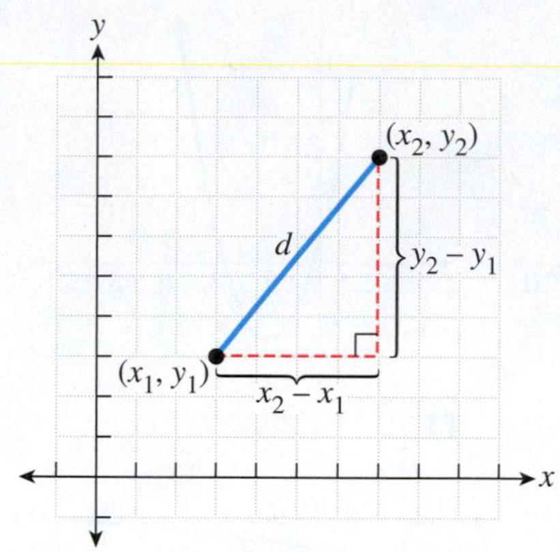

The Pythagorean theorem says that $d^2 = (x_2 - x_1)^2 + (y_2 - y_1)^2$. If we take the square root of both sides of this equation, we obtain a formula for the distance in terms of the coordinates of the two points: $d = \sqrt{(x_2 - x_1)^2 + (y_2 - y_1)^2}$. (We omit the negative square root because a distance is a nonnegative number.)

The Distance Formula

The distance, d, between two points (x_1, y_1) and (x_2, y_2) is given by the formula

$$d = \sqrt{(x_2 - x_1)^2 + (y_2 - y_1)^2}.$$

EXAMPLE 1 Find the distance between the points (3, 9) and (6, 5).

SOLUTION Let (3, 9) be the point (x_1, y_1) and (6, 5) be the point (x_2, y_2). Now use the distance formula.

$d = \sqrt{(x_2 - x_1)^2 + (y_2 - y_1)^2}$

$d = \sqrt{(6 - 3)^2 + (5 - 9)^2}$ Substitute for x_2, x_1, y_2, and y_1.

$d = \sqrt{(3)^2 + (-4)^2}$ Simplify the expressions in the radicand.

$d = \sqrt{25}$ Simplify the radicand.

$d = 5$ Simplify the radical.

The distance between the two points is five units.

The choice of which point will be (x_1, y_1) and which point will be (x_2, y_2) is completely arbitrary. Verify that if we let (6, 5) be the point (x_1, y_1) and (3, 9) be the point (x_2, y_2), d still would have been equal to 5.

EXAMPLE 2 Find the distance between the points (2, −6) and (−5, −11). Round to the nearest tenth.

SOLUTION We will let (2, −6) be the point (x_1, y_1) and (−5, −11) be the point (x_2, y_2).

$d = \sqrt{(x_2 - x_1)^2 + (y_2 - y_1)^2}$

$d = \sqrt{(-5 - 2)^2 + (-11 - (-6))^2}$ Substitute for x_2, x_1, y_2, and y_1.

$d = \sqrt{74}$ Simplify the radicand.

$d \approx 8.6$ Approximate using a calculator.

The distance between the two points is approximately 8.6 units.

Quick Check 1

Find the distance between the given points. Round to the nearest tenth if necessary.

a) (1, 2) and (9, 8)
b) (3, −6) and (−1, 6)

The Midpoint Formula

Suppose we are given two points, (x_1, y_1) and (x_2, y_2), and are asked to find their **midpoint**, which is the point on the line segment that connects the two points and is located exactly halfway between the two points. The x-coordinate of the midpoint must be equal to the average of the x-coordinates of the two points, and the y-coordinate must be equal to the average of the y-coordinates of the two points.

The Midpoint Formula

The midpoint of a line segment connecting two points (x_1, y_1) and (x_2, y_2) is the point whose coordinates are $\left(\frac{x_1 + x_2}{2}, \frac{y_1 + y_2}{2}\right)$.

EXAMPLE 3 Find the midpoint of the line segment that connects the points $(2, -9)$ and $(6, 3)$.

SOLUTION Use the midpoint formula, treating $(2, -9)$ as (x_1, y_1) and $(6, 3)$ as (x_2, y_2). This choice is arbitrary.

***x*-coordinate**	***y*-coordinate**
$\frac{x_1 + x_2}{2} = \frac{2 + 6}{2}$	$\frac{y_1 + y_2}{2} = \frac{-9 + 3}{2}$
$= 4$	$= -3$

The midpoint is $(4, -3)$.

Quick Check 2

Find the midpoint of the line segment that connects the points $(-4, -3)$ and $(5, 7)$.

Circles Centered at the Origin

Objective 2 Graph circles centered at the origin.

A **circle** is defined as the collection of all points (x, y) in a plane that are a fixed distance from a point called its **center**. The distance from the center to each point on the circle is called the **radius** of the circle.

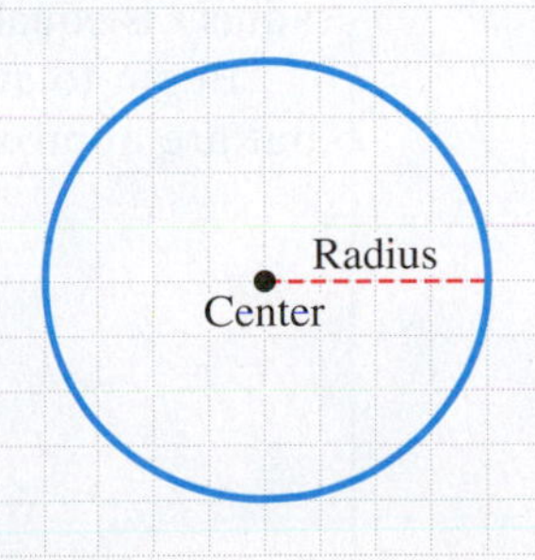

We will begin by learning to graph circles centered at the origin.

EXAMPLE 4 Graph the circle centered at the origin whose radius is 5.

SOLUTION Begin by plotting the center at the origin (0, 0). From there, move five units to the right of the origin and plot a point. Repeat this process for points that are five units to the left of the center and five units above and below the center.

Finish by drawing the circle that passes through these points.

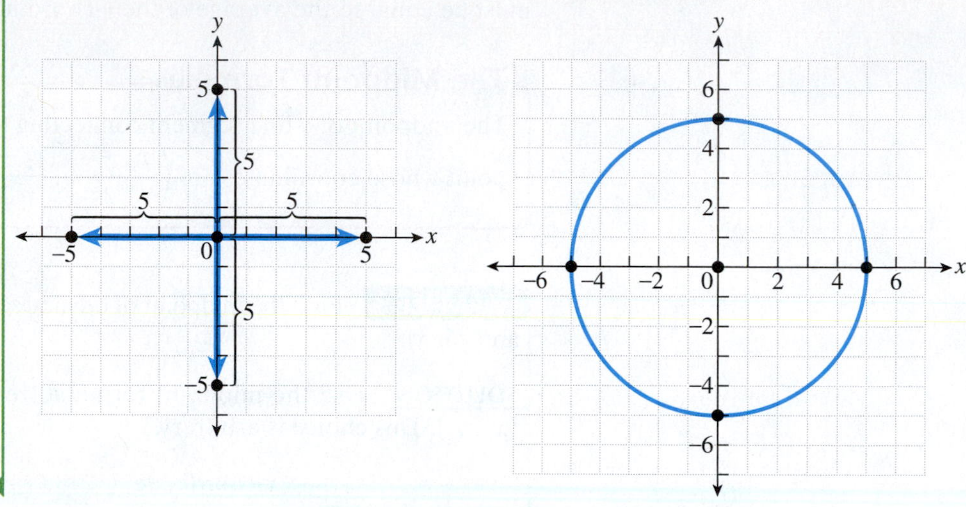

The circle in the previous example is the collection of all points (x, y) that are a distance of five units from the origin (0, 0). We can use this fact and the distance formula to find the equation for this circle.

$$d = \sqrt{(x_2 - x_1)^2 + (y_2 - y_1)^2}$$

$$5 = \sqrt{(x - 0)^2 + (y - 0)^2}$$ Substitute 5 for d, x for x_2, 0 for x_1, y for y_2, and 0 for y_1.

$$5 = \sqrt{x^2 + y^2}$$ Simplify.

$$25 = x^2 + y^2$$ Square both sides.

The equation of the circle centered at the origin with a radius of 5 can be written as $x^2 + y^2 = 25$.

Equation of a Circle Centered at the Origin (Standard Form)

The equation for a circle with radius r centered at the origin is $x^2 + y^2 = r^2$.

EXAMPLE 5 Graph the circle $x^2 + y^2 = 10$. State the center and radius.

SOLUTION This circle also has its center at the origin, and because $r^2 = 10$, the radius r is equal to $\sqrt{10}$. Note that $\sqrt{10} \approx 3.2$.

Begin to graph the circle by plotting the center (0, 0). Next, plot the points that are approximately 3.2 units to the right and left of the center as well as the

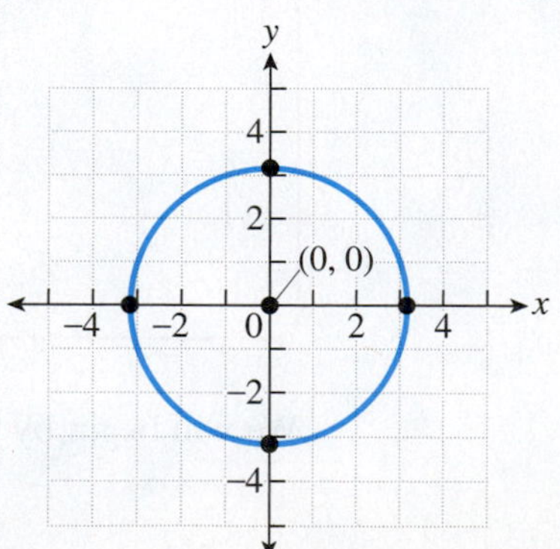

points that are approximately 3.2 units above and below the center. Once these points have been plotted, draw the circle that passes through them.

Quick Check 3

Graph the circle $x^2 + y^2 = 12$. State the center and radius.

Circles Centered at a Point Other Than the Origin

Objective 3 Graph circles centered at a point (h, k). The equation for a circle centered at a point (h, k) can be derived by the distance formula, just as the equation for a circle centered at the origin was derived.

$$\sqrt{(x-h)^2 + (y-k)^2} = r \quad \text{The distance from } (x, y) \text{ to the center } (h, k) \text{ is } r.$$
$$(x-h)^2 + (y-k)^2 = r^2 \quad \text{Square both sides of the equation.}$$

Standard Form of the Equation of a Circle

The equation for a circle with radius r centered at the point (h, k) is

$$(x-h)^2 + (y-k)^2 = r^2.$$

EXAMPLE 6 Graph the circle $(x-3)^2 + (y-4)^2 = 4$. State the center and radius.

SOLUTION Because the equation of the circle is of the form $(x-h)^2 + (y-k)^2 = r^2$, the center of this circle is the point $(3, 4)$.

Because $r^2 = 4$, the radius r is equal to $\sqrt{4}$, or 2.

Begin by plotting the center $(3, 4)$. Next, plot the points that are two units to the right and left of the center as well as the points that are two units above and below the center. Once these points have been plotted, draw the circle that passes through them.

Quick Check 4

Graph the circle $(x-6)^2 + (y-5)^2 = 9$. State the center and radius.

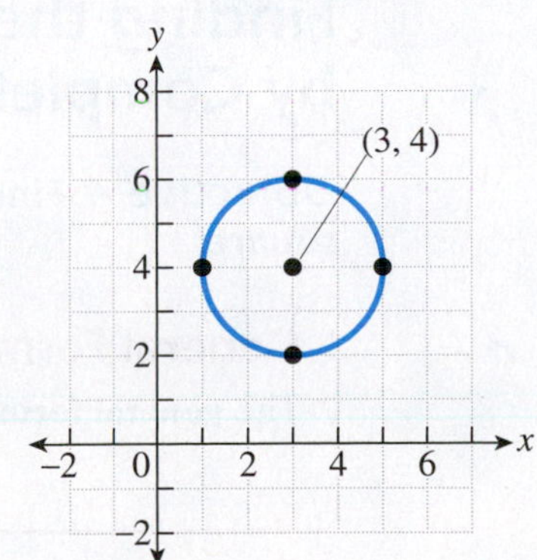

EXAMPLE 7 Graph the circle $(x + 3)^2 + (y - 6)^2 = 25$. State the center and radius.

SOLUTION The center of this circle is the point $(-3, 6)$. To find the x-coordinate of the center, we can rewrite $x + 3$ as $x - (-3)$. This shows that h is -3. Alternatively, we could have set the expression containing x that was being squared, $x + 3$, equal to 0 and solved for x. A similar approach shows that the y-coordinate of the center is $y = 6$.

Because $r^2 = 25$, the radius r is equal to $\sqrt{25}$, or 5.

We begin to graph the circle by plotting the center $(-3, 6)$. Next, we plot the points that are five units to the right and left of the center as well as the points that are five units above and below the center. Once these points have been plotted, we draw the circle that passes through them.

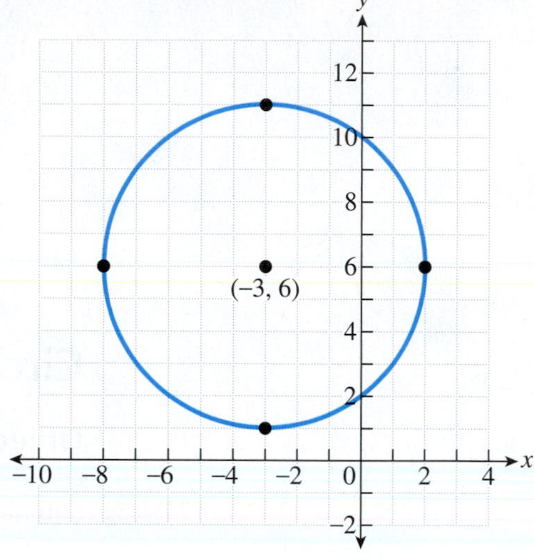

Quick Check 5

Graph the circle. State the center and radius.

a) $(x - 1)^2 + (y + 6)^2 = 16$ **b)** $(x + 2)^2 + (y - 3)^2 = 64$

Finding the Center and Radius by Completing the Square

Objective 4 **Find the center of a circle and its radius by completing the square.**

General Form of the Equation of a Circle

The **general form** of the equation of a circle is

$$Ax^2 + Ay^2 + Bx + Cy + D = 0, A \neq 0.$$

To graph a circle in this form, we must rewrite the equation in the standard form, which will allow us to determine the center and radius of the circle. This is done by completing the square, which must be done for both x and y.

EXAMPLE 8 Graph the circle $x^2 + y^2 - 8x + 10y + 25 = 0$. State the center and radius.

SOLUTION Rewrite the equation in standard form by completing the square for x and y.

$$x^2 + y^2 - 8x + 10y + 25 = 0$$

$$(x^2 - 8x) + (y^2 + 10y) = -25$$

Collect terms containing x. Collect terms containing y. Subtract 25 from both sides.

$$(x^2 - 8x + 16) + (y^2 + 10y + 25) = -25 + 16 + 25$$

Add 16 to both sides to complete the square for x. Add 25 to both sides to complete the square for y.

$$(x - 4)^2 + (y + 5)^2 = 16$$

Factor the two quadratic expressions on the left side of the equation.

Because the equation of this circle has been written in standard form, the center of the circle is the point $(4, -5)$. Because $r^2 = 16$, the radius r is equal to $\sqrt{16}$, or 4. Begin to graph the circle by plotting the center $(4, -5)$. Next, plot the points that are four units to the right and left of the center as well as the points that are four units above and below the center. Finish by drawing the circle that passes through these four points.

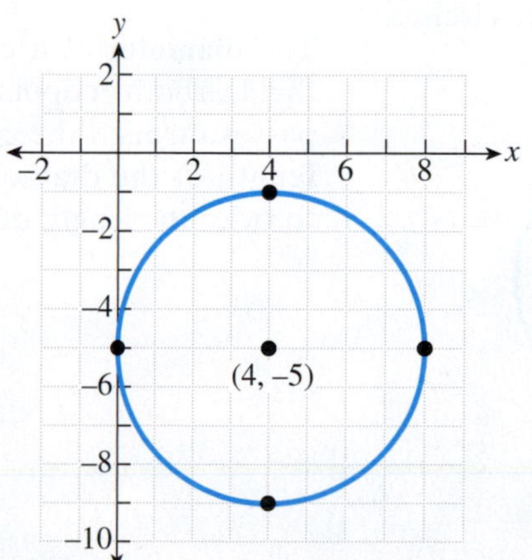

Quick Check 6

Graph the circle $x^2 + y^2 + 14x - 6y + 57 = 0$. State the center and radius.

Finding the Equation of a Circle Whose Center and Radius Are Known

Objective 5 Find the equation of a circle that meets the given conditions.

EXAMPLE 9 Find the equation of the circle.

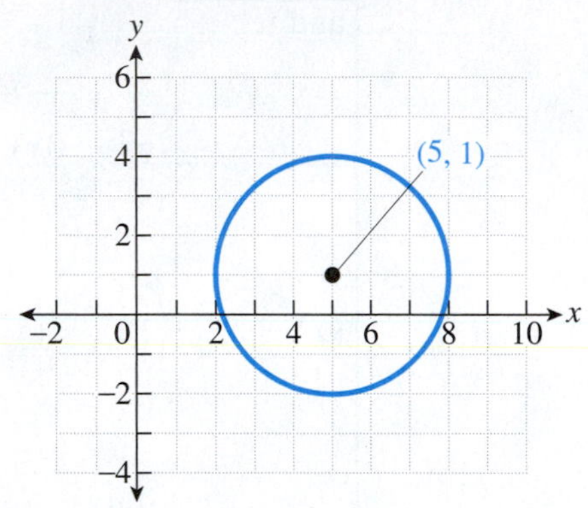

SOLUTION Begin by finding the center of the circle, which is the point (5, 1). Next, measure the distance from the center of the circle to a point on the circle. This distance is the radius. In this case, the radius is 3.

Now write the equation in standard form.

$$(x - h)^2 + (y - k)^2 = r^2$$

$$(x - 5)^2 + (y - 1)^2 = 3^2 \quad \text{Substitute 5 for } h, 1 \text{ for } k, \text{ and 3 for } r.$$

$$(x - 5)^2 + (y - 1)^2 = 9 \quad \text{Simplify.}$$

The equation of this circle is $(x - 5)^2 + (y - 1)^2 = 9$.

Quick Check 7

Find the equation of the circle.

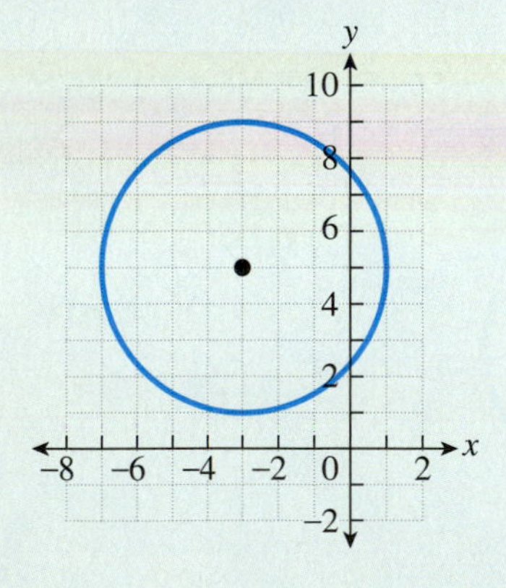

The **diameter** of a circle is a line segment that has both endpoints on the circle and that passes through the center of the circle. The length of the diameter of a circle is equal to twice the length of the radius of the circle.

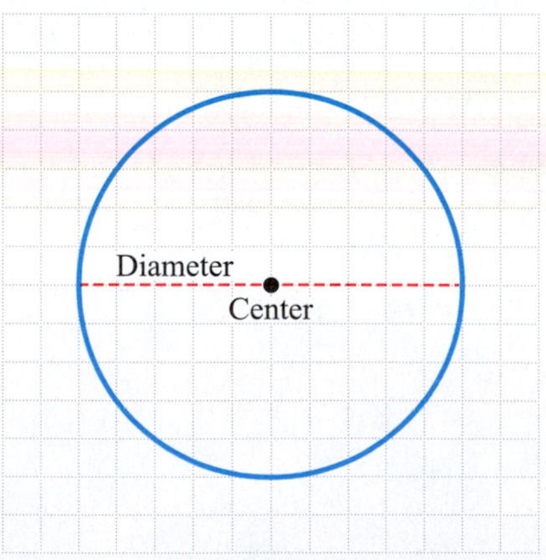

EXAMPLE 10 Find the equation in standard form of the circle that has a diameter with endpoints (−1, −5) and (7, 1).

SOLUTION Begin by finding the center of the circle, which is the midpoint of the given endpoints of a diameter.

$$x = \frac{x_1 + x_2}{2} \qquad y = \frac{y_1 + y_2}{2}$$

$$x = \frac{-1 + 7}{2} \qquad y = \frac{-5 + 1}{2}$$

$$= 3 \qquad = -2$$

The center of the circle is $(3, -2)$. Next, find the radius by calculating the distance from the center of the circle to a point on the circle. Use the point $(7, 1)$.

$$d = \sqrt{(x_2 - x_1)^2 + (y_2 - y_1)^2}$$
$$d = \sqrt{(7 - 3)^2 + (1 - (-2))^2} \quad \text{Substitute for } x_2, x_1, y_2, \text{ and } y_1.$$
$$d = \sqrt{25} \quad \text{Simplify the radicand.}$$
$$d = 5 \quad \text{Simplify.}$$

The radius is 5. Now write the equation in standard form.

$$(x - h)^2 + (y - k)^2 = r^2$$
$$(x - 3)^2 + (y - (-2))^2 = 5^2 \quad \text{Substitute 3 for } h, -2 \text{ for } k, \text{ and 5 for } r.$$
$$(x - 3)^2 + (y + 2)^2 = 25$$

The equation of this circle is $(x - 3)^2 + (y + 2)^2 = 25$.

Quick Check 8

Find the equation in standard form of the circle that has a diameter with endpoints $(4, -7)$ and $(-2, 3)$.

BUILDING YOUR STUDY STRATEGY

Preparing for a Final Exam (Part 1), 2 Availability of Resources The end of the semester is a hectic time for students. The same can be said for employees of your college. Talk to your instructor and find out when he or she will be available on campus to answer questions in the time leading up to the final exam. Also check with the tutorial center or math lab to find out if there will be any change in its hours of operation.

For more advice on making effective use of all of your resources, consult the Study Strategies in Chapter 3.

Exercises 13.2

PRACTICE

WATCH

DOWNLOAD

READ

REVIEW

Vocabulary

1. The distance, d, between two points (x_1, y_1) and (x_2, y_2) is given by the formula ________________.
2. The ________ of a line segment connecting two points (x_1, y_1) and (x_2, y_2) is the point whose coordinates are $\left(\frac{x_1 + x_2}{2}, \frac{y_1 + y_2}{2}\right)$.
3. The collection of all points in a plane that are a fixed distance from a specified point is called a(n) ______.
4. The point that is equidistant from each point on a circle is the ______ of the circle.
5. The distance from the center to each point on the circle is called the ______ of the circle.
6. The ________ of a circle is a line segment that has both endpoints on the circle and that passes through the center of the circle.
7. The equation for a circle with radius r centered at the origin is __________.
8. The equation for a circle with radius r centered at the point (h, k) is ________________.

Find the distance between the given points. Round to the nearest tenth if necessary.

9. $(7, 6)$ and $(4, 2)$
10. $(16, 1)$ and $(4, 10)$
11. $(15, -53)$ and $(54, 27)$
12. $(-32, -15)$ and $(-77, 13)$
13. $(-4, -6)$ and $(-4, 5)$
14. $(7, 3)$ and $(-9, 3)$
15. $(2, -3)$ and $(-5, 4)$
16. $(-8, -6)$ and $(-4, 7)$
17. $(-13, -9)$ and $(-18, 1)$
18. $(16, 11)$ and $(-14, -19)$

Find the midpoint of the line segment that connects the given points.

19. $(0, 0)$ and $(6, 4)$
20. $(3, 15)$ and $(7, 1)$
21. $(6, -9)$ and $(8, -25)$
22. $(-5, 10)$ and $(-12, 32)$

Graph the circle. State the center and radius of the circle.

23. $x^2 + y^2 = 16$

24. $x^2 + y^2 = 9$

25. $x^2 + y^2 = 100$

26. $x^2 + y^2 = 144$

27. $x^2 + y^2 = 6$

28. $x^2 + y^2 = 18$

29. $(x - 2)^2 + (y - 7)^2 = 16$

30. $(x - 4)^2 + (y - 2)^2 = 9$

31. $(x + 5)^2 + (y - 1)^2 = 25$

32. $(x - 6)^2 + (y + 5)^2 = 4$

33. $(x + 2)^2 + (y - 1)^2 = 49$

34. $(x + 6)^2 + (y + 4)^2 = 25$

35. $(x + 5)^2 + y^2 = 25$

36. $x^2 + (y - 9)^2 = 36$

37. $(x - 2)^2 + (y + 6)^2 = 18$

38. $(x + 8)^2 + (y + 3)^2 = 8$

Find the center and radius of each circle by completing the square.

39. $x^2 + y^2 + 4x + 12y - 41 = 0$

40. $x^2 + y^2 + 6x + 18y + 65 = 0$

41. $x^2 + y^2 - 10x + 2y - 23 = 0$

42. $x^2 + y^2 - 8x - 40y + 127 = 0$

43. $x^2 + y^2 + 16x - 57 = 0$

44. $x^2 + y^2 - 10y - 11 = 0$

45. $x^2 + y^2 + 4x + 14y - 27 = 0$

46. $x^2 + y^2 - 8x + 6y + 14 = 0$

Find the equation of each circle with the given center and radius.

47. Center $(0, 0)$, radius 6

48. Center $(0, 0)$, radius 9

49. Center $(4, 2)$, radius 7

50. Center $(1, 8)$, radius 4

51. Center $(-9, -10)$, radius 5

52. Center $(-6, 0)$, radius 2

Find the equation of each circle.

53.

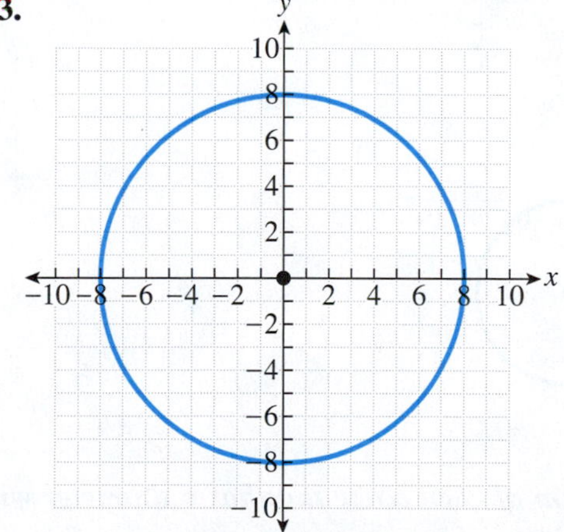

54.

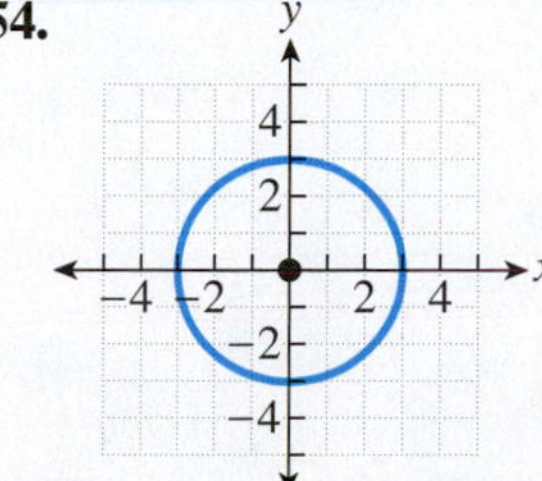

55.

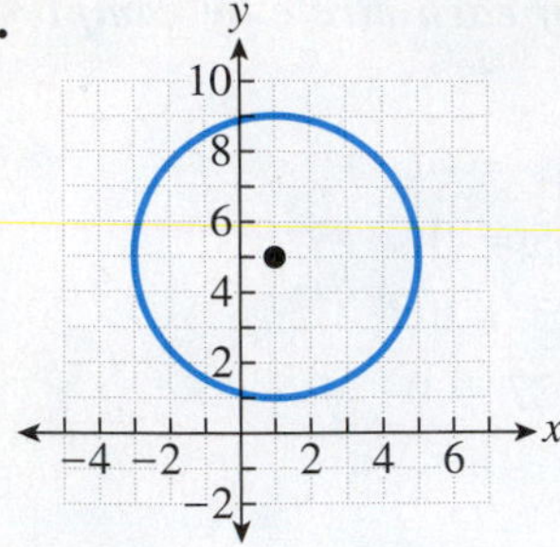

56.

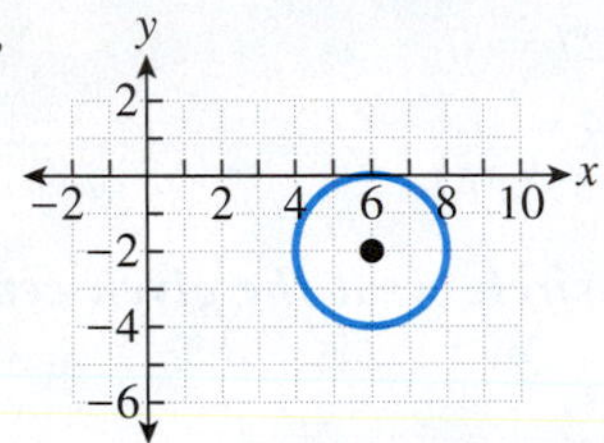

57.

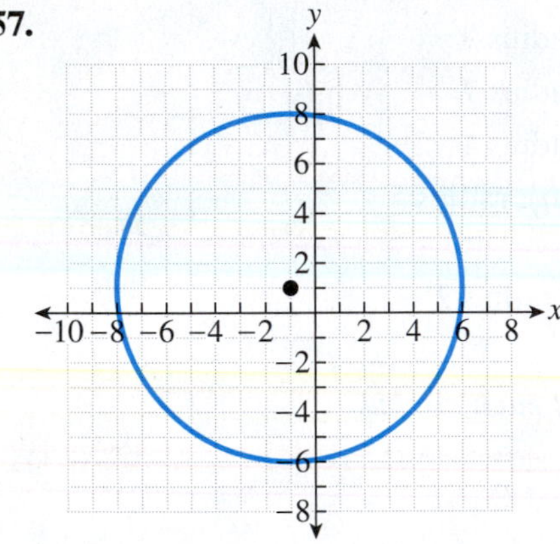

58.

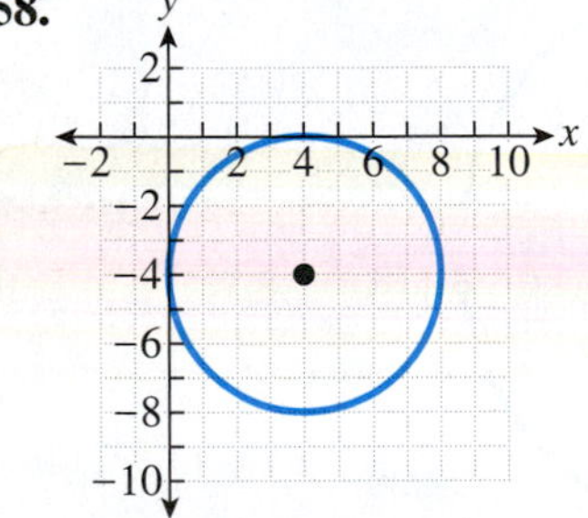

Find the equation of each circle that has a diameter with the given endpoints.

59. (1, 7) and (9, 7)

60. (−6, 3) and (4, 3)

61. (−2, 10) and (6, 4)

62. (8, −17) and (−8, 13)

63. (4, 1) and (12, −3)

64. (−3, 9) and (−15, −11)

Mixed Practice, 65–76

Graph. For graphs that are parabolas, label the vertex and any intercepts. For graphs that are circles, label the center.

65. $x^2 + (y + 6)^2 = 16$

66. $(x + 4)^2 + (y - 6)^2 = 25$

67. $x^2 + y^2 + 18x - 6y - 10 = 0$

68. $y = -(x - 5)^2 + 9$

69. $y = x^2 + 6x - 7$

70. $(x + 1)^2 + (y + 5)^2 = 25$

71. $x = -(y + 1)^2 - 5$

72. $x^2 + y^2 - 6x + 8y - 39 = 0$

73. $(x + 8)^2 + y^2 = 1$

74. $x = y^2 + 6y + 5$

75. $x^2 + (y - 7)^2 = 25$

76. $x^2 + y^2 = 4$

77. Bart built a rectangular barbecue area that measures 20 feet by 50 feet. He plans to draw a circle around the barbecue area, as shown, and fill in this area with grass.

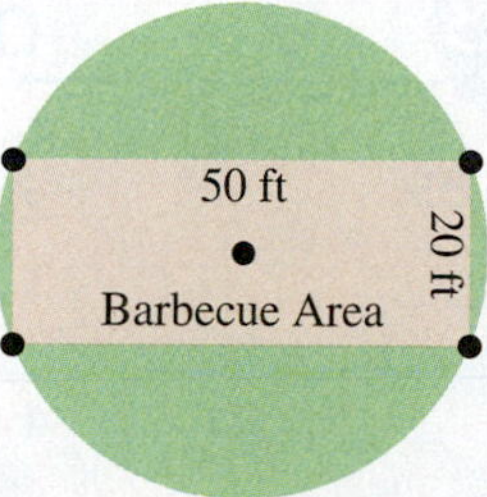

a) Find the equation of the circle assuming that the point in the center of the barbecue area is the origin.

b) How many square feet of sod need to be ordered to cover the grass area? Round to the nearest tenth.

78. The London Eye is a gigantic wheel built to celebrate the millennium. The diameter of the wheel is 122 meters, and its height is 131 meters.

The London Eye can be represented on a rectangular coordinate plane, plotted as shown.

a) Find an equation for the circle.

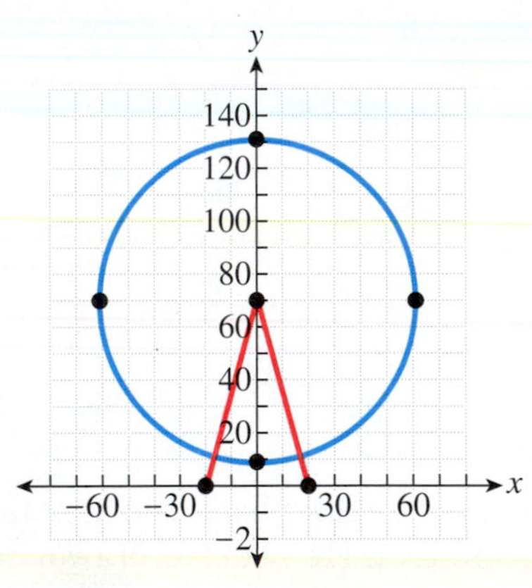

b) How far to the left or right of center is a car that is 100 meters above the ground? Round to the nearest tenth.

Writing in Mathematics

Answer in complete sentences.

79. Explain how to determine whether the graph of an equation will be a parabola that opens up, a parabola that opens down, a parabola that opens to the right, a parabola that opens to the left, or a circle. Give an example of each type of equation.

80. ***Solutions Manual*** Write a solutions manual page for the following problem.
Graph $x^2 + y^2 + 6x - 8y - 11 = 0$. *State the center and radius of the circle.*

13.3 Ellipses

OBJECTIVES

1. **Graph ellipses centered at the origin.**
2. **Graph ellipses centered at a point (h, k).**
3. **Find the center of an ellipse and the lengths of its axes by completing the square.**
4. **Find the equation of an ellipse that meets the given conditions.**

In this section, we will investigate a conic section called an ellipse. An **ellipse** is the collection of all points (x, y) in the plane for which the sum of the distances d_1 and d_2 between the point (x, y) and two fixed points, F_1 and F_2, called foci, is a positive constant. Each fixed point is called a **focus**. The graph is similar to a circle, but is more oval in shape. For any point (x, y) on the ellipse, the sum of the distances d_1 and d_2 remains the same.

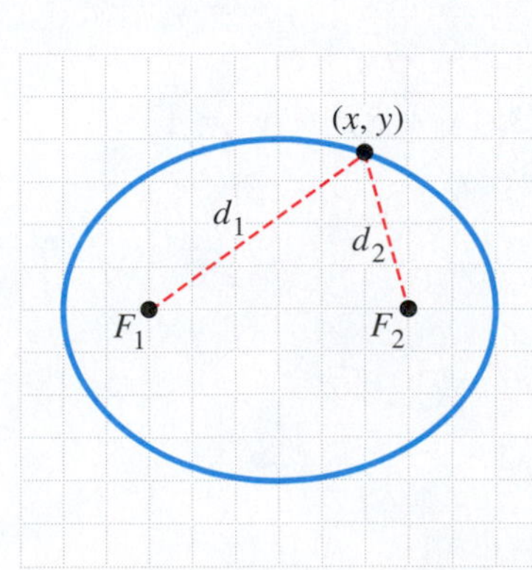

In this text, the foci will be on a horizontal line or on a vertical line. Like a circle, an ellipse has a **center** and it is the point located midway between the two foci.

A horizontal line passing through the center of an ellipse intersects the ellipse at two points, as does a vertical line passing through the center. The horizontal and vertical line segments connecting these points are called the **axes** of the ellipse. The longer line segment is called the **major axis**, and its endpoints are the **vertices** of the ellipse. The other line segment is called the **minor axis**, and its endpoints are the **co-vertices** of the ellipse.

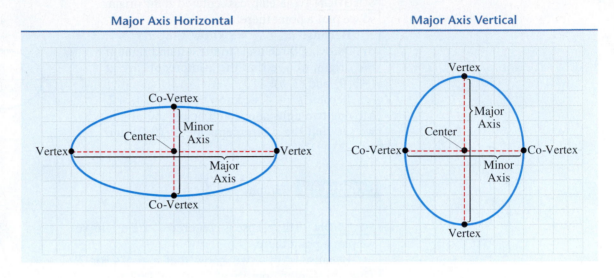

Major Axis Horizontal	Major Axis Vertical
Co-Vertex, Minor Axis, Center, Vertex, Vertex, Major Axis, Co-Vertex	Vertex, Major Axis, Center, Co-Vertex, Co-Vertex, Minor Axis, Vertex

Ellipses Centered at the Origin

Objective 1 Graph ellipses centered at the origin.

Standard Form of the Equation of an Ellipse Centered at the Origin

The equation of an ellipse that is centered at the origin is of the form $\frac{x^2}{a^2} + \frac{y^2}{b^2} = 1$.

This is the **standard form** of the equation of an ellipse.

The ellipse will have x-intercepts at $(a, 0)$ and $(-a, 0)$ and y-intercepts at $(0, b)$ and $(0, -b)$. If $a > b$, the major axis is horizontal and the minor axis is vertical. If $b > a$, the major axis is vertical and the minor axis is horizontal.

To graph an ellipse centered at the origin, we begin by plotting a point at the origin. We then plot points that are a units to the left and right of the origin as well as points that are b units above and below the origin. We finish by drawing the ellipse that passes through these four points.

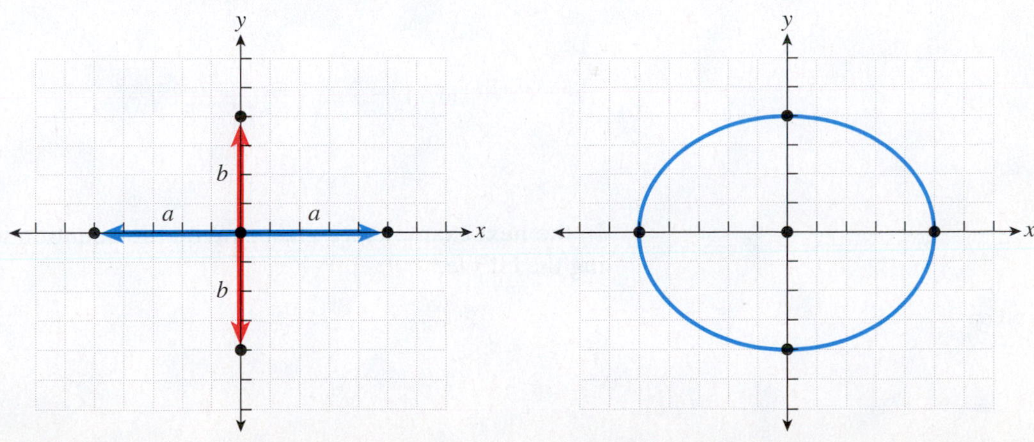

A WORD OF CAUTION When drawing the ellipse that passes through the four points, make sure your graph looks like an oval, not a diamond.

EXAMPLE 1 Graph the ellipse $\frac{x^2}{9} + \frac{y^2}{4} = 1$.

SOLUTION This ellipse is centered at the origin, so we plot a point there.

Because $a^2 = 9$, we know that $a = \sqrt{9}$, or 3. Moving three units to the left and right of the origin, we see that the x-intercepts are at $(-3, 0)$ and $(3, 0)$.

Because $b^2 = 4$, we know that $b = \sqrt{4}$, or 2. Moving two units above and below the origin, we see that the y-intercepts are at $(0, 2)$ and $(0, -2)$.

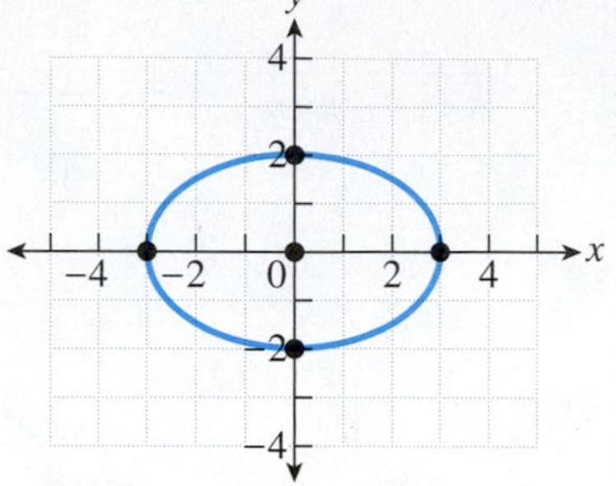

EXAMPLE 2 Graph $\frac{x^2}{4} + \frac{y^2}{36} = 1$.

SOLUTION This ellipse is centered at the origin, so we plot a point there.

Because $a^2 = 4$, we know that $a = \sqrt{4}$, or 2. Thus, the x-intercepts are at $(2, 0)$ and $(-2, 0)$.

Because $b^2 = 36$, we know that $b = \sqrt{36}$, or 6. It follows that the y-intercepts are at $(0, 6)$ and $(0, -6)$.

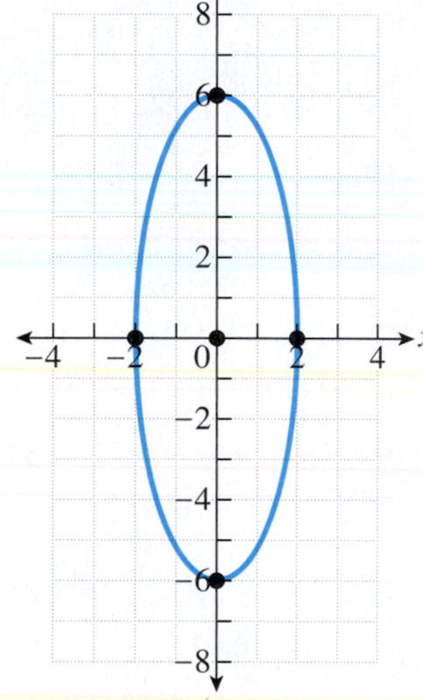

Quick Check 1

Graph.

a) $\frac{x^2}{25} + \frac{y^2}{9} = 1$ **b)** $\frac{x^2}{9} + \frac{y^2}{16} = 1$

In the next example, we must rewrite the equation in standard form before graphing the ellipse.

EXAMPLE 3 Graph $25x^2 + 16y^2 = 400$.

SOLUTION For the equation of an ellipse to be in standard form, the sum of the variable terms must be equal to 1. To rewrite this equation in standard form, we divide both sides of the equation by 400 and simplify.

$$25x^2 + 16y^2 = 400$$

$$\frac{25x^2 + 16y^2}{400} = \frac{400}{400} \quad \text{Divide both sides by 400.}$$

$$\frac{\overset{1}{\cancel{25}}x^2}{\underset{16}{\cancel{400}}} + \frac{\overset{1}{\cancel{16}}y^2}{\underset{25}{\cancel{400}}} = 1 \quad \text{Rewrite the left side of the equation as the sum of two fractions and simplify.}$$

$$\frac{x^2}{16} + \frac{y^2}{25} = 1 \quad \text{Simplify.}$$

The center of this ellipse is the origin. Because $a^2 = 16$, we know that $a = 4$. The x-intercepts are at $(4, 0)$ and $(-4, 0)$. Because $b^2 = 25$, we know that $b = 5$ and that the y-intercepts are at $(0, 5)$ and $(0, -5)$.

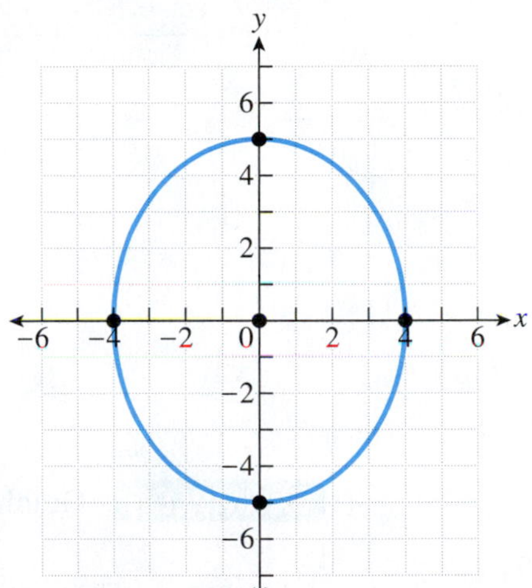

Quick Check 2

Graph $4x^2 + 36y^2 = 144$.

Ellipses Centered at a Point Other Than the Origin

Objective 2 Graph ellipses centered at a point (h, k).

Standard Form of the Equation of an Ellipse with Center (h, k)

The equation for an ellipse centered at the point (h, k) whose horizontal axis has length $2a$ and whose vertical axis has length $2b$ is $\frac{(x - h)^2}{a^2} + \frac{(y - k)^2}{b^2} = 1$.

To graph an ellipse centered at the point (h, k), we will begin by plotting a point at the center. The endpoints of the horizontal axis can be found by moving a units to the left and right of the center. The endpoints of the vertical axis can be found by moving b units above and below the center. Once these four endpoints have been plotted, we will draw the ellipse that passes through them.

EXAMPLE 4 Graph $\frac{(x-3)^2}{4} + \frac{(y-5)^2}{9} = 1$.

SOLUTION Because the equation has the form $\frac{(x-h)^2}{a^2} + \frac{(y-k)^2}{b^2} = 1$, the center of this ellipse is $(3, 5)$.

Because $a^2 = 4$, we know that $a = 2$. The endpoints of the horizontal axis are two units to the left and right of the center. Because $b^2 = 9$, we know that $b = 3$. The endpoints of the vertical axis are three units above and below the center. We finish by drawing the ellipse that passes through these four endpoints.

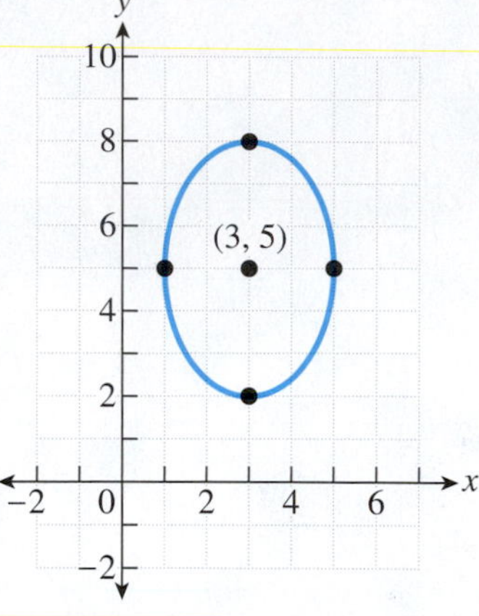

Quick Check 3

Graph $\frac{(x-8)^2}{49} + \frac{(y-7)^2}{25} = 1$.

EXAMPLE 5 Graph $(x+3)^2 + \frac{(y+2)^2}{10} = 1$.

SOLUTION The center of this ellipse is $(-3, -2)$.

When we do not see a denominator under the squared term containing x, $a^2 = 1$; so $a = 1$, and we plot the endpoints of the horizontal axis one unit to the left and right of the center. Because $b^2 = 10$, we know that $b = \sqrt{10}$. Because $\sqrt{10} \approx 3.2$, we plot the endpoints of the vertical axis approximately 3.2 units above and below the center. We finish by drawing the ellipse that passes through the four endpoints.

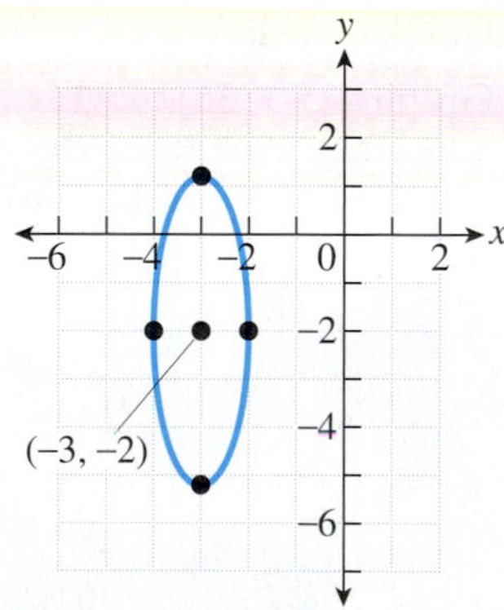

Quick Check 4

Graph $\frac{(x-4)^2}{20} + (y+5)^2 = 1$.

Finding the Center of an Ellipse and the Lengths of Its Axes by Completing the Square

Objective 3 **Find the center of an ellipse and the lengths of its axes by completing the square.**

General Form of the Equation of an Ellipse

The **general form** of the equation of an ellipse is $Ax^2 + By^2 + Cx + Dy + E = 0$, $A \neq 0$, $B \neq 0$, and $A \neq B$.

Notice that the coefficient of the x^2 term is not equal to the coefficient of the y^2 term. If those two coefficients are equal, the graph of the equation is more specifically defined as a circle than an ellipse. To graph an ellipse in general form, we must rewrite the equation in standard form. This allows us to determine the center, as well as a and b. This is done by completing the square, which must be done for both x and y.

EXAMPLE 6 Graph the ellipse $25x^2 + 4y^2 + 150x - 16y + 141 = 0$. Give the center.

SOLUTION We convert the equation to standard form by completing the square for x and y.

$$25x^2 + 4y^2 + 150x - 16y + 141 = 0$$

$$(25x^2 + 150x) + (4y^2 - 16y) = -141$$

Collect terms containing x. Collect terms containing y. Subtract 141 from both sides.

$$25(x^2 + 6x) + 4(y^2 - 4y) = -141$$

Factor 25 from the terms containing x so that the coefficient of the x^2 term is 1. Factor 4 from the terms containing y.

$$25(x^2 + 6x + 9) + 4(y^2 - 4y + 4) = -141 + 225 + 16$$

Add 9 inside the parentheses containing x to complete the square for x. Add $25 \cdot 9$, or 225, to the right side of the equation. Add 4 inside the parentheses containing y to complete the square for y. Add $4 \cdot 4$, or 16, to the right side of the equation.

$$25(x + 3)^2 + 4(y - 2)^2 = 100$$

Factor the two quadratic expressions on the left side of the equation.

$$\frac{\overset{1}{\cancel{25}}(x + 3)^2}{\underset{4}{\cancel{100}}} + \frac{\overset{1}{\cancel{4}}(y - 2)^2}{\underset{25}{\cancel{100}}} = 1$$

Divide both sides of the equation by 100. Rewrite the left side of the equation as the sum of two fractions. Divide out common factors.

$$\frac{(x + 3)^2}{4} + \frac{(y - 2)^2}{25} = 1$$

Simplify.

Because the equation is now in standard form, the center of this ellipse is $(-3, 2)$. Because $a^2 = 4$, $a = 2$ and we plot the endpoints of the horizontal axis two units to the left and right of the center. Because $b^2 = 25$, $b = 5$ and we plot the endpoints of the vertical axis five units above and below the center. Draw the ellipse that passes through the four endpoints.

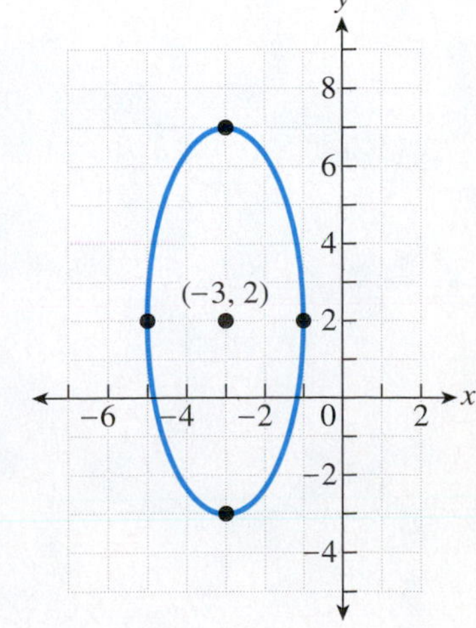

Quick Check 5

Graph the ellipse $9x^2 + 16y^2 - 36x + 160y + 292 = 0$. Give the center.

Finding the Equation of an Ellipse

Objective 4 Find the equation of an ellipse that meets the given conditions. To find the equation of an ellipse, we will begin by determining the center (h, k), a, and b. Then we will write the equation of the ellipse in standard form $\frac{(x - h)^2}{a^2} + \frac{(y - k)^2}{b^2} = 1$.

EXAMPLE 7 Find the standard form equation of the ellipse.

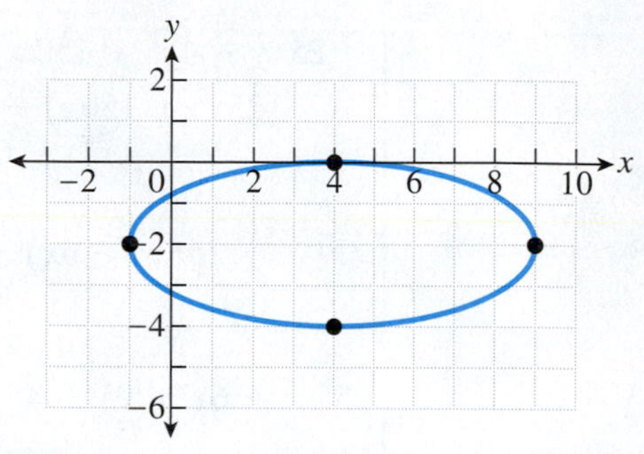

SOLUTION The center of the ellipse is $(4, -2)$.

Measure the distance from the center of the ellipse to an endpoint of its horizontal axis. This distance is a. In this case, $a = 5$.

Next, measure the distance from the center of the ellipse to an endpoint of its vertical axis. This distance is b. In this case, $b = 2$.

Now write the equation in standard form.

$$\frac{(x - 4)^2}{5^2} + \frac{(y - (-2))^2}{2^2} = 1$$ Substitute 4 for h, -2 for k, 5 for a, and 2 for b in the standard form of the equation of an ellipse.

$$\frac{(x - 4)^2}{25} + \frac{(y + 2)^2}{4} = 1$$ Simplify.

The equation of this ellipse is $\frac{(x - 4)^2}{25} + \frac{(y + 2)^2}{4} = 1$.

Quick Check 6

Find the standard form equation of the ellipse.

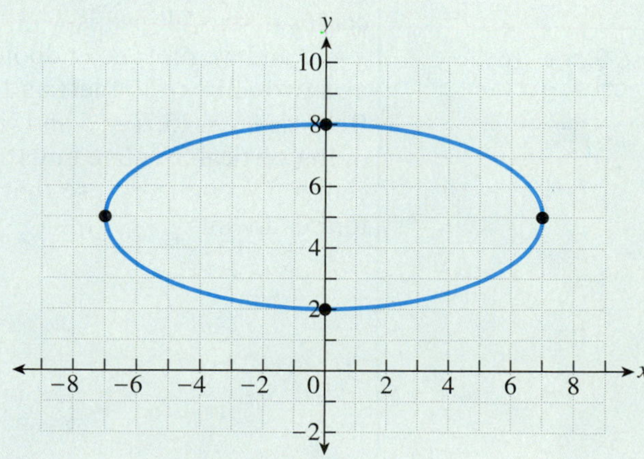

BUILDING YOUR STUDY STRATEGY

Preparing for a Final Exam (Part 1), 3 Time Management As you begin to prepare for a final exam, understand that you have only a finite amount of time before the exam; so you must use it wisely. Arrange your schedule so that you include extra time to study math. It is important that you prepare for the final exam every day; do not take any days off from your preparation.

If you have a job, try to work fewer hours. Ask for days off if possible.

For further suggestions on effective time management, read the Study Strategies in Chapter 10.

Exercises 13.3

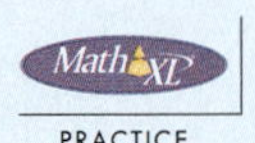
PRACTICE

WATCH

DOWNLOAD

READ

REVIEW

Vocabulary

1. The collection of all points (x, y) in the plane for which the sum of the distances d_1 and d_2 between the point (x, y) and two fixed points, F_1 and F_2, is a positive constant is called a(n) ______.
2. The two fixed points, F_1 and F_2, referred to in the definition of an ellipse are called ______.
3. The ______ of an ellipse is the point located midway between the two foci.
4. The longer of an ellipse's two axes is called the ______ axis, and the shorter of an ellipse's two axes is called the ______ axis.
5. The endpoints of the major axis of an ellipse are called ______.
6. The endpoints of the minor axis of an ellipse are called ______.
7. The equation of an ellipse that is centered at the origin is of the form ______.
8. The equation of an ellipse that is centered at the point (h, k) is of the form ______.

Graph the ellipse. Give the coordinates of the center as well as the values of a and b.

9. $\frac{x^2}{4} + \frac{y^2}{9} = 1$

10. $\frac{x^2}{16} + \frac{y^2}{25} = 1$

11. $\frac{x^2}{25} + \frac{y^2}{4} = 1$

12. $\frac{x^2}{49} + \frac{y^2}{9} = 1$

13. $\frac{x^2}{36} + \frac{y^2}{12} = 1$

14. $\dfrac{x^2}{25} + \dfrac{y^2}{32} = 1$

15. $81x^2 + 4y^2 = 324$

16. $25x^2 + 49y^2 = 1225$

17. $18x^2 + 32y^2 = 288$

18. $27x^2 + 108y^2 = 972$

19. $\dfrac{(x - 6)^2}{25} + \dfrac{(y - 8)^2}{36} = 1$

20. $\dfrac{(x - 4)^2}{4} + \dfrac{(y - 7)^2}{16} = 1$

21. $\dfrac{(x + 3)^2}{36} + \dfrac{(y - 5)^2}{4} = 1$

22. $\dfrac{(x + 1)^2}{16} + \dfrac{(y + 4)^2}{25} = 1$

23. $\dfrac{(x + 5)^2}{4} + \dfrac{y^2}{36} = 1$

24. $\dfrac{x^2}{36} + \dfrac{(y + 3)^2}{25} = 1$

25. $\dfrac{(x + 2)^2}{25} + (y - 5)^2 = 1$

26. $(x + 7)^2 + \dfrac{(y - 1)^2}{25} = 1$

27. $\dfrac{(x - 6)^2}{20} + \dfrac{(y + 6)^2}{4} = 1$

28. $\dfrac{(x - 3)^2}{8} + \dfrac{(y + 8)^2}{25} = 1$

29. $4x^2 + 9y^2 - 24x - 90y + 225 = 0$

30. $25x^2 + 4y^2 - 300x + 56y + 996 = 0$

31. $2x^2 + 8y^2 + 16x - 64y + 88 = 0$

32. $49x^2 + 9y^2 + 98x + 36y - 356 = 0$

Find the standard-form equation of the ellipse.

33.

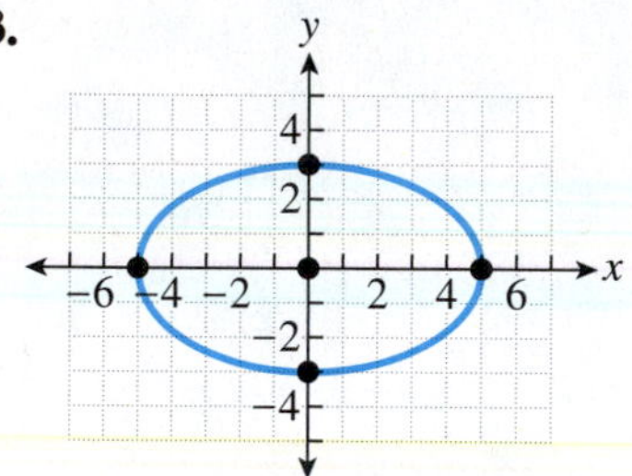

34.

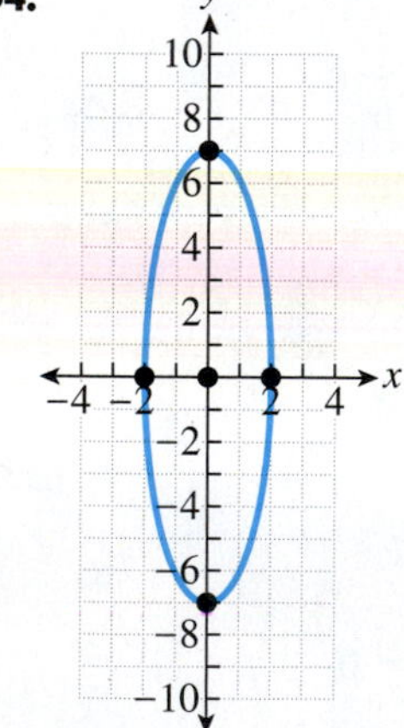

35.

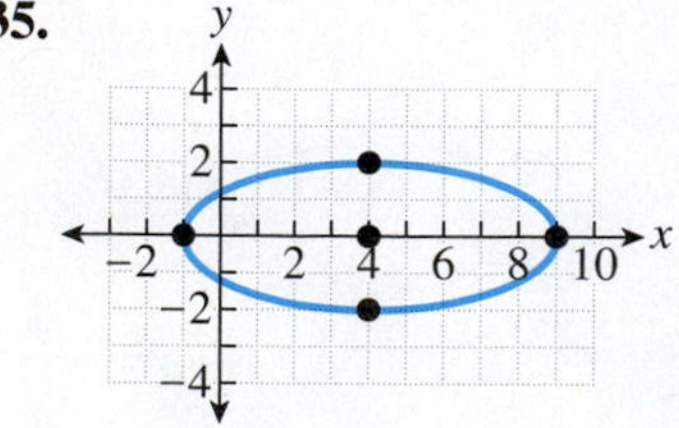

36.

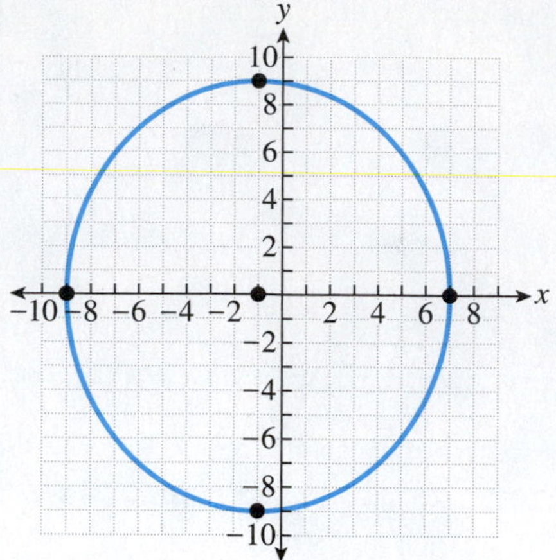

37.

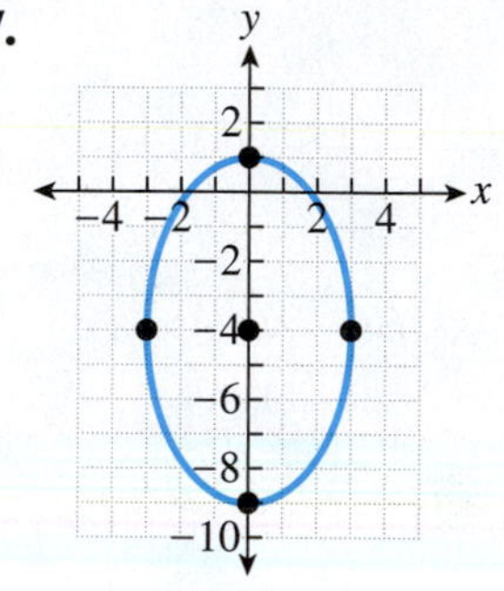

38.

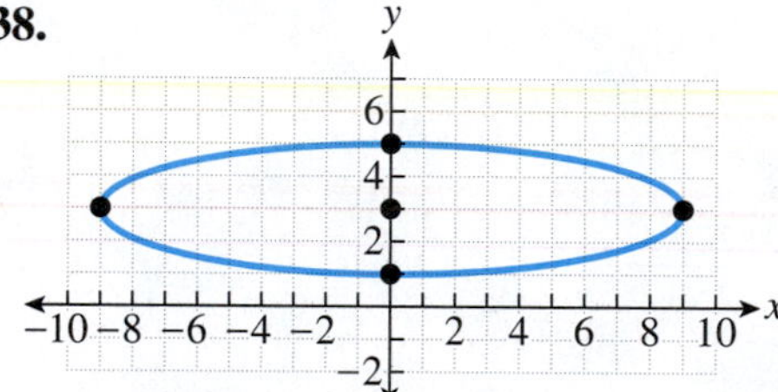

39.

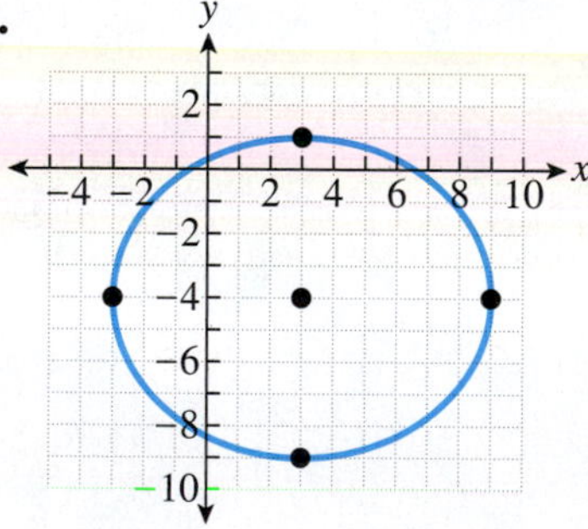

40.

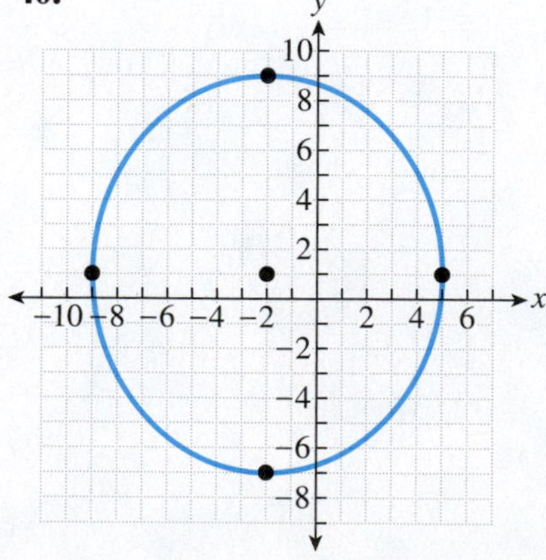

Mixed Practice, 41–58

Graph each equation. For graphs that are parabolas, label the vertex and any intercepts. For graphs that are circles or ellipses, label the center.

41. $(x + 4)^2 + (y - 6)^2 = 9$

42. $y = -(x - 2)^2 + 5$

43. $x = y^2 + 6y + 10$

44. $\dfrac{x^2}{36} + \dfrac{y^2}{4} = 1$

45. $3x^2 + 5y^2 = 45$

46. $x = -y^2 + 2y + 7$

47. $\dfrac{(x - 1)^2}{49} + \dfrac{(y + 3)^2}{36} = 1$

48. $\dfrac{(x - 1)^2}{9} + \dfrac{(y + 5)^2}{16} = 1$

49. $9x^2 + 25y^2 + 54x - 100y - 44 = 0$

50. $\dfrac{x^2}{36} + (y - 4)^2 = 1$

51. $x^2 + y^2 - 12x - 14y - 59 = 0$

52. $y = x^2 + 6x + 2$

53. $x^2 + y^2 = 32$

54. $x^2 + 49y^2 - 12x + 196y + 183 = 0$

55. $x = -(y + 4)^2 - 9$

56. $x^2 + y^2 = 81$

57. $\dfrac{(x + 6)^2}{25} + \dfrac{(y + 1)^2}{4} = 1$

58. $y = (x - 3)^2 + 4$

59. An elliptical track can be described by the equation $1089x^2 + 2500y^2 = 2{,}722{,}500$, where x and y are in feet. Assume that the major axis runs from west to east.

a) Find the width of the track from west to east.

b) Find the width of the track from north to south.

60. A tunnel through a hillside is in the shape of a semi-ellipse. The base of the tunnel is 80 feet across, and at its highest point, the tunnel is 30 feet high. The tunnel is plotted as follows on a rectangular coordinate plane (not to scale):

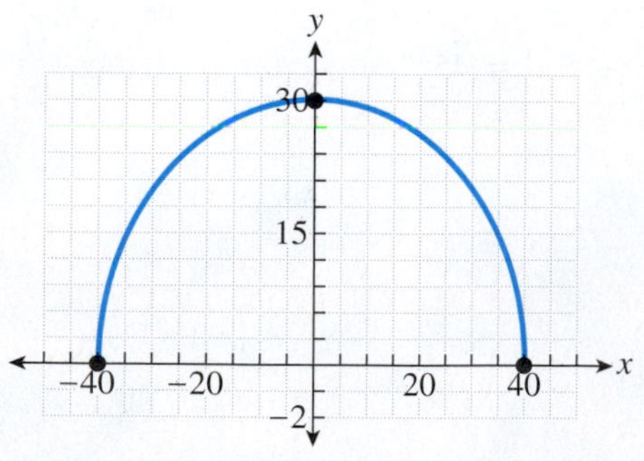

a) Find the equation of the ellipse.

b) How far from the center of the tunnel is the height 16 feet? Round to the nearest tenth.

61. The area of an ellipse given by the equation $\frac{(x-h)^2}{a^2} + \frac{(y-k)^2}{b^2} = 1$ is πab. A homeowner is building an elliptical swimming pool that is 30 feet from end to end at its widest point and 20 feet from side to side at its widest point. Find the area covered by the swimming pool. Round to the nearest tenth.

62. A small park in the shape of an ellipse covers an area of approximately 1372.88 square feet. If the park measures 46 feet from west to east, as shown, what is the distance from south to north?

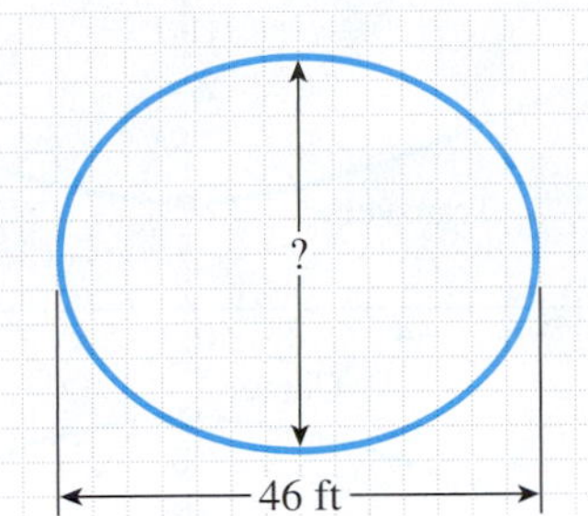

Writing in Mathematics

Answer in complete sentences.

63. Explain the difference between an ellipse and a circle. Are there any similarities between the two?

64. ***Newsletter*** Write a newsletter explaining how to graph an ellipse whose equation is of the form $\frac{(x-h)^2}{a^2} + \frac{(y-k)^2}{b^2} = 1$.

13.4 Hyperbolas

OBJECTIVES

1. **Graph hyperbolas centered at the origin.**
2. **Graph hyperbolas centered at a point (h, k).**
3. **Find the center of a hyperbola and the lengths of its axes by completing the square.**
4. **Find the equation of a hyperbola that meets the given conditions.**

In this section, we will investigate a conic section called a hyperbola. Here are two examples of graphs of hyperbolas.

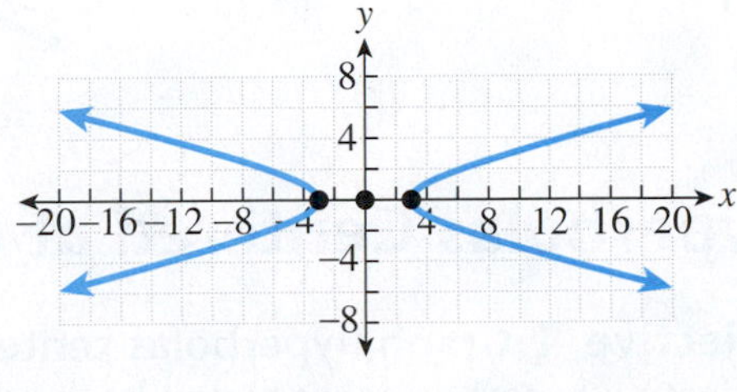

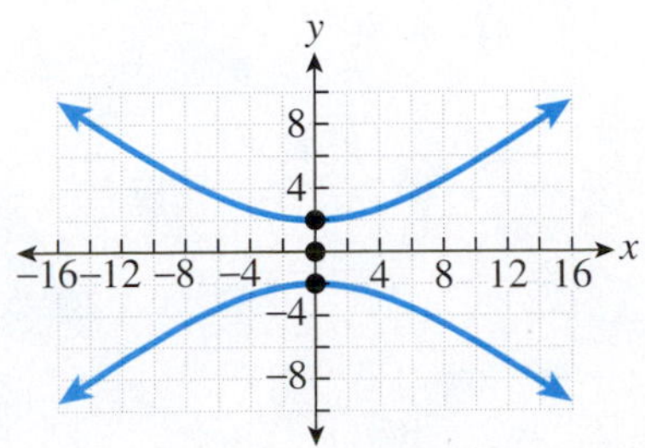

Notice that the graph of a hyperbola is different from other graphs we have drawn in that it has two parts, called **branches**.

Hyperbola

A **hyperbola** is the collection of all points (x, y) in the plane for which the *difference* of the distances, d_1 and d_2, between the point and two fixed points, F_1 and F_2, called foci, is a constant.

This definition is similar to that of an ellipse, but for an ellipse, the *sum* of the distances remains constant, not the difference. For any point (x, y) on the hyperbola, $|d_1 - d_2|$ remains the same.

Like a circle and an ellipse, a hyperbola has a center. The **center** of a hyperbola is the point that is located midway between the two foci.

The line that passes through the two foci intersects the hyperbola at two points. These points are the **vertices** of the hyperbola, and the line segment between them is called the **transverse axis** of the hyperbola.

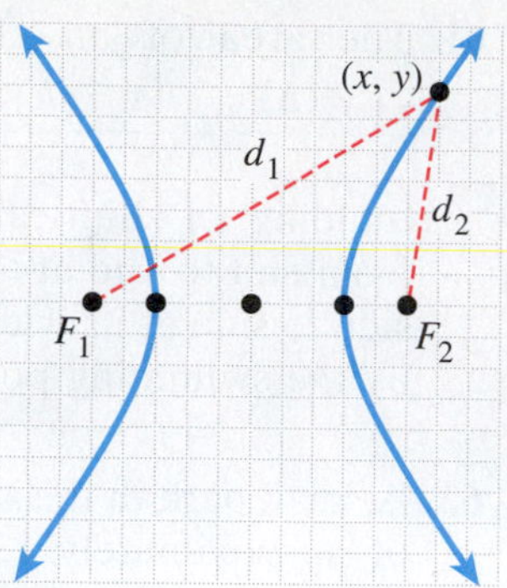

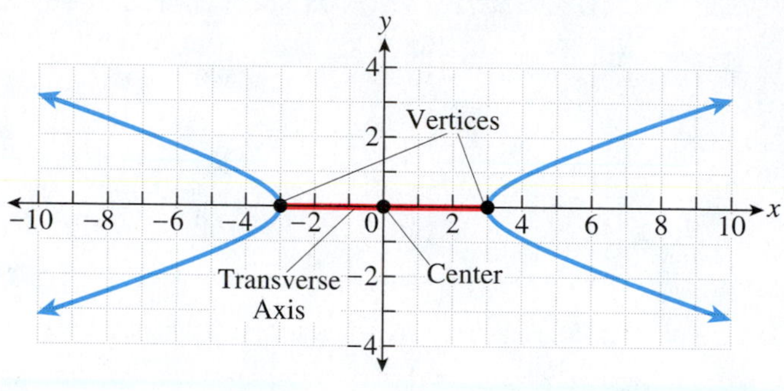

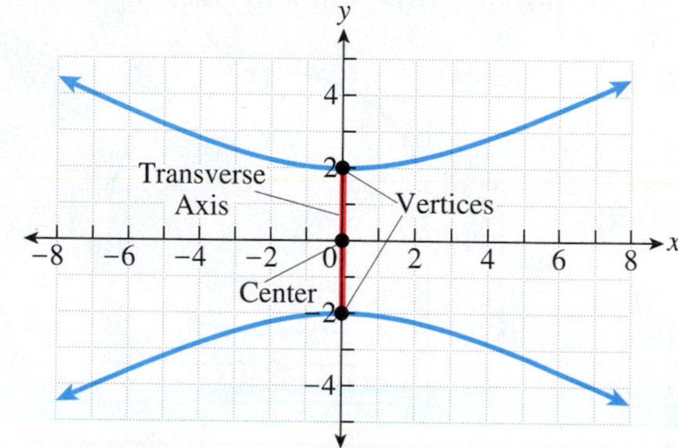

Each hyperbola has a pair of asymptotes. Each **asymptote** is a line that passes through the center of the hyperbola, showing the behavior of the branches of the hyperbola as we move farther out from the center of the hyperbola.

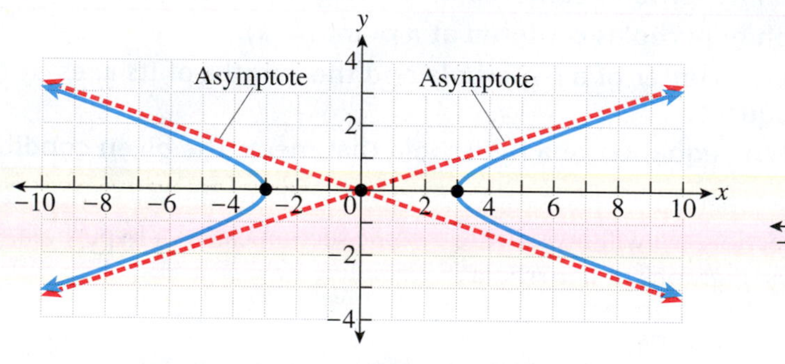

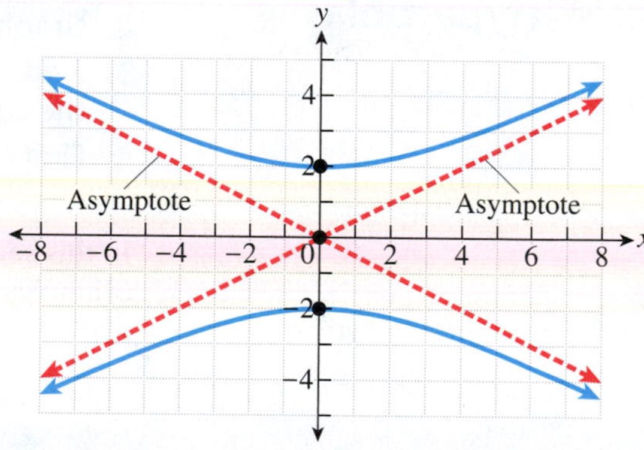

Hyperbolas Centered at the Origin

Objective 1 Graph hyperbolas centered at the origin. We will begin by investigating hyperbolas that have a horizontal transverse axis, whose branches open to the left and to the right.

Standard Form of the Equation of a Hyperbola with a Horizontal Transverse Axis (Opening to the Left and Right)

The standard-form equation of a hyperbola with a horizontal transverse axis centered at the origin has the form $\frac{x^2}{a^2} - \frac{y^2}{b^2} = 1$.

The hyperbola will have vertices at $(a, 0)$ and $(-a, 0)$. The asymptotes of the hyperbola will be the lines $y = \frac{b}{a}x$ and $y = -\frac{b}{a}x$.

To graph a hyperbola, we will begin by plotting the center and each vertex. Then we will graph the asymptotes, using dashed lines to indicate that they are not part of the hyperbola. We will finish the graph by drawing a branch through each vertex so that the branch approaches the asymptotes.

To graph the asymptotes, a rectangle is constructed around the center of the hyperbola that extends a units to the left and right of the center and b units above and below the center. Once this rectangle has been constructed, the asymptotes are the lines that pass through the diagonals of the rectangle.

Notice that four points are labeled on the rectangle. Two of the points are vertices and the endpoints of the transverse axis. The two points that are not vertices are the endpoints of what is known as the **conjugate axis**. If the branches of the hyperbola open to the left and the right, the conjugate axis will be vertical, and its length is $2b$.

EXAMPLE 1 Graph the hyperbola $\frac{x^2}{4} - \frac{y^2}{25} = 1$. Give the center, the vertices, and the equations of the asymptotes.

SOLUTION Because the equation is of the form $\frac{x^2}{a^2} - \frac{y^2}{b^2} = 1$, the hyperbola is centered at the origin and its branches open to the left and to the right. We begin by plotting a point at the origin and graphing the asymptotes.

Because $a^2 = 4$ and $b^2 = 25$, we know that $a = 2$ and $b = 5$. The rectangle used to graph the asymptotes extends two units to the left and to the right of the center and five units above and below the center. The equations for these asymptotes are $y = \frac{5}{2}x$ and $y = -\frac{5}{2}x$.

The vertices are located at the points $(-2, 0)$ and $(2, 0)$ on the rectangle that we drew to graph the asymptotes. Next, we draw the branches of the hyperbola opening to the left and to the right in such a way that they approach the asymptotes.

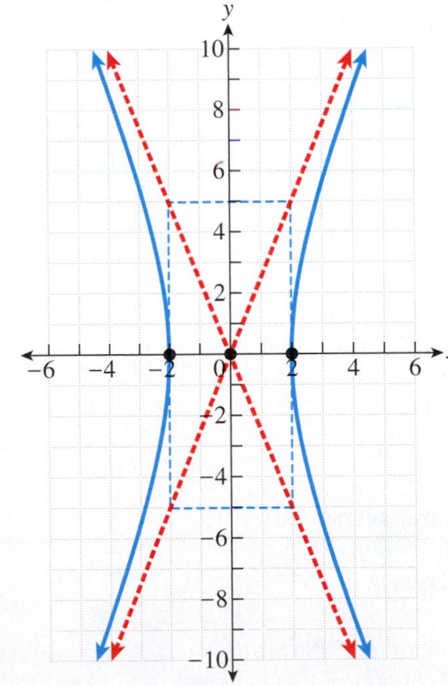

Quick Check 1

Graph the hyperbola $\frac{x^2}{16} - \frac{y^2}{9} = 1$. Give the center, the vertices, and the equations of the asymptotes.

We now turn to hyperbolas that have a vertical transverse axis. For this type of hyperbola, the branches open upward and downward.

Standard Form of the Equation of a Hyperbola with a Vertical Transverse Axis (Opening Up and Down)

The equation of a hyperbola with a vertical transverse axis that is centered at the origin has the form $\frac{y^2}{b^2} - \frac{x^2}{a^2} = 1$.

The hyperbola will have vertices at $(0, b)$ and $(0, -b)$. The asymptotes of the hyperbola will be the lines $y = \frac{b}{a}x$ and $y = -\frac{b}{a}x$. The conjugate axis will be horizontal, and its length will be $2a$.

EXAMPLE 2 Graph $\frac{y^2}{9} - \frac{x^2}{16} = 1$. Give the center, the vertices, and the equations of the asymptotes.

SOLUTION This hyperbola is centered at the origin, and its branches open upward and downward. We begin by plotting a point at the origin and graphing the asymptotes.

Because $a^2 = 16$ and $b^2 = 9$, we know that $a = 4$ and $b = 3$. The rectangle used to graph the asymptotes extends four units to the left and to the right of the center and three units above and below the origin. The equations for these asymptotes are $y = \frac{3}{4}x$ and $y = -\frac{3}{4}x$.

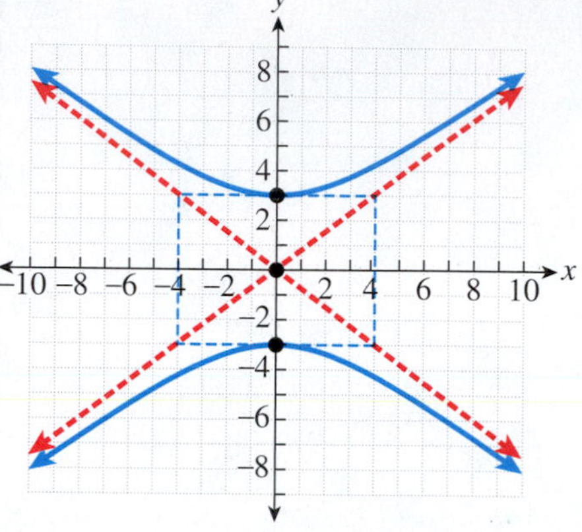

The vertices are located at the points $(0, 3)$ and $(0, -3)$. After plotting the vertices, we draw the branches of the hyperbola opening upward and downward in such a way that they approach the asymptotes.

Quick Check 2

Graph $\frac{y^2}{4} - \frac{x^2}{36} = 1$. Give the center, the vertices, and the equations of the asymptotes.

Here is a brief summary of hyperbolas centered at the origin:

Transverse Axis	Horizontal	Vertical
Equation	$\frac{x^2}{a^2} - \frac{y^2}{b^2} = 1$	$\frac{y^2}{b^2} - \frac{x^2}{a^2} = 1$
Sample Graph		
Vertices	$(a, 0), (-a, 0)$	$(0, b), (0, -b)$
Asymptotes	$y = \frac{b}{a}x, y = -\frac{b}{a}x$	$y = \frac{b}{a}x, y = -\frac{b}{a}x$

It is crucial for us to be able to tell what type of hyperbola we are graphing from the equation. For a hyperbola whose equation is of the form $\frac{x^2}{a^2} - \frac{y^2}{b^2} = 1$, the branches of the hyperbola open to the left and to the right. For a hyperbola whose equation is of the form $\frac{y^2}{b^2} - \frac{x^2}{a^2} = 1$, the branches of the hyperbola open upward and downward.

We also must be able to differentiate between the equation of a hyperbola and the equation of an ellipse. The major difference between the two equations is that the equation of a hyperbola involves the difference of the terms containing x and y, while the equation of an ellipse involves their sum.

Hyperbolas Centered at a Point Other Than the Origin

Objective 2 Graph hyperbolas centered at a point (h, k). We will now learn how to graph hyperbolas centered at any point (h, k).

Standard Form Equation of a Hyperbola with Center (h, k)

- The equation of a hyperbola centered at the point (h, k) with a horizontal transverse axis (opening to the left and right) has the standard form

$$\frac{(x-h)^2}{a^2} - \frac{(y-k)^2}{b^2} = 1.$$

The vertices are located a units to the left and to the right of the center. The slopes of the asymptotes are $m = \frac{b}{a}$ and $m = -\frac{b}{a}$.

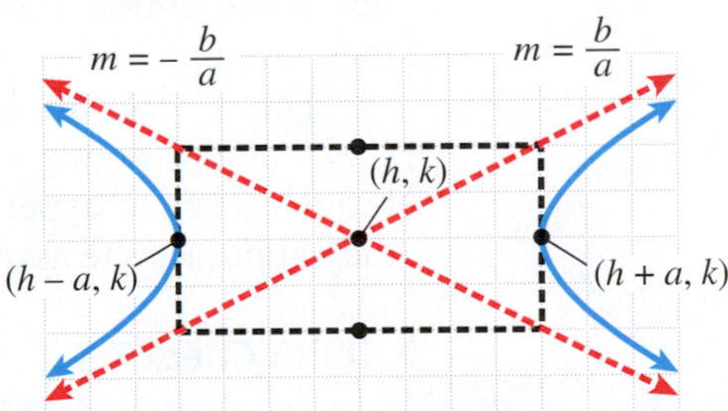

- The equation for a hyperbola centered at the point (h, k) with a vertical transverse axis (opening up and down) has the standard form

$$\frac{(y-k)^2}{b^2} - \frac{(x-h)^2}{a^2} = 1.$$

The vertices are located b units above and below the center. The slopes of the asymptotes are $m = \frac{b}{a}$ and $m = -\frac{b}{a}$.

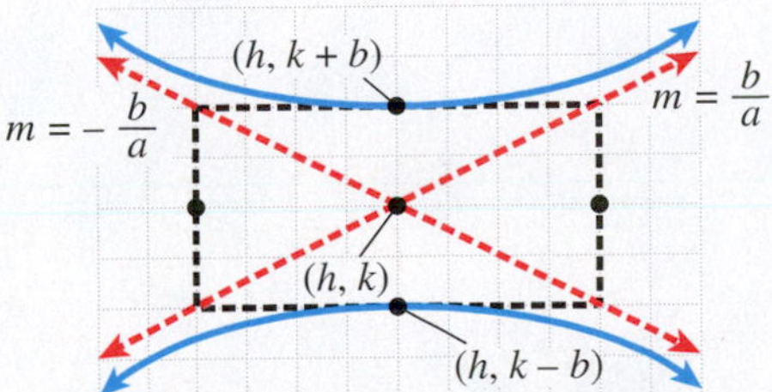

To graph a hyperbola centered at a point (h, k), we will begin by plotting the center of the hyperbola and its vertices. If the branches of the hyperbola open to the left and to the right, the vertices can be found by moving a units to the left and right of the center. If the branches of the hyperbola open upward and downward, the vertices can be found by moving b units above and below the center.

Once the center and vertices have been plotted, we will graph the asymptotes of the hyperbola by using dashed lines. Although we can graph these asymptotes from their equations, we will continue to use a rectangle as we did for hyperbolas centered at the origin.

Then we will draw the two branches of the hyperbola in such a way that each branch passes through a vertex and approaches the asymptotes.

EXAMPLE 3 Graph the hyperbola $\frac{(x-4)^2}{9} - \frac{(y-2)^2}{4} = 1$. Give the center, the vertices, and the slopes of the asymptotes.

SOLUTION Because the equation is of the form $\frac{(x-h)^2}{a^2} - \frac{(y-k)^2}{b^2} = 1$, the hyperbola is centered at $(4, 2)$ and its branches open to the left and to the right. We begin by plotting the center and graphing the asymptotes.

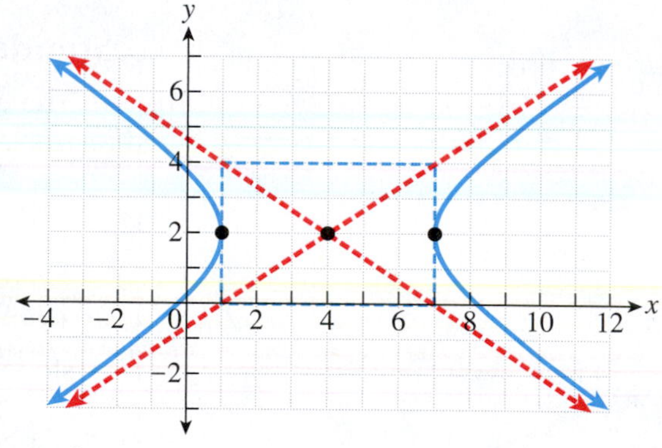

Because $a^2 = 9$ and $b^2 = 4$, we know that $a = 3$ and $b = 2$. The rectangle used to graph the asymptotes extends three units to the left and to the right of the center and two units above and below the center. The slopes for these asymptotes are $m = \frac{2}{3}$ and $m = -\frac{2}{3}$.

The vertices are located three units to the left and to the right of the center $(4, 2)$ at the points $(1, 2)$ and $(7, 2)$. After plotting the vertices, we draw the branches of the hyperbola opening to the left and to the right in such a way that they approach the asymptotes.

Quick Check 3

Graph the hyperbola $\frac{(x-3)^2}{4} - \frac{(y-5)^2}{16} = 1$. Give the center, the vertices, and the slopes of the asymptotes.

EXAMPLE 4 Graph the hyperbola $\frac{(y+5)^2}{16} - (x-1)^2 = 1$. Give the center, the vertices, and the slopes of the asymptotes.

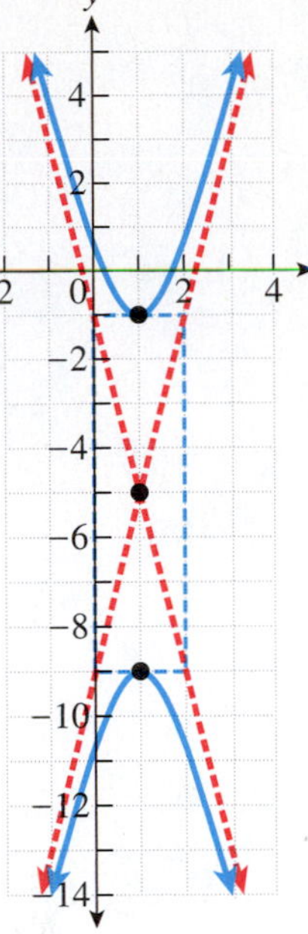

SOLUTION Because the equation is of the form $\frac{(y-k)^2}{b^2} - \frac{(x-h)^2}{a^2} = 1$, the hyperbola is centered at $(1, -5)$ and its branches open upward and downward. We begin by plotting a point at the center and graphing the asymptotes.

There is no denominator under the squared term containing x, so $a^2 = 1$. Therefore, $a = 1$. Because $b^2 = 16$, $b = 4$. The rectangle used to graph the asymptotes extends one unit to the left and to the right of the center and four units above and below the center. The slopes for these asymptotes are $m = 4$ and $m = -4$.

The vertices are located four units above and below the center at the points $(1, -1)$ and $(1, -9)$. After plotting the vertices, we draw the branches of the hyperbola opening upward and downward in such a way that they approach the asymptotes.

Quick Check 4

Graph the hyperbola $\frac{(y-2)^2}{49} - \frac{(x+3)^2}{9} = 1$. Give the center, the vertices, and the slopes of the asymptotes.

Finding the Center of a Hyperbola and the Lengths of Its Axes by Completing the Square

Objective 3 **Find the center of a hyperbola and the lengths of its axes by completing the square.**

General Form of the Equation of a Hyperbola

The **general form** of the equation of a hyperbola is $Ax^2 + By^2 + Cx + Dy + E = 0$, where $A \neq 0$, $B \neq 0$, and A and B have opposite signs.

If both A and B have the same sign, the graph will be a circle or an ellipse, not a hyperbola. To graph a hyperbola in general form, we must rewrite the equation in standard form. This will allow us to determine the center as well as a and b. This is done by completing the square, which must be done for both x and y.

EXAMPLE 5 Graph the hyperbola $-4x^2 + 25y^2 + 24x + 50y - 111 = 0$. Give the center, the vertices, and the equations of the asymptotes.

SOLUTION Convert the equation to standard form by completing the square for x and y.

$$-4x^2 + 25y^2 + 24x + 50y - 111 = 0$$

$$(-4x^2 + 24x) + (25y^2 + 50y) = 111$$

Collect terms containing x. Collect terms containing y. Add 111 to both sides.

$$-4(x^2 - 6x) + 25(y^2 + 2y) = 111$$

Factor -4 from the terms containing x so that the coefficient of the x^2 term is 1. In the same fashion, factor 25 from the terms containing y.

$$-4(x^2 - 6x + 9) + 25(y^2 + 2y + 1) = 111 - 36 + 25$$

Add 9 inside the parentheses containing x to complete the square for x. Because $-4 \cdot 9 = -36$, subtract 36 on the right side of the equation. Add 1 inside the parentheses containing y to complete the square for y. Add $25 \cdot 1$, or 25, to the right side of the equation.

$$-4(x - 3)^2 + 25(y + 1)^2 = 100$$

Factor the two quadratic expressions on the left side of the equation.

$$25(y + 1)^2 - 4(x - 3)^2 = 100$$

Rewrite so that the term containing x is being subtracted from the term containing y.

$$\frac{(y + 1)^2}{4} - \frac{(x - 3)^2}{25} = 1$$

Divide both sides by 100 and simplify.

Quick Check 5

Graph the hyperbola $16x^2 - 9y^2 - 64x - 54y - 161 = 0$. Give the center, the vertices, and the equations of the asymptotes.

Begin by plotting the center, $(3, -1)$, and graphing the asymptotes.

For this hyperbola, $a = 5$ and $b = 2$. The equations for the asymptotes are $y = \frac{2}{5}(x - 3) - 1$ and $y = -\frac{2}{5}(x - 3) - 1$.

Because the equation has the form $\frac{(y - k)^2}{b^2} - \frac{(x - h)^2}{a^2} = 1$, the branches of this hyperbola open upward and downward.

The vertices are $(3, 1)$ and $(3, -3)$. After plotting the vertices, draw the branches of the hyperbola opening upward and downward in such a way that they approach the asymptotes.

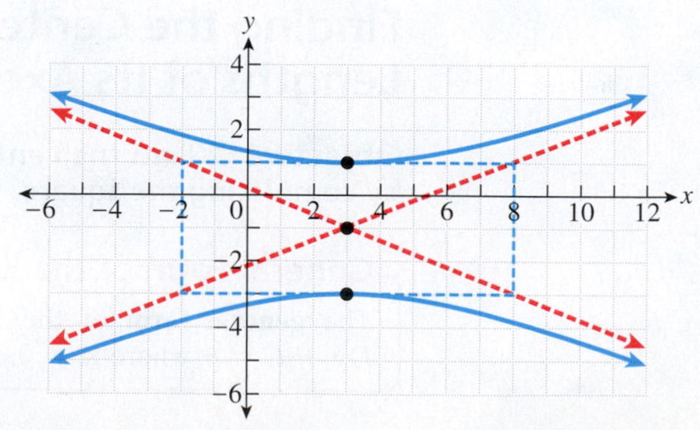

Finding the Standard-Form Equation of a Hyperbola

Objective 4 Find the equation of a hyperbola that meets the given conditions. To find the equation of a hyperbola in standard form, we must find the center (h, k), a, and b.

EXAMPLE 6 Find the equation of the hyperbola whose graph is as follows:

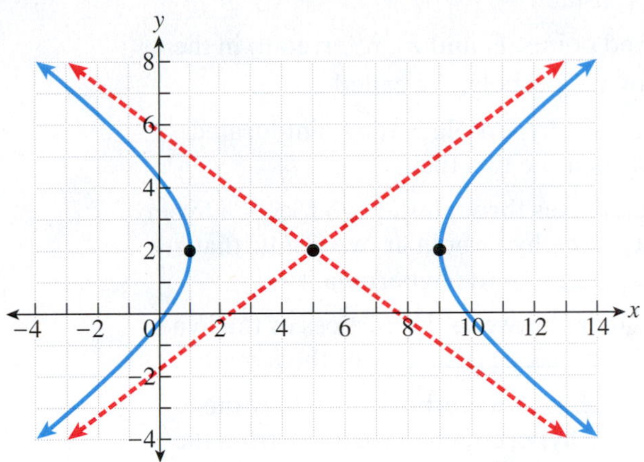

SOLUTION Because the branches of the hyperbola open to the left and to the right, its equation is of the form $\frac{(x-h)^2}{a^2} - \frac{(y-k)^2}{b^2} = 1$.

We begin by finding the center of the hyperbola, which is at the point $(5, 2)$.

Then we measure the distance a from the center of the hyperbola to one of its vertices and find that $a = 4$.

Next, we measure the vertical distance b from a vertex of the hyperbola to one of its asymptotes and find that $b = 3$.

We can now write the equation in standard form.

$$\frac{(x-h)^2}{a^2} - \frac{(y-k)^2}{b^2} = 1$$

$$\frac{(x-5)^2}{4^2} - \frac{(y-2)^2}{3^2} = 1 \quad \text{Substitute 5 for } h, \text{2 for } k, \text{4 for } a, \text{and 3 for } b.$$

$$\frac{(x-5)^2}{16} - \frac{(y-2)^2}{9} = 1 \quad \text{Simplify.}$$

The equation of this hyperbola is $\frac{(x-5)^2}{16} - \frac{(y-2)^2}{9} = 1$.

Quick Check 6

Find the equation of the hyperbola whose graph is shown.

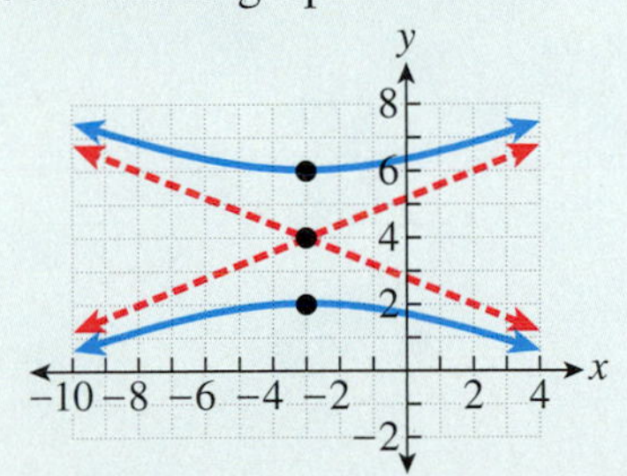

BUILDING YOUR STUDY STRATEGY

Preparing for a Final Exam (Part 1), 4 Study Environment Choosing the location where you will study for the final exam is not a trivial matter. You should study in a distraction-free zone; you cannot afford to waste time and effort at this point in the course. Study in a well-lit area that allows you easy access to all of your materials.

See the Study Strategies in Chapter 11 for further advice on creating a positive study environment.

Exercises 13.4

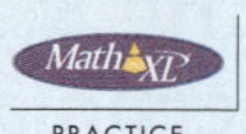
PRACTICE

WATCH

DOWNLOAD

READ

REVIEW

Vocabulary

1. A(n) ________ is the collection of all points (x, y) in the plane for which the difference of the distances, d_1 and d_2, between the point and two fixed points, F_1 and F_2, is a constant.
2. The two fixed points, F_1 and F_2, referred to in the definition of a hyperbola, are called _____.
3. The _____ of a hyperbola is the point located midway between the two foci.
4. The line that passes through the two foci of a hyperbola intersects the hyperbola at two points that are called the ______ of the hyperbola.
5. The line segment between the vertices of the hyperbola is called the ____________ of the hyperbola.
6. The ________ of a hyperbola show how the branches of the hyperbola behave away from the center of the hyperbola.
7. The standard form equation of a hyperbola centered at the origin with a horizontal transverse axis has the form __________, while a hyperbola centered at the origin with a vertical transverse axis has the form __________.
8. The standard form equation of a hyperbola centered at the point (h, k) with a horizontal transverse axis has the form ______________________, while a hyperbola centered at the point (h, k) with a vertical transverse axis has the form ______________________.

Graph the hyperbola. Give the coordinates of the center as well as the values of a and b.

9. $\dfrac{x^2}{9} - \dfrac{y^2}{4} = 1$

10. $\dfrac{x^2}{16} - \dfrac{y^2}{25} = 1$

11. $\dfrac{y^2}{49} - \dfrac{x^2}{9} = 1$

12. $\dfrac{y^2}{16} - \dfrac{x^2}{36} = 1$

13. $\dfrac{y^2}{25} - x^2 = 1$

14. $\dfrac{x^2}{16} - y^2 = 1$

15. $\dfrac{x^2}{12} - \dfrac{y^2}{9} = 1$

16. $\dfrac{y^2}{6} - \dfrac{x^2}{25} = 1$

17. $49x^2 - 4y^2 = 196$

18. $9x^2 - 36y^2 = 324$

19. $y^2 - 9x^2 = 9$

20. $y^2 - 4x^2 = 4$

21. $\dfrac{(y-5)^2}{16} - \dfrac{(x-2)^2}{9} = 1$

22. $\dfrac{(y-6)^2}{25} - \dfrac{(x-3)^2}{9} = 1$

23. $\dfrac{(x+1)^2}{9} - \dfrac{(y+4)^2}{4} = 1$

24. $\dfrac{(x+3)^2}{4} - \dfrac{(y-3)^2}{25} = 1$

25. $\dfrac{(y-1)^2}{4} - \dfrac{x^2}{4} = 1$

26. $\dfrac{(x+1)^2}{9} - \dfrac{(y+5)^2}{9} = 1$

27. $\dfrac{(x-4)^2}{49} - \dfrac{(y+2)^2}{16} = 1$

28. $\dfrac{(y-2)^2}{9} - \dfrac{(x+4)^2}{49} = 1$

29. $\dfrac{(x+2)^2}{25} - \dfrac{(y+3)^2}{28} = 1$

30. $\dfrac{(x+1)^2}{18} - \dfrac{(y-2)^2}{4} = 1$

31. $9x^2 - 4y^2 + 18x - 24y - 63 = 0$

32. $-16x^2 + 9y^2 - 160x + 72y - 400 = 0$

33. $-25x^2 + 16y^2 - 100x - 224y + 284 = 0$

34. $x^2 - 64y^2 - 10x + 768y - 2343 = 0$

Find the standard-form equation of the hyperbola whose graph is shown.

35.

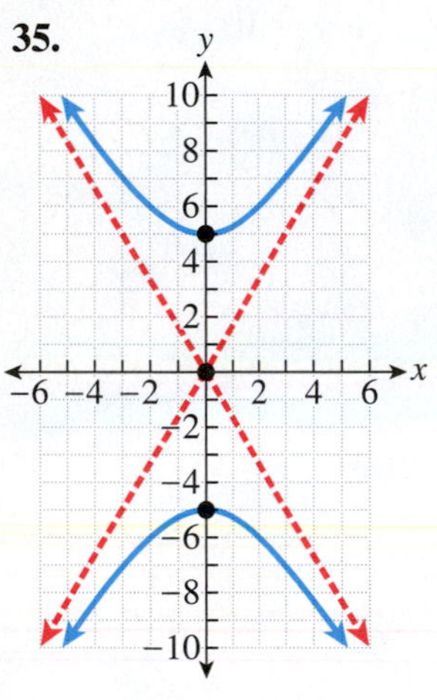

36.

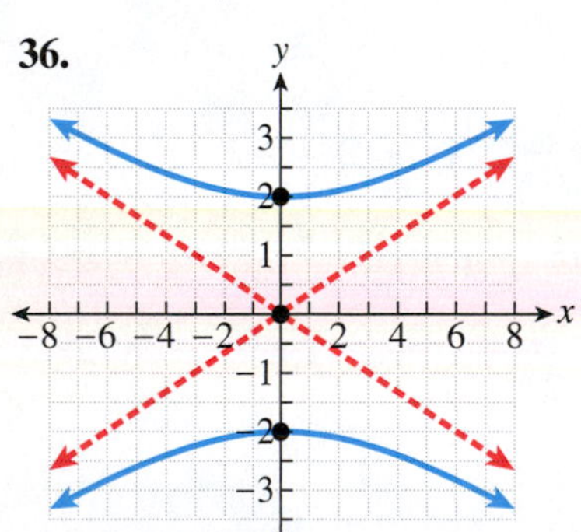

37.

38.

39.

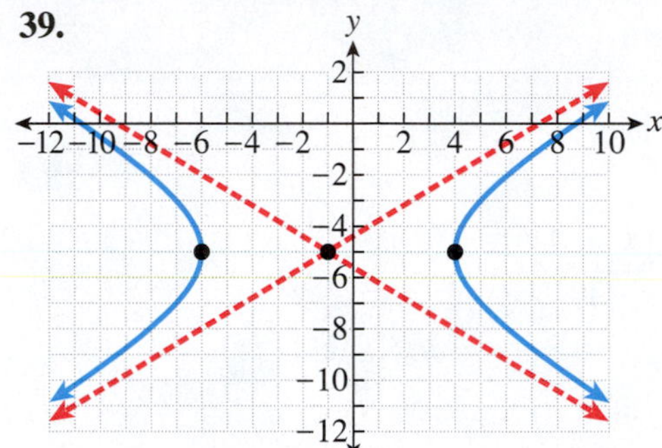

40.

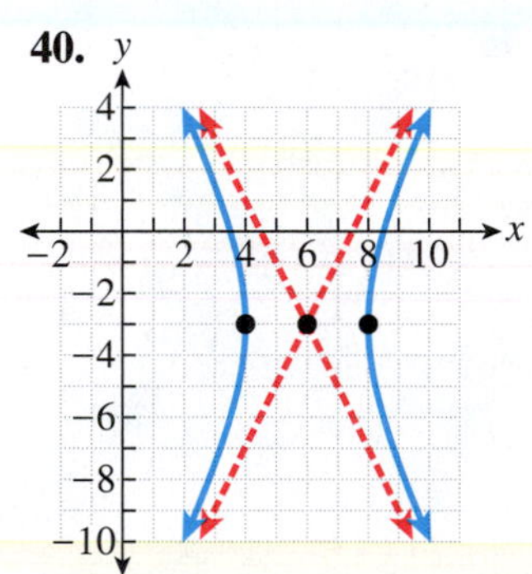

Mixed Practice, 41–54

Graph each equation. For graphs that are parabolas, label the vertex and any intercepts. For graphs that are circles, ellipses, or hyperbolas, label the center.

41. $y = -x^2 + 6x - 17$

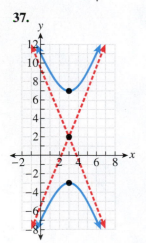

42. $\dfrac{y^2}{25} - x^2 = 1$

43. $\dfrac{(x-6)^2}{9} + \dfrac{y^2}{49} = 1$

44. $y = (x-3)^2 - 4$

45. $\dfrac{(x+2)^2}{25} + \dfrac{(y-5)^2}{9} = 1$

46. $\dfrac{(y-3)^2}{25} - \dfrac{(x+7)^2}{36} = 1$

47. $x^2 + y^2 = 64$

48. $\dfrac{x^2}{36} + \dfrac{y^2}{4} = 1$

49. $4x^2 - 9y^2 + 40x + 54y - 17 = 0$

50. $(x-4)^2 + (y+7)^2 = 9$

51. $x^2 + y^2 - 8x + 18y + 48 = 0$

53. $\dfrac{(x + 6)^2}{9} - \dfrac{(y + 2)^2}{16} = 1$

52. $16x^2 + 9y^2 - 256x + 90y + 1105 = 0$

54. $x = y^2 + 10y + 16$

For Exercises 55–60, determine which graph is associated with each given equation.

a)

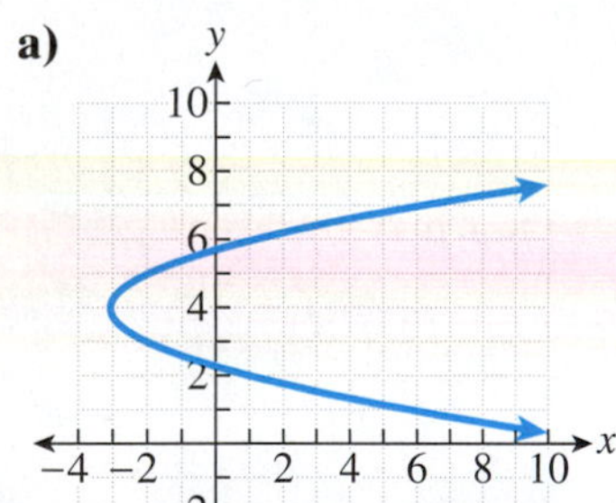

b)

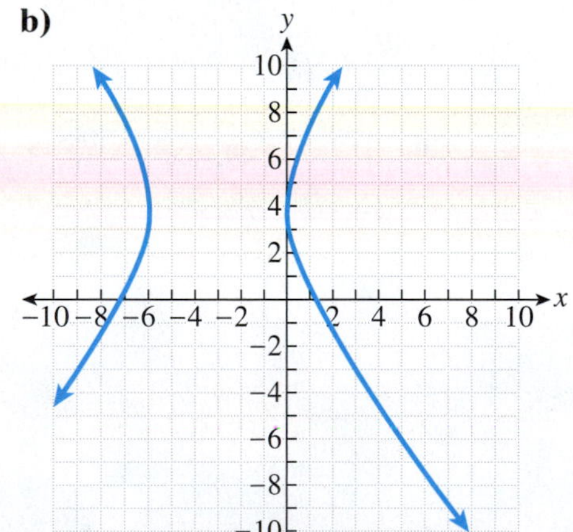

c)

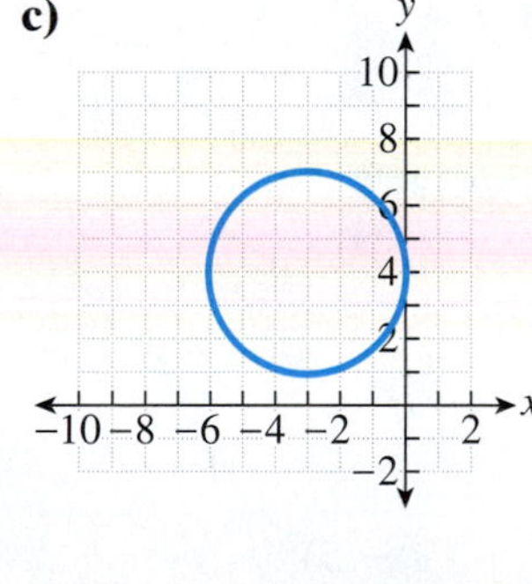

d)

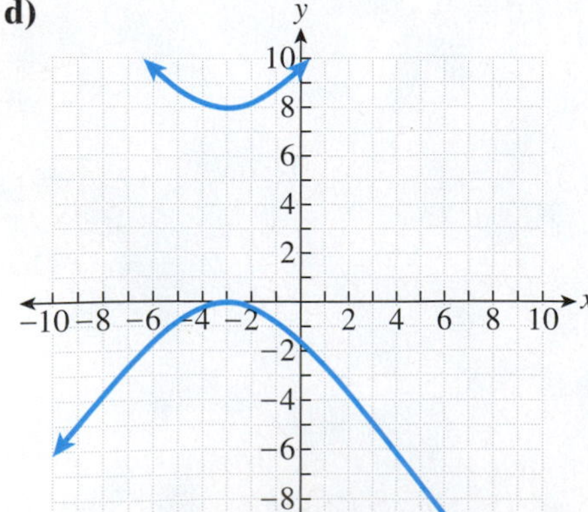

e)

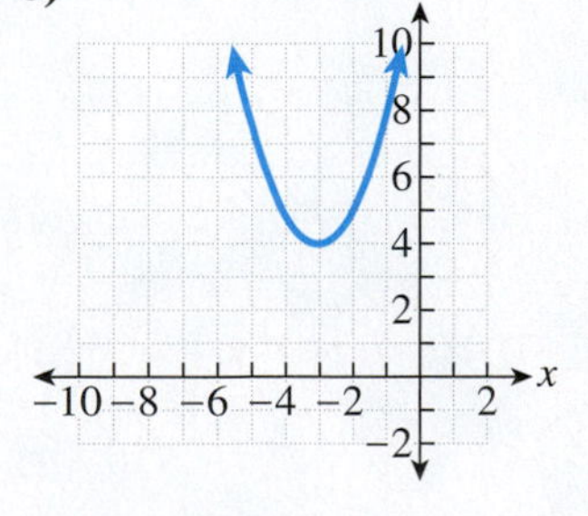

f)

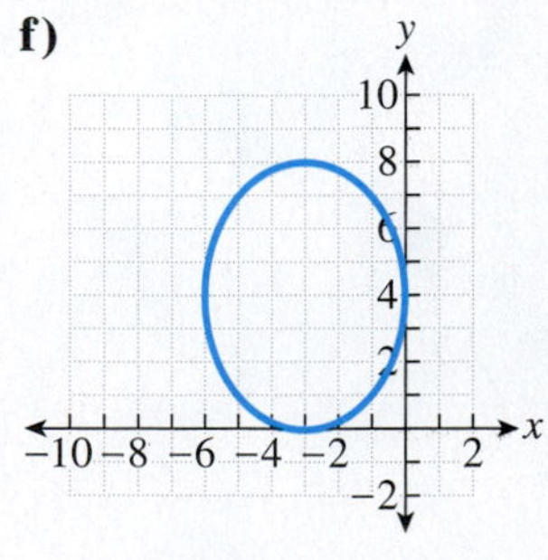

55. $y = (x + 3)^2 + 4$

56. $x = (y - 4)^2 - 3$

57. $(x + 3)^2 + (y - 4)^2 = 9$

58. $\dfrac{(x + 3)^2}{9} + \dfrac{(y - 4)^2}{16} = 1$

59. $\dfrac{(x + 3)^2}{9} - \dfrac{(y - 4)^2}{16} = 1$

60. $\dfrac{(y - 4)^2}{16} - \dfrac{(x + 3)^2}{9} = 1$

Writing in Mathematics

Answer in complete sentences.

61. Explain how to determine whether the graph of an equation is a hyperbola whose branches open to the left and right. Also explain how to determine whether the graph of an equation is a hyperbola whose branches open up and down.

Quick Review Exercises

Section 13.4

Solve.

1. $x + 8y = 19$
$7x - 5y = 11$

2. $y = 5x + 6$
$-3x + 4y = -27$

3. $2x - 3y = 11$
$5x + 6y = 14$

4. $4x - 5y = 15$
$-2x + 15y = -30$

13.5 Nonlinear Systems of Equations

OBJECTIVES

1. Solve nonlinear systems by using the substitution method.
2. Solve nonlinear systems by using the addition method.
3. Solve applications of nonlinear systems.

A **nonlinear system of equations** is a system of equations in which the graph of at least one of the equations is not a line. For example, if the graph of one equation is a circle and the graph of the other equation is a line, the system is a nonlinear system of equations. For a system of two equations in two unknowns, an ordered pair (x, y) is a solution of the system if it is a solution of each equation. Graphically, a solution of a system of two equations in two unknowns is a point of intersection of the graphs of the two equations. Some nonlinear systems can have one or more solutions. Following are examples of nonlinear systems of equations and their solutions, which are points of intersection for the two graphs:

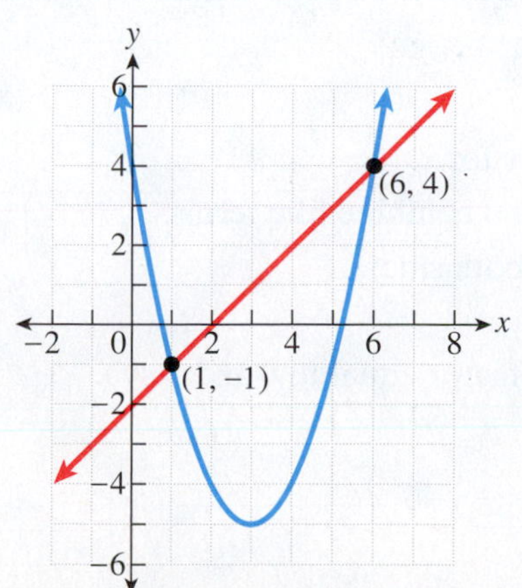

Two Solutions

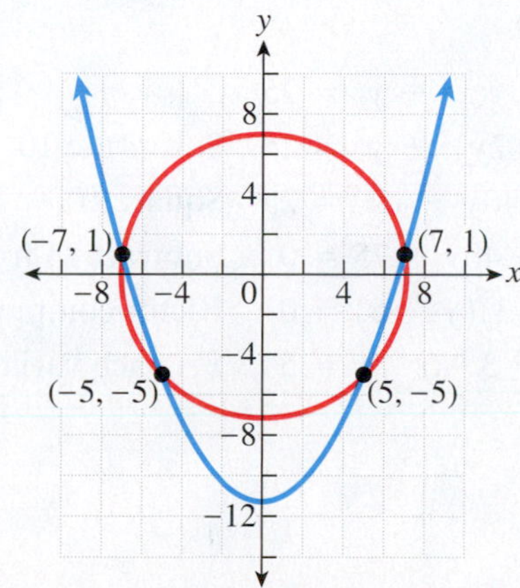

Four Solutions

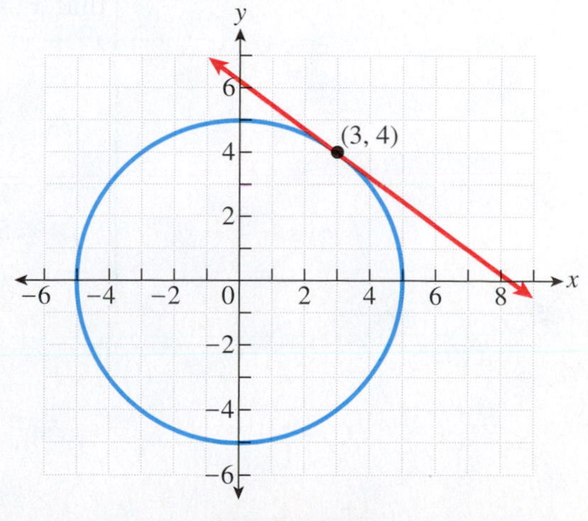

One Solution

If the graphs of the two equations do not intersect, the system of equations has no solution.

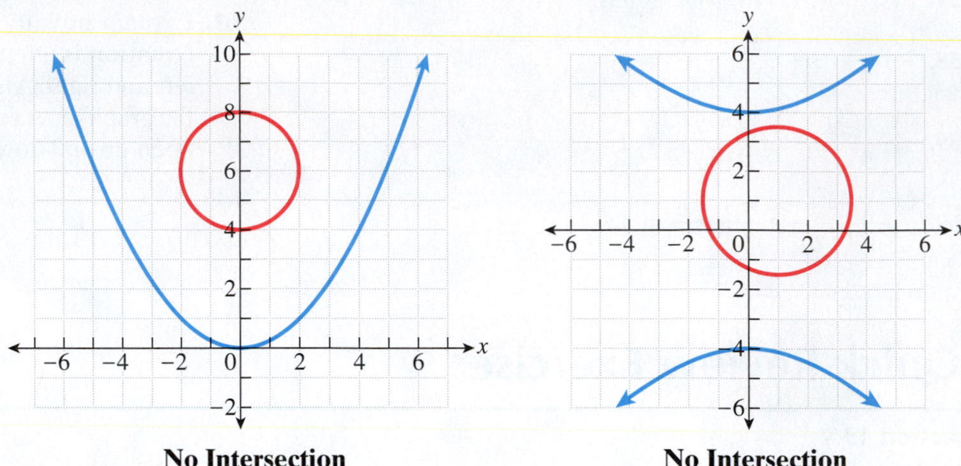

No Intersection **No Intersection**

The Substitution Method

Objective 1 Solve nonlinear systems by using the substitution method. One effective technique for solving nonlinear systems of equations is the substitution method. As in Chapter 4, we will solve one of the equations for one of the variables in terms of the other variable and then substitute this expression for that variable in the other equation. This will result in an equation with only one variable, which we can then solve.

Solving a System of Equations by the Substitution Method

1. Solve one of the equations for either variable.
2. Substitute this expression for the variable in the other equation.
3. Solve this equation.
4. Substitute each value for the variable in the equation from Step 1.

EXAMPLE 1 Solve the nonlinear system: $\begin{aligned} x^2 + y^2 &= 25 \\ x + 2y &= 10 \end{aligned}$

SOLUTION We begin by solving the second equation for x because the coefficient of that term is 1. If we subtract $2y$ from both sides of the equation, we find that $x = 10 - 2y$. This expression can then be substituted for x in the equation $x^2 + y^2 = 25$.

$$
\begin{aligned}
x^2 + y^2 &= 25 \\
(10 - 2y)^2 + y^2 &= 25 && \text{Substitute } 10 - 2y \text{ for } x. \\
5y^2 - 40y + 100 &= 25 && \text{Square } 10 - 2y \text{ and combine like terms.} \\
5y^2 - 40y + 75 &= 0 && \text{Subtract 25 from both sides.} \\
5(y - 3)(y - 5) &= 0 && \text{Factor completely.} \\
y = 3 \quad \text{or} \quad y &= 5 && \text{Set each variable factor equal to 0 and solve.}
\end{aligned}
$$

We can now substitute these values for y in the equation $x = 10 - 2y$ to find the corresponding x-coordinates of the solutions.

$y = 3$	$y = 5$	
$x = 10 - 2(3)$	$x = 10 - 2(5)$	Substitute for y.
$x = 4$	$x = 0$	Simplify.
$(4, 3)$	$(0, 5)$	Simplify.

The two solutions are $(4, 3)$ and $(0, 5)$.

The graph of the first equation in the system is a circle, and the graph of the second equation is a line. Here are the graphs of the two equations, showing their points of intersection.

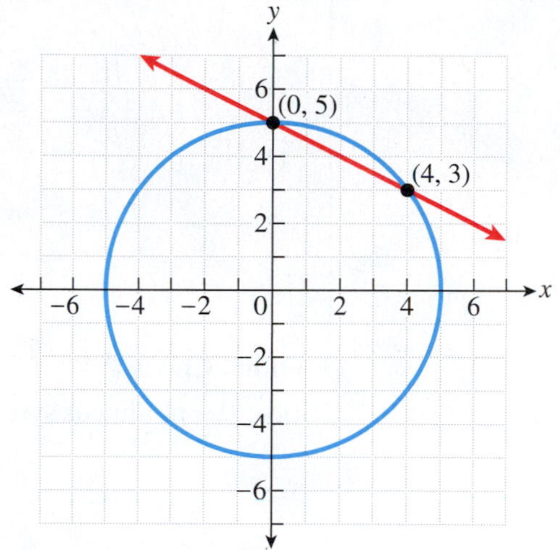

Quick Check 1

Solve the nonlinear system: $\begin{aligned} 4x^2 + y^2 &= 41 \\ x + 2y &= -8 \end{aligned}$

Although it is not necessary, it is a good idea to graph both equations in the system. This will give us an idea of the number of solutions we are looking for as well as the approximate coordinates of the solutions.

EXAMPLE 2 Solve the nonlinear system: $\begin{aligned} y &= x + 6 \\ x^2 + y^2 &= 16 \end{aligned}$

SOLUTION Because the first equation is already solved for y, we can substitute $x + 6$ for y in the equation $x^2 + y^2 = 16$.

$$x^2 + y^2 = 16$$

$$x^2 + (x + 6)^2 = 16 \quad \text{Substitute } x + 6 \text{ for } y.$$

$$2x^2 + 12x + 36 = 16 \quad \text{Square } x + 6 \text{ and combine like terms.}$$

$$2x^2 + 12x + 20 = 0 \quad \text{Subtract 16 from both sides.}$$

$$2(x^2 + 6x + 10) = 0 \quad \text{Factor out a common factor of 2.}$$

Because $x^2 + 6x + 10$ cannot be factored, we use the quadratic formula.

$$x = \frac{-6 \pm \sqrt{6^2 - 4(1)(10)}}{2(1)} \quad \text{Substitute 1 for } a\text{, 6 for } b\text{, and 10 for } c \text{ in the quadratic formula.}$$

$$x = \frac{-6 \pm \sqrt{-4}}{2} \quad \text{Simplify the discriminant and the denominator.}$$

Because the discriminant was negative, this equation has no real-number solutions. Therefore, this system of equations has no solution. We could have determined this from the graphs of the two equations. The graph of the equation $y = x + 6$ is a line with a slope of 1 and a y-intercept at $(0, 6)$, while the graph of $x^2 + y^2 = 16$ is a circle centered at the origin with a radius of 4. Notice that the two graphs do not intersect.

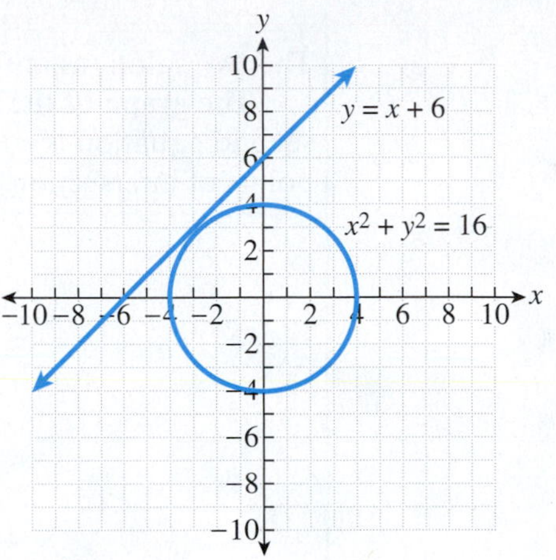

Quick Check 2

Solve the nonlinear system:

$$y = 2x - 8$$
$$x^2 + y^2 = 9$$

EXAMPLE 3 Solve the nonlinear system: $\begin{aligned} y &= x^2 - 3 \\ x^2 + y^2 &= 9 \end{aligned}$

SOLUTION The graph on the left of the first equation is a parabola that opens upward, while the graph of the second equation is a circle. Based on the graph, there appear to be three solutions.

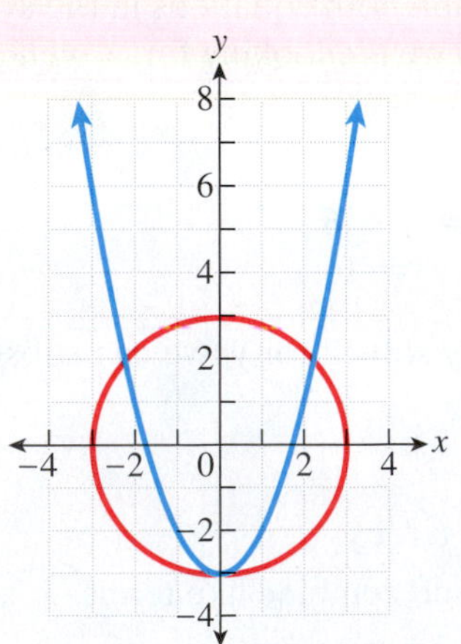

Solve the first equation for x^2, giving $x^2 = y + 3$, and substitute that expression for x^2 in the second equation. This is a useful technique when the graph of neither equation is a line.

$$\begin{aligned} x^2 + y^2 &= 9 & & \\ (y + 3) + y^2 &= 9 & & \text{Substitute } y + 3 \text{ for } x^2. \\ y^2 + y - 6 &= 0 & & \text{Collect all terms on the left side.} \\ (y + 3)(y - 2) &= 0 & & \text{Factor.} \\ y = -3 \quad \text{or} \quad y &= 2 & & \text{Set each factor equal to 0 and solve.} \end{aligned}$$

We can now substitute these values for y in the equation $x^2 = y + 3$ to find the corresponding x-coordinates of the solutions.

$$\begin{aligned} &y = -3 & &y = 2 & & \\ &x^2 = (-3) + 3 & &x^2 = (2) + 3 & &\text{Substitute for } y. \\ &x = 0 & &x = \pm\sqrt{5} & &\text{Simplify and take the square root of each side.} \end{aligned}$$

The solutions are $(0, -3)$, $(\sqrt{5}, 2)$, and $(-\sqrt{5}, 2)$.

Quick Check 3

Solve the nonlinear system: $\begin{aligned} y &= x^2 \\ 5x^2 + y^2 &= 24 \end{aligned}$

The Addition Method

Objective 2 Solve nonlinear systems by using the addition method. Another effective technique for solving nonlinear systems of equations is the addition method, which was introduced in Chapter 4. By rewriting the two equations in such a way that the coefficients of like terms are opposites, we can add the two equations to obtain a single equation with only one variable, which we can then solve.

EXAMPLE 4 Solve the nonlinear system: $\begin{aligned} 16x^2 + 25y^2 &= 400 \\ x^2 + y^2 &= 16 \end{aligned}$

SOLUTION If we multiply both sides of the second equation by -16, the coefficients of the two terms containing x^2 will be opposites. Then we can eliminate terms containing x^2, leaving an equation whose only variable is y.

$$\begin{aligned} 16x^2 + 25y^2 &= 400 \\ x^2 + y^2 &= 16 \end{aligned} \quad \xrightarrow{\text{Multiply by } -16.} \quad \begin{aligned} 16x^2 + 25y^2 &= 400 \\ -16x^2 - 16y^2 &= -256 \end{aligned}$$

$$\begin{aligned} 16x^2 + 25y^2 &= 400 \\ -16x^2 - 16y^2 &= -256 \quad \text{Add.} \\ \hline 9y^2 &= 144 \end{aligned}$$

Now we can solve the equation $9y^2 = 144$ for y.

$$\begin{aligned} 9y^2 &= 144 \\ y^2 &= 16 \quad \text{Divide both sides by 9.} \\ y &= \pm 4 \quad \text{Take the square root of each side.} \end{aligned}$$

We can now substitute 4 and -4 for y in either equation to find the corresponding x-coordinates of the solutions. We will use the equation of the circle $x^2 + y^2 = 16$.

$y = 4$	$y = -4$	
$x^2 + (4)^2 = 16$	$x^2 + (-4)^2 = 16$	Substitute for y.
$x = 0$	$x = 0$	Solve for x.

The solutions are $(0, 4)$ and $(0, -4)$.

Quick Check 4

Solve the nonlinear system: $\begin{aligned} 4x^2 + 7y^2 &= 211 \\ x^2 + y^2 &= 34 \end{aligned}$

We can attempt to solve nonlinear systems by graphing, but the solutions will depend on the accuracy of the graph, not to mention the estimate of solutions that do not have integer coordinates. However, technology such as a graphing calculator or computer software have built-in functions for finding the points of intersection of two graphs.

Applications of Nonlinear Systems of Equations

Objective 3 Solve applications of nonlinear systems. Now we will turn our attention to applied problems requiring solving a nonlinear system of equations to solve the problem.

EXAMPLE 5 A rectangular dorm room has a perimeter of 40 feet and an area of 91 square feet. Find its dimensions.

SOLUTION The unknowns in this problem are the length and width of the room. We will let x represent the length of the room, and we will let y represent the width of the room.

Unknowns
Length: x
Width: y

Because the perimeter of the rectangle is 40 feet, we know that $2x + 2y = 40$. Also, because the area of the rectangle is 91 square feet, we know that $xy = 91$. If we solve the perimeter equation for x or y, we can substitute the expression for that variable in the area equation.

$$2x + 2y = 40$$
$$2x = 40 - 2y \quad \text{Subtract } 2y.$$
$$x = 20 - y \quad \text{Divide both sides by 2.}$$

We can now substitute $20 - y$ for x in the equation $xy = 91$.

$$xy = 91$$
$$(20 - y)y = 91 \quad \text{Substitute } 20 - y \text{ for } x.$$
$$0 = y^2 - 20y + 91 \quad \text{Simplify and collect all terms on the right side of the equation.}$$
$$0 = (y - 7)(y - 13) \quad \text{Factor the quadratic expression.}$$
$$y = 7 \quad \text{or} \quad y = 13 \quad \text{Set each factor equal to 0 and solve.}$$

We can find the corresponding x-coordinates of the solutions by substituting these values for y in one of the two equations and solving for x. We will use the perimeter equation.

$y = 7$	$y = 13$	
$2x + 2(7) = 40$	$2x + 2(13) = 40$	Substitute for y.
$x = 13$	$x = 7$	Solve for x.

We see that the length is 13 feet if the width is 7 feet and the length is 7 feet if the width is 13 feet. In either case, the dimensions of the room are 7 feet by 13 feet.

Quick Check 5

A rectangle has a perimeter of 30 feet and an area of 36 square feet. Find its dimensions.

EXAMPLE 6 A rectangular computer monitor has a perimeter of 46 inches and a diagonal that is 17 inches long. Find the dimensions of the monitor.

SOLUTION The unknowns in this problem are the length and width of the monitor. We let x represent the length of the monitor, and we let y represent the width of the monitor.

Unknowns
Length: x
Width: y

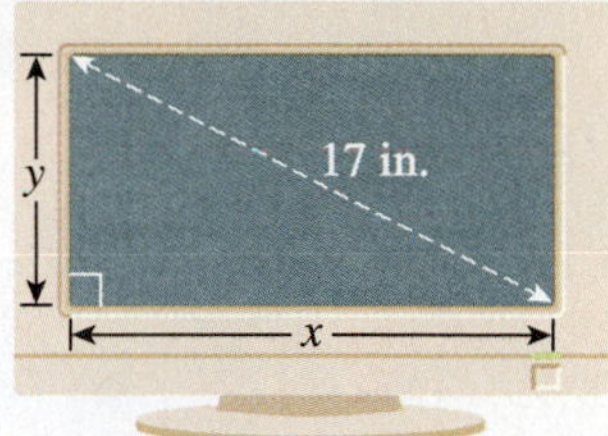

Because the perimeter of the monitor is 46 inches, we know that $2x + 2y = 46$. Also, because the diagonal of the monitor is 17 inches, we know from the Pythagorean theorem that $x^2 + y^2 = 17^2$, or $x^2 + y^2 = 289$.

Solve the perimeter equation for x, giving $x = 23 - y$; then substitute that expression for x into the equation $x^2 + y^2 = 289$.

$$x^2 + y^2 = 289$$

$$(23 - y)^2 + y^2 = 289 \quad \text{Substitute } 23 - y \text{ for } x.$$

$$2y^2 - 46y + 240 = 0 \quad \text{Square } 23 - y \text{ and collect all terms on the left side.}$$

$$2(y - 8)(y - 15) = 0 \quad \text{Factor completely.}$$

$$y = 8 \quad \text{or} \quad y = 15 \quad \text{Set each variable factor equal to 0 and solve.}$$

We can find the corresponding x-coordinates of the solutions by substituting these values for y in one of the two equations and solving for x. We will use the perimeter equation.

$y = 8$	$y = 15$	
$2x + 2(8) = 46$	$2x + 2(15) = 46$	Substitute for y.
$x = 15$	$x = 8$	Solve for x.

We see that the length is 15 inches if the width is 8 inches and that the length is 8 inches if the width is 15 inches. In either case, the dimensions of the rectangle are 8 inches by 15 inches.

Quick Check 6

A rectangle has a perimeter of 42 inches and a diagonal 15 inches long. Find the dimensions of the rectangle.

We conclude with an example involving the height of a projectile. Recall that if an object is launched with an initial velocity of v_0 feet per second from an initial height of s feet, its height in feet after t seconds is given by the function $h(t) = -16t^2 + v_0t + s$.

EXAMPLE 7 A rock is thrown upward from ground level with an initial velocity of 100 feet per second. At the same instant, a water balloon is dropped from a helicopter that is hovering 300 feet above the ground. At what time will the two objects be the same height above the ground? What is the height?

SOLUTION We begin by finding the time it takes for the two objects to be at the same height. The function describing the height of the rock is $h(t) = -16t^2 + 100t$, while the function describing the height of the water balloon dropped from the helicopter is $h(t) = -16t^2 + 300$. If we let the variable y represent the height of each object, we get the following nonlinear system of equations:

$$y = -16t^2 + 100t$$
$$y = -16t^2 + 300$$

Because each equation is already solved for y, we can set $-16t^2 + 100t$ equal to $-16t^2 + 300$ and solve for t.

$$-16t^2 + 100t = -16t^2 + 300$$

$$100t = 300 \quad \text{Add } 16t^2 \text{ to both sides of the equation. The resulting equation is linear.}$$

$$t = 3 \quad \text{Divide both sides by 100.}$$

The two objects will be at the same height after 3 seconds. We can substitute 3 for t in either function to find the height.

$$h(t) = -16t^2 + 100t$$

$$h(3) = -16(3)^2 + 100(3) \quad \text{Substitute 3 for } t.$$

$$= 156 \quad \text{Simplify.}$$

After 3 seconds, each object will be at a height of 156 feet.

Quick Check 7

A boy throws a ball upward from the ground with an initial velocity of 40 feet per second. At the same instant, a boy standing on a platform 10 feet above the ground throws another ball upward with an initial velocity of 35 feet per second. At what time will the two balls be the same height above the ground?

BUILDING YOUR STUDY STRATEGY

Preparing for a Final Exam (Part 1), 5 What to Review

Old Exams and Quizzes

When preparing for a final exam, begin by reviewing old exams and quizzes. Make sure you understand any mistakes you made on them.

Once you understand your errors, rework all of the problems on the exam or quiz without referring to your notes or textbook. Doing this will provide feedback on those topics you need to review and those topics you have under control.

Old Homework Assignments

When studying a particular topic, look back at your old homework for that topic. If you struggled with a particular type of problem, your homework will reflect that. Your homework may contain notes you made to yourself about how to do certain problems or mistakes to avoid.

Notes and Note Cards

Look over your class notes from that topic. Your notes should include examples of the problems in that section of the textbook, as well as pointers from your instructor.

The note cards you created during the semester should focus on problems you considered difficult at the time and contain strategies for solving these types of problems. A quick glance at these note cards will speed up your review.

Materials from Your Instructor

Materials from your instructor, such as a review sheet for the final or a practice final, are very helpful.

As you work through your study schedule, try to solve your instructor's problems when you review that section or chapter.

If you are having trouble with a certain problem, use a note card to write down the steps you need to follow to solve the problem. Review your note cards each day before the exam.

Textbook

Each chapter in the textbook contains a chapter review assignment and a chapter test. Also, cumulative review assignments are provided at the end of Chapters 4, 7, 11, and 14. Use the cumulative review exercises to prepare for a final exam. These exercise sets contain problems representative of the material covered up to that point in the text.

Try doing these exercises without referring to your notes, note cards, or the text. In this way, you can use these problems to determine which topics require more study. If you made a mistake while solving a problem, make note of the mistake and how to avoid it in the future. If there are problems you do not recognize or know how to begin, ask your instructor or a tutor for help.

Applied Problems

Many students have a difficult time with applied problems on a final exam. You will find the following strategy helpful:

- Make a list of the applied problems you have covered this semester.
- Create a study sheet for each type of problem.
- Write down an example or two of each type of problem.
- List the steps necessary to solve each type of problem.

Review these study sheets frequently as the exam approaches. This should help you identify the applied problems on the exam and to remember how to solve them.

Exercises 13.5

MyMathLab

PRACTICE

WATCH

DOWNLOAD
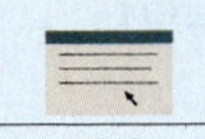
READ

REVIEW

Vocabulary

1. A(n) __________ system of equations is a system of equations in which the graph of at least one of the equations is not a line.
2. An ordered pair (x, y) is a(n) __________ of a nonlinear system of two equations if it is a solution of each equation.

Solve by the substitution method.

3. $x^2 + y^2 = 100$
$x - 7y = 50$

4. $x^2 + y^2 = 25$
$x - y = -1$

5. $x^2 + y^2 = 85$
$3x + y = -29$

6. $x^2 + y^2 = 50$
$-x + 2y = 15$

7. $4x^2 + y^2 = 16$
$y = 2x - 4$

8. $2x^2 + 3y^2 = 14$
$x = -3y + 7$

9. $y = x^2 - 3x + 8$
$-2x + y = 4$

10. $y = x^2 + 6x$
$-4x + y = 3$

11. $x^2 - y^2 = 8$
$y = x - 2$

12. $y^2 - x^2 = 7$
$y = x + 7$

13. $y = 3x - 9$
$x^2 + y^2 = 4$

14. $y = x - 2$
$(x + 2)^2 + (y - 1)^2 = 9$

15. $x^2 + y^2 = 25$
$y = x^2 - 5$

16. $x^2 + y^2 = 16$
$y = x^2 - 4$

17. $x^2 + y^2 = 30$
$x = y^2$

18. $x^2 + y^2 = 42$
$x = -y^2$

19. $x^2 + 4y^2 = 4$
$y = x^2 + 1$

20. $25x^2 + 9y^2 = 225$
$x = y^2 + 3$

21. $x^2 - y^2 = 1$
$x = y^2 + 11$

22. $x^2 - y^2 = 15$
$x = \frac{1}{3}y^2 - 1$

23. $x^2 - 4y^2 = 16$
$x = y^2 - 11$

24. $y^2 - 9x^2 = 9$
$y = x^2 + 1$

25. $y = x^2 - 7$
$y = -x^2 + 11$

26. $y = x^2 - 8$
$y = -x^2$

27. $y = x^2 - 6$
$x^2 + (y - 4)^2 = 8$

28. $y = x^2 + 1$
$x^2 + \frac{(y - 5)^2}{4} = 1$

Solve by the addition method.

29. $x^2 + y^2 = 5$
$x^2 - y^2 = 3$

30. $x^2 + y^2 = 20$
$y^2 - x^2 = 12$

31. $3x^2 + y^2 = 14$
$x^2 - y^2 = 2$

32. $5x^2 + y^2 = 45$
$x^2 - y^2 = 9$

33. $x^2 + 2y^2 = 18$
$y^2 - x^2 = 6$

34. $x^2 + 9y^2 = 9$
$y^2 - x^2 = 1$

35. $9x^2 + 4y^2 = 87$
$x^2 - 2y^2 = 6$

36. $8x^2 + 6y^2 = 44$
$4x^2 - 2y^2 = 12$

37. $x^2 + y^2 = 24$
$x^2 + 5y^2 = 60$

38. $x^2 + y^2 = 19$
$10x^2 + y^2 = 100$

39. $x^2 + y^2 = 6$
$4x^2 + 9y^2 = 39$

40. $x^2 + y^2 = 13$
$3x^2 + 18y^2 = 99$

41. $x^2 + 12y^2 = 117$
$x^2 + 4y^2 = 45$

42. $3x^2 + y^2 = 21$
$10x^2 + y^2 = 49$

43. $3x^2 + 2y^2 = 21$
$15x^2 + 4y^2 = 51$

44. $3x^2 + 4y^2 = 84$
$9x^2 + 21y^2 = 333$

45. A rectangular flower garden has a perimeter of 44 feet and an area of 105 square feet. Find the dimensions of the garden.

46. A rectangular classroom has a perimeter of 78 feet and an area of 360 square feet. Find the dimensions of the classroom.

47. The cover for a rectangular swimming pool measures 400 square feet. If the perimeter of the pool is 82 feet, find the dimensions of the pool.

48. The perimeter of a soccer field is 320 yards, and the area of the field is 6000 square yards. Find the dimensions of the soccer field.

49. Juan is remodeling his home office by replacing the molding around the floor of the room, as well as the carpet. The door to the office is 4 feet wide, and no molding is required there. If Juan uses 60 feet of molding and 192 square feet of carpet, find the dimensions of the office.

50. Rosalba is painting a rectangular wall in her home. The perimeter of the wall is 60 feet. The wall has two windows, each with an area of 15 square feet. If the painting area is 170 square feet, find the dimensions of the wall.

51. George's vegetable garden is in the shape of a rectangle and has 210 feet of fencing around it. Diagonally, the garden measures 75 feet from corner to corner. Find the dimensions of the garden.

52. A rectangular lawn has a perimeter of 100 feet and a diagonal of $10\sqrt{13}$ feet. Find the dimensions of the lawn.

53. A rectangular sports court has an area of 800 square feet. If the diagonal of the court measures $20\sqrt{5}$ feet, find the dimensions of the court.

54. Laura's horses live in a rectangular pasture that has an area of 10,000 square feet. The diagonal of the pasture measures $50\sqrt{17}$ feet. Find the dimensions of the pasture.

55. An arrow is fired upward with an initial velocity of 120 feet per second. At the same instant, a ball is dropped from a helicopter that is hovering 540 feet above the ground. At what time will the arrow and ball be the same height above the ground?

56. A ball is thrown upward from a beach with an initial velocity of 32 feet per second. At the same instant, another ball is dropped from a cliff 48 feet above the beach. At what time will the two balls be the same height above the beach?

57. A model rocket is fired upward from the ground with an initial velocity of 60 feet per second. At the same instant, another model rocket is fired upward with an initial velocity of 32 feet per second from a roof 84 feet above the ground.

a) At what time will the two rockets be the same height above the ground?

b) How high above the ground are the two rockets at that time?

58. A slingshot fires a rock upward from the ground with an initial velocity of 96 feet per second. At the same instant, a cannonball is fired upward with an initial velocity of 66 feet per second from a roof 120 feet above the ground.

a) At what time will the rock and cannonball be the same height above the ground?

b) How high above the ground are the two objects at that time?

Writing in Mathematics

Answer in complete sentences.

59. Write a word problem leading to a system of nonlinear equation whose solution is *The rectangle is 50 feet by 30 feet.*

CHAPTER 13 SUMMARY

Section 13.1 Parabolas

Graphing Equations of the Form $y = ax^2 + bx + c$, pp. 758–759

- Determine whether the parabola opens upward or downward.
- Find the vertex of the parabola. Use the formula $x = \frac{-b}{2a}$ to find the x-coordinate of the vertex.
- Find the y-intercept of the parabola by substituting 0 for x.
- If there are any x-intercepts, find them by letting $y = 0$ and solving for x.
- Find the axis of symmetry $\left(x = \frac{-b}{2a}\right)$ and use it to find other points on the parabola.

Graph: $y = x^2 - 4x - 2$

Opens upward; vertex: $(2, -6)$; y-intercept: $(0, -2)$; x-intercepts: $(2 - \sqrt{6}, 0)$, $(2 + \sqrt{6}, 0)$

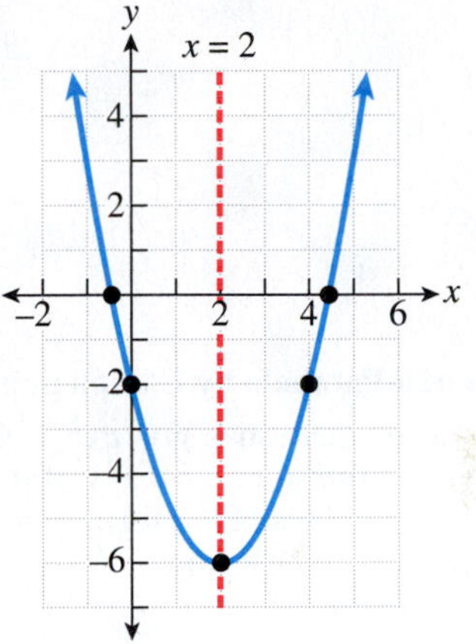

Graphing Equations of the Form $y = a(x - h)^2 + k$, pp. 759–760

- Determine whether the parabola opens upward or downward.
- Find the vertex of the parabola: (h, k).
- Shift the graph of $y = x^2$ (or $y = -x^2$ if the graph opens downward) by h units horizontally and by k units vertically.
- Find the y-intercept of the parabola by substituting 0 for x.
- If there are any x-intercepts, find them by letting $y = 0$ and solving for x.

Graph: $y = -(x + 1)^2 - 3$

Opens downward; vertex: $(-1, -3)$; shift one unit to the left and three units down; y-intercept: $(0, -4)$; x-intercepts: none

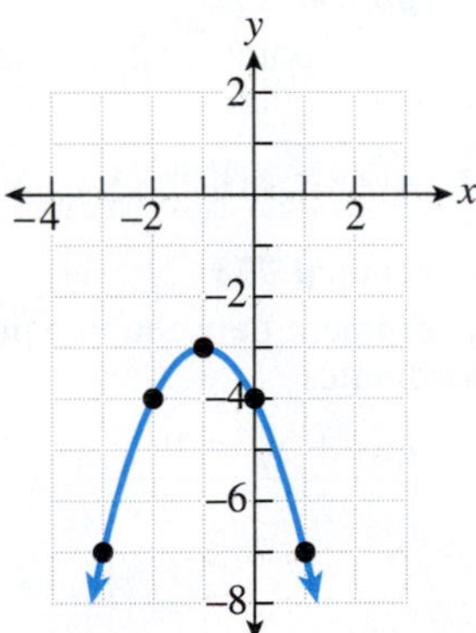

Graphing Equations of the Form $x = ay^2 + by + c$, pp. 761–762

- Determine whether the parabola opens to the right or to the left.
- Find the vertex of the parabola. Use the formula $y = \frac{-b}{2a}$ to find the y-coordinate of the vertex.
- Find the x-intercept of the parabola by substituting 0 for y.
- If there are any y-intercepts, find them by letting $x = 0$ and solving for y.
- Find the axis of symmetry $\left(x = \frac{-b}{2a}\right)$ and use it to find other points on the parabola.

Graph: $x = -y^2 - 2y + 15$

Opens left; vertex: $(16, -1)$; x-intercept: $(15, 0)$; y-intercepts: $(0, -5)$, $(0, 3)$

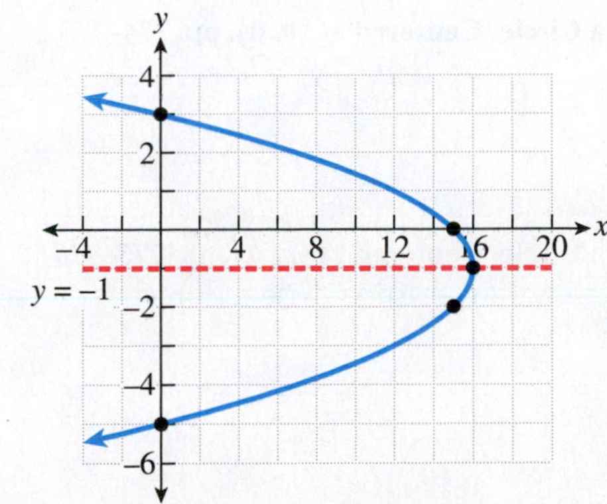

Graphing Equations of the Form $x = a(y - k)^2 + h$, pp. 763–764

- Determine whether the parabola opens to the right or to the left.
- Find the vertex of the parabola: (h, k).
- Shift the graph of $x = y^2$ (or $x = -y^2$ if the graph opens to the left) by h units horizontally and by k units vertically.
- Find the x-intercept of the parabola by substituting 0 for y.
- If there are any y-intercepts, find them by letting $x = 0$ and solving for y.

Graph: $x = (y + 3)^2 - 4$

Opens right; vertex: $(-4, -3)$; shift four units to left and three units down; x-intercept: $(5, 0)$; y-intercepts: $(0, -5)$, $(0, -1)$

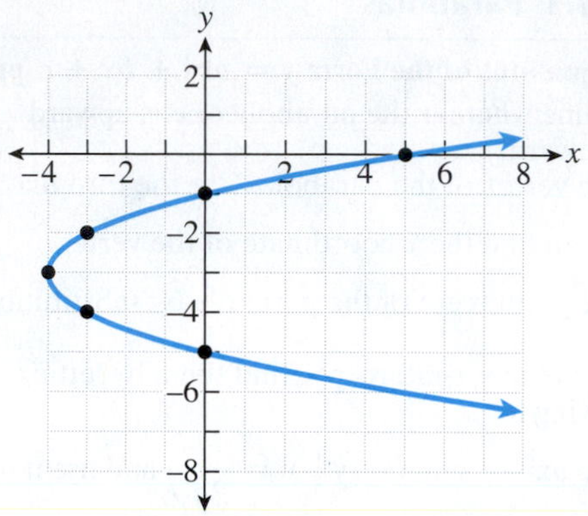

Finding the Vertex of a Parabola by Completing the Square, pp. 764–765

For the parabola whose equation is $y = ax^2 + bx + c$, find its vertex by completing the square for x: $y = a(x - h)^2 + k$.

If the equation of a parabola is of the form $x = ay^2 + by + c$, complete the square for y: $x = a(y - k)^2 + h$.

Find the vertex by completing the square: $x = y^2 - 18y + 75$

$$x = y^2 - 18y + 75$$
$$x - 75 = y^2 - 18y$$
$$x - 75 + 81 = y^2 - 18y + 81$$
$$x + 6 = (y - 9)^2$$
$$x = (y - 9)^2 - 6$$

Vertex: $(-6, 9)$

Section 13.2 Circles

The Distance Formula, pp. 771–772

The distance, d, between two points (x_1, y_1) and (x_2, y_2) is given by the formula

$$d = \sqrt{(x_2 - x_1)^2 + (y_2 - y_1)^2}.$$

Find the distance between the points: $(2, 7)$, $(11, -5)$

$$d = \sqrt{(11 - 2)^2 + (-5 - 7)^2} = \sqrt{225} = 15$$

Midpoint of a Line Segment, p. 773

The midpoint of a line segment between two points (x_1, y_1) and (x_2, y_2) has the coordinates

$$\left(\frac{x_1 + x_2}{2}, \frac{y_1 + y_2}{2}\right).$$

Find the midpoint of the points: $(5, -3)$, $(8, -11)$

$$\left(\frac{5 + 8}{2}, \frac{-3 + (-11)}{2}\right)$$

$$\left(\frac{13}{2}, -7\right)$$

Circle, pp. 773

A circle is the collection of points in a plane that are a fixed distance from a point called its center. The distance from the center to each point on the circle is called the radius of the circle.

Equation of a Circle, Centered at (0, 0), pp. 774–775

$$x^2 + y^2 = r^2$$

Find the center and radius:

$$x^2 + y^2 = 49$$

Center: $(0, 0)$; $r = \sqrt{49} = 7$

Equation of a Circle, Centered at (h, k), pp. 775–776

$$(x - h)^2 + (y - k)^2 = r^2$$

Find the center and radius:

$$(x - 6)^2 + (y + 3)^2 = 25$$

Center: $(6, -3)$; $r = \sqrt{25} = 5$

Graphing a Circle, pp. 774–776

Plot a point at its center.
Plot two points that are r units to the left and right of the center as well as two other points that are r units above and below the center.
Graph the circle that passes through these four points.

Graph: $(x + 2)^2 + (y - 4)^2 = 36$
Center: $(-2, 4)$; $r = \sqrt{36} = 6$

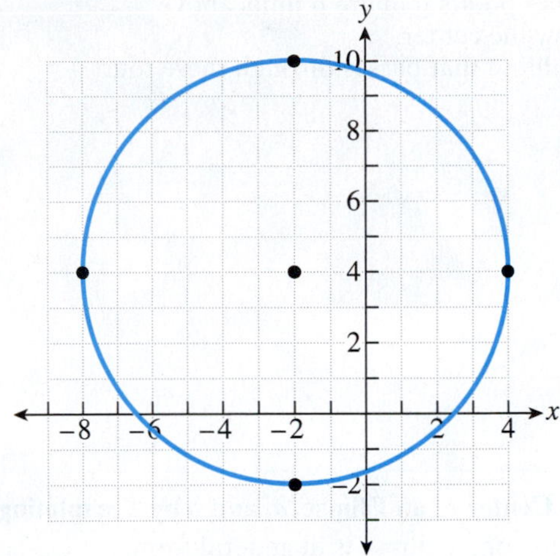

Finding the Center of a Circle and r by Completing the Square, pp. 776–777

If the equation of a circle is in general form, find the center as well as r by completing the square for both x and y.

Find the center and radius: $x^2 + y^2 - 6x + 18y - 10 = 0$

$$x^2 + y^2 - 6x + 18y - 10 = 0$$
$$x^2 - 6x + y^2 + 18y = 10$$
$$x^2 - 6x + 9 + y^2 + 18y + 81 = 10 + 9 + 81$$
$$(x - 3)^2 + (y + 9)^2 = 100$$

Center: $(3, -9)$; $r = \sqrt{100} = 10$

Section 13.3 Ellipses

Ellipse, pp. 784–785

An ellipse is the collection of all points in a plane for which the sum of the distances between the point and two fixed points called foci is a constant.

Equation of an Ellipse, Centered at (0, 0), pp. 785–787

$$\frac{x^2}{a^2} + \frac{y^2}{b^2} = 1$$

Find the center, a, and b:

$$\frac{x^2}{36} + \frac{y^2}{49} = 1$$

Center: $(0, 0)$; $a = \sqrt{36} = 6$; $b = \sqrt{49} = 7$

Equation of an Ellipse, Centered at (h, k), pp. 787–788

$$\frac{(x - h)^2}{a^2} + \frac{(y - k)^2}{b^2} = 1$$

Find the center, a, and b:

$$\frac{(x - 1)^2}{25} + \frac{(y + 7)^2}{4} = 1$$

Center: $(1, -7)$; $a = \sqrt{25} = 5$; $b = \sqrt{4} = 2$

Graphing an Ellipse, pp. 785–788

Plot a point at its center.
Plot two points that are a units to the left and right of the center.
Plot two other points that are b units above and below the center.
Graph the ellipse that passes through these four points.

Graph: $\dfrac{(x-2)^2}{16}+\dfrac{(y-3)^2}{9}=1$

Center: $(2, 3)$; $a=\sqrt{16}=4$; $b=\sqrt{9}=3$

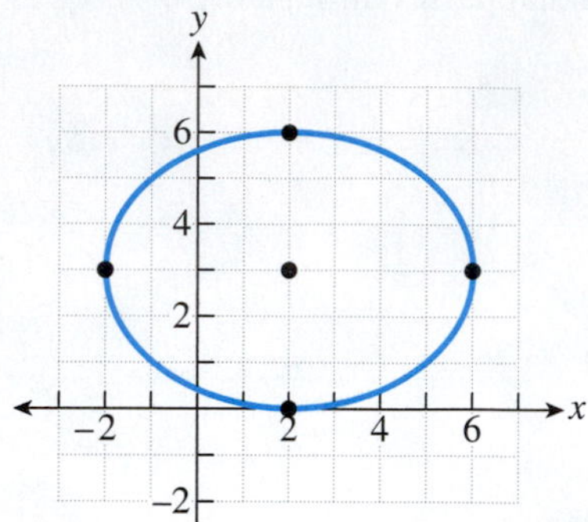

Finding the Center of an Ellipse, a, and b by Completing the Square, pp. 788–789

If the equation of an ellipse is in general form, find the center as well as a and b by completing the square for both x and y.

Find the center, a, and b: $16x^2+25y^2-160x+200y+400=0$

$$16x^2+25y^2-160x+200y+400=0$$
$$16(x^2-10x+25)+25(y^2+8y+16)$$
$$=-400+16\cdot25+25\cdot16$$
$$16(x-5)^2+25(y+4)^2=400$$
$$\frac{(x-5)^2}{25}+\frac{(y+4)^2}{16}=1$$

Center: $(5, -4)$; $a=\sqrt{25}=5$; $b=\sqrt{16}=4$

Section 13.4 Hyperbolas

Hyperbola, pp. 797–798

A hyperbola is the collection of all points in a plane for which the difference of the distances between the point and two foci is a constant. The graph of a hyperbola has two branches.

Equation of a Hyperbola, Centered at (0, 0), pp. 798–801

Centered at origin, branches opening to the left and right:

$$\frac{x^2}{a^2}-\frac{y^2}{b^2}=1$$

Centered at origin, branches opening upward and downward:

$$\frac{y^2}{b^2}-\frac{x^2}{a^2}=1$$

Opening to the left and right:

Find the center, a, and b: $\dfrac{x^2}{9}-\dfrac{y^2}{64}=1$

Center: $(0, 0)$; $a=\sqrt{9}=3$; $b=\sqrt{64}=8$

Opening upward and downward:

Find the center, a, and b: $\dfrac{y^2}{25}-\dfrac{x^2}{121}=1$

Center: $(0, 0)$; $a=\sqrt{121}=11$; $b=\sqrt{25}=5$

Equation of a Hyperbola, Centered at (h, k), pp. 801–803

Centered at (h, k), branches opening to the left and right:

$$\frac{(x-h)^2}{a^2}-\frac{(y-k)^2}{b^2}=1$$

Centered at (h, k), branches opening upward and downward:

$$\frac{(y-k)^2}{b^2}-\frac{(x-h)^2}{a^2}=1$$

Opening to the left and right:

Find the center, a, and b: $\dfrac{(x-4)^2}{49}-\dfrac{(y-3)^2}{16}=1$

Center: $(4, 3)$; $a=\sqrt{49}=7$; $b=\sqrt{16}=4$

Opening upward and downward:

Find the center, a, and b: $\dfrac{(y-6)^2}{4}-\dfrac{(x+2)^2}{81}=1$

Center: $(-2, 6)$; $a=\sqrt{81}=9$; $b=\sqrt{4}=2$

Graphing a Hyperbola Whose Branches Open to the Left and Right, pp. 798–803

Plot a point at its center.
Plot the vertices, which are a units to the left and right of the center.
Graph the asymptotes using dashed lines.
Graph the branches in such a way that each branch passes through a vertex and approaches the asymptotes.

Graph: $\frac{x^2}{25} - \frac{y^2}{4} = 1$
Center: $(0, 0)$; $a = \sqrt{25} = 5$; $b = \sqrt{4} = 2$
Vertices: $(-5, 0)$, $(5, 0)$

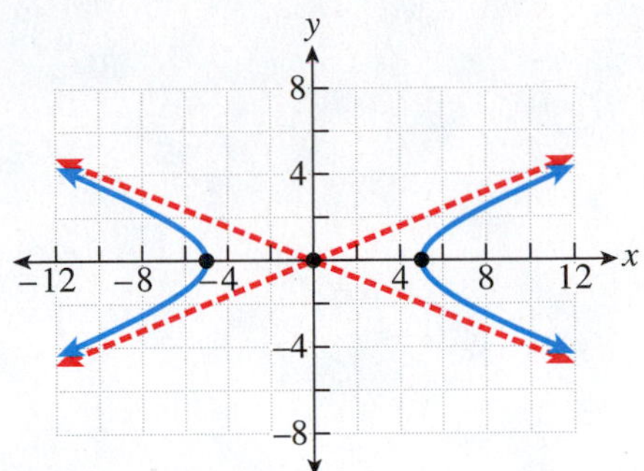

Graphing a Hyperbola Whose Branches Open Upward and Downward, pp. 798–803

Plot a point at its center.
Plot the vertices, which are b units above and below the center.
Graph the asymptotes using dashed lines.
Graph the branches in such a way that each branch passes through a vertex and approaches the asymptotes.

Graph: $\frac{(y+3)^2}{16} - \frac{(x-4)^2}{4} = 1$
Center: $(4, -3)$; $a = \sqrt{4} = 2$; $b = \sqrt{16} = 4$
Vertices: $(4, 1)$, $(4, -7)$

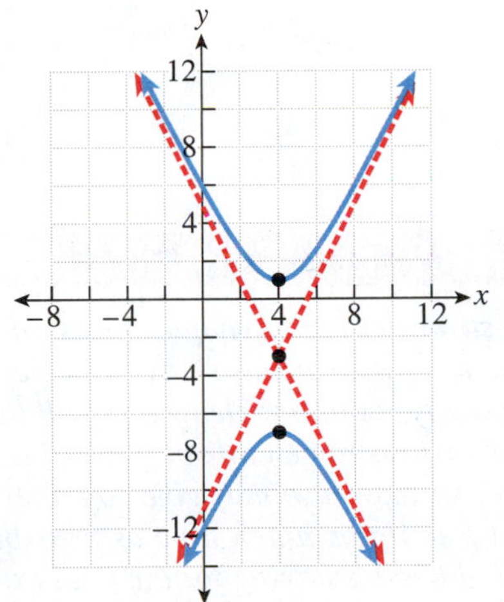

Finding the Center of a Hyperbola, a, and b by Completing the Square, pp. 803–804

If the equation of a hyperbola is in general form, find the center as well as a and b by completing the square for both x and y.

Find the center, a, and b: $9x^2 - 4y^2 - 72x - 24y + 72 = 0$

$$9(x^2 - 8x + 16) - 4(y^2 + 6y + 9) = -72 + 9 \cdot 16 - 4 \cdot 9$$
$$9(x-4)^2 - 4(y+3)^2 = 36$$
$$\frac{(x-4)^2}{4} - \frac{(y+3)^2}{9} = 1$$

Center: $(4, -3)$; $a = \sqrt{4} = 2$; $b = \sqrt{9} = 3$

Section 13.5 Nonlinear Systems of Equations

Nonlinear System of Equations, pp. 813–814

A nonlinear system of equations is a system of equations in which the graph of at least one of the equations is not a line.

$$x^2 + y^2 = 25$$
$$y = 3x - 5$$

Solving a Nonlinear System of Two Equations in Two Unknowns, pp. 814–817

Nonlinear systems of equations can be solved by using the substitution method or the addition method.

Solve: $\begin{aligned} x^2 + y^2 &= 25 \\ y &= x - 1 \end{aligned}$

Substitution method

$$\begin{aligned} x^2 + (x - 1)^2 &= 25 \\ x^2 + x^2 - 2x + 1 &= 25 \\ 2x^2 - 2x - 24 &= 0 \\ 2(x^2 - x - 12) &= 0 \\ 2(x + 3)(x - 4) &= 0 \end{aligned}$$

$x = -3$: $y = (-3) - 1 = -4$ $x = 4$: $y = (4) - 1 = 3$

$(-3, -4), (4, 3)$

Solve: $\begin{aligned} x^2 + y^2 &= 26 \\ 2x^2 + 3y^2 &= 53 \end{aligned}$

Addition method

$$\begin{aligned} x^2 + y^2 &= 26 \xrightarrow{\text{Multiply by } -2.} & -2x^2 - 2y^2 &= -52 \\ 2x^2 + 3y^2 &= 53 & 2x^2 + 3y^2 &= 53 \\ & & y^2 &= 1 \end{aligned}$$

$y = -1$:
$$\begin{aligned} x^2 + (-1)^2 &= 26 \\ x^2 + 1 &= 26 \\ x^2 &= 25 \\ x &= \pm 5 \end{aligned}$$

$y = 1$:
$$\begin{aligned} x^2 + (1)^2 &= 26 \\ x^2 + 1 &= 26 \\ x^2 &= 25 \\ x &= \pm 5 \end{aligned}$$

$(-5, -1), (5, -1), (-5, 1), (5, 1)$

SUMMARY OF CHAPTER 13 STUDY STRATEGIES

The study strategies in this chapter focus on how to prepare to take a final exam and serve as a summary of the various study strategies presented throughout the textbook. After determining the types of problems that will be on the exam, you can look for problems to practice in various reviews throughout this textbook, in your notes, and on your old exams. Your instructor also may give you a review assignment for the final exam.

Try to free up as much time as possible in your daily schedule to maximize your studying and begin studying for the final exam at least two weeks before the actual exam.

Study strategies for preparing for a final exam will continue in Chapter 14.

CHAPTER 13 REVIEW

Graph. Label the vertex and any intercepts. **[13.1]**

1. $y = x^2 - 8x - 9$

2. $y = -x^2 + 6x - 3$

3. $y = (x - 4)^2 - 1$

4. $y = -(x + 3)^2 - 4$

5. $x = y^2 - 10y + 15$

6. $x = y^2 + 3y - 28$

7. $x = -y^2 + y + 20$

8. $x = -y^2 - 6y - 14$

9. $x = (y + 5)^2 - 4$

10. $x = -(y + 1)^2 + 10$

Worked-out solutions to Review Exercises marked with ● can be found on page AN-59.

11. $x = -(y - 4)^2 + 9$

12. $x = (y - 3)^2 + 6$

***Find the distance between the given points. Round to the nearest tenth if necessary.* [13.2]**

13. $(-7, -8)$ and $(1, 7)$

14. $(3, -5)$ and $(-6, -2)$

***Graph the circle. Label the center of the circle. Give its radius as well.* [13.2]**

15. $x^2 + y^2 = 36$

16. $x^2 + y^2 = 20$

17. $(x + 3)^2 + (y - 4)^2 = 4$

18. $(x - 8)^2 + (y + 3)^2 = 9$

19. $(x + 4)^2 + (y - 5)^2 = 25$

20. $(x - 2)^2 + y^2 = 16$

21. $x^2 + y^2 + 12x - 2y - 12 = 0$

***Find the center and radius of the circle by completing the square.* [13.2]**

22. $x^2 + y^2 + 6x - 18y - 10 = 0$

23. $x^2 + y^2 - 8x - 20y + 99 = 0$

***Find the equation of the circle.* [13.2]**

24.

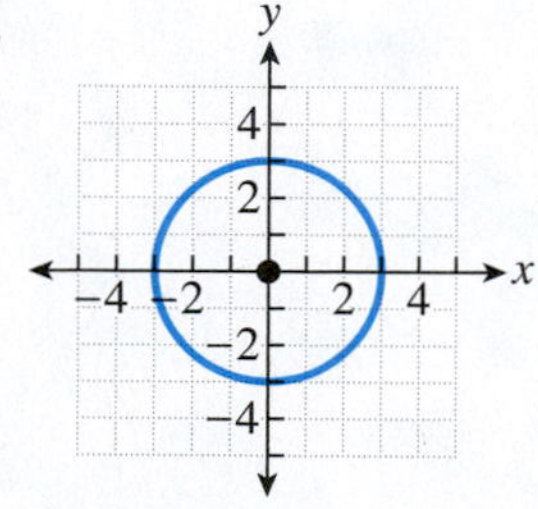

25.

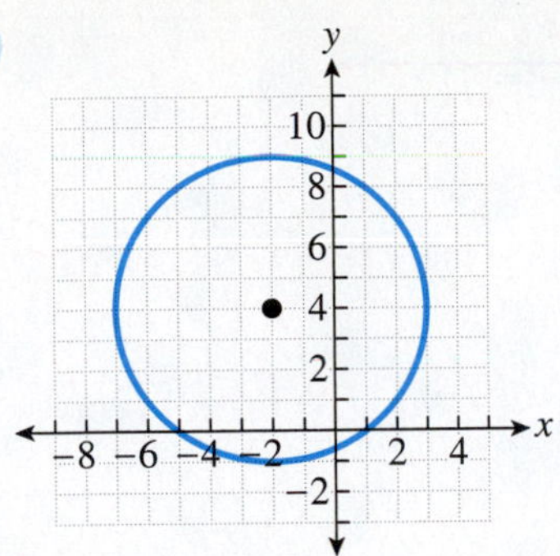

***Graph the ellipse. Label the center. Give the values of a and b.* [13.3]**

26. $\dfrac{x^2}{25} + \dfrac{y^2}{9} = 1$

27. $x^2 + \dfrac{y^2}{16} = 1$

28. $\dfrac{(x + 3)^2}{9} + \dfrac{(y - 4)^2}{16} = 1$

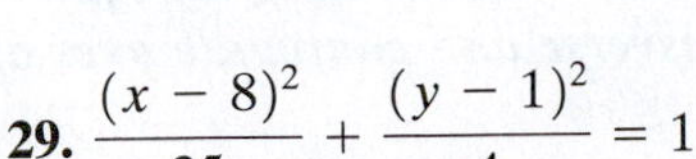

29. $\dfrac{(x - 8)^2}{25} + \dfrac{(y - 1)^2}{4} = 1$

30. $\dfrac{(x + 6)^2}{9} + \dfrac{y^2}{36} = 1$

31. $9x^2 + 16y^2 + 90x - 64y + 145 = 0$

***Find the center of the ellipse by completing the square. Find the values of a and b as well.* [13.3]**

32. $4x^2 + 49y^2 - 48x - 98y - 3 = 0$

33. $81x^2 + 16y^2 + 324x + 128y - 716 = 0$

***Find the equation of the ellipse.* [13.3]**

34.

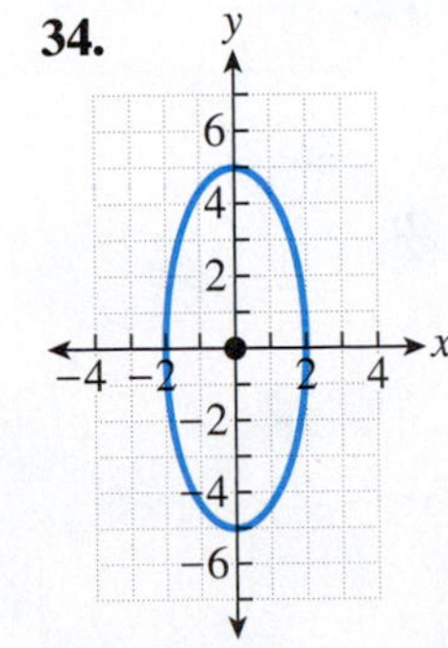

35.

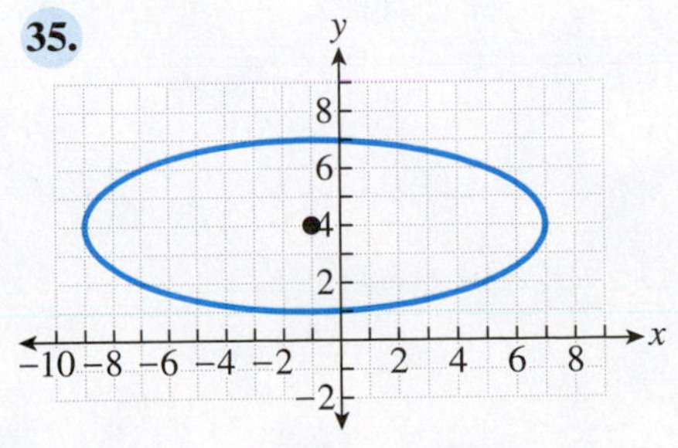

***Graph the hyperbola. Label the center. Give the values of a and b.* [13.4]**

36. $\dfrac{x^2}{9} - \dfrac{y^2}{25} = 1$

37. $\dfrac{y^2}{16} - \dfrac{x^2}{16} = 1$

38. $\dfrac{(x - 1)^2}{4} - \dfrac{(y - 3)^2}{9} = 1$

39. $\dfrac{(x + 4)^2}{25} - \dfrac{(y - 2)^2}{36} = 1$

40. $\dfrac{(y + 3)^2}{4} - \dfrac{(x + 2)^2}{81} = 1$

41. $-4x^2 + 25y^2 - 24x + 300y + 764 = 0$

***Find the center of the hyperbola by completing the square. Find the lengths of its transverse and conjugate axes as well.* [13.4]**

42. $-9x^2 + 4y^2 + 90x - 56y - 65 = 0$

43. $49x^2 - 36y^2 + 784x + 288y + 796 = 0$

Find the equation of the hyperbola. **[13.4]**

44.

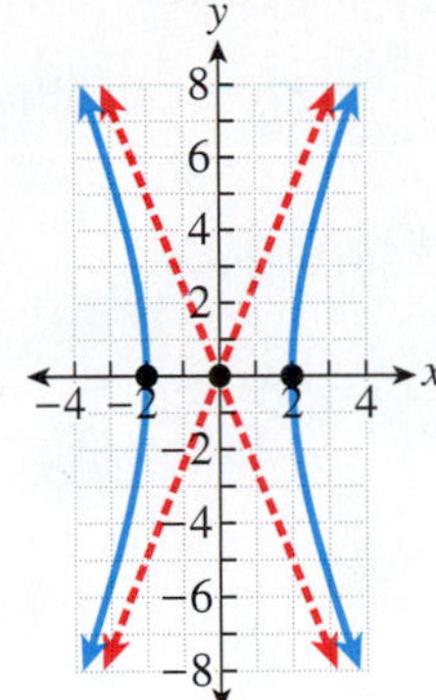

45.

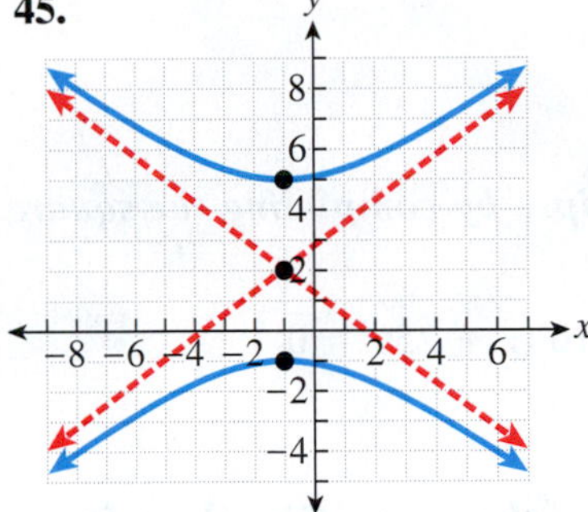

Solve by substitution. **[13.5]**

46. $x^2 + y^2 = 50$
$x + 2y = 15$

47. $4x^2 + 3y^2 = 48$
$y = 2x - 4$

48. $y = x^2 - 8x + 23$
$2x - y = 1$

49. $4y^2 - x^2 = 96$
$6y - x = 32$

50. $x^2 + y^2 = 17$
$y = x^2 + 3$

51. $3x^2 + 5y^2 = 72$
$y = x^2 - 6$

52. $9x^2 - y^2 = 9$
$y = x^2 - 1$

53. $y = -x^2 + 2x + 12$
$y = x^2$

Solve by the addition method. **[13.5]**

54. $x^2 + y^2 = 100$
$y^2 - x^2 = 62$

55. $3x^2 + 4y^2 = 31$
$x^2 - 2y^2 = 7$

56. $x^2 + y^2 = 8$
$2x^2 + 15y^2 = 68$

57. $6x^2 + y^2 = 159$
$2x^2 + 9y^2 = 131$

58. A rectangular flower garden has a perimeter of 70 feet and an area of 300 square feet. Find the dimensions of the garden. [13.5]

59. A rectangle has a perimeter of 42 inches and a diagonal of 15 inches. Find the dimensions of the rectangle. [13.5]

60. A projectile is fired upward from the ground with an initial velocity of 76 feet per second. At the same instant, another projectile is fired upward with an initial velocity of 41 feet per second from a roof 140 feet above the ground. [13.5]

a) After how long will the two projectiles be the same height above the ground?

b) How high above the ground are the two projectiles at that time?

CHAPTER 13 TEST

For Extra Help

Step-by-step test solutions are found on the Chapter Test Prep Videos available via the Video Resources on DVD, in *MyMathLab*, and on YouTube (search "WoodburyElemIntAlg" and click on "Channels").

Graph. Label the vertex and any intercepts.

1. $y = x^2 - 4x + 7$

2. $y = -(x + 3)^2 + 9$

3. $x = -y^2 + 9y - 14$

4. $x = (y + 1)^2 - 8$

Find the distance between the given points. Round to the nearest tenth if necessary.

5. $(7, -3)$ and $(3, 5)$

Graph the circle. Identify the center and radius of the circle.

6. $(x + 4)^2 + (y - 2)^2 = 25$

Find the center and radius of the circle by completing the square.

7. $x^2 + y^2 - 12x - 6y + 13 = 0$

Find the equation of the circle with the given center and radius.

8. Center $(3, -8)$, radius 16

Graph the ellipse. Label the center. Give the values of a and b.

9. $\dfrac{(x - 4)^2}{16} + \dfrac{(y + 7)^2}{9} = 1$

Find the center of the ellipse by completing the square. Give the values of a and b.

10. $4x^2 + 25y^2 - 64x + 100y + 256 = 0$

Graph the hyperbola. Label the center. Give the values of a and b.

11. $\dfrac{x^2}{16} - y^2 = 1$

Find the equation of the conic section that has been graphed.

12.

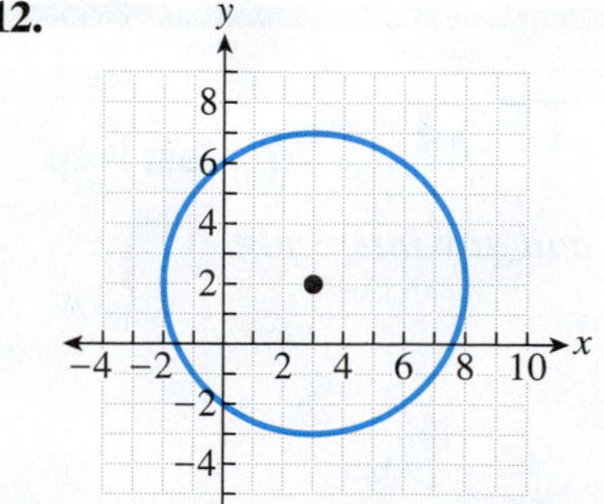

13. 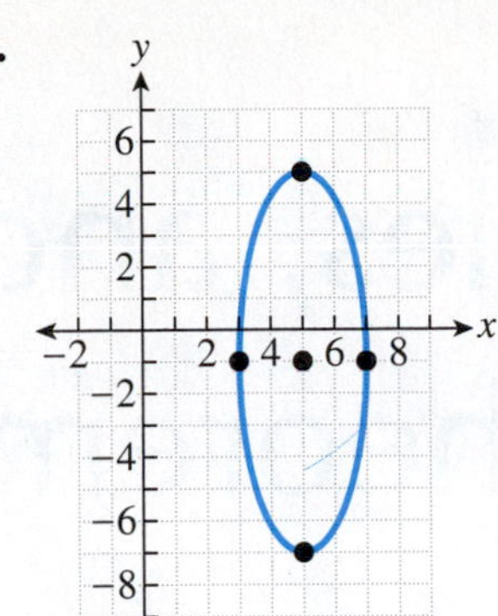

Solve the nonlinear system.

14. $x^2 + y^2 = 100$
$x + 3y = 10$

15. $x^2 + 6y^2 = 49$
$x - 2y = 1$

16. $y^2 - x^2 = 10$
$y = x^2 - 32$

17. $x^2 + 4y^2 = 85$
$3x^2 + 2y^2 = 165$

18. A rectangular corral has a perimeter of 340 feet and an area of 6000 square feet. Find the dimensions of the corral.

Mathematicians in History

Sofia Kovalevskaya The story of 19th-century Russian mathematician Sofia Kovalevskaya is one of genius and determination. At a time when women were discouraged from studying mathematics, she became one of the most respected mathematicians of her day.

Write a one-page summary (*or* make a poster) of the life of Sofia Kovalevskaya and her accomplishments.

Interesting issues:

- Where and when was Sofia Kovalevskaya born?
- With what were the walls of her bedroom papered when she was 11 years old?
- Who convinced Kovalevskaya's parents to allow her to study mathematics?
- Whom did Kovalevskaya marry? What was his fate?
- By the spring of 1874, Kovalevskaya had completed three papers while studying with which prominent mathematician in Berlin?
- What job did Kovalevskaya hold after earning her doctorate?
- In what European city did Kovalevskaya finally obtain a position?
- When did Kovalevskaya die? What were the circumstances of her death?
- Kovalevskaya once said, "It is impossible to be a mathematician without being a poet in soul." Explain what you think Kovalevskaya meant by that statement.

CHAPTER 14

Sequences, Series, and the Binomial Theorem

In this chapter, we will learn about *sequences*, which are ordered lists of numbers. We also will investigate *series*, which are the sums of the numbers in a sequence. In particular, we will discuss two specific types of sequences and series: the arithmetic sequence and series and the geometric sequence and series.

The chapter concludes with a section on the binomial theorem, which provides a method for raising a binomial to a power that is a natural number, such as $(2x + 3y)^9$.

STUDY STRATEGY

Using the Previous Study Strategies to Prepare for a Final Exam (Part 2)
The study strategies in this chapter continue to focus on how to incorporate the study strategies from previous chapters into your preparation for a final exam.

14.1 Sequences and Series

OBJECTIVES

1. Find the terms of a sequence, given its general term.
2. Find the general term of a sequence.
3. Find partial sums of a sequence.
4. Use summation notation to evaluate a series.
5. Use sequences to solve applied problems.

Sequences

Suppose you deposited \$100 in a bank account that pays 6% interest, compounded monthly. Here is the amount of money in the account after each of the first 5 years if no further deposits or withdrawals are made.

At the end of year . . .	Principal Balance
1	\$106.17
2	\$112.72
3	\$119.67
4	\$127.05
5	\$134.89

Using this information, we can create a function whose input is a natural number from 1 to 5 and whose output is the balance of the account after that number of years. This type of function is called a sequence.

Sequence

A **sequence** is a function whose domain is a set of consecutive natural numbers beginning with 1. If the domain is the entire set of natural numbers $\{1, 2, 3, \ldots\}$, the sequence is an **infinite sequence**. If the domain is the set $\{1, 2, 3, \ldots, n\}$, for some natural number n, the sequence is a **finite sequence**.

The input for the function gives the order of the term in the sequence, while the output of the function is the actual term. We denote the first number in a sequence as a_1, the second number in a sequence as a_2, and so on. In the example of the bank account, $a_1 = \$106.17$, $a_2 = \$112.72$, and so on.

Finding the Terms of a Sequence

Objective 1 Find the terms of a sequence, given its general term. A sequence is often denoted by its **general term**, a_n. The general term of a sequence provides a formula for finding the nth term of a sequence.

EXAMPLE 1 Find the first five terms of the sequence whose general term is $a_n = n^2 - 10$.

SOLUTION Substitute the first five natural numbers for n in the general term.

$$a_1 = (1)^2 - 10 = -9 \qquad a_2 = (2)^2 - 10 = -6 \qquad a_3 = (3)^2 - 10 = -1$$
$$a_4 = (4)^2 - 10 = 6 \qquad a_5 = (5)^2 - 10 = 15$$

The first five terms of the sequence are $-9, -6, -1, 6$, and 15. The entire sequence could be represented as $-9, -6, -1, 6, 15, \ldots$.

Quick Check 1

Find the first four terms of the sequence whose general term is $a_n = \dfrac{1}{3n + 4}$.

A sequence whose terms alternate between being positive and negative is called an **alternating sequence**. An alternating sequence has a factor of $(-1)^n$ or $(-1)^{n+1}$ in its general term. Each time the value of n increases by 1, both $(-1)^n$ and $(-1)^{n+1}$ alternate between -1 and 1.

EXAMPLE 2 Find the first four terms of the sequence whose general term is $a_n = \frac{(-1)^n}{3^{n+2}}$.

SOLUTION Substitute the first four natural numbers for n in the general term.

$$a_1 = \frac{(-1)^1}{3^{1+2}} = \frac{-1}{3^3} = -\frac{1}{27} \qquad a_2 = \frac{(-1)^2}{3^{2+2}} = \frac{1}{3^4} = \frac{1}{81}$$

$$a_3 = \frac{(-1)^3}{3^{3+2}} = \frac{-1}{3^5} = -\frac{1}{243} \qquad a_4 = \frac{(-1)^4}{3^{4+2}} = \frac{1}{3^6} = \frac{1}{729}$$

The first four terms of the sequence are $-\frac{1}{27}, \frac{1}{81}, -\frac{1}{243}$, and $\frac{1}{729}$.

Quick Check 2

Find the first four terms of the sequence whose general term is $a_n = \frac{(-1)^{n+1}}{2n}$.

Finding the General Term of a Sequence

Objective 2 Find the general term of a sequence. To find the general term of a sequence, we must be able to detect the pattern of the terms of the sequence; consequently, we must know enough terms of the sequence. Suppose we were trying to find the general term of the sequence 2, 4, ... There are not enough terms to determine the pattern. It could be the sequence whose general term is $a_n = 2^n$ (2, 4, 8, 16, 32, ...), or it could be the sequence whose general term is $a_n = 2n$ (2, 4, 6, 8, 10, ...). There are simply not enough terms to know for sure.

EXAMPLE 3 Find the next three terms of the sequence 3, 6, 9, 12, 15, ..., and find its general term, a_n.

SOLUTION All of the terms are multiples of 3; so continuing the pattern gives the terms 18, 21, and 24. The general term is $a_n = 3n$.

When examining the terms of a sequence, determine whether the successive terms increase by the same amount each time. If they do, the general term, a_n, will contain a multiple of n.

EXAMPLE 4 Find the next three terms of the sequence 9, 13, 17, 21, 25, ... and find its general term, a_n.

SOLUTION First, we should notice that each successive term increases by 4. So the next three terms are 29, 33, and 37. When each successive term increases by 4, this tells us that the general term will contain the expression $4n$. However, $4n$ is not the general term of this sequence, as its terms would be 4, 8, 12, 16, 20, ... Notice that each term is five less than the corresponding term in the sequence 9, 13, 17, 21, 25, ... The general term can be found by adding 5 to $4n$; so the general term is $a_n = 4n + 5$.

Now we will turn to finding the general term of an alternating sequence.

EXAMPLE 5 Find the next three terms of the sequence −1, 4, −9, 16, −25, ... and find its general term, a_n.

SOLUTION Notice that if we disregard the signs, the first five terms are the squares of 1, 2, 3, 4, and 5. Because the signs of the terms are alternating, the next term will be positive. The next three terms are 36, −49, and 64.

When an alternating sequence begins with a negative term, the general term will contain the factor $(-1)^n$. [If the first term had been positive, the general term would have contained the factor $(-1)^{n+1}$.] The general term for this sequence is $a_n = (-1)^n \cdot n^2$.

Quick Check 3

Find the next three terms of the sequence $7, -13, 19, -25, 31, \ldots$ and find its general term a_n.

On occasion, we represent a sequence by giving its first term and explaining how to find any term after that from the preceding term. In this case, we are defining the general term **recursively**. For example, recall the sequence $9, 13, 17, 21, 25, \ldots$ from Example 4. We found its general term to be $a_n = 4n + 5$. Because each term in the sequence is four more than the previous term, we could define the general term recursively as $a_n = a_{n-1} + 4$, where a_{n-1} is the term that precedes the term a_n. A recursive formula for a general term, a_n, defines it in terms of the preceding term, a_{n-1}.

EXAMPLE 6 Find the first five terms of the sequence whose first term is $a_1 = 19$ and whose general term is $a_n = a_{n-1} - 7$.

SOLUTION We know that the first term is $a_1 = 19$. We can now use the recursive formula $a_n = a_{n-1} - 7$ to find the next four terms.

$$a_2 = a_{2-1} - 7 = a_1 - 7 = 19 - 7 = 12$$
$$a_3 = a_{3-1} - 7 = a_2 - 7 = 12 - 7 = 5$$
$$a_4 = a_{4-1} - 7 = a_3 - 7 = 5 - 7 = -2$$
$$a_5 = a_{5-1} - 7 = a_4 - 7 = -2 - 7 = -9$$

The first five terms of the sequence are 19, 12, 5, -2, and -9.

Quick Check 4

Find the first five terms of the sequence whose first term is $a_1 = 2$ and whose general term is $a_n = (a_{n-1})^2 + 1$.

Series

We now turn our attention to series, which are sums of the terms in a sequence.

Infinite Series, Finite Series

An **infinite series**, s, is the sum of the terms in an infinite sequence.

$$s = a_1 + a_2 + a_3 + \cdots + a_n + \cdots$$

A **finite series**, s_n, is the sum of the first n terms of a sequence.

$$s_n = a_1 + a_2 + a_3 + \cdots + a_n$$

The finite series s_n is also called the ***n*th partial sum** of the sequence.

Finding Partial Sums

Objective 3 Find partial sums of a sequence.

EXAMPLE 7 For the sequence whose general term is $a_n = (-1)^n \cdot (2n + 1)$, find s_6.

SOLUTION The first six terms of the sequence are $a_1 = -3$, $a_2 = 5$, $a_3 = -7$, $a_4 = 9$, $a_5 = -11$, and $a_6 = 13$. Now find the partial sum.

$$s_6 = (-3) + 5 + (-7) + 9 + (-11) + 13 = 6$$

Quick Check 5

For the sequence whose general term is $a_n = n^2 + 3n$, find s_4.

EXAMPLE 8 For the sequence whose first term is $a_1 = -4$ and whose general term is $a_n = \frac{a_{n-1}}{2}$, find s_5.

SOLUTION We know that $a_1 = -4$. Because each term in the sequence is half of the preceding term, $a_2 = -2$, $a_3 = -1$, $a_4 = -\frac{1}{2}$, and $a_5 = -\frac{1}{4}$. We can now find the partial sum.

$$s_5 = (-4) + (-2) + (-1) + \left(-\frac{1}{2}\right) + \left(-\frac{1}{4}\right) = -7\frac{3}{4}$$

Quick Check 6

For the sequence whose first term is $a_1 = 1$ and whose general term is $a_n = 3a_{n-1} + 4$, find s_4.

Summation Notation

Objective 4 Use summation notation to evaluate a series. The nth partial sum of a finite sequence, $s_n = a_1 + a_2 + a_3 + \cdots + a_n$, can be represented as $\sum_{i=1}^{n} a_i$. This notation for a partial sum is called **summation notation**. The number n is the **upper limit of summation**, while the number 1 is the **lower limit of summation**. The variable i is called the **index of summation**, or simply the **index**. The symbol Σ is the capital Greek letter sigma that is often used in mathematics to represent a sum. This notation is also called **sigma notation**.

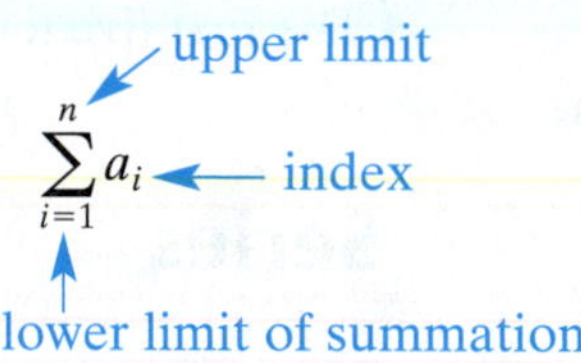

EXAMPLE 9 Find the sum $\sum_{i=1}^{8}(3i + 5)$.

SOLUTION In this example, we need to find the sum of the first eight terms of this sequence. The first term of the sequence is 8, and each successive term increases by 3.

$$\sum_{i=1}^{8}(3i + 5) = 8 + 11 + 14 + 17 + 20 + 23 + 26 + 29$$ Rewrite as the sum of the first eight terms of the sequence.

$$= 148$$ Add.

Quick Check 7

Find the sum $\sum_{i=1}^{4} 6i$.

Applications

Objective 5 Use sequences to solve applied problems. We will conclude this section with an applied problem involving sequences.

EXAMPLE 10 Erin found a new job with a starting salary of \$40,000 per year. Each year she will be given a 5% raise. Write a sequence showing Erin's annual salary for the first 5 years and find the total amount she will earn during those 5 years.

SOLUTION Because Erin earned \$40,000 in the first year, $a_1 = \$40,000$. To find her salary for the second year, we calculate her 5% raise and add it to her salary. Five percent of \$40,000 can be calculated by multiplying 0.05 by \$40,000.

$$0.05(\$40,000) = \$2000$$

With her \$2000 raise, Erin's salary for the second year is \$42,000; so $a_2 = \$42,000$. Continuing in the same fashion, we find that the sequence is \$40,000, \$42,000, \$44,100, \$46,305, \$48,620.25. During the first 5 years, Erin will earn \$221,025.25.

Quick Check 8

A copy machine was purchased for \$8000. Its value decreases by 20% each year. Write a sequence showing the value of the copy machine at the end of each of the first four years.

BUILDING YOUR STUDY STRATEGY

Preparing for a Final Exam (Part 2), 1 Practice Quizzes Creating and taking practice quizzes is an effective way to prepare for a final exam. Begin by creating a practice test for an entire chapter. For problems you struggle with, review the material before trying another practice test. Once you feel comfortable with one chapter, create a practice test for another chapter. Repeat until you have covered all of the chapters.

For further information on practice quizzes, see the Study Strategies in Chapter 12.

Exercises 14.1

PRACTICE

WATCH

DOWNLOAD

READ

REVIEW

Vocabulary

1. A(n) ________ is a function whose domain is a set of consecutive natural numbers beginning with 1.
2. If the domain of a sequence is the entire set of natural numbers, the sequence is a(n) ________ sequence.
3. The ________ term of a sequence is often denoted as a_n.
4. A sequence whose terms alternate between being positive and negative is called a(n) ________ sequence.
5. The sum of the terms of a sequence is called a(n) ________.
6. A(n) ________ series, s, is the sum of the terms in an infinite sequence.
7. A(n) ________ series, s_n, is the sum of the first n terms of a sequence.
8. A finite series, s_n, is also called the nth ________ of the sequence.

9. Find the fifth term of the sequence whose general term is $a_n = 3n - 4$.
10. Find the third term of the sequence whose general term is $a_n = (n + 3)^2 - 5n$.
11. Find the sixth term of the sequence whose general term is $a_n = (-1)^n \cdot 3^{n+1}$.
12. Find the ninth term of the sequence whose general term is $a_n = (-1)^n \cdot \dfrac{1}{2n - 1}$.

Find the first five terms of the sequence with the given general term.

13. $a_n = 4n$
14. $a_n = -3n$
15. $a_n = 2n + 9$
16. $a_n = 5n - 8$
17. $a_n = (-1)^n \cdot (2n + 3)$
18. $a_n = \dfrac{(-1)^{n+1}}{n^2}$
19. $a_n = 3 \cdot 4^n$
20. $a_n = 6 \cdot 2^{n-1}$

Find the next three terms of the given sequence and find its general term a_n.

21. 5, 10, 15, 20, 25, . . .
22. 3, 10, 17, 24, 31, . . .
23. 1, 3, 9, 27, 81

24. 4, 9, 16, 25, 36

25. $-\frac{1}{5}, \frac{1}{10}, -\frac{1}{15}, \frac{1}{20}, -\frac{1}{25}$

26. $\frac{1}{2}, \frac{1}{4}, \frac{1}{6}, \frac{1}{8}, \frac{1}{10}, \ldots$

27. 1, 8, 27, 64, 125, . . .

28. 1, −4, 16, −64, 256

Find the first five terms of the sequence with the given first term and general term.

29. $a_1 = 6, a_n = a_{n-1} + 3$

30. $a_1 = 10, a_n = 2a_{n-1}$

31. $a_1 = -5, a_n = 3a_{n-1} + 8$

32. $a_1 = -9, a_n = a_{n-1} - 10$

33. $a_1 = 16, a_n = -3a_{n-1}$

34. $a_1 = 8, a_n = \frac{-a_{n-1}}{2}$

Find the indicated partial sum for the given sequence.

35. s_8, 13, 20, 27, 34, . . .

36. s_9, 25, 16, 7, −2, . . .

37. s_7, −18, −31, −44, −57, . . .

38. s_6, −19, −2, 15, 32, . . .

39. s_6, 7, 14, 28, 56, . . .

40. s_6, 5, −15, 45, −135, . . .

41. s_6, 25, −36, 49, −64, . . .

42. s_{40}, 1, −1, 1, −1, . . .

Find the indicated partial sum for the sequence with the given general term.

43. $s_5, a_n = 3n + 10$

44. $s_4, a_n = 2n - 19$

45. $s_6, a_n = -7n$

46. $s_7, a_n = -n + 8$

47. $s_5, a_n = n^2$

48. $s_5, a_n = 2n - 1$

49. $s_{10}, a_n = (-1)^n \cdot (5n)$

50. $s_9, a_n = (-1)^n \cdot n^2$

Find the indicated partial sum for the sequence with the given first term and general term.

51. $s_6, a_1 = 2, a_n = a_{n-1} + 7$

52. $s_7, a_1 = 9, a_n = a_{n-1} + 4$

53. $s_8, a_1 = 5, a_n = a_{n-1} - 9$

54. $s_5, a_1 = -11, a_n = a_{n-1} - 14$

55. $s_4, a_1 = 3, a_n = 20a_{n-1}$

56. $s_5, a_1 = -4, a_n = -3a_{n-1} + 13$

Find the sum.

57. $\sum_{i=1}^{15} i$

58. $\sum_{i=1}^{7} (i - 3)$

59. $\sum_{i=1}^{8} (2i - 1)$

60. $\sum_{i=1}^{5} (21i + 35)$

61. $\sum_{i=1}^{10} i^2$

62. $\sum_{i=1}^{6} i^3$

63. $\sum_{i=1}^{15} \left(\frac{1}{i} - \frac{1}{i+1}\right)$

64. $\sum_{i=1}^{8} \frac{1}{2^i}$

Rewrite the sum using summation notation.

65. 7 + 14 + 21 + 28

66. 2 + 4 + 6 + 8 + 10

67. 5 + (−10) + 15 + (−20) + 25 + (−30)

68. (−3) + 6 + (−9) + 12 + (−15)

69. 1 + 4 + 9 + 16 + 25 + 36 + 49

70. $1 + \frac{1}{2} + \frac{1}{3} + \frac{1}{4} + \frac{1}{5}$

71. Katy spent $30,000 on a new delivery van for her catering business. If the value of the van decreases by 20% each year, construct a sequence showing the value of the van after each of the first 4 years.

72. The president of a company with 10,000 employees announced a plan to increase the number of employees by 10% a year for the next 4 years. Construct a sequence showing the projected number of employees at the end of each of the next 4 years.

73. Tina started a new job with an annual salary of $50,000. Each year she will be given a 4% raise.

a) Write a sequence showing Tina's annual salary for the first 5 years.

b) Find the total amount that Tina earns during those 5 years.

74. Shane started a new job with an annual salary of \$50,000. Each year he will be given a \$2000 raise.

a) Write a sequence showing Shane's annual salary for the first 5 years.

b) Find the total amount that Shane earns during those 5 years.

Writing in Mathematics

Answer in complete sentences.

75. Describe a real-world application involving a sequence.

76. Describe a real-world application involving a series.

14.2 Arithmetic Sequences and Series

OBJECTIVES

1. Find the general term of an arithmetic sequence.
2. Find partial sums of an arithmetic sequence.
3. Use an arithmetic sequence to solve an applied problem.

Arithmetic Sequences

In this section, we will examine a particular type of sequence called an arithmetic sequence.

Arithmetic Sequence

An **arithmetic sequence** is a sequence in which each term after the first term differs from the preceding term by a constant amount d. This means that the general term a_n can be defined recursively by $a_n = a_{n-1} + d$, for $n \neq 1$. The number d is called the **common difference** of the arithmetic sequence.

If a sequence is an arithmetic sequence, successive terms must increase or decrease by the same amount for each term in the sequence after the first term. For example, the sequence 4, 13, 22, 31, 40, ... is an arithmetic sequence with $d = 9$ because each term is nine greater than the preceding term. The sequence $\frac{5}{4}, \frac{3}{4}, \frac{1}{4}, -\frac{1}{4}, -\frac{3}{4}, \ldots$ is also an arithmetic sequence with $d = -\frac{1}{2}$, as each term is $\frac{1}{2}$ less than the preceding term.

The sequence 2, 4, 8, 16, 32, ... is not an arithmetic sequence. The second term is two greater than the first term, but the third term is four greater than the second term. Because the terms are not increasing by the same amount for the entire sequence, the sequence is not an arithmetic sequence.

EXAMPLE 1 Find the first five terms of the arithmetic sequence whose first term is 5 and whose common difference is $d = 4$.

SOLUTION Because the first term of this sequence is 5 and each term is four greater than the previous term, the first five terms of the sequence are 5, 9, 13, 17, and 21.

Quick Check 1

Find the first six terms of the arithmetic sequence whose first term is -8 and whose common difference is $d = 3$.

Finding the General Term of an Arithmetic Sequence

Objective 1 Find the general term of an arithmetic sequence. For any arithmetic sequence whose first term is a_1 and whose common difference is d, we know that the second term a_2 is equal to $a_1 + d$. Because we could add d to the second term to find the third term of the sequence, we know that $a_3 = a_1 + 2d$.

$$\begin{aligned} a_3 &= a_2 + d \\ &= (a_1 + d) + d \\ &= a_1 + 2d \end{aligned}$$

Continuing this pattern, we have $a_4 = a_1 + 3d$, $a_5 = a_1 + 4d$, and so on. This leads to the following result for the general term, a_n, of an arithmetic sequence:

General Term of an Arithmetic Sequence

If an arithmetic sequence whose first term is a_1 has a common difference of d, the general term of this sequence is $a_n = a_1 + (n - 1)d$.

EXAMPLE 2 Find the general term, a_n, of the arithmetic sequence 5, 11, 17, 23, 29, ...

SOLUTION The first term, a_1, of this sequence is 5. Subtract the first term of the sequence from the second term to find the common difference d.

$$d = 11 - 5 = 6$$

Now find the general term of this sequence by using the formula $a_n = a_1 + (n - 1)d$.

$$\begin{aligned} a_n &= 5 + (n - 1)6 && \text{Substitute 5 for } a_1 \text{ and 6 for } d. \\ &= 6n - 1 && \text{Simplify.} \end{aligned}$$

The general term for this arithmetic sequence is $a_n = 6n - 1$.

Quick Check 2

Find the general term, a_n, of the arithmetic sequence 2, 14, 26, 38, 50, ...

After we find the general term, a_n, of an arithmetic sequence, we can go on to find any particular term of the sequence.

EXAMPLE 3 Find the 45th term, a_{45}, of the arithmetic sequence 3, 7, 11, 15, 19, ...

SOLUTION Begin by finding the general term, a_n, of this arithmetic sequence. Then find a_{45}.

The first term, a_1, of this sequence is 3, and the common difference d is 4. Now find the general term of this sequence by using the formula $a_n = a_1 + (n - 1)d$.

$$\begin{aligned} a_n &= 3 + (n - 1)4 && \text{Substitute 3 for } a_1 \text{ and 4 for } d. \\ &= 4n - 1 && \text{Simplify.} \end{aligned}$$

The general term for this arithmetic sequence is $a_n = 4n - 1$. Now find the 45th term.

$$\begin{aligned} a_{45} &= 4(45) - 1 && \text{Substitute 45 for } n \text{ in } a_n = 4n - 1. \\ &= 179 && \text{Simplify.} \end{aligned}$$

The 45th term of this arithmetic sequence is 179.

Quick Check 3

Find the 61st term, a_{61}, of the arithmetic sequence 57, 59, 61, 63, 65, ...

Arithmetic Series

Objective 2 Find partial sums of an arithmetic sequence. An **arithmetic series** is the sum of the terms of an arithmetic sequence. A finite arithmetic series consisting of only the first n terms of an arithmetic sequence is also known as the ***n*th partial sum** of the arithmetic sequence and is denoted as s_n.

EXAMPLE 4 Find s_7 for the arithmetic sequence 4, 8, 12, 16, 20, ...

SOLUTION Because only the first five terms are given, begin by finding the sixth and seventh terms. Because each term in this arithmetic sequence is four greater than the preceding term, the next two terms of the sequence are 24 and 28. Now find the seventh partial sum of this arithmetic sequence.

$$
\begin{aligned}
s_7 &= 4 + 8 + 12 + 16 + 20 + 24 + 28 \\
&= 112 \qquad \text{Add.}
\end{aligned}
$$

Quick Check 4

Find s_9 for the arithmetic sequence 20, 26, 32, 38, 44, ...

We now present a formula for finding partial sums of an arithmetic sequence. This formula will be quite helpful, as it depends on only the first and nth terms of the arithmetic sequence (a_1 and a_n), as well as on the number of terms being added.

*n*th Partial Sum of an Arithmetic Sequence

The sum of the first n terms of an arithmetic sequence is given by

$$s_n = \frac{n}{2}(a_1 + a_n),$$

where a_1 is the first term and a_n is the nth term of the sequence.

Before we apply this formula, let's investigate where the formula comes from. We will let a_1 be the first term of the arithmetic sequence, a_n be its nth term, and d be the common difference. Because s_n is the sum of the first n terms of the sequence, it can be expressed as

$$s_n = a_1 + (a_1 + d) + (a_1 + 2d) + (a_1 + 3d) + \cdots + (a_1 + (n - 1)d).$$

The first n terms of an arithmetic sequence also can be generated by starting with the nth term a_n and subtracting d to find the preceding term: $a_n, a_n - d, a_n - 2d, \ldots$ $a_n - (n - 1)d$. So the nth partial sum can also be expressed as

$$s_n = a_n + (a_n - d) + (a_n - 2d) + (a_n - 3d) + \cdots + (a_n - (n - 1)d).$$

We now have two different expressions for s_n. If we add the two expressions term by term, we obtain the following result:

$$
\begin{aligned}
s_n &= a_1 + (a_1 + d) + (a_1 + 2d) + (a_1 + 3d) + \cdots + (a_1 + (n - 1)d) \\
s_n &= a_n + (a_n - d) + (a_n - 2d) + (a_n - 3d) + \cdots + (a_n - (n - 1)d) \\
\hline
2s_n &= (a_1 + a_n) + (a_1 + a_n) + (a_1 + a_n) + (a_1 + a_n) + \cdots + (a_1 + a_n)
\end{aligned}
$$

So $2s_n = n(a_1 + a_n)$ because the expression $(a_1 + a_n)$ appears n times on the right side of the equation. Dividing both sides of the equation $2s_n = n(a_1 + a_n)$ by 2 produces the result $s_n = \frac{n}{2}(a_1 + a_n)$.

The seventh term of the arithmetic sequence 4, 8, 12, 16, 20, . . . is 28; so s_7 can be found as follows:

$$
\begin{aligned}
s_7 &= \frac{7}{2}(a_1 + a_7) \\
s_7 &= \frac{7}{2}(4 + 28) \qquad \text{Substitute 7 for } n, 4 \text{ for } a_1, \text{ and 28 for } a_7. \\
&= 112 \qquad \text{Simplify.}
\end{aligned}
$$

To find the nth partial sum of an arithmetic sequence by using the formula $s_n = \frac{n}{2}(a_1 + a_n)$, we must know the values of the first term and the last term that are being added. In the next example, we will find the general term of the arithmetic sequence, $a_n = a_1 + (n - 1)d$, to help us find the last term in the sum. This becomes particularly useful as n becomes larger.

EXAMPLE 5 Find s_{20} for the arithmetic sequence 98, 93, 88, 83, 78, ...

SOLUTION We already know that $n = 20$ and $a_1 = 98$. To find the value of a_{20}, begin by finding the general term of this arithmetic sequence using the formula $a_n = a_1 + (n - 1)d$. The common difference for this sequence is $d = -5$.

$$a_n = 98 + (n - 1)(-5) \quad \text{Substitute 98 for } a_1 \text{ and } -5 \text{ for } d.$$
$$= -5n + 103 \quad \text{Simplify.}$$

Now use the fact that $a_n = -5n + 103$ to find the value of a_{20}.

$$a_{20} = -5(20) + 103 \quad \text{Substitute 20 for } n.$$
$$= 3 \quad \text{Simplify.}$$

Now apply the formula $s_n = \frac{n}{2}(a_1 + a_n)$ to find s_{20}.

$$s_{20} = \frac{20}{2}(98 + 3) \quad \text{Substitute 20 for } n\text{, 98 for } a_1\text{, and 3 for } a_{20}.$$
$$= 1010 \quad \text{Simplify.}$$

Quick Check 5

Find s_{36} for the arithmetic sequence 690, 671, 652, 633, 614, ...

EXAMPLE 6 Find the sum of the first 147 odd positive integers.

SOLUTION The first 147 odd positive integers make up the sequence that begins 1, 3, 5, 7, 9, ... This is an arithmetic sequence with $a_1 = 1$ and $d = 2$. We find the general term of this sequence by using the formula $a_n = a_1 + (n - 1)d$ to help us find a_{147}.

$$a_n = 1 + (n - 1)2 \quad \text{Substitute 1 for } a_1 \text{ and 2 for } d.$$
$$= 2n - 1 \quad \text{Simplify.}$$

We can now use $a_n = 2n - 1$ to find the value of a_{147}.

$$a_{147} = 2(147) - 1 \quad \text{Substitute 147 for } n.$$
$$= 293 \quad \text{Simplify.}$$

We can now apply the formula $s_n = \frac{n}{2}(a_1 + a_n)$ to find s_{147}.

$$s_{147} = \frac{147}{2}(1 + 293) \quad \text{Substitute 147 for } n\text{, 1 for } a_1\text{, and 293 for } a_{147}.$$
$$= 21{,}609 \quad \text{Simplify.}$$

The sum of the first 147 odd positive integers is 21,609.

Quick Check 6

Find the sum of the first 62 even positive integers.

Applications

Objective 3 Use an arithmetic sequence to solve an applied problem.

EXAMPLE 7 Joyce found a new job with a starting salary of \$40,000 per year. Each year she will be given a \$2500 raise.

a) Write a sequence showing Joyce's annual salary for each of the first 6 years.
b) Find the general term, a_n, for the sequence from part a.
c) What will Joyce's total earnings be if she stays with the company for 30 years?

SOLUTION

a) Because the starting salary is \$40,000 and it increases by \$2500 each year, the sequence is \$40,000, \$42,500, \$45,000, \$47,500, \$50,000, \$52,500.

b) The first term, a_1, of this sequence is \$40,000, and the common difference d is \$2500. We now can find the general term of this sequence by using the formula $a_n = a_1 + (n - 1)d$.

$$a_n = 40{,}000 + (n - 1)2500 \quad \text{Substitute 40,000 for } a_1 \text{ and 2500 for } d.$$
$$= 2500n + 37{,}500 \quad \text{Simplify.}$$

The general term for this arithmetic sequence is $a_n = 2500n + 37{,}500$.

c) To find Joyce's total earnings over 30 years, we need to find s_{30}. We already know that $n = 30$ and $a_1 = 40{,}000$. To apply the formula $s_n = \frac{n}{2}(a_1 + a_n)$, we must find the value of a_{30}.

$$a_{30} = 2500(30) + 37{,}500 \quad \text{Substitute 30 for } n \text{ in } a_n = 2500n + 37{,}500.$$
$$= 112{,}500 \quad \text{Simplify.}$$

We can now apply the formula $s_n = \frac{n}{2}(a_1 + a_n)$.

$$s_{30} = \frac{30}{2}(40{,}000 + 112{,}500) \quad \text{Substitute 30 for } n, \text{ 40,000 for } a_1, \text{ and 112,500 for } a_{30}.$$
$$= 2{,}287{,}500 \quad \text{Simplify.}$$

During the first 30 years, Joyce will earn \$2,287,500.

BUILDING YOUR STUDY STRATEGY

Preparing for a Final Exam (Part 2), 2 Study Groups If you have participated in a study group throughout the semester, now is not the time to begin studying on your own. Suggest that all of the students in the group bring problems they are struggling with or problems they believe are important and work through them as a group. A group study session is a great place to complete difficult problems.

At the end of your group study session, as a group, try to write a practice exam. In determining which problems to include, focus on the types of problems most likely to appear on your final exam.

For further information on working with a study group, see the Study Strategies in Chapter 1.

Exercises 14.2

Vocabulary

1. A(n) __________ sequence is a sequence in which each term after the first term differs from the preceding term by a constant amount d.
2. The difference, d, between consecutive terms in an arithmetic sequence is called the __________ difference of the sequence.
3. The general term, a_n, of an arithmetic sequence can be defined recursively by __________ for $n \neq 1$.
4. If an arithmetic sequence whose first term is a_1 has a common difference of d, the general term of this sequence is given by the formula __________.
5. A(n) __________ is the sum of the terms of an arithmetic sequence.
6. The sum, s_n, of the first n terms of an arithmetic sequence is given by the formula __________.

Find the common difference, d, of each given arithmetic sequence.

7. 6, 10, 14, 18, 22, ...

8. 9, 28, 47, 66, 85, ...

9. $\frac{7}{2}, \frac{29}{6}, \frac{37}{6}, \frac{15}{2}, \frac{53}{6}, \ldots$

10. $13, \frac{59}{4}, \frac{33}{2}, \frac{73}{4}, 20, \ldots$

11. −13, −10, −7, −4, −1, ...

12. −4, −8, −12, −16, −20, ...

13. $\frac{9}{4}, \frac{41}{4}, \frac{73}{4}, \frac{105}{4}, \frac{137}{4}, \ldots$

14. $\frac{31}{6}, \frac{17}{2}, \frac{71}{6}, \frac{91}{6}, \frac{37}{2}, \ldots$

Find the first five terms of the arithmetic sequence with each given first term and common difference.

15. $a_1 = 1, d = 5$

16. $a_1 = 1, d = -2$

17. $a_1 = -9, d = 12$

18. $a_1 = -13, d = -8$

19. $a_1 = -10, d = -21$

20. $a_1 = 9, d = \frac{5}{2}$

Find the general term, a_n, of each given arithmetic sequence.

21. 18, 29, 40, 51, 62, ...

22. −3, 3, 9, 15, 21, ...

23. 3, 1, −1, −3, −5, ...

24. 47, 62, 77, 92, 107, ...

25. −7, 13, 33, 53, 73, ...

26. −27, −35, −43, −51, −59, ...

27. $-\frac{3}{2}, \frac{5}{2}, \frac{13}{2}, \frac{21}{2}, \frac{29}{2}, \ldots$

28. $8, \frac{16}{3}, \frac{8}{3}, 0, -\frac{8}{3}, \ldots$

29. Find the 19th term, a_{19}, of the arithmetic sequence 16, 21, 26, 31, 36, ...

30. Find the 14th term, a_{14}, of the arithmetic sequence −9, 29, 67, 105, 143, ...

31. Which term of the arithmetic sequence 23, 29, 35, 41, 47, ... is equal to 389?

32. Which term of the arithmetic sequence −99, −94, −89, −84, −79, ... is equal to 336?

33. Which term of the arithmetic sequence 611, 603, 595, 587, 579, ... is equal to −365?

34. Which term of the arithmetic sequence −327, −336, −345, −354, −363, ... is equal to −3981?

Find the partial sum of the arithmetic sequence with each given first term and common difference.

35. $s_7, a_1 = 8, d = 5$

36. $s_8, a_1 = 10, d = 7$

37. $s_{15}, a_1 = -6, d = 11$

38. $s_{12}, a_1 = -19, d = -15$

39. $s_{36}, a_1 = 65, d = -8$

40. $s_{25}, a_1 = 725, d = 75$

41. $s_{32}, a_1 = 109, d = 99$

42. $s_{19}, a_1 = 62, d = -45$

Find the partial sum for each given arithmetic sequence.

43. s_{10}, 1, 21, 41, 61, ...

44. s_{15}, 9, 22, 35, 48, ...

45. s_{17}, −17, −1, 15, 31, ...

46. s_9, −38, −31, −24, −17, ...

47. s_{20}, 86, 47, 8, −31, ...

48. s_{27}, −3, −16, −29, −42, ...

49. Find the sum of the first 1000 positive integers.

50. Find the sum of the first 418 positive integers.

51. Find the sum of the first 67 odd positive integers.

52. Find the sum of the first 109 odd positive integers.

53. Find the sum of the first 200 even positive integers.

54. Find the sum of the first 82 even positive integers.

55. Andrea finds a job that pays $10 per hour. Each year her pay rate will increase by $1 per hour.

a) Write a sequence showing Andrea's hourly pay rate for the first 5 years.

b) Find the general term a_n for the sequence from part a.

56. During its first year of existence, a minor league baseball team sold 2800 season tickets. In each year after that, the number of season tickets sold decreased by 120.

a) Write a sequence showing the number of season tickets sold in the first 4 years.

b) Find the general term, a_n, for the number of season tickets sold in the team's nth year.

57. Jon starts a job with an annual salary of $38,500. Each year he will be given a raise of $750.

a) Write a sequence showing Jon's annual salary for the first 6 years.

b) Find the general term, a_n, for Jon's annual salary in his nth year on the job.

c) How much money will Jon earn from this job over the first 15 years?

58. A teacher has a retirement savings plan. In the first year the teacher contributes \$100 per month. Each year the teacher will increase her contributions by \$10 per month so that she contributes \$110 per month in the second year \$120 per month in the third year and so on.

a) Find the general term, a_n, of an arithmetic sequence showing the annual contribution of the teacher in the nth year.

b) How much money will the teacher contribute over a period of 30 years?

Writing in Mathematics

Answer in complete sentences.

59. Describe a real-world application involving an arithmetic sequence.

60. Describe a real-world application involving an arithmetic series.

61. ***Solutions Manual*** Write a solutions manual page for the following problem.

Find s_{22} for an arithmetic sequence with first term $a_1 = 44$ and common difference $d = 13$.

14.3 Geometric Sequences and Series

OBJECTIVES

1. Find the common ratio of a geometric sequence.
2. Find the general term of a geometric sequence.
3. Find partial sums of a geometric sequence.
4. Find the sum of an infinite geometric series.
5. Use a geometric sequence to solve an applied problem.

Geometric Sequences

Consider the sequence 4, 12, 36, 108, 324, ... Each term in this sequence is equal to three times the preceding term. This is an example of a geometric sequence, which is the focus of this section.

Geometric Sequence

A **geometric sequence** is a sequence in which each term after the first term is a constant multiple of the preceding term. This means that the general term, a_n, can be defined recursively by $a_n = r \cdot a_{n-1}$, for $n \neq 1$. The number r is called the **common ratio** of the geometric sequence.

A sequence is a geometric sequence if the quotient of any term after the first term and its preceding term, $\frac{a_n}{a_{n-1}}$, is a constant. For example, the sequence 1, 5, 25, 125, 625, ... is a geometric sequence because $\frac{a_n}{a_{n-1}} = 5$ for each term after the first term in the sequence. In other words, each term is five times greater than the preceding term. The value 5 is the common ratio, r, of this geometric sequence.

The sequence 2, 6, 10, 14, 18, ... is not a geometric sequence because there is no common ratio. The second term is three times greater than the first term, but the third term is not three times greater than the second term.

Finding the Common Ratio of a Geometric Sequence

Objective 1 Find the common ratio of a geometric sequence. To find the common ratio, r, of a geometric sequence, take any term of the sequence after the first term and divide it by the preceding term.

EXAMPLE 1 Find the common ratio, r, of the geometric sequence 2, 8, 32, 128, 512, ...

SOLUTION To find r, divide the second term of the sequence by the first term.

$$r = \frac{8}{2} = 4$$

Quick Check 1

Find the common ratio, r, of the geometric sequence 256, 64, 16, 4, 1, . . .

A geometric sequence can be an alternating sequence. In this case, the common ratio r will be a negative number.

EXAMPLE 2 Find the common ratio, r, of the geometric sequence 5, −10, 20, −40, 80, ...

SOLUTION To find r, divide the second term of the sequence by the first term.

$$r = \frac{-10}{5} = -2$$

The common ratio for this geometric sequence is $r = -2$.

Quick Check 2

Find the common ratio, r, of the geometric sequence −120, 60, −30, 15, $-\frac{15}{2}$, ...

EXAMPLE 3 Find the first five terms of the geometric sequence whose first term is 7 and whose common ratio is $r = 2$.

SOLUTION Because the first term of this sequence is 7 and each term is two times greater than the previous term, the first five terms of the sequence are 7, 14, 28, 56, and 112.

Quick Check 3

Find the first five terms of the geometric sequence whose first term is 5 and whose common ratio is $r = -\frac{3}{4}$.

Finding the General Term of a Geometric Sequence

Objective 2 Find the general term of a geometric sequence. The general term of a geometric sequence can be determined from the first term of the sequence, a_1, and the sequence's common ratio r. We know that the second term a_2 is equal to $a_1 \cdot r$. Continuing this pattern, we obtain the following:

$$a_3 = a_2 \cdot r = (a_1 \cdot r) \cdot r = a_1 \cdot r^2$$
$$a_4 = a_3 \cdot r = (a_1 \cdot r^2) \cdot r = a_1 \cdot r^3$$
$$a_5 = a_4 \cdot r = (a_1 \cdot r^3) \cdot r = a_1 \cdot r^4$$

This leads to the following result about the general term, a_n, of a geometric sequence:

General Term of a Geometric Sequence

If a geometric sequence whose first term is a_1 has a common ratio of r, the general term of the sequence is $a_n = a_1 \cdot r^{n-1}$.

EXAMPLE 4 Find the general term, a_n, of the geometric sequence 3, 30, 300, 3000, ...

SOLUTION The first term, a_1, of this sequence is 3. Divide the second term of the sequence by the first term to find the common ratio, r.

$$r = \frac{30}{3} = 10$$

Now find the general term of this sequence using the formula $a_n = a_1 \cdot r^{n-1}$.

$$a_n = 3 \cdot 10^{n-1} \quad \text{Substitute 3 for } a_1 \text{ and 10 for } r.$$

The general term for this geometric sequence is $a_n = 3 \cdot 10^{n-1}$.

Quick Check 4

Find the general term, a_n, of the geometric sequence 9, 18, 36, 72, . . .

EXAMPLE 5 Find the general term, a_n, of the geometric sequence −1, 4, −16, 64, ...

SOLUTION The first term, a_1, of this sequence is −1. Divide the second term of the sequence by the first term to find the common ratio, r.

$$r = \frac{4}{-1} = -4$$

Now find the general term of this sequence using the formula $a_n = a_1 \cdot r^{n-1}$.

$$a_n = -1 \cdot (-4)^{n-1} \quad \text{Substitute } -1 \text{ for } a_1 \text{ and } -4 \text{ for } r.$$

The general term for this geometric sequence is $a_n = -1 \cdot (-4)^{n-1}$.

Quick Check 5

Find the general term, a_n, of the geometric sequence 11, −55, 275, −1375, . . .

Geometric Series

Objective 3 Find partial sums of a geometric sequence. A **geometric series** is the sum of the terms of a geometric sequence. We will begin by examining finite geometric series consisting of only the first n terms of a geometric sequence. Again, the sum of the first n terms of a sequence is known as the nth partial sum of the sequence and is denoted by s_n. In the case of a finite geometric series,

$$s_n = a_1 + a_1 r + a_1 r^2 + a_1 r^3 + \cdots + a_1 r^{n-1}.$$

EXAMPLE 6 Find s_7 for the geometric sequence 1, 2, 4, 8, 16, ...

SOLUTION Because only the first five terms are given, begin by finding the sixth and seventh terms. Because each term in this geometric sequence is equal to two times the preceding term, the next two terms of the sequence are 32 and 64. Now find the seventh partial sum of this geometric sequence.

$$\begin{aligned} s_7 &= 1 + 2 + 4 + 8 + 16 + 32 + 64 \\ &= 127 \qquad \text{Add.} \end{aligned}$$

Quick Check 6

Find s_8 for the geometric sequence 12, 36, 108, 324, ...

Now we will present a formula for finding partial sums of a geometric sequence. This formula will be quite helpful, as it depends only on the first term of the geometric sequence (a_1) and the common ratio, r, of the sequence.

*n*th Partial Sum of a Geometric Sequence

The sum of the first n terms of a geometric sequence is given by

$$s_n = a_1 \cdot \frac{1 - r^n}{1 - r},$$

where a_1 is the first term and r is the common ratio of the sequence.

Before we apply this formula, let's investigate where the formula comes from. Suppose we have a geometric sequence whose first term is a_1 and whose common ratio is r. Because s_n is the sum of the first n terms of the sequence, it can be expressed as follows:

$$s_n = a_1 + a_1 r + a_1 r^2 + a_1 r^3 + \cdots + a_1 r^{n-1}$$

If we multiply both sides of this equation by r, the resulting equation is

$$r \cdot s_n = a_1 r + a_1 r^2 + a_1 r^3 + a_1 r^4 + \cdots + a_1 r^n.$$

Now we can subtract corresponding terms of these two equations as follows:

$$\begin{aligned} s_n &= a_1 + a_1 r + a_1 r^2 + a_1 r^3 + \cdots + a_1 r^{n-1} \\ r \cdot s_n &= a_1 r + a_1 r^2 + a_1 r^3 + \cdots + a_1 r^{n-1} + a_1 r^n \\ \hline s_n - r \cdot s_n &= a_1 - a_1 r^n \\ s_n(1 - r) &= a_1(1 - r^n) \end{aligned}$$

Factor out the common factor s_n on the left side of the equation and the common factor a_1 on the right side of the equation.

$$\frac{s_n(1 - r)}{1 - r} = \frac{a_1(1 - r^n)}{1 - r}$$ Divide both sides by $1 - r$ to isolate s_n.

$$s_n = a_1 \cdot \frac{1 - r^n}{1 - r}$$ Simplify.

For the geometric sequence 1, 2, 4, 8, 16, . . . in the previous example, $a_1 = 1$ and $r = 2$.

Now we can apply the formula $s_n = a_1 \cdot \frac{1 - r^n}{1 - r}$ to find s_7.

$$s_7 = 1 \cdot \frac{1 - 2^7}{1 - 2}$$ Substitute 1 for a_1, 2 for r, and 7 for n.

$$= 127$$ Simplify.

Note that this matches the result found in the previous example.

In the next example, we will find the partial sum of a geometric sequence whose common ratio r is negative.

EXAMPLE 7 Find s_9 for the geometric sequence 6, −42, 294, −2058, ...

SOLUTION For this sequence, $a_1 = 6$ and $r = \frac{-42}{6} = -7$. Now we can apply the formula $s_n = a_1 \cdot \frac{1 - r^n}{1 - r}$ to find s_9.

$$s_9 = 6 \cdot \frac{1 - (-7)^9}{1 - (-7)}$$ Substitute 6 for a_1, −7 for r, and 9 for n.

$$= 30{,}265{,}206$$ Simplify.

Quick Check 7

Find s_{10} for the geometric sequence 1, −3, 9, −27, ...

Infinite Geometric Series

Objective 4 Find the sum of an infinite geometric series. Now we will turn our attention to **infinite geometric series**, $s = a_1 + a_1 r + a_1 r^2 + \cdots$. Consider the geometric sequence $\frac{1}{2}, \frac{1}{4}, \frac{1}{8}, \frac{1}{16}, \frac{1}{32}, \ldots$, where $a_1 = \frac{1}{2}$ and $r = \frac{1}{2}$. The first partial sum of this series is $s_1 = \frac{1}{2}$. Here are the next four partial sums.

$$s_2 = \frac{1}{2} + \frac{1}{4} = \frac{2}{4} + \frac{1}{4} = \frac{3}{4}$$

$$s_3 = \frac{1}{2} + \frac{1}{4} + \frac{1}{8} = \frac{4}{8} + \frac{2}{8} + \frac{1}{8} = \frac{7}{8}$$

$$s_4 = \frac{1}{2} + \frac{1}{4} + \frac{1}{8} + \frac{1}{16} = \frac{8}{16} + \frac{4}{16} + \frac{2}{16} + \frac{1}{16} = \frac{15}{16}$$

$$s_5 = \frac{1}{2} + \frac{1}{4} + \frac{1}{8} + \frac{1}{16} + \frac{1}{32} = \frac{16}{32} + \frac{8}{32} + \frac{4}{32} + \frac{2}{32} + \frac{1}{32} = \frac{31}{32}$$

As n increases, the partial sums of this sequence are getting closer and closer to 1. In such a case, we say that the **limit** of the infinite geometric series $\frac{1}{2} + \frac{1}{4} + \frac{1}{8} + \frac{1}{16} + \frac{1}{32} + \cdots$ is equal to 1.

Limit of an Infinite Geometric Series

Suppose a geometric sequence has a common ratio, r. If $|r| < 1$, the corresponding infinite geometric series has a limit, and this limit is given by the formula

$$s = \frac{a_1}{1 - r}.$$

If $|r| \geq 1$, as it is for the series $\frac{1}{2} + \frac{3}{4} + \frac{9}{8} + \cdots \left(r = \frac{3}{2}\right)$, no limit exists.

EXAMPLE 8 For the geometric sequence $16, 4, 1, \frac{1}{4}, \ldots$, does the infinite series have a limit? If so, what is that limit?

SOLUTION Begin by determining whether $|r| < 1$ for this sequence.

$$r = \frac{a_2}{a_1} = \frac{4}{16} = \frac{1}{4}$$

Because the absolute value of r is less than 1, the infinite series does have a limit.

To find the limit, use the formula $s = \frac{a_1}{1 - r}$.

$$s = \frac{16}{1 - \frac{1}{4}} = \frac{16}{\frac{3}{4}} = \frac{64}{3} \quad \text{Substitute 16 for } a_1 \text{ and } \tfrac{1}{4} \text{ for } r. \text{ Simplify.}$$

The limit of this infinite geometric series is $\frac{64}{3}$, or $21\frac{1}{3}$.

Quick Check 8

For the geometric sequence $3, 2, \frac{4}{3}, \frac{8}{9}, \ldots$, does the infinite series have a limit? If so, what is that limit?

EXAMPLE 9 For the geometric sequence $4, 6, 9, \frac{27}{2}, \ldots$, does the infinite series have a limit? If so, what is that limit?

SOLUTION First, determine whether $|r| < 1$.

$$r = \frac{a_2}{a_1} = \frac{6}{4} = \frac{3}{2}$$

Because the absolute value of r is greater than 1, the infinite series does not have a limit. In this case, the partial sums increase without limit as n continues to increase.

Quick Check 9

For the geometric sequence $-2, 8, -32, 128, \ldots$, does the infinite series have a limit? If so, what is that limit?

In the next example, we will find the limit of an infinite geometric series, given the general term of the sequence.

EXAMPLE 10 For the geometric sequence whose general term is $a_n = 8 \cdot \left(\frac{1}{3}\right)^{n-1}$, does the infinite series have a limit? If so, what is that limit?

SOLUTION For this sequence, $r = \frac{1}{3}$. Because the absolute value of r is less than 1, the infinite series does have a limit. The first term of this series is $a_1 = 8$, and the limit can be found by $s = \frac{a_1}{1 - r}$.

$$s = \frac{8}{1 - \frac{1}{3}} \quad \text{Substitute 8 for } a_1 \text{ and } \tfrac{1}{3} \text{ for } r.$$

$$= 12 \quad \text{Simplify.}$$

The limit of this infinite geometric series is 12.

Applications

Objective 5 Use a geometric sequence to solve an applied problem. We will finish this section with an application involving a geometric sequence.

EXAMPLE 11 Donna's parents have established a trust fund for their daughter. On her 21st birthday, Donna receives a payment of \$10,000. On each successive birthday, she receives a payment that is 80% of the previous payment.

a) Write a sequence showing Donna's payments for the first 4 years.

b) Find the total amount of Donna's first 25 payments.

SOLUTION

a) Because the first payment is \$10,000 and the second payment is 80% of that amount, she will receive 0.8(\$10,000), or \$8000, on her 22nd birthday. Repeating this process, we find that the sequence is \$10,000, \$8000, \$6400, \$5120.

b) To find the total of Donna's payments for the first 25 years, we need to find s_{25} for this geometric sequence. We already know that $n = 25$ and $a_1 = 10{,}000$. To apply the formula $s_n = a_1 \cdot \frac{1 - r^n}{1 - r}$, we must find the common ratio r.

$$r = \frac{a_2}{a_1} = \frac{8000}{10{,}000} = 0.8$$

It should be no surprise that $r = 0.8$ for this sequence, as each term is 80% of the preceding term. We can now apply the formula $s_n = a_1 \cdot \frac{1 - r^n}{1 - r}$.

$$s_{25} = 10{,}000 \cdot \frac{1 - (0.8)^{25}}{1 - 0.8} \quad \text{Substitute 25 for } n\text{, 10,000 for } a_1\text{, and 0.8 for } r.$$

$$\approx 49{,}811.11 \quad \text{Evaluate using a calculator.}$$

During the first 25 years, Donna will receive \$49,811.11.

Quick Check 10

A car was purchased for \$25,000. Its value decreases by 30% each year.

a) Write a sequence showing the value of the car at the end of each of the first 6 years.

b) Find the value of the car at the end of year 10.

BUILDING YOUR STUDY STRATEGY

Preparing for the Final Exam (Part 2), 3 Taking the Test and Anxiety When you arrive to take your final exam, bring along a positive attitude. Be confident. If you come well prepared, trust your ability.

Once you sit down to take the final exam, begin by writing down any information you may find useful during the exam, such as formulas and notes to yourself. Briefly read through the test to get an idea of its length and difficulty. Begin by working problems about which you are most confident.

For further information on dealing with the anxiety that accompanies test taking, see the Study Strategies in Chapter 6.

Exercises 14.3

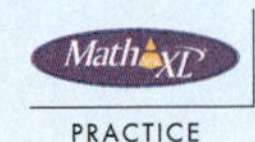

PRACTICE

WATCH

DOWNLOAD

READ

REVIEW

Vocabulary

1. A(n) ________________ is a sequence in which each term after the first term is a constant multiple of the preceding term.
2. The ratio $r = \frac{a_n}{a_{n-1}}$ between consecutive terms in a geometric sequence is called the ____________ of the sequence.
3. The general term, a_n, of a geometric sequence can be defined recursively by ____________, for $n \neq 1$.
4. A(n) ______________ is the sum of the terms of a geometric sequence.
5. The sum, s_n, of the first n terms of a geometric sequence is given by the formula ______________.
6. For an infinite geometric sequence with $|r| < 1$, the corresponding infinite geometric series has a limit s, and this limit is given by the formula __________.

Find the common ratio, r, of the given geometric sequence.

7. $1, 8, 64, 512, 4096, \ldots$
8. $1, 5, 25, 125, 625, \ldots$
9. $16, 4, 1, \frac{1}{4}, \frac{1}{16}, \ldots$
10. $1458, 486, 162, 54, 18, \ldots$
11. $-3, 21, -147, 1029, -7203, \ldots$
12. $6, -12, 24, -48, 96, \ldots$
13. $-6, 8, -\frac{32}{3}, \frac{128}{9}, -\frac{512}{27}, \ldots$
14. $32, -12, \frac{9}{2}, -\frac{27}{16}, \frac{81}{128}, \ldots$

Find the first five terms of the geometric sequence with the given first term and common ratio.

15. $a_1 = 1, r = 10$
16. $a_1 = 1, r = -3$
17. $a_1 = 4, r = -2$
18. $a_1 = 2, r = 5$
19. $a_1 = 8, r = \frac{3}{4}$
20. $a_1 = 12, r = \frac{1}{2}$
21. $a_1 = -20, r = -\frac{7}{4}$
22. $a_1 = \frac{1}{3}, r = -\frac{4}{3}$

Find the general term, a_n, of the given geometric sequence.

23. $1, \frac{2}{3}, \frac{4}{9}, \frac{8}{27}, \ldots$
24. $1, 6, 36, 216, \ldots$
25. $-1, 4, -16, 64, \ldots$
26. $-1, -5, -25, -125, \ldots$
27. $2, 14, 98, 686, \ldots$
28. $10, 30, 90, 270, \ldots$

29. $16, -12, 9, -\frac{27}{4}, \ldots$

30. $1000, 700, 490, 343, \ldots$

Find the partial sum of the geometric sequence with the given first term and common ratio.

31. $s_7, a_1 = 5, r = 4$

32. $s_8, a_1 = 15, r = 3$

33. $s_{20}, a_1 = 48, r = -2$

34. $s_5, a_1 = -6, r = -10$

35. $s_9, a_1 = 256, r = \frac{1}{2}$

36. $s_7, a_1 = 4096, r = 0.25$

Find the indicated partial sum for the given geometric sequence.

37. $s_6, 1, -8, 64, -512, \ldots$

38. $s_8, 1, 10, 100, 1000, \ldots$

39. $s_7, 4, 12, 36, 108, \ldots$

40. $s_9, 50, 100, 200, 400, \ldots$

41. $s_6, 13, 156, 1872, 22{,}464, \ldots$

42. $s_7, 9, -27, 81, -243, \ldots$

For the given geometric sequence, does the infinite series have a limit? If so, find that limit.

43. $30, 10, \frac{10}{3}, \frac{10}{9}, \ldots$

44. $432, 72, 12, 2, \ldots$

45. $1, -\frac{4}{5}, \frac{16}{25}, -\frac{64}{125}, \ldots$

46. $\frac{1}{9}, -\frac{5}{27}, \frac{25}{81}, -\frac{125}{243}, \ldots$

47. $\frac{81}{100}, \frac{243}{200}, \frac{729}{400}, \frac{2187}{800}, \ldots$

48. $108, 90, 75, \frac{125}{2}, \ldots$

49. $\frac{1}{2}, -\frac{1}{4}, \frac{1}{8}, -\frac{1}{16}, \ldots$

50. $1, -1, 1, -1, \ldots$

For the geometric sequence whose general term is given, does the infinite series have a limit? If so, what is that limit?

51. $a_n = 5 \cdot \left(\frac{5}{6}\right)^{n-1}$

52. $a_n = 200 \cdot \left(\frac{1}{4}\right)^{n-1}$

53. $a_n = -8 \cdot \left(\frac{3}{10}\right)^{n-1}$

54. $a_n = 6 \cdot \left(-\frac{5}{3}\right)^{n-1}$

55. $a_n = \frac{5}{7} \cdot \left(\frac{11}{8}\right)^{n-1}$

56. $a_n = \frac{4}{25} \cdot \left(-\frac{5}{12}\right)^{n-1}$

57. A copy machine purchased for \$8000 decreases in value by 20% each year.

a) Write a geometric sequence showing the value of the copy machine at the end of each of the first 3 years after it was purchased.

b) Find the value of the copy machine at the end of the eighth year.

58. A farmer purchased a new tractor for \$30,000. The tractor's value decreases by 10% each year.

a) Write a geometric sequence showing the value of the tractor at the end of each of the first 6 years after it was purchased.

b) Find the value of the tractor at the end of year 10.

59. A motor home valued at \$40,000 decreases in value by 50% each year. Find the general term, a_n, of a geometric sequence showing the value of the motor home n years after it was purchased.

60. A diamond ring valued at \$4000 increases in value by 5% each year. Find the general term, a_n, of a geometric sequence showing the value of the diamond ring n years after it was purchased.

61. A professional baseball player signed a 10-year contract. The salary for the first year was \$10,000,000, with a 10% raise for each of the remaining 9 years. At the end of the contract, what is the total amount of money earned by the player?

62. Cathy takes a job as a high school teacher. The starting salary is $32,000 per year, with a 2% raise each year. If Cathy teaches for 30 years, what is the total amount she will be paid?

Writing in Mathematics

Answer in complete sentences.

63. Describe a real-world application involving a geometric sequence.

64. Describe a real-world application involving a geometric series.

65. Explain the difference between an arithmetic sequence and a geometric sequence.

66. Explain how to find the general term of a geometric sequence. Use an example to illustrate the process.

67. ***Newsletter*** Write a newsletter explaining how to find the limit of an infinite geometric series.

Quick Review Exercises

Section 14.3

Simplify.

1. $(x + 5)^2$

2. $(x - y)^2$

3. $(2x + 9y)^2$

4. $(5a^3 - 4b^4)^2$

14.4 The Binomial Theorem

OBJECTIVES

1. Evaluate factorials.
2. Calculate binomial coefficients.
3. Use the binomial theorem to expand a binomial raised to a power.
4. Use Pascal's triangle and the binomial theorem to expand a binomial raised to a power.

Recall that a binomial is a polynomial with two terms, such as $2x - 3$ and $5x + 8y$. In this section, we will learn a method for raising a binomial to a power, such as $(x + y)^9$.

Factorials

Objective 1 Evaluate factorials.

Factorials

The product of all positive integers from 1 through some positive integer n, $1 \cdot 2 \cdot 3 \cdots (n - 1) \cdot n$, is called ***n* factorial** and is denoted as $n!$

For example, $5! = 1 \cdot 2 \cdot 3 \cdot 4 \cdot 5$, or 120. By definition, $0! = 1$.

EXAMPLE 1 Evaluate: 7!

SOLUTION To evaluate 7!, multiply the successive integers from 1 through 7.

$$7! = 1 \cdot 2 \cdot 3 \cdot 4 \cdot 5 \cdot 6 \cdot 7 \quad \text{Multiply the positive integers from 1 through 7.}$$
$$= 5040 \quad \text{Simplify.}$$

Quick Check 1

Evaluate: 9!

Most calculators have a built-in function for calculating factorials; consult your calculator manual for directions. Most nongraphing calculators have a button labeled [n!] or [x!], while many graphing calculators store the factorial function in the [MATH] menu.

Binomial Coefficients

Objective 2 Calculate binomial coefficients. We now will use factorials to calculate numbers called binomial coefficients.

Binomial Coefficients

For any two nonnegative integers n and r, $n \geq r$, the number ${}_nC_r$ is a **binomial coefficient** and is given by the formula ${}_nC_r = \dfrac{n!}{r! \cdot (n-r)!}$.

EXAMPLE 2 Evaluate the binomial coefficient ${}_9C_4$.

SOLUTION Begin by substituting 9 for n and 4 for r in the formula ${}_nC_r = \dfrac{n!}{r! \cdot (n-r)!}$.

$${}_9C_4 = \frac{9!}{4! \cdot (9-4)!} \quad \text{Substitute 9 for } n \text{ and 4 for } r.$$

$$= \frac{9!}{4! \cdot 5!} \quad \text{Subtract } 9 - 4.$$

$$= \frac{362{,}880}{24 \cdot 120} \quad \text{Evaluate each factorial.}$$

$$= \frac{362{,}880}{2880} \quad \text{Simplify the denominator.}$$

$$= 126 \quad \text{Simplify the fraction.}$$

Quick Check 2

Evaluate the binomial coefficient ${}_6C_2$.

Using Your Calculator To find binomial coefficients on the TI–84, access the PRB menu. To evaluate the binomial coefficient ${}_9C_4$, begin by keying the first number, 9. Then tap [MATH] to access the MATH menu. Use the right arrow to scroll over to the PRB menu. Select option **3: nCr**. When you return to the main screen, key the second number, 4, and tap [ENTER]. Here is the calculator screen you should see.

```
9 nCr 4
                126
```

EXAMPLE 3 Evaluate the binomial coefficient ${}_8C_0$.

SOLUTION Begin by substituting 8 for n and 0 for r in the formula ${}_nC_r = \dfrac{n!}{r! \cdot (n-r)!}$.

$${}_8C_0 = \frac{8!}{0! \cdot 8!} \quad \text{Substitute 8 for } n \text{ and 0 for } r.$$

$$= \frac{8!}{8!} \quad 0! = 1$$

$$= 1 \quad \text{Simplify the fraction.}$$

Quick Check 3

Evaluate the binomial coefficient ${}_{20}C_{20}$.

For any nonnegative integer n, the binomial coefficient ${}_nC_0 = 1$.
The binomial coefficient ${}_nC_n = 1$ as well because ${}_nC_n = \dfrac{n!}{n! \cdot 0!}$.

We will finish our exploration of binomial coefficients with an example that saves time in the calculation of several binomial coefficients. It also demonstrates an interesting property of binomial coefficients.

EXAMPLE 4 Evaluate the binomial coefficients ${}_7C_0, {}_7C_1, {}_7C_2, \ldots, {}_7C_7$.

SOLUTION Verify the following results using your calculator's built-in function for calculating binomial coefficients or applying the formula:

$${}_7C_0 = 1 \quad {}_7C_1 = 7 \quad {}_7C_2 = 21 \quad {}_7C_3 = 35 \quad {}_7C_4 = 35 \quad {}_7C_5 = 21 \quad {}_7C_6 = 7 \quad {}_7C_7 = 1$$

Notice the following properties:

- Both the first and last coefficients are equal to 1.
- The second coefficient is equal to 7. For any positive integer n, ${}_nC_1 = n$.
- The coefficients follow a symmetric pattern. For any nonnegative integers n and a, $a \le n$, ${}_nC_a = {}_nC_{n-a}$. This means that once we calculate ${}_7C_2$, we know that ${}_7C_5$ is equal to the same value.

These properties will be helpful when expanding binomials using the binomial theorem.

Quick Check 4

Complete the accompanying table using the properties of binomial coefficients that were introduced in Example 4. You should not have to calculate any of these binomial coefficients.

${}_9C_0 =$	${}_9C_1 =$	${}_9C_2 = 36$	${}_9C_3 =$	${}_9C_4 = 126$
${}_9C_5 =$	${}_9C_6 = 84$	${}_9C_7 =$	${}_9C_8 =$	${}_9C_9 =$

The Binomial Theorem

Objective 3 Use the binomial theorem to expand a binomial raised to a power. We use the **binomial theorem** to raise a binomial expression to a power that is a positive integer without having to repeatedly multiply the binomial expression by itself. For instance, we can use the binomial theorem to expand $(x + y)^5$ rather than multiplying $(x + y)(x + y)(x + y)(x + y)(x + y)$.

The Binomial Theorem

For any positive integer n,

$$(x + y)^n = {}_nC_0x^ny^0 + {}_nC_1x^{n-1}y^1 + {}_nC_2x^{n-2}y^2 + \cdots + {}_nC_{n-1}x^1y^{n-1} + {}_nC_nx^0y^n.$$

This can be expressed by summation notation as

$$(x + y)^n = \sum_{r=0}^{n} {}_nC_r \cdot x^{n-r}y^r.$$

EXAMPLE 5 Expand $(x + y)^5$ using the binomial theorem.

SOLUTION We apply the binomial theorem with $n = 5$.

$$(x + y)^5 = {}_5C_0 \cdot x^5y^0 + {}_5C_1 \cdot x^4y^1 + {}_5C_2 \cdot x^3y^2 + {}_5C_3 \cdot x^2y^3 + {}_5C_4 \cdot x^1y^4 + {}_5C_5 \cdot x^0y^5$$

After evaluating the binomial coefficients and simplifying factors whose exponent is 0 or 1, we obtain the following:

$$(x + y)^5 = x^5 + 5x^4y + 10x^3y^2 + 10x^2y^3 + 5xy^4 + y^5$$

Quick Check 5

Expand $(x + y)^3$ using the binomial theorem.

If the binomial being raised to a power is a difference rather a sum, such as $(x - y)^4$, the signs of the terms in the binomial expansion alternate between positive and negative.

EXAMPLE 6 Expand $(x - y)^4$ using the binomial theorem.

SOLUTION Apply the binomial theorem with $n = 4$, alternating the signs of the terms.

$$(x - y)^4 = {}_4C_0 \cdot x^4y^0 - {}_4C_1 \cdot x^3y^1 + {}_4C_2 \cdot x^2y^2 - {}_4C_3 \cdot x^1y^3 + {}_4C_4 \cdot x^0y^4$$

$$= x^4y^0 - 4x^3y^1 + 6x^2y^2 - 4x^1y^3 + x^0y^4$$ Evaluate the binomial coefficients.

$$= x^4 - 4x^3y + 6x^2y^2 - 4xy^3 + y^4$$ Simplify factors whose exponents are 0 or 1.

Quick Check 6

Expand $(x - 4)^4$ using the binomial theorem.

Pascal's Triangle

Objective 4 Use Pascal's triangle and the binomial theorem to expand a binomial raised to a power. One convenient way for finding binomial coefficients is through the use of a triangular array of numbers called **Pascal's triangle**, which is named in honor of the 17th-century French mathematician Blaise Pascal. Here is the portion of Pascal's triangle showing the binomial coefficients for values of n through 7.

$n = 0$ 1
$n = 1$ 1 1
$n = 2$ 1 2 1
$n = 3$ 1 3 3 1
$n = 4$ 1 4 6 4 1
$n = 5$ 1 5 10 10 5 1
$n = 6$ 1 6 15 20 15 6 1
$n = 7$ 1 7 21 35 35 21 7 1

Consider the row labeled $n = 5$, which contains the values 1, 5, 10, 10, 5, and 1. These are the binomial coefficients ${}_5C_0$, ${}_5C_1$, ${}_5C_2$, ${}_5C_3$, ${}_5C_4$, and ${}_5C_5$, respectively. Each value in Pascal's triangle is the sum of the two entries diagonally above it.

$n = 4$ 1 4 6 4 1

$n = 5$ 1 5 10 10 5 1

To construct Pascal's triangle, begin with a single 1 in the first row ($n = 0$), with two 1's written diagonally underneath it in the next row ($n = 1$).

$n = 0$ 1
$n = 1$ 1 1

The next row will have one more value than the previous row, and each number in the row is the sum of the values diagonally above it.

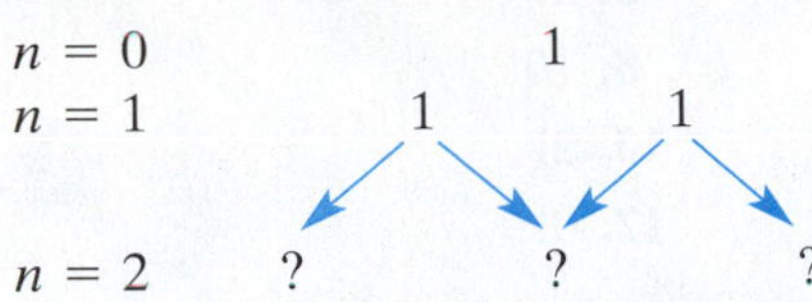

The values for this next row will be 1, 2, and 1. Then proceed to the next row, repeating this process until you have reached the desired row.

EXAMPLE 7 Expand $(x + y)^6$ using the binomial theorem and Pascal's triangle.

SOLUTION We apply the binomial theorem with $n = 6$. Here is the row of Pascal's triangle associated with $n = 6$.

$$n = 6 \qquad 1 \qquad 6 \qquad 15 \qquad 20 \qquad 15 \qquad 6 \qquad 1$$

This says that the coefficients of the seven terms are 1, 6, 15, 20, 15, 6, and 1, respectively. We know that the exponent of x in the first term will be 6 and will decrease by 1 in each successive term. We also know that the exponent of y in the first term will be 0 and will increase by 1 in each successive term.

$$(x + y)^6 = 1x^6y^0 + 6x^5y^1 + 15x^4y^2 + 20x^3y^3 + 15x^2y^4 + 6x^1y^5 + 1x^0y^6$$

Apply the binomial theorem, obtaining the binomial coefficients from the $n = 6$ row of Pascal's triangle.

$$= x^6 + 6x^5y + 15x^4y^2 + 20x^3y^3 + 15x^2y^4 + 6xy^5 + y^6$$

Simplify each term.

Quick Check 7

Expand $(x + y)^7$ using the binomial theorem and Pascal's triangle.

BUILDING YOUR STUDY STRATEGY

Preparing for the Final Exam (Part 2), 4 Taking the Test and Pacing Set a schedule for yourself. For example, establish a goal to be finished with at least 25 percent of the exam by the time one-fourth of the time has expired. This will help you determine whether you are working quickly enough.

Save time at the end of the exam period to check your work. Make sure you have attempted every problem. Some students, in a rush, may skip an entire page.

For further information on test taking strategies, see the Study Strategies in Chapter 5.

Exercises 14.4

Vocabulary

1. The product of all positive integers from 1 through some positive integer n is called __________ and is denoted as ___.

2. For any two nonnegative integers n and r, $n \geq r$, the number ${}_nC_r$ is a(n) ______________ and is

given by the formula ______________.

3. The __________ theorem is used to raise a binomial expression to a power that is a positive integer without the need to repeatedly multiply the binomial expression by itself.

4. One convenient way for finding binomial coefficients is through the use of a triangular array of numbers called ______________.

Evaluate.

5. 6!
6. 4!
7. 8!
8. 10!
9. 3!
10. 2!
11. 1!
12. 0!

Rewrite each given product as a factorial.

13. $1 \cdot 2 \cdot 3 \cdots 17$
14. $1 \cdot 2 \cdot 3 \cdots 12$
15. $1 \cdot 2 \cdot 3 \cdots 33$
16. $1 \cdot 2 \cdot 3 \cdots 19$

Simplify.

17. $\frac{11!}{4!}$
18. $\frac{10!}{8!}$
19. $\frac{9!}{3! \cdot 6!}$
20. $\frac{11!}{7! \cdot 4!}$
21. $8! + 7!$
22. $(5!)^2$

Evaluate each given binomial coefficient.

23. ${}_7C_2$
24. ${}_{10}C_8$
25. ${}_8C_3$
26. ${}_9C_4$
27. ${}_{12}C_{11}$
28. ${}_{14}C_3$
29. ${}_{13}C_4$
30. ${}_{13}C_9$
31. ${}_8C_0$
32. ${}_{15}C_{15}$

Complete the table by using the properties of binomial coefficients. Do not calculate any of binomial coefficients.

33. ${}_{11}C_0 =$ ${}_{11}C_1 =$ ${}_{11}C_2 =$
${}_{11}C_3 = 165$ ${}_{11}C_4 =$ ${}_{11}C_5 = 462$
${}_{11}C_6 =$ ${}_{11}C_7 = 330$ ${}_{11}C_8 =$
${}_{11}C_9 = 55$ ${}_{11}C_{10} =$ ${}_{11}C_{11} =$

34. ${}_{10}C_0 =$ ${}_{10}C_1 =$ ${}_{10}C_2 = 45$
${}_{10}C_3 = 120$ ${}_{10}C_4 =$ ${}_{10}C_5 = 252$
${}_{10}C_6 = 210$ ${}_{10}C_7 =$ ${}_{10}C_8 =$
${}_{10}C_9 =$ ${}_{10}C_{10} =$

Expand using the binomial theorem.

35. $(x + y)^3$
36. $(x + a)^5$
37. $(x + 2)^3$
38. $(x + y)^4$
39. $(x + 6)^4$
40. $(x + 1)^3$
41. $(x - y)^5$
42. $(x - y)^2$
43. $(x - 3)^4$
44. $(x - 1)^3$
45. $(x + 3y)^3$
46. $(4x + y)^4$
47. $(4x - y)^3$
48. $(x - 2y)^6$

List the coefficients in the row of Pascal's triangle for the given value of n.

49. $n = 10$
50. $n = 12$

51. Complete the table with the sum of the binomial coefficients of Pascal's triangle for $n = 1$ through $n = 6$.

n	1	2	3	4	5	6
Sum						

52. a) Find a formula for the sum of the binomial coefficients associated with row $n = k$ of Pascal's triangle.

b) Use the sum to predict the sum for the row associated with $n = 16$.

Expand using the binomial theorem and Pascal's triangle.

53. $(x + y)^5$
54. $(x + 2y)^3$
55. $(x + 2)^7$
56. $(x + 4)^6$
57. $(x - y)^3$
58. $(x - y)^8$
59. $(x - 7)^4$
60. $(x - 5)^5$

Writing in Mathematics

Answer in complete sentences.

61. If $a + b = n$, explain why ${}_nC_a = {}_nC_b$.

CHAPTER 14 SUMMARY

Section 14.1 Sequences and Series

Sequences, pp. 836–837

A sequence is a function whose domain is a set of consecutive natural numbers beginning with 1.

If the domain is the entire set of natural numbers $\{1, 2, 3, \ldots\}$, the sequence is an infinite sequence.

If the domain is the set $\{1, 2, 3, \ldots, n\}$ for some natural number n, the sequence is a finite sequence.

100, 120, 140, 160, 180, 200

5, 20, 80, 320, 1280, . . .

General Term of a Sequence, pp. 837–839

The general term of a sequence is denoted a_n. The general term is usually defined in terms of its index n or in terms of the preceding term a_{n-1}.

Find the first five terms of the sequence: $a_n = 3n - 4$

$-1, 2, 5, 8, 11$

Alternating Sequence, p. 837

A sequence whose terms alternate between being positive and negative is called an alternating sequence. An alternating sequence frequently has a factor of $(-1)^n$ or $(-1)^{n+1}$ in its general term.

Find the first five terms of the sequence: $a_n = (-1)^n(4n + 5)$

$-9, 13, -17, 21, -25$

Sequences Defined Recursively, p. 839

If the general term of a sequence is defined in terms of the preceding term a_{n-1}, the sequence is defined recursively.

Find the first five terms of the sequence: $a_1 = -6, a_n = -2a_{n-1} + 5$

$$a_2 = -2(-6) + 5 = 17$$
$$a_3 = -2(17) + 5 = -29$$
$$a_4 = -2(-29) + 5 = 63$$
$$a_5 = -2(63) + 5 = -121$$

$-6, 17, -29, 63, -121$

Series, pp. 839–840

An infinite series, s, is the sum of the terms in an infinite sequence.

$$s = a_1 + a_2 + a_3 + \cdots + a_n + \cdots$$

A finite series, s_n, is the sum of the first n terms of a sequence.

$$s_n = a_1 + a_2 + a_3 + \cdots + a_n$$

The finite series s_n is also called the nth partial sum of the sequence.

Find s_6 for the sequence: $a_n = 2n + 23$

First six terms: 25, 27, 29, 31, 33, 35

$s_6 = 180$

Summation Notation, p. 840

The nth partial sum of a sequence,

$$s_n = a_1 + a_2 + a_3 + \cdots + a_n,$$

can be represented in summation notation as follows:

$$\sum_{i=1}^{n} a_i$$

Find the sum: $\sum_{i=1}^{5}(4i - 1)$

$$\sum_{i=1}^{5}(4i - 1) = 3 + 7 + 11 + 15 + 19$$
$$= 55$$

Section 14.2 Arithmetic Sequences and Series

Arithmetic Sequences, p. 843

An arithmetic sequence is a sequence in which each term after the first term differs from the preceding term by a constant amount d.

This means that the general term is $a_n = a_{n-1} + d$ for $n \neq 1$.

The number d is called the common difference of the arithmetic sequence.

Find the first five terms of an arithmetic sequence whose first term is 13 and whose common difference is 9.

13, 22, 31, 40, 49

General Term of an Arithmetic Sequence, pp. 843–844

If an arithmetic sequence whose first term is a_1 has a common difference of d, the general term of this sequence is $a_n = a_1 + (n - 1)d$.

Find the general term of the arithmetic sequence: 17, 23, 29, 35, 41, . . .

$$d = 23 - 17 = 6$$

$$\begin{aligned} a_n &= a_1 + (n-1)d \\ a_n &= 17 + (n-1)6 \\ &= 17 + 6n - 6 \\ &= 6n + 11 \end{aligned}$$

Arithmetic Series, pp. 844–846

An arithmetic series is the sum of the terms of an arithmetic sequence. A finite arithmetic series consisting of only the first n terms of an arithmetic sequence is also known as the nth partial sum of the arithmetic sequence and is denoted as s_n.

The sum of the first n terms of an arithmetic sequence is given by

$$s_n = \frac{n}{2}(a_1 + a_n),$$

where a_1 is the first term of the sequence and a_n is the nth term of the sequence.

Find s_{20} for the arithmetic sequence: 5, 14, 23, 32, 41, . . .

$$d = 9$$

$$\begin{aligned} a_n &= 5 + (n-1)9 \\ &= 9n - 4 \end{aligned} \qquad \begin{aligned} a_{20} &= 9(20) - 4 \\ &= 176 \end{aligned}$$

$$s_n = \frac{20}{2}(5 + 176) = 10(181) = 1810$$

Section 14.3 Geometric Sequences and Series

Geometric Sequences, pp. 849–850

A geometric sequence is a sequence in which each term after the first term is a constant multiple of the preceding term. This means that the general term is $a_n = r \cdot a_{n-1}$ for $n \neq 1$. The number r is called the common ratio of the geometric sequence.

Find the first five terms of a geometric sequence whose first term is 8 and whose common ratio is 3.

8, 24, 72, 216, 648

General Term of a Geometric Sequence, pp. 850–851

If a geometric sequence whose first term is a_1 has a common ratio of r, the general term of this sequence is $a_n = a_1 \cdot r^{n-1}$.

Find the general term of the geometric sequence: 8, 20, 50, 125, . . .

$$r = \frac{20}{8} = \frac{5}{2} \qquad a_n = 8 \cdot \left(\frac{5}{2}\right)^{n-1}$$

Geometric Series, pp. 851–852

A geometric series is the sum of the terms of a geometric sequence. A finite geometric series consists of only the first n terms of a geometric sequence and is denoted as s_n.

The sum of the first n terms of a geometric sequence is given by

$s_n = a_1 \cdot \dfrac{1 - r^n}{1 - r}$, where a_1 is the first term of the sequence and r is the common ratio of the sequence.

Find s_8 for the geometric sequence: 3, 12, 48, 192, . . .

$$r = 4$$

$$s_8 = 3 \cdot \frac{1 - 4^8}{1 - 4} = 3 \cdot \frac{-65{,}535}{-3} = 65{,}535$$

Infinite Geometric Series, pp. 853–854

If a geometric sequence has a common ratio r such that $|r| < 1$, the infinite geometric series has a limit, and this limit is given by the formula $s = \frac{a_1}{1 - r}$.
If $|r| \geq 1$, no limit exists.

Find s for the geometric sequence: $3, \frac{6}{5}, \frac{12}{25}, \frac{24}{125}, \ldots$

$$r = \frac{\frac{6}{5}}{3} = \frac{2}{5}$$

$$s = \frac{3}{1 - \frac{2}{5}} = \frac{3}{\frac{3}{5}} = 3 \cdot \frac{5}{3} = 5$$

Section 14.4 The Binomial Theorem

Factorials, pp. 857–858

The product of all positive integers from 1 through some positive integer n, $1 \cdot 2 \cdot 3 \cdot \cdots \cdot n$, is called n factorial and is denoted $n!$.

Find: 7!

$$7! = 1 \cdot 2 \cdot 3 \cdot 4 \cdot 5 \cdot 6 \cdot 7 = 5040$$

Binomial Coefficients, pp. 858–859

For any two nonnegative integers n and r, $n > r$, the number ${}_nC_r$ is a binomial coefficient and is given by the formula ${}_nC_r = \frac{n!}{r! \cdot (n - r)!}$.

Find: ${}_9C_3$

$${}_9C_3 = \frac{9!}{3! \cdot 6!} = \frac{362{,}880}{6 \cdot 720} = 84$$

The Binomial Theorem, pp. 859–860

For any positive integer n,

$$(x + y)^n = \sum_{r=0}^{n} {}_nC_r \cdot x^{n-r}y^r$$
$$= {}_nC_0x^ny^0 + {}_nC_1x^{n-1}y^1 + \cdots + {}_nC_nx^0y^n.$$

If a binomial being raised to a power is a difference rather than a sum, the signs of the terms in the binomial expansion alternate between positive and negative.

Simplify: $(x + 3)^4$

$${}_4C_0x^4 \cdot 3^0 + {}_4C_1x^3 \cdot 3^1 + {}_4C_2x^2 \cdot 3^2 + {}_4C_3x^1 \cdot 3^3 + {}_4C_4x^0 \cdot 3^4$$
$$= 1x^4 \cdot 1 + 4x^3 \cdot 3 + 6x^2 \cdot 9 + 4x \cdot 27 + 1 \cdot 81$$
$$= x^4 + 12x^3 + 54x^2 + 108x + 81$$

Pascal's Triangle, pp. 860–861

This triangular array of numbers is a convenient way for finding binomial coefficients. Each value in Pascal's triangle is the sum of the two entries directly above it.

Here is the portion of Pascal's triangle showing the binomial coefficients for values of n through 7.

$n = 0$	1
$n = 1$	1 1
$n = 2$	1 2 1
$n = 3$	1 3 3 1
$n = 4$	1 4 6 4 1
$n = 5$	1 5 10 10 5 1
$n = 6$	1 6 15 20 15 6 1
$n = 7$	1 7 21 35 35 21 7 1

SUMMARY OF CHAPTER 14 STUDY STRATEGIES

- The study tips in this chapter focused on preparing for and doing well on a final exam. One effective way to prepare for a final exam is by studying with a small group of students. In a small group, you can help each other to ensure that everyone is prepared for the exam. It is a great way to get answers to any questions you may have.
- Create a series of practice tests to determine whether you are prepared for an exam. Start with practice tests that cover a wide range of material; then move on to more specialized practice tests for problems and concepts you find difficult.
- Bring confidence and a positive attitude to your final exam. Students who arrive fearing for the worst often realize their fears.
- Finally, even though a final exam may seem more important than a chapter exam, do not abandon the test-taking strategies that have got you to this point in the course.

CHAPTER 14 REVIEW

***Find the first five terms of the sequence with each given general term.* [14.1]**

1. $a_n = n + 7$
2. $a_n = 5n - 19$

***Find the next three terms of each given sequence and find its general term, a_n.* [14.1]**

3. 13, 26, 39, 52, 65, . . .
4. 12, 19, 26, 33, 40, . . .

***Find the first five terms of the sequence with each given first term and general term.* [14.1]**

5. $a_1 = -9, a_n = a_{n-1} + 17$
6. $a_1 = -4, a_n = 2a_{n-1} - 23$

***Find a recursive formula for the general term, a_n, of each given sequence.* [14.1]**

7. 3, 17, 31, 45, 59, . . .
8. 5, −10, 20, −40, 80, . . .

***Find the indicated partial sum for each given sequence.* [14.1]**

9. s_6, 4, 14, 24, 34, . . .
10. s_9, 2, 12, 72, 432, . . .

***Find the indicated partial sum for the sequence with each given general term.* [14.1]**

11. $s_7, a_n = 2n - 5$
12. $s_8, a_n = 5n + 6$

***Find the indicated partial sum for the sequence with each given general term.* [14.1]**

13. $s_7, a_1 = 50, a_n = a_{n-1} - 8$
14. $s_5, a_1 = 20, a_n = 4a_{n-1} + 5$

***Find each sum.* [14.1]**

15. $\sum_{i=1}^{12} i$
16. $\sum_{i=1}^{11} (5i + 6)$

***Find the common difference, d, of each given arithmetic sequence.* [14.2]**

17. 7, 31, 55, 79, 103, . . .
18. −3, −9, −15, −21, −27, . . .

***Find the first five terms of the arithmetic sequence with each given first term and common difference.* [14.2]**

19. $a_1 = 16, d = -9$
20. $a_1 = -18, d = 11$

***Find the general term, a_n, of each given arithmetic sequence.* [14.2]**

21. 4, 11, 18, 25, 32, . . .
22. 8, 1, −6, −13, −20, . . .
23. −15, −13, −11, −9, −7, . . .
24. $\frac{5}{6}, \frac{19}{12}, \frac{7}{3}, \frac{37}{12}, \frac{23}{6}, \ldots$
25. Find the 39th term, a_{39}, of the arithmetic sequence 25, 46, 67, 88, 109, ... [14.2]
26. Find the 27th term, a_{27}, of the arithmetic sequence −9, −32, −55, −78, −101, ... [14.2]
27. Which term of the arithmetic sequence −12, −5, 2, 9, 16, . . . is equal to 1633? [14.2]
28. Which term of the arithmetic sequence 40, 21, 2, −17, −36, . . . is equal to −9346? [14.2]

***Find the indicated partial sum of each given arithmetic sequence.* [14.2]**

29. s_8, 10, 17, 24, 31, . . .
30. s_{15}, −102, −84, −66, −48, . . .
31. s_{24}, 16, 13, 10, 7, . . .
32. s_{30}, −10, −3, 4, 11, . . .
33. Find the sum of the first 340 positive integers. [14.2]
34. The new CEO of a company started the job with an annual salary of $20,000,000. Her contract states that she will be given a raise of $2,000,000 each year. [14.2]
    - **a)** Write a sequence showing the CEO's annual salary for the first 5 years.
    - **b)** Find the general term, a_n, for the CEO's annual salary in her nth year in this job.
    - **c)** How much money will the CEO earn from this job over the first 10 years?

Worked-out solutions to Review Exercises marked with ● can be found on page AN-63.

***Find the common ratio, r, of each given geometric sequence.* [14.3]**

35. 1, 10, 100, 1000, 10000, . . .

36. 5, 30, 180, 1080, 6480, . . .

***Find the general term, a_n, of each given geometric sequence.* [14.3]**

37. 1, 12, 144, 1728, . . .

38. $8, 6, \frac{9}{2}, \frac{27}{8}, \ldots$

39. −4, −12, −36, −108, . . .

40. $-24, 8, -\frac{8}{3}, \frac{8}{9}, \ldots$

41. Find the eighth term, a_8, of the geometric sequence 9, 18, 36, 72, ... [14.3]

***Find the partial sum of the geometric sequence with each given first term and common ratio.* [14.3]**

42. $s_9, a_1 = 1, r = 3$

43. $s_{11}, a_1 = 10, r = -2$

***Find the partial sum for each given geometric sequence.* [14.3]**

44. s_7, 1, 6, 36, 216, . . .

45. $s_8, \frac{3}{2}, 12, 96, 768, \ldots$

46. s_{10}, 320, 160, 80, 40, . . .

47. s_9, 35, −140, 560, −2240, . . .

***For each given geometric sequence, does the infinite series have a limit? If so, find that limit.* [14.3]**

48. 64, 80, 100, 125, . . .

49. $\frac{36}{7}, \frac{24}{7}, \frac{16}{7}, \frac{32}{21}, \ldots$

***For each geometric sequence whose general term is given, does the infinite series have a limit? If so, find that limit.* [14.3]**

50. $a_n = -21 \cdot \left(\frac{2}{7}\right)^{n-1}$

51. $a_n = 45 \cdot \left(-\frac{7}{8}\right)^{n-1}$

52. A community service organization started a homeless shelter with a budget of \$150,000 for the first year. Their plans call for an annual increase in the budget of 10% over the previous year. How much will be budgeted for the shelter over the first 8 years of operation? [14.3]

***Evaluate.* [14.4]**

53. 9!

54. 7!

***Evaluate each given binomial coefficient.* [14.4]**

55. ${}_8C_3$

56. ${}_{11}C_8$

***Expand using the binomial theorem.* [14.4]**

57. $(x + y)^6$

58. $(x + 6)^4$

59. $(x - 9)^3$

***Expand using the binomial theorem and Pascal's triangle.* [14.4]**

60. $(x + y)^9$

CHAPTER 14 TEST

For Extra Help
CHAPTER Test Prep VIDEOS

Step-by-step test solutions are found on the Chapter Test Prep Videos available via the Video Resources on DVD, in MyMathLab, and on YouTube (search "WoodburyElemIntAlg" and click on "Channels").

1. Find the next three terms of the sequence 11, 26, 41, 56, 71, . . . and find its general term, a_n.

2. Find the sum $\sum_{i=1}^{9}(4i + 11)$.

3. Find the first five terms of the arithmetic sequence whose first term is $a_1 = 9$ and whose common difference is $d = 8$.

4. Find the general term, a_n, of the given arithmetic sequence 20, 13, 6, −1, −8, ...

5. Find the 15th partial sum, s_{15}, of the arithmetic sequence $14, \frac{35}{2}, 21, \frac{49}{2}, \ldots$.

6. Find the sum of the first 200 positive even integers.

7. Find the first five terms of the geometric sequence whose first term is $a_1 = 5$ and whose common ratio is $r = 8$.

8. Find the general term, a_n, of the geometric sequence $4, 28, 196, 1372, \ldots$.

9. Find the tenth partial sum, s_{10}, of the geometric sequence $3, 15, 75, 375, \ldots$.

10. For the geometric sequence $10, -6, \frac{18}{5}, -\frac{54}{25}, \ldots$, does the infinite series have a limit? If so, find that limit.

11. Jonah's parents have decided to start a college fund for him. On Jonah's first birthday, his parents deposit \$2500 in the fund. On each birthday after that, they increase the amount they deposit by \$500.

a) Write a sequence showing the amounts that Jonah's parents deposit on his first six birthdays.

b) Find the general term, a_n, for the amount deposited on Jonah's nth birthday.

c) What is the total amount deposited in the fund for Jonah's first 18 birthdays?

12. A new car valued at \$18,000 decreases in value by 25% each year.

a) Write a geometric sequence showing the value of the car at the end of each of the first 4 years after it was purchased.

b) Find the general term, a_n, for the value of the car n years after it was purchased.

13. Evaluate the binomial coefficient ${}_{12}C_5$.

Expand using the binomial theorem.

14. $(x + 1)^4$

Expand using the binomial theorem and Pascal's triangle.

15. $(4x + y)^7$

Mathematicians in History

Fibonacci was a mathematician in the 12th and 13th centuries who introduced the world to an important sequence of numbers known as the Fibonacci sequence.

Write a one-page summary (*or* make a poster) of the life of Fibonacci and his accomplishments.

Interesting issues:

- Where and when was Fibonacci born?
- Where was Fibonacci educated?
- What was the occupation of Fibonacci's father?
- What was Fibonacci's real name?
- What is the Fibonacci sequence? How is it created?
- Which problem in Fibonacci's book *Liber abaci* introduced the sequence that came to be known as the Fibonacci sequence?
- List at least three occurrences of the Fibonacci sequence in nature.
- Fibonacci also wrote problems involving perfect numbers. What is a perfect number?

CUMULATIVE REVIEW CHAPTERS 12–14

***Evaluate each given function. Round to the nearest thousandth.* [12.1–12.2]**

1. $f(x) = e^{x+3} - 21, f(2)$

2. $f(x) = \ln(x + 7) + 32, f(25)$

***Simplify.* [12.2]**

3. $\log_3 81$

4. $\ln e^8$

***Rewrite as a single logarithmic expression. Simplify if possible. Assume that all variables represent positive real numbers.* [12.3]**

5. $2 \ln x - 4 \ln y + 5 \ln z$

***Rewrite as the sum or difference of logarithmic expressions whose arguments have an exponent of 1. Simplify if possible. Assume that all variables represent positive real numbers.* [12.3]**

6. $\log\left(\frac{b^4c^5}{a}\right)$

***Evaluate using the change-of-base formula. Round to the nearest thousandth.* [12.3]**

7. $\log_{13} 200$

8. $\log_9 3224$

***Solve. Round to the nearest thousandth.* [12.4]**

9. $5^{x-6} = 25$

10. $7^x = 219$

11. $e^{x+7} = 6205$

***Solve. Round to the nearest thousandth.* [12.4]**

12. $\log(x^2 + 7x - 9) = \log(3x + 23)$

13. $\log_3(x - 9) = 4$

14. $\log_2(x + 3) + \log_2(x + 9) = 4$

***For the given function f(x), find its inverse function $f^{-1}(x)$.* [12.4]**

15. $f(x) = e^{x-2} + 3$

16. $f(x) = \ln(x + 9) - 16$

17. Jean deposited $2000 in an account that pays 6% annual interest, compounded quarterly. How long will it take until the balance of the account is $10,000? [12.5]

18. The population of a city increased from 120,000 in 1992 to 150,000 in 2002. If the population continues to increase exponentially at the same rate, what will the population be in 2022? [12.5]

***Graph. Label any intercepts and asymptotes. State the domain and range of the function.* [12.6]**

19. $f(x) = 2^x - 1$

20. $f(x) = \ln(x + 5) - 3$

***Graph each parabola. Label each vertex and any intercepts.* [13.1]**

21. $y = x^2 + 6x - 8$

22. $y = (x + 3)^2 - 4$

23. $x = -y^2 + y + 6$

24. $x = (y - 2)^2 + 3$

Graph each circle. Label the center of the circle. Identify the radius of the circle. **[13.2]**

25. $x^2 + y^2 = 64$

26. $(x + 3)^2 + (y - 2)^2 = 25$

Graph each ellipse. Label the center. Give the values of a and b. **[13.3]**

27. $\dfrac{x^2}{9} + \dfrac{y^2}{49} = 1$

28. $36x^2 + 25y^2 - 288x - 300y + 576 = 0$

Graph each hyperbola. Label the center. Give the values of a and b. **[13.4]**

29. $\dfrac{x^2}{25} - \dfrac{y^2}{4} = 1$

30. $(y - 2)^2 - \dfrac{(x + 4)^2}{9} = 1$

Find the equation of each conic section. **[13.2–13.4]**

31.

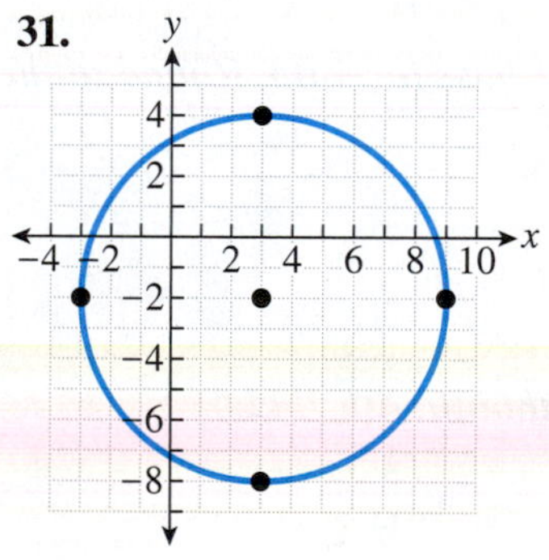

32.

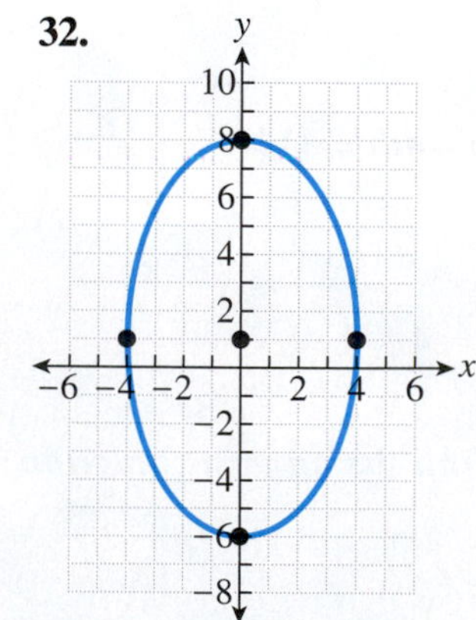

***Solve by substitution.* [13.5]**

33. $x^2 + y^2 = 40$
$2x + y = 10$

34. $x^2 + 2y^2 = 18$
$y = x - 3$

***Solve by the addition method.* [13.5]**

35. $x^2 + y^2 = 100$
$x^2 - y^2 = 28$

36. $2x^2 + 5y^2 = 38$
$4x^2 - 3y^2 = 24$

37. A rectangular lawn has a perimeter of 100 feet and an area of 600 square feet. Find the dimensions of the lawn. [13.5]

***Find the next three terms of the given sequence and find its general term, a_n.* [14.1]**

38. $-4, 9, 22, 35, 48, \ldots$

***Find the first five terms of the sequence with the given first term and general term.* [14.1]**

39. $a_1 = 13, a_n = 2a_{n-1} + 5$

***Find the general term, a_n, of the given arithmetic sequence.* [14.2]**

40. $3, 12, 21, 30, 39, \ldots$

***Find the partial sum for the given arithmetic sequence.* [14.2]**

41. $s_{12}, -6, -1, 4, 9, \ldots$

42. Find the sum of the first 200 positive integers. [14.2]

***Find the general term, a_n, of the given geometric sequence.* [14.3]**

43. $12, 9, \frac{27}{4}, \frac{81}{16}, \ldots$

***Find the partial sum of the geometric sequence of the given first term and common ratio.* [14.3]**

44. $s_8, a_1 = 1, r = 4$

***Find the partial sum for the given geometric sequence.* [14.3]**

45. $s_7, 2, 14, 98, 686, \ldots$

***For the geometric sequence whose general term is given, does the infinite series have a limit? If so, what is that limit?* [14.3]**

46. $a_n = 6 \cdot \left(\frac{3}{8}\right)^{n-1}$

***Evaluate the given binomial coefficient.* [14.4]**

47. ${}_{10}C_4$

***Expand using the binomial theorem.* [14.4]**

48. $(x + y)^4$

49. $(x - y)^6$

***Expand using the binomial theorem and Pascal's triangle.* [14.4]**

50. $(x + y)^8$

Synthetic Division

Synthetic division is a shorthand alternative to polynomial long division when the divisor is of the form $x - b$. When we use synthetic division, we do not write any variables, only the coefficients.

Suppose we wanted to divide $\frac{2x^2 - x - 25}{x - 4}$. Note that $b = 4$ in this example. Before we begin, we check the dividend to make sure it is written in descending order and no terms are missing. If there are missing terms, we add placeholders. On the first line, we write the value for b as well as the coefficients for the dividend.

$$\underline{4|} \quad 2 \quad -1 \quad -25$$

Notice that we set the value of b apart from the coefficients of the divisor. After we use synthetic division, we find the quotient and the remainder in the last of the three rows. To begin the actual division, we bring the first coefficient down to the bottom row as follows:

$$\begin{array}{r|rrr} 4 & 2 & -1 & -25 \\ & \downarrow & & \\ \hline & 2 & & \end{array}$$

This number is the leading coefficient of the quotient. In the next step, we multiply 4 by this first coefficient and write it under the next coefficient in the dividend (-1). Totaling this column gives the next coefficient in the quotient.

$$\begin{array}{r|rrr} 4 & 2 & -1 & -25 \\ & \downarrow & 8 & \\ \hline & 2 & 7 & \end{array}$$

We repeat this step until we total the last column. We multiply 4 by 7 and write the product under -25, then total the last column.

$$\begin{array}{r|rrr} 4 & 2 & -1 & -25 \\ & \downarrow & 8 & 28 \\ \hline & 2 & 7 & \boxed{3} \end{array}$$

We are now ready to read the quotient and remainder from the bottom row. The remainder is the last number in the bottom row, which in the example is 3. The numbers in front of the remainder are the coefficients of the quotient, written in descending order. In this example, the quotient is $2x + 7$. So $\dfrac{2x^2 - x - 25}{x - 4} = 2x + 7 + \dfrac{3}{x - 4}$.

EXAMPLE 1 Use synthetic division to divide $\dfrac{x^3 - 11x^2 - 13x + 6}{x + 2}$.

SOLUTION The divisor is of the correct form, and $b = -2$. This is because $x + 2$ can be rewritten as $x - (-2)$. Because the dividend is in descending form with no missing terms, we may begin.

$$\begin{array}{r|rrrr} -2 & 1 & -11 & -13 & 6 \\ & \downarrow & -2 & 26 & -26 \\ \hline & 1 & -13 & 13 & \boxed{-20} \end{array}$$

So $\dfrac{x^3 - 11x^2 - 13x + 6}{x + 2} = x^2 - 13x + 13 - \dfrac{20}{x + 2}$.

Now we turn to an example that has a dividend with missing terms, requiring the use of placeholders.

EXAMPLE 2 Use synthetic division to divide $\dfrac{3x^4 - 8x - 5}{x - 3}$.

SOLUTION The dividend, written in descending order with placeholders, is $3x^4 + 0x^3 + 0x^2 - 8x - 5$. The value for b is 3.

$$\begin{array}{r|rrrrr} 3 & 3 & 0 & 0 & -8 & -5 \\ & \downarrow & 9 & 27 & 81 & 219 \\ \hline & 3 & 9 & 27 & 73 & \boxed{214} \end{array}$$

So $\dfrac{3x^4 - 8x - 5}{x - 3} = 3x^3 + 9x^2 + 27x + 73 + \dfrac{214}{x - 3}$.

Exercises A-1

MyMathLab

PRACTICE

WATCH

DOWNLOAD

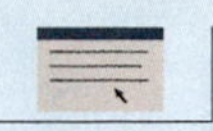
READ

REVIEW

Divide using synthetic division.

1. $\dfrac{x^2 - 18x + 77}{x - 7}$

2. $\dfrac{x^2 + 4x - 45}{x - 5}$

3. $(x^2 + 13x - 48) \div (x - 3)$

4. $(2x^2 - 43x + 221) \div (x - 13)$

5. $\dfrac{7x^2 - 11x - 12}{x + 2}$

6. $\dfrac{3x^2 + 50x - 30}{x + 17}$

7. $\dfrac{12x^2 - 179x - 154}{x - 14}$

8. $\dfrac{x^2 + 14x - 13}{x + 1}$

9. $\dfrac{x^3 - 34x + 58}{x - 5}$

10. $\dfrac{x^3 - 59x - 63}{x + 7}$

11. $\dfrac{x^4 + 8x^2 - 33}{x + 8}$

12. $\dfrac{5x^5 - 43x^3 - 8x^2 + 9x - 9}{x - 3}$

APPENDIX

A-2 Using Matrices to Solve Systems of Equations

Matrix

A **matrix** is a rectangular array of numbers, such as $\begin{bmatrix} 1 & 5 & 7 & 20 \\ 0 & 4 & -3 & 5 \\ 0 & 0 & 6 & 6 \end{bmatrix}$. The numbers in a matrix are called **elements**. A matrix is often described by the number of **rows** and **columns** it contains. The matrix in this example is a three by four, or 3×4, matrix. (The plural form of matrix is **matrices**.) We refer to the first row of a matrix as R_1, the second row as R_2, and so on.

We can use matrices to solve systems of equations in a manner that is similar to using the addition method. For example, to solve the system of equations $\begin{aligned} 2x - 5y &= -9 \\ 4x + 3y &= -5 \end{aligned}$, we begin by creating the **augmented matrix** $\left[\begin{array}{cc:c} 2 & -5 & -9 \\ 4 & 3 & -5 \end{array}\right]$. The first column of the matrix contains the coefficients of the variable x from the two equations. The second column contains the coefficients of the variable y. The dashed vertical line separates these coefficients from the constants in each equation.

We will use **row operations** to convert a matrix to a form in which we can easily solve the system of equations.

Row Operations

1. Any two rows of a matrix can be interchanged.
2. The elements in any row can be multiplied by any nonzero number.
3. The elements in any row can be changed by adding a nonzero multiple of another row's elements to them.

To solve a system of two equations in two unknowns, our goal is to transform the augmented matrix associated with that system to one of the form $\left[\begin{array}{cc|c} a & b & c \\ 0 & d & e \end{array}\right]$.

This matrix corresponds to the system of equations $\begin{aligned} ax + by &= c \\ dy &= e \end{aligned}$. We can then easily solve the second equation for y and substitute that value in the first equation to solve for x. So the goal is to transform the matrix into one whose first element in the second row is 0.

EXAMPLE 1 Solve $\begin{aligned} 2x - 5y &= -9 \\ 4x + 3y &= -5 \end{aligned}$.

SOLUTION The augmented matrix associated with this system is $\left[\begin{array}{cc|c} 2 & -5 & -9 \\ 4 & 3 & -5 \end{array}\right]$. We can change the first element in row 2 to 0 by adding -2 times the first row to the second row.

$$\left[\begin{array}{cc|c} 2 & -5 & -9 \\ 4 & 3 & -5 \end{array}\right]$$

$$\left[\begin{array}{cc|c} 2 & -5 & -9 \\ 0 & 13 & 13 \end{array}\right] \quad \text{Multiply } R_1 \text{ by } -2 \text{ and add to } R_2.$$

The second row of this matrix is equivalent to the equation $13y = 13$, which can be solved by dividing both sides of the equation by 13. The solution of this equation is $y = 1$.

The first row of this matrix is equivalent to the equation $2x - 5y = -9$. If we substitute 1 for y, we can solve this equation for x.

$$\begin{aligned} 2x - 5(1) &= -9 \\ 2x - 5 &= -9 \\ 2x &= -4 \\ x &= -2 \end{aligned}$$

The solution of the system of equations is $(-2, 1)$.

EXAMPLE 2 Solve $\begin{aligned} 3x + 4y &= 6 \\ 2x - y &= -7 \end{aligned}$.

SOLUTION The augmented matrix associated with this system is $\left[\begin{array}{cc|c} 3 & 4 & 6 \\ 2 & -1 & -7 \end{array}\right]$. We begin by multiplying row 2 by 3 so that the first element in row 2 is a multiple of the first element in row 1.

$$\left[\begin{array}{cc|c} 3 & 4 & 6 \\ 2 & -1 & -7 \end{array}\right]$$

$$\left[\begin{array}{cc|c} 3 & 4 & 6 \\ 6 & -3 & -21 \end{array}\right] \quad \text{Multiply } R_2 \text{ by 3.}$$

$$\left[\begin{array}{cc|c} 3 & 4 & 6 \\ 0 & -11 & -33 \end{array}\right] \quad \text{Multiply } R_1 \text{ by } -2 \text{ and add to } R_2.$$

This matrix is equivalent to the system of equations $\begin{array}{r} 3x + 4y = 6 \\ -11y = -33 \end{array}$. We solve the second equation for y.

$$\begin{aligned} -11y &= -33 \\ y &= 3 \quad \text{Divide both sides by } -11. \end{aligned}$$

Substitute 3 for y in the equation $3x + 4y = 6$ and solve for x.

$$\begin{aligned} 3x + 4(3) &= 6 \\ 3x + 12 &= 6 \\ 3x &= -6 \\ x &= -2 \end{aligned}$$

The solution of the system of equations is $(-2, 3)$.

To solve a system of three equations in three unknowns, we begin by setting up a 3×4 augmented matrix. Using row operations, our goal is to transform the matrix to one of the form $\left[\begin{array}{ccc|c} a & b & c & d \\ 0 & e & f & g \\ 0 & 0 & h & i \end{array}\right]$.

EXAMPLE 3 Solve the system $\begin{array}{r} x + y + z = 1 \\ 2x - 3y + 4z = 29 \\ -x - 2y + 3z = 8 \end{array}$.

SOLUTION The augmented matrix associated with this system is

$\left[\begin{array}{ccc|c} 1 & 1 & 1 & 1 \\ 2 & -3 & 4 & 29 \\ -1 & -2 & 3 & 8 \end{array}\right]$. We begin by changing the first element in row 2 and row 3 to 0.

$$\left[\begin{array}{ccc|c} 1 & 1 & 1 & 1 \\ 2 & -3 & 4 & 29 \\ -1 & -2 & 3 & 8 \end{array}\right]$$

$$\left[\begin{array}{ccc|c} 1 & 1 & 1 & 1 \\ 0 & -5 & 2 & 27 \\ 0 & -1 & 4 & 9 \end{array}\right] \quad \text{Multiply } R_1 \text{ by } -2 \text{ and add to } R_2. \text{ Add } R_1 \text{ to } R_3.$$

Now we need to change the second element of row 3 to a 0. We begin to do that by changing that element to a multiple of the second element of row 2.

$$\left[\begin{array}{ccc|c} 1 & 1 & 1 & 1 \\ 0 & -5 & 2 & 27 \\ 0 & 5 & -20 & -45 \end{array}\right] \quad \text{Multiply } R_3 \text{ by } -5.$$

$$\left[\begin{array}{ccc|c} 1 & 1 & 1 & 1 \\ 0 & -5 & 2 & 27 \\ 0 & 0 & -18 & -18 \end{array}\right] \quad \text{Add } R_2 \text{ to } R_3.$$

The last row is equivalent to the equation $-18z = -18$, whose solution is $z = 1$. We substitute this value for z in the equation that is equivalent to row 2, $-5y + 2z = 27$.

$$\begin{aligned} -5y + 2(1) &= 27 \\ -5y + 2 &= 27 \\ -5y &= 25 \\ y &= -5 \end{aligned}$$

Now we substitute -5 for y and 1 for z in the equation that is equivalent to row 1, $x + y + z = 1$.

$$\begin{aligned} x + (-5) + (1) &= 1 \\ x - 4 &= 1 \\ x &= 5 \end{aligned}$$

The solution of this system is $(5, -5, 1)$.

Exercises A-2

PRACTICE

WATCH

DOWNLOAD

READ

REVIEW

Solve the system of equations by using matrices.

1. $2x + 3y = 13$
$5x - 4y = -2$

2. $3x - 2y = 16$
$7x + 5y = 18$

3. $-4x + 9y = 75$
$3x - 7y = -58$

4. $x + 2y = -23$
$6x + y = -39$

5. $11x + 7y = 42$
$19x - 13y = -78$

6. $15x + 8y = 9$
$-10x + 9y = -6$

7. $x + 3y + 2z = 11$
$3x - 2y - z = 4$
$-2x + 4y + 3z = 5$

8. $x - y + z = 1$
$4x + 3y + 2z = 10$
$4x - 2y - 5z = 31$

9. $x + y + z = 3$
$2x + 3y + 4z = 16$
$3x - y + 4z = 23$

10. $x + y + z = 2$
$-2x - 3y + 8z = -26$
$5x - 4y - 9z = 20$

11. $3x + 7y - 8z = 60$
$5x - 4y + 7z = -44$
$-4x + y + 10z = -36$

12. $6x + y + 3z = -40$
$-5x + \frac{3}{2}y - z = 20$
$\frac{9}{2}x + \frac{1}{5}y + \frac{3}{4}z = 19$

ANSWERS TO SELECTED EXERCISES

CHAPTER 1

Quick Check 1.1 **1.** 0 1 2 3 4 5 6 7 8 9 10 **2. a)** $>$ **b)** $<$ **3. a)** 13 **b)** -8 **4. a)** $<$ **b)** $>$ **5.** 9

Section 1.1 **1.** empty set **3.** infinity **5.** 0 1 2 3 4 5 6 7 8 9 10 **7.** 0 1 2 3 4 5 6 7 8 9 10 **9.** 0 1 2 3 4 5 6 7 8 9 10 **11.** $<$ **13.** $>$ **15.** $>$ **17.** -10 -8 -6 -4 -2 0 **19.** 0 2 4 6 8 10 **21.** -15 -12 -9 -6 -3 0 **23.** 7 **25.** -22 **27.** 0 **29.** $>$ **31.** $<$ **33.** $<$ **35.** $>$ **37.** 15 **39.** 0 **41.** 7 **43.** -7 **45.** -29 **47.** $>$ **49.** $<$ **51.** $<$ **53.** A, B, C, D **55.** B, C, D **57.** C, D **59.** $-5, 5$ **61.** No number possible **63.** $-14, 14$ **65.** No, the absolute value of 0 is not positive. **67.** Negative; explanations will vary.

Quick Check 1.2 **1.** 8 **2.** -13 **3.** -11 **4.** 18 **5.** -4 **6.** -60 **7.** 63 **8.** -400 **9.** -9

Section 1.2 **1.** larger **3.** positive **5.** positive **7.** divisor **9.** -5 **11.** -27 **13.** 1 **15.** 8 **17.** -11 **19.** 2 **21.** -6 **23.** -8 **25.** -22 **27.** 61 **29.** -9 **31.** -11 **33.** -6 **35.** 27 **37.** 1 **39.** \$8 **41.** 4°C **43.** 2150 ft **45.** -42 **47.** -40 **49.** 180 **51.** -82 **53.** 0 **55.** 90 **57.** 180 **59.** -9 **61.** -6 **63.** 4 **65.** -14 **67.** 0 **69.** Undefined **71.** 132 **73.** -8 **75.** -21 **77.** -144 **79.** -216 **81.** 0 **83.** \$92 **85.** $-\$200$ **87.** 19 **89.** -128 **91.** True **93.** True **95.** Taking away the negative number -5 from 11 is equivalent to adding the positive number 5 to 11. **97.** It is the opposite of the product of a positive integer and a negative integer.

Quick Check 1.3 **1.** $\{1, 2, 3, 4, 6, 9, 12, 18, 36\}$ **2. a)** Composite **b)** Prime **c)** Composite **3.** $3 \cdot 3 \cdot 7$ **4. a)** $\frac{3}{14}$ **b)** $\frac{1}{16}$ **5.** $9\frac{4}{13}$ **6.** $\frac{49}{6}$

Section 1.3 **1.** factor set **3.** composite **5.** top **7.** common factors **9.** improper **11.** No **13.** Yes **15.** Yes **17.** $\{1, 2, 3, 4, 6, 8, 12, 16, 24, 48\}$ **19.** $\{1, 3, 9, 27\}$ **21.** $\{1, 2, 4, 5, 10, 20\}$ **23.** $\{1, 3, 9, 27, 81\}$ **25.** $\{1, 31\}$ **27.** $\{1, 7, 13, 91\}$ **29.** $2 \cdot 3 \cdot 3$ **31.** $2 \cdot 3 \cdot 7$ **33.** $3 \cdot 13$ **35.** $3 \cdot 3 \cdot 3$ **37.** $5 \cdot 5 \cdot 5$ **39.** Prime **41.** $3 \cdot 3 \cdot 11$ **43.** Prime **45.** $2 \cdot 2 \cdot 2 \cdot 3 \cdot 5$ **47.** $\frac{5}{8}$ **49.** $\frac{1}{5}$ **51.** $\frac{4}{9}$ **53.** $\frac{27}{64}$ **55.** $\frac{10}{11}$ **57.** $\frac{7}{13}$ **59.** $\frac{19}{5}$ **61.** $\frac{50}{17}$ **63.** $\frac{151}{11}$ **65.** $7\frac{4}{5}$ **67.** $14\frac{3}{7}$ **69.** $7\frac{8}{19}$ **71.** Answers will vary. One possible answer is $\frac{6}{8}, \frac{9}{12}, \frac{12}{16}$, and $\frac{15}{20}$. **73.** Answers will vary. One possible answer is 30, 42, 70, and 105. **75.** Answers will vary; some possible answers include cooking and construction.

Quick Check 1.4 **1.** $\frac{5}{56}$ **2.** 23 **3.** $\frac{8}{105}$ **4.** $\frac{5}{14}$ **5.** $\frac{3}{5}$ **6.** $11\frac{3}{4}$ **7.** 126 **8.** $\frac{26}{45}$ **9.** $1\frac{11}{30}$

Section 1.4 **1.** improper fraction **3.** Invert the divisor and multiply. **5.** least common multiple (LCM) **7.** $\frac{1}{18}$ **9.** $-\frac{22}{27}$ **11.** $2\frac{2}{5}$ **13.** $31\frac{1}{2}$ **15.** $\frac{16}{27}$ **17.** $\frac{27}{20}$ **19.** $-\frac{7}{4}$ **21.** $\frac{17}{20}$ **23.** $\frac{5}{44}$ **25.** $-\frac{21}{50}$ **27.** $\frac{11}{15}$ **29.** 1 **31.** $-\frac{2}{5}$ **33.** $\frac{3}{8}$ **35.** 30 **37.** 84 **39.** 80 **41.** 280 **43.** $\frac{31}{20}$ **45.** $\frac{53}{40}$ **47.** $11\frac{1}{5}$ **49.** $11\frac{5}{6}$ **51.** $\frac{1}{5}$ **53.** $\frac{13}{28}$ **55.** $4\frac{1}{4}$ **57.** $3\frac{3}{10}$ **59.** $-\frac{41}{36}$ **61.** $-\frac{17}{48}$ **63.** $\frac{146}{105}$ **65.** $-\frac{89}{90}$ **67.** $-\frac{67}{240}$ **69.** $\frac{8}{15}$ **71.** $\frac{8}{25}$ **73.** $-\frac{17}{24}$ **75.** $\frac{56}{45}$ **77.** 104 **79.** $2\frac{6}{35}$ **81.** $-\frac{83}{168}$ **83.** $-\frac{77}{468}$ **85.** 8 **87.** 14 **89.** $1\frac{7}{12}$ cups **91.** $\frac{9}{20}$ fluid ounce **93.** $\frac{7}{10}$ of a foot **95.** No, the craftsperson needs 3 feet of wood. **97.** $\frac{2}{3}$ cup **99.** You cannot divide a common factor from two numerators.

Quick Check 1.5 **1.** 0.75 **2.** $0.2\overline{7}$ **3.** $\frac{17}{40}$ **4. a)** 70% **b)** $52\frac{1}{2}$% **5.** 42% **6. a)** $\frac{7}{20}$ **b)** $\frac{7}{60}$ **7. a)** 0.08 **b)** 2.4

Section 1.5 **1.** tenths **3.** numerator, denominator **5.** divide **7.** 7.85 **9.** -36.4 **11.** 4.7 **13.** 33.92 **15.** -8.03 **17.** 35.032 **19.** 55.2825 **21.** 94 **23.** 0.9 **25.** -2.875 **27.** 0.52 **29.** 24.58 **31.** $\frac{1}{5}$ **33.** $\frac{17}{20}$ **35.** $-\frac{37}{50}$ **37.** $\frac{3}{8}$ **39.** 101.4° F **41.** \$224.33 **43.** \$257.64 **45.** 75% **47.** 80% **49.** $87\frac{1}{2}$% **51.** 675% **53.** 40% **55.** 15% **57.** 9% **59.** 320% **61.** $\frac{21}{25}$ **63.** $\frac{7}{100}$ **65.** $\frac{1}{9}$ **67.** $5\frac{1}{5}$ **69.** 0.54 **71.** 0.16 **73.** 0.07 **75.** 0.003 **77.** 4 **79.** Answers will vary. Possible answers include $\frac{3}{8}, \frac{6}{16}, \frac{9}{24}$. **81.** $\frac{1}{5}$, 20% **83.** $\frac{8}{25}$, 0.32 **85.** 1.625, 162.5% **87.** Answers will vary. They now match U.S. currency—dollars and cents.

Quick Check 1.6 **1.** \$854 **2. a)** 30, **b)** 43 **3.** Mode: 65; midrange: 61 **4.** 23 centimeters **5.**

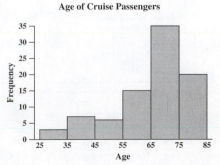

Age of Cruise Passengers

Frequency: 0, 5, 10, 15, 20, 25, 30, 35

Age: 25, 35, 45, 55, 65, 75, 85

Section 1.6 **1.** center **3.** mean **5.** mode **7.** range **9.** 59 **11.** 46.75 **13.** 76 **15.** 59 **17.** 56 **19.** 32, 41 **21.** none **23.** 129, 134 **25.** 435, 162 **27. a)** 23, **b)** 22, **c)** 30, **d)** 22.5, **e)** 19 **29. a)** 52, **b)** 49, **c)** 63, **d)** 54.5, **e)** 37 **31. a)** 63, **b)** 67, **c)** 44, 76, **d)** 58, **e)** 56 **33. a)** 264, **b)** 266, **c)** 210, **d)** 275, **e)** 412 **35.** 100.25 **37.** 133 **39.** 115.2 **41.** 109 **43.** 95 **45.** 16.5 **47. a)** \$217.80, **b)** \$219, **c)** \$199, \$219, \$259, **d)** \$213, **e)** \$172 **49. a)** 295.375, **b)** 270.5, **c)** none, **d)** 377, **e)** 394 **51. a)** 1003, **b)** 955, **c)** none, **d)** 1067.5, **e)** 1359 **53.** 76 **55.** 105 **57.** 10 **59.**

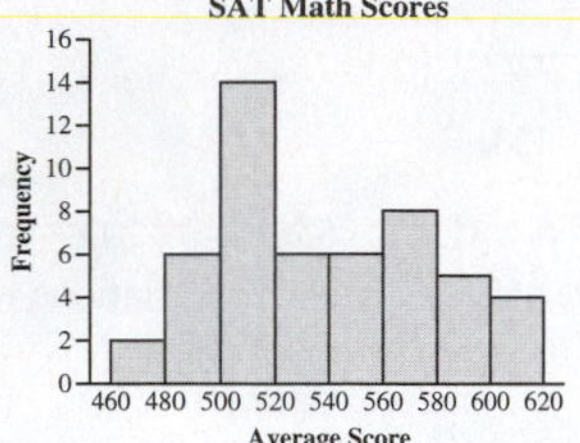

61.

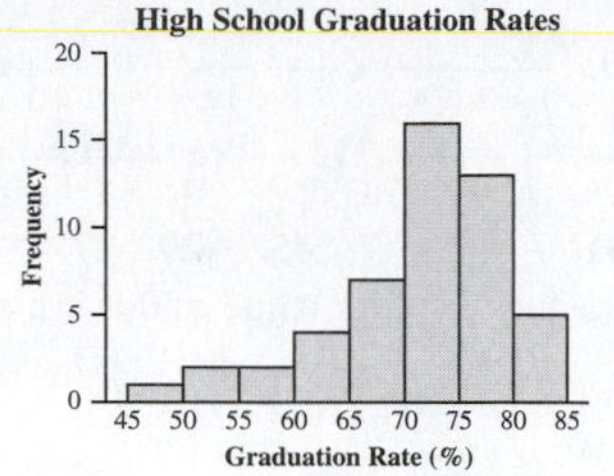

63. Frequency: 2, 11, 30, 15, 2

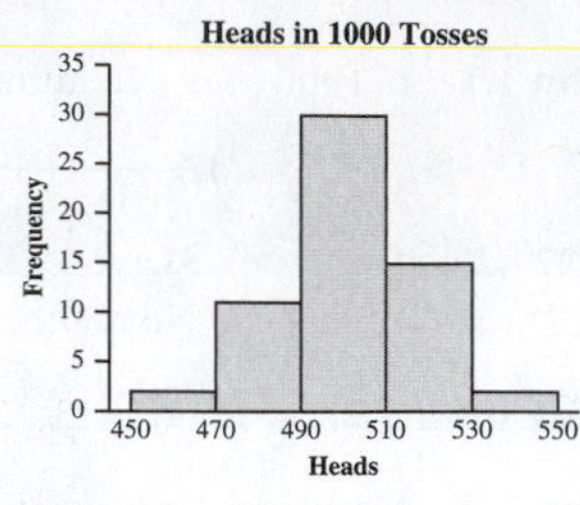

65. Frequency: 2, 9, 10, 5, 9, 1

67.

Score	Frequency
50 to 59	9
60 to 69	8
70 to 79	11
80 to 89	8
90 to 99	9

69.

Height	Frequency
150 to 154	3
155 to 159	5
160 to 164	7
165 to 169	7
170 to 174	8
175 to 179	4

Quick Check 1.7 **1.** 64 **2.** $\frac{1}{4096}$ **3.** -84 **4.** 1 **5.** 360 **6.** -84

Section 1.7 **1.** base **3.** squared **5.** grouping **7.** 2^3 **9.** $(-2)^4$ **11.** -3^5 **13.** 5^3 **15.** 81 **17.** 16,807 **19.** 1,000,000 **21.** 1 **23.** $\frac{27}{64}$ **25.** 0.00032 **27.** 81 **29.** -128 **31.** 200 **33.** -144 **35.** 21 **37.** 16 **39.** -23 **41.** -14.44 **43.** 14.9 **45.** -71 **47.** 25 **49.** 51 **51.** $\frac{11}{14}$ **53.** $\frac{11}{125}$ **55.** $\frac{1}{21}$ **57.** $-\frac{1}{4}$ **59.** $\frac{10}{3}$ **61.** 151 **63.** -100 **65.** -64 **67.** -1995 **69.** $7 + 4 + 5 \cdot 6$ **71.** $5 - 11 + 2 - 3^2$ **73.** 1024 **75.** 1,073,741,824 **77.** 3 **79.** 9 **81.** 8 **83.** $3 \cdot 5 - 9 + 8 = 14$ **85.** $(3 + 7 \cdot 9) \div 2 = 33$ **87.** Explanations will vary. $(-2)^6 = 64, -2^6 = -64$

Quick Check 1.8 **1.** $x + 9$ **2.** $x - 25$ **3.** $2x$ **4.** $\frac{x}{20}$ **5.** 57 **6.** 128 **7.** 97 **8.** $35x - 28$ **9.** $12x - 24y + 36z$ **10.** $-24x - 66$ **11. Four terms:** $x^3, -x^2, 23x, -59$; **coefficients:** $1, -1, 23, -59$ **12.** $-2x + 8y$ **13.** $19x + 7$ **14.** $-2x + 51$

Section 1.8 **1.** variable **3.** evaluate **5.** associative **7.** term **9.** like terms **11.** $x + 15$ **13.** $x - 24$ **15.** $3x$ **17.** $2x + 19$ **19.** $x + y$ **21.** $7(x - y)$ **23.** $325c$ **25.** $25{,}000 + 22a$ **27.** Nine less than a number **29.** Seven times a number **31.** Ten less than eight times a number **33.** 79 **35.** -87 **37.** 46 **39.** 88 **41.** -20 **43.** 0 **45.** 49 **47.** 44 **49.** -31.95 **51.** 8.91 **53. a)** $7 - 5 \neq 5 - 7$ **b)** Yes, a and b must be equal. **55.** $3x - 27$ **57.** $15 - 35x$ **59.** $-10x - 14$ **61.** $-54x + 30$ **63.** $16x$ **65.** $-5x$ **67.** $7x + 4$ **69.** $-2x - 22y$ **71.** $10x - 8$ **73.** $8x - 29$ **75.** $-9y + 85$ **77.** $30z + 6$ **79.** $-10a - 5b + 26c$ **81. a)** 4 **b)** $5x^3, 3x^2, -7x, -15$ **c)** $5, 3, -7, -15$ **83. a)** 2 **b)** $3x, -17$ **c)** $3, -17$ **85. a)** 3 **b)** $15x^2, -41x, 55$ **c)** $15, -41, 55$ **87. a)** 4 **b)** $-35a, -33b, 52c, 69$ **c)** $-35, -33, 52, 69$ **89.** Answers will vary. Items are worth \$60 each.

Chapter 1 Review **1.** < **2.** > **3.** 8 **4.** 13 **5.** 6 **6.** -12 **7.** -7 **8.** -3 **9.** -41 **10.** 27 **11.** -37 **12.** 20 **13.** -54 **14.** 16 **15.** $\{1, 2, 3, 6, 7, 14, 21, 42\}$ **16.** $\{1, 2, 3, 4, 6, 9, 12, 18, 27, 36, 54, 108\}$ **17.** $2 \cdot 2 \cdot 2 \cdot 2 \cdot 2$ **18.** $2 \cdot 2 \cdot 3 \cdot 5$ **19.** $\frac{4}{7}$ **20.** $\frac{1}{8}$ **21.** $\frac{20}{117}$ **22.** 14 **23.** $\frac{17}{3}$ **24.** $\frac{307}{25}$ **25.** $3\frac{4}{5}$ **26.** $9\frac{1}{6}$ **27.** $\frac{21}{100}$ **28.** $2\frac{1}{3}$ **29.** $\frac{14}{25}$ **30.** $\frac{32}{75}$ **31.** $\frac{29}{24}$ **32.** $\frac{89}{60}$ **33.** $\frac{1}{6}$ **34.** $-\frac{29}{252}$ **35.** $\frac{25}{77}$ **36.** $\frac{5}{24}$ **37.** $\frac{21}{104}$ **38.** $\frac{5}{12}$ **39.** 12.62 **40.** 8.818 **41.** 30.24 **42.** 8.8 **43.** 0.32 **44.** 0.9375 **45.** $\frac{3}{4}$ **46.** $\frac{7}{25}$ **47.** 40% **48.** 28% **49.** 90% **50.** 45% **51.** $\frac{3}{10}$ **52.** $\frac{11}{20}$ **53.** 0.9 **54.** 0.04 **55.** \$17 **56.** $-$\$47 **57.** \$12,600 **58.** $4\frac{1}{6}$ cups **59.** 76.5 inches. **60.** 126.5 **61.** 88, 89, 20

62.

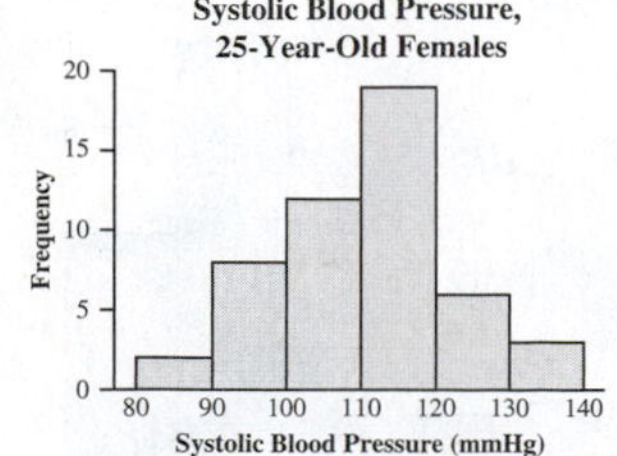

63. 64 **64.** $\frac{8}{125}$ **65.** -64 **66.** 6075 **67.** 37 **68.** 22 **69.** 48 **70.** 60 **71.** 51 **72.** $\frac{87}{100}$ **73.** $x + 14$ **74.** $x - 20$ **75.** $2x - 8$ **76.** $6x + 9$ **77.** $3.55c$ **78.** $20 + 0.15m$ **79.** 44 **80.** 25 **81.** 56 **82.** -69 **83.** 0 **84.** 84 **85.** $5x + 35$ **86.** $27x$ **87.** $5x - 8$ **88.** $-16y + 126$ **89.** $5k - 48$ **90.** $3x - 303$ **91. a)** 4 **b)** $x^3, -4x^2, -10x, 41$ **c)** $1, -4, -10, 41$ **92. a)** 3 **b)** $-x^2, 5x, -30$ **c)** $-1, 5, -30$

Chapter 1 Review Exercises: Worked-Out Solutions

12. $8 - (-19) - 7$
$= 8 + 19 - 7$
$= 27 - 7$
$= 20$

15. $\underline{42}$
$1 \cdot 42$
$2 \cdot 21$
$3 \cdot 14$
$6 \cdot 7$
$\{1, 2, 3, 6, 7, 14, 21, 42\}$

18. 60 → (2), 30; 30 → (2), 15; 15 → (3), (5)
$60 = 2 \cdot 2 \cdot 3 \cdot 5$

19. $\frac{24}{42} = \frac{\overset{1}{\cancel{2}} \cdot 2 \cdot 2 \cdot \overset{1}{\cancel{3}}}{\underset{1}{\cancel{2}} \cdot \underset{1}{\cancel{3}} \cdot 7} = \frac{4}{7}$

24. $12 \cdot 25 + 7 = 307$
$12\frac{7}{25} = \frac{307}{25}$

26. $6\overline{)55}$ = 9, -54, remainder 1
$\frac{55}{6} = 9\frac{1}{6}$

27. $\frac{\overset{3}{\cancel{9}}}{\underset{4}{\cancel{16}}} \cdot \frac{\overset{7}{\cancel{28}}}{\underset{25}{\cancel{75}}} = \frac{21}{100}$

31. $\frac{5}{8} + \frac{7}{12}$
$= \frac{15}{24} + \frac{14}{24}$
$= \frac{29}{24}$

43. $25\overline{)8.00}$ = 0.32
$-7\,5$
50
-50
0
$\frac{8}{25} = 0.32$

45. $0.75 = \frac{75}{100} = \frac{3}{4}$

47. $\frac{2}{\underset{1}{\cancel{5}}} \cdot \overset{20}{\cancel{100}}\% = 40\%$

51. $30\% = \frac{30}{100} = \frac{3}{10}$

71. $54 - 27 \div 3^2$
$= 54 - 27 \div 9$
$= 54 - 3$
$= 51$

83. $x^2 - 7x - 30$
$(-3)^2 - 7(-3) - 30$
$= 9 + 21 - 30$
$= 0$

88. $8y - 6(4y - 21)$
$= 8y - 24y + 126$
$= -16y + 126$

Chapter 1 Test **1.** $>$ **2.** 17 **3.** -6 **4.** 63 **5.** $\{1, 3, 5, 9, 15, 45\}$ **6.** $2 \cdot 2 \cdot 3 \cdot 3 \cdot 3$ **7.** $\frac{5}{7}$ **8.** $3\frac{13}{18}$ **9.** $\frac{5}{84}$ **10.** $\frac{7}{12}$ **11.** $\frac{91}{60}$ **12.** $-\frac{17}{72}$ **13.** 18.2735 **14.** $\frac{9}{25}$ **15.** \$6487 **16.** \$795.51 **17.** $\frac{18}{25}$ **18.** 0.06 **19.** 55, 45 **20.** -24 **21.** 25 **22.** $\frac{5}{11}$ **23.** $4n - 7$ **24.** $50 + 20h$ **25.** 61 **26.** -1 **27.** $10x - 65$ **28.** $-9y + 240$

CHAPTER 2

Quick Check 2.1 **1.** Yes **2.** $\{-16\}$ **3.** $\{14\}$ **4.** $\{-16\}$ **5.** $\left\{\frac{14}{3}\right\}$ **6.** $\{-30\}$ **7.** $\{15\}$ **8.** \$65

Section 2.1 **1.** equation **3.** solution set **5.** For any algebraic expressions A and B and any number n, if $A = B$, then $A + n = B + n$ and $A - n = B - n$. **7.** No, there is a term containing x^2. **9.** Yes **11.** Yes **13.** Yes **15.** No, there is a variable in a denominator. **17.** Yes **19.** No **21.** Yes **23.** No **25.** $\{-13\}$ **27.** $\{14\}$ **29.** $\left\{\frac{11}{4}\right\}$ **31.** $\{0\}$ **33.** $\{-7\}$ **35.** $\{14\}$ **37.** $\{-45\}$ **39.** $\{21\}$ **41.** $\{-96\}$ **43.** $\{8\}$ **45.** $\{-10\}$ **47.** $\{2.17\}$ **49.** $\{7\}$ **51.** $\{-8\}$ **53.** $\{43\}$ **55.** $\{2.5\}$ **57.** $\{9\}$ **59.** $\{20\}$ **61.** $\{3\}$ **63.** $\{23\}$ **65.** $\{-9\}$ **67.** $\{-4\}$ **69.** $\{-4\}$ **71.** $\{12\}$ **73.** $\{16\}$ **75.** $\{30\}$ **77.** $\{-41\}$ **79.** $\{-77\}$ **81.** $\{-56\}$ **83.** $\{-18\}$ **85.** $\left\{-\frac{21}{10}\right\}$ **87.** $\{0\}$ **89.** $x - 3 = 4$ **91.** $2n = 5$ **93.** $0.05x = 1.35$; 27 nickels **95.** $3x = 477$; 159 people **97.** $x + 8 = 174$; 166 employees **99.** $x - 19 = 37$; 56°F **101.** Explanations will vary; cannot divide by 0.

Quick Check 2.2 **1.** $\{7\}$ **2.** $\{-8\}$ **3.** $\left\{\frac{35}{12}\right\}$ **4.** $\{-9\}$ **5.** $\{2\}$ **6.** $\{-3\}$ **7.** $\{-6\}$ **8.** $\varnothing$ **9.** $\mathbb{R}$ **10.** $y = \frac{-x + 5}{2}$ **11.** $x = \frac{5z}{y}$

Section 2.2 **1.** LCM **3.** the empty set **5.** b **7.** $\{-3\}$ **9.** $\left\{\frac{7}{2}\right\}$ **11.** $\{-7\}$ **13.** $\{0\}$ **15.** $\{5.7\}$ **17.** $\{-10\}$ **19.** $\left\{-\frac{15}{2}\right\}$ **21.** $\mathbb{R}$ **23.** $\{3\}$ **25.** $\{6\}$ **27.** $\{-2\}$ **29.** $\{125\}$ **31.** $\{26\}$ **33.** $\{18\}$ **35.** $\varnothing$ **37.** $\{-5\}$ **39.** $\{1200\}$ **41.** $\{8200\}$ **43.** $\{30\}$ **45.** $\{3\}$ **47.** $\{37\}$ **49.** $\left\{\frac{104}{25}\right\}$ **51.** $\left\{-\frac{87}{2}\right\}$ **53.** Answers will vary; $5x - 6 = 5x - 7$ **55.** Answers will vary; $2(x - 4) - 3 = 2x - 11$ **57.** Answers will vary; $x + 3 = -8$ **59.** $y = -5x - 2$ **61.** $y = \frac{-7x + 4}{2}$ **63.** $y = \frac{4x + 10}{3}$ **65.** $b = P - a - c$ **67.** $t = \frac{d}{r}$ **69.** $r = \frac{C}{2\pi}$ **71.** 20°C **73.** 87 **75.** **a)** 80 kites **b)** \$6.50 **c)** \$12.75 **77.** 4 **79.** 15 **81.** $\left\{-\frac{15}{2}\right\}$ **83.** $\mathbb{R}$ **85.** $\left\{\frac{13}{2}\right\}$ **87.** $x = \frac{8y + 34}{11}$ **89.** $\left\{\frac{1}{8}\right\}$ **91.** $\{1800\}$ **93.** Answers/explanations will vary; $x = \frac{23}{2}, x = \frac{y + 9}{2}$

Quick Check 2.3 **1.** 8 **2.** Length: 24 feet; width: 16 feet **3.** 14 inches, 48 inches, 50 inches **4.** 70, 71, 72 **5.** 7 hours **6.** 17 and 86 **7.** 22 nickels and 9 quarters

Section 2.3 **1.** b **3.** b **5.** $x + 1$, $x + 2$ **7.** $x + 2$, $x + 4$ **9.** 46 **11.** 145 **13.** 8.9 **15.** Length: 9 meters; width: 4 meters **17.** Length: 18 feet; width: 10 feet **19.** Length: 52 centimeters; width: 13 centimeters **21.** 55 feet **23.** 45 centimeters **25.** 8 inches, 13 inches, 15 inches **27.** 11 inches **29.** A: 60°; B: 30° **31.** 20°, 70° **33.** 42°, 138° **35.** 50°, 60°, 70° **37.** 667 square inches **39.** 52 inches **41.** 152, 153, 154 **43.** 80, 82, 84 **45.** 73, 75, 77, 79 **47.** 15, 16, 17 **49.** 500 miles **51.** $5\frac{1}{2}$ hours **53.** $1\frac{11}{14}$ meters per second **55.** **a)** 15 hours **b)** 560 miles **57.** 6 hours **59.** 5 hours **61.** 27, 44 **63.** 44, 52 **65.** 13, 18 **67.** 18 quarters **69.** 90 students **71.** Answers will vary. Width is 30 feet less than length; perimeter is 140 feet.

Quick Check 2.4 **1.** 11 **2.** 65 **3.** 70% **4.** 140 **5.** 54% **6.** \$60.90 **7.** \$5000 at 3%; \$7000 at 5% **8.** \$850 at 4%; \$1050 at 5% **9.** 20 mL of 30% and 60 mL of 42% **10.** $\{33\}$ **11.** 68 dentists **12.** 79.45 kilograms

Section 2.4 **1.** Amount = Percent · Base **3.** original value **5.** $I = P \cdot r \cdot t$ **7.** proportion **9.** 36 **11.** 230 **13.** 87.5% **15.** 39 **17.** 163.2 **19.** 44 **21.** 84 **23.** 1025 **25.** 300 mL **27.** \$4095 **29.** \$3.25 **31.** 7.5% **33.** **a)** 400 **b)** 100 points **c)** 25% **35.** 13.6% **37.** 122.2%

39. \$700 at 2%, \$1400 at 5% **41.** \$2500 at 4%, \$4000 at 5% **43.** \$5000 **45.** \$3000 **47.** 24 gallons of 70%, 36 gallons of 40% **49.** 400 gallons of 5%, 600 gallons of 2% **51.** 11 milliliters 27%, 33 milliliters 43% **53.** 1.28 liters 40%, 0.32 liters pure alcohol **55.** 240 milliliters **57.** {32} **59.** {36} **61.** {81.75} **63.** {62.1} **65.** {9} **67.** {3} **69.** 5520 females **71.** 22.5 tablespoons **73.** 144 teachers **75.** 23.4 in. or 23.6 in. **77.** 4.0 gal **79.** 176 lb or 176.2 lb **81.** Answers will vary; \$10,000 more at 5% than 3%; total interest: \$2100.

Quick Review Exercises **1.** {6} **2.** {−16} **3.** {13} **4.** {16}

Quick Check 2.5 **1.** (graph) **2.** $x < 7$, (graph), $(-\infty, 7)$ **3.** $x \ge 5$, (graph), $[5, \infty)$ **4.** $x < -3$, (graph), $(-\infty, -3)$ **5.** $x \ge -6$, (graph), $[-6, \infty)$ **6.** $x \le 4$ or $x \ge 7$, (graph), $(-\infty, 4] \cup [7, \infty)$ **7.** $-3 < x < \frac{3}{2}$, (graph), $\left(-3, \frac{3}{2}\right)$ **8.** $x < 130$ **9.** lower than 48

Section 2.5 **1.** linear inequality **3.** interval **5.** compound **7.** (graph), $(-\infty, 3)$ **9.** (graph), $[-1, \infty)$ **11.** (graph), $(-2, 8)$ **13.** (graph), $\left(\frac{9}{2}, \infty\right)$ **15.** (graph), $(-\infty, 2] \cup (8, \infty)$ **17.** $x > 5$ **19.** $-4 \le x \le 7$ **21.** $x < 2$ or $x > 9$ **23.** $x < -4$ **25.** $x < -4$, (graph), $(-\infty, -4)$ **27.** $x > 6$, (graph), $(6, \infty)$ **29.** $x < -3$, (graph), $(-\infty, -3)$ **31.** $x \ge -3.5$, (graph), $[-3.5, \infty)$ **33.** $x < 4$, (graph), $(-\infty, 4)$ **35.** $x \ge -10$, (graph), $[-10, \infty)$ **37.** $x > 7$, (graph), $(7, \infty)$ **39.** $x > \frac{5}{3}$, (graph), $\left(\frac{5}{3}, \infty\right)$ **41.** $x > -8$, (graph), $(-8, \infty)$ **43.** $x > -\frac{11}{2}$, (graph), $\left(-\frac{11}{2}, \infty\right)$ **45.** $x < -4$ or $x > -2$, (graph), $(-\infty, -4) \cup (-2, \infty)$ **47.** $x \le 2$ or $x \ge 5$, (graph), $(-\infty, 2] \cup [5, \infty)$ **49.** $x > 6$ or $x < 3$, (graph), $(-\infty, 3) \cup (6, \infty)$ **51.** $x < 7$, (graph), $(-\infty, 7)$ **53.** $x \le \frac{3}{2}$ or $x \ge 3$, (graph), $\left(-\infty, \frac{3}{2}\right] \cup [3, \infty)$ **55.** $-1 < x < 4$, (graph), $(-1, 4)$ **57.** $1 < x < 7$, (graph), $(1, 7)$ **59.** $-3 < x \le 6$, (graph), $(-3, 6]$ **61.** $\frac{16}{15} \le x \le \frac{25}{6}$, (graph), $\left[\frac{16}{15}, \frac{25}{6}\right]$ **63.** Answers will vary. 8 is not less than 3. **65.** $x \ge 2$ **67.** $x < 30$ **69.** $0.10x \ge 30$, at least 300 **71.** $0.07x + 19.95 \le 85$, at most 929 **73.** $x \ge 0.15 \cdot 37{,}500$, at least \$5625 **75.** $(92 + 93 + 85 + 96 + x)/5 \ge 80$, at least 34 **77.** Explanations will vary; produces opposite relationship between two quantities

Chapter 2 Review **1.** No **2.** Yes **3.** {16} **4.** {−6} **5.** {−6} **6.** {−6} **7.** $8x = 944$, 118 people **8.** $x - 23 = 189$, 212 points **9.** {7} **10.** {−9} **11.** {2} **12.** {36} **13.** {10} **14.** $\left\{\frac{35}{4}\right\}$ **15.** {−33} **16.** $\left\{-\frac{3}{5}\right\}$ **17.** {16} **18.** {−7} **19.** $y = -7x + 8$ **20.** $y = \frac{-15x + 30}{7}$ **21.** $d = \frac{C}{\pi}$ **22.** $a = \frac{P - b - 2c}{3}$ **23.** 36, 62 **24.** Length: 16 feet; width: 9 feet **25.** 66, 68, 70 **26.** $4\frac{3}{4}$ hours **27.** 41 dimes **28.** 19 **29.** $37\frac{1}{2}\%$ **30.** 4250 **31.** 5786 **32.** $63\frac{3}{4}\%$ **33.** \$12,300 **34.** \$47.96 **35.** \$1700 at 6%, \$300 at 5% **36.** 32 milliliters 17%, 48 milliliters 42% **37.** {4} **38.** {32} **39.** {78} **40.** {50} **41.** 6408 students **42.** 15 **43.** $x < -3$, $(-\infty, -3)$ (graph) **44.** $x > -5$, $(-5, \infty)$ (graph) **45.** $x \ge 5$, $[5, \infty)$ (graph) **46.** $x \le -4$, $(-\infty, -4]$ (graph) **47.** $x < -6$ or $x > 7$, $(-\infty, -6) \cup (7, \infty)$ (graph) **48.** $x \le -8$ or $x \ge -3$, $(-\infty, -8] \cup [-3, \infty)$ (graph) **49.** $-5 \le x \le 10$, $[-5, 10]$ (graph) **50.** $-1 < x < 10$, $(-1, 10)$ (graph) **51.** $x \ge 7$ **52.** At least \$34.75 **53.** 79 to 100

Chapter 2 Review Exercises: Worked-Out Solutions

1.
$$4x + 9 = 19$$
$$4(7) + 9 = 19$$
$$28 + 9 = 19$$
$$37 = 19$$
False

$x = 7$ is not a solution.

4.
$$-9x = 54$$
$$\frac{-9x}{-9} = \frac{54}{-9}$$
$$x = -6$$
{−6}

6.
$$a + 14 = 8$$
$$a + 14 - 14 = 8 - 14$$
$$a = -6$$
{−6}

9.
$$5x - 8 = 27$$
$$5x - 8 + 8 = 27 + 8$$
$$5x = 35$$
$$\frac{5x}{5} = \frac{35}{5}$$
$$x = 7$$
{7}

13.
$$\frac{x}{6} - \frac{5}{12} = \frac{5}{4}$$
$$12 \cdot \frac{x}{6} - 12 \cdot \frac{5}{12} = 12 \cdot \frac{5}{4}$$
$$2x - 5 = 15$$
$$2x - 5 + 5 = 15 + 5$$
$$2x = 20$$
$$\frac{2x}{2} = \frac{20}{2}$$
$$x = 10$$
{10}

16.
$$2m + 23 = 17 - 8m$$
$$2m + 23 + 8m = 17 - 8m + 8m$$
$$10m + 23 = 17$$
$$10m + 23 - 23 = 17 - 23$$
$$10m = -6$$
$$\frac{10m}{10} = \frac{-6}{10}$$
$$m = -\frac{3}{5}$$
$\left\{-\frac{3}{5}\right\}$

22.
$$P = 3a + b + 2c$$
$$P - b - 2c = 3a$$
$$\frac{P - b - 2c}{3} = \frac{3a}{3}$$
$$a = \frac{P - b - 2c}{3}$$

24. Length: x

Width: $x - 7$
$$2x + 2(x - 7) = 50$$
$$2x + 2x - 14 = 50$$
$$4x - 14 = 50$$
$$4x = 64$$
$$x = 16$$
Length: $x = 16$

Width: $x - 7 = 16 - 7 = 9$

30.
$$1360 = 0.32x$$
$$\frac{1360}{0.32} = \frac{0.32x}{0.32}$$
$$4250 = x$$

31. $x = 0.55 \cdot 10{,}520$
$x = 5786$

37. $\frac{n}{6} = \frac{18}{27}$
$n \cdot 27 = 6 \cdot 18$
$27n = 108$
$n = 4$

41. $\frac{3}{7} = \frac{n}{14{,}952}$
$3 \cdot 14{,}952 = 7 \cdot n$
$44{,}856 = 7n$
$6408 = n$

46. $3x + 4 \leq -8$
$3x \leq -12$
$x \leq -4$
$(-\infty, -4]$

49. $-21 \leq 3x - 6 \leq 24$
$-15 \leq 3x \leq 30$
$-5 \leq x \leq 10$
$[-5, 10]$

52. $2x \geq 69.5$
$\frac{2x}{2} \geq \frac{69.5}{2}$
$x \geq 34.75$
The price needs to be at least \$34.75.

Chapter 2 Test **1.** $\{-7\}$ **2.** $\{-39\}$ **3.** $\{8\}$ **4.** $\left\{-\frac{21}{4}\right\}$ **5.** $\left\{\frac{55}{2}\right\}$ **6.** $\{-3\}$ **7.** $\{12\}$ **8.** $\{55\}$ **9.** $y = \frac{-4x + 28}{3}$ **10.** 28 **11.** 15%
12. $x \leq -\frac{5}{2}, \left(-\infty, -\frac{5}{2}\right]$ **13.** $x \leq -9$ or $x \geq -\frac{3}{2}, (-\infty, -9] \cup \left[-\frac{3}{2}, \infty\right)$
14. $-10 < x < \frac{16}{3}, \left(-10, \frac{16}{3}\right)$ **15.** Length: 23 feet; width: 13 feet **16.** 17 \$5 bills **17.** \$6600
18. 37.5 milliliters 13%, 62.5 milliliters 21% **19.** 88 professors **20.** At least 34

CHAPTER 3

Quick Check 3.1 **1.** No **2.**

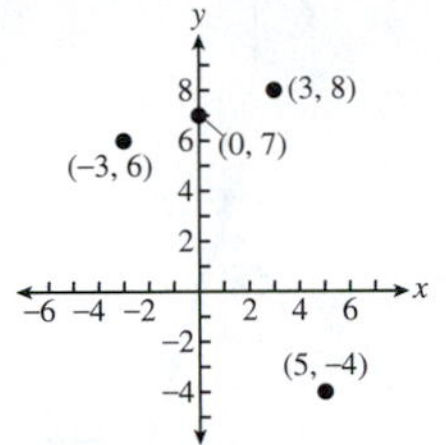

3. $A(-6, 2), B(-2, 6), C(0, -5), D(-4, -2), E(8, 1)$

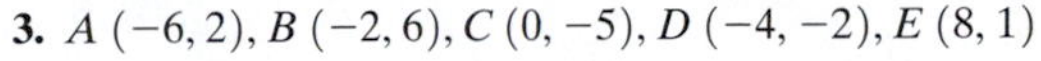

4.

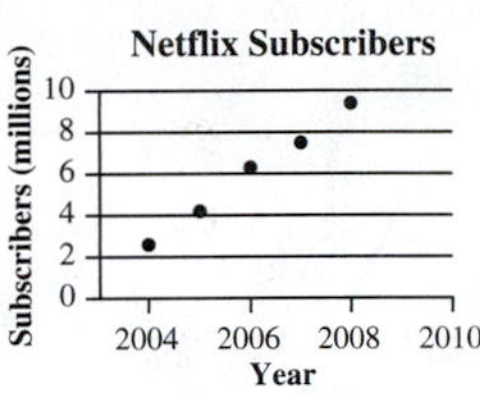

5. $y = -4$ **6.**

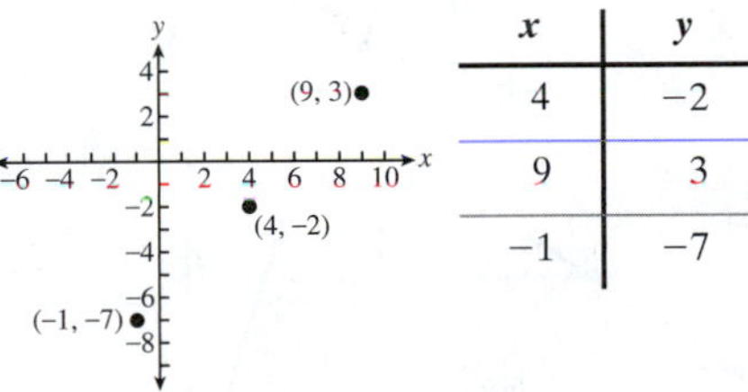

x	y
4	−2
9	3
−1	−7

7.

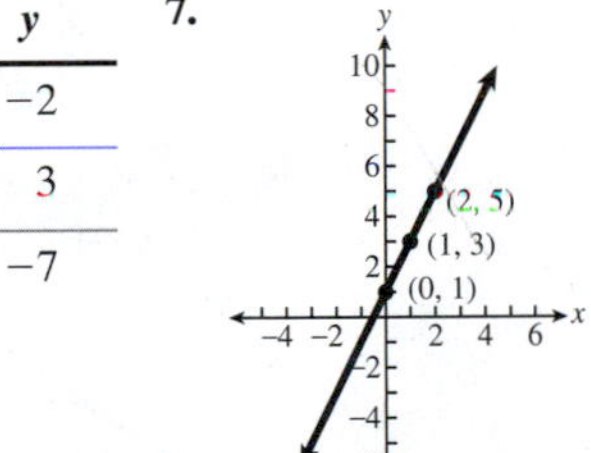

Section 3.1 **1.** $Ax + By = C$ **3.** x-axis **5.** origin **7.** Yes **9.** No **11.** Yes **13.** Yes **15.** Yes **17.**

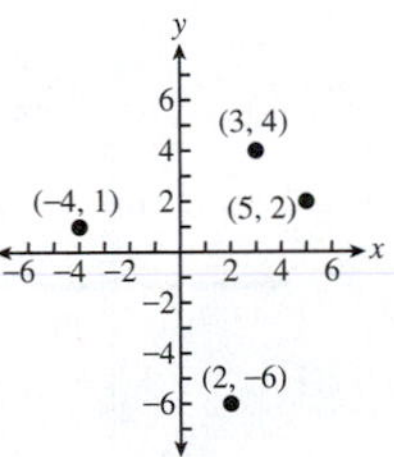

19. **21.**

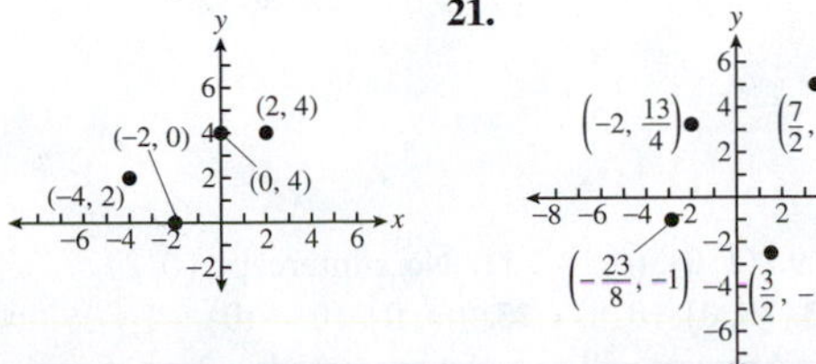

23. $A(-3, 1), B(2, -5), C(-6, -4), D(0, 3)$

25. $A(30, -15), B(-40, 0), C(-10, -30), D(-15, 15)$ **27.** IV **29.** I **31.** I **33.** IV **35.**

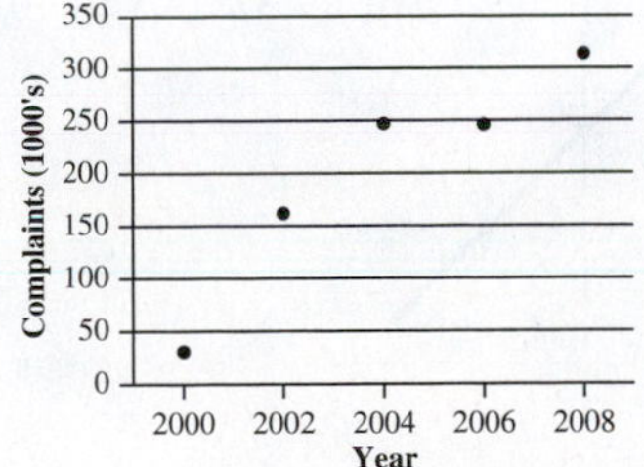

37.

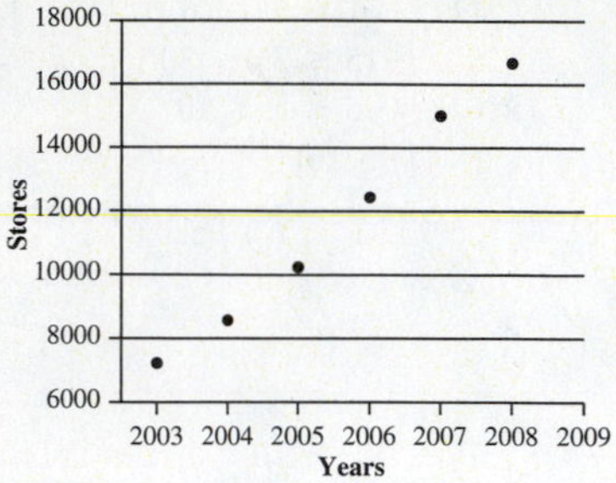

39.

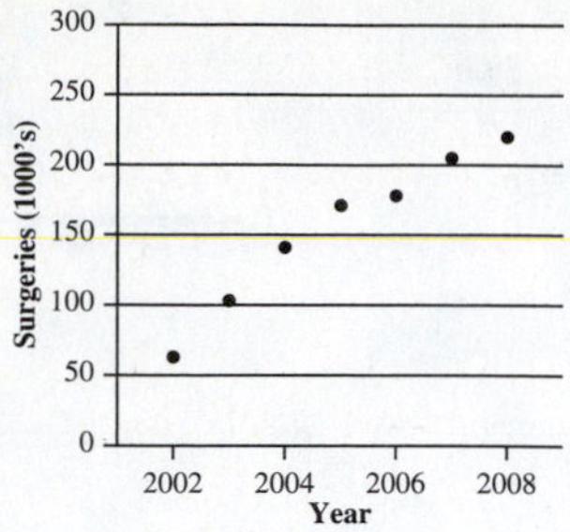

41. -1 **43.** -5 **45.** 11 **47.** 2 **49.** -2 **51.** $\frac{5}{2}$ **53.** y: $-1, 4, 9$

55. y: $-9, -5, -3$ **57.** y: $-4, -2, 2$ **59.** Example: $x + y = 7$ **61.** Example: $5x + 4y = 24$ **63.**

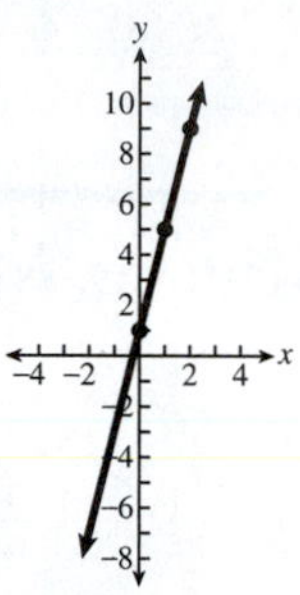

65.

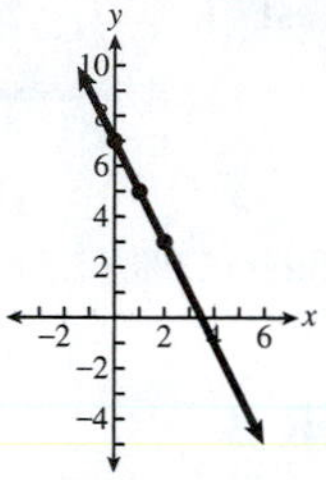

67.

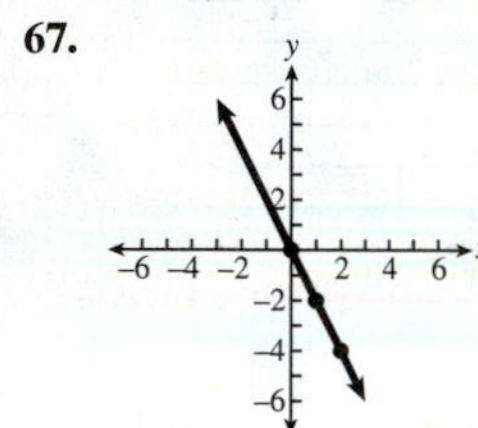

69.

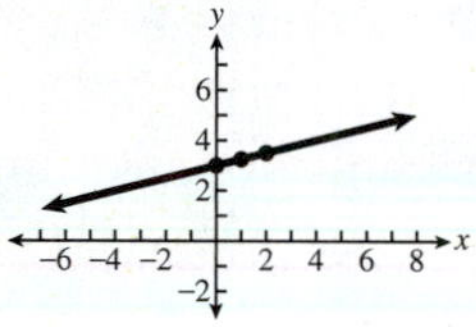

71.

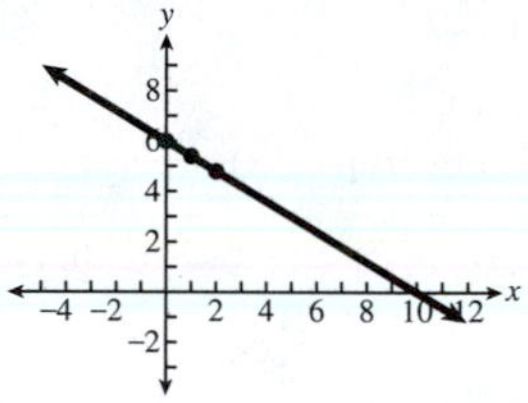

73.

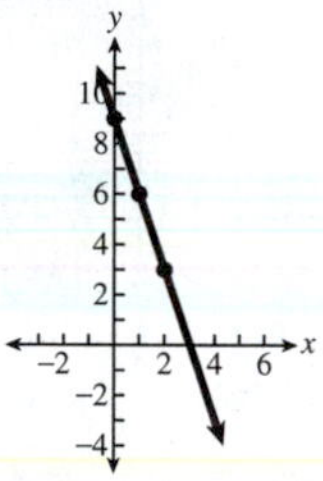

75. Explanations will vary. The first coordinate gives the horizontal location.

Quick Check 3.2 **1.** $(-4, 0), (0, 5)$ **2.** $(4, 0), (0, -3)$ **3.** $\left(\frac{9}{2}, 0\right), (0, -9)$ **4.**

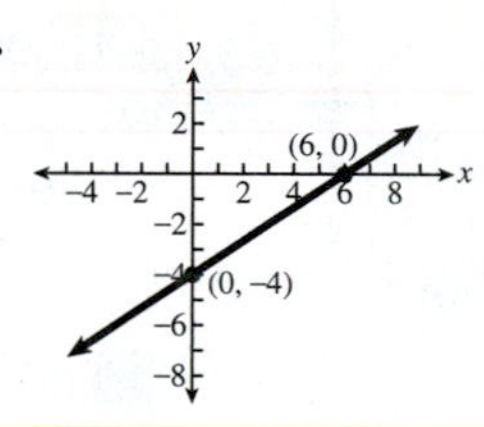

5.

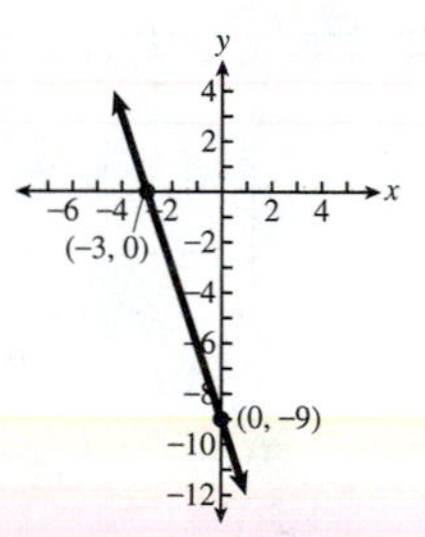

6.

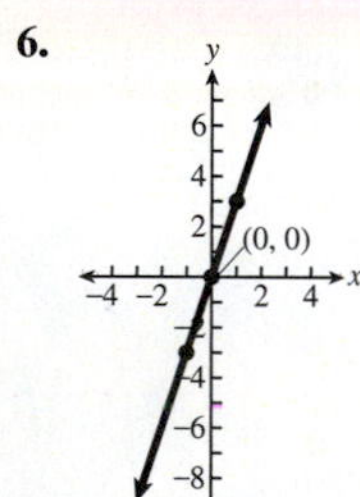

7.

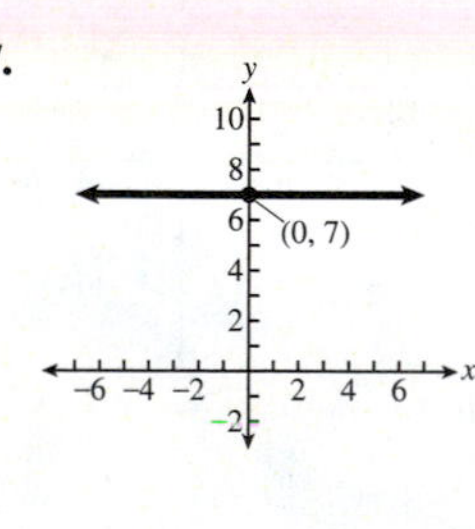

8.

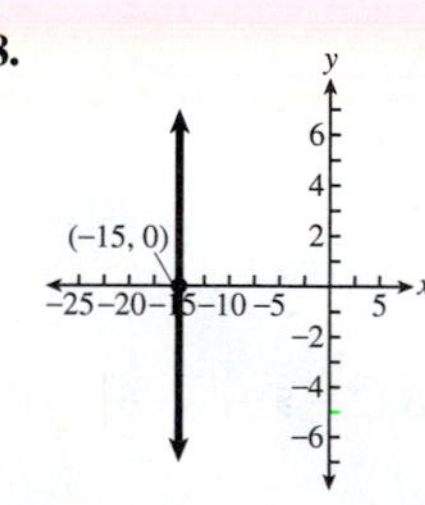

9. a) Initially Nancy was $10,000 in debt.

b) It will take five months before Nancy breaks even.

Section 3.2 **1.** x-intercept **3.** vertical **5.** Substitute 0 for y and solve for x. **7.** $(-3, 0), (0, 6)$ **9.** $(1, 0), (0, 5)$ **11.** No x-intercept, $(0, 2)$ **13.** $(6, 0), (0, 6)$ **15.** $(-8, 0), (0, 8)$ **17.** $(2, 0), (0, 3)$ **19.** $(10, 0), (0, -6)$ **21.** $(5, 0), (0, 6)$ **23.** $(8, 0), (0, 8)$ **25.** $\left(\frac{10}{3}, 0\right), (0, -10)$ **27.** $(0, 0), (0, 0)$ **29.** No x-intercept, $(0, 2)$ **31.** Answers will vary. An example is $x + y = 4$. **33.** Answers will vary. An example is $-2x + 5y = -50$.

35. $(4, 0), (0, 4)$

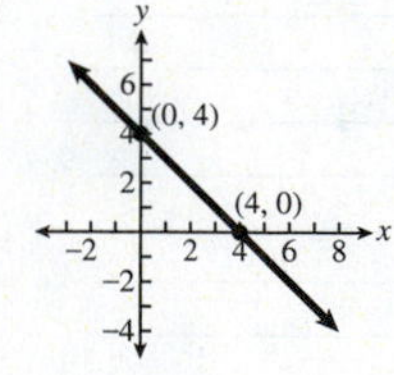

37. $(-3, 0), (0, 3)$

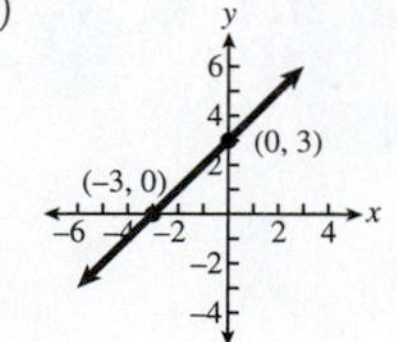

39. $(5, 0), (0, -2)$

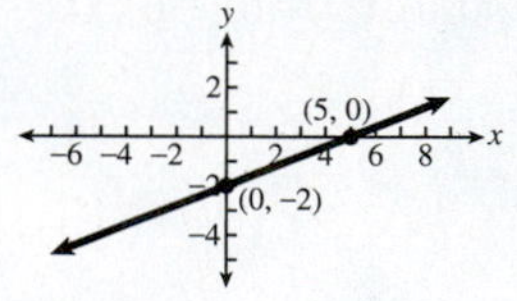

41. $(-8, 0), (0, -6)$

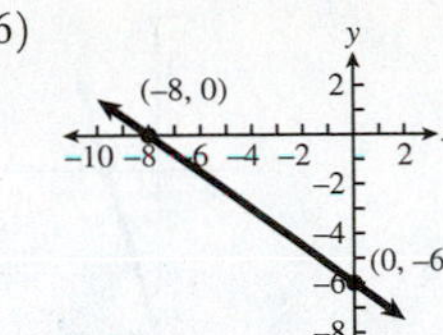

43. $\left(-\frac{7}{2}, 0\right), (0, 2)$

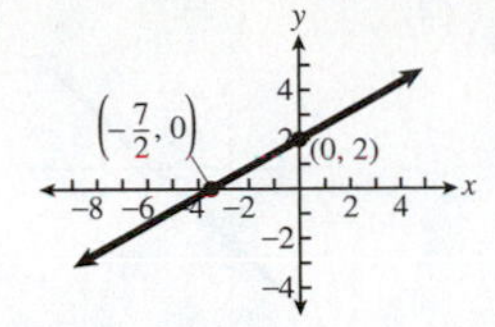

45. $(4, 0), (0, -4)$

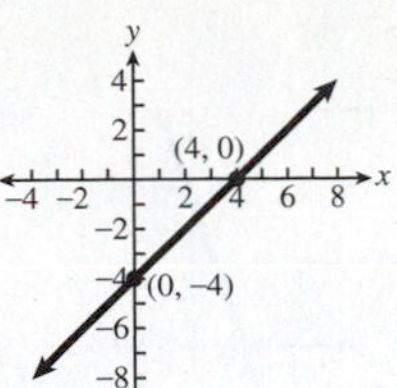

47. $\left(\frac{7}{2}, 0\right), (0, 7)$

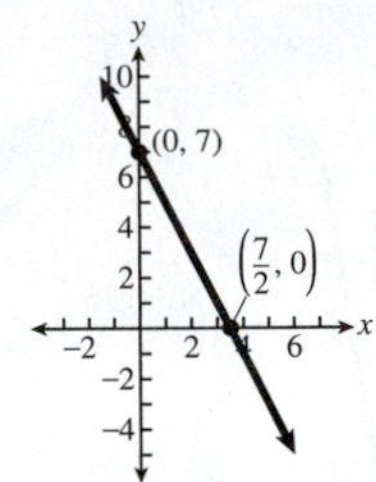

49. $(0, 0), (0, 0)$

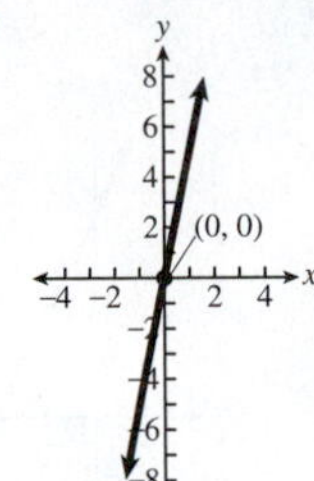

51. $(0, 0), (0, 0)$

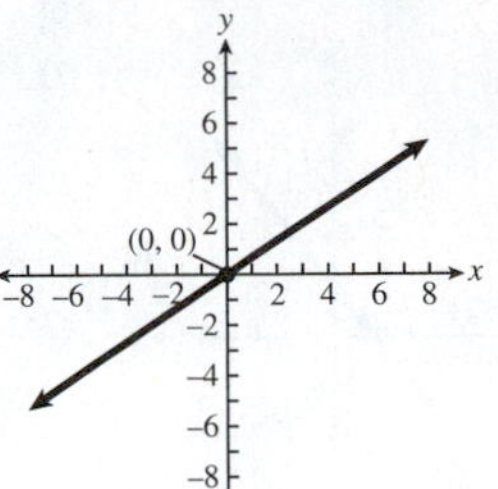

53. No x-intercept, $(0, 4)$

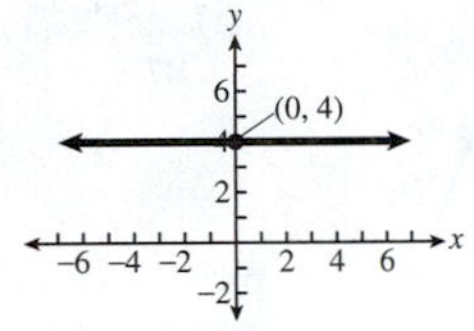

55. $(3, 0)$, no y-intercept

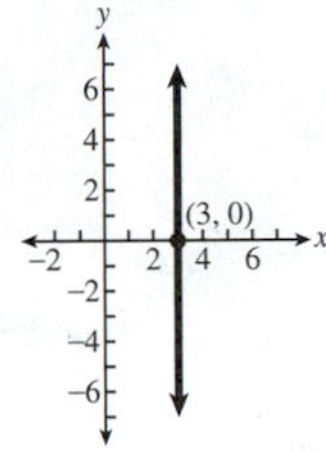

57. a) $(0, 12{,}000)$; the original value of the copy machine is \$12,000. **b)** $(8, 0)$; the copy machine has no value after eight years. **c)** \$7500

59. a) $(0, 300)$; the cost to belong to the club is \$300. **b)** Because the minimum cost is \$300, the cost to a member cannot be \$0. **c)** \$1500

Quick Review Exercises **1.** $y = -2x + 8$ **2.** $y = 4x - 6$ **3.** $y = 4x + 5$ **4.** $y = -\frac{2}{3}x + 2$

Quick Check 3.3 **1.** -2 **2.** 3 **3.** $-\dfrac{5}{2}$ **4.** Slope is undefined. **5.** $m = 0$ **6.** $m = \frac{5}{2}, (0, -6)$ **7.** $m = -3, (0, 5)$ **8.** $y = \frac{1}{2}x - \frac{4}{7}$

9.

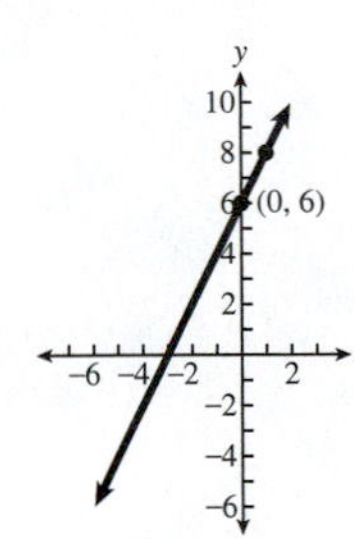

10.

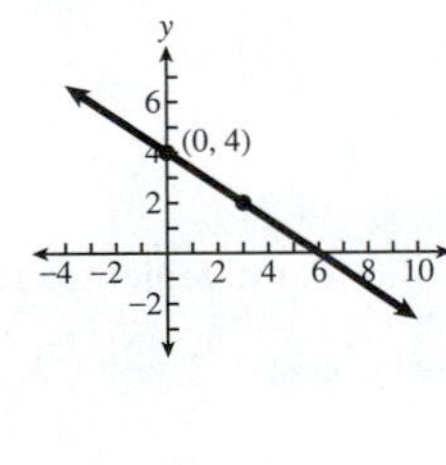

11. Slope is 218; number of women accepted to medical school increases by 218 per year. y-intercept is $(0, 7485)$; approximately 7485 women were accepted to medical school in 1997.

Section 3.3 **1.** slope **3.** negative **5.** horizontal **7.** slope–intercept **9.** Negative **11.** Positive **13.** $-\dfrac{1}{2}$ **15.** $-\dfrac{3}{4}$ **17.** 0 **19.** 2 **21.** 3 **23.** $-\dfrac{4}{3}$ **25.** $-\dfrac{9}{4}$ **27.** 0 **29.** Undefined **31.** y-intercept: $(0, 4)$; m: 0

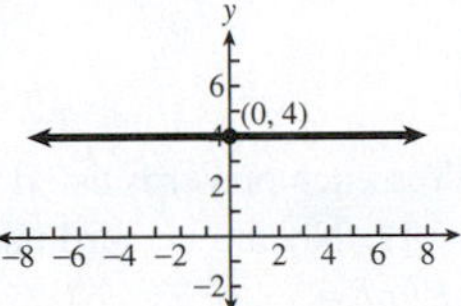

33. x-intercept: $(3, 0)$; m: undefined

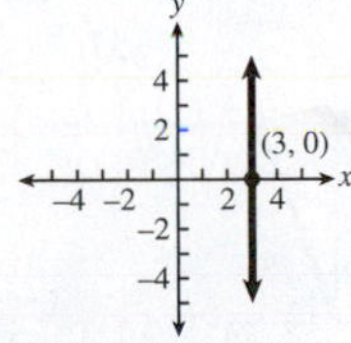

35. x-intercept: $(8, 0)$; m: undefined

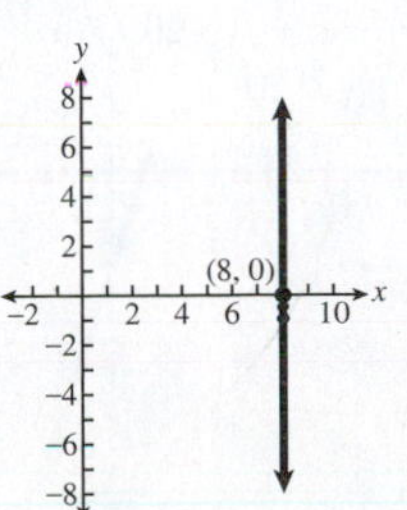

37. $x = 6$; m: undefined

39. $y = 2; m{:}0$ **41.** $6, (0, -7)$ **43.** $-2, (0, 3)$ **45.** $-\dfrac{3}{2}, \left(0, -\dfrac{5}{2}\right)$ **47.** $\dfrac{5}{8}, \left(0, -\dfrac{5}{4}\right)$ **49.** $\dfrac{1}{5}, \left(0, -\dfrac{8}{5}\right)$ **51.** $y = -2x + 5$ **53.** $y = 3x - 6$

55. $y = -4$ **57.**

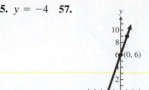

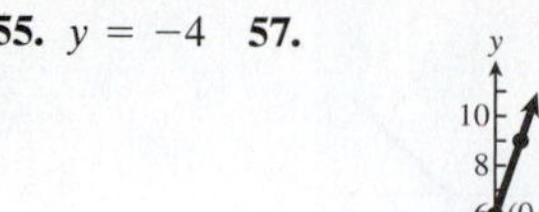

59.

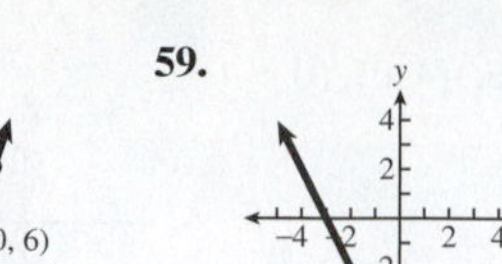

61.

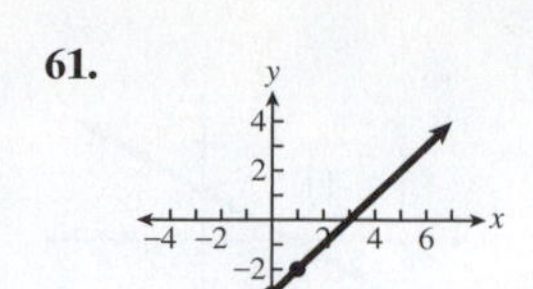

63.

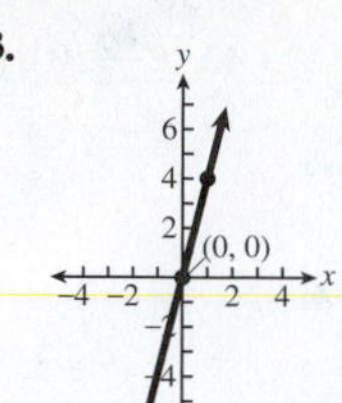

65.

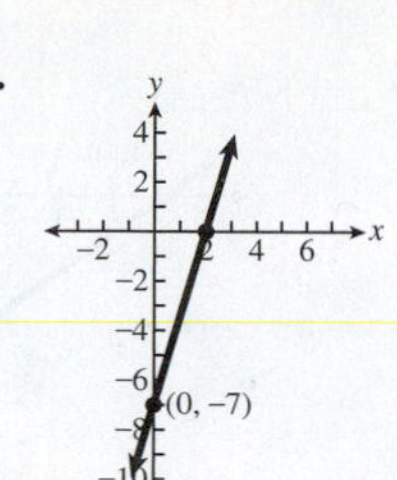

67.

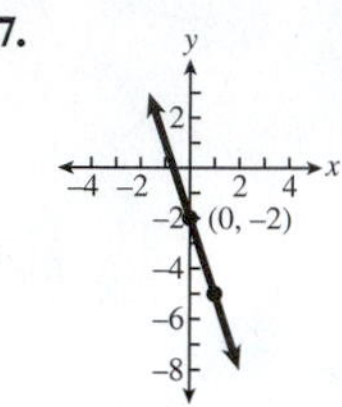

69.

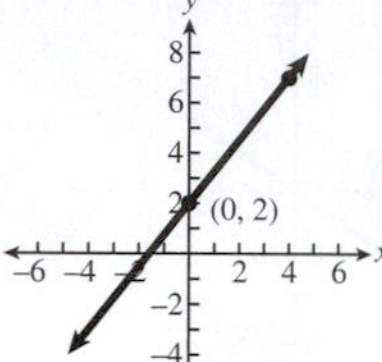

71.

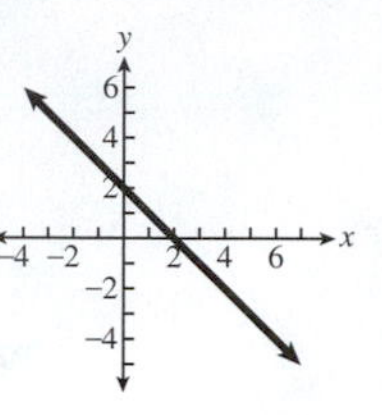

73.

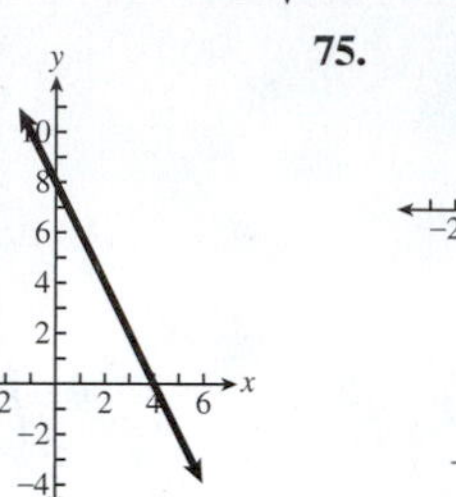

75.

77.

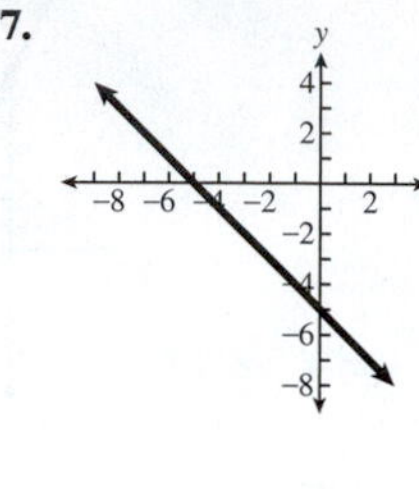

79.

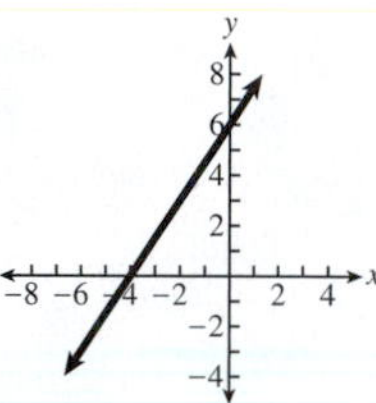

81.

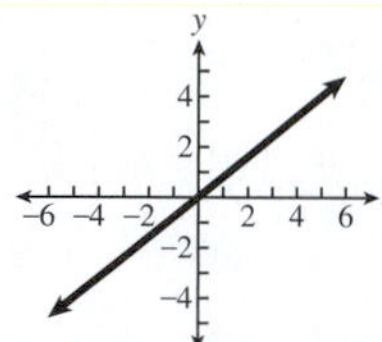

83.

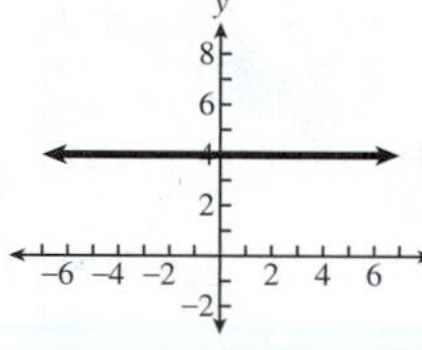

85.

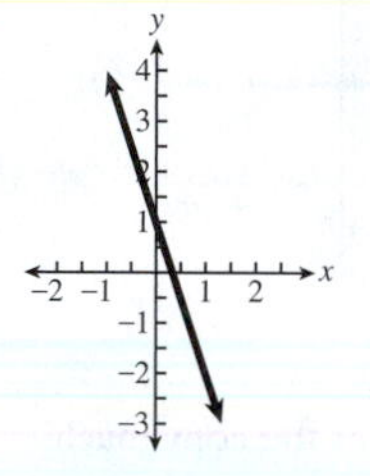

87.

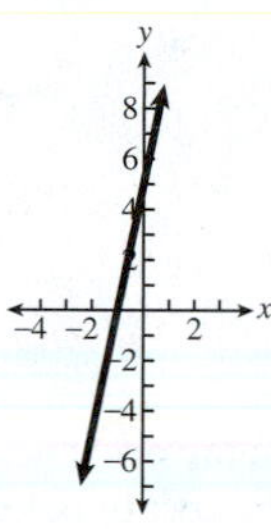

89. (1, −3)

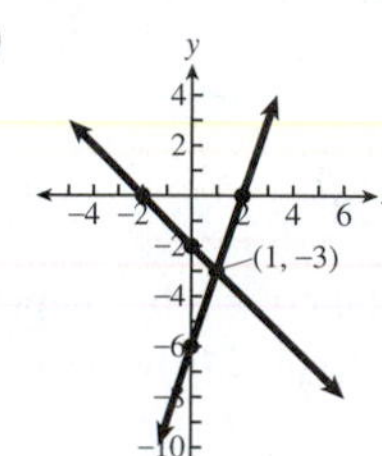

91. a) −3000; the value decreases by \$3000 per year. **b)** (0, 25,000); the original value of the car in 2008 was \$25,000. **c)** \$7000 **93. a)** $\frac{1}{3}$ **b)** $\frac{3}{2}$ **95.** $\frac{1}{20}$ **97.** Positive; the unemployment rate (y) increases as the time after the statement (x) increases.

Quick Check 3.4 **1.** Cost $= 19.95 + 0.15x$; domain: set of all possible miles; range: set of all possible costs **2.** −3 **3.** $7a + 44$

4.

5.

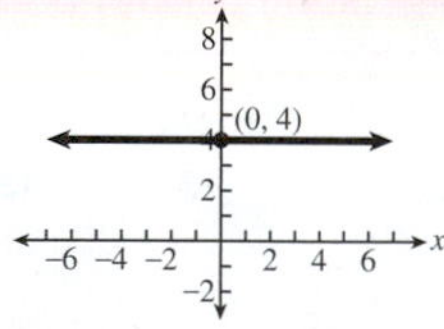

6. a) −6 **b)** 3

Section 3.4 **1.** function **3.** range **5.** linear **7.** Yes, each player is listed with only one team. **9.** Yes, each person has only one mother. **11. a)** Yes, **b)** No **13. a)** Yes, **b)** No **15.** Yes **17.** No, the x-coordinate 5 is associated with two y-coordinates. **19.** No, the x-coordinate 2 is associated with five different y-coordinates. **21.** a) $F(x) = \frac{9}{5}x + 32$ b) 32° F, 212° F, 86° F, 14° F, −40° F **23. a)** $f(x) = 7x + 36$ **b)** \$120 **25.** $f(x) = 4x + 3$ **27.** $f(x) = -3x - 4$ **29.** $f(x) = 6$ **31.** −22 **33.** −77 **35.** 13 **37.** −25 **39.** 1 **41.** $3a + 4$ **43.** $7a + 19$ **45.** $-6a + 31$ **47.** $6x + 6h + 4$ **49.**

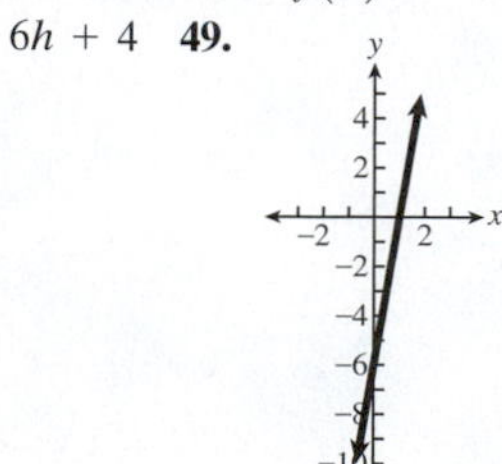

51.

53.

55.

57.

59. a) 4 **b)** −1 **c)** Domain: $(-\infty, \infty)$; range: $(-\infty, \infty)$ **61. a)** 3 **b)** −2 **c)** Domain: $(-\infty, \infty)$; range: $(-\infty, \infty)$
63. a) 5 **b)** Domain: $(-\infty, \infty)$; range: $\{5\}$ **65.** Answers will vary; A: people, B: $\{1, \ldots, 12\}$, person to birth month, birth month to person.

Quick Review Exercises **1.** $m = -3$ **2.** $m = \frac{5}{4}$ **3.** $m = \frac{1}{5}$ **4.** $m = -\frac{7}{5}$

Quick Check 3.5 **1.** No **2.** Yes **3.** $\frac{2}{9}$ **4.** Yes **5.** Yes **6.** $-\frac{3}{7}$ **7.** Neither **8.** Parallel **9.** Perpendicular

Section 3.5 **1.** parallel **3.** vertical **5.** parallel **7.** Yes **9.** No **11.** Yes **13.** No **15.** Yes **17.** Yes **19.** Neither **21.** Perpendicular **23.** Parallel **25.** Neither **27.** Parallel **29.** Perpendicular **31.** Parallel **33.** Neither **35.** Parallel **37.** -6 **39.** 5 **41.** -3 **43.** $-\frac{1}{8}$ **45.** $\frac{3}{5}$ **47.** $\frac{7}{4}$ **49.** $-\frac{A}{B}$ **51.** Yes **53.** Yes **55.** One example: rails on railroad tracks **57.** Find the slope of each line by solving for y. If the two slopes are equal, the lines are parallel.

Quick Review Exercises **1.** $m = -2$ **2.** $m = \frac{4}{3}$ **3.** $m = 5$ **4.** $m = \frac{3}{4}$

Quick Check 3.6 **1.** $y = 2x + 3$ **2.** $y = -3x + 6$ **3.** $f(x) = -4x + 5$ **4.** $y = -\frac{3}{2}x + 4$ **5.** $y = 4$ **6.** $y = 316x + 3620$ **7.** $y = \frac{2}{5}x + 2$ **8.** $y = -\frac{4}{3}x + 6$

Section 3.6 **1.** $y - y_1 = m(x - x_1)$. **3.** $y = 3x + 5$ **5.** $y = \frac{1}{5}x - 2$ **7.** $y = 7x - 3$ **9.** $y = -3x + 5$ **11.** $y = \frac{2}{3}x - 3$ **13.** $y = 5x + 8$ **15.** $y = 3x - 17$ **17.** $y = -4x + 22$ **19.** $y = \frac{3}{2}x$ **21.** $y = 5$ **23.** $f(x) = -2x + 15$ **25.** $f(x) = \frac{3}{5}x - 8$ **27.** $y = x - 5$ **29.** $y = 3x + 6$ **31.** $y = \frac{6}{5}x$ **33.** $x = -2$ **35.** $y = -\frac{1}{3}x + 2$ **37.** $y = x - 4$ **39.** $y = x - 6$ **41. a)** $y = 10x + 36$ **b)** \$36 **43. a)** $y = 12.50x + 176$ **b)** \$176 **c)** \$12.50 **45. a)** $y = 30x + 500$ **b)** \$500 **c)** \$30 **d)** \$6500 **47.** $y = -2x - 15$ **49.** $y = 3x - 18$ **51.** $y = 7$ **53.** $y = -\frac{1}{3}x + 7$ **55.** $y = \frac{5}{4}x + 7$ **57.** $x = -5$ **59.** $y = x + 3$ **61.** $y = -\frac{1}{4}x + 2$ **63.** $y = -3x + 4$ **65.** $y = \frac{5}{3}x + 23$ **67.** $y = 11$ **69.** $y = 4x + 21$ **71.** $y = 14x - 37$ **73.** Explanations will vary. Vertical line: each x-coordinate is the same value.

Quick Review Exercises **1.**

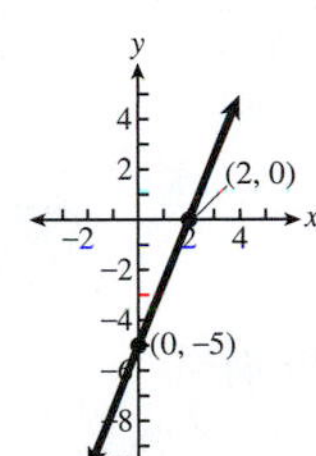

2.

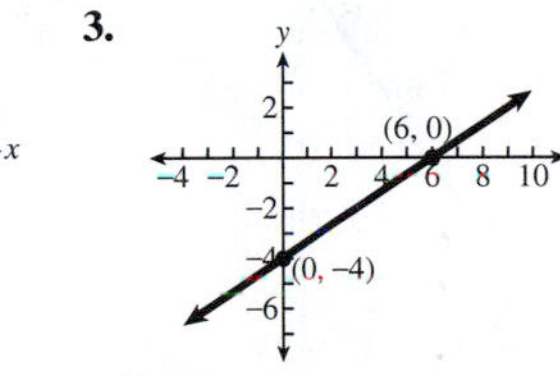

3.

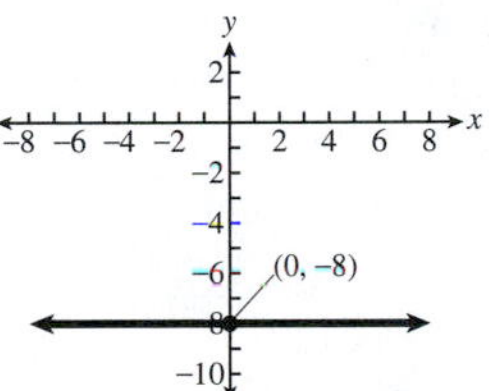

4.

Quick Check 3.7 **1.**

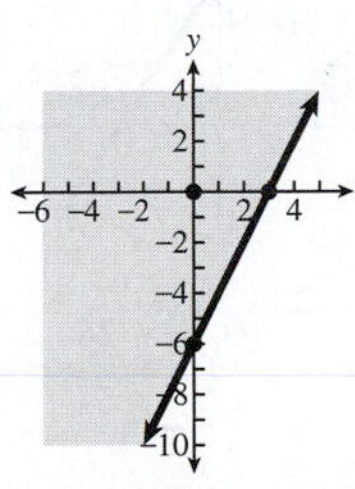

2.

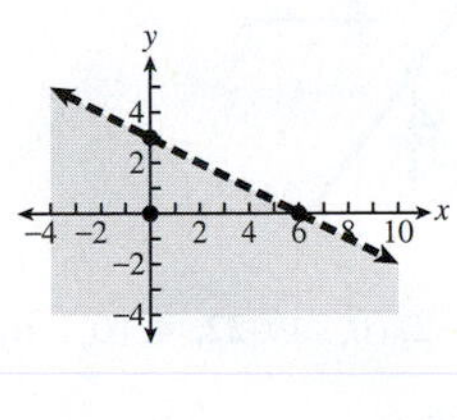

3.

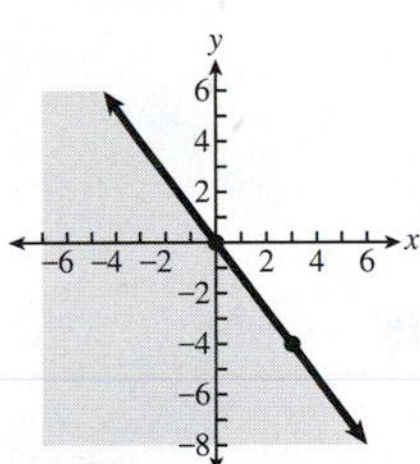

4. $x + y \geq 60$

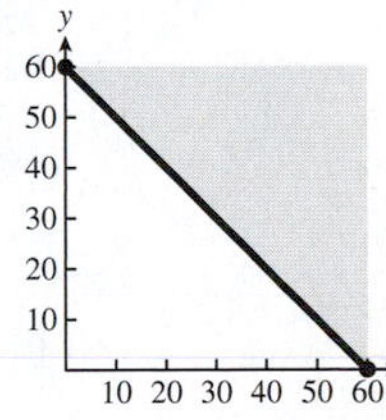

Section 3.7 **1.** solution **3.** dashed **5. a)** Yes, **b)** No, **c)** Yes, **d)** Yes **7. a)** Yes, **b)** No, **c)** No, **d)** No **9. a)** Yes, **b)** Yes, **c)** No, **d)** No **11.**

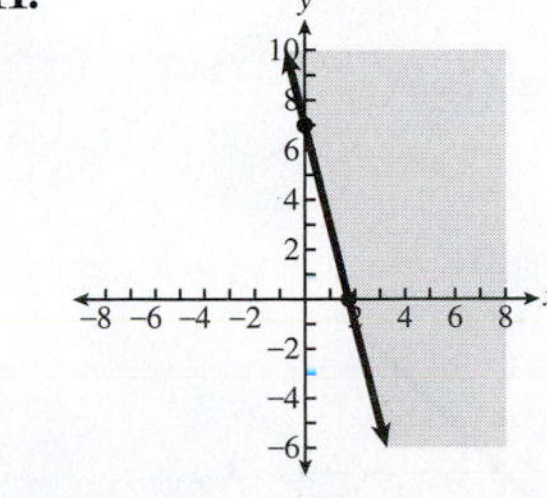

13.

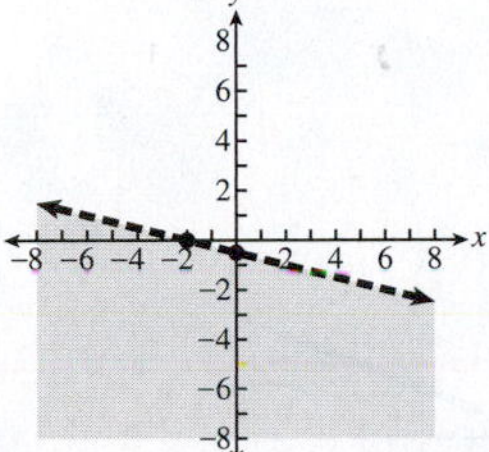

15. A **17.** A **19.** $\leq$ **21.** $<$ **23.**

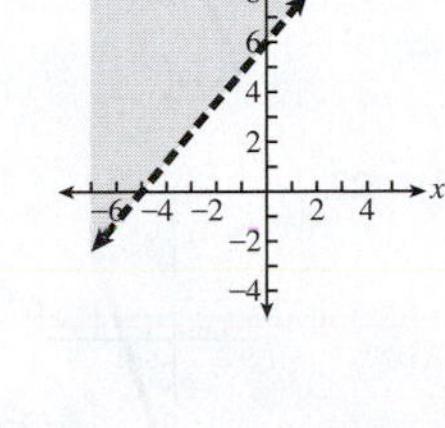

25.

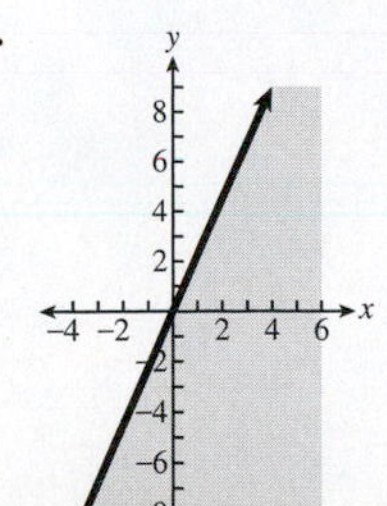

27.

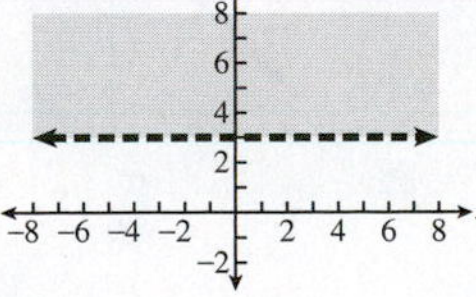

29.

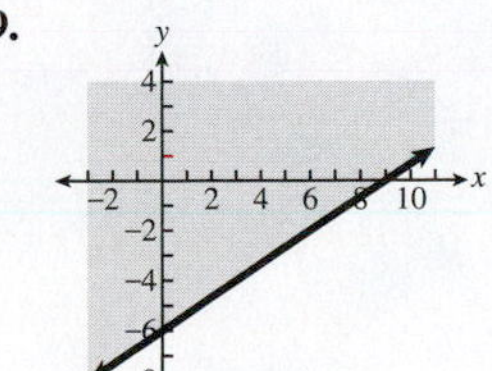

31.

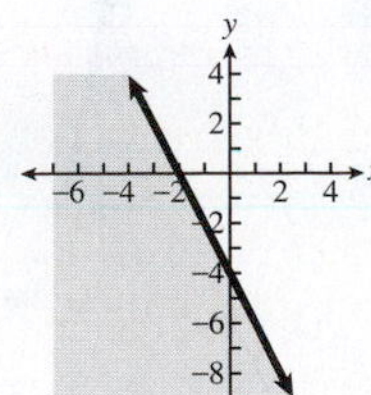

33.

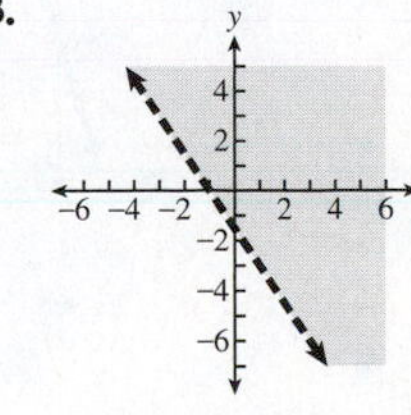

35.

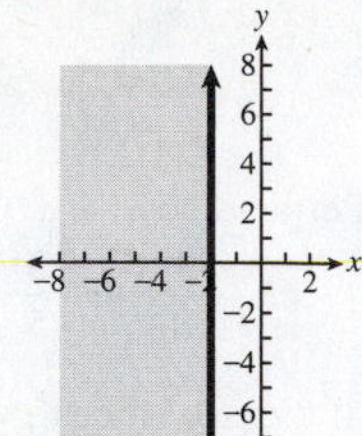

37. $y \le -x - 5$ **39.** $y > 8x$ **41.**

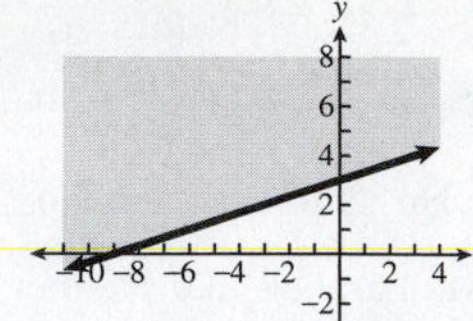

43.

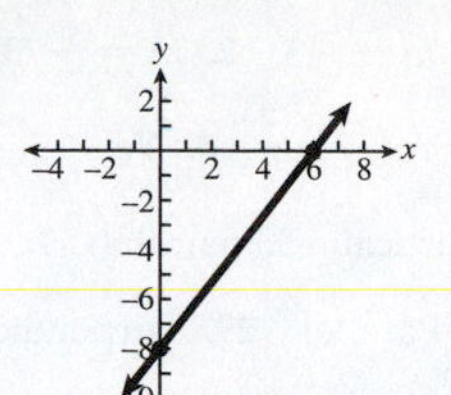

45.

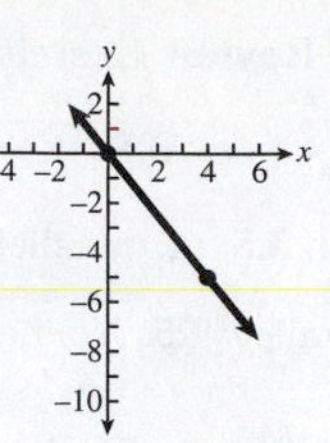

47.

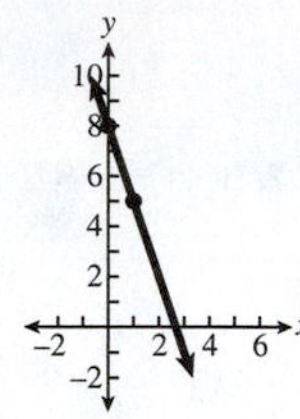

49.

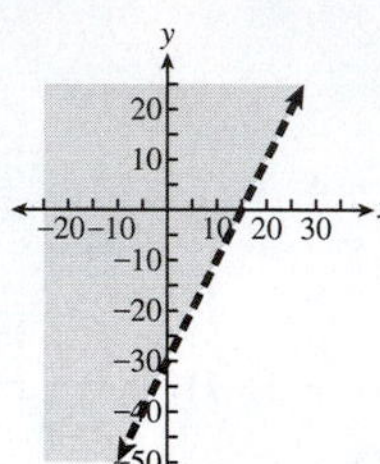

51. $x + y > 50$

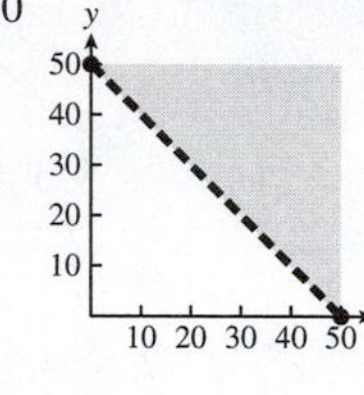

53.

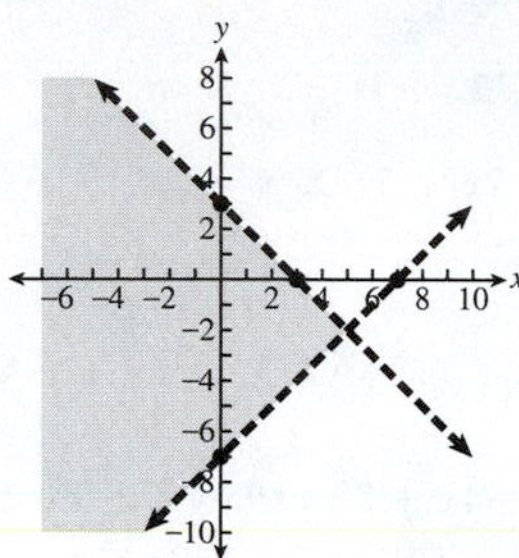

55. Explanations will vary. The line contains points for which the two sides are equal.

Chapter 3 Review **1.** $A\,(7, -2)$, $B\,(-5, -1)$, $C\,(-1, -5)$, $D\,(0, 8)$ **2.** III **3.** II **4.** No **5.** Yes **6.** Yes **7.** $(-20, 0), (0, 5)$ **8.** $(-3, 0), (0, 15)$ **9.** $(0, 0)$ **10.** $\left(\frac{7}{2}, 0\right), (0, -2)$ **11.** $(-4, 0), (0, 6)$

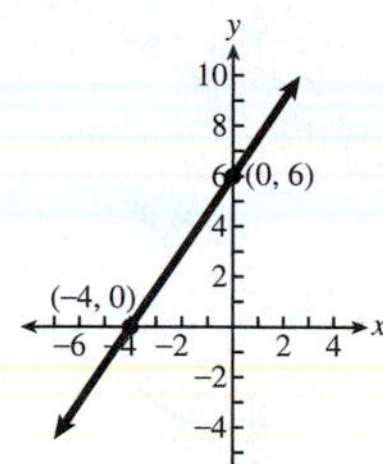

12. $(-8, 0), (0, 2)$

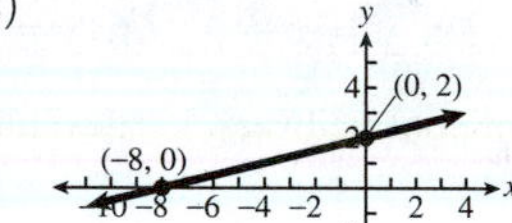

13. $(0, 0), (0, 0)$

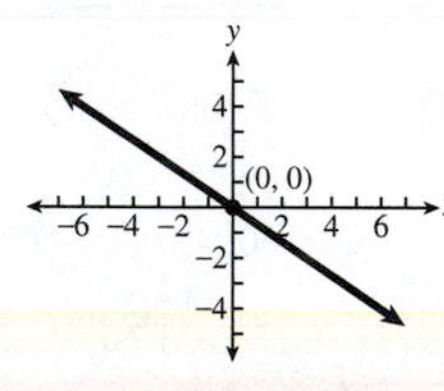

14. $\left(-\frac{3}{2}, 0\right), (0, -2)$

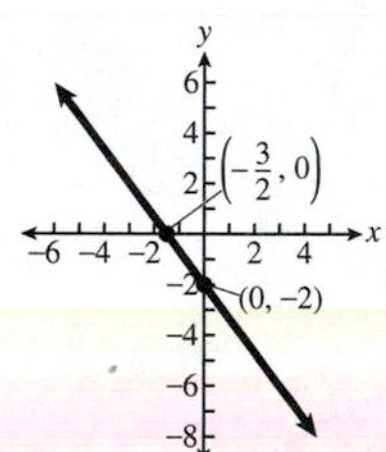

15. $(4, 0), (0, 6)$

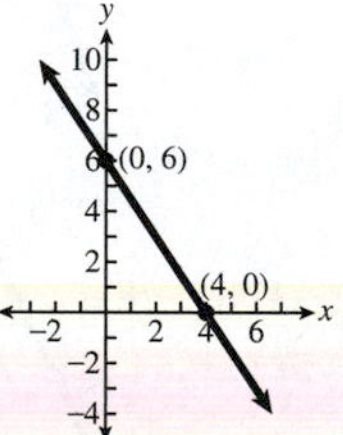

16. $(2, 0), (0, -4)$

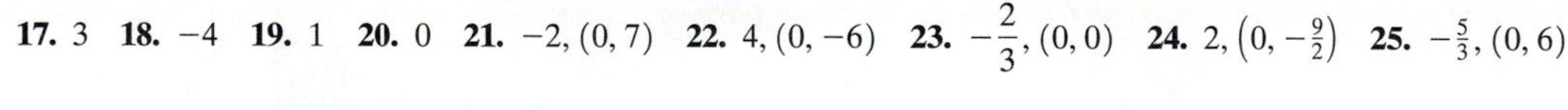

17. 3 **18.** -4 **19.** 1 **20.** 0 **21.** $-2, (0, 7)$ **22.** $4, (0, -6)$ **23.** $-\frac{2}{3}, (0, 0)$ **24.** $2, \left(0, -\frac{9}{2}\right)$ **25.** $-\frac{5}{3}, (0, 6)$

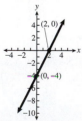

26. $0, (0, -7)$ **27.**

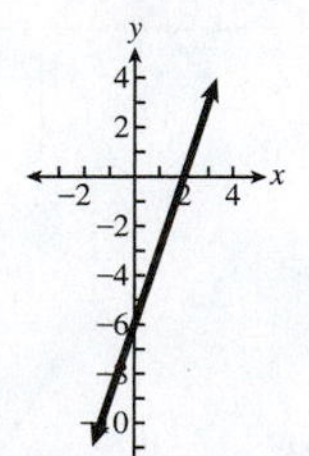

28.

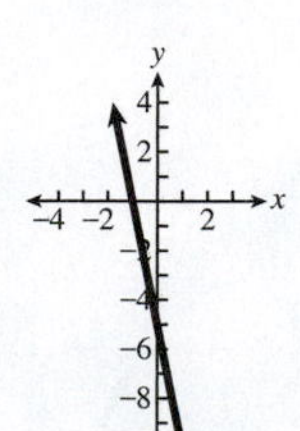

29.

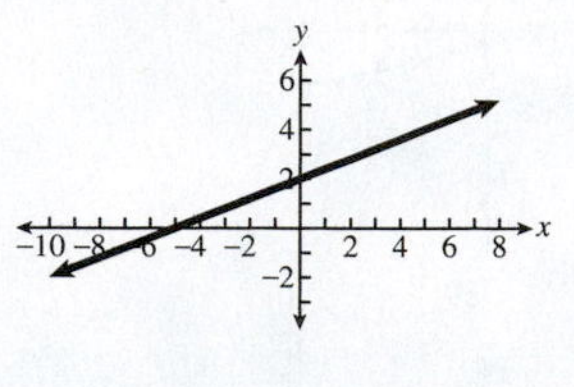

30.

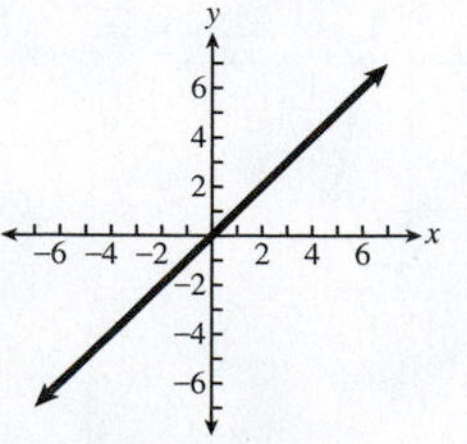

31. 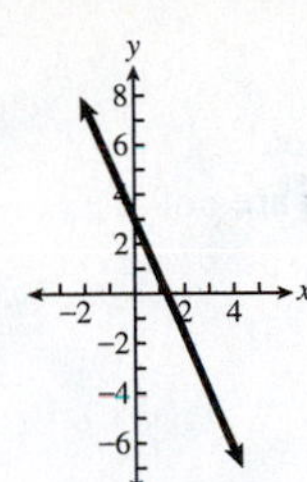 **32.** 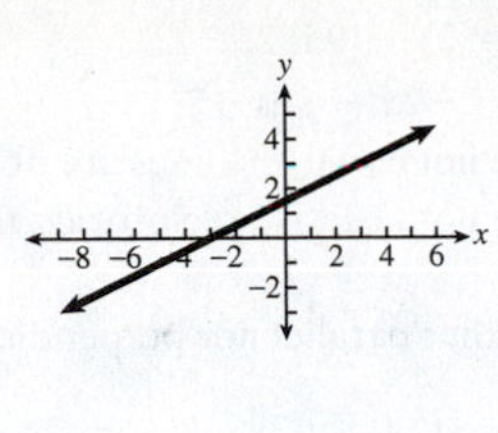 **33.** Neither **34.** Perpendicular **35.** Parallel **36.** $y = -4x - 2$ **37.** $y = \frac{2}{5}x - 6$

38. a) $f(x) = 720 + 50x$ **b)** $1270 **c)** 41 months **39.** -11 **40.** 43 **41.** $15a + 4$ **42. a)** -7 **b)** 6 **c)** $(-\infty, \infty)$ **d)** $(-\infty, \infty)$ **43.** $y = x - 6$ **44.** $y = -5x + 8$ **45.** $y = -\frac{3}{2}x + 3$ **46.** $y = -2x - 3$ **47.** $y = \frac{3}{2}x + 4$ **48.** $y = 3$ **49.** $y = \frac{3}{2}x + 6$ **50.** $y = 3x$ **51.** **52.** 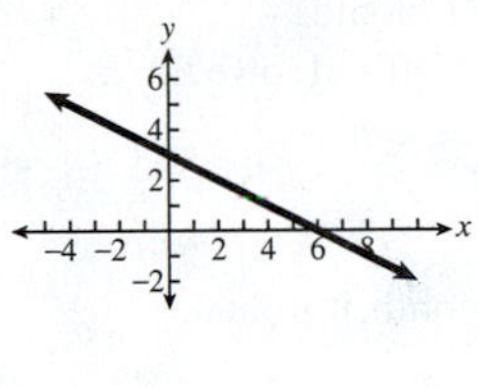

53. 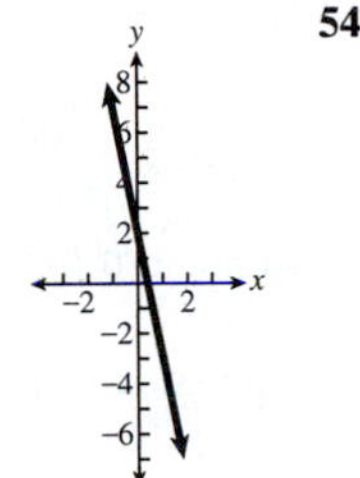 **54.** 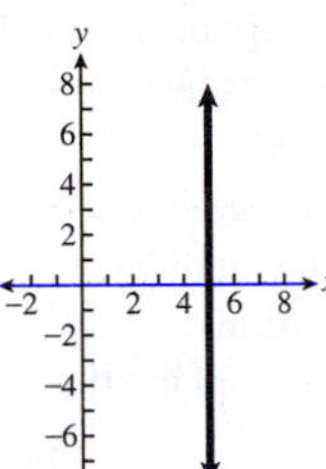 **55.** 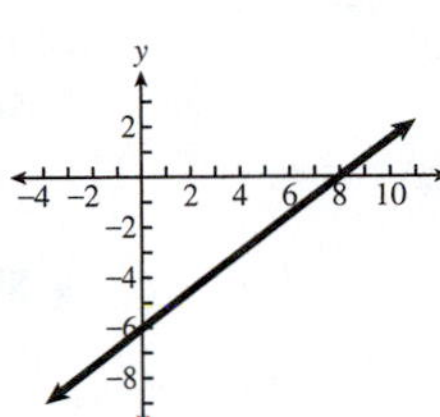 **56.** 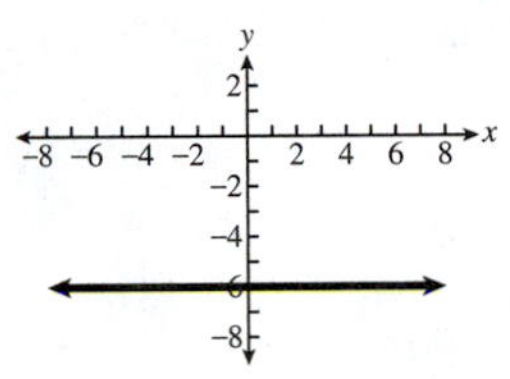 **57.** 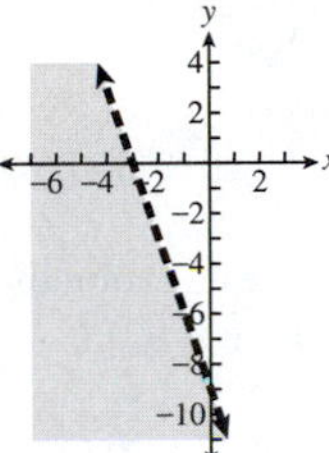

58. 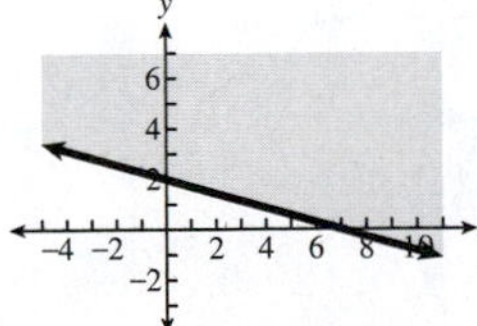 **59.** 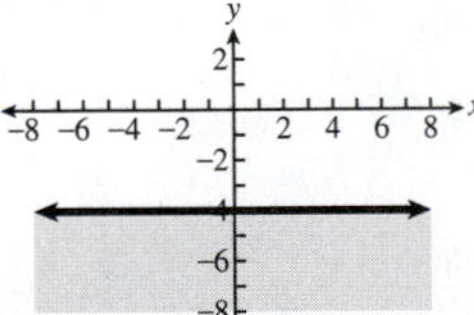 **60.** 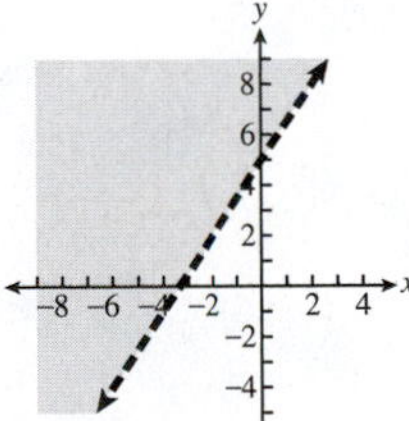

Chapter 3 Review Exercises: Worked-Out Solutions

4.
$$4x - 2y = 16$$
$$4(5) - 2(-2) = 16$$
$$20 + 4 = 16$$
$$24 = 16$$
False, so $(5, -2)$ is not a solution

7. x-intercept
$$2x - 8(0) = -40$$
$$2x = -40$$
$$\frac{2x}{2} = \frac{-40}{2}$$
$$x = -20$$
$(-20, 0)$

y-intercept
$$2(0) - 8y = -40$$
$$-8y = -40$$
$$\frac{-8y}{-8} = \frac{-40}{-8}$$
$$y = 5$$
$(0, 5)$

11. x-intercept
$$-3x + 2(0) = 12$$
$$-3x = 12$$
$$x = -4$$
$(-4, 0)$

y-intercept
$$-3(0) + 2y = 12$$
$$2y = 12$$
$$y = 6$$
$(0, 6)$

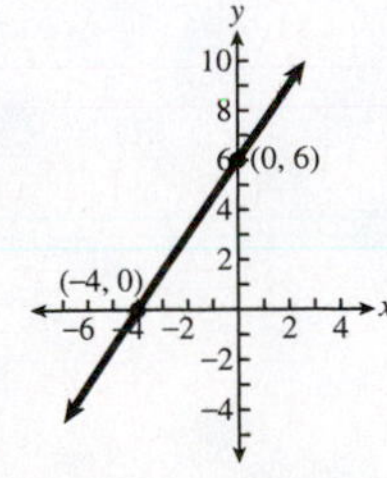

17. (x_1, y_1): $(-5, 3)$ (x_2, y_2): $(-3, 9)$
$$m = \frac{y_2 - y_1}{x_2 - x_1}$$
$$m = \frac{9 - 3}{-3 - (-5)}$$
$$m = \frac{6}{2}$$
$$m = 3$$

24.
$$4x - 2y = 9$$
$$-2y = -4x + 9$$
$$\frac{-2y}{-2} = \frac{-4x}{-2} + \frac{9}{-2}$$
$$y = 2x - \frac{9}{2}$$
$m = 2$, y-intercept: $\left(0, -\frac{9}{2}\right)$

27. Put the y-intercept $(0, -6)$ on the graph
The slope is 3.
Beginning at $(0, -6)$, move up 3 units and 1 unit to the right.
Plot a point at the location $(1, -3)$.
Graph the line that goes through these two points.

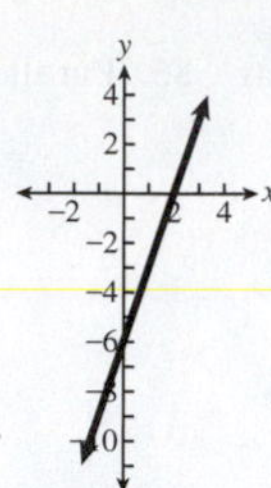

33. The slope of the line $y = 2x - 9$ is 2.
The slope of the line $y = -2x + 9$ is -2.
Since the two slopes are not equal, the lines are not parallel.
Since the two slopes are not negative reciprocals, the lines are not perpendicular.
So, the two lines are neither parallel nor perpendicular.

38. a) The fund started with \$720 and added \$50 for each month (x), so the function is $f(x) = 720 + 50x$
b) Substitue 11 for the number of months.
$f(x) = 720 + 50(11) = 1270$
There will be \$1270 after 11 months.
c) Set the function equal to \$2750 and solve for x.
$720 + 50x = 2750$
$50x = 2030$
$x = 40.6$
Rounding up to the next month, it will take 41 months.

39.
$f(x) = 9x + 7$
$f(-2) = 9(-2) + 7$
$f(-2) = -18 + 7$
$f(-2) = -11$

44.
$y - y_1 = m(x - x_1)$
$y - 3 = -5(x - 1)$
$y - 3 = -5x + 5$
$y = -5x + 8$

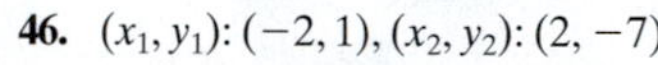

46. (x_1, y_1): $(-2, 1)$, (x_2, y_2): $(2, -7)$
First, find the slope of the line.
$m = \frac{y_2 - y_1}{x_2 - x_1}$
$m = \frac{-7 - 1}{2 - (-2)}$
$m = \frac{-8}{4}$
$m = -2$
Then find the equation of the line.
$y - y_1 = m(x - x_1)$
$y - 1 = -2(x - (-2))$
$y - 1 = -2(x + 2)$
$y - 1 = -2x - 4$
$y = -2x - 3$

49. The line passes through $(-2, 3)$ and $(2, 9)$.
$m = \frac{9 - 3}{2 - (-2)} = \frac{6}{4} = \frac{3}{2}$
$y - 3 = \frac{3}{2}(x - (-2))$
$y - 3 = \frac{3}{2}x + 3$
$y = \frac{3}{2}x + 6$

57. Graph the dashed line $3x + y = -9$
The x-intercept of the line is $(-3, 0)$
The y-intercept of the line is $(0, -9)$.
Select $(0, 0)$ as a test point.
$3(0) + (0) < -9$
$0 < -9$
Since this inequality is false, the test point $(0, 0)$ is not a solution.
Shade the half-plane that does not contain $(0, 0)$

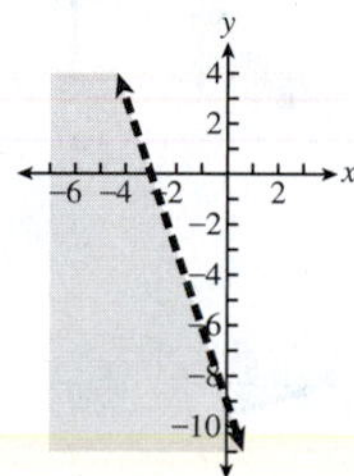

Chapter 3 Test **1.** II **2.** Yes **3.** $(-3, 0), \left(0, \frac{5}{2}\right)$ **4.** $(6, 0), (0, 21)$ **5.** $\left(\frac{3}{2}, 0\right), (0, -6)$

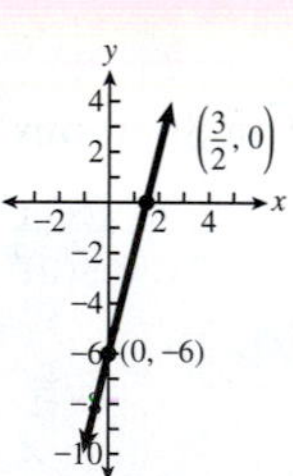

6. $(6, 0), (0, -4)$

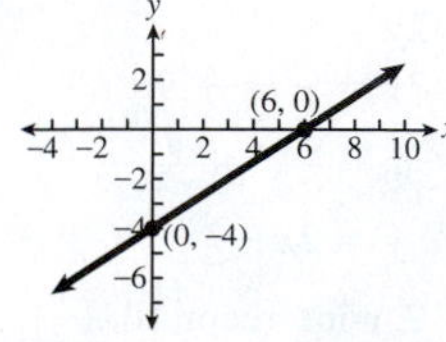

7. $-\frac{1}{5}$ **8.** $7, (0, -8)$ **9.** $-\frac{3}{5}, (0, 9)$ **10.**

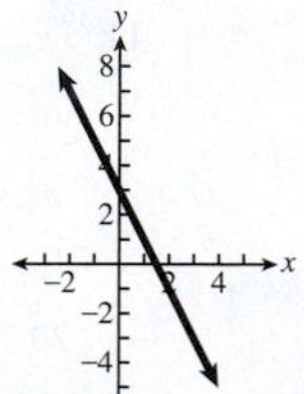

11.

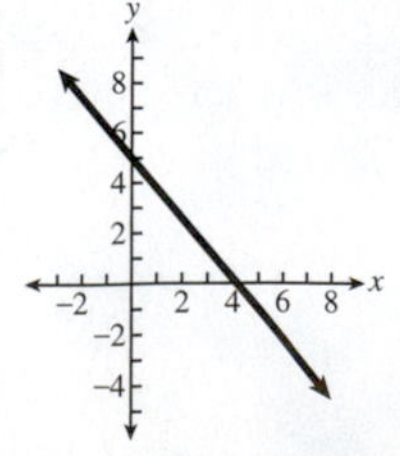

12. Parallel **13.** $y = -6x + 5$ **14.** 13 **15.** $y = -\frac{5}{2}x + 4$ **16.** $y = -2x - 8$

17. **18.**

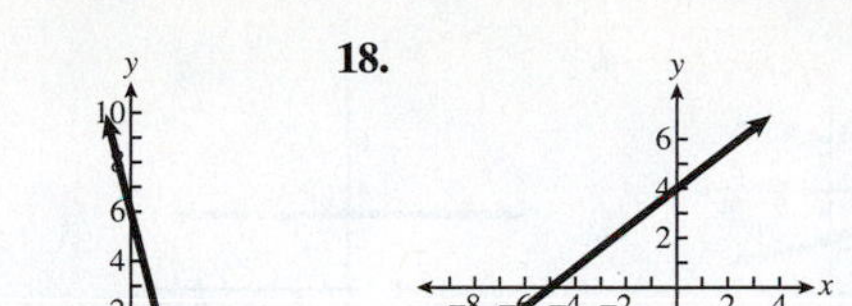

19.

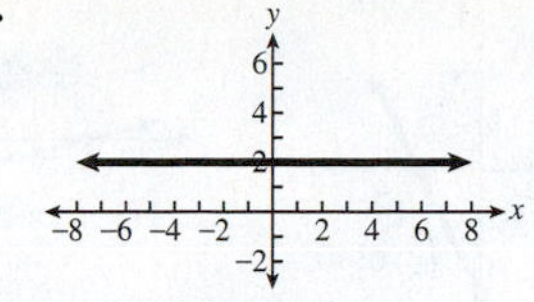

20.

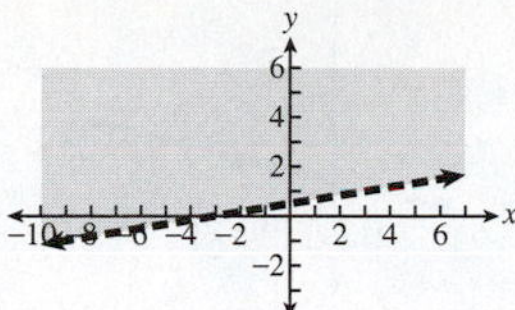

CHAPTER 4

Quick Check 4.1 **1.** Yes **2.** $(5, 0)$ **3.** $(2, 4)$ **4.** $(-5, 2)$ **5.** No solution **6.** $\left(x, -\frac{2}{3}x + 2\right)$

Section 4.1 **1.** system of linear equations **3.** solution **5.** inconsistent **7.** Yes **9.** No **11.** Yes **13.** No **15.** Yes **17.** $(2, 4)$ **19.** $(-3, 6)$ **21.** $(4, 3)$ **23.** $(5, -2)$ **25.** Dependent, $(x, 2x + 4)$ **27.** $(-2, 0)$ **29.** $(-3, -10)$ **31.** $(0, 4)$ **33.** Inconsistent, $\varnothing$ **35.** $(-4, -9)$ **37.** Answers will vary. Following is a correct example.

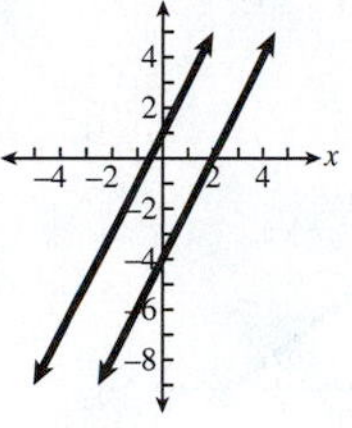

39. Answers will vary. Following is a correct example.

41. Answers will vary. Following is a correct example.

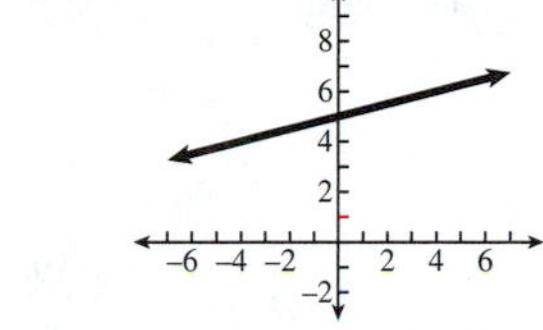

43. Explanations will vary. The ordered pair is a solution of each equation.

Quick Review Exercises **1.** $\{3\}$ **2.** $\{-4\}$ **3.** $\{-5\}$ **4.** $\{-7\}$

Quick Check 4.2 **1.** $(5, 2)$ **2.** $(2, 6)$ **3.** $\varnothing$ **4.** $\left(x, \frac{1}{6}x - \frac{2}{3}\right)$ **5.** $(1, -4)$ **6.** 11 desktops and 4 laptops

Section 4.2 **1.** one of the equations for one of the variables **3.** dependent **5.** $(3, 4)$ **7.** $(17, -3)$ **9.** Inconsistent, $\varnothing$ **11.** Dependent, $(x, -3x - 2)$ **13.** $(12, -49)$ **15.** $(4, 2)$ **17.** $(2, 9)$ **19.** $(-3, -15)$ **21.** $\left(3, \frac{11}{2}\right)$ **23.** $(1, 6)$ **25.** $(4, -10)$ **27.** $(-8, -2)$ **29.** $\left(\frac{5}{4}, \frac{7}{2}\right)$ **31.** $(35, 10)$ **33.** $(10, 30)$ **35.** $(3000, 2000)$ **37.** $(450, 550)$ **39.** $(48, 32)$ **41.** Dependent, $(x, -6x + 21)$ **43.** $(-5, 3)$ **45.** $(-8, 5)$ **47.** $(2, 1)$ **49.** $(-2, 4)$ **51.** 250 students, 950 nonstudents **53.** 47 chicken, 95 steak **55.** 2037 **57.** Explanations will vary. Solve one equation for one variable; then substitute.

Quick Check 4.3 **1.** $(7, 1)$ **2.** $(3, -3)$ **3.** $(-8, -1)$ **4.** $\left(x, \frac{2}{3}x - 4\right)$ **5.** $\varnothing$ **6.** $(-6, 4)$ **7.** $(275, 425)$ **8.** 43 nickels, 32 dimes

Section 4.3 **1.** LCM **3.** dependent **5.** $(-8, -4)$ **7.** $(-6, 3)$ **9.** $(4, -5)$ **11.** Dependent, $\left(x, \frac{1}{2}x - \frac{11}{4}\right)$ **13.** $(7, 1)$ **15.** $(-2, -5)$ **17.** $(3, -4)$ **19.** $(3, 16)$ **21.** $(1, -4)$ **23.** $\left(\frac{3}{2}, 4\right)$ **25.** Inconsistent, $\varnothing$ **27.** $\left(\frac{5}{2}, \frac{1}{2}\right)$ **29.** $(7, 2)$ **31.** $(3, -2)$ **33.** $(2, 1)$ **35.** $(3, 12)$ **37.** $(2700, 900)$ **39.** $(50, 70)$ **41.** $(25, 75)$ **43.** 4 roses, 11 carnations **45.** Tree: \$40; shrub: \$8 **47.** 49 dimes, 32 nickels **49.** 29 nickels, 22 quarters **51.** $(9, -1)$ **53.** $(-7, 6)$ **55.** $(12, 10)$ **57.** Dependent, $(x, 4x - 8)$ **59.** $(-4, -9)$ **61.** $\left(\frac{3}{2}, -5\right)$

Quick Check 4.4 **1.** 40 **2.** 75 feet by 40 feet **3.** \$3200 at 5%, \$1000 at 4% **4.** 32 pounds of almonds and 16 pounds of cashews **5.** 160 milliliters of 60% solution and 240 milliliters of 50% solution **6.** $\frac{1}{2}$ hour **7.** Plane: 475 mph; wind: 25 mph

Section 4.4 **1.** $P = 2L + 2W$ **3.** $d = r \cdot t$ **5.** Algebra: 38; statistics: 55 **7.** Andre: 35; father: 63 **9.** 55, 24 **11.** 15 field goals, 32 extra points **13.** 21 **15.** 35 **17.** Pizza: \$11; soda: \$4 **19.** Length: 27 inches; width: 19 inches **21.** Length: 135 feet; width: 30 feet **23.** Length: 175 centimeters; width: 75 centimeters **25.** Length: 40 feet; width: 25 feet **27.** \$800 at 6%, \$1200 at 3% **29.** \$20,000 at 4.25%, \$5000 at 3.5% **31.** \$6000 at 4% profit, \$1500 at 13% loss **33.** 16 pounds with fruit, 24 pounds without fruit **35.** 30 pounds Kona, 50 pounds regular **37.** 80 milliliters of 80%, 240 milliliters of 44% **39.** 12 ounces of 20%, 6 ounces of 50% **41.** 2.2 liters of vodka, 1.8 liters of tonic water **43.** George: 2 hours; Tina: 4 hours **45.** After 1.5 hours **47.** Kayak: 4 mph; current: 2 mph **49.** Plane: 550 mph; wind: 50 mph **51.** Answers will vary. Length is 7 meters more than the width; perimeter is 46 meters. **53.** Answers will vary. Combine 40% and 58% to make 90 milliliters of 42%.

Quick Review Exercises **1.**

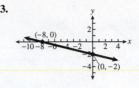

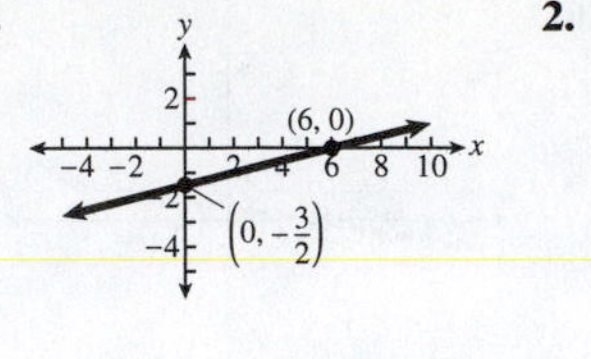

2. **3.**

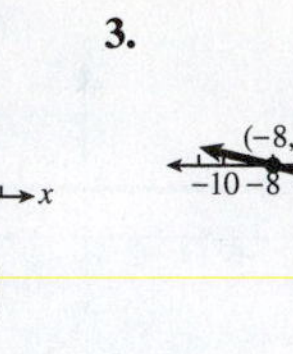

4.

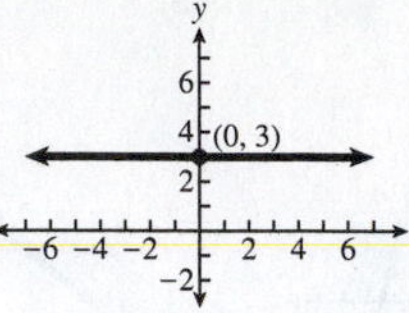

Quick Check 4.5 **1.**

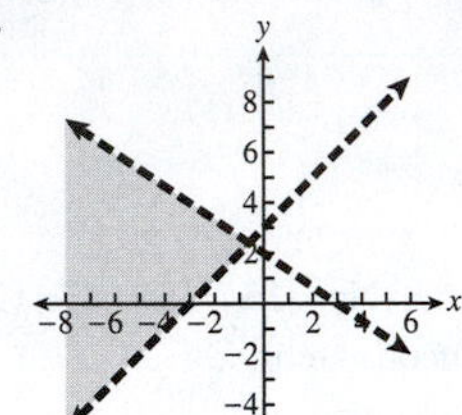

2.

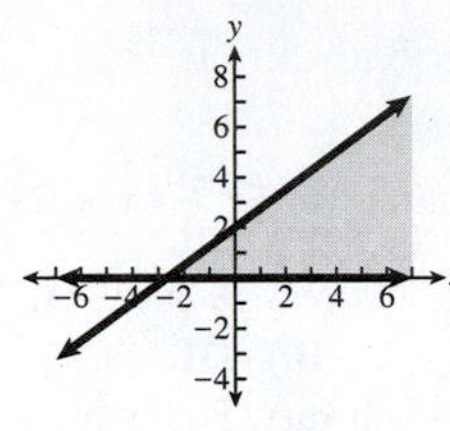

3.

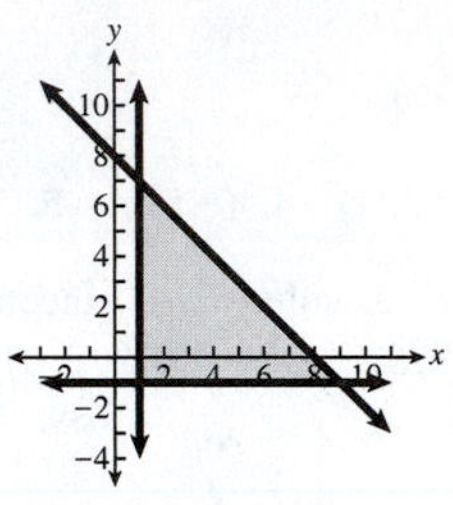

Section 4.5 **1.** system of linear inequalities **3.** solid **5.**

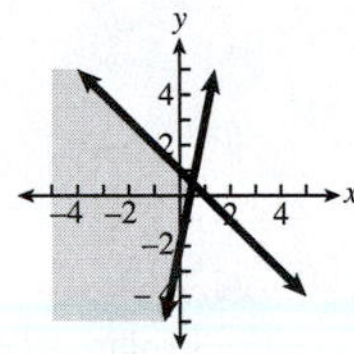

7.

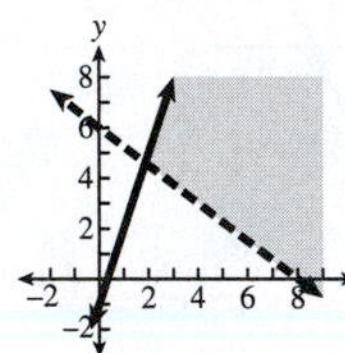

9. **11.**

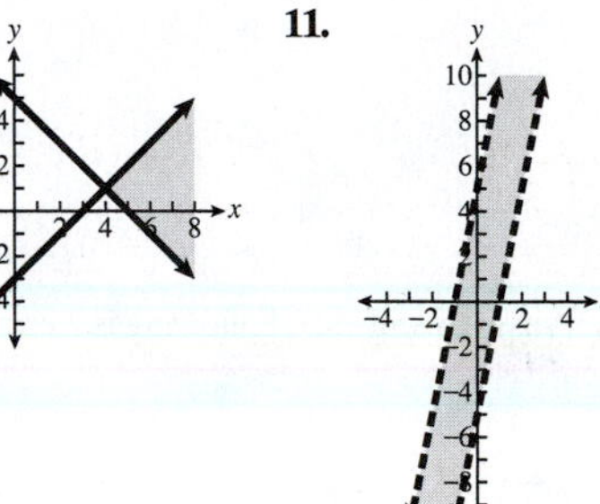

13.

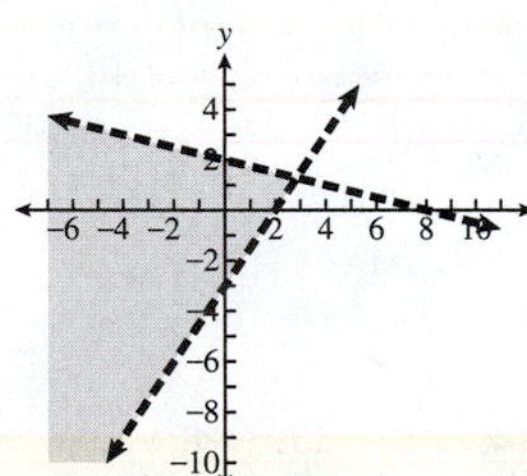

15.

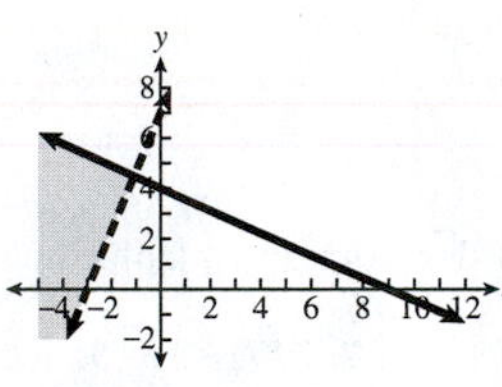

17.

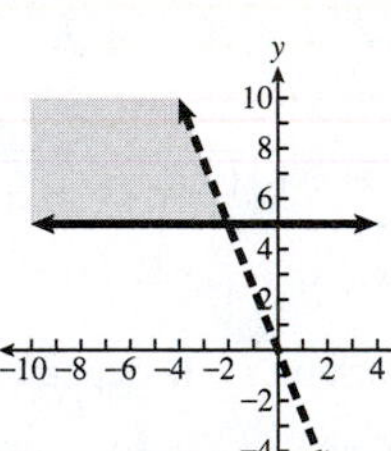

19.

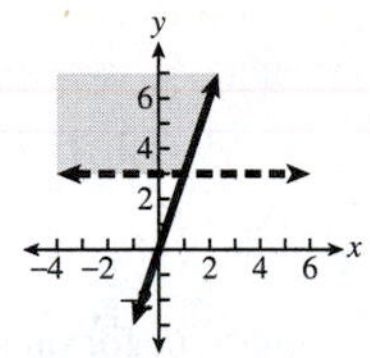

21.

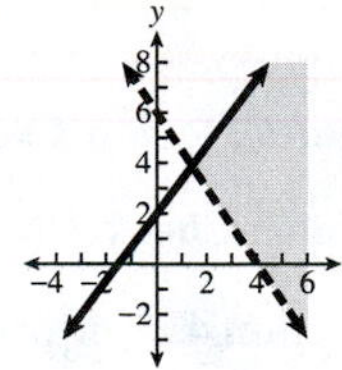

23.

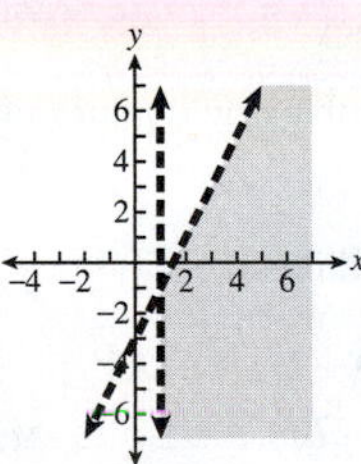

25.

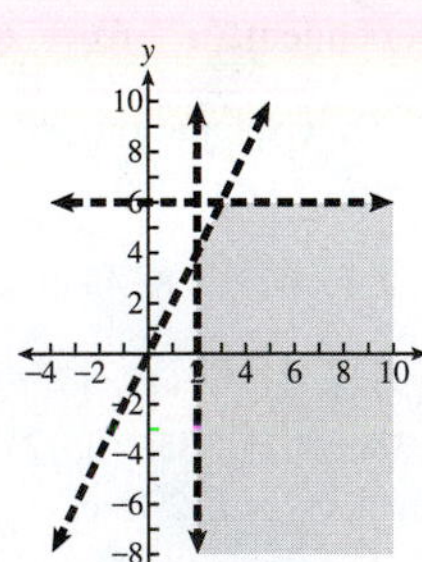

27.

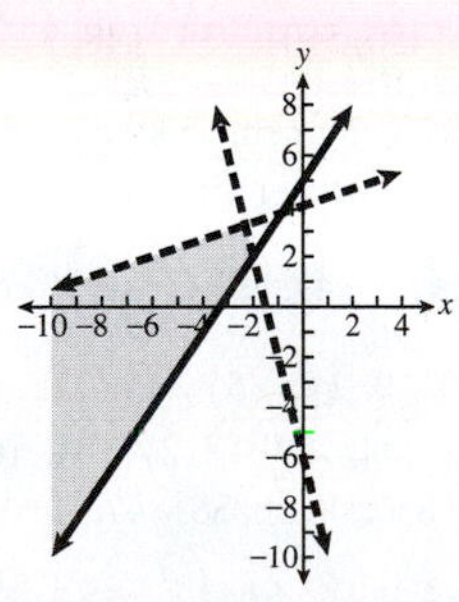

29. $x > 0$ $y > 0$ **31.** 18 square units

Chapter 4 Review **1.** $(-4, -5)$ **2.** $(6, -1)$ **3.** $(-3, 6)$ **4.** $(-7, 0)$ **5.** $(8, 3)$ **6.** $(3, -3)$ **7.** $(2, 4)$ **8.** $(-6, -4)$ **9.** $(-3, -2)$ **10.** $(-7, 4)$ **11.** $(0, -5)$ **12.** $\varnothing$ **13.** $(9, 1)$ **14.** $\left(4, \frac{3}{2}\right)$ **15.** $\left(-\frac{5}{3}, -8\right)$ **16.** $(x, 2x - 4)$ **17.** $(-2, -3)$ **18.** $(-4, 6)$ **19.** $(2, -4)$ **20.** $(-3, 7)$ **21.** $(-2, -5)$ **22.** $(5, -3)$ **23.** $(x, 3x - 2)$ **24.** $\left(-\frac{5}{2}, -\frac{3}{2}\right)$ **25.** $(0, -6)$ **26.** $(27, -16)$ **27.** $\varnothing$ **28.** $(-3, 5)$ **29.** $(5, 3.9)$ **30.** $(6000, 2000)$ **31.** 23, 64 **32.** 1077 **33.** 834 **34.** 46 **35.** 400 feet by 260 feet **36.** 76 inches **37.** \$1200 at 5%, \$2800 at 4.25% **38.** \$1000 at 20%, \$1500 at 8% **39.** 15 pounds 4%, 30 pounds 13% **40.** 5 mph **41.**

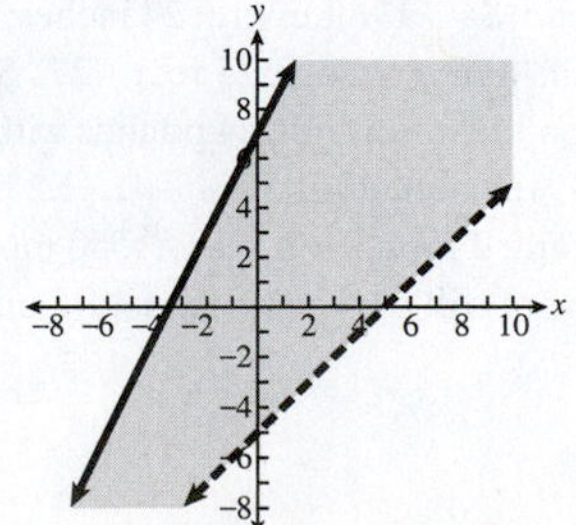

42.

43.

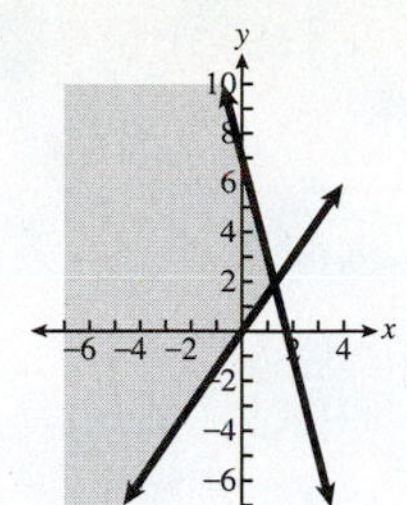

44.

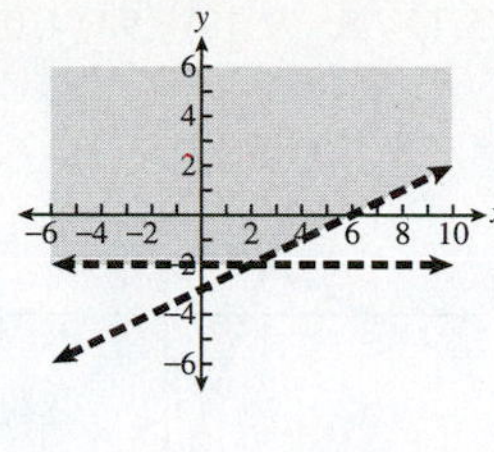

Chapter 4 Review Exercises: Worked-Out Solutions **1.** Look for the point where the two lines intersect. The solution to the system is $(-4, -5)$

5.

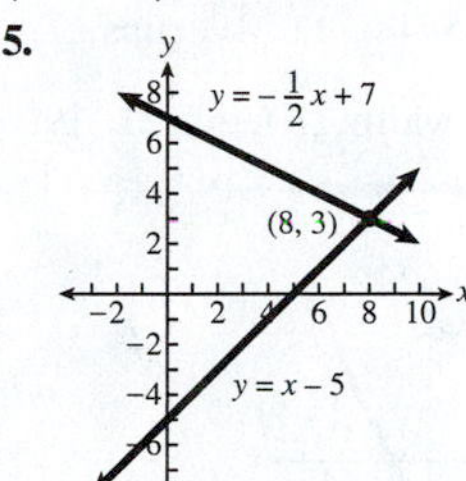

$(8, 3)$

9.
$$\begin{aligned} x &= 2y + 1 \\ 2x - 5y &= 4 \\ 2(2y + 1) - 5y &= 4 \\ 4y + 2 - 5y &= 4 \\ -y + 2 &= 4 \\ -y &= 2 \\ y &= -2 \end{aligned}$$

$$\begin{aligned} x &= 2(-2) + 1 \\ x &= -4 + 1 \\ x &= -3 \end{aligned}$$

$(-3, -2)$

15.
$$\begin{aligned} 6x + 7y &= -66 \\ -3x + 4y &= -27 \end{aligned} \xrightarrow{\text{Multiply by 2}} \begin{aligned} 6x + 7y &= -66 \\ -6x + 8y &= -54 \end{aligned}$$

$$\begin{aligned} 6x + 7y &= -66 \\ -6x + 8y &= -54 \\ \hline 15y &= -120 \\ y &= -8 \end{aligned}$$

$$\begin{aligned} 6x + 7(-8) &= -66 \\ 6x - 56 &= -66 \\ 6x &= -10 \\ x &= -\frac{10}{6} \\ x &= -\frac{5}{3} \end{aligned}$$

$\left(-\frac{5}{3}, -8\right)$

33. Women: x
Men: y

$$\begin{aligned} x + y &= 1764 \\ 5x + 10y &= 13470 \end{aligned} \xrightarrow{\text{Multiply by } -5} \begin{aligned} -5x - 5y &= -8820 \\ 5x + 10y &= 13470 \end{aligned}$$

$$\begin{aligned} -5x - 5y &= -8820 \\ 5x + 10y &= 13470 \\ \hline 5y &= 4650 \\ y &= 930 \end{aligned}$$

$$\begin{aligned} x + 930 &= 1764 \\ x &= 834 \end{aligned}$$

There were 834 ladies.

35. Length: l
Width: w

$$\begin{aligned} 2l + 2w &= 1320 \\ l &= w + 140 \end{aligned}$$

$$\begin{aligned} 2l + 2w &= 1320 \\ 2(w + 140) + 2w &= 1320 \\ 2w + 280 + 2w &= 1320 \\ 4w + 280 &= 1320 \\ 4w &= 1040 \\ w &= 260 \end{aligned}$$

$$\begin{aligned} l &= 260 + 140 \\ l &= 400 \end{aligned}$$

The length is 400 feet and the width is 260 feet.

37. First account (5%): x
Second account (4.25%): y

$$\begin{aligned} x + y &= 4000 \\ 0.05x + 0.0425y &= 179 \end{aligned} \xrightarrow{\text{Multiply by 10,000}} \begin{aligned} x + y &= 4000 \\ 500x + 425y &= 1{,}790{,}000 \end{aligned}$$

$$\begin{aligned} x + y &= 4000 \\ 500x + 425y &= 1{,}790{,}000 \end{aligned} \xrightarrow{\text{Multiply by } -425} \begin{aligned} -425x - 425y &= -1{,}700{,}000 \\ 500x + 425y &= 1{,}790{,}000 \end{aligned}$$

$$\begin{aligned} -425x - 425y &= -1{,}700{,}000 \\ 500x + 425y &= 1{,}790{,}000 \\ 75x &= 90{,}000 \end{aligned}$$

$x = 1200$

$$\begin{aligned} 1200 + y &= 4000 \\ y &= 2800 \end{aligned}$$

He invested \$1200 at 5% interest and \$2800 at 4.25% interest.

41.

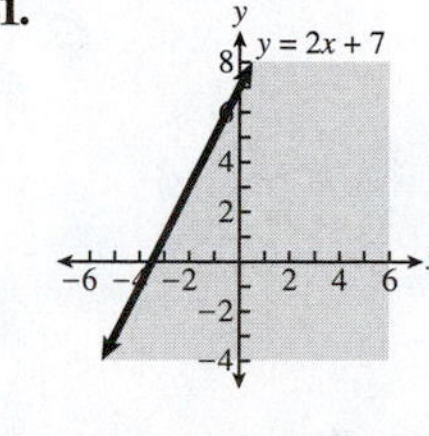

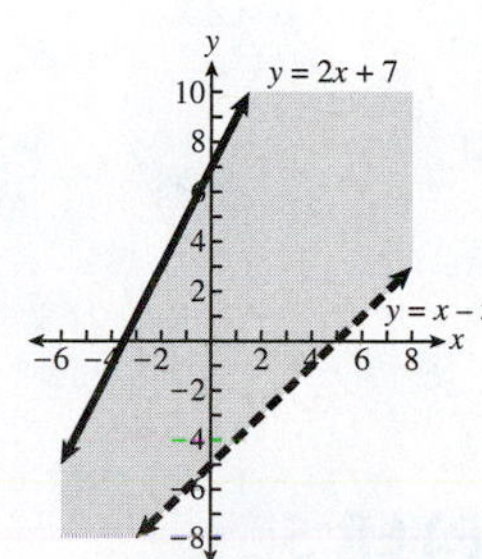

Chapter 4 Test **1.** $(-5, 2)$ **2.** $(4, -5)$ **3.** $(3, -3)$ **4.** $(-4, -1)$ **5.** $(-3, 2)$ **6.** $\varnothing$ **7.** $(5, 6)$ **8.** $(9, 10)$ **9.** $(4, 0)$ **10.** $\left(x, \frac{2}{3}x - 3\right)$ **11.** 180 **12.** 5400 **13.** Length: 25 feet; width: 8 feet **14.** \$20,000 at 7%, \$12,000 at 3% **15.**

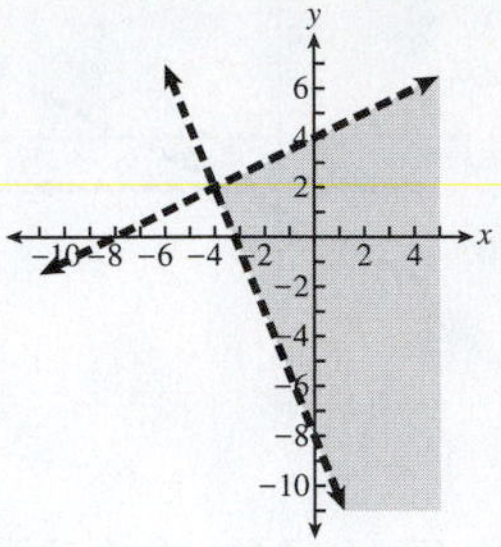

Cumulative Review Chapters 1–4 **1.** $>$ **2.** -34 **3.** -165 **4.** 13 **5.** $2^3 \cdot 3^2$ **6.** $\frac{2}{49}$ **7.** $\frac{33}{20}$ **8.** 6.08 **9.** 17.86 **10.** \$448 **11.** $4\frac{1}{12}$ cups **12.** -43 **13.** 15 **14.** -28 **15.** $5x - 3$ **16.** $-11x - 69$ **17.** $\{-9\}$ **18.** $\left\{\frac{7}{2}\right\}$ **19.** $\{14\}$ **20.** $\{-14\}$ **21.** Length: 22 feet; width: 13 feet **22.** 19 **23.** 25 **24.** 24% **25.** \$248,600 **26.** $\{6\}$ **27.** 4615 **28.** $x \le 3$

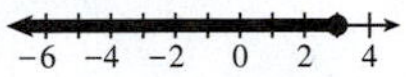

$(-\infty, 3]$ **29.** $-5 \le x \le 1$

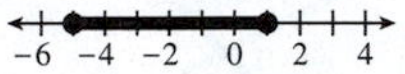

$[-5, 1]$

30. $(3, 0), (0, -6)$

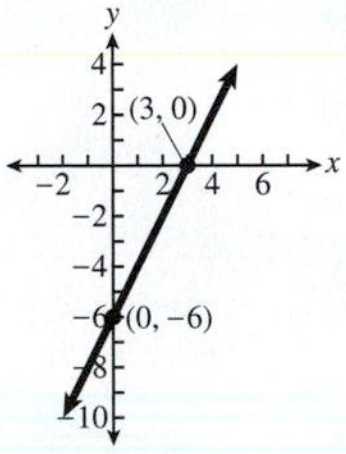

31. $(8, 0), (0, -6)$

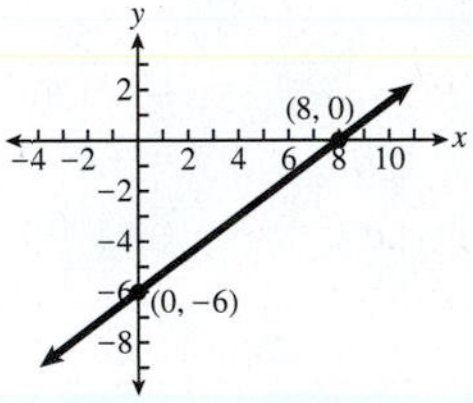

32. 2 **33.** $m = -\frac{2}{3}$, $(0, 7)$ **34.**

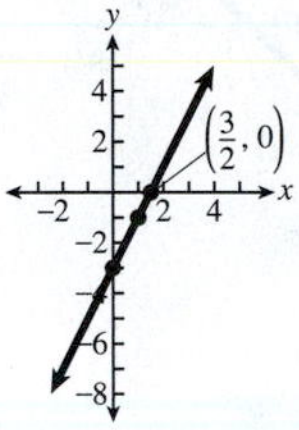

35.

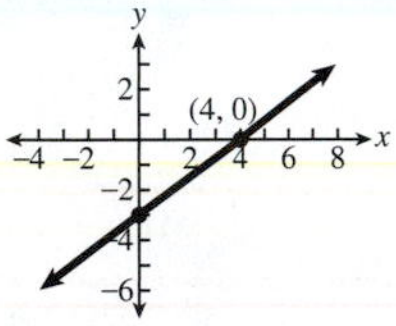

36. Perpendicular **37. a)** $f(x) = 220 + 40x$ **b)** \$540 **c)** 15 months **38.** 391 **39.** 1129

40. $y = -2x + 2$ **41.** $y = -3x + 7$ **42.**

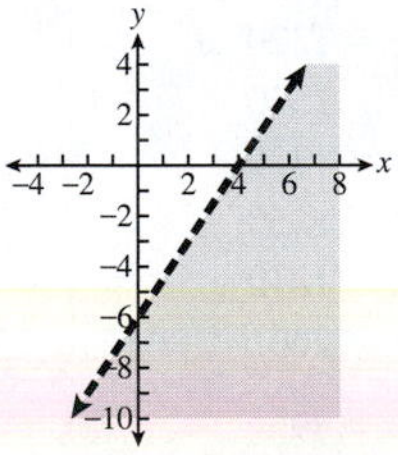

43. $(5, 2)$ **44.** $(3, 7)$ **45.** $(1, 2)$ **46.** $(-1, 10)$ **47.** 250 children

48. Length: 40 feet; width: 25 feet **49.** \$11,000 at 3%, \$4000 at 2.5% **50.**

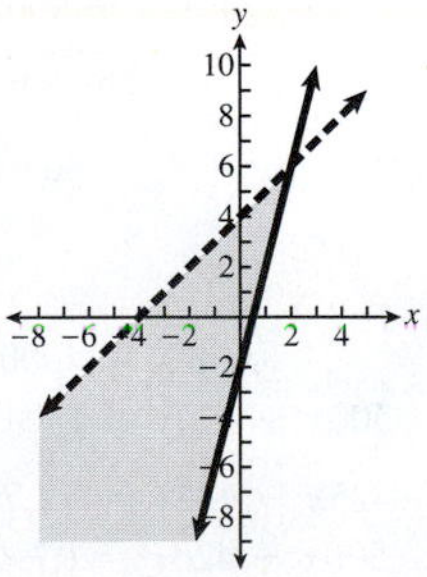

CHAPTER 5

Quick Check 5.1 **1. a)** x^9 **b)** 729 **2.** $(a - 6)^{27}$ **3.** $a^{15}b^{10}$ **4.** x^{63} **5.** x^{34} **6.** $a^{40}b^{24}$ **7.** $64a^{26}b^{19}c^{35}$ **8. a)** x^{16} **b)** x^{16} **9.** $4a^8b^6c^6$ **10. a)** 1 **b)** 1 **c)** 9 **11.** $\frac{a^5b^{35}}{c^{15}d^{20}}$ **12.** 4096 **13.** a^{56}

Section 5.1 **1.** $x^m \cdot x^n = x^{m+n}$ **3.** $(xy)^n = x^ny^n$ **5.** $x^0 = 1$ **7.** 64 **9.** x^{13} **11.** m^{40} **13.** b^{14} **15.** $(2x - 3)^{14}$ **17.** x^8y^{13} **19.** x^5 **21.** b **23.** x^{24} **25.** a^{63} **27.** 1024 **29.** x^{29} **31.** 7 **33.** 5 **35.** $81x^4$ **37.** $64x^6$ **39.** $m^{56}n^{35}$ **41.** $81x^{16}y^4$ **43.** $x^{14}y^{67}z^{61}$ **45.** 7 **47.** 6, 13 **49.** x^9 **51.** r^{15} **53.** $8x^{14}$ **55.** $(a + 5b)^2$ **57.** a^5b^5 **59.** 12 **61.** 30 **63.** 1 **65.** 1 **67.** 3 **69.** 1 **71.** $\frac{27}{64}$ **73.** $\frac{a^8}{b^8}$ **75.** $\frac{x^{20}}{y^{15}}$ **77.** $\frac{8a^{18}}{b^{21}}$ **79.** $\frac{a^{45}b^{18}}{c^{63}}$ **81.** 4 **83.** 12

85. 16 **87.** 16 **89.** a^{24} **91.** a^{42} **93.** $\frac{125x^{12}}{8y^{21}}$ **95.** x^{12} **97.** $\frac{a^{40}b^{45}}{32c^{10}}$ **99.** b^{20} **101.** x^{90} **103.** $\frac{81a^8b^{36}}{2401c^4d^{16}}$ **105.** 144 feet **107.** 8100 square feet **109.** 615.44 square feet **111.** $x^5 \cdot x^4$; explanations will vary.

Quick Check 5.2 **1.** $\frac{1}{64}$ **2.** $\frac{a^5}{b^4}$ **3. a)** $x^{12}y^7$ **b)** $\frac{y^4zw^9}{x^2}$ **4.** $\frac{1}{x^7}$ **5.** x^{42} **6.** $\frac{b^6}{81a^2c^4}$ **7.** $\frac{1}{x^6}$ **8.** $\frac{1}{x^{20}y^{16}}$ **9. a)** 4.6×10^{-3} **b)** 3.57×10^6 **10. a)** 3,200,000 **b)** 0.000721 **11.** 6.96×10^{13} **12.** 3.0×10^{12} **13.** 8000 seconds

Section 5.2 **1.** $\frac{1}{x^n}$ **3.** $\frac{1}{25}$ **5.** $\frac{1}{64}$ **7.** $-\frac{1}{169}$ **9.** $\frac{1}{b^{12}}$ **11.** $\frac{12}{x^6}$ **13.** $-\frac{5}{m^{19}}$ **15.** x^5 **17.** $3y^4$ **19.** $\frac{y^2}{x^{11}}$ **21.** $-\frac{8b^5}{a^6c^7}$ **23.** $\frac{5c^4}{a^6b^9d^5}$ **25.** $\frac{1}{x^5}$ **27.** a^6 **29.** $\frac{1}{m^{21}}$ **31.** $\frac{1}{x^{18}}$ **33.** x^{42} **35.** $\frac{z^{12}}{x^{10}y^8}$ **37.** $\frac{a^{15}b^{24}}{64z^6}$ **39.** $\frac{1}{x^{15}}$ **41.** x^{17} **43.** $\frac{1}{a^3}$ **45.** $\frac{1}{x^{11}}$ **47.** 1 **49.** $\frac{y^{20}}{x^{15}}$ **51.** $\frac{9c^{10}}{a^8b^{14}d^{16}}$ **53.** 0.000000307 **55.** 8,935,000,000 **57.** 90,210 **59.** 2.7×10^{-4} **61.** 8.6×10^6 **63.** 4.2×10^{11} **65.** 9.43×10^{19} **67.** 3.4×10^{-22} **69.** 4.1496×10^{-11} **71.** 1.74×10^5 **73.** 3.36×10^4 seconds (33,600 seconds) **75.** 1.116×10^8 miles **77.** 1.1×10^{45} grams **79.** $\$9.487 \times 10^{10}$ **81.** 2,310,000 **83.** Answers will vary; negative exponent versus negative base

Quick Review Exercises **1.** $11x$ **2.** $-7a$ **3.** $-6x^2 + 11x$ **4.** $12y^3 + 5y^2 - 13y - 37$

Quick Check 5.3 **1. a)** Trinomial; 2, 1, 0 **b)** Binomial; 2, 0 **c)** Monomial; 7 **2.** 1, −6, −11, 32 **3.** $-x^3 + 4x^2 + x + 9$; Leading term: $-x^3$; Leading coefficient: −1; Degree: 3 **4.** 111 **5.** 280 **6.** $x^3 + 3x + 169$ **7.** $-3x^3 - x^2 + 15x + 5$ **8.** $6x^2 + 13x - 97$ **9.** 289 **10.** Each term: 5, 8, 10; polynomial: 10 **11.** $13x^4y^2 - 4x^3y^3$

Section 5.3 **1.** polynomial **3.** monomial **5.** trinomial **7.** descending order **9.** 4, 2, 1, 0 **11.** 1, 7, 4 **13.** 7, 1, −15 **15.** 10, −17, 6, −1, 2 **17.** trinomial **19.** binomial **21.** monomial **23.** $3x^2 + 8x - 7, 3x^2, 3, 2$ **25.** $2x^4 + 6x^2 - 11x + 10, 2x^4, 2, 4$ **27.** 30 **29.** 32 **31.** −8 **33.** −2 **35.** −591 **37.** $8x^2 - 6x + 3$ **39.** $2x^2 - 17x + 80$ **41.** $2x^3 + 8x^2 - 14x - 11$ **43.** $x^3 - x^2 + 5x + 3$ **45.** $2x^9 + 7x^6 - 4x^5 - 5x^4 + 7x^2 + 12$ **47.** $4x^2 - 3x - 13$ **49.** $2x^4 - 3x^3 + x^2 + 11x + 11$ **51.** $12x^2 + x + 9, 2x^2 + 19x - 3$ **53.** $5x^3 + x^2 - 5x - 6, -3x^3 - x^2 - 11x - 56$ **55.** 184 **57.** 148 **59.** 75,680 **61.** 10, 6, 3; polynomial: 10 **63.** 8, 9, 11; polynomial: 11 **65.** $7x^2y + 11xy^2 - 3x^2y^3$ **67.** $-14a^3b^2 + 13a^5b + 5a^2b^3$ **69.** $2x^3yz^2 - 15xy^4z^3 - 7x^2y^2z^5 - 4x^2yz^3$ **71.** \$1,297,200 **73.** 5331 **75.** Explanations will vary; exponent operations

Quick Check 5.4 **1.** $56x^2$ **2.** $60x^{14}yz^{13}$ **3.** $28x^4 - 24x^3 + 20x^2 - 32x$ **4.** $6x^9 - 3x^8 + 3x^7 + 45x^6 - 63x^5$ **5.** $x^2 + 20x + 99$ **6.** $10x^2 - 49x + 18$ **7.** $3x^3 + 30x^2 + 46x - 16$ **8.** $x^2 - 100$ **9.** $4x^2 - 49$ **10.** $25x^2 - 80x + 64$ **11.** $x^2 + 12x + 36$

Section 5.4 **1.** add **3.** multiply **5.** $36x^5$ **7.** $-54m^{18}$ **9.** $91a^{12}b^{11}$ **11.** $-70x^4y^2z^9w^4$ **13.** $84x^{10}$ **15.** $6x^{25}$ **17.** $-12x^9$ **19.** $15x - 20$ **21.** $-12x + 18$ **23.** $18x^2 + 30x$ **25.** $3x^5 - 4x^4 + 7x^3$ **27.** $6x^3y^2 - 12x^2y^3 + 14xy^4$ **29.** $x^2 - 2x - 63$ **31.** $x^2 - 12x + 27$ **33.** $4x^2 - 21x - 18$ **35.** $x^2 + x - 72$ **37.** $4x^2 - 20x - 39$ **39.** $3x^2 + 13x - 30$ **41.** $3x^3 - 13x^2 - 37x - 18$ **43.** $x^4 - 4x^3 - 51x^2 + 310x - 400$ **45.** $x^2 - 2xy - 8y^2$ **47.** $20x^2y^2 + 39xy + 18$ **49.** $3x^3$ **51.** $3x^3 - 5x^2 - 12$ **53.** 3 **55.** 3, 6 **57.** $x^2 - 7x - 18$ **59.** $-18x^9$ **61.** $-4x^8 + 20x^7 - 32x^6 - 12x^5$ **63.** $x^2 - 81$ **65.** $9x^2 - 49$ **67.** $x^2 + 14x + 49$ **69.** $16x^2 - 24x + 9$ **71.** $x - 9$ **73.** $x - 6$ **75.** $x^2 + 6x - 7$ **77.** $5x^3 - 40x^2 - 45x$ **79.** $-28x^6y^9$ **81.** $20x^{14}$ **83.** $25x^2 - 9$ **85.** $x^2 + 26x + 169$ **87.** $6x^4y - 2x^6y^4 - 14x^2y^2$ **89.** $16x^2 - 72x + 81$ **91.** $x^4 - 2x^3 - 7x^2 - 8x + 16$ **93.** Answers will vary; combining like terms versus multiplying

Quick Check 5.5 **1.** $8x^5$ **2.** $5x^2y^9$ **3.** $x^2 - 3x - 7$ **4.** $8x^8 + 2x^5 + 5x^3$ **5.** $x + 9$ **6.** $x - 10 + \frac{62}{x + 7}$ **7.** $4x + 3 + \frac{16}{3x - 4}$ **8.** $x^2 - x - 2 - \frac{13}{x - 2}$

Section 5.5 **1.** term **3.** factor **5.** $8x^{21}$ **7.** $-5n^7$ **9.** $13a^6b^2$ **11.** $5x^4$ **13.** $\frac{3x^2}{2}$ **15.** $\frac{x^8}{3}$ **17.** $10x^{11}$ **19.** $-4x^5$ **21.** $3x^2 - 5x - 8$ **23.** $8x^3 + 10x - 9$ **25.** $2x^6 + 3x^5 + 5x^4$ **27.** $-2x^4 + 3x^2$ **29.** $x^5y^4 - x^3y^3 + xy^2$ **31.** $6x$ **33.** $6x^7 - 8x^5 - 18x^3$ **35.** $x + 5$ **37.** $x + 6$ **39.** $x - 9$ **41.** $x + 4 + \frac{3}{x - 8}$ **43.** $x^2 - 12x - 25 + \frac{39}{x + 1}$ **45.** $2x - 7 + \frac{3}{x + 5}$ **47.** $3x + 2 - \frac{4}{2x + 7}$ **49.** $x - 13$ **51.** $x^3 + 3x^2 + 11x + 18 + \frac{86}{x - 3}$ **53.** $x^2 + 5x + 25$ **55.** $x^2 + 5x - 24$ **57.** $x + 13$ **59.** Yes **61.** No **63.** $x^2 + 9x - 27 + \frac{72}{x + 3}$ **65.** $x^2 - 4x + 24 - \frac{115}{x + 4}$ **67.** $3x^4 - 2x^2 + 5x - 1$ **69.** $8x - 6 + \frac{11}{x + 6}$ **71.** $5x - 9 - \frac{33}{4x - 3}$ **73.** $-9x^7 + 6x^5 + 2x^4 + x$ **75.** $4x^2 - 6x - 10 - \frac{9}{2x + 3}$ **77.** $3x^2 - 5x - 17$ **79.** $-3x^2yz^5$

Chapter 5 Review **1.** x^6 **2.** $64x^{21}$ **3.** 7 **4.** $\frac{625x^{36}}{16y^8}$ **5.** $28x^{19}$ **6.** $24x^{16}y^8$ **7.** $36a^{10}b^6c^{12}$ **8.** −2 **9.** $x^{40}y^{52}z^{16}$ **10.** x^{10} **11.** $\frac{1}{25}$ **12.** $\frac{1}{1000}$ **13.** $\frac{7}{x^4}$ **14.** $-\frac{6}{x^6}$ **15.** $-8y^5$ **16.** $\frac{27}{x^{12}}$ **17.** $\frac{x^{10}}{4}$ **18.** x^{15} **19.** $\frac{b^{21}}{a^{12}}$ **20.** $\frac{1}{m^6}$ **21.** $\frac{x^{15}z^{21}}{64y^{12}}$ **22.** $\frac{y^{36}}{x^{66}}$ **23.** $\frac{1}{a^{20}}$ **24.** $\frac{1}{x^{11}}$ **25.** 1.4×10^9 **26.** 2.1×10^{-12} **27.** 5.002×10^{-6} **28.** 0.000123 **29.** 40,750,000 **30.** 6,127,500,000,000 **31.** 1.5×10^{14} **32.** 2.5×10^{-18} **33.** 2.49×10^{-12} grams **34.** 1.5×10^4 seconds **35.** 80 **36.** −47 **37.** 30 **38.** −112 **39.** 27 **40.** 371 **41.** $2x^2 - 6x - 9$ **42.** $8x^2 - 19x - 50$ **43.** $8x - 21$ **44.** $x^2 + 9x - 22$ **45.** $3x^3 - 5x^2 + 13x - 30$ **46.** $-4x^3 + 4x^2 - 6x - 115$ **47.** −9 **48.** 37 **49.** $a^8 - 7a^4 + 10$ **50.** $4a^6 + 6a^3 - 30$ **51.** $f(x) + g(x) = 10x^2 - 2x + 31, f(x) - g(x) = -4x^2 - 14x - 49$ **52.** $f(x) + g(x) = -7x^2 - 10x - 24, f(x) - g(x) = 9x^2 + 20x - 36$

53. $x^2 - 10x + 25$ **54.** $x^2 + 9x - 52$ **55.** $12x^3 - 28x^2 - 64x$ **56.** $4x^2 - 28x + 49$ **57.** $15x^{13}$ **58.** $x^2 - 64$ **59.** $6x^2 - 19x - 130$ **60.** $-18x^6$ **61.** $36x^2 - 25$ **62.** $x^2 + 20x + 100$ **63.** $16x^2 + 56x + 49$ **64.** $-24x^7 + 56x^6 + 48x^5$ **65.** $x^3 + 3x^2 - 42x + 108$ **66.** $3x^3 - 11x^2 + x + 18$ **67.** $-120x^{18}$ **68.** $10x^6 + 22x^5 - 24x^4$ **69.** $f(x)\cdot g(x) = x^2 + 3x - 70$ **70.** $f(x)\cdot g(x) = 24x^8 + 54x^7 - 120x^6$ **71.** $3x^2 - 4x + 10$ **72.** $3x + 4 - \frac{9}{x-4}$ **73.** $-2y^6$ **74.** $x^2 - 2x + 10 - \frac{65}{x+5}$ **75.** $x^3 - 3x^2 + 9x - 27 + \frac{162}{x+3}$ **76.** $6yz^5$ **77.** $8x + 15 + \frac{36}{x+9}$ **78.** $6x + 12$ **79.** $3x - 17$ **80.** $2x + 1 + \frac{23}{8x-3}$

Chapter 5 Review Exercises: Worked-Out Solutions

1. $\frac{x^9}{x^3} = x^{9-3} = x^6$

6. $3x^8y^5 \cdot 8x^8y^3 = 24x^{8+8}y^{5+3} = 24x^{16}y^8$

7. $(-6a^5b^3c^6)^2 = (-6)^2a^{5\cdot2}b^{3\cdot2}c^{6\cdot2} = 36a^{10}b^6c^{12}$

18. $\frac{x^{10}}{x^{-5}} = x^{10-(-5)} = x^{15}$

21. $(4x^{-5}y^4z^{-7})^{-3} = 4^{-3}x^{15}y^{-12}z^{21} = \frac{x^{15}z^{21}}{4^3y^{12}} = \frac{x^{15}z^{21}}{64y^{12}}$

23. $a^{-15}\cdot a^{-5} = a^{-15+(-5)} = a^{-20} = \frac{1}{a^{20}}$

32. $(3.0 \times 10^{-13}) \div (1.2 \times 10^5) = \frac{3.0}{1.2} \times \frac{10^{-13}}{10^5} = 2.5 \times 10^{-13-5} = 2.5 \times 10^{-18}$

33. $1.66 \times 10^{-24}\cdot 1.5 \times 10^{12} = 1.66\cdot1.5 \times 10^{-24}\cdot10^{12} = 2.49 \times 10^{-24+12} = 2.49 \times 10^{-12}$

2.49×10^{-12} grams

39. $5x^2 + 10x + 12$
$5(-3)^2 + 10(-3) + 12 = 5(9) + 10(-3) + 12 = 45 - 30 + 12 = 27$

41. $(x^2 + 3x - 15) + (x^2 - 9x + 6) = x^2 + 3x - 15 + x^2 - 9x + 6 = 2x^2 - 6x - 9$

43. $(x^2 - 6x - 13) - (x^2 - 14x + 8) = x^2 - 6x - 13 - x^2 + 14x - 8 = 8x - 21$

47. $f(x) = x^2 - 25$
$f(-4) = (-4)^2 - 25 = 16 - 25 = -9$

51. $f(x) + g(x) = (3x^2 - 8x - 9) + (7x^2 + 6x + 40) = 3x^2 - 8x - 9 + 7x^2 + 6x + 40 = 10x^2 - 2x + 31$

$f(x) - g(x) = (3x^2 - 8x - 9) - (7x^2 + 6x + 40) = 3x^2 - 8x - 9 - 7x^2 - 6x - 40 = -4x^2 - 14x - 49$

54. $(x - 4)(x + 13) = x^2 + 13x - 4x - 52 = x^2 + 9x - 52$

55. $4x(3x^2 - 7x - 16) = 4x\cdot3x^2 - 4x\cdot7x - 4x\cdot16 = 12x^3 - 28x^2 - 64x$

56. $(2x - 7)^2 = (2x - 7)(2x - 7) = 4x^2 - 14x - 14x + 49 = 4x^2 - 28x + 49$

57. $5x^7\cdot3x^6 = 15x^{7+6} = 15x^{13}$

69. $f(x)\cdot g(x) = (x + 10)(x - 7) = x^2 - 7x + 10x - 70 = x^2 + 3x - 70$

71. $\frac{6x^4 - 8x^3 + 20x^2}{2x^2} = \frac{6x^4}{2x^2} - \frac{8x^3}{2x^2} + \frac{20x^2}{2x^2} = 3x^2 - 4x + 10$

72.

```
            3x + 4
x - 4 )3x² - 8x - 25
       3x² - 12x   ↓
             4x - 25
             4x - 16
                  -9
```

$\frac{3x^2 - 8x - 25}{x - 4} = 3x + 4 - \frac{9}{x-4}$

74.

```
             x² - 2x + 10
x + 5 )x³ + 3x² + 0x - 15
       x³ + 5x²    ↓    ↓
           -2x² + 0x - 15
           -2x² - 10x   ↓
                  10x - 15
                  10x + 50
                       -65
```

$\frac{x^3 + 3x^2 - 15}{x + 5} = x^2 - 2x + 10 - \frac{65}{x+5}$

Chapter 5 Test **1.** x^8 **2.** $\frac{x^{30}}{729y^{24}}$ **3.** $x^{72}y^{40}z^{64}$ **4.** $\frac{1}{256}$ **5.** $\frac{32}{x^{35}}$ **6.** x^{17} **7.** x^7 **8.** $\frac{y^{12}}{9x^2z^{10}}$ **9.** 2.35×10^7 **10.** 0.000000047 **11.** 7.74×10^4 **12.** -70 **13.** $-4x^2 - 14x + 1$ **14.** $2x^2 + 14x - 64$ **15.** 14 **16.** $30x^5 + 40x^4 - 85x^3$ **17.** $x^2 - 36$ **18.** $20x^2 - x - 63$ **19.** $7x^5 + 11x^4 - 5x^3$ **20.** $6x - 29 + \frac{125}{x+3}$

CHAPTER 6

Quick Check 6.1 **1.** 4 **2.** 12 **3. a)** $4x^3$ **b)** x^3z^4 **4.** $6x^2(x^2 - 7x - 15)$ **5.** $3x^4(5x^3 - 10x + 1)$ **6. a)** $(x - 9)(5x + 14)$ **b)** $(x - 8)(7x - 6)$ **7.** $(x + 4)(x^2 + 7)$ **8.** $(x - 5)(2x - 9)$ **9.** $(x - 9)(4x^2 + 1)$ **10.** $3(x + 8)(x - 4)$

Section 6.1 **1.** factored **3.** smallest **5.** 2 **7.** 6 **9.** 4 **11.** x^3 **13.** a^2b^2 **15.** $2x^3$ **17.** $5a^2c$ **19.** $7(x - 2)$ **21.** $x(5x + 4)$ **23.** $4x(2x^2 + 5)$ **25.** $6(10x^2 + 6x + 1)$ **27.** $5x^3(4x^3 - 7x - 10)$ **29.** $m^5n^3(m^2 - n + mn^3)$ **31.** $10x^2y^4(x^3y - 3x^5 + 8y^6)$ **33.** $4x(-2x^2 + 3x - 4)$ **35.** $(2x - 7)(5x + 8)$ **37.** $(3x - 4)(x - 9)$ **39.** $(4x + 7)(5x - 1)$ **41.** $(x + 10)(x + 3)$ **43.** $(x - 9)(x^2 + 6)$ **45.** $(x - 5)(x - 12)$ **47.** $(x + 5)(3x + 4)$ **49.** $(x + 4)(7x - 6)$ **51.** $(x + 7)(3x + 1)$ **53.** $(x - 5)(2x - 1)$ **55.** $(x + 8)(x^2 + 6)$ **57.** $(x^2 + 3)(4x + 3)$ **59.** $2(x + 5)(x + 3)$ **61.** Example: $x^2 + 3x - 40$ **63.** Example: $2x^2 + x - 36$ **65.** Answers will vary; multiplication

Quick Review Exercises **1.** $x^2 + 11x + 28$ **2.** $x^2 - 20x + 99$ **3.** $x^2 - 3x - 40$ **4.** $x^2 - 4x - 60$

Quick Check 6.2 **1. a)** $(x + 2)(x + 5)$ **b)** $(x - 5)(x - 6)$ **2. a)** $(x + 12)(x - 3)$ **b)** $(x + 6)(x - 7)$ **3. a)** $(x + 6)(x - 1)$ **b)** $(x + 2)(x + 3)$ **c)** $(x - 2)(x - 3)$ **d)** $(x - 6)(x + 1)$ **4.** $(x + 5)^2$ **5.** Prime **6.** $5(x + 2)(x + 6)$ **7.** $-(x - 6)(x - 8)$ **8.** $(x - 8y)(x + 4y)$ **9.** $(xy + 12)(xy - 2)$

Section 6.2 **1.** leading coefficient **3.** prime **5.** $(x - 10)(x + 2)$ **7.** $(x + 9)(x - 4)$ **9.** Prime **11.** $(x + 6)(x + 8)$ **13.** $(x + 15)(x - 2)$ **15.** Prime **17.** $(x - 20)(x + 3)$ **19.** $(x + 12)(x - 1)$ **21.** $(x + 7)^2$ **23.** $(x + 7)(x + 13)$ **25.** $(x - 12)^2$ **27.** $(x - 5)(x - 8)$ **29.** $(x - 15)(x + 2)$ **31.** $(x - 4)(x - 5)$ **33.** $(x - 6)(x - 10)$ **35.** $6(x - 4)(x - 5)$ **37.** $5(x - 3)(x + 10)$ **39.** $-(x + 7)(x - 10)$ **41.** $x(x + 4)^2$ **43.** $-3x^3(x + 10)(x - 8)$ **45.** $10x^6(x + 11)(x - 9)$ **47.** $(x + 7y)(x - 6y)$ **49.** $(x + 4y)(x - 9y)$ **51.** $(xy + 8)(xy + 9)$ **53.** $5x^3(x + 3y)(x + 7y)$ **55.** $(x + 9)(x^3 + 5)$ **57.** Prime **59.** $(x + 2)(x + 20)$ **61.** $(x + 20)^2$ **63.** $4(x + 6)(x - 4)$ **65.** $x^3(x - 15)$ **67.** $(x - 7)(x - 9)$ **69.** Prime **71.** $(x - 25)(x + 4)$ **73.** 16 **75.** 60

Quick Review Exercises **1.** $2x^2 + 13x + 20$ **2.** $6x^2 - 37x + 56$ **3.** $3x^2 + 14x - 24$ **4.** $8x^2 + 30x - 27$

Quick Check 6.3 **1.** $(3x + 4)(x + 3)$ **2.** $(4x - 9)(x - 4)$ **3.** $(2x - 3)(8x - 7)$ **4.** $(x + 4)(2x - 7)$ **5.** $(4x - 3)(5x + 8)$ **6.** $9(x + 6)(x - 3)$

Section 6.3 **1.** $ax^2 + bx + c$, where $a \neq 1$ **3.** $(x + 3)(5x + 1)$ **5.** $(2x + 5)(2x + 7)$ **7.** $(3x - 2)(2x - 5)$ **9.** $(7x - 3)(2x + 3)$ **11.** $(2x + 3)(6x - 5)$ **13.** $(3x - 4)(4x - 3)$ **15.** Prime **17.** $(3x - 2)(x + 5)$ **19.** $(3x + 8)(4x - 7)$ **21.** $(4x - 3)^2$ **23.** $3(5x + 2)(x - 2)$ **25.** $8(2x + 5)(x - 1)$ **27.** $16(x - 2)^2$ **29.** $5(3x + 16)(x - 7)$ **31.** $(x + 1)(x + 13)$ **33.** $9(2x - 1)$ **35.** $(2x - 3)(x - 2)$ **37.** Prime **39.** $x^4(x - 6)$ **41.** $6x(x - 8)$ **43.** $-(x + 11)(x - 7)$ **45.** Prime **47.** $(x^2 - 7)(x^3 + 4)$ **49.** $(x + 6)(x^2 - 15)$ **51.** $(x - 12)^2$ **53.** $9(x - 6)(x + 1)$ **55.** $-(5x + 2)(x - 6)$ **57.** $(x - 9)(2x + 3)$ **59.** $(x + 15)(x - 3)$ **61.** $-3(x + 8)(x - 4)$ **63.** $(5x - 3)(3x + 5)$ **65.** $(2x + 1)(x + 4)$ **67.** Answers will vary.

Quick Review Exercises **1.** $x^2 - 64$ **2.** $16x^2 - 9$ **3.** $x^3 + 512$ **4.** $8x^3 - 125$

Quick Check 6.4 **1.** $(x + 5)(x - 5)$ **2.** $(9x^6 + 10y^8)(9x^6 - 10y^8)$ **3.** $7(x + 5)(x - 5)$ **4.** $(x^2 + y^2)(x + y)(x - y)$ **5.** $(x - 5)(x^2 + 5x + 25)$ **6.** $(x^4 - 2)(x^8 + 2x^4 + 4)$ **7.** $(x + 6)(x^2 - 6x + 36)$

Section 6.4 **1.** difference of squares **3.** multiples of 2 **5.** difference of cubes **7.** $(x + 8)(x - 8)$ **9.** $(x + 10)(x - 10)$ **11.** $(5a + 7)(5a - 7)$ **13.** Prime **15.** $(a + 4b)(a - 4b)$ **17.** $(5x + 8y)(5x - 8y)$ **19.** $(4 + x)(4 - x)$ **21.** $4(x + 7)(x - 7)$ **23.** Prime **25.** $5(x^2 + 4)$ **27.** $(x^2 + 4)(x + 2)(x - 2)$ **29.** $(x^2 + 9y)(x^2 - 9y)$ **31.** $(x - 1)(x^2 + x + 1)$ **33.** $(x - 2)(x^2 + 2x + 4)$ **35.** $(10 - y)(100 + 10y + y^2)$ **37.** $(a - 4b)(a^2 + 4ab + 16b^2)$ **39.** $(5x - 2y)(25x^2 + 10xy + 4y^2)$ **41.** $(x + y)(x^2 - xy + y^2)$ **43.** $(x + 2)(x^2 - 2x + 4)$ **45.** $(x^2 + 3)(x^4 - 3x^2 + 9)$ **47.** $(4m + n)(16m^2 - 4mn + n^2)$ **49.** $3(x - 5)(x^2 + 5x + 25)$ **51.** $5(4a + 3b)(16a^2 - 12ab + 9b^2)$ **53.** $(x - 8)^2$ **55.** $(x + 2)(x + 14)$ **57.** $11x^2(2x^2 - 5)$ **59.** $(2x - 5)(3x + 5)$ **61.** $x^3(x - 10)(x - 6)$ **63.** $-9(x^2 + 25)$ **65.** $-6(x - 4)(x - 6)$ **67.** $(x - 13)(x + 3)$ **69.** $(5x + 6y^4)(5x - 6y^4)$ **71.** Prime **73.** $(x + 8y^3)(x - 8y^3)$ **75.** $4x^2(2x^3 - 5x^2 - 13)$ **77.** $5(x^2 + 9)$ **79.** $(x + 5)(x + 7)$ **81.** $(x + 2)(3x - 19)$ **83.** $(x + 12)(x^2 - 6)$ **85.** $(x + 5y)(x^2 - 5xy + 25y^2)$ **87.** $x^5(2x + 3)^2$ **89.** d **91.** f **93.** g **95.** b **97.** Answers will vary; two terms, even exponents, coefficients are perfect squares

Quick Check 6.5 **1.** $-6(x - 10)(x + 1)$ **2.** $(6x^4y^7 + 5)(36x^8y^{14} - 30x^4y^7 + 25)$ **3.** $(x + 5)(x - 2)(x^2 + 2x + 4)$ **4.** $(x^2 + y)(x^4 - x^2y + y^2)(x^2 - y)(x^4 + x^2y + y^2)$ **5.** $(x^2 + 4)(x^4 - 4x^2 + 16)$ **6.** $2(4x - 3)(x + 6)$

Section 6.5 **1.** common factors **3.** sum of squares **5.** sum of cubes **7.** $x(x^6 + 18x^4 - 36)$ **9.** $4(21x^2 + x - 5)$ **11.** $a^2b^6(a^3b - 3a^2 + 8b^3)$ **13.** $(x + 6)(x^2 + 5)$ **15.** $(x - 7)(x^2 - 3)$ **17.** $(x - 13)(x + 5)$ **19.** $(x + 6)(x + 7)$ **21.** $(x + 15)(x - 11)$ **23.** Prime **25.** $(x - 2)(x - 21)$ **27.** $(x + 10)^2$ **29.** $(x + 12y)(x - 5y)$ **31.** $(xy + 1)(xy + 16)$ **33.** $4(x + 6)(x + 8)$ **35.** $3(x - 9)(x - 5)$ **37.** $-10(x - 9)(x + 2)$ **39.** $(3x - 8)(x + 6)$ **41.** $(3x + 4)(7x + 1)$ **43.** Prime **45.** $(4x - 11)(3x - 2)$ **47.** $(3x - 5)^2$ **49.** $6(3x - 7)(x - 1)$ **51.** $(x + 8)(x - 8)$ **53.** $(x^4 + 4y^3)(x^4 - 4y^3)$ **55.** $x^2(x + 6)(x - 6)$ **57.** Prime **59.** $(x - 10)(x^2 + 10x + 100)$ **61.** $(3x - 2y)(9x^2 + 6xy + 4y^2)$ **63.** $(x + 6)(x^2 - 6x + 36)$ **65.** $(x^2 + 9y)(x^4 - 9x^2y + 81y^2)$ **67.** $7(x + 3)(x^2 - 3x + 9)$ **69.** $(x - 9)(x + 2)(x - 2)$ **71.** $(x - 7)(2x + 1)(2x - 1)$ **73.** $(x + 16)(x + 1)(x^2 - x + 1)$ **75.** $(x^4 + 1)(x^2 + 1)(x + 1)(x - 1)$ **77.** $(x + y)(x^2 - xy + y^2)(x - y)(x^2 + xy + y^2)$ **79.** $(x + 4 + y)(x + 4 - y)$ **81.** $(x - 5 + y)(x - 5 - y)$ **83.** $(x + 3 + y)(x + 3 - y)$ **85.** $(2x + 5)(2x - 1)$ **87.** $(x - 6)(x^2 + 8)$ **89.** $(x + 15)(x - 4)$ **91.** Prime **93.** $(2x + 7)(4x^2 - 14x + 49)$ **95.** $(x + 7)(x + 14)$ **97.** $(4x^2 + 9)(2x + 3)(2x - 3)$ **99.** $6(x + 5)(x + 8)$ **101.** $(3x + 11y^3)(3x - 11y^3)$ **103.** $(x + 4)(x + 5)(x - 5)$ **105.** $(9x - 8y)(81x^2 + 72xy + 64y^2)$ **107.** $(xy - 6)(xy - 9)$ **109.** $(3x + 2)(x - 10)$ **111.** Prime **113.** $5x^2(x^3 + 4x - 7)$ **115.** $(8x + 13)(8x - 13)$

Quick Review Exercises **1.** $\{7\}$ **2.** $\{-3\}$ **3.** $\{\frac{12}{5}\}$ **4.** $\{-\frac{29}{2}\}$

Quick Check 6.6 **1. a)** $\{-8, 2\}$, **b)** $\{-\frac{9}{4}, 0\}$ **2.** $\{-4, 5\}$ **3.** $\{-8, -7\}$ **4.** $\{\frac{1}{3}, \frac{7}{2}\}$ **5.** $\{-9, 5\}$ **6.** $\{-7, 10\}$ **7.** $\{-9, 3\}$ **8.** $x^2 - 2x - 48 = 0$

Section 6.6 **1.** quadratic equation **3.** zero-factor **5.** $\{-7, 100\}$ **7.** $\{0, 12\}$ **9.** $\{-8, 6\}$ **11.** $\left\{-7, -\frac{4}{3}\right\}$ **13.** $\{2, 9\}$ **15.** $\{-11, -4\}$ **17.** $\{-10, 4\}$ **19.** $\{-3, 15\}$ **21.** $\{8\}$ **23.** $\{-9, 9\}$ **25.** $\{-7, 0\}$ **27.** $\{0, \frac{22}{5}\}$ **29.** $\{-9, 10\}$ **31.** $\{5, 7\}$ **33.** $\{-8, \frac{2}{3}\}$ **35.** $\{-\frac{2}{9}, \frac{1}{3}\}$ **37.** $\{-5, 7\}$ **39.** $\{4\}$ **41.** $\{-6, -3\}$ **43.** $\{-2, 9\}$ **45.** $\{-3, 3\}$ **47.** $\{-6, 6\}$ **49.** $\{-8, 8\}$ **51.** $\{-3, 10\}$ **53.** $\{-4, 3\}$ **55.** $\{-17\}$ **57.** $\{3, 6\}$ **59.** $\left\{-\frac{3}{2}, 9\right\}$ **61.** $\left\{\frac{5}{4}, 4\right\}$ **63.** $x^2 - 9x + 20 = 0$ **65.** $x^2 - 6x = 0$ **67.** $x^2 - 25 = 0$ **69.** $10x^2 - x - 2 = 0$ **71.** $x = -\frac{14}{3}$ **73.** $x = -29$ **75.** $x = \frac{28}{5}$ **77.** Answers will vary. One factor must equal 0; set each factor equal to 0.

Quick Review Exercises **1.** 23 **2.** 101 **3.** $-14b - 40$ **4.** 58

Quick Check 6.7 **1.** 495 **2.** 8, 9 **3.** −6, 12 **4. a)** 48 feet **b)** 3 seconds **c)** 1 second **d)** 64 feet

Section 6.7 **1.** quadratic function **3.** 104 **5.** −45 **7.** 0 **9.** 0 **11.** 147 **13.** 17 **15.** 160 **17.** $16a^2 + 16a - 20$ **19.** $a^2 + 3a - 35$ **21.** −7, 6 **23.** 4, 5 **25.** 9, 14 **27.** −7, 7 **29.** −8, 2 **31.** −8, −5 **33. a)** 5 seconds **b)** 336 feet **35. a)** 192 feet **b)** 5 seconds **c)** 1 second **d)** 256 feet **37. a)** \$70 **b)** 5 mugs **c)** 20 mugs **d)** \$80

Quick Check 6.8 **1.** 11 and 12 **2.** 9 and 12 **3.** Length: 15 feet; width: 7 feet **4.** 11 feet by 14 feet **5. a)** $h(t) = -16t^2 + 144t$ **b)** 9 seconds **c)** [0, 9] **6. a)** 5050, **b)** 9

Section 6.8 **1.** $x, x + 1, x + 2$ **3.** Area = Length · Width **5.** 11, 12 **7.** 22, 24 **9.** 15, 17 **11.** 8, 9 **13.** 20, 22 **15.** 6, 11 **17.** 6, 14 **19.** 14, 15 **21.** 3, 22 **23.** 8, 11 **25.** 14 years old **27.** 21 years old **29.** Length: 15 inches; width: 8 inches **31.** Length: 30 centimeters; width: 14 centimeters **33.** Length: 15 meters; width: 7 meters **35.** Length: 40 feet; width: 24 feet **37.** Length: 8 inches; width: 5 inches **39.** Length: 9 feet; width: 5 feet **41.** 8 meters **43.** Base: 8 feet; height: 21 feet **45.** 10 seconds **47.** 5 seconds **49.** 1 second, 3 seconds **51.** 8 seconds **53.** 80 feet/second **55.** 3 seconds **57.** 8 **59.** 20 **61.** Answers will vary. Length is 4 feet more than width, and each is increased by 5 feet. Area is 192 square feet.

Chapter 6 Review **1.** $3(x - 7)$ **2.** $x(x - 20)$ **3.** $5x^3(x^4 + 3x - 4)$ **4.** $a^3b(6a^2b - 10b^2 + 15a)$ **5.** $(x + 6)(x^2 + 4)$ **6.** $(x + 9)(x^2 - 7)$ **7.** $(x - 3)(x^2 - 11)$ **8.** $(x - 3)^2(x + 3)$ **9.** $(x - 4)^2$ **10.** $(x + 7)(x - 4)$ **11.** $(x - 3)(x - 8)$ **12.** $-5(x + 8)(x - 4)$ **13.** $(x + 5y)(x + 9y)$ **14.** Prime **15.** $(x + 2)(x + 15)$ **16.** $(x - 7)(x + 6)$ **17.** $(x - 8)(10x - 3)$ **18.** $(4x + 5)(x - 2)$ **19.** $6(3x - 1)(x + 1)$ **20.** $(2x - 3)^2$ **21.** $3(x + 5)(x - 5)$ **22.** $(2x + 5y)(2x - 5y)$ **23.** $(x + 2y)(x^2 - 2xy + 4y^2)$ **24.** Prime **25.** $(x - 6)(x^2 + 6x + 36)$ **26.** $7(x - 3)(x^2 + 3x + 9)$ **27.** $\{-7, 5\}$ **28.** $\{-\frac{1}{2}, \frac{8}{3}\}$ **29.** $\{-5, 5\}$ **30.** $\{4, 10\}$ **31.** $\{-8\}$ **32.** $\{-12, 9\}$ **33.** $\{-5, 14\}$ **34.** $\{0, 16\}$ **35.** $\{-15, -4\}$ **36.** $\{8, 10\}$ **37.** $\{-1, 16\}$ **38.** $\{-4, \frac{15}{2}\}$ **39.** $\{-6, 6\}$ **40.** $\{3, 12\}$ **41.** $\{-7, -3\}$ **42.** $\{-13, 4\}$ **43.** $x^2 - 13x + 42 = 0$ **44.** $x^2 - 16 = 0$ **45.** $5x^2 - 7x - 6 = 0$ **46.** $28x^2 + 27x - 10 = 0$ **47.** 9 **48.** −16 **49.** 8 **50.** 96 **51.** −6, 6 **52.** −9, 6 **53.** −10, −2 **54.** −4, 11 **55. a)** 80 feet **b)** 4 seconds **c)** 1 second **d)** 144 feet **56. a)** \$20.85 **b)** 400 **c)** \$4.85 **57.** 16, 18 **58.** Rosa: 7, Dale: 15 **59.** Length: 13 meters; width: 8 meters **60. a)** 8 seconds **b)** 2 seconds, 3 seconds

Chapter 6 Review Exercises: Worked-Out Solutions

3. GCF: $5x^3$

$$5x^7 + 15x^4 - 20x^3 = 5x^3(x^4 + 3x - 4)$$

5.
$$\begin{aligned} x^3 + 6x^2 + 4x + 24 &= x^2(x + 6) + 4(x + 6) \\ &= (x + 6)(x^2 + 4) \end{aligned}$$

10. $m \cdot n = -28, m + n = 3$
$$x^2 + 3x - 28 = (x + 7)(x - 4)$$

11. $m \cdot n = 24, m + n = -11$
$$x^2 - 11x + 24 = (x - 3)(x - 8)$$

15. $m \cdot n = 30, m + n = 17$
$$x^2 + 17x + 30 = (x + 2)(x + 15)$$

17. By grouping:
$$\begin{aligned} 10x^2 - 83x + 24 &= 10x^2 - 80x - 3x + 24 \\ &= 10x(x - 8) - 3(x - 8) \\ &= (x - 8)(10x - 3) \end{aligned}$$

21.
$$\begin{aligned} 3x^2 - 75 &= 3(x^2 - 25) \\ &= 3(x + 5)(x - 5) \end{aligned}$$

23.
$$\begin{aligned} x^3 + 8y^3 &= (x)^3 + (2y)^3 \\ &= (x + 2y)(x^2 - 2xy + 4y^2) \end{aligned}$$

25.
$$\begin{aligned} x^3 - 216 &= (x)^3 - (6)^3 \\ &= (x - 6)(x^2 + 6x + 36) \end{aligned}$$

30.
$$\begin{aligned} &x^2 - 14x + 40 = 0 \\ &(x - 4)(x - 10) = 0 \\ &x - 4 = 0 \quad \text{or} \quad x - 10 = 0 \\ &x = 4 \quad \text{or} \quad x = 10 \\ &\{4, 10\} \end{aligned}$$

41.
$$\begin{aligned} x(x + 10) &= -21 \\ x^2 + 10x &= -21 \\ x^2 + 10x + 21 &= 0 \\ (x + 3)(x + 7) &= 0 \\ x + 3 = 0 \quad &\text{or} \quad x + 7 = 0 \\ x = -3 \quad &\text{or} \quad x = -7 \\ \{-7, -3\} \end{aligned}$$

43. $\{6, 7\}$
$$\begin{aligned} &x = 6 \quad \text{or} \quad x = 7 \\ &x - 6 = 0 \quad \text{or} \quad x - 7 = 0 \\ &(x - 6)(x - 7) = 0 \\ &x^2 - 13x + 42 = 0 \end{aligned}$$

49.
$$\begin{aligned} f(-8) &= (-8)^2 + 10(-8) + 24 \\ &= 64 - 80 + 24 \\ &= 8 \end{aligned}$$

51.

$$\begin{aligned} f(x) &= 0 \\ x^2 - 36 &= 0 \\ (x+6)(x-6) &= 0 \\ x + 6 = 0 \quad &\text{or} \quad x - 6 = 0 \\ x = -6 \quad &\text{or} \quad x = 6 \end{aligned}$$

56. a)

$$\begin{aligned} f(800) &= 0.0001(800)^2 - 0.08(800) + 20.85 \\ &= 0.0001(640{,}000) - 0.08(800) + 20.85 \\ &= 64 - 64 + 20.85 \\ &= 20.85 \end{aligned}$$

b) The lowest point on the graph occurs when the number of toys is 400.

c)

$$\begin{aligned} f(400) &= 0.0001(400)^2 - 0.08(400) + 20.85 \\ &= 0.0001(160{,}000) - 0.08(400) + 20.85 \\ &= 16 - 32 + 20.85 \\ &= 4.85 \end{aligned}$$

57. First: x

Second: $x + 2$

$$\begin{aligned} x(x+2) &= 288 \\ x^2 + 2x &= 288 \\ x^2 + 2x - 288 &= 0 \\ (x+18)(x-16) &= 0 \\ x + 18 = 0 \quad &\text{or} \quad x - 16 = 0 \\ x = -18 \quad &\text{or} \quad x = 16 \end{aligned}$$

Omit the negative solution.

First: $x = 16$

Second: $x + 2 = 16 + 2 = 18$

The two numbers are 16 and 18.

59. Length: $2x - 3$

Width: x

$$\begin{aligned} x(2x-3) &= 104 \\ 2x^2 - 3x &= 104 \\ 2x^2 - 3x - 104 &= 0 \\ (2x+13)(x-8) &= 0 \\ 2x + 13 = 0 \quad &\text{or} \quad x - 8 = 0 \\ x = -\tfrac{13}{2} \quad &\text{or} \quad x = 8 \end{aligned}$$

Omit the negative solution.

Length: $2x - 3 = 2(8) - 3 = 13$

Width: $x = 8$

The length is 13 meters and the width is 8 meters.

Chapter 6 Test **1.** $x(x - 25)$ **2.** $(x - 3)(x^2 - 5)$ **3.** $(x - 4)(x - 5)$ **4.** $(x - 18)(x + 4)$ **5.** $3(x + 1)(x + 19)$ **6.** $(6x + 5)(x - 1)$ **7.** $(x + 5y)(x - 5y)$ **8.** $(x - 6y)(x^2 + 6xy + 36y^2)$ **9.** $\{-10, 10\}$ **10.** $\{4, 9\}$ **11.** $\{-12, 5\}$ **12.** $\{-6, 13\}$ **13.** $x^2 - 4 = 0$ **14.** $5x^2 + 21x - 54 = 0$ **15.** 120 **16.** -18 **17.** $-8, -6$ **18. a)** \$2.245 **b)** 8000 **c)** \$2.11 **19.** Length: 20 feet; width: 5 feet **20. a)** 5 seconds **b)** 3 seconds

CHAPTER 7

Quick Check 7.1 **1.** $\frac{1}{3}$ **2.** $\frac{7}{2}$ **3.** $-1, 9$ **4.** -2 **5.** All real numbers except 9 and -5: $(-\infty, -5) \cup (-5, 9) \cup (9, \infty)$ **6.** $\frac{5x}{7}$ **7.** $\frac{x+4}{x-8}$ **8.** $\frac{3x+1}{x+3}$ **9.** $-\frac{x+9}{1+x}$

Section 7.1 **1.** rational expression **3.** rational function **5.** lowest terms **7.** $\frac{3}{4}$ **9.** $\frac{2}{3}$ **11.** $\frac{26}{7}$ **13.** $-\frac{75}{26}$ **15.** 5 **17.** $-8, 7$ **19.** $-5, \frac{2}{3}$ **21.** $-6, 6$ **23.** $\frac{1}{3}$ **25.** $-\frac{8}{7}$ **27.** $\frac{21}{16}$ **29.** All real numbers except -10 and 0: $(-\infty, -10) \cup (-10, 0) \cup (0, \infty)$ **31.** All real numbers except -3 and 5: $(-\infty, -3) \cup (-3, 5) \cup (5, \infty)$ **33.** All real numbers except -6 and 3: $(-\infty, -6) \cup (-6, 3) \cup (3, \infty)$ **35.** Quadratic **37.** Rational **39.** Linear **41.** $\frac{1}{3x^3}$ **43.** $\frac{1}{x+6}$ **45.** $\frac{x+4}{x-3}$ **47.** $\frac{x}{x-10}$ **49.** $\frac{x-9}{x^2-5x+25}$ **51.** $\frac{x+6}{x-4}$ **53.** $\frac{3x+5}{x+7}$ **55.** $\frac{3(x+5)}{x-4}$ **57.** Not opposites **59.** Opposites **61.** Opposites **63.** $-\frac{7+x}{x-5}$ **65.** $-\frac{x}{x-11}$ **67.** $-\frac{x^2+2x+4}{2+x}$ **69. a)** 4 **b)** -1 **c)** 4 **71.** Answers will vary. Set denominator equal to 0 and solve. **73.** Answers will vary. Two differences are written in opposite order.

Quick Review Exercises **1.** $\frac{41}{60}$ **2.** $\frac{49}{120}$ **3.** $\frac{1}{6}$ **4.** $\frac{68}{231}$

Quick Check 7.2 **1.** $\frac{(x-6)(x+2)}{(x-4)(x+10)}$ **2.** $-\frac{6+x}{x+7}$ **3.** $-\frac{1}{7+x}$ **4.** $\frac{(x-2)(x+7)}{(x-1)(x-8)}$ **5.** $-(x+5)$ **6.** $\frac{x-6}{(x-1)(x-10)(x-2)}$

Section 7.2 **1.** factoring **3.** factored **5.** $\frac{x+6}{x-3}$ **7.** $-\frac{(11+x)(x-14)}{(x-12)(x-7)}$ **9.** $\frac{(x-11)(2x+1)}{(x-4)(x+2)}$ **11.** $\frac{(x-11)(x+9)}{(x+4)(x+6)}$ **13.** $\frac{x^2+2x+4}{x-6}$ **15.** $\frac{(3-x)(x+4)}{(x-10)(x+7)}$ **17.** 1 **19.** $-\frac{x(2x-9)}{(x+10)(x-6)}$ **21.** $\frac{2}{x+4}$ **23.** $\frac{(x-11)(x+5)}{(x+12)(x-4)}$ **25.** $\frac{(x+7)(x+5)}{(x-6)(x-9)}$ **27.** $\frac{(x-3)(x+12)}{(x-9)(x+3)}$ **29.** $-\frac{(x-4)(x+11)}{x(x-2)}$ **31.** $\frac{(x-1)(x-4)}{(x+5)(x+6)}$ **33.** 1 **35.** $-\frac{(x+7)(x-10)}{(x-7)(9+x)}$ **37.** $\frac{x+4}{4x+1}$ **39.** $-\frac{(x+7)(x-3)}{x(x-8)}$ **41.** $\frac{x-1}{(x-5)^2}$ **43.** $\frac{x(x-9)}{(x-1)(x+3)}$ **45.** $\frac{(x-3)(x+11)}{(2-x)(x+2)}$ **47.** $\frac{(x-9)(x+4)}{x(x-12)}$ **49.** $-\frac{x+1}{x-1}$ **51.** $\frac{3x+2}{2x-7}$ **53.** $\frac{x-12}{x-8}$ **55.** $-\frac{x-8}{12+x}$ **57.** $\frac{(x+12)(x+8)}{(x-4)(2x-1)}$ **59.** $\frac{(x-6)(x-8)}{x(x-3)}$ **61.** Answers will vary. Invert divisor and rewrite as multiplication.

Quick Review Exercises **1.** $9x + 3$ **2.** $2x^2 - 6x - 29$ **3.** $3x - 42$ **4.** $-2x^2 - 2x - 29$

Quick Check 7.3 **1.** $\frac{21}{2x+7}$ **2.** $\frac{5}{x-4}$ **3.** $\frac{x+4}{x-3}$ **4.** $\frac{4}{3(x+6)}$ **5.** $-\frac{x-3}{8+x}$ **6.** $\frac{1}{5}$ **7.** $\frac{x+9}{x+4}$

Section 7.3 **1.** numerators **3.** first term **5.** $\frac{13}{x+3}$ **7.** $\frac{x+10}{x-5}$ **9.** $\frac{1}{x-4}$ **11.** $\frac{x-1}{x-5}$ **13.** $\frac{x+9}{x-8}$ **15.** $\frac{2(x-12)}{x-6}$ **17.** $\frac{8}{x}$ **19.** $\frac{x+2}{x-9}$ **21.** $\frac{4}{x+1}$ **23.** $\frac{x-7}{x+7}$ **25.** 2 **27.** $\frac{1}{x-10}$ **29.** $\frac{2}{x-4}$ **31.** $\frac{x+2}{x+8}$ **33.** $\frac{x+12}{3x+2}$ **35.** $\frac{(x-3)(x-9)}{(x-5)(x+3)}$ **37.** $-\frac{11}{x+8}$ **39.** $\frac{(x-7)(x+2)}{(x-2)(x-9)}$ **41.** 2 **43.** $x+11$ **45.** 4 **47.** $\frac{x+3}{2}$ **49.** $\frac{(x-7)(x-1)}{3(x+6)(x+4)}$ **51.** $\frac{x-18}{x+10}$ **53.** $\frac{2(x+4)}{x-9}$ **55.** $-\frac{(11+x)(x+3)}{(x-4)(x+8)}$ **57.** $\frac{3(x-10)}{x+4}$ **59.** $\frac{2(3x+2)}{3x-2}$ **61.** $\frac{x+9}{x-9}$ **63.** $\frac{(x+10)(x+3)}{x(x-2)}$ **65.** $\frac{x-3}{x+2}$ **67.** $3x$ **69.** $4x+7$ **71.** Answers will vary. Multiply one denominator by -1.

Quick Check 7.4 **1.** $36r^2s^6$ **2.** $(x-8)(x-5)(x+4)$ **3.** $(x+9)^2(x-9)$ **4.** $\frac{11x+18}{(x-2)(x+6)}$ **5.** $\frac{x+5}{(x-5)(x+3)}$ **6.** $\frac{2(x-1)}{(x+2)(x-6)}$

Section 7.4 **1.** LCD **3.** $6a$ **5.** $(x-9)(x+5)$ **7.** $(x+3)(x-3)(x-6)$ **9.** $(x+2)^2(x+5)$ **11.** $\frac{19x}{24}$ **13.** $\frac{1}{20n}$ **15.** $\frac{10n+13m^2}{m^7n^3}$ **17.** $\frac{2(4x+13)}{(x+2)(x+4)}$ **19.** $\frac{6(x-7)}{(x+3)(x-3)}$ **21.** $\frac{11x+31}{(x+2)(x+5)(x-3)}$ **23.** $\frac{-4x+37}{(x-6)(x+1)(x+7)}$ **25.** $\frac{14}{(x-10)(x+4)}$ **27.** $-\frac{1}{(x-5)(x-4)}$ **29.** $\frac{6}{(x+2)(x-4)}$ **31.** $\frac{x+6}{(x+3)(x+4)}$ **33.** $\frac{x^2-3x-12}{(x-5)(x+4)(x-6)}$ **35.** $\frac{x^2+11x+110}{(x+10)(x-5)(x+5)}$ **37.** $\frac{x-1}{x(x+7)}$ **39.** $\frac{(x-5)(x+8)}{(x+7)(x+4)(x-8)}$ **41.** $\frac{2x}{(x-2)(x+6)}$ **43.** $\frac{13}{(x-6)(x+7)}$ **45.** $\frac{12}{(x-7)(x-3)}$ **47.** $\frac{2(x+4)}{(x+2)(x+5)}$ **49.** $\frac{2x-9}{x-1}$ **51.** $\frac{x+3}{(x-4)(x+10)}$ **53.** $\frac{x-7}{x}$ **55.** $\frac{2x+7}{6-x}$ **57.** $\frac{(3x-4)(x-7)}{x(x^2-2x+4)}$ **59.** $\frac{x-10}{x+9}$ **61.** Answers will vary. Factor each denominator and list all factors.

Quick Check 7.5 **1.** $\frac{31}{38}$ **2.** $\frac{x+5}{x}$ **3.** $\frac{x+8}{8x}$ **4.** $\frac{11}{x+7}$ **5.** $\frac{(x-8)(x-2)}{(x+10)(x+1)}$

Section 7.5 **1.** complex fraction **3.** $\frac{3}{68}$ **5.** $\frac{39}{38}$ **7.** $\frac{7(5x+3)}{5(7x+4)}$ **9.** $\frac{6}{x}$ **11.** $\frac{2}{3}$ **13.** $\frac{3x}{x-5}$ **15.** $\frac{6}{x+6}$ **17.** $\frac{15}{x+5}$ **19.** $\frac{x-4}{x+5}$ **21.** $-\frac{2(x-2)}{2+x}$ **23.** $\frac{x+5}{x-9}$ **25.** $-\frac{(2x-9)(x+7)}{x(2x-1)}$ **27.** $\frac{x-1}{x-7}$ **29.** $\frac{x^2+6x+14}{(x-6)(x+4)(x-2)}$ **31.** $\frac{x(x+2)}{(3x-4)(x+5)}$ **33.** $\frac{x+6}{x(x+7)}$ **35.** 5 **37.** $\frac{5}{x-3}$ **39.** $\frac{2x+3}{x}$ **41.** $-\frac{8x}{8+x}$ **43.** $\frac{12(x+5)}{(x+3)(x+4)(x+6)}$ **45.** $\frac{x^2-4x+16}{x+7}$ **47.** $\frac{x+11}{x-8}$ **49.** $\frac{x+6}{x+1}$ **51.** $\frac{(x^2+10x+100)(x+5)}{(x+11)(3x-7)}$
53. Answers will vary. The numerator and denominator are not polynomials; they contain fractions.

Quick Check 7.6 **1.** $\{20\}$ **2.** $\{-3, 8\}$ **3.** $\varnothing$ **4.** $\{-3, 6\}$ **5.** $\{4\}$ **6.** $x = \frac{2yz}{4y-3z}$ **7.** $x = \frac{3y}{4y-5}$

Section 7.6 **1.** rational equation **3.** extraneous solution **5.** $\{5\}$ **7.** $\{12\}$ **9.** $\{28\}$ **11.** $\{3, 4\}$ **13.** $\{2, 6\}$ **15.** $\{2, 5\}$ **17.** $\{-5, 5\}$ **19.** $\{3\}$ **21.** $\varnothing$ **23.** $\{2\}$ **25.** $\{7\}$ **27.** $\{16\}$ **29.** $\{-5, 12\}$ **31.** $\{2\}$ **33.** $\varnothing$ **35.** $\{-2, 5\}$ **37.** $\left\{\frac{7}{2}, 10\right\}$ **39.** $\{1\}$ **41.** $\{-1, 15\}$ **43.** $\{-7, -6\}$ **45.** $\{-6, 20\}$ **47.** $\{5\}$ **49.** $\left\{-\frac{1}{2}, 7\right\}$ **51.** $\left\{-5, \frac{1}{6}\right\}$ **53.** $\{4\}$ **55.** $\{-5, -2\}$ **57.** $\{-10\}$ **59.** $W = \frac{A}{L}$ **61.** $x = -\frac{5y}{2y-1}$ **63.** $x = \frac{8y-9}{3y-2}$ **65.** $r = \frac{2x+y}{2}$ **67.** $x = \frac{y-y_1+mx_1}{m}$ **69. a)** 116 **b)** $x = -15$ **71.** Answers will vary. The solution causes the denominator to equal 0.

Quick Check 7.7 **1.** 10 **2.** 3 and 12 **3.** $12\frac{8}{11}$ minutes **4.** 24 hours **5.** 75 mph **6.** 6 mph **7.** $360 **8.** $163\frac{3}{4}$ minutes **9.** 864

Section 7.7 **1.** reciprocal **3.** divided **5.** vary directly **7.** 15 **9.** 20 **11.** 2 **13.** 6, 9 **15.** 10, 20 **17.** 6 minutes **19.** $10\frac{2}{7}$ hours **21.** $17\frac{1}{7}$ minutes **23.** 28 hours **25.** 18 hours **27.** 6 hours **29.** 6 hours **31.** 10 mph **33.** 6 mph **35.** 40 kilometers/hour **37.** 150 mph **39.** 80 **41.** 14 **43.** 294 **45.** 108 **47.** $190 **49.** 3 amperes **51.** $15 **53.** 18 amperes **55.** $7\frac{1}{2}$ foot-candles **57.** Answers will vary. One person takes 8 hours, and another person takes 10 hours. How long does it take the two of them together?

Chapter 7 Review **1.** $-\frac{1}{5}$ **2.** $\frac{3}{8}$ **3.** $\frac{5}{2}$ **4.** $\frac{3}{2}$ **5.** -9 **6.** $-6, 9$ **7.** $\frac{1}{x-3}$ **8.** $\frac{x+10}{x-5}$ **9.** $-\frac{6+x}{x-1}$ **10.** $\frac{x+3}{4x+3}$ **11.** $\frac{5}{9}$ **12.** $\frac{17}{22}$
13. All real numbers except -6 and 0; $(-\infty, -6) \cup (-6, 0) \cup (0, \infty)$ **14.** All real numbers except 1 and 4; $(-\infty, 1) \cup (1, 4) \cup (4, \infty)$ **15.** $\frac{x}{x+5}$ **16.** $\frac{(x+9)(x+8)}{(x-7)(x-2)}$ **17.** $\frac{x-4}{x-6}$ **18.** $-\frac{(4+x)(7x-1)}{x+6}$ **19.** $\frac{(x+1)(x+5)}{(x-1)(x+3)}$ **20.** $\frac{(x+8)(7-x)}{(x-6)(x+3)}$ **21.** $\frac{x+6}{x-4}$ **22.** $\frac{(x-3)(x+12)}{(x-11)(x+8)}$ **23.** $\frac{(x+1)(x-1)}{x(3x+2)}$ **24.** $-\frac{(3+x)(x+2)}{x-5}$ **25.** $\frac{x(x-6)}{(x+5)(x-9)}$ **26.** $\frac{(x+2)(x+3)}{(x+1)(x-10)}$ **27.** $\frac{12}{x+6}$ **28.** $\frac{x-3}{x+7}$ **29.** $\frac{x+9}{x+4}$ **30.** 6 **31.** $x+6$ **32.** $\frac{x-6}{x-5}$ **33.** $\frac{7}{(x+4)(x-3)}$ **34.** $\frac{5}{(x-2)(x-7)}$ **35.** $\frac{x+2}{(x+5)(x-4)}$ **36.** $\frac{x-3}{(x+3)(x+9)}$ **37.** $\frac{2(x+4)}{(x+6)(x-2)}$ **38.** $\frac{x+5}{(x+1)(x-5)}$

39. $\frac{x+3}{(x-1)(x+7)}$ **40.** $\frac{x}{x+9}$ **41.** $\frac{x-11}{x+11}$ **42.** $\frac{x+7}{x-2}$ **43.** $\frac{(x-3)(x+3)}{(x-2)(x+6)}$ **44.** $\left\{\frac{72}{13}\right\}$ **45.** $\{3,4\}$ **46.** $\{16\}$ **47.** $\{-8,5\}$ **48.** $\{2\}$ **49.** $\{-3\}$ **50.** $\{-2,5\}$ **51.** $\{-2,-1\}$ **52.** $h=\frac{2A}{b}$ **53.** $x=\frac{7y}{4y-3}$ **54.** $r=\frac{15x-10y}{6}$ **55.** 8 **56.** 5, 10 **57.** 20 minutes **58.** 84 minutes **59.** 60 mph **60.** 6 mph **61.** 130 calories **62.** 225 feet **63.** 675 pounds **64.** 80 foot-candles

Chapter 7 Review Exercises: Worked-Out Solutions

4.
$$\frac{(-4)^2-5(-4)-15}{(-4)^2+17(-4)+66}=\frac{16+20-15}{16-68+66}=\frac{21}{14}=\frac{3}{2}$$

6.
$$x^2-3x-54=0$$
$$(x+6)(x-9)=0$$
$$x=-6 \quad\text{or}\quad x=9$$

8.
$$\frac{x^2+6x-40}{x^2-9x+20}=\frac{(x+10)(x-4)}{(x-4)(x-5)}=\frac{x+10}{x-5}$$

11.
$$r(3)=\frac{(3)+7}{(3)^2+8(3)-15}=\frac{3+7}{9+24-15}=\frac{10}{18}=\frac{5}{9}$$

13.
$$x^2+6x=0$$
$$x(x+6)=0$$
$$x=0 \quad\text{or}\quad x=-6$$

All real numbers except -6 and 0.

16.
$$\frac{x^2+14x+45}{x^2-6x-7}\cdot\frac{x^2+9x+8}{x^2+3x-10}=\frac{(x+5)(x+9)}{(x-7)(x+1)}\cdot\frac{(x+1)(x+8)}{(x+5)(x-2)}=\frac{(x+9)(x+8)}{(x-7)(x-2)}$$

19.
$$f(x)\cdot g(x)=\frac{x-6}{x-1}\cdot\frac{x^2+6x+5}{x^2-3x-18}=\frac{x-6}{x-1}\cdot\frac{(x+1)(x+5)}{(x-6)(x+3)}=\frac{(x+1)(x+5)}{(x-1)(x+3)}$$

22.
$$\frac{x^2-x-6}{x^2-19x+88}\div\frac{x^2+10x+16}{x^2+4x-96}=\frac{(x-3)(x+2)}{(x-11)(x-8)}\cdot\frac{(x+12)(x-8)}{(x+8)(x+2)}=\frac{(x-3)(x+12)}{(x-11)(x+8)}$$

28.
$$\frac{x^2+10x}{x^2+15x+56}-\frac{5x+24}{x^2+15x+56}=\frac{x^2+10x-5x-24}{x^2+15x+56}=\frac{x^2+5x-24}{x^2+15x+56}=\frac{(x+8)(x-3)}{(x+7)(x+8)}=\frac{x-3}{x+7}$$

32.
$$\frac{x^2-3x-12}{x^2-25}-\frac{2x-18}{25-x^2}=\frac{x^2-3x-12}{x^2-25}+\frac{2x-18}{x^2-25}=\frac{x^2-3x-12+2x-18}{x^2-25}=\frac{x^2-x-30}{x^2-25}=\frac{(x-6)(x+5)}{(x+5)(x-5)}=\frac{x-6}{x-5}$$

33.
$$\frac{5}{x^2+3x-4}+\frac{2}{x^2-4x+3}=\frac{5}{(x+4)(x-1)}+\frac{2}{(x-3)(x-1)}$$
$$=\frac{5}{(x+4)(x-1)}\cdot\frac{x-3}{x-3}+\frac{2}{(x-3)(x-1)}\cdot\frac{x+4}{x+4}$$
$$=\frac{5x-15+2x+8}{(x+4)(x-1)(x-3)}=\frac{7x-7}{(x+4)(x-1)(x-3)}=\frac{7(x-1)}{(x+4)(x-1)(x-3)}=\frac{7}{(x+4)(x-3)}$$

39.
$$f(x)-g(x)=\frac{x+2}{x^2+4x-5}-\frac{1}{x^2+12x+35}=\frac{x+2}{(x+5)(x-1)}-\frac{1}{(x+5)(x+7)}$$
$$=\frac{x+2}{(x+5)(x-1)}\cdot\frac{x+7}{x+7}-\frac{1}{(x+5)(x+7)}\cdot\frac{x-1}{x-1}$$
$$=\frac{(x^2+9x+14)-(x-1)}{(x+5)(x-1)(x+7)}=\frac{x^2+9x+14-x+1}{(x+5)(x-1)(x+7)}=\frac{x^2+8x+15}{(x+5)(x-1)(x+7)}$$
$$=\frac{(x+3)(x+5)}{(x+5)(x-1)(x+7)}=\frac{(x+3)}{(x-1)(x+7)}$$

42.
$$\frac{1+\frac{2}{x}-\frac{35}{x^2}}{1-\frac{7}{x}+\frac{10}{x^2}}=\frac{1+\frac{2}{x}-\frac{35}{x^2}}{1-\frac{7}{x}+\frac{10}{x^2}}\cdot\frac{x^2}{x^2}=\frac{1\cdot x^2+\frac{2}{x}\cdot x^2-\frac{35}{x^2}\cdot x^2}{1\cdot x^2-\frac{7}{x}\cdot x^2+\frac{10}{x^2}\cdot x^2}=\frac{x^2+2x-35}{x^2-7x+10}=\frac{(x+7)(x-5)}{(x-5)(x-2)}=\frac{x+7}{x-2}$$

45.
$$x-5+\frac{12}{x}=2$$
$$x\cdot x-5\cdot x+\frac{12}{x}\cdot x=2\cdot x$$
$$x^2-5x+12=2x$$
$$x^2-7x+12=0$$
$$(x-3)(x-4)=0$$
$$x=3 \quad\text{or}\quad x=4$$
$$\{3,4\}$$

49.
$$\frac{x}{x+5}+\frac{3}{x-7}=\frac{36}{x^2-2x-35}$$
$$\frac{x}{x+5}+\frac{3}{x-7}=\frac{36}{(x+5)(x-7)}$$
$$(x+5)(x-7)\cdot\frac{x}{x+5}+(x+5)(x-7)\cdot\frac{3}{x-7}=(x+5)(x-7)\cdot\frac{36}{(x+5)(x-7)}$$
$$x(x-7)+3(x+5)=36$$
$$x^2-7x+3x+15=36$$
$$x^2-4x+15=36$$
$$x^2-4x-21=0$$
$$(x+3)(x-7)=0$$
$$x=-3 \quad\text{or}\quad x=7$$

$x=7$ is an extraneous solution.

$\{-3\}$

53.
$$y=\frac{3x}{4x-7}$$
$$y(4x-7)=3x$$
$$4xy-7y=3x$$
$$4xy-3x=7y$$
$$x(4y-3)=7y$$
$$x=\frac{7y}{4y-3}$$

58.

Pipe	Time alone	Work-rate	Time Working	Portion Completed
Smaller	$t + 42$ minutes	$\frac{1}{t+42}$	28 minutes	$\frac{28}{t+42}$
Larger	t minutes	$\frac{1}{t}$	28 minutes	$\frac{28}{t}$

$$\frac{28}{t+42} + \frac{28}{t} = 1$$
$$t(t+42)\cdot\frac{28}{t+42} + t(t+42)\cdot\frac{28}{t} = 1\cdot t(t+42)$$
$$28t + 28(t+42) = t(t+42)$$
$$28t + 28t + 1176 = t^2 + 42t$$
$$56t + 1176 = t^2 + 42t$$
$$0 = t^2 - 14t - 1176$$
$$0 = (t-42)(t+28)$$

$t = 42$ or $t = -28$ (Omit negative solution.)
It would take the smaller pipe $42 + 42 = 84$ minutes.

60. Rate upstream: $r - 2$, Rate downstream: $r + 2$

$$\frac{10}{r+2} = \frac{5}{r-2}$$
$$(r-2)(r+2)\cdot\frac{10}{r+2} = (r-2)(r+2)\cdot\frac{5}{r-2}$$
$$10(r-2) = 5(r+2)$$
$$10r - 20 = 5r + 10$$
$$5r = 30$$
$$r = 6$$

Dominique can paddle 6mph in still water.

Chapter 7 Test **1.** $-\frac{5}{6}$ **2.** $-3, -7$ **3.** $\frac{x-5}{x+10}$ **4.** -13 **5.** $\frac{x-1}{x-3}$ **6.** $-\frac{(x-5)^2}{(x+3)(5+x)}$ **7.** 3 **8.** $x+8$ **9.** $\frac{11}{(x+7)(x-4)}$ **10.** $\frac{x-6}{(x-5)(x-2)}$ **11.** $\frac{x+9}{x+7}$ **12.** $\frac{x-5}{(x+9)(x+11)}$ **13.** $\frac{(x+5)(x+9)}{x(x+6)}$ **14.** $\frac{x+6}{x+2}$ **15.** $\frac{(x-10)(x+3)}{(x-1)(x+8)(x+2)}$ **16.** $\{36\}$ **17.** $\{-2, 9\}$ **18.** $y = \frac{5x}{7x-2}$ **19.** 90 minutes **20.** 20 mph

Cumulative Review Chapters 5–7 **1.** x^7 **2.** $108m^{16}n^{21}$ **3.** $\frac{625a^{24}b^{44}}{c^{28}}$ **4.** t^{21} **5.** $\frac{a^{48}}{b^{56}}$ **6.** $\frac{1}{x^{35}}$ **7.** 3.3×10^{11} **8.** 7.14×10^{-13} **9.** 0.001 second **10.** -44 **11.** $2x^2 - 7x - 59$ **12.** $-x^2 - 12x + 22$ **13.** 67 **14.** $15x^3 - 24x^2 + 36x$ **15.** $12x^2 - 53x + 56$ **16.** $5x^7 + 12x^4 - 9x$ **17.** $x + 7 - \frac{3}{x+6}$ **18.** $(x-4)(x^2+6)$ **19.** $2(x-5)(x-8)$ **20.** $(x+9)(x-4)$ **21.** $(2x-5)(x+2)$ **22.** $(x-5)(x^2+5x+25)$ **23.** $(x+9)(x-9)$ **24.** $\{-7, 7\}$ **25.** $\{3, 8\}$ **26.** $\{-7\}$ **27.** $\{-10, 2\}$ **28.** $x^2 - 5x - 14 = 0$ **29.** $-10, 10$ **30.** $-7, 5$ **31. a)** 240 feet **b)** 6 seconds **c)** 2 seconds **d)** 256 feet **32.** 12, 14 **33.** Length: 19 meters; width: 7 meters **34.** 5, 9 **35.** $\frac{x-6}{x+7}$ **36.** $\frac{3}{7}$ **37.** $\frac{(x-3)(x-2)}{(x+11)(x+2)}$ **38.** $\frac{(x+10)(x-2)}{(x+7)(x+1)}$ **39.** $\frac{3}{x+2}$ **40.** $\frac{2(4x-13)}{(x+2)(x-1)(x-5)}$ **41.** $\frac{x-2}{(x-7)(x+3)}$ **42.** $\frac{x+8}{(x+6)(x+5)}$ **43.** $\frac{x-9}{x+8}$ **44.** $\frac{x-4}{x+8}$ **45.** $\{5\}$ **46.** $\{0, 18\}$ **47.** $\{-3\}$ **48.** $x = \frac{y}{2y-1}$ **49.** 10 **50.** 12 hours

CHAPTER 8

ick Check 8.1 **1.** $\{-3\}$ **2.** $\{3\}$ **3.** $\left\{\frac{31}{5}\right\}$ **4.** $\mathbb{R}$ **5.** $\varnothing$ **6.** $\{-7, 7\}$ **7.** $\{-\frac{13}{3}, -1\}$ **8.** $\{\frac{1}{2}, \frac{9}{2}\}$ **9.** $\varnothing$ **10.** $\{-22, -\frac{4}{3}\}$

.1 **1.** equation **3.** LCM **5.** the empty set **7.** $X = a, X = -a$ **9.** $\{-4\}$ **11.** $\{-\frac{8}{5}\}$ **13.** $\{-4\}$ **15.** $\{-11\}$ **17.** $\{\frac{23}{3}\}$ **19.** $\{-18\}$ $\{-\frac{15}{4}\}$ **25.** $\{-2\}$ **27.** $\{\frac{59}{20}\}$ **29.** $\{-\frac{1}{3}\}$ **31.** $\{-2\}$ **33.** $\{12\}$ **35.** $\left\{\frac{9}{4}\right\}$ **37.** $\mathbb{R}$ **39.** $\varnothing$ **41.** $\mathbb{R}$ **43.** $\{-2, 2\}$ **45.** $\{-19, 7\}$ $\varnothing$ **51.** $\{-22, 14\}$ **53.** $\{-2, 8\}$ **55.** $\{0, \frac{11}{2}\}$ **57.** $\{-5, \frac{11}{3}\}$ **59.** $\{-9, 3\}$ **61.** $\{-18, 4\}$ **63.** $\{-1, 21\}$ **65.** $\{-18, -8\}$ **67.** $\varnothing$ **73.** $\left\{-\frac{75}{4}\right\}$ **75.** $\{\frac{5}{11}, \frac{17}{3}\}$ **77.** $\{-\frac{29}{3}, -1\}$ **79.** $\{-2\}$ **81.** $|x| = 2$; answers will vary. **83.** $|x + 1| = 4$; answers will vary. $- 7|$ cannot equal -2.

-32 **2.** -120 **3.** $\frac{17}{13}$ **4.** 10

Quick Check 8.2 **1.** $x \le 2$, $(-\infty, 2]$ **2.** $x \ge -2$, $[-2, \infty)$ **3.** $x < 5$, $(-\infty, 5)$ **4.** $-5 \le x \le -2$, $[-5, -2]$ **5.** $x < -6$ or $x > \frac{9}{2}$, $(-\infty, -6) \cup \left(\frac{9}{2}, \infty\right)$ **6.** $-4 \le x \le -2$, $[-4, -2]$ **7.** $-2 < x < 3$, $(-2, 3)$ **8.** $\varnothing$ **9.** $x < -2$ or $x > 6$, $(-\infty, -2) \cup (6, \infty)$ **10.** $x < -6$ or $x > -1$, $(-\infty, -6) \cup (-1, \infty)$ **11.** $\mathbb{R}$, $(-\infty, \infty)$

Section 8.2 **1.** linear inequality **3.** interval **5.** compound **7.** $x \le 8$, $(-\infty, 8]$ **9.** $x \ge -4$, $[-4, \infty)$ **11.** $x > -5$, $(-5, \infty)$ **13.** $x > -\frac{7}{4}$, $\left(-\frac{7}{4}, \infty\right)$ **15.** $x \ge 18$, $[18, \infty)$ **17.** $8 \le x \le 14$, $[8, 14]$ **19.** $-2 \le x \le 7$, $[-2, 7]$ **21.** $-7 < x < -\frac{9}{2}$, $\left(-7, -\frac{9}{2}\right)$ **23.** $-\frac{32}{3} < x < -4$, $\left(-\frac{32}{3}, -4\right)$ **25.** $x < 2$ or $x > 8$, $(-\infty, 2) \cup (8, \infty)$ **27.** $x \le -7$ or $x \ge 9$, $(-\infty, -7] \cup [9, \infty)$ **29.** $x < -4$ or $x > 6$, $(-\infty, -4) \cup (6, \infty)$ **31.** $-5 < x < 1$, $(-5, 1)$ **33.** $-1 \le x \le 11$, $[-1, 11]$ **35.** $-\frac{7}{2} \le x \le -1$, $\left[-\frac{7}{2}, -1\right]$ **37.** $-1 < x < \frac{5}{3}$, $\left(-1, \frac{5}{3}\right)$ **39.** $\varnothing$, **41.** $x < 3$ or $x > 5$, $(-\infty, 3) \cup (5, \infty)$ **43.** $x \le -9$ or $x \ge -3$, $(-\infty, -9] \cup [-3, \infty)$ **45.** $\mathbb{R}$, $(-\infty, \infty)$ **47.** $x < \frac{3}{2}$ or $x > 7$, $\left(-\infty, \frac{3}{2}\right) \cup (7, \infty)$ **49.** $x < -5$ or $x > 2$, $(-\infty, -5) \cup (2, \infty)$ **51.** $-10 \le x \le 2$, $[-10, 2]$ **53.** $x \le -8$ or $x \ge 7$, $(-\infty, -8] \cup [7, \infty)$ **55.** $\mathbb{R}$, $(-\infty, \infty)$ **57.** $x \ge -8$, $[-8, \infty)$ **59.** $\varnothing$, **61.** $x \le -9$ or $x \ge 4$, $(-\infty, -9] \cup [4, \infty)$ **63.** $4 \le x < 8$, $[4, 8)$ **65.** $|x| < 2$ **67.** $|x - 5| > 4$ **69.** Answers will vary. An absolute value cannot be negative.

Quick Check 8.3 **1.** $(-2, 0)$, $(0, -5)$

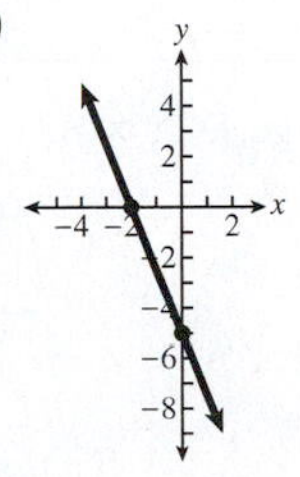

2.

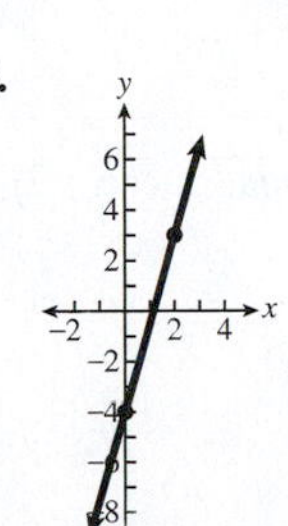

3.

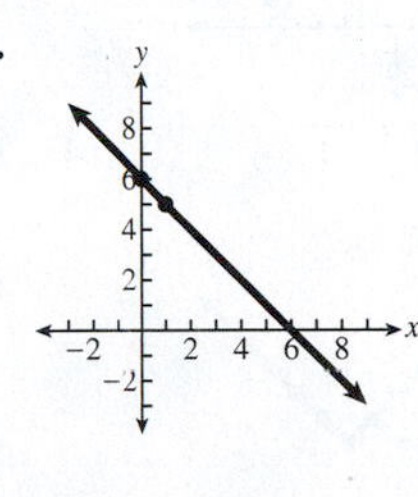

4.

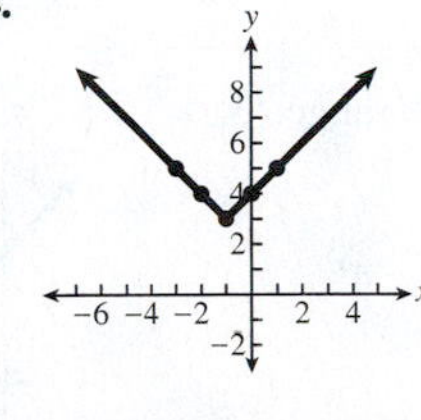

5. Domain: $(-\infty, \infty)$; range: $[-4, \infty)$

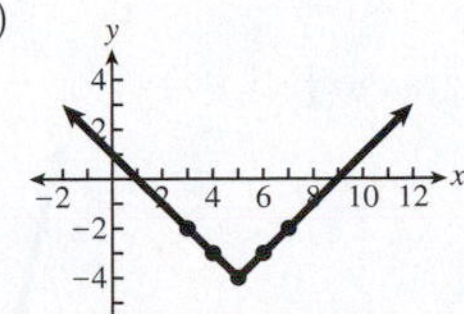

6. Domain: $(-\infty, \infty)$; range: $(-\infty, -5]$

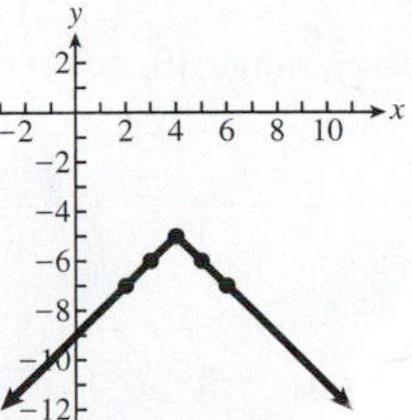

7. $f(x) = |x + 6| - 3$

Section 8.3 **1.** absolute value function **3.** y-intercept **5.** $(-6, 0), (0, -8)$

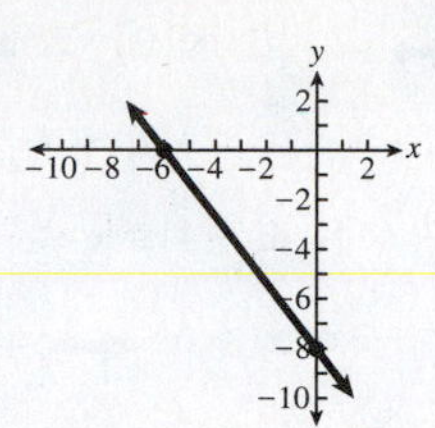

7. $(5, 0), (0, -2)$

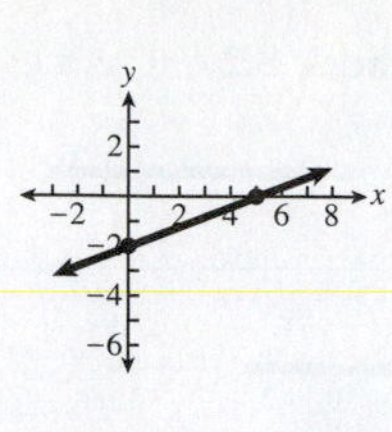

9. $(9, 0), (0, 6)$

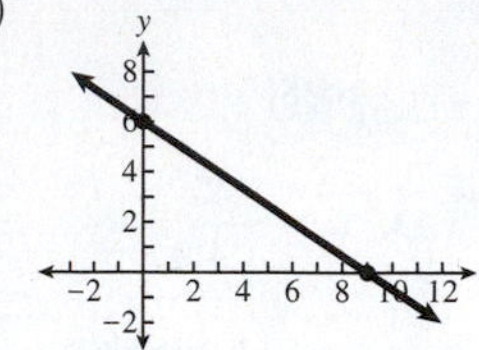

11. $\left(\frac{3}{2}, 0\right), (0, -4)$

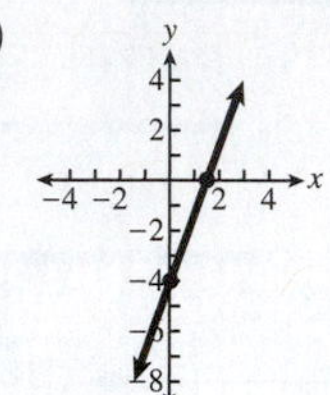

13. $\left(\frac{4}{3}, 0\right), (0, 4)$ **15.**

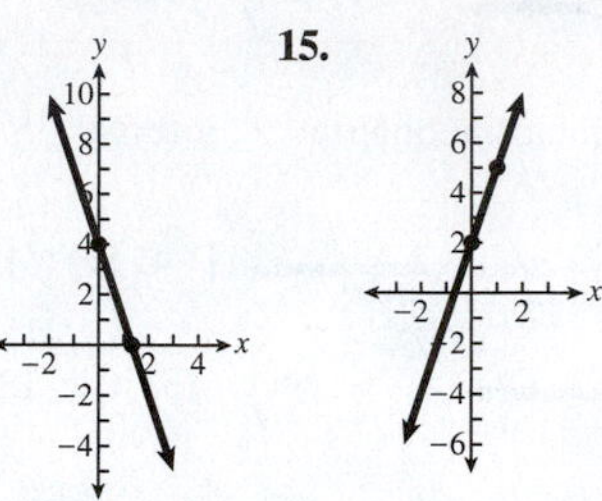

17.

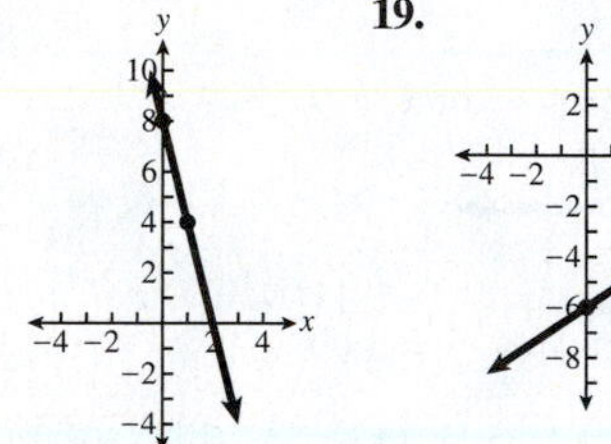

19.

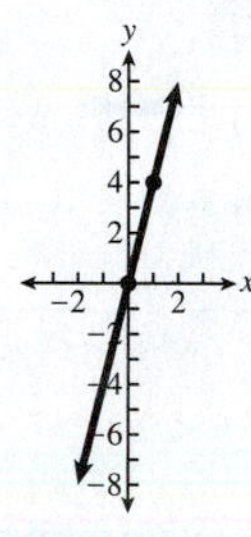

21. **23.**

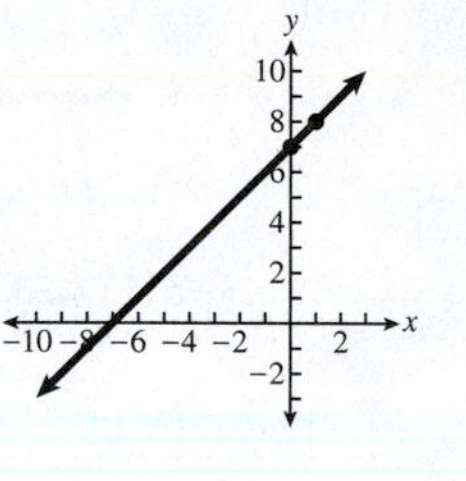

25. **27.**

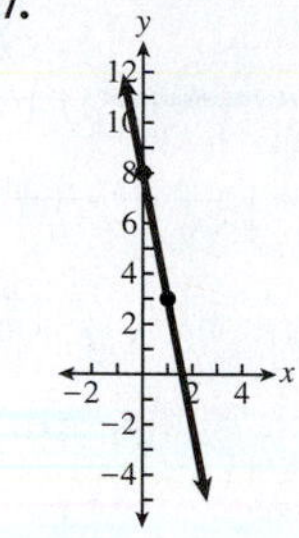

29. **31.**

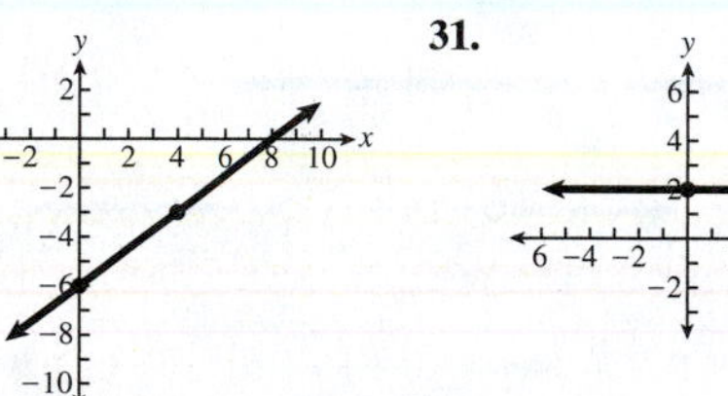

33. Domain: $(-\infty, \infty)$; range: $[0, \infty)$

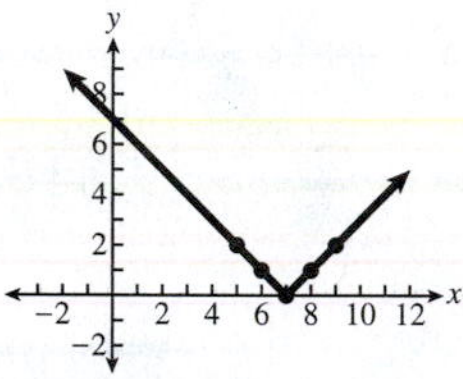

35. Domain: $(-\infty, \infty)$; range: $[3, \infty)$

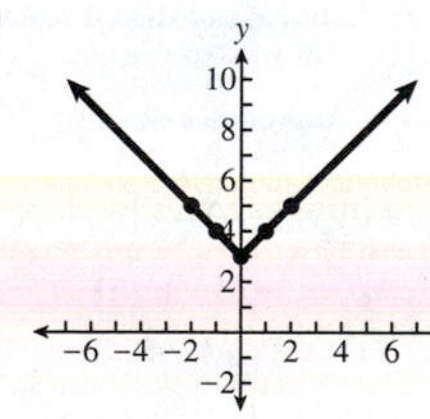

37. Domain: $(-\infty, \infty)$; range: $[-4, \infty)$

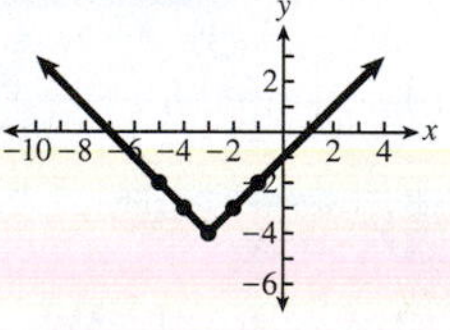

39. Domain: $(-\infty, \infty)$; range: $[3, \infty)$

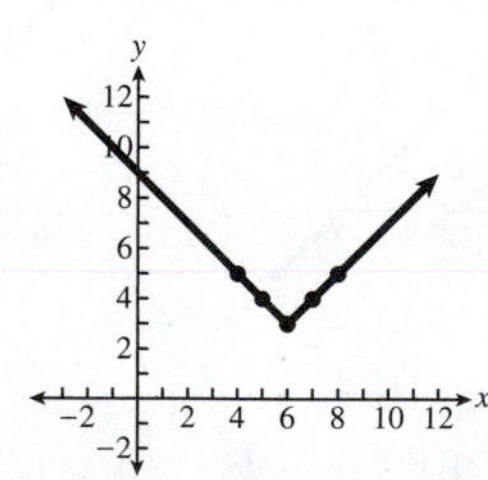

41. Domain: $(-\infty, \infty)$; range: $[-2, \infty)$

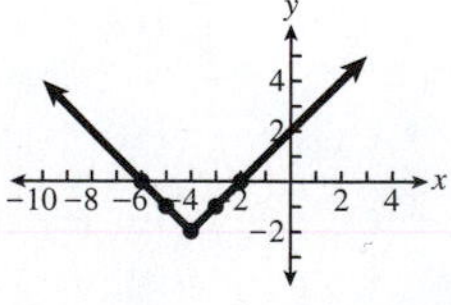

ain: $(-\infty, \infty)$; range: $[0, \infty)$

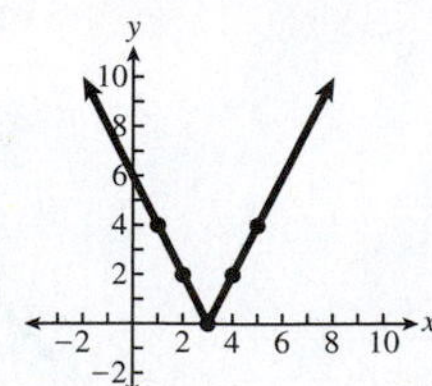

45. Domain: $(-\infty, \infty)$; range: $[-5, \infty)$

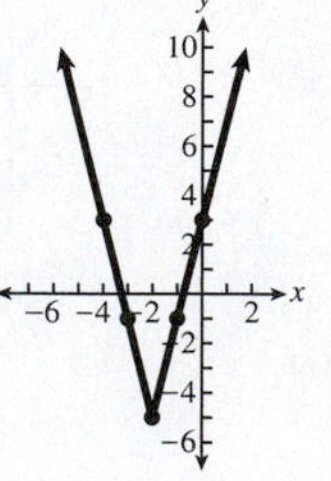

47. Domain: $(-\infty, \infty)$; range: $[3, \infty)$

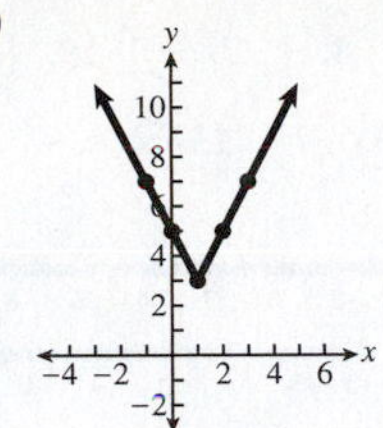

49. Domain: $(-\infty, \infty)$; range: $(-\infty, 0]$

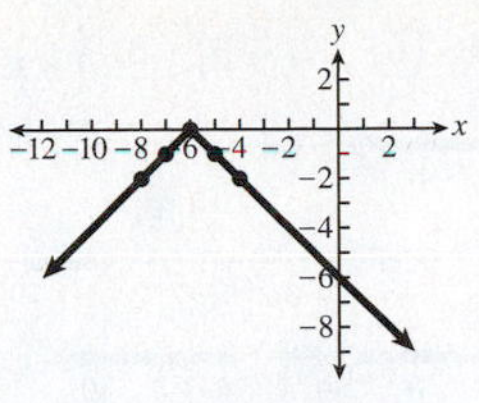

51. Domain: $(-\infty, \infty)$; range: $(-\infty, -4]$

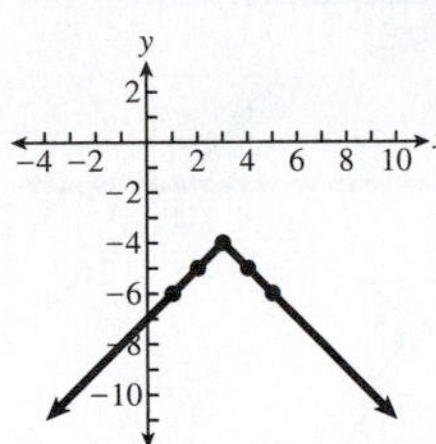

53. Domain: $(-\infty, \infty)$; range: $(-\infty, 9]$

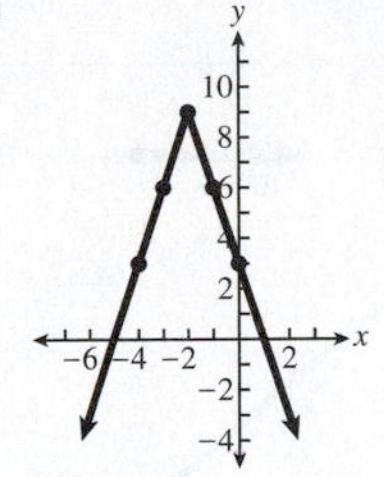

55. $(-4, 0), (10, 0), (0, -4)$ **57.** $(-6, 0), (6, 0), (0, -6)$ **59.** $(0, 6)$ **61.** $(1, 0), (7, 0), (0, -1)$ **63.** $f(x) = |x + 6|$ **65.** $f(x) = |x + 1| + 3$ **67.** $f(x) = -|x + 2| + 5$ **69.**

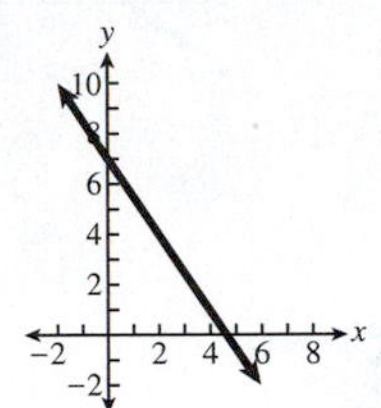

71. **73.**

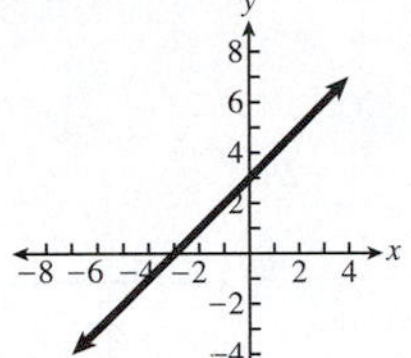

75.

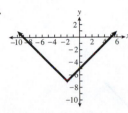

77. **79.**

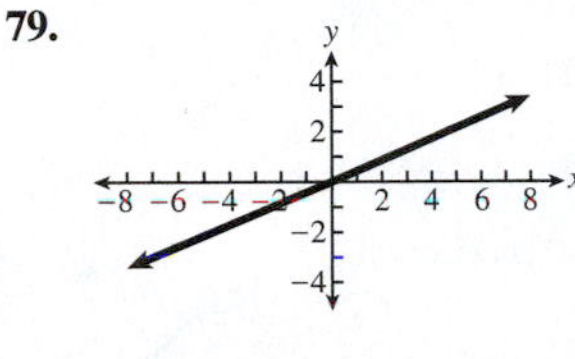

81. No, explanations will vary. Turning point is not on the graph.

Quick Check 8.4 **1.** $12x^3(3x^3 + 5x^2 - 1)$ **2.** $(x - 8)(x^2 + 4)$ **3. a)** $(x + 4)(x + 7)$ **b)** $(x - 3)(x - 10)$ **4.** $2(x - 9)(x + 2)$ **5.** $-(x - 7)(x - 10)$ **6.** $(2x - 5)(3x + 4)$ **7.** $(3x + 10)(3x - 10)$ **8.** $(3x - 4y)(9x^2 + 12xy + 16y^2)$ **9.** $\{-4, 8\}$ **10.** $\{-2, \frac{5}{2}\}$ **11.** $\{5\}$ **12.** $\{1, 3\}$

Section 8.4 **1.** factored **3.** common factors **5.** difference of cubes **7.** zero-factor **9.** $2x^3(18x^5 - 8x + 25)$ **11.** $4a^2b(5a - 7b^3 - 4a^3b)$ **13.** $(n + 4)(n^2 + 5)$ **15.** $(x + 7)(x^2 - 3)$ **17.** $(x - 5)(x - 8)$ **19.** $2(x - 8)(x + 4)$ **21.** $(a + 4b)(a + 9b)$ **23.** $4(x + 2)(x - 7)$ **25.** $(2x + 9)(x - 4)$ **27.** $(2x + 3)(3x + 8)$ **29.** $(x + 6)(x - 6)$ **31.** $(4b + 3)(4b - 3)$ **33.** Prime; cannot be factored **35.** $(x - 2)(x^2 + 2x + 4)$ **37.** $(3x + 5)(9x^2 - 15x + 25)$ **39.** $(x^2 + 9)(x + 3)(x - 3)$ **41.** $(x - 7)(x + 3)(x - 3)$ **43.** $(n + 4)(n + 2)(n^2 - 2n + 4)$ **45.** $\{-4, 1\}$ **47.** $\{-9, -8\}$ **49.** $\{-6, 8\}$ **51.** $\{2, 9\}$ **53.** $\{-2, -\frac{3}{2}\}$ **55.** $\{-5, 5\}$ **57.** $\{-\frac{2}{3}, \frac{2}{3}\}$ **59.** $\{0, 25\}$ **61.** $\{-7, 3\}$ **63.** $\{-6, -1\}$ **65.** $\{-3, -\frac{1}{2}\}$ **67.** $\{-2, 3, 10, 15\}$ **69.** $\{-12, -6, -4, 2\}$ **71.** $x^2 + 5x - 36 = 0$ **73.** $3x^2 - 16x + 5 = 0$ **75.** $x^2 - 144 = 0$ **77.** $\{16\}$ **79.** $\{-2, 5\}$ **81.** $\{-11, 4\}$ **83.** $\{-3, 2\}$ **85.** $\{-4\}$ **87.** $\{-5, 5\}$ **89.** $\varnothing$ **91.** $\{-15, 3\}$ **93.** $\{-12, -8, 4, 6\}$ **95.** Answers will vary. The solution causes a denominator to equal 0.

Quick Check 8.5 **1.** $(-2, 4)$ **2.** $(3, 2)$ **3.** 80 milliliters of 60% solution and 320 milliliters of 50% solution **4.** No **5.** $(1, 2, -3)$ **6.** $(\frac{1}{2}, -4, 5)$ **7.** 16 \$1's, 9 \$5's, 5 \$10's

Section 8.5 **1.** system of linear equations **3.** independent **5.** dependent **7.** ordered triple **9.** $(7, -2)$ **11.** $(3, -5)$ **13.** $(-6, 7)$ **15.** $(x, 3x + 7)$ **17.** $(8, 2)$ **19.** $(5, -\frac{1}{2})$ **21.** $(-4, -4)$ **23.** $\varnothing$ **25.** $(4, -6)$ **27.** Length: 19 inches; width: 8 inches **29.** 21 dimes, 44 quarters **31.** \$5400 at 6%, \$2100 at 3% **33.** 20 milliliters 15%, 40 milliliters 30% **35.** No **37.** Yes **39.** $(2, 6, 3)$ **41.** $(5, -2, -1)$ **43.** $\left(\frac{2}{3}, 5, -4\right)$ **45.** $(-30, 20, 5)$ **47.** $(-8, -4, 2)$ **49.** $\left(\frac{3}{4}, -\frac{9}{2}, 1\right)$ **51.** $(6, -3, 1)$ **53.** $(-7, 4, 7)$ **55.** 19 \$20 bills **57.** 5 bacon-and-egg sandwiches **59.** Abraham: 45; Belen: 37; Celeste: 56 **61.** A: 60°; B: 40°; C: 80° **63.** \$8000 at 3%, \$20,000 at 5%, \$12,000 at 6% **65.** \$5000 at 2%, \$2000 at 10 \$1000 at 25% **67.** \$13,000 at 5% profit, \$6000 at 7% profit, \$6000 at 40% loss **69.** 60 milliliters of 10%, 30 milliliters of 15%, 10 milliliters of 3 **71.** Answers will vary; 300 people, \$2/child, \$5/adult, \$3/senior, 20 more seniors than children; total: \$1260

Chapter 8 Review **1.** $\{6\}$ **2.** $\left\{\frac{13}{4}\right\}$ **3.** $\left\{\frac{5}{2}\right\}$ **4.** $\{7\}$ **5.** $\{-6, 6\}$ **6.** $\varnothing$ **7.** $\{-15, 1\}$ **8.** $\left\{-7, -\frac{5}{3}\right\}$ **9.** $\{-3, -5\}$ **10.** $\{-6, 3\}$
11. $x < -2$, $(-\infty, -2)$ **12.** $x \le 7$, $(-\infty, 7]$ **13.** $x \ge 9$, $[9, \infty)$
14. $17 \le x \le \frac{39}{2}$, $\left[17, \frac{39}{2}\right]$ **15.** $-8 < x < 9$, $(-8, 9)$
16. $x < 2$ or $x \ge 6$, $(-\infty, 2) \cup [6, \infty)$ **17.** $-10 \le x \le 10$, $[-10, 10]$
18. $\frac{4}{3} < x < 6$, $\left(\frac{4}{3}, 6\right)$ **19.** $-8 < x < 7$, $(-8, 7)$
20. $x < -9$ or $x > 9$, $(-\infty, -9) \cup (9, \infty)$
21. $x \le -6$ or $x \ge -1$, $(-\infty, -6] \cup [-1, \infty)$ **22.** $\mathbb{R}$, (∞, ∞)

23. $(6, 0), (0, 5)$

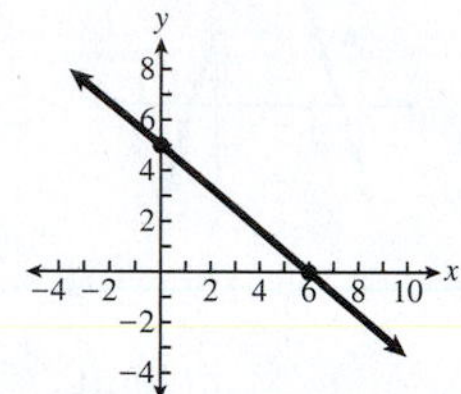

24. $\left(-\frac{3}{2}, 0\right), (0, 4)$

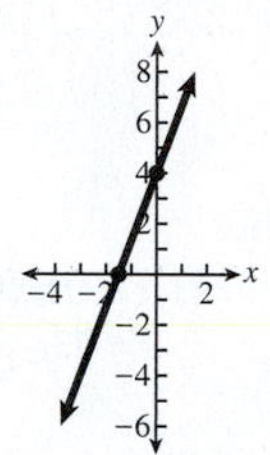

25.

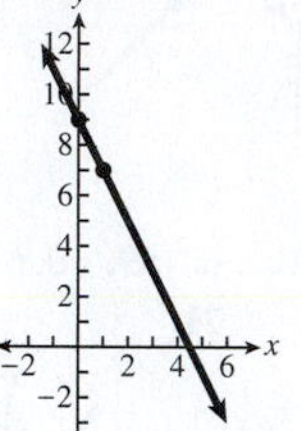

26.

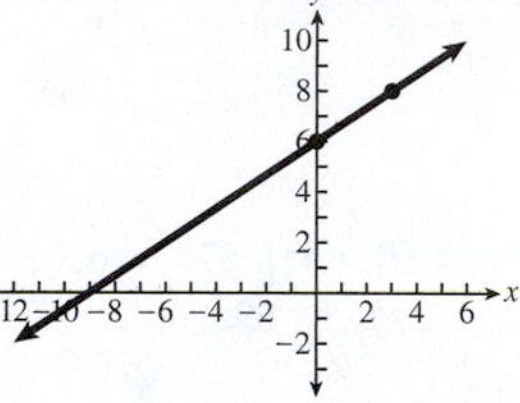

27.

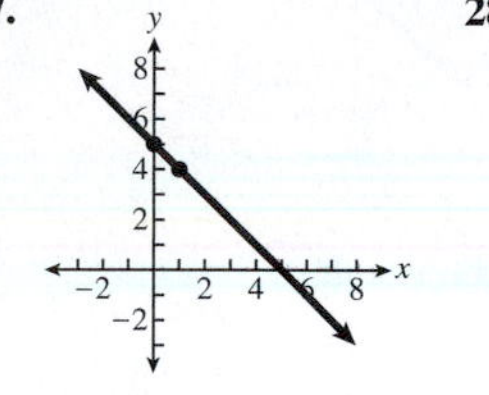

28.

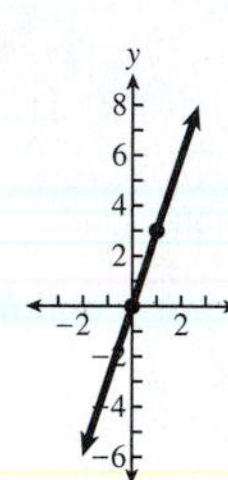

29.

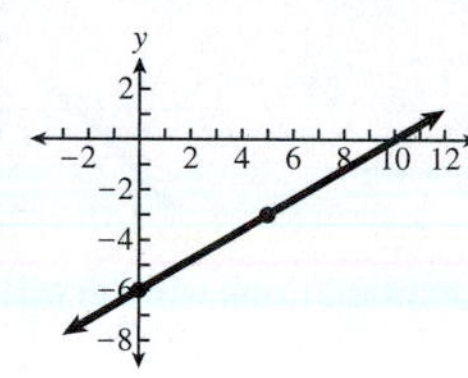

30.

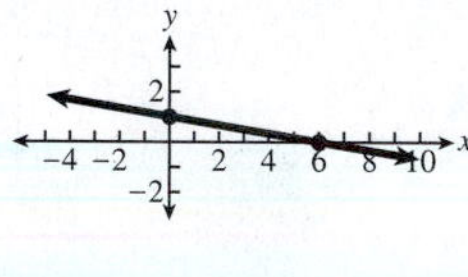

31. Domain: $(-\infty, \infty)$; range: $[3, \infty)$

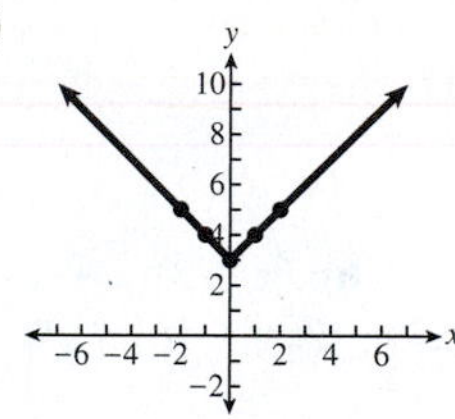

32. Domain: $(-\infty, \infty)$; range: $[0, \infty)$

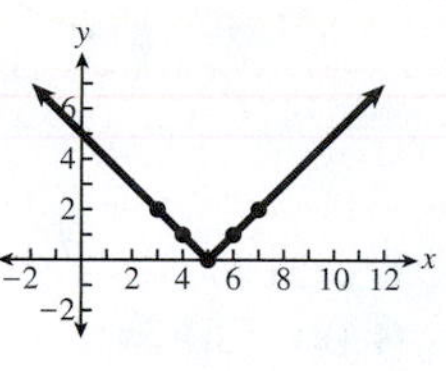

33. Domain: $(-\infty, \infty)$; range: $[-4, \infty)$

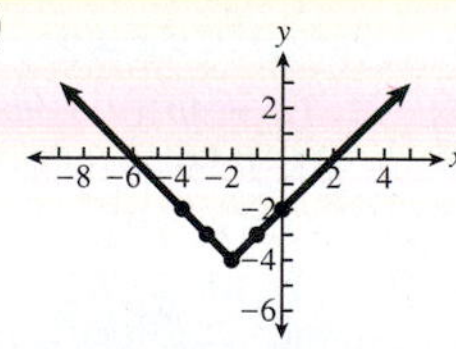

34. Domain: $(-\infty, \infty)$; range: $(-\infty, 2]$

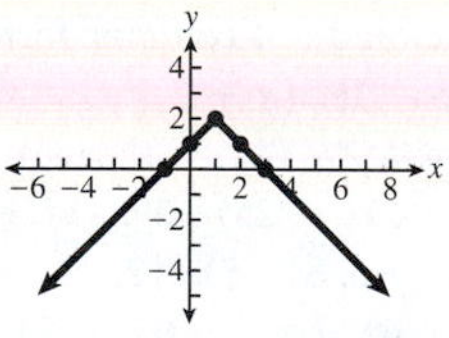

35. $2x(3x^8 - 2x^6 + 5x^2 - 9x + 1)$ **36.** $5a^2b(a^2b^2 - 2b^4 - 5a)$ **37.** $(x - 8)(x^2 + 11)$ **38.** $(x + 9)(x + 2)(x - 2)$ **39.** $(x - 7)(x + 2)$
40. $3(x - 3)(x - 8)$ **41.** $(x + 8y)(x + 10y)$ **42.** $(x + 15)(x - 2)$ **43.** $4(x + 6)(x - 1)$ **44.** $(x - 9)(x + 6)$ **45.** $(2x - 9)(3x - 1)$
46. $(2x + 3)(2x - 7)$ **47.** $(x + 6)(x - 6)$ **48.** $(3x + 1)(3x - 1)$ **49.** $(2x + 7)(2x - 7)$ **50.** $(5x + 8y)(5x - 8y)$ **51.** $(x + 3)(x^2 - 3x + 9)$
52. $(x - 9)(x^2 + 9x + 81)$ **53.** $\{-2, 3\}$ **54.** $\{-6, -4\}$ **55.** $\{-9, 2\}$ **56.** $\{-12, 12\}$ **57.** $\{13\}$ **58.** $\{-3, 8\}$ **59.** $\varnothing$ **60.** $\{5\}$ **61.** $(4, -1)$
$(-5, 3)$ **63.** $(1, 7)$ **64.** $(-7, 8)$ **65.** $(5, 4, -2)$ **66.** $(3, -8, 4)$ **67.** $(0, 6, -6)$ **68.** $\left(6, -\frac{5}{3}, \frac{7}{2}\right)$ **69.** 50 senior citizens **70.** A: 72°; B: 48°; C: 60°

r 8 Review Exercises: Worked-Out Solutions

$$- \frac{1}{6} = \frac{2}{15}x + \frac{1}{3}$$
$$) = 30 \cdot \left(\frac{2}{15}x + \frac{1}{3}\right)$$
$$) \cdot \frac{2}{15}x + 30 \cdot \frac{1}{3}$$
10

%

%

7.
$$|x + 7| - 14 = -6$$
$$|x + 7| = 8$$
$$x + 7 = 8 \quad \text{or} \quad x + 7 = -8$$
$$x = 1 \quad \text{or} \quad x = -15$$
$$\{-15, 1\}$$

11.
$$3x + 16 < 10$$
$$3x < -6$$
$$x < -2$$

$(-\infty, -2)$

15. $-14 < 3x + 10 < 37$
$-24 < 3x < 27$
$-8 < x < 9$

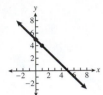

$(-8, 9)$

18. $|3x - 11| < 7$
$-7 < 3x - 11 < 7$
$4 < 3x < 18$
$\frac{4}{3} < x < 6$

$\left(\frac{4}{3}, 6\right)$

20. $|x| + 4 < 13$
$|x| > 9$
$x > 9$ or $x < -9$

$(-\infty, -9) \cup (9, \infty)$

23.

x-intercept	y-intercept
$5x + 6(0) = 30$	$5(0) + 6y = 30$
$5x = 30$	$6y = 30$
$x = 6$	$y = 5$
$(6, 0)$	$(0, 5)$

25. The line has slope $m = -2$ and its y-intercept is $(0, 9)$. Plot the y-intercept. A second point on the line can be found by moving down 2 units from the y-intercept and 1 unit to the right.

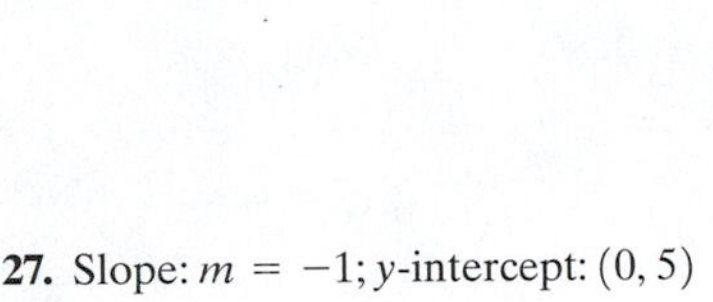

27. Slope: $m = -1$; y-intercept: $(0, 5)$

33. Select values of x centered around $x = -2$.

x	$f(x) = \|x + 2\| - 4$
-4	$f(-4) = \|-2\| - 4 = 2 - 4 = -2$
-3	$f(-3) = \|-1\| - 4 = 1 - 4 = -3$
-2	$f(-2) = \|0\| - 4 = 0 - 4 = -4$
-1	$f(-1) = \|1\| - 4 = 1 - 4 = -3$
0	$f(0) = \|2\| - 4 = 2 - 4 = -2$

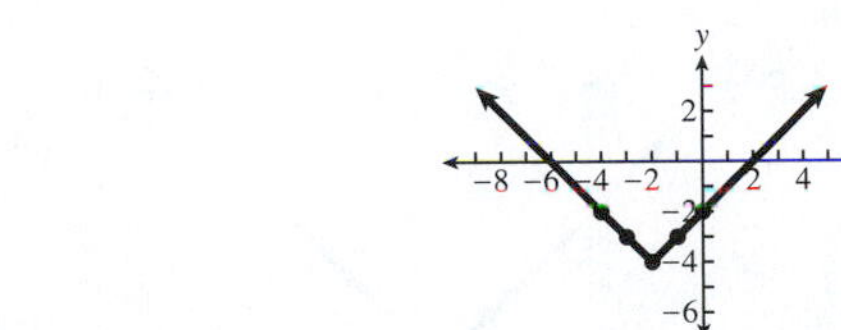

The domain of this function is $(-\infty, \infty)$. The function's minimum value is -4, so its range is $[-4, \infty)$.

36. The GCF is $5a^2b$.
$5a^4b^3 - 10a^2b^5 - 25a^3b = 5a^2b(a^2b^2 - 2b^4 - 5a)$

37. Factoring by Grouping
$x^3 - 8x^2 + 11x - 88$
$= x^2(x - 8) + 11(x - 8)$
$= (x - 8)(x^2 + 11)$

39. Find m and n such that $m \cdot n = -14$ and $m + n = -5$.
$x^2 - 5x - 14 = (x - 7)(x + 2)$

40. Factor out the GCF 3 first.
$3x^2 - 33x + 72$
$= 3(x^2 - 11x + 24)$
$= 3(x - 3)(x - 8)$

47. Difference of Squares
$x^2 - 36$
$= (x)^2 - (6)^2$
$= (x + 6)(x - 6)$

53. $x^2 - x - 6 = 0$
$(x + 2)(x - 3) = 0$
$x + 2 = 0$ or $x - 3 = 0$
$x = -2$ or $x = 3$
$\{-2, 3\}$

58. $x - 5 - \frac{24}{x} = 0$
$x \cdot \left(x - 5 - \frac{24}{x}\right) = x \cdot 0$
$x \cdot x - x \cdot 5 - x \cdot \frac{24}{x} = 0$
$x^2 - 5x - 24 = 0$
$(x + 3)(x - 8) = 0$
$x + 3 = 0$ or $x - 8 = 0$
$x = -3$ or $x = 8$
$\{-3, 8\}$

62. Substitution Method
$6(7 - 4y) - 5y = -45$
$42 - 24y - 5y = -45$
$42 - 29y = -45$
$-29y = -87$
$y = 3$
$x = 7 - 4(3)$
$x = 7 - 12$
$x = -5$
$(-5, 3)$

63. Addition Method
$2x + 3y = 23 \xrightarrow{\text{Multiply by 3.}} 6x + 9y = 69$
$-6x + 2y = 8 \qquad\qquad -6x + 2y = 8$
$11y = 77$

$\frac{11y}{11} = \frac{77}{11}$
$y = 7$

$2x + 3(7) = 23$
$2x + 21 = 23$
$2x = 2$
$x = 1$

$(1, 7)$

67. $x + y + z = 0$
$2x + 3y - z = 24$
$3x - 2y + 4z = -36$

Eliminate z using Eq. 1 and Eq. 2, as well as Eq. 2 and Eq. 3.

$x + y + z = 0$	(Eq. 1)	$8x + 12y - 4z = 96$	4(Eq. 2)
$2x + 3y - z = 24$	(Eq. 2)	$3x - 2y + 4z = -36$	(Eq. 3)
$3x + 4y = 24$	(Eq. 4)	$11x + 10y = 60$	(Eq. 5)

Solve for x using Eq. 4 and Eq. 5.

$3x + 4y = 24 \xrightarrow{\text{Multiply by 5.}} 15x + 20y = 120$
$11x + 10y = 60 \xrightarrow{\text{Multiply by -2.}} -22x - 20y = -120$
$-7x = 0$

$-7x = 0$
$x = 0$

Back substitute.

Equation 4	Equation 1
$3(0) + 4y = 24$	$(0) + (6) + z = 0$
$0 + 4y = 24$	$6 + z = 0$
$4y = 24$	$z = -6$
$y = 6$	

$(0, 6, -6)$

Chapter 8 Test **1.** $\{-16\}$ **2.** $\{-16, -2\}$ **3.** $\{-3, 13\}$ **4.** $x > -4$, $(-4, \infty)$ **5.** $-6 \le x \le 5$, $[-6, 5]$ **6.** $x < 3$ or $x > 5$,  $(-\infty, 3) \cup (5, \infty)$ **7.** $-9 \le x \le 1$, $[-9, 1]$ **8.** $x < 0$ or $x > 5$, $(-\infty, 0) \cup (5, \infty)$

9. 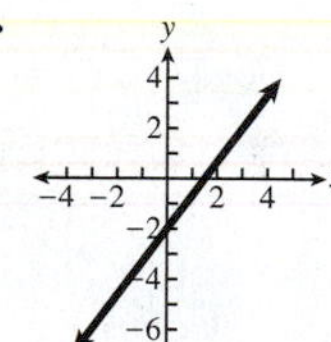

10. 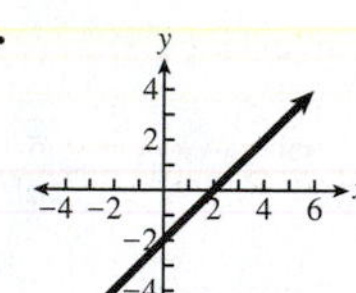

11. Domain: $(-\infty, \infty)$; range: $[3, \infty)$

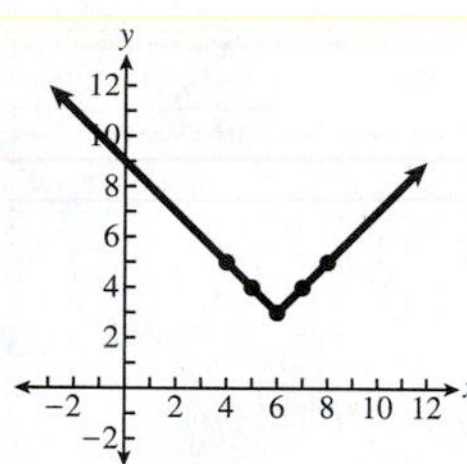

12. $(x + 8)(x - 5)$

13. $4(x - 2)(x - 5)$ **14.** $(2x + 5)(3x + 2)$ **15.** $(3x + 10)(3x - 10)$ **16.** $\{-9, 3\}$ **17.** $\{-7, 1\}$ **18.** $(-3, 4)$ **19.** $(8, 2, 1)$ **20.** A: 115°; B: 45°; C: 20°

CHAPTER 9

Quick Check 9.1 **1. a)** 7 **b)** $\frac{2}{5}$ **c)** -6 **2. a)** $|a^7|$ **b)** $5|b^{11}|$ **3.** $8x^{12}$ **4.** 10.440 **5. a)** 3 **b)** 4 **c)** -7 **6.** x^7 **7.** $-3x^7y^8$ **8.** $x + 5$ **9.** $30b^4$ **10.** 144 **11. a)** 5 **b)** $3a^3$ **12.** 35 **13.** $[-18, \infty)$ **14.** $(-\infty, \infty)$

…ion 9.1 **1.** square root **3.** nth root **5.** radicand **7.** radical function **9.** 6 **11.** 2 **13.** $\frac{1}{9}$ **15.** $\frac{5}{6}$ **17.** Not a real number **19.** -7 **21.** $|a^7|$ … **25.** $3|x^3|$ **27.** $\frac{1}{7}x^2$ **29.** $|a^9b^{11}|$ **31.** $x^4|y^5z^7|$ **33.** 81 **35.** $9x^2$ **37.** 7.416 **39.** 18.055 **41.** 0.728 **43.** 5 **45.** -3 **47.** x^5 **49.** a^7b^4 … x^3y^5 **55.** $x + 3$ **57.** 9 **59.** 30 **61.** a^{13} **63.** x^{19} **65.** 6 **67.** 10 **69.** x^9 **71.** a^6 **73.** 6 **75.** 6 **77.** x^5 **79.** a^3b^2 **81.** 3 **83.** x^5 … **89.** $108x$ **91.** 5 **93.** 392 **95.** 4 **97.** 3 **99.** 12.385 **101.** 5 **103.** $[4, \infty)$ **105.** $\left[\frac{8}{3}, \infty\right)$ **107.** $\left(-\infty, \frac{13}{2}\right]$ **109.** $(-\infty, \infty)$ … 4, $\sqrt[3]{-64}$; not real: $\sqrt{-64}$ Explanations will vary.

… $36 = 6$ **b)** $\sqrt[5]{-32x^{15}} = -2x^3$ **2.** $\sqrt[4]{x^{32}y^4z^{24}} = x^8yz^6$ **3.** $a^{1/9}$ **4.** $(\sqrt[6]{4096x^{18}})^5 = 1024x^{15}$ **5.** $x^{5/12}$ **6.** $x^{29/24}$ **7.** $x^{14/45}$

… $^{1/4}$ **7.** $7^{1/2}$ **9.** $9d^{1/4}$ **11.** 8 **13.** -7 **15.** x^6 **17.** $4a^{11}b^7$ **19.** $-3x^{11}y^5z$ **21.** $x^{3/5}$ **23.** $y^{8/9}$ **25.** $(3x^2)^{7/3}$ … **35.** $64a^6b^{18}$ **37.** $625x^{12}y^{20}z^4$ **39.** $x^{4/5}$ **41.** $x^{11/6}$ **43.** $a^{31/12}$ **45.** $x^{1/2}y^{11/12}$ **47.** $x^{3/14}$ **49.** $a^{9/25}$ **51.** $x^{5/3}$ … $\frac{1}{7776}$ **65.** $\frac{1}{3125}$ **67.** $\frac{1}{6}$ **69.** $\sqrt[20]{x^9}$ **71.** $\sqrt[3]{a}$ **73.** $\sqrt[5]{m}$ **75.** $\frac{1}{\sqrt[6]{x}}$ **77.** Yes; explanations will vary.

Quick Check 9.3 **1.** $x^3\sqrt[5]{x^2}$ **2.** $a^4b^7c^3\sqrt{bcd}$ **3.** $3\sqrt{10}$ **4.** $2\sqrt[3]{35}$ **5.** $x^6yz^{10}w^3\sqrt[5]{x^3yw^3}$ **6.** $15\sqrt{10}-5\sqrt{5}$ **7.** $11\sqrt[5]{a^4b^3}-11\sqrt[7]{a^4b^3}$ **8.** $4\sqrt{7}$ **9.** $6\sqrt{2}$

Section 9.3 **1.** like radicals **3.** $2\sqrt{3}$ **5.** $3\sqrt{5}$ **7.** $12\sqrt{3}$ **9.** $4\sqrt[3]{3}$ **11.** $6\sqrt[3]{6}$ **13.** $2\sqrt[4]{75}$ **15.** $x^4\sqrt{x}$ **17.** $m^4\sqrt[3]{m}$ **19.** $x^{14}\sqrt[5]{x^4}$ **21.** $x^7y^6\sqrt{x}$ **23.** $y^3\sqrt{xy}$ **25.** $x^{16}y^8z^8\sqrt{xy}$ **27.** $x^4y^3\sqrt[3]{xy^2z}$ **29.** $a^3b^8c^2\sqrt[4]{a^2b^3c}$ **31.** $3x^5y^6\sqrt{3x}$ **33.** $3x^5yz^4\sqrt[4]{2x^3y^2}$ **35.** $27\sqrt{5}$ **37.** $-6\sqrt[3]{4}$ **39.** $15\sqrt{15}$ **41.** $5\sqrt{7}-21\sqrt{3}$ **43.** $2\sqrt{2}-13\sqrt{3}$ **45.** $-2\sqrt{x}$ **47.** $11\sqrt[5]{a}$ **49.** $22\sqrt{x}$ **51.** $5\sqrt{x}-17\sqrt{y}$ **53.** $21\sqrt[3]{x}-3\sqrt[4]{x}$ **55.** $13\sqrt{2}$ **57.** $-31\sqrt{6}$ **59.** $46\sqrt{3}$ **61.** $35\sqrt[4]{2}$ **63.** $-51\sqrt{3}-75\sqrt{2}$ **65.** $7\sqrt{3}-11\sqrt{2}$ **67.** $-6\sqrt{2}+17\sqrt{5}$ **69.** Answers will vary. Factor; then look for squares.

Quick Check 9.4 **1. a)** 60 **b)** $x^4yz\sqrt[3]{z^2}$ **c)** $144\sqrt{3}$ **2.** $6+6\sqrt{5}$ **3.** $x^6y^6-x^8y^6\sqrt{xy}$ **4. a)** $57-2\sqrt{6}$, **b)** $17-2\sqrt{70}$ **5. a)** 9, **b)** 529 **6.** $\frac{\sqrt{210}}{3}$ **7.** $\frac{\sqrt{5}}{5}$ **8.** $\frac{2y^2\sqrt{xy}}{5x^2z^5}$ **9.** $\frac{5\sqrt{3}-3\sqrt{5}}{2}$ **10.** $\frac{33-20\sqrt{3}}{3}$

Section 9.4 **1.** x **3.** rationalizing **5.** 30 **7.** 216 **9.** $280\sqrt{3}$ **11.** $72\sqrt[3]{2}$ **13.** x^{11} **15.** $2a^6b^7\sqrt{15b}$ **17.** $x^6y^3z^5\sqrt[4]{z^3}$ **19.** $4-2\sqrt{3}$ **21.** $42+14\sqrt{5}$ **23.** $162\sqrt{2}-72\sqrt{5}$ **25.** $m^4n^6\sqrt{n}+m^6n^8\sqrt{m}$ **27.** $a^6b^{15}c^4\sqrt[3]{c}+a^{10}b^3c^7\sqrt[3]{b}$ **29.** $2-\sqrt{6}+\sqrt{10}-\sqrt{15}$ **31.** $58+13\sqrt{6}$ **33.** $154-26\sqrt{21}$ **35.** $24+10\sqrt{15}$ **37.** $98-24\sqrt{10}$ **39.** -3 **41.** 414 **43.** -178 **45.** $\frac{2\sqrt{3}}{x^2y^3}$ **47.** $3\sqrt{2}$ **49.** $\frac{3\sqrt{3}}{2}$ **51.** $\frac{2\sqrt{6}}{3}$ **53.** $\frac{\sqrt{2}}{2}$ **55.** $\frac{3\sqrt{33a}}{11a^4}$ **57.** $\frac{s^2\sqrt[3]{4rs^2t^2}}{2rt}$ **59.** $\frac{7\sqrt{3}}{3}$ **61.** $\frac{4\sqrt{2a}}{3a}$ **63.** $\frac{x^2z^3\sqrt{xy}}{y^3}$ **65.** $\frac{3\sqrt{13}+3\sqrt{7}}{2}$ **67.** $-\frac{3-\sqrt{33}}{2}$ **69.** $\frac{3\sqrt{39}+9\sqrt{3}}{2}$ **71.** $\frac{59-11\sqrt{15}}{34}$ **73.** $\frac{328+43\sqrt{22}}{126}$ **75.** $-\frac{7+10\sqrt{6}}{19}$ **77.** $\frac{3\sqrt{5}}{7}$ **79.** $573+80\sqrt{35}$ **81.** $5a^3b^2c^4\sqrt[3]{5c^2}$ **83.** $-\frac{99-17\sqrt{35}}{2}$ **85.** $6\sqrt[3]{5}$ **87.** $120+135\sqrt{2}-56\sqrt{5}-63\sqrt{10}$ **89.** $-14x^5\sqrt{2}+7x^{11}$ **91.** $a^4b^8c^{11}\sqrt[4]{ab^2}$ **93.** $\frac{a^2b^4\sqrt{3ac}}{6c^5}$ **95.** $\frac{5\sqrt[3]{6x}}{3x}$ **97.** 2257 **99.** $a+2\sqrt{ab}+b$ **101.** Answers will vary. One term is the same; the other term is the opposite.

Quick Review Exercises **1.** $2^4\cdot 3^2$ **2.** $2^3\cdot 5^2\cdot 7$ **3.** 2^{10} **4.** $2\cdot 3^3\cdot 7^2\cdot 11$

Quick Check 9.5 **1. a)** $\{47\}$ **b)** $\{-73\}$ **2.** $\varnothing$ **3.** 11 **4. a)** $\{9\}$ **b)** $\{-220\}$ **5.** $\{2, 10\}$ **6.** $\{8\}$ **7.** $\{-22\}$ **8.** $\{6\}$ **9.** 2.72 seconds **10.** 7.30 feet **11.** 70 miles per hour

Section 9.5 **1.** radical **3.** isolate **5.** pendulum **7.** $\{97\}$ **9.** $\{58\}$ **11.** $\varnothing$ **13.** $\{20\}$ **15.** $\{10\}$ **17.** $\{-10, 5\}$ **19.** 45 **21.** 13 **23.** $-2, 7$ **25.** $\{4\}$ **27.** $\{39\}$ **29.** $\{42\}$ **31.** $\{8\}$ **33.** $\{9\}$ **35.** $\{7\}$ **37.** $\varnothing$ **39.** $\{3\}$ **41.** $\{2\}$ **43.** $\{26\}$ **45.** $\{9\}$ **47.** $\{5\}$ **49.** $\{-1, 3\}$ **51.** $\{-4\}$ **53.** $\{5\}$ **55.** 2.48 seconds **57.** 1.49 seconds **59.** 2.6 feet **61.** 40 mph **63.** 73 mph **65.** 2.09 square meters **67.** 1.45 square meters **69.** 68 kilograms **71.** 138.6 gallons/minute **73.** 71.4 gallons/minute **75.** 2.3 feet **77.** 1.7 miles **79.** 12.2 miles **81.** Answers will vary. The problem needs skid distance and drag factor; then solve for speed.

Quick Check 9.6 **1. a)** $6i$ **b)** $3i\sqrt{7}$ **c)** $-3i\sqrt{6}$ **2. a)** $-5+27i$ **b)** $11-28i$ **3.** -60 **4.** $-10\sqrt{6}$ **5.** $48-42i$ **6.** $26-13i$ **7.** $65-72i$ **8.** 65 **9.** 137 **10.** $\frac{6}{5}-\frac{9}{10}i$ **11.** $\frac{57}{58}+\frac{41}{58}i$ **12.** $-\frac{2}{3}-\frac{5}{4}i$ **13.** -1

Section 9.6 **1.** imaginary **3.** -1 **5.** real **7.** $2i$ **9.** $9i$ **11.** $-13i$ **13.** $5i\sqrt{2}$ **15.** $-6i\sqrt{7}$ **17.** $11+14i$ **19.** $4-17i$ **21.** $4i$ **23.** $26i$ **25.** $-2-24i$ **27.** $6+9i$ **29.** $2+9i$ **31.** -98 **33.** 20 **35.** -56 **37.** $-6\sqrt{6}$ **39.** -30 **41.** $-35+56i$ **43.** $24-54i$ **45.** $33+17i$ **47.** $62+10i$ **49.** $-13+84i$ **51.** $-24-10i$ **53.** 29 **55.** 208 **57.** 61 **59.** $\frac{15}{13}-\frac{3}{13}i$ **61.** $-\frac{35}{26}+\frac{45}{26}i$ **63.** $\frac{14}{53}+\frac{57}{53}i$ **65.** $\frac{43}{85}+\frac{36}{85}i$ **67.** $-8i$ **69.** $-\frac{7}{3}-\frac{5}{3}i$ **71.** $-\frac{1}{4}+\frac{3}{2}i$ **73.** $\frac{3}{2}-\frac{7}{2}i$ **75.** $-\frac{13}{10}-\frac{31}{10}i$ **77.** $3+8i$ **79.** $31+33i$ **81.** i **83.** -1 **85.** -1 **87.** $2i$ **89.** $48-9i$ **91.** $\frac{4}{5}-\frac{9}{5}i$ **93.** $45-108i$ **95.** $-\frac{6}{5}+\frac{12}{5}i$ **97.** 29 **99.** $-18\sqrt{2}$ **101.** $-\frac{6}{17}+\frac{27}{17}i$ **103.** $6i\sqrt{7}$ **105.** $72-54i$ **107.** $23+10i$ **109.** 126 **111.** $-i$ **113.** $-24+16i$ **115.** Answers will vary; similar to combining like terms.

Chapter 9 Review **1.** 5 **2.** 13 **3.** -4 **4.** x^3 **5.** $9x^6$ **6.** $3x^3y^2z^5$ **7.** 4.123 **8.** 6.403 **9.** 9 **10.** 2.828 **11.** $[-3, \infty)$ **12.** $\left[-\frac{25}{3}, \infty\right)$ **13.** $a^{1/5}$ **14.** $x^{3/4}$ **15.** $x^{13/2}$ **16.** $n^{8/5}$ **17.** 2 **18.** $49x^{14}$ **19.** $x^{5/3}$ **20.** $x^{7/16}$ **21.** $x^{1/4}$ **22.** $\frac{1}{7}$ **23.** $\frac{1}{1024}$ **24.** $\frac{1}{81}$ **25.** $2\sqrt{7}$ **26.** $x^3\sqrt[3]{x}$ **27.** $x^2yz^3\sqrt[5]{xy^3}$ **28.** $3a^8b^4c^5\sqrt[3]{15ac}$ **29.** $6rs^5t^6\sqrt{5rs}$ **30.** $5r^3s^4t^2\sqrt{7}$ **31.** $12\sqrt{2}$ **32.** $6\sqrt[3]{x}$ **33.** $25\sqrt{5}$ **34.** $57\sqrt{3}-40\sqrt{2}$ **35.** $2x^2\sqrt[3]{9}$ **36.** $30\sqrt{2}-60$ **37.** $3\sqrt{2}-\sqrt{6}+4\sqrt{3}-4$ **38.** $296-43\sqrt{55}$ **39.** $303-108\sqrt{5}$ **40.** -648 **41.** $4x^2\sqrt{7}$ **42.** $\frac{5a^3b^3\sqrt{b}}{2c^5}$ **43.** $\frac{2\sqrt[3]{50}}{5}$ **44.** $\frac{4\sqrt{14}}{7}$ **45.** $\frac{\sqrt{22}}{6}$ **46.** $\frac{a^3c^2\sqrt{bc}}{b^2}$ **47.** $\frac{4(\sqrt{11}+\sqrt{5})}{3}$ **48.** $\frac{5\sqrt{5}+3}{116}$ **49.** $\frac{44+13\sqrt{14}}{10}$ **50.** $\frac{-53+27\sqrt{5}}{209}$ **51.** $\{12\}$ **52.** $\{20\}$ **53.** $\{9\}$ **54.** $\{2\}$ **55.** $\{9\}$ **57.** 4 **58.** $-2, 8$ **59.** 2.48 seconds **60.** 67 mph **61.** $7i$ **62.** $2i\sqrt{10}$ **63.** $6i\sqrt{7}$ **64.** $-15i\sqrt{3}$ **65.** $11+5i$ **66.** $16-2i$ **67.** $1+5i$ **69.** -16 **70.** $-10\sqrt{3}$ **71.** $28+21i$ **72.** $114+106i$ **73.** $72-54i$ **74.** 205 **75.** $\frac{21}{29}+\frac{9}{29}i$ **76.** $\frac{6}{5}+\frac{9}{10}i$ **77.** $\frac{13}{106}+\frac{19}{106}i$ **78.** -16 **80.** -1

Chapter 9 Review Exercises: Worked-Out Solutions

4. $\sqrt[3]{x^9} = \sqrt[3]{(x^3)^3}$
$= x^3$

9. $f(x) = \sqrt{x - 10}$
$f(91) = \sqrt{91 - 10}$
$= \sqrt{81}$
$= 9$

11. $2x + 6 \geq 0$
$2x \geq -6$
$x \geq -3$
$[-3, \infty)$

18. $(343x^{21})^{2/3} = (\sqrt[3]{343x^{21}})^2$
$= (7x^7)^2$
$= 49x^{14}$

19. $x^{3/2} \cdot x^{1/6} = x^{3/2+1/6}$
$= x^{9/6+1/6}$
$= x^{10/6}$
$= x^{5/3}$

27. $\sqrt[5]{x^{11}y^8z^{15}} = \sqrt[5]{x^{10}y^5z^{15}} \cdot \sqrt[5]{xy^3}$
$= x^2yz^3\sqrt[5]{xy^3}$

33. $5\sqrt{20} + 3\sqrt{125} = 5\sqrt{4 \cdot 5} + 3\sqrt{25 \cdot 5}$
$= 5 \cdot 2\sqrt{5} + 3 \cdot 5\sqrt{5}$
$= 10\sqrt{5} + 15\sqrt{5}$
$= 25\sqrt{5}$

38. $(10\sqrt{5} - 7\sqrt{11})(9\sqrt{5} + 2\sqrt{11}) = 10\sqrt{5} \cdot 9\sqrt{5} + 10\sqrt{5} \cdot 2\sqrt{11} - 7\sqrt{11} \cdot 9\sqrt{5} - 7\sqrt{11} \cdot 2\sqrt{11}$
$= 90 \cdot 5 + 20\sqrt{55} - 63\sqrt{55} - 14 \cdot 11$
$= 450 - 43\sqrt{55} - 154$
$= 296 - 43\sqrt{55}$

45. $\sqrt{\frac{11}{18}} = \frac{\sqrt{11}}{\sqrt{18}} \cdot \frac{\sqrt{2}}{\sqrt{2}}$
$= \frac{\sqrt{22}}{\sqrt{36}}$
$= \frac{\sqrt{22}}{6}$

48. $\frac{2\sqrt{3}}{10\sqrt{15} - 3\sqrt{12}} = \frac{2\sqrt{3}}{10\sqrt{15} - 3\sqrt{12}} \cdot \frac{10\sqrt{15} + 3\sqrt{12}}{10\sqrt{15} + 3\sqrt{12}}$
$= \frac{20\sqrt{45} + 6\sqrt{36}}{100 \cdot 15 - 9 \cdot 12}$
$= \frac{20 \cdot 3\sqrt{5} + 6 \cdot 6}{1500 - 108}$
$= \frac{60\sqrt{5} + 36}{1392}$
$= \frac{5\sqrt{5} + 3}{116}$

51. $\sqrt{4x + 1} = 7$
$(\sqrt{4x + 1})^2 = 7^2$
$4x + 1 = 49$
$4x = 48$
$x = 12$
$\{12\}$

54. $\sqrt{17 - 4x} - x = 1$
$\sqrt{17 - 4x} = x + 1$
$(\sqrt{17 - 4x})^2 = (x + 1)^2$
$17 - 4x = (x + 1)(x + 1)$
$17 - 4x = x^2 + 2x + 1$
$0 = x^2 + 6x - 16$
$0 = (x - 2)(x + 8)$
$x = 2$ or $x = -8$
$x = -8$ is an extraneous solution.
$\{2\}$

60. $s = \sqrt{30df}$
$s = \sqrt{30(200)(0.75)}$
$s = \sqrt{4500}$
$s \approx 67$

62. $\sqrt{-40} = \sqrt{-1} \cdot \sqrt{40}$
$= i \cdot \sqrt{4 \cdot 10}$
$= i \cdot 2\sqrt{10}$
$= 2i\sqrt{10}$

65. $(4 + 2i) + (7 + 3i) = 4 + 7 + 2i + 3i$
$= 11 + 5i$

72. $(10 + 2i)(13 + 8i) = 130 + 80i + 26i + 16i^2$
$= 130 + 106i - 16$
$= 114 + 106i$

75. $\frac{6}{7 - 3i} = \frac{6}{7 - 3i} \cdot \frac{7 + 3i}{7 + 3i}$
$= \frac{42 + 18i}{49 - 9i^2}$
$= \frac{42 + 18i}{49 + 9}$
$= \frac{42 + 18i}{58}$
$= \frac{21 + 9i}{29}$ or $\frac{21}{29} + \frac{9}{29}i$

78. $\frac{16}{i} = \frac{16}{i} \cdot \frac{i}{i}$
$= \frac{16i}{i^2}$
$= \frac{16i}{-1}$
$= -16i$

79. $i^{53} = (i^4)^{13} \cdot i^1$
$= 1 \cdot i$
$= i$

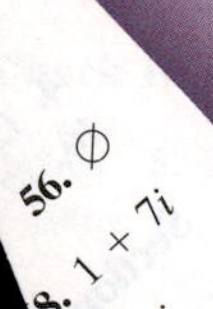

...pter 9 Test **1.** $5x^5y^7$ **2.** 8.718 **3.** $a^{1/2}$ **4.** $b^{11/24}$ **5.** $\frac{1}{25}$ **6.** $6\sqrt{2}$ **7.** $a^5b^{10}c\sqrt[4]{ab^2}$ **8.** $5\sqrt{2}$ **9.** $140\sqrt{2} - 40\sqrt{7}$ **10.** $226 - 168\sqrt{3}$ ... $\sqrt{xy}$ **12.** $\frac{-564 + 167\sqrt{14}}{50}$ **13.** $\{20\}$ **14.** $\{8\}$ **15.** 13.0 ft **16.** $3i\sqrt{11}$ **17.** $-18 - 11i$ **18.** 85 **19.** $-\frac{33}{34} + \frac{47}{34}i$

...0

... **a)** $\{-3, 5\}$ **b)** $\{-11, 11\}$ **2. a)** $\{-2\sqrt{7}, 2\sqrt{7}\}$ **b)** $\{-4i\sqrt{2}, 4i\sqrt{2}\}$ **3. a)** $\left\{-\frac{11}{3}, 3\right\}$ **b)** $\left\{\frac{3 - 3i}{2}, \frac{3 + 3i}{2}\right\}$ ...meters **6. a)** $\{-6, -2\}$ **b)** $\{3 - i, 3 + i\}$ **7.** $\{-9, 2\}$ **8.** $\{8 - 3\sqrt{2}, 8 + 3\sqrt{2}\}$ **9.** $3x^2 + x - 10 = 0$

...oots **3.** completing the square **5.** $\{-2, 5\}$ **7.** $\{-7, 0\}$ **9.** $\left\{\frac{1}{3}, 4\right\}$ **11.** $\{-4, 6\}$ **13.** $\{-3, 3\}$ ... $2\sqrt{6}\}$ **21.** $\{-6i\sqrt{2}, 6i\sqrt{2}\}$ **23.** $\{-2i\sqrt{5}, 2i\sqrt{5}\}$ **25.** $\{7 - 3\sqrt{2}, 7 + 3\sqrt{2}\}$ ...$5 + 3i\sqrt{2}\}$ **31.** $\{9 - 5\sqrt{2}, 9 + 5\sqrt{2}\}$ **33.** $\left\{-1, \frac{9}{5}\right\}$ **35.** $x^2 + 12x + 36 = (x + 6)^2$ ... $8x + 16 = (x - 4)^2$ **41.** $x^2 + \frac{3}{4}x + \frac{9}{64} = \left(x + \frac{3}{8}\right)^2$ **43.** 9 meters **45.** 11.3 inches **47.** 2.8 inches ...$3 - 2\sqrt{5}, -3 + 2\sqrt{5}\}$ **55.** $\{2 - 2i\sqrt{2}, 2 + 2i\sqrt{2}\}$ **57.** $\{7 - \sqrt{19}, 7 + \sqrt{19}\}$

59. $\{-4 - 2i\sqrt{7}, -4 + 2i\sqrt{7}\}$ **61.** $\{-8, 2\}$ **63.** $\{-6, 9\}$ **65.** $\left\{\frac{-7 - \sqrt{89}}{2}, \frac{-7 + \sqrt{89}}{2}\right\}$ **67.** $\left\{\frac{9 - 3i\sqrt{7}}{2}, \frac{9 + 3i\sqrt{7}}{2}\right\}$ **69.** $\left\{-2, -\frac{1}{2}\right\}$
71. $\{-8, 9\}$ **73.** $\left\{\frac{1}{2}, 4\right\}$ **75.** $\left\{\frac{3 - i}{2}, \frac{3 + i}{2}\right\}$ **77.** $x^2 + 3x - 10 = 0$ **79.** $x^2 - 14x + 48 = 0$ **81.** $4x^2 + 5x - 6 = 0$ **83.** $x^2 - 9 = 0$
85. $x^2 + 25 = 0$ **87.** $\{-10, -3\}$ **89.** $\{-3 - \sqrt{26}, -3 + \sqrt{26}\}$ **91.** $\{-1, 7\}$ **93.** $\left\{\frac{5}{2}, 5\right\}$ **95.** $\{4 - 3\sqrt{3}, 4 + 3\sqrt{3}\}$
97. $\{2 - 2\sqrt{6}, 2 + 2\sqrt{6}\}$ **99.** $\{-1, 6\}$ **101.** $\{4 - i\sqrt{3}, 4 + i\sqrt{3}\}$ **103.** $\{-3 - 2i\sqrt{2}, -3 + 2i\sqrt{2}\}$ **105.** $\left\{\frac{-3 - 3i\sqrt{3}}{2}, \frac{-3 + 3i\sqrt{3}}{2}\right\}$
107. $\{-16, 0\}$ **109.** $\{-5, 7\}$ **111.** $\{-1, 5\}$ **113.** $\{-5, 1\}$ **115.** Answers will vary. Using ± prevents losing a solution.

Quick Check 10.2 **1. a)** $a = 1, b = 11$, and $c = -13$ **b)** $a = 5, b = -9$, and $c = -11$ **c)** $a = 1, b = 0$, and $c = -30$ **2. a)** $\{-10, 3\}$ **b)** $\left\{\frac{-21 - 3\sqrt{21}}{14}, \frac{-21 + 3\sqrt{21}}{14}\right\}$ **3.** $\left\{\frac{7 - 3i\sqrt{3}}{2}, \frac{7 + 3i\sqrt{3}}{2}\right\}$ **4.** $\left\{\frac{-2 - i\sqrt{14}}{3}, \frac{-2 + i\sqrt{14}}{3}\right\}$ **5.** $\left\{-\frac{5}{2}, \frac{4}{3}\right\}$ **6. a)** Two real solutions **b)** Two nonreal complex solutions **c)** One real solution **7. a)** Not factorable **b)** Factorable **8.** 5 seconds **9.** 0.32 seconds, 1.93 seconds **10.** No

Section 10.2 **1.** quadratic **3.** zero **5.** two **7.** $\{-4, 9\}$ **9.** $\{2 - \sqrt{2}, 2 + \sqrt{2}\}$ **11.** $\left\{\frac{-1 - 3i\sqrt{3}}{2}, \frac{-1 + 3i\sqrt{3}}{2}\right\}$ **13.** $\{-2i\sqrt{3}, 2i\sqrt{3}\}$
15. $\left\{\frac{-7 - \sqrt{5}}{2}, \frac{-7 + \sqrt{5}}{2}\right\}$ **17.** $\{-8, 11\}$ **19.** $\left\{-\frac{3}{2}, 4\right\}$ **21.** $\{2 - 2i\sqrt{7}, 2 + 2i\sqrt{7}\}$ **23.** $\left\{\frac{3 - 3\sqrt{5}}{2}, \frac{3 + 3\sqrt{5}}{2}\right\}$ **25.** $\left\{-\frac{1}{5}, 4\right\}$ **27.** $\{3\}$
29. $\{-2\sqrt{6}, 2\sqrt{6}\}$ **31.** $\{-4, 5\}$ **33.** $\left\{\frac{1 - i\sqrt{74}}{10}, \frac{1 + i\sqrt{74}}{10}\right\}$ **35.** $\{3, 4\}$ **37.** Two real **39.** Two nonreal complex **41.** Two nonreal complex
43. One real **45.** Prime **47.** Factorable **49.** Factorable **51.** Prime **53.** $\left\{\frac{5 - \sqrt{85}}{2}, \frac{5 + \sqrt{85}}{2}\right\}$ **55.** $\left\{-1, \frac{1}{3}\right\}$ **57.** $\left\{-\frac{2}{5}, 2\right\}$ **59.** $\left\{\frac{-4 - i\sqrt{2}}{2}, \frac{-4 + i\sqrt{2}}{2}\right\}$
61. $\{-18i, 18i\}$ **63.** $\left\{\frac{4}{3}, \frac{7}{2}\right\}$ **65.** $\left\{\frac{-1 - \sqrt{10}}{2}, \frac{-1 + \sqrt{10}}{2}\right\}$ **67.** $\left\{\frac{9 - \sqrt{165}}{2}, \frac{9 + \sqrt{165}}{2}\right\}$ **69.** $\left\{\frac{3}{4}\right\}$ **71.** $\left\{\frac{-1 - i\sqrt{79}}{2}, \frac{-1 + i\sqrt{79}}{2}\right\}$ **73.** $\{2 - \sqrt{6}, 2 + \sqrt{6}\}$
75. $\{3 - i, 3 + i\}$ **77.** $\left\{\frac{-4 - \sqrt{70}}{9}, \frac{-4 + \sqrt{70}}{9}\right\}$ **79.** $\left\{-\frac{7}{3}, \frac{5}{2}\right\}$ **81.** $\left\{\frac{9 - 4\sqrt{2}}{2}, \frac{9 + 4\sqrt{2}}{2}\right\}$ **83.** $\{-4, 1\}$ **85.** $\{5 - \sqrt{7}, 5 + \sqrt{7}\}$ **87.** $\{7, 9\}$
89. 8 seconds **91.** 5.4 seconds **93.** 1 second, 4 seconds **95.** 1.5 seconds, 6.0 seconds **97.** No, the discriminant is negative. **99.** Yes, 0.9 second **101.** 96 feet per second **103.** Answers will vary. The square root of a positive number produces two real solutions; ±0 does not change the numerator; the square root of a negative number is imaginary.

Quick Check 10.3 **1.** $\{-2, 2, -i\sqrt{3}, i\sqrt{3}\}$ **2. a)** $u = \sqrt{x}$, **b)** $u = x^{1/3}$, **c)** $u = x^2 - 4x$ **3. a)** $\{1, 25\}$, **b)** $\{1, 27\}$ **4.** $\{-5\}$ **5.** $\{0, 3\}$ **6.** 71.5 minutes

Section 10.3 **1.** u-substitution **3.** $\{-1, 1, -2, 2\}$ **5.** $\{-\sqrt{3}, \sqrt{3}\}$ **7.** $\{-3i, 3i, -\sqrt{2}, \sqrt{2}\}$ **9.** $\{-2, 2, -3, 3\}$ **11.** $\{16, 81\}$ **13.** $\{36\}$ **15.** $\varnothing$
17. $\{1, 49\}$ **19.** $\{\varnothing\}$ **21.** $\{100\}$ **23.** $\{8, 27\}$ **25.** $\{-216, 1\}$ **27.** $\{-125, -64\}$ **29.** $\{-1, 2\}$ **31.** $\{3, 9\}$ **33.** $\{-8, 2\}$ **35.** $\{2, 5\}$ **37.** $\{1\}$
39. $\{2, 10\}$ **41.** $\{-10, 4\}$ **43.** $\{-12, 5\}$ **45.** $\{1\}$ **47.** $\left\{-\frac{13}{5}, 5\right\}$ **49.** Small: 120 minutes; large: 60 minutes **51.** 6.5 hours **53.** 88.2 minutes
55. $\{4 - \sqrt{29}, 4 + \sqrt{29}\}$ **57.** $\{64\}$ **59.** $\left\{\frac{2 - 4\sqrt{2}}{3}, \frac{2 + 4\sqrt{2}}{3}\right\}$ **61.** $\left\{\frac{5 - i\sqrt{31}}{2}, \frac{5 + i\sqrt{31}}{2}\right\}$ **63.** $\{-7, 13\}$ **65.** $\{-10, -2\}$ **67.** $\{6\}$ **69.** $\left\{-\frac{8}{5}, 0\right\}$
71. $\{-8 - 12i, -8 + 12i\}$ **73.** $\{-i, i, -2, 2\}$ **75.** $\{-343, -125\}$ **77.** $\{-5\}$ **79.** After replacing u with x^2, take the square root of each side.

Quick Review Exercises **1.** $\{-4, 10\}$ **2.** $\{-3, 12\}$ **3.** $\{3 - \sqrt{22}, 3 + \sqrt{22}\}$ **4.** $\left\{\frac{-11 - i\sqrt{39}}{2}, \frac{-11 + i\sqrt{39}}{2}\right\}$

Quick Check 10.4 **1.**

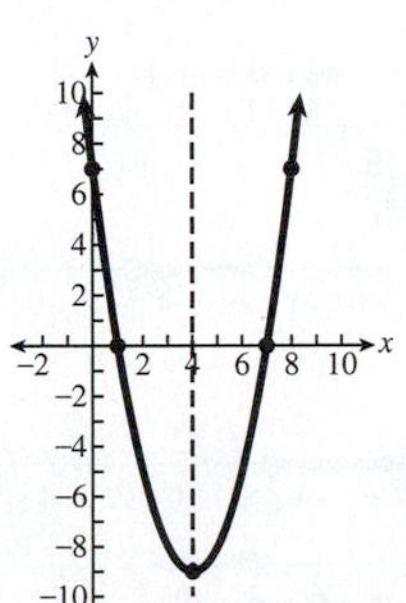

2.

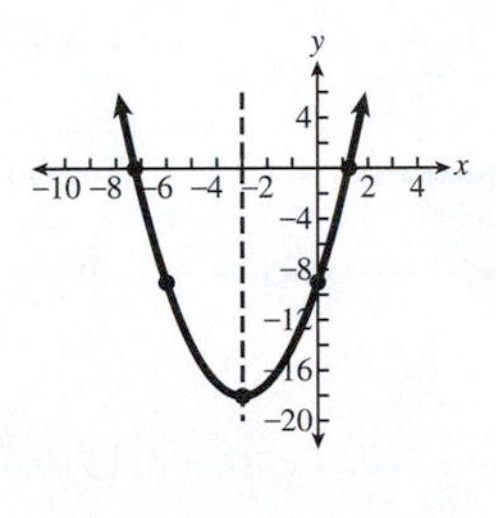

3.

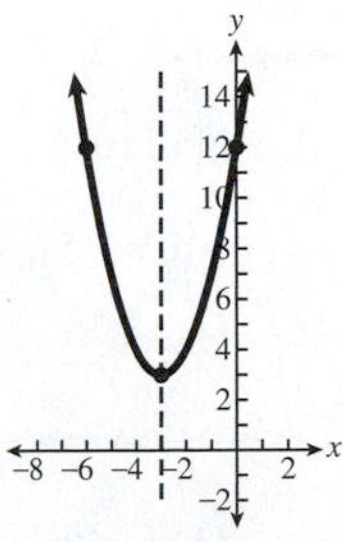

4.

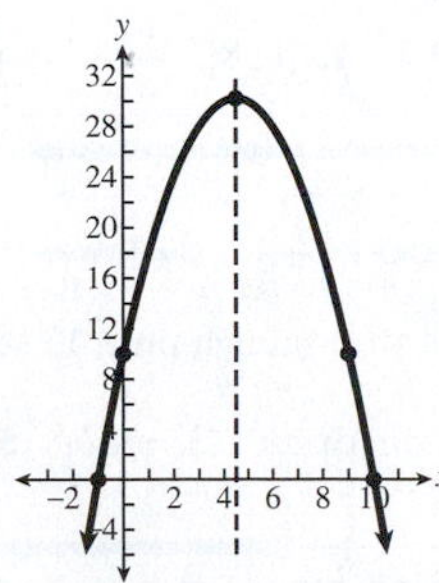

5.

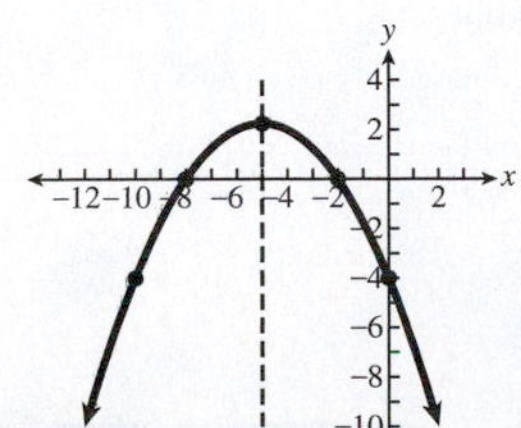

6. a) $(-2, -8), x = -2$ **b)** $(8, 7), x = 8$ **7.**

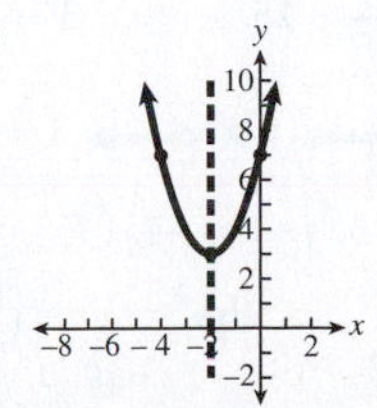

8.

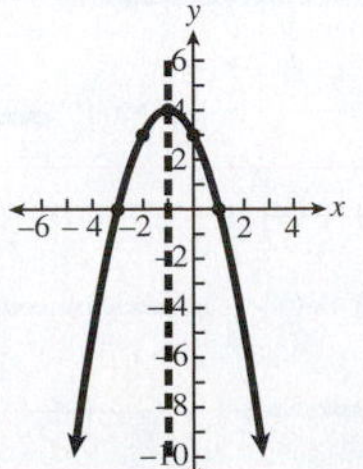

Section 10.4 **1.** parabola **3.** negative **5.** y-intercept **7.** $(-4, -38), x = -4$ **9.** $\left(\frac{7}{2}, -\frac{9}{4}\right), x = \frac{7}{2}$ **11.** $(6, 98), x = 6$ **13.** $(-2, -27), x = -2$
15. $(3, 25), x = 3$ **17.** $\left(-\frac{3}{4}, \frac{119}{16}\right), x = -\frac{3}{4}$ **19.** $(-5, -225), x = -5$ **21.** $(7, 12), x = 7$ **23.** $(-1, -5), x = -1$ **25.** $(9, -7), x = 9$
27. $(-8, 0), (5, 0), (0, -40)$ **29.** $(0.7, 0), (4.3, 0), (0, 3)$ **31.** No x-intercepts, $(0, 15)$ **33.** $(-4.6, 0), (1.6, 0), (0, -15)$ **35.** $(-6, 0), (1, 0), (0, 6)$
37. $(-2.4, 0), (8.4, 0), (0, 20)$ **39.** No x-intercepts, $(0, -15)$ **41.** $(4, 0), (0, 16)$ **43.** $(4, 0), (10, 0), (0, 40)$ **45.** No x-intercepts, $(0, 29)$

47. $(-2.5, 0), (4.5, 0), (0, 11)$ **49.** **51.** **53.**

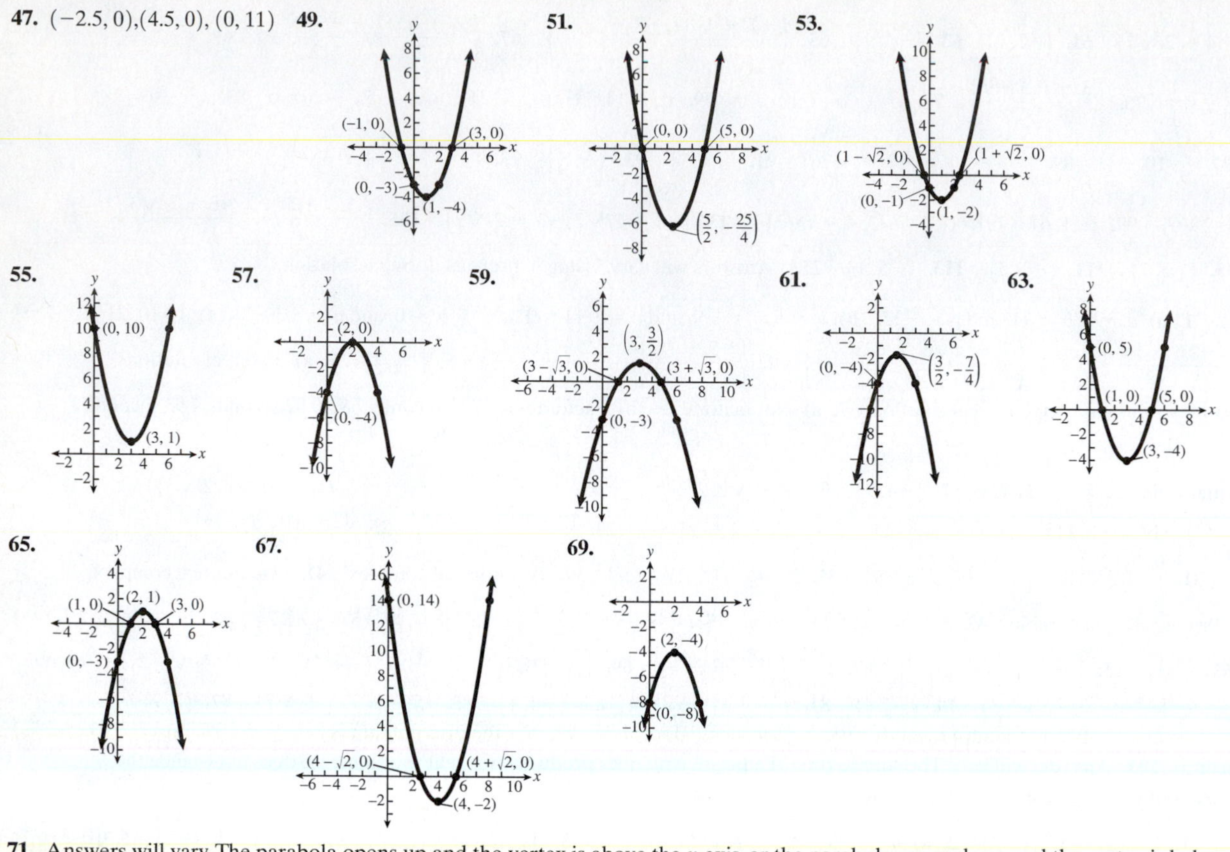

71. Answers will vary. The parabola opens up and the vertex is above the x-axis, or the parabola opens down and the vertex is below the x-axis.

Quick Check 10.5 **1.** Base: 20 inches; height: 6 inches **2.** Length: 23.1 inches; width: 15.1 inches **3.** 9.80 inches **4.** 7.7 feet **5.** 250 feet

Section 10.5 **1.** $A = \frac{1}{2}bh$ **3.** legs **5.** Length: 11 inches; width: 6 inches **7.** Length: 15.4 meters; width: 7.7 meters **9.** Length: 9.9 feet; width: 6.8 feet **11.** Base: 12 inches; height: 7 inches **13.** Base: 3.6 inches; height: 8.2 inches **15.** Width: 8 inches; height: 10 inches **17.** Length: 61.0 inches; width: 35.5 inches **19.** Length: 32.6 inches; width: 16.6 inches **21.** Height: 40 inches; width: 32 inches **23.** Base: 32 inches; height: 16 inches **25.** Base: 6.4 feet; height: 14.4 feet **27.** 18.9 inches **29.** 12 feet **31.** 20.8 meters **33.** 100 miles **35.** 6 feet **37.** 30.7 feet **39.** 16.4 inches **41.** Length: 5.1 feet; width: 4.1 feet **43. a)** Base: 12 feet; height: 9 feet **b)** 15 feet **45. a)** 69.3 centimeters **b)** 2772 sq cm **47.** Answers will vary. The length is 5 feet more than the width; the area is 126 sq.ft.

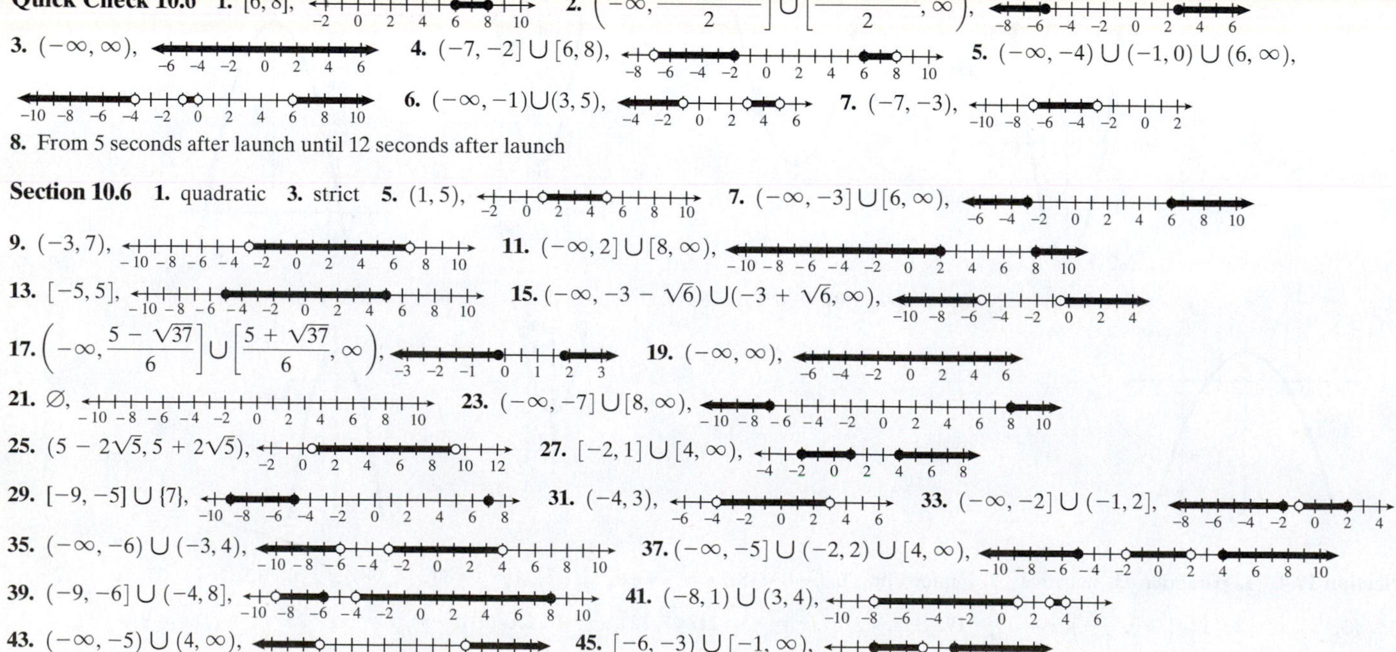

Quick Check 10.6 **1.** $[6, 8]$, **2.** $\left(-\infty, \frac{-3 - \sqrt{69}}{2}\right] \cup \left[\frac{-3 + \sqrt{69}}{2}, \infty\right)$, **3.** $(-\infty, \infty)$, **4.** $(-7, -2] \cup [6, 8)$, **5.** $(-\infty, -4) \cup (-1, 0) \cup (6, \infty)$, **6.** $(-\infty, -1) \cup (3, 5)$, **7.** $(-7, -3)$, **8.** From 5 seconds after launch until 12 seconds after launch

Section 10.6 **1.** quadratic **3.** strict **5.** $(1, 5)$, **7.** $(-\infty, -3] \cup [6, \infty)$, **9.** $(-3, 7)$, **11.** $(-\infty, 2] \cup [8, \infty)$, **13.** $[-5, 5]$, **15.** $(-\infty, -3 - \sqrt{6}) \cup (-3 + \sqrt{6}, \infty)$, **17.** $\left(-\infty, \frac{5 - \sqrt{37}}{6}\right] \cup \left[\frac{5 + \sqrt{37}}{6}, \infty\right)$, **19.** $(-\infty, \infty)$, **21.** $\varnothing$, **23.** $(-\infty, -7] \cup [8, \infty)$, **25.** $(5 - 2\sqrt{5}, 5 + 2\sqrt{5})$, **27.** $[-2, 1] \cup [4, \infty)$, **29.** $[-9, -5] \cup \{7\}$, **31.** $(-4, 3)$, **33.** $(-\infty, -2] \cup (-1, 2]$, **35.** $(-\infty, -6) \cup (-3, 4)$, **37.** $(-\infty, -5] \cup (-2, 2) \cup [4, \infty)$, **39.** $(-9, -6] \cup (-4, 8]$, **41.** $(-8, 1) \cup (3, 4)$, **43.** $(-\infty, -5) \cup (4, \infty)$, **45.** $[-6, -3) \cup [-1, \infty)$,

47. $[-2, 2)$, **49.** 1 second to 3 seconds **51.** 0.5 second to 1.5 seconds **53. a)** $h(t) = -16t^2 + 125t + 48$ **b)** 1.5 seconds to 6.3 seconds **55.** $[2, 6]$ **57.** $(-\infty, 2] \cup [3, \infty)$ **59.** $[-9, 6)$ **61.** $(-\infty, -7) \cup (-6, -3) \cup (11, \infty)$ **63.** $x^2 - 2x - 3 < 0$ **65.** $x^2 - 4 \geq 0$ **67.** $\frac{x-5}{x+4} \leq 0$ **69.** $\frac{x^2 - 2x - 3}{x^2 + 10x + 24} \leq 0$

Chapter 10 Review **1.** $\{-7, -4\}$ **2.** $\{-10, 3\}$ **3.** $\{-4, 0\}$ **4.** $\{\frac{3}{2}, 5\}$ **5.** $\{-5, 1\}$ **6.** $\{7, 11\}$ **7.** $\{-7, 7\}$ **8.** $\{-6\sqrt{2}, 6\sqrt{2}\}$ **9.** $\{-2i, 2i\}$ **10.** $\{6 - 6i, 6 + 6i\}$ **11.** $\{8 - 3i, 8 + 3i\}$ **12.** $\{-9, 2\}$ **13.** $\{-2, 4\}$ **14.** $\{3 - \sqrt{6}, 3 + \sqrt{6}\}$ **15.** $\{-7 - \sqrt{2}, -7 + \sqrt{2}\}$ **16.** $\left\{\frac{-5 - i\sqrt{15}}{2}, \frac{-5 + i\sqrt{15}}{2}\right\}$ **17.** $\left\{\frac{-7 - 3\sqrt{5}}{2}, \frac{-7 + 3\sqrt{5}}{2}\right\}$ **18.** $\{-2 - \sqrt{6}, -2 + \sqrt{6}\}$ **19.** $\left\{\frac{3 - 3i\sqrt{3}}{2}, \frac{3 + 3i\sqrt{3}}{2}\right\}$ **20.** $\left\{\frac{5 - 2i}{2}, \frac{5 + 2i}{2}\right\}$ **21.** $\{-7, 3\}$ **22.** $\{6 - i, 6 + i\}$ **23.** $\{-1, 1, -2, 2\}$ **24.** $\{64\}$ **25.** $\{1\}$ **26.** $\{-27, 343\}$ **27.** $\{-3, 2\}$ **28.** $\{-6\}$ **29.** $\{-6, -5\}$ **30.** $\{6\}$ **31.** $\{9, 10\}$ **32.** $\{-2, 2, -3i, 3i\}$ **33.** $\{-6 - 2i\sqrt{2}, -6 + 2i\sqrt{2}\}$ **34.** $\{-2 - 2i\sqrt{3}, -2 + 2i\sqrt{3}\}$ **35.** $\{-10, 0\}$ **36.** $\{3 - \sqrt{29}, 3 + \sqrt{29}\}$ **37.** $\{5\}$ **38.** $\{20\}$ **39.** $\{-4, 5\}$ **40.** $\left\{\frac{4 - \sqrt{10}}{3}, \frac{4 + \sqrt{10}}{3}\right\}$ **41.** $\{9\}$ **42.** $\{-9 - 8i, -9 + 8i\}$ **43.** $\{-16, -8\}$ **44.** $\left\{\frac{7 - \sqrt{129}}{2}, \frac{7 + \sqrt{129}}{2}\right\}$ **45.** $x^2 + 5x - 36 = 0$ **46.** $x^2 - 8x + 16 = 0$ **47.** $6x^2 - 47x + 52 = 0$ **48.** $x^2 - 8 = 0$ **49.** $x^2 + 25 = 0$ **50.** $x^2 + 64 = 0$ **51.** $(4, -36), x = 4$ **52.** $\left(-\frac{5}{2}, -\frac{89}{4}\right), x = -\frac{5}{2}$ **53.** $(-1, 36), x = -1$ **54.** $(3, -4), x = 3$ **55.** No x-intercepts; y-int.: $(0, 11)$ **56.** x-int.: $(2 - \sqrt{7}, 0), (2 + \sqrt{7}, 0)$; y-int.: $(0, -3)$ **57.** x-int.: $(5, 0)$; y-int.: $(0, -25)$ **58.** x-int.: $(-4 - 2\sqrt{2}, 0), (-4 + 2\sqrt{2}, 0)$; y-int.: $(0, 8)$

59.

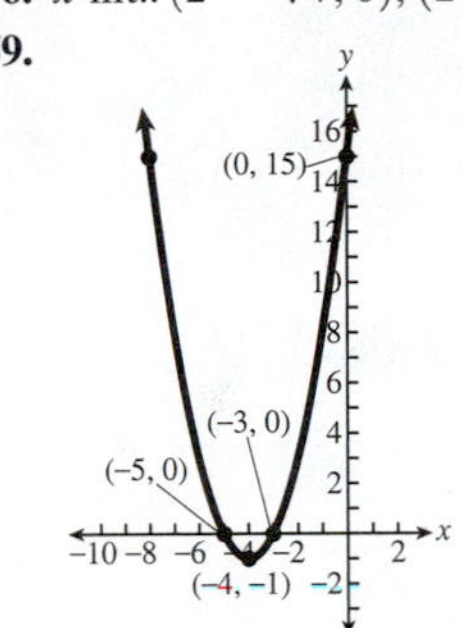

60.

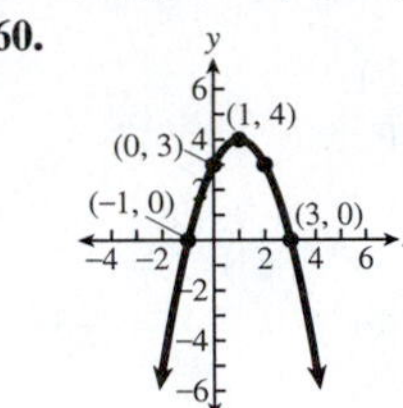

61.

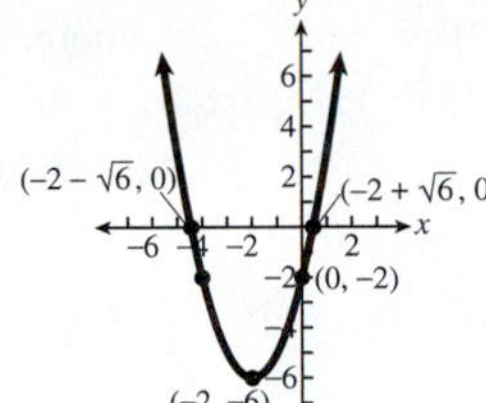

62.

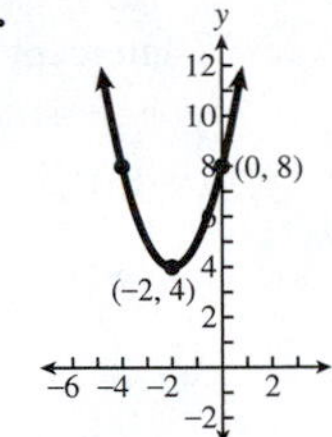

63.

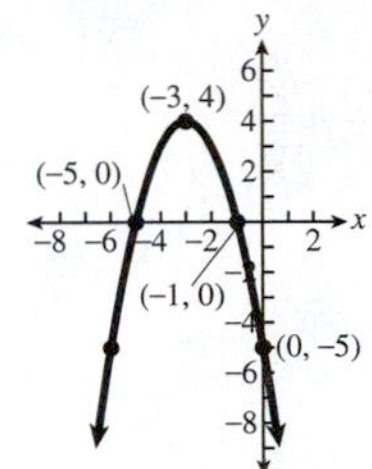

64.

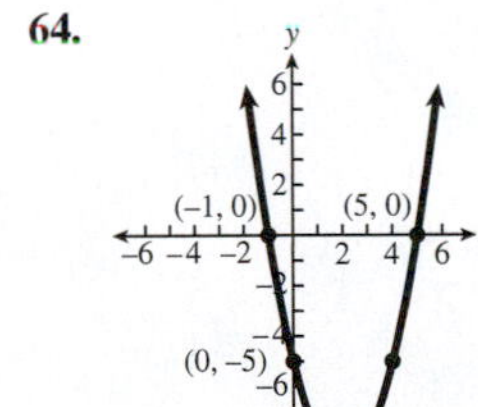

65. $[2, 7]$, **66.** $(-\infty, -8) \cup (-1, \infty)$,

67. $(-\infty, -2 - \sqrt{15}) \cup (-2 + \sqrt{15}, \infty)$ **68.** $(-\infty, \infty)$,

69. $(-\infty, 6] \cup (7, 9]$, **70.** $(-\infty, -10) \cup [-2, 2] \cup (8, \infty)$,

71. $(-\infty, -10] \cup [-5, \infty)$, **72.** $(-\infty, -4] \cup (1, 4] \cup (9, \infty)$,

73. $6\frac{1}{2}$ seconds **74.** 4.54 seconds **75.** 9.1 hours **76.** 61.1 hours **77.** Length: 18 centimeters; width: 16 centimeters **78.** Length: 29.8 feet; width: 16.8 feet **79.** 1280.6 miles **80.** Length: 9.9 feet; width: 6.9 feet

Chapter 10 Review Exercises: Worked-Out Solutions

1.
$$x^2 + 11x + 28 = 0$$
$$(x + 7)(x + 4) = 0$$
$$x + 7 = 0 \quad \text{or} \quad x + 4 = 0$$
$$x = -7 \quad \text{or} \quad x = -4$$
$$\{-7, -4\}$$

11.
$$(x - 8)^2 + 30 = 21$$
$$(x - 8)^2 = -9$$
$$\sqrt{(x - 8)^2} = \pm\sqrt{-9}$$
$$x - 8 = \pm 3i$$
$$x = 8 \pm 3i$$
$$\{8 - 3i, 8 + 3i\}$$

13.
$$x^2 - 2x - 8 = 0$$
$$x^2 - 2x = 8$$
$$x^2 - 2x + 1 = 8 + 1$$
$$(x - 1)^2 = 9$$
$$\sqrt{(x - 1)^2} = \pm\sqrt{9}$$
$$x - 1 = \pm 3$$
$$x = 1 \pm 3$$
$$x = 1 + 3 \quad \text{or} \quad x = 1 - 3$$
$$x = 4 \quad \text{or} \quad x = -2$$
$$\{-2, 4\}$$

17.
$$x = \frac{-7 \pm \sqrt{7^2 - 4(1)(1)}}{2(1)}$$
$$= \frac{-7 \pm \sqrt{49 - 4}}{2}$$
$$= \frac{-7 \pm \sqrt{45}}{2}$$
$$= \frac{-7 \pm 3\sqrt{5}}{2}$$
$$\left\{\frac{-7 - 3\sqrt{5}}{2}, \frac{-7 + 3\sqrt{5}}{2}\right\}$$

23.
$$\begin{aligned} u &= x^2 \\ u^2 - 5u + 4 &= 0 \\ (u-1)(u-4) &= 0 \\ u - 1 = 0 \quad &\text{or} \quad u - 4 = 0 \\ u = 1 \quad &\text{or} \quad u = 4 \\ x^2 = 1 \quad &\text{or} \quad x^2 = 4 \\ \sqrt{x^2} = \pm\sqrt{1} \quad &\text{or} \quad \sqrt{x^2} = \pm\sqrt{4} \\ x = \pm 1 \quad &\text{or} \quad x = \pm 2 \end{aligned}$$
$\{-1, 1, -2, 2\}$

27.
$$\begin{aligned} \sqrt{x^2 + x - 2} &= 2 \\ (\sqrt{x^2 + x - 2})^2 &= 2^2 \\ x^2 + x - 2 &= 4 \\ x^2 + x - 6 &= 0 \\ (x+3)(x-2) &= 0 \\ x + 3 = 0 \quad &\text{or} \quad x - 2 = 0 \\ x = -3 \quad &\text{or} \quad x = 2 \end{aligned}$$
$\{-3, 2\}$

29.
$$\begin{aligned} 1 + \tfrac{11}{x} + \tfrac{30}{x^2} &= 0 \\ x^2\left(1 + \tfrac{11}{x} + \tfrac{30}{x^2}\right) &= x^2 \cdot 0 \\ x^2 \cdot 1 + x^2 \cdot \tfrac{11}{x} + x^2 \cdot \tfrac{30}{x^2} &= 0 \\ x^2 + 11x + 30 &= 0 \\ (x+6)(x+5) &= 0 \\ x + 6 = 0 \quad &\text{or} \quad x + 5 = 0 \\ x = -6 \quad &\text{or} \quad x = -5 \end{aligned}$$
$\{-6, -5\}$

43.
$$\begin{aligned} f(x) &= 0 \\ (x+12)^2 - 16 &= 0 \\ (x+12)^2 &= 16 \\ \sqrt{(x+12)^2} &= \pm\sqrt{16} \\ x + 12 &= \pm 4 \\ x &= -12 \pm 4 \\ x = -12 + 4 \quad &\text{or} \quad x = -12 - 4 \\ x = -8 \quad &\text{or} \quad x = -16 \end{aligned}$$
$\{-16, -8\}$

45.
$$\begin{aligned} x = -9 \quad &\text{or} \quad x = 4 \\ x + 9 = 0 \quad &\text{or} \quad x - 4 = 0 \\ (x+9)(x-4) &= 0 \\ x^2 + 5x - 36 &= 0 \end{aligned}$$

59. Vertex

$x = \frac{-8}{2(1)} = -4$

$y = (-4)^2 + 8(-4) + 15 = -1$

$(-4, -1)$

y-intercept $(x = 0)$

$y = 0^2 + 8(0) + 15 = 15$

$(0, 15)$

x-intercept $(y = 0)$

$$\begin{aligned} x^2 + 8x + 15 &= 0 \\ (x+5)(x+3) &= 0 \\ x + 5 = 0 \quad &\text{or} \quad x + 3 = 0 \\ x = -5 \quad &\text{or} \quad x = -3 \end{aligned}$$
$(-5, 0), (-3, 0)$

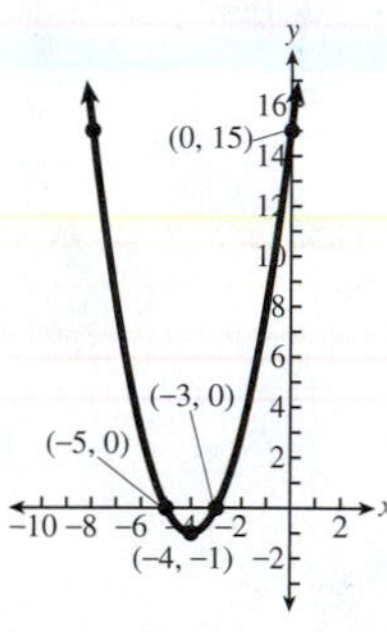

63. Vertex (h, k)

$(-3, 4)$

y-intercept $(x = 0)$

$y = -(0+3)^2 + 4$

$y = -9 + 4$

$y = -5$

$(0, -5)$

x-intercept $(y = 0)$

$$\begin{aligned} 0 &= -(x+3)^2 + 4 \\ (x+3)^2 &= 4 \\ \sqrt{(x+3)^2} &= \pm\sqrt{4} \\ x + 3 &= \pm 2 \\ x &= -3 \pm 2 \end{aligned}$$
$(-5, 0), (-1, 0)$

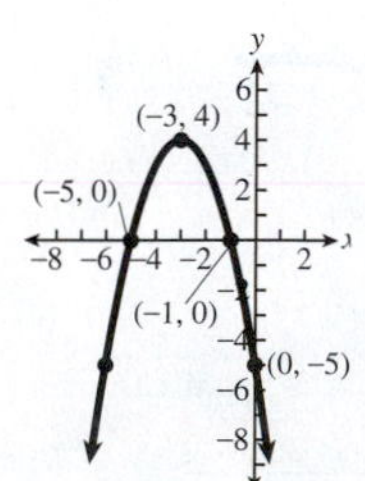

65.
$$\begin{aligned} x^2 - 9x + 14 &= 0 \\ (x-2)(x-7) &= 0 \end{aligned}$$
Critical Values: $x = 2, 7$

Test Point	$x - 2$	$x - 7$	$(x-2)(x-7)$
0	$-$	$-$	$+$
3	$+$	$-$	$-$
8	$+$	$+$	$+$

$[2, 7]$

69.
$$\begin{aligned} x^2 - 15x + 54 &= 0 \\ (x-6)(x-9) &= 0 \\ x = 6 \quad &\text{or} \quad x = 9 \end{aligned}$$

$x - 7 = 0$

$x = 7$

Critical Values: $x = 6, 7, 9$

Test Point	$x - 6$	$x - 9$	$x - 7$	$\frac{(x-6)(x-9)}{x-7}$
0	$-$	$-$	$-$	$-$
6.5	$+$	$-$	$-$	$+$
8	$+$	$-$	$+$	$-$
10	$+$	$+$	$+$	$+$

$(-\infty, 6] \cup (7, 9]$

74. $-16t^2 + 66t + 30 = 0$
$-2(8t^2 - 33t - 15) = 0$
$t = \frac{-(-33) \pm \sqrt{(-33)^2 - 4(8)(-15)}}{2(8)}$
$= \frac{33 \pm \sqrt{1569}}{16}$
$\frac{33 + \sqrt{1569}}{16} \approx 4.54, \frac{33 - \sqrt{1569}}{16} \approx -0.41$
Omit the negative solution.
4.54 seconds.

80. Length: x, Width: $x - 3$
$x^2 + (x - 3)^2 = 12^2$
$x^2 + x^2 - 6x + 9 = 144$
$2x^2 - 6x - 135 = 0$
$x = \frac{-(-6) \pm \sqrt{(-6)^2 - 4(2)(-135)}}{2(2)}$
$= \frac{6 \pm \sqrt{1116}}{4}$
$\frac{6 + \sqrt{1116}}{4} \approx 9.9, \frac{6 - \sqrt{1116}}{4} \approx -6.9$
Omit the negative solution.
Length: 9.9 feet, Width: $9.9 - 3 = 6.9$ feet

Chapter 10 Test **1.** $\{-6, -3\}$ **2.** $\left\{0, \frac{16}{3}\right\}$ **3.** $\{7 - 2\sqrt{13}, 7 + 2\sqrt{13}\}$ **4.** $\{3 - i\sqrt{3}, 3 + i\sqrt{3}\}$ **5.** $\{-2\sqrt{3}, 2\sqrt{3}, -2i, 2i\}$ **6.** $\{1\}$ **7.** $\left\{-1, \frac{3}{2}\right\}$ **8.** $\left\{-\frac{3}{2}, 9\right\}$ **9.** $\{4, 10\}$ **10.** $\{-4, 14\}$ **11.** $\{5 - 3i, 5 + 3i\}$ **12.** $7x^2 - 39x + 20 = 0$ **13.** $x^2 + 16 = 0$ **14.** $(-4, -55)$ **15.** $(7, 20)$ **16.** x-int.: $(2.8, 0), (7.2, 0)$; y-int.: $(0, 20)$ **17.**

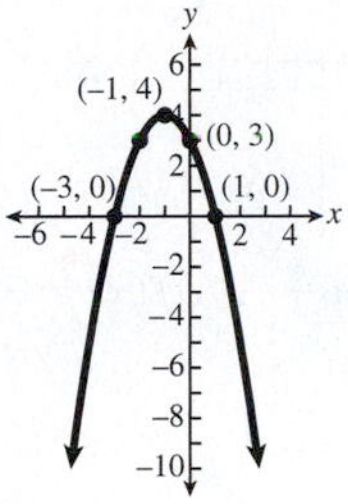

18.

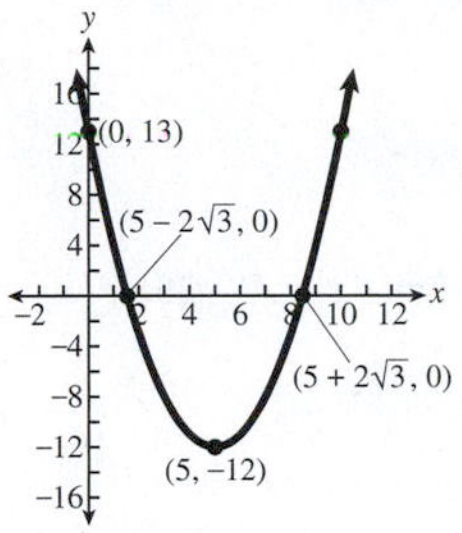

19. $(-\infty, -7) \cup (4, \infty)$,

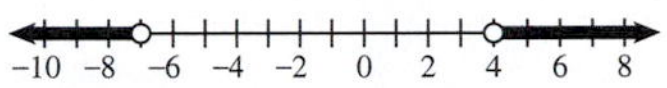

20. $[-8, -3] \cup (-2, 7)$,

21. Length: 97.1 yards; width: 37.1 yards **22.** 1700 miles

CHAPTER 11

Quick Check 11.1 **1. a)** Yes **b)** No **2.** No **3.** No **4.** 20 **5.** $-12m + 57$

Section 11.1 **1.** function **3.** evaluating **5.** Yes **7.** Yes **9.** No **11.** Function; domain: {1960, 1964, 1968, 1972, 1976, 1980, 1984}; range: {Kennedy, Johnson, Nixon, Carter, Reagan} **13.** Function; domain: $\{-5, -3, -1, 1, 3, 5\}$; range: $\{-5, -3, -1, 1, 3, 5\}$ **15.** Function; domain: $\{-6, -3, 0, 3, 6\}$; range: $\{5\}$ **17.** Yes **19.** No **21.** No **23.** No **25.** -21 **27.** -5 **29.** 4 **31.** $7a + 38$ **33.** $-10n - 21$ **35.** 52 **37.** 128 **39.** 118 **41.** 31 **43.** Domain: $(-\infty, \infty)$; range: $[-1, \infty)$, x-intercept: $(2, 0), (4, 0)$; y-intercept: $(0, 8)$ **45.** Domain: $[-4, \infty)$; range: $[2, \infty)$, x-intercept: none; y-intercept: $(0, 4)$ **47.** 4 **49.** 8 **51.** 5 **53.** $-1, 9$ **55.**

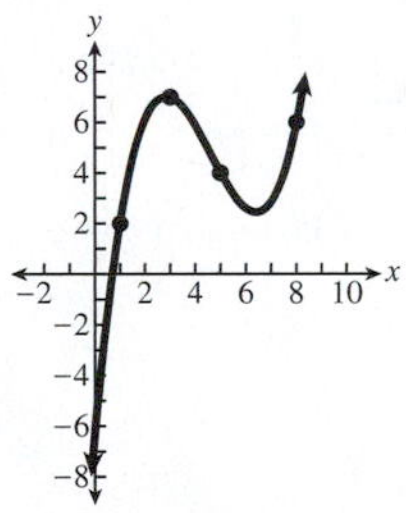

57.

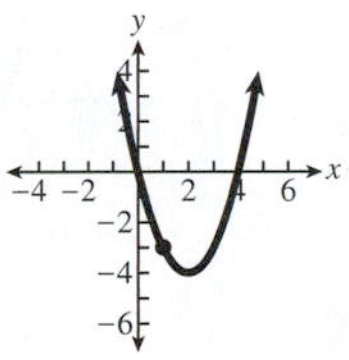

59. Answers will vary. Two y values are associated with one x value. **61.** Answers will vary. Find x-coordinates of all points with a y-coordinate of 2.

Quick Review Exercises **1.**

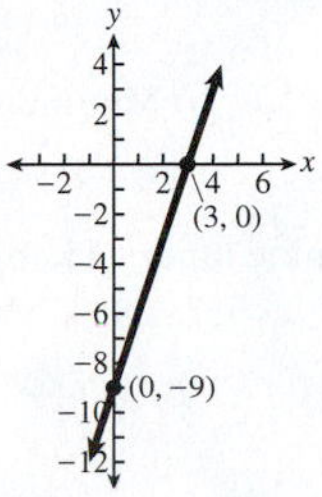

2.

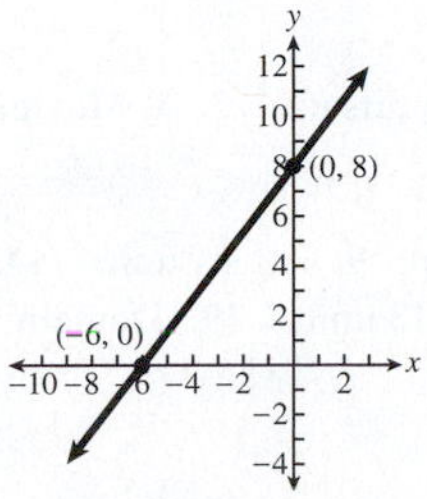

3.

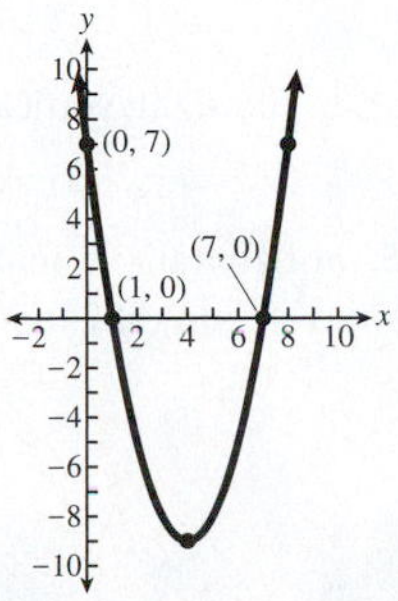

4.

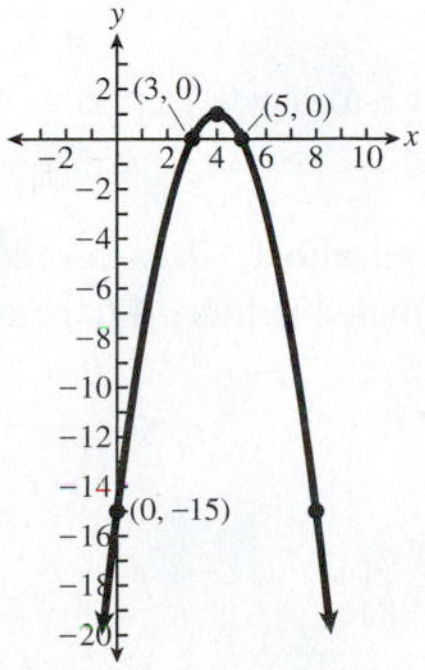

Quick Check 11.2 **1.**

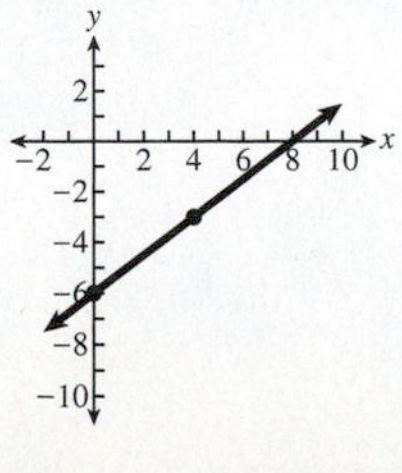

2.

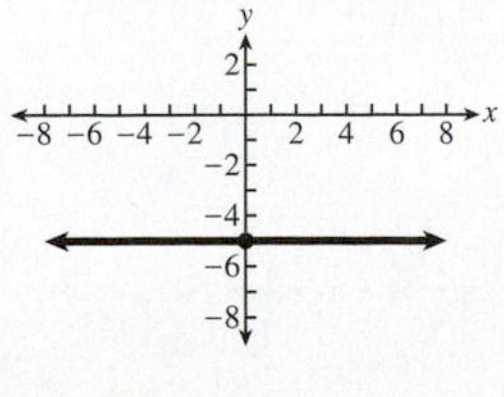

3. $f(x) = 2x - 6$ **4.** $f(x) = 20x + 500$
5. a) $C(x) = 50 + 1.25x, R(x) = 9.95x, P(x) = 8.70x - 50$
b) 121.

Section 11.2 **1.** linear **3.** revenue **5.**

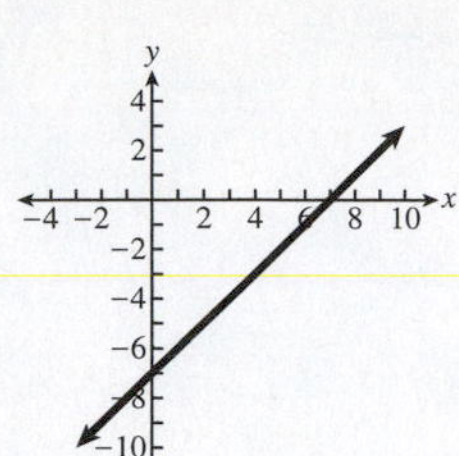

7.

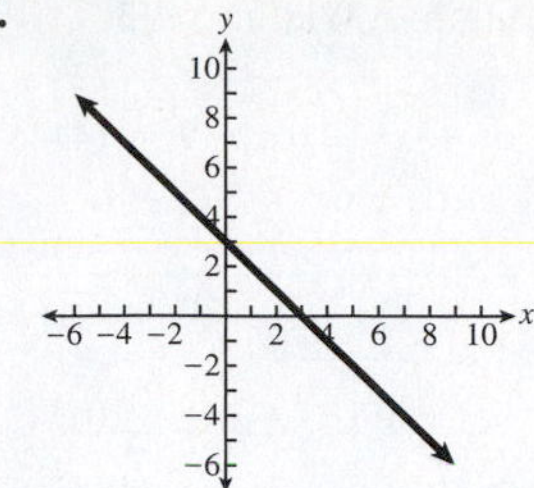

9.

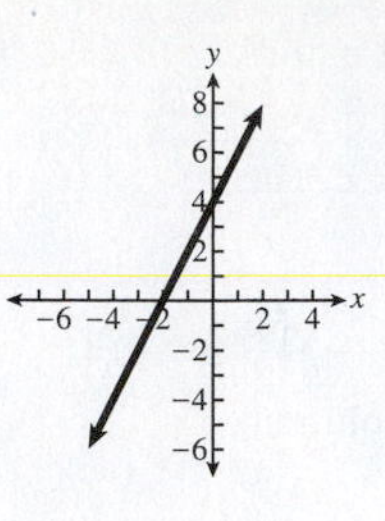

11.

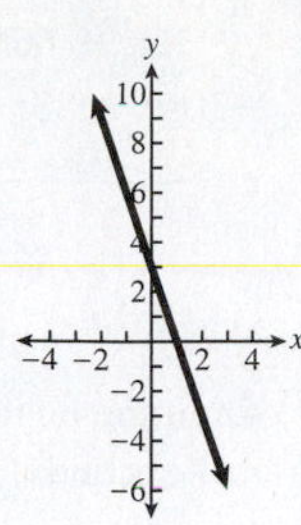

13.

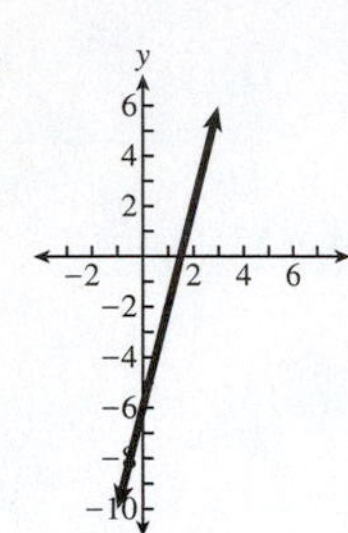

15.

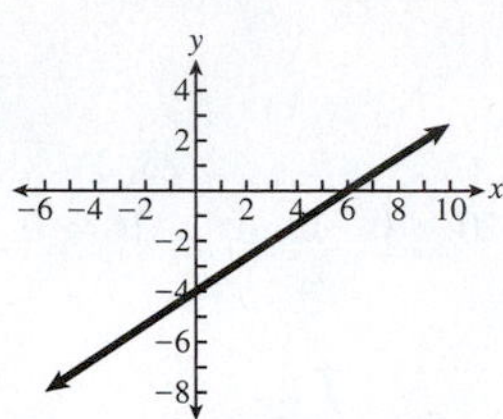

17.

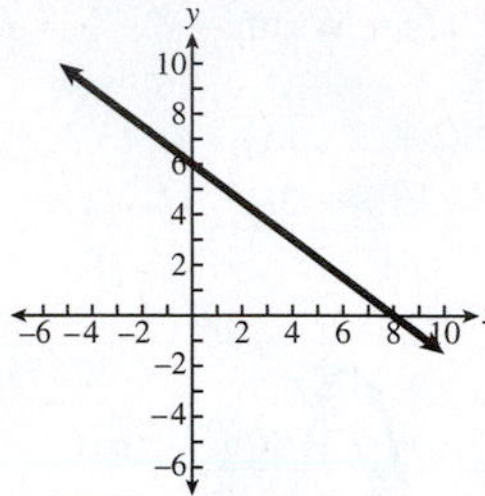

19.

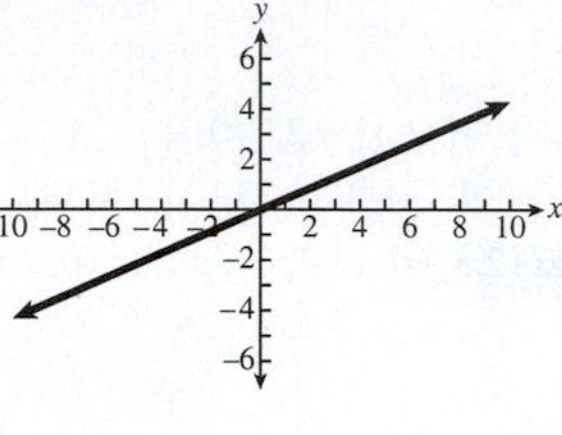

21.

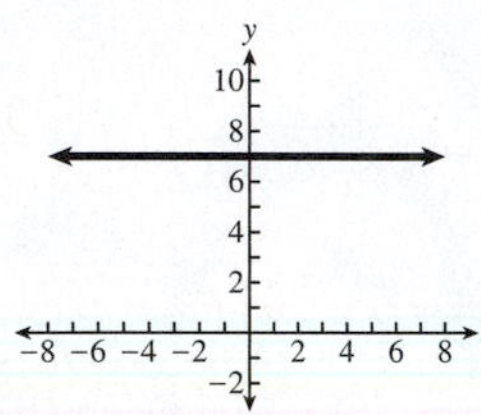

23. $f(x) = 3x + 7$ **25.** $f(x) = -\frac{4}{3}x + \frac{7}{3}$ **27.** $f(x) = 3x$ **29.** $f(x) = -6$ **31. a)** $f(x) = 0.03x + 0.39$
b) Slope: \$0.03/minute; y-intercept: \$0.39 connection fee **c)** \$1.71 **33. a)** $f(x) = 0.15x + 20$ **b)** Slope: \$0.15/mile; y-intercept: \$20 base fee **c)** \$59
35. a) $C(x) = 125$, $R(x) = 4x$, $P(x) = 4x - 125$ **b)** \$175 **37. a)** $C(x) = 0.40x + 12$, $R(x) = 3x$, $P(x) = 2.6x - 12$ **b)** \$144
39. a) $C(x) = 7.50x + 1200$, $R(x) = 29.95x$, $P(x) = 22.45x - 1200$ **b)** At least 499. **41.** $m = 3$ **43.** $m = 2$ **45.** Answers will vary. Plot the y-intercept; then use the slope. **47.** Answers will vary. Fixed cost is \$9540 and cost/item is \$3; sell the item for \$7.

Quick Check 11.3 **1.**

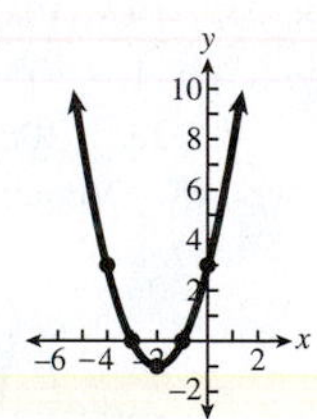

Domain: $(-\infty, \infty)$; range: $[-1, \infty)$ **2.**

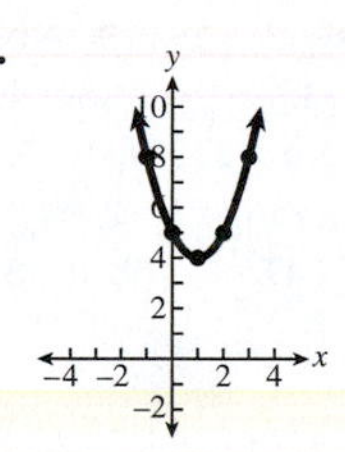

Domain: $(-\infty, \infty)$; range: $[4, \infty)$

3.

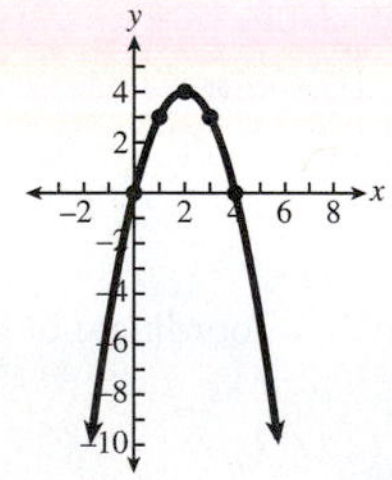

Domain: $(-\infty, \infty)$; range: $(-\infty, 4]$ **4. a)** Rotation about x-axis: no; horizontal: 13 units to the left; vertical: 7 units down
b) Rotation about x-axis: yes; horizontal: 4 units to the right; vertical: 30 units up **5. a)** Maximum: -10 **b)** Minimum: 47 **6.** 961
7. 4356 square meters **8.** $2x + h + 3$

Section 11.3 **1.** parabola **3.** y-intercept **5.** horizontal **7.** minimum **9.** left six units **11.** up nine units **13.** right five units, down nine units **15.** left 10 units, down 40 units **17.** rotate about x-axis, right 2 units, up 13 units **19.** Domain: $(-\infty, \infty)$; range: $[7, \infty)$

(0, 7)

21. 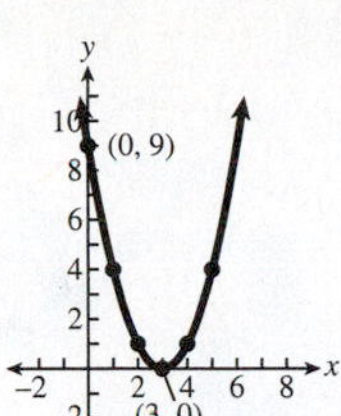

Domain: $(-\infty, \infty)$; range: $[0, \infty)$

23. 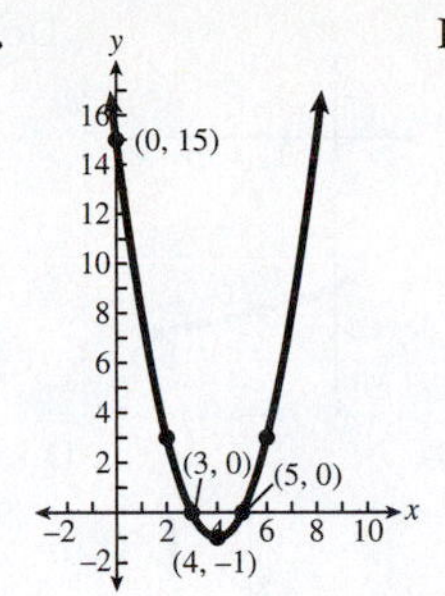

Domain: $(-\infty, \infty)$; range: $[-1, \infty)$

25. 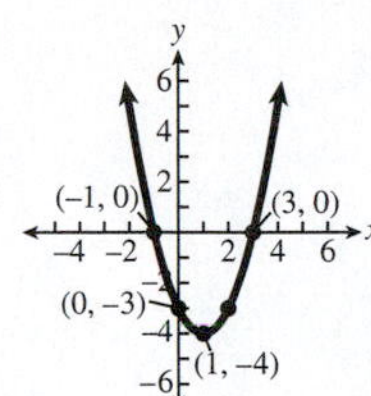

Domain: $(-\infty, \infty)$; range: $[-4, \infty)$

27. 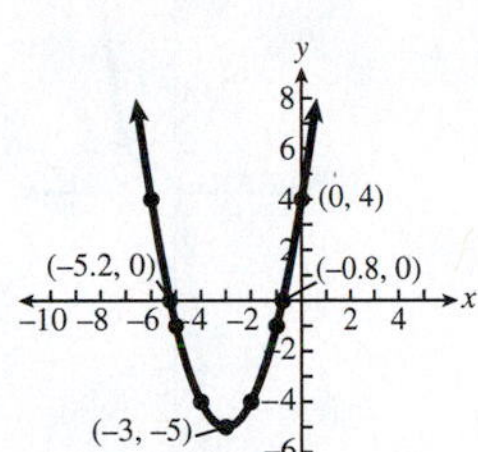

Domain: $(-\infty, \infty)$; range: $[-5, \infty)$

29. 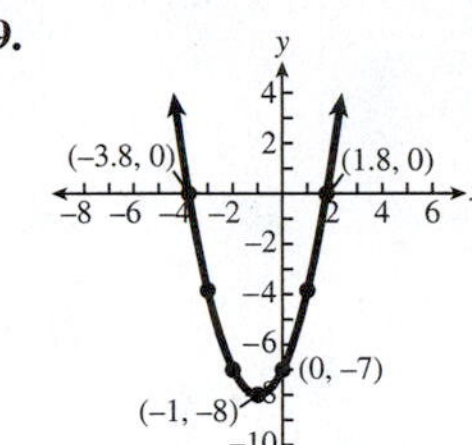

Domain: $(-\infty, \infty)$; range: $[-8, \infty)$

31. 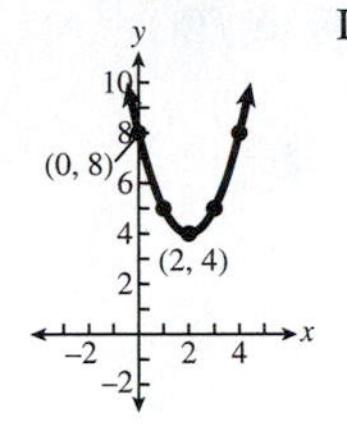

Domain: $(-\infty, \infty)$; range: $[4, \infty)$

33. 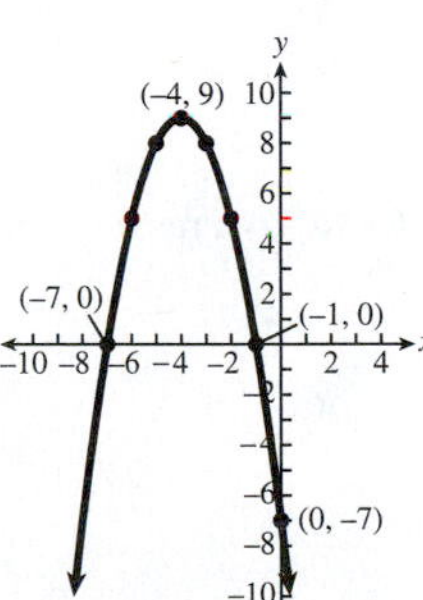

Domain: $(-\infty, \infty)$; range: $(-\infty, 9]$

35. 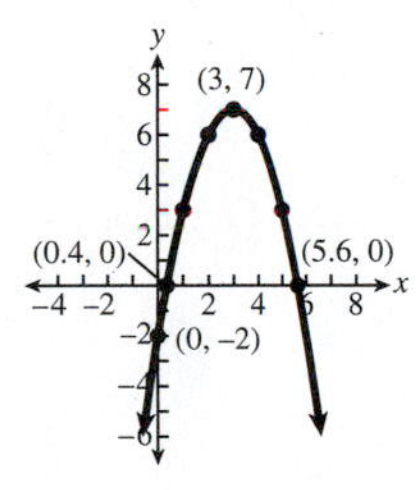

Domain: $(-\infty, \infty)$; range: $(-\infty, 7]$

37. 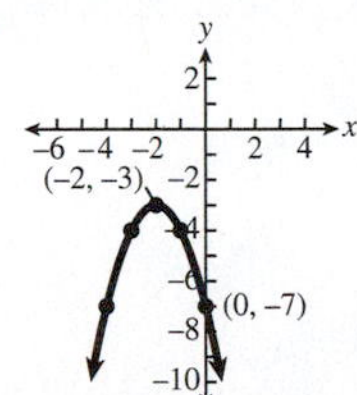

Domain: $(-\infty, \infty)$; range: $(-\infty, -3]$ **39.** Minimum, 4 **41.** Maximum, -26 **43.** Minimum, 18 **45.** Maximum, -33

47. Minimum, 46 **49.** Minimum, $\frac{263}{3}$ **51.** 21 feet **53.** 1.725 meters **55.** 144 **57.** 75 feet by 75 feet, 5625 square feet **59.** 70,000 pans, \$3.4375/pan **61.** \$3 increase, \$37,750 revenue **63.** $f(x) = (x - 6)^2$ **65.** $f(x) = (x + 3)^2 - 11$ **67.** $f(x) = -(x - 7)^2 + 13$ **69.** $2x + h$ **71.** $2x + h - 9$ **73.** $6x + 3h - 11$ **75.** Answers will vary. The parabola opens up and the vertex is above the x-axis, or the parabola opens down and the vertex is below x-axis.

Quick Check 11.4 **1.** 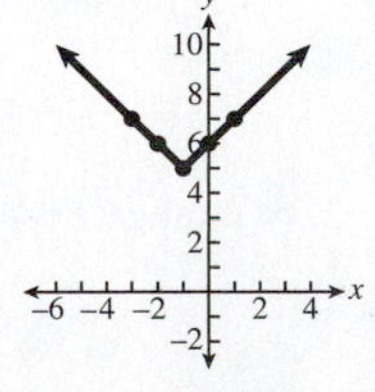
Domain: $(-\infty, \infty)$; range: $[5, \infty)$

2. 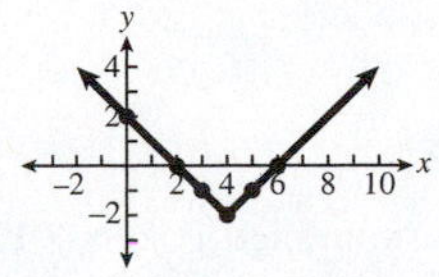
Domain: $(-\infty, \infty)$; range: $[-2, \infty)$

3. 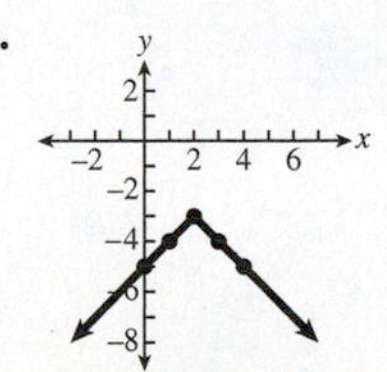
Domain: $(-\infty, \infty)$; range: $(-\infty, -3]$

4. 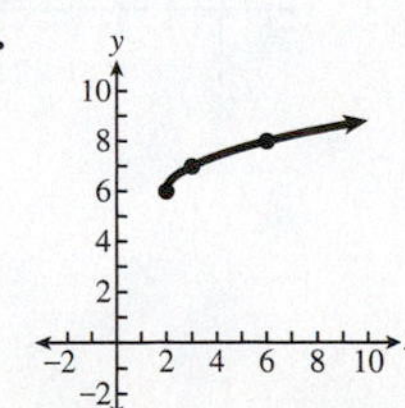
Domain: $[2, \infty)$; range: $[6, \infty)$

5. 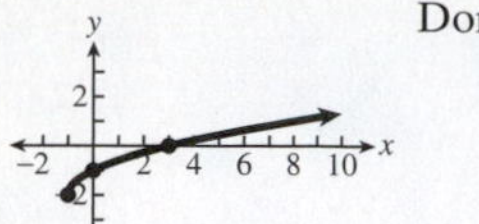 Domain: $[-1, \infty)$; range: $[-2, \infty)$ **6.** 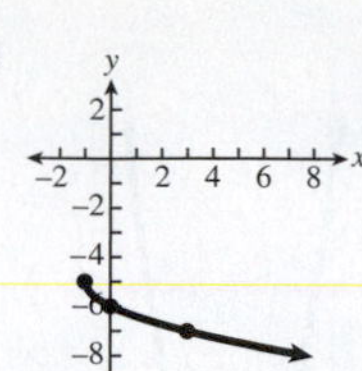 Domain: $[-1, \infty)$; range: $(-\infty, -5]$

7. Domain: $(-\infty, -4]$; range: $[-2, \infty)$ **8.** 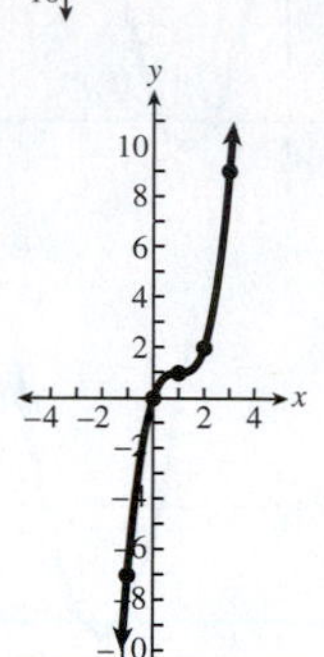 Domain: $(-\infty, \infty)$; range: $(-\infty, \infty)$

9. Domain: $(-\infty, \infty)$; range: $(-\infty, \infty)$ **10. a)** $f(x) = \sqrt{x - 2} + 5$ **b)** $f(x) = |x + 4| - 2$

Section 11.4 **1.** radicand **3.** 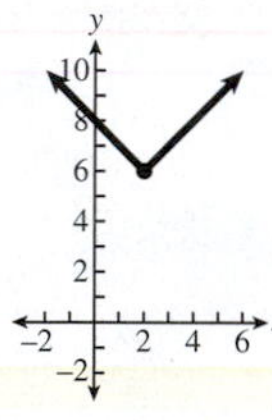 Domain: $(-\infty, \infty)$; range: $[6, \infty)$ **5.** 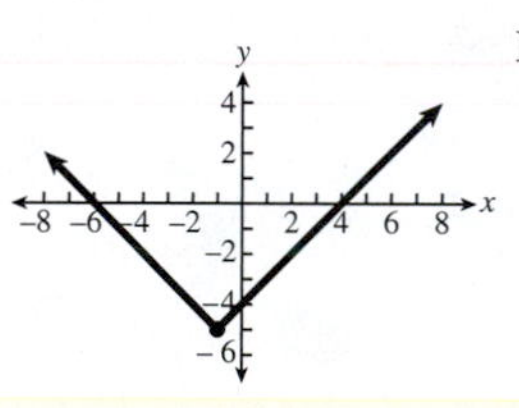 Domain: $(-\infty, \infty)$; range: $[-5, \infty)$

7. 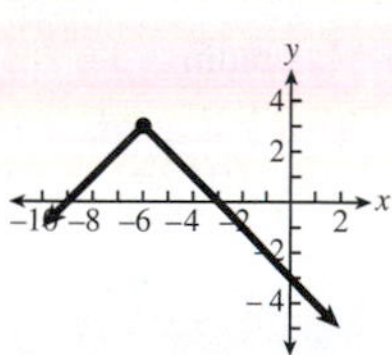 Domain: $(-\infty, \infty)$; range: $(-\infty, 3]$ **9.** 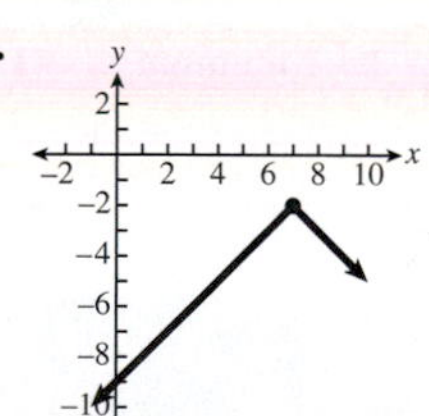 Domain: $(-\infty, \infty)$; range: $(-\infty, -2]$

11. 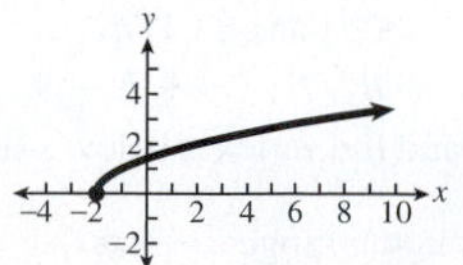 Domain: $[-2, \infty)$; range: $[0, \infty)$ **13.** 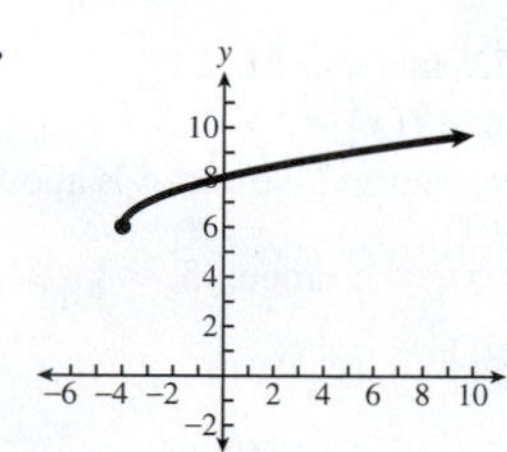 Domain: $[-4, \infty)$; range: $[6, \infty)$

15. Domain: $[5, \infty)$; range: $[-2, \infty)$ **17.** 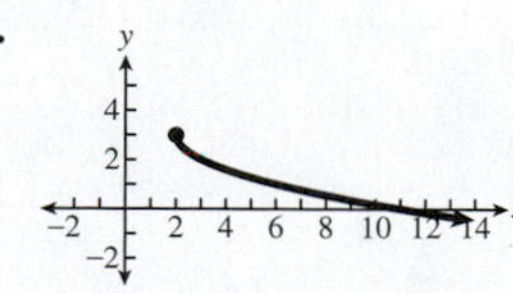 Domain: $[2, \infty)$; range: $(-\infty, 3]$

19. 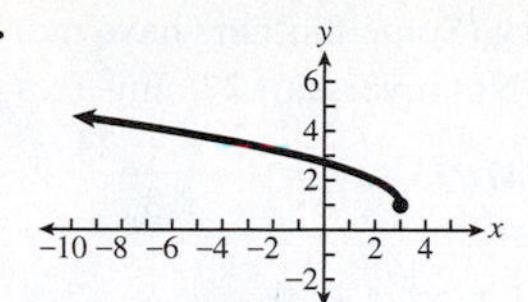 Domain: $(-\infty, 3]$; range: $[1, \infty)$ **21.** 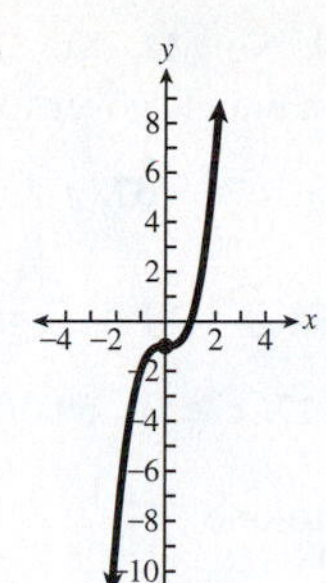 Domain: $(-\infty, \infty)$; range: $(-\infty, \infty)$

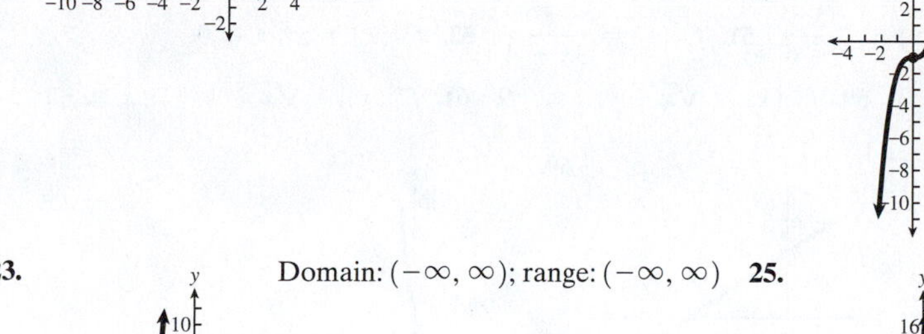

23. 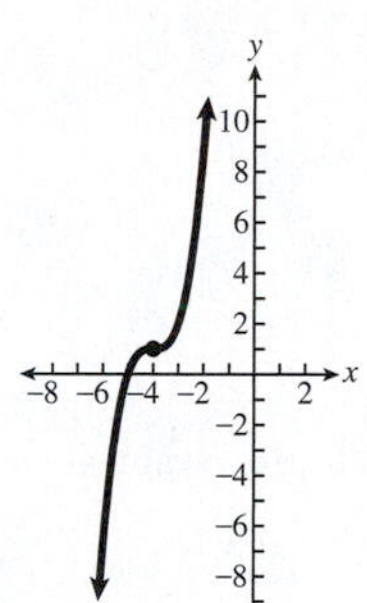 Domain: $(-\infty, \infty)$; range: $(-\infty, \infty)$ **25.** 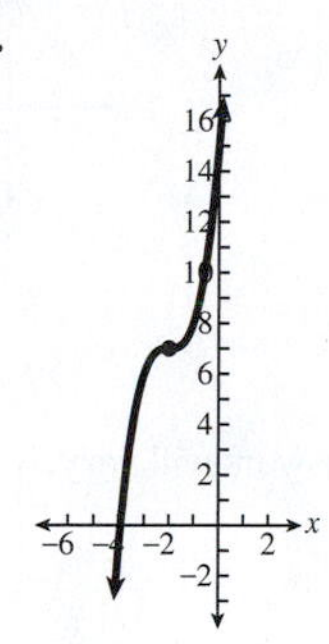 Domain: $(-\infty, \infty)$; range: $(-\infty, \infty)$

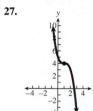

27. Domain: $(-\infty, \infty)$; range: $(-\infty, \infty)$ **29.** $f(x) = |x - 4| - 5$ **31.** $f(x) = \sqrt{x - 4} + 3$ **33.** $f(x) = \sqrt{-(x - 3)} - 1$

35. $f(x) = x^3 + 5$ **37.** Answers will vary. The graph of a square-root function is half of a parabola opening to the left or right.

Quick Check 11.5 **1. a)** $x^2 - 6x - 24$ **b)** $x^2 - 12x + 64$ **c)** -17 **d)** 92 **2.** $12x^2 - 5x - 28$ **3.** $\frac{x - 3}{x + 7}$; $x \neq -6, -7$ **4. a)** $155{,}889x + 12{,}119{,}000$; total number of students enrolled in a public or private college x years after 1990 **b)** $67{,}943x + 6{,}937{,}000$; how many more students are enrolled at public colleges than at private colleges x years after 1990 **5.** $8x + 19$ **6. a)** $x^2 - 3x - 25$ **b)** $x^2 + 3x - 28$ **7.** $\frac{2x - 3}{x + 3}$; domain: all real numbers except $x = -3$

Section 11.5 **1.** sum **3.** $(f \cdot g)(x)$ **5.** composition **7. a)** $7x - 6$ **b)** 50 **c)** -27 **9. a)** $x^2 + 4x - 11$ **b)** 85 **c)** -14 **11. a)** $8x - 6$ **b)** 42 **c)** -46 **13. a)** $-x^2 + 15x + 49$ **b)** 103 **c)** -51 **15. a)** $2x^2 + 5x - 12$ **b)** 63 **c)** -9 **17. a)** $x^3 - 3x^2 - 22x + 24$ **b)** -36 **c)** 36 **19. a)** $\frac{3x + 1}{x + 7}$ **b)** $\frac{11}{7}$ **c)** -1 **21. a)** $\frac{x - 3}{x + 4}$ **b)** $\frac{4}{11}$ **c)** $-\frac{5}{2}$ **23.** 64 **25.** 22 **27.** 1421 **29.** $-\frac{5}{11}$ **31.** $7x - 26$ **33.** $10x^2 - 103x + 153$ **35.** $-x^2 + 9x + 10$ **37.** $\frac{1}{x + 2}$ **39. a)** $(f + g)(x) = 19.5x + 447$; this is the total number of doctors, in thousands, in the United States x years after 1980. **b)** 1227; there will be 1,227,000 doctors in the United States in the year 2020. **41. a)** $(f - g)(x) = 5.8x + 20.9$; this tells how many more females than males, in thousands, will earn a master's degree in the United States x years after 1990. **b)** 194.9; there will be 194,900 more master's degrees earned by females than by males in the United States in the year 2020. **43.** 15 **45.** 0 **47. a)** $3x^2 + 15x - 16$ **b)** 134 **49. a)** $x^2 + 9x + 20$ **b)** 90 **51. a)** $6x + 17$ **b)** $6x + 14$ **53. a)** $x^2 + 3x - 36$ **b)** $x^2 + 11x - 12$ **55.** $\frac{10x + 65}{2x + 20}$, all real numbers except -10 **57.** $\sqrt{2x + 12}$, $[-6, \infty)$ **59.** $9x - 40$ **61.** $x^4 + 12x^3 + 66x^2 + 180x + 228$ **63.** $x = -4$ **65.** $x = -3$ **67.** $f(x) = x - 8$ **69.** $f(x) = \frac{x - 8}{3}$ **71.** Yes, explanations will vary. Addition is commutative.

Quick Check 11.6 **1.** Not one-to-one **2.** Not one-to-one **3.** Inverse functions **4.** Inverse functions **5.** $f^{-1}(x) = 2x - 20$ **6.** $f^{-1}(x) = \frac{3}{x - 5}$, $x \neq 5$ **7.** $f^{-1}(x) = x^2 - 16x + 73$; domain: $[8, \infty)$ **8.**

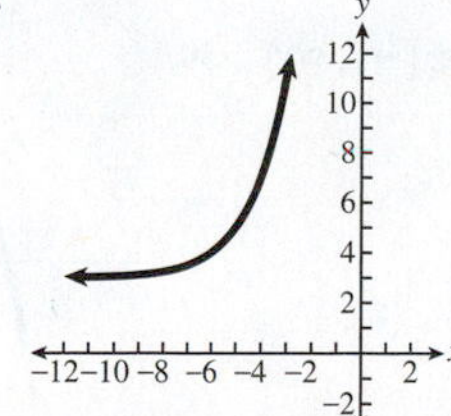

Section 11.6 **1.** one-to-one **3.** inverse **5.** $f^{-1}(x)$ **7.** Yes **9.** No **11.** No **13.** No **15.** No; explanations will vary. (Some mothers have more than one child.) **17.** Yes; explanations will vary. (Each state has a unique governor.) **19.** No **21.** Yes **23.** No **25.** Not inverses **27.** Inverses **29.** Inverses **31.** Not inverses **33.** $f^{-1}(x) = x - 9$ **35.** $f^{-1}(x) = 3x$ **37.** $f^{-1}(x) = \frac{x - 14}{3}$ **39.** $f^{-1}(x) = \frac{x + 6}{7}$ **41.** $f^{-1}(x) = -\frac{4x - 24}{3}$ **43.** $f^{-1}(x) = \frac{x}{m}$ **45.** $f^{-1}(x) = \frac{1}{x - 3}$ **47.** $f^{-1}(x) = \frac{1 - 3x}{x}$ **49.** $f^{-1}(x) = \frac{5}{x + 6}$ **51.** $f^{-1}(x) = \frac{8}{3x - 4}$ **53.** $f^{-1}(x) = x^2, x \geq 0$ **55.** $f^{-1}(x) = x^2 + 14x + 49, x \geq -7$ **57.** $f^{-1}(x) = x^2 - 6x + 17, x \geq 3$ **59.** $f^{-1}(x) = \sqrt{x + 9}, x \geq -9$ **61.** $f^{-1}(x) = \sqrt{x + 1} + 3, x \geq -1$ **63.** $\{(-17, -5), (-9, -1), (-5, 1), (1, 4), (7, 7)\}$ **65.** Not one-to-one **67.**

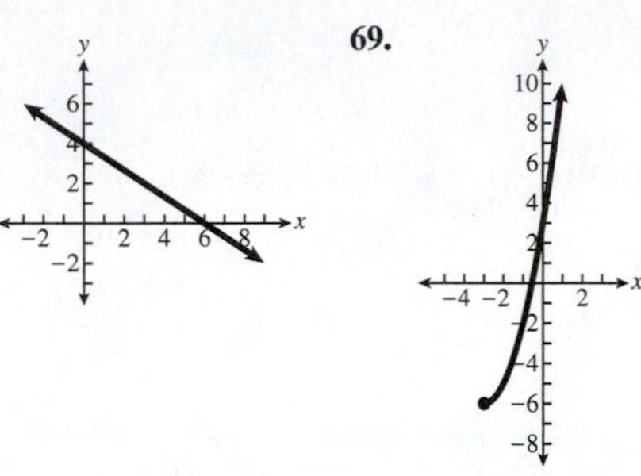

69.

71.

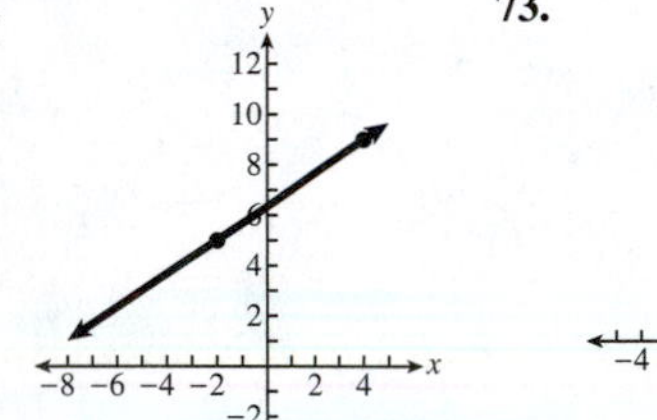

73.

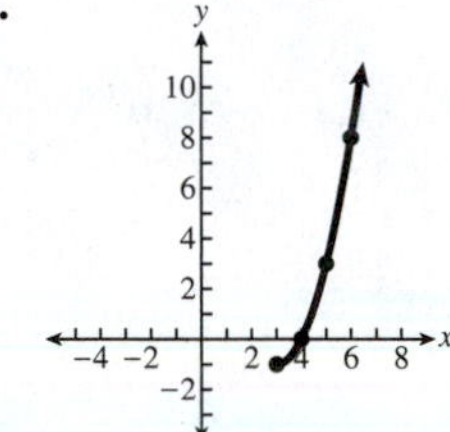

75. Answers will vary. It shows one y-value associated with two or more x-values.

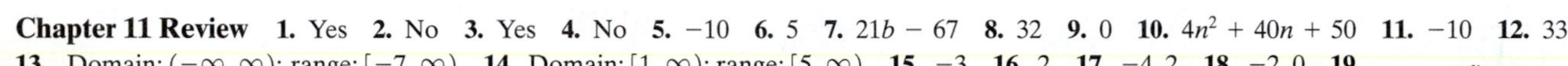

Chapter 11 Review **1.** Yes **2.** No **3.** Yes **4.** No **5.** -10 **6.** 5 **7.** $21b - 67$ **8.** 32 **9.** 0 **10.** $4n^2 + 40n + 50$ **11.** -10 **12.** 33 **13.** Domain: $(-\infty, \infty)$; range: $[-7, \infty)$ **14.** Domain: $[1, \infty)$; range: $[5, \infty)$ **15.** -3 **16.** 2 **17.** $-4, 2$ **18.** $-2, 0$ **19.**

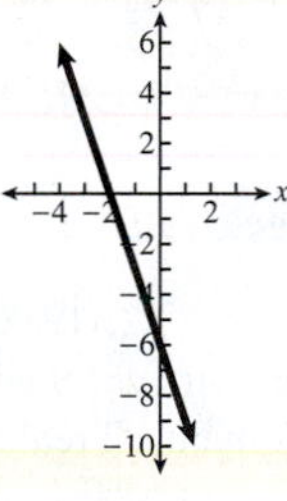

20.

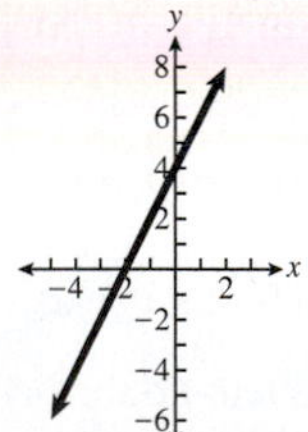

21.

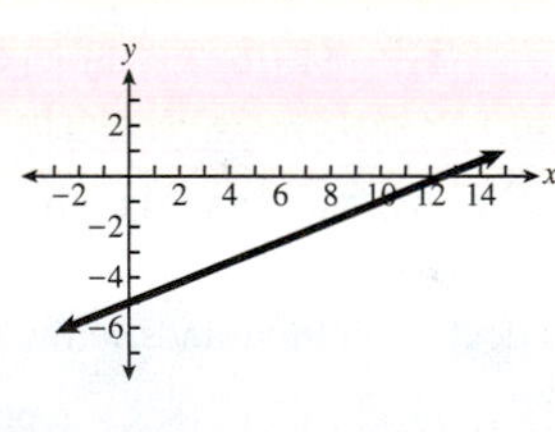

22.

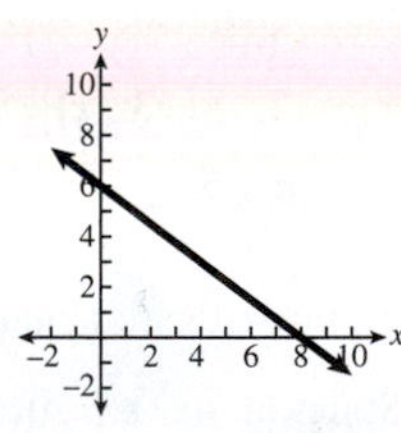

23. $f(x) = 2x + 6$ **24.** $f(x) = -\frac{1}{4}x$

25. a) $f(x) = 0.05x + 30$ **b)** \$33.75 **26. a)** $C(x) = 0.10x + 100$ **b)** $R(x) = 1.50x$

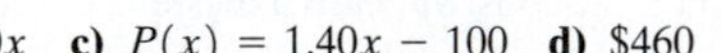

c) $P(x) = 1.40x - 100$ **d)** \$460

27.

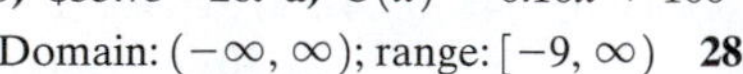

Domain: $(-\infty, \infty)$; range: $[-9, \infty)$

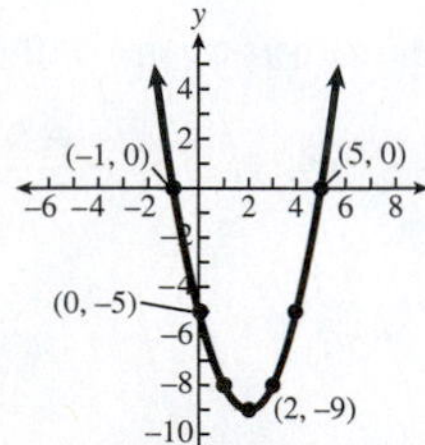

28. Domain: $(-\infty, \infty)$; range: $[3, \infty)$

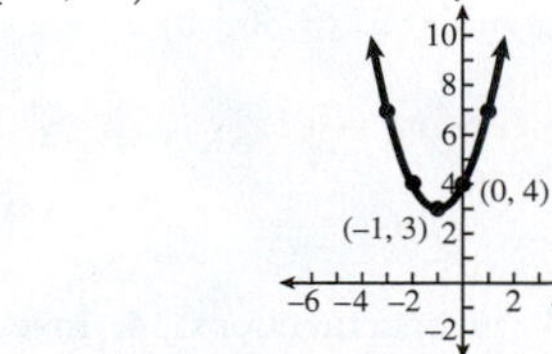

29.

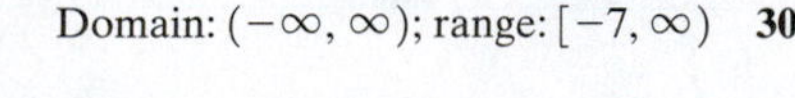

Domain: $(-\infty, \infty)$; range: $[-7, \infty)$

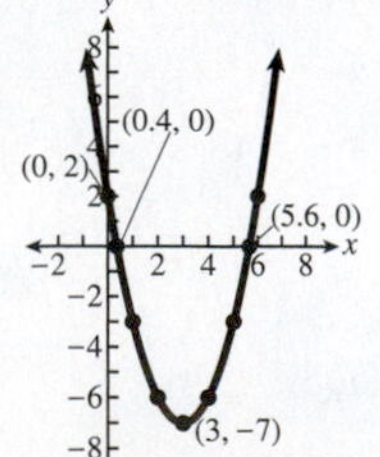

30. Domain: $(-\infty, \infty)$; range: $(-\infty, 4]$

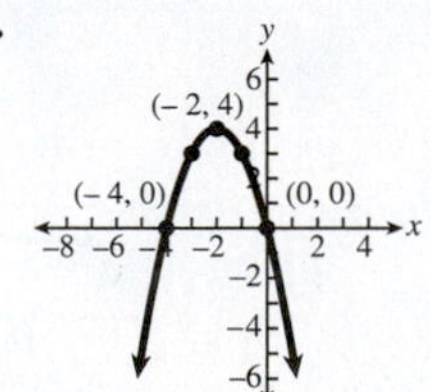

31. Minimum, 24

32. Maximum, −6 **33.** Minimum, −25 **34.** Maximum, −39 **35.** 526 feet **36.** 36 feet by 36 feet, 1296 square feet **37.** $2x + h$ **38.** $2x + h - 6$

39. Domain: $(-\infty, \infty)$; range: $[-4, \infty)$

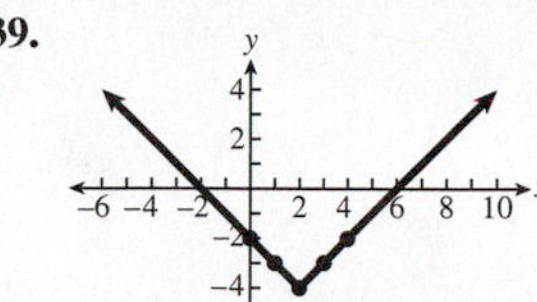

40. Domain: $(-\infty, \infty)$; range: $(-\infty, -2]$

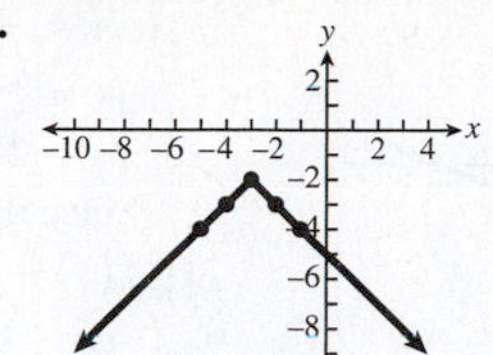

41. Domain: $[2, \infty)$; range: $[7, \infty)$

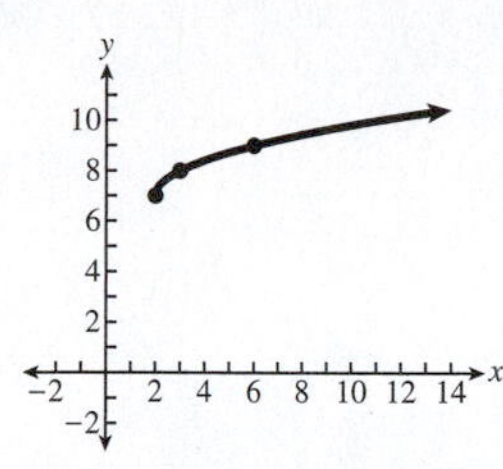

42. Domain: $[-5, \infty)$; range: $[-3, \infty)$

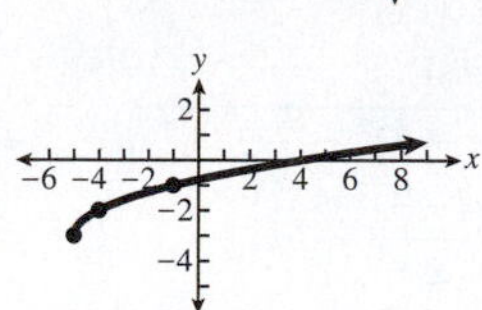

43. Domain: $[-1, \infty)$; range: $(-\infty, 2]$

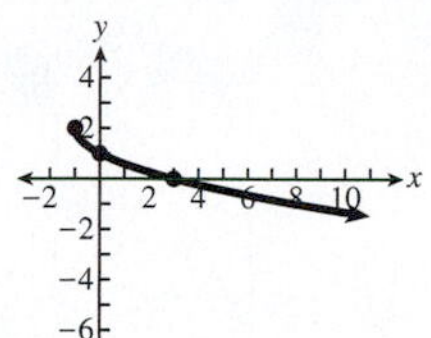

44. Domain: $(-\infty, 2]$; range: $[5, \infty)$

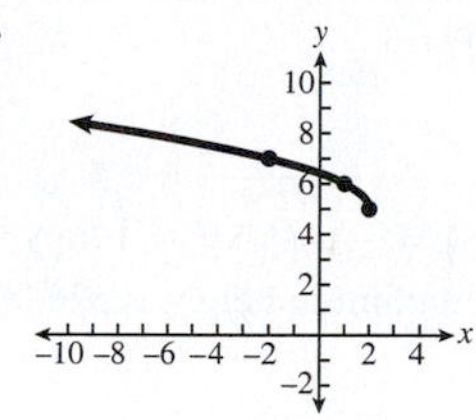

45. Domain: $(-\infty, \infty)$; range: $(-\infty, \infty)$

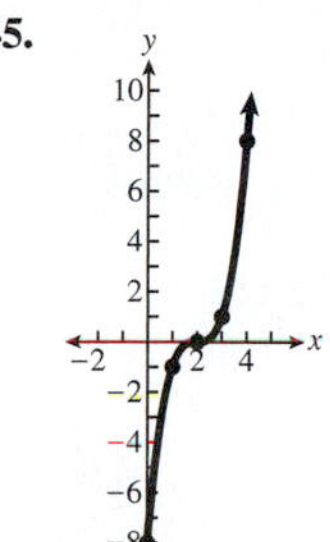

46. Domain: $(-\infty, \infty)$; range: $(-\infty, \infty)$

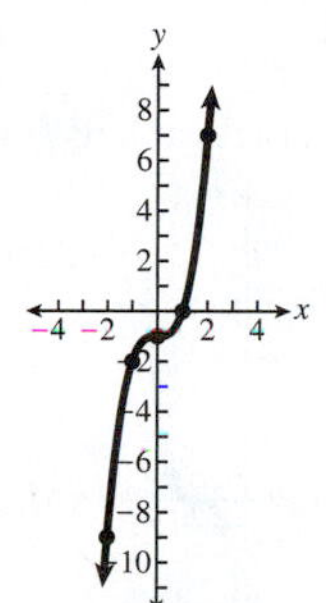

47. $f(x) = \sqrt{x - 1} + 6$

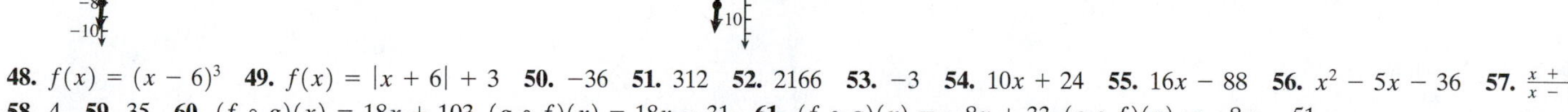

48. $f(x) = (x - 6)^3$ **49.** $f(x) = |x + 6| + 3$ **50.** −36 **51.** 312 **52.** 2166 **53.** −3 **54.** $10x + 24$ **55.** $16x - 88$ **56.** $x^2 - 5x - 36$ **57.** $\frac{x+2}{x-7}$
58. 4 **59.** 35 **60.** $(f \circ g)(x) = 18x + 103, (g \circ f)(x) = 18x - 31$ **61.** $(f \circ g)(x) = -8x + 33, (g \circ f)(x) = -8x - 51$
62. $(f \circ g)(x) = x^2 + 7x - 64, (g \circ f)(x) = x^2 - 9x - 48$ **63.** $(f \circ g)(x) = 3x^2 - 27x + 130, (g \circ f)(x) = 9x^2 + 15x + 27$
64. No, two people could have the same favorite. **65.** Yes, each player has a unique number. **66.** Yes **67.** No **68.** No **69.** Yes **70.** No
71. Yes **72.** $f^{-1}(x) = x + 10, (-\infty, \infty)$ **73.** $f^{-1}(x) = \dfrac{x + 15}{2}, (-\infty, \infty)$ **74.** $f^{-1}(x) = -x + 19, (-\infty, \infty)$ **75.** $f^{-1}(x) = \dfrac{-x + 27}{8}, (-\infty, \infty)$
76. $f^{-1}(x) = \dfrac{2}{3x - 9}$, all real numbers except 3 **77.** $f^{-1}(x) = \dfrac{21x}{7 - 4x}$, all real numbers except $\frac{7}{4}$ **78.**

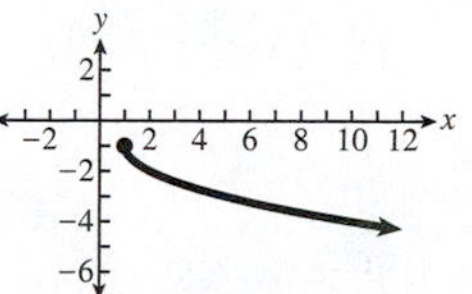

79.

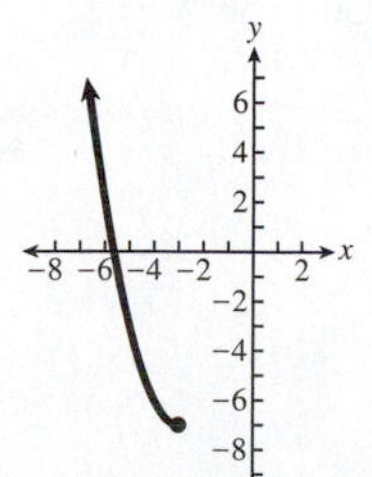

Chapter 11 Review Exercises: Worked-Out Solutions

8. $f(7) = (7)^2 - 7(7) + 32$
$= 49 - 49 + 32$
$= 32$

14. The domain, read from left to right on the graph, is $[1, \infty)$. The range, read from the bottom of the graph to the top, is $[5, \infty)$.

15. The point on the graph with an x-coordinate of 4 is $(4, -3)$, so $f(4) = -3$.

18. There are two points with a y-coordinate of -5: $(-2, -5)$ and $(0, -5)$. So $x = -2$ and $x = 0$ are solution to the equation $f(x) = -5$.

21. Slope: $m = \frac{2}{5}$; y-intercept: $(0, -5)$

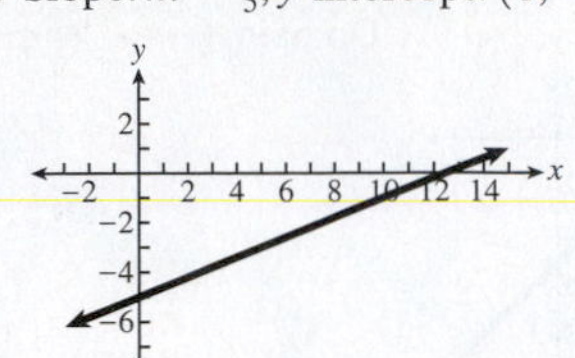

23. The points $(-3, 0)$ and $(0, 6)$ are on the line.

$m = \frac{6 - 0}{0 - (-3)}$

$m = \frac{6}{3}$

$m = 2$

Since the y-intercept is $(0, 6)$, $b = 6$.

$f(x) = 2x + 6$

29.

Vertex (h, k)	y-intercept $(x = 0)$	x-intercept $(y = 0)$
$(3, -7)$	$y = (0 - 3)^2 - 7$	$0 = (x - 3)^2 - 7$
	$y = 9 - 7$	$7 = (x - 3)^2$
	$y = 2$	$\pm\sqrt{7} = \sqrt{(x - 3)^2}$
	$(0, 2)$	$\pm\sqrt{7} = x - 3$
		$x = 3 \pm \sqrt{7}$
		$3 - \sqrt{7} \approx 0.4,\ 3 + \sqrt{7} \approx 5.6$
		$(0.4, 0), (5.6, 0)$

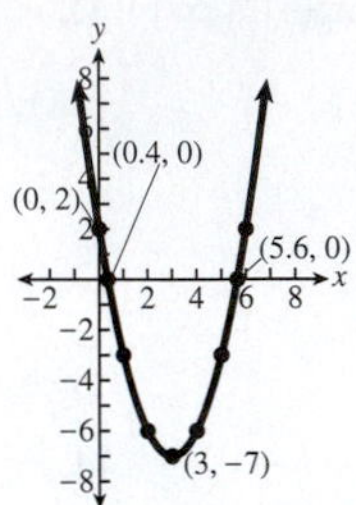

31. Since $a > 0$, the function has a minimum.

$x = \frac{-(-12)}{2(1)} = 6$

$f(6) = 6^2 - 12(6) + 60$
$= 36 - 72 + 60$
$= 24$

The minimum value is 24.

35. $t = \frac{-176}{2(-16)} = 5.5$

$h(5.5) = -16(5.5)^2 + 176(5.5) + 42 = 526$

The maximum height is 526 feet.

41. Shift the basic graph of $f(x) = \sqrt{x}$ by 2 units to the right and up by 7 units.

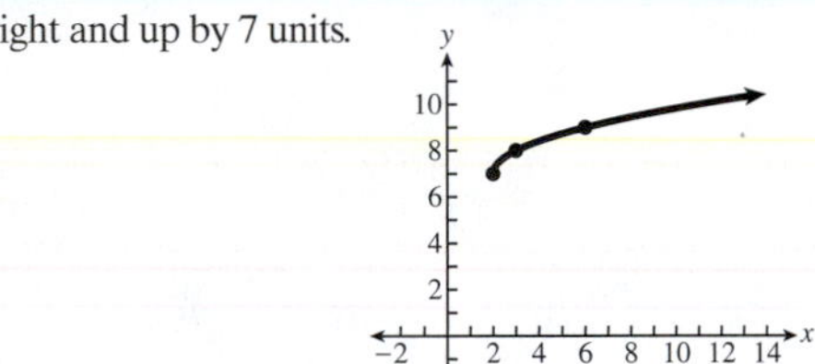

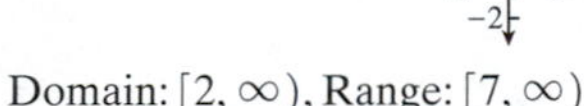

Domain: $[2, \infty)$, Range: $[7, \infty)$

45. Shift the basic graph of $f(x) = x^3$ by 2 units to the right.

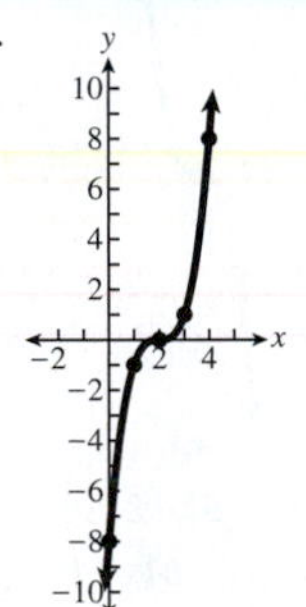

Domain: $(-\infty, \infty)$, Range: $(-\infty, \infty)$

54. $(f + g)(x) = f(x) + g(x)$
$= (6x - 11) + (4x + 35)$
$= 10x + 24$

58. $g(8) = (8)^2 - 14(8) + 48$
$= 64 - 112 + 48$
$= 0$

$(f \circ g)(8) = f(g(8))$
$= f(0)$
$= (0) + 4$
$= 4$

62. $(f \circ g)(x) = f(g(x))$
$= f(x^2 + 7x - 56)$
$= (x^2 + 7x - 56) - 8$
$= x^2 + 7x - 64$

$(g \circ f)(x) = g(f(x))$
$= g(x - 8)$
$= (x - 8)^2 + 7(x - 8) - 56$
$= x^2 - 16x + 64 + 7x - 56 - 56$
$= x^2 - 9x - 48$

69. $(f \circ g)(x) = f(g(x))$
$= f(\frac{x + 10}{2})$
$= 2(\frac{x + 10}{2}) - 10$
$= x + 10 - 10$
$= x$

$(g \circ f)(x) = g(f(x))$
$= g(2x - 10)$
$= \frac{(2x - 10) + 10}{2}$
$= \frac{2x}{2}$
$= x$

Yes, the functions are inverse functions.

73. $y = 2x - 15$
$x = 2y - 15$
$x + 15 = 2y$
$\frac{x + 15}{2} = y$
$f^{-1}(x) = \frac{x + 15}{2}$

Domain: $(-\infty, \infty)$

Chapter 11 Test **1.** Yes **2.** No **3.** 51 **4.** $30a - 9$ **5.** 270 **6.** Domain: $(-\infty, \infty)$; range: $(-\infty, 5]$ **7.**

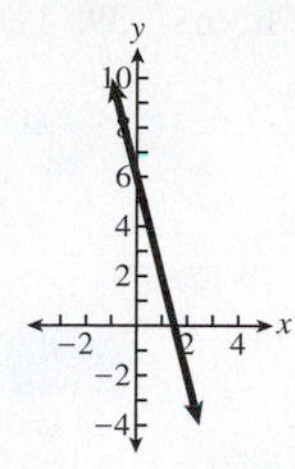

8.

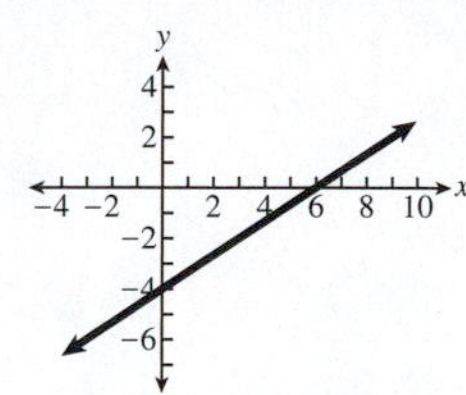

9. $f(x) = 7x + 35$ **10.**

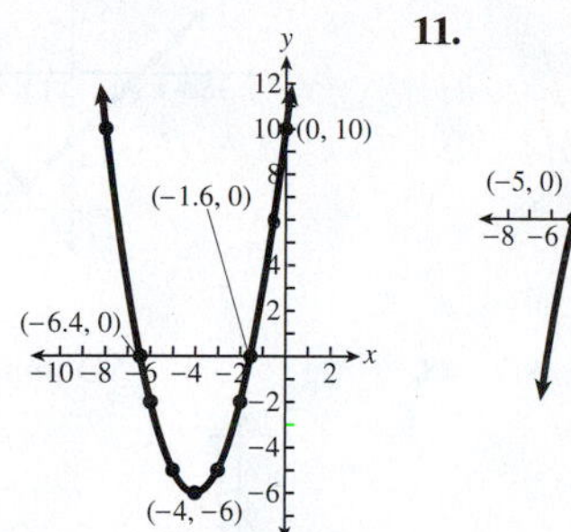

11. (−3, 4), (−5, 0), (−1, 0), (0, −5)

12. 9.25 feet

13.

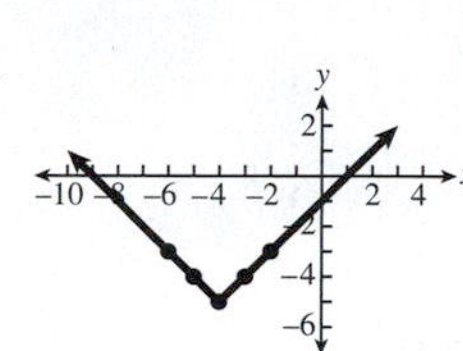

Domain: $(-\infty, \infty)$; range: $[-5, \infty)$ **14.**

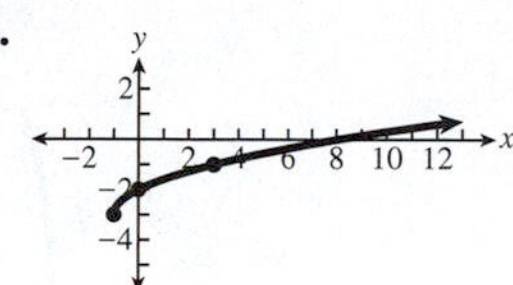

Domain: $[-1, \infty)$; range: $[-3, \infty)$

15.

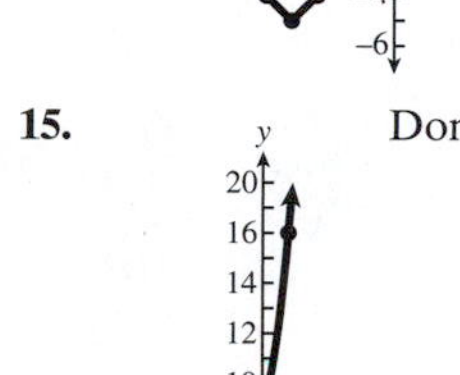

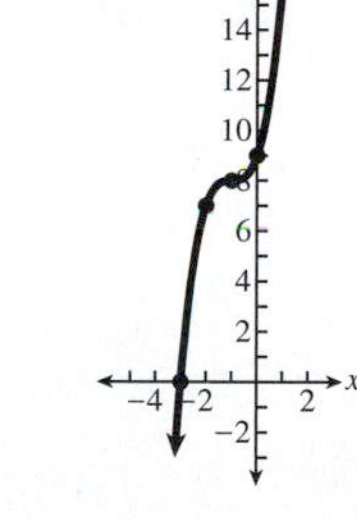

Domain: $(-\infty, \infty)$; range: $(-\infty, \infty)$ **16.** $f(x) = \sqrt{x + 6}$ **17.** -13 **18.** $x^3 - 216$ **19.** 11 **20.** 18 **21.** $-8x + 105$

22. $x^2 + 15x + 6$ **23.** No **24.** Yes **25.** $f^{-1}(x) = \dfrac{-x + 20}{4}$ **26.** $f^{-1}(x) = \dfrac{-9}{4x - 1}, x \neq \dfrac{1}{4}$

Cumulative Review Chapters 8–11 **1.** $\{-9, 3\}$ **2.** $x \leq 4$, $(-\infty, 4]$ **3.** $\frac{1}{2} \leq x \leq 13$, $\left[\frac{1}{2}, 13\right]$ **4.** $x < -3$ or $x > 2$, $(-\infty, -3) \cup (2, \infty)$ **5.** $-5 < x < 5$, $(-5, 5)$ **6.** $(x - 9)(x + 6)$ **7.** $4(x - 1)(x - 8)$ **8.** $(x + 5)(x + 6)$ **9.** $(7x + 8)(7x - 8)$ **10.** $(6, -2)$ **11.** $(3, 5, -2)$ **12.** 7.483 **13.** $64x^{15}$ **14.** $x^{9/14}$ **15.** $5r^6s^5t^4\sqrt{3t}$ **16.** $34\sqrt{3}$ **17.** $20 + 6\sqrt{15}$ **18.** $\dfrac{c\sqrt{ac}}{ab}$ **19.** $\dfrac{136 + 39\sqrt{10}}{62}$ **20.** $\{21\}$ **21.** $\{3\}$ **22.** 17 **23.** $6i\sqrt{10}$ **24.** $28 + 41i$ **25.** $\dfrac{4}{5} + \dfrac{3}{5}i$ **26.** $\{-2 - 3i\sqrt{2}, -2 + 3i\sqrt{2}\}$ **27.** $\left\{\dfrac{5 - 2\sqrt{22}}{3}, \dfrac{5 + 2\sqrt{22}}{3}\right\}$ **28.** $\{-\sqrt{2}, \sqrt{2}, -3, 3\}$ **29.** $\{4\}$ **30.** $\{2\}$ **31.** $\{3, 9\}$ **32.**

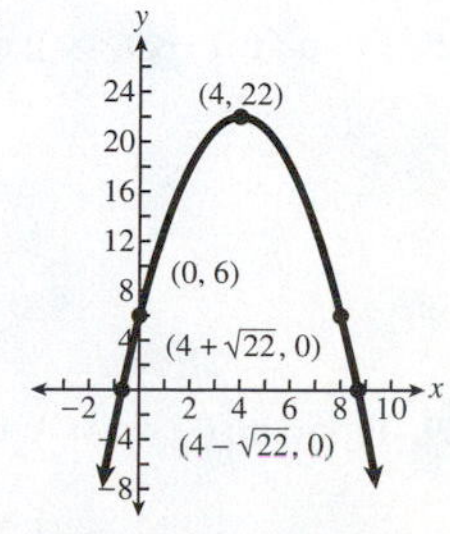

33.

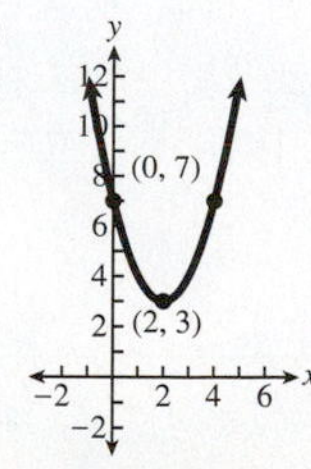

34. $[4, 6]$, **35.** $(-\infty, -6] \cup (-5, 8]$,

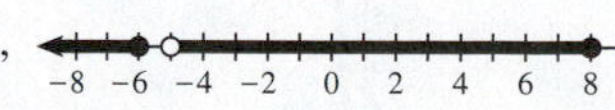

36. Length: 33.9 feet; width: 28.9 feet

37. 5.15 seconds **38.** 11.7 hours **39.** $12n - 43$ **40.** 126 **41.**

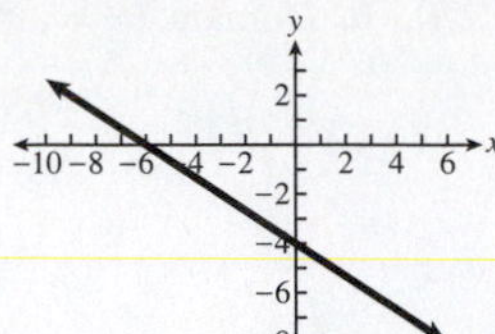
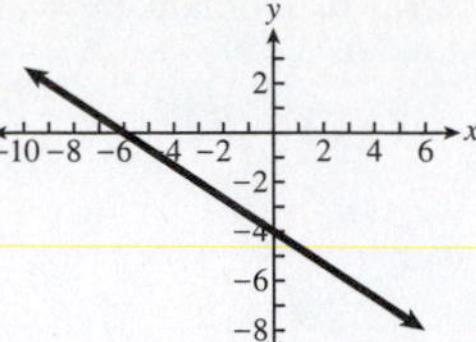

42. a) $f(x) = 0.15x + 20$ **b)** \$65 **43.** Minimum, -96

44. 374 feet **45.**

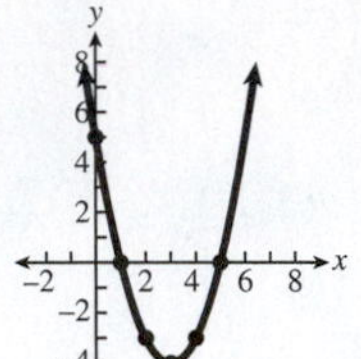

Domain: $(-\infty, \infty)$; range: $[-4, \infty)$ **46.**

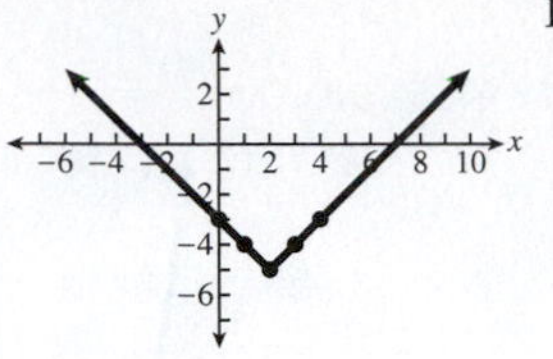

Domain: $(-\infty, \infty)$; range: $[-5, \infty)$

47.

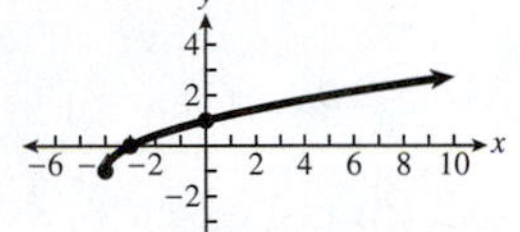

Domain: $[-4, \infty)$; range: $[-1, \infty)$ **48.**

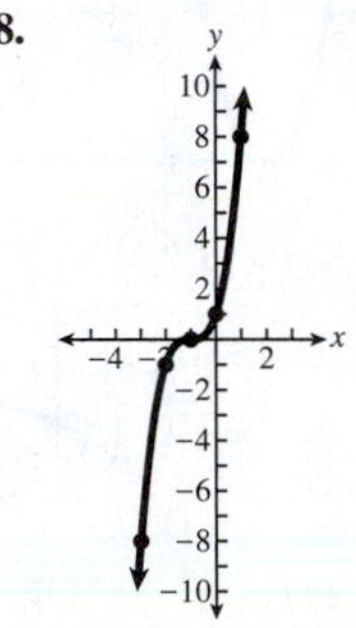

Domain: $(-\infty, \infty)$; range: $(-\infty, \infty)$ **49.** $10x - 31$

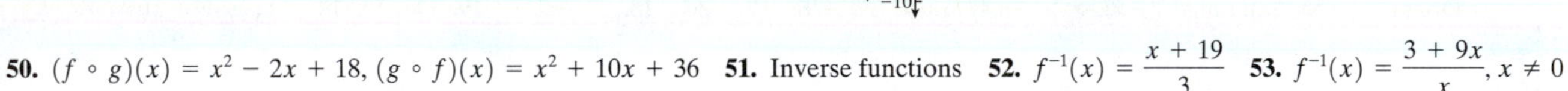

50. $(f \circ g)(x) = x^2 - 2x + 18, (g \circ f)(x) = x^2 + 10x + 36$ **51.** Inverse functions **52.** $f^{-1}(x) = \dfrac{x + 19}{3}$ **53.** $f^{-1}(x) = \dfrac{3 + 9x}{x}, x \neq 0$

CHAPTER 12

Quick Check 12.1 **1.** $f(5) = 1024, f(-3) = \frac{1}{64}$ **2.** 79 **3.**

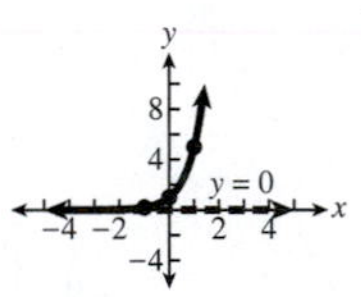

Domain: $(-\infty, \infty)$; range: $(0, \infty)$

4.

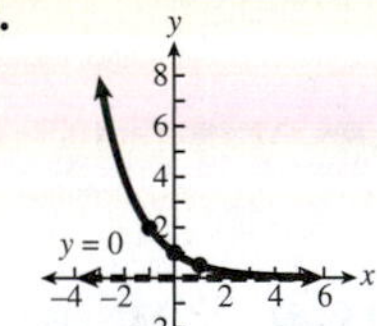
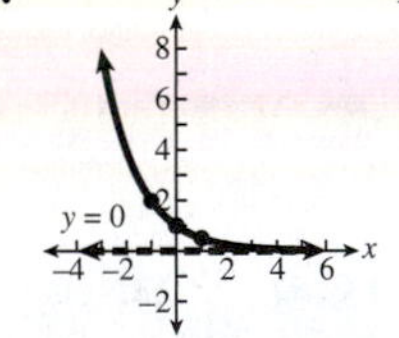

Domain: $(-\infty, \infty)$; range: $(0, \infty)$ **5.** $f(-1) \approx 0.368, f(4.3) \approx 73.700$ **6. a)** $\{4\}$ **b)** $\{9\}$ **7.** Approximately \$54.7 billion

Section 12.1 **1.** exponential **3.** increasing **5.** $r = s$ **7.** 16 **9.** 1 **11.** $\frac{4}{25}$ **13.** $\frac{625}{16}$ **15.** 16 **17.** 710 **19.** 3525 **21.** $\dfrac{1}{15{,}625}$ **23.** 34

25. Domain: $(-\infty, \infty)$; range: $(0, \infty)$

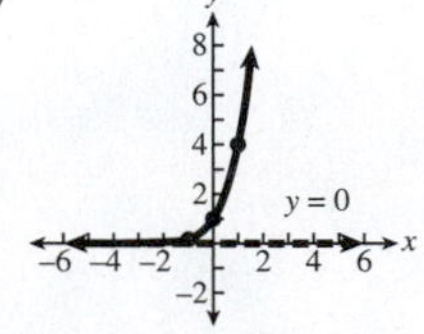

27. Domain: $(-\infty, \infty)$; range: $(0, \infty)$

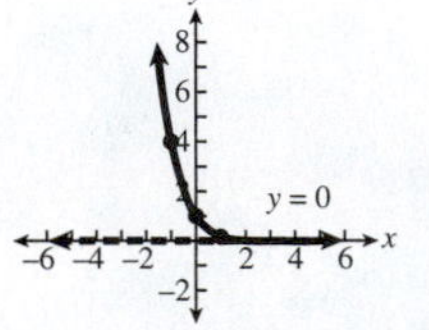

29. Domain: $(-\infty, \infty)$; range: $(0, \infty)$

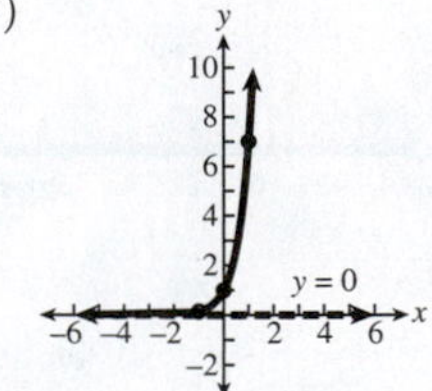

31. Domain: $(-\infty, \infty)$; range: $(0, \infty)$

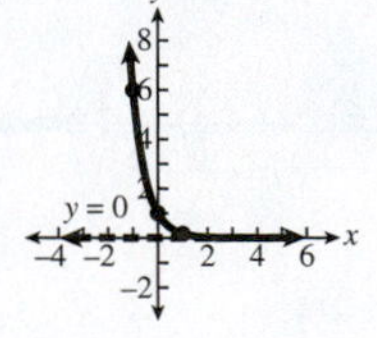
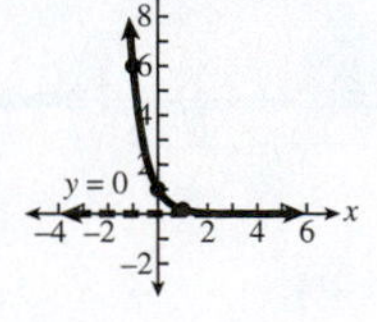

33. 7.389 **35.** 0.006

37. 20.086 **39.** 0.001 **41.** $\{4\}$ **43.** $\{3\}$ **45.** $\{-1\}$ **47.** $\{11\}$ **49.** $\{-6\}$ **51.** $\{5\}$ **53.** $\{8\}$ **55.** $\{-2\}$ **57.** $\{4\}$ **59.** $\{-5, 2\}$ **61.** \$663,907.76 **63.** \$15,605.09 **65.** \$99.33 **67.** \$47.3 billion **69.** 36.8°C **71.** \$1.78 **73.** Answers will vary. For $b > 1$, the function is increasing, and for $0 < b < 1$, the function is decreasing. **75.** Answers will vary. $2^{x+1} = 32$; rewrite 32 as a power of 2.

Quick Check 12.2 **1. a)** 2 **b)** -3 **c)** 1 **d)** 0 **2.** -4 **3. a)** 4 **b)** 1.875 **4. a)** -3 **b)** 6.405 **5.** $\log_7 20 = x$ **6.** $3^6 = x$ **7.** $\{1024\}$ **8.** $\{3\}$ **9. a)**

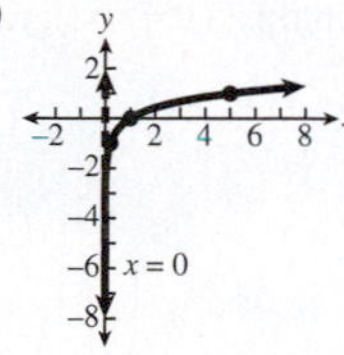

Domain: $(0, \infty)$; range: $(-\infty, \infty)$ **b)**

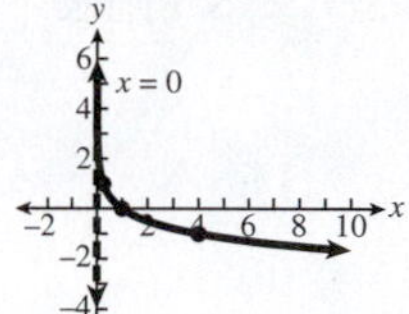

Domain: $(0, \infty)$; range: $(-\infty, \infty)$

10. Approximately \$46.54 billion **11.** 3.4

Section 12.2 **1.** $\log_b x$ **3.** common **5.** exponential **7.** vertical **9.** 2 **11.** 3 **13.** 3 **15.** 0 **17.** -2 **19.** -3 **21. a)** 2 **b)** 5 **23. a)** 0 **b)** 7 **25.** 3 **27.** -2 **29.** 2.049 **31.** 0.870 **33.** 1 **35.** -6 **37.** 3.466 **39.** 2.787 **41.** $\log_4 64 = 3$ **43.** $\log_8 \frac{1}{64} = -2$ **45.** $\log_5 42 = x$ **47.** $\log 321 = x$ **49.** $2^7 = 128$ **51.** $9^{-4} = \frac{1}{6561}$ **53.** $e^6 = x$ **55.** $10^7 = x$ **57.** $\{32\}$ **59.** $\{\frac{1}{81}\}$ **61.** $\{10{,}000\}$ **63.** $\{7.389\}$ **65.** $\{8\}$ **67.** $\{57.598\}$ **69.** $\{41\}$ **71.** $\{108\}$ **73.** $\{3\}$ **75.** $\{5\}$ **77.** $\{\frac{3}{2}\}$ **79.**

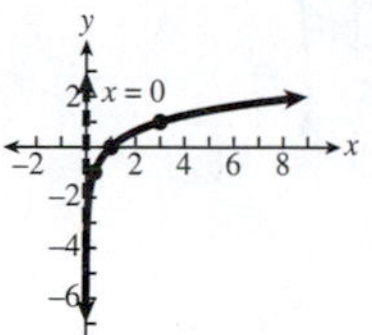

Domain: $(0, \infty)$; range: $(-\infty, \infty)$

81.

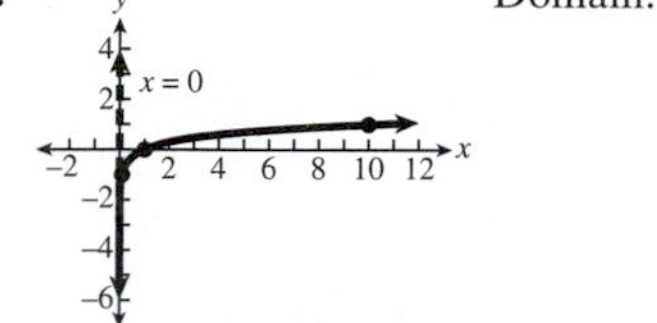

Domain: $(0, \infty)$; range: $(-\infty, \infty)$ **83.**

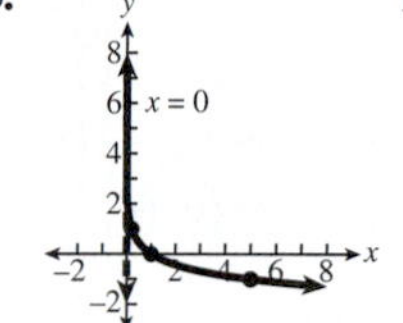

Domain: $(0, \infty)$; range: $(-\infty, \infty)$ **85.** 19,934

87. 33.8 inches **89.** 4.2 **91.** 630,957,345 times larger **93.** Answers will vary; exponent to which a base must be raised to produce a given result.

Quick Review Exercises **1.** x^{16} **2.** b^{12} **3.** n^{56} **4.** 1

Quick Check 12.3 **1.** $\log(35x)$ **2.** $\log_2 5 + \log_2 b$ **3.** 2 **4.** $\log_b x - 1$ **5.** $2 + 2\log_3 x + 4\log_3 y$ **6.** $\frac{1}{7}\ln x$ **7.** $\log_8 x^8$ **8.** -4 **9.** 16 **10.** $\log_2\left(\frac{acd}{b}\right)$ **11.** $\log\left(\frac{a^6\sqrt[5]{b}}{c^9d^{12}}\right)$ **12.** $3\log_5 z - 9\log_5 x - 4\log_5 y$ **13.** 4.395

Section 12.3 **1.** $\log_b(xy)$ **3.** $r \cdot \log_b x$ **5.** $\log_4(3a)$ **7.** $\log(xy)$ **9.** $\log_9 81 = 2$ **11.** $\log_3 10 + \log_3 x$ **13.** $\ln 3 + 1$ **15.** $\log_4 a + \log_4 b + \log_4 c$ **17.** $\log_3 7$ **19.** $\ln\left(\frac{x}{3}\right)$ **21.** $\log_4\left(\frac{1}{64}\right) = -3$ **23.** $\log_8 10 + \log_8 a - \log_8 b$ **25.** $\log_4 6 + \log_4 x + \log_4 y + \log_4 z - \log_4 w$ **27.** $\log_5 x + \log_5 y - 2 - \log_5 z$ **29.** $7\log_2 x$ **31.** $4\log_5 x + 7\log_5 y$ **33.** $\frac{1}{2}\log_3 x$ **35.** $4 + 5\log_2 x + 7\log_2 y$ **37.** $\log_5 x^8$ **39.** $\ln\sqrt[3]{x}$ **41.** $\log_3 x^{10}$ **43.** 12 **45.** 9 **47.** 7 **49.** b **51.** $\log_7\left(\frac{20}{3}\right)$ **53.** $\log_9 2000$ **55.** $\ln\left(\frac{4}{5}\right)$ **57.** $\log_2\left(\frac{x}{yzw}\right)$ **59.** $\ln\left(\frac{x^4y^9}{z^{13}}\right)$ **61.** $\log_b(x^2 - 2x - 48)$ **63.** $\log(3x^2 + 26x - 40)$ **65.** $\ln\left(\frac{a^3\sqrt[5]{b}}{c^{10}}\right)$ **67.** $A + B$ **69.** $C + D$ **71.** $3A$ **73.** $3C$ **75.** $\log_3 a + \log_3 b - \log_3 c - \log_3 d$ **77.** $3\log_2 a - 3 - 4\log_2 b$ **79.** $6\ln a + \frac{1}{2}\ln b - 7\ln c$ **81.** $-3\ln a - 2\ln d + \ln b + 7\ln c$ **83.** 3.459 **85.** 1.256 **87.** 8.601 **89.** 0.774 **91.** -0.981 **93.** 1.829 **95.** 1.701 **97.** Examples will vary.

Quick Check 12.4 **1. a)** $\{2\}$ **b)** $\{12\}$ **2.** $\left\{\frac{\log 12}{\log 4} + 5\right\}, \frac{\log 12}{\log 4} + 5 \approx 6.792$ **3.** $\left\{\frac{\log 11}{\log 9} - 3\right\}, \frac{\log 11}{\log 9} - 3 \approx -1.909$ **4.** $\{\ln 28 - 2\}, \ln 28 - 2 \approx 1.332$ **5.** $\{2\}$ **6.** $\varnothing$ **7.** $\{-2, 5\}$ **8.** $\{333{,}338\}$ **9. a)** $\{4\}$, **b)** $\{3\}$ **10.** $\{11\}$ **11.** $f^{-1}(x) = \ln(x - 9) - 6$ **12.** $f^{-1}(x) = e^{x+18} - 10$

Section 12.4 **1.** extraneous **3.** $\{3\}$ **5.** $\{9\}$ **7.** $\{9\}$ **9.** $\{\frac{3}{2}\}$ **11.** $\{\frac{9}{2}\}$ **13.** $\{-12\}$ **15.** $\{2.044\}$ **17.** $\{2.111\}$ **19.** $\{9.755\}$ **21.** $\{-1.277\}$ **23.** $\{1.062\}$ **25.** $\{6.947\}$ **27.** $\{-2.785\}$ **29.** $\{4\}$ **31.** $\{9\}$ **33.** $\varnothing$ **35.** $\{-2, 7\}$ **37.** $\{4\}$ **39.** $\{243\}$ **41.** $\{\frac{1}{625}\}$ **43.** $\{407.429\}$ **45.** $\{24\}$ **47.** $\{-9, 4\}$ **49.** $\{4, 5\}$ **51.** $\{8\}$ **53.** $\{-4\}$ **55.** $\{36\}$ **57.** $\{2\}$ **59.** $\{8\}$ **61.** $\{3\}$ **63.** $\{13\}$ **65.** $\{9\}$ **67.** $\{-10, 7\}$ **69.** $\{-2, 15\}$ **71.** $\{-5.267\}$ **73.** $\{5.577\}$ **75.** $\{-5\}$ **77.** $\{347\}$ **79.** $\{-3\}$ **81.** $\{10.069\}$ **83.** $\{\frac{5}{3}\}$ **85.** $\{4\}$ **87.** $\{37\}$ **89.** $\{111\}$ **91.** $\{-7\}$ **93.** $\{12\}$ **95.** $\{-2.432\}$ **97.** $\{-6\}$ **99.** $\{41\}$ **101.** $\{-4, 8\}$ **103.** $\{\frac{1}{4}, 32\}$ **105.** $\{\frac{1}{81}, 9\}$ **107.** $f^{-1}(x) = \ln x + 2$ **109.** $f^{-1}(x) = \ln(x - 4) - 5$ **111.** $f^{-1}(x) = \log_2(x + 9) + 6$ **113.** $f^{-1}(x) = 10^x - 6$ **115.** $f^{-1}(x) = e^{x+9} + 4$ **117.** $f^{-1}(x) = 3^{x-8} - 1$ **119.** Answers will vary. Logarithms have restricted domains. **121.** Answers will vary. Interchange x and y and solve for y.

Quick Check 12.5 **1.** \$464.65 **2.** Approximately 14.0 years **3.** Approximately 15.35 years **4.** Approximately \$1.45 **5.** Approximately 117.5 days **6.** In 2017 **7.** 2.8, acid **8.** 5.01×10^{-1} moles/liter **9.** 60 db **10.** 10^{-4} watts per square meter

Section 12.5 **1.** compound **3.** continuously **5.** negative **7.** \$8452.44 **9.** \$27,198.41 **11.** 4.5 years **13.** 23.1 years **15.** \$2588.68 **17.** 5.8 years **19.** 12,583 **21.** 13.7 billion **23.** 2016 **25.** 86.4 minutes **27.** 278 **29.** 22.6 milligrams **31.** 167 years **33.** 229,919.7 years **35.** 2094

37. 2016 **39.** 83,652 **41.** 2020 **43.** 2047 **45.** 7 years **47. a)** 90.3° F **b)** 2.5 hours **49.** 2018 **51.** 12.6, base **53.** 2.3, acid **55.** 1.0×10^{-5} moles per liter **57.** 6.31×10^{-11} moles per liter **59.** 80 db **61.** 110 db **63.** 10^{-1} watts per square meter **65.** $10^{-2.5}$ watts per square meter **67.** 80 **69.** 2016 **71.** Answers will vary. $1000 is invested at 4% interest, compounded quarterly. When will the balance reach $1967? **73.** Answers will vary. It is the time it takes for half of a radioactive substance to decay.

Quick Check 12.6 **1.** Domain: $(-\infty, \infty)$; range: $(-9, \infty)$ **2.** 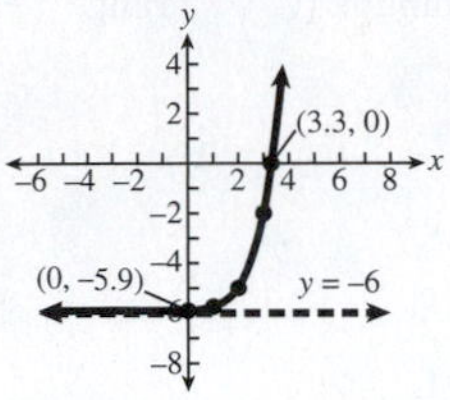 Domain: $(-\infty, \infty)$; range: $(-6, \infty)$ **3.** Domain: $(-\infty, \infty)$; range: $(-9, \infty)$ **4.** 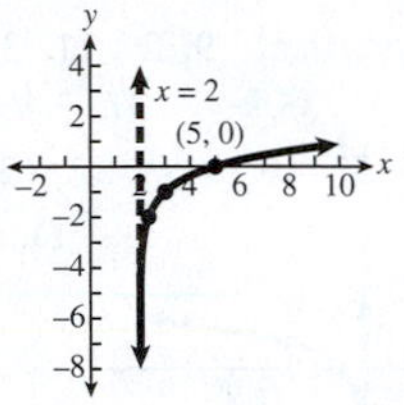 Domain: $(2, \infty)$; range: $(-\infty, \infty)$ **5.** Domain: $(-7, \infty)$; range: $(-\infty, \infty)$ **6.** 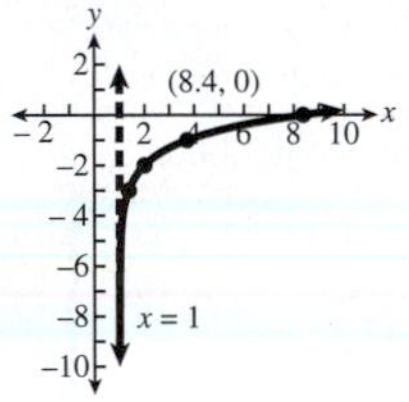 Domain: $(1, \infty)$; range: $(-\infty, \infty)$

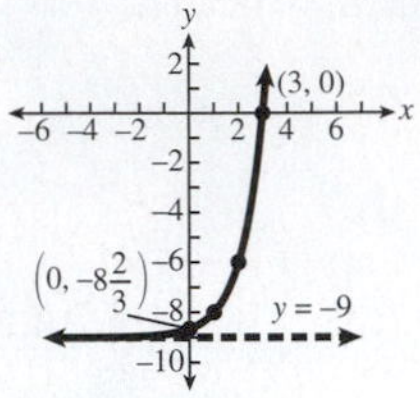

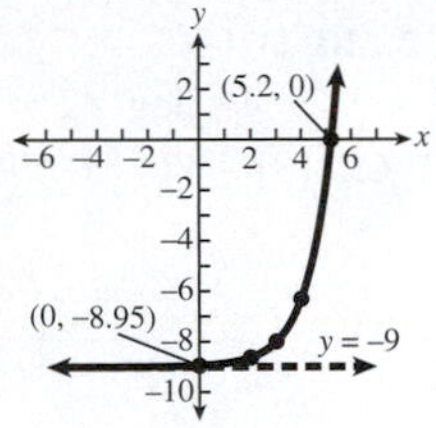

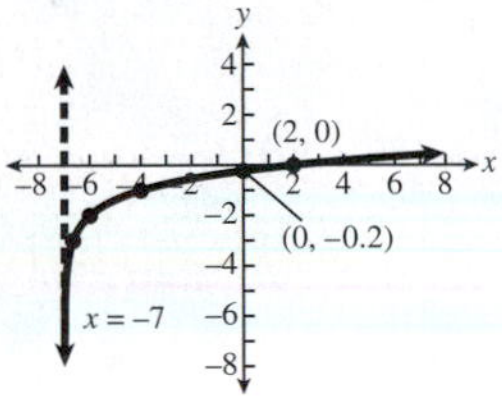

Section 12.6 **1.** horizontal **3.** exponential **5.** y- **7.** x-int: none; y-int: (0, 22) **9.** x-int: (−0.3, 0); y-int: (0, 15) **11.** x-int: (−0.7, 0); y-int: (0, 10.1) **13.** x-int: (13, 0); y-int: none **15.** x-int: (51.6, 0); y-int: (0, −2.9) **17.** Asymptote: $y = 8$; domain: $(-\infty, \infty)$; range: $(8, \infty)$ **19.** Asymptote: $y = -9$; domain: $(-\infty, \infty)$; range: $(-9, \infty)$ **21.** Asymptote: $x = 5$; domain: $(5, \infty)$; range: $(-\infty, \infty)$ **23.** Asymptote: $x = \frac{1}{2}$; domain: $\left(\frac{1}{2}, \infty\right)$; range: $(-\infty, \infty)$ **25.** Domain: $(-\infty, \infty)$; range: $(4, \infty)$ **27.** 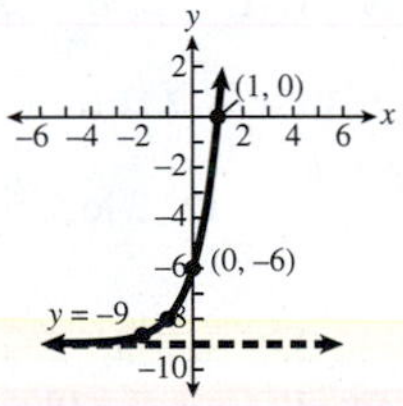 Domain: $(-\infty, \infty)$; range: $(-9, \infty)$ **29.** Domain: $(-\infty, \infty)$; range: $(-4, \infty)$ **31.** 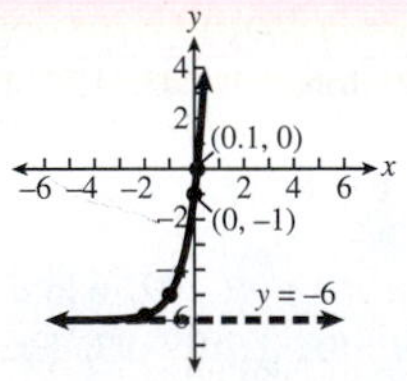 Domain: $(-\infty, \infty)$; range: $(-6, \infty)$ **33.** Domain: $(-\infty, \infty)$; range: $(3, \infty)$ **35.** 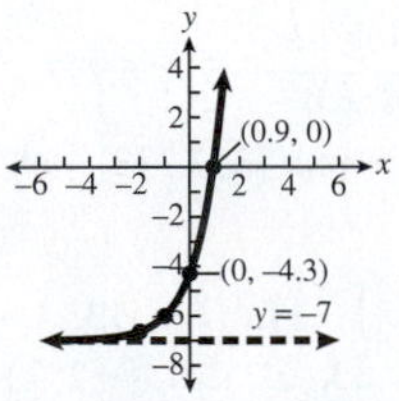 Domain: $(-\infty, \infty)$; range: $(-7, \infty)$ **37.** Domain: $(2, \infty)$; range: $(-\infty, \infty)$ **39.** 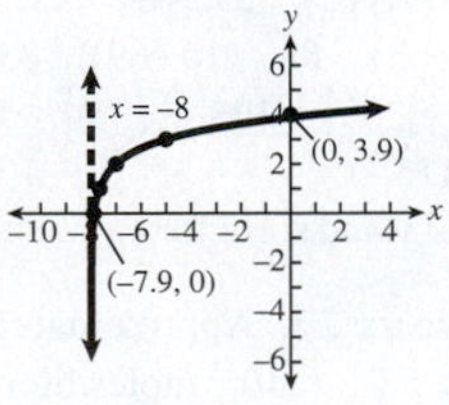 Domain: $(-8, \infty)$; range: $(-\infty, \infty)$

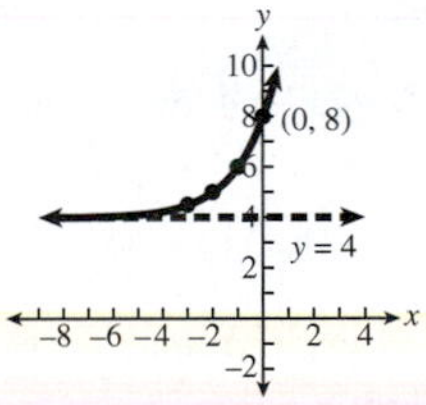

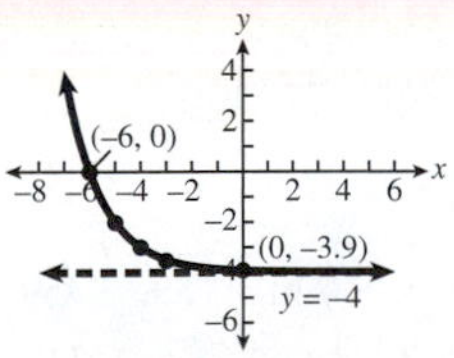

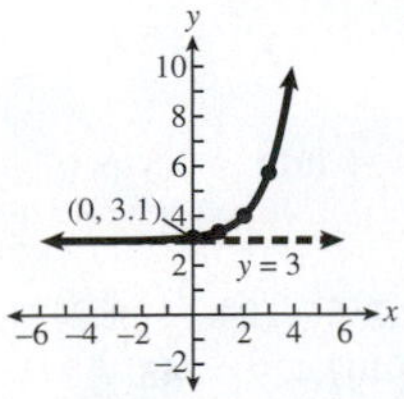

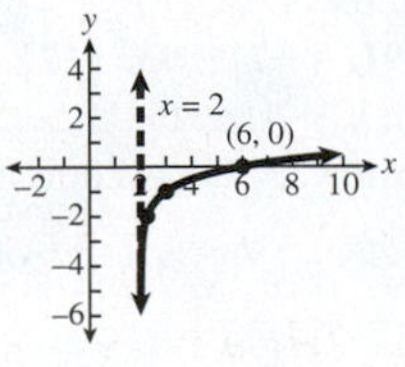

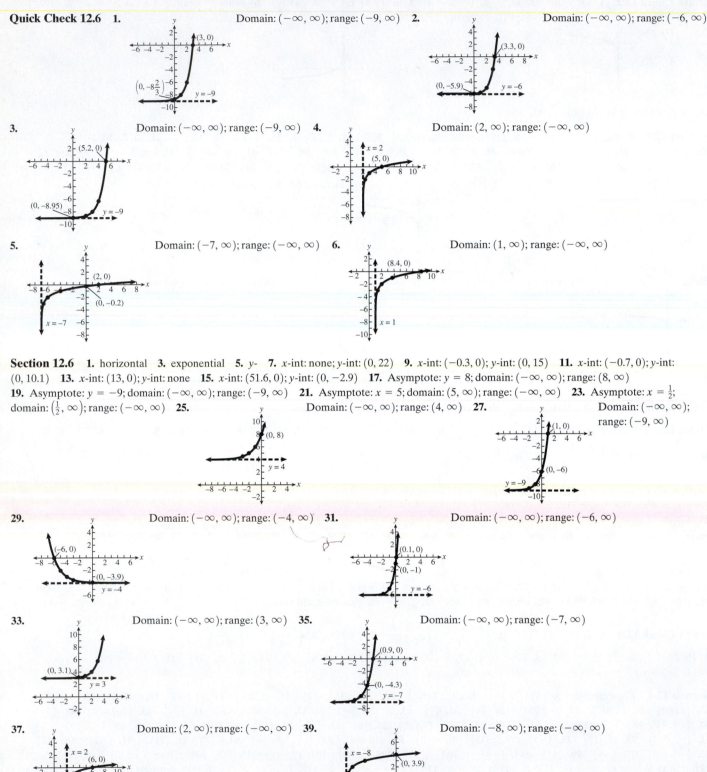

41. 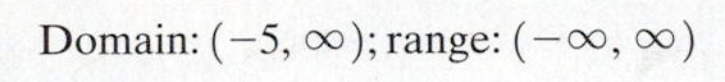
Domain: $(-5, \infty)$; range: $(-\infty, \infty)$

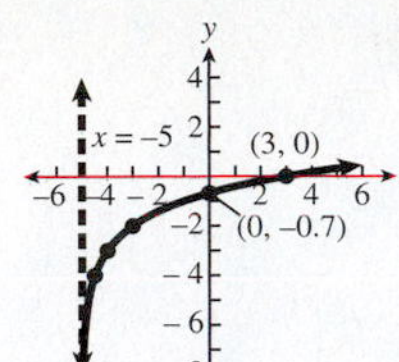

43. Domain: $(3, \infty)$; range: $(-\infty, \infty)$

x = 3
(3.4, 0)

45. 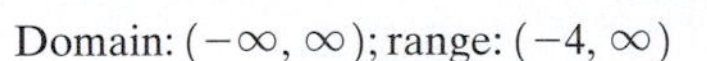
Domain: $(-1, \infty)$; range: $(-\infty, \infty)$

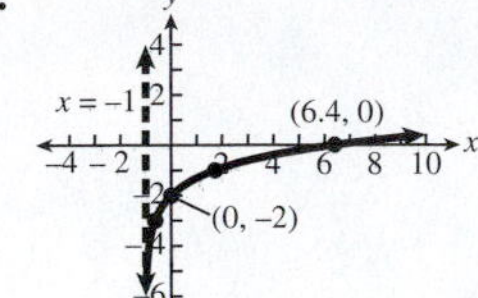

47. Domain: $(-\infty, \infty)$; range: $(-4, \infty)$

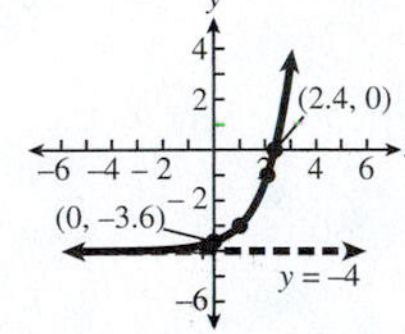

49. Domain: $(-4, \infty)$; range: $(-\infty, \infty)$

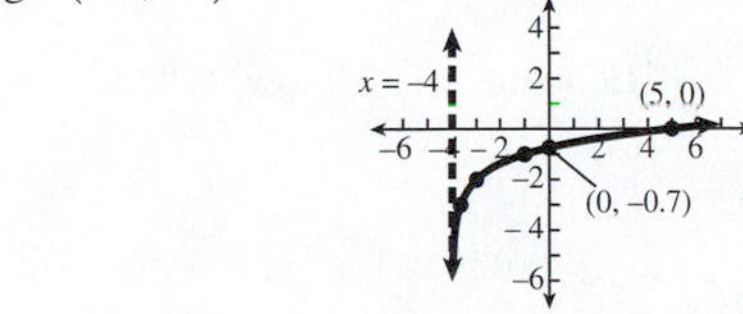

51. Domain: $(-6, \infty)$; range: $(-\infty, \infty)$ **53.** Domain: $(-\infty, \infty)$; range: $(-8, \infty)$

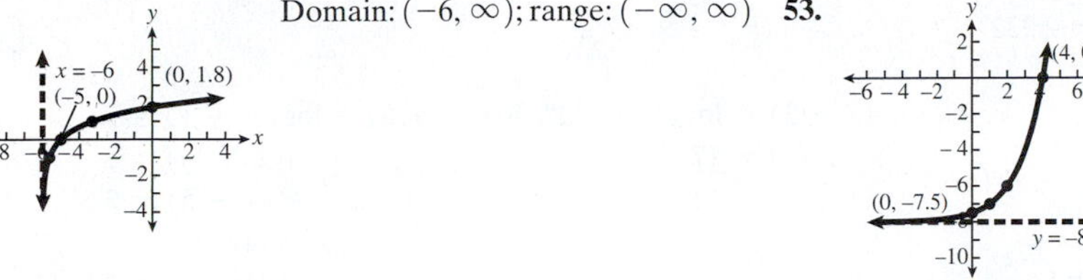

55. Domain: $(-\infty, \infty)$; range: $(-3, \infty)$

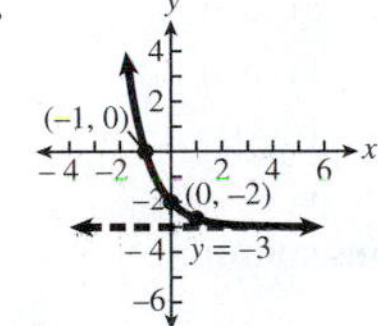

57.

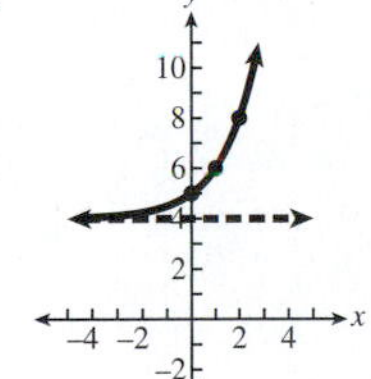

59.

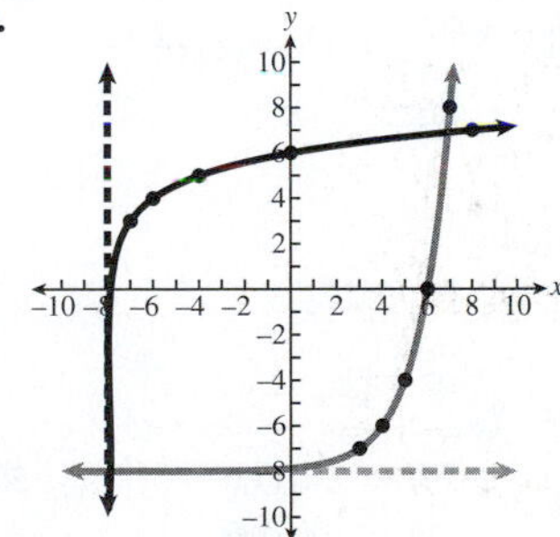

61. b **63.** h **65.** a **67.** f **69.** Answers will vary. Shift basic graph; x-int: set equal to 0 and solve; y-int: substitute 0 for x

Chapter 12 Review **1.** 1015 **2.** 7.086 **3.** −2 **4.** 5 **5.** 0 **6.** 12 **7.** 23.761 **8.** −2.756 **9.** $\log_4 1024 = x$ **10.** $\ln 20 = x$ **11.** $10^{3.2} = x$ **12.** $e^{11} = x$ **13.** $\log_2 104$ **14.** $\log\left(\frac{a^2}{b^5}\right)$ **15.** $\log_2\left(\frac{x^4z^9}{y^3}\right)$ **16.** $\log_b \frac{1}{972}$ **17.** $2 + \log_8 x$ **18.** $\log_3 n - 3$ **19.** $6 \log_b a + 3 \log_b c - 2$ **20.** $3 + 9 \log_2 x - \log_2 y - 7 \log_2 z$ **21.** 2.322 **22.** −8.205 **23.** 10.641 **24.** 4.340 **25.** $\{3\}$ **26.** $\{7\}$ **27.** $\{\frac{5}{4}\}$ **28.** $\{12\}$ **29.** $\{1.232\}$ **30.** $\{0.468\}$ **31.** $\{12.127\}$ **32.** $\{5.806\}$ **33.** $\{14\}$ **34.** $\{1\}$ **35.** $\{148.413\}$ **36.** $\{3, 10\}$ **37.** $\{2\}$ **38.** $\{10\}$ **39.** $\{4\}$ **40.** $\{2\}$ **41.** $\{3.107\}$ **42.** $\{54.598\}$ **43.** $f^{-1}(x) = \ln(x - 4)$ **44.** $f^{-1}(x) = \log_2(x + 1) + 5$ **45.** $f^{-1}(x) = e^x - 8$ **46.** $f^{-1}(x) = 3^{x+10} - 6$ **47.** 13.3 years **48.** 334,133 **49.** 6.25 grams **50.** 167 years **51.** 2007 **52.** 78.5° F **53.** 4.3 **54.** 3.16×10^{-4} moles/liter **55.** Domain: $(-\infty, \infty)$; range: $(2, \infty)$

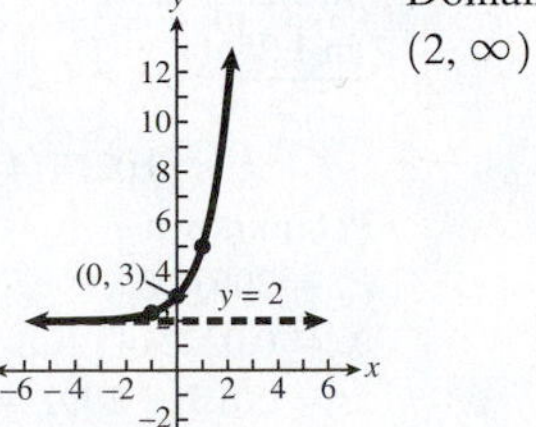

56. Domain: $(-\infty, \infty)$; range: $(-8, \infty)$

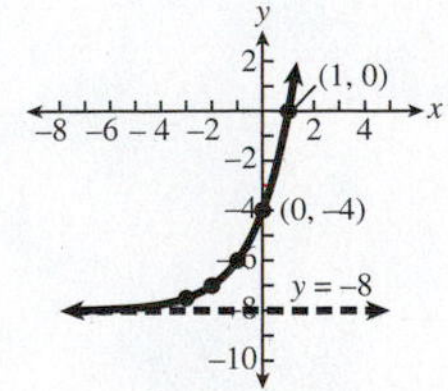

57. Domain: $(-\infty, \infty)$; range: $(-6, \infty)$

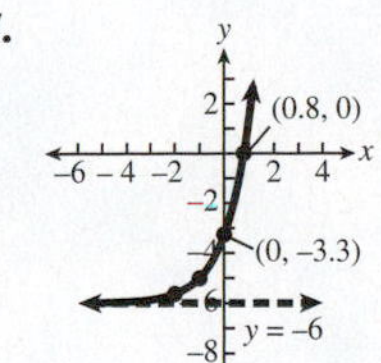

58.

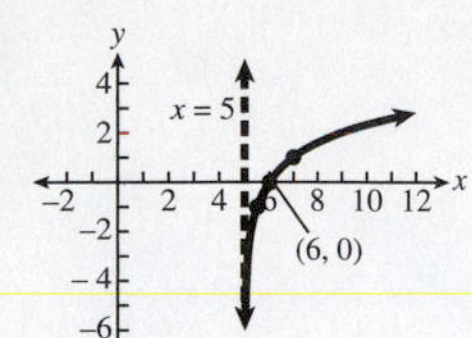

Domain: $(5, \infty)$; range: $(-\infty, \infty)$

59.

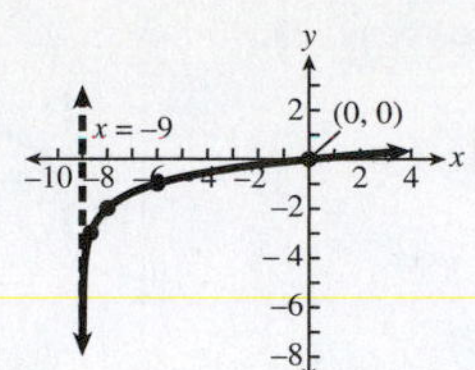

Domain: $(-9, \infty)$; range: $(-\infty, \infty)$

60.

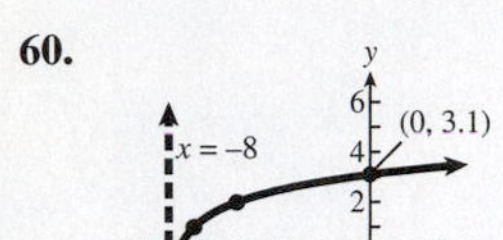
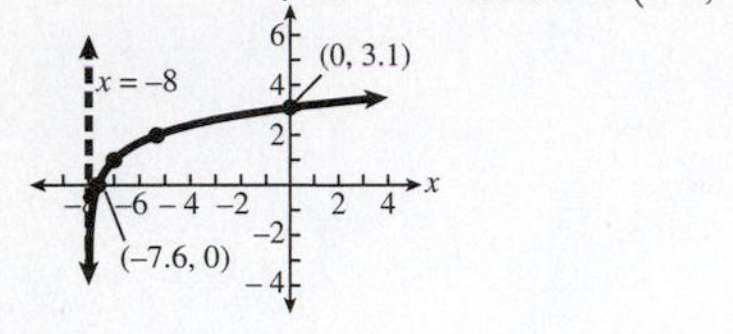

Domain: $(-8, \infty)$; range: $(-\infty, \infty)$

Chapter 12 Review Exercises: Worked-Out Solutions

2. $f(11) = e^{11-8} - 13$
$= e^3 - 13$
≈ 7.086

3. $\log_6 \frac{1}{36} = -2$ because $6^{-2} = \frac{1}{36}$.

15. $4 \log_2 x - 3 \log_2 y + 9 \log_2 z = \log_2 x^4 - \log_2 y^3 + \log_2 z^9$
$= \log_2 \left(\frac{x^4 z^9}{y^3}\right)$

19. $\log_b\left(\frac{a^6 c^3}{b^2}\right) = \log_b a^6 + \log_b c^3 - \log_b b^2$
$= 6 \log_b a + 3 \log_b c - 2 \log_b b$
$= 6 \log_b a + 3 \log_b c - 2$

21. $\log_4 25 = \frac{\log 25}{\log 4}$
≈ 2.322

26. $8^{x-5} = 64$
$8^{x-5} = 8^2$
$x - 5 = 2$
$x = 7$
$\{7\}$

28. $9^{x+6} = 27^x$
$(3^2)^{x+6} = (3^3)^x$
$3^{2x+12} = 3^{3x}$
$2x + 12 = 3x$
$12 = x$
$\{12\}$

31. $e^{x-8} = 62$
$\ln e^{x-8} = \ln 62$
$x - 8 = \ln 62$
$x = \ln 62 + 8$
$x \approx 12.127$
$\{12.127\}$

33. $\log_4(x + 3) = \log_4 17$
$x + 3 = 17$
$x = 14$
$\{14\}$

39. $\log_3(x + 5) + \log_3(x - 1) = 3$
$\log_3((x + 5)(x - 1)) = 3$
$\log_3(x^2 + 4x - 5) = 3$
$x^2 + 4x - 5 = 3^3$
$x^2 + 4x - 5 = 27$
$x^2 + 4x - 32 = 0$
$(x + 8)(x - 4) = 0$
$x + 8 = 0$ or $x - 4 = 0$
$x = -8$ or $x = 4$
$x = -8$ is an extraneous solution.
$\{4\}$

43. $y = e^x + 4$
$x = e^y + 4$
$x - 4 = e^y$
$\ln(x - 4) = \ln e^y$
$\ln(x - 4) = y$
$f^{-1}(x) = \ln(x - 4)$

48. $P = 250{,}000$
$P_0 = 200{,}000$
k: Unknown
$t = 2002 - 1992 = 10$
$250{,}000 = 200{,}000e^{k(10)}$
$\frac{250{,}000}{200{,}000} = \frac{200{,}000e^{10k}}{200{,}000}$
$1.25 = e^{10k}$
$\ln 1.25 = \ln e^{10k}$
$\ln 1.25 = 10k$
$\frac{\ln 1.25}{10} = k$
$k \approx 0.022314$
P: Unknown
$P_0 = 200{,}000$
$k = 0.022314$
$t = 2015 - 1992 = 23$
$P = 200{,}000e^{0.022314(23)}$
$P = 200{,}000e^{0.513222}$
$P \approx 334{,}133$

50. $0.5 = 1 \cdot e^{5730k}$
$\ln 0.5 = \ln e^{5730k}$
$\ln 0.5 = 5730k$
$\frac{\ln 0.5}{5730} = k$
$k \approx -0.000121$
$0.98 = 1 \cdot e^{-0.000121t}$
$\ln 0.98 = \ln e^{-0.000121t}$
$\ln 0.98 = -0.000121t$
$\frac{\ln 0.98}{-0.000121} = t$
$t \approx 167$

51. $30 = 11.5 \cdot 1.057^x$
$\frac{30}{11.5} = 1.057^x$
$\log \frac{30}{11.5} = \log 1.057^x$
$\log \frac{30}{11.5} = x \cdot \log 1.057$
$\frac{\log\left(\frac{30}{11.5}\right)}{\log 1.057} = x$
$x \approx 17.3$
It reached 30% in $1990 + 17 = 2007$.

55. Shift graph of $f(x) = 3^x$ up by 2 units. Domain: $(-\infty, \infty)$, Range: $(2, \infty)$

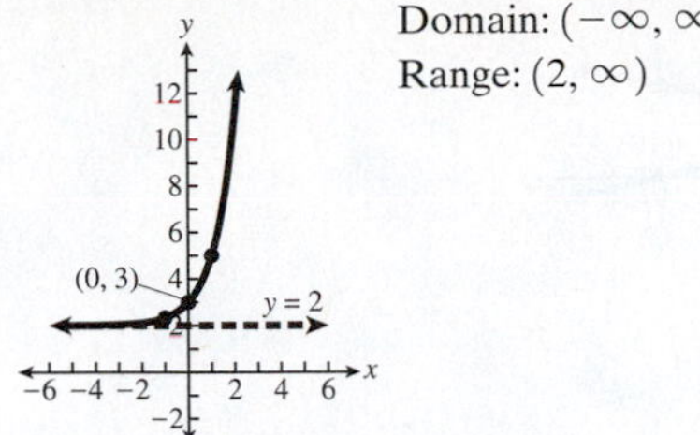

Horizontal Asymptote: $y = 2$
No x-intercepts ($0 = 3^x + 2$ has no solution.)
y-intercept ($x = 0$): $(0, 3)$

58. Shift graph of $f(x) = \log_2 x$ by 5 units to the right. Domain: $(5, \infty)$, Range: $(-\infty, \infty)$

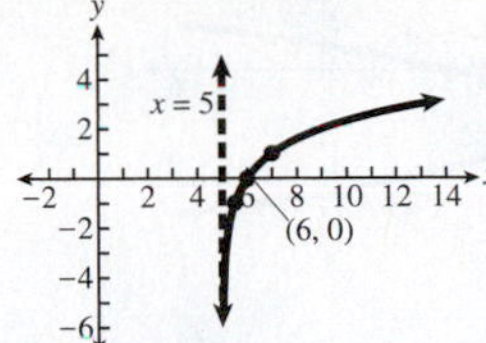

Vertical Asymptote: $x = 5$
x-intercept ($y = 0$): $(6, 0)$
No y-intercepts ($0 = 2^y + 5$ has no solution.)

Chapter 12 Test **1.** -2.611 **2.** 6.890 **3.** 2 **4.** -3 **5.** 3 **6.** $\log_b\left(\frac{x^2y^5}{z^3}\right)$ **7.** $4 \ln a + 6 \ln b - 8 \ln c - \ln d$ **8.** 1.878 **9.** $\{-6\}$ **10.** $\{12.124\}$ **11.** $\{4.207\}$ **12.** $\{8\}$ **13.** $\{60\}$ **14.** $\{8\}$ **15.** $f^{-1}(x) = \ln(x + 17) - 9$ **16.** 6 years **17.** \$80,946.56 **18.** 870.7 years

19. Domain: $(-\infty, \infty)$; range: $(-9, \infty)$

(4.2, 0)
(0, −8.9)
y = −9

20. Domain: $(-4, \infty)$; range: $(-\infty, \infty)$

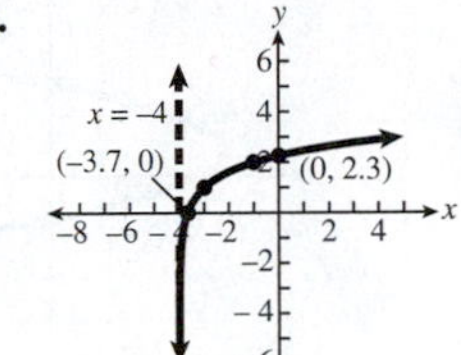

CHAPTER 13

Quick Check 13.1 **1.**

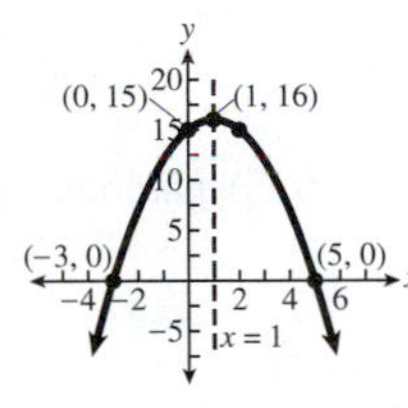

2.

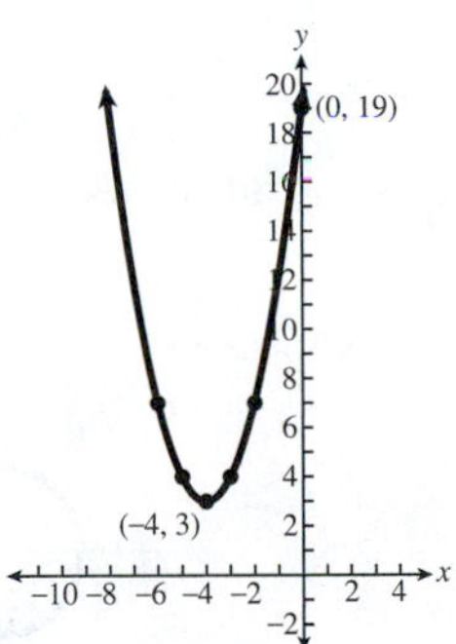

3.

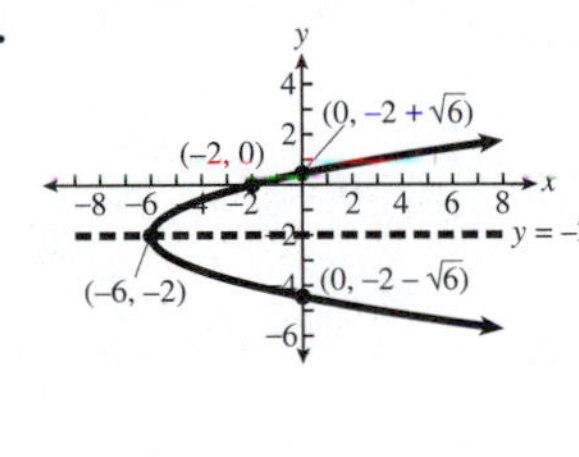

4.

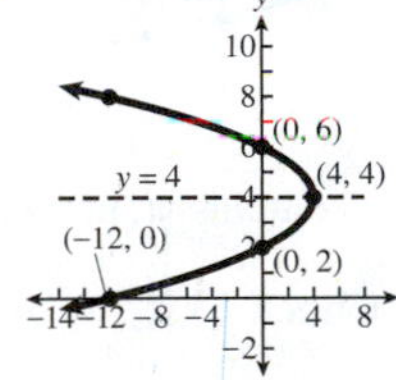

5.

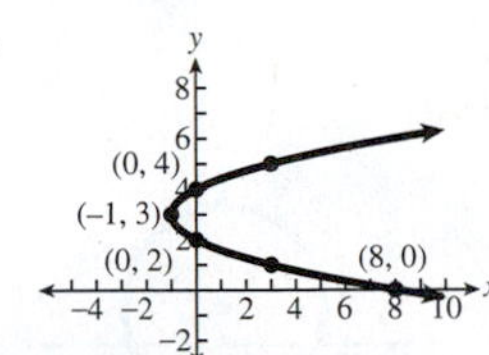

6.

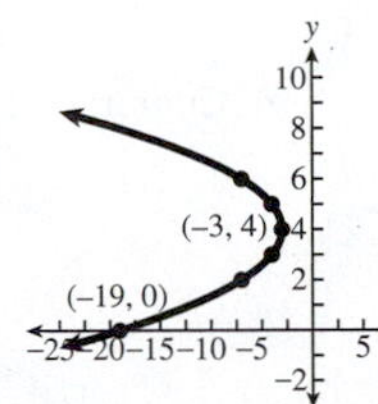

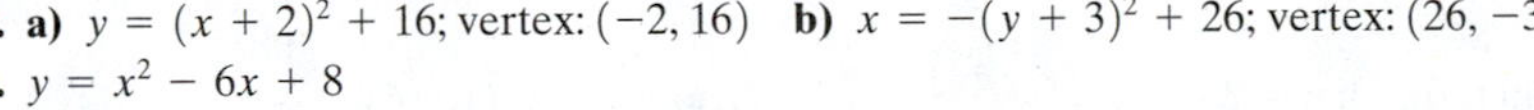

7. a) $y = (x + 2)^2 + 16$; vertex: $(-2, 16)$ **b)** $x = -(y + 3)^2 + 26$; vertex: $(26, -3)$ **8.** $y = x^2 - 6x + 8$

Section 13.1 **1.** upward, downward **3.** right, left **5.** (h, k) **7.**

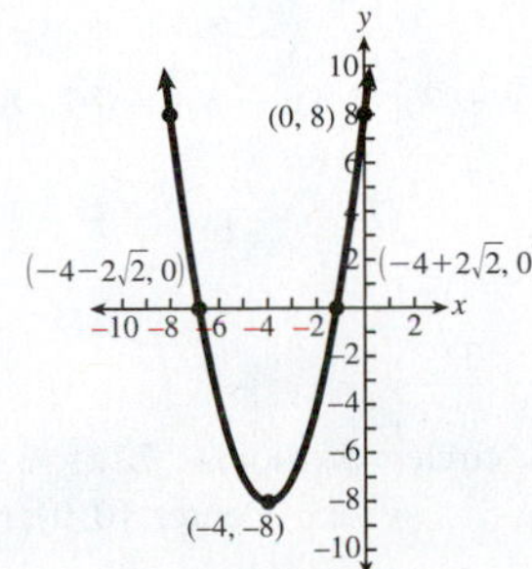

9.

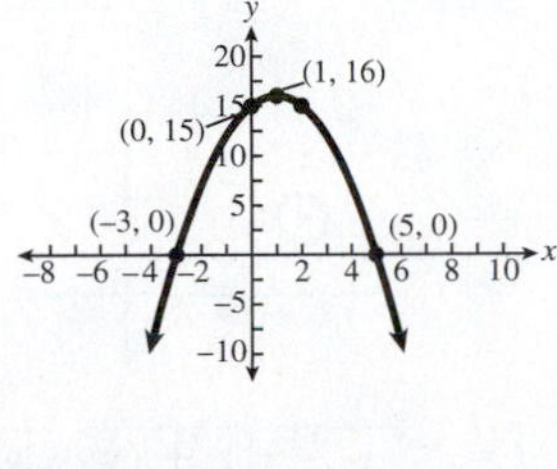

11.
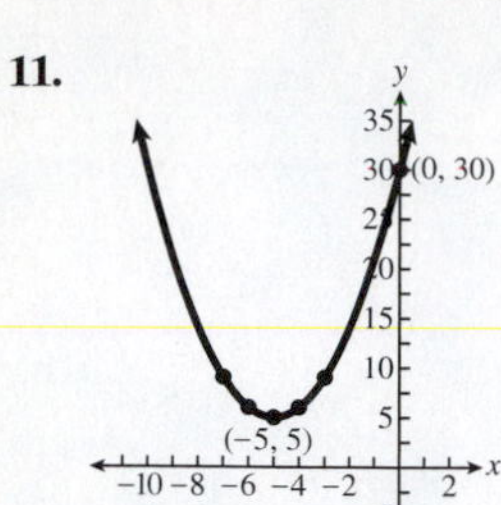
13.
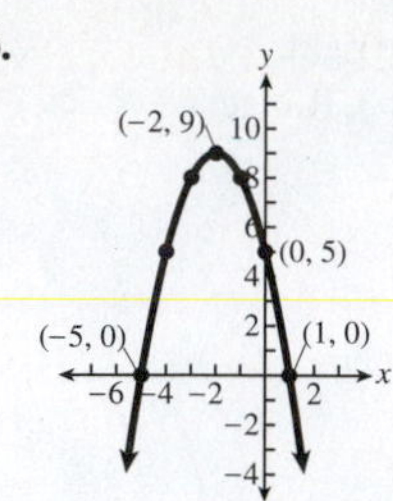
15.
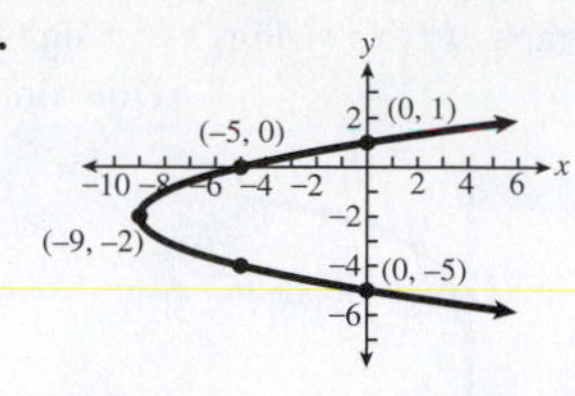
17.
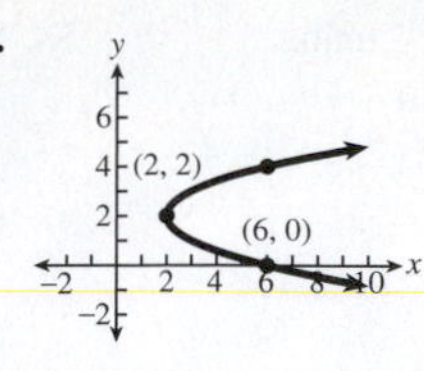

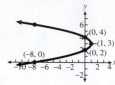

19.

21.
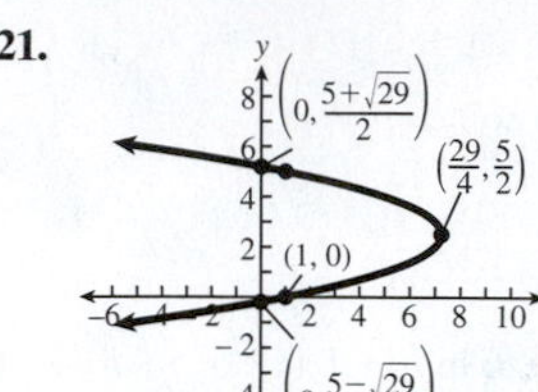
23.
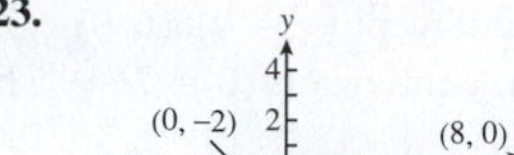
25.
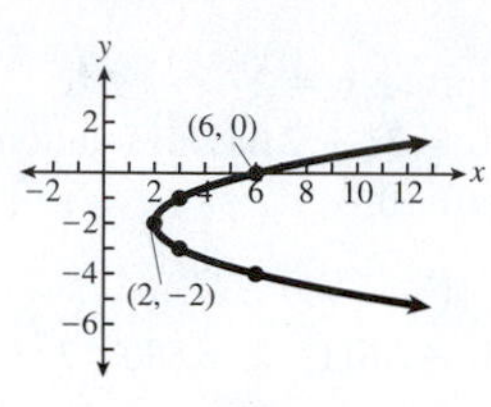

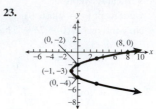

27.
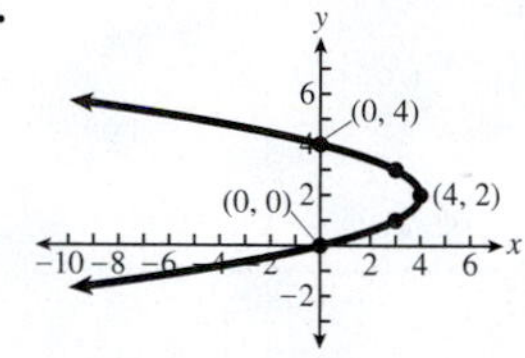
29.
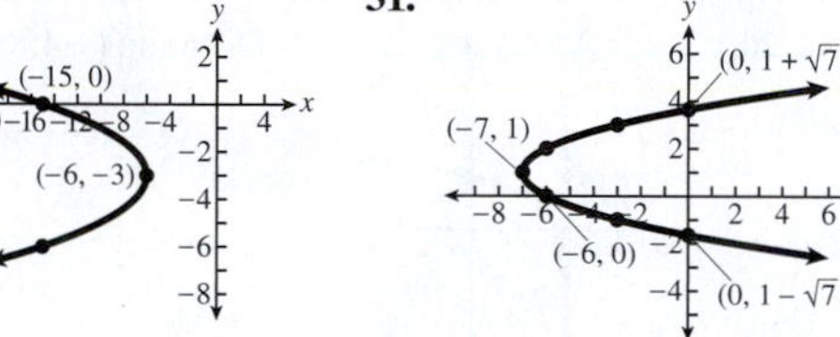
31.
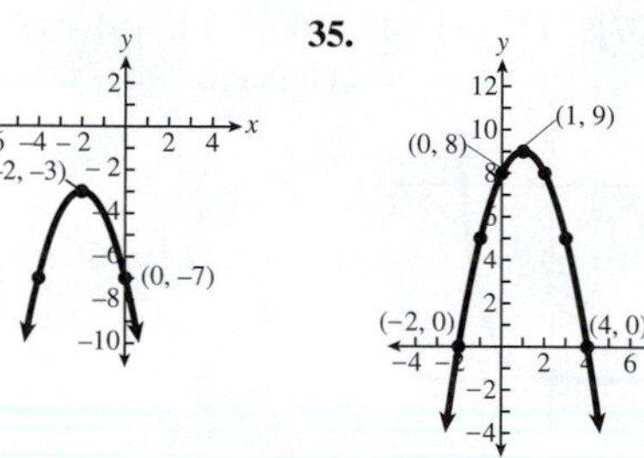
33.
35.

37.
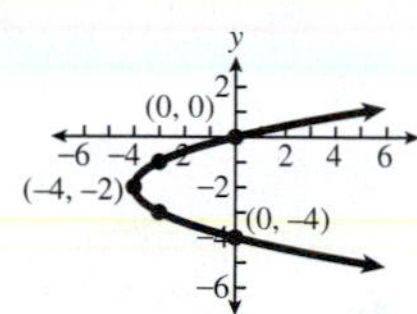
39.
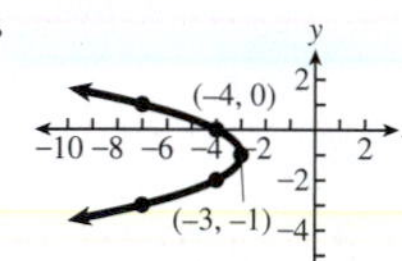

41. $(1, 14)$ **43.** $(7, 125)$ **45.** $(-49, -3)$ **47.** $(-47, 2)$ **49.** $y = x^2 - 6x + 5$
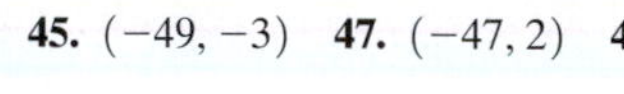

51. $x = y^2 - 4y - 5$ **53. a)** $y = \frac{1}{500}x^2 - 5$ **b)** 4.2 feet **55.** b **57.** h **59.** g **61.** c **63.** Answers will vary. A parabola opens left or right if the equation contains y^2, not x^2.

Quick Check 13.2 **1. a)** 10 **b)** 12.6 **2.** $\left(\frac{1}{2}, 2\right)$ **3.** Center: $(0, 0)$; $r = 2\sqrt{3} \approx 3.5$

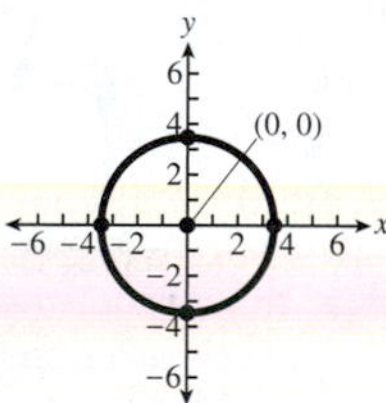

4. Center: $(6, 5)$; $r = 3$

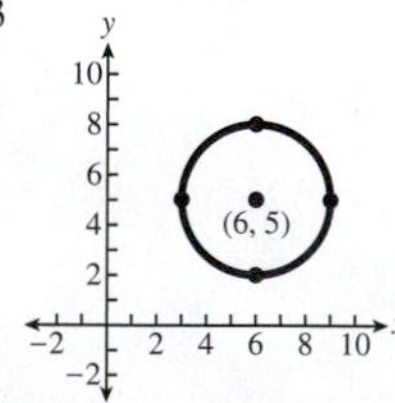

5. a) Center: $(1, -6)$; $r = 4$

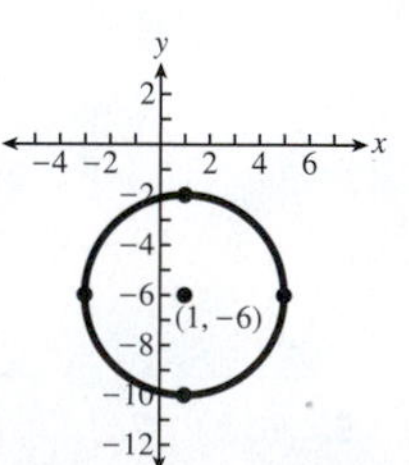

b) Center: $(-2, 3)$; $r = 8$

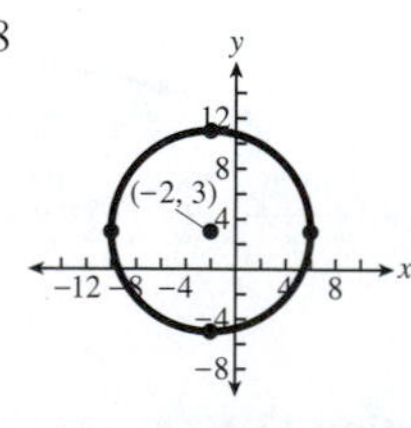

6. Center: $(-7, 3)$; $r = 1$

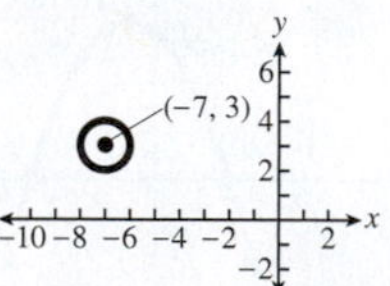
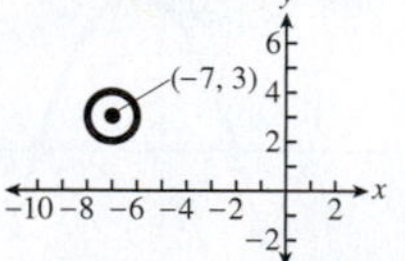

7. $(x + 3)^2 + (y - 5)^2 = 16$ **8.** $(x - 1)^2 + (y + 2)^2 = 34$

Section 13.2 **1.** $d = \sqrt{(x_2 - x_1)^2 + (y_2 - y_1)^2}$ **3.** circle **5.** radius **7.** $x^2 + y^2 = r^2$ **9.** 5 **11.** 89 **13.** 11 **15.** 9.9 **17.** 11.2 **19.** $(3, 2)$ **21.** $(7, -17)$ **23.** Center: $(0, 0)$; $r = 4$

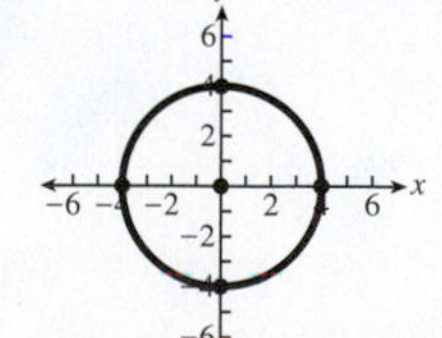

25. Center: $(0, 0)$; $r = 10$

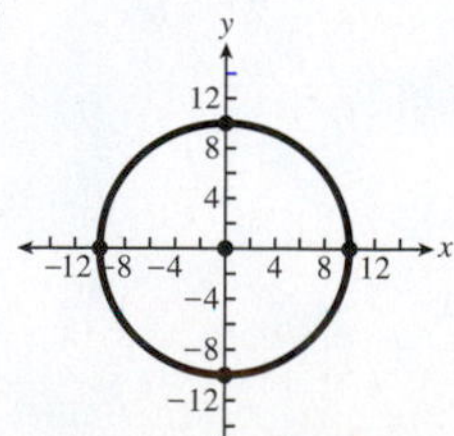

27. Center: $(0, 0)$; $r = \sqrt{6}$

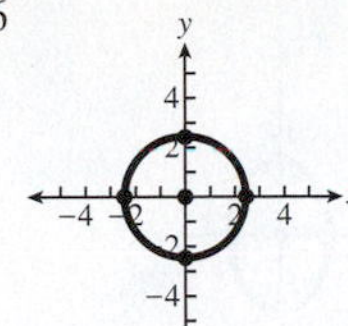

29. Center: $(2, 7)$; $r = 4$

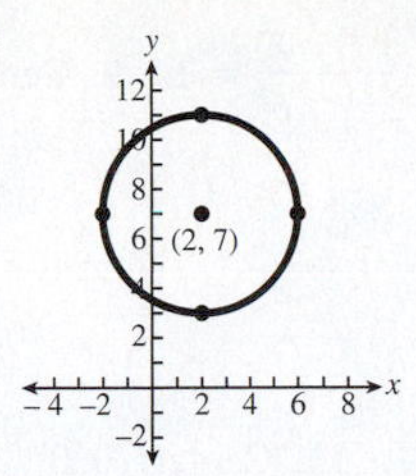

31. Center: $(-5, 1)$; $r = 5$

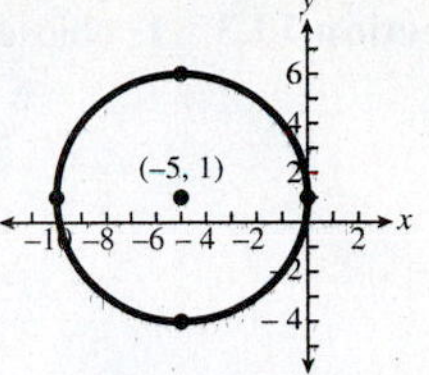

33. Center: $(-2, 1)$; $r = 7$

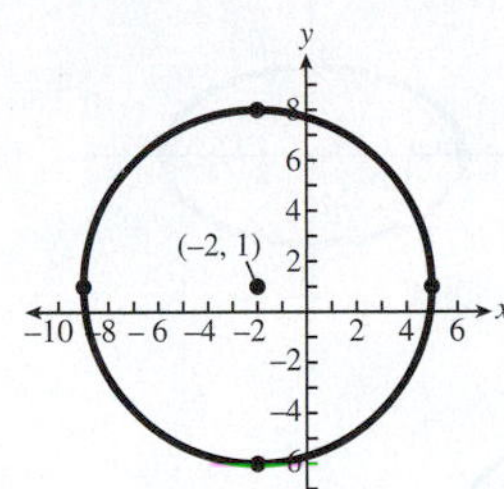

35. Center: $(-5, 0)$; $r = 5$

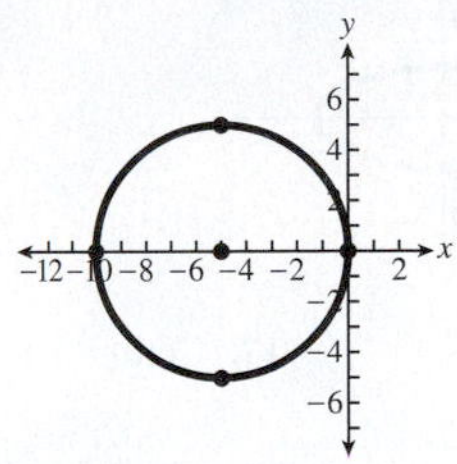

37. Center: $(2, -6)$; $r = 3\sqrt{2}$

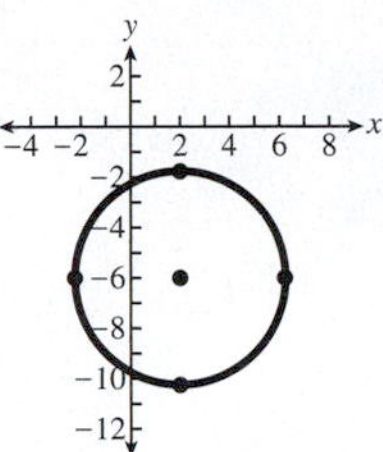
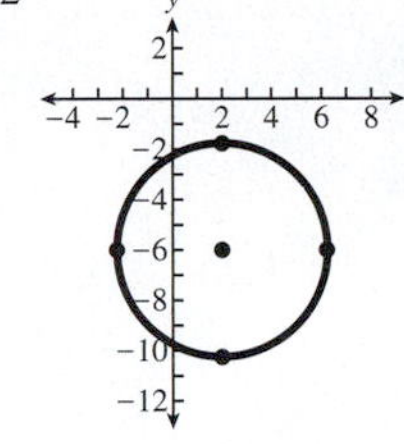

39. $(-2, -6), 9$ **41.** $(5, -1), 7$ **43.** $(-8, 0), 11$ **45.** $(-2, -7), 4\sqrt{5}$ **47.** $x^2 + y^2 = 36$

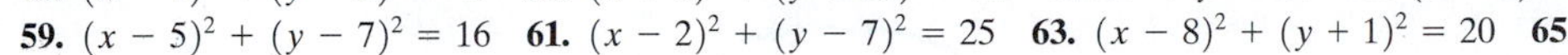
49. $(x - 4)^2 + (y - 2)^2 = 49$ **51.** $(x + 9)^2 + (y + 10)^2 = 25$ **53.** $x^2 + y^2 = 64$ **55.** $(x - 1)^2 + (y - 5)^2 = 16$ **57.** $(x + 1)^2 + (y - 1)^2 = 49$
59. $(x - 5)^2 + (y - 7)^2 = 16$ **61.** $(x - 2)^2 + (y - 7)^2 = 25$ **63.** $(x - 8)^2 + (y + 1)^2 = 20$ **65.**

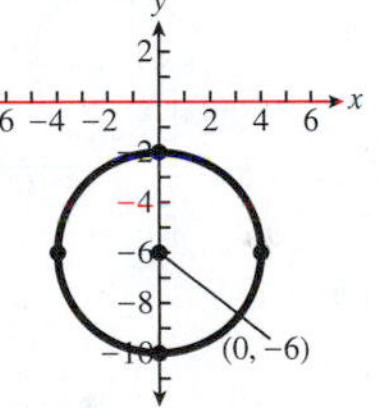

67.

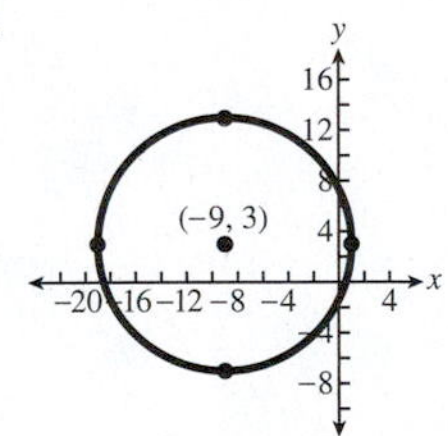

69.

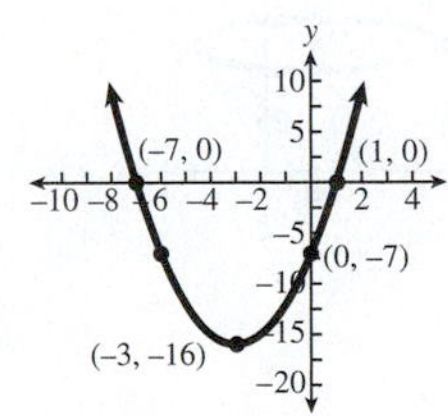

71.

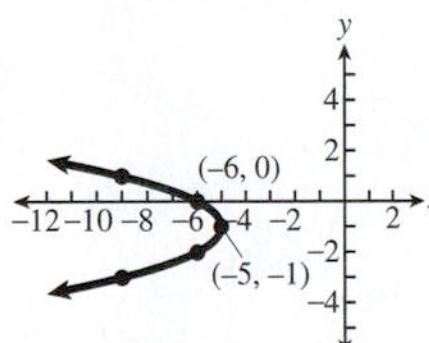

73.

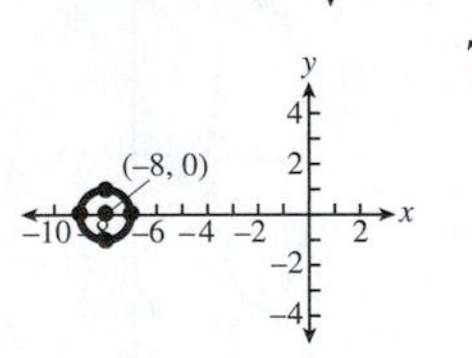

75.

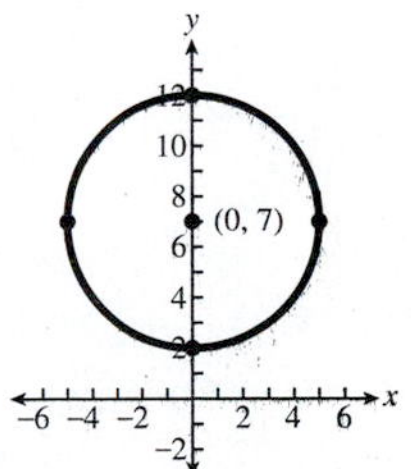

77. a) $x^2 + y^2 = 725$ **b)** 1277.7 square feet **79.** Answers will vary. Only x^2: up/down; only y^2: left/right; $x^2 + y^2$: circle

Quick Check 13.3 **1. a)**

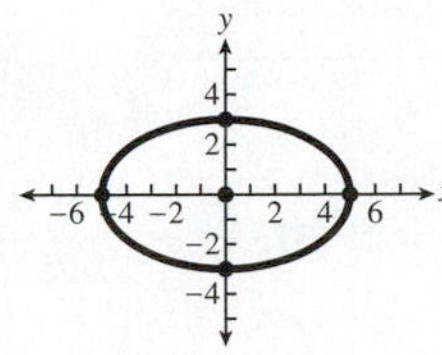

b)

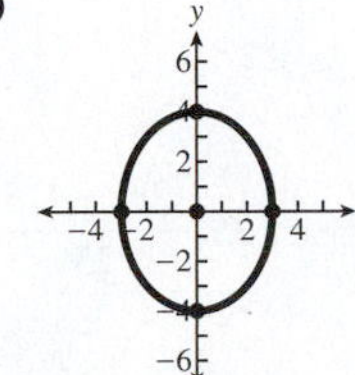

2.

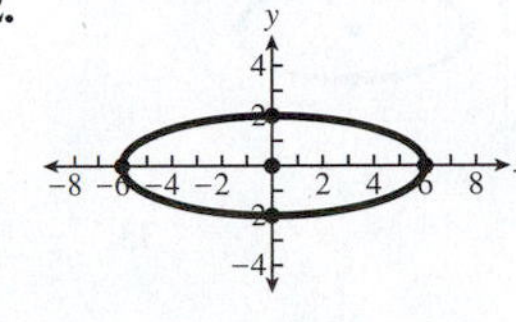

3.

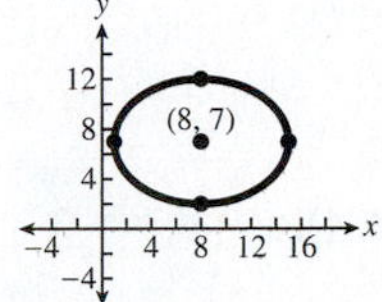

4.

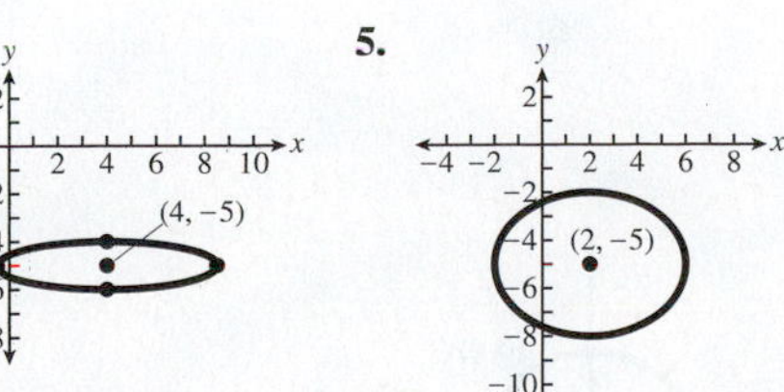

5.

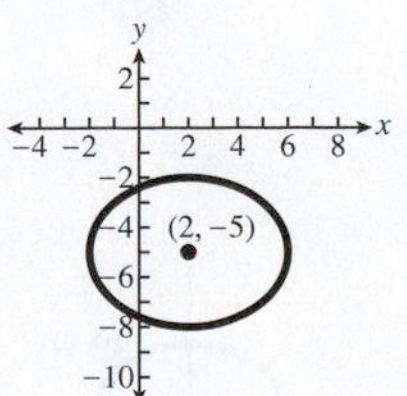

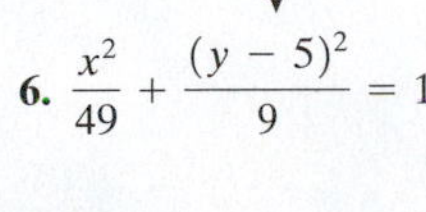
6. $\dfrac{x^2}{49} + \dfrac{(y - 5)^2}{9} = 1$

Section 13.3 **1.** ellipse **3.** center **5.** vertices **7.** $\frac{x^2}{a^2} + \frac{y^2}{b^2} = 1$ **9.** Center: $(0, 0)$; $a = 2$; $b = 3$

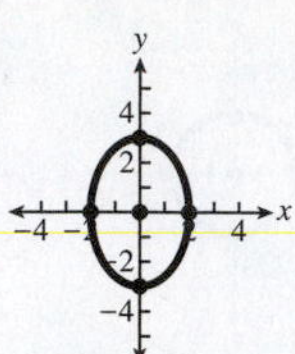

11. Center: $(0, 0)$; $a = 5$; $b = 2$

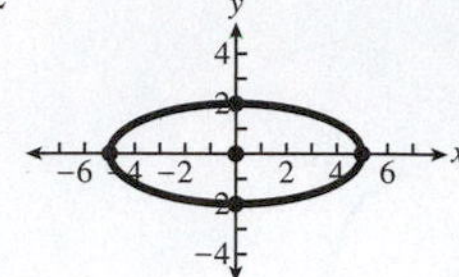

13. Center: $(0, 0)$; $a = 6$; $b = 2\sqrt{3}$

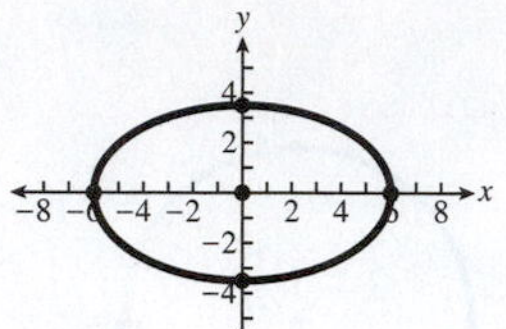

15. Center: $(0, 0)$; $a = 2$; $b = 9$

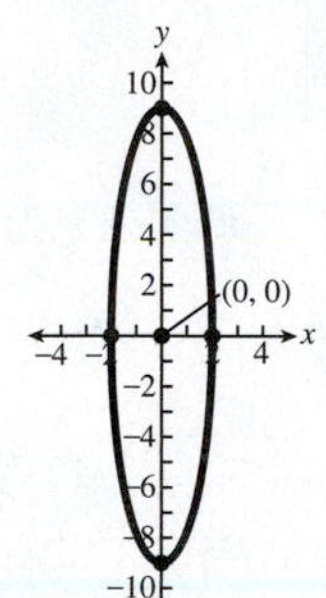

17. Center: $(0, 0)$; $a = 4$; $b = 3$

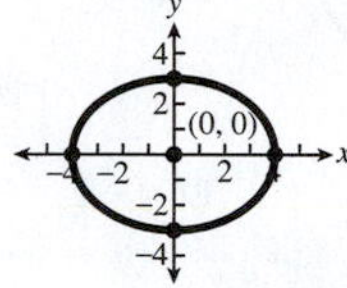

19. Center: $(6, 8)$; $a = 5$; $b = 6$

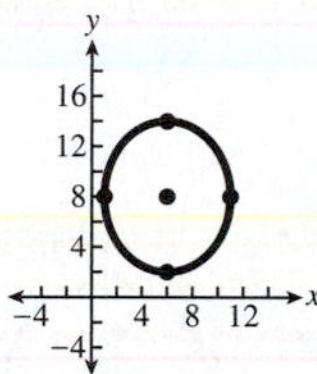

21. Center: $(-3, 5)$; $a = 6$; $b = 2$

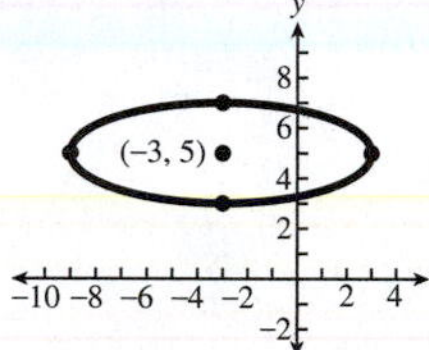

23. Center: $(-5, 0)$; $a = 2$; $b = 6$

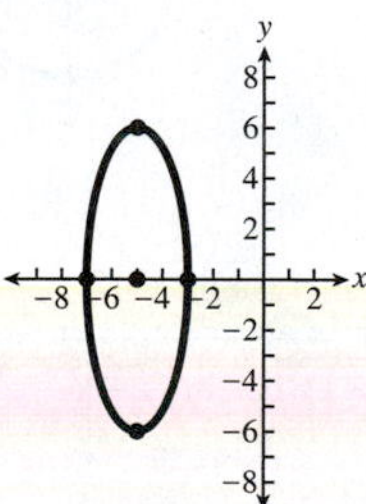

25. Center: $(-2, 5)$; $a = 5$, $b = 1$

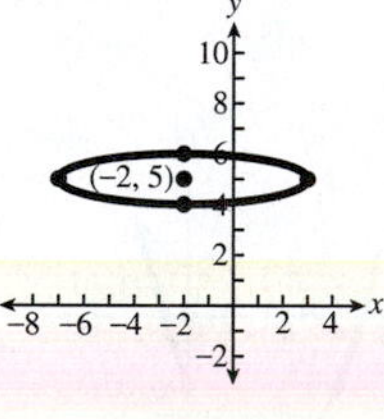

27. Center: $(6, -6)$; $a = 2\sqrt{5}$; $b = 2$

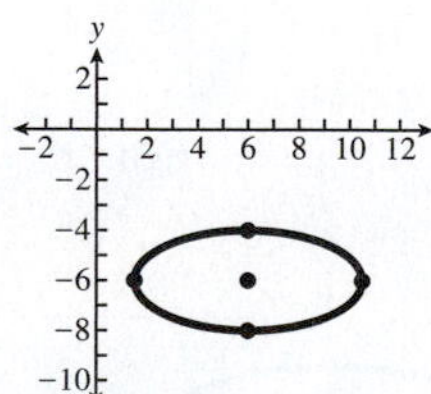

29. Center: $(3, 5)$; $a = 3$; $b = 2$

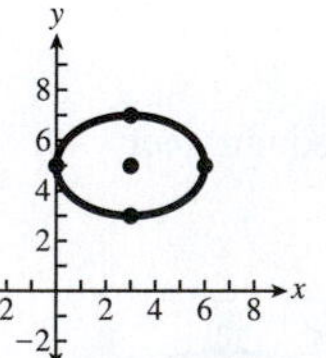

31. Center: $(-4, 4)$; $a = 6$; $b = 3$

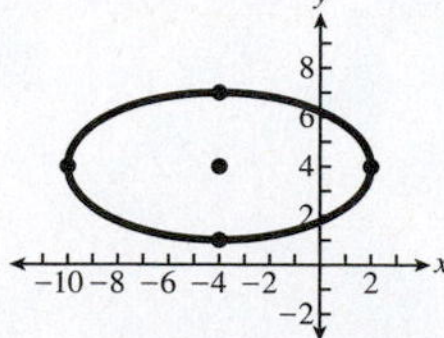

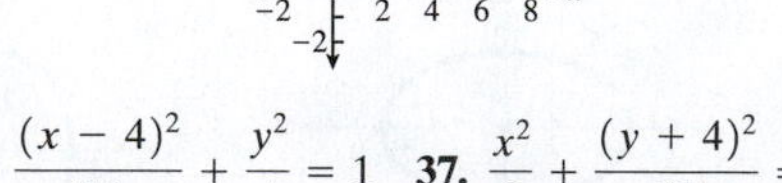

33. $\frac{x^2}{25} + \frac{y^2}{9} = 1$ **35.** $\frac{(x-4)^2}{25} + \frac{y^2}{4} = 1$ **37.** $\frac{x^2}{9} + \frac{(y+4)^2}{25} = 1$

39. $\frac{(x-3)^2}{36} + \frac{(y+4)^2}{25} = 1$ **41.**

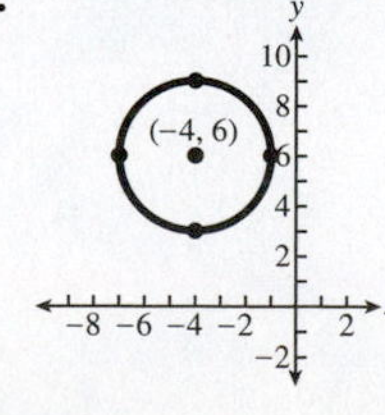

43.

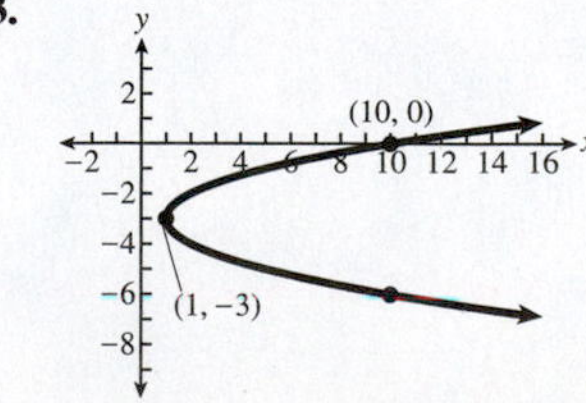

45.

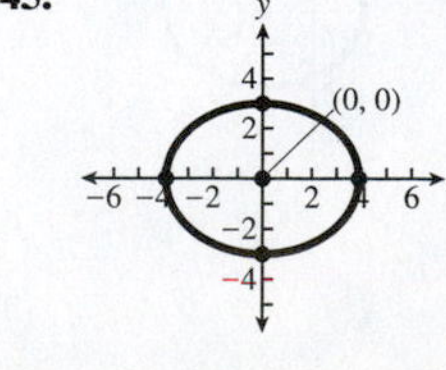

47.

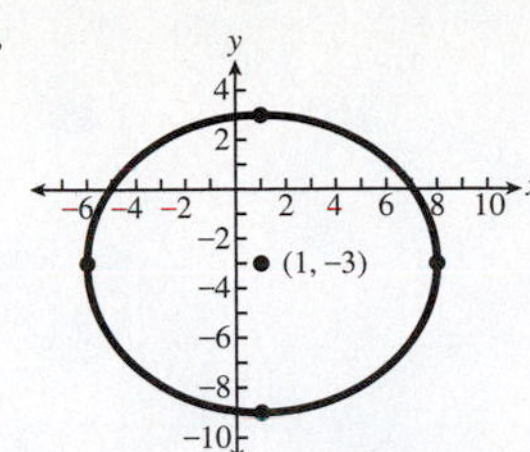

49.

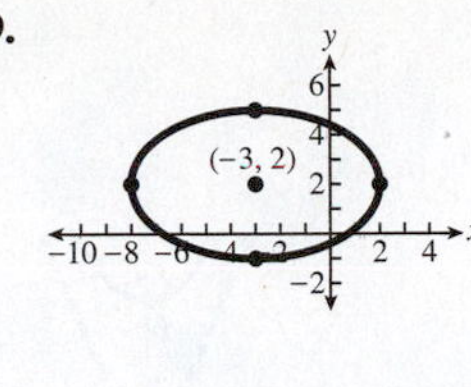

51.

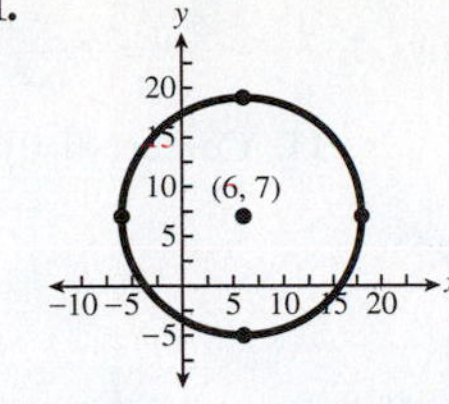

53.

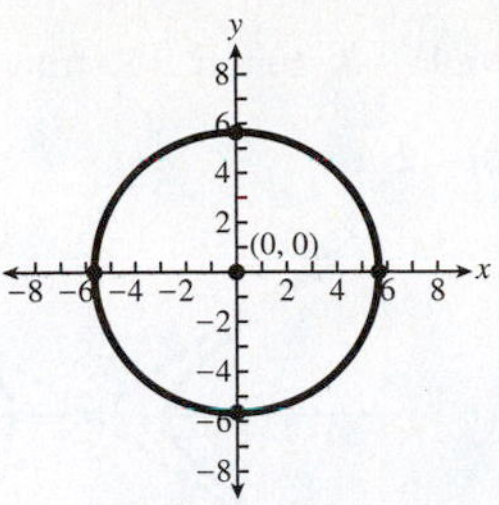

55.

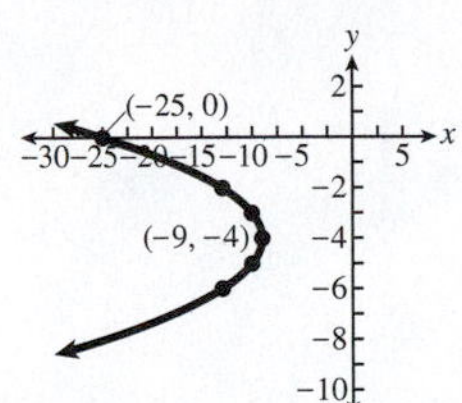

57.

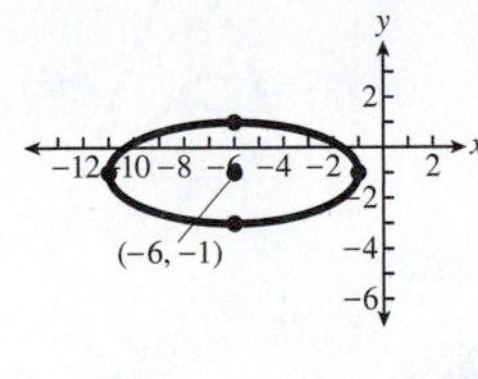

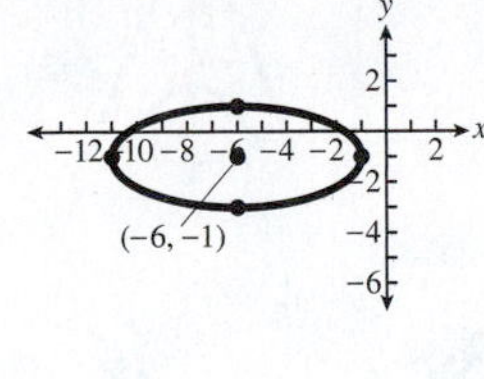

59. a) 100 feet **b)** 66 feet **61.** 471.2 square feet **63.** Answers will vary. The graph of an ellipse is more oval in shape; they have similar equations.

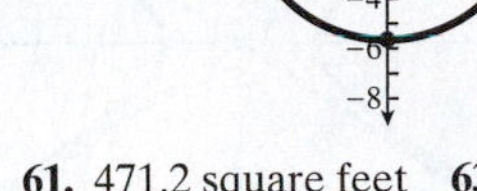

Quick Check 13.4 **1.** Center: (0, 0); vertices: (−4, 0); (4, 0); asymptotes: $y = \frac{3}{4}x$, $y = -\frac{3}{4}x$

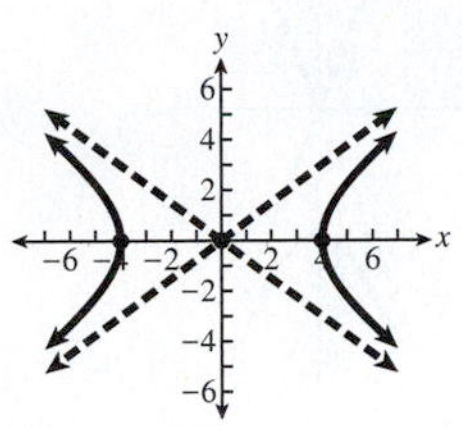

2. Center: (0, 0); vertices: (0, −2), (0, 2); asymptotes: $y = \frac{1}{3}x$, $y = -\frac{1}{3}x$

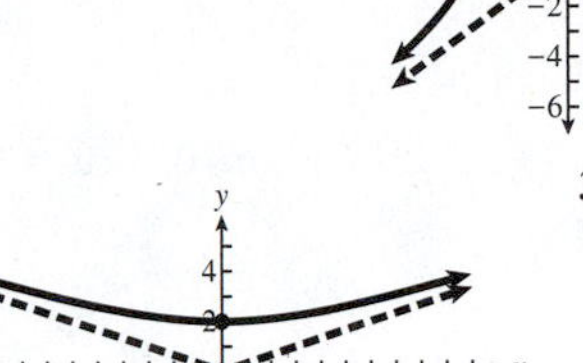

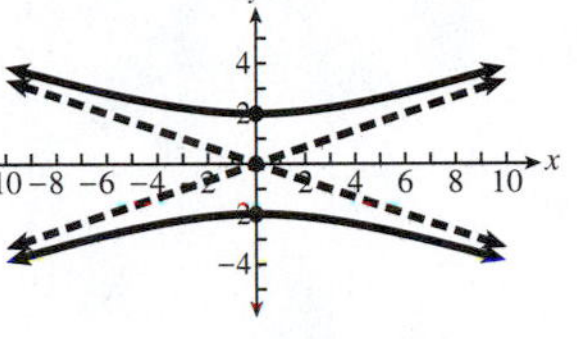

3. Center: (3, 5); vertices: (1, 5), (5, 5); asymptotes: $m = 2, m = -2$

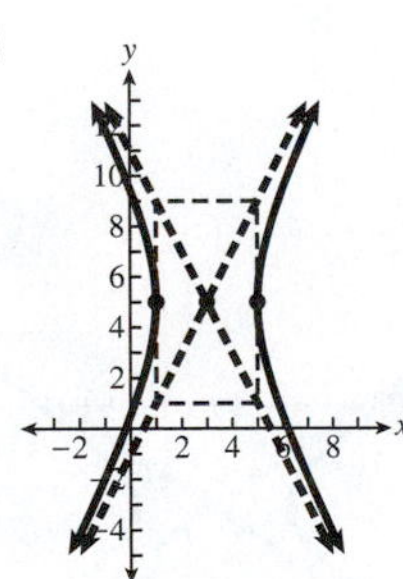

4. Center: (−3, 2); vertices: (−3, 9), (−3, −5); asymptotes: $m = \frac{7}{3}$, $m = -\frac{7}{3}$

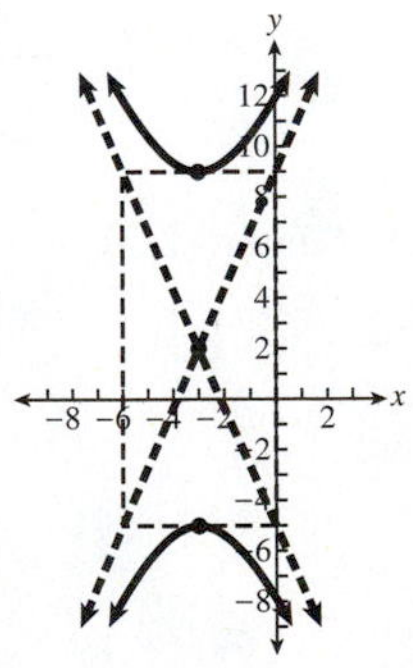

5. Center: (2, −3); vertices: (−1, −3), (5, −3); Asymptotes: $y = \frac{4}{3}(x - 2) - 3$, $y = -\frac{4}{3}(x - 2) - 3$

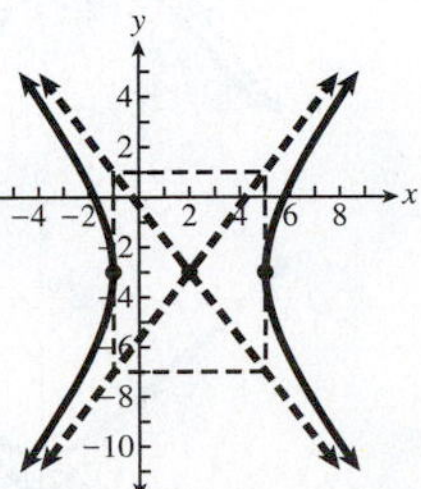

6. $\frac{(y-4)^2}{4} - \frac{(x+3)^2}{25} = 1$

Section 13.4 **1.** hyperbola **3.** center **5.** transverse axis **7.** $\dfrac{x^2}{a^2} - \dfrac{y^2}{b^2} = 1, \dfrac{y^2}{b^2} - \dfrac{x^2}{a^2} = 1$

9. Center: $(0, 0)$; $a = 3$; $b = 2$

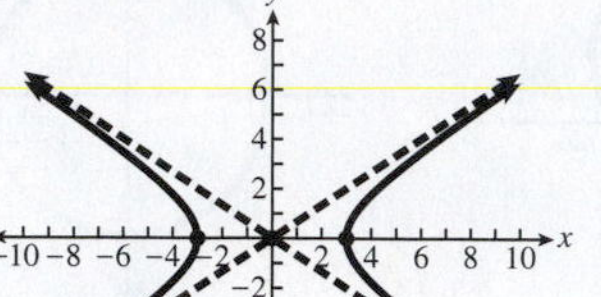

11. Center: $(0, 0)$; $a = 3$; $b = 7$

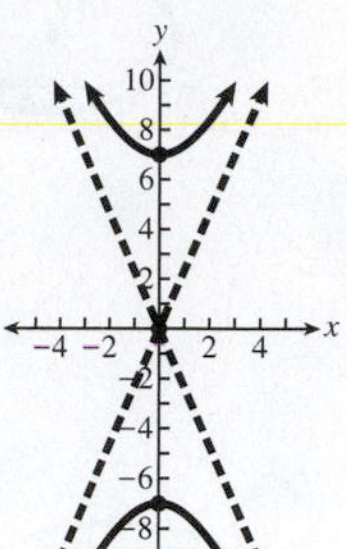

13. Center: $(0, 0)$; $a = 1$; $b = 5$

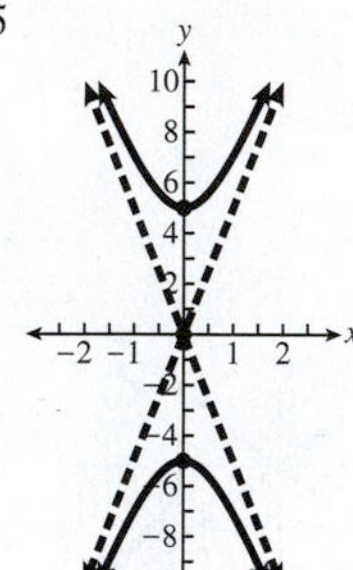

15. Center: $(0, 0)$; $a = 2\sqrt{3}$; $b = 3$

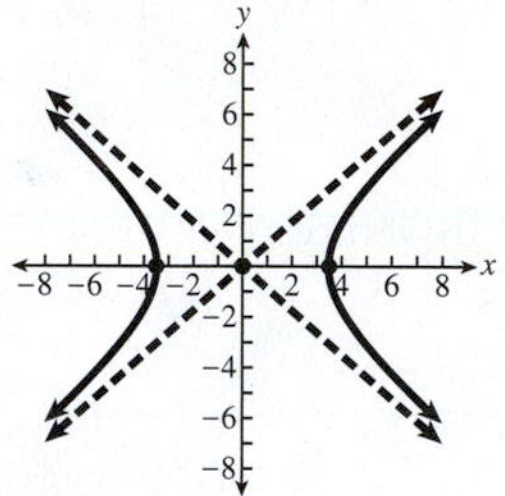

17. Center: $(0, 0)$; $a = 2$; $b = 7$

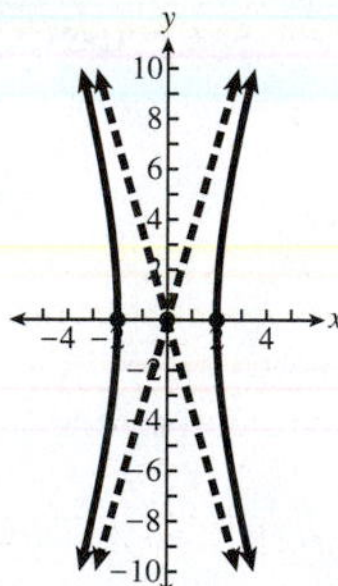

19. Center: $(0, 0)$; $a = 1$; $b = 3$

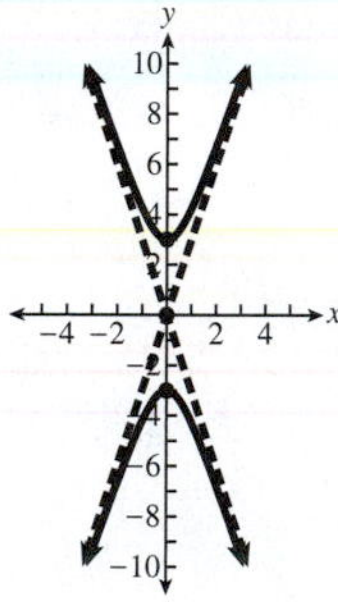

21. Center: $(2, 5)$; $a = 3$; $b = 4$

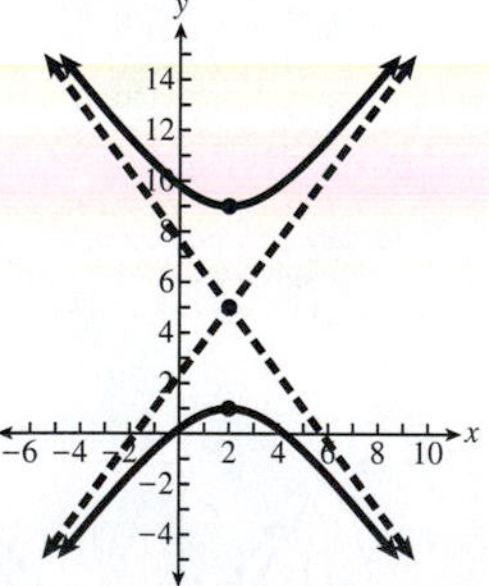

23. Center: $(-1, -4)$; $a = 3$; $b = 2$

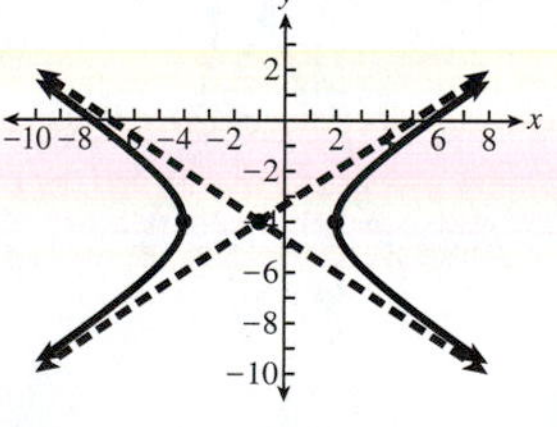

25. Center: $(0, 1)$; $a = 2$; $b = 2$

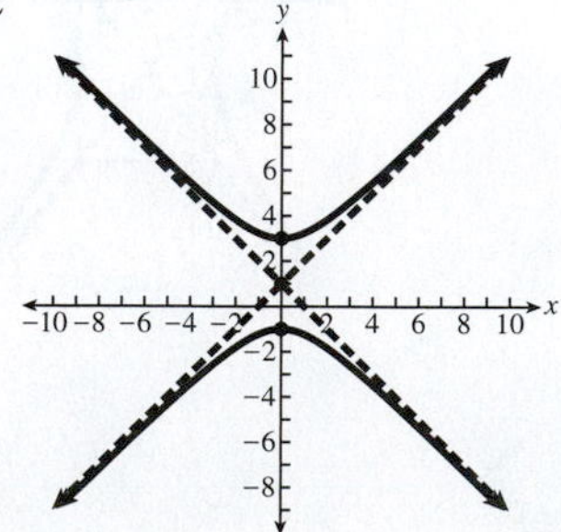

27. Center: $(4, -2)$; $a = 7$; $b = 4$

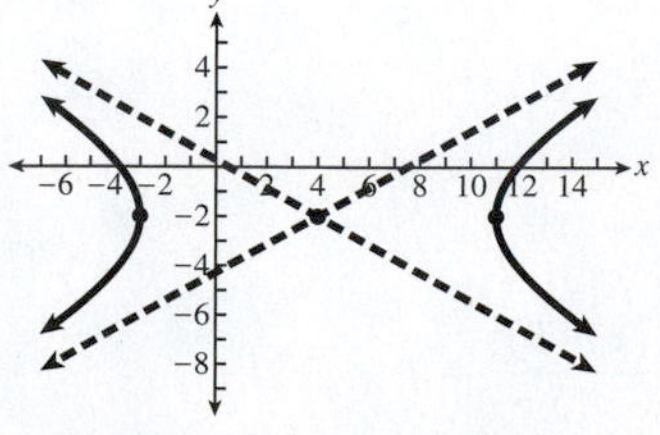

29. Center: $(-2, -3)$; $a = 5$; $b = 2\sqrt{7}$

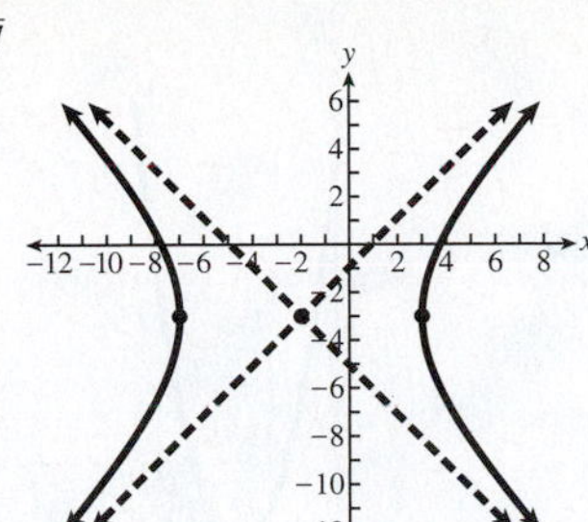

31. Center: $(-1, -3)$; $a = 2$; $b = 3$

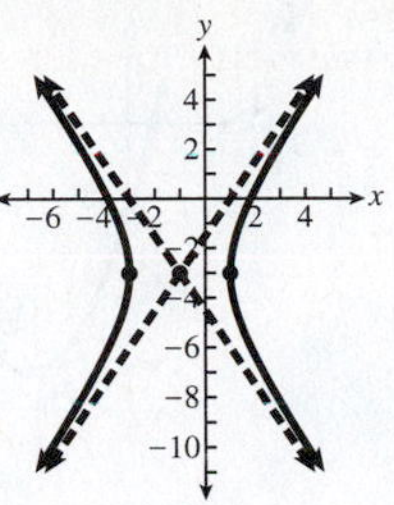

33. Center: $(-2, 7)$; $a = 4$; $b = 5$

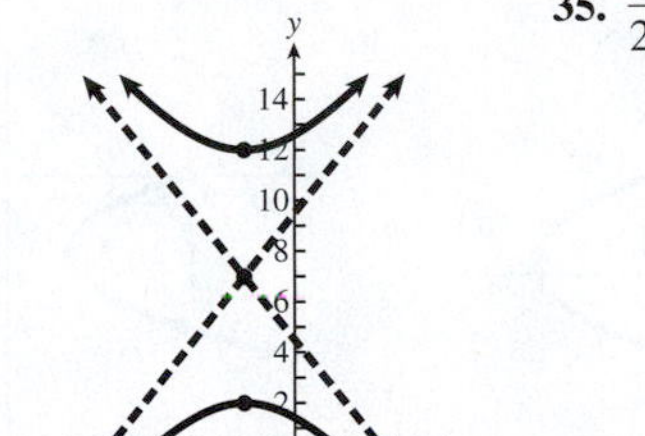

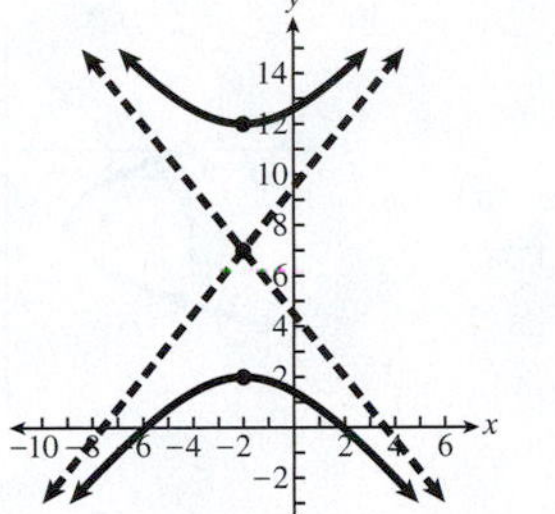

35. $\frac{y^2}{25} - \frac{x^2}{9} = 1$ **37.** $\frac{(y-2)^2}{25} - \frac{(x-3)^2}{4} = 1$ **39.** $\frac{(x+1)^2}{25} - \frac{(y+5)^2}{9} = 1$

41.

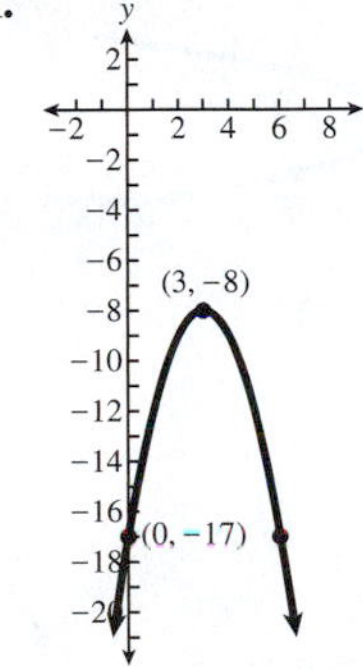

43.

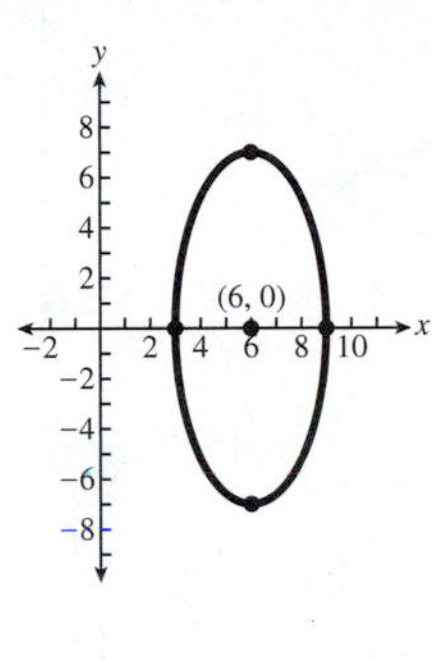

45.

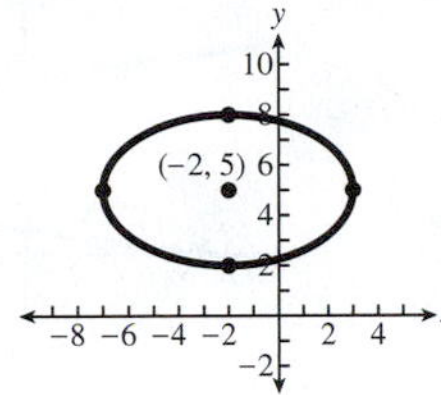

47.

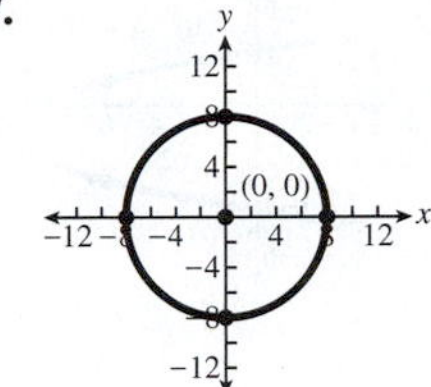

49.

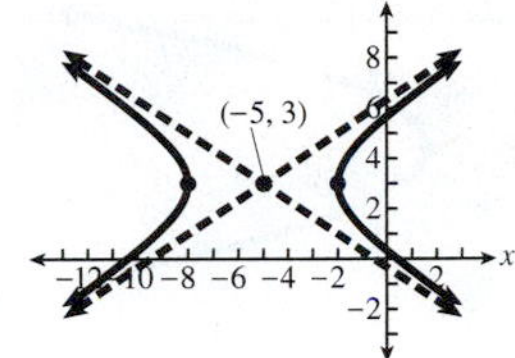

51.

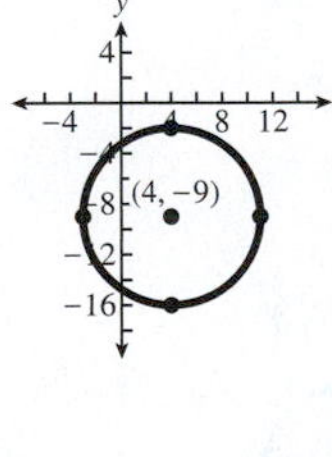

53.

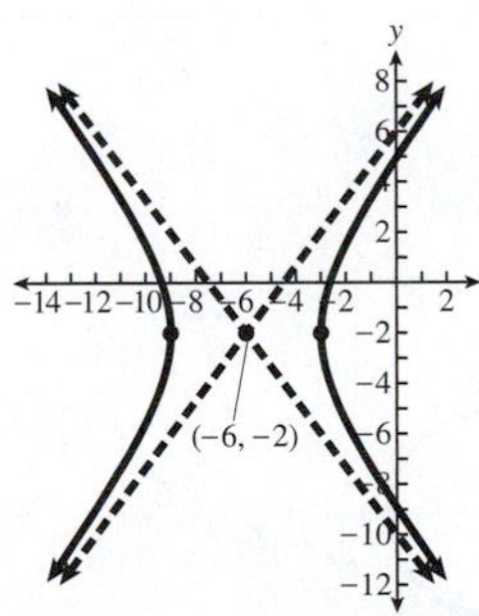

55. e **57.** c **59.** b **61.** Answers will vary. Left/right: $x^2 - y^2 =$ positive number; up/down: $y^2 - x^2 =$ positive number

Quick Review Exercises **1.** $(3, 2)$ **2.** $(-3, -9)$ **3.** $(4, -1)$ **4.** $\left(\frac{3}{2}, -\frac{9}{5}\right)$

Quick Check 13.5 **1.** $(2, -5)$, $\left(-\frac{50}{17}, -\frac{43}{17}\right)$ **2.** $\varnothing$ **3.** $(-\sqrt{3}, 3)$, $(\sqrt{3}, 3)$ **4.** $(3, 5)$, $(-3, 5)$, $(3, -5)$, $(-3, -5)$ **5.** 3 feet by 12 feet **6.** 9 inches by 12 inches **7.** 2 seconds

Section 13.5 **1.** nonlinear **3.** $(8, -6)$, $(-6, -8)$ **5.** $(-9, -2)$, $\left(-\frac{42}{5}, -\frac{19}{5}\right)$ **7.** $(0, -4)$, $(2, 0)$ **9.** $(1, 6)$, $(4, 12)$ **11.** $(3, 1)$ **13.** $\varnothing$ **15.** $(0, -5)$, $(3, 4)$, $(-3, 4)$ **17.** $(5, \sqrt{5})$, $(5, -\sqrt{5})$ **19.** $(0, 1)$ **21.** $\varnothing$ **23.** $(10, \sqrt{21})$, $(10, -\sqrt{21})$, $(-6, \sqrt{5})$, $(-6, -\sqrt{5})$ **25.** $(3, 2)$, $(-3, 2)$ **27.** $\varnothing$ **29.** $(2, 1)$, $(2, -1)$, $(-2, 1)$, $(-2, -1)$ **31.** $(2, \sqrt{2})$, $(2, -\sqrt{2})$, $(-2, \sqrt{2})$, $(-2, -\sqrt{2})$ **33.** $(\sqrt{2}, 2\sqrt{2})$, $(\sqrt{2}, -2\sqrt{2})$, $(-\sqrt{2}, 2\sqrt{2})$, $(-\sqrt{2}, -2\sqrt{2})$ **35.** $\left(3, \frac{\sqrt{6}}{2}\right)$, $\left(3, -\frac{\sqrt{6}}{2}\right)$, $\left(-3, \frac{\sqrt{6}}{2}\right)$, $\left(-3, -\frac{\sqrt{6}}{2}\right)$ **37.** $(\sqrt{15}, 3)$, $(\sqrt{15}, -3)$, $(-\sqrt{15}, 3)$, $(-\sqrt{15}, -3)$ **39.** $(\sqrt{3}, \sqrt{3})$, $(\sqrt{3}, -\sqrt{3})$, $(-\sqrt{3}, \sqrt{3})$, $(-\sqrt{3}, -\sqrt{3})$ **41.** $(3, 3)$, $(3, -3)$, $(-3, 3)$, $(-3, -3)$ **43.** $(1, 3)$, $(1, -3)$, $(-1, 3)$, $(-1, -3)$ **45.** 7 feet by 15 feet **47.** 16 feet by 25 feet **49.** 8 feet by 24 feet **51.** 45 feet by 60 feet **53.** 40 feet by 20 feet **55.** 4.5 seconds **57. a)** 3 seconds **b)** 36 feet **59.** Answers will vary. The area of a rectangle is 1500 square feet; its perimeter is 160 feet. Find the dimensions of the rectangle.

Chapter 13 Review **1.**

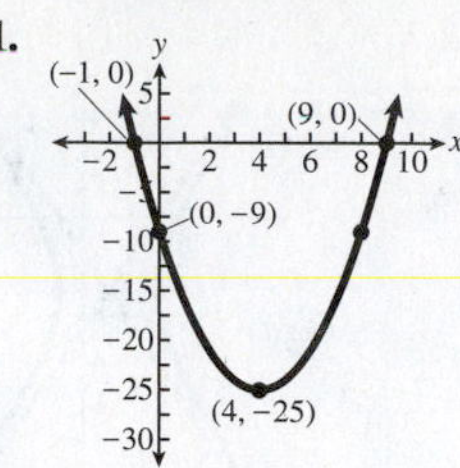

2.

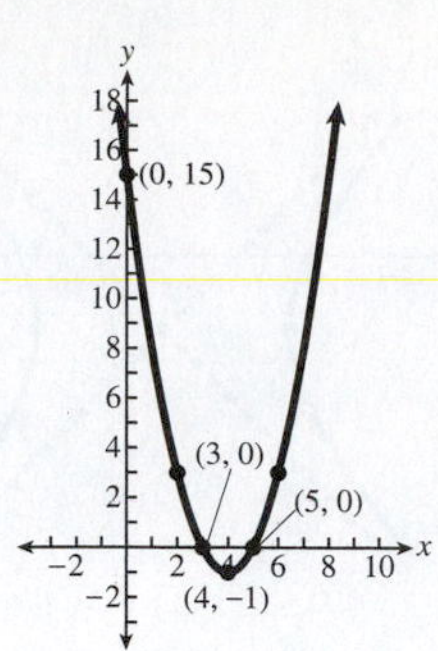

3.

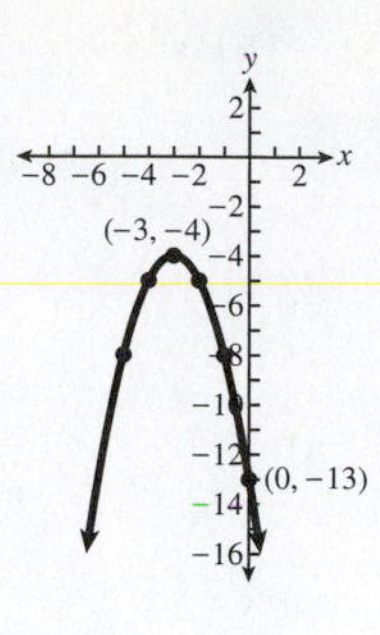

4.

5.

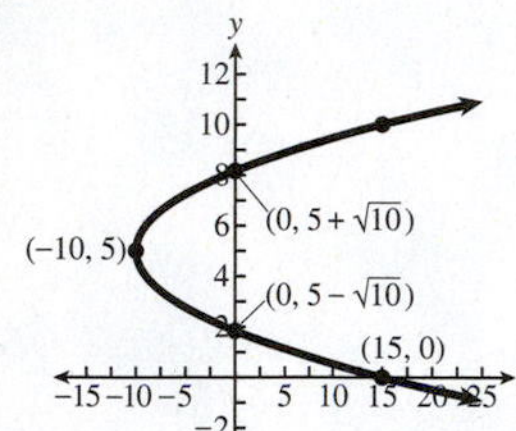

6.

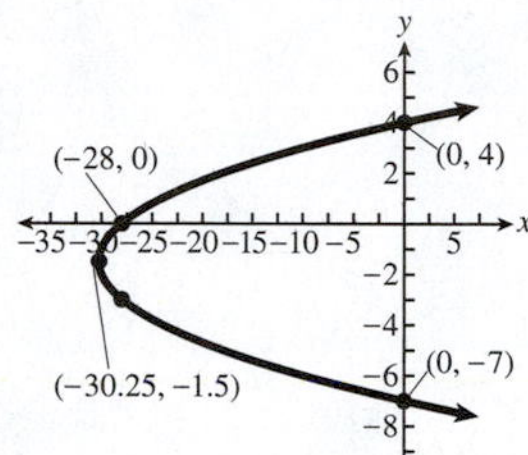

7.

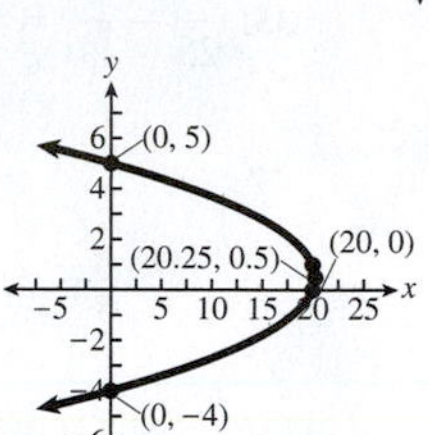

8.

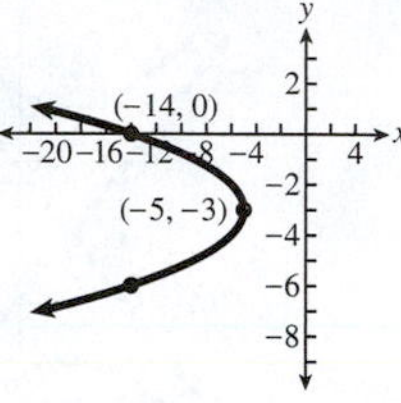

9.

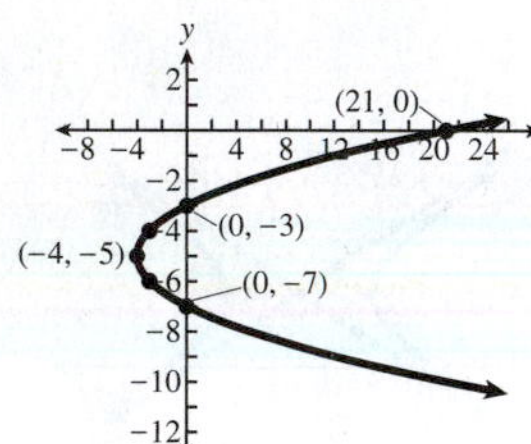

10.

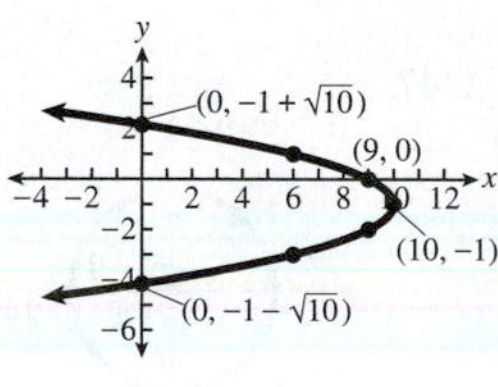

11.

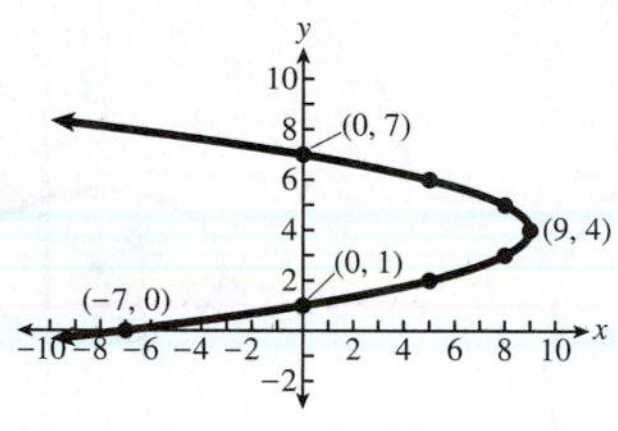

12.

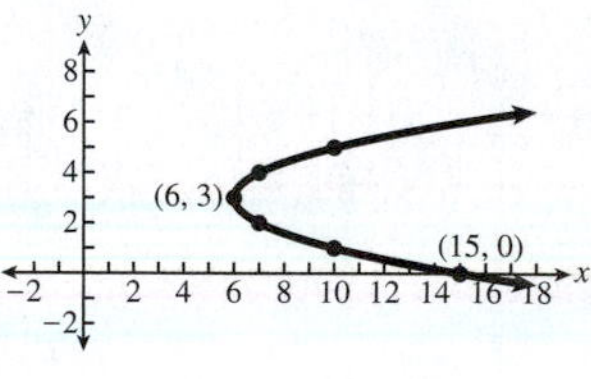

13. 17 **14.** 9.5 **15.** $r = 6$

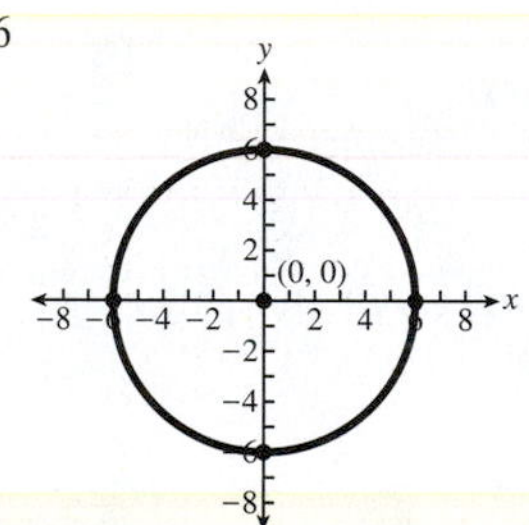

16. $r = 2\sqrt{5}$

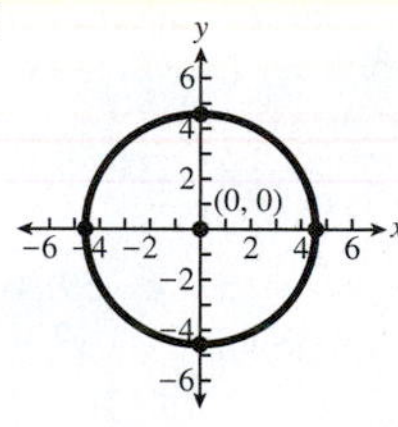

17. $r = 2$

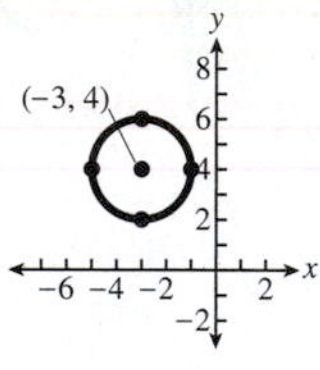

18. $r = 3$

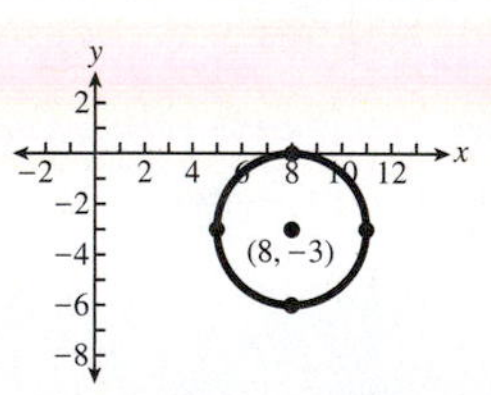

19. $r = 5$

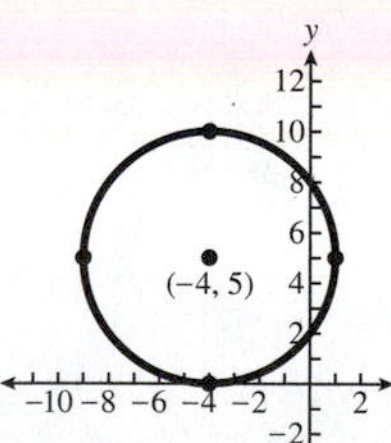

20. $r = 4$

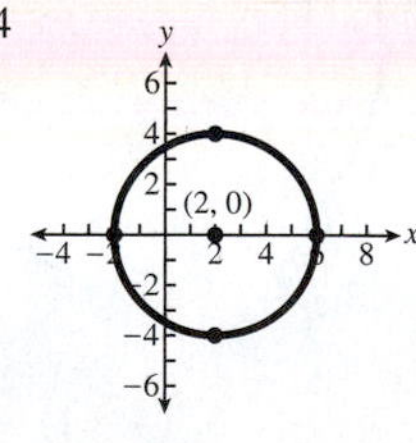

21. $r = 7$

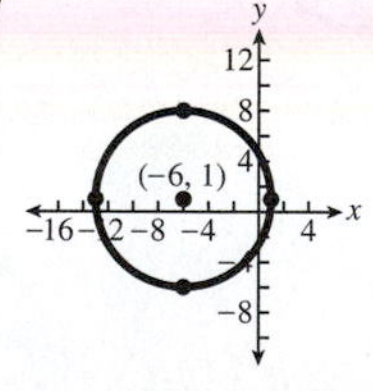

22. $(-3, 9), r = 10$ **23.** $(4, 10), r = \sqrt{17}$ **24.** $x^2 + y^2 = 9$ **25.** $(x + 2)^2 + (y - 4)^2 = 25$ **26.** $a = 5, b = 3$

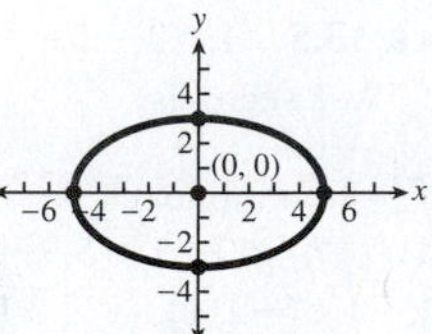

27. $a = 1, b = 4$

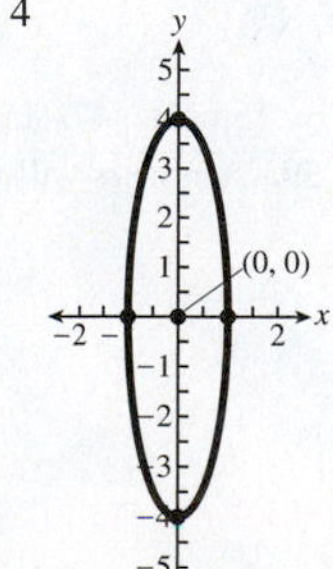

28. $a = 3, b = 4$

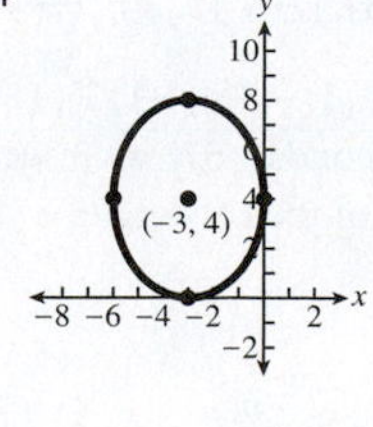

29. $a = 5, b = 2$

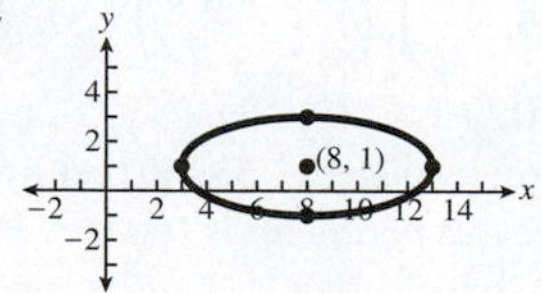

30. $a = 3, b = 6$

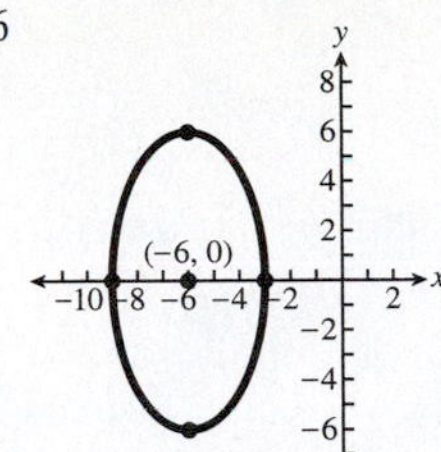

31. $a = 4, b = 3$

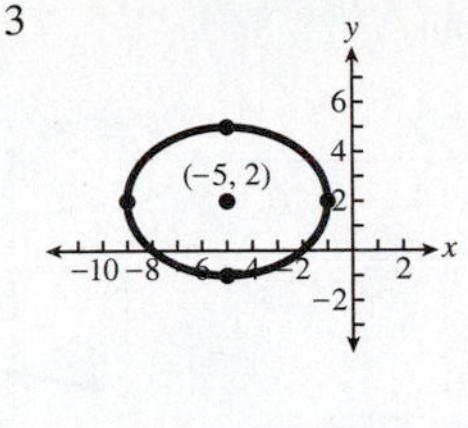

32. $(6, 1), a = 7, b = 2$ **33.** $(-2, -4), a = 4, b = 9$

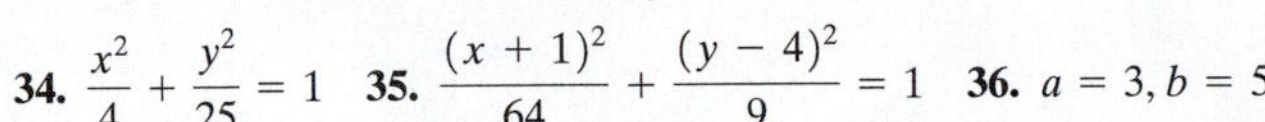

34. $\dfrac{x^2}{4} + \dfrac{y^2}{25} = 1$ **35.** $\dfrac{(x+1)^2}{64} + \dfrac{(y-4)^2}{9} = 1$ **36.** $a = 3, b = 5$

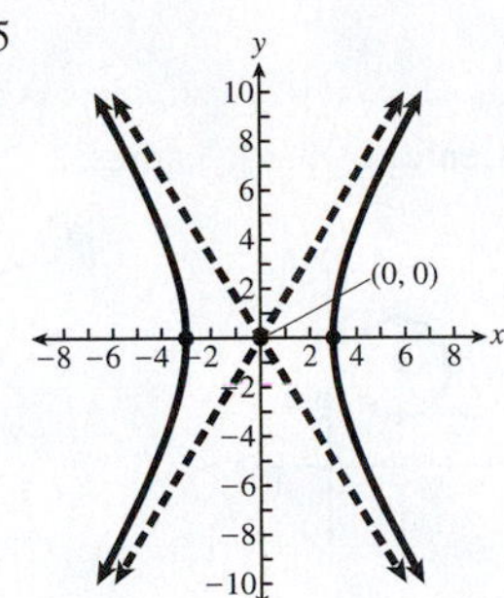

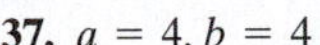

37. $a = 4, b = 4$

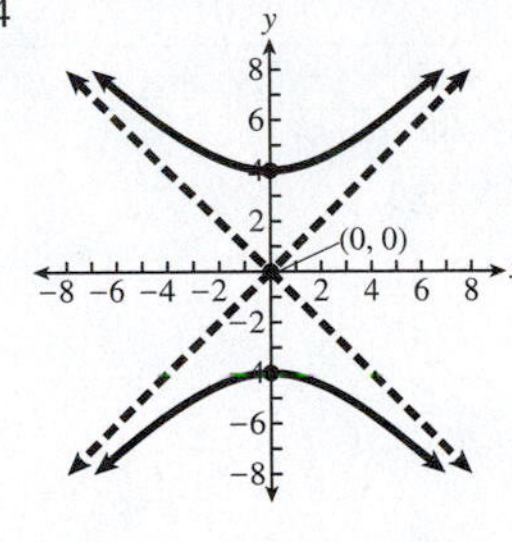

38. $a = 2, b = 3$

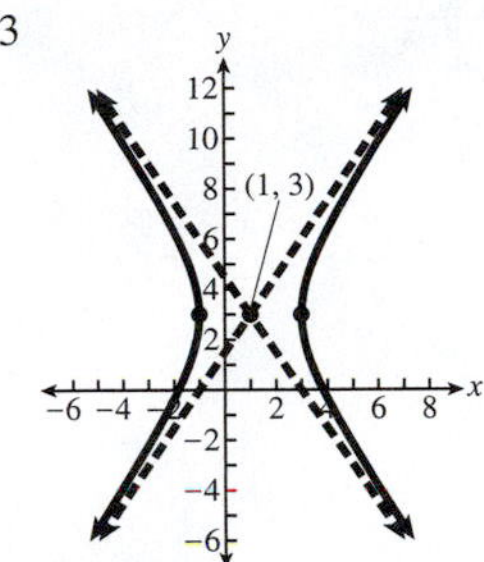

39. $a = 5, b = 6$

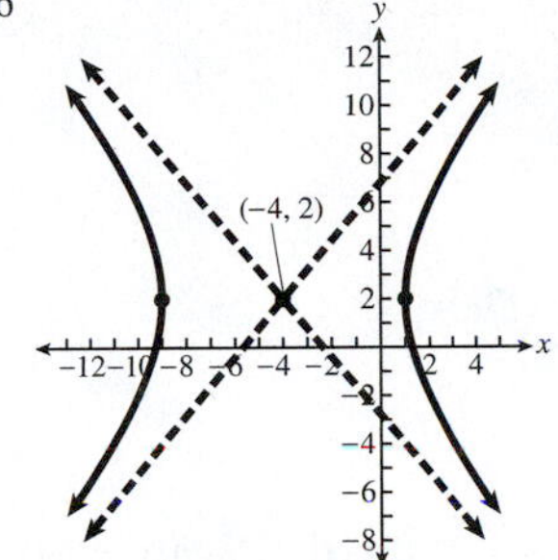

40. $a = 9, b = 2$

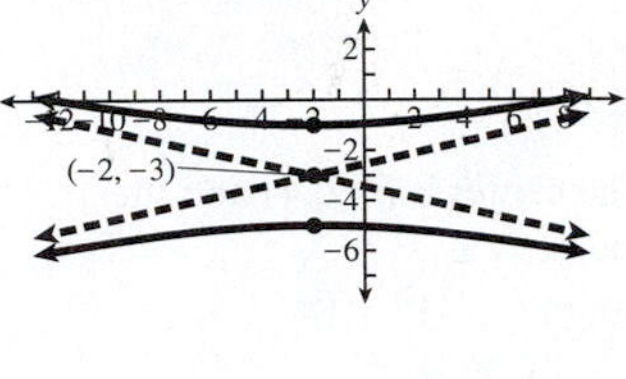

41. $a = 5, b = 2$

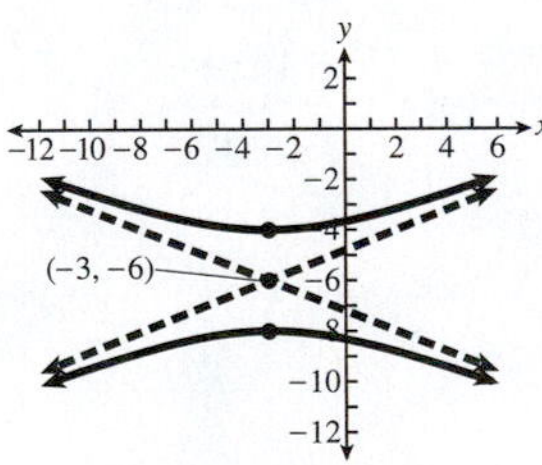

42. Center: $(5, 7)$; transverse axis: $2b = 6$; conjugate axis: $2a = 4$

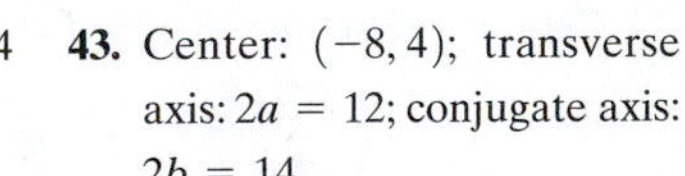

43. Center: $(-8, 4)$; transverse axis: $2a = 12$; conjugate axis: $2b = 14$

44. $\dfrac{x^2}{4} - \dfrac{y^2}{25} = 1$ **45.** $\dfrac{(y-2)^2}{9} - \dfrac{(x+1)^2}{16} = 1$ **46.** $(5, 5), (1, 7)$ **47.** $(3, 2), (0, -4)$ **48.** $(6, 11), (4, 7)$ **49.** $(10, 7), (-2, 5)$ **50.** $(1, 4), (-1, 4)$
51. $\left(\dfrac{2\sqrt{15}}{5}, -\dfrac{18}{5}\right), \left(-\dfrac{2\sqrt{15}}{5}, -\dfrac{18}{5}\right), (3, 3), (-3, 3)$ **52.** $(\sqrt{10}, 9), (-\sqrt{10}, 9), (1, 0), (-1, 0)$ **53.** $(3, 9), (-2, 4)$
54. $(\sqrt{19}, 9), (-\sqrt{19}, 9), (\sqrt{19}, -9), (-\sqrt{19}, -9)$ **55.** $(3, 1), (-3, 1), (3, -1), (-3, -1)$ **56.** $(2, 2), (-2, 2), (2, -2), (-2, -2)$
57. $(5, 3), (-5, 3), (5, -3), (-5, -3)$ **58.** 15 feet by 20 feet **59.** 9 inches by 12 inches **60. a)** 4 seconds **b)** 48 feet

Chapter 13 Review Exercises: Worked-Out Solutions

6. Vertex	x-intercept ($y = 0$)	y-intercept ($x = 0$)
$y = \frac{-3}{2(1)} = -1.5$ $x = (-1.5)^2 + 3(-1.5) - 28 = -30.25$ $(-30.25, -1.5)$	$x = 0^2 + 3(0) - 28 = -28$ $(-28, 0)$	$y^2 + 3y - 28 = 0$ $(y + 7)(y - 4) = 0$ $y + 7 = 0$ or $y - 4 = 0$ $y = -7$ or $y = 4$ $(0, -7), (0, 4)$

(-28, 0) (0, 4) (-30.25, -1.5) (0, -7)

10. Shift the graph of $x = -y^2$ by 10 units to the right and down by 1 unit.

Vertex (h, k)	x-intercept $(y = 0)$	y-intercept $(x = 0)$
$(10, -1)$	$x = -(0 + 1)^2 + 10$ $x = -1 + 10$ $x = 9$ $(9, 0)$	$0 = -(y + 1)^2 + 10$ $(y + 1)^2 = 10$ $\sqrt{(y + 1)^2} = \pm\sqrt{10}$ $y + 1 = \pm\sqrt{10}$ $y = -1 \pm \sqrt{10}$ $(0, -1 \pm \sqrt{10})$

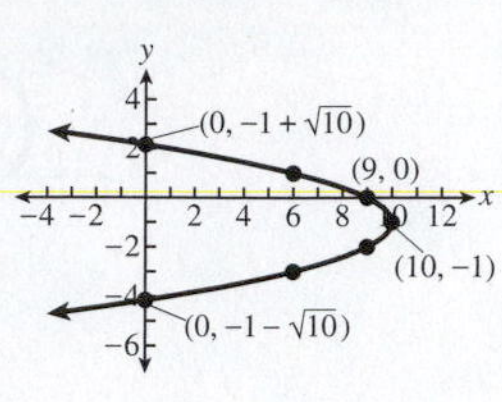

13. $d = \sqrt{(1 - (-7))^2 + (7 - (-8))^2}$
$d = \sqrt{8^2 + 15^2}$
$d = \sqrt{289}$
$d = 17$

17. 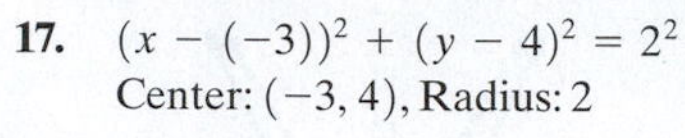
$(x - (-3))^2 + (y - 4)^2 = 2^2$
Center: $(-3, 4)$, Radius: 2

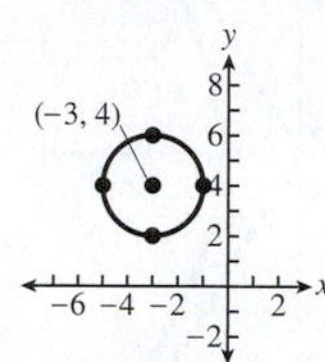

21.
$$x^2 + y^2 + 12x - 2y - 12 = 0$$
$$x^2 + 12x + 36 + y^2 - 2y + 1 = 12 + 36 + 1$$
$$(x + 6)^2 + (y - 1)^2 = 49$$
$$(x - (-6))^2 + (y - 1)^2 = 7^2$$
Center: $(-6, 1)$, Radius: 7

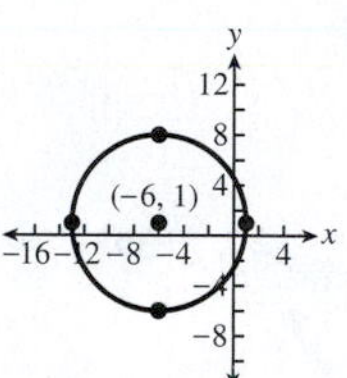

25. The center is $(-2, 4)$ and the radius is 5.
$$(x - (-2))^2 + (y - 4)^2 = 5^2$$
$$(x + 2)^2 + (y - 4)^2 = 25$$

28. $\frac{(x - (-3))^2}{3^2} + \frac{(y - 4)^2}{4^2} = 1$
Center: $(-3, 4)$, $a = 3$, $b = 4$

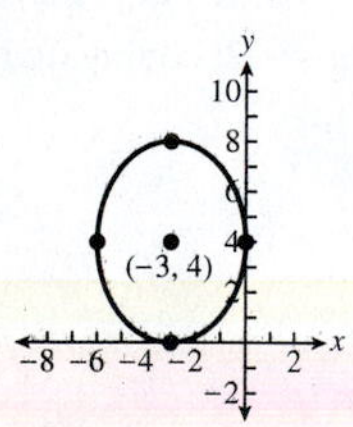

31.
$$9x^2 + 16y^2 + 90x - 64y + 145 = 0$$
$$9(x^2 + 10x) + 16(y^2 - 4y) = -145$$
$$9(x^2 + 10x + 25) + 16(y^2 - 4y + 4) = -145 + 9 \cdot 25 + 16 \cdot 4$$
$$9(x + 5)^2 + 16(y - 2)^2 = 144$$
$$\frac{9(x + 5)^2}{144} + \frac{16(y - 2)^2}{144} = \frac{144}{144}$$
$$\frac{(x - (-5))^2}{4^2} + \frac{(y - 2)^2}{3^2} = 1$$
Center: $(-5, 2)$, $a = 4$, $b = 3$

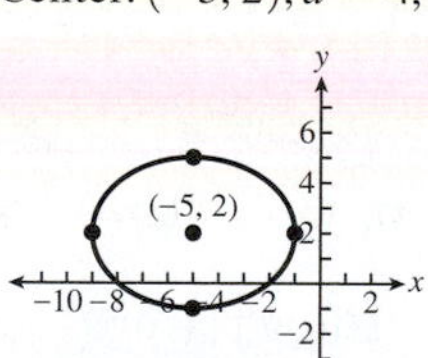

35. The center is $(-1, 4)$, $a = 8$, and $b = 3$.
$$\frac{(x - (-1))^2}{8^2} + \frac{(y - 4)^2}{3^2} = 1$$
$$\frac{(x + 1)^2}{64} + \frac{(y - 4)^2}{9} = 1$$

36. Center: $(0, 0)$, $a = 3$, $b = 5$

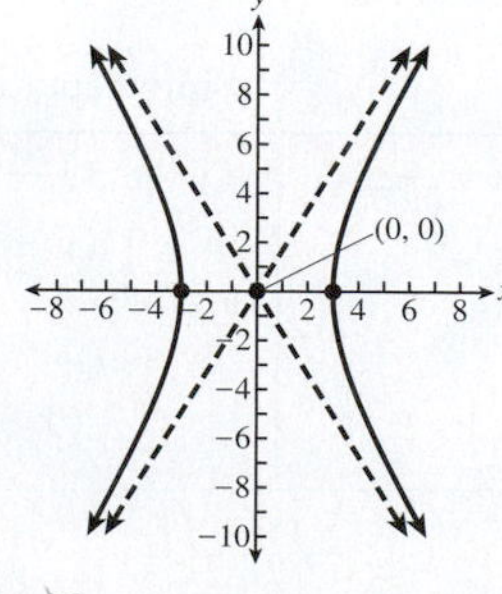

40. $\frac{(y - (-3))^2}{2^2} - \frac{(x - (-2))^2}{9^2} = 1$
Center: $(-2, -3)$, $a = 9$, $b = 2$

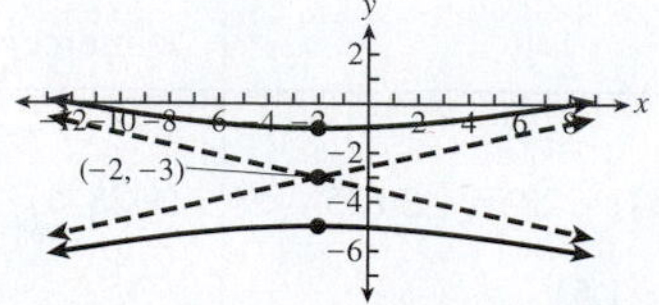

43.

$$\begin{aligned}
49x^2 - 36y^2 + 784x + 288y + 796 &= 0\\
49(x^2 + 16x) - 36(y^2 - 8y) &= -796\\
49(x^2 + 16x + 64) - 36(y^2 - 8y + 16) &= -796 + 49\cdot 64 - 36\cdot 16\\
49(x + 8)^2 - 36(y - 4)^2 &= 1764\\
\frac{49(x + 8)^2}{1764} - \frac{36(y - 4)^2}{1764} &= \frac{1764}{1764}\\
\frac{(x - (-8))^2}{6^2} - \frac{(y - 4)^2}{7^2} &= 1
\end{aligned}$$

Center: $(-8, 4)$, $a = 6$, Transverse Axis: $2a = 12$, $b = 7$, Conjugate Axis: $2b = 14$

46.

$$\begin{aligned}
x &= 15 - 2y\\
(15 - 2y)^2 + y^2 &= 50\\
225 - 60y + 4y^2 + y^2 &= 50\\
5y^2 - 60y + 175 &= 0\\
5(y^2 - 12y + 35) &= 0\\
5(y - 5)(y - 7) &= 0
\end{aligned}$$

$$\begin{array}{lcl}
y - 5 = 0 & \text{or} & y - 7 = 0\\
y = 5 & \text{or} & y = 7\\
x = 15 - 2(5) & & x = 15 - 2(7)\\
x = 5 & & x = 1\\
(5, 5) & & (1, 7)
\end{array}$$

50.

$$\begin{aligned}
x^2 + (x^2 + 3)^2 &= 17\\
x^2 + x^4 + 6x^2 + 9 &= 17\\
x^4 + 7x^2 - 8 &= 0\\
(x^2 + 8)(x^2 - 1) &= 0
\end{aligned}$$

$$\begin{array}{lcl}
x^2 + 8 = 0 & \text{or} & x^2 - 1 = 0\\
x^2 = -8 & \text{or} & x^2 = 1\\
\varnothing & & x = \pm 1\\
 & & y = 1^2 + 3 \quad y = (-1)^2 + 3\\
 & & y = 4 \qquad\quad y = 4\\
 & & (1, 4) \qquad\ (-1, 4)
\end{array}$$

55. $3x^2 + 4y^2 = 31$, $x^2 - 2y^2 = 7$ $\xrightarrow{\text{Multiply by 2}}$ $3x^2 + 4y^2 = 31$, $2x^2 - 4y^2 = 14$

$$\begin{aligned}
3x^2 + 4y^2 &= 31\\
2x^2 - 4y^2 &= 14\\
\hline
5x^2 \qquad &= 45\\
5x^2 &= 45\\
x^2 &= 9\\
x &= \pm 3
\end{aligned}$$

$$\begin{array}{ll}
3(3)^2 + 4y^2 = 31 & 3(-3)^2 + 4y^2 = 31\\
4y^2 = 4 & 4y^2 = 4\\
y^2 = 1 & y^2 = 1\\
y = \pm 1 & y = \pm 1\\
(3, 1), (3, -1) & (-3, 1), (-3, -1)
\end{array}$$

59.

$$\begin{aligned}
2x + 2y &= 42\\
2x &= 42 - 2y\\
x &= 21 - y\\
(21 - y)^2 + y^2 &= 15^2\\
441 - 42y + y^2 + y^2 &= 225\\
2y^2 - 42y + 216 &= 0\\
2(y^2 - 21y + 108) &= 0\\
2(y - 9)(y - 12) &= 0
\end{aligned}$$

$$\begin{array}{lcl}
y - 9 = 0 & \text{or} & y - 12 = 0\\
y = 9 & \text{or} & y = 12\\
x = 21 - 9 & & x = 21 - 12\\
x = 12 & & x = 9\\
(12, 9) & & (9, 12)
\end{array}$$

The dimensions are 9 inches by 12 inches.

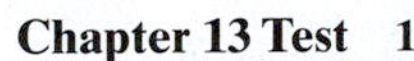

Chapter 13 Test **1.**

2.

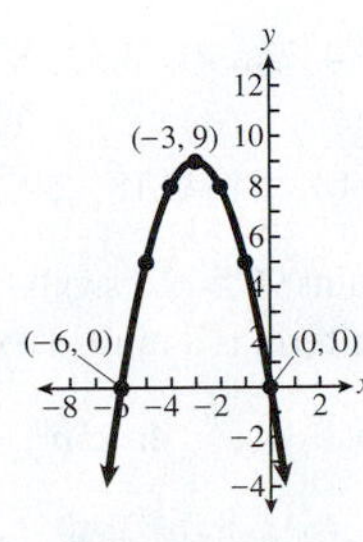

3.

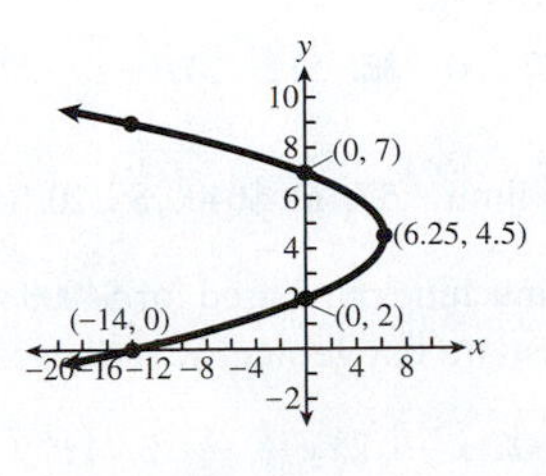

4.

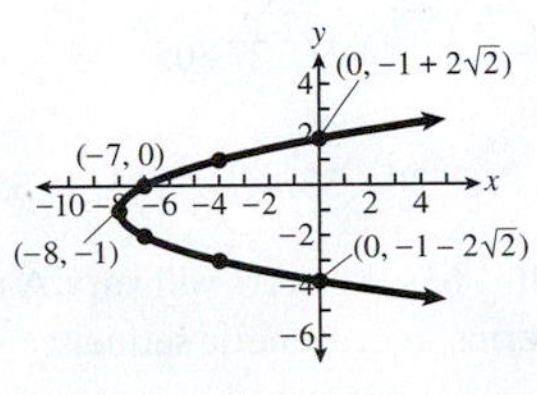

5. 8.9

6.

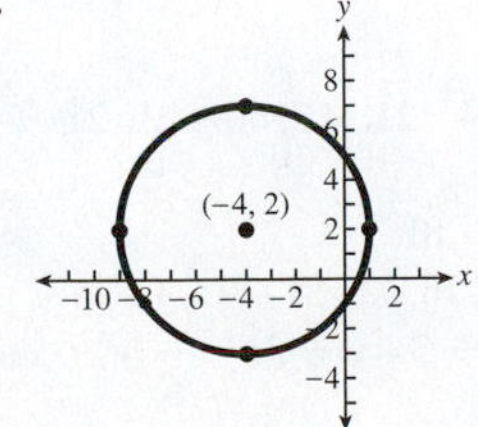

$r = 5$ **7.** $(6, 3), r = 4\sqrt{2}$ **8.** $(x - 3)^2 + (y + 8)^2 = 256$ **9.** $a = 4, b = 3$

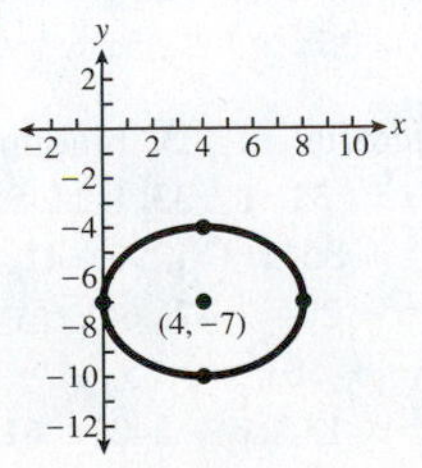

10. $(8, -2), a = 5, b = 2$ **11.** $a = 4, b = 1$

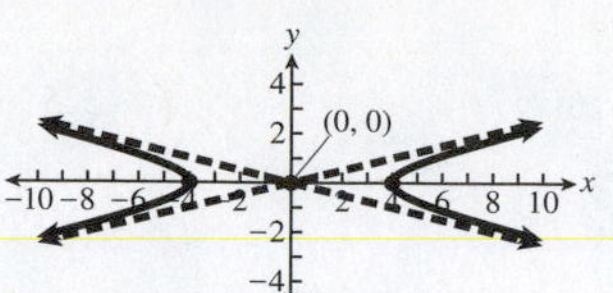

12. $(x - 3)^2 + (y - 2)^2 = 25$ **13.** $\frac{(x-5)^2}{4} + \frac{(y+1)^2}{36} = 1$

14. $(10, 0), (-8, 6)$ **15.** $(5, 2), \left(-\frac{19}{5}, -\frac{12}{5}\right)$ **16.** $(\sqrt{39}, 7), (-\sqrt{39}, 7), (\sqrt{26}, -6), (-\sqrt{26}, -6)$ **17.** $(7, 3), (-7, 3), (7, -3), (-7, -3)$ **18.** 50 feet by 120 feet

CHAPTER 14

Quick Check 14.1 **1.** $\frac{1}{7}, \frac{1}{10}, \frac{1}{13}, \frac{1}{16}$ **2.** $\frac{1}{2}, -\frac{1}{4}, \frac{1}{6}, -\frac{1}{8}$ **3.** $-37, 43, -49; a_n = (-1)^{n+1}(6n + 1)$ **4.** 2, 5, 26, 677, 458330 **5.** 60 **6.** 112 **7.** 60 **8.** \$6400, \$5120, \$4096, \$3276.80

Section 14.1 **1.** sequence **3.** general **5.** series **7.** finite **9.** 11 **11.** 2187 **13.** 4, 8, 12, 16, 20 **15.** 11, 13, 15, 17, 19 **17.** −5, 7, −9, 11, −13 **19.** 12, 48, 192, 768, 3072 **21.** $30, 35, 40; a_n = 5n$ **23.** $243, 729, 2187; a_n = 3^{n-1}$ **25.** $\frac{1}{30}, -\frac{1}{35}, \frac{1}{40}; a_n = \frac{(-1)^n}{5n}$ **27.** $216, 343, 512; a_n = n^3$ **29.** 6, 9, 12, 15, 18 **31.** −5, −7, −13, −31, −85 **33.** 16, −48, 144, −432, 1296 **35.** 300 **37.** −399 **39.** 441 **41.** −45 **43.** 95 **45.** −147 **47.** 55 **49.** 25 **51.** 117 **53.** −212 **55.** 25,263 **57.** 120 **59.** 64 **61.** 385 **63.** $\frac{15}{16}$ **65.** $\sum_{i=1}^{4} 7i$ **67.** $\sum_{i=1}^{6} (-1)^{i+1} \cdot 5i$ **69.** $\sum_{i=1}^{7} i^2$ **71.** \$24,000, \$19,200, \$15,360, \$12,288 **73. a)** \$50,000, \$52,000, \$54,080, \$56,243.20, \$58,492.93 **b)** \$270,816.13 **75.** Answers will vary; decreasing the value of a car by year

Quick Check 14.2 **1.** −8, −5, −2, 1, 4, 7 **2.** $a_n = 12n - 10$ **3.** 177 **4.** 396 **5.** 12,870 **6.** 3906

Section 14.2 **1.** arithmetic **3.** $a_n = a_{n-1} + d$ **5.** arithmetic series **7.** 4 **9.** $\frac{4}{3}$ **11.** 3 **13.** 8 **15.** 1, 6, 11, 16, 21 **17.** −9, 3, 15, 27, 39 **19.** −10, −31, −52, −73, −94 **21.** $a_n = 11n + 7$ **23.** $a_n = -2n + 5$ **25.** $a_n = 20n - 27$ **27.** $a_n = 4n - \frac{11}{2}$ **29.** 106 **31.** 62nd **33.** 123rd **35.** 161 **37.** 1065 **39.** −2700 **41.** 52,592 **43.** 910 **45.** 1887 **47.** −5690 **49.** 500,500 **51.** 4489 **53.** 40,200 **55. a)** \$10, \$11, \$12, \$13, \$14 **b)** $a_n = n + 9$ **57. a)** \$38,500, \$39,250, \$40,000, \$40,750, \$41,500, \$42,250 **b)** $a_n = 750n + 37{,}750$ **c)** \$656,250 **59.** Answers will vary. A gym membership starts at \$300/year and increases \$25 each year.

Quick Check 14.3 **1.** $\frac{1}{4}$ **2.** $-\frac{1}{2}$ **3.** $5, -\frac{15}{4}, \frac{45}{16}, -\frac{135}{64}, \frac{405}{256}$ **4.** $a_n = 9 \cdot 2^{n-1}$ **5.** $a_n = 11 \cdot (-5)^{n-1}$ **6.** 39,360 **7.** −14,762 **8.** Yes, 9 **9.** No **10. a)** \$17,500, \$12,250, \$8575, \$6002.50, \$4201.75, \$2941.23 **b)** \$706.19

Section 14.3 **1.** geometric sequence **3.** $a_n = r \cdot a_{n-1}$ **5.** $s_n = a_1 \cdot \frac{1 - r^n}{1 - r}$ **7.** 8 **9.** $\frac{1}{4}$ **11.** −7 **13.** $-\frac{4}{3}$ **15.** 1, 10, 100, 1000, 10000 **17.** 4, −8, 16, −32, 64 **19.** $8, 6, \frac{9}{2}, \frac{27}{8}, \frac{81}{32}$ **21.** $-20, 35, -\frac{245}{4}, \frac{1715}{16}, -\frac{12{,}005}{64}$ **23.** $a_n = \left(\frac{2}{3}\right)^{n-1}$ **25.** $a_n = -1 \cdot (-4)^{n-1}$ **27.** $a_n = 2 \cdot 7^{n-1}$ **29.** $a_n = 16 \cdot \left(-\frac{3}{4}\right)^{n-1}$ **31.** 27,305 **33.** −16,777,200 **35.** 511 **37.** −29,127 **39.** 4372 **41.** 3,528,889 **43.** Yes, 45 **45.** Yes, $\frac{5}{9}$ **47.** No limit **49.** Yes, $\frac{1}{3}$ **51.** Yes, 30 **53.** Yes, $-11\frac{3}{7}$ **55.** No limit **57. a)** \$6400, \$5120, \$4096 **b)** \$1342.18 **59.** $a_n = 20{,}000(0.5)^{n-1}$ or $40{,}000(0.5)^n$ **61.** \$159,374,246 **63.** Answers will vary. A copy machine purchased for \$4000 maintains 90% of its value each year. **65.** Answers will vary. The difference between terms in arithmetic sequence is constant. For geometric sequence, each term is a multiple of previous term.

Quick Review Exercises **1.** $x^2 + 10x + 25$ **2.** $x^2 - 2xy + y^2$ **3.** $4x^2 + 36xy + 81y^2$ **4.** $25a^6 - 40a^3b^4 + 16b^8$

Quick Check 14.4 **1.** 362,880 **2.** 15 **3.** 1 **4.** 1, 9, 84, 126, 36, 9, 1 **5.** $x^3 + 3x^2y + 3xy^2 + y^3$ **6.** $x^4 - 16x^3 + 96x^2 - 256x + 256$ **7.** $x^7 + 7x^6y + 21x^5y^2 + 35x^4y^3 + 35x^3y^4 + 21x^2y^5 + 7xy^6 + y^7$

Section 14.4 **1.** n factorial, $n!$ **3.** binomial **5.** 720 **7.** 40,320 **9.** 6 **11.** 1 **13.** 17! **15.** 33! **17.** 1,663,200 **19.** 84 **21.** 45,360 **23.** 21 **25.** 56 **27.** 12 **29.** 715 **31.** 1 **33.** 1, 11, 55, 330, 462, 165, 11, 1 **35.** $x^3 + 3x^2y + 3xy^2 + y^3$ **37.** $x^3 + 6x^2 + 12x + 8$ **39.** $x^4 + 24x^3 + 216x^2 + 864x + 1296$ **41.** $x^5 - 5x^4y + 10x^3y^2 - 10x^2y^3 + 5xy^4 - y^5$ **43.** $x^4 - 12x^3 + 54x^2 - 108x + 81$ **45.** $x^3 + 9x^2y + 27xy^2 + 27y^3$ **47.** $64x^3 - 48x^2y + 12xy^2 - y^3$ **49.** 1, 10, 45, 120, 210, 252, 210, 120, 45, 10, 1 **51.** 2, 4, 8, 16, 32, 64 **53.** $x^5 + 5x^4y + 10x^3y^2 + 10x^2y^3 + 5xy^4 + y^5$ **55.** $x^7 + 14x^6 + 84x^5 + 280x^4 + 560x^3 + 672x^2 + 448x + 128$ **57.** $x^3 - 3x^2y + 3xy^2 - y^3$ **59.** $x^4 - 28x^3 + 294x^2 - 1372x + 2401$ **61.** Answers will vary. Both denominators are $a! \cdot b!$.

Chapter 14 Review **1.** 8, 9, 10, 11, 12 **2.** −14, −9, −4, 1, 6 **3.** $78, 91, 104; a_n = 13n$ **4.** $47, 54, 61; a_n = 7n + 5$ **5.** −9, 8, 25, 42, 59 **6.** −4, −31, −85, −193, −409 **7.** $a_n = a_{n-1} + 14$ **8.** $a_n = -2a_{n-1}$ **9.** 174 **10.** 4,031,078 **11.** 21 **12.** 228 **13.** 182 **14.** 7380 **15.** 78 **16.** 396 **17.** 24 **18.** −6 **19.** 16, 7, −2, −11, −20 **20.** −18, −7, 4, 15, 26 **21.** $a_n = 7n - 3$ **22.** $a_n = -7n + 15$ **23.** $a_n = 2n - 17$ **24.** $a_n = \frac{3}{4}n + \frac{1}{12}$

25. 823 **26.** −607 **27.** 236th **28.** 495th **29.** 276 **30.** 360 **31.** −444 **32.** 2745 **33.** 57,970 **34. a)** \$20,000,000, \$22,000,000, \$24,000,000, \$26,000,000, \$28,000,000 **b)** $a_n = 2{,}000{,}000n + 18{,}000{,}000$ **c)** \$290,000,000 **35.** 10 **36.** 6 **37.** $a_n = 1 \cdot 12^{n-1}$ **38.** $a_n = 8\left(\frac{3}{4}\right)^{n-1}$ **39.** $a_n = -4 \cdot 3^{n-1}$ **40.** $a_n = -24\left(-\frac{1}{3}\right)^{n-1}$ **41.** 1152 **42.** 9841 **43.** 6830 **44.** 55,987 **45.** 3,595,117.5 **46.** 639.375 **47.** 1,835,015 **48.** No limit **49.** Yes, $15\frac{3}{7}$ **50.** Yes, $-29\frac{2}{5}$ **51.** Yes, 24 **52.** \$1,715,383.22 **53.** 362,880 **54.** 5040 **55.** 56 **56.** 165 **57.** $x^6 + 6x^5y + 15x^4y^2 + 20x^3y^3 + 15x^2y^4 + 6xy^5 + y^6$ **58.** $x^4 + 24x^3 + 216x^2 + 864x + 1296$ **59.** $x^3 - 27x^2 + 243x - 729$ **60.** $x^9 + 9x^8y + 36x^7y^2 + 84x^6y^3 + 126x^5y^4 + 126x^4y^5 + 84x^3y^6 + 36x^2y^7 + 9xy^8 + y^9$

Chapter 14 Review Exercises: Worked-Out Solutions

2. $a_1 = 5(1) - 19 = -14$
$a_2 = 5(2) - 19 = -9$
$a_3 = 5(3) - 19 = -4$
$a_4 = 5(4) - 19 = 1$
$a_5 = 5(5) - 19 = 6$

5. $a_1 = -9$
$a_2 = -9 + 17 = 8$
$a_3 = 8 + 17 = 25$
$a_4 = 25 + 17 = 42$
$a_5 = 42 + 17 = 59$

9. $s_6 = 4 + 14 + 24 + 34 + 44 + 54 = 174$

11. $s_7 = (-3) + (-1) + 1 + 3 + 5 + 7 + 9 = 21$

15. $\sum_{i=1}^{12} i = 1 + 2 + \cdots + 12 = 78$

21. $d = 11 - 4 = 7$
$a_n = 4 + (n-1)7$
$a_n = 4 + 7n - 7$
$a_n = 7n - 3$

25. $a_n = 25 + (n-1)21 = 21n + 4$
$a_{39} = 21(39) + 4 = 823$

29. $a_n = 10 + (n-1)7 = 7n + 3$
$a_1 = 10, a_8 = 7(8) + 3 = 59$
$s_8 = \frac{8}{2}(10 + 59) = 4(69) = 276$

36. $r = \frac{30}{5} = 6$

38. $r = \frac{6}{8} = \frac{3}{4}$
$a_n = 8(\frac{3}{4})^{n-1}$

43. $s_{11} = 10 \cdot \frac{1 - (-2)^{11}}{1 - (-2)} = 10 \cdot \frac{2049}{3} = 6830$

44. $r = \frac{6}{1} = 6$
$s_7 = 1 \cdot \frac{1 - 6^7}{1 - 6} = 1 \cdot \frac{-279{,}935}{-5} = 55{,}987$

49. $r = \dfrac{(\frac{24}{7})}{(\frac{36}{7})} = \frac{2}{3}$
Since $|r| < 1$, the series has a limit.
$s = \dfrac{\frac{36}{7}}{1 - \frac{2}{3}} = \dfrac{\frac{36}{7}}{\frac{1}{3}} = \frac{36}{7} \cdot \frac{3}{1} = \frac{108}{7} = 15\frac{3}{7}$

53. $9! = 1 \cdot 2 \cdot \ldots \cdot 9 = 362{,}880$

55. ${}_8C_3 = \frac{8!}{3! \cdot 5!} = \frac{40{,}320}{6 \cdot 120} = 56$

58. $(x + 6)^4 = {}_4C_0 \cdot x^4 \cdot (6)^0 + {}_4C_1 \cdot x^3 \cdot (6)^1 + {}_4C_2 \cdot x^2 \cdot (6)^2 + {}_4C_3 \cdot x^1 \cdot (6)^3 + {}_4C_4 \cdot x^0 \cdot (6)^4$
$= 1x^4 \cdot 1 + 4x^3 \cdot 6 + 6x^2 \cdot 36 + 4x \cdot 216 + 1 \cdot 1296$
$= x^4 + 24x^3 + 216x^2 + 864x + 1296$

Chapter 14 Test **1.** 86, 101, 116; $a_n = 15n - 4$ **2.** 279 **3.** 9, 17, 25, 33, 41 **4.** $a_n = -7n + 27$ **5.** 577.5 **6.** 40,200 **7.** 5, 40, 320, 2560, 20480 **8.** $a_n = 4 \cdot 7^{n-1}$ **9.** 7,324,218 **10.** Yes, $6\frac{1}{4}$ **11. a)** \$2500, \$3000, \$3500, \$4000, \$4500, \$5000 **b)** $a_n = 500n + 2000$ **c)** \$121,500 **12. a)** \$13,500, \$10,125, \$7593.75, \$5695.31 **b)** $a_n = 18{,}000(0.75)^{n-1}$ **13.** 792 **14.** $x^4 + 4x^3 + 6x^2 + 4x + 1$ **15.** $16{,}384x^7 + 28{,}672x^6y + 21{,}504x^5y^2 + 8960x^4y^3 + 2240x^3y^4 + 336x^2y^5 + 28xy^6 + y^7$

Cumulative Review Chapters 12–14 **1.** 127.413 **2.** 35.466 **3.** 4 **4.** 8 **5.** $\ln\left(\frac{x^2z^5}{y^4}\right)$ **6.** $4 \log b + 5 \log c - \log a$ **7.** 2.066 **8.** 3.677 **9.** {8} **10.** {2.769} **11.** {1.733} **12.** {4} **13.** {90} **14.** {−1} **15.** $f^{-1}(x) = \ln(x - 3) + 2$ **16.** $f^{-1}(x) = e^{x+16} - 9$ **17.** 27 years **18.** Approximately 234,373 **19.** Domain: $(-\infty, \infty)$; range: $(-1, \infty)$

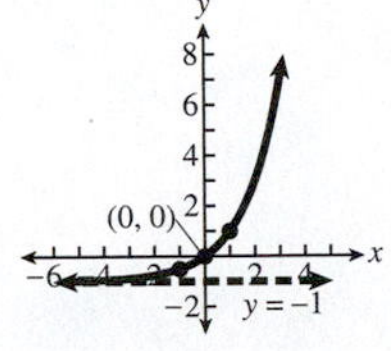

20. Domain: $(-5, \infty)$; range: $(-\infty, \infty)$

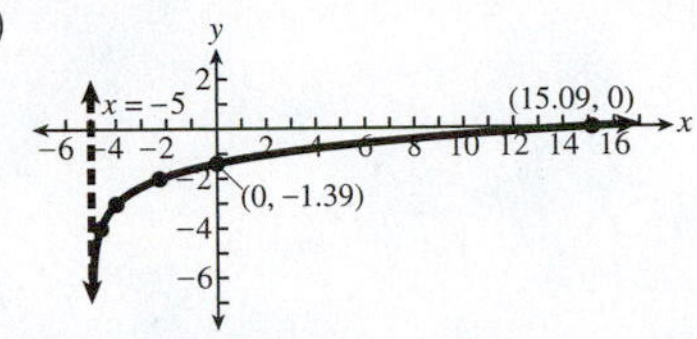

21.

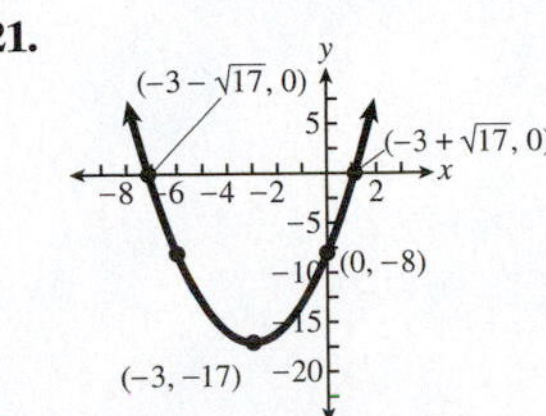

22.

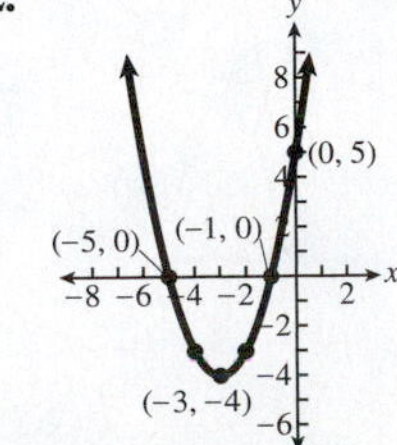

23.

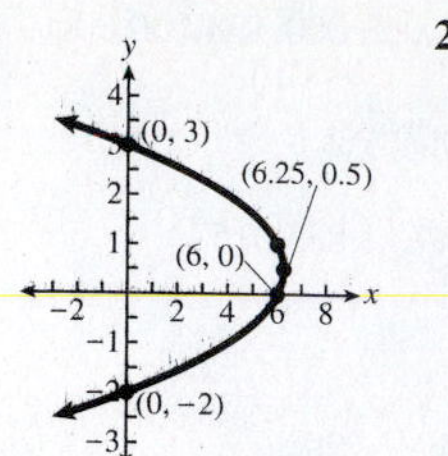

24.

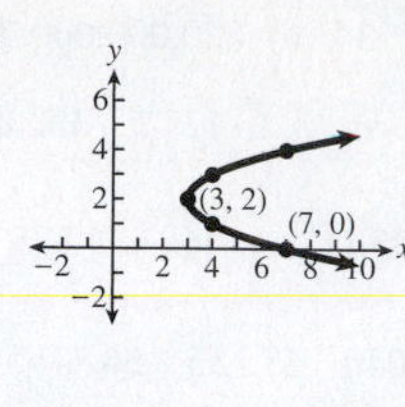

25. $r = 8$

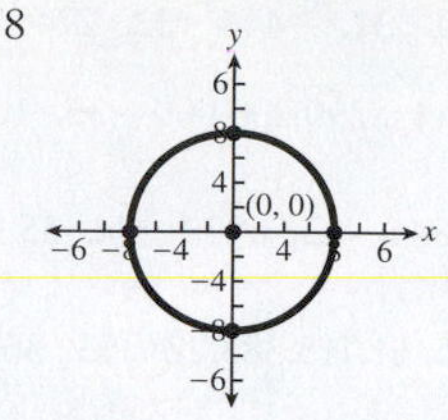

26. $r = 5$

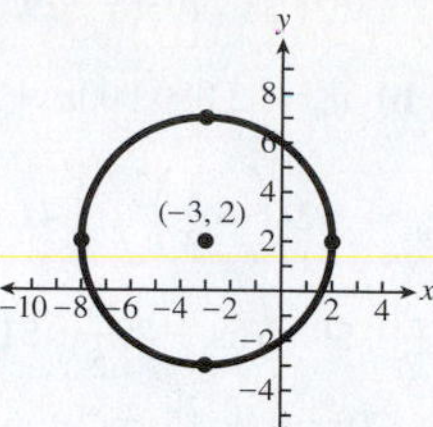

27. $a = 3, b = 7$

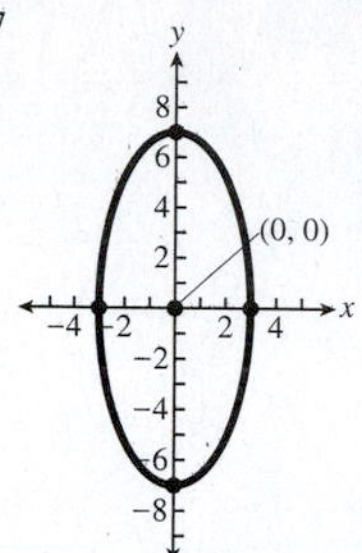

28. $a = 5, b = 6$

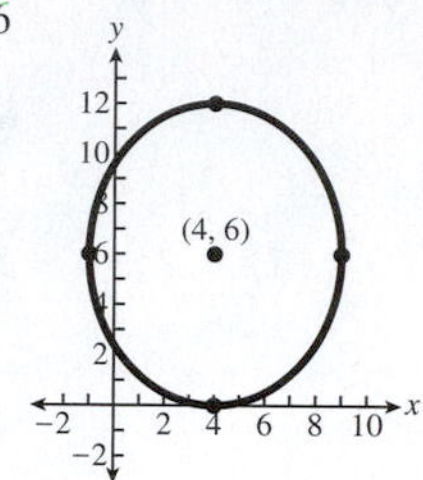

29. $a = 5, b = 2$

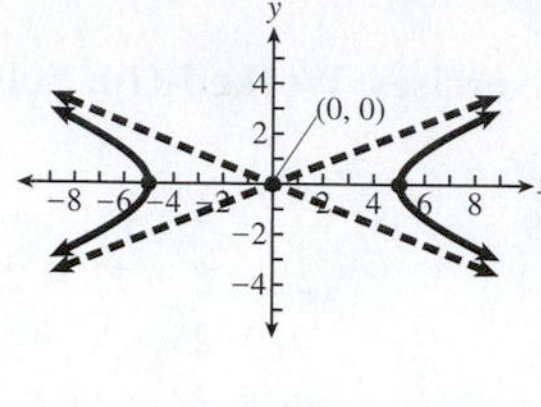

30. $a = 3, b = 1$

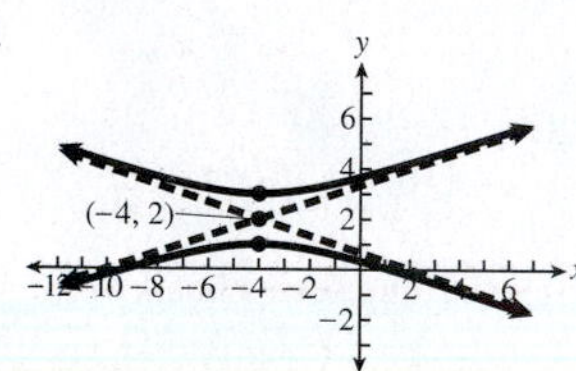

31. $(x - 3)^2 + (y + 2)^2 = 36$ **32.** $\frac{x^2}{16} + \frac{(y - 1)^2}{49} = 1$ **33.** $(2, 6), (6, -2)$ **34.** $(0, -3), (4, 1)$

35. $(8, 6), (-8, 6), (8, -6), (-8, -6)$ **36.** $(3, 2), (-3, 2), (3, -2), (-3, -2)$ **37.** 30 feet by 20 feet **38.** $61, 74, 87; a_n = 13n - 17$

39. 13, 31, 67, 139, 283 **40.** $a_n = 9n - 6$ **41.** 258 **42.** 20,100 **43.** $a_n = 12 \cdot \left(\frac{3}{4}\right)^{n-1}$ **44.** 21,845 **45.** 274,514 **46.** Yes, $9\frac{3}{5}$ **47.** 210

48. $x^4 + 4x^3y + 6x^2y^2 + 4xy^3 + y^4$ **49.** $x^6 - 6x^5y + 15x^4y^2 - 20x^3y^3 + 15x^2y^4 - 6xy^5 + y^6$

50. $x^8 + 8x^7y + 28x^6y^2 + 56x^5y^3 + 70x^4y^4 + 56x^3y^5 + 28x^2y^6 + 8xy^7 + y^8$

Appendix A-1 **1.** $x - 11$ **3.** $x + 16$ **5.** $7x - 25 + \frac{38}{x + 2}$ **7.** $12x - 11 - \frac{308}{x - 14}$ **9.** $x^2 + 5x - 9 + \frac{13}{x - 5}$

11. $x^3 - 8x^2 + 72x - 576 + \frac{4575}{x + 8}$

Appendix A-2 **1.** $(2, 3)$ **3.** $(-3, 7)$ **5.** $(0, 6)$ **7.** $(3, 2, 1)$ **9.** $(-1, -2, 6)$ **11.** $(0, 4, -4)$

INDEX OF APPLICATIONS

Education

Entertainment/Sports

Environment

General Interest

Geometry

Government

Health Care

Hobbies

Real Estate

Science/Technology

Statistics

INDEX

E

F

G

M

N

O

S

T

U

V

W

X

Y

Z

PHOTO CREDITS

Page 14 Rubberball Productions **Page 14** Thinkstock/Getty Images **Page 26** Digital vision **Page 26** PhotoDisc Red/Getty Images **Page 26** PhotoDisc/Getty Images **Page 31** PhotoDisc/Getty Images **Page 47** PhotoDisc/Getty Images **Page 53** Kobal Collection **Page 62** Murleen Ray/Courtesy of author **Page 68** Philip Gould/Corbis **Page 70** Rubberball Productions/Getty Images **Page 78** Comstock/Getty Images **Page 80** Stockbyte Platinum/Getty Images **Page 85** Brand X Pictures/Getty Images **Page 85** Kobal Collection **Page 89** Bettmann/Corbis **Page 92** *Eldorado: Aristide Bruant* (1892), Henri de Toulouse-Lautrec. Lithograph. Photo by Peter Harholdt/Corbis **Page 95** iStockphoto **Page 107** Digital Vision/Getty Images **Page 110** PhotoDisc Red/Getty Images **Page 117** AP Photos **Page 139** Digital Vision/Getty Images **Page 152** PhotoDisc Red/Getty Images **Page 154** PhotoDisc Blue/Getty Images **Page 174** PhotoDisc Blue/Getty Images **Page 181** PhotoDisc/Getty Images **Page 197** Murleen Ray/Courtesy of author **Page 211** Robbie Jack/Corbis **Page 213** David Bergman/Corbis **Page 227** PhotoDisc/Getty Images **Page 229** Reuters/Corbis **Page 246** PhotoDisc Blue/Getty Images **Page 248** Murleen Ray/ Courtesy of author **Page 266** NASA **Page 268** NASA **Page 294** Photo Researchers, Inc. **Page 334** Getty Images/Stone Allstock **Page 336** Thinkstock/Getty Images **Page 337** Image Source/Getty Images **Page 352** PhotoDisc Red/Getty Images **Page 353** Library of Congress Prints and Photographs Division [LC-USZ62-101363] **Page 397** Stockbyte Platinum/getty Images **Page 400** PhotoDisc Blue/Getty Images **Page 405** Brand X Pictures/Getty Images **Page 416** Najlah Feanny/Saba/Corbis **Page 466** Robert Maass/Corbis **Page 470** PhotoDisc **Page 479** PhotoDisc/Getty Images **Page 480** Photos.com/Thinkstock **Page 517** James Shaffer/PhotoEdit **Page 535** Photos.com/Thinkstock **Page 562** Brand X Pictures/Getty Images **Page 563** PhotoDisc/Getty Images **Page 580** Brand X Pictures/Getty Images **Page 603** Bettmann/Corbis **Page 605** Yannis Behrakis/Reuters/Corbis **Page 611** Mike Blake/Reuters/Corbis **Page 617** José Rodriguez **Page 650** PhotoDisc **Page 655** PhotoDisc Red **Page 681** the Granger Collection, New York **Page 692** PhotoDisc **Page 695** PhotoDisc Red **Page 703** Comstock **Page 725** PhotoDisc Red **Page 729** PhotoDisc Blue **Page 733** Corbis **Page 734** Brand X Pictures **Page 756** Lucien Aigner/Corbis **Page 784** Thinkstock **Page 822** Paul Erickson/iStockphoto **Page 822** PhotoDisc **Page 835** Michael Nicholson/Corbis **Page 841** PhotoDisc **Page 854** Digital Vision **Page 856** Comstock **Page 868** Roberto A. Sanchez/iStockphoto

Geometry Formulas and Definitions

- Perimeter of a triangle with sides s_1, s_2, and s_3: $P = s_1 + s_2 + s_3$

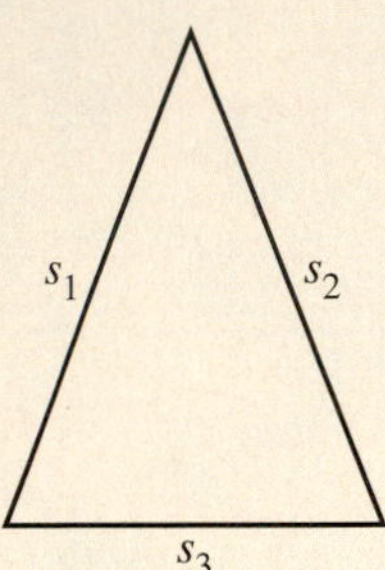

- Perimeter of a square with side s: $P = 4s$

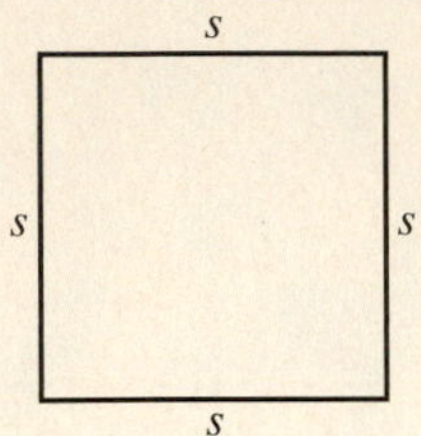

- Perimeter of a rectangle with length L and width W: $P = 2L + 2W$

- **Circumference** (distance around the outside) of a circle with radius r: $C = 2\pi r$

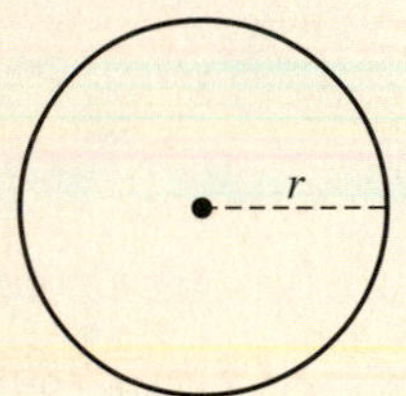

- Area of a triangle with base b and height h: $A = \frac{1}{2}bh$

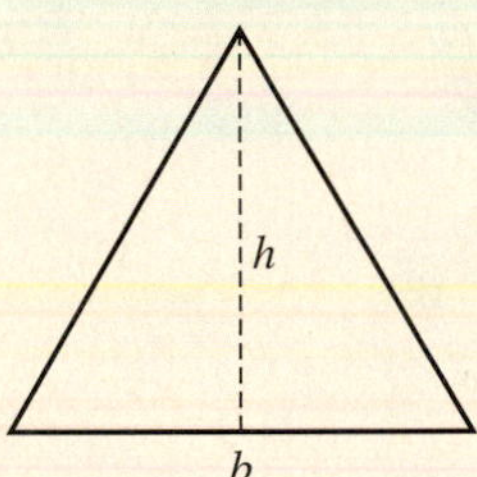

- Area of a square with side s: $A = s^2$

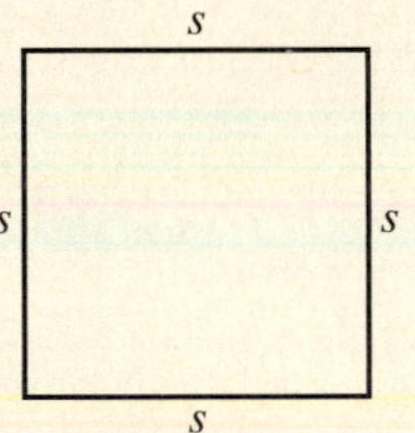

- Area of a rectangle with length L and width W: $A = LW$

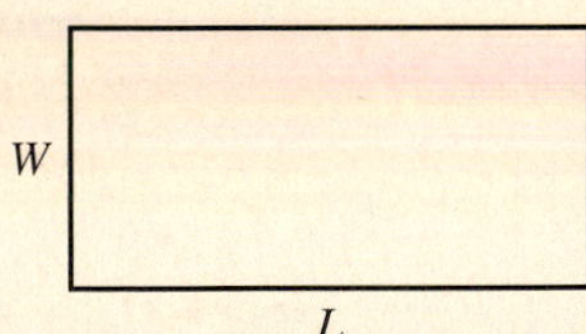

- An **equilateral triangle** is a triangle that has three equal sides.

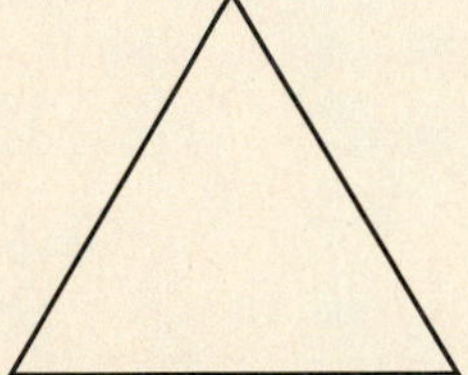

- Area of a circle with radius r: $A = \pi r^2$

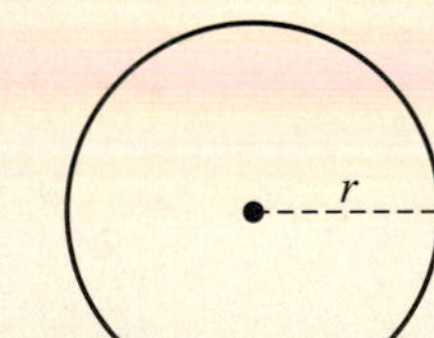

- An **isosceles triangle** is a triangle that has at least two equal sides.

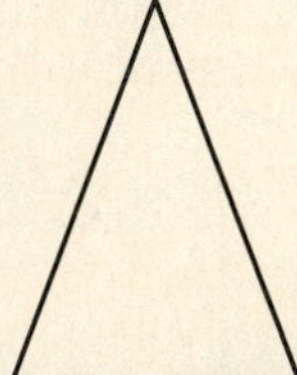

Formulas and Equations

Distance Formula (see page 84) $d = r \cdot t$

Basic Percent Equation (see page 90)

Amount = Percent · Base

Simple Interest (see page 93) $I = p \cdot r \cdot t$

Horizontal Line (see page 134) $y = b$

Vertical Line (see page 134) $x = a$

Slope Formula (see page 142) $m = \dfrac{y_2 - y_1}{x_2 - x_1}$

Slope–Intercept Form of a Line (see page 145)

$y = mx + b$

Point–Slope Form of a Line (see page 168)

$y - y_1 = m(x - x_1)$

Rules for Exponents (see pages 251–254, 256, 260)

$x^m \cdot x^n = x^{m+n}$

$(x^m)^n = x^{m \cdot n}$

$(xy)^n = x^n y^n$

$\dfrac{x^m}{x^n} = x^{m-n}\ (x \neq 0)$

$x^0 = 1 (x \neq 0)$

$\left(\dfrac{x}{y}\right)^n = \dfrac{x^n}{y^n}\ (y \neq 0)$

$x^{-n} = \dfrac{1}{x^n}\ (x \neq 0)$

Special Products (see page 278)

$(a + b)(a - b) = a^2 - b^2$

$(a + b)^2 = a^2 + 2ab + b^2$

$(a - b)^2 = a^2 - 2ab + b^2$

Factoring Formulas

Difference of Squares:

$a^2 - b^2 = (a + b)(a - b)$ (see page 313)

Difference of Cubes:

$a^3 - b^3 = (a - b)(a^2 + ab + b^2)$ (see page 315)

Sum of Cubes:

$a^3 + b^3 = (a + b)(a^2 - ab + b^2)$ (see page 317)

Height, in feet, of a projectile after *t* seconds (see page 342)

$h(t) = -16t^2 + v_0 t + s$

Direct Variation (see page 401) $y = kx$

Inverse Variation (see page 402) $y = \dfrac{k}{x}$

Properties of Radicals (see page 486)

$\sqrt[n]{a} \cdot \sqrt[n]{b} = \sqrt[n]{ab}, \dfrac{\sqrt[n]{a}}{\sqrt[n]{b}} = \sqrt[n]{\dfrac{a}{b}}$

Fractional Exponents (see page 491) $a^{m/n} = (\sqrt[n]{a})^m$

Period of a Pendulum (see page 514) $T = 2\pi\sqrt{\dfrac{L}{32}}$

Imaginary Numbers (see page 519) $i = \sqrt{-1}, i^2 = -1$

Quadratic Formula (see page 547) $x = \dfrac{-b \pm \sqrt{b^2 - 4ac}}{2a}$

Pythagorean Theorem (see page 578) $a^2 + b^2 = c^2$

Properties of Logarithms (see pages 707, 708, 709)

$\log_b x + \log_b y = \log_b(xy)\ (x > 0, y > 0, b > 0, b \neq 1)$

$\log_b x - \log_b y = \log_b\left(\dfrac{x}{y}\right)\ (x > 0, y > 0, b > 0, b \neq 1)$

$\log_b x^r = r \cdot \log_b x\ (x > 0, b > 0, b \neq 1)$

Change-of-Base Formula (see page 712)

$\log_b x = \dfrac{\log_a x}{\log_a b}\ (x > 0, a > 0, b > 0, a \neq 1, b \neq 1)$

Compound Interest Formula (see page 725)

$A = P\left(1 + \dfrac{r}{n}\right)^{nt}$

Continuous Compound Interest (see page 727) $A = Pe^{rt}$

Exponential Growth and Decay (see pages 728, 729)

$P = P_0 e^{kt}$

pH of a Liquid (see page 731) $\text{pH} = -\log[\text{H}^+]$

Volume of a Sound (see page 732) $L = 10 \log\left(\dfrac{I}{10^{-12}}\right)$

Distance Formula (see page 772)

$d = \sqrt{(x_2 - x_1)^2 + (y_2 - y_1)^2}$

General Term of an Arithmetic Sequence (see page 844)

$a_n = a_1 + (n - 1)d$

***n*th Partial Sum of an Arithmetic Sequence** (see page 845)

$s_n = \dfrac{n}{2}(a_1 + a_n)$

General Term of a Geometric Sequence (see page 850)

$a_n = a_1 \cdot r^{n-1}$

***n*th Partial Sum of a Geometric Sequence** (see page 852)

$s_n = a_1 \cdot \dfrac{1 - r^n}{1 - r}$

Limit of an Infinite Geometric Series (If It Exists) (see page 853)

$s = \dfrac{a_1}{1 - r}$

Binomial Theorem (see page 859)

$(x + y)^n = \sum_{r=0}^{n} {}_nC_r \cdot x^{n-r} y^r$

Functions, Equations, and Their Graphs

Linear Equations and Functions (see pages 123, 157)

$Ax + By = C, y = mx + b, f(x) = mx + b$

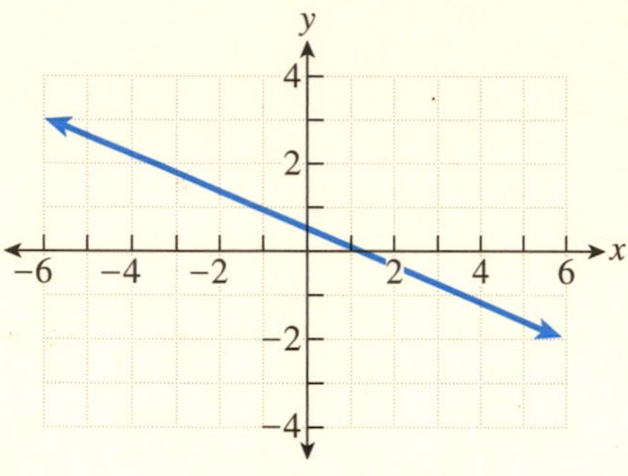

Square Root Functions (see page 637)

$f(x) = a\sqrt{x - h} + k$

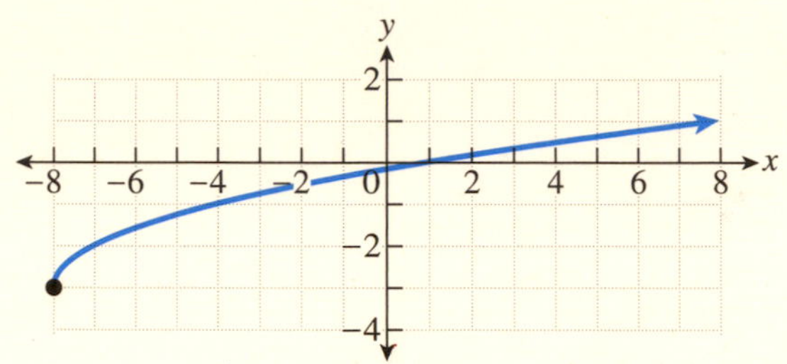

Absolute Value Functions (see pages 422, 439)

$f(x) = a|x - h| + k$

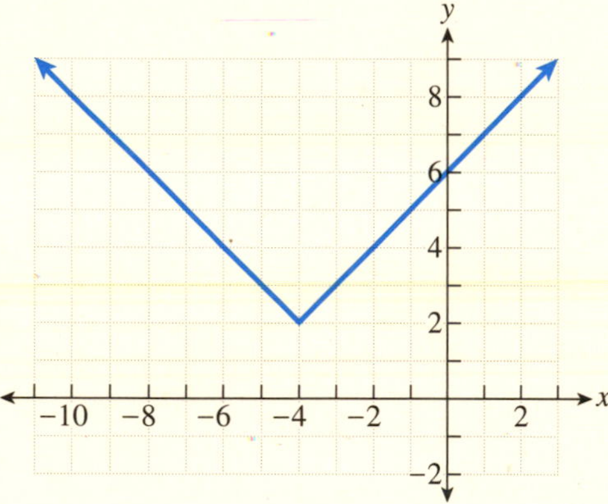

Cubic Functions (see page 641)

$f(x) = a(x - h)^3 + k$

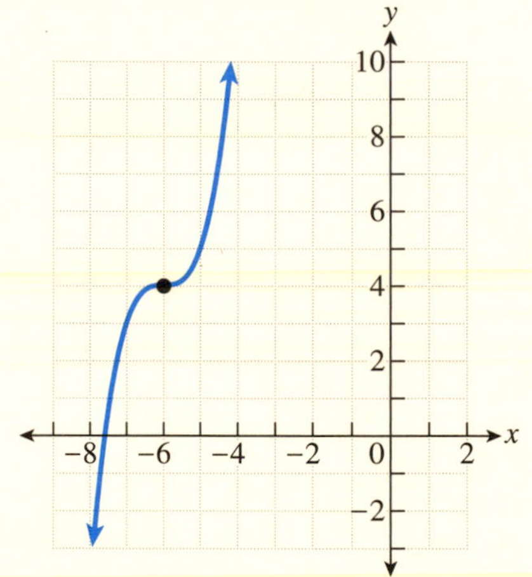

Quadratic Equations and Functions (see pages 325, 332)

$y = ax^2 + bx + c, \quad y = a(x - h)^2 + k,$

$f(x) = ax^2 + bx + c, \quad f(x) = a(x - h)^2 + k$

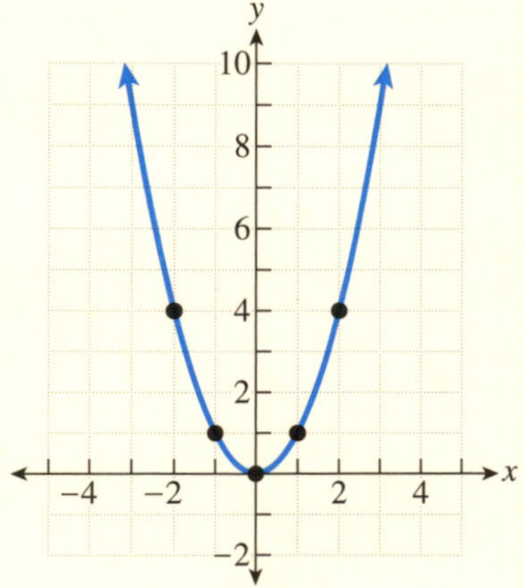

Exponential Functions (see pages 686, 736)

$f(x) = b^{x-h} + k$

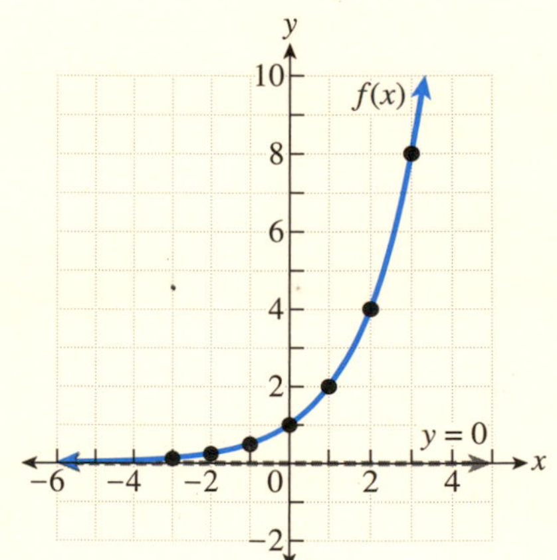